Principles of Microbiology

Principles of Microbiology

Editor
Michael A. Buratovich, PhD

SALEM PRESS
A Division of EBSCO Information Services, Inc.
Ipswich, Massachusetts

GREY HOUSE PUBLISHING

Cover photo: Clean culture of aerobic bacteria on agar plate. Photo by AndreasReh, iStock.

Principles of Microbiology, published by Grey House Publishing, Inc., Amenia, NY, under exclusive license from EBSCO Information Services, Inc.

∞ The paper used in these volumes conforms to the American National Standard for Permanence of Paper for Printed Library Materials, Z39.48 1992 (R2009).

Publisher's Cataloging-In-Publication Data
(Prepared by The Donohue Group, Inc.)

Names: Buratovich, Michael A., editor.
Title: Principles of microbiology / editor, Michael A. Buratovich, PhD.
Description: Ipswich, Massachusetts : Salem Press, a division of EBSCO Information Services, Inc. ; Amenia, NY : Grey House Publishing, [2022] | Series: Principles of science | Includes bibliographical references and index.
Identifiers: ISBN 9781637000953
Subjects: LCSH: Microbiology. | LCGFT: Reference works.
Classification: LCC QR41.2 .P75 2022 | DDC 579—dc23

FIRST PRINTING
PRINTED IN THE UNITED STATES OF AMERICA

Contents

Publisher's Note

Microbiology is the next volume in Salem's *Principles of Science* series, which includes *Energy*, *Marine Science*, *Geology*, *Information Technology*, and *Mathematics*.

This new resource explores the study of the invisible world of microorganisms, introducing readers to the main principles of microbiology, its importance, and its many real-world applications. Microorganisms are all around us; many live on or in the human body. They play a crucial role in oxygenating the atmosphere, in optimizing agricultural soil, in pharmaceutical development, in genetic engineering, and, last but not least, in gaining control over deadly infectious disease outbreaks. Topics covered in this volume include: major groups of microorganisms; adaptive immunity; microorganisms and human disease; environmental microbiology; applied and industrial microbiology; and microbiomes.

This work begins with a comprehensive Editor's Introduction to the topic of microbiology written by Michael A. Buratovich, PhD.

Following the Introduction, *Principles of Microbiology* includes 167 entries arranged in 12 broad categories:

The Microbes surveys types of microbes along with how they can be viewed. A range of unicellular organisms, including bacteria, archaea, protists, and fungi, are discussed. Microbes often define the limits of life in a particular biosphere, creating conditions necessary to the survival and evolution of other life forms. Entries in this section include algae, eukaryotes and prokaryotes, and bacteria.

Microbial Processes examines the metabolic pathways of microbes, which are much more diverse than those of animals. The study of these pathways are ideal models for biochemists. Topics discussed include fermentation, photosynthesis, DNA and RNA synthesis, vitamins and minerals, and nitrogen fixation.

Microbial Genetics explores gene expression as it applies to bacterial gene expression, one of the first explorations in the field of molecular genetics and consequently a well understood branch of this complicated topic. Microbial genetics have been exploited in genetic engineering and in developing antibodies. Recombinant DNA technology, plasmids, and antibiotic resistance are among the topics discussed.

Geochemical Cycles discusses the role of microbes in geochemical cycles. Microbes play a major part in nutrient recycling, helping to extract specific minerals—including carbon, nitrogen, phosphorous, and sulfur—and making them available to living sources.

Microbial Exploitation covers the unique metabolic capabilities of microbes and their use in making food and drinks, in treating sewage, and in creating biosynthetics. Entries in this section—which include anaerobic digestion and lactic acid—discuss each of these uses.

Microbial Symbioses examines the symbiosis between microbes and multicellular organisms, such as lichens' mutualistic relationship with photosynthetic algae. Topics include mycorrhizae, ruminants, and termites.

The Bacteria discusses some of the better-known bacteria and the often devastating diseases they cause. Among the topics discussed are food poisoning, tetanus, cholera, typhoid fever, and pneumonia.

Fungi covers specific fungi and the diseases they can cause, which can be particularly harmful to immunocompromised individuals. Also discussed is yeast, which is used to make food and beverages. Topics include coccidioidomycosis, mycotoxins, ringworm, and yeasts.

Protozoans describes pathogenic protozoans that cause human and animal diseases, which cause major problems in the Third World. This section includes entries on amoebic dysentery, malaria, trypanosomiasis, toxoplasmosis, and others.

Viruses introduces the primary viral diseases that infect human populations, helping to differentiate between common childhood diseases and those that are potentially debilitating or even deadly. Virus types and the structure and life cycle of viruses are discussed, as well as viral genetics, hepatitis B and hepatitis C, retroviruses, polio, influenza, measles, rubella, and rabies.

Immunology discusses the complexities of the human immune system, both innate and acquired. This section discusses the components of each branch and how they interact with microbes. Topics include antibodies, autoimmune disorders, infection control, immunization, sepsis, and steroids.

Vaccines explains the vaccines that protect us from childhood and adult diseases, discussing both commonly available vaccines as well as a few experimental ones. Among the vaccines covered are those for cancer, chickenpox, COVID-19, hepatitis, typhoid, and rabies.

Entries begin by specifying the Category an entry falls into and include a brief Definition and a list of Key Terms summarizing important points; all entries end with a helpful Further Reading section.

This work also includes helpful appendices, including:

- Bibliography;
- Glossary;
- Organizations;
- Subject Index

Salem Press extends appreciation to all involved in the development and production of this work. Names and affiliations of contributors to this work follow the Editor's Introduction.

Principles of Microbiology, as well as all Salem Press reference books, is available in print and as an e-book. Please visit www.salempress.com for more information.

Introduction

Microbes are everywhere. Even our bodies, be it our skin, mouths, throats, or guts, are awash in microbes. The human body is more an ecosystem than an organism. Our microbes keep infectious microbes at bay, help us digest and absorb complex molecules, and produce small compounds that keep our bodies humming.

Microbes have been found living deep inside the Earth's crust at the bottom of the sea. They live throughout the Arctic. Heat-living (thermophiles) and acid-living (acidophiles) microbes live near Brother's Volcano, approximately 200 miles northeast of New Zealand and 6,000 feet underwater. Almost anywhere we look, we find microbes.

Human beings have co-opted microbes to make food (bread, sauerkraut, yogurt, kimchi, pickles, kefir, tempeh, miso, natto, and cheese) and drinks (wine, beer, sake, kvass, vinegar, kombucha, pulque, and others). More recently, genetic engineering has extended the uses for microbes to make vaccines, insulin, growth factor, and other medicines. Other uses include environmental clean-up, biological control, and preventing frost damage in crops.

Given the importance of microbes and their ubiquity, no "Principles of" series would be complete without a volume on microbiology, the study of microbes.

We begin this volume by introducing our subjects, the microbes. These tiny, living creatures are not a monolithic group but a highly diverse, weird, and wonderful collection of organisms. The Archaea live in extreme environments, and their basic molecular biology is more akin to human cells than that of bacteria. The bacteria or eubacteria cause diseases and occupy most of the environmental niches we see. Fungi are integral decomposers, particularly in forests. A few fungi cause diseases, typically in immunosuppressed individuals, and unicellular fungi, yeasts, ferment juices to make wine, beers, and dough to make bread. This section includes the photosynthetic algae and techniques for observing and staining microbes.

After that, our volume covers the metabolic processes that make microbes unique and immensely useful. Many microbes use the same metabolic processes as plants, animals, and others. Therefore, the bacterium Escherichia coli was the primary model organism for metabolic biochemistry research for decades. However, others have unique metabolic pathways (e.g., ethanol fermentation, lactic acid fermentation, butanol/acetone fermentation). Their biochemistry makes them useful for industrial purposes. This section focuses on those molecules synthesized and used by microbes, including DNA and RNA, amino acids, lipids, polysaccharides, and biofilms. Other articles elucidate how our body's native microbes help us assimilate various vitamins. Nitrogen fixation, a uniquely microbial capability that is the cornerstone of the global nitrogen cycle, is the final entry in this section.

The next section examines microbial genetics. Bacteria were the first model system for molecular genetics. The 1961 *Journal of Molecular Biology* paper by François Jacob and Jacques Monod (for which they won the 1965 Nobel Prize in Physiology or Medicine) proposed the operon model for gene regulation. This work became the launchpad for decades of fruitful investigations of gene expression in prokaryotes (microorganisms without internal compartments like a nucleus) and eukaryotes (organisms whose cells contain multiple internal compartments like a nucleus). Our section examines the nature of the operon, how microbes transfer genes between them, recombinant DNA technology, and the genetics of antibiotic resistance.

The following section examines the vital geochemical cycles and the integral roles microbes play in them. Without microbes, these cycles would grind to a crawl and make complex life on Earth impossible.

After that, our volume examines how human food and industrial technologies have exploited microbes and their metabolic idiosyncrasies to digest and make complex molecules, treat our sewage, make useful organic acids, and create unique foods like beer, wine, bread, and the like.

Next, we catalog and describe some of the special relationships between microbes and other multicellular organisms. Symbiosis simply means living together, and microbes often live as a unique part of multicellular organisms in a way that benefits both. Lichens, for example, are a unique community consisting of a fungus and an alga. The algae photosynthesize sunlight to make energy and fix atmospheric carbon dioxide, and fungi keep the algae moist and protected. Other symbiotic relationships include microbes and termites, bovines, plants, and humans.

The following section examines a selection of the bacteria and the diseases they cause. While only a minority of bacteria cause disease, that minority is the best-known of the bacteria. Our volume examines everything from sinusitis to syphilis. Readers can find helpful material about most of the more common and interesting bacterial infections.

The next section examines the fungi, particularly the pathogenic fungi. Fungi deserve a volume of their own, but we focused on those fungi that cause human disease in this admittedly brief survey. We end with an entry about yeasts, unicellular fungi that cause disease and help make our alcoholic drinks and bread.

The fungal section is followed by examining protozoans and the diseases they cause. Protozoans also deserve a volume of their own, but this survey examines the more common protozoan diseases. Protozoan diseases are interesting and worth their own category.

Viruses are the subject of the next section. Since viruses are obligate intracellular parasites, they all cause disease. All multicellular organisms are infected by viruses. Plant, animal, and microbial viruses play vital roles in biological evolution and gene transfer. Our viral section has an excellent survey of the more well-known human diseases.

The penultimate section is a survey of the human immune system and how it influences human interactions with those microbes that do and those that do not cause disease. (The immune system is the subject of another Salem Press volume, *Principles of Health: Allergies & Immune Disorders.*) This survey hits the primary concepts of immunity and constitutes an excellent introduction to this topic.

The final section examines vaccines that protect us against the previously discussed diseases. Vaccine production requires a knowledge of the microbes from which they are made and their genetics. Vaccine technology is rapidly changing with the advent of messenger RNA vaccines. This section will keep our readers up to date with the latest vaccines for travel and general health.

This volume will leave the reader better informed and hopefully with a sense of wonder about the universe contained on human skin or in a drop of water that you cannot see but that profoundly influence life.

—Michael A. Buratovich, PhD

Contributors

Erika A. Abrahamsen
Gordon College

Sarah Acker
Yale University

Christine Adamec, MBA
Independent Scholar

Richard Adler, PhD
University of Michigan-Dearborn

E. Victor Adlin, MD
Temple University School of Medicine

Anubhav Agarwal, MD
South Nassau Communities Hospital

Patricia A. Ainsa, MPH, PhD
University of Texas, El Paso

Saeed Akhter, MD
Texas Technological University Health Science Center

Rick Alan
Medical Writer and Editor

Bruce Ambuel, PhD
Medical College of Wisconsin

John J. B. Anderson, PhD
University of North Carolina, Chapel Hill

Wendell Anderson
Northstar Writing & Editing, American Medical Writers Association

Earl R. Andresen, PhD
University of Texas, Arlington

Daruenie Andujar
Gordon College

Thessicar Antoine-Reid, PhD
Vanderbilt University Medical Center

Walter Appleton
Grosse Pointe Woods, Michigan

Michele Arduengo, PhD
Milton, Wisconsin

Samar Aslam, MD
South Nassau Communities Hospital

Bryan C. Auday, PhD
Gordon College

Catherine Avelar, BS
Cornell University of Veterinary Medicine

Mihaela Avramut, MD, PhD
Verlan Medical Communications, American Medical Writers Association

Michelle Badash
Wakefield, MA

Gloria Reyes Báez, MD
University of Puerto Rico School of Medicine

Dana K. Bagwell
Memory Health and Fitness Institute

Jimmy Bajaj, DO
South Nassau Communities Hospital

Pamela J. Baker, PhD
Bates College

Anita Baker-Blocker, MPH, PhD
Ann Arbor, Michigan

Iona C. Baldridge
Lubbock Christian University

Beverly Ballarlo
Northeastern University

Nancy Banasiak, MSN, PNP-BC, APRN
Yale University School of Nursing

Veronica N. Baptista, MD
Stanford Medical Group

Adriel Barrios-Anderson
Brown University

Lawrence W. Bassett, MD
University of California, Los Angeles, School of Medicine

Robert J. Baumann, MD
University of Kentucky

John A. Bavaro, EdD, RN
Slippery Rock University

Barbara C. Beattie
Sarasota, Florida

Tanja Bekhuis, PhD
TCB Research

Paul F. Bell, PhD
Heritage Valley Health System

Gregory A. Benitz, PsyD
San Francisco, California

Alvin K. Benson, PhD
Utah Valley State College

Cynthia Breslin Beres
Glendale, California

Carol D. Berkowitz, MD
Harbor-UCLA Medical Center

Leonard Berkowitz, DO
South Nassau Communities Hospital

Milton Berman, PhD
University of Rochester

Matthew Berria, PhD
Weber State University

Silvia M. Berry, MSc, RVT
Englewood Hospital and Medical Center, New Jersey

Leah M. Betman, MS, CGC
Centreville, Virginia

Massimo D. Bezoari, MD
Huntingdon College

Jigna Bhalla, PharmD
American Medical Writers Association

Dawn Bielawski, PhD
Wayne State University

Anna Binda, PhD
American Medical Writers Association

Jennifer Birkhauser, MS, MD
University of California, Irvine

Tyler Biscontini
Independent Scholar

Virginiae Blackmon
Fort Worth, Texas

Stephanie McCallum Blake, MSN
Duke University Medical Center

Robert W. Block, MD
University of Oklahoma

Jane Blood-Siegfried, RN, DNSc, CPNP
Duke University School of Nursing

Paul R. Boehlke, PhD
Wisconsin Lutheran College

Prodromos G. Borboroglu, MD
Navy Medical Center, San Diego

Wanda Todd Bradshaw, MSN, NNP, PNP, CCRN
Duke University School of Nursing

Barbara Brennessel, PhD
Wheaton College

Peter N. Bretan, MD
University of California Medical Center, San Francisco

Douglas H. Brown
Wellesley College

Kenneth H. Brown, PhD
Northwestern Oklahoma State University

Thomas L. Brown, PhD
Wright State University School of Medicine

Carolynn Bruno, PhD, APRN, CNS, FNP-C
New York University, Rory Meyers College of Nursing

Mitzie L. Bryant, BSN, MEd
St. Louis Board of Education

Faith Hickman Brynie, PhD
Bigfork, Montana

Fred Buchstein
John Carrol University

Suzette Buhr, RTR, CDA
American Medical Writers Association

Amy Webb Bull, DSN, APN
Tennessee State University School of Nursing

Michael A. Buratovich, PhD
Spring Arbor University

Edmund C. Burke, MD
University of California Medical Center, San Francisco

John T. Burns, PhD
Bethany College

Rosslynn S. Byous, DPA, PA-C
University of Southern California, Keck School of Medicine

Jeffrey R. Bytomski, DO
Duke University Medical Center

Cait Caffrey
Independent Scholar

Lauren M. Cagen, PhD
University of Tennessee, Memphis

David Caldwell, PhD
Indianapolis, IN

James J. Campanella, PhD
Montclair State University

William J. Campbell
Independent Scholar

Edmund J. Campion, PhD
University of Tennessee

Brian Campos
Brown University

Louis A. Cancellaro, MD
Veterans Affairs Medical Center, Mountain Home, Tennessee

Byron D. Cannon, PhD
University of Utah

Shiliang Alice Cao
Brown University

Richard P. Capriccioso, MD
University of Phoenix

Mary Allen Carey, PhD
University of Oklahoma

Cheryl Carrao, DO
South Nassau Communities Hospital

Christine M. Carroll, RN, BSN, MBA
American Medical Writers Association

Culley C. Carson III, MD
University of North Carolina School of Medicine

Rosalyn Carson-DeWitt, MD
Everyday Health

Donatella M. Casirola, PhD
University of Medicine and Dentistry, New Jersey Medical School

Cristina Cesaro, DO
South Nassau Communities Hospital

Anne Lynn S. Chang, MD
Stanford University

Karen Chapman-Novakofski, RD, LDN, PhD
University of Illinois

Kathleen A. Chara, MS
Roseville, Minnesota

Paul J. Chara, Jr., PhD
Northwestern College

Kerry L. Cheesman, PhD
Capital University

Richard W. Cheney, Jr., PhD
Christopher Newport University

Derrick Cheng
Brown University

David L. Chesemore, PhD
California State University, Fresno

Valentina Chiarelli, PhD
Università degli Studi di Trieste

Francis P. Chinard, MD
New Jersey Medical School

Leland J. Chinn, PhD
Biola University

Ariel Choi
Brown University

Kathleen Chung
Brown University

Rose Ciulla-Bohling, PhD
Independent Scholar

David A. Clark, MD
Louisiana State University Medical School, New Orleans

Nancy Handshaw Clark, PhD
American University of the Caribbean School of Medicine/Kingston Hospital, Surrey, England

Julien M. Cobert
Duke University

Jaime S. Colomé, PhD
California Polytechnic, San Luis Obispo

Pam Conboy
American Medical Writers Association

Arlene R. Courtney, PhD
Western Oregon State College

Mark S. Coyne
University of Kentucky

Sarah Crawford, PhD
Southern Connecticut State University

James Danckert, PhD
University of Waterloo

Tish Davidson, AM
Fremont, California

LeAnna DeAngelo, PhD
Arizona State University

Roy L. DeHart, MD, MPH
University of Oklahoma

Anna Delamerced
Brown University

Patrick J. DeLuca, PhD
Mount St. Mary College

Cynthia L. De Vine
American Medical Writers Association

Thomas E. DeWolfe, PhD
Hampden-Sydney College

Shawkat Dhanani, MD, MPH
VA Greater Los Angeles Healthcare System

Jackie Dial, PhD
MedicaLink, LLC

Mary Dietmann, EdD, APRN, CNS
Sacred Heart University College of Nursing

Kenneth Dill, MD
South Nassau Communities Hospital

Stephen DiMaria
Brown University

Sandra Ripley Distelhorst
Vashon, Washington

Suzanne Dixon, MPH, MS, RD
Verywell Health

Matthew Doiron, BA
Spaulding Rehabilitation Hospital

Katherine Hoffman Doman
Greene County, Tennessee

Mark R. Doman, MD
Veterans Affairs Medical Center, Mountain Home, Tennessee

Lillian Dominguez
Brown University

Desiree Dreeuws
Sunland, California

Cherie H. Dunphy, MD
University of North Carolina, Chapel Hill

Allison Dussault, RN
Yale University

Mark Dziak
Northeast Editing

Patricia Stanfill Edens, PhD, RN, FACHE
The Oncology Group, LLC

Miriam Ehrenberg, PhD
City University of New York, John Jay College

Ebrahim Elahi, MD
Mount Sinai School of Medicine

Ophelia Empleo-Frazier, MSN, GNP-BC, WCC, DCP
Yale University School of Nursing

Benjamin Estrada, MD
University of South Alabama

Elicia Estrella, MS, CGC, LGC
Children's Hospital, Boston

Renée Euchner, RN
American Medical Writers Association

Merrill Evans, MA
Tucson, AZ

C. Richard Falcon
Roberts and Raymond Associates, Philadelphia

L. Fleming Fallon, Jr., MD, PhD, MPH
Bowling Green State University

Meika A. Fang, MD
Veterans Affairs Medical Center, West Los Angeles

Phillip A. Farber, PhD
Bloomsburg University

Elizabeth Farrington, AGPCNP- BC
Yale University

Frank J. Fedel
Henry Ford Hospital, Detroit

Nicholas Feo
South Nassau Communities Hospital

Caroline Taylor Ferguson
International Neuroethics Society

Jill Ferguson, PhD
National Writers Union

Marisela Fermin-Schon, RN
Yale University

Adi R. Ferrara, BS, ELS
Bellevue, WA

Mary C. Fields, MD
Collin County Community College

K. Thomas Finley, PhD
State University of New York, Brockport

Jesse Fishman, PharmD
Children's Healthcare of Atlanta

Ryan Fogle
Independent Scholar

M. A. Foote, PhD
M. A. Foote Associates

Kimberly Y. Z. Forrest, PhD
Slippery Rock University of Pennsylvania

Ronald B. France, PhD
LDS Hospital, Salt Lake City

Katherine B. Frederich, PhD
Eastern Nazarene College

Paul Freudigman, Jr., MD
Pennsylvania State University, Hershey Medical Center

Paul J. Frisch
Nanuet, New York

Daniel R. Gallie, PhD
University of California, Riverside

Joanne R. Gambosi, BSN, MA
Gold Canyon, Arizona

Christi N. Gandham, DO
Oceanside, New York

Laura Garasimowicz, MS
Prenatal Diagnosis of Northern California

Bianca Garcia, MD
South Nassau Communities Hospital

Frances García, MD
Universidad Central del Caribe School of Medicine

Keith Garebian, PhD
Mississauga, Ontario

Jason Georges
Glendale, California

Soraya Ghayourmanesh, PhD
City University of New York

Sibdas Ghosh, PhD
University of Wisconsin, Whitewater

Joseph T. Giacino, PhD
Spaulding Rehabilitation Hospital, Massachusetts

Jennifer L. Gibson, PharmD
Marietta, Georgia

Margaret Ring Gillock, MS
Libertyville, IL

Lenela Glass-Godwin, MWS
Texas A&M University

Wallace A. Gleason, Jr., MD
University of Texas, Houston

James S. Godde, PhD
Monmouth College

D.R. Gossett
Louisiana State University, Shreveport

Amanda Grannis, BA
Binghamton University

Daniel G. Graetzer, PhD
University of Montana

Hans G. Graetzer, PhD
South Dakota State University

James E. Grant, MPH
Michigan Department of Community Health, Lansing, MI

Carly A. Gray
Yale University

David A. Gremse, MD
University of South Alabama College of Medicine

Dennis W. Grogan, PhD
University of Cinncinati

Frank Guerra, MD
University of Colorado School of Medicine

Yelena Guller, PhD
Spaulding Rehabilitation Hospital, Massachusetts

Lonnie J. Guralnick, PhD
Western Oregon State College

L. Kevin Hamberger, PhD
Medical College of Wisconsin

Ronald C. Hamdy, MD
James H. Quillen College of Medicine

Emad Hanna, MD, MSc
South Nassau Communities Hospital

Beth M. Hannan, MS, CGC
Albany Medical Center

Robert J. Harmon, MD
University of Colorado School of Medicine

Linda Hart, MS, MA
University of Wisconsin-Madison

Peter M. Hartmann, MD
York Hospital, Pennsylvania

Robin Hasslen, PhD
St. Cloud State University

Katherine Hauswirth, MSN, RN
Hauswirth Writing Solutions

H. Bradford Hawley, MD
Boonshoft School of Medicine, Wright State University

Robert M. Hawthorne, Jr., PhD
Marlboro, Vermont

Carol A. Heintzelman, DSW
Millersville University of Pennsylvania

Peter B. Heller, PhD
Manhattan College

Collette Bishop Hendler, RN, MS, CIC
Abington Memorial Hospital

Jerald D. Hendrix
Kennesaw State University

Diane Andrews Henningfeld, PhD
Adrian College

Stephen Henry, DO
South Nassau Communities Hospital

Martha M. Henze, MS, RD
Boulder Community Hospital, Colorado

Michelle L. Herdman, PhD
University of Charleston

David Hernandez
Brown University

Margaret Trexler Hessen, MD
Drexel University College of Medicine

Jane F. Hill, PhD
Bethesda, Maryland

Trevor Hinshaw
Gordon College

Carl W. Hoagstrom, PhD
Ohio Northern University

David Wason Hollar, Jr., PhD
Rockingham Community College

Jenna Hollenstein
Genzyme

Carol A. Holloway
Grosse Point Woods, Michigan

Christine G. Holzmueller
Glen Rock, Pennsylvania

Elaine Hong
Gordon College

Robert M. Hordon
Rutgers University

David L. Horn
Medical Writer

David Hornung, PhD
St. Lawrence University

Ryan C. Horst
Eastern Mennonite University

Howard L. Hosick, PhD
Washington State University

Katherine H. Houp, PhD
Midway College

Carina Endres Howell, PhD
Lock Haven University of Pennsylvania

Shih-Wen Huang, MD
University of Florida

Jason A. Hubbart, MS
University of Idaho, Moscow

Larry Hudgins, MD
Veterans Affairs Medical Center, Mountain Home, Tennessee

Mary Hurd
East Tennessee State University

Christopher Iliades, MD
Centerville, Massachusetts

Peter Iltis, PhD
Gordon College

April D. Ingram
Kelowna, British Columbia, Canada

Tracy Irons-Georges
Glendale, California

Vicki J. Isola, PhD
Hope College

Micah Issitt
St. Louis, MO

Louis B. Jacques, MD
Wayne State University School of Medicine

Samar Post Jamali
Yale University

Ilia Jbankov
Yale University

Thomas C. Jefferson, MD
University of Arkansas for Medical Sciences

Krystal Jenkins
New York Institute of Technology

Albert C. Jensen, MS
Central Florida Community College

Cheryl Pokalo Jones
Townsend, Delaware

Daniel Jones, DDS, PhD
Indiana Wesleyan University

Claire Joseph, MLS, MA, AHIP
South Nassau Communities Hospital

Roushig Grace Kalebjian, MA, RN
Yale University

Karen E. Kalumuck, PhD
The Exploratorium, San Francisco

Ahmad Kamal, MD
Stanford University Medical School

Clair Kaplan, APRN, MSN
Yale University School of Nursing

Susan J. Karcher, PhD
Purdue University

Armand M. Karow, PhD
Xytex Corporation

Laurence Katz, MD
University of North Carolina, Chapel Hill

Rupinder P. Kaur, DO
South Nassau Communities Hospital

Allee Keener
Gordon College

Gerald W. Keister, MA
American Medical Writers Association

Mara Kelly-Zukowski, PhD
Felician College

Ing-Wei Khor, PhD
Oceanside, California

Camillia King, MPH
Huntsville, AL

Michael R. King, PhD
University of Rochester

Cassandra Kircher, PhD
Elon University

Vernon N. Kisling, Jr., PhD
University of Florida

Samuel V. A. Kisseadoo, PhD
Hampton University

Hillar Klandorf, PhD
West Virginia University

Robert T. Klose, PhD
University College of Bangor

Walter Klyce
Brown University

Jeffrey A. Knight, PhD
Mount Holyoke College

Marylane Wade Koch, RN, MSN
University of Memphis, Loewenberg School of Nursing

Nicholas Koen
Brown University

Anita P. Kuan, PhD
Woonsocket, Rhode Island

Steven A. Kuhl, PhD
Lander University

Jeanne L. Kuhler, MS, PhD
Auburn University, Montgomery

Joyce W. Lacy, PhD
Azusa Pacific University

David J. Ladouceur, PhD
University of Notre Dame

Joshua Lampert, MS-III
Miami, Florida

Nicholas Lanzieri
Pace University

Jeffrey P. Larson, PT, ATC
Northern Medical Informatics

Joshua Lampert, MS-III
Miami, Florida

Jack Lasky
Independent Scholar

Victor R. Lavis, MD
University of Texas, Houston

David M. Lawrence
J. Sargeant Reynolds Community College

Christie Leal, DO
Long Beach, New York

Charles T. Leonard, PhD, PT
University of Montana

Lorraine Lica, PhD
San Diego, California

Lauren Lichten, MS, CGC
Tufts Medical Center

Stan Liu, MD
University of California, Los Angeles, School of Medicine

Martha Oehmke Loustaunau, PhD
New Mexico State University

Eric V. D. Luft, PhD, MLS
SUNY, Upstate Medical University

Arthur J. Lurigio, PhD
Loyola University Chicago

Courtney H. Lyder, MD
Yale University School of Nursing

Elizabeth Lynn, MD
South Nassau Communities Hospital

Nancy E. Macdonald, PhD
University of South Carolina, Sumter

Marianne M. Madsen, MS
University of Utah

Janet Mahoney, RN, PhD, APRN
Monmouth University

Laura Gray Malloy, PhD
Bates College

Nancy Farm Mannikko, PhD
Centers for Disease Control and Prevention

Katia Marazova, MD, PhD
Paris, France

Cherie Marcel, BS
Independent Scholar

Debra Ellen Margolis, DO
St. Joseph's Hospital and Medical Center, Paterson, New Jersey

Mary E. Markland, MA
Argosy University

Sergei A. Markov, PhD
Austin Peay State University

Bonita L. Marks, PhD
University of North Carolina, Chapel Hill

Geraldine F. Marrocco, EdD, APRN, CNS, ANP-BC
Yale University School of Nursing

Charles C. Marsh, PharmD
University of Arkansas for Medical Sciences

Robert L. Martone
Wyeth Neuroscience

Julie J. Martin
Medical Writer

Amber M. Mathiesen, MS
Saint Luke's Boise Medical Center

Maki Matsumura
Yale University

Karen A. Mattern
Visiting Nurse Association Home Health Services

Grace D. Matzen
Molloy College

Maura S. McAuliffe
Austin, Texas

Daniel E. McCallus, PhD
Nucleonics, Inc.

Krisha McCoy
Independent Scholar

Mary Frances McGibbon, DNP, RN, MSN, FNP-BC
City University of New York

Jeffrey A. McGowan
West Virginia University

Mary Beth McGranaghan
Chestnut Hill College

James P. McKenna, MD
Heritage Valley Health System

Wayne R. McKinny, MD
University of Hawaii School of Medicine

Trudy Mercadal, PhD
Plantation, FL

Ralph R. Meyer, PhD
University of Cincinnati

Robert D. Meyer, PhD
Chestnut Hill College

Elva B. Miller, OD
Harrisonburg, Virginia

George Miller, Jr., MD
South Nassau Communities Hospital

Roman J. Miller, PhD
Eastern Mennonite College

Shari Parsons Miller, MA
Accenture

Randall L. Milstein, PhD
Oregon State University

Eli C. Minkoff, PhD
Bates College

Vicki J. Miskovsky
Walter Reed Army Medical Center

Nicole Mitchell
Yale University

Beatriz Manzor Mitrzyk
Mitrzyk Medical Communications, LLC

Briana Moglia
Fort Lauderdale, Florida

Michael Moglia
Glen Cove, New York

Paul Moglia, PhD
South Nassau Communities Hospital

Robin Kamienny Montvilo, RN, PhD
Rhode Island College

Sharon Moore, MD
Veterans Affairs Medical Center, Mountain Home, Tennessee

George A. Morgan, PhD
Colorado State University

Marvin Morris, LAc, MPA
American Medical Writers Association

Sheila J. Mosee, MD
Howard University

Judy Mouchawar, MD, MSPH
University of Colorado Health Sciences Center

Rodney C. Mowbray, PhD
University of Wisconsin, LaCrosse

William L. Muhlach, PhD
Southern Illinois University

Karen M. Nagel, RPh, PhD
Midwestern University

John Panos Najarian, PhD
William Patterson College

Kimberly A. Napoli, MS
Kan Com Biomedical Communications

Donald J. Nash, PhD
Colorado State University

Victor H. Nassar, MD
Emory University

Mary A. Nastuk, PhD
Wellesley College

Nicole Negbenebor
Brown University

Elizabeth Marie McGhee Nelson, PhD
Christian Brothers University

Cindy Nesci, DC
East Pointe, Michigan

Bryan Ness, PhD
Pacific Union College

Marsha M. Neumyer
Pennsylvania State University College of Medicine

Derek Nhan
University of Colorado, Denver

William D. Niemi, PhD
Russell Sage College

Christopher J. Norman, BA, BSN, RN-BC, BC
Yale University

Jane C. Norman, PhD, RN, CNE
Tennessee State University

Kathleen O'Boyle
Wayne State University

Annette O'Connor, PhD
La Salle University

Oladele A. Ogunseitan, PhD, MPH
University of California, Irvine

David A. Olle, MS
Eastshire Communications Inner Medicine Publishing

Colm A. Ó'Moráin, MA, MD, MSc, DSc
University of Dublin, Trinity College

Kristin S. Ondrak, PhD
University of North Carolina at Chapel Hill

Gwenelle S. O'Neal, DSW
Rutgers University

J. Timothy O'Neill, PhD
Uniformed Services University of the Health Sciences

Oliver Oyama, PhD
Duke/Fayetteville Area Health Education Center

Oladayo Oyelola, PhD, SC(ASCP)
American Medical Writers Association

Maria Pacheco, PhD
Buffalo State College

Barbara Pahud, MD, MPH
Children's Mercy Kansas City, University of Missouri, Kansas City, University of Kansas Medical Center

Robert J. Paradowski, PhD
Rochester Institute of Technology

Gowri Parameswaran
Southwest Missouri State University

David Parr
Gordon College

Matthew Parsons, MA
Tulane University

RoseMarie Pasmantier, MD
State University of New York Health Science Center, Brooklyn

Paul M. Paulman, MD
University of Nebraska Medical Center

Cheryl Pawlowski, PhD
University of Northern Colorado

Joseph G. Pelliccia, PhD
Bates College

Angel Perez, MD
South Nassau Communities Hospital

Fortunato Perez-Benavides, MD
Texas Tech University

Madeline Pesec
Brown University

Carol Moore Pfaffly, PhD
Fort Collins Family Medicine Center

Kenneth A. Pidcock, PhD
Wilkes University

Heather E. Pierce, LSW
Unison Behavioral Health Group, Toledo, Ohio

Linda L. Pierce, PhD, RN
Medical College of Ohio

Scott W. Pierce, LSW
Firelands Counseling and Recovery Services, Sandusky, Ohio

Kathryn Pierno, MS
American Medical Writers Association

Diego Pineda, MS
National Network for Immunization Information

Nancy A. Piotrowski, PhD
University of California, Berkeley

Luzanna Plancarte, MD
South Nassau Communities Hospital

George R. Plitnik, PhD
Frostburg State University

Connie Pollock
Glendale, California

Melanie Porter, PhD
Macquarie University

Darleen Powars, MD
Los Angeles County-USC Medical Center

Frank J. Prerost, PhD
Midwestern University

Layne A. Prest, PhD
University of Nebraska Medical Center

Nancy E. Price, PhD
American Medical Writers Association

Victoria Price, PhD
Lamar University

Cynthia F. Racer, MA, MPH
American Medical Writers Association

Rashmi Ramasubbaiah, MD
Western Kentucky University

Lillian M. Range, PhD
University of Southern Mississippi

Dandamudi V. Rao, PhD
University of Medicine and Dentistry of New Jersey

C. Mervyn Rasmussen, MD
Bremerton Naval Hospital, Washington

Darrell L. Ray
University of Tennessee

Diane C. Rein, PhD, MLS
Purdue University

Andrew J. Reinhart, MS
Washington University School of Medicine

Douglas Reinhart, MD
University of Utah

Jerome L. Rekart, PhD
Rivier University

Andrew Ren, MD
Kaiser Permanente Los Angeles Medical Center

Richard M. Renneboog, MSc
Independent Scholar

Wendy E. S. Repovich, PhD
Eastern Washington University

Peter D. Reuman, MD, MPH
University of Florida

Mindy Rice, MSN, RN, PhD
Spring Arbor University

Alex K. Rich
Lititz, PA

Betty Richardson, PhD
Southern Illinois University, Edwardsville

Alice C. Richer, RD, MBA, LD
Norwood, Massachusetts

Brad Rikke, PhD
University of Colorado at Boulder

John L. Rittenhouse
Eastern Mennonite College

Felix Rivera, MD
South Nassau Communities Hospital

Connie Rizzo, MD, PhD
Columbia University

Jeffrey B. Roberts, MD
Duke University Medical Center

Larry M. Roberts, JD
Pasadena, California

James L. Robinson, PhD
University of Illinois, Urbana-Champaign

Hilda Velez Rodriguez, MS
San Juan, Puerto Rico

Ana Maria Rodriguez-Rojas, MS
GXP Medical Writing, LLC

Linda Roethel, MD
South Nassau Family Medicine

Charles W. Rogers, PhD
Southwestern Oklahoma State University

Eugene J. Rogers, MD
Chicago Medical School

Laurie Rosenblum, MPH
Education Development Center, Massachusetts

John Alan Ross, PhD
Eastern Washington University

Lynne T. Roy
Cedars-Sinai Medical Center, Los Angeles

Nadja Rozovsky, PhD
Somerville, Massachusetts

Claudia Daileader Ruland, MA
Johns Hopkins University

Irene Struthers Rush
Boise, Idaho

Virginia L. Salmon
Northeast State Community College

Pamela Rose V. Samonte
Villanova University

Susan L. Sandel, PhD
MidState Behavioral Health System

Robert Sandlin, PhD
San Diego State University

Sarit Sandowski, OMS-II
New York Institute of Technology

Alexander Sandra, MD
University of Iowa, Carver College of Medicine

Tulsi B. Saral, PhD
University of Houston, Clear Lake

Andrea Sarchi, OMS-IV
New York Institute of Technology

Lisa M. Sardinia, PhD
Pacific University

David K. Saunders, PhD
Emporia State University

Jacob S. Sawyer, PhM
Columbia University

Zachary Sax
South Nassau Communities Hospital

Panagiotis D. Scarlatos
Florida Atlantic University

Elaine M. Schaefer, DO
South Nassau Communities Hospital

Elizabeth D. Schafer, PhD
Loachapoka, Alabama

Rosemary Scheirer, EdD
Chestnut Hill College

David M. Schlom
Independent Scholar

Steven A. Schonefeld, PhD
Tri-State University

Kathleen Schongar, MS, MA
The May School

John Richard Schrock, PhD
Emporia State University

Tanja Schub, BS
Binghampton University

Jay D. Schvaneveldt, PhD
Utah State University

Jason J. Schwartz, PhD, JD
Los Angeles, California

Marie Schwartz
Yale University

Miriam E. Schwartz, MD, MA, PhD
University of California, Los Angeles

Tom E. Scola
University of Wisconsin, Whitewater

Rebecca Lovell Scott, PhD, PA-C
Sandwich, Massachusetts

Rose Secrest
Chattanooga, Tennessee

John J. Seidl, MD
Medical College of Wisconsin

Carol A. Selden, RD
Allegiance Health, Jackson, Michigan

Sibani Sengupta, PhD
American Medical Writers Association

Thomas L. Sevier, MD, FACSM, FACP
Performance Dynamics, Inc., Astym Program

Gregory B. Seymann, MD
University of California, San Diego School of Medicine

Frank E. Shafer, MD
Allegheny University of the Health Sciences

Uzma Shahzad, MD
South Nassau Communities Hospital

Christopher D. Sharp, MD
Stanford University School of Medicine

Holly K. Shaw, PhD, RN
Adelphi University

John M. Shaw
Education Systems

Richard Sheposh
Penn State University

Martha Sherwood, PhD
University of Oregon

George C. Shields, PhD
Lake Forest College

Lisa J. Shientag, VMD
University of Massachusetts

Sonia Shubert
Yale University

R. Baird Shuman, PhD
University of Illinois, Urbana-Champaign

Mel Siegel, MA
Fordham University

Sanford S. Singer, PhD
University of Dayton

Virginia Slaughter, PhD
University of Queensland, Australia

Jane A. Slezak, PhD
Fulton Montgomery Community College

Julie M. Slocum, RN, MS, CDE
Women and Infants' Hospital

Genevieve Slomski, PhD
New Britain, Connecticut

Caroline M. Small
Silver Spring, Maryland

Dwight G. Smith, PhD
Southern Connecticut State University

H. David Smith, PhD
University of Michigan

Jane Marie Smith
Butler County Community College

Roger Smith, PhD
Portland, Oregon

Lisa Levin Sobczak, RNC
Santa Barbara, California

Angela Spano
Huntingdon College

Nancy Sprague
Independent Scholar

Carly Stabb
Yale University

Claire L. Standen, PhD
University of Massachusetts Medical School

Sharon W. Stark, RN, APRN, DNSc
Monmouth University

William D. Stark, DDS
Monrovia, California

Toby R. Stewart, PhD
University of Nebraska Medical Center, Omaha

Glenn Ellen Starr Stilling, MA, MLS
Appalachian State University

Bruce L. Stinchcomb
St. Louis Community College

James R. Stubbs, MD
University of South Alabama

Wendy L. Stuhldreher, PhD, RD
Slippery Rock University

Giri Sulur, PhD
University of California, Los Angeles

Pavel Svilenov
Wisconsin Lutheran College

Rena Christina Tabata
Think Tank Innovations

Steven R. Talbot, RVT
University of Utah Vascular Laboratory

Sue Tarjan
Santa Cruz, California

Billie M. Taylor, MSE, MLS
Arkadelphia, Arkansas

William F. Taylor
Detroit, Michigan

Gerald T. Terlep, PhD
Wayne State University

Bethany Thivierge, MPH, ELS
Technicality Resources

Susan E. Thomas, MLS
Indiana University South Bend

Nicole Thomasian
Brown University

Roberta Tierney, MSN, JD, APN
Indiana University

Venkat Raghavan Tirumala, MD, MHA
Western Kentucky University

Leslie V. Tischauser, PhD
Prairie State College

Winona Tse, MD
Mount Sinai School of Medicine

Mary S. Tyler, PhD
University of Maine

Jane Ungvarsky
Independent Scholar

John V. Urbas, PhD
Kennesaw State College

Maxine M. Urton, PhD
Xavier University

Rhea U. Vallente, PhD
University of Washington

Nicole M. Van Hoey, PharmD
Arlington, Virginia

Anju Varanasi, MD
South Nassau Communities Hospital

Mikhail Varshavski, OMS-IV
New York Institute of Technology

Charles L. Vigue, PhD
University of New Haven

James Waddell, PhD
University of Minnesota

Peter J. Waddell, PhD
University of South Carolina

Anthony J. Wagner, PhD
Medical University of South Carolina

Edith K. Wallace, PhD
Heartland Community College

C. J Walsh, PhD
Mote Marine Laboratory

Peter J. Walsh, PhD
Fairleigh Dickinson University

Marc H. Walters, MD
Portland Community College

Annita Marie Ward, EdD
Salem-Teikyo University

John F. Ward, MD
Navy Medical Center, San Diego

Marcia Watson-Whitmyre, PhD
University of Delaware

Judith Weinblatt, MA, MS
New York, New York

Marcia J. Weiss, MA, JD
Point Park College

Barry A. Weissman, OD, PhD
University of California, Los Angeles, School of Medicine

David J. Wells, Jr., PhD
University of South Alabama Medical Center

John B. Welsh, MD, PhD
San Diego, CA

Zhongqi Weng
Yale University

Mark Wengrovitz, MD
Pennsylvania State University, Hershey Medical Center

Althea Williams, MBA
South Nassau Communities Hospital

Beth Williams
Yale University

Lee Williams
New York City, New York

Russell Williams, MSW
University of Arkansas for Medical Sciences

S.M. Willis, MS, MA
Independent Scholar

Lindsey L. Wilner, PsyD
Gansevoort, New York

Bradley R. A. Wilson, PhD
University of Cincinnati

Michael Windelspecht, PhD
Appalachian State University

Barbara Woldin
American Medical Writers Association

Stephen L. Wolfe, PhD
University of California, Davis

Bonnie L. Wolff
Pacific Coast Cardiac and Vascular Surgeons

Debra Wood, RN
Orlando, FL

Paul Y. K. Wu, MD
University of Southern California

Robin L. Wulffson, MD
American Medical Writers Association

Geetha Yadav, PhD
Bio-Rad Laboratories Inc.

Lynda J.-S. Yang, MD, PD, FAANS
University of Michigan

Daniel L. Yazak, DED
Montana State University, Billings

Claudine Yee
Brown University

Jay R. Yett
Independent Scholar

Garry Young, PhD
Nottingham Trent University

Rachel Zahn, MD
Solana Beach, California

Kathleen Zanolli, PhD
University of Kansas

W. Michael Zawada, PhD
University of Colorado Health Sciences Center

Mark Zelman, PhD
Aurora University

George D. Zgourides, MD, PsyD
John Peter Smith Hospital

Ming Y. Zheng, PhD
Gordon College

Susan M. Zneimer, PhD
US Labs

Nillofur Zobairi, PhD
Southern Illinois University

The Microbes

In this section, we survey the representatives of microbes and the main methods to view them. Microbial diversity is the range of different kinds of unicellular organisms, including bacteria, archaea, protists, and fungi. Various microbes thrive throughout distinct strata of the biosphere. The resident microbes in any given biosphere often define the limits of life and create conditions conducive to the survival and evolution of other living beings. Entries include algae, eukaryotes and prokaryotes, and bacteria.

Microbes

Microbial Methods

Microbes

Algae

Category: Microbes

Specialties and related fields: Botany, ecology; evolutionary biology, limnology, marine biology; phycology, taxonomy

Definition: a varied and multifaceted group of simple, nonflowering, and typically aquatic plants that includes seaweeds and many single-celled forms; however, they lack true stems, roots, leaves, and vascular tissue

KEY TERMS

chloroplast: any chlorophyll-containing organelle in plant and algal cells

clade: a group of organisms that probably evolved from a common ancestor

endosymbiosis: a type of symbiotic relationship between two organisms in one organism lives inside the cells of the other organism, usually in a mutualistic relationship that benefits both organisms

flagellum (pl: *flagella*): relatively long, delicate, whiplike structures on the surface of cells, used to drive motion

heterotrophic: organisms that cannot produce their food but must acquire it from other organic carbon sources, such as bacterial, plant, or animal matter

photoautotrophy: the ability of algae and other organisms to derive energy directly from sunlight through the process of photosynthesis and assimilate carbon from atmospheric carbon dioxide

prokaryotes: microorganisms that lack internal compartmentalization (i.e., their cells lack an organized nucleus and other internal organelles)

ALGAL CLASSIFICATION

The study of algae is known as "phycology" ("phyco" from the ancient Greek word *phûkos*, meaning "seaweed"). In the past, biologists thought algae were lower plants because some forms looked like plants. As in plants, the primary photosynthetic pigment in algae is chlorophyll a, and oxygen is produced during photosynthesis. Detailed studies of algae's variety and complex evolutionary history drove scientists to move algae from the Plantae kingdom (in the Eukarya domain) and into the Protist kingdom. Comparative molecular studies of algal nuclear and chloroplast genomes have recently established that algae have much closer affinities to plants than any other group of organisms. Therefore, algae were regrouped once again with plants.

Cyanobacteria often referred to as blue-green algae, are not a type of algae but bacteria. However, cyanobacteria are photosynthetic and occupy many of the same ecological niches as algae.

All green algae and plants are grouped in a clade called "Viridiplantae." This clade contains about 450,000-500,000 species that play integral roles in terrestrial and aquatic ecosystems. The Viridiplantae is split into two lineages: Chlorophyta and Streptophyta. The Streptophyta consists of the charophyte green algae from which the land plants evolved. Members of the Viridiplantae originated from organisms that engulfed cyanobacteria. Instead of ingesting these photosynthetic bacteria, they became, over time, an integrated part of the cell. The ingested cyanobacteria became modern chloroplasts. This handily explains why chloroplasts possess bacterial genomes and ribosomes that bear a remarkable resemblance to those from cyanobacteria. This explanation of the evolutionary origin of plants is known as the "endosymbiont theory."

Other groups of algae result from "secondary endosymbiosis," in which green or red algae were taken up by other organisms and integrated into those cells. Several modern organisms have endosymbionts that were recently acquired and are in

A variety of algae growing on the sea bed in shallow waters. Photo by Toby Hudson, via Wikimedia Commons.

the process of becoming organelles. For example, the freshwater flagellate *Cyanophora paradoxa* has an internal organelle called a "cyanelle" that consists of cyanobacteria that still have their cell walls and most of their bacterial genomes. Another example of endosymbiosis in progress is the fungus *Geosiphon pyriforme*, that has an internal cyanobacterium, *Nostoc*, that also fixes atmospheric nitrogen.

THE RANGE OF ALGAL LIFE

Algae can be found nearly everywhere on earth: oceans, rivers, lakes, in the snow of mountaintops, on forest and desert soils, on rocks, on plants and animals (such as within the hollow hair of polar bears living in a zoo and the pelts of sloths in the wilds), or even on other algae. They are involved in diverse interactions with other organisms, including symbiosis, parasitism, and epiphytism (growing while attached to another organism).

Lichens are symbiotic associations between cyanobacteria or green algae and fungi. Atmospheric nitrogen-fixing cyanobacteria occur in symbiotic associations with plants such as bryophytes, water ferns, gymnosperms (such as cycads), and angiosperms. The aquatic fern *Azolla*, commonly used as a biofertilizer in rice fields in Asian countries, harbors the symbiotic cyanobacterium *Anabaena azollae*. *Gunnera*, the only flowering plant to house symbiotic cyanobacterium *Nostoc*, is widely distributed in the tropics.

Symbiotic dinoflagellates known as "zooxanthellae" live within the tissues of corals. Corals get their colors and obtain energy from their photosynthetic symbionts. About 15 percent of red algae occur as parasites of other red algae. Parasitic algae may even transfer nuclei into host cells and transform them. After transformation, the reproductive cells of the host algae carry the parasite's genes. Various algae live on the surfaces of plants and other algae as epiphytes. Sometimes algae can be found in strange places—the pink color of flamingos, for example, comes from a pigment in the algae these birds consume.

ALGAL STRUCTURE AND PROPERTIES

A cell wall bounds algal cells. Cyanobacteria, which used to be considered prokaryotic algae, lacks both a nucleus and complex membrane-bound organelles, namely chloroplasts and mitochondria. Photosynthesis occurs in cyanobacteria in thylakoid membranes like those of plants. However, there is no double membrane surrounding the thylakoids of cyanobacteria.

All algal groups are eukaryotic. Eukaryotic algae differ from cyanobacteria in that they possess chloroplasts and flagella with associated structures and their cell wall composition. According to the endosymbiont hypothesis, some eukaryotic algae (red and green algae) obtained their chloroplasts by acquiring symbiotic prokaryotic cyanobacteria. This is known as "primary endosymbiosis." Other eukaryotic algae probably obtained their chloroplasts by taking up eukaryotic endosymbiotic algae, a process known as "secondary endosymbiosis." The existence of secondary endosymbiosis is indicated by the occurrence of more than two membranes around the chloroplasts of some algae, such as haptophytes, euglenophytes, dinoflagellates and cryptomonads.

Pigments found in algae include chlorophylls, phycobilins, and carotenoids. All algae contain chlorophyll a. Accessory pigments vary among different algal groups. For example, green algae also contain chlorophyll b.

Photoautotrophy is the principal mode of nutrition in algae; in other words, they are "self-feeders," using light energy and a photosynthetic apparatus to produce their food (organic carbon) from carbon dioxide and water. Most algal groups contain heterotrophic species, which obtain their organic food molecules by consuming other organisms. Numerous algae are mixotrophs; that is, they use different modes of nutrition (such as autotrophy and heterotrophy), depending on the availability of resources. The molecules used as food reserves differ among and are characteristic for algal groups. Food reserve molecules are polymers of glucose with different links between monomers.

Many algae are capable of movement. Movement is accomplished through flagellar action and by extrusion of mucilage. There is also peristaltic and amoeba-like algal movement. Within algal cells, movement of the cytoplasm, plastids, and nucleus also occurs. Advantages conferred by mobility include achieving optimal light conditions for photosynthesis, avoiding damage caused by excess light, and obtaining inorganic nutrients.

ALGAL REPRODUCTION AND LIFE CYCLES

Algae may reproduce either asexually or sexually. Asexual reproduction among algae includes the production of unicellular spores that germinate without fusing with other cells, fragmentation of filamentous forms, and cell division by splitting. In sexual reproduction, parent cells release gametes, which then fuse to form a zygote. Zygotes may either develop into new filaments or produce haploid spores by meiotic division.

Algae exhibit different types of life cycles. Some algal life cycles are characterized by an alternation of generations like that of plants. Two phases occur: sporophyte (usually diploid) and gametophyte (usually haploid). The sporophyte produces haploid spores through meiosis, and the haploid gametophyte produces male or female gametes by mitosis. Gametophyte and sporophyte may be structurally identical or dissimilar, depending on the algal group.

ECOLOGICAL ROLES OF ALGAE

Algae and the semirelated cyanobacteria have played significant roles in earth's ecosystems since the origin of cyanobacteria more than three billion years ago. Early cyanobacteria were responsible for developing significant amounts of free oxygen in the atmosphere, which then made aerobic respiration possible. More than 70 percent of all photosynthetic activity on earth is carried out by phytoplankton— floating microscopic algae—rather than plants. Phytoplankton recharge the atmosphere with oxygen and simultaneously absorb carbon dioxide, helping to support the complex web of aquatic biota.

Algae are also essential in the global cycling of other elements, such as carbon, nitrogen, phosphorus, and silicon. Several algal groups—such as green algae, red algae, and the haptophyte algae—can generate calcium carbonate. Sedimented algae are the major contributors to deep-sea carbonate deposits (sand), covering about half of the world's ocean floor. Calcified coralline red algae contribute to coral reefs in tropical waters. Silica sediments in oceans (sand) are based on the abundant growth of another algal group, the diatoms, which contain silica in their cell walls.

Some cyanobacteria can fix atmospheric nitrogen and convert it to ammonia. Ammonia, in turn, can be a nitrogen source for plants and animals. On the other hand, high levels of nitrogen and phosphorus in rivers and lakes owing to pollution can cause the rapid and uncontrollable growth of algae, known as "algal blooms." An algal bloom threatens human and marine health, both directly and indirectly. It clogs fishes' gills, interferes with water filters, and ruins recreation sites. More than 50 percent of algal blooms produce toxins. Cases of human respiratory, skin and gastrointestinal disorders associated with algal toxins have been reported. Specific algal blooms are called "red tides." The water appears to be red or brown because of the color of algal bodies, mainly dinoflagellates that contain the pigment xanthophyll.

TECHNOLOGICAL APPLICATIONS

Algae have been used as food, medicine, and fertilizer for centuries. The earliest known reference to algae as food occurs in Chinese poetic literature dated about 600 BCE. More recently, algae have played important roles in specific biotechnological processes, including algaculture.

Red, brown, and green algae and cyanobacteria are used for food in Pacific and Asian countries, especially Japan. The annual harvest of the red alga *Porphyra* worldwide is worth several billion dollars. *Porphyra* (laver, Japanese *nori*, Chinese *zicai*) is used as a wrapper for sushi or eaten alone. It is also sometimes added to miso soup, a traditional Japanese soup made from a miso paste, a fermented mixture of soybeans, sea salt, and rice koji. Another edible alga with a high iodine content is the brown alga *Laminaria* (kelp, Japanese *kombu*). The cyanobacterium *Spirulina*, with a protein level of 50 to 70 percent, was cultivated for centuries by indigenous Central Americans at Lake Texcoco near modern-day Mexico City for use as human food. Its modern-day use began in the late twentieth century, and it has become a popular dietary supplement.

Several gelling agents are produced from red and brown algae. Agar from red algae is used as a medium for culturing microorganisms, including algae, as a food gel, and pharmaceutical capsules. Red algal carrageenan is used in toothpaste, cosmetics, and food such as ice cream and chocolate milk. Alginates from brown algae have extensive applications in the cosmetics, soap, and detergent industries. Sources of alginates are *Laminaria*, some *Fucus* species, and the giant kelp *Macrocystis*, which can grow to more than 60 meters long. Algae are also used as feed in the culture of commercially important fish and shrimp.

Mass cultivation of algae (microalgae)—in open ponds and photobioreactors to produce fuels (such as biomass) and biochemicals (such as carotenoids, amino acids, and carbohydrates) and for water purification—is a rapidly developing area based on the use of solar energy as an energy source. The green alga *Dunaliella* is used in the industrial production of carotene. In wastewater treatment plants, algae are used to remove nutrients and heavy metals and add oxygen to the water.

Algaculture, or algae farming, has been considered a replacement for when land is not suitable for agriculture. Algal fuel (also called "algal biofuel") may include biodiesel, biogasoline, bioethanol, biomethanol, biobutanol, and vegetable oil. The production of algae is expensive, but algal fuel can produce more biomass per unit each year than any other type of biomass. Its yield has been reported to be between ten and one hundred times more than other biofuel crops. Algae release carbon dioxide when burned, but this is offset by the growth of algal and other biofuels that take carbon dioxide out of the atmosphere. Algae can also be grown in freshwater, ocean saltwater, and even in wastewater. Algaculture has minimal impact on water resources and is considered harmless to the environment if spilled.

Algae are used worldwide as indicators (biomonitors) of water quality, helping detect the presence of toxic compounds in water samples. Several fast-growing algae are used, including the green alga *Selenastrum capricornutum*. Many algae are widely employed as research tools because they are easy to culture and manipulate. Danish biologist Joachim Hammerling's experiments with the green alga *Acetabularia* identified the nucleus as the likely storage site of genetic information.

DIVERSITY

Estimates for how many species of algae exist vary widely. Taxonomists believe that there could be anywhere between thirty thousand to one million species of algae, or possibly ten million or more. Molecular comparisons using nucleotide sequences in ribosomal RNA (ribonucleic acid) suggest that algae do not fall within a single group linked by a common ancestor but evolved independently. Algae are divided into approximately six supergroup affiliations, depending on their endosymbiont. Other features important for classification include their photosynthetic pigments, food reserves, cell structure, and reproduction. These groups include euglenoids, cryptomonads, dinoflagellates, haptophytes, and red algae, among others.

The green algae are a large, informal group of algae that consists of the Chlorophyta, an ancient, morphologically and ecologically diverse algal lineage that includes three major classes: Ulvophyceae, Trebouxiophyceae, and Chlorophyceae and other more primitive algal groups. This group includes unicellular and colonial cells with two whiplike flagella, filamentous and spherical "coccoid" forms, and larger seaweeds. Green algae have chlorophyll a and b, the accessory pigments beta-carotene (red-orange), and yellow xanthophylls. Their cells walls are composed of cellulose, and their storage carbohydrate is starch. There are about 22,000 species of green algae. The original endosymbionts of the Chlorophyta were cyanobacteria.

The Rhodophyta, or the red algae, has between 4,000 and 6,000 species. Red algae lack any flagellated stages. The photosynthetic pigments include chlorophyll a as well as accessory phycobilins and carotenoids. Two membranes surround each chloroplast. The food reserve is a floridean starch. A red algal cell is encircled by a wall composed of cellulose. Asexual and sexual reproduction, as well as alteration of generations, are widespread among *Rhodophyta*. A triphasic life cycle is unique for this group of algae. The original endosymbiont of the Rhodophyta were cyanobacteria.

The Euglenophyta contains mostly unicellular forms with one or two flagella. Only one-third of this group possess chlorophyll-containing chloroplasts. Other euglenoids are strictly heterotrophic. The phylum contains more than 900 primarily freshwater species. The food reserve is the carbohydrate paramylon, a polymer of glucose. Euglenophytes have chlorophyll a and b as well as carotenoids as their photosynthetic pigments. There is no cell wall. Cells have several small chloroplasts; three membranes surround each. A relative of euglenophytes is the protozoan *Trypanosoma*, which causes the human disease African sleeping sickness. Reproduction in the euglenophytes occurs by the mitotic division of cells. The original endosymbiont of euglenoids was green algae.

A group related to the euglenoids are the Chlorarachniophytes, a small group of marine algae with wide distribution in tropical and temperate waters. These amoebae contain endogenous green algae that have become integrated into the rest of the cell. Their chloroplasts, however, contain a remnant of the green algal endosymbiont known as a "nucleomorph." Chlorarachniophytes are typically mixotrophic and ingest bacteria and smaller protozoa in addition to conducting photosynthesis.

The Cryptophyta includes unicellular, biflagellate cryptomonads. The original endosymbiont of the cryptomonads was red algae. There are about 220 species of cryptomonads that occur in freshwater, marine, and brackish water environments. In addition to chlorophyll a, chloroplasts can contain chlorophyll c, carotenoids, and phycobilins. The carotenoid pigment alloxanthin is unique to Cryptophytes. Four membranes surround each chloroplast. Chloroplast endoplasmic reticulum borders the chloroplasts. The principal food reserve is starch. Instead of a typical cell wall, a periplast composed of protein plates occurs beneath the cell membrane. Reproduction is primarily asexual. A distinguishing feature of cryptomonads is an "extrusome." This subcellular structure has two connected spiral ribbons held under tension that deploys if the cryptomonad is irritated by mechanical, chemical, or light stress. Once discharged, the extrusome propels the cell in a zig-zag course away from the source of the disturbance.

Members of the Dinophyta, or dinoflagellates, have unicellular forms with two different flagella. There are an estimated two thousand to four thousand marine species and about 200 freshwater forms. Many have chlorophylls a and c, as well as the unique carotenoid peridinin. Some members of Dinophyta have fucoxanthin. Chloroplasts have three closely associated membranes. The primary food reserve is starch, but lipids are also important storage molecules. A dinoflagellate cell is not surrounded by a cell wall but has a theca (a sort of armor) made of cellulose. Dinoflagellates can reproduce asexually and sexually. The original endosymbiont of dinoflagellates was red algae.

The Haptophyta includes primarily marine unicellular biflagellated algae. A haptophyte cell also has a flagellum-like haptonema, used to capture food. There are about 300 species. The photosynthetic pigments include chlorophyll a, and accessory pigments chlorophyll c, and the carotenoid fucoxanthin. Each chloroplast has four membranes. The food reserve is a polymer of glucose called "chrysolaminarin." Several layers of scales, or coccoliths, composed primarily of calcium carbonate may cover the haptophyte cell. Asexual and sexual reproduction is widespread.

The Heterokonta is a vast group of photosynthetic and nonphotosynthetic organisms. The paired flagella in these organisms are different sizes, causing the scientist Alexander Luther in 1899 to refer to them as "Heterokontae" or unequal flagella. Many nonphotosynthetic Heterokonta have genes from photosynthetic organisms in their genomes, suggesting that these organisms secondarily lost their ability to perform photosynthesis (e.g., Oomycetes, the water molds). Many algae in this group have chloroplasts surrounded by four membranes: the two membranes are continuous with the endoplasmic reticulum of the chloroplast, and the inner two membranes are those of the chloroplast. The chloroplast has endoplasmic reticular membranes. This is evidence that the original endosymbiont of the Heterokonts was a eukaryotic organism, and in this case, a red alga. The Heterokonta include the diatoms, brown algae, golden algae, yellow-green algae, and several other lesser-known algal groups.

The Phaeophytes or brown algae contain most of the main kelps or seaweeds. Phaeophytes serve vital ecological roles in marine environments as food sources for animals and significant consumers of atmospheric carbon dioxide. Phaeophytes include such well-known species as *Laminaria*, *Macrocystis*, which may grow to lengths of 200 feet, and *Fucus*, which grows on rocks along the seashore. The photosynthetic pigments of Phaeophytes include chlorophyll a and chlorophyll c, ß-carotene, fucoxanthin, violaxanthin, diatoxanthin, and other xanthophylls. Interestingly, brown algae have an excess of carotenoid over chlorophyll pigments. Their storage molecular is a glucose polymer called "laminarin."

The Xanthophyta or yellow-green algae are closely related to the Phaeophyta and live in marine and soil environments. These algae consist of single-celled flagellated and colonial forms and filamentous organisms. Their photosynthetic pigments include chlorophyll a and c, ß-carotene, and the carotenoid diadinoxanthin. Their storage polysaccharide is chrysolaminarin, a linear polymer of glucose linked by ß-linkages.

The Chrysophytes or golden algae live mainly in freshwater. Ecologists use Chrysophytes to assess the health of freshwater ecosystems in areas where acid rain degrades the environment. The golden algae have a characteristic siliceous cyst within their cells called a "statospore" that can appear smooth, ornamented, round, or ellipsoid. Most Chrysophytes are unicellular flagellates, but some are amoeboid, and others are nonmotile. The golden algae use chlorophyll a and c and fucoxanthin as their photosynthetic pigments. The storage polymer of Chrysophytes is chrysolaminarin.

—Sergei A. Markov and Michael A. Buratovich, PhD

Further Reading

Barsanti, Laura, and Paolo Gualtieri. *Algae: Anatomy, Biochemistry, and Biotechnology*. 2nd ed., CRC Press, 2014.

Bellinger, Edward G., and David C. Singee. *Freshwater Algae: Identification and Use as Bioindicators*. Wiley, 2010.

Graham, Linda E., and Lee W. Wilcox. *Algae*. 2nd ed., Benjamin Cummings, 2009.

Lee, Robert Edward. *Phycology*. 5th ed., Cambridge UP, 2018.

Pereira, Leonel, and Joao Magalhaes Neto, editors. *Marine Algae: Biodiversity, Taxonomy, Environmental Assessment, and Biotechnology*. CRC Press, 2014.

Phillips, Julie A. *The Lives of Seaweeds: A Natural History of Our Planet's Seaweeds and Other Algae*. Princeton UP, 2022.

Piganeau, Gwenaël. *Genomic Insights into the Biology of Algae*. Academic, 2012.

Raven, Peter H., Ray F. Evert, and Susan E. Eichhorn. *Biology of Plants*. 8th ed., W. H. Freeman, 2013.

Spellman, Frank R., and Melissa L. Stoudt. *Environmental Science: Principles and Practices*. Scarecrow, 2013.

Archaea

Category: Prokaryotic diversity

Also known as: Archaebacteria

Anatomy or system affected: Gastrointestinal tract, skin

Specialties and related fields: Bacteriology, biochemistry, biotechnology, microbial ecology, microbiology, molecular biology, taxonomy

Definition: a group of highly diverse prokaryotic organisms distinct from the historically familiar bacteria that have molecular properties previously thought to occur only in eukaryotes and others commonly associated with bacteria; many archaea require extreme conditions for growth, and their genetic processes have adapted to these extreme conditions in ways that are not fully understood

KEY TERMS

conjugation: the process by which one bacterial cell transfers deoxyribonucleic acid (DNA) directly to another

domain: the highest-level division of life, sometimes called a "superkingdom"

extreme halophiles: microorganisms that require extremely high salt concentrations for optimal growth

insertion sequence: a small, independently transposable genetic element

methanogens: microorganisms that derive energy from the production of methane

prokaryotes: unicellular organisms with simple ultrastructures lacking nuclei and other intracellular organelles

small subunit ribosomal RNA (ssu rRNA): the ribonucleic acid (RNA) molecule found in the small subunit of the ribosome; also called 16S rRNA (in prokaryotes) or 18S rRNA (in eukaryotes)

GENE SEQUENCES MEASURE THE DIVERSITY OF PROKARYOTES

Prokaryotic microorganisms have been on earth for as many as 3.5 billion years. They have diverged tremendously in genetic and metabolic terms. Unfortunately, the magnitude of this divergence has made it difficult to measure the relatedness of prokaryotes to one another. In the 1970s, Carl R. Woese addressed this problem by reading short sequences of ribonucleotides from a highly conserved ribonucleic acid (RNA) molecule, the small subunit ribosomal RNA (ssu rRNA) or 16S rRNA (18S rRNA in eukaryotes). Because this RNA is present in all organisms and has sections that evolve very slowly, others that evolve less slowly, and nine hypervariable regions. It is an ideal molecule for determining the degree of relatedness between two microorganisms. The proportion of shared sequences thus provide a quantitative index of similarity by which all cellular organisms could, in principle, be compared.

When Woese and others used the nucleotide sequence data to construct an evolutionary tree, eukaryotes (plants, animals, fungi, and protozoa) formed a cluster separated from the common bacteria. Unexpectedly, however, a third cluster emerged that was equally distinct from both eukaryotes and common bacteria. This cluster consisted of prokaryotes that lacked biochemical features of most bacteria (such as a cell wall composed of peptidoglycan); possessed other features not found in any other organisms (such as membranes composed of isoprenoid ether lipids); and occurred in unusual, typically harsh, environments. Woese and his coworkers eventually designated the three divisions of life represented by these clusters, or "domains," naming the nonbacterial prokaryotes the domain *Archaea*.

ARCHAEA CLASSIFICATION

The Archaea differ from the Bacteria (also known as the Eubacteria) in several significant ways. First, the majority of the Bacteria have cell walls composed of peptidoglycan, but the Archaea lack peptidoglycan in their cell walls. Second, the rRNAs of the Archaea differ distinctly from those of Bacteria. Third, the Archaea have unusual cofactors not found in members of the Bacteria.

Archaea classification is based on the enormous quantities of 16S rRNA sequence data compiled from them. Unfortunately, the riches of sequence data have generated substantial disagreement as well. Bacterial taxonomists agree that the Archaea are divided into a group called "DPANN" and two other groups called the "Euryarchaeota" and the "Proteoarchaeota."

DPANN contains ultrasmall archaea or nanoarchaea. The name DPANN is an acronym made from the first letters of the first five groups discovered: *Diapherotrites*, *Parvarchaeota*, *Aenigmarchaeota*, *Nanoarchaeota*, and *Nanohaloarchaeota*. Besides being tiny compared to other Archaea, many members of the DPANN have small genomes devoid of genes for primary biosynthetic pathways. Consequently, many DPANN members depend on other microbes and live as episymbionts with other microorganisms. These organisms live in oxygen-free condition (anaerobes) and extreme environments such as high salt, heat, acid, or metal concentrations.

The Euryarchaeota are incredibly diverse and include the methanogens, the halobacteria, and the extreme thermophiles. Methanogens metabolize carbon dioxide and hydrogen gas to form methane gas. They are one of the most metabolically unique organisms on the earth. Halogens live and survive at extreme concentrations of salt. The extreme thermophiles live at temperatures between 41° to 122° Celsius.

The Proteoarchaeota are the Archaea most closely related to eukaryotes. This wide-ranging group includes the Korarchaeota that live in hydrothermal environments, the Crenarchaeota that live in marine environments, high temperature and sulfur-dependent environments, and, among others, the Asgardarchaeota, which have the high similarity to eukaryotes.

THE GENETIC MACHINERY OF ARCHAEA

Because bacteria and eukaryotes differ significantly concerning gene and chromosome structure and gene expression details, molecular biologists have examined the same properties in archaea and have found a mixture of "bacterial" and "eukaryotic" features. The organization of deoxyribonucleic acid (DNA) within archaeal cells is bacterial because archaeal chromosomes are circular DNAs of between 2 million and 4 million base pairs having single origins of replication, typically replicated bidirectionally. As in bacteria, the genes are densely packed and often grouped into clusters of related genes transcribed from a common promoter called "operons." However, the promoters themselves resemble the TATA box/BRE (B recognition element) combination of eukaryotic DNA polymerase II (Pol II) promoters. The RNA polymerases have the complex subunit composition of eukaryotes rather than the simple composition found in bacteria. Furthermore, archaea initiate transcription by a simplified version of the process seen in eukaryotic cells. Transcription factors (TATA-binding protein and a TFIIB) first bind to regions ahead of the promoter, then recruit RNA polymerase to attach and begin transcription. However, introns are rare in archaea and do not interrupt protein-encoding genes but have been found to interrupt RNA-coding genes. Also, the regulation of transcription in archaea seems to depend heavily on the types of repressor and activator proteins found in bacteria; however, regulatory proteins of the eukaryotic type, and those unique to archaea, have also been found.

GENOMES OF ARCHAEA

The availability of complete DNA sequences now enables archaeal genomes to be compared to the genomes of bacteria and eukaryotes. One pattern that emerges from these comparisons is that most

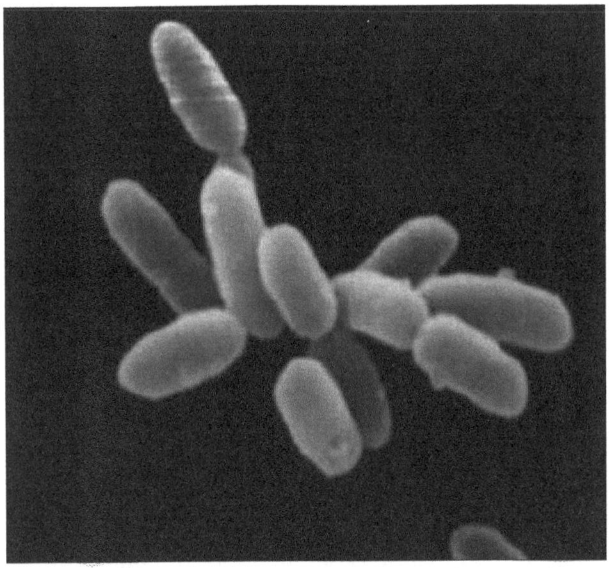

Cluster of cells of Halobacterium sp. *strain NRC-1. Image via NASA/Wikimedia Commons. [Public domain.]*

archaeal genes responsible for processing genetic information (synthesis of DNA, RNA, and proteins) resemble their eukaryotic counterparts. In contrast, most archaeal genes for metabolic functions (biosynthetic pathways, for example) resemble their bacterial counterparts. The genomes of archaea also reveal probable cases of gene acquisition from distant relatives, a process called "lateral gene transfer."

A third pattern to emerge from genome comparisons is that some archaea are missing genes thought to be essential. For example, at least two methanogenic archaea genomes do not encode an enzyme that charges transfer RNA (tRNA) with cysteine. These archaea instead use a novel strategy for making cysteinyl tRNA. Some of the seryl tRNA made by these cells is converted to cysteinyl tRNA by a specialized enzyme. Nanoarchaeum equitans provides a more extreme example of gene deficiency, the first reported parasitic or symbiotic archaea that grow attached to an *Ignicoccus*, another hyperthermophile. *N. equitans* has been reported to have a volume approximating 1 percent of an *Escherichia coli* cell, the smallest nonviral cellular genome (0.49 Mbp), and numerous 16S rRNA nucleotide base substitutions even in regions normally conserved in other archaeal species. This extremely small genome lacks genes necessary for numerous metabolic functions, including genes coding for lipid, amino acid, nucleotide, and enzyme cofactor biosynthesis. This includes genes coding for vital catabolic pathways, including glycolysis. It has been suggested but not proven that its Ignicoccus host supplies these functions. Even more intriguing is the much longer list of archaea— all of which happen to be hyperthermophiles, which grow optimally at 80° C (176° Fahrenheit) or above— that lack genes for the DNA mismatch repair proteins found in all other organisms.

UNIQUE GENETIC PROPERTIES

This last observation raises an important question: Has an evolutionary history distinct from all conventional genetic systems, combined with the unique demands of life in unusual environments, resulting in unique genetic properties in archaea? Although basic genetic assays can be performed in only a few species, the results help identify which genetic properties of cellular organisms are genuinely universal and which may have unusual features in archaea.

The methanogen *Methanococcus voltae* transfers short chromosome pieces from one cell to another, using particles that resemble bacterial viruses (bacteriophages). This means of gene transfer has been seen in only a few bacteria. In other methanogens, researchers have used more conventional genetic phenomena, such as antibiotic-resistance genes, plasmids, and transposable elements, to develop tools for cloning or inactivating genes. As a result, new details about the regulation of gene expression in archaea and the genetics of methane formation are now coming to light.

The extreme halophile *Halobacterium salinarum* exhibits extremely high rates of spontaneous mutation of the genes for its photosynthetic pigments and gas vacuoles. This genetic instability reflects the fact that insertion sequences transpose very frequently into these and other genes. A distantly related species, *Haloferax volcanii*, can transfer chromosomal genes through conjugation. Although many bacteria engage in conjugation, the mechanism used by *H. volcanii* does not resemble the typical bacterial system since no plasmid seems to be involved. There is no apparent distinction between donor strain and recipient strain in the transfer of DNA.

Genetic tools for the archaea from geothermal environments are less well developed. Still, specific selections have made it possible to study spontaneous mutation and homologous recombination in some species of *Sulfolobus*. At the normal growth temperatures of these aerobic archaea, 75 to 80° C (167 to 176° F), spontaneous chemical decomposition of DNA is calculated to be about one thousand times more frequent than in the organisms previously studied by geneticists. Despite this, *Sulfolobus acidocaldarius* exhibits the same spontaneous mutation frequency as *E. coli* and significantly lower proportions of base-pair substitutions and deletions. This indicates effective mechanisms for avoiding or accurately repairing DNA damage, including mismatched bases, even though no mismatch repair genes have been found in *Sulfolobus* species. Also, *S. acidocaldarius*, like *H. volcanii*, has a mechanism of conjugation that does not require a plasmid or distinct donor and recipient genotypes. The transferred DNA recombines efficiently into the resident chromosome, as indicated by frequent recombination between mutations spaced only a few base pairs apart.

Finding two similar and unusual conjugation mechanisms in two different and distantly related archaea (*H. volcanii* and *S. acidocaldarius*) raises questions regarding the possible advantages of this capability. Population genetic theory predicts that organisms that reproduce clonally (as bacteria and archaea do) would benefit from occasional exchange and recombination of genes because this accelerates the production of beneficial combinations of alleles. Such recombination may be significant for archaea such as *Haloferax* and *Sulfolobus* species, whose extreme environments are like islands separated by vast areas that cannot support growth. For these organisms, frequent DNA transfer between cells of the same species may provide an efficient way to enhance genetic diversity within small, isolated populations.

IMPACT

Woese's monumental discovery that two very different prokaryotic groups (bacteria and archaea) exist based on DNA sequencing of a conserved macromolecule (rRNA) led to a complete reevaluation of the evolution of not only bacteria (previously including archaea) but also eukaryotes. Before, biologists thought that eukaryotes evolved from prokaryotes. His data definitively showed that all three were derived from a common ancestor. Even more surprising was the fact that all eukaryotes were more closely related to each other and distantly related to both bacteria and archaea. Finally, his data showed that archaea were more closely related to eukaryotes than bacteria and were the most ancient organisms derived first from the common ancestor.

Since then, the DNA sequences of numerous archaeal isolates of different groups have been compared to each other and those of bacteria and eukaryotes, further delineating the evolution of different diverse archaeal groups. These data suggest that a hyperthermophile was probably the common ancestor of *Archaea*. Also, the properties of numerous types of archaea have been studied. Many are found in extremely harsh environments that normal bacteria and eukaryotes cannot tolerate. This has led to molecular and genetic studies of the macromolecules, including proteins and lipids that allow the survival of these organisms so that their mechanisms can be elucidated. Finally, various DNA polymerases naturally found in these organisms have been employed in modern DNA analysis techniques that require enzymes to be heat-stable, thus providing a practical result of studying these important organisms.

—Dennis W. Grogan, PhD, Steven A. Kuhl, PhD, and Michael A. Buratovich, PhD

Further Reading

Albers, Sonja, et al. "Archaea." *Essentials of Glycobiology*, edited by Ajit Varki et al., 3rd ed., Cold Spring Harbor Laboratory Press, 2017, pp. 283-292, doi:10.1101/glycobiology.3e.022.

Barker, David M. *Archaea: Salt-Lovers, Methane-Makers, Thermophiles, and Other Archaeans*. Crabtree Pub. Co., 2010.

Cavicchioii, Richard, editor. *Archaea: Molecular and Cellular Biology*. ASM Press, 2007.

Garrett, Roger A., and Hans-Peter Klenk, editors. *Archaea: Evolution, Physiology, and Molecular Biology*. Wiley-Blackwell, 2008.

Madigan, Michael T., and John M. Martinko. *Brock Biology of Micro-organisms*. 11th ed., Prentice-Hall, 2006.

Olsen, Gary, and Carl R. Woese. "Archaeal Genomics: An Overview." *Cell*, vol. 89, 1997, pp. 991-94.
Woese, Carl R. "Archaebacteria." *Scientific American*, vol. 244, 1981, pp. 98-122.

EUKARYOTES

Category: Biology
Anatomy or system affected: Cells
Specialties and related fields: Microbiology, evolutionary biology
Definition: organisms whose constituent cells are characterized by the presence of internal compartments

KEY TERMS

chloroplast: the organelle in eukaryotes in which photosynthesis is performed by algae and green plants
deoxyribonucleic acid (DNA): a stable molecule that contains most of the genetic information of a cell
endosymbiotic theory: the concept that eukaryotes arose from prokaryotes by incorporating free-living microbes into symbiotic relationships
eukaryotic cell: a cell that contains a nucleus and other membrane-bounded organelles
mitochondrion: the eukaryotic organelle in which energy is generated by aerobic respiration
organelles: subcellular membrane-bounded compartments that perform specific functions within the eukaryotic cell
Precambrian: the interval of geologic time from the formation of the earth (4.6 billion years ago) to the beginning of the Cambrian period (544 million years ago)
prokaryotic cell: a cell that does not contain a nucleus or other membrane-bounded organelles

CHARACTERISTICS

All the commonly seen organisms on the earth are eukaryotes, or organisms built up of eukaryotic cells. These organisms evolved from a prokaryotic ancestor and developed over the last 1.4 billion years into highly diverse and successful groups of invertebrates, fish, amphibians, reptiles, and mammals. Today, eukaryotes live in a vast array of different environments in almost all areas of the earth's surface.

Eukaryotes are organisms whose constituent cells are characterized by a nucleus and other membrane-bounded organelles such as mitochondria and, in plants, chloroplasts. Prokaryotic cells do not have a nucleus or membrane-bounded organelles. Biologists now recognize that the most significant discontinuity in life is not between animals and plants but between the prokaryotes and the eukaryotes. All organisms on the earth except viruses, bacteria, archaea, and cyanobacteria (blue-green algae) are eukaryotes. The vast majority of fossil species of the last 700 million years have also been eukaryotes. The prokaryotes dominated the fossil record of the Precambrian, and the transition from prokaryotic to eukaryotic cells represents one of the most extraordinary steps in the development of life on the earth. This transition probably occurred between 2 billion years ago and 1.4 billion years ago, leading to the evolution of complex multicellular plants and animals. The evolution of highly successful groups such as fish, reptiles, and mammals depended on the formation and elaboration of the eukaryotic cell.

The primary distinctive trait of the eukaryotic cell is the nucleus, an organelle that houses deoxyribonucleic acid (DNA). In prokaryotic cells, the DNA molecules are arranged in a single loose strand within the cell's cytoplasm. Prokaryotes generally reproduce by simple splitting (binary fission). In contrast, the reproduction of cells and organisms is much more highly organized in eukaryotes. Eukaryotic cells reproduce by the complicated processes of mitosis and meiosis. In mitosis, the DNA is copied, and each new cell receives an exact copy of the original DNA. However, in meiosis, a second division of the genetic information occurs, forming sperm and egg cells. These sex cells from separate organisms then fuse to form offspring with a combination of genetic information from each parent. This sexual reproduction, which is not found in prokaryotes, has led to a high level of diversity among the eukaryotes.

Eukaryotic cells also contain other organelles that perform specific metabolic functions more tightly organized than they occur in prokaryotes. Mitochondria are small organelles found in almost all eukaryotic cells and are the sites of aerobic respiration. Carbon-rich molecules are broken down into smaller molecules during aerobic respiration, and the cell releases and captures chemical energy. This process requires oxygen. The amount of energy released in aerobic respiration is much greater than the energy released by fermentation, which occurs in prokaryotes.

Algae and green-plant cells also have chloroplasts, organelles that are the sites of photosynthesis. During photosynthesis, chemical energy in the form of adenosine triphosphate (ATP) is harvested by photosystems, with the release of oxygen (from the water molecule. That energy is used to fix carbon dioxide into aldehyde-containing compounds used to make sugars. Several prokaryotes are photosynthetic, but photosynthesis occurs in extensions of the cell membrane in prokaryotes, not chloroplasts as in eukaryotes. Eukaryotic photosynthesis is efficient enough to supply most of the cellular carbon for all other eukaryotes on the earth. Many other organelles are found within eukaryotic cells. They perform such functions as waste removal, transport of materials, oxidation of unusual lipids, and cell movement.

The advantage organelles provide to eukaryotes is the spatially ordered arrangement of sequential biochemical reactions that increases the efficiency of metabolic processes. This increased efficiency permitted the development of larger and more complex multicellular organisms. Although a few prokaryotes are loosely assembled into multicellular organisms, it is the eukaryotes that defined separation of functions between cells.

EUKARYOTE FOSSILS

The origins of eukaryotes are largely unknown as a result of the notorious selectivity of the fossil record. Advanced eukaryotes such as clams, fish, birds, and mammals have shells, bones, or teeth commonly preserved in sedimentary rocks. These eukaryotes have left behind a decipherable if sporadic, fossil record over the last 700 million years. The older single-cell eukaryotes were rarely fossilized, and the evidence for the first eukaryotes probably will never be found. However, the eukaryotes arose after the first accumulation of significant amounts of oxygen in the atmosphere about 2 billion years ago and the appearance of the first multicellular animals about 680 million years ago.

However rare an occurrence, single cells are occasionally fossilized in the rocks. The oldest fossils are from sedimentary rocks in the Pilbara Shield of Australia. They are dated at about 3.5 billion years old. All these fossils and all the other Archean and early Proterozoic fossils are prokaryotes, showing that only prokaryotes lived during the first two-thirds of the earth's history.

Fossil eukaryotic cells are rarely preserved with intact organelles, but they may differ from prokaryotes by size. Eukaryotic cells range in size from 5 microns to 1 millimeter (one micron equals 0.001 millimeters). In contrast, prokaryotes are generally 1 to 20 microns in diameter. Relatively large fossilized cells have been found in the Precambrian rocks in the Death Valley region of California and Australia. These rocks have been dated at about 1.4 billion years old. Although some paleontologists are not convinced that these fossils are eukaryotes, a statistical analysis of more than eight thousand fossil cell sizes from Precambrian rocks throughout the world shows that the eukaryotic stage of evolution had probably been achieved by about 1.75 billion years ago. Large

Eukaryotes and some examples of their diversity. Clockwise from top left: Red mason bee, Boletus edulis, *chimpanzee,* Isotricha intestinalis, Ranunculus asiaticus, *and* Volvox carteri. *Photos by Karwath/Hillewaert/Nedelcu/Logan/Stridvall/Dostál, via Wikimedia Commons.*

sheets of algae have been found in some of the sedimentary rocks of the Northwest Territory of Canada. They have been dated at about 1 billion years old. Almost undoubtedly, eukaryotic cells have been found in the 900-million-year-old rocks of Australia. It appears that eukaryotes developed in the interval between 2 billion years, and the appearance of definite eukaryotes at about 1 billion years ago, or about 1.75 billion years ago. How complex eukaryotic cells arose from a prokaryotic ancestor could not be recorded in the fossil record, and theories attempting to describe the evolution of the early eukaryotes must be based upon the biochemical relationships between present-day organisms.

EVOLUTIONARY THEORIES

Two general theories have been proposed to explain the evolution of eukaryotes. The traditional view, direct filiation, suggests that the nucleus, mitochondria, chloroplasts, and other organelles arose by mutations with the prokaryotes. Although most mutations are deleterious, some may have been beneficial to the ancestral prokaryotes and were retained within the organisms. Through a sequential accumulation of mutations, each organelle developed over a long interval in the Precambrian eon. This view is supported by the very complex interrelationships between the organelles within the eukaryotic cell.

A competing theory was proposed in the early twentieth century by the Russian biologist R. C. Mereschkowsky. It was revised by some modern biologists, particularly Lynn Margulis of Boston University. This theory, known as the endosymbiotic theory, suggests that mitochondria and chloroplasts were sequentially incorporated into symbiotic relationships within the ancestral prokaryotes. The mitochondria were, in this theory, originally free-living aerobic bacteria that arose after the initial oxygenation of the atmosphere. These bacteria were capable of normal living processes. Still, they invaded a large prokaryotic cell and continued to live and respire within the host cell. Both the invader and the host cell benefited from this arrangement. The host provided food to the invader in exchange for some of the energy derived from that food. In addition, the invader may have supplied some protection from oxygen to the host cell. The two continued to live together and, over a long time in the later Precambrian, became closely related until they grew utterly dependent upon each other. The same general scenario is used to explain the origin of chloroplasts from cyanobacteria.

There is biochemical evidence to support the endosymbiotic theory as interpreted by Margulis and others. First, separate and different DNA genomes are in mitochondria and chloroplasts. These genomes are circular molecules, as in bacteria, and the organization of their genes into operons is unmistakably bacterial. Second, mitochondrial and chloroplast DNAs are not complexed with histone proteins, as is the nuclear DNA. Second, some mitochondria can replicate independently of the main cell. Finally, the proteins and ribonucleic acids (RNA)s encoded by mitochondrial genomes are more closely related to bacterial proteins and RNAs from the alpha-proteobacteria than any other group of organisms. Likewise, the proteins and RNAs encoded by chloroplast genomes resemble those of blue-green bacteria more closely than any other group.

However, the earliest eukaryotic cells arose, they rapidly diversified. This diversification in the eukaryotes is in the arrangement and the organization of the cells, not in the internal biochemistry of the cells, which remains remarkably consistent throughout all the eukaryotes. The eukaryotes joined together and differentiated into tissues, organs, and systems to form multicellular organisms. Soft-bodied metazoans were present 680 million years ago, and all the major phyla of animals were present 500 million years ago. The eukaryotes have developed different forms, abilities, and behaviors over the last 500 million years. They can live in most of the environments of the world. Eukaryotes are found in polar to tropical regions, from the depths of the oceans to the tops of mountain ranges, and from rain forests to deserts. It is estimated that there are between 3 and 10 million living species of eukaryotes today and perhaps one hundred to one thousand times as many eukaryotic species that have lived in the geologic past. All those organisms are based upon the eukaryotic cell that developed about 1.75 billion years ago from some prokaryotic ancestor.

—*Jay R. Yett*

Further Reading

Gray, Michael W. "Lynn Margulis and the Endosymbiont Hypothesis: 50 Years Later." *Molecular Biology of the Cell*, vol. 28, no.10, 2017, pp. 1285-87, doi:10.1091/mbc.E16-07-0509.

Gunde-Cimerman, Nina, Aharon Oren, and Ana Plemenita. *Adaptation to Life at High Salt Concentrations in Archaea, Bacteria, and Eukarya*. Springer, 2011.

Margulis, Lynn, and Michael J. Chapman. *Kingdoms and Domains: An Illustrated Guide to the Phyla of Life on Earth*. 4th ed., Elsevier, 2009.

McMenamin, Mark. *Discovering the First Complex Life: The Garden of the Ediacara*. Columbia UP, 1998.

Ruggiero, Michael A., et al. "A Higher Level Classification of All Living Organisms." *PloS One*, vol. 10, no. 4, 2015, p. e0119248, doi:10.1371/journal.pone.0119248.

Scheckenbach, Frank, et al. "Large-Scale Patterns in Biodiversity of Microbial Eukaryotes from the Abyssal Sea Floor." *Proceedings of the National Academy of Sciences of the United States of America*, vol. 107, 2010, pp. 115-20.

Schopf, J. William. *Major Events in the History of Life*. Jones and Bartlett, 1992.

Starr, Cecie, et al. *Biology: The Unity and Diversity of Life*. 15th ed., Cengage Learning, 2018.

Wernegreen, Jennifer J. "In It for the Long Haul: Evolutionary Consequences of Persistent Endosymbiosis." *Current Opinion in Genetics & Development*, vol. 47, 2017, pp. 83-90, doi:10.1016/j.gde.2017.08.006.

Yaacov, Davidov, and Eduard Jurkevitch. "Predation Between Prokaryotes and the Origin of Eukaryotes." *BioEssays*, vol. 31, 2009, pp. 738-57.

Prokaryotes

Category: Biology
Anatomy or system affected: None
Specialties and related fields: Bacteriology, biology, biotechnology, cell biology, environmental biotechnology, evolutionary biology, geology, microbiology
Definition: microscopic single-celled organisms that have neither a distinct nucleus with a membrane nor other specialized organelles

KEY TERMS

archaea: a group of single-celled prokaryotic organisms with distinct molecular characteristics separating them from bacteria and eukaryotes

bacteria: ubiquitous, mostly free-living organisms often consisting of one biological cell that constitute a large domain of prokaryotic microorganisms; typically, a few micrometers in length, bacteria were among the first life-forms to appear on Earth and are present in most of its habitat

Monera: a biological kingdom that is made up of prokaryotes

Prokaryotes are primitive, one-celled organisms that have left an extensive fossil record in sedimentary structures produced by the physiological activity of cell communities. For 80 percent of the Earth's history, communities of prokaryotes made up the biosphere of the Earth. They are a well-defined group of organisms and occupy a highly diverse variety of habitats.

CHARACTERISTICS OF PROKARYOTES

From about 3.5 to 1 billion years ago, life on Earth, as determined from the fossil record, consisted entirely of one-celled organisms with cell morphology and a metabolism different from other life-forms. These organisms, the prokaryotes, are characterized by their lack of a cell nucleus, lack of sexual reproduction (meiosis), the small size of the prokaryotic cell, and distinctive biochemistry. Prokaryotes are neither plants nor animals, although the aerobic photosynthetic forms, often called "blue-green algae", have been placed with the plants in the past. Eukaryotes, organisms with a cell nucleus and a larger, more complex cell, make up animals, plants, fungi, and protists. Prokaryotes are thus quite separate from all other life-forms in terms of their cell biology. Prokaryotes and eukaryotes are the two most basic categories of living things, exhibiting basic differences in their biological processes that are greater than those between animals or plants or between any of the other kingdoms.

Prokaryotes are mainly single-celled organisms, usually found living together in "colonies" consisting of immense numbers of cells. Their deoxyribonucleic acid (DNA) is distributed throughout the cell; it is not, as in the case of the eukaryotes, localized in a cell nucleus surrounded by a nuclear membrane. The prokaryotic cell is ten times smaller than the average eukaryotic cell. It lacks chloroplasts and mitochondria and consequently is considered primitive when compared with the eukaryotic cell.

Prokaryotes constitute the kingdom Monera, one of five kingdoms in modern taxonomy. (The other kingdoms are the protists, fungi, animals, and plants, which have more complex eukaryotic-type cells.) Phyla, or categories, within the kingdom Monera include the bacteria, the cyanobacteria (or blue-green algae), the archaebacteria, and the prochlorophytes. The bacteria, as well as the other moneran phyla, are further subdivided into several classes. Bacterial classes of the Monera include the eubacteria, photosynthetic bacteria, myxobacteria (slime bacteria), actinomycetes (mold-like bacteria), and other groups, each characterized by its unique metabolism and biochemistry. The bacteria consist of obligate or strict anaerobes and facultative anaerobes; the former include the photosynthetic bacteria, which differ from the cyanobacteria in their ability to function, if re-

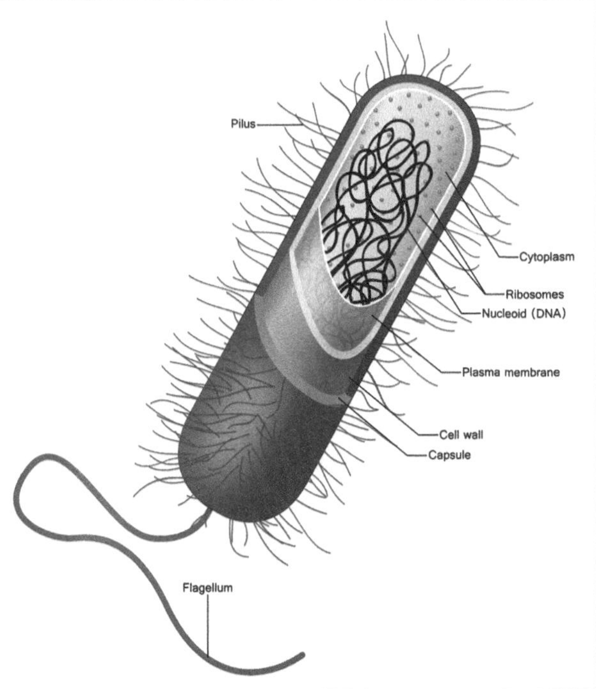

Prokaryotes are microscopic organisms that have been on Earth for millions of years. They differ from eukaryotes in that they are very simple and do not have a nucleus to enclose DNA or specialized cell organelles. Image by Ali Zifan, own work, via Wikimedia Commons.

quired under anaerobic conditions and low light levels, but also in their different photosynthetic pigment.

The archaebacteria are considered to be the most primitive and ancient of the monerans. Archaebacteria have many biochemical and metabolic characteristics that allow them to live under very adverse conditions—conditions such as those that appear to have existed during the Earth's early history. The archaebacteria are defined from their ribosomal ribonucleic acid (RNA), which in sequencing is quite different from all other monerans. The archaebacteria differ fundamentally from the other bacteria classes in structural and biochemical aspects.

GEOLOGICAL SIGNIFICANCE

Fossil prokaryote cells of great antiquity have become widely known from the fossil record. They were first reported in the 1910s by C. D. Walcott from 1.5-billion-year-old strata of western Montana (Belt series). However, the authenticity of these fossils was doubted until the discovery, in 1954, of one of the oldest known paleontological "windows" on the life of the past, the 2-billion-year-old Gunflint biota. Since then, many prokaryote cell fossils have been reported, most from very fine-textured flinty cherts associated with stromatolites of the Proterozoic (latter part of the Precambrian) eon and dating from as far back as 3.5 billion years.

The geologic significance of the prokaryotes is great: They (at present and in the geologic past) play an important part in the recycling of many chemical elements, but they also have a role in basic geologic processes as weathering and other rock alterations. For example, prokaryotes are involved in the formation of stromatolites. Stromatolites are layered organosedimentary structures, frequently found fossilized in rock strata of many different geologic ages. Stromatolites come in various shapes and sizes; the

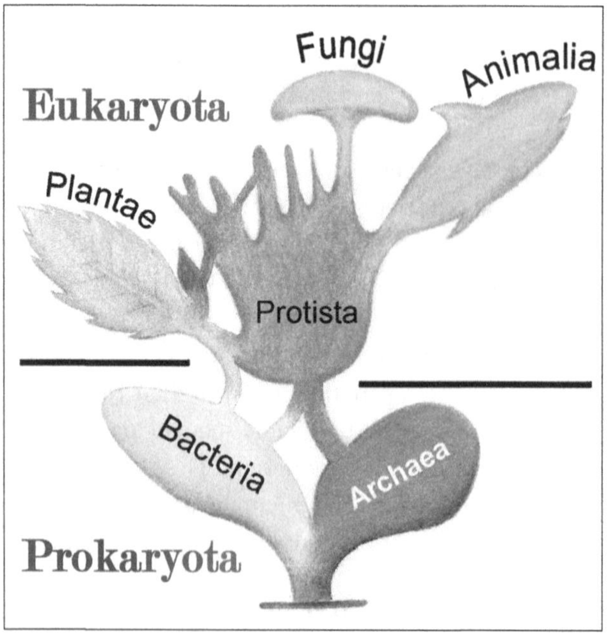

Symbiogenetic and phylogenetic diagram of living organisms, showing the origins of eukaryotes and prokaryotes. Image by Maulucioni y Doridí, own work, via Wikimedia Commons.

different types have often been given Linnaean biological names. They were thought to be fossil organisms like corals or sponges when originally discovered. Most stromatolites are dome-shaped, finger-shaped, or laminar structures with a characteristic "signature." They can form significant parts of rock strata, particularly in limestone and dolomites. Stromatolites are found in rock strata as ancient as the Archean (former part of the Precambrian) eon. They are particularly diverse and abundant in strata of the Proterozoic. Locally, they can be quite common in early Paleozoic marine strata.

The origin of stromatolites was debated for many years; as late as the 1950s, many paleontologists seriously doubted their biogenic origin. This doubt stemmed, at least in part, from the fact that stromatolites occur so much further in the geologic past than do any other fossils. Through thousands and thousands of meters of Precambrian strata, they are the only fossils that can be found. Early workers on stromatolites, such as Walcott, suggested a cyanobacterial origin for them. The discovery of the well-preserved cells of the prokaryotic type in digitate (fingerlike) stromatolites of the Gunflint Chert of Ontario in the 1950s led to a gradual acceptance by most geologists and paleontologists of the organic origin of most stromatolites. It became clear that, under the right conditions, small, fragile cells could be preserved in very ancient strata.

During the 1970s and 1980s, studies on Precambrian stromatolites and the prokaryotic organisms responsible for them became widespread. Stromatolite occurrences going as far back as 3.5 billion years have been documented. These ancient stromatolites yield morphological information and carbon isotope ratios indicative of a biogenic origin. They sometimes supply biochemical information in the form of hydrocarbons, amino acids, and porphyrins (the latter is a degradation product of original photosynthetic pigment). In 1999, studies of Proterozoic rocks from western Australia confirmed the presence of chemicals that could only have been synthesized by cyanobacteria.

The morphology of a prokaryotic organism is simple. Unlike fossils of eukaryotic organisms, fossils of most prokaryotes provide little specific information about the actual living organism. Prokaryotic cells can be single coccoid (spherical) forms, or they can be long chains of cells, like the filaments or trichomes of the cyanobacteria.

STUDY OF PROKARYOTES

A standard petrographic thin section mounted on a glass slide is the standard mount for observing cells preserved in a stromatolite. Oil immersion is usually required if fossil prokaryote cells are to be observed. Thin slivers of a stromatolite can also be examined under oil immersion; however, the best results are generally with well-made thin sections. Often considerable trial and error are involved in finding stromatolites that preserve cells and then, in actuality locating those cells; different parts of a particular stromatolite specimen usually have varying degrees of cell preservation. Very fine-grained sediments, such as those that occur with stromatolites preserved by black cherts or finely crystalline limestones, generally give the best results.

In this section, a stromatolite may exhibit fossil cyanobacterial cells as either filaments or rod-shaped forms under high optical magnification. If preservation of these small prokaryote cells is excellent, as in the stromatolites of the Gunflint Formation, the biogenic origin of the cells will be clear, and distinct cell types can be observed. When most stromatolites are examined in thin sections under high magnification, however, the biogenicity of the small objects seen is usually not so certain. Often, small black globules of carbon, suggestive of macerated cells, are evident, but their origin usually cannot be proved. Contaminants such as spores, pollen grains, bacteria cells, and fungi fragments can be a problem, particularly in examining suspected fossiliferous rocks when thin sections are not used. Even with most thin sections, the unequivocal verification of a biogenic origin for fossil cells is rare. In the case of the Gunflint prokaryotes, the detail preserved in these fossil cells is highly remarkable; some of them show internal cell structure and cells in the process of division.

The earliest stromatolites that yield these fossil cells are generally broad domes or laminar forms. Associated cells either are single-cell coccoid forms or consist of probable chains of photosynthetic bacteria. Chains of cells of filamentous cyanobacteria generally first appeared about 2.3 billion years ago. This appearance of filaments agrees fairly well with the first appearance of branched or digitate stromatolites, for which filamentous cyanobacteria seem to be responsible.

CHEMICAL SIGNATURES

Often more significant than single-cell morphology or the megascopic morphology of a stromatolite is the chemical signature left by a group of prokaryotes due to their metabolic activity. Prokaryotes are classified according to their type of metabolism; some prokaryotes have a metabolism that enables them to occupy a wider variety of ecological niches than do eukaryotic organisms. Anaerobic and aerobic forms are the two fundamental forms of prokaryotic metabolism. These two categories are the autotrophs and the heterotrophs; heterotrophic prokaryotes require previously formed organic material on which to live, while autotrophs do not. The autotrophs obtain their energy from their environment either in sunlight (photoautotrophs) or through chemical reactions such as oxidation, as in the sulfur-oxidizing bacteria; such bacteria are called "chemoautotrophs." This type of metabolism is unique in the organic world. All other life-forms obtain their energy from photosynthesis or by utilizing the chemical energy contained in previously formed organic compounds. The cyanobacteria are photoautotrophs and are responsible for forming the various types of stromatolites.

The process of photosynthesis changes the microenvironment around the photosynthesizing prokaryote; the mineral precipitation that results is responsible for the formation of stromatolite layers.

Some stromatolites contain oxidized manganese, cobalt, or other "transitional" elements, possibly incorporated into these fossil communities by the oxidative metabolism of bacteria. Chemoautotrophic prokaryotes, which are various types of bacteria, may leave a chemical signature in the form of these oxides and precipitate their production of a layered stromatolite-like structure containing these oxidized metals. Several bacteria oxidize manganese to higher oxidation states so that it is precipitated; deep-sea manganese nodules presently being formed are believed to have such an origin. Sectioning of these nodules shows a finely layered, stromatolite-like structure. Some of the heavy-metal-bearing stromatolites of the early Precambrian may reflect a similar chemoautotrophic metabolism. Analysis of organic residues present in many stromatolites in small quantities can sometimes shed light upon the specific organisms responsible for forming them. This technique, however, has met with only limited success, although degradation products of the photosynthetic pigment present in cyanobacteria have been identified, supporting the cyanobacterial origin of many ancient stromatolites.

The earliest stromatolites from the Archean age (about 3.5 billion years old) exhibit certain distinctive morphological and chemical aspects. Some of these early stromatolites may be products of photosynthetic bacteria's anaerobic assemblages rather than cyanobacteria communities. Geochemical evidence suggests that the atmosphere in the Archean may have been anoxygenic (oxygen-free). Not being obligate aerobes, the photosynthetic bacteria would have been favored by such an environment.

—*Bruce L. Stinchcomb*

Further Reading

Broadhead, T. W., editor. *Fossil Prokaryotes and Protists: Notes for a Short Course.* U of Tennessee, 1987.

Fedonkin, Mikhail A., James G. Gehling, Kathleen Grey, et al. *The Rise of Animals: Evolution and Diversification of the Kingdom Animalia.* Johns Hopkins UP, 2007.

Gunde-Cimerman, Nina, Aharon Oren, and Ana Plemenita. *Adaptation to Life at High Salt Concentrations in Archaea, Bacteria, and Eukarya.* Springer, 2011.

Margulis, Lynn, and Michael J. Chapman. *Kingdoms and Domains: An Illustrated Guide to the Phyla of Life on Earth.* 4th ed., Elsevier, 2009.

McMenamin, Mark. *Discovering the First Complex Life: The Garden of the Ediacara.* Columbia UP, 1998.

Nisbet, Evan G. *The Young Earth: An Introduction to Archean Geology.* Unwin Hyman, 1987.

Schopf, J. William, editor. *Major Events in the History of Life.* Jones and Bartlett, 1992.

Yaacov, Davidov, and Eduard Jurkevitch. "Predation Between Prokaryotes and the Origin of Eukaryotes." *BioEssays,* vol. 31, 2009, pp. 738-57.

BACTERIA

Category: Biology

Also known as: Eubacteria

Anatomy or system affected: Anus, gastrointestinal tract, oral cavity, pharynx, skin, teeth

Specialties and related fields: Bacteriology, microbiology

Definition: single-celled organisms lacking a nucleus, found in and on humans and widespread in the environment

KEY TERMS

capsule: an outer layer common to many bacteria, made of polysaccharides outside the cell envelope. It is regarded as part of the outer envelope of bacterial cells, is a well-organized layer, not easily washed off, and contributes to disease causation in many pathogenic bacteria

gram-negative: a group of bacteria that have a thin peptidoglycan cell wall with an outer membrane extender to it, and stain pink with a Gram stain
gram-positive: a group of bacteria that have a thick peptidoglycan cell wall without an outer membrane and stain purple with a Gram stain
lipopolysaccharide: a complex sugar-lipid molecule situated in the outer leaf of the outer membrane of gram-negative bacteria
pathogenic: able to cause disease in living organisms
peptidoglycan: a complex sugar-amino acid polymer that is the main component of bacterial cell walls

SIGNIFICANCE

Bacteria are ubiquitous on Earth, and some species can cause disease in humans. An understanding of the classification of bacteria and how bacterial populations grow and reproduce is helpful in the identification, diagnosis, and treatment of bacterial diseases.

The tiny unicellular organisms known as bacteria define the biosphere on Earth. That is, if bacteria do not inhabit a particular environment, no living things reside there. Bacteria are incredibly adaptable and have managed to exploit a wide variety of habitats successfully. One niche exploited by bacteria is the human body. Humans support a population of more than two hundred species of bacteria in numbers greater than the cells that make up an individual human host. These members of the normal flora are found on the skin and in the digestive, urinary, reproductive, and upper respiratory tracts of humans.

Although some species of bacteria can cause disease in humans, other animals, and plants, most bacterial species are not pathogenic (disease-causing). Bacteria are vital players in the ecology of the Earth, functioning in essential roles in global chemical cycles. Perhaps most importantly, bacteria are the only organisms on Earth that can fix nitrogen—that is, to convert the nitrogen gas in the atmosphere to a form usable by other organisms.

Disease-causing bacteria have attracted the most interest and study since the confirmation of the germ

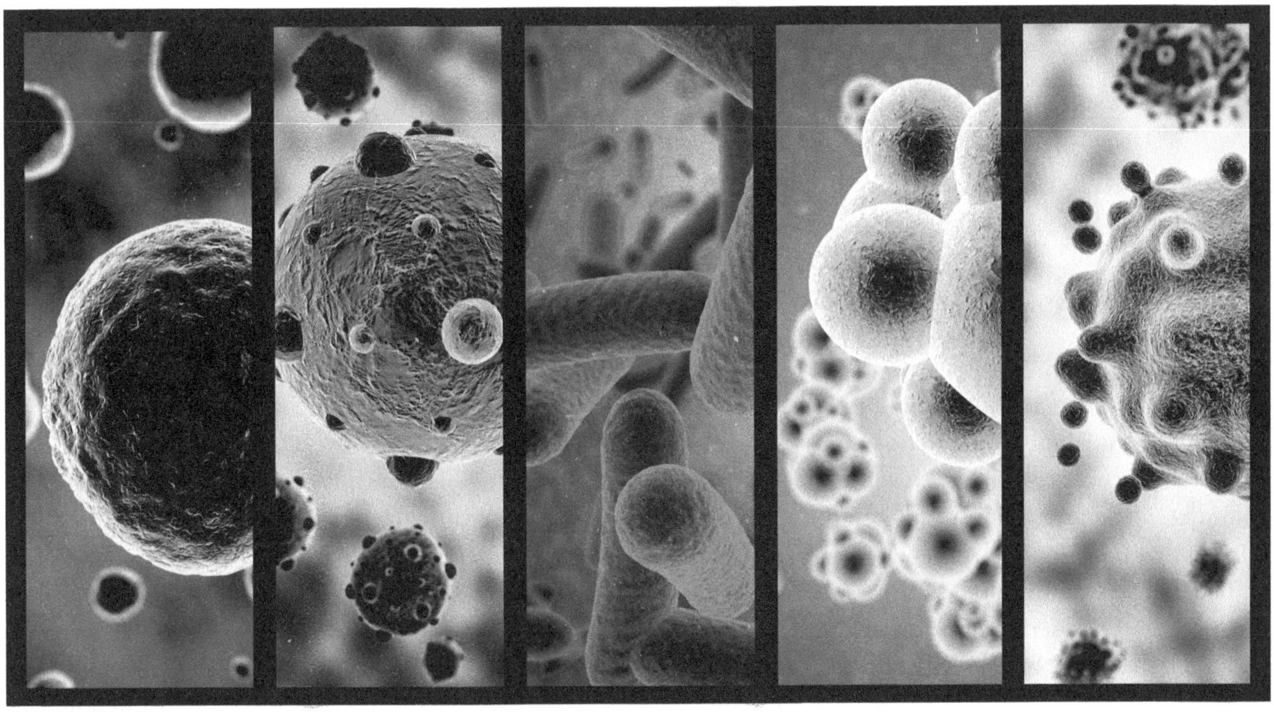

Various microorganisms. Image via iStock/Yuri_Arcurs. [Used under license.]

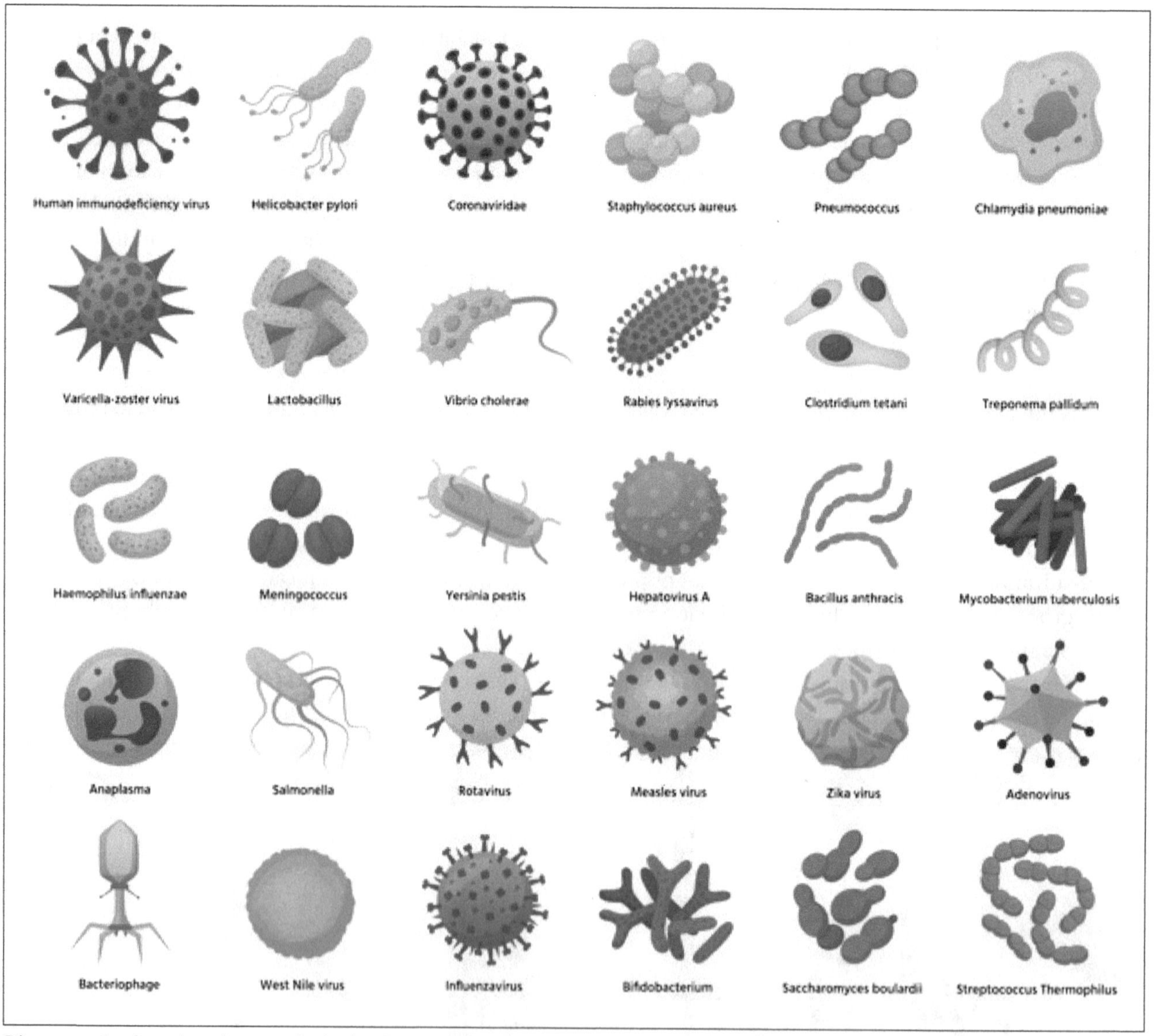

Disease-causing bacterias, viruses and microbes. Image via iStock/Tetiana Lazunova. [Used under license.]

theory of disease by Louis Pasteur and Robert Koch in the 1870s. It is interesting to note that Koch's proof that germs cause disease involved the bacterium *Bacillus anthracis*, which causes anthrax, an organism that has been used as a biological weapon.

The first sixty years of the study of medical bacteriology focused on identification and diagnosis, with little attention to the basic biology of bacteria. The discovery and development of antibiotics led to an overly optimistic view that modern medicine had conquered infectious diseases. The emergence of antibiotic-resistant strains of bacteria and outbreaks of previously unknown pathogens stimulated a renewed interest in bacteriology.

CLASSIFICATION AND TAXONOMY

Bacteria are classified as prokaryotic cells—that is, the genetic material of a bacterium is not enclosed in a nucleus. This lack of a nucleus distinguishes bacterial cells from eukaryotic cells that compose plants and animals. Additional differences between bacterial and eukaryotic cells include the types of molecules found in the cell walls, organization and expression of genes, and sensitivity to certain antibiotics.

The most basic division of bacteria relies upon the structure of their cell walls. One group of bacteria has a thick cell wall consisting of a carbohydrate-amino acid polymer called "peptidoglycan" outside their cell membrane. These bacteria live on the skin, fur, and clothing, indoor environments, and in soil. Because members of this bacterial group stain purple in a Gram stain, they are called "gram-positive." A second, highly diverse group of bacteria has a cell wall with a thin peptidoglycan layer outside the cell membrane and an outer membrane external to the peptidoglycan layer. The outer membrane has a molecule called "lipopolysaccharide" in its outer leaf. A lipopolysaccharide is a complicated molecule with a carbohydrate polymer attached to multiple fatty acids. Because these bacteria stain pink with a Gram stain, they are called "gram-negative."

The International Committee on Systematics of Prokaryotes (ICSP) is the organization that oversees the nomenclature of prokaryotes and issues opinions concerning related taxonomic matters. When a researcher discovers a previously undescribed bacterium, the ICSP must approve the researcher's proposed name for the newly described species and the taxonomic classification of the species.

Bacteria are presently distinguished from the Archaea, which are also prokaryotes but have basic molecular biology more akin to eukaryotes than the bacteria. Additionally, Archaea have cell walls and membrane lipids, unlike those of the bacteria. The bacteria are divided into the Terrabacteria, the Gracilicutes, an odd group of tiny cells with tiny genomes called the "Ultramicrobacteria," and the "Thermotogae."

The Thermotogae contains gram-negative, anaerobic bacteria that thrive at high temperatures and metabolize complex carbohydrates to produce hydrogen gas. Ultramicrobacteria are small than a tenth of a micron, which is ten times smaller than most bacteria. These organisms are parasites on larger organisms, have minimal genomes, and were initially isolated from soil.

The Gracilicutes contains most gram-negative bacteria. Within the Gracilicutes are the Proteobacteria (α, ß, γ, δ, and ε proteobacteria), the acidobacteria, *Chlamydia*/Planctomycetes bacteria, the FCB group (Fibrobacteres, Chlorobi, and Bacteroidetes), the spirochaetes, and the fusobacteria.

The Terrabacteria contains all the gram-positive bacteria, the cyanobacteria (blue-green bacteria), and a wonderfully diverse bacterial group called the "Chloroflexi." Some of the Chloroflexi are heat-loving organisms, others are anaerobic photosynthetic

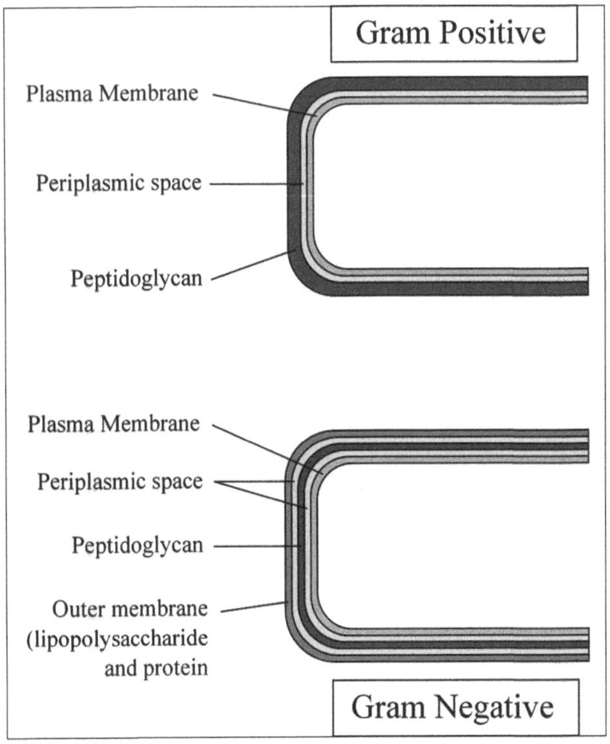

Gram-positive and gram-negative cell wall structure. Image by Graevemoore, via Wikimedia Commons.

bacteria, and others degrade complex organic compounds (halorespirers). The Chloroflexi stain gram-negative with a Gram stain, but they have one cell wall surrounding their cells. The gram-positive bacteria include the Firmicutes (low G + C gram-positive bacteria), the Actinobacteria (high G + C gram-positive bacteria), and the *Deinococcus-Thermus* group that stains gram-positive but have an outer membrane like gram-negative bacteria. The Firmicutes includes most gram-positive bacteria and bacteria that lack cell walls, a small group called the "Mollicutes."

Clinically, the classification of bacteria significantly aids the diagnosis of specific diseases. Identification of bacteria in a clinical specimen can be accomplished through direct microscopic examination, isolation, the culture of the responsible bacteria, and biochemical and immunological tests. Researchers have developed and marketed several automated microbial diagnosis systems that allow rapid diagnosis without isolating the organisms of interest.

CELL AND POPULATION GROWTH

In discussing the growth of living organisms, one can focus on the growth of an individual or the growth of a population. Because bacteria are single-celled organisms, the growth of an individual bacterium does not include the development of organs or other body parts but rather just the enlargement of the cell itself.

Discussion of the growth of bacterial species is usually concerned with the growth of a population of cells. Because almost all bacteria reproduce through the division of one cell into two, the growth of a population of bacterial cells is geometric—that is, the population doubles in size with each round of cell division. The length of time required for a population of bacterial cells to double varies depending on the species and strain of bacteria and the environmental conditions, including temperature, pH, nutrient availability, and waste accumulation.

Some bacteria, such as *Escherichia coli*, have a maximum doubling rate of fewer than thirty minutes. At this rate, a single cell could generate a population of one million cells in less than ten hours. If the environmental conditions remained optimal, with ready nutrients and regular waste removal, a culture of maximally reproducing *E. coli* bacteria would equal the mass of the planet Earth within one week. Other bacteria, such as *Mycobacterium tuberculosis*, divide much more slowly, taking twelve to eighteen hours under optimal conditions for one round of binary fission. The optimal growth rates estimated for many bacteria are merely speculative because most species have not yet been cultured on defined or artificial media.

Even slowly dividing bacteria can reproduce in far less time than nearly every other type of organism. Bacteria can rapidly overwhelm any unpreserved biological sample because of their rapid reproductive rates and omnipresence in the living world. Unrefrigerated food, blood and tissue samples, and other biological specimens can quickly become host to a diverse, rapidly growing population of bacteria.

REPRODUCTION

Most bacteria reproduce by binary fission. One cell grows by manufacturing more cellular components. The genome is replicated, and the single cell divides into two essentially identical cells. This type of reproduction is termed asexual because it does not involve recombining genetic material from two parents. Because the cells that result from binary fission are virtually identical genetically, the individual cells in a group or colony of bacteria all descended from a single ancestral cell could well be clones of the original cell.

The cellular machinery involved in replicating the genetic material does not perform this replication with perfect fidelity. At each round of replication, there is a finite probability of errors occurring. These errors lead to changes in the genetic material known as mutations. These mutations may result in cells with characteristics different from those of the other cells in the population. These altered characteristics may lead to better-adapted cells to a particular environment—perhaps the ability to metabolize a new nutrient or survive in the presence of an antibiotic. Because bacterial cells reproduce by simple cell division, altered characteristics are transmitted to all offspring of the altered cell (barring further mutation).

Although bacteria do not reproduce sexually by recombination of genetic material from two parents, many bacteria can obtain genetic material from other cells through various methods. Some bacteria can take up DNA (deoxyribonucleic acid) from the environment (probably released from decomposing cells), receive DNA through viral infections, and transfer DNA directly from one living cell to another. These genetic recombination processes allow genes (such as those that confer antibiotic resistance) to rapidly spread throughout a bacterial population.

—Lisa M. Sardinia and Michael A. Buratovich, PhD

Further Reading

Anderson, Denise G., et al. ISE Nester's Microbiology: A Human Perspective. 10th ed., McGraw-Hill Education, 2021.

Betsy, Tom, and Jim Keogh. *Microbiology DeMYSTiFieD*. 2nd ed., McGraw-Hill Education, 2012.

Forterre, Patrick. *Microbes from Hell*. Translated by Teresa Lavender Fagan. U of Chicago P, 2016.

Madigan, Michael, et al. Brock Biology of Microorganisms. 15th ed., Pearson, 2017.

Pommerville, Jeffrey C. *Alcamo's Fundamentals of Microbiology*. 8th ed., Jones & Bartlett, 2007.

Tortora, Gerard, et al. Microbiology: An Introduction. 13th ed., Pearson, 2018.

Willey, Joanne, et al. *Prescott's Microbiology*. 11th ed., McGraw-Hill Education, 2019.

BACTERIA: STRUCTURE AND GROWTH

Category: Pathogen
Anatomy or system affected: Cells
Specialties and related fields: Bacteriology, microbiology
Definition: bacteria are single-celled organisms that reside in every habitat, including the human body; bacteria are a necessary part of the human body's normal flora; very few species cause illness, and many are beneficial; and they are also the smallest known organisms that can reproduce independently

KEY TERMS

cell wall: a rigid structure that surrounds cells and gives them their shape

cell membrane: a fatty layer made mostly of phospholipids that surrounds cells and delimits them

eukaryote: organisms made of cells with extensive internal compartmentalization

organelles: tiny compartments within eukaryotic cells that carry out specific functions

phospholipids: lipid molecules with phosphate in their structure that are the main component of cell membranes

prokaryotes: single-celled organisms that lack intracellular compartmentalization

GENERAL STRUCTURE

Bacteria are the most common life-form on Earth. These single-celled organisms come in a variety of shapes and sizes. The millions of known species of bacteria live in a wide range of environments, from vents deep in the ocean floor to the recesses of the human digestive tract. Most bacteria are harmless to humans; some are helpful and necessary for human health, while a small fraction is pathogenic. Despite these various features, all types of bacteria have fundamental characteristics in common.

Bacteria have a simpler cell structure than plant and animal cells, which are higher life-forms called "eukaryotes." Eukaryotes have cells that are divided into smaller compartments by membranes. Each compartment, or organelle, carries out specialized functions. Bacteria are prokaryotes, which have no organelles. They consist of just one compartment separated from the outside world by a cell membrane and a cell wall. The interior of the cell, called the "cytoplasm," contains a solution of sugars, salts, vitamins, enzymes, and other substances dissolved in water. Suspended in the cytoplasm are many ribosomes and a nucleoid made of deoxyribonucleic acid (DNA).

The cell membrane is a semipermeable barrier that separates the inside of the cell from the outside. This thin structure is vital to the survival of the cell. The membrane is created by the assembly of phospholipids and proteins into a bilayer. The inner and outer surfaces of the bilayer are charged and, thus, are attracted to the water molecules inside and outside the cell. The center layer of this structure is composed of fatty acids, which repel water. These chemical properties of the cell membrane ensure that the watery contents of the cell cannot leak through.

The structure of cell membranes also allows for the selective passage of specific molecules. This important feature ensures that necessary nutrients are allowed to enter the cell, and waste products can exit.

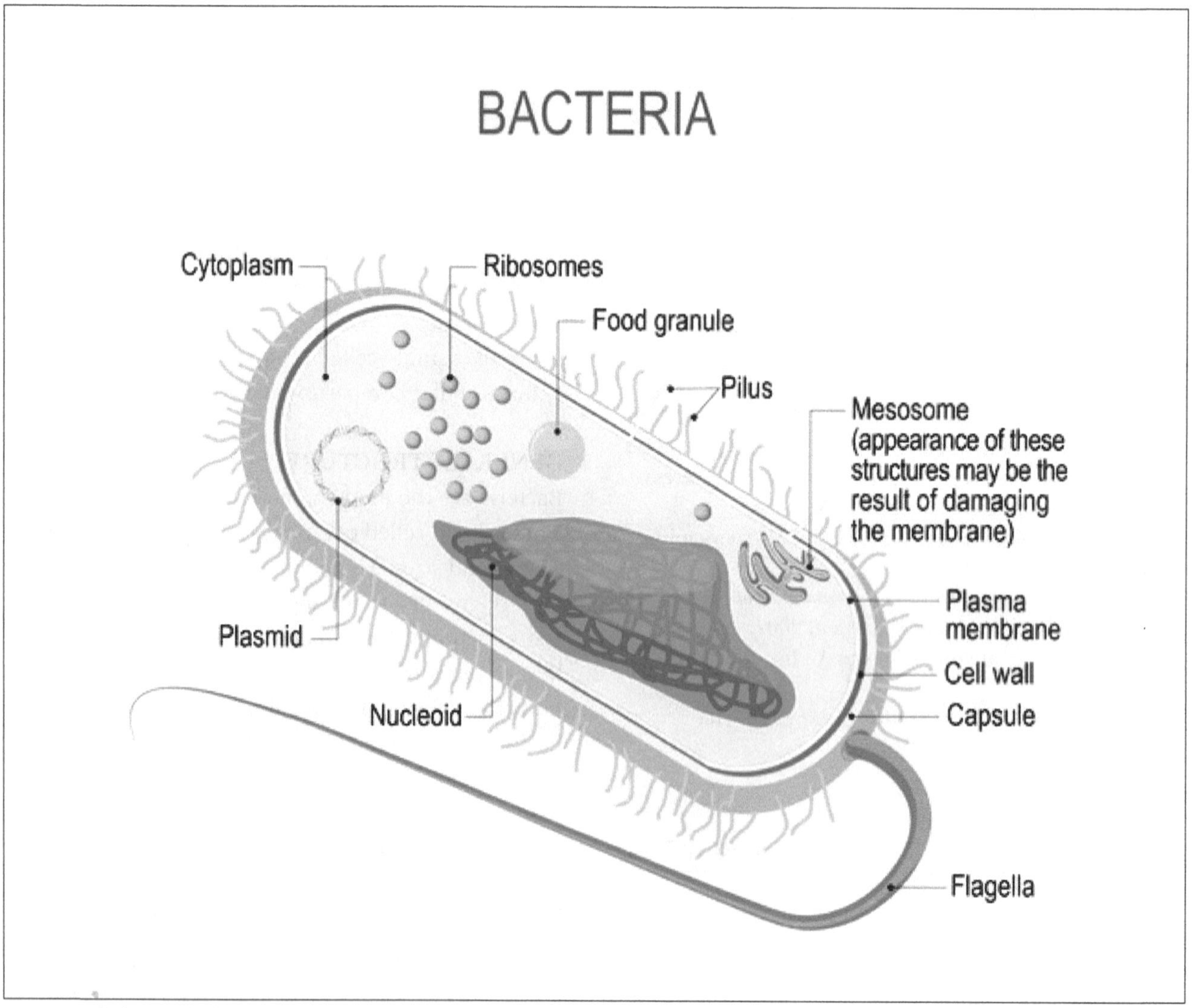

Structure of a bacterial cell. Image via iStock/ttsz. [Used under license.]

While some substances cross the membrane through passive diffusion, most are transported actively by processes that require energy. The active transport of molecules across the membrane is mediated by proteins embedded in the cell membrane.

The cell membrane also serves as a site for the attachment of proteins involved in essential biochemical reactions. One example is the electron transport system, which generates adenosine triphosphate (ATP), the cell's energy currency. In bacteria, ATP is generated by a chain of proteins bound to the inner side of the cell membrane. In eukaryotes, this process occurs on the inner membranes of mitochondria. The bacterial cell membrane thus provides some of the functions carried out by organelles in eukaryotes.

The cell wall is a rigid network of fibers that encloses and protects the bacterial cell. The substance that makes up the cell wall is a unique polymer called "peptidoglycan," which is not found in eukaryotes. Peptidoglycan is made of long sugar molecules that are connected by short peptides. Bacteria can be divided into two major groups based on the structure of their cell walls. Gram-positive bacteria have a thicker peptidoglycan cell wall that will turn purple when treated with a Gram's stain. The cell walls of gram-negative bacteria are surrounded by an outer membrane, which prevents the adhesion of a Gram's stain. The extra protection provided by the more complex cell wall of gram-negative bacteria makes them less sensitive to some antibiotics, which can penetrate the cell walls of only the gram-positive bacteria.

Several classes of antibiotics target the cell walls of bacteria. Penicillins, cephalosporins, and glycopeptides, including vancomycin and other antibiotics, interfere with cell-wall construction, causing the bacteria to rupture and die. The goal in treating bacterial infections with antibiotics is to kill the intended organisms without damaging the host's cells. Because human and animal cells lack cell walls, they are not affected by such drugs.

The internal components of bacteria use nutrients in the environment to allow the organisms to grow and reproduce. The bacterial cytoplasm is rich with ribosomes. As in eukaryotic cells, bacterial ribosomes carry out protein synthesis and are made of ribonucleic acid (RNA) and ribosomal proteins. Differences in the structure of eukaryotic and prokaryotic ribosomes make the ribosome a target for antibiotic action. Multiple classes of antibiotics, including aminoglycosides (e.g., streptomycin, neomycin, and gentamicin), tetracyclines (e.g., doxycycline and minocycline), and macrolides (e.g., azithromycin erythromycin, and clarithromycin), disrupt protein synthesis in bacteria but not in the cells of the host.

Bacterial DNA is organized into one large ring-shaped chromosome. In contrast to eukaryotes, the bacterial chromosome is not encased in a nucleus. The bacterial chromosome contains all the information needed to provide for the basic functions of the organism. Bacteria may also contain circular DNA structures called "plasmids." The genes on plasmids are not usually necessary for survival, but they may become so in certain environments; plasmids can carry genes for antibiotic resistance, allowing the host bacteria to survive in the presence of a drug that is usually deadly to its species.

SPECIALIZED FEATURES

The variety of specialized features found in bacteria reflects their adaptation to the broadest range of environments of any organism on Earth. Bacteria are diverse in their size and morphology. Although the average size of a bacterial cell is 1 to 5 micrometers (μm) in diameter, they range from 0.1 to 750 μm in diameter. One of the most distinguishing features of bacterial cells is their shapes, which can be used diagnostically. The most common shapes are spheres (cocci), rods (bacilli), comma shapes (vibrios), and spirals (spirochetes and spirillum).

Many bacteria have developed specialized structures that allow them to move in their environment.

Some have flagella, which are long filaments that protrude from the cell wall and are used to produce a swimming motion. The arrangement of flagella on the bacterial cell depends on the species. A cell can have a single flagellum or multiple flagella, either clumped at one end of the cell or spread over the entire surface. Some bacteria exhibit a gliding motion, which is created by structures known as pili. These cell surface projections can extend and retract, causing the bacteria to move. Bacteria also use pili to attach to surfaces and each other. Some aquatic bacteria use gas vesicles to adjust their position in their environment. Gas vesicles are hollow structures made of protein. When present, they increase the buoyancy of the organism, making it rise to the water surface. Gas vesicles disintegrate and reassemble according to the concentration of nutrients in the cell.

Capsules are specialized structures that add an extra layer of protection to the exterior of some bacterial cells. The capsule is made of a polysaccharide-containing material that forms rigid layers on the cell wall's exterior. Species with capsules are highly resistant to the action of phagocytes, cells of the host immune system that engulf and kill bacteria. Capsule-bearing strains of *Streptococcus pneumoniae*, for example, cause a particularly invasive and dangerous form of pneumonia.

Some species of bacteria can survive harsh conditions by forming endospores, which allow the bacteria to become dormant. Endospores, tiny cells that develop within bacterial cells, contain DNA and a portion of the cytoplasm. A strong wall surrounds and protects the endospore. Once the bacteria die, the endospores are released into the environment, where they can survive indefinitely. These tough structures are resistant to heat, radiation, chemicals, and desiccation. When environmental conditions improve, the endospore rapidly germinates and develops into a bacterial cell. Endospore-forming bacteria include *Bacillus anthracis*, which causes anthrax, and *Clostridium botulinum*, responsible for a serious form of food poisoning called "botulism."

BACTERIAL GROWTH

Bacteria possess all the machinery necessary to grow and reproduce independently of other cells. They are the smallest creatures on Earth that have this capacity. While they may use a host organism as a habitat, nearly all bacteria can reproduce without invading host cells. This feature sets them apart from viruses, which carry their genetic material but require host-cell components for reproduction. The small size and relatively simple structure of bacteria allow them to grow and reproduce much faster than eukaryotic cells.

Bacteria reproduce asexually by dividing in half in a process called "binary fission." Individual bacterial cells grow continuously, making copies of their components and duplicating their DNA. The two copies of the chromosome move toward opposite ends of the cell, ensuring that each "daughter" cell will receive this essential DNA. When enough new material is present to sustain two cells, the cell membrane pinches inward at the center. A cell wall grows to form a partition that divides the cell into two daughter cells. Because bacterial reproduction is asexual, each daughter will be identical to the parent cell.

Populations of bacteria increase at a rate determined by the time it takes individual cells to grow and divide, creating the next generation. With each generation, the bacterial population doubles in size. The time required for a population of cells to double is known as the doubling time. Bacterial doubling times vary with the species, ranging from a few minutes to several hours. The nearly explosive growth rate of bacteria is about one hundred times faster than that of eukaryotic cells. Rapid binary fission allows bacteria to become extremely numerous in a short amount of time. If one bacterium with a doubling time of twenty minutes were allowed to grow for forty-four hours, the resulting mass of bacteria produced would equal the mass of the Earth.

FACTORS AFFECTING BACTERIAL GROWTH RATES

The actual occurrence of exponential bacterial growth is greatly limited by environmental factors, both in natural habitats and laboratories. Long before a bacterial population could grow to match the Earth's mass, the bacteria's supply of nutrients in the environment would be depleted. Bacterial growth rates are highly dependent on many factors, including temperature, the availability of nutrients, pH (acidity), and oxygen concentrations. Measures that reduce the rate of bacterial growth can be used to prevent illnesses caused by bacteria; most pathogenic bacteria must be present in large numbers to cause illness.

The optimal temperature for bacterial growth depends upon the species. Bacteria that live inside humans, including those of medical significance, thrive at an optimal temperature of about 98.6° Fahrenheit (37° Celsius). They can survive at temperatures generally ranging from 50° to 118.4° F (10° to 48° C). However, their growth rates significantly decrease at lower temperatures. Their ability to survive below the optimal temperature may allow them to live outside a host for short periods until they enter a new host. This temperature tolerance facilitates the spread of bacteria from one host to another.

Controlling the environmental temperature reduces bacterial growth rates. Refrigeration of food slows the growth of bacteria, keeping their numbers low enough to prevent illness. Aqueous solutions heated to boiling 212° F (100° C) for thirty minutes will kill all bacteria in the solution. Medical instruments and solutions can be sterilized in an autoclave by heating above 248° F (120° C), killing bacteria and heat-tolerant endospores.

Bacteria take in nutrients from their environment, but the available nutrients will vary depending on the habitat of a given species. Most bacteria's general nutritional requirements include a carbon source for energy, such as sugar; a nitrogen source, such as ammonia or nitrate; a variety of minerals and salts; vitamins; and other growth factors.

Bacteria are sensitive to the pH of their environment and can live only within a relatively narrow pH range. Most species of bacteria grow optimally in neutral environments, with a pH level between 6 and 8. Some species are specially adapted to live in highly acidic or basic environments. The optimal pH of a given species will determine where it thrives, even within the human body. With a pH of 2, the stomach is home to low numbers of acid-tolerant species of lactobacilli and streptococci. With a neutral pH of 7, the large intestine is a much more popular residence; enormous numbers of bacteria from a minimum of ten different species live in the large intestine. The sensitivity of most bacteria to low pH inhibits bacterial growth during food preservation, as occurs when foods are pickled in vinegar.

The presence of oxygen in the environment is another factor that affects bacterial growth. Most species, the aerobes, require oxygen for growth. For these species, low oxygen will cause a decrease in growth rate; if oxygen levels fall too low, they will not survive. For other species, the anaerobes, oxygen is not necessary for growth. Oxygen is toxic to some species; these obligate anaerobes cannot survive in environments where oxygen is present. Oxygen tolerance is an attribute used to identify bacterial species.

IMPACT

Bacteria are ubiquitous, and they will remain so. They have developed diverse traits that allow them to thrive in an amazing variety of habitats, including unimaginably harsh conditions. Their demonstrated adaptability should give pause and guide future scientific and medical strategies for preventing and treating bacterial illnesses.

—*Kathryn Pierno, MS*

Further Reading

Brooker, Robert J., et al. *Biology*. 5th ed., McGraw-Hill Higher Education, 2019.

Gladwin, Mark T., et al. *Clinical Microbiology Made Ridiculously Simple*. 8th ed., MedMaster, 2021.

Koch, Arthur L. *The Bacteria: Their Origin, Structure, Function, and Antibiosis*. Springer, 2006.

Madigan, Michael T., et al. *Brock Biology of Microorganisms*. 16th ed., Pearson, 2020.

Riedel, Stefan, et al., *Jawetz Melnick & Adelbergs Medical Microbiology*, 28th ed. McGraw-Hill Education/Medical, 2019.

Yong, Ed. *I Contain Multitudes: The Microbes Within Us and a Grander View of Life*. Ecco, 2016.

FUNGI CLASSIFICATION AND TYPES

Category: Microbial classification
Also known as: Fungal systematics
Anatomy or system affected: All systems
Specialties and related fields: Ecology, evolutionary biology, microbiology, mycology, taxonomy
Definition: eukaryotic, heterotrophic organisms that digest organic matter by secreting enzymes into the extracellular environment, assimilating nutrients, including fixed carbon, by osmosis

KEY TERMS

ascus: a cylindrical sac in which the spores of ascomycete fungi develop

asexual reproduction: organismal production used by many eukaryotes and all prokaryotes that involves a single parent, resulting in genetically identical offspring; includes binary fission, fragmentation, and budding

basidium: a microscopic club-shaped spore-bearing structure produced by certain fungi

meiosis: a type of cell division that results in four daughter cells, each with half the number of chromosomes of the parent cell, as in the production of gametes and plant spores

mutualism: symbiosis that is beneficial to both organisms involved

parasitism: the relationship between two species of plants or animals in which one benefits at the expense of the other

saprophytes: an organism that feeds on nonliving organic matter known as detritus at a microscopic level

spores: a reproductive cell capable of developing into a new individual without fusion with another reproductive cell

symbiosis: any of several living arrangements between members of two different species, including mutualism, commensalism, and parasitism

zygospores: the thick-walled resting cell of certain fungi and algae, arising from the fusion of two similar gametes

GENERAL CHARACTERISTICS

Except in unicellular forms, the basic growth pattern of fungi consists of filamentous hyphae (slender tubes that are the basic building blocks of fungi), with little cellular differentiation in vegetative tissues. Reproduction, both sexual and asexual, is through spores, usually microscopic.

Biologists formerly included fungus as the phylum Mycota within the plant kingdom. In the 1960s, they adopted a five-kingdom classification; one of the kingdoms was reserved for fungi. With increasing knowledge of ultrastructure and physiological processes at the molecular level, it became evident that the fungi were not monophyletic. Sequencing of ribosomal deoxyribonucleic acid (DNA) confirmed this. It also helped clarify the probable taxonomic position of fungal species (including some important human pathogens) without diagnostic morphological features. Sequencing also aided taxonomists in constructing a more natural system reflecting phylogeny and actual biological affinity. From a practical point of view, the better a taxonomic system, the more useful it is for making identifications critical to diagnosis and predicting what therapies are most promising.

Ribonucleic acid (RNA) sequencing and a form of mathematical analysis known as "cladistics" have identified dozens of distinct evolutionary lines among the eukaryotes, most of them consisting of obscure groups of protozoa. The Myxomycetes (slime molds), included with fungi in older classifications, are now considered animals closely related to one group of amoeboid protozoa. A significant group of aquatic organisms, the Oomycetes or Oomycota, were initially grouped with the fungi, but molecular and phylogenetic studies revealed significant differences between fungi and oomycetes. Consequently, the oomycetes are now placed in the clade Stramenopiles with several phyla of algae, including kelp and diatoms, with the same flagellar structure and cell wall chemistry. The remaining fungi fall on that portion of the tree of life, including Metazoa (multicellular animals), vascular plants and green algae, and some protozoa.

The clade (a group consisting of an organism and all its descendants), including the Zygomycota, Ascomycota, and Basidiomycota, sometimes termed "Eumycota," consists of overwhelmingly terrestrial organisms that lack motile spores or any vestige of a flagellar base. Most are haploid except for the zygote, which immediately undergoes meiosis before the spore formation. Several orders formerly included in the Zygomycetes may be distinct enough to warrant recognition at a higher taxonomic level. The Microsporidia, a group of obligate animal parasites formerly regarded as primitive or degenerate protozoa, groups with the Eumycota. The Chytridiomycota encompasses aquatic forms with zoospores equipped with a whiplash 9+2 flagellum, cell walls containing chitin, and, in several species, regular alternation between haploid and diploid generations. Many species are parasitic on algae.

EUMYCOTA: PHYLUM CHYTRIDIOMYCOTA

In older classifications, Chytridiomycetes were grouped with the Oomycetes in a general category, Phycomycetes, or algal-like fungi. The diagnostic feature is a zoospore with a single 9+2 flagellum, the same type found in plants and multicellular animals. Sexual reproduction is through the fusion of zoospores to form a zygote. Some chytrids form extensive mycelium (mass of hyphae). In contrast, in others, the vegetative body consists of a single cell anchored to the host by rhizoids that converts to a zoosporangium. Growth and asexual reproduction can take place in both the haploid and diploid phases.

Based on DNA sequencing, there are three distinct phyletic groups of fungi with uniflagellate zoospores: the Chytridiomycota, including the Chytridiales, Spizellomycetales, and Rhizophydiales; the Blastocladiomycota, which are mainly parasites on soil and freshwater invertebrates; and the Neocallimastigomycota, a small group of anaerobes inhabiting the stomachs of ruminants. There have been no known reports of human disease caused by Chytridiomycota. However, *Batrachochytrium dendrobalis* causes a devastating disease in frogs and other amphibians.

EUMYCOTA: PHYLUM GLOMEROMYCOTA

Based on DNA sequencing and morphology and host relationships, the Glomeromycota replaced the older phylum Zygomycota in 2001. This group contains fungi that form symbiotic relationships with the roots of green plants. The arbuscular mycorrhizae infect plant roots and form fine, hairlike extensions that significantly increase the absorptive capacity of plant roots. This mutualistic relationship benefits the plants and fungi since plants benefit from increased nutrient absorption and fungi benefit from protection and nutritional access.

Other fungi in this group include the traditional Zygomycota, the Mucoromycotina, Kickxellomycotina, Zoopagomycotina, and insect pathogens, the Entomophthoromycotina. These fungi are characterized by the absence of motile stages in the life cycle and by non septate hyphae, cell walls containing chitin, production of nonmotile sexual spores in sporangia, and fusion of hyphal outgrowths of equal size to form a diploid zygospore, often thick-walled and ornamented, that serves as a resting spore. About nine hundred species are known, amounting to approximately 1 percent of the described fungi.

The Mucormycotina, comprising the Mucorales, Endogonales, and Mortierellales, includes fast-growing saprophytes (organisms that live on dead or decaying matter) with abundant asexual reproduction, including the familiar black bread mold *Rhizopus stolonifera*. Species of *Mucor* and several other genera cause rare but perilous fulminating infections in immunocompromised persons.

Most members of the Kickxellomycotina and Zoopagomycotina are specialized parasites or commensals on invertebrate animals and other fungi. They have not been implicated in human disease. The Entomophthoromycota, which includes several insect pathogens used as agents of biological control, is characterized by conidia (a type of asexually produced spores) that are actively discharged. Several species of *Conidiobolus* and *Basidiobolus ranarum* infect humans, usually immunocompromised persons, producing chronic skin ulcers and polyps.

Microsporidia, which cause chronic infections in many vertebrate animals, including humans, were thought initially to be protozoa and near the base of the eukaryotic family tree based on small cell size, small genomes, and a lack of mitochondria. A unique feature of the cell is a triggered filament that aids in the penetration of host cells. DNA analysis suggests the simple structure is not primitive but instead evolved in the parasitic habitat.

EUMYCOTA: PHYLUM BLASTOCLADIOMYCOTA

The Blastocladiomycota were originally considered in the Chytridiomycota. However, molecular and ultrastructural data put these fungi into their own phylum as a sister group to the Glomeromycota. These fungi are saprophytic, meaning that they feed on decomposing plant and animal material. However, some members of this group live as parasites of eukaryotes. The blastocladiomycetes are closely related to the chytrids. However, the chytrids undergo meiosis immediately after the nuclei of mating hyphae fuse (zygotic meiosis). The blastocladiomycetes, however, undergo sporic meiosis in which the zygote divides by mitosis to produce a multicellular, diploid sporophyte.

Examples of blastocladiomycetes include *Coelomomyces*, a pathogen of mosquitoes, *Allomyces*, a common laboratory fungus, and *Blastocladiella*, an aquatic fungus that grows on submerged twigs and fruit.

EUMYCOTA: SUBKINGDOM DIKARYA

The subkingdom Dikarya is a well-defined clade composed of the Ascomycota and Basidiomycota and includes more than 90 percent of the species described as fungi. These predominantly terrestrial fungi lack motile spores and have chitinous cell walls. An extensive mycelium composed of regularly septate hyphae is usually present. The distinctive feature defining this clade is a life-cycle stage between plasmogamy (fusion of cells) and karyogamy (fusion of nuclei), during which the cells are binucleate, with a complete set of chromosomes from each parent. In Ascomycota, the binucleate stage is confined to the actual sexual fruit body. However, in the Basidiomycota, it constitutes the primary vegetative thallus—persisting, in some genera of wood-destroying fungi, for decades or even centuries.

Older classifications sometimes formally recognized a third class, the Deuteromycetes or Fungi Imperfecti, for fungi with septate hyphae and no known sexual cycle. Manuals for identifying these fungi still group them in a form-class for convenience. Some morphologically defined form-genera of asexually reproducing fungi, such as *Penicillium* and *Alternaria*, represent distinct biological entities connecting to genera defined by the sexual stage, while others do not. The trend in recent years has been to use biochemistry of metabolites to classify yeasts, which are very simple morphologically and represent a growth phase of many human pathogens. With the advent of DNA sequencing, it has become possible to classify any organism of interest correctly.

Phylum Ascomycota. These organisms have a vegetative thallus, except in yeasts, which comprises haploid septate hyphae, cells that are generally uninucleate. Asexual reproduction is through conidia unicellular to multicellular spores, typically airborne and often produced abundantly. Sexual reproduction in most families is initiated by fertilizing a specialized enlarged cell, the ascogonium, with small airborne conidia known as spermatia, followed by limited proliferation of binucleate cells. Karyogamy and meiosis occur inside a sac-like cell called an "ascus," within which ascospores (usually eight) are delimited.

The Ascomycota is the largest and most diverse natural phylum of Eumycota. It includes the majority of species of medically important fungi and the majority of plant pathogens. Between one-quarter and one-third of the known species form symbiotic lichen associations with algae and cyanobacteria.

For the most part, the division of the Ascomycota into classes and orders, proposed in the mid-twentieth century and based on the structure and development of ascocarps (mature fruiting bodies) and asci (spore sacs), agrees with the division based on DNA sequencing. However, the old subclass Hemiascomycetes is defined by the absence of fruit bodies and includes the Taphrinales, mainly obligate biotrophic parasites of higher plants. The Saccharomycetales (ascomycetous yeasts with no or limited mycelial growth) becomes the subphyla Taphrinomycota and Saccharomycota.

Molecular studies have shown that the important human pathogen *Pneumocystis carinii*, which occurs as undifferentiated yeastlike cells in host tissue and has not been successfully cultured, is a member of the Taphrinomycota. Another important pathogen, *Candida albicans*, is a representative member of the Saccharomycota, which also includes brewer's yeast. The yeast growth form is characterized by single cells that bud off multiple daughter cells from undifferentiated loci on the cell surface. Some yeasts, including *C. albicans*, have a diploid vegetative state.

The old class Euascomycetes, renamed subphylum Pezizomycotina, includes fifty-eight recognized orders of Ascomycetes producing asci in distinct fruiting bodies. They are grouped into seven well-defined classes, plus four orders in classes by themselves. The most important divisions are Pezizomycetes, Eurotiomycetes, Laboulbeniomycetes, Lecanoromycetes, Leotiomycetes, and Sordariomycetes.

Pezizomycetes are fungi with disc-shaped fruit bodies (apothecia) and operculate asci, related hypogeous gastroid forms, true truffles, Dothideomycetes, fungi with enclosed fruit bodies (perithecia), ascostromatic development, and bitunicate asci. This group includes many plant pathogens. Abundantly sporulating asexual stages are common allergens, and a few species are opportunistic human pathogens.

Eurotiomycetes are fungi with enclosed, aporate, often reduced fruit bodies and simple thin-walled asci. Asexual stages of Eurotiales include the genera *Penicillium* and *Aspergillus*. This class includes the majority of true human pathogens and agents of food spoilage that produce toxins and carcinogens. The dermatophyte genera *Trichophyton* and *Microsporon* and the serious pathogens *Histoplasma* and *Paracoccidiodes* belong to the order Onygenales.

Laboulbeniomycetes are specialized ectoparasites of arthropods, with significantly reduced thalli. Lecanoromycetes are Lichen mycobionts and saprophytes with apothecia and complex (but not functionally bitunicate) asci. Leotiomycetes are plant parasites that have unitunicate asci, apothecia, and ascohymenial development. Sordariomycetes have a perithecium fruit body, ascohymenial development, and unitunicate asci. This diverse group includes many important plant pathogens. *Fusarium* and *Sporothrix* are conidial stages of Sordariomycetes.

Phylum Basidiomycota. These organisms have a vegetative thallus, except in yeasts, comprised of haploid dikaryotic septate hyphae. Production of asexual spores is infrequent in Hymenomycetes but a regular part of the life cycles of Uredinomycetes (rusts) and Ustilagomycetes (smuts). Sexual reproduction is initiated by fusing undifferentiated haploid mycelial cells (Hymenomycetes) or pycniospores functioning as spermatia (Uredinomycetes). Karyogamy and meiosis take place inside a club-like structure called a "basidium," Basidiospores, produced externally, are forcibly discharged. Basidiomycetes are most important as wood decomposers, plant pathogens (rusts and smuts), and mycorrhizal symbionts of forest trees. Subphyla are Puccinomycotina, Ustilagomycotina, and Agaricomycotina.

From a human perspective, the most important groups in the subphylum Puccinomycotina are the Pucciniales (rusts), obligate biotrophic plant parasites with complex life cycles, and the Sporidiobolales, the main group of basidiomycetous yeasts. Of particular interest to medical mycologists is *Cryptococcus (Filobasidiella) neoformans*, in which multiple mitotic divisions follow meiosis in the basidium, and basidiospores are budded off in chains.

The subphylum Ustilagomycotina consists of mainly obligate plant parasites with complex life cycles, divided into the Ustilagomycetes (smuts) and Exobasidiomycetes (leaf curl diseases and some smuts). There is one human parasite, the dermatophyte yeast *Melasseza*.

As the name implies, the subphylum Agaricomycotina includes the familiar edible mushroom *Agaricus campestris*. Also called "Hymenomycetes," members of this group have a life cycle including a limited undifferentiated mycelial haploid phase followed by hyphal fusion establishing a dikaryon. Dikaryotic hyphae have characteristic clamp connections and complex dolipore septa. Basidia are typically borne in a layer (the hymenium) on complex fruit bodies. Basidia are septate in the Tremellomycetes (jelly fungi) and unicellular in the Agaricales (mushrooms), Polyporales (woody pore fungi), and Phallales (stinkhorns). The orders of Agaricomycotina have long been defined by microanatomy and chemistry rather than gross fruit-body form, and present classifications based on DNA analysis closely approximate older treatments. This subphylum contains no important human pathogens.

IMPACT

Having an accurate classification system for fungi or any other group of organisms that have a significant impact on humans is critical to identifying species and devising methods of control tailored to the particular organism. In a clinical setting, health-care providers need to identify the agent causing the disease to initiate appropriate therapies. In research laboratories, the development of effective therapies depends on understanding the biochemistry and life cycle of the target pathogen, a process greatly aided by being able to classify it with a biologically related species.

Human pathogenic fungi have always presented a challenge to medicine because fungi are more closely related to humans than are bacteria and most protozoa. Drugs that inhibit fungal growth are therefore likely to be toxic to humans. Most common fungal infections are superficial or localized and can be treated topically. However, growing populations of immunocompromised persons, including those with human immunodeficiency virus (HIV) infection, transplant recipients, and persons undergoing chemotherapy, have led to the emergence of several systemic, life-threatening mycoses.

DNA sequencing is an excellent aid in identifying and treating fungal diseases. It has established the taxonomic position and, therefore, the most promising avenues for therapy for morphologically ambiguous pathogens such as Microsporidia and

Pneumocystis. DNA sequencing is becoming available as a clinical diagnostic tool for establishing the identity of a pathogen in tested persons.

—*Martha A. Sherwood, PhD*

Further Reading

Gnanam, Chelin Rani. *Introduction To Mycology*. MJP Publishers, 2013.

Hibbet, David S., et al. "A Higher-Level Phylogenetic Classification of the Fungi." *Mycological Research*, vol. 111, 2007, pp. 509-47.

Hoenigl, Martin, and Alida Fe Talento, editors. *Fungal Infections Complicating COVID-19*. Mdpi AG, 2021.

Priest, Fergus G., and Michael Goodfellow, editors. *Applied Microbial Systematics*. Kluwer Academic, 2000.

Walsh, Thomas J., et al. *Larone's Medically Important Fungi: A Guide to Identification*. 6th ed., ASM Press, 2018.

Flagella and cilia

Category: Cell biology

Anatomy or system affected: Cells, fallopian tubes, respiratory tract, reproductive tract, spermatozoa,

Specialties and related fields: Bacteriology, cell biology, microbiology, obstetrics and gynecology, otolaryngology, protozoology, pulmonary medicine

Definition: extensions of cells that propel them through their environment (flagella) or generate currents at the cell surface (cilia)

KEY TERMS

cilia: short, microscopic hairlike vibrating structures found in large numbers on the surface of specific cells, either causing currents in the surrounding fluid or, in some protozoans and other small organisms, providing propulsion

dynein: a family of cytoskeletal motor proteins that move along microtubules in cells

flagella: a slender threadlike structure, especially a microscopic appendage that enables many protozoa, bacteria, spermatozoa, and other organisms

flagellin: the structural protein of bacterial flagella

microtubules: a microscopic tubular structure present in numbers in the cytoplasm of cells, sometimes aggregating to form more complex structures

variable region: the N-terminal portion of the antibody, which possesses antigen-binding sites and has variable amino acid sequences

INTRODUCTION

Although the term "flagellum" is used for both prokaryotes (archaea and bacteria) and eukaryotes (fungi, protists, plants, and animals), the structure and mechanism of action of this structure in prokaryotes are quite different from the structure and mechanism of action in eukaryotes. Eukaryotic flagella and cilia, however, are structurally and functionally identical. The differences between them are in their number, length, and position. Flagella are less numerous, longer, and usually polar, while cilia are more numerous and shorter, covering much of the cell's surface. Because the dividing line between eukaryotic flagella and cilia is not precise, many scientists use *undulipodia* as a collective word for both eukaryotic flagella and cilia. In some algae, other protists and the gametes of certain plants with motile sperm, flagella, and cilia occur.

BACTERIAL FLAGELLA

Bacterial flagella are made from a protein called "flagellin." The flagellum is a hollow tube, about 20 nanometers (1 billionth of a meter) in diameter. The hollow tube forms a delicate, helical filament. Just outside the cell wall, the flagellin shaft is attached to the "hook." The hook causes the shaft to take a sharp bend just outside the cell wall. The hook attaches to a basal body that passes through the cell wall and membrane with a series of protein rings that act as bearings. Embedded in the cell membrane is a motor that rotates between 200 to 1,000 times per minute. The motor works through hydrogen ions pumped into the cell from the outside (known as a "proton motive force"). The shaft passes through the motor and rotates with it.

Bacteria have many different flagellar arrangements. Monotrichous bacteria, like *Vibrio cholera*, have only one flagellum. Lophotrichous bacteria, like *Helicobacter pylori*, have multiple flagella attached to the same spot. Amphitrichous bacteria, like the photosynthetic bacterium *Rhodospirillum rubrum*, have a flagellum at opposite ends of the cell. However, only one is operational at a time. Peritrichous bacteria, like *Proteus mirabilis*, have flagella all over the cell. Like the syphilis organism *Treponema pallidum*, spirochetes have flagella attached to the cell membrane that do not extend outside the cell wall but stay within the cell wall. These so-called endoflagella cause the cell to corkscrew through the medium.

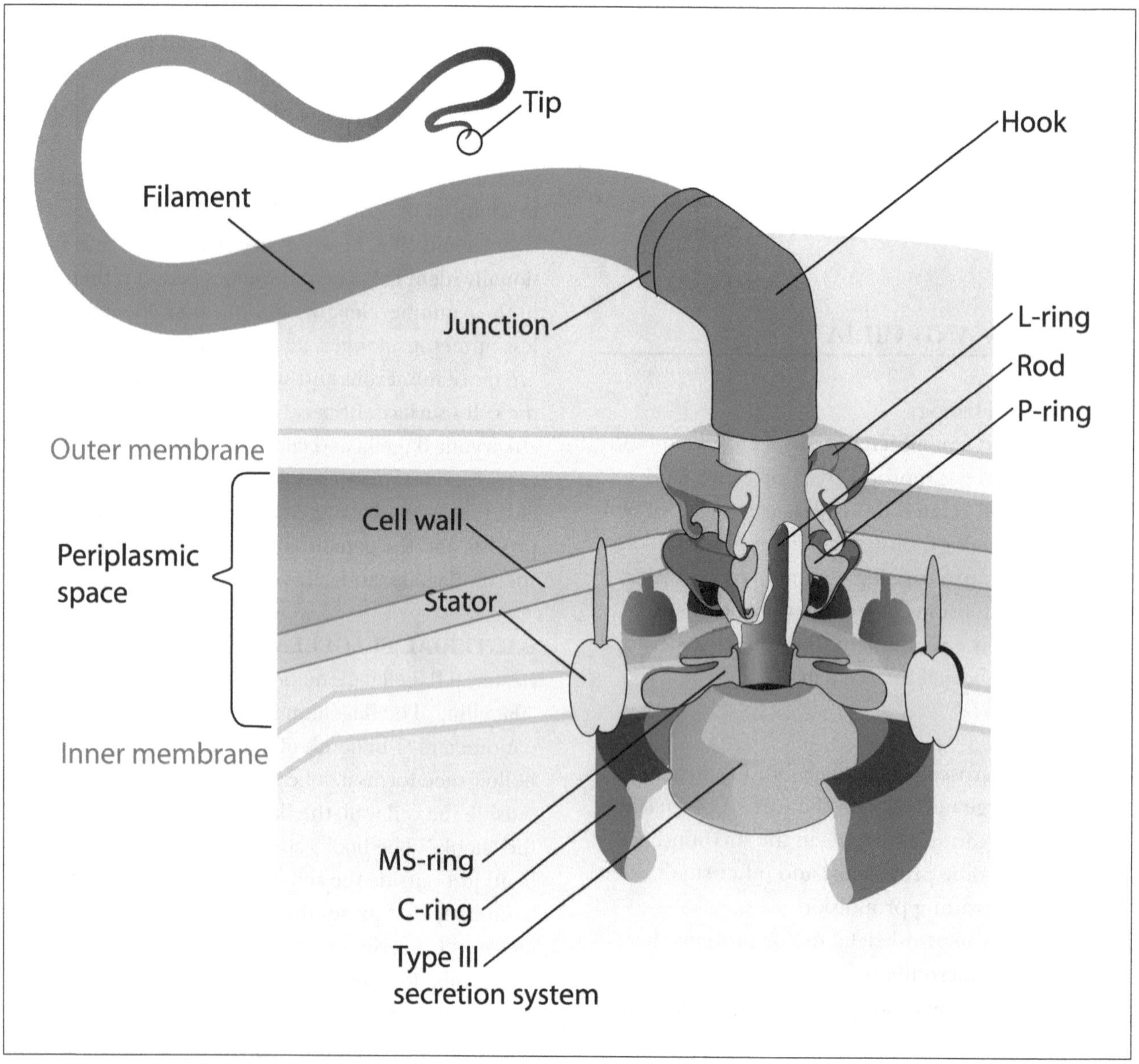

A gram-negative bacterial flagellum. By LadyofHats, via Wikimedia Commons. [Public domain.]

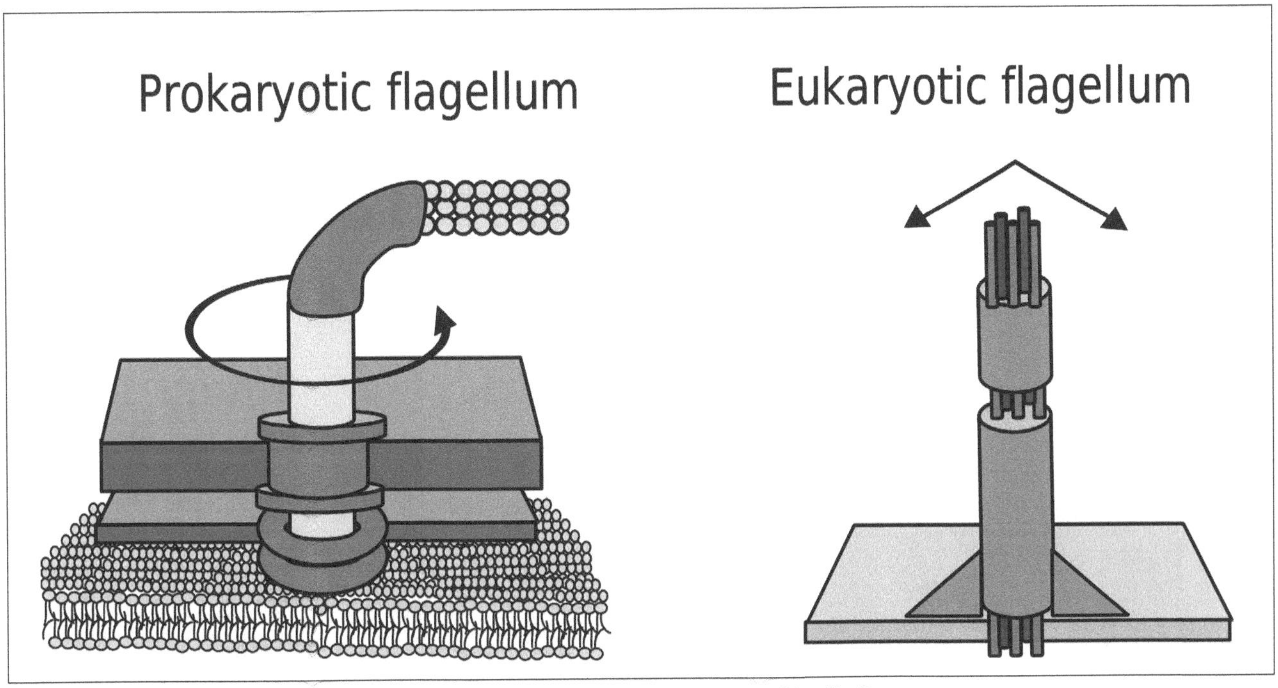

Difference between prokaryotic flagellum and eukaryotic flagellum. By Mgaetani, via Wikimedia Commons.

EUKARYOTIC FLAGELLA AND CILIA

Eukaryotic flagella and cilia, or undulipodia, are more complex and larger (approximately 0.25 micrometer in diameter) than their prokaryotic counterparts. The main component of these eukaryotic structures is the "microtubule," a long, cylindrical structure composed of tubulin proteins. In eukaryotic flagella and cilia, two central microtubules are surrounded by a circular arrangement of nine microtubule pairs. Eukaryotic flagella and cilia also contain more than five hundred other proteins, including dynein and kinesin, motor proteins that use cell energy to slide the microtubules past each other, causing an undulating motion (hence, the name undulipodium). Unlike bacterial flagella, eukaryotic flagella and cilia are intracellular structures because a continuation of the plasma membrane covers them.

Although absent from most fungi (except in some gametes), undulipodia are found in many protists and some plants. Unicellular algae (such as *Chlamydomonas* and *Euglena*) and colonial algae (*Volvox*) use undulipodia for locomotion. Multicellular algae (*Phaeophyta*, *Rhodophyta*) produce flagellated sperm. Among the true plants, bryophytes (*Hepatophyta*, *Anthocerotophyta*, and *Bryophyta*), ferns, and their allies (*Psilotophyta*, *Lycophyta*, *Sphenophyta*, and *Pterophyta*), and some gymnosperms (*Cycadophyta* and *Ginkgophyta*) also produce flagellated sperm. Other gymnosperms (*Coniferophyta*) and angiosperms (*Anthophyta*) do not produce cells with flagella or cilia.

Mutations that affect undulipodia in humans cause Kartagener's syndrome, sometimes known as Kartagener's triad. The symptom triad includes recurrent respiratory infections, situs inversus, and male infertility. Situs inversus is a condition in which the heart and other organs are transposed to the opposite side. People with Kartagener's syndrome have frequent respiratory, ear, and sinus infections, male infertility, and chronic nasal congestion.

—Richard W. Cheney Jr. and Michael A. Buratovich, PhD

Further Reading

Amos, W. Bradshaw, and J. G. Duckett, editors. *Prokaryotic and Eukaryotic Flagella*. Cambridge UP, 1982.

Grognot, Marianne, and Katja M. Taute. "More Than Propellers: How Flagella Shape Bacterial Motility Behaviors." *Current Opinion in Microbiology*, vol. 61, 2021, pp. 73-81, doi:10.1016/j.mib.2021.02.005.

Raven, Peter H., Ray F. Evert, and Susan E. Eichhorn. *Biology of Plants*. 6th ed., W. H. Freeman/Worth, 1999.

Satir, Peter. *Structure and Function in Cilia and Flagella*. Springer, 1965.

Sleigh, Michael A., editor. *Cilia and Flagella*. Academic Press, 1974.

Wilson, Leslie, William Dentler, and Paul T. Matsudaira, editors. *Cilia and Flagella*. Methods in Cell Biology 47. Academic Press, 1995.

Zariwala, Maimoona A., et al. "Primary Ciliary Dyskinesia." *GeneReviews*, edited by Margaret P. Adam, et al., University of Washington, Seattle, 24 Jan. 2007.

MICROBIAL METHODS

KOCH'S POSTULATES

Category: Epidemiology
Anatomy or system affected: All systems
Specialties and related fields: Bacteriology, epidemiology, microbiology, preventive medicine, public health
Definition: a set of experimental guidelines used to determine if a particular microorganism is the causative agent of a particular disease

KEY TERMS

hypothesis: a proposition made as a basis for reasoning, without any assumption of its truth
postulate: a thing suggested or assumed as true as the basis for reasoning, discussion, or belief

HISTORICAL OVERVIEW

In the nineteenth century, Robert Koch, a German physician, and bacteriologist played a significant role in determining an infectious disease's etiology (cause). Through his work with *Bacillus anthracis* (the causative agent of anthrax), he linked a specific microorganism to a specific infectious disease. Koch conducted experiments showing that *B. anthracis* was always present in diseased animals, that healthy animals inoculated with the bacterium would develop the disease, and that cultivation of the bacterium in artificial media followed by inoculation resulted in the disease.

Koch also discovered the causative organisms for several other diseases, including tuberculosis and cholera. In describing the etiology of tuberculosis, Koch proposed a set of guidelines for establishing a cause and effect relationship between a given microorganism and a specific disease. These scientific criteria are known as Koch's postulates.

THE POSTULATES

Koch's postulates are a set of four experimental criteria used to establish the etiology of a disease. The first criterion states that the pathogen must be present in all infected persons and absent in healthy persons. The second criterion states that the pathogen must be isolated from the diseased person and cultivated in the laboratory. The third criterion states that the cultivated pathogen must cause the disease in a healthy person after inoculation. The fourth criterion states that the pathogen must be isolated again from the infected person and identified as identical to the original isolate.

Robert Koch. Photo via Wikimedia Commons. [Public domain.]

EXCEPTIONS

There are some exceptions to Koch's postulates. Specific pathogens and fastidious microorganisms have complex and unusual growth requirements and survive only within living host cells. Such microorganisms cannot be cultured on artificial media. Numerous pathogens infect a specific species only, while others become transformed in vitro. Some infectious diseases have unclear origins, while others cause multiple disease conditions. Many infections develop from the combined effects of several different microorganisms. Various diseases do not originate from a microorganism. They may be the result of poor nutrition, chromosomal abnormality, organ failure, or environmental influences. These exceptions have stimulated the need for modifications to Koch's postulates.

IMPACT

Koch's contributions were invaluable in advancing medical microbiology and the understanding of the nature of a disease. Koch's postulates still provide the essential principles for determining the causative agents of emerging infectious diseases and the foundation to address disease and public health.

—Rose Ciulla-Bohling, PhD

Further Reading

Byrd, Allyson L., and Julia A. Segre. "Infectious Disease: Adapting Koch's Postulates." *Science*, vol. 351, no. 6270, 2016, pp. 224-26, doi:10.1126/science.aad6753.

Daniel, Wayne W. *Biostatistics: A Foundation for Analysis in the Health Sciences*. 9th ed., John Wiley & Sons, 2009.

Engelkirk, Paul G., and Gwendolyn R. W. Burton. *Burton's Microbiology for the Health Sciences, 11th ed.* 8th ed., Jones & Bartlett Learning, 2020.

Hardy, Simon P. *Human Microbiology*. Taylor and Francis, 2003.

Murray, Patrick R., et al. *Medical Microbiology*. 9th ed., Elsevier, 2020.

Rabins, Peter V. *The Why of Things: Causality in Science, Medicine, and Life*. Columbia UP, 2013.

Straus, Eugene, and Alex Straus. *Medical Marvels: The One Hundred Greatest Advances in Medicine*. Prometheus Books, 2006.

Tortora, Gerard J., et al. *Microbiology: An Introduction*. 13th ed., Pearson, 2018.

MICROSCOPY

Category: Biotechnology
Anatomy or system affected: Cells
Specialties and related fields: Bacteriology, microbiology, pathology
Definition: the examination of minute objects through a microscope, an instrument that provides an enlarged image of an object not visible with the naked eye

KEY TERMS

compound microscope: a microscope that uses multiple lenses to enlarge the image of a sample

objective lens: the lens on the microscope closest to the sample

resolution: the minimum distance at which two distinct points of a specimen can still be seen as separate entities

simple microscope: essentially a magnifying glass made of a single convex lens with a short focal length that magnifies the object through angular magnification

DEFINITION

The word "microscopy" defines the technique wherein microscopes are used to study organisms and cells that are too small to be seen by the unaided eye. When first invented, microscopes comprised simply a series of magnifying lenses that made the object or specimen under study appear much bigger than its actual size. Moreover, these early inventions relied on sunlight as the source of illumination, a feature that has been modified over the years and now includes

Photo via iStock/PeopleImages. [Used under license.]

microscopes with a diverse collection of illuminators, from visible light and ultraviolet light and laser to sound waves, electron beams, and thin metal probes. The spectrum of microscopy also has changed since the technique that began with two-dimensional images of protists now offers, for example, three-dimensional colored imaging that can be used to study molecular processes in atomic detail.

HISTORICAL BACKGROUND

As early as the first century, the Romans discovered lenses while experimenting with different kinds of glass and their ability to enhance the visibility of objects seen through them. In the late sixteenth century, Dutch spectacle maker Zaccharias Jansen created the world's first compound microscope by placing several lenses inside a tube. Encouraged by their findings, Galileo started building his microscope. The word "microscope," though, was coined by Giovanni Faber in the seventeenth century to describe his friend Galileo's invention. "Microscope" was derived from two Greek words, *micron*, meaning "small," and *skopein*, meaning "to look at."

Soon after that, in the mid-seventeenth century, British scientist Robert Hooke made several important contributions in microscopy and documented them in his famous book *Micrographia* (1665). Hooke studied cork tissue sections and discovered tiny chambers in the tissue that he called "cells."

In the late seventeenth and early eighteenth centuries, several pioneering discoveries (influenced by Hooke) in biology were made using microscopes made by a Dutch tradesman, Antoni van Leeuwenhoek. Despite a lack of formal training as a scientist,

Antoni van Leeuwenhoek, via Wikimedia Commons. [Public domain.].

Leeuwenhoek discovered bacteria, protists, and many other microbes, thus opening up the field of microbiology. He is often regarded as the founder of microscopy, which now includes microscopes such as the scanning probe and atomic probe, allowing one to visualize structures at the atomic and molecular levels.

THE PROCESS OF MICROSCOPY

An important goal in all kinds of microscopy is to enhance the contrast between the specimen and the medium (also called the "background"). This is done to provide a sharp and detailed image because most cells and organisms have very slight coloration. Common methods to enhance contrast include stains, dyes, and alternative illumination sources, such as ultraviolet and laser.

The process of staining will be discussed here as it applies to the widely used compound light microscope. There are two broad categories of stains: acidic stain, wherein the chromophore (coloring unit) is an anion, and basic stain, wherein the chromophore is a cation. Stains also are classified as simple or differential. Simple stains will color all microbes in a nonspecific manner; thus, they are typically helpful for studying cell shapes, morphology, and arrangements. Differential stains are specific and will therefore stain only certain cells; they are often used in microbial identification.

The staining process typically starts with creating a smear, composed of simply two to three drops of the bacterial suspension spread out on a clean glass slide. The next step is fixing, which helps attach the specimen to the glass slide. Typically, bacterial specimens are fixed by quickly passing the air-dried smear over a flame. Once the specimen has been fixed, the actual staining process begins. It typically involves adding the stain or dye, then waiting a few minutes before washing off the stain, adding a mordant (color enhancer) if required, and counterstaining (with the secondary stain).

Once the staining process is finished, depending upon the specimen size, the slide is covered with a square piece of thin glass called the "coverslip" and then observed under the microscope. A drop of immersion oil is placed over the specimen if greater magnification is required with the brightfield microscope (a compound light microscope is commonly used). The specimen is then viewed using a special objective called the "immersion lens."

IMPACT

With microscopy's evolution and ongoing advances in the field, microscopy is now an integral part of clinical diagnostics. For instance, several imaging tools, such as confocal, multiphoton, and widefield microscopes, have been integrated to allow studies of tumor cell migration. Such imaging allows one to see intricate details, such as how the tumor cells interact with the extracellular matrix and if there is any likelihood of metastasis. These microscopic studies, which provide insight into the molecular basis of tumor cell migration, can, in the long run, help scientists develop anticancer therapies.

Malaria diagnosis is another example of advances in microscopy and their utility. Malaria is an infectious disease prevalent primarily in tropical regions. According to the World Health Organization, each year, it causes more than 1 million deaths worldwide. Traditionally, the standard method to confirm the presence of the malarial parasite in red blood cells relied on manual microscopy, which, in addition to being error-prone, is tedious. Scientists now combine computer vision with imaging tools to allow for malarial parasite diagnosis in thin blood smears. This feature will allow clinicians to treat malaria in the early stages.

Furthermore, the development of superresolution microscopes, such as those that incorporate multiphoton techniques in fiber optic microscopy and automated image analysis for high throughput screens, have allowed scientists and the pharmaceutical industry to improve current drug assays and also to equip them with newer and better disease models. Microscopy and imaging tools thus continue to play a critical role not only in traditional cell biology but also in more recent clinical diagnosis and drug discovery.

—Sibani Sengupta, PhD

Further Reading

Boray Tek, F., A. G. Dempster, and I. Kale. "Computer Vision for Microscopy Diagnosis of Malaria." *Malaria Journal,* vol. 8, 2009, p. 153.

Croft, William J. *Under the Microscope: A Brief History of Microscopy*. World Scientific, 2006.

Le Dévédec, S. E., et al. "Systems Microscopy Approaches to Understand Cancer Cell Migration and Metastasis." *Cellular and Molecular Life Sciences,* vol. 67, 2010, pp. 3219-40.

Tortora, Gerard J., Berdell R. Funke, and Christine L. Case. *Microbiology: An Introduction*. 10th ed., Benjamin Cummings, 2010.

Confocal microscopy

Categories: Microscopy, biotechnology
Also known as: Confocal laser scanning microscopy, laser confocal scanning microscopy
Anatomy or system affected: Cells
Specialties and related fields: Bacteriology, biochemistry, biotechnology, cell biology, pathology
Definition: optical imaging technique used when a high degree of contrast or reconstruction of a three-dimensional image is desired

KEY TERMS

fluorescence: the visible or invisible radiation emitted by certain substances because of incident radiation of a shorter wavelength such as X-rays or ultraviolet light

focus: the adjustment of a lens to produce a clear image

laser: a device that generates an intense beam of coherent monochromatic light (or other electromagnetic radiation) by stimulated emission of photons from excited atoms or molecules

resolution: the degree of detail visible in a microscopic image

SIGNIFICANCE

Confocal microscopy has rapidly gained popularity in forensic science as a method of choice for imaging evidence samples because confocal microscopes produce images of a quality superior to what can be achieved with conventional fluorescence microscopes.

Forensic scientists can use various microscopic methods to examine samples obtained from accident or crime scenes. The choice of technique is determined in part by the size of the target. Confocal microscopy utilizes point illumination and a pinhole in an optically conjugate plane to eliminate light flare, producing high-quality images.

Three types of confocal microscopes are available: confocal laser scanning microscopes (CLSMs), spinning-disk (Nipkow disk) confocal microscopes, and programmable array microscopes (PAMs). Modern instruments are highly evolved compared with the earliest versions. Still, the principles of confocal imaging established by Marvin Minsky in 1957 are shared by all confocal microscopes. The image formation method in confocal microscopes is fundamentally different from that of wide-field microscopes, which light entire specimens. Confocal microscopes produce in-focus images of thick specimens through optical sectioning (also known as Z-sections) using focused light beams. Using digital image-processing technology, serial (consecutive) images can be reassembled to construct three-dimensional representations of the sample or structures studied.

Before imaging with confocal microscopy, specimens are usually fixed and stained. The preparatory protocols (cutting, fixing, and staining of specimens) are primarily derived from those used in conventional microscopy. Specific regions of specimens (such as specific organelles) can be labeled with antibodies conjugated with fluorescent probes during the staining stage. By examining the relative distribution of epitopes of interest, investigators can ascertain many details about a sample, including the type of specimen, pathological condition, and phase in the cell cycle.

GFP fusion protein being expressed in Nicotiana benthamiana. *The fluorescence is visible by confocal microscopy. Photo by I, Synapomorphy, via Wikimedia Commons.*

Live-cell imaging and time-lapse imaging can be achieved with confocal microscopy, and inert and nonbiological specimens can also be examined using this technique. Forensic scientists can use confocal microscopes to examine evidence samples that are not easily visualized with conventional microscopes, such as the marks on bullets and cartridge cases and gunshot residue that is expelled when a firearm is discharged.

Another application of confocal microscopes in forensic science is in the analysis of paper documents. Specifically, confocal microscopy can enable an analyst to determine the sequence of two crossing strokes in different colors or different types of inks. Because confocal microscopes can capture serial images in various depths, scientists can identify the sequence in which marks were made on a given document with computer reconstruction imaging techniques.

—*Rena Christina Tabata*

Further Reading

Brzostowski, Joseph, and Haewon Sohn, editors. *Confocal Microscopy: Methods and Protocols* (Methods in Molecular Biology, 2304). Humana, 2021.

Matsumoto, Brian, editor. *Cell Biological Applications of Confocal Microscopy*. 2nd ed., Academic Press, 2002.

Paddock, Stephen W., editor. *Confocal Microscopy Methods and Protocols*. Humana Press, 1999.

Pawley, James B., editor. *Handbook of Biological Confocal Microscopy*. 3rd ed., Springer, 2006.

IMMUNOCYTOCHEMISTRY AND IMMUNOHISTOCHEMISTRY

Category: Medical specialties, procedures
Also known as: Immunostaining, immunocytochemistry, immunohistochemistry
Anatomy or system affected: Cells
Specialties and related fields: immunology, microbiology, oncology, pathology
Definition: immunocytochemistry (ICC) refers to using antibody-based protocols on cells to identify and localize visually specific molecules (antigens); immunohistochemistry (IHC) refers to similar procedures carried out on tissue sections

KEY TERMS

biopsy: an examination of tissue removed from a living body to discover the presence, cause, or extent of a disease

horseradish peroxidase (HRP): an enzyme found in the roots of horseradish used extensively in biochemistry applications

primary antibody: an antibody that binds directly to the target; the variable region of the primary antibody recognizes an epitope on the target; it is produced by a host organism that is of a different species than the specimen

secondary antibody: an antibody that binds to the primary antibody-target complex to capture the complex and to deliver a means of detecting the complex

tissue sectioning: cleanly and consistently cutting paraffin-embedded or frozen tissue into thin slices

INTRODUCTION

Immunocytochemistry and immunohistochemistry (ICC) uses antibody-based protocols on cells to identify and localize visually specific molecules (antigens). Immunohistochemistry (IHC) utilizes identical procedures, except that they are carried out on tissue sections. ICC and IHC help diagnose undifferentiated or metastatic cancers, leukemias, lymphomas, breast cancers.

WHY PERFORMED

IHC and ICC identify specific antigens on or in tissues or cells. These procedures are valuable in several aspects of the evaluation of human cancers. In cases of undifferentiated malignant tumors, the presence of antigens (such as tyrosinase) can confirm the origin of the cells, identifying the tumor (in this example, as a melanoma). In cases of metastatic tumors of unknown origin, the presence of antigens with highly restricted specificity can identify the primary site, as in the case of prostate-specific antigen (PSA) expression by a skeletal lesion. On occasion, subclassification of histologically similar tumors has prognostic implications, as in the case of embryonal carcinomas, which are distinguished from seminomas by the presence of keratin.

Leukemias and lymphomas are especially amenable to IHC-assisted diagnosis and classification because of the abundant and diverse surface antigens present on white blood cells. However, flow cytometry is sometimes favored in this class of malignancies. Several antigens of therapeutic importance can be detected by IHC, including estrogen and progesterone receptors in breast cancer. Finally, IHC can identify antigens associated with tumor cells' growth rate, such as Ki67. While primarily performed after the sample collection procedure, immunostains can also be performed rapidly as part of intraoperative consultations.

PATIENT PREPARATION

Samples taken as part of more extensive surgeries (such as mastectomy) require general anesthesia preceded by an overnight fast on the part of the patient. More straightforward biopsy procedures (such as bone marrow biopsy) may require only local anesthesia with or without sedation.

STEPS OF THE PROCEDURE

The first task of the pathologist is to construct a differential diagnosis and decide whether immunostaining is necessary. If so, the antigen or antigens to be interrogated must be selected based on specific hypotheses. Sample acquisition, immunostaining, and diagnostic interpretation are the next steps.

Individual cells for ICC can be scraped or brushed from surfaces, aspirated from cavities or compartments via a fine needle, or concentrated from body fluids by centrifugation. For IHC, tissue samples are usually obtained by standard biopsy or open procedures. Routine tissue processing involving formalin fixation, paraffin embedding, and sectioning is suitable for many antigen-antibody interactions. Still, in some cases, the interaction requires special sample handling such as frozen sectioning (to preserve the antigen) or heating (to increase exposure or "retrieve" the antigen).

Immunostaining itself requires exposure of the sample to a specific (primary) antibody under conditions that allow antigen-antibody binding to take place. Hundreds of primary antibodies against tumor-specific molecules are commercially available. In some cases, the primary antibody is conjugated to an enzyme such as horseradish peroxidase (HRP). This format is rapid, but this direct conjugate-labeled method has low sensitivity and requires high antibody concentrations. A more typical procedure involves a different (secondary) antibody that binds to the primary antibody's constant (Fc) region and is linked to HRP or another enzyme. Since multiple secondary antibodies can attach to a single primary antibody, sensitivity is improved. Other methods are possible, but all result in localizing a detectable "tag" molecule near the antigen(s) of interest. Typically used tags include enzymes such as HRP, alkaline phosphatase, glucose oxidase, and beta-galactosidase. Antibody-conjugated fluorescent molecules are sometimes used as tags. Yet, visualization requires specialized light sources and filters, and fluorescent tags degrade quickly. If higher magnification is needed, antibody-conjugated gold particles can also serve as tags and are visualized with the electron microscope. Gold particle sizes can be tightly controlled, facilitating double-labeling in individual sections.

Signal generation on samples with attached enzyme tags is achieved by incubation with a color-producing (chromogenic) substrate system that produces insoluble precipitates locally. Diaminobenzidine or aminoethylcarbazole are popular substrates for HRP because the catalysis products are stable and easily identified in the microscope as brown or red deposits, respectively. Glucose oxidase and its substrate tetrazolium are popular for double-labeling techniques since the blue reaction product is easily distinguished from products of HRP.

The final processing step is counterstaining and mounting slides. Counterstaining is necessary to visualize cells and structures around the immunostain. Hematoxylin, which stains nuclei blue, is a popular choice since many immunostains identify cell surface antigens. Appropriate positive and negative controls should always be stained and evaluated in parallel with the patient sample.

AFTER THE PROCEDURE

Samples collected as part of more extensive surgical procedures require standard postoperative care. More straightforward biopsy collection procedures require that the biopsy site be kept clean and dry until fully healed. Patients and caregivers are instructed to monitor for signs of infections.

RISKS

Immunostaining itself poses no additional risks to the patient. However, biopsies may cause some patients immense discomfort or pain under certain circumstances.

RESULTS

Meaningful results from the patient sample are impossible without appropriate staining of parallel positive and negative control samples. The most important attribute of an adequate positive control is the heterogeneous distribution of the stain within and among groups of cells. Results for patient samples are usually given in semiquantitative terms. They include estimates of staining intensity, cellular distribution (membranous, cytoplasmic, or nuclear), and abundance of positively staining cells. The report includes an interpretation that is either favored or ruled out by the observed staining pattern.

—John B. Welsh, MD, PhD

Further Reading

Al-Nafussi, Awatif. *Tumor Diagnosis: Practical Approach and Pattern Analysis*. Oxford UP, 2005.

Dabbs, David J., editor. *Diagnostic Immunohistochemistry: Theranostic and Genomic Applications*. 5th ed., Elsevier, 2018.

Leong A. S.-Y., and T. Y.-M. Leong. "Newer Developments in Immunohistology." *Journal of Clinical Pathology*, vol. 59, 2006, pp. 1117-26.

Leong, Anthony S.-Y., Kumarasen Cooper, and F. Joel W.-M. Leong. *Manual of Diagnostic Antibodies for Immunohistology*. 2nd ed., Greenwich Medical Media, 2002.

Microbial Processes

Microbes use many of the same metabolic pathways humans do. However, the metabolic diversity of microbes far outstrips that of animals. Consequently, microbes are model systems for biochemists who investigate metabolic pathways. This section examines the pathways and molecules made and used by microbes. Topics discussed include fermentation, photosynthesis, DNA and RNA synthesis, vitamins and minerals, and nitrogen fixation.

Chemotaxis

Category: Cell biology
Anatomy or system affected: Immune system
Specialties and related fields: Bacteriology, biochemistry, biotechnology, immunology, microbiology
Definition: the movement of a motile cell or organism, or part of an organism, in a direction corresponding to a gradient of increasing or decreasing concentrations of particular substances

KEY TERMS

chemoattractants: molecules that attract cells towards them
chemorepellents: molecules that drive cells away from them
flagella: a whiplike extension of cells that either rotates in bacteria or moves back and forth as in eukaryotic cells to propel the cell through a medium
methylation: the attachment of methyl (-CH_3) groups to molecules, usually proteins.
neutrophils: white blood cells filled with granules that phagocytose and destroy invading bacteria and cell debris

INTRODUCTION

Many microorganisms possess the ability to move toward a chemical environment favorable for growth. They will move toward a region that is rich in nutrients and other growth factors and away from chemical irritants that might damage them. Among the organisms that display this "chemotactic behavior," none is simpler than bacteria. Bacteria are single-celled "prokaryotic" microorganisms, which means that their deoxyribonucleic acid (DNA) is not contained within a well-defined nucleus surrounded by a nuclear membrane, as in "eukaryotic" (plant and animal) cells. "Prokaryotes" lack many cellular structures associated with more complex eukaryotic cells; nevertheless, many species of bacteria are capable of sensing chemicals in their environment and responding by movement.

BACTERIAL FLAGELLA

Bacteria capable of movement are called "motile bacteria." Not all bacteria are motile, but most species possess some form of motility. Although bacteria can move in three different ways, the most common means is by long, whiplike structures called "flagella."

Bacterial flagella are attached to cell surfaces and rotate like propellers to push the cells forward. A bacterial cell must overcome much resistance from the water through which it swims. Despite this, some bacteria can move at a velocity of almost 90 micrometers per second, equivalent to more than one hundred bacterial cell lengths per second.

A flagellum comprises three major structural components: the filament, the hook, and the basal body. The filament is a hollow cylinder composed of a protein called "flagellin." A single filament contains several thousand spherically shaped flagellin molecules bound in a spiral pattern, forming a long, thin cylinder. A typical filament is between 15 and 20 micrometers long but only 0.02 micrometers thick. The filament is attached to the cell through the hook and basal body. The hook is an L-shaped structure composed of protein and slightly wider than the filament. One hook end is connected to the filament, and the other is attached to the basal body. The basal body, also known as the "rotor," consists of protein rings embedded in the cell wall and plasma membrane. Inside these rings is a central rod attached to the hook. The central rod of the basal body rotates inside the rings, much like the shaft of a motor. As it rotates, it causes the hook and the filament to turn.

BACTERIA IN MOTION

While they are moving, bacteria change direction by reversing the rotation of their flagella. As a bacterium swims forward in a straight line, its flagella spin in a counterclockwise direction. Because of their structure, the flagella twist together when they rotate counterclockwise and act cooperatively to push the cell forward. The forward movement is referred to as a "run." Every few seconds, a chemical change in the basal body of each flagellum causes it to reverse its spin from counterclockwise to clockwise. When the flagella spin clockwise, they fly apart and can no longer work together to move the cell forward. The cell stops and tumbles randomly until the flagella reverse again, returning to counterclockwise spin and a forward run. This type of movement, in which the cell swims forward for a short distance and then randomly changes its direction, is called "run-and-tumble" movement.

Certain eukaryotic microorganisms, such as Euglena and some other protozoa, are also motile using flagella. The structure and activity of eukaryotic flagella are, however, completely different from those of bacteria. Eukaryotic flagella are composed of protein fibers called "microtubules," which move back and forth in a wavelike fashion to achieve movement. The rotation of bacterial flagella and the run and tumble movement they produce are unique to bacteria.

ATTRACTANTS AND REPELLANTS

Bacteria respond by chemotaxis to two broad classes of substances, "attractants" and "repellants." They move toward high concentrations of attractants (positive chemotaxis) and away from high concentrations of repellants (negative chemotaxis). Attractants are nutrients and growth factors most often, such as monosaccharides (simple sugars), amino acids (the building blocks of protein), and specific vitamins required for bacterial metabolism. Repellants include waste products given off by bacteria and other toxic substances found in the environment.

Bacteria respond to attractants and repellants by altering the time between tumbles in their run and tumble movement. When a bacterial cell detects an attractant, the time between tumbles and the runs' time increases. If the cell moves toward a higher concentration of an attractant, its runs will be longer. The opposite effect occurs when a cell encounters a repellant. A repellant causes the time between tumbles to decrease, resulting in shorter runs as the cell changes direction more frequently while avoiding the repellant. The net result is that the cell tends to move toward a lower concentration of the repellant.

CHEMOTACTIC RECEPTORS

Bacteria recognize attractants and repellants through specialized proteins called "chemotactic receptors," also called "methyl-accepting chemotactic proteins" (MCPs), which are embedded in their plasma membranes just inside the cell wall. Biologists have identified roughly twenty different receptors for attractants and some ten for repellants. Each receptor protein is believed to respond to only a single type of attractant or repellant.

When an attractant molecule binds to its chemotactic receptor, two separate events occur. First, there is a rapid activation of the receptor. The attractant molecule binds to a particular site on the receptor protein to form an activated receptor. However, this binding is not permanent, so a cell must remain in an area with attractant molecules to activate its receptors. The activated receptor sends a chemical signal to the basal bodies of flagella, which causes them to spin in a counterclockwise direction, producing continuous swimming in one direction.

At the same time, there is an adaptation of the activated receptors to the attractant. Adaptation is crucial because it keeps the cell from swimming too long in one direction. It is accomplished by "methylation" of the receptors. Methyl groups are attached to the protein by an enzyme in the cell. A methyl group consists of an atom of carbon attached to three atoms of hydrogen ($-CH_3$). Methylated receptors do not stimulate the basal bodies for counterclockwise rotation as effectively as nonmethylated receptors. After a cell has been in the presence of an attractant for a short while, its receptors adapt to the attractant, and it returns to the original pattern of run and tumble movement. Adaptation is reversed by "demethylation," the removal of methyl groups from the receptor by a separate enzyme. Together, the balance between methylation and demethylation makes the receptors very sensitive to small changes in attractant concentration so that cells remain in the region with the greatest concentration of attractant.

The action of repellants appears to be like that of attractants. Repellant molecules bind to sites on their chemotactic receptors, activating the receptors. The activated receptors signal the flagella to spin clockwise instead of counterclockwise, causing the cell to tumble and change direction. Repellant receptors also adapt through methylation and demethylation, much like attractant receptors.

It is not entirely understood how an activated chemotactic receptor can signal flagella to rotate. Four different proteins inside the bacterial cell have been identified as a possible link between the chemotactic receptors and the basal bodies of flagella. These proteins are believed to regulate flagellar rotation using a process called "phosphorylation." Phosphorylation, the attachment of phosphate molecules to a protein, is used in all types of cells as a kind of "on and off" switch to regulate protein activity.

EUKARYOTIC CHEMOTAXIS

The chemotactic mechanism employed by eukaryotic cells differs substantially from those employed by bacteria. The chemotactic receptors of eukaryotic cells are embedded in their plasma membranes. Some receptors bind chemoattractants and activate a signal transduction pathway that induces movement toward greater attractant concentrations. Chemorepellents also bind receptors that induce different signal transduction pathways that drive the cell away from large repellent concentrations toward lower concentrations of it. Many eukaryotic cells are polarized—they have a front side and a backside. Activation of chemoattractive receptors on one side of the cell changes that side into the front. The cell moves its new front side toward the increased concentration gradient of the chemoattractant. Conversely, when chemorepellents bind to receptors on the surface of a cell, the side of the cell where the receptors are activated becomes the new rear of the cell. The cell moves away from high concentrations of the chemorepellent.

In the immune system, white blood cells called "neutrophils" vigorously move toward molecules produced by inflammation, bacterial infection, and cell damage. Neutrophils respond to several molecules, including lipids like leukotrienes and platelet activation factor (PAF), peptides like N-formylated peptides from bacteria, and LL-37 (neutrophil granule- and epithelial cell-derived cathelicidin), protein fragments such as complement protein C5a, and proteins like chemokines. Some molecules, like leukotrienes and chemokines, are released by damaged tissues and summon neutrophils to come and clean up cell debris. Others, such as formylated bacterial peptides, result from damaged bacterial cells and summon neutrophils to sites of infection to phagocytose and kill invading bacteria. Still, others, like LL-37 and C5a, result from inflammation and summon neutrophils to breaches in the skin or other barriers to fight potential infections.

Most chemotactic receptors in the immune system are members of the G protein-coupled receptor family. These receptors pass through the cell membrane seven times and associate with a three-part protein called a "G protein." When the receptor binds its chemoattractant, the G protein inside the cell is activated. The activated G protein sends signals into the cell to remodel its cytoskeleton and reorient it towards the chemoattractant source. Cell movement and the direction it moves depend on its cytoskeleton. Therefore, intracellular signals that restructure the cytoskeleton, set in motion by chemoattractant binding to its receptor, change the direction the cell moves.

—*Jerald D. Hendrix and Michael A. Buratovich, PhD*

Further Reading

Gerhardt, Philipp, et al., editors. *Methods for General Bacteriology*. American Society for Microbiology, 1994.

Lodish, Harvey, et al. *Molecular Cell Biology*. 9th ed., WH Freeman, 2021.

Murray, Patrick R., et al. *Medical Microbiology*. 9th ed., Elsevier, 2020.

Prescott, Lansing, John P. Harley, and Donald A. Klein. *Microbiology*. 5th ed., McGraw-Hill, 2002.

Willey, Joanne, et al. *Prescott's Microbiology*. 11th ed., McGraw-Hill Education, 2019.

GLYCOLYSIS

Category: Biochemistry

Anatomy or system affected: Blood, cells, muscles, musculoskeletal system

Specialties and related fields: Biochemistry, cytology, exercise physiology, pharmacology, sports medicine

Definition: the chemical process of splitting a glucose molecule to obtain energy for other cellular processes; at times of intense activity, glycolysis produces most of the energy used by muscles

KEY TERMS

adenosine triphosphate (ATP): an important biological molecule that represents the energy currency of the cell; the energy in a special high-energy bond in ATP is used to drive almost all cellular processes that require energy

aerobic: occurring in the presence of oxygen

anaerobic: occurring in the absence of oxygen

cellular respiration: a complex series of chemical reactions by which chemical energy stored in the bonds of food molecules is released and used to form ATP

chemical energy: the energy locked up in the chemical bonds that hold the atoms of a molecule together; food molecules, such as glucose, contain much energy in their bonds

creatine phosphate: an energy-containing molecule present in significant quantities in muscle tissue; energy is stored in a high-energy bond like that of ATP

enzyme: a biological catalyst that speeds up a chemical reaction without itself being used up; enzymes are made of protein, and a single enzyme can usually only catalyze a single chemical reaction

nicotinamide adenine dinucleotide (NAD): a molecule used to hold pairs of electrons when they have been removed from a molecule by some biological process; the empty molecule is denoted by NAD+, while it is denoted as NADH when it is carrying electrons

STRUCTURE AND FUNCTIONS

Glycolysis is the first step in the process that cells use to extract energy from food molecules. Although energy can be extracted from most types of food molecules, glycolysis is usually considered to begin with glucose. The term "glycolysis" actually means the splitting (*lysis*) of glucose (*glyco*). This is a good description of the process since the glucose molecule is split into two halves. The glucose molecule consists of a backbone of six carbon atoms attached, in various ways, to twelve hydrogen atoms and six oxygen atoms. The glucose molecule is inherently stable and unlikely to split spontaneously at any appreciable rate.

When the energy is extracted from a glucose molecule, it is stored, for the short term, in a much less stable molecule called "adenosine triphosphate" (ATP). The ATP molecule consists of a complex organic molecule (adenosine), to which are attached three simple phosphate groups (see figure).

ATP consists of a five-carbon sugar called "ribose," linked on one side to the nitrogenous base adenine and on the other side to a linear chain of phosphate groups. The molecule formed by the attachment of adenine to ribose is called "adenosine," and the linkage of three phosphates generates adenosine triphosphate. The first phosphate is attached to the ribose sugar through a chemical bond whose energy is no greater than those bonds found anywhere else in the molecule. While the first phosphate is attached by what one could call a "normal" chemical bond, the second and third phosphates are attached by high-energy bonds. These are chemical bonds that require a considerable amount of energy to create. Thus, ATP is an ideal energy storage molecule that provides readily available energy for the biosynthetic reactions of the cell and other energy-requiring processes.

When one of the high-energy bonds of ATP is broken, a large amount of energy is released. Usually, only the bond holding the last phosphate is broken, producing a molecule of adenosine diphosphate (ADP) and a free phosphate group. The phosphate group is only split from ATP at the precise moment when some other process in the cell requires energy. This breaking of ATP provides the energy to drive cellular processes. The processes include activities such as the synthesis of molecules, the movement of molecules, and muscle contraction. The third phosphate can be reattached to ADP using the energy released from glycolysis or by other components of cellular respiration. The production of ATP can be diagrammed as follows: "energy from glycolysis + ADP + phosphate → ATP." Similarly, the breakdown of ATP can be diagrammed as "ATP → ADP + phosphate + usable energy." With this understanding of how ATP works, one can look at how glycolysis generates it in the cell.

The first step in the production of energy from sugar is an energy-consuming process. Since glucose is inherently a stable molecule, it must be activated before it will split. It is activated by attaching a phosphate group to each end of the six-carbon backbone. ATP supplies these phosphate groups. Therefore, glycolysis begins by using the energy from two ATP molecules. The atoms of the glucose molecule are also rearranged during the activation process so that it is changed into a very similar sugar, fructose. A fructose molecule with a phosphate group on either end is called "fructose 1,6-diphosphate." Thus, one can summarize the activation process as "glucose + 2 ATP → fructose 1,6-diphosphate + 2 ADP."

Fructose 1,6-diphosphate is a much more reactive molecule. It can be readily split by an enzyme called "aldolase" into two three-carbon compounds called "dihydroxyacetone phosphate" (DHAP) and "glyceraldehyde 3-phosphate" (G3P). DHAP is converted into G3P by an enzyme called "triose phosphate isomerase," making G3P the starting point for all the following steps of glycolysis. Each G3P undergoes several reactions, but only the more consequential reactions will be mentioned. G3P undergoes an oxidation reaction catalyzed by an enzyme called "glyceraldehyde 3-phosphate dehydrogenase." Oxidation reactions involve the loss of high-energy electrons. Electrons are highly energetic and have a negative electrical charge. They are picked up and carried by molecules specially designed for this purpose.

These energy-carrying molecules are called "nicotinamide adenine dinucleotide" (NAD). Biologists have agreed on a conventional notation for this molecule to allow the reader to know whether the molecule is carrying electrons or is empty. Since the empty molecule has a net positive charge, it is denoted as NAD+. When full, it holds a pair of electrons. One electron would neutralize the positive charge, while two results in a negative charge. The negative charge attracts one of the many hydrogen ions (H+) in the cell. Thus when carrying electrons, the molecule is denoted NADH. G3P surrenders two high-energy electrons to NAD+. The G3P molecule also picks up a free phosphate group at the end opposite from where one is already attached to form

1,3-bisphosphoglycerate. One can summarize the reaction as "2 Glyceraldehyde 3-phosphate + 2 NAD^+ + 2 inorganic phosphates → 2 1,3-bisphosphoglycerate + NADH + H^+." The following reactions merely transfer the energy in these chemical bonds to high-energy bonds by transferring these phosphate groups to ADP molecules to produce ATP. Since each G3P eventually produces two ATPs, and two G3Ps are produced from each original glucose molecule, glycolysis produces four ATP molecules altogether. However, since two ATPs were used to activate the glucose, the cell has a net gain of two ATP molecules for each glucose molecule used.

The rearrangement of the atoms leaves them in a form called "pyruvate." Pyruvate still contains much energy locked up in its chemical bonds. In most of the body's cells and most of the time, pyruvate will be further broken down and all of its energy released. This further breakdown of pyruvate requires oxygen and is beyond the scope of this topic. However, the complete breakdown of two molecules of pyruvate can produce more than thirty additional ATP molecules. With the addition of oxygen, the end products are the simple molecules of carbon dioxide and water.

The oxidative pathways that completely break down pyruvate are limited by the lack of oxygen in very active muscles. The ability to deal with electrons from NADH is also drastically reduced. Glycolysis can continue even in the absence of oxygen, but the electrons produced by glycolysis must be dealt with.

There is a minimal amount of NAD+ in each cell. NAD+ is designed to hold electrons briefly while they are transferred to some other system. In the absence of oxygen, the electrons are transferred to pyruvate. Since pyruvate cannot be broken down without oxygen, there is an ample supply. Transferring electrons from NADH to pyruvate allows the empty NAD+ to pick up more electrons produced by glycolysis. Therefore, glycolysis can continue producing two ATP molecules from each glucose molecule used. While two ATPs per glucose molecule is small compared to the more than thirty ATPs produced by oxidative metabolism, it is better than none.

The process of generating energy (ATPs) in the absence of oxygen is referred to as fermentation. Most people are familiar with the fermentation of grapes to produce wine. Yeast has the enzymes to transfer electrons from NADH to a pyruvate derivative and convert the resulting molecule into alcohol and carbon dioxide. No further energy is obtained from this process, and alcohol still contains much of the energy that was in glucose. Humans and other mammals have different enzymes than yeast cells. These enzymes transfer the electrons from NADH to pyruvate, producing lactate.

GLYCOLYSIS AND MUSCLE ACTIVITY

When yeast is fermented anaerobically (without oxygen), it will continue producing alcohol until it poisons itself. Most yeast cannot tolerate more than about 12 percent alcohol, the concentration found in most wine. The lactate produced by fermentation in humans is also poisonous. People, however, do not respire completely anaerobically. The two ATPs produced per glucose molecule used are not enough to supply the energy needs of most human cells. Muscle cells have to be somewhat of an exception. One asks the muscle cells to use energy much faster than one can supply them with oxygen. One may consider a muscle working under various physical activity levels and examine its oxygen requirements and waste products.

At rest, a muscle requires very little ATP energy. For an individual sitting on the couch watching television, energy demands are minimal. The lungs inhale and exhale slowly and take in enough oxygen to keep their concentration in the blood high. A relatively slow heart rate can pump enough of this oxygen-rich blood to the muscles to supply their very minimal needs. As soon as one uses a muscle, however, its ATP consumption increases dramatically. Even if an individual walks as far as the refrigerator, large quantities of ATP are required to cause the leg muscles to contract. Muscle cells maintain a constant level of ATP so that, as soon as one asks a muscle to contract, it can do so. The broken-down ATP is almost instantly regenerated from an additional energy store peculiar to muscle cells. Creatine phosphate is a molecule similar to ATP in that a high-energy bond attaches the phosphate group. There is more creatine phosphate in muscle cells than ATP. As soon as ATP is broken down, phosphates, and their high-energy bonds, are transferred from creatine phosphate. Within the first few seconds of activity, the ATP concentration in a muscle cell remains almost constant. However, the creatine phosphate level begins to drop.

As soon as the creatine phosphate concentration drops, the aerobic (oxygen-requiring) respiratory processes speed up. These processes completely break down glucose to carbon dioxide and water and release plenty of ATP. This ATP can then be used for muscle contraction. If the muscle has now stopped contracting, the new ATP produced will be used to rebuild the store of creatine phosphate.

Within the first minute or so of muscle contraction, the use of oxygen can be pretty high. The circulatory system has not yet responded to this increased oxygen demand. Muscle tissue, however, has a reserve of oxygen. The red color of most mammalian muscles is attributable to myoglobin, which is similar to hemoglobin in that it has a strong affinity for oxygen. The myoglobin stores oxygen directly in the muscle so that the muscle can operate aerobically. At the same time, the circulatory and respiratory systems adjust to the increased oxygen demand.

At low or moderate muscle activity, the carbon dioxide produced by aerobic respiration in muscles will increase the activity of both the circulatory and the respiratory systems. An increased blood flow supplies the increased demand for oxygen by the muscles. Jogging around a track or participating in aerobic exercises would be considered low to moderate muscular activity. Respiration rate and pulse rate both increase with jogging. This increase in oxygen supply to the muscles provides all that they need. The level of creatine phosphate will be lower than that in resting muscles. However, it will soon be replenished when the activity is stopped. The muscle cells have a good supply of food molecules in the form of glycogen. Glycogen is simply a long string of glucose molecules connected for convenient storage. The glycogen supply can last for hours at a rate of activity such as that created by jogging. Even after it is used up, glycogen stored in the liver can be broken down to glucose and carried to the muscles by the blood. An individual will probably want to stop jogging before their muscles will want to quit.

High levels of muscular activity pose a different set of problems. After more than a minute of vigorous exercise, the muscles begin to use ATP faster than oxygen can be supplied to regenerate it. The additional ATP is supplied by lactic acid fermentation. Glucose is only broken down as far as pyruvate, then converted to lactate by adding electrons from NADH. Lactate begins to accumulate in the muscle tissue. Since the body is still using large amounts of ATP but not taking in enough oxygen, it is said to enter a state of oxygen debt. When the muscular activity ends, the oxygen debt is repaid.

One can use an example of someone running to catch a bus, sprinting for fifty yards at full speed. That is not enough time for the circulation and lungs to respond to the increased demand for oxygen. The muscles have made up the difference between supply and demand with lactic acid fermentation. The individual now sits down in the bus and pants to repay their oxygen debt.

Some of the oxygen will go to replenish the store in muscle myoglobin. Some of it will be used in oxidative metabolism in the muscle to replenish creatine phosphate reserves. The rest deals with the accumulated lactate. The lactate is not all dealt with in the muscle where it was produced. Being a small molecule, it quickly enters the bloodstream. In muscles throughout the body, it is converted back to pyruvate. Pyruvate can then re-enter the oxidative pathway and be used to generate ATP with the use of oxygen. The lactate, then, is being used as a food molecule to supply the needs of resting muscle. Much of the lactate is metabolized in the liver. Some of it will be metabolized with oxygen to produce the energy to convert the rest of it back to glucose. The glucose can then be circulated in the blood or stored in the liver or muscles as glycogen. A minimal amount of lactate is excreted in the urine or sweat.

If the subject of the preceding example kept running at full speed, having missed the bus and run to the office, lactate would build up in the muscles and the blood. If the office were far enough away, the subject would eventually reach the point of exhaustion and stop running. At that point, the level of lactate in the leg muscles would be high enough to inhibit the enzymes of glycolysis. Glycolysis would slow down so that lactate would not become any more concentrated. The muscles' supply of creatine phosphate would be almost exhausted. Still, the ATP supply would be only slightly lower than in a resting muscle. The body is protected from damaging itself: Too much lactate would lower the pH to dangerous levels. The absolute lack of ATP causes muscles to lock, as in rigor mortis. The body's self-protection mechanisms force one to stop before either of these conditions exists. Once the subject stops running and pants long enough, they can continue. The additional oxygen taken in by increased respiration will have metabolized a sufficient amount of lactate to allow the muscles to start working again.

In cases where an individual has an inherited deficiency of particular enzymes of glycolysis, the consequences for muscle tissue are rather dire. Muscles, which depend heavily on glycolysis when operating under conditions of oxygen debt, fail to perform well if any of the glycolytic enzymes are defective. Symptoms include frequent muscle cramps, easy fatigability, and evidence of heavy muscle damage after strenuous exertion.

GLYCOLYSIS AND RED BLOOD CELL FUNCTION

Red blood cells are the oxygen-ferrying units of the bloodstream. They are filled with an iron-containing protein called "hemoglobin." Hemoglobin binds oxygen tightly when oxygen concentrations are high and releases oxygen when oxygen concentrations are low. Red blood cells must maintain the health and functionality of their hemoglobin stores to perform their task successfully, and glycolysis helps them do that. Approximately 90 to 95 percent of the glucose that enters the cell is metabolized to lactate through glycolysis and lactate dehydrogenase in red blood cells. The ATP generated by glycolysis brings charged atoms into the cell, such as calcium, potassium, and others. The NADH generated by glycolysis is also used to maintain the iron found in hemoglobin in a state that allows it to bind oxygen. Glycolysis is also used to form the metabolite 2,3-DPG (2,3-Diphosphoglycerate). 2,3-DPG binds to hemoglobin and forces it to release oxygen more readily when oxygen concentrations are low. Thus 2,3-DPG aids hemoglobin delivery of oxygen to the tissues.

Abnormalities in the enzymes that catalyze the reactions of glycolysis are inherited. Individuals who inherit two copies of a gene that encodes a glycolytic enzyme's mutant form experience uncontrolled red blood cells' uncontrolled destruction (hemolysis). The red blood cell destruction that results from defects in glycolytic enzymes is chronic and not eased by drugs. An enlarged spleen is a typical symptom of glycolytic enzyme abnormalities. The spleen tends to fill with dying red blood cells. The red blood cell destruction can be so severe that blood transfusions might be necessary. Removal of the spleen reduces red blood cell destruction.

INSULIN, DIABETES, AND GLYCOLYSIS

The hormones insulin and glucagon heavily regulate glycolysis. Insulin, a hormone made and released by the beta cells of the pancreatic islets, stimulates the insertion of the GLUT4 glucose transporter into the membranes of cells. People with type 1 diabetes mellitus, who are incapable of making sufficient quantities of insulin, tend to have very high blood sugar readings since their cells cannot receive the signal to insert the glucose transporter into their membranes and take up glucose from the blood. This prevents the removal of glucose from the blood. In type 1 diabetics, the blood glucose level climbs to abnormally high levels. GLUT4 allows the uptake of glucose without the input of energy. Therefore, glycolysis occurs as fast as the cells can take up glucose.

Insulin also stimulates the synthesis of a metabolite called "fructose 2,6-bisphosphate." Fructose 2,6-bisphosphate is a potent activator of phosphofructokinase, and activation of this enzyme ensures the activation of glycolysis. Insulin also activates the expression of genes that encode the protein involved in glycolysis. During uncontrolled diabetes, reduced glucose transport in muscle inhibits muscle cell glycolysis. In liver cells, reduced glycolytic gene expression and attenuation of the levels of fructose 2,6-bisphosphate reduce glycolysis. This contributes to voluntary muscle weakness, liver dysfunction, and heart problems sometimes observed in diabetics.

GLYCOLYSIS AND CANCER

Glucose uptake and its degradation by glycolysis occur ten times faster in tumor cells than in nontumor cells. This phenomenon, called the "Warburg effect," seems to benefit tumor cells since they lack an extensive capillary network to feed them oxygen and rely on anaerobic glycolysis to generate ATP.

Oxygen-poor conditions also induce the synthesis of a protein called "hypoxia-inducible factor" (HIF). HIF is a transcription factor that helps turn on the expression of specific genes that help cells survive oxygen-poor conditions. HIF activates the synthesis of at least eight glycolytic enzymes. These fundamental observations of cancer cells have shown that glycolytic enzymes are excellent potential drug targets for anticancer agents.

PERSPECTIVE AND PROSPECTS

Cellular respiration is how organisms harvest usable energy in the form of ATP molecules from food molecules. Lactic acid fermentation is used by human muscles when oxygen is in a limited supply. Glycolysis is the energy-producing component of lactic acid fermentation, which is much less efficient than aerobic cellular respiration. Fermentation harvests only two molecules of ATP for every glucose molecule used. In comparison, aerobic respiration produces a yield of more than thirty molecules of ATP. Most forms of life will resort to fermentation only when oxygen is absent or in short supply. While higher forms of life, such as humans, can obtain energy by fermentation for short periods, they incur an oxygen debt that must eventually be repaid. The yield of two molecules of ATP for each glucose molecule used is not enough to sustain their high energy demand.

Nevertheless, lactic acid fermentation is an essential source of ATP for humans during strenuous physical exercise. Even though it is an inefficient use of glucose, it can provide enough ATP for a short burst of activity. After the activity is over, the lactate produced must be dealt with, which usually requires oxygen.

Most popular exercise programs focus on aerobic activity. Aerobic exercises do not stress muscles to the point where the blood cannot supply enough oxygen. These exercises are designed to improve the efficiency of the oxygen delivery system so that there is less need for anaerobic metabolism. Training programs in general attempt to tune the body to reduce the need for lactic acid fermentation. They concentrate on improving the delivery of oxygen to the muscles, storing oxygen in the muscles, or increasing muscular contraction efficiency.

Insulin signaling activates glycolysis, whereas another pancreatic peptide hormone, glucagon, inhibits glycolysis. People with diabetes can suffer from inadequate glycolytic activity in particular organs, resulting in organ dysfunction. The expression of mutant forms of various glycolytic enzymes or supporting enzymes in transgenic mice has elucidated the link between abnormalities in glycolysis and the pathology of diabetes mellitus.

In the 1920s, the German biochemist Otto Warburg demonstrated that cancer cells voraciously take up glucose and metabolize it to lactate. Glycolysis is very active in cancer cells and helps them flourish under low-oxygen conditions. The development of new glycolytic inhibitors may constitute a new class of anticancer drugs with wide-ranging therapeutic applications.

—*James Waddell, PhD and Michael A. Buratovich, PhD*

Further Reading

Da Poian, Andrea T., and Miguel A. R. B. Castanho. *Integrative Human Biochemistry: A Textbook for Medical Biochemistry*. 2nd ed., Springer, 2021.

Denniston, Katherine, et al. *General, Organic, and Biochemistry*. 10th ed., McGraw-Hill Education, 2019.

Fox, Stuart Ira, and Krista Rompolski. *Human Physiology*. 16th ed., McGraw-Hill, 2021.

Lodish, Harvey, et al. *Molecular Cell Biology*. 9th ed., W. H. Freeman, 2021.

Nelson, David L., and Michael A. Cox. *Lehninger Principles of Biochemistry*. 8th ed., New York: W. H. Freeman, 2021.

Reece, Jane B., et al. *Campbell Biology: Concepts & Connections*. 8th ed., Pearson, 2020.

Wu, Chaodong, et al. "Regulation of Glycolysis: The Role of Insulin." *Experimental Gerontology*, vol. 40, 2005, pp. 894-99.

Fermentation

Category: Biochemistry, metabolism

Anatomy or system affected: Cells

Specialties and related fields: Bacteriology, biochemistry, biotechnology, microbiology

Definition: an oxygen-independent metabolic process that degrades complex organic compounds to carbon dioxide, organic acids, and alcohols

KEY TERMS

ATP: adenosine triphosphate; a triphosphorylated nucleotide that is the energy currency within cells

ethanol: C_2H_5OH, organic alcohol that is the end product of some types of fermentation

enzymes: biological catalysts that increase the rate of chemical reactions in living systems without being ultimately changed in the process

glycolysis: the breakdown of glucose by enzymes, releasing energy in the form of ATP and pyruvic acid.

isotypes: the different classes of antibodies

light chain: the smaller subunits of an antibody

monoclonal antibodies: antibodies derived from a single parent clonal cell

variable region: the N-terminal portion of the antibody, which possesses antigen-binding sites and has variable amino acid sequences

INTRODUCTION

Fermentation is a biological process in which a microorganism converts carbohydrates, typically starch or sugar, into simpler components such as alcohols, acids, and gases. More broadly, the term "fermentation" is also used to refer to any transformation of organic matter by enzymes. The most common groups of microorganisms involved in food fermentation include yeasts, bacteria, and molds that produce enzymes that catalyze the fermentation process.

In common usage, "fermentation" usually refers to an anaerobic process, which does not require oxygen. However, oxygen may be present in certain types of fermentation processes. Humans have been taking advantage of fermentation products for culinary purposes since the Neolithic period. However, the science behind this phenomenon was not explored in earnest until the mid-nineteenth century. The advances that this exploration set in motion enabled the practical applications of fermentation to expand into other fields, such as pharmaceuticals, biotechnology, and nutrition.

BACKGROUND

Fermentation has been used for culinary applications since at least several thousand years BCE. The earliest known example comes from China ca. 7000 BCE, where scientists have discovered evidence in pottery remnants of a beverage made from fermented rice, honey, and fruit. Fermented beverages have since become vital in societies across historical and geographical boundaries. Due to their once-mysterious combination of anesthetic, antiseptic, and intoxicating properties, alcoholic beverages, in particular, have played important social roles, both secular and religious, throughout the history of civilization.

OVERVIEW

Around the same time fermentation was beginning to be used in China, alcoholic beverages were also being fermented in the Middle East. Both beer and wine were essential offerings to the gods in ancient Egypt. Later, in ancient Greece and Rome, wine was the fermented beverage of choice. Indigenous South Americans produced a wide variety of fermented beverages using readily available ingredients, such as corn and cassava. As with the Egyptians, these beverages often played important roles in religious rituals. In medieval Europe, beer and wine were of particular importance, serving as sanitized versions of drinking water to prevent the rampant spread of waterborne infectious diseases.

Throughout the Middle Ages and beyond, fermented beverages continued to play important social, medical, and religious roles. Across Europe, beer brewing became a crucial part of supporting monasteries run by Trappists, a religious order of the Roman Catholic Church. Trappist monks brewed for their consumption, the community's consumption, and eventually as a primary means of funding their works through the sale of beer. On a social level, the community was often centered on tavern life and the consumption of beer. This tradition was brought to North America by European settlers in the seventeenth and eighteenth centuries.

The culinary applications of fermentation have by no means been limited to beverages. Since prehistoric times, fermented foods have been a dietary staple for most of the human race. Many kinds of bread and cheese are produced using fermentation, and bread is leavened using yeast fermentation. Many kinds of cheese are produced through mold fermentation by adding an enzyme complex called "rennet."

In the West, fermentation, except for cheeses, has always been more closely associated with yeast. Historically, East Asian cuisine has made a very different, more extensive use of fermented food. A crucial component of the East Asian diet since at least 300 BCE has been *Aspergillus oryzae*, also known as koji, a fungus used to ferment grains and soybeans. Koji has been used in East Asian cultures to produce such foods and beverages as miso, sake, and amazake for thousands of years.

In the nineteenth century, several advances were made in understanding the science behind fermentation, called "zymology" or "zymurgy." French microbiologist and chemist Louis Pasteur (1822-95) made the most important of these in the 1850s and 1860s. While studying fermentation processes, Pasteur proved that fermentation is brought on by living organisms rather than by spontaneous generation, as was previously believed. Further investigation showed that specific microorganisms cause specific types of fermentation—this knowledge allowed for increased control over the fermentation process.

The second significant advancement in the technology of fermentation was the result of experiments by German chemist Eduard Buchner (1860-1917), who discovered that living yeast cells do not need to be present for fermentation to occur, as the enzymes produced by the yeast cells are the agents responsible for fermentation. Buchner also discovered that fermentation could occur in the presence of oxygen. For his contributions to the science of fermentation, Buchner received the 1907 Nobel Prize in chemistry.

In addition to bread and cheese, many other foods are produced using fermentation. The processes used to make chocolate, yogurt, bread, sauerkraut, soy sauce, kimchi, sour cream, tempeh, poi, kombucha, traditional pickles, and vinegar involve fermentation in some capacity. Coffee is sometimes produced via a process called the "ferment and wash" method, in which the pulp of coffee cherries is removed from the bean by breaking down the cellulose through fermentation. Still, this process was predated by forms of dry processing and is being steadily replaced by machine-assisted wet-processing methods.

The twenty-first century has seen increased recognition of the nutritional value of fermentation. Proponents of incorporating fermented foods into one's diet cite numerous health benefits, including increased vitamin levels and improved digestibility. The fermentation process creates probiotics that make foods more easily digestible and enriches them with vitamins and enzymes. In addition, fermentation changes the texture and flavors of some foods, making them more palatable to human tastes. It also acts as a natural preservative and provides far more health benefits than chemical preservatives.

Increased knowledge of the nature of fermentation has also had a profound effect on the modern beer industry. The resurgence of the craft-beer industry in the twenty-first century would not have been possible without the work of scientists such as Pasteur. Expanded knowledge of the effect that different agents of fermentation, such as yeast and bacteria, have on the properties of the finished product is a crucial tool used by craft brewers.

Studies in zymology have led to significant developments in nonculinary fields as well. In the pharmaceutical and medical fields, fermentation is used to synthesize antibiotics, vitamins, and other drugs. A variety of valuable products can be created by way of industrial fermentation. Because one of the byproducts of fermentation is alcohol, often butanol and ethanol, the process can produce biofuel. Acetone, used as a cleaning solvent, is another byproduct of certain types of fermentation. Fermentation is also used to treat wastewater, a process that has the added benefit of producing usable biofuel and breaking down the waste product.

—Matthew Parsons, MA and Shari Parsons Miller, MA

Further Reading

Banschbach, Valerie S., and Robert Letovsky. "The Use of Corn versus Sugarcane to Produce Ethanol Fuel: A Fermentation Experiment for Environmental Studies." *American Biology Teacher*, vol. 72, no. 1, 2010, pp. 31-36.

Behme, Stefan. *Manufacturing of Pharmaceutical Proteins: From Technology to Economy*. Wiley, 2009.

Boulton, Christopher, and David Quain. *Brewing Yeast and Fermentation*. Blackwell, 2001.

El-Mansi, E. M. T., et al., editors. *Fermentation Microbiology and Biotechnology*. 3rd ed., CRC, 2012.

Katz, Sandor Ellix. *The Art of Fermentation: An In-Depth Exploration of Essential Concepts and Processes from around the World*. Chelsea Green, 2012.

Mitchell, David A., et al., editors. *Solid-State Fermentation Bioreactors: Fundamentals of Design and Operation*. Springer, 2006.

Sheppard, John, editor. *Introduction to Brewing and Fermentation Science: Essential Knowledge for Those Dedicated to Brewing Better Beer*. World Scientific, 2021.

OXIDATIVE PHOSPHORYLATION

Category: Biochemistry
Also known as: Cellular respiration
Anatomy or system affected: All systems
Specialties and related fields: Bacteriology, biochemistry, biotechnology, microbiology
Definition: the synthesis of ATP by phosphorylation of ADP for which energy is obtained by electron

transport and takes place in the mitochondria during aerobic respiration in eukaryotes and the cell membrane in bacteria

KEY TERMS

ATP: Adenosine 5'-triphosphate, or ATP, is the principal molecule for storing and transferring energy in cells

ATP synthase: an enzyme that catalyzes the synthesis of ADP or makes a phosphoanhydride bond to form ATP from ADP and inorganic phosphate.

electron transport chain (ETC): a series of protein complexes that transfer electrons from electron donors to electron acceptors via redox reactions and couples this electron transfer with the transfer of protons across a membrane

mitochondria: an organelle found in large numbers in most cells, in which the biochemical processes of respiration and energy production occur. It has a double membrane; the inner layer being folded inward to form layers (cristae)

OVERVIEW OF CELLULAR RESPIRATION

Oxidative phosphorylation is one of four processes that collectively comprise the cellular energy metabolism system. Cells break down macromolecules from food and use them to power cellular activity. Cellular metabolism is a series of chemical reactions that convert the energy from degraded macronutrients into adenosine triphosphate, or ATP. ATP is a biological molecule that stores energy in molecular bonds. ATP consists of a nucleotide base, adenosine, bonded to a sugar (ribose), and three phosphates linked by high-energy phosphodiesterase bonds. When phosphates are hydrolyzed from the molecule, energy is released, which can power cellular activities. By storing energy in a semistable state, ATP can move energy within a cell or a cellular system. ATP, therefore, serves as the basic unit of biological energy.

The breakdown of macronutrients begins anaerobically, which means that the process does not initially utilize oxygen. The first step in cellular respiration is glycolysis, which converts one molecule of glucose, a six-carbon sugar, into two pyruvate molecules, a three-carbon molecule. ATP is produced directly by glycolysis, and the process reduces nicotinamide adenine dinucleotide (NAD+), a cofactor molecule found in living cells. The reduction of NAD+ results in the conversion of NAD+ to NADH. NADH is one of the molecules needed for oxidative phosphorylation, and two molecules of NADH are produced during glycolysis.

The second stage in cellular respiration is pyruvate oxidation. The pyruvate produced by glycolysis enters the mitochondrion and is converted into a two-carbon molecule known as acetate. Acetate is linked to a carrier molecule called "coenzyme-A" during the oxidative process to generate acetyl CoA. Pyruvate oxidation produces another NADH as a byproduct, later used in oxidative phosphorylation. Carbon dioxide is another byproduct of pyruvate oxidation, and this carbon dioxide is exhaled from the body during respiration.

The third stage of cellular respiration is the Krebs cycle, citric acid cycle, or the tricarboxylic acid (TCA) cycle. This more complex process involves a cyclic series of chemical reactions that utilize acetyl CoA to produce byproducts used in the final stage, oxidative phosphorylation. The Krebs cycle has eight basic steps: Oxaloacetate is combined with acetyl CoA to create citrate, which is converted to isocitrate, which is oxidized to form alpha-ketoglutarate, which is then oxidized again to create succinyl CoA, which is converted to succinate, which is converted to fumarate, and then converted to malate, which is finally converted into oxaloacetate. The oxaloacetate produced in the eighth step of the process then recombines with another acetyl CoA produced by pyruvate oxidation to begin the cycle again. The Krebs cycle produces two molecules of carbon dioxide, three molecules of NADH, three hydrogen ions, one molecule of flavin adenine dinucleotide ($FADH_2$), and one molecule of guanosine triphosphate (GTP). Each molecule of glucose absorbed by the body requires two rotations of the cycle. So each molecule of glucose gives rise to two carbon dioxide, six NADH, six hydrogen ions, two $FADH_2$, and two GTP. GTP can be used directly by cells to form ribonucleic acid (RNA) or other processes. Still, the NADH and $FADH_2$ molecules are the most important products of the Krebs cycle in terms of the continuation of cellular respiration, as these are the basic inputs for oxidative phosphorylation.

THE FINAL STAGE OF RESPIRATION

Without oxygen, aerobic organisms quickly lose energy and eventually die. Aerobic cells deprived of energy can no longer "respire" or make ATP. Some energy is produced during glycolysis, pyruvate oxidation, and the Krebs cycle. Still, most of the ATP produced is derived from oxidative phosphorylation, which consists of two coupled processes: the electron transport chain (ETC) and chemiosmosis.

Two membranes surround the organelle within cells known as a mitochondrion. The outer membrane encloses an area known as the "intermembrane space," while the inner membrane surrounds the matrix. The membranes, made up of phospholipids, are similar to the membranes surrounding a bacterial cell. These membranes separate the intermembrane space from the matrix and create two compartments with strikingly distinct properties. There are a variety of proteins and other molecules embedded in the mitochondrial membranes, and these molecules are involved in the respiratory process. These proteins and molecules are organized into four complexes, called "complexes I to IV.:"

The ETC utilizes a series of chemical reactions made possible by membrane complexes that transport electrons across the inner membrane by transferring electrons between electron donor molecules and electron receptors. The energy created by transporting electrons shuttles protons across the membrane. As this occurs, the pH value of the intermembrane space becomes more acidic than the pH value within the matrix, resulting in what is called an "electrochemical gradient." The differential in pH values between the membrane spaces essentially means potential energy across the membrane as the system attempts to return to a balanced state. The primary purpose of the electron transport chain is to create this electrochemical gradient, and it is the potential energy from this gradient that drives cellular energy production and keeps cells and organisms alive.

Once the gradient is created, protons flow back through the channel through a special protein complex known as ATP synthase. This process is known as "chemiosmosis" because it is a form of osmosis, which is the movement of atoms or molecules from an area of high concentration to one of lower concentration, which ultimately equalizes the concentration on both sides of a membrane. As protons pass through ATP synthase, ADP molecules are phosphorylated, which means an additional phosphate group is added. ADP phosphorylation transforms adenosine diphosphate (ADP) into adenosine triphosphate (ATP), producing the energy needed by the cell.

The respiration process is slightly different when it occurs in an aerobic environment (one with oxygen) than in an anaerobic environment. In an aerobic environment, a molecule of oxygen acts as the final electron receptor at the end of the ETC. When this oxygen molecule accepts electrons and picks up protons from the chain, it produces water, which is one of the byproducts of cellular respiration. In aerobic systems, the system will shut down without oxygen. Some anaerobic organisms can use a molecule other than oxygen as the final electron receptor, such as nitrate, sulfate, or hydrogen ions. All processes that reduce compounds other than oxygen only occur in strictly anaerobic conditions since these processes are exquisitely oxygen sensitive. The electron carriers that provide electrons to the ETC are NADH and $FADH_2$, produced by glycolysis and the Krebs cycle.

During this process, NADH and $FADH_2$ are oxidized back into NAD+ and FAD, both of which then flow back into the respiration process, where they are again reduced by glycolysis and the Krebs cycle.

—*Micah Issitt*

Further Reading

Berg, J. M., et al. *Biochemistry*. 9th ed., W. H. Freeman, 2019.

Nelson, David L., and Michael M. Cox. *Lehninger Principles of Biochemistry*. 8th ed., W. H. Freeman, 2021.

"Oxidative Phosphorylation." *Science Direct*. Accessed 18 May 2020.

Zimmerman, Jerry J., et al. "Cellular Respiration." *Science Direct*. Accessed 18 May 2020.

Photosynthesis

Category: Plant biochemistry
Anatomy or system affected: None
Specialties and related fields: Biochemistry, biology, botany, cellular biology, photochemistry
Definition: the process by which green plants and some other organisms use sunlight to synthesize foods from carbon dioxide and water

KEY TERMS

photolysis: the splitting of molecules into component atoms and parts of molecules through the action of light energy

rubisco: a short-form name for the enzyme ribulose bisphosphate carboxylase/oxygenase, reportedly the most abundant protein on earth

Photosynthesis is the process by which organic sugar molecules (glucose) are synthesized from water and an inorganic carbon source (carbon dioxide or bicarbonate), using sunlight as the energy source to drive the process. Although most often associated with plants (in which the reactions of photosynthesis occur within compartments called chloroplasts), algae and certain types of bacteria are also capable of photosynthesis.

A RATHER VALUABLE PROCESS

From an ecological and evolutionary perspective, photosynthesis is significant because converting inorganic carbon to organic carbon represents the entry point of carbon atoms into biological systems. Photosynthesis is also significant because it is the means whereby oxygen is released into the atmosphere. The atmospheric concentration of oxygen is approximately 21 percent, and most of this oxygen originates from photosynthesis. In addition, solar energy absorbed during photosynthesis serves as the ultimate source of energy for almost all nonphotosynthetic organisms.

NATURE OF LIGHT

Light from the sun is composed of various types of electromagnetic radiation. Plants can use only a portion of this solar radiation for photosynthesis. This photosynthetically active radiation (PAR) ranges from 400 to 700 nanometers. It corresponds approximately to the visible light perceived by the human eye. The energy content of light depends on its wavelength, with shorter wavelengths having a higher energy content than longer wavelengths.

ROLE OF PIGMENTS

For light energy to drive photosynthesis, it first must be absorbed. Several types of photosynthetic pigments are found in plants. When these pigments absorb light, some of the pigments' electrons are elevated to a higher energy level. These high-energy electrons are used to drive the reactions of photosynthesis, thus converting light energy into chemical energy. The most common photosynthetic pigment in plants is the green-colored chlorophyll. Two types of chlorophyll are found in plants, chlorophyll 1, and chlorophyll b, with other types of chlorophyll found in various types of algae and photosynthetic bacteria.

Additional plant accessory pigments, such as carotenoids, which are yellow or orange, play a minor role in absorbing wavelengths of light not absorbed by chlorophyll. Carotenoids also help protect chlorophyll from damage that may occur due to absorbing excess light energy. As the most abundant plant pigment, chlorophyll gives plants an intense green color that typically masks the other colored pigments. However, in deciduous trees and shrubs, chlorophyll is degraded during the autumn, revealing a spectacular display of colors from carotenoids and other pigments.

REACTIONS OF PHOTOSYNTHESIS

The process of photosynthesis is complex, involving many biochemical reactions. Historically, the reactions of photosynthesis have been divided into light reactions and dark reactions. The light reactions include the absorption of light and converting light energy to chemical energy. The dark reactions use the chemical energy produced in the light reactions to incorporate (or fix) carbon dioxide molecules into organic molecules (sugars). The light reactions are localized in the internal membrane network called "thylakoid membranes" within the chloroplast.

The dark reactions occur in the aqueous region of the chloroplast called the "stroma." The term "dark reactions" is somewhat misleading because several photosynthetic enzymes are not active in the dark, so these reactions will not occur without light. Although it is common to separate the light and dark reactions when describing photosynthesis, they are tightly coupled and occur simultaneously in the plant.

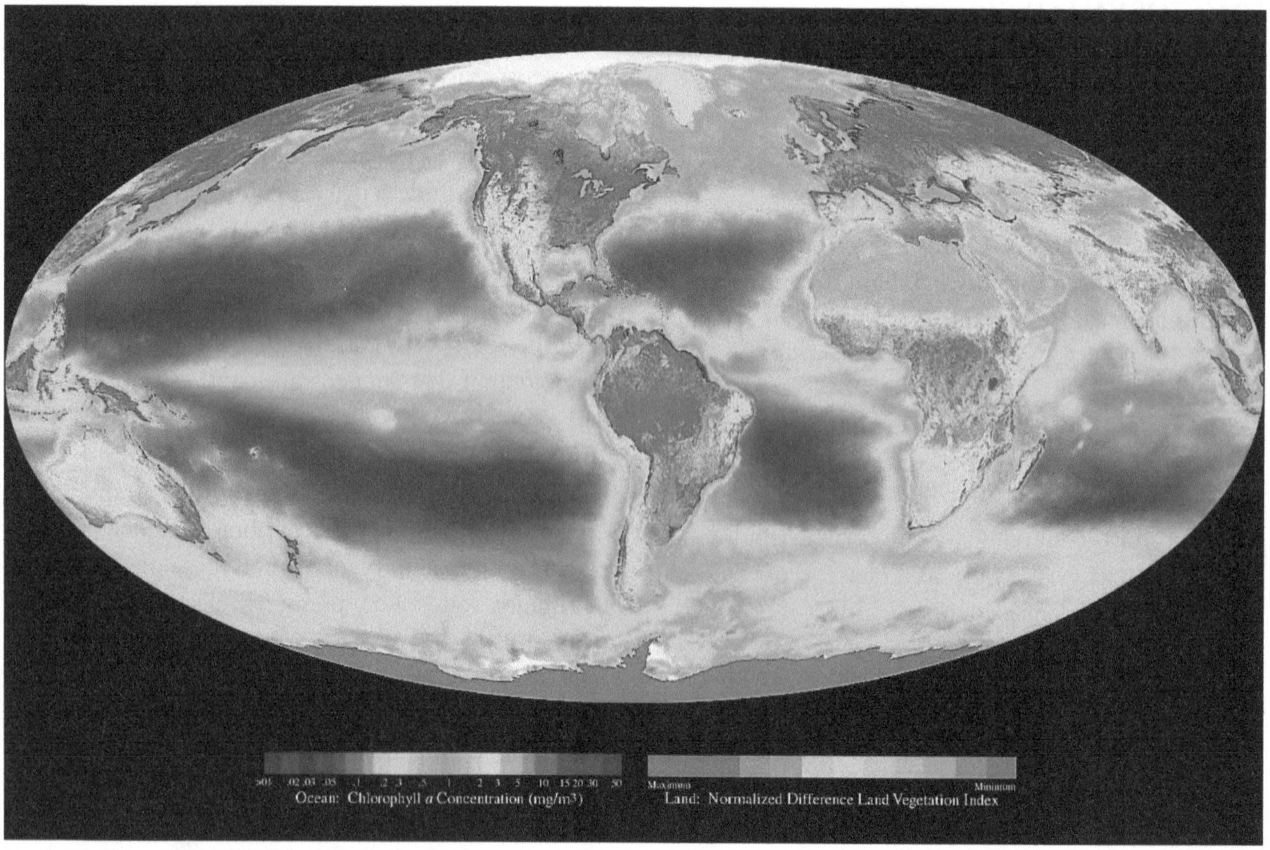

Composite image showing the global distribution of photosynthesis, including both oceanic phytoplankton and terrestrial vegetation. Image via Wikimedia Commons. [Public domain.]

LIGHT REACTIONS

Chlorophyll and other accessory pigments that absorb light energy, along with certain proteins, are organized into structures called photosystems. Two types of photosystems occur in plants, photosystem I and photosystem II, and both are embedded in the thylakoid membranes. When photosystems absorb light, the energy is transferred to special chlorophyll molecules, called "reaction center chlorophylls." The energy is transferred to electrons. High-energy electrons are released from the reaction centers. They are passed along the thylakoid membranes by electron transport molecules. Energy is extracted from the electrons as they are passed along. The energy transports protons (H^+) across the thylakoid membrane to the thylakoid interior. This process establishes a proton concentration gradient that is used to make ATP (adenosine triphosphate) from ADP (adenosine diphosphate) and inorganic phosphate (P_i) in a process called "photophosphorylation." The acceptor molecule $NADP^+$ (nicotinamide adenine dinucleotide phosphate, oxidized form) finally accepts the high-energy electrons and combines them with protons (H^+) to form the high-energy molecule NADPH (nicotinamide adenine dinucleotide phosphate, reduced form). This process is called "noncyclic electron flow." The ATP and NADPH produced are forms of chemical energy utilized by the dark reactions.

Many of the functions of photosystems I and II are similar. However, only photosystem II can split apart water molecules in the thylakoid interior into electrons, protons, and oxygen in a process called "photolysis." The electrons released from water during photolysis replace electrons lost by chlorophyll molecules during the electron transport reactions. The protons released from water accumulate in the thylakoid interior, adding to the concentration gradient established by noncyclic electron flow. Oxygen produced from the photolysis of water is released as a gas to the atmosphere. Accordingly, oxygen gas is considered to be a byproduct of plant photosynthesis. Algae and some bacteria (cyanobacteria) also release oxygen during photosynthesis. Still, other photosynthetic bacteria do not split water molecules and thus do not release oxygen.

The proton gradient across the thylakoid membranes represents a potential energy source used in making ATP. Protons cannot diffuse across the thylakoid membranes unless permitted to do so by a special enzyme complex called the "ATP synthase." Protons provide the energy required to generate ATP as they move through the ATP synthase from the thylakoid interior, where there is a high concentration of protons, to the stroma, where there is a lower concentration of protons. The energy associated with the proton gradient is analogous to a reservoir of water held back by a dam. Water may be allowed to pass through the dam through a turbine, thus using water to produce electrical power. The use of a proton gradient across a membrane as the energy source for the synthesis of ATP by ATP synthase is called "chemiosmosis." It occurs in both the chloroplast and the mitochondria. In chloroplasts, during photosynthesis, the process is known as photophosphorylation. In mitochondria, ATP is synthesized during oxidative phosphorylation (often referred to as "ox-phos"), a component of cellular respiration. ATP, like NADPH, is a high-energy molecule produced by light reactions and is consumed during dark reactions.

In a cyclic electron flow process, electrons can travel within the electron transport system as described above but are diverted to an acceptor in the electron transport chain between photosystems I and II. Passing through the chain back to photosystem I, the electrons enable the transport of protons across the thylakoid membrane, thus supplying power for the generation of ATP.

DARK REACTIONS

Carbon dioxide gas is a normal but minor component of the atmosphere. It enters leaves when air diffuses through stomata, small pores on the plant surfaces. The first reaction in converting carbon dioxide to sugar molecules occurs when the enzyme Rubisco (also known as ribulose bisphosphate carboxylase/oxygenase and reportedly the most abundant protein on earth) combines ribulose bisphosphate (RuBP) containing five carbon atoms with a carbon dioxide molecule to produce two identical molecules of simple sugar, each containing three carbon atoms. These three-carbon sugar molecules are then subsequently metabolized through a series of reactions leading to producing a three-carbon sugar called "glyceraldehyde 3-phosphate" (G3P). G3P is used to make glucose and other organic molecules, and the remaining G3P is used to regenerate RuBP so the process can continue. This cyclic pathway is known as the "Calvin cycle" (or the Calvin-Benson cycle).

Sugar products may be stored within the chloroplast as starch or transported as sucrose to other parts of the plant as needed. ATP and NADPH from the light reactions are required for several of the Calvin cycle reactions. Because the first molecule produced by this pathway is a simple sugar with three carbon atoms, this is known as the C_3 pathway. Plants using this pathway are known as C_3 plants, and common examples include many trees and the majority of crops.

In addition to catalyzing the uptake of carbon dioxide during the Calvin cycle, the Rubisco enzyme can use oxygen, initiating another metabolic pathway called "photorespiration." In photorespiration, when oxygen is attached to RuBP instead of carbon dioxide, a product results that cannot be used in the Calvin cycle, and that product must go through a different set of complex reactions. For this reason, photorespiration is often described as a wasteful process that competes with photosynthesis. The relative amounts of carbon dioxide and oxygen gases inside the chloroplast determine the relative rates of photosynthesis and photorespiration. Experiments in which the carbon dioxide concentration of air has been altered have demonstrated that the rate of photosynthesis increases and photorespiration decreases when the concentration of carbon dioxide increases. In some agricultural and horticultural greenhouse operations, carbon dioxide amounts in the atmosphere are elevated to stimulate photosynthesis, leading to plant production and yield increases.

Some plants have an adaptation whereby carbon dioxide is initially fixed by a pathway other than the Calvin cycle. This adaptation involves the enzyme phosphoenolpyruvate carboxylase (PEP carboxylase), an enzyme that lacks the oxygenase activity of Rubisco. PEP carboxylase attaches bicarbonate to phosphoenolpyruvate (PEP) in mesophyll cells that contact air spaces in the leaf. PEP is then converted into a series of organic acids and, in the process, is transported into a specialized set of cells called "bundle sheath cells" that are separated from the air spaces in the leaf. In the bundle sheath cells, carbon dioxide is released from the last organic acid in the series, which raises the carbon dioxide level in these cells where the carbon dioxide is used in the C_3 cycle. Raising the carbon dioxide concentration within the chloroplast increases photosynthesis while reducing photorespiration. The initial product of the pathway in the mesophyll cells is an organic acid with four carbon atoms. Thus, the pathway is called the "C_4 pathway." Plants possessing this pathway are known as C_4 plants. Examples include most grasses and a few crops, including corn and sugarcane.

A second adaptation that circumvents photorespiration is the CAM (crassulacean acid metabolism) pathway, named after the family of plants where it was discovered, *Crassulaceae*, or the stonecrop family. CAM photosynthesis is similar to C_4 photosynthesis. It is another adaptation for raising the concentration of carbon dioxide inside the chloroplast. CAM plants accomplish photosynthesis using a biochemical process essentially the same as C_4 plants. Still, instead of carrying out these reactions in separate cells, they carry out certain reactions at night. CAM plants open their stomata only at night. The carbon dioxide is transferred to PEP, converted into another organic acid stored throughout the night. During the day, the stomata remain closed, and the carbon dioxide needed for the C_3 cycle is supplied by releasing carbon dioxide from the last organic acid in the CAM cycle. Examples of CAM plants include cactus and pineapple. Both C_4 and CAM plants typically require less water than C_3 plants and may be found in warmer and drier environments. C_4 plants tend to have high rates of photosynthesis. In contrast, CAM plants have low photosynthetic rates because the CAM cycle is less efficient than the C_4 cycle. C_3 plants typically have intermediate photosynthetic rates under optimal conditions.

PHOTOSYNTHESIS AND THE ENVIRONMENT

Several environmental factors affect the rate of photosynthesis. For example, temperature extremes and water stress inhibit photosynthesis. As light intensity increases, so do photosynthetic rates. However, when photosynthesis becomes light-saturated, further increases in light intensity will not result in greater rates of photosynthesis. Leaves of plants that grow in full-sun conditions are smaller and thicker, with more extensive vascular systems than shade plants. Although so-called sun leaves and shade leaves have similar photosynthetic rates in low light, shade leaves have much lower rates of photosynthesis at high light intensities. They can be damaged when exposed to such conditions.

As mentioned above, atmospheric carbon dioxide concentrations can also regulate photosynthesis. At present, the concentration of carbon dioxide in the atmosphere is 0.0415 percent. Still, scientific data show that this concentration is increasing. Higher concentrations of atmospheric carbon dioxide may stimulate plant photosynthesis and plant growth but may have undesirable climatic effects.

—William J. Campbell

Further Reading

Blankenship, Robert E. *Molecular Mechanisms of Photosynthesis.* 2nd ed., Wiley-Blackwell, 2014.
Buchanan, Bob B., et al. *Biochemistry and Molecular Biology of Plants.* American Society of Plant Physiologists, 2000.
Hopkins, William G., and Normal P. A. Hüner. *Introduction to Plant Physiology.* 4th ed., John Wiley & Sons, 2008.
Rao, K. K., and D. O. Hall. *Photosynthesis.* 6th ed., Cambridge UP, 1999.

DNA AND RNA SYNTHESIS

Category: Biochemistry, genetics, molecular biology
Anatomy or system affected: All systems
Specialties and related fields: Bacteriology, biochemistry, biotechnology, genetics; molecular biology
Definition: the fundamental process of DNA and RNA synthesis

KEY TERMS

complementary strand: one of the two strands of nucleotides that make up a DNA molecule, with each nucleotide in one strand corresponding to the position of its complementary nucleotide (cytosine for guanine, adenine for thymine, and vice versa) in the other.

deoxyribonucleic acid (DNA): a large molecule formed by two complementary strands of nucleotides that encode all living organisms' genetic information

gene expression: the process by which RNA copies genes, which are specific segments of the DNA molecule, and uses the information to synthesize either proteins or other types of RNA

nucleotide: the fundamental structural component of DNA and RNA, consisting of a ribose (in RNA) or deoxyribose (in DNA) sugar molecule bonded to a phosphate group and one of five nucleobases: cytosine, adenine, guanine, thymine (DNA only), or uracil (RNA only) *polymerase chain reaction*: a laboratory method in which a minimal amount of DNA can be replicated thousands or even millions of times, using free nucleotides and an enzyme called "DNA polymerase"

ribonucleic acid (RNA): a category of large molecules, typically consisting of a single strand of nucleotides, that perform various functions in cells, including the transcription of DNA molecules and the transfer of specific genetic information for protein synthesis

UNDERSTANDING DNA AND RNA SYNTHESIS

It is tempting to oversimplify deoxyribonucleic acid (DNA) formation by likening it to zipping up a zipper, but that is perhaps the easiest way to visualize the process. Indeed, DNA synthesis is diagrammed in virtually all biochemistry texts as such. The analogy is further simplified by associating the "teeth" of the zipper with the purine and pyrimidine nucleotides from which the molecular structures of DNA and ribonucleic acid (RNA) are formed.

A nucleotide is formed when a purine or pyrimidine base and a phosphate group are chemically bonded to a sugar molecule. Only five different purine and pyrimidine bases are utilized in constructing

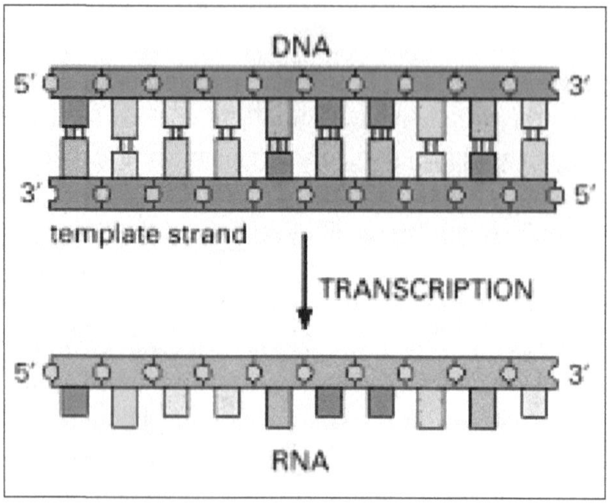

DNA trranscription produces a single-stranded RNA molecule that is complementary to one strand of DNA. Image via NCBI (ncbi.nlm.nih.gov).

DNA or RNA nucleotides. In DNA nucleotides, these are the bases adenine, cytosine, guanine, and thymine. At the same time, RNA uses the base uracil instead of thymine in its nucleotides. The different bases are indicated by the first letter of their names: A, C, T, G, and U. There are different mnemonic devices for recalling the complementarity of the different bases. One is that the curved letters C and G go together, as do the pointed letters A and T. Another is an easily remembered phrase such as "Cary Grant Ate Tacos." Any number of such devices can be used to suit an individual's personal preference.

The second significant difference between DNA and RNA is the nature of the sugars with which the nucleotides are constructed. In RNA nucleotides, the sugar molecule is ribose, a five-carbon simple sugar related to fructose. Sugar molecules are carbohydrates, indicating that each carbon atom is chemically bonded to both an H atom and an -OH group. These are the components of the water molecule, so the term indicates that each carbon (*carbo-*) atom in the molecule is hydrated (*-hydrate*). In DNA nucleotides, however, the sugar is deoxyribose. The name indicates that the molecule lacks an oxygen atom that is part of the ribose sugar molecular structure. This difference in the structure of the sugar portion of the nucleotide is subtle but of essential importance because it alone determines whether the molecule and its role in the biochemical process of life are DNA or RNA. In both DNA and RNA, the sugar molecules form a five-member ring structure made up of one oxygen atom and four of the five carbon atoms.

The third component of DNA/RNA nucleotides is the phosphate group, PO_4^{3-}, or P_i, generally referred to as "inorganic phosphate" when in that form and "phosphate" when bonded to another biomolecule, such as adenosine in adenosine triphosphate (ATP). The bonds between an oxygen atom in the phosphate group and other molecules are stable, even though they are deemed "high energy" bonds. In respiration and glycolysis (the decomposition of glucose), the bond between the third and second phosphate group in the triphosphate component is utilized to store energy by its formation and release energy when that bond is cleaved. Both DNA and RNA are large molecules. Their respective molecular structures consist of a long, biopolymer chain of alternating sugar molecules and phosphate groups. Each sugar component has a purine or pyrimidine base molecule bonded to the carbon atom on one side of the ring oxygen atom in both the DNA and RNA structures. The phosphate group bonded to a carbon atom on the other side of the ring oxygen atom. It is at this point that the difference between ribose and deoxyribose sugar becomes vitally important. The RNA molecule consists of a single strand made of nucleotides with the bases adenine, cytosine, guanine, and uracil. However, DNA molecules consist of two complementary strands of adenine, cytosine, guanine, and thymine nucleotides. Bases from each strand match up and connect to the corresponding bases in the other strand; adenine pairs with thymine, and cytosine pairs with guanine. The result is that a DNA molecule is considerably bigger than an RNA molecule. It has the form of a double helix as the two-component strands coil around each other. The RNA molecule consists of a single nucleotide strand that assumes different shapes according to its role in transcription and gene expression.

DNA/RNA SYNTHESIS SAMPLE PROBLEM

In 2005, a fossilized leg bone of Tyrannosaurus rex was found to contain viable tissue. Suppose a fragment of a DNA strand was recovered from such material and found to have the nucleotide order AGTTCGCGGAACTATTCG. What is the nucleotide order of the complementary strand in duplex DNA?

What is its RNA complement?

Answer: Recall the mnemonic device "Cary Grant Ate Tacos" (or whatever mnemonic you wish to use to signify that C/G and A/T are complementary nucleotides):

The order of the "found" DNA fragment is

AGTTCGCGGAACTATTCG

Place the complementary nucleotide below each one in the series, as

AGTTCGCGGAACTATTCG

TCAAGCGCCTTGATAAGC

This is the complementary sequence that would exist in a duplex DNA molecule. To generate the RNA complement, recall that RNA uses uracil (U) instead of thymine (T). The RNA complement is then easily defined by substituting U for T in the DNA complement, yielding

UCAAGCGCCUUGAUAAGC

(Note that listing the complete sequence of a human DNA molecule in this form would require a book of approximately one million closely printed pages. It is rather unlikely that T. rex could be cloned from the above fragment.)

DISCOVERY AND ANALYSIS OF DNA AND RNA MOLECULES

DNA was isolated from cell nuclei as early as 1869. Still, the fact that it bears genetic information was not known until 1943, when Oswald Avery, Colin MacLeod, and Maclyn McCarty demonstrated that introducing DNA from a virulent strain of the bacterium *Streptococcus pneumoniae* into a nonvirulent strain could transform the nonvirulent pneumococcal strain into a virulent strain. In 1953, James D. Watson and Francis Crick, with significant assistance from King's College biophysicist Maurice Wilkins, and the analytical and theoretical work of Rosalind Franklin, first published the discovery that the DNA molecule has a double helix form, an image now so well recognized. Since then, methods and techniques for manipulating and analyzing DNA have advanced, permitting biochemists and geneticists to understand better the role of genes and chromosomes in the DNA molecule. In 2001, the journal *Nature* published the first complete analysis of the human genome, which demonstrated, among other things, that all humans alive today are descended from a very few human populations that originated in Africa in the distant past. As the genome was deciphered over time, an understanding of the mechanisms by which DNA and RNA are synthesized was also obtained.

FORMATION OF DNA AND RNA IN LIVING CELLS: FOUR BASIC RULES

Both DNA and RNA are produced by copying a preexisting DNA strand according to the base pairings of adenine to thymine and cytosine to guanine. The DNA molecule must first "unzip" before it can replicate. Strand separation allows nucleotide fragments to form a complementary RNA strand in which uracil nucleotides replace thymine nucleotides. From this complementary RNA strand, a duplicate of the original DNA strand is assembled from other fragments. This process can be thought of as making a mold from one-half of the DNA molecule and then using it to cast a copy of the original.

Second, both RNA and DNA strands grow in one direction only. The phosphate group of each nucleotide is situated at the 5 position of the sugar molecule, the number indicating a specific location in the molecular structure according to the conventions for naming organic molecules. At the 3 position of each sugar molecule, there is a free hydroxyl (-OH) substituent that can form the phosphate ester bond with another nucleotide. The synthesis of a DNA or an RNA strand always proceeds from the

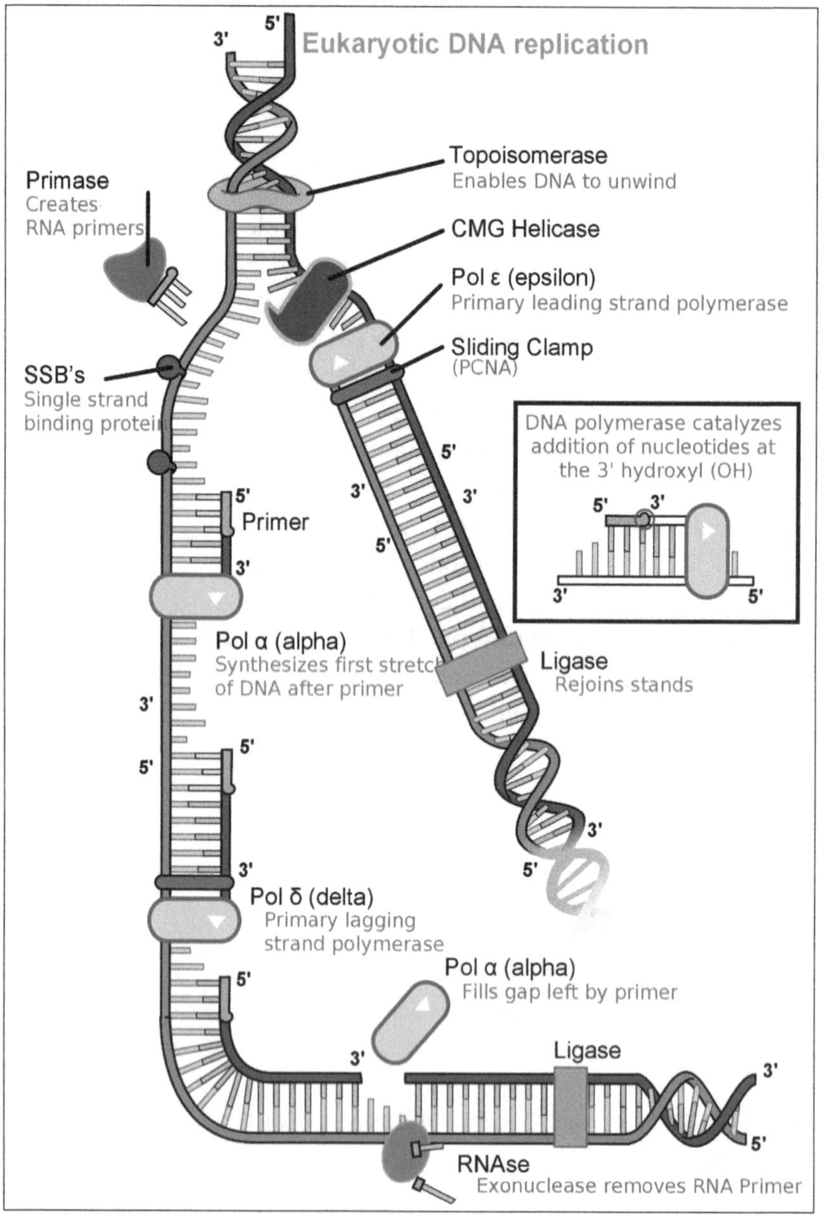

The enzyme DNA helicase unwinds the double-stranded DNA molecule. DNA polymerases only make DNA on single-stranded DNA. Single-stranded DNA-binding proteins bind non-specifically to single-stranded DNA to prevent them from re-annealing. Enzymes called "topoisomerases" work ahead of DNA helicase to unwind potential knots in the DNA. A DNA polymerase called "Pol ?" synthesizes DNA on the leading strand where synthesis is continuous toward the replication fork. On the lagging strand, DNA synthesis proceeds away from the replication fork and must constantly restart. DNA synthesis on the lagging strand is discontinuous. A DNA polymerase called "Pol ?" synthesizes DNA on the lagging strand. An RNA polymerase called "primase" synthesizes the RNA primer used by DNA polymerases to initiate DNA synthesis. After DNA synthesis, the RNA primer is removed by RNA endonucleases. The hole within the DNA strand is replaced by DNA polymerase ?, and the enzyme DNA ligase seals the gap in the DNA backbone. By LadyofHats, own work, via Wikimedia Commons.

5 -position in one nucleotide to the 3 -position in the next nucleotide as nucleotides are added in sequence.

Third, both DNA and RNA are synthesized by particular enzymes in polymerase chain reactions. RNA polymerases produce new strands of RNA, and DNA polymerases produce new DNA strands in DNA/RNA polymerase reactions. Through the transcription process, a new RNA strand is produced as RNA polymerases transcribe the nucleotide pattern of the parent DNA strand. RNA polymerase enzymes can initiate the formation of a new strand by coordinating to an appropriate site on a duplex strand of DNA (the double helix form of the molecule), where they temporarily separate the two strands of the DNA molecule and begin the process of assembling a new RNA strand from the corresponding nucleotides. DNA polymerases are not able to initiate the formation of a new strand directly. Instead, the process requires the formation of a primer, a DNA, or an RNA segment that is bound to the parent DNA strand and acts as the template for the new DNA strand. Both RNA and DNA polymerases contain several protein subunits, each of which carries out a specific function.

Fourth, synthesis of a new duplex DNA strand proceeds only from a particular formation known as a "replication fork." Specific enzymes function to open the duplex strand, allowing other enzymes to assemble the matching complementary strands from the appropriate nucleotides. As the new strands are formed, other enzymes still function to rejoin the strands as the growing fork progresses along the length of the template duplex DNA molecule. An essential aspect of duplex DNA replication is that, because the strands only grow in one direction, the growth directions on the two branches of the growing fork are opposite. The new strand on the "leading" branch of the fork grows continuously, nucleotide by nucleotide. The new strand on the "trailing" branch of the fork is assembled instead in bits and pieces from various nucleotide segments. A replication fork in eukaryotic cells is shown below.

AMPLIFYING DNA FOR ANALYSIS

Among the many techniques and methods that have been developed for the manipulation of DNA samples—and that are especially important for the science of DNA "fingerprinting"—probably the most important is DNA amplification. By this method, a minute sample of DNA, as might be obtained from just a few hair follicles found at a crime scene, for example, undergoes repeated replications so that enough of the DNA is present to produce a clear fragmentation pattern that is the "fingerprint" of that particular DNA. This methodology has been used to convict criminals who might otherwise have gone free, as well as to free individuals from prison who had been wrongfully convicted of crimes they did not commit.

—Richard M. Renneboog, MSc

Further Reading

Berg, Jeremy M., et al. *Biochemistry*. 9th ed., W. H. Freeman, 2019.

"The Human Genome." *Nature, vol.* 409, no. 6822, 2001, pp. 813-958.

Lafferty, Peter, and Julian Rowe, editors. *The Hutchinson Dictionary of Science*. 2nd ed., Helicon, 1998.

Lodish, Harvey, et al. *Molecular Cell Biology*. 9th ed., Freeman, 2021.

Mitra, Sandhya. *Genetic Engineering: Principles and Practice*. 2nd ed., McGraw Hill Education, 2015.

Nelson, David L., and Michael M. Cox. *Lehninger Principles of Biochemistry*. 8th ed., W. H. Freeman, 2021.

Pelczar, Michael J., Jr., E. C. S. Chan, and Noel R. Krieg. *Microbiology: Concepts and Applications*. McGraw, 1993.

Watson, James D., et al. *Molecular Biology of the Gene*. 7th ed., Pearson, 2013.

Protein synthesis

Category: Molecular genetics
Anatomy or system affected: All
Specialties and related fields: Biochemistry, microbiology, molecular genetics

Definition: a core biological process, occurring inside cells, balancing the loss of cellular proteins through the production of new proteins

KEY TERMS

amino acid: the basic subunit of a protein; there are twenty commonly occurring amino acids, any of which may join together by chemical bonds to form a complex protein molecule

peptide bond: the chemical bond between amino acids in protein

polypeptide: a linear molecule composed of amino acids joined together by peptide bonds; all proteins are functional polypeptides

ribonucleic acid (RNA): a nucleic acid (chain of nucleotides) that serves various functions concerning deoxyribonucleic acid (DNA), including protein synthesis and gene expression, and regulation

translation: the process of forming proteins according to instructions contained in an RNA molecule

SIGNIFICANCE

Cellular proteins can be grouped into two general categories: proteins with a structural function that contribute to the three-dimensional organization of cells and proteins with an enzymatic function that catalyze the biochemical reactions required for cell growth and function. Understanding how proteins are synthesized provides insight into how a cell organizes itself and how defects in this process can lead to disease.

THE FLOW OF INFORMATION FROM STORED TO ACTIVE FORM

The cell can be viewed as a unit that assembles resources from its environment into biochemically functional molecules, then organizes these molecules in three-dimensional space to allow for cell growth and replication. A cell must have the biosynthetic means to assemble resources into molecules to carry out this process. It must contain the information required to produce the biosynthetic and structural machinery. Deoxyribonucleic acid, or DNA, represents the stored form of this information, whereas proteins are the end product. There are thousands of different proteins in cells, either serving a structural role or acting as enzymes that catalyze the biosynthetic reactions of the cell. Following the discovery of the structure of DNA in 1953 by James Watson and Francis Crick, scientists began to study the process by which the information stored in DNA is converted into protein molecules.

BACTERIAL PROTEIN SYNTHESIS

Bacterial DNA is transcribed into messenger ribonucleic acid (mRNA) and then translated into protein. Proteins are linear, functional molecules composed of a unique sequence of amino acids. About twenty different amino acids are used to synthesize proteins. Although each protein sequence of amino acids is present in each DNA molecule, DNA cannot synthesize proteins by itself. Another similar molecule, called "ribonucleic acid" (RNA), is necessary to decode the information contained in DNA and do the work of assembling proteins.

There are various types of RNA, each one distinguished by its function. The process of protein synthesis involves three types of RNA. Messenger RNA (mRNA) copies the information in a cell's DNA—the sequence of amino acids that makes up a particular protein—and carries it to a part of the cell called the "ribosome," where protein synthesis occurs. Transfer RNA (tRNA) decodes the information carried by the mRNA and then transports the required amino acids to the appropriate location during synthesis. Ribosomal RNA (rRNA) acts as the engine that carries out most of the steps during protein synthesis. With a specific set of proteins, rRNA forms ribosomes that bind the mRNA, serve as the platform for tRNA to decode the mRNA, and catalyze the formation of peptide bonds between amino acids. Each ribosome is composed of two subunits: a small (40S) and a large

(60S) subunit, each of which has its function. The "S" in the 40S and 60S is an abbreviation for Svedberg units, which measure how quickly a large molecule or complex molecular structure sediments (or sinks) to the bottom of a centrifuge tube while being centrifuged. The larger the number, the larger the molecule.

Like all RNA, mRNA is composed of just four types of nucleotides: adenine (A), guanine (G), cytosine (C), and uracil (U). Each type bonds only with one other type: guanine with cytosine and adenine with uracil (or, in DNA, thymine). (DNA contains thymine in the place of uracil, but the other three bases are the same.) therefore, when an mRNA molecule temporarily bonds with a DNA molecule, it forms a mirror image of the nucleotide sequence contained in the DNA. This step is called "transcription."

Transcription is the first step in the process. A DNA molecule composed of a linear sequence of nucleotides gives rise to a protein molecule composed of a linear sequence of amino acids. This process is called "translation" since it converts the "language" of nucleotides that make up DNA into the "language" of amino acids that make up a protein. This is achieved through three-nucleotide sequences called "codons." The four nucleotides—A, G, C, and U—mean sixty-four possible codons. Each codon (save for some exceptions) corresponds to a specific amino acid. Only twenty amino acids are coded for in DNA, and most amino acids correspond to several different codons. For example, six codons (UCU, UCC, UCA, UCG, AGU, and AGC) specify the amino acid serine, whereas only one (AUG) specifies the amino acid methionine. A molecule of mRNA, therefore, is simply a linear array of codons (that is, three-nucleotide "words" that are "read" by tRNAs together with ribosomes). The region within an mRNA containing this sequence of codons is called the "coding region."

Before translation can occur in eukaryotic cells, mRNA molecules undergo processing steps at both ends to add features that will be necessary for translation. (These processing steps do not occur in prokaryotic cells.) Nucleotides are structured such that they have two ends, a 5 and a 3 end, that are available to form chemical bonds with other nucleotides. Each nucleotide present in an mRNA has a 5-to-3 orientation that gives directionality to the mRNA so that the RNA begins with a 5 end and finishes in a 3 end. The ribosome reads the coding region of an mRNA in a 5-to-3 direction. Following the synthesis of an mRNA from its DNA template, one guanine is added to the 5 end of the mRNA in an in-

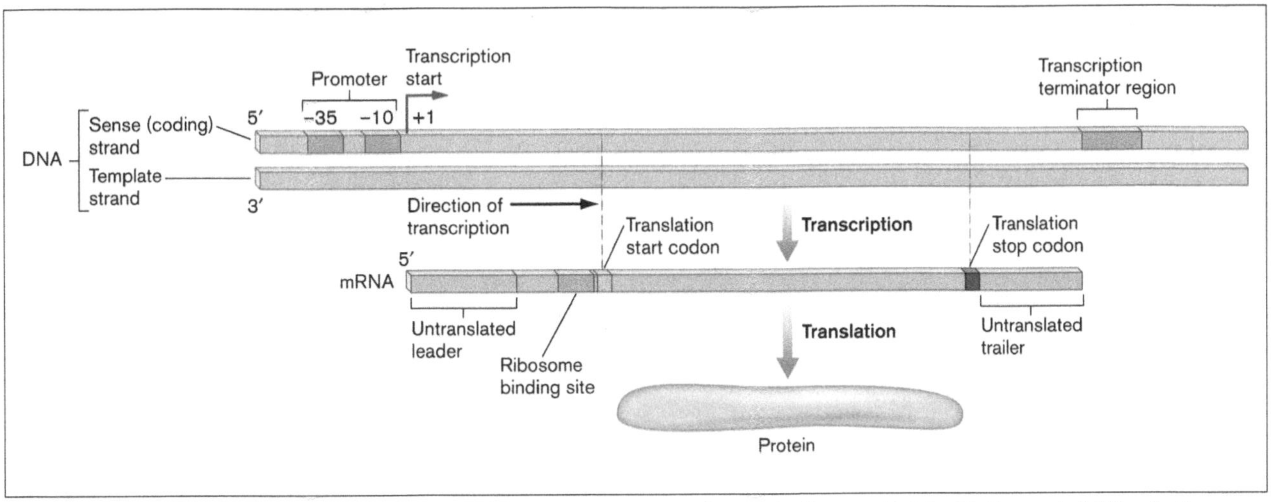

Bacterial DNA is transcribed into mRNA and then translated into protein. Image by Joan L. Slonczewski, John W. Foster, via Wikimedia Commons.

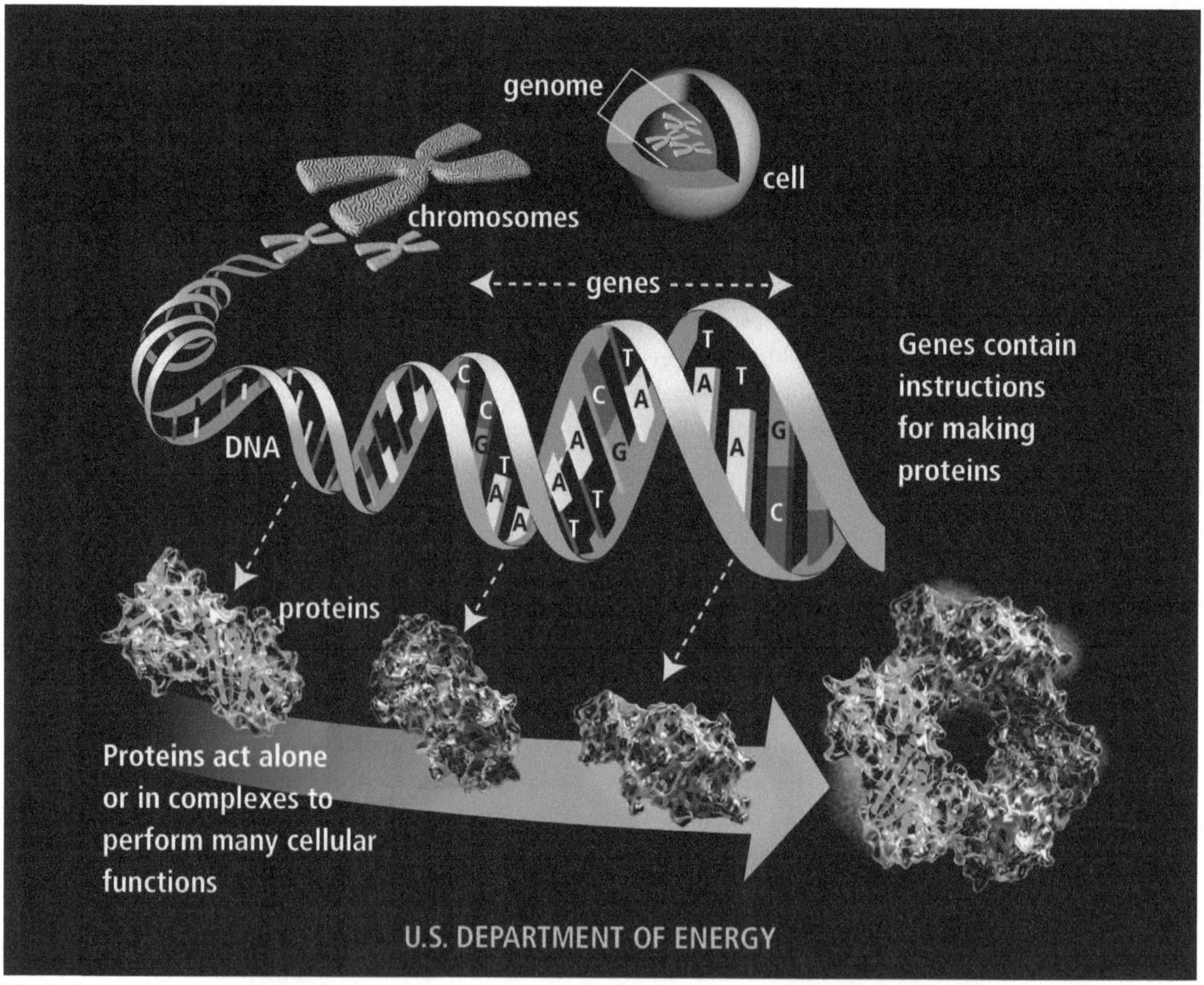

The fundamental steps in protein synthesis. Image by US Department of Energy Human Genome Program. [Public domain.]

verted orientation. It is the only nucleotide in the entire mRNA present in a 3 to 5 orientation and is referred to as the cap. A long stretch of adenosine is added to the 3 end of the mRNA to make the poly-A tail.

Typically, mRNAs have a stretch of nucleotide sequence that lies between the cap and the coding region. This is referred to as the leader sequence and is not translated. Therefore, a signal is necessary to indicate where the coding region initiates. The codon AUG usually serves as this initiation codon; however, other AUG codons may be present in the coding region. Any one of three possible codons (UGA, UAG, or UAA) can serve as stop codons that signal the ribosome to terminate translation. Several accessory proteins assist ribosomes in binding mRNA and help carry out the required steps during translation.

THE TRANSLATION PROCESS: INITIATION

Translation occurs in three phases: initiation, elongation, and termination. The function of the 40S ribosomal subunit is to bind to an mRNA and locate the correct AUG as the initiation codon. It does this by binding close to the cap at the 5 end of the mRNA and scanning the nucleotide sequence in its 5 to 3 direction in search of the initiation codon. Marilyn Kozak identified a certain nucleotide sequence surrounding the initiator AUG of eukaryotic mRNAs that indicate that this AUG is the initiation codon to the ribosome. She found that the presence of an A or G three nucleotides before the AUG and a G in the position immediately following the AUG were critical in identifying the correct AUG as the initiation codon. This is referred to as the "sequence context" of the initiation codon. Therefore, as the 40S ribosomal subunit scans the leader sequence of an mRNA in a 5 to 3 direction, it searches for the first AUG and may bypass other AUGs, not in this context.

Nahum Sonenberg demonstrated that the scanning process by the 40S subunit could be impeded by the presence of stem-loop structures present in the leader sequence. These form from a base pairing between complementary nucleotides in the leader sequence. Two nucleotides are complementary when they join together by hydrogen bonds. For instance, the nucleotide (or base) A is complementary to U, and these two can form what is called a "base pair." Likewise, the nucleotides C and G are complementary. Several accessory proteins, called "eukaryotic initiation factors" (eIFs), aid the binding and scanning of 40S subunits. The first of these, eIF4F, consists of three subunits called "eIF4E, eIF4A, and eIF4G." The protein eIF4E is the subunit responsible for recognizing and binding to the cap of the mRNA. The eIF4A subunit of eIF4F, together with another factor called "eIF4B," removes the presence of stem-loop structures in the leader sequence through the disruption of the base pairing between nucleotides in the stem-loop. The protein eIF4G is the large subunit of eIF4F, and it serves to interact with several other proteins, one of which is eIF3. This latter initiation factor is that the 40S subunit first associates with during its initial binding to an mRNA.

Through the combined action of eIF4G and eIF3, the 40S subunit is bound to the mRNA, and through the action of eIF4A and eIF4B, the mRNA is prepared for 40S subunit scanning. As the cellular concentration of eIF4E is very low, mRNAs must compete for this protein. Those that do not compete well for eIF4E will not be translated efficiently. This represents one means by which a cell can regulate protein synthesis. One class of mRNA that competes poorly for eIF4E encodes growth-factor proteins. Growth factors are required in small amounts to stimulate cellular growth. Sonenberg has shown that the overproduction of eIF4E in animal cells reduces the competition for this protein. Messenger RNAs such as growth-factor mRNAs that were previously poorly translated when the concentration of eIF4E was low are now translated at a higher rate when eIF4E is abundant. This, in turn, results in the overproduction of growth factors, which leads to uncontrolled growth, a characteristic typical of cancer cells.

A protein that specifically binds to the poly-A tail at the 3 end of an mRNA is called the "poly-A-binding protein" (PABP). Discovered in the 1970s, the only function of this protein was thought to be to protect the mRNA from attack at its 3 end by enzymes that degrade RNA. Daniel Gallie demonstrated another function for PABP by showing that the PABP-poly-A-tail complex was required for the function of the eIF4F-cap complex during translation initiation. The idea that a protein located at the 3 end of an mRNA should participate in events occurring at the opposite end of an mRNA seemed strange initially. However, RNA is quite flexible and is rarely present in a straight, linear form in the cellular environment. Consequently, the poly-A tail can easily approach the cap at the 5' end. Gallie showed that PABP interacts with eIF4G and eIF4B, two initiation factors closely associated with the cap, through protein-to-protein contacts. The consequence of this interaction is that the 3 end of an mRNA is held in close physical proximity to its cap. The interaction between these proteins stabilizes their binding to the mRNA, promoting protein synthesis. Therefore, mRNAs can be thought of as adopting a circular form during translation that looks similar to a snake biting its tail. This idea is now widely accepted by scientists.

One additional factor, called "eIF2," is needed to bring the first tRNA to the 40S subunit. Along with the initiator tRNA (which decodes the AUG codon specifying the amino acid methionine), eIF2 aids the 40S subunit in identifying the AUG initiation. Once the 40S subunit has located the initiation codon, the 60S ribosomal subunit joins the 40S subunit to form the intact 80S ribosome. (Svedberg units are not additive; therefore, the 40S and 60S units joined together do not make a 100s unit.) This marks the end of the initiation phase of translation.

THE TRANSLATION PROCESS: ELONGATION AND TERMINATION

During the elongation phase, tRNAs bind to the 80S ribosome as it passes over the codons of the mRNA, and the amino acids attached to the tRNAs are transferred to the growing polypeptide. The binding of the tRNAs to the ribosome is assisted by an accessory protein called "eukaryotic elongation factor 1" (eEF1). The appropriate tRNA decodes a codon through base pairing between the three nucleotides that make up the codon in the mRNA and three complementary nucleotides within a specific region (called the "anticodon") within the tRNA. The tRNA binding sites in the 80s ribosome are located in the 60S subunit. In a process known as translocation, the ribosome moves over the coding region one codon at a time or in steps of three nucleotides. When the ribosome moves to the next codon to be decoded, the tRNA containing the appropriate anticodon will bind tightly in the open site in the 60S subunit (the A site). The tRNA bound to the previous codon is present in a second site in the 60S subunit (the P site). Once a new tRNA has bound to the A site, the ribosomal RNA itself catalyzes the formation of a peptide bond between the growing polypeptide and the new amino acid. This results in the transfer of the polypeptide attached to the tRNA present in the P site to the amino acid on the tRNA present in the A site. A second elongation factor, eEF2, catalyzes the ribosome's movement to the next codon to be decoded. This process is repeated one codon at a time until a stop codon is reached.

The termination phase of translation begins when the ribosome reaches one of the three termination or stop codons. These are also called "nonsense" codons, as the cell does not produce any tRNAs that can decode them. Accessory factors, called "release factors," are also required to assist this stage of translation. They bind to the empty A site where the stop codon is present. This triggers the cleavage of the bond between the completed protein from the last tRNA in the P site, releasing the protein. The ribosome then dissociates into its 40S and 60S subunits, the latter of which diffuse away from the mRNA. The close physical proximity of the cap and poly-A tail of an mRNA maintained by the interaction between PABP and the initiation factors (eIF4G and eIF4B) is thought to assist the recycling of the 40S subunit back to the 5 end of the mRNA to participate in a subsequent round of translation.

IMPACT AND APPLICATIONS

The elucidation of the process and control of protein synthesis provides a ready means by which scientists can manipulate these processes in cells. In addition to infectious diseases, insufficient dietary protein represents one of the greatest challenges to world health. The majority of people now living are limited to obtaining their dietary protein solely through the consumption of plant matter. Knowledge of protein synthesis may allow molecular biologists to increase the amount of protein in important crop species. Moreover, most plants contain an imbalance in the amino acids needed in the human diet that can lead to disease. For example, protein from corn is poor in the amino acid lysine. In contrast, the protein from soybeans is poor in methionine and cysteine. Molecular biologists may be able to correct this imbalance by changing the codons present in plant genes, thus improving this source of protein for those people who rely on it for life.

—*Daniel R. Gallia, PhD*

Further Reading

Atkins, John F., Raymond F. Gesteland, and Thomas R. Cech, editors. *RNA Worlds: From Life's Origins to Diversity in Gene Regulation*. Cold Spring Harbor Laboratory, 2011.

Bethaz, Carlo, and Vito Li Puma, editors. *New Research on Protein Synthesis*. Nova, 2014.

Crick, Francis H. C. "The Genetic Code: III." *Scientific American*, Oct. 1966, pp. 55-62.

Keiler, Kenneth C., editor. *Bacterial Regulatory RNA: Methods and Protocols*. Humana, 2012.

Krebs, Jocelyn E., Elliott S. Goldstein, and Stephen T. Kilpatrick, eds. *Lewin's Genes XI*. 11th ed., Jones, 2014.

Lake, James A. "The Ribosome." *Scientific American*, Aug. 1981, pp. 84-97.

Liljas, Anders, and Mans Ehrenberg. *Structural Aspects of Protein Synthesis*. 2nd ed., World Scientific, 2013.

Rich, Alexander, and Sung Hou Kim. "The Three-Dimensional Structure of Transfer RNA." *Scientific American*, Jan. 1978, pp. 52-62.

Tropp, Burton E. *Molecular Biology: Genes to Proteins*. 4th ed., Jones, 2012.

Whitford, David. *Proteins: Structure and Function*. Wiley, 2005.

LIPIDS

Category: Biochemistry

Anatomy or system affected: All systems

Specialties and related fields: Biochemistry, cytology, nutrition, vascular medicine

Definition: organic compounds found in the tissues of plants and animals that serve as energy-storage molecules, function as solvents for water-insoluble vitamins, provide insulation against the loss of body heat, act as a protective cushion for vital organs, and are structural components of cell membranes

KEY TERMS

alcohol: an organic compound containing a hydroxyl (-OH) group attached to a carbon atom

carboxylic acid: an organic compound that contains the carboxyl (-CO2H) group

ester: the relatively non-water-soluble compound formed when an alcohol reacts with a carboxylic acid

fatty acid: an organic compound that is composed of a long hydrocarbon chain with a carboxyl group at one end

glycerol: three-carbon alcohol that has one hydroxyl compound on each carbon atom

hydrocarbon: an organic compound composed of only hydrogen and carbon atoms that does not dissolve in water (water-insoluble)

hydrophilic: "water-loving" or "water-attracting"; a term given to molecules or regions of molecules that interact favorably with water

hydrophobic: "water-hating" or "water-repelling"; a term given to molecules or regions of molecules that do not interact favorably with water

saponification: a reaction in which a strong basic solution splits a molecule into a carboxylic acid unit and an alcohol unit

STRUCTURE AND FUNCTIONS

Lipids are a class of bio-organic compounds, typically insoluble in water and relatively soluble in organic solvents such as alcohols, ethers, and hydrocarbons. Unlike the other classes of organic molecules found in biological systems (carbohydrates, proteins, and nucleic acids), lipids possess a unifying physical property-solubility behavior rather than a unifying structural feature. Fats, oils, some vitamins and hormones, and most of the nonprotein components of cell membranes are lipids.

There are two categories of lipids—those that undergo saponification and those that are nonsaponifiable. The saponifiable lipids can be divided into simple and complex lipids. Simple lipids, composed of carbon, hydrogen, and oxygen, yield fatty acids and an alcohol upon saponification. Complex lipids contain additional elements, such as phosphorus, nitrogen, and sulfur, yielding fatty acids, alcohol, and other compounds on saponification.

The fatty acid building blocks of saponifiable lipids may be either saturated, which means that as many hydrogen atoms as possible are attached to the carbon chain, or unsaturated, which means that at least two hydrogen atoms are missing. Saturated fatty acids are white solids at room temperature. In contrast, unsaturated ones are liquids at room temperature because of a geometrical difference in the long carbon chains. The carbon atoms of a saturated fatty acid are arranged in a zigzag or accordion configuration. These chains are stacked on top of one another in a very orderly and efficient fashion, making it difficult to separate the chains from one another. When carbons in the chain are missing hydrogen atoms, the regular zigzag of the chain is disrupted, leading to

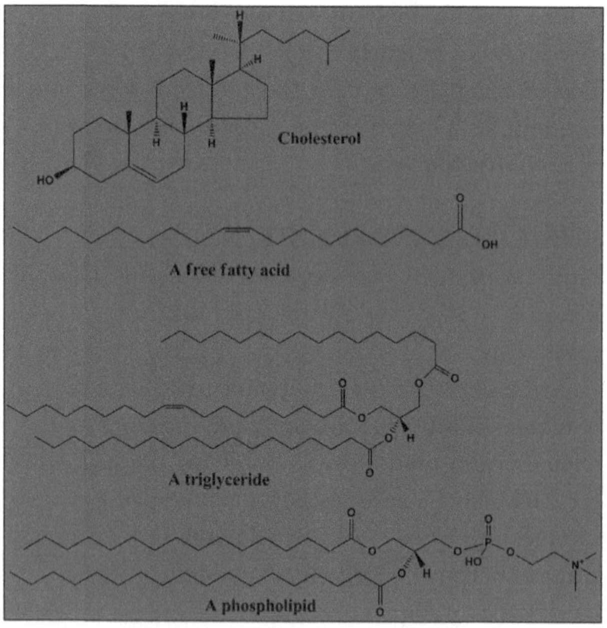

Structures of some common lipids: cholesterol and oleic acid (top); a triglyceride composed of oleoyl, stearoyl, and palmitoyl chains attached to a glycerol backbone (middle); the common phospholipid phosphatidylcholine (bottom). Image by Lmaps at the English Wikipedia, via Wikimedia Commons.

less efficient packing, which allows the chains to be separated more easily. Saturated fatty acids have a higher melting temperature because they require more energy to separate their chains than unsaturated fatty acids. Unsaturated fatty acids can be converted into saturated ones by adding hydrogen atoms through a process called "hydrogenation."

Simple lipids can be divided into triglycerides and waxes. Waxes such as beeswax, lanolin (from lamb's wool), and carnauba wax (from a palm tree) are esters formed from an alcohol with a long carbon chain and a fatty acid. These compounds, which are solids at room temperature, serve as protective coatings. Most plant leaves are coated with a wax film to prevent attack by microorganisms and loss of water through evaporation. Animal fur and bird feathers have a wax coating. For example, the wax coating on their feathers is what allows ducks to stay afloat.

Edible fats and oils such as lard (pig fat), tallow (beef fat), corn oil, and butter are triglycerides. Triglyceride molecules are fatty acid esters in which three fatty acids (all saturated, all unsaturated, or mixed) combine with one molecule of the alcohol glycerol. Oils are triglycerides that are liquid at room temperature, while fats are solid at room temperature. A triglyceride's fluidity depends on the nature of its fatty acid chains; the more unsaturated the triglyceride, the more fluid its structure. The triglycerides found in animals tend to have more saturated fatty acids than do those found in plants. Vegetable oils and fish oils are frequently polyunsaturated.

Complex lipids are classified as phospholipids or glycolipids. Structurally, phospholipids are composed of fatty acids and a phosphate group. Glycerol-based lipids called "phosphoglycerides" contain glycerol, two fatty acids, and a phosphate group. The phosphoglyceride structure contains a hydrophilic (polar) head, the phosphate unit, and two hydrophobic (nonpolar) fatty acid tails. The polar head can interact strongly with water, while the nonpolar tails interact strongly with organic solvents and avoid water.

Egg yolks contain a large amount of the phosphoglyceride phosphatidylcholine (also called "lecithin"). This lipid is used to form the emulsion mayonnaise from oil and vinegar. Usually, oil and water do not mix. The hydrophobic oil forms a separate layer on top of the water. Since lecithin's structure contains both a hydrophobic and a hydrophilic region, it can attach to the water with its polar head and the oil with its nonpolar tail, preventing the two materials from separating. Lipids derived from the alcohol sphingosine are called "sphingolipids." They contain one fatty acid, one long hydrocarbon chain, and a phosphate group. Like the phosphoglycerides, sphingolipids have a head-and-two-tail structure. Sphingolipids are important components in the protective and insulating coating that surrounds nerves.

Glycolipids differ from phospholipids in that they possess a sugar group in place of the phosphate group. Their structure is again the polar head and dual tail arrangement in which the sugar is the hydrophilic unit. Cerebrosides, sphingosine-based glycolipids containing a simple sugar such as galactose or glucose, are found in large amounts in the brain's white matter and the myelin sheath. Gangliosides, which are found in the brain's gray matter, in neural tissue, and the receptor sites for neurotransmitters, contain a more complex sugar component.

Nonsaponifiable lipids do not contain esters of fatty acids as their primary structural feature. Steroids are an important class of nonsaponifiable lipids. All steroids possess an identical four-ring framework called the "steroid nucleus." Still, they differ in the groups that are attached to their ring systems. Examples of steroids are cholesterol, the bile acids secreted by the liver, the sex hormones, corticosteroids secreted by the adrenal cortex, and digitoxin from the digitalis plant, which is used to treat heart disease.

Lipids constitute about 50 percent of the mass of most animal cell membranes. Biological membranes control the chemical environment of the space they enclose. They are selective filters controlling what substances enter and exit the cell since they constitute a relatively impermeable barrier against most water-soluble molecules. The three types of lipids involved are phospholipids (most abundant), glycolipids, and cholesterol. When surrounded by an aqueous environment, phospholipids tend to organize into a double layer of lipid molecules, a bilayer, allowing their hydrophobic tails to be buried internally and their hydrophilic heads to be exposed to water. These phospholipids have one saturated and one unsaturated tail. Differences in tail length and saturation influence the packing efficiency of the molecules and affect the membrane's fluidity. Short, unsaturated tails increase the fluidity of the membrane. Cholesterol is essential in maintaining the mechanical stability of the lipid bilayer, thereby preventing a change from the fluid state to a rigid crystalline state. It also decreases the permeability of small water-soluble molecules.

The lipid bilayer provides the basic structure of the membrane and serves as a two-dimensional solvent for protein molecules. Protein molecules are responsible for most membrane functions; for example, they can provide receptor sites, catalyze reactions, or transport molecules across the membrane. These proteins may extend across the bilayer (transmembrane proteins) or be associated with only one face of the bilayer. Cell membranes also have carbohydrates attached to the outer face of the bilayer. These carbohydrates are bound to membrane proteins or part of a glycolipid. Typically, 2 to 10 percent of a membrane's total weight is carbohydrate. Evidence exists that cell-surface carbohydrates are used as recognition sites for chemical processes.

Lipids play an important role in health and well-being. The body acquires lipids directly from dietary lipids and indirectly by converting other nutrients into lipids. There are two fatty acids, linoleic and linolenic acids, which are called "essential fatty acids." Since these fatty acids cannot be synthesized in the body in sufficient amounts, their supply must come directly from dietary sources. Fortunately, these acids are widely found in foodstuffs, so deficiency is rarely observed in adults.

About 95 percent of the lipids in foods are triglycerides, which provide 30 to 50 percent of the calories in an average diet. Triglycerides produce 4,000 calories per pound, compared to the 1,800 calories per pound produced by carbohydrates or proteins. Since triglyceride is such an efficient energy source, the body converts carbohydrates and proteins into adipose (reserve fatty) tissue for storage to be used when extra fuel is required.

While carbohydrates and proteins undergo significant degradation in the stomach, triglycerides remain intact, forming large globules that float to the top of the mixture. Fats spend longer than other nutrients in the stomach, slowing molecular activity before continuing into the intestines. Thus, a fat-laden meal gives more prolonged satiety than a low-fat one.

In the small intestine, bile salts split fat globules into smaller droplets, allowing enzymes called "lipases" to saponify the triglycerides. The fatty acids at the two ends are removed in some instances, leaving one attached as a monoglyceride. About 97 percent of dietary triglycerides are absorbed into the bloodstream; the remainder is excreted. Although glycerol and fatty acids with short carbon chains are water-soluble enough to dissolve in the blood, the long-chain fatty acids and monoglycerides are not. These insoluble materials recombine to form new triglycerides. Since these hydrophobic triglycerides would form large globules if excreted directly into the blood, small triglyceride droplets are surrounded with a protective protein coat that can dissolve in water, taking the encapsulated triglyceride with it. This structure is an example of a lipoprotein.

Cholesterol is found in relatively small (milligram) quantities in foods, compared to triglycerides. Cholesterol supplies raw materials for bile salt production and to be used as a structural constituent of brain and nerve tissue. Since these functions are essential to animals but serve no purpose in plants, cholesterol is found only in animals. Only about 50 percent of dietary cholesterol is absorbed into the blood; the rest is excreted. Much of the body's supply of cholesterol is produced in the liver. For most individuals, the amount of cholesterol synthesized in the body is larger than the amount absorbed directly from the diet.

Digested lipids released from the intestine and those synthesized in the liver compose the lipid content of the blood. The fatty acids required by the liver are obtained directly from the bloodstream or by synthesis from sources such as glucose, amino acids, and alcohol. Liver-synthesized triglycerides are incorporated into lipoprotein packages before entering the bloodstream. There are three types of lipoprotein packages that transport lipids to and from the liver. Very-low-density lipoproteins (VLDLs) transport triglycerides to tissues; low-density lipoproteins (LDLs) transport the cholesterol from the liver to other cells; and high-density lipoproteins (HDLs) transport cholesterol from other tissues to the liver for destruction.

DISORDERS AND DISEASES

Lipid consumption is an important dietary concern. Lipid deficiency is rarely observed in adults but can occur in infants who are fed nonfat formulas. Since fatty acids are essential for growth, lipid consumption should not be restricted in individuals under two years of age. Excess lipid consumption is associated with health problems such as obesity and cardiovascular disease. Although excess calories from any dietary source can lead to obesity, the body must expend less energy to store dietary fat than to store dietary carbohydrates as body fat. Thus, high-fat diets produce more body fat than do high-carbohydrate, low-fat diets.

Atherosclerosis, or "hardening of the arteries," is the leading cause of cardiovascular disease. A strong correlation exists between diets high in saturated fats and the incidence of atherosclerosis. In this condition, deposits called "plaques," which have a high cholesterol content, form on artery walls. Over time, these deposits narrow the artery and decrease its elasticity, resulting in reduced blood flow. Blockages can occur, resulting in a heart attack or stroke. High serum cholesterol levels (total blood cholesterol content) often result in increased plaque formation. Since dietary cholesterol is not efficiently absorbed into the bloodstream and the serum cholesterol level is primarily determined by the amount of cholesterol synthesized in the liver, high serum cholesterol levels are frequently related to high saturated fat intake.

Since the serum cholesterol level measurement gives the total cholesterol concentration of the blood, it can be a somewhat misleading predictor of atherosclerosis risk; cholesterol is not free in the blood but is encapsulated in lipoproteins. Since the cholesterol packaged in the LDL, cholesterol that can be deposited in plaques ("bad" cholesterol), has a very different fate from that in the HDL, which is transporting cholesterol for destruction ("good" cholesterol), measuring the ratio of LDL cholesterol to HDL cholesterol is a better indicator of atherosclerosis risk. Decreasing dietary intake of cholesterol and saturated fats, increasing water-soluble fibers in the diet, removing excess body weight, and increasing aerobic exercise will improve the LDL-C/HDL-C ratio.

Several hereditary diseases are known that result from abnormal accumulation of the complex lipids utilized in membranes. These diseases are called "lipid" (or "lysosomal") storage diseases or lipidoses. In normal individuals, the amount of each complex lipid present in the body is relatively constant; in other words, the formation rate equals the rate of destruction. The lipids are broken down by enzymes that attack specific bonds in the lipid structure. Lipid storage diseases occur when a lipid-degrading enzyme is defective or absent. In these cases, the lipid synthesis usually proceeds, but the degradation is impaired, causing the lipid or a partial degradation product to accumulate, with consequences such as an enlarged liver and spleen, mental disability, blindness, and death.

Niemann-Pick, Gaucher's, and Tay-Sachs's diseases are examples of lipidoses. Niemann-Pick disease is caused by a defect in an enzyme that breaks down sphingomyelin. The disease becomes apparent in infancy, causing mental retardation and death, usually by age four. Gaucher's disease, a more common disease involving the accumulation of a glycolipid, produces two different syndromes. The acute cerebral form affects infants, causing severe nervous system abnormalities, retardation, and death before age one. The chronic form, which may become evident at any age, causes enlargement of the spleen, anemia, and erosion of the bones. In Tay-Sachs disease, a partially degraded lipid accumulates in the tissues of the central nervous system. Symptoms include progressive loss of vision, paralysis, and death at three or four years of age. Although Tay-Sachs disease is relatively rare (1 in 300,000 births), it has a high incidence in individuals of Eastern European Jewish descent (1 in 3,600 births). This defect is a recessive genetic trait found in one of every twenty-eight members of this population. There is a one in four chance that their child will develop Tay-Sachs disease for two parents who are both carriers of this trait. Tests have been developed to detect the presence of the defective gene in the parent, and the amniotic fluid of a developing fetus can be sampled using a technique called "amniocentesis" to detect Tay-Sachs disease. Lipid storage diseases have no known cures; however, they can be prevented through genetic counseling.

PERSPECTIVE AND PROSPECTS

The ability of a cell to discriminate in its chemical exchanges with the environment is fundamental to life. How the cell membrane accomplishes this feat has been a subject of intense biochemical research since the beginning of the twentieth century.

In 1895, Ernst Overton observed that lipid-soluble substances enter cells more quickly than those that are lipid-insoluble. He reasoned that the membrane must be composed of lipids. About twenty years later, chemical analysis showed that membranes also contain proteins. Irwin Langmuir prepared the first artificial membrane in 1917 by mixing a phospholipid-containing hydrocarbon solution with water. Evaporation of the hydrocarbon left a phospholipid film on the water's surface, which showed that only the hydrophilic heads contacted the water. When the Dutch biologists E. Gorter and F. Grendel deposited the lipids from red blood cell membranes on a water surface and decreased the occupied surface area with a movable barrier, a continuous film occupied an area approximately twice the surface area of the original red blood cells. In 1935, all these observations, along with the fact that the surfaces of artificial membranes containing only phospholipids are less water-absorbent than the surfaces of true biological membranes, were combined by Hugh Davson and James Danielli into a membrane model in which a phospholipid bilayer was sandwiched between two water-absorbent protein layers.

The technological advances of the 1950s in X-ray diffraction and electron microscopy allowed the structures of membranes to be probed directly. Such studies revealed that membranes are indeed composed of parallel orderly arrays of lipids. However, many of the proteins are attached to one of the faces of the bilayer: The Davson-Danielli model was too simplistic. The freeze-fracture technique of preparing cells for electron microscopy has provided the most information about the nature of membrane proteins. In this technique, the two layers are separated so that the inner topography can be studied. Instead of the smooth surface predicted by the Davson-Danielli model, a cobblestone-like surface was observed, resulting from proteins penetrating the membrane's interior. All experimental evidence supports the fluid mosaic model for biological membranes, a model first proposed by Seymour Singer and Garth Nicholson in 1972. In this model, proteins are dispersed and embedded in a phospholipid bilayer in a fluid state. How membranes function was the following question to be considered.

Although most of the small molecules needed by cells cross the barrier via protein channels, some essential nutrients, such as cholesterol in its LDL package, are too large to pass through a small channel. In 1986, Michael Brown and Joseph Goldstein received the Nobel Prize for discovering specific protein receptors on the membranes of liver cells to which LDL molecules attach. These receptors move across the surface until they encounter a shallow indentation or pit. As the pit deepens, the membrane closes behind the LDL, forming a coating allowing transport across the hydrophobic membrane interior. The presence of insufficient numbers of these receptors causes abnormal LDL-cholesterol buildup in the blood.

Many questions remain unanswered concerning the roles of proteins and glycolipids in membranes. Membranes are involved in the movement, growth, and development of cells. How the membrane is involved in the uncontrolled multiplication and migration in cancer is one medically important question. Experiments that will answer questions about how membrane structure affects functioning should lead to new medical treatments.

—Arlene R. Courtney, PhD

Further Reading

Bettelheim, Frederick A., et al. *Introduction to General, Organic, and Biochemistry.* 10th ed., Brooks/Cole Cengage Learning, 2013.

Bloomfield, Molly M., and Lawrence J. Stephens. *Chemistry and the Living Organism.* 6th ed., John Wiley, 1996.

Brown, Michael S., and Joseph L. Goldstein. "How LDL Receptors Influence Cholesterol and Atherosclerosis." *Scientific American,* vol. 251, Nov. 1984, pp. 58-66.

Carlson, Emily. "The Big, Fat World of Lipids." *NIH National Institute of General Medical Sciences: Inside Life Science,* 9 Aug. 2012.

Christian, Janet L., and Janet L. Greger. *Nutrition for Living.* 4th ed., Benjamin/Cummings, 1994.
Cornatzer, W. E. *Role of Nutrition in Health and Disease.* Thomas, 1989.
MedlinePlus. "Dietary Fats." *MedlinePlus*, 28 June 2013.
National Institute of General Medical Sciences. "You Are What You Eat." *NIH National Institute of General Medical Sciences: ChemHealthWeb*, 9 Aug. 2012.
Sikorski, Zdzisław E., and Anna Kołakowska, editors. *Chemical and Functional Properties of Food Lipids.* CRC Press, 2002.
Vance, Dennis E., and Jean E. Vance, editors. *Biochemistry of Lipids, Lipoproteins, and Membranes.* 5th ed., Elsevier, 2008.

AMINO ACIDS

Category: Biochemistry, organic chemistry
Anatomy or system affected: Every system
Specialties and related fields: Biochemistry, microbiology
Definition: simple organic compounds that contain a carboxyl (-COOH) and an amino (-NH_2) group that are the building blocks of proteins

KEY TERMS

amino group: a functional group containing a nitrogen atom bonded to two hydrogen atoms (-NH^2)

carboxyl group: a functional group containing a carbon atom double-bonded to an oxygen atom and single bonded to a hydroxyl group (-OH); has the formula CO_2H, typically written -COOH

catalyst: a chemical species that initiates or speeds up a chemical reaction but is not consumed in the reaction

peptide bond: a covalent bond that links the carboxyl group of one amino acid to the amine group of another, enabling the formation of proteins and other polypeptides

protein: a biological polymer consisting of one or more long chains of amino acids linked by peptide bonds in a sequence specified by an organism's deoxyribonucleic acid (DNA)

THE NATURE OF AMINO ACIDS

Strictly speaking, an amino acid is any compound whose molecular structure contains both an amino group and a carboxyl group, also called a "carboxylic acid group"—hence the term "amino acid." However, the term in general use refers to the specific group of amino acids relevant to the genetic code in the deoxyribonucleic acid (DNA) molecule. These twenty (or sometimes twenty-three, depending on how they are classified) amino acids are called "proteinogenic amino acids," which refers to the fact that they are the only amino acids used to create proteins, enzymes, and other biomolecules. The three disputed amino acids are selenocysteine and pyrrolysine, which are not directly coded for in the genetic code but rather synthesized by other means and incorporated later, and *N*-formylmethionine, which initiates protein creation in some prokaryotes but is typically removed afterward. Selenocysteine is the only one of the three found in eukaryotes.

Of the standard twenty proteinogenic amino acids, nine are deemed "essential" because they are not synthesized in human metabolism but must be acquired through diet. All proteinogenic amino acids are also called "α-amino acids," which means that their amino and carboxyl groups are both bonded to the same carbon atom, known as the α-carbon (alpha carbon). This same carbon atom is also bonded to a hydrogen atom. The fourth atom or group bonded to the α-carbon determines the identity of the amino acid.

Chemically, amino acids have unique properties due to the presence of both a base and an acid in the same molecule. Self-neutralization, in which the acid transfers a proton to the base, readily takes place to produce a zwitterion, an electrically neutral molecule in which both a positive and a negative charge exist in separate parts of the molecule at the same time. In addition, each amino acid has a unique isoelectric point, which is the specific degree of acidity or basicity (pH) at which the amino acid has no net electrical charge. These particular characteristics are responsi-

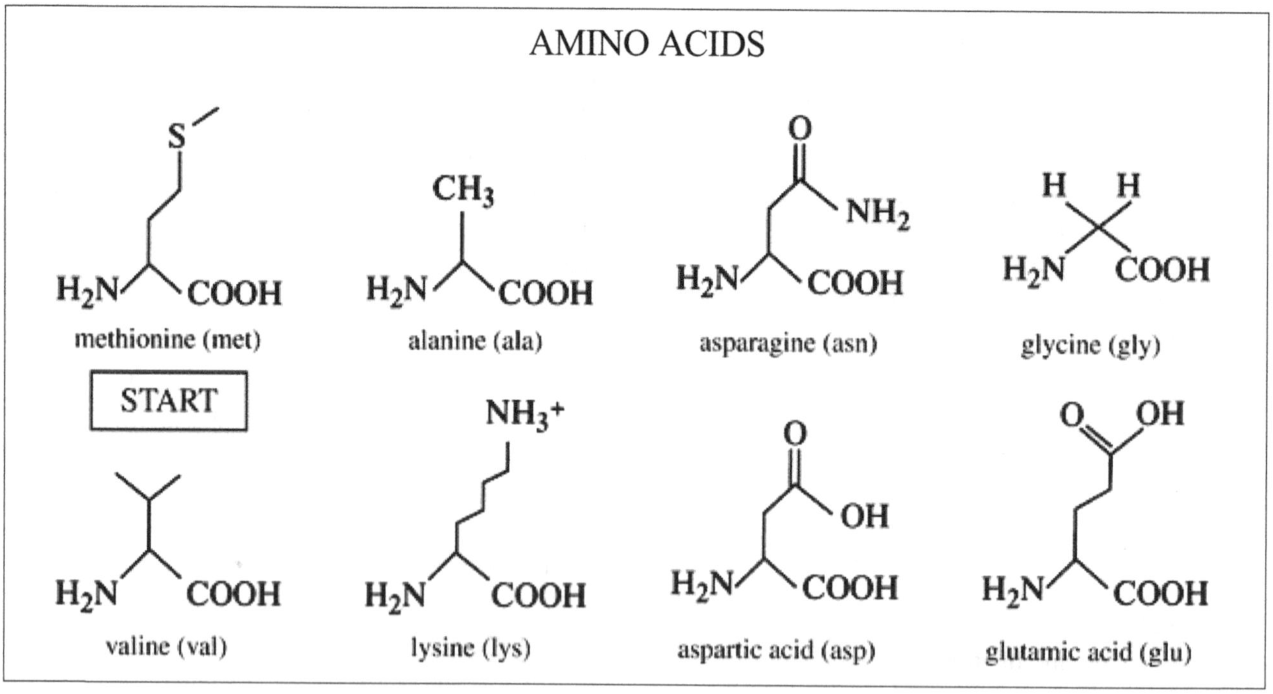

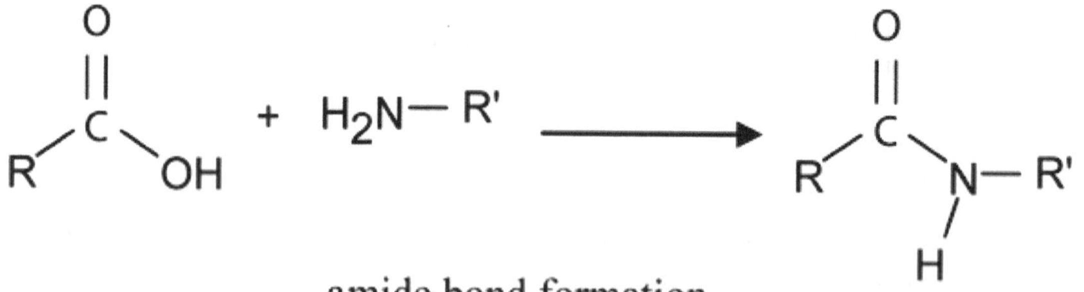

amide bond formation

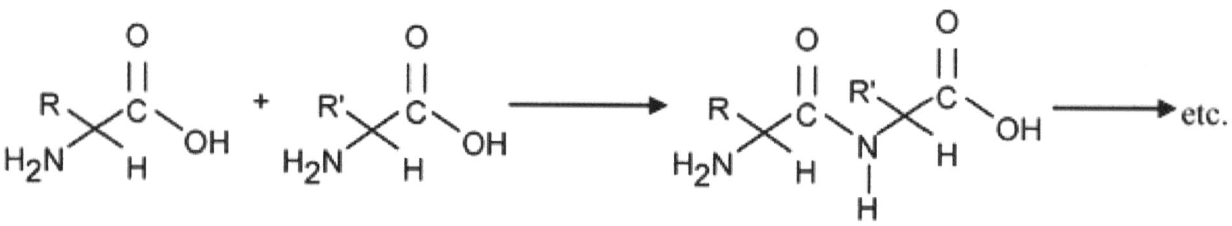

peptide bond formation

Amino Acid	Symbol	mRNA Codon
Alanine	A	GCA, GCC, GCG, GCU
Arginine	R	AGA, AGG, CGA, CGG, CGC, CGU
Asparagine	N	AAC, AAU
Aspartic acid	D	GAC, GAU
Cysteine	C	UGC, UGU
Glutamic acid	E	GAA, GAG
Glutamine	Q	CAG, CAA
Glycine	G	GGA, GGC, GGG, GGU
Histidine	H	CAC, CAU
Isoleucine	I	AUA, AUC, AUU
Leucine	L	CUA, CUC, CUG, CUU, UUA, UUG
Lysine	K	AAA, AAG
Methionine	M	AUG
Phenylalanine	F	UUU, UUC
Proline	P	CCA, CCC, CCG, CCU
Serine	S	AGC, AGU, UCA, UCG, UCC, UCU
Threonine	T	ACA, ACG, ACC, ACU
Tryptophan	W	UGG
Tyrosine	Y	UAC, UAU
Valine	V	GUA, GUG, GUC, GUU
Start codon	M	AUG
Stop codon		UGA, UAA, UAG

ble for most, if not all, of the behavior of amino acids and the much larger compounds they form as proteins and enzymes.

FORMATION OF PROTEINS AND ENZYMES

The structures of all proteins are determined by the sequence of amino acids encoded in the DNA molecule. With just one each of the twenty standard amino acids, there are thousands of trillions of possible ways to arrange them in what is called a "polypeptide chain." Most proteins and enzymes contain far more than just twenty amino acids.

The synthesis of polypeptides and their formation into proteins is rapid, taking as little as six minutes, according to various tracer studies using radioactively labeled amino acids. The process begins with transcription, during which specific enzymes open up the double-stranded structure of a DNA molecule and assemble copies of the nucleotide sequence using RNA segments. Each segment, called "messenger RNA" (mRNA), carries specific sequences of three nucleotides called "codons." During the next step, translation, the mRNA translates the genetic code from DNA to structures called "ribosomes," composed of ribosomal RNA (rRNA), where they match up with the rRNA sequence of nucleotides. When this occurs, the codons are exposed. In the cell's cytosol (intracellular fluid), a third type of RNA called "transfer RNA" (tRNA) transfers the specific amino acid corresponding to a particular codon to the mRNA strand in the ribosome. The anticodon on the tRNA segment matches the codon on the mRNA strand.

Specific enzymes there act as a catalyst to form a peptide bond between two neighboring amino acids. A peptide bond is just the standard amide structure that forms between a carboxylic acid and an amine:

The term is used to refer specifically to an amide bond formed between amino acids in a polypeptide.

TRANSLATING THE GENETIC CODE INTO PROTEINS

The DNA and RNA molecules use only four different nucleotide bases to specify the entire genetic code. Yet, the number of possible three-nucleotide codons formed from these is more than sufficient to differentiate the twenty standard amino acids. The system is quite redundant, with several different codons signifying the same amino acid. Specific nucleotide sequences also designate the starting and ending points of a particular sequence of amino acids and hence the protein structure that derives from that sequence. DNA was first identified in the late 1860s by Swiss chemist Friedrich Miescher. Still, it was not thought to be related to genetic information until Oswald Avery and colleagues established its connection in 1943. Subsequent research eventually revealed the structure and function of DNA and RNA. By preparing synthetic sequences of mRNA codons, researchers determined which codons encoded for each specific amino acid in transcription and translation. The code is translated in the accompanying chart.

Interestingly, the same codon, AUG, indicates both methionine and the start sequence, while the three stop codons are unique. The context in which the AUG codon appears determines whether or not it functions as a start codon or the codon for methionine.

AMINO ACIDS AND PROTEINS

The sequence of amino acids in a protein molecule defines its primary structure. Since each amino acid group has a specific geometry dictated by molecular structure and bond formation rules, no polypeptide chain or protein can be just a linear molecule. The angles of the bonds at each atom create all sorts of twists and turns along the entire length of the polypeptide molecule, moving the various functional groups on each amino acid into positions that allow them to interact with each other. Some segments of the protein molecule form larger physical shapes, such as spirals or flattened sheets. These constitute the secondary structure of the protein. A third, or tertiary, structure results from the interaction of the various functional groups as they form bonds due to their proximity to each other. A fourth, or quaternary, structure results when two or more protein molecules combine to form a larger reactive complex.

—Richard M. Renneboog, MSc

Further Reading

Berg, Jeremy M., et al. *Biochemistry*. 9th ed., W.H. Freeman, 2019.

Klein, David R. *Organic Chemistry and a Second Language*. 5th ed., Wiley, 2019.

Lodish, Harvey, et al. *Molecular Cell Biology*. 9th ed., W.H. Freeman, 2021.

Nelson, David L, and Michael M. Cox. *Lehninger Principles of Biochemistry*. 8th ed., W.H. Freeman, 2021.

Winter, Arthur. *Organic Chemistry for Dummies*. 2nd ed., For Dummies, 2016.

POLYSACCHARIDE

Category: Biochemistry

Also known as: Complex carbohydrates

Anatomy or system affected: All systems

Specialties and related fields: Bacteriology, biochemistry, biotechnology, gastroenterology, microbiology

Definition: a carbohydrate (e.g., starch, cellulose, or glycogen) whose molecules consist of several sugar molecules bonded together

KEY TERMS

glycogen: a substance deposited in bodily tissues as a store of carbohydrates; a polysaccharide forms glucose on hydrolysis

heterosaccharide: any saccharide composed of more than one simple sugar

homosaccharide: any saccharide composed of one simple sugar

monosaccharide: any class of sugars (e.g., glucose) that cannot be hydrolyzed to give a simpler sugar

starch: an odorless, tasteless white substance that occurs widely in plant tissue and is obtained chiefly from cereals and potatoes; it is a polysaccharide that functions as a carbohydrate store and is an important constituent of the human diet

INTRODUCTION

Polysaccharide is a substance composed of long chains of simple sugars connected by special bonds. Polysaccharides are the most complex members of a broader family of compounds known as carbohydrates or saccharides. The carbohydrate family also includes monosaccharides and disaccharides, which are the building blocks of polysaccharides. Polysaccharides serve as carbohydrate storage units in both plants and animals and play several other key structural roles. The three main types of polysaccharides include starch, glycogen, and cellulose, which are important components of living organisms. In plants, most polysaccharides are produced through photosynthesis or how plants create energy using sunlight, carbon dioxide, and water. Certain types of bacteria also can produce polysaccharides.

BACKGROUND

Carbohydrates are molecular compounds made from carbon, hydrogen, and oxygen. In plants and animals, carbohydrates are structurally and functionally vital. Plants can create carbohydrates through photosynthesis. In addition to providing energy for plants, this process is indirectly responsible for the continued existence of animal life because animals depend on plant carbohydrates as the source of their food supply. As a result, carbohydrates are essential to all living organisms.

The three types of carbohydrates differ primarily based on their construction. Monosaccharides are the smallest and least complex type of carbohydrate, consisting of only a single carbohydrate unit. They are referred to as simple sugars. Because of their small size and ability to dissolve in water, monosaccharides can easily pass through cell membranes by diffusion. Different types of monosaccharides include glucose, galactose, fructose, ribose, and deoxyribose. These substances play key roles in the normal functioning of living organisms.

Disaccharides are slightly more complex carbohydrates formed when a pair of monosaccharides join through a condensation reaction. The three main types of disaccharides include sucrose, lactose, and maltose. Although they are soluble in water, disaccharides are too large to pass through cell membranes via diffusion. During the digestion process, disaccharides are broken into monosaccharides through a hydrolysis reaction that releases energy.

Polysaccharides are created when monosaccharides undergo condensation reactions resulting in an extended chain of molecules. This process is known as condensation polymerization. The three main types of polysaccharides include starch, glycogen, and cellulose. The characteristics of its molecule chain determine the properties of a polysaccharide. Some of the factors involved are the length of the chain, the extent to which it branches out, the presence of folds in the chain, and whether the chain is straight or coiled. Depending on their specific properties, polysaccharides serve different purposes in plants and animals.

OVERVIEW

Polysaccharides can be classified into one of two categories: homopolysaccharides or heteropolysaccharides. Homopolysaccharides are polysaccharides made of one type of monosaccharide. Heteropolysaccharides are made from two or more different types of monosaccharides. Polysaccharides are sometimes categorized according to their primary function as energy storage or structural components of cells. This latter form of categorization is an easy way to understand the similarities and differences between the three main types of polysaccharides.

Starches, made of glucose chains held together by glycosidic bonds, primarily provide short-term energy storage. Starches serve as plants' main energy reserve. When starch chains are broken into glucose during the digestive process, the energy held within is released. Humans consume more starches than any other kind of carbohydrate. Foods such as potatoes, rice, corn, and wheat are common sources of dietary starches.

Glycogen is the main energy reserve in animals. Glycogen is a fuel reserve stored mostly in the liver and skeletal muscles that the body can tap as needed. When a quick burst of energy is required, the pancreas releases a hormone called "glucagon," which causes the liver to convert some of its stored glycogen into glucose. Glycogen that is stored in the muscles is used internally when converted into glucose. In both the liver and muscles, this conversion can also be triggered by the abrupt release of adrenaline in response to stress.

Unlike starches or glycogen, cellulose does not function as an energy store. Cellulose is a key structural component found in the cell walls of plants, and it gives different plants their unique shape and structure. Cellulose is not a source of energy for humans. Although some animals can digest cellulose, humans cannot because they lack the necessary enzymes. As a result, grasses and other plants that are made primarily of cellulose are inedible for humans. Cellulose is still useful to humans because of its important role in producing items such as cotton and paper.

In addition to starches, glycogen, and cellulose, other less common polysaccharides exist. Chitin is a glucose-derived polysaccharide found in the exoskeletons of crabs and other similar animals. It is used as a fertilizer in agricultural settings, a binding agent in various industrial applications, and dissolvable surgical thread in the medical field. Pectin is a polysaccharide found in the cell walls of various plants that is used as a gelling agent for jams, jellies, preserves, and a thickener in other food products.

One of the most important modern applications of polysaccharides is in the production of vaccines. Pneumococcal polysaccharide vaccine (PPSV) can protect people from bacterial infections such as pneumonia, bacteremia, and meningitis. Usually recommended for elderly individuals and patients with weakened immune systems or other health issues, PPSV can protect against twenty-three different types of pneumococcal bacteria.

—*Jack Lasky*

Further Reading

Aspinall, Gerald O. *Polysaccharides.* Pergamon Press, 1970.

"Carbohydrates." *Royal Society of Chemistry,* www.rsc.org/Education/Teachers/Resources/cfb/Carbohydrates.htm. Accessed 19 Apr. 2017.

George, Helga. "What Are Polysaccharides?" *WiseGeek,* 12 Apr. 2017, www.wisegeek.org/what-are-polysaccharides.htm. Accessed 19 Apr. 2017.

Group, Edward. "Understanding Your Nutrition: What Are Polysaccharides?" *Global Healing Center,* 18 Feb. 2016, www.globalhealingcenter.com/natural-health/understanding-nutrition-polysaccharides. Accessed 19 Apr. 2017.

Kroll, Jess. "The Differences between Monosaccharides & Polysaccharides." *Sciencing,* sciencing.com/differences-between-monosaccharides-polysaccharides-8319130.html. Accessed 19 Apr. 2017.

Nall, Rachel. "In Which Foods Are Polysaccharides Found?" *Livestrong,* 22 June 2011, www.livestrong.com/article/477021-polysaccharides-are-found-in-which-foods. Accessed 19 Apr. 2017.

"Pneumococcal Polysaccharide Vaccine." *MedlinePlus,* 15 Nov. 2016, medlineplus.gov/druginfo/meds/a607022.html. Accessed 19 Apr. 2017.

"What Are the Different Polysaccharides?" *InnovateUS,* www.innovateus.net/science/what-are-different-polysaccharides. Accessed 19 Apr. 2017.

Biofilm

Category: Biology
Also known as: Gamma globulin proteins, immunoglobulin (Ig)
Anatomy or system affected: All
Specialties and related fields: All
Definition: a consortium of microbes that stick to each other and a surface while embedded within a slimy extracellular polymer made by the cells

KEY TERMS

extracellular polymer: a polymer like a protein, polysaccharide, glycoprotein, lipids, or polynucleotides secreted by microbes that take up large quantities of water and form a slime layer outside the cell
glycocalyx: a collective term for the biofilm in which the microorganisms are embedded
quorum-sensing: the regulation of gene expression in response to fluctuations in cell-population density

INTRODUCTION

A biofilm is a collection of microbial cells that secrete a substance that enables them to stick together to create a film, or thin layer, on a liquid or surface. The film can be limited to one microbial species but most commonly comprises multiple types of cells, including bacteria, algae, protozoa, and yeast. One type of film can include hundreds of different microbial cells.

Liquid is required for the formation of a biofilm, which can form on natural surfaces, including parts of the human body and plants, rocks, metals, plastics, and other materials. Some familiar examples of biofilms include the sludge that forms in drains, dental plaque, and the slippery slime-type substance that sometimes forms on rocks along a body of water.

FORMATION

Biofilms can form anywhere a microbe encounters a surface in a liquid environment. Within minutes of settling on the surface—for example, an underwater rock—the previously free-floating cells start making extracellular polymeric substances (EPS). These slippery, slimy substances serve as glue to help individual microbes form a complex community or colony. Within hours of the first microbe landing on a surface, the colony is formed. It is ready to propagate by releasing either individual cells or sections of the developed colony.

The cells that become a biofilm have specific patterns of behavior regarding the ways they attach, develop into a colony, and detach from that colony, which differs from other microbial forms.

SIGNIFICANCE

As a natural occurrence, biofilms have existed for as long as microbes and liquids have been present. Researchers were aware that films such as dental plaque were teeming with tiny microorganisms since Dutch scientist Antony van Leeuwenhoek (1632-1723) first used his microscope to observe tiny organisms moving in a sample collected from between someone's teeth in 1684. Leeuwenhoek is credited with discovering biofilm, but he and other scientists lacked the tools to study the full formation of biofilm for several centuries. In the 1940s, scientists began to realize the difference between planktonic or floating microbes and sessile cells, or those that had attached to surfaces and could form a biofilm. In the latter part of the twentieth century, researchers finally were able to study biofilms in enough detail to understand the magnitude of their effect and growth.

Biofilms can be helpful. They are part of the food chain, providing nutrient sources for some insects and other species. They can also help break down soil contaminants from accidental spills and are helpful to the nutrient absorption process of some plants. The powers of biofilms can also be harnessed to help with such processes as water and wastewater treatment and to extract some metal ores through a process known as microbial leaching. While some biofilms that form on humans, including dental plaque, can cause infections or decay, others—such as those in the digestive tract—can have beneficial properties and present few if any problems.

There are also many harmful effects of biofilms. They can foul water supplies by depleting oxygen and cause the formation of harmful algal blooms on the surfaces of bodies of water. They can help spread toxic chemicals like mercury and arsenic and diseases such as cholera, plague, and West Nile virus. Biofilms can also produce gases that affect the atmosphere.

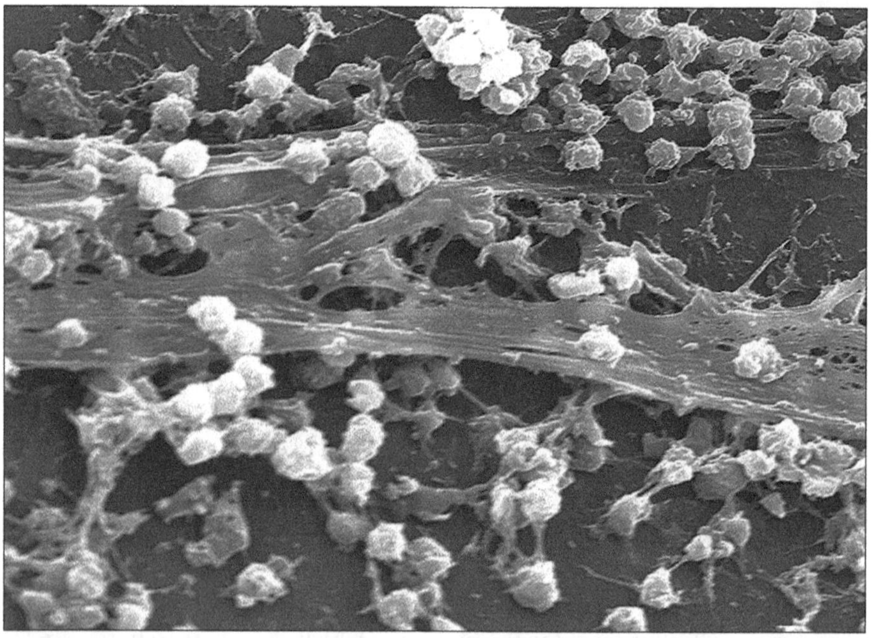

Staphylococcus aureus *biofilm on an indwelling catheter. Photo via CDC/Wikimedia Commons. [Public domain.]*

Biofilm colonies demonstrate behaviors, properties, and abilities that the individual cells do not possess alone. For example, an antibiotic dose that would kill an individual cell might need to be hundreds of times stronger to kill the biofilm colony. Understanding how biofilms form and how they interact with the human body is vital to medical professionals. Biofilms can contain microbes such as bacteria that cause illness and infection. For example, biofilms can be responsible for chronic otitis media or ear infections and wounds that will not heal. Pneumonia in cystic fibrosis patients is sometimes the result of a biofilm infection as well. The solid surfaces of catheters and implants, which are often close to bodily fluids such as urine and blood, provide a surface for microbes to begin forming a biofilm and causing a related infection. These biofilms are often identified when the infection cannot be fought off by the body's normal defenses and is also resistant to treatment with antibiotics.

The infectious biofilms can also form on equipment used in medical procedures. Their tenacious nature makes them difficult to remove by the usual disinfection processes. Biofilm-forming microorganisms can spread within a medical facility and force the replacement of thousands of dollars of medical equipment. These infections can take hold on implanted medical devices such as artificial joints, pacemakers, and stents, subjecting patients who already have a health issue to new problems, increased need for surgery and antibiotics, and the risk of further infections.

MEDICAL RESEARCH

Researchers are very interested in learning how these biofilm colonies develop to be stronger than the sum of the individual parts to eliminate these costly and dangerous problems. In 2015, researchers at the University of Maryland and University of Michigan, working under the auspices of the US National Institutes of Health and National Science Foundation, conducted studies into how one common hospital bacteria, *Pseudomonas aeruginosa,* forms a biofilm.

Scientists had previously determined that when microbes detect a change in their environment that makes it favorable to form a biofilm, they use signaling molecules to start this process. They also determined that a molecule called "Cyclic-di-GMP," also known as "c-di-GMP," is used by many forms of bacteria to initiate the biofilm-making process. The c-di-GMP is then converted into another molecule, known as "pGpG," which continues the signaling process.

What was not known previously was what signaled biofilms to stop the formation process. The researchers from the Universities of Maryland and Michigan were the first to identify an enzyme known as oligoribonuclease as the off-switch for the biofilm process. It works by disrupting and pulling apart the pGpG, putting an end to the signal, and stopping biofilm formation. While the study focused on the biofilms formed by *P. aeruginosa*, oligoribonuclease is also known to be a signaling trigger for ending several other types of bacterial biofilms. Researchers believe this discovery can be applied to developing disinfectants, antibiotics, and other treatments and inhibitors of biofilm formation. They are also searching for similar shut-off triggers in other infective microbes.

—*Janine Ungvarsky*

Further Reading

"Biofilm Basics." *Montana State University Center for Biofilm Engineering*, www.biofilm.montana.edu/biofilm-basics.html. Accessed 4 Feb. 2016.

Bjarnshot, T. "The Role of Bacterial Biofilms in Chronic Infections." *Acta Pathologica, Microbiologica et Immunologica Scandinavica*, May 2013, www.ncbi.nlm.nih.gov/pubmed/23635385. Accessed 4 Feb. 2016.

Donlan, Rodney M. "Biofilms: Microbial Life on Surface." *Emerging Infectious Diseases*. US Centers for Disease Control, vol. 8, no. 9, Sept. 2002, wwwnc.cdc.gov/eid/article/8/9/02-0063_article. Accessed 4 Feb. 2016.

"Off Switch for Biofilm Formation Discovered." *Science Daily*, 24 Aug. 2015, www.sciencedaily.com/releases/2015/08/150824163001.htm. Accessed 4 Feb. 2016.

Sanders, Robert. "Discovery Opens Door to Attacking Biofilms that Cause Chronic Infections." *Berkeley News*. University of Berkeley, 12 July 2012, news.berkeley.edu/2012/07/12/discovery-opens-door-to-attacking-biofilms-that-cause-chronic-infections/. Accessed 4 Feb. 2016.

Vickery, Karen, editor. *Microbial Biofilms in Healthcare: Formation, Prevention and Treatment*. Mdpi AG, 2020.

Porphyrin

Category: Biochemistry
Also known as: Heme
Anatomy or system affected: All systems
Specialties and related fields: Biochemistry, biotechnology
Definition: any class of pigments (including heme and chlorophyll) whose molecules contain a flat ring of four linked heterocyclic groups, sometimes with a central metal atom

KEY TERMS

chlorophyll: a green pigment, present in all green plants and in cyanobacteria, responsible for absorbing light to provide energy for photosynthesis with a magnesium atom held in a porphyrin ring.

cobalamin: a water-soluble vitamin involved in metabolism that acts as a cofactor in deoxyribonucleic acid (DNA) synthesis, in both fatty acid and amino acid metabolism

cytochrome: compounds consisting of heme bonded to a protein that function as electron transfer agents in many metabolic pathways, especially cellular respiration

INTRODUCTION

Porphyrins are specialized molecules that capture metal ions. These molecules are essential for a wide variety of chemical processes. In plants, they are associated with photosynthesis. In humans and other mammals, they are associated with transporting oxygen throughout the body and absorbing the B_{12} vitamin. They are also responsible for bright pigments in a wide variety of organisms.

Incorrectly functioning porphyrins can cause severe physical and mental illness. Most porphyrin disorders are genetic. For this reason, treating porphyrin disorders can be difficult.

BACKGROUND

Atoms are microscopic particles that make up all matter. They are made up of a single element, which can all be found on the periodic table of elements. Microscopic objects cannot be seen with the naked eye. Instead, an observer must use high levels of magnification to view them. Atoms are made of three types of subatomic particles: protons, neutrons, and electrons. Protons and neutrons are both found in the center of the atom, called the "nucleus." This is where most of the mass of the atom is located. Protons are positively charged, while neutrons are neutrally charged. Electrons, negatively charged subatomic particles, travel around the nucleus in defined orbits. The ratio of protons to electrons determines the charge of the atom. If an atom has more protons than electrons, the atom is positively charged. If the reverse is true, the atom is negatively charged. The charge of an atom determines how easily it attracts and bonds with other atoms.

Atoms do not usually exist alone, and in most cases, they bond together to form molecules. One commonly used example of a molecule is hydrogen dioxide, which we know as water. Water molecules are composed of a single hydrogen atom bonded to two oxygen atoms. These molecules bond together under different conditions, manifesting as liquid water, ice, or steam.

Atoms can bond together in several ways. Ionic bonds are formed when atoms gain or lose electrons. Suppose one atom is positively charged, having too few electrons, and another atom is negatively charged, having too many electrons. In that case, the two atoms may bond together. The positive atom takes the appropriate amount of electrons from the negative atom. Both atoms are left as close to neutrally charged as possible. Ionic bonds are relatively weak compared to other atomic bonds. They are usually weak enough to be dissolved in water.

Covalent bonds are also formed when one atom is negatively charged and another atom is positively charged. However, in a covalent bond, the two atoms share the electron in question. The electron orbits both atoms, binding them together. Covalent bonds are much more powerful than ionic bonds.

OVERVIEW

Porphyrins are rigid, square molecules with a precisely sized gap in the middle. They are made up of four five-membered rings centered around nitrogen atoms. These structures are called "pyrroles." They keep the porphyrin rigid and allow it to trap a metal ion in its center. Once a metal ion has been captured, the structure is known as a "complex." Complexes are bound tightly together and use the captured metal to attract other atoms for various chemical processes.

Porphyrin complexes are responsible for several of the most important molecular structures on earth. For example, chlorophyll is formed when a magnesium ion bonds with four nitrogen pyrroles. Chlorophyll is a green-colored pigment, and it allows plants to absorb sunlight, a key part of the photosynthesis process. Photosynthesis fuels plants, allowing them to grow, reproduce, and carry out many other functions.

Porphyrin complexes are essential to the function of mammals, including humans. Porphyrin complexes called "hemoglobin" are found in the red blood cells of all mammals and the blood cells of several arthropods. Hemoglobin is formed when an iron ion is attached to a set of pyrroles. It is responsible for carrying oxygen through the bloodstream. Porphyrins are also necessary for the development of the B_{12} vitamin. During its creation, the metal cobalt is bound to a porphyrin structure. This vitamin is essential to health; a lack of B_{12} vitamin can cause severe health problems, including anemia and nervous system malfunctions.

Incorrectly functioning porphyrins or a deficiency in porphyrins can have severe side effects. Malfunctioning genes can cause porphyria, which causes the porphyrins in the body to function at a reduced capacity. One famous historical figure who experts believe suffered from porphyria was King George III. The monarch was a highly gifted military tactician, as well as a skilled head of domestic policy. Under his rule, England flourished. However, physicians of the early nineteenth century believed that a mysterious illness beset King George III that they could not cure. The disease caused the king to experience prolonged bouts of insanity, culminating in uncharacteristically terrible government decisions. Some of these bouts lasted more than a year. The disease also caused the king to produce strangely discolored urine, lose control of his limbs, and suffer severe abdominal pain. These symptoms and the periods of mental instability incapacitated King George III before his death. His symptoms closely match the symptoms of variegate porphyria. This genetic disorder could not have been treated by the methods of the time. It makes the sufferer hypersensitive to poisoning from heavy metals, which can cause bouts of acute mental instability.

Porphyrins can also be found in the exoskeletons of a variety of invertebrates. They cause the animals' exoskeletons to show a variety of vibrant colors. In some rare cases, as with some millipedes of the genus *Motyxia*, porphyrins can cause the exoskeletons to glow in the dark. Additionally, porphyrins from the decomposed remains of animals can commonly be found in fossil fuels, such as coal and oil.

—*Tyler Biscontini*

Further Reading

"About Porphyria." *American Porphyria Foundation*, 2015, www.porphyriafoundation.com/about-porphyria. Accessed 9 May 2017.

"Chemistry of Porphyrins." *The Museum of Organic Chemistry*, www.org-chem.org/yuuki/porphyrin/porphyrin.html. Accessed 9 May 2017.

"Chemistry I: Atoms and Molecules." *Estrella Mountain*, www2.estrellamountain.edu/faculty/farabee/biobk/BioBookCHEM1.html. Accessed 9 May 2017.

"Discovery of a Glowing Millipede in California and the Gradual Evolution of Bioluminescence in Diplopoda." *Crossmark*, 2015, www.pnas.org/content/112/20/6419.full.pdf. Accessed 9 May 2017.

"Heme and Bilirubin Metabolism." *The Medical Biochemistry Page*, 2021, themedicalbiochemistrypage.org/heme-and-bilirubin-metabolism/. Accessed 1 Dec 2021.

"Porphyrins." *TCI America*, www.tcichemicals.com/eshop/en/us/category_index/10825/. Accessed 9 May 2017.

"Porphyrins: One Ring in the Colors of Life." *American Scientist*, May/June 2011, www.americanscientist.org/issues/pub/porphyrins-one-ring-in-the-colors-of-life/2. Accessed 9 May 2017.

"Porphyrins & Porphyria Diagnosis." *American Porphyria Foundation*, 2015, www.porphyriafoundation.com/testing-and-treatment/testing-for-porphyria/porphyrins-and-porphyria-diagnosis. Accessed 9 May 2017.

VITAMIN A

Category: Dietary supplements

Also known as: Retinol, retinoic acid, retinal, ß-carotene

Anatomy or system affected: All systems

Specialties and related fields: Biochemistry, biotechnology, dermatology, immunology, internal medicine, oncology, obstetrics, ophthalmology, pharmacology, preventive medicine, public health

Definition: a group of unsaturated nutritional, organic compounds that include retinol, retinal, and several carotenoids; vitamin A has multiple functions: it is important for growth and development, for the maintenance of the immune system, and good vision

KEY TERMS

11-cis-retinal: a cofactor of opsins in the retina that are light activated

beta-carotene: an organic, intensely colored red-orange pigment abundant in fungi, plants, and fruits *provitamin*: a substance that may be converted within the body to a vitamin

vitamin: any of a group of organic compounds which are essential for normal growth and nutrition and are required in small quantities in the diet because the body cannot synthesize them

WHAT WE KNOW

Vitamin A is the generic term used to describe a subgroup of related fat-soluble compounds in a larger group known as retinoids; compounds in the subgroup include retinol, retinal (i.e., preformed vitamin A that the body converts as needed into retinoic acid, which affects cellular gene expression), retinoic acid, and related compounds. Provitamin A carotenoid is a generic term for any of the red, orange, or yellow phytonutrients (i.e., plant-derived nutrients) found in fruits and vegetables that can be converted within the body into retinal, a biologically active form of vitamin A. Examples of provitamin A carotenoids include beta-carotene, alpha-carotene, and beta-cryptoxanthin.

Although vitamin A is toxic if ingested in large amounts, beta-carotene and most other vitamin A precursor molecules are generally considered nontoxic. Vitamin A is obtained from foods of animal origin, particularly liver and dairy products high in fat. Vitamin A precursors (e.g., provitamin A carotenoids), including beta-carotene, are found in plant foods, especially orange and dark green leafy vegetables.

Approximately 10 percent of the nearly 600 carotenoids identified are converted to vitamin A in the body as required to meet physiologic needs for vitamin A. Beta-carotene is the carotenoid converted into vitamin A most efficiently. Other than beta-carotene, alpha-carotene and beta-cryptoxanthin contribute the most dietary vitamin A; they are converted to vitamin A with approximately half the efficiency of beta-carotene.

Vitamin A is fat-soluble and is stored in the body for significant periods. Due to its fat-soluble nature and the fact that body stores of vitamin A are significant in well-nourished populations, several weeks of insufficient vitamin A intake and/or absorption are unlikely to cause vitamin A deficiency in otherwise healthy individuals. About 50 percent of consumed vitamin A is stored, primarily in the liver; 10 percent is not absorbed; 20 percent is transported via bile to the feces and excreted; 17 percent is excreted in the urine, and 3 percent is released through respiration.

The history of human awareness of vitamin A dates back thousands of years to recognizing night blindness as the first nutritional-deficiency disease. Ancient Egyptian writings from 1500 BCE recommend topical application to the eyes of cooked liver juice to treat night blindness. Greek writings of the era include information on liver juice ingestion and topical application to treat night blindness. Assuming permanent damage is not present, the component of the liver that can cure night blindness is vitamin A.

In the United States in 1913, a growth factor called "fat-soluble A" was identified in butter, eggs, and cod liver oil but not in lard. It was not yet called "vitamin" A because the term *vitamin* is reserved for nutrients definitively identified as vital for life. This nutrient was classified as a growth factor because rats given a purified diet without fat-soluble A failed to grow properly. Subsequent research in the 1920s confirmed the results of earlier studies that showed that deficiency of the nutrient vitamin A has profound adverse effects on growth and tissue differentiation. In 1930, the structures of vitamin A, now recognized as a vitamin, and beta-carotene, which the body converts to vitamin A as required to meet physiologic needs, were identified.

VITAMIN A ACTION

Vitamin A has numerous essential functions in the body, including roles in vision, growth, cell division and differentiation, and immune function.

In the eye, the 11-cis-retinal form of vitamin A binds to a protein called "opsin" to form rhodopsin in eye cells. The presence of rhodopsin allows eye cells to detect very small light levels, making the nutrient important for night vision. Vitamin A deficiency leads to impaired adaptation to darkness, commonly called "night blindness."

Regulation of cell differentiation is among the most crucial roles of vitamin A in the body; vitamin A promotes proper development of specialized tissues, including fetal cell development into brain, muscle, liver, pancreas, kidney, blood, and other specialized tissues during gestation.

Vitamin A plays a crucial role in immune status by maintaining the integrity of skin and mucous membranes of the eyes and the respiratory, urinary, and intestinal tracts. Vitamin A is required for the growth and differentiation of white blood cells (WBCs) and activation of T lymphocytes, a subset of WBCs that are critical components of normal immune function.

Vitamin A is critical to fetal development and proper limbs, eyes, ears, and heart formation. Both

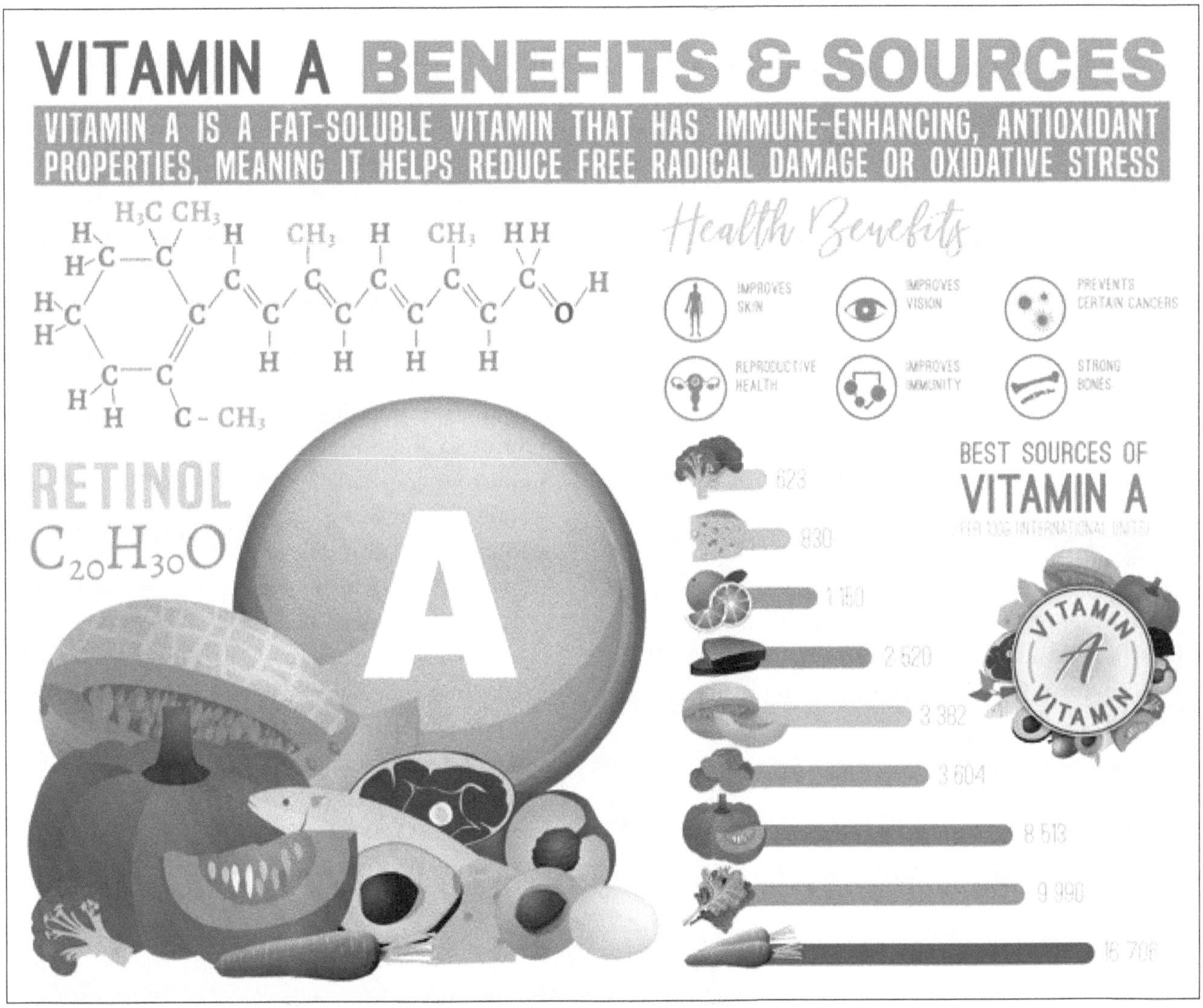

Image via iStock/newannyart. [Used under license.]

vitamin A deficiency and vitamin A excess can cause congenital disabilities. During fetal development, vitamin A is required to regulate growth hormone gene expression.

Vitamin A is needed for red blood cell (RBC) formation and assists in the movement of stored iron to the site of RBC development for incorporation into hemoglobin.

SOURCES OF VITAMIN A

Preformed vitamin A is found only in foods of animal origin, certain dietary supplements, and fortified foods. Provitamin A carotenoids are found in plant foods, notably dark green leafy vegetables, and orange-yellow vegetables and fruits.

Excellent sources of preformed vitamin A include the following:

- Beef liver
- Turkey and chicken giblets (e.g., liver, heart, gizzard, neck)
- Pork liver
- Whole milk and dairy products high in fat
- Vitamin A-fortified foods (e.g., skim milk, certain cereals)
- Eggs
- Fish, especially fatty, cold-water fish such as cod, halibut, and salmon, and oils from fatty cold-water fish

Excellent sources of provitamin A carotenoids, particularly beta-carotene, alpha-carotene, and beta-cryptoxanthin, include the following foods and their juices:

- Carrots
- Pumpkin
- Spinach
- Beet, collard, dandelion, and turnip greens
- Squash
- Cantaloupe
- Butterhead, romaine, and green leaf lettuce
- Cabbage
- Red bell peppers
- Apricots
- Tomato soup and juice and canned tomatoes

Lycopene, the key provitamin A carotenoid found in tomatoes, is not bioavailable from raw tomatoes; the body cannot absorb lycopene from tomatoes unless they are heated and processed:

- Broccoli
- Papayas
- Tangerines
- Plantains

RECOMMENDED INTAKE OF VITAMIN A

The daily recommended dietary intake (RDI) for vitamin A has been established for various age groups. The recommended intake for vitamin A is reported as micrograms (mcg) of retinol activity equivalents (RAE) to account for the varying biologic activities of retinol and provitamin A carotenoids. RAEs can be converted to International Units (IU) of vitamin A (1 RAE = 3.3 IU). IU are listed on food and dietary supplement labels.

- Birth to six months: 400 mcg (1,320 IU)
- Seven to twelve months: 500 mcg (1,650 IU)
- One to three years: 300 mcg (1,000 IU)
- Four to eight years: 400 mcg (1,320 IU)
- Nine to thirteen years: 600 mcg (2,000 IU)
- Fourteen to eighteen years
    - Males: 900 mcg (3,000 IU)
    - Females: 700 mcg (2,310 IU)
    - During pregnancy: 750 mcg (2,500 IU)
    - While lactating: 1,200 mcg (4,000 IU)
- Nineteen-plus years
    - Males: 900 mcg (3,000 IU)
    - Females: 700 mcg (2,310 IU)
    - During pregnancy: 770 mcg (2,565 IU)
    - While lactating: 1,300 mcg (4,300 IU)

Current guidelines recommend that preterm infants be given 700-1,500IU/kg/day vitamin A; those with significant lung disease may require a dose of 2,000-3,000 IU/kg/day.

VITAMIN A DEFICIENCY

Plasma vitamin A concentrations lower than ten ug/dL (0.35 umol/L) indicate vitamin A deficiency.

Those at risk for vitamin A deficiency include residents of the developing world. Vitamin A deficiency is uncommon in the United States and the rest of the developed world. Still, it is a leading cause of night blindness in the developing world, particularly in Africa and Southeast Asia. Vitamin A deficiency in the developing world typically begins after conception and early in life with insufficient intake during nutritionally demanding periods, including pregnancy, lactation, infancy, and childhood. Insufficient breastfeeding and maternal nutritional deficiencies that lead to poor quality breast milk contribute to vitamin A deficiency in infants and children.

In adulthood, lack of preformed vitamin A in the diet from liver, milk, cheese, eggs, or fortified food products coupled with an inadequate intake of provitamin A carotenoids contribute to widespread vitamin A deficiency in the developing world. In the absence of preformed vitamin A, which is commonly found in the developing world, dietary intake of modest amounts of vegetables and fruits as the sole source of vitamin A may not deliver adequate amounts of provitamin A carotenoids to prevent deficiency.

Up to half a million malnourished children in the developing world develop night blindness and eventually become completely blind each year due to vitamin A deficiency; approximately half of these children will die within one year of the onset of blindness due to vitamin A deficiency and general malnutrition. It is estimated that a total of 670,000 children die each year solely due to vitamin A deficiency. Vitamin A deficiency is common in:

- persons who abuse alcohol.
- preterm and low birth weight infants, especially in the developing world.
- persons with malabsorptive conditions, including Crohn's disease, ulcerative colitis (UC), celiac disease, cystic fibrosis, chronic liver disease, and pancreatic insufficiency.
- persons who have undergone gastrointestinal surgery that affects nutrient absorption (e.g., gastrectomy, small intestinal resection, bariatric surgery).
- persons with liver disease.

Signs and symptoms of vitamin A deficiency include night blindness, technically referred to as impaired dark adaptation; loss of appetite; growth retardation; impaired fertility; decreased immune function; frequent infections; follicular hyperkeratosis (i.e., excessive development of keratin in hair follicles resulting in rough, elevated skin bumps); xerophthalmia (i.e., parched eyes); Bitot's spots (i.e., a condition characterized by abnormal changes in the conjunctiva and cornea); and complete blindness. Impaired iron utilization leading to iron-deficiency anemia is associated with vitamin A deficiency. Vitamin A deficiency also increases infection risk, with infections leading to diarrhea and depressed appetite, which further decreases vitamin A intake.

Vitamin A deficiency increases the risk of pregnancy-related complications, including maternal anemia and preterm delivery.

Treatment of vitamin A deficiency includes dietary intake and supplementation. Dietary intake of vitamin A or beta-carotene supplements can correct vitamin A deficiency; supplements of preformed vitamin A are rarely prescribed due to the high toxicity of vitamin A.

Persons who have not received a diagnosis of vitamin A deficiency should not take preformed vitamin A supplements above the recommended dietary intake because there is a low threshold for vitamin A toxicity.

VITAMIN A TOXICITY AND MEDICATION INTERACTIONS

Daily safe and tolerable upper intake levels (ULs) for vitamin A have been established as follows:

- Birth to three years: 600 mcg (2,000 IU)
- Four to eight years: 900 mcg (3,000 IU)
- Nine to thirteen years: 1,700 mcg (5,610 IU)
- Fourteen to eighteen years (all individuals, including pregnant and lactating females): 2,800 mcg (9,240 IU)
- Nineteen-plus years (all individuals, including pregnant and lactating females): 3,000 mcg (10,000 IU)

Normal serum vitamin A concentrations are 30-95 ug/dL (1.05-3.32umol/L).

In many clinical settings, plasma vitamin A values can be challenging to interpret because retinol-binding protein (RBP, i.e., the carrier of vitamin A in blood) is an acute phase response protein (i.e., change in response to inflammation); production of RBP is increased in response to inflammation or infection, making plasma vitamin A levels an unreliable indicator of vitamin A status in patients with inflammation or infection.

Preformed vitamin A is highly toxic in large amounts and is particularly toxic for pregnant women, even in moderate doses; moderate amounts of preformed vitamin A, especially when taken in early pregnancy, can lead to spontaneous abortion, congenital disabilities, and learning disabilities.

Provitamin A carotenoids are generally nontoxic, except for canthaxanthin; ingestion of large quantities of canthaxanthin for prolonged periods can lead to retinopathy; the quantity of canthaxanthin required to induce retinopathy in humans has not been conclusively determined, but studies in nonhuman primates suggest that doses of 200-500 mg of canthaxanthin per kilogram of body weight, which would be approximately 15-40 grams per day for a person weighing 175 pounds, taken for several months can cause retinopathy.

Vitamin A toxicity is classified as acute, chronic, or teratogenic.

ACUTE VITAMIN A TOXICITY

A vitamin A dose of 200 mg (660,000 IU) or greater in adults or 100 mg (330,000 IU) or greater in children can lead to acute vitamin A toxicity.

Signs and symptoms of acute vitamin A toxicity include nausea, vomiting, headache, increased cerebrospinal fluid pressure, bulging of the fontanelles in infants, vertigo, double vision, loss of muscular coordination, drowsiness, loss of appetite, and skin itching and flaking.

A single vitamin A dose of 11,800 mg (11.8 g) or more in adults and 500 mg or more in children can be lethal.

CHRONIC VITAMIN A TOXICITY

Chronic vitamin A toxicity can be caused by recurrent intake of approximately ten times the recommended dietary allowance (RDA); the margin between safe and toxic intake is much narrower for vitamin A than any other nutrient.

Signs and symptoms of chronic vitamin A toxicity include alopecia, ataxia, bone and muscle pain, cheilitis (i.e., inflammation and cracking of the lips), conjunctivitis, headaches, liver damage and abnormalities of liver-related blood markers, hyperlipidemia, hyperostosis (i.e., excessive bone growth), drying of the mucous membranes, pruritus, increased intracranial pressure, skin abnormalities, and vision impairment; signs and symptoms are often, although not always, reversible when vitamin A intake is discontinued.

TERATOGENIC VITAMIN A TOXICITY

All retinoids, including vitamin A, are potent teratogens (i.e., causative of congenital disabilities).

A single huge dose or high daily or weekly intake of preformed vitamin A, especially during early pregnancy, can lead to spontaneous abortion and fetal malformations. The lowest teratogenic dose of preformed vitamin A in humans has not been determined. Although the US Food and Drug Administration (FDA) reports that 8,000 IU of supplemental preformed vitamin A should be considered the maximum safe dosage before or during pregnancy, most health experts advise against any supplementation with preformed vitamin A during pregnancy.

Vitamin A toxicity during pregnancy can lead to fetal defects, including craniofacial malformations, congenital heart disease, kidney and thymus defects, permanent learning disabilities, and central nervous system (CNS) disorders.

When combined with other teratogens (e.g., alcohol), synergistic effects can lead to the occurrence of fetal defects at lower-than-expected doses of vitamin A.

Pregnant women should be advised to avoid vitamin A supplements and limit intake of vitamin A-rich foods (e.g., liver).

Prescription formulations of retinoids (e.g., tretinoin [Retin-A], which is not available over the counter) are long-acting; adverse effects and congenital disabilities are reported to occur months after prescribed retinoid therapy is discontinued.

Vitamin A is hepatotoxic (i.e., toxic to the liver) and can cause severe adverse effects that range from elevated liver enzymes to liver failure; hepatotoxic effects can increase when vitamin A is taken with certain medications, including acetaminophen, carbamazepine, isoniazid, methotrexate, and prescription retinoids (e.g., tretinoin [Retin-A]).

Concurrent use of tetracyclines (i.e., antibiotics) and vitamin A supplements can increase the risk of intracranial hypertension (e.g., rise in the pressure of brain fluid), a potentially life-threatening condition.

Long-term use or taking a single high dose of vitamin A can increase the risk of bleeding in persons receiving blood thinners because vitamin A can antagonize (i.e., interfere with, lessen, or oppose) vitamin K activity.

Medications that reduce fat absorption, including orlistat and bile acid sequestrants (e.g., cholestyramine), can reduce vitamin A absorption; vitamin A supplementation can increase absorption to a sufficient degree to avoid deficiency.

Adequate vitamin A intake can improve iron-deficiency anemia in persons with adequate iron intake and low serum retinol levels. Chronic alcohol ingestion can potentiate the hepatotoxic effects of vitamin A. Without adequate zinc, a vitamin A deficiency can develop despite vitamin A supplementation because zinc deficiency interferes with vitamin A breakdown.

RESEARCH FINDINGS

Authors of a recent systematic review and meta-analysis concluded that vitamin A supplementation in children aged six months to five years living in low- and middle-income countries is associated with a 24 percent reduction in all-cause mortality, a 15 percent reduction in the incidence of diarrhea, a 50 percent reduction in the incidence of measles, a 68 percent reduction in the prevalence of night blindness, a 55 percent reduction in the prevalence of Bitot's spots (i.e., the buildup of keratin in the eye), and a 69 percent reduction in the prevalence of xerophthalmia (i.e., dryness of eyes).

Data are mixed regarding whether intake of vitamin A alone within the safe range of intake and in the presence of adequate vitamin D status increases the risk of osteoporosis; however, the combination of high vitamin A intake and low vitamin D intake can increase osteoporosis risk.

Although several liver diseases are associated with oxidative stress, investigators who conducted a recent meta-analysis concluded that persons with liver disease should not take vitamin A supplements unless they have verifiable vitamin A deficiency.

Vitamin A deficiency may be more common in older persons than previously believed; assessment of vitamin A status in older persons (age seventy years and older) can help identify persons at risk for vitamin A deficiency.

Researchers conducted a recent study of twenty-two neonates with congenital diaphragmatic hernia (CDH) and thirty-four healthy neonates. They concluded that low retinol in newborns, independent of maternal retinol status, is associated with an increased risk of CDH.

In a study of 261 long-term hemodialysis patients, investigators found that low retinol levels were associated with 27 and 31 percent increases in the risk of all-cause and cardiovascular mortality, respectively.

Although carotenoids are generally considered nontoxic, researchers warn that current and former smokers should not take supplemental beta-carotene. Results of controlled clinical trials show that beta-carotene supplements may increase the risk of lung, head and neck, and possibly other cancers in this population and may increase the risk of cancer recurrence in both smokers and non-smokers who have been treated for head and neck cancer.

In contrast, researchers evaluated the relationship between the consumption of fruits and vegetables and lung cancer risk among 61,491 Chinese men from the Shanghai Men's Health Study, a population-based, prospective cohort study. Study results revealed that the combined intake of fruits and vegetables was marginally associated with a lower risk for lung cancer. In contrast, consumption of vegetables rich in carotenoids was significantly inversely associated with lung cancer risk. Further studies conducted on animals and cells have indicated that treatment with carotenoids can inhibit cancer development. Researchers suggest that while high dietary intake of beta-carotene, as found in food sources, can be chemo-preventive, high doses of beta-carotene might interfere with the antioxidant reactions of other compounds in individuals who smoke.

Other carotenoids have not been well studied regarding cancer risk in smokers; in the absence of definitive studies establishing the safety of other carotenoids in smokers, carotenoid supplements should be avoided.

SUMMARY

Consumers should become knowledgeable about vitamin A. Vitamin A maintains eye health, develops tissues, and is critical in fetal development. A well-balanced diet includes good sources of vitamin A, such as sources of preformed vitamin A including fortified milk, eggs, and fish, and sources of provitamin A, including carrots, pumpkin, and spinach. Symptoms of vitamin A deficiency include night blindness, loss of appetite, and decreased immune function. Consumers should be aware of the interactions between vitamin A and certain medications. Research suggests that vitamin A may help prevent certain cancers; however, vitamin A supplementation may increase the risk of developing lung cancer in smokers.

—*Suzanne Dixon, MPH, MS, RD and Tanja Schub, BS*

Further Reading

Bjelakovic, G., L. L. Gluud, D. Nikolova, et al. "Antioxidant Supplements for Liver Diseases." *Cochrane Database of Systematic Reviews*, vol. 3, 2011, CD007749.

Black, R. E., L. H. Allen, Z. A. Bhutta, et al. "Maternal and Child Undernutrition: Global and Regional Exposures and Health Consequences." *The Lancet*, vol. 371, no. 9608, 2008, pp. 243-60.

Eckert, M. J., J. T. Perry, V. Y. Sohn, et al. "Incidence of Low Vitamin A Levels and Ocular Symptoms after Roux-en-Y Gastric Bypass." *Surgery for Obesity and Related Diseases*, vol. 6, no. 6, 2010, pp. 653-57.

Investigative Ophthalmology & Visual Science, vol. 41, no. 6, pp. 1513-22.

Gudas, L. J., and J. A. Wagner. "Retinoids Regulate Stem Cell Differentiation." *Journal of Cellular Physiology*, vol. 226, no. 2, 2011, pp. 322-30.

Mactier, H., M. M. Mokaya, L. Farrell, and C. A. Edwards. "Vitamin A Provision for Preterm Infants: Are We Meeting Current Guidelines?" *Archives of Disease in*

Childhood: Fetal and Neonatal Edition, vol. 96, no. 4, 2011, pp. F286-89, doi:10.1136/adc.2010.190017.
Mark, M., N. B. Ghyselinck, and P. Chambon. "Function of Retinoic Acid Receptors during Embryonic Development." *Nuclear Receptor Signaling Atlas*, vol. 7, 2009, p. e002.
National Institutes of Health. "Vitamin A." *Medline Plus*, 12 Jan. 2015, www.nlm.nih.gov/medlineplus/ency/article/002400.htm. Accessed 3 Feb. 2015.
———. "Dietary Supplement Fact Sheet: Vitamin A." *Office of Dietary Supplements*, 5 June 2013, ods.od.nih.gov/factsheets/VitaminA-HealthProfessional/. Accessed 3 Feb. 2015.
Olson, J. A. "Vitamin A, Retinoids, and Carotenoids." *Modern Nutrition in Health and Disease*, edited by M. E. Shils, J. A. Olson, and M. Shike, 8th ed., Lea & Febiger, pp. 287-307.
Oregon State University. "Vitamin A." *Linus Pauling Institute*. Micronutrient Information Center, 2007, lpi.oregonstate.edu/infocenter/vitamins/vitaminA/. Accessed 3 Feb. 2015.
Rock, C. L. "Carotenoid Update." *Journal of the American Dietetic Association*, vol. 103, no. 4, 2003, pp. 423-25.
Sathe, M. N., and A. S. Patel. "Update in Pediatrics: Focus on Fat-Soluble Vitamins." *Nutrition in Clinical Practice*, vol. 25, no. 4, 2010, pp. 340-46, doi:10.1177/0884533610374198.
The Teratology Society. "Teratology Society Position Paper: Recommendations for Vitamin A Use during Pregnancy." *Teratology*, vol. 35, no. 2, 1987, pp. 269-75.
US Food and Drug Administration. News & Events. "FDA Warns Consumers to Stop Using Soladek Vitamin Solution." *FDA, News & Events*, 28 Mar. 2011, www.fda.gov/NewsEvents/Newsroom/PressAnnouncements/ucm248588.htm. Accessed 3 Feb. 2015.
USDA National Nutrient Database for Standard Reference, Release 24. "Vitamin A, IU Content of Selected Foods Per Common Measure, Sorted by Nutrient Content." *USDA*, 31 May 2013, www.docstoc.com/docs/30903437/USDA-National-Nutrient-Database-for-Standard-Reference_-Release-22. Accessed 3 Feb. 2015.
World Health Organization. "Global Prevalence of Vitamin A Deficiency in Populations at Risk 1995-2005." *Global Database on Vitamin A Deficiency, WHO*, 2009.

Vitamin B12

Category: Dietary supplements
Also known as: Cobalamin, cyanocobalamin, hydroxocobalamin, methylcobalamin
Anatomy or system affected: All systems
Specialties and related fields: Biochemistry, dietetics, internal medicine, hematology, preventive medicine, public health
Definition: a water-soluble that is one of the eight B vitamins and a cofactor in deoxyribonucleic acid (DNA) synthesis, fatty acid, and amino acid metabolism

KEY TERMS

cyanide poisoning: poisoning from exposure to any form of cyanide
gastric bypass surgery: surgical procedures performed on the stomach or intestines to induce weight loss.
megaloblastic anemia: a condition in which the bone marrow produces unusually large, structurally abnormal, immature red blood cells called "megaloblasts"
S-adenosylmethionine: an essential cofactor in the body made from the amino acid methionine and adenosine that participates in single carbon transfer reactions

OVERVIEW

Vitamin B_{12}, an essential nutrient, is also known as cobalamin. The "cobal" in the name refers to the metal cobalt contained in B_{12}. Vitamin B_{12} is required for the regular activity of nerve cells. It works with folate and vitamin B_6 to lower blood levels of homocysteine, a chemical in the blood that might contribute to heart disease. B_{12} also plays a role in the body's manufacture of S-adenosylmethionine (SAMe).

Anemia is usually (but not always) the first sign of B_{12} deficiency. Early in the twentieth century, doctors coined the name "pernicious anemia" for a stubborn form of anemia that did not improve even when the patient was given iron supplements. Experts now know that pernicious anemia occurs when the stomach fails to excrete a particular intrinsic factor. The body needs the intrinsic factor for efficient absorption of vitamin B_{12}. In 1948, vitamin B_{12} was identified as the cure for pernicious anemia. B_{12} deficiency also causes nerve damage, and this may, in some cases, occur without anemia first developing.

Vitamin B_{12} has also been proposed as a treatment for numerous other conditions. Still, there is no definitive evidence that it is effective for any purpose other than correcting the deficiency.

REQUIREMENTS AND SOURCES

Extraordinarily small amounts of vitamin B12 suffice for daily nutritional needs. The official US and Canadian recommendations for daily intake (in micrograms or mcg) are as follows:

Infants aged zero to six months (0.4 mcg) and seven to twelve months (0.5 mcg); children aged one to three years (0.9 mcg), four to eight years (1.2 mcg), and nine to thirteen years (1.8 mcg); males and females aged fourteen years and older (2.4 mcg); pregnant women (2.6 mcg); and nursing women (2.8 mcg).

Vitamin B_{12} deficiency is rare in the young. However, it is not unusual in older people: 10 to 20 percent of the elderly are deficient in B_{12}. Older people have lower levels of stomach acid, and vitamin B_{12} in food comes attached to proteins. It must be released by acid in the stomach to be absorbed. When stomach acid levels are low, people do not absorb as much vitamin B_{12} from their food. Vitamin B_{12} supplements do not need acid for absorption and should therefore get around this problem. However, for unclear reasons, one study found that Buble absorb-deficient seniors need very high dosages of the supplements to normalize their levels, as high as 600 to 1,000 mcg daily. Similarly, people who take medications that significantly reduce stomach acids, such as omeprazole (Prilosec) or ranitidine (Zantac), may have trouble absorbing B_{12} from food and could benefit from supplementation.

Stomach surgery and other conditions affecting the digestive tract can also lead to B_{12} deficiency. Vitamin B_{12} absorption or levels in the blood may also be impaired by colchicine (for gout), metformin and phenformin (for diabetes), and azidothymidine ([AZT] for acquired immunodeficiency syndrome [AIDS]). Exposure to nitrous oxide, such as may be experienced by dentists and dental hygienists, might cause B_{12} deficiency, but studies disagree. Slow-release potassium supplements might also impair B_{12} absorption.

Vitamin B_{12} is found in most animal foods; it is also found only in animal food. Beef, liver, clams, and lamb provide a whopping 80 to 100 mcg of B_{12} per 3.5-ounce serving, at least forty times the dietary requirement. Sardines, chicken liver, beef kidney, and calf liver are good sources, providing between 25 and

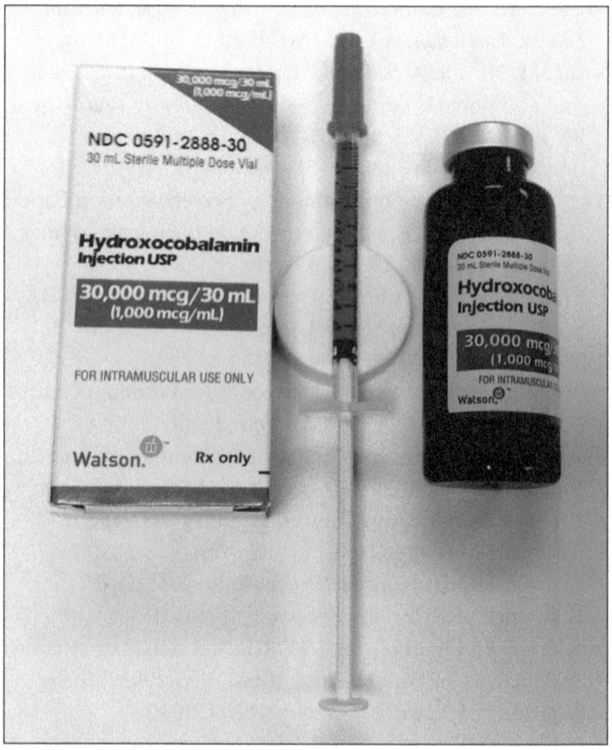

A Vitamin B_{12} solution (hydroxocobalamin) with a single dose drawn up into a syringe. Photo by Sbharris, via Wikimedia Commons.

60 mcg per serving. Trout, salmon, tuna, eggs, whey, and many kinds of cheese provide at least the recommended daily intake.

Total vegetarians (vegans) must take vitamin B_{12} supplements or consume B_{12}-fortified foods, or they will eventually become deficient. Contrary to some reports, seaweed and tempeh do not provide B_{12}. Some forms of blue-green algae, such as spirulina, contain B_{12}, but it is not absorbable.

Vitamin B_{12} is available in three forms: cyanocobalamin, hydroxocobalamin, and methylcobalamin. The first is the most widely available and least expensive. Still, some experts think that the other two forms are preferable.

Severe B_{12} deficiency can cause anemia and, potentially, nerve damage. The latter may become permanent if the deficiency is not corrected in time. Anemia most often develops first, leading to treatment before permanent nerve damage develops. However, folate supplements can get in the way of this "early warning system." This is why people are cautioned against taking high doses of folate without medical supervision. When taken at a dosage higher than 400 mcg daily, folate can prevent anemia caused by B_{12} deficiency, thereby allowing permanent nerve damage to develop without any warning. More mild deficiencies of vitamin B_{12} may cause elevated levels of homocysteine in the blood, potentially increasing the risk of heart disease. Mild B_{12} deficiency, too slight to cause anemia, may also impair brain function.

THERAPEUTIC DOSAGES

For correcting absorption problems caused by medications, taking vitamin B_{12} at the level of dietary requirements should suffice. Enormously higher daily doses—ranging from 100 to 2,000 mcg—are sometimes recommended for other purposes.

THERAPEUTIC USES

It appears that individuals who take medications that dramatically lower stomach acid, such as H2 blockers or proton pump inhibitors, would benefit by taking B_{12} supplements. Other individuals likely to be deficient in B_{12}, such as the elderly and those taking the medications listed above in Requirements and Sources, might well benefit from a daily B_{12} supplement to prevent B_{12} deficiency.

For pernicious anemia, B_{12} injections are traditionally used. However, research has shown that oral B_{12} works just as well, provided people take enough of it: between 300 and 1,000 mcg daily.

Vitamin B_{12} is a standard treatment for cyanide poisoning. In people with cyanide poisoning, large hydroxocobalamin doses are given intravenously with sodium thiosulfate. The central cobalt atom in hydroxocobalamin binds cyanide ions to form cyanocobalamin (vitamin B_{12}), which is nontoxic. The kidneys efficiently excrete nontoxic cyanocobalamin is excreted in the urine.

Various types of gastric bypass or gastric restriction surgeries treat morbid obesity. One procedure, specifically the Roux-en-Y gastric bypass surgery (RYGB) but not sleeve gastric bypass surgery or gastric banding, increases the risk of vitamin B_{12} deficiency. Patients who have had the RYGB procedure require postoperative B_{12}. Because absorption of this vitamin is diminished in people who have had RYGB, such individuals usually take sublingual or injected forms of vitamin B_{12} (1000 μg/day).

Weak evidence suggests that B_{12} supplements may improve sperm activity and sperm count; on this basis, they could be helpful for male infertility. Some cases of recurrent miscarriage might be due to vitamin B_{12} deficiency.

One placebo-controlled, double-blind study, enrolling forty-nine people with eczema, found benefit with a cream containing vitamin B_{12} at a concentration of 0.07 percent. Topical B_{12} is hypothesized to work for eczema by reducing local levels of the substance nitric oxide (not related to nitrous oxide).

Based on weak and sometimes contradictory evidence, vitamin B_{12} has been suggested for human immunodeficiency virus (HIV), amyotrophic lateral sclerosis, carpal tunnel syndrome, diabetic neuropathy, multiple sclerosis, restless legs syndrome, and tinnitus. Some evidence suggests that people with vitiligo (splotchy loss of skin pigmentation) might be deficient in vitamin B_{12}, and supplementation along with folate may be helpful. However, the evidence is feeble, and not all studies agree.

Some alternative practitioners recommend the use of injected vitamin B_{12} for Bell's palsy. However, the only scientific support for this approach comes from one study that was not double-blind.

Vitamin B_{12} is sometimes recommended for numerous other problems, including asthma, osteoporosis, periodontal disease, and depression. However, there is not much evidence that it works.

A double-blind trial of vitamin B_{12} for seasonal affective disorder, a type of depression related to lack of light during the winter, failed to find evidence of benefit. In addition, a randomized trial involving older adults with mild depression found that taking folate (400 mcg) and vitamin B_{12} (100 mcg) daily for two years was no better than a placebo for reducing depressive symptoms.

One double-blind, placebo-controlled study of 140 people with mildly low B_{12} levels failed to find the supplement to improve mental function and mood. Another study failed to find evidence that vitamin B_{12} improved the general sense of well-being among seniors with signs of mild B_{12} deficiency.

Although vitamin B_{12} has been proposed as a treatment for Alzheimer's disease, this recommendation is based solely on the results of one small, poorly designed study. More recent and better-designed studies found little to no benefit.

SCIENTIFIC EVIDENCE

Vitamin B_{12} deficiencies in men can lead to reduced sperm counts and lowered sperm mobility. For this reason, B_{12} supplements have been tried to improve fertility in men with abnormal sperm production. In one double-blind study of 375 infertile men, supplementation with vitamin B_{12} produced no benefits on average in the group as a whole. However, in a particular subgroup of men with sufficiently low sperm count and sperm motility, B_{12} appeared to be helpful. Such "dredging" of the data is suspect from a scientific point of view. Consequently, this study cannot be taken as proof of effectiveness.

SAFETY ISSUES

Vitamin B_{12} appears to be extremely safe. However, in some cases, very high doses of the vitamin can cause or worsen acne symptoms.

IMPORTANT INTERACTIONS

People who are taking colchicine; the anti-HIV drug azidothymidine; medications that reduce stomach acid, such as the H2 blocker ranitidine (Zantac) or the proton pump inhibitor omeprazole (Prilosec); oral hypoglycemics, such as metformin; and slow-release potassium supplements; and people who are exposed to nitrous oxide anesthesia may need extra B_{12}. Another option is to take extra calcium, which may, in turn, improve B_{12} absorption.

—EBSCO CAM Review Board

Further Reading

Hvas, A. M., et al. "No Effect of Vitamin B_{12} Treatment on Cognitive Function and Depression." *Journal of Affective Disorders,* vol. 81, 2004, pp. 269-73.

___ "Vitamin B12 Treatment Has Limited Effect on Health-Related Quality of Life Among Individuals with Elevated Plasma Methylmalonic Acid." *Journal of Internal Medicine,* vol. 253, 2003, pp. 146-52.

Malouf, R., and J. Grimley Evans. "Folic Acid with or Without Vitamin B_{12} for the Prevention and Treatment of Healthy Elderly and Demented People." *Cochrane Database of Systematic Reviews,* vol. 4, 2008, CD004514.

Reznikoff-Etievant, M. F., et al. "Low Vitamin B_{12} Level as a Risk Factor for Very Early Recurrent Abortion." *European Journal of Obstetrics, Gynecology, and Reproductive Biology,* vol. 104, 2002, pp. 156-59.

Sato, Y., et al. "Amelioration by Mecobalamin of Subclinical Carpal Tunnel Syndrome Involving Unaffected Limbs in Stroke Patients." *Journal of Neurological Sciences*, vol. 231, 2005, pp. 13-18.
Seussen, S. J., et al. "Oral Cyanocobalamin Supplementation in Older People with Vitamin B_{12} Deficiency." *Archives of Internal Medicine*, vol. 165, 2005, pp. 1167-72.
Stucker, M., et al. "Topical Vitamin B, a New Therapeutic Approach in Atopic Dermatitis: Evaluation of Efficacy and Tolerability in a Randomized Placebo-Controlled Multicentre Clinical Trial." *British Journal of Dermatology*, vol. 150, 2005, pp. 977-83.
Ting, R. Z., et al. "Risk Factors of Vitamin B_{12} Deficiency in Patients Receiving Metformin." *Archives of Internal Medicine*, vol. 166, 2006, pp. 1975-99.
Walker, J. G., et al. "Mental Health Literacy, Folic Acid and Vitamin B_{12}, and Physical Activity for the Prevention of Depression in Older Adults." *British Journal of Psychiatry*, vol. 197, no. 1, 2020, pp. 45-54.

VITAMIN C

Also known as: Ascorbate, ascorbic acid
Category: Dietary supplements
Anatomy or system affected: All systems
Specialties and related fields: Biochemistry, biotechnology, dietetics, preventive medicine, public health
Definition: a vitamin found in various foods and sold as a dietary supplement that prevents and treats scurvy, is essential for tissue repair, collagen formation, and the production of certain neurotransmitters

KEY TERMS

ascorbic acid: a vitamin found particularly in citrus fruits and green vegetables; it is essential in maintaining healthy connective tissue and is also thought to act as an antioxidant; and severe deficiency causes scurvy

vitamin: any of a group of organic compounds essential for normal growth and nutrition and required in small dietary quantities because the body cannot synthesize them

OVERVIEW

Although most animals can make vitamin C, humans have lost the ability to do so through evolution. Because of this, humans must get the vitamin from food, chiefly fresh fruits and vegetables. One of this vitamin's primary functions is helping the body manufacture collagen, an essential protein in connective tissues, cartilage, and tendons.

From ancient times through the early nineteenth century, sailors and others deprived of fresh fruits and vegetables developed scurvy. Scurvy involves scorbutic symptoms, which include nonhealing wounds, bleeding gums, bruising, and overall weak-

Image via iStock/stevezmina1. [Used under license.]

ness. It is now known that scurvy is nothing more than vitamin C deficiency. Scurvy was successfully treated with citrus fruit during the mid-eighteenth century. In 1931-32, Hungarian physiologist Albert Szent-Györgyi isolated the active ingredient of citrus fruits, calling it the antiscorbutic principle, or ascorbic acid (vitamin C).

Vitamin C is a powerful antioxidant that neutralizes damaging natural substances called "free radicals." It works in water, both inside and outside cells. Vitamin C complements another antioxidant vitamin, vitamin E, which works in fatty parts of the body.

Vitamin C is the single most popular vitamin supplement in the United States and perhaps the most controversial. In the 1960s, two-time Nobel Prize winner Linus Pauling claimed that vitamin C could effectively treat both cancer and the common cold. Subsequent research has mostly discounted these claims but has not dampened enthusiasm for this essential nutrient. The vitamin C "movement" has led to hundreds of clinical studies testing the vitamin on dozens of illnesses; at present, however, no dramatic benefits have been discerned.

Linus Pauling, a Nobel Prize winner, recommended taking vitamin C for the common cold in a 1970 book. Photo byNobel Foundation, via Wikimedia Commons. [Public domain.]

REQUIREMENTS AND SOURCES

Vitamin C is an essential nutrient obtained from food or supplements; the body cannot manufacture it. The official US and Canadian recommendations for daily intake (in milligrams) are as follows:

- Infants to six months of age (40) and seven to twelve months of age (50);
- children one to three years of age (15), four to eight years of age (25), and nine to thirteen years of age (45);
- boys aged fourteen to eighteen years (75) and girls aged fourteen to eighteen years (65);
- men (90) and women (75); pregnant girls (80) and pregnant women (85);
- nursing girls (115) and nursing women (120).

Smoking cigarettes significantly reduces levels of vitamin C in the body. The recommended daily intake for smokers is 35 mg higher across all age groups.

Scurvy, the classic vitamin C deficiency disease, is now a rarity in the developed world. However, a more subtle deficiency of vitamin C is relatively common. According to one study, 40 percent of Americans do not get enough vitamin C. Also, vitamin C deficiency significant enough to cause bleeding problems during surgery turns out to be more common than previously thought.

Aspirin and other anti-inflammatory drugs might lower body levels of vitamin C, as might oral contraceptives. Supplementation may be helpful if one is taking any of these medications.

Most people think of orange juice as the definitive source of vitamin C. However, many vegetables are even richer sources. Red chili peppers, sweet peppers, kale, parsley, collard, and turnip greens are full of vitamin C, as are broccoli, Brussels sprouts, watercress, cauliflower, cabbage, and strawberries. (Oranges and other citrus fruits are good sources, too.)

One great advantage of getting vitamin C from foods rather than from supplements is that many other potentially healthful nutrients are obtained simultaneously, nutrients such as bioflavonoids and carotenes. However, vitamin C in food is partially destroyed by cooking and exposure to air. For maximum nutritional benefit, one could try eating freshly made salads rather than dishes that require a lot of cooking.

Vitamin C supplements are available in two forms: ascorbic acid and ascorbate. The latter is less intensely sour.

THERAPEUTIC DOSAGES

Since the time of Pauling, proponents have recommended taking vitamin C in enormous doses, as high as 20,000 to 30,000 mg daily. However, some evidence suggests that there might be no reason to take more than 200 mg of vitamin C daily (ten to one hundred times less than the amount recommended by vitamin C proponents). The reason is that if a person consumes more than 200 mg daily (researchers have tested up to 2,500 mg), the kidneys begin to excrete the excess at a steadily increasing rate, matching the increased dose. The digestive tract also stops absorb-

Albert Szent-Györgyi won the Nobel Prize for discovering how to mass-produce Vitamin C while living in Szeged, which was the center of the paprika (red pepper) industry. Image by Douglas RM, Hemilä H. National Centre for Epidemiology and Population Health, Australian National University, via Wikimedia Commons.

ing it well. The net effect is that no matter how much is taken, the blood levels of vitamin C do not increase very much.

However, there are some flaws in this research. Vitamin C levels might rise in other tissues even if they remain constant in the blood. Furthermore, this study did not evaluate the possible effects of taking vitamin C several times daily rather than once daily.

Many nutritional experts recommend 500 mg of vitamin C daily, and this dose is almost undoubtedly safe. Others recommend taking as much vitamin C as possible, up to 30,000 mg daily, cutting back only when one starts to develop stomach cramps and diarrhea. This recommendation seems based more on a semireligious enthusiasm for vitamin C than on any evidence that such massive doses of the vitamin are beneficial.

Intravenous vitamin C can easily raise vitamin C levels to 140 times higher than the maximum achievable with oral vitamin C. However, there is no meaningful evidence that intravenous vitamin C provides any medical benefits.

THERAPEUTIC USES

According to numerous double-blind, placebo-controlled studies, the regular use of vitamin C supplements can slightly reduce symptoms of colds and modestly shorten the length of the illness. However, taking vitamin C at the onset of a cold probably will not work.

Regular use of vitamin C does not seem to help prevent colds. One exception is the "postmarathon sniffle," colds that develop after heavy exercise. Vitamin C may help prevent this condition, although not all studies agree.

Two double-blind studies suggest that the use of vitamin C with vitamin E might slightly reduce the risk of developing preeclampsia, a complication of pregnancy. However, a much larger follow-up study failed to find benefits. Two studies conducted by a single research group have found that vitamin C at a dose of 500 mg daily might help prevent reflex sympathetic dystrophy. This poorly understood condition can follow injuries such as fractures.

Over time, the body develops tolerance to drugs in the nitrate family (such as nitroglycerin). Some evidence suggests that the use of vitamin C can help maintain the effectiveness of these medications.

Other small double-blind trials suggest that vitamin C might be helpful for anterior uveitis (when taken with vitamin E), autism, easy bruising, minor injuries, protecting the liver in nonalcoholic steatohepatitis, speeding recovery from bedsores, treating female infertility (specifically, a condition called "luteal phase defect"), and preventing early rupture of the chorioamniotic membranes in pregnancy. Vitamin C might also improve the effectiveness of antibiotic treatment for Helicobacter pylori, the cause of most peptic ulcers.

Preliminary evidence suggests that cream containing vitamin C may improve the appearance of aging or sun-damaged skin. Inconsistent evidence suggests that oral or topical vitamin C, taken by itself or with vitamin E, may also help protect the skin from sun damage.

Double-blind studies of vitamin C for the following conditions have yielded mixed results: asthma, male infertility, reducing the muscle soreness that typically develops after exercise, and hypertension. Unexpectedly, one study found that a combination of vitamin C (500 mg daily) and grape seed oligomeric proanthocyanidins (1,000 mg daily) slightly increased blood pressure. Whether this was a fluke of statistics or a real combined effect remains unclear.

Limited and, in some cases, contradictory evidence suggests possible benefit in the prevention or treatment of allergies, atrial fibrillation following coronary artery bypass grafting, bladder infections during pregnancy, gallbladder disease (in women), glaucoma, gout, obesity, and vascular dementia. Also, the intravaginal use of vitamin C tablets might be helpful for nonspecific vaginitis.

Observational studies indicate that people with a higher intake of vitamin C have a lower incidence of cataracts, macular degeneration, heart disease, cancer, and osteoarthritis. However, these findings do not indicate that vitamin C supplements will help prevent or treat these conditions. Observational studies are notoriously unreliable for showing the efficacy of treatments; only double-blind studies can do that. Only one has been performed that directly examined vitamin C's potential benefits for preventing these conditions. Two large double-blind trials exploring the effectiveness of vitamin C for heart disease prevention, one in women at high risk and the other in men at low risk, failed to find any benefit. Vitamin C has been proposed as a treatment for cancer. However, this claim is very controversial, and there is no scientifically meaningful evidence that it works.

Massive doses of vitamin C have at times been popular among people with human immunodeficiency virus (HIV) infection, based on preliminary evidence. An observational study linked high doses of vitamin C with slower progression to acquired immunodeficiency syndrome (AIDS). However, a double-blind study of forty-nine people with HIV who took combined vitamins C and E or placebo for three months did not significantly affect the amount of HIV detected or the number of opportunistic infections. Furthermore, one study found that vitamin C at a dose of 1 gram (g) daily substantially reduced blood levels of the drug indinavir, a protease inhibitor used to treat HIV infection. This could potentially cause the drug to fail.

In a study of eighty women with Chlamydia trachomatis infection, adding vitamin C to doxycycline and triple sulfa vaginal cream reduced discharge and pain associated with intercourse. According to a double-blind, placebo-controlled study of 141 women with cervical dysplasia (early cervical cancer), vitamin C, taken at a dosage of 500 mg daily, does not help reverse the dysplasia.

One substantial study failed to find vitamin C helpful in improving high cholesterol. Vitamin C also does not appear to help treat Raynaud's phenomenon caused by scleroderma.

SCIENTIFIC EVIDENCE

Colds. As the best known of all-natural treatments for the common cold, vitamin C has been subjected to irresponsible hype from both proponents and opponents. Enthusiasts claim that if one takes vitamin C daily, one will never get sick. At the same time, critics of the treatment insist that vitamin C has no benefit.

However, a reasoned evaluation of the research indicates something in between. Numerous studies have found that vitamin C supplements taken at a dose of 1,000 mg daily or more throughout the cold season can modestly reduce symptoms of colds and help a person get over a cold faster. Still, they do not generally help prevent colds.

Reducing cold symptoms. Most studies on vitamin C have evaluated the potential benefits of taking vitamin C throughout the cold season. A review of twenty-nine placebo-controlled trials involving more than eleven thousand people found that the use of vitamin C in this way can reduce symptoms and decrease the duration of colds. Other studies have found similar results.

Many people begin taking vitamin C only when cold symptoms start. Vitamin C is probably not effective when used in this way. One double-blind trial enrolled four hundred persons with new-onset cold symptoms. It divided them into four different daily vitamin-C-dosage groups: 30 mg daily (a dose lower than the minimum daily requirement and used by the researchers as a placebo), 1,000 mg, 3,000 mg, or 3,000 mg with bioflavonoids. Participants were instructed to take the vitamin at the onset of symptoms and for the following two days. The results showed no difference in the duration or severity of cold symptoms among the groups. High-dose vitamin C taken at the onset of a cold, in other words, did not help. A review of seven randomized and nonrandomized trials found that taking vitamin C at the start of a cold did not offer any benefits. Indeed, there are numerous other natural treatments for the common cold, some of which may be more helpful than vitamin C.

Preventing colds. Although two relatively late studies suggest that regular use of vitamin C throughout the cold season can help prevent colds, these studies had a variety of flaws. Most other studies have found little to no benefit along these lines. However, people who are truly deficient in vitamin C, such as older adults in nursing homes, may show increased resistance to infection if they take vitamin C (or other nutrients).

In addition, vitamin C might help prevent respiratory infections that can follow heavy endurance exercise. Marathon running and similar forms of exertion can temporarily weaken the immune system, leading to infections. Vitamin C may be helpful. According to a double-blind, placebo-controlled study involving ninety-two runners, taking 600 mg of vitamin C for twenty-one days before a race made a significant difference in the incidence of sickness afterward. Within two weeks of the race, 68 percent of the runners taking placebo developed cold symptoms, versus only 33 percent of those taking the vitamin C supplement. As part of the same study, nonrunners of similar age and gender to those running were also given vitamin C or a placebo. The supplement had no apparent effect on the incidence of upper respiratory infections in this group. Vitamin C seemed to be effective in this capacity only for those who exercised intensively.

Two other studies found that vitamin C could reduce the number of colds experienced by groups of people involved in rigorous exercise in extremely cold environments. One study involved 139 children attending a skiing camp in the Swiss Alps. At the same time, the other enrolled 56 military men engaged in a training exercise in northern Canada during the winter months. In both cases, the participants took either 1 g of vitamin C or a placebo daily when their training program began. Cold symptoms were monitored for one to two weeks following training and found significant differences in favor of vitamin C.

However, one large study of 674 US Marine Corps recruits in basic training found no such benefit. The results showed no difference in the number of colds between the treatment and placebo groups.

There are many possibilities for this discrepancy. Perhaps basic training in the Marine Corps is significantly different from the other forms of exercise studied. Another point to consider is that the Marine recruits did not start taking vitamin C right at the beginning of training but waited three weeks. The study also lasted a bit longer than the positive studies mentioned above, continuing for two months; maybe vitamin C is more effective at preventing colds in the short term. Another possibility is that vitamin C does not work, and more research is needed to know for sure.

Preeclampsia prevention. Preeclampsia is a dangerous complication of pregnancy that involves high blood pressure, swelling of the whole body, and improper kidney function. A double-blind, placebo-controlled study of 283 women at increased risk for preeclampsia found that supplementation with vitamin C (1,000 mg daily) and vitamin E (400 international units daily) significantly reduced the chances of developing this disease.

While this research is promising, more extensive studies are necessary to confirm whether vitamins C and E will work. The authors of this study point out that similarly sized studies found benefits with other treatments, such as aspirin, that later proved to be ineffective when large-scale studies were performed. Furthermore, it is unknown whether such high dosages of these vitamins are safe for pregnant women.

Cancer treatment. Cancer treatment is one of the more controversial proposed uses of vitamin C. An early study tested vitamin C in eleven hundred terminally ill people with cancer. One hundred participants received 10,000 mg daily of vitamin C, while one thousand other participants (the control group) received no treatment. Those taking the vitamin survived more than four times longer on average (210 days) than those in the control group (50 days). A large (1,826-participant) follow-up study by the same researchers found a nearly doubled survival rate (343 days versus 180 days) in vitamin-C-treated participants whose cancers were deemed incurable compared with untreated controls. However, these studies were poorly designed. Other generally better-constructed studies have found no benefit of vitamin C in cancer. Vitamin C cannot be regarded as a proven treatment for cancer.

Reflex sympathetic dystrophy. Reflex sympathetic dystrophy (RSD) is a set of symptoms that can develop in the legs, arms, feet, and hands after fractures and other injuries. It is also called "complex regional pain syndrome." Symptoms include skin temperature and color changes over the affected area, accompanied by burning pain, sensitivity to touch, sweating, and a limited range of motion. The cause of RSD is unknown, and the condition is tough to treat.

Two studies performed by a single research group reported evidence that vitamin C can help prevent RSD after wrist fractures. In one of these studies, 123 adults with wrist fractures were enrolled and followed for one year. All were given 500 mg of vitamin C or a placebo daily for fifty days. The results showed significantly fewer cases of RSD in the treated group.

A subsequent study conducted by the same research group compared placebo with three dosages of vitamin C in 416 people who had a wrist fracture. Again, treatment continued for fifty days. The results indicated that approximately 10 percent of those given a placebo developed RSD. In comparison, less than 2 percent of those given either 500 or 1,500 mg of vitamin C daily did so. According to the statistical analysis used by the authors, this relative benefit was statistically significant. The 200-mg dose of vitamin C did appear to offer some protection too, but not as much.

Vitamin C for Preventing Reflex Sympathetic Dystrophy

Reflex sympathetic dystrophy (RSD) is a set of symptoms that can develop in the legs, arms, feet, and hands after fractures and other injuries. Also called "complex regional pain syndrome," its symptoms include skin temperature and color changes over the affected area, accompanied by burning pain, sensitivity to touch, sweating, and limitation of range of motion. The cause of RSD is unknown, and it is challenging to treat.

Two studies performed by a single research group suggest that RSD might be preventable by the timely use of vitamin C following a fracture. The most recent studies compared placebo with three different dosages of vitamin C in 416 people with wrist fractures.

For fifty days, participants received either placebo or vitamin C at a dose of 200 milligrams (mg), 500 mg, or 1,500 mg daily. They were then followed to see how many developed RSD. The results indicated that approximately 10 percent of those given placebo developed RSD, while less than 2 percent of those given vitamin C in the 500 mg or the 1,500 mg daily dose did so. (This difference was statistically significant.) The 200-mg dose of vitamin C did appear to offer some protection too, but not as much.

Based on these findings, the researchers concluded that people who have been injured, placing them at risk for RSD, should take vitamin C at a dose of 500 mg daily. However, it should be noted that confirmation by an independent research group is still lacking.

—Steven Bratman, MD

Easy bruising. A two-month, double-blind study of ninety-four older adults with marginal vitamin C deficiency found that vitamin C supplements decreased their bruising tendency.

Hypertension. According to a thirty-day, double-blind study of thirty-nine persons taking medications for hypertension (high blood pressure), treatment with 500 mg of vitamin C daily can reduce blood pressure by about 10 percent. Smaller benefits were seen in studies of persons with normal blood pressure or borderline hypertension. However, other studies have failed to find any significant blood-pressure-lowering effect. This mixed evidence suggests, on balance, that if vitamin C does have any blood-pressure-lowering effect, it is at most quite small.

Maintaining the effectiveness of nitrate drugs. Nitroglycerin and related nitrate medications are used for the treatment of angina. However, the effectiveness of these medications tends to diminish over time. According to a double-blind study of forty-eight people, the use of vitamin C at a dose of 2,000 mg three times daily helped maintain the effectiveness of nitroglycerin. These findings are supported by other studies too.

Angina is too serious a disease for self-treatment. Persons with angina should not take vitamin C (or any other supplement) except a physician's advice.

SAFETY ISSUES

The US government has issued recommendations regarding tolerable upper intake levels (ULs) for vitamin C. The UL can be thought of as the highest daily intake over a prolonged time known to pose no risks to most members of a healthy population. The ULs for vitamin C are as follows:

- children one to three years of age (400), four to eight years of age (650), and nine to thirteen years of age (1,200);
- boys and girls aged fourteen to eighteen years (1,800);
- men and women (2,000);
- pregnant girls (1,800) and pregnant women (2,000);
- nursing girls (1,800) and nursing women (2,000).

Even within the safe intake range for vitamin C, some persons may develop diarrhea. This side effect will likely go away with the continued use of vitamin C. However, one might have to cut down the dosage for a while and then gradually build up again.

Chronic intake of large quantities of vitamin C is associated with an increased risk of kidney stones. However, large-scale observational studies have shown that kidney stone risk varies between individuals. Some people have a higher risk for vitamin-C-induced kidney stones. People with a history of kidney stones and kidney failure who have a defect in vitamin C or oxalate metabolism should probably restrict vitamin C intake to approximately 100 mg daily. Persons with glucose-6-phosphate dehydrogenase deficiency, iron overload, or a history of intestinal surgery should also avoid high-dose vitamin C.

Interestingly, the type of microbes within the digestive system affects blood oxalate levels. Calcium oxalate is the component of most kidney stones. The bacterium *Oxalobacter formigenes* degrades oxalate in the bowel, decreases blood oxalate levels, and reduces the risk of kidney stones. Consequently, someone's microbial flora can influence their kidney stone risk.

Vitamin C supplements increase the absorption of iron. Because it is not good to get more iron than needed, persons using iron supplements should not take vitamin C simultaneously as the iron supplements except under a physician's supervision.

One study from the 1970s suggests that high doses of vitamin C (3 g daily) might increase the levels of acetaminophen (such as Tylenol) in the body. This could potentially put a person at higher risk for acetaminophen toxicity. This interaction is probably unimportant when acetaminophen is taken in single doses for pain and fever or a few days during a cold. However, if one uses acetaminophen daily or has kidney or liver problems, simultaneous use of high-dose vitamin C is probably not advisable.

Weak evidence suggests that vitamin C, when taken in high doses, might reduce the blood-thinning effects of warfarin (Coumadin) and heparin. One study found that vitamin C at a dose of 1 g daily substantially reduced blood levels of the drug indinavir, a protease inhibitor used to treat HIV infection.

Heated disagreement exists regarding whether it is safe or appropriate to combine antioxidants such as vitamin C with standard chemotherapy drugs. The reasoning behind the concern is that some chemotherapy drugs may work in part by creating free radicals that destroy cancer cells, and antioxidants might interfere with this beneficial effect. However, there is no good evidence that antioxidants interfere with chemotherapy drugs; there is growing evidence that they do not. Finally, the maximum safe dosages of vitamin C for people with severe liver or kidney disease have not been determined.

IMPORTANT INTERACTIONS

Persons taking aspirin, other anti-inflammatory drugs, or oral contraceptives may need more vitamin C. The risk of liver damage from high doses of acetaminophen may be increased if one also takes large doses of vitamin C. High-dose vitamin C might reduce the effectiveness of warfarin and heparin.

High-dose vitamin C can cause a person to absorb too much iron. This is especially a problem for people with diseases that cause them to store too much iron. Vitamin C may help maintain the effectiveness of medications in the nitrate family.

Persons with angina should not take vitamin C (or any other supplement) except on a physician's advice. High-dose vitamin C may reduce the effectiveness of protease inhibitors for HIV infection. Finally, persons undergoing cancer chemotherapy should not use vitamin C except on a physician's advice.

—EBSCO CAM Review Board

Further Reading

Bryer, S. C., and A. H. Goldfarb. "Effect of High Dose Vitamin C Supplementation on Muscle Soreness, Damage, Function, and Oxidative Stress to Eccentric Exercise." *International Journal of Sport Nutrition and Exercise Metabolism,* vol. 16, 2006, pp. 270-80.

Chuang, C. H., et al. "Adjuvant Effect of Vitamin C on Omeprazole-Amoxicillin-Clarithromycin Triple Therapy for *Helicobacter pylori* Eradication." *Hepatogastroenterology,* vol. 54, 2007, pp. 320-24.

Connolly, D. A., et al. "The Effects of Vitamin C Supplementation on Symptoms of Delayed Onset Muscle Soreness." *Journal of Sports Medicine and Physical Fitness,* vol. 46, 2006, pp. 462-67.

Cook, N. R., et al. "A Randomized Factorial Trial of Vitamins C and E and Beta Carotene in the Secondary Prevention of Cardiovascular Events in Women: Results from the Women's Antioxidant Cardiovascular Study." *Archives of Internal Medicine,* vol. 167, 2007, pp. 1610-18.

Dosedìl, Martin et al. "Vitamin C-Sources, Physiological Role, Kinetics, Deficiency, Use, Toxicity, and Determination." *Nutrients,* vol. 13, no. 2, 2021, p. 615, doi:10.3390/nu13020615.

Hemila, H., E. Chalker, and B. Douglas. "Vitamin C for Preventing and Treating the Common Cold." *Cochrane Database of Systematic Reviews,* 2010, CD000980.

Rumbold, A. R., et al. "Vitamins C and E and the Risks of Preeclampsia and Perinatal Complications." *New England Journal of Medicine,* vol. 354, 2006, pp. 1796-1806.

Sesso, H. D., et al. "Vitamins E and C in the Prevention of Cardiovascular Disease in Men." *Journal of the American Medical Association,* vol. 300, 2008, pp. 2123-33.

Tecklenburg, S. L., et al. "Ascorbic Acid Supplementation Attenuates Exercise-Induced Bronchoconstriction in Patients with Asthma." *Respiratory Medicine,* vol. 101, 2007, pp. 1770-78.

Zollinger, P. E., et al. "Can Vitamin C Prevent Complex Regional Pain Syndrome in Patients with Wrist Fractures?" *Journal of Bone and Joint Surgery: American Volume,* 89, 2007, pp. 1424-31.

VITAMIN D

Category: Dietary supplements

Related terms: Cholecalciferol (vitamin D_3), ergocalciferol (vitamin D_2)

Anatomy or system affected: All systems

Specialties and related fields: Biochemistry, biotechnology, dietetics, pharmacology, preventive medicine, public health

Definition: a group of vitamins, including calciferol (vitamin D_2) and cholecalciferol (vitamin D_3), found in liver and fish oils, essential for calcium absorption and the prevention of rickets in children and osteomalacia in adults

KEY TERMS

calcium: An essential element for the proper functioning of the body, including bone formation, muscle and heart contraction, and nervous system functioning

cholecalciferol: another name for the mature form of vitamin D

osteoporosis: a medical condition in which the bones become brittle and fragile from loss of tissue, typically because of hormonal changes, or deficiency of calcium or vitamin D

parathyroid gland: a gland next to the thyroid which secretes a hormone (parathyroid hormone) that regulates calcium levels in a person's body

vitamin: an organic molecule that is an essential micronutrient which an organism needs in small quantities for the proper functioning of its metabolism

Image via iStock/StudioBarcelona. [Used under license.]

OVERVIEW

Vitamin D is both a vitamin and a hormone. It is a vitamin because the body cannot absorb calcium without it. It is a hormone because the body manufactures it in response to the skin's exposure to sunlight. Although vitamin D helps the body absorb calcium, it is also crucial to the muscles for movement, nerves to the brain and throughout the body, and the immune system to ward off infection by bacteria and viruses. Vitamin D deficiency is a public health concern, with about 35 percent of adults in the United States having a deficiency.

There are two major forms of vitamin D, and both have the word "calciferol" in their names. In Latin, *calciferol* means "calcium carrier." Vitamin D_3 (cholecalciferol) is made by the body and is found in some foods. Vitamin D_2 (ergocalciferol) is the form most often added to milk and other foods. Both are available as supplements, but D_3 may raise the blood level more effectively.

Strong evidence suggests that using both vitamin D and calcium supplements can help prevent and treat osteoporosis. Other research areas include conditions such as cancer, multiple sclerosis, osteomalacia, psoriasis, rickets, and cognitive health.

REQUIREMENTS AND SOURCES

Dosages of vitamin D are often expressed in terms of international units (IU) rather than milligrams. The US recommendations for daily (IU) intake of vitamin D are as follows: infants to twelve months of age (400); ages one year to seventy years (600); adults aged seventy-one years and older (800) pregnant and nursing females (600).

In a study of military personnel in submarines, 400 IU of vitamin D daily was inadequate to maintain bone health. At the same time, six days of sun exposure proved capable of supplying enough vitamin D for forty-nine sunless days. In addition, a study of veiled Islamic women living in Denmark found that 600 IU of vitamin D daily was insufficient to raise vitamin D levels in the blood to normal levels. The authors of this study recommend that sun-deprived persons should receive 1,000 IU of vitamin D daily.

The level of vitamin D a person's skin will make is related to the time of day for sun exposure, the season, the latitude of the site of sun exposure, the amount of clothes covering the skin, and the person's skin pigmentation. One's lifestyle can influence the amount of vitamin D the skin will create. For example, some persons stay more indoors for various reasons and have less exposure to the sun. In contrast, others may be outside more and use sunscreen to minimize the adverse effects of the sun. As indicated by the study of submarine personnel, by far the best source of vitamin D is sunlight. However, current recommendations that stress sun avoidance and the use of sunblock may have the unintended effect of increasing the prevalence of vitamin D deficiency. Severe vitamin D deficiency was common in England in the nineteenth century because coal smoke often obscured the sun. During that time, cod liver oil, which is high in vitamin D, became popular as a supplement for children to help prevent rickets, a disease caused by vitamin D deficiency in which developing bones soften and curve because they are not receiving enough calcium.

Little vitamin D is found naturally in certain foods. Some dietary sources include fatty fish like tuna, mackerel, salmon, and beef liver, cheese, and egg yolks. In many countries, vitamin D is added to fortify milk, juices, and other foods like breakfast cereals contributing to daily intake. Cod liver oil is a good source of vitamin D.

Vitamin D deficiency occurs in older adults who often receive less sun exposure and in people who live in northern latitudes and do not drink vitamin-D-enriched milk. The consequences of this deficiency may be an increased risk of hypertension, osteoporosis, and several forms of cancer.

Additionally, phenytoin (Dilantin), primidone (Mysoline), and phenobarbital for seizures; corticosteroids; cimetidine (Tagamet) for ulcers; the blood-thinning drug heparin; and the antituberculosis drugs isoniazid (INH) and rifampin may interfere with vitamin D absorption or activity.

THERAPEUTIC DOSAGES

For therapeutic purposes, vitamin D is taken at the nutritional doses described in the preceding Requirements and Sources section (and sometimes in even higher amounts). Persons who wish to exceed nutritional levels of vitamin D intake should consult their health-care provider.

THERAPEUTIC USES

Persons concerned about osteoporosis should take calcium and vitamin D. The combination appears to help prevent bone loss. This is true even if one is taking other treatments for osteoporosis. One cannot build bone without calcium. One cannot properly absorb and utilize calcium without adequate vitamin D. Studies suggest vitamin D may also help prevent the falls that lead to osteoporotic fractures. Some evi-

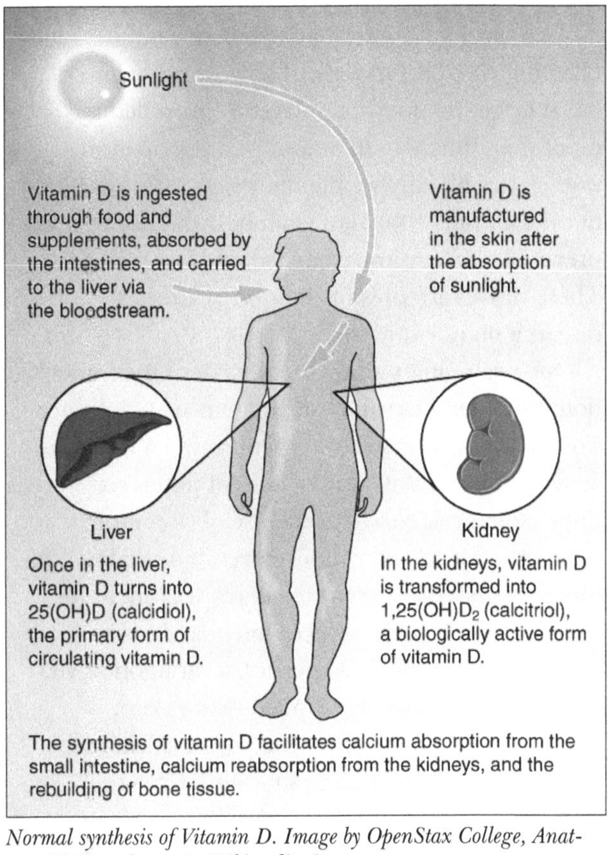

Normal synthesis of Vitamin D. Image by OpenStax College, Anatomy & Physiology, via Wikimedia Commons.

dence suggests that getting adequate vitamin D may be connected to health issues such as diabetes, hypertension, and autoimmune conditions, including multiple sclerosis.

Vitamin D may play a role in preventing cancer of the breast, colon, pancreas, prostate, and skin. However, research on this question has yielded mixed results. One study suggests that the combined use of calcium plus vitamin D, but not either supplement separately, can help reduce the risk of colon cancer. However, an extensive study involving more than thirty-six thousand postmenopausal women found that supplementing the diet with 1,000 milligrams (mg) of calcium plus 400 IU of vitamin D daily did not lower the risk of breast cancer in seven years. Based on the results of this placebo-controlled study, there does not appear to be a connection between vitamin D and breast cancer risk.

Weak evidence hints that adequate vitamin D intake might reduce the risk of hypertension and diabetes. A large, randomized, placebo-controlled trial of more than thirty-six thousand postmenopausal women found daily supplementation with 1,000 mg of calcium plus 400 IU of vitamin D did not reduce or prevent hypertension during seven years of follow-up. These results are possibly limited by the lack of research on calcium use.

One preliminary study suggests that supplementation with vitamin D and calcium may be helpful for women with polycystic ovary syndrome. A meta-analysis (formal statistical review) of published studies found some evidence that the use of vitamin D at recommended levels may reduce overall mortality. Vitamin D is sometimes mentioned as a treatment for psoriasis. However, this recommendation is based on Danish studies using calcipotriol, a variation of vitamin D_3 used externally (applied to the skin). Calcipotriol does not affect the body's calcium absorption, so it is a different substance from the vitamin D one can purchase at a store.

It has been suggested that because vitamin D levels in the body drop in the wintertime, vitamin D supplements might be helpful for seasonal affective disorder. A small, double-blind, placebo-controlled trial conducted in winter with forty-four people found that vitamin D supplements produced improvements in various measures of mood. However, a double-blind, placebo-controlled study of 2,217 women older than seventy years failed to find benefit. It has been hypothesized that light therapy (used successfully for seasonal affective disorder) works by raising vitamin D levels. Still, there is some evidence that this is not the case. Finally, vitamin D supplements also do not appear to help enhance growth in healthy children.

SCIENTIFIC EVIDENCE

Osteoporosis. Persons with severe osteoporosis often have low levels of vitamin D. Supplementing with vitamin D alone is probably no more than minimally helpful. Still, the combination of calcium and vitamin D is probably more effective.

Vitamin D may offer another benefit for osteoporosis in older adults. Some studies have found that vitamin D supplementation improves balance (especially in women) and reduces the risk of falling. Because the most common adverse consequence of osteoporosis is a fracture caused by a fall, this could be a meaningful benefit. Supplementation with vitamin D plus calcium may also aid healing after a fracture has occurred.

SAFETY ISSUES

When taken at recommended dosages, vitamin D appears to be safe. However, vitamin D can build up in the body when used in considerable excess and cause toxic symptoms such as nausea and vomiting: constipation, weight loss, weakness, disorientation, and irregular heart rhythm. At an intake level of about 40,000 IU daily (about one hundred times the recommended daily intake), vitamin D can cause dangerous elevations in blood calcium levels. A few persons consumed doses five times higher because of a manufacturing error; the resulting toxicity was severe and may have caused one death.

However, short of these vastly excessive dosages, it is not clear at what level vitamin D becomes toxic. The official safe upper limits for vitamin D daily supplement intake limit for persons aged nine and older and pregnant and nursing females is 4,000 IU. Vitamin D created by the skin by sunshine is self-regulated by the body to proper levels. People with sarcoidosis or hyperparathyroidism should use caution to never take vitamin D without consulting their health-care provider due to potential complications. Also, taking vitamin D and calcium supplements might interfere with some of the effects of drugs in the calcium-channel blocker family. One must consult a health-care provider before trying this combination.

The combination of calcium, vitamin D, and thiazide diuretics could potentially lead to excessive calcium levels in the body. Persons taking thiazide diuretics should consult with a health-care provider about the proper doses of vitamin D and calcium.

IMPORTANT INTERACTIONS

The Mayo Clinic warns readers through their website that Vitamin D supplements can interact with anticonvulsants, aluminum, steroids, stimulant laxatives, cytochrome P450 3A4 (CYP3A4) substrates, and thiazide diuretics. They suggest caution when taking specific medications such as verapamil, orlistat, diltiazem, digoxin, atorvastatin, calcipotriene, and cholestyramine.

Persons who may need extra vitamin D include those who are taking antiseizure drugs, such as phenobarbital, primidone (Mysoline), valproic acid (Depakene), phenytoin (Dilantin), corticosteroids, cimetidine (Tagamet), heparin, isoniazid, (INH), and rifampin. Persons taking calcium-channel blockers should not take high-dose vitamin D (with calcium) except under physician supervision. Finally, persons taking thiazide diuretics should not take calcium and vitamin D supplements unless under a doctor's supervision.

—Marylane Wade Koch, RN, MSN

Further Reading

Chlebowski, R. T., et al. "Calcium Plus Vitamin D Supplementation and the Risk of Breast Cancer." *Journal of the National Cancer Institute,* vol. 100, 2008, pp. 1581-91

Fosnight, S. M., W. J. Zafirau, and S. E. Hazelett. "Vitamin D Supplementation to Prevent Falls in the Elderly: Evidence and Practical Considerations." *Pharmacotherapy,* vol. 28, 2008, pp. 225-34.

Gunners, Kris. "Vitamin D 101—A Detailed Beginner's Guide." *Healthline,* Mar. 2019, www.healthline.com/nutrition/vitamin-d-101.

Margolis, K. L., et al. "Effect of Calcium and Vitamin D Supplementation on Blood Pressure: The Women's Health Initiative Randomized Trial." *Hypertension,* vol. 52, 2008, pp. 847-55.

The Mayo Staff. "Vitamin D." *The Mayo Clinic,* 18 Oct. 2017, www.mayoclinic.org/drugs-supplements-vitamin-d/art-20363792.

National Center for Complementary and Integrative Health: Fact Sheet Vitamin D. https://ods.od.nih.gov/search.aspx?zoom_query=Vitamin%20D.

"Vitamin D: The Nutrition Source." *Harvard T. H. Chan, School of Public Health,* Mar. 2020, www.hsph.harvard.edu/nutritionsource/vitamin-d.

VITAMIN E

Category: Dietary supplements
Anatomy or system affected: All systems
Specialties and related fields: Biochemistry, biotechnology, dietetics, pharmacology, preventive medicine, public health
Definition: a fat-soluble nutrient needed by the body in small amounts to stay healthy; it is found in seeds, nuts, leafy green vegetables, and vegetable oils, boosts the immune system, and helps keep blood clots from forming

KEY TERMS

alpha-tocopherol: another name for vitamin E

blood clotting: the coagulation of the blood to prevent blood loss after damage to a blood vessel

dosage: the size or frequency of a dose of a medicine or drug

vitamin: an organic molecule that is an essential micronutrient which an organism needs in small quantities for the proper functioning of its metabolism

OVERVIEW

Vitamin E is an antioxidant that fights damaging natural substances known as free radicals. It works in lipids (fats and oils), which complements vitamin C that fights free radicals dissolved in water. As an antioxidant, vitamin E has been widely advocated for preventing heart disease and cancer. However, the results of large, well-designed trials have generally not been encouraging. Many other proposed benefits of vitamin E have also failed to prove useful in studies. There are no medicinal uses for vitamin E with solid scientific support.

REQUIREMENTS AND SOURCES

Vitamin E dosage recommendations are a bit complex because the vitamin exists in many forms. New vitamin E recommendations are in milligrams (mg) of alpha-tocopherol. Alpha-tocopherol can come from either natural vitamin E (called, somewhat incorrectly, "d-alpha-tocopherol") or synthetic vitamin E (called, also somewhat incorrectly, "dl-alpha-tocopherol"). However, much of the alpha-tocopherol in synthetic vitamin E is inactive. For this reason, one has to take about twice as much of it to get the same effect.

There are other forms of vitamin E, such as beta-, delta-, and gamma-tocopherols, all of which occur in food. These other forms may be necessary; for example, preliminary evidence hints that gamma-tocopherol may be the most important (or, perhaps, the only) form of vitamin E for preventing prostate cancer. On this basis, it has been suggested that the best vitamin E supplement would be a mixture of all these.

Vitamin E dosages are commonly listed on labels as international units (IU). One IU natural vitamin E equals 0.67 mg alpha-tocopherol; one IU synthetic vitamin E equals 0.45 mg alpha-tocopherol. Therefore, to meet the new dietary recommendations for vitamin E (15 mg per day), one needs to get either 22 IU natural vitamin E (22 IU x 0.67 = 15 mg) or 33 IU synthetic vitamin E (33 IU x 0.45 = 15 mg). The official US and Canadian recommendations for daily intake (in milligrams) of vitamin E are as follows:

- Infants to six months of age (4) and seven to twelve months of age (5);
- children one to three years of age (6), four to eight years of age (7), and nine to thirteen years of age (11);
- males and females aged fourteen years and older (15);
- pregnant females (15);
- and nursing females (19).

In developed countries, mild dietary deficiency of vitamin E is relatively common. The best food sources of vitamin E are polyunsaturated vegetable oils, seeds, nuts, and whole grains. To get a therapeutic dosage, though, one needs to take a supplement.

THERAPEUTIC DOSAGES

The optimal therapeutic dosage of vitamin E has not been established. Most studies have used between 50 and 800 IU daily, and some have used even higher doses. This would correspond to about 50 to 800 mg of synthetic vitamin E (dl-alpha-tocopherol) or 25 to 400 mg of natural vitamin E (d-alpha-tocopherol or mixed tocopherols).

In purchasing natural vitamin E, one should look for a label that reads "mixed tocopherols." However, some manufacturers use this term to mean the synthetic dl-alpha-tocopherol, so the contents need to be read closely. Natural tocopherols come as d-alpha-, d-gamma-, d-delta-, and d-beta-tocopherol.

THERAPEUTIC USES

Observational studies raised hopes that vitamin E supplements could help prevent various forms of cancer and heart disease. However, observational studies are notoriously unreliable for determining the effectiveness of treatments. Only double-blind trials can do that, and such studies have, on balance, found vitamin E ineffective for preventing heart disease or any common form of cancer other than, possibly, prostate cancer. The use of high-dose vitamin E for a long time might slightly increase the death rate. Other potential uses of vitamin E have limited supporting evidence.

Intriguing but far from definitive studies suggest that vitamin E might improve immune response to vaccinations, control symptoms of restless leg syndrome, help prevent deep venous thrombosis, reduce symptoms of premenstrual syndrome, and decrease symptoms of menstrual pain. Vitamin E, combined with evening primrose, has also been studied to alleviate premenstrual breast pain (mastalgia).

While there is weak evidence that vitamin E supplements can reduce discomfort in rheumatoid arthritis, there is strong evidence that it does not prevent it. Although preliminary studies hinted that the use of vitamin E might prevent or slow the progression of cataracts, in a ten-year study of almost forty thousand female health-care professionals, the use of natural vitamin E at a dose of 600 mg every other day failed to have any effect on cataract development.

Evidence regarding whether vitamin E can slow the progression of Alzheimer's disease is inconsistent. A large study failed to find vitamin E helpful for preventing mental decline (resulting from any cause) in women older than sixty-five years of age. Studies of vitamin E in combination with vitamin C for the prevention of preeclampsia have yielded inconsistent results.

Vitamin E has also shown equivocal promise in diabetes. One double-blind trial found benefits for cardiac autonomic neuropathy, a complication of diabetes. Weaker evidence hints at possible benefits for diabetic peripheral neuropathy. However, the best-designed study of all, a long-term trial involving 3,654 people with diabetes, found that the use of vitamin E did not protect against diabetes-induced kidney or heart damage. Similarly, while a few studies performed by one research group suggested that vitamin E might help improve glucose control in people with diabetes, subsequent evidence showed that the benefits are limited to the short term. In addition, in an extensive double-blind study, the use of vitamin E at a dose of 600 IU every other day failed to reduce the risk of participants developing type 2 diabetes. Finally, a study unexpectedly found that their blood pressure increased when people with diabetes took 500 mg of vitamin E daily (either as natural alpha-tocopherol or as a mixture of alpha and gamma-tocopherol). Similarly, studies on whether vitamin E is helpful for allergic rhinitis (hay fever) have produced conflicting results.

A small double-blind study in Iran reported that vitamin E (400 IU daily) was more effective than a placebo for treating menopausal hot flashes. However, a larger study in the United States failed to find vitamin E significantly helpful for hot flashes associated with breast cancer treatment.

Vitamin E might help reduce the lung-related side effects caused by the drug amiodarone, which is used to prevent abnormal heart rhythms. A trial of 108 persons undergoing chemotherapy with cisplatin found that vitamin E supplementation (extended three months past chemotherapy) reduced cisplatin-related neurotoxicity (damage to nerves common during treatment with cisplatin). Studies have yielded mixed results on whether vitamin E helps control seizures in people with epilepsy, reducing symptoms of tardive dyskinesia, aiding recovery during heavy exercise, and treating male infertility.

When combined with vitamin C, vitamin E may protect against sunburn to a small extent. The same combination has also shown promise for acute anterior uveitis. A separate study failed to find vitamin E alone (at the high dose of 1,600 mg daily) helpful for macular edema (swelling of the retina's center) associated with uveitis.

Vitamin E has been tried for amyotrophic lateral sclerosis (Lou Gehrig's disease). However, the results in the first reported double-blind study showed questionable benefits, if any. Some vitamin E proponents felt that the dose of vitamin E used in this study might have been too low. Accordingly, they conducted another study using ten times the dose, lasting eighteen months and enrolling 160 people. Once again, vitamin E failed to prove significantly more effective than placebo.

In one observational study, the high intake of vitamin E was linked to decreased risk of progression to acquired immunodeficiency syndrome in people with human immunodeficiency virus (HIV) infection. However, a double-blind study of forty-nine people with HIV who took combined vitamins C and E or placebo for three months did not significantly affect the amount of HIV detected or the number of opportunistic infections. It has been suggested that vitamin E may enhance the antiviral effects of azidothymidine, but this is minimal evidence.

Vitamin E has been suggested for preventing the cardiac toxicity caused by the drug doxorubicin. However, while it has shown promise in animal studies when studied in people, vitamin E has persistently failed to prove effective for this purpose.

Vitamin E is sometimes recommended for osteoarthritis. However, a two-year, double-blind, placebo-controlled study of 136 people with osteoarthritis of the knee failed to find any benefit in symptom control or slowing disease progression. An earlier six-month, double-blind, placebo-controlled trial of seventy-seven people with osteoarthritis also failed to find benefit.

A four-year, double-blind, placebo-controlled trial of 1,193 people with macular degeneration failed to find vitamin E alone helpful for preventing or treating macular degeneration. Vitamin E has also so far failed to prove helpful for preventing or treating alcoholic hepatitis, asthma, congestive heart failure, fibrocystic breast disease, or Parkinson's disease.

In an extensive study involving more than 29,000 male smokers, researchers failed to find a benefit to alpha-tocopherol (50 IU per day), beta-carotene (20 mg per day), or the two taken together for the prevention of type 2 diabetes in a five-to-eight-year period.

SCIENTIFIC EVIDENCE

Cancer prevention. The results of observational trials have been mixed. Still, on balance, they suggest that a high intake of vitamin E and other antioxidants is associated with a reduced risk of lung cancer and many other forms of cancer, including bladder, stomach, mouth, throat, laryngeal, liver, and prostate. Based on these and other results, researchers developed the hypothesis that antioxidants can help prevent cancer and set in motion large, long-term, double-blind, placebo-controlled studies to verify it. However, these studies generally failed to find vitamin E helpful for preventing cancer in people at high risk for it.

The one positive note came in a double-blind trial of 29,133 smokers. In this study, 50 mg of synthetic vitamin E (dl-alpha-tocopherol) daily for five to eight years led to a 32 percent reduction in the incidence of prostate cancer and a 41 percent drop in prostate cancer deaths.

Results were seen soon after the beginning of supplementation, and this was unexpected because prostate cancer grows very slowly. The fact that vitamin E almost immediately lowered the incidence of prostate cancer suggests that it somehow blocks the step at which a hidden prostate cancer leaps to being detectable.

Nonetheless, the negative results regarding most other types of cancer have made scientists hesitant to place too much hope in these findings. It has been suggested that alpha-tocopherol alone is less effective than the multiple forms of tocopherol that occur in nature; in particular, it has been suggested that gamma-tocopherol rather than alpha-tocopherol might be the most relevant form of vitamin E for cancer prevention. The use of alpha-tocopherol supplements may deplete both gamma- and delta-tocopherol levels, potentially producing a negative effect. However, gamma-tocopherol has not been tested in meaningful controlled trials. It is quite possible that were one to be performed, and the results would prove as disappointing as those for other forms of vitamin E. In addition, vitamin E may have a pro-oxidant effect (the reverse of what is desired) under certain circumstances.

Cardiovascular disease. Most observational studies have found associations between high intake of vitamin E and reduced risk of cardiovascular disease (heart disease and strokes). However, observational studies by themselves cannot be relied upon to identify helpful treatments. Double-blind studies, which provide much more convincing evidence of effectiveness, have generally failed to find vitamin E supplements effective.

The Heart Outcomes Prevention Evaluation trial found that natural vitamin E (d-alpha-tocopherol) at a dose of 400 IU daily did not reduce the number of heart attacks, strokes, or deaths from heart disease any more than placebo. The trial followed more than nine thousand men and women who had existing heart disease or high risk for it.

Negative results were seen in numerous other large trials too. When the results of these studies began to come in, some antioxidant proponents suggested that the people enrolled in these trials already had diseases too advanced for vitamin E to help. However, a subsequent large trial found vitamin E ineffective for slowing the progression of heart disease also in healthy people. Moreover, in an extensive placebo-controlled trial involving more than fourteen thousand male physicians in the United States at low risk for heart disease, 400 IU of vitamin E every other day failed to lower the risk of major cardiovascular events or mortality in eight years. On the contrary, vitamin E was associated with a slightly increased risk of stroke.

As with preventing cancer, critics have suggested that the form of vitamin E used in these studies (alpha-tocopherol) was not the best choice and that gamma-tocopherol might be more helpful. Gamma-tocopherol is present in the diet much more abundantly than alpha-tocopherol, and it could be that the studies showing benefits with dietary vitamin E tracked the influence of gamma-tocopherol. However, an observational study specifically examining if gamma-tocopherol levels were associated with the risk of heart attack found no relationship between the two.

In addition, under certain circumstances, vitamin E may have a pro-oxidant effect, which could explain the negative outcomes. One study found that vitamin E might help prevent serious cardiovascular events in persons with diabetes who also have the particular genetic marker Hp 2. It has been hypothesized that people with the Hp 2 gene have an inadequate endogenous (built-in) antioxidant defense system. For this reason, they might be particularly benefited from taking antioxidant supplements such as vitamin E. However, this concept remains highly preliminary.

Preeclampsia prevention. Preeclampsia is a dangerous complication of pregnancy that involves high blood pressure, swelling of the whole body, and improper kidney function. A double-blind, placebo-controlled study of 283 women at increased risk for preeclampsia found that supplementation with vitamin E (400 IU daily of natural vitamin E) and vitamin

C (1,000 mg daily) significantly reduced the chances of developing this disease.

While this research is promising, larger studies are necessary to confirm whether vitamins E and C will actually work. The authors of this study point out that studies of similar size found benefits with other treatments, such as aspirin, that later proved to be ineffective when large-scale studies were performed. Furthermore, it is unknown if such high dosages of these vitamins are safe for pregnant women.

Tardive dyskinesia. Between 1987 and 1998, several double-blind studies indicated vitamin E was beneficial in treating tardive dyskinesia (TD). Although most of these studies were small and lasted only four to twelve weeks, one thirty-six-week study enrolled forty people. Three small double-blind studies reported that vitamin E was not helpful. Nonetheless, a statistical analysis of the double-blind studies done before 1999 found good evidence that vitamin E was more effective than placebo. Most studies found that vitamin E worked best for TD of more recent onset.

However, in 1999, the picture on vitamin E changed with one more study, the largest and longest to date. This double-blind study included 107 participants from nine different research sites who took 1,600 IU of vitamin E or placebo daily for a minimum of one year. In contrast to most previous studies, this trial did not find vitamin E effective in decreasing TD symptoms.

Researchers proposed several possible explanations for the discrepancy. One explanation was that the earlier studies were too small or too short to be accurate and that vitamin E did not help. Another was the most complicated: that vitamin E might help only a subgroup of people who have TD (those with milder TD symptoms of more recent onset) and that fewer of these people had participated in the latest study. They also pointed to changes in schizophrenia treatment since the last study was done, including the growing use of antipsychotic medications that do not cause TD.

The effectiveness of vitamin E for a person is simply not known. Given the lack of other suitable treatments for TD and the general safety of the vitamin, it may be worth discussing with one's physician.

Immune support. The elderly often do not respond adequately to vaccinations. One double-blind study suggests that vitamin E may be able to strengthen the immune response to vaccines. In this trial, eighty-eight people over sixty-five years were given either placebo or vitamin E at 60, 200, or 800 IU dl-alpha-tocopherol daily. The researchers then gave all participants immunizations against hepatitis B, tetanus, diphtheria, and pneumonia. They looked at participants' immune responses to these vaccinations. The researchers also used a skin test that evaluates the overall strength of the immune response.

The results were promising. Vitamin E at 200 mg per day and, to a lesser extent, at 800 mg per day significantly increased the strength of the immune response. However, it is not clear whether vitamin E has a general "immune support" effect.

One study in the elderly found that vitamin E did not help prevent colds and other respiratory infections and even seemed to increase slightly the severity of infections that did occur. In a similar-sized double-blind study of long-term-care residents, the use of vitamin E at 200 IU daily failed to reduce the incidence or number of days of respiratory infection or antibiotic use. The researchers found some evidence of benefit by breaking down the respiratory infections by type. Still, such after-the-fact analysis is questionable from a statistical perspective. The same researchers repeated the study with a larger group and did find a reduction in the frequency of colds. Another researcher found evidence that vitamin E can have either a harmful or a beneficial effect, depending on who takes it (the exact differences remaining undefined).

Alzheimer's disease. Evidence is conflicting regarding whether high-dose vitamin E can slow the progression of Alzheimer's disease. In a double-blind, placebo-controlled study, 341 people with Alzheimer's disease received either 2,000 IU daily of vitamin E (dl-alpha-tocopherol), the antioxidant drug selegiline, or a placebo. Those given vitamin E took nearly two hundred days longer to reach a severe state of the disease than the placebo group. (Selegiline was even more effective.)

However, negative results were seen in a study of 769 people at high risk of developing Alzheimer's disease (judging based on early symptoms). Participants were given 2,000 IU of vitamin E, the drug donepezil, or a placebo for three years. Neither treatment reduced the percentage of people who went on to develop Alzheimer's disease. Such high dosages of vitamin E should not be taken except under a doctor's supervision.

Dysmenorrhea. In a double-blind, placebo-controlled trial, one hundred young women with significant dysmenorrhea (menstrual pain) were given 500 IU of vitamin E or a placebo for five days. Treatment began two days before and continued for three days after the expected onset of menstruation. While both groups showed significant improvement in pain in the two months of the study (presumably because of the power of the placebo), pain reduction was greater in the treatment group than in the placebo group.

Benefits were also seen in a four-month, double-blind, placebo-controlled study of 278 adolescent girls in Iran. The dose used in this study was 200 IU twice daily.

Mastalgia. Eight-five women with premenstrual mastalgia (breast pain) were randomized to receive one of four treatments for six months: vitamin E (1,200 IU) and placebo, evening primrose (3,000 mg) and placebo, vitamin E and evening primrose, or placebo alone. In this small study, none of the treatment groups experienced better results than the placebo group.

Male infertility. In a double-blind, placebo-controlled study of 110 men whose sperm showed subnormal activity, treatment with 100 IU of vitamin E daily resulted in improved sperm activity and higher actual fertility (measured in pregnancies). However, a smaller double-blind trial found no benefit.

Cardiac autonomic neuropathy. People with diabetes sometimes develop cardiac autonomic neuropathy, irregular heartbeats. A four-month, double-blind, placebo-controlled trial found that vitamin E at a dose of 600 mg daily might improve these symptoms.

SAFETY ISSUES

The safe upper intake level (UL) for vitamin E for adults is set at 1,000 mg daily. The equivalent amounts are 1,500 IU of natural vitamin E and 1,100 IU of synthetic vitamin E. For pregnant girls (females eighteen years old and younger), the upper limit is 800 mg.

Vitamin E has a blood-thinning effect that could lead to problems in certain situations. In one study of 28,519 men, vitamin E supplementation at the low dose of about 50 IU synthetic vitamin E per day caused an increase in fatal hemorrhagic strokes, the kind of stroke caused by bleeding. (However, it reduced the risk of a more common type of stroke, and the two effects were essentially canceled out.) Based on its blood-thinning effects, there are concerns that vitamin E could cause problems if it is combined with medications that also thin the blood, such as warfarin (Coumadin), heparin, clopidogrel (Plavix), pentoxifylline (Trental), and aspirin. Theoretically, the net result could be to thin the blood too much, causing bleeding problems. A study that evaluated vitamin E plus aspirin did find an additive effect. In contrast, the results of a study on vitamin E and Coumadin found no evidence of an interaction. However, it would still not be advisable to combine these treatments except under a physician's supervision.

There is also a remote possibility that vitamin E could also interact with supplements that possess a mild blood-thinning effect, such as garlic, policosanol, and ginkgo. Persons with bleeding disorders, such as hemophilia and those about to undergo surgery or labor and delivery, should also approach vitamin E with caution.

In addition, vitamin E might temporarily enhance the body's sensitivity to its insulin in persons with type 2 diabetes. This could lead to a risk of blood sugar levels falling too low. In addition, one study found that vitamin E can raise blood pressure in people with diabetes. Persons with diabetes should not take high-dose vitamin E without first consulting a physician.

When all major vitamin E studies are statistically combined through meta-analysis, some evidence suggests that the long-term usage of vitamin E at high doses might increase the overall death rate for unclear reasons. The results of one large study involving 29,000 males indicate that vitamin E supplementation may increase the risk of tuberculosis in heavy smokers. Curiously, however, this was true only in those participants who also consumed high levels of vitamin C (a minimum of 90 mg daily) in their diet. Consuming high levels of vitamin C without supplemental vitamin E led to a reduction in tuberculosis risk.

Finally, considerable controversy exists regarding whether it is safe or appropriate to combine vitamin E with standard chemotherapy drugs. The reasoning behind this concern is that some chemotherapy drugs may work in part by creating free radicals that destroy cancer cells. Antioxidants like vitamin E might interfere with this beneficial effect. However, there is no good evidence that antioxidants interfere with chemotherapy drugs, growing evidence that they do not, and some evidence of potential benefit under certain circumstances. Nonetheless, given the high stakes involved, it is strongly recommended that persons should not take any supplements while undergoing cancer chemotherapy, except on the advice of a physician.

IMPORTANT INTERACTIONS

One should seek medical advice before taking vitamin E if also taking blood-thinning drugs, such as Coumadin, heparin, Plavix, Trental, and aspirin. Vitamin E may help protect from lung-related side effects if one is taking amiodarone. Vitamin E may help reduce side effects if one is taking phenothiazine drugs. One should seek medical advice before taking vitamin E if also taking chemotherapy drugs. High-dose vitamin E might cause blood sugar levels to fall too low, requiring an adjustment in medication dosage if one takes oral hypoglycemic medications.

—EBSCO CAM Review Board

Further Reading

Christen, W. G., et al. "Vitamin E and Age-Related Cataract in a Randomized Trial of Women." *Ophthalmology,* vol. 115, 2008, pp. 822-29..

Kang, J. H., et al. "A Randomized Trial of Vitamin E Supplementation and Cognitive Function in Women." *Archives of Internal Medicine,* vol. 166, 2006, pp. 2462-68.

Karlson, E. W., et al. "Vitamin E in the Primary Prevention of Rheumatoid Arthritis." *Arthritis and Rheumatism,* vol. 59, 2008, pp. 1589-95.

Kataja-Tuomola, M., et al. "Effect of Alpha-tocopherol and Beta-carotene Supplementation on Macrovascular Complications and Total Mortality from Diabetes." *Annals of Medicine,* vol. 42, 2010, pp. 178-86.

Keith, M. E., et al. "A Controlled Clinical Trial of Vitamin E Supplementation in Patients with Congestive Heart Failure." *American Journal of Clinical Nutrition,* vol. 73. 2001, pp. 219-24.

Manning, P. J., et al. "Effect of High-Dose Vitamin E on Insulin Resistance and Associated Parameters in Overweight Subjects." *Diabetes Care,* vol. 27, 2004, pp. 2166-71.

Meydani, S. N., et al. "Vitamin E and Respiratory Infection in the Elderly." *Annals of the New York Academy of Sciences,* vol. 1031, 2005, pp. 214-22.

Pace, A., et al. "Vitamin E Neuroprotection for Cisplatin Neuropathy." *Neurology,* vol. 74, 2010, pp. 762-66.

Peters, U., et al. "Vitamin E and Selenium Supplementation and Risk of Prostate Cancer in the Vitamins and Lifestyle (VITAL) Study Cohort." *Cancer Causes and Control,* vol. 19, 2008, pp. 75-87.

Pruthi, S., et al. "Vitamin E and Evening Primrose Oil for Management of Cyclical Mastalgia." *Alternative Medicine Review*, vol. 15, 2010, pp. 59-67.
Sesso, H. D., et al. "Vitamins E and C in the Prevention of Cardiovascular Disease in Men." *Journal of the American Medical Association*, vol. 300, 2008, pp. 2123-33.
Shahar, E., G. Hassoun, and S. Pollack. "Effect of Vitamin E Supplementation on the Regular Treatment of Seasonal Allergic Rhinitis." *Annals of Allergy, Asthma, and Immunology*, vol. 92, 2004, pp. 654-58.
Ziaei, S., A. Kazemnejad, and M. Zareai. "The Effect of Vitamin E on Hot Flashes in Menopausal Women." *Gynecologic and Obstetric Investigation*, vol. 64, 2007, pp. 204-7.
Ziaei, S., et al. "A Randomised Controlled Trial of Vitamin E in the Treatment of Primary Dysmenorrhoea." *BJOG: An International Journal of Obstetrics and Gynaecology*, vol. 112, 2005, pp. 466-69.

VITAMIN K

Category: Herbs and supplements
Also known as: Vitamin K_1 (phylloquinone), vitamin K_2 (menaquinone), vitamin K_3 (menadione)
Anatomy or system affected: Blood, cells, bones, circulatory system, gastrointestinal system, immune system, liver
Specialties and related fields: Biochemistry, dietetics, hematology, preventive medicine, public health
Definition: any group of vitamins found mainly in green leaves and essential for the blood-clotting process; they include phylloquinone (vitamin K_1), menaquinone (vitamin K_2), and menadione (vitamin K_3)

KEY TERMS

blood clotting: an important process that prevents excessive bleeding when a blood vessel is injured
clotting factors: several plasma components (such as fibrinogen, prothrombin, thromboplastin, and factor VIII) involved in the clotting of blood
warfarin: a drug used to treat blood clots (such as in deep vein thrombosis-DVT or pulmonary embolus-PE) and prevent new clots from forming

OVERVIEW

Vitamin K plays a significant role in the body's blood-clotting system. There are three forms of vitamin K: K_1 (phylloquinone), found in plants; K_2 (menaquinone), produced by bacteria in the intestines; and K_3 (menadione), a synthetic form.

Vitamin K is used medically to reverse the effects of "blood-thinning" drugs, such as warfarin (Coumadin). Growing evidence suggests that it may also be helpful for osteoporosis.

REQUIREMENTS AND SOURCES

Vitamin K is an essential nutrient, but a person needs only a tiny amount of it. The official US recommendations for daily intake (in micrograms or mcg) have been set as follows:

Infants aged zero to six months (2 mcg) and seven to twelve months (2.5 mcg); children aged one to three years (30 mcg) and four to eight years (55 mcg); boys aged nine to thirteen years (60 mcg) and aged fourteen to eighteen years (75 mcg); men (120 mcg); girls aged nine to thirteen years (60 mcg) and aged fourteen to eighteen years (75 mcg); women (90 mcg); pregnant girls (75 mcg); pregnant women (90 mcg, preferably the K_1 variety [phylloquinone]); and nursing girls (75 mcg) and women (90 mcg, preferably phylloquinone).

Vitamin K (in the form of K_1) is found in green leafy vegetables. Kale and turnip greens are the best food sources, providing about ten times the daily adult requirement in a single serving. Spinach, broccoli, lettuce, and cabbage are also very rich sources. People can get perfectly respectable amounts of vitamin K in such common foods as oats, green peas, whole wheat, and green beans, as well as watercress and asparagus.

Bacteria in the intestines also manufacture vitamin K (in the form of K_2). The most important vitamin K producers include *Enterobacter agglomerans*, *Serratia marcescens*, and *Enterococcus faecium*. Intestinal synthesis is a major source of vitamin K. Long-term use of antibiotics can cause a vitamin K deficiency by killing these bacteria. However, this effect seems to be significant only in people deficient in vitamin K, to begin with. Pregnant and postmenopausal women are also sometimes deficient in this vitamin. In addition, children born to women taking anticonvulsants while pregnant may be significantly deficient in vitamin K, causing bleeding problems and facial bone abnormalities. Vitamin K supplementation during pregnancy may help prevent this.

The blood-thinning drug warfarin (Coumadin) works by antagonizing the effects of vitamin K. Conversely, vitamin K supplements or intake of foods containing high levels of vitamin K block the action of this medication and can be used as an antidote. Specific cephalosporin antibiotics and possibly other antibiotics may also interfere with vitamin-K-dependent blood clotting. Cephalosporin antibiotics with an N-methylthiotetrazole (NMTT) side chain interfere with vitamin K metabolism and increase clotting times. These antibiotics include the following second-generation cephalosporin, cefamandole, cefbuperazone, cefmetazole, cefminox, cefotetan, and the following third-generation cephalosporins, cefmenoxime, cefoperazone, and moxalactam.

However, this interaction seems to be significant only in people who have diets poor in vitamin K.

People with digestive tract disorders, such as chronic diarrhea, celiac sprue, ulcerative colitis, or Crohn's disease, may become deficient in vitamin K. Alcoholism can also lead to vitamin K deficiency.

What Is the Best Source of Vitamin K?

Researchers with the US Department of Agriculture (USDA) studied vitamin K to see the supplement's effects on volunteers who consumed a vegetable or fortified oil, both rich in vitamin K. A discussion of the study was presented in the USDA magazine Agricultural Research in January 2000. The discussion is excerpted here.

Worldwide, only a handful of researchers study vitamin K—long known for its critical role in blood clotting. But with the aging of the US population, this vitamin may command a more extensive following as its importance to the integrity of bones becomes increasingly evident. It activates at least three proteins involved in bone health, says Sarah Booth, from the Vitamin K Laboratory at the Jean Mayer USDA Human Nutrition Research Center on Aging at Tufts University in Boston.

Vegetables provide the lion's share of this vitamin [K] in the diet. Still, nutritionists have assumed that people absorb more from oil or oil-based supplements than from vegetables. To find out, Booth led a study with colleagues at Yale University School of Medicine to compare the absorption and use—known as bioavailability—of vitamin K from broccoli and oil fortified with the vitamin. For five days each, volunteers consumed a helping of broccoli or fortified oil along with a base diet. This increased their phylloquinone [vitamin K_1] intake to around 400 micrograms per day—five to six times the recommended dietary allowance.

"What's really exciting," Booth says, "is to look at the functional markers for vitamin K status. There were no differences between vitamin K from broccoli and vitamin K from oil overall. That's good because green leafy vegetables contain so many other nutrients." For instance, when the volunteers ate broccoli, blood levels of an important carotenoid—lutein—increased compared to when they ate the base diet only.

THERAPEUTIC DOSAGES

In one osteoporosis study, vitamin K was taken at the high dose of 1 milligram (mg) daily, more than ten times the necessary nutritional intake.

THERAPEUTIC USES

Growing but not definitive, evidence suggests that vitamin K should be added to the list of nutrients that prevent osteoporosis. Based on its ability to help blood clot normally, vitamin K has also been proposed to treat excessive menstrual bleeding. However, the last actual study testing this idea was carried out more than fifty-five years ago. Vitamin K has also been recommended for nausea, although there is no meaningful evidence that it works.

Preliminary evidence suggests that vitamin K supplementation may help prevent liver cancer. Very high doses of intravenous vitamin K have also been used to treat advanced liver cancer, with, perhaps, marginal benefits.

SCIENTIFIC EVIDENCE

Vitamin K plays a known biochemical role in the formation of bone. This has led researchers to look for relationships between vitamin K intake and osteoporosis.

Observational studies have found that people with osteoporosis often have low levels of vitamin K and that people with a higher intake of vitamin K have a lower incidence of osteoporosis. Research also suggests that supplemental vitamin K can reduce the amount of calcium lost in the urine. This is indirect evidence of a beneficial effect on bone.

However, while these studies are interesting, only double-blind, placebo-controlled trials can prove a treatment effective. Several such studies have been performed on vitamin K for osteoporosis, with generally positive results.

One of these was a three-year, double-blind, placebo-controlled trial of 181 women; it found that vitamin K significantly enhanced the effectiveness of supplementation with calcium, vitamin D, and magnesium. Postmenopausal women between the ages of fifty and sixty were divided into three groups: receiving either placebo, calcium plus vitamin D plus magnesium, or calcium plus vitamin D plus magnesium plus vitamin K_1 (at the high dose of 1 mg daily). Researchers monitored bone loss by using a standard dual-energy X-ray (DEXA) bone density scan. The results showed that the study participants using vitamin K and the other nutrients lost less bone than those in the other two groups.

Benefits were also seen in other studies. However, another placebo-controlled trial involving 452 older men and women with normal calcium and vitamin D levels failed to demonstrate any beneficial effects of 500 micrograms (mcg) per day of vitamin K supplementation on bone density and other measures of bone health over three years.

If there is a favorable effect, it appears to be relatively modest. Vitamin K may show its influence most strongly when, instead of a DEXA scan alone, more complex tests of bone strength are used. Some evidence hints that vitamin K works by reducing bone breakdown rather than by enhancing bone formation.

SAFETY ISSUES

Vitamin K is relatively safe at the recommended therapeutic dosages. The vitamin directly counters the effects of the anticoagulant warfarin (Coumadin). Persons taking warfarin should not take vitamin K supplements or alter their dietary intake of vitamin K without doctor supervision.

One study suggests a novel way of using this effect deliberately. Researchers gave people on warfarin a fixed daily dose of vitamin K to override the changes in warfarin action caused by the natural variation in day-to-day dietary vitamin K consumption. The results were positive: international normalized ratio (INR) values (the standard measurement of warfarin's blood-thinning effect) became more stable. However, this method should not be used except under close physician supervision.

Newborns are commonly given vitamin K_1 injections to prevent bleeding problems. Although some have suggested that this practice may increase the risk of cancer, enormous observational studies have found no such connection (one such trial involved more than one million participants).

IMPORTANT INTERACTIONS

People who are taking warfarin (Coumadin) should not take vitamin K supplements or eat foods high in vitamin K except under the supervision of a physician since they will need to have their medication dosages adjusted. People taking cephalosporins or other antibiotics may need more vitamin K if they are already deficient in this nutrient. People taking anticonvulsants, such as phenytoin (Dilantin), carbamazepine, phenobarbital, and primidone (Mysoline), and are pregnant may also need more vitamin K.

—*EBSCO CAM Review Board*

Further Reading

Bolton-Smith, C., et al. "A Two-Year Randomized Controlled Trial of Vitamin K_1 (Phylloquinone) and Vitamin D_3 Plus Calcium on the Bone Health of Older Women." *Journal of Bone and Mineral Research,* vol. 22, no. 4, 2007, pp. 509-19.

Booth, S. L., et al. "Dietary Vitamin K Intakes Are Associated with Hip Fracture but Not with Bone Mineral Density in Elderly Men and Women." *American Journal of Clinical Nutrition,* vol. 71, 2000, pp. 1201-8.

___ "Effect of Vitamin K Supplementation on Bone Loss in Elderly Men and Women." *Journal of Clinical Endocrinology and Metabolism,* vol. 93, no. 4, 2008, pp. 1217-23.

Braam, L. A., et al. "Vitamin K_1 Supplementation Retards Bone Loss in Postmenopausal Women Between Fifty and Sixty Years of Age." *Calcified Tissue International,* vol. 73, 2003, pp. 21-26.

Cockayne, S., et al. "Vitamin K and the Prevention of Fractures." *Archives of Internal Medicine,* vol. 166, 2006, pp. 1256-61.

Habu, D., et al. "Role of Vitamin K_2 in the Development of Hepatocellular Carcinoma in Women with Viral Cirrhosis of the Liver." *JAMA: The Journal of the American Medical Association,* vol. 292, 2004, pp. 358-61.

Knapen, M. H., L. J. Schurgers, and C. Vermeer. "Vitamin K_2 Supplementation Improves Hip Bone Geometry and Bone Strength Indices in Postmenopausal Women." *Osteoporosis International,* vol. 18, no. 7, 2007, pp. 963-72.

Martini, L. A., et al. "Dietary Phylloquinone Depletion and Repletion in Postmenopausal Women: Effects on Bone and Mineral Metabolism." *Osteoporosis International,* vol. 17, no. 6, 2006, pp. 929-35.

Purwosunu, Y., et al. "Vitamin K Treatment for Postmenopausal Osteoporosis in Indonesia." *Journal of Obstetrics and Gynaecology Research,* vol. 32, 2006, pp. 230-34.

Rombouts, E. K., F. R. Rosendaal, and F. J. van der Meer. "Daily Vitamin K Supplementation Improves Anticoagulant Stability." *Journal of Thrombosis and Haemostasis,* vol. 5, no. 10, 2007, pp. 2043-48.

Sarin, S. K., et al. "High Dose Vitamin K_3 Infusion in Advanced Hepatocellular Carcinoma." *Journal of Gastroenterology and Hepatology,* vol. 21, 2006, pp. 1478-82.

Vitamins and minerals

Category: Biology

Anatomy or system affected: All

Specialties and related fields: Dietetics, endocrinology, family medicine, internal medicine, nutrition

Definition: chemicals that supply the body with the means of metabolizing (extracting and using the energy from) the macronutrients (fats, carbohydrates, and proteins) it ingests; essential diet ingredients

KEY TERMS

fat-soluble vitamins: vitamins that, because of their structure and solubility, migrate to fatty tissues in the body, where they are stored

macronutrients: materials ingested in large amounts to supply the energy and materials for physical bodies

megadose: ten or more times the recommended daily allowance of a nutrient

micronutrients: substances of which only milligrams are needed in the daily diet, such as vitamins and minerals

mineral: an inorganic salt of particular metals or elements needed for good health

recommended daily (or dietary) allowance (RDA): the in-take levels of the essential nutrients that are considered adequate to meet the known nutritional needs of most healthy persons

trace elements: elements needed in the diet at levels of less than 100 milligrams per day *vitamin*: an organic compound constituent of food that is consumed in relatively small amounts (less than 0.1 gram per kilogram of body weight per day) and that is essential to the maintenance of life

water-soluble vitamins: vitamins that, because of their structure, show strong solubility in water; they usually pass through the body in a relatively short time

STRUCTURE AND FUNCTIONS

Vitamins are organic compounds (that is, compounds made up of carbon, oxygen, nitrogen, sulfur, or hydrogen) that are food constituents and crucial to the maintenance of life and good health. They make possible the production of energy and the formation of coherent body tissues from the macronutrients usually consumed in a regular diet. They are, among other things, coenzymes that serve as oxidizing and reducing agents and carriers for other chemical groups at the active sites of enzymes. Vitamins are part of the approximately one hundred organic compounds that are of the proper size and stability to be absorbed from the digestive tract into the bloodstream without digestion or breakdown. Nevertheless, they are not produced in the body in large amounts to keep a person healthy; they have always been available in food, and there was probably no need for the human metabolism to produce them. Plants synthesize vitamins, and therefore plants constitute the principal natural source of these compounds.

Vitamins are divided into two main groups: water-soluble and fat-soluble vitamins. Structural differences account for the two types of solubility. Fat-soluble vitamins (vitamins A, D, E, and K) consist mainly of hydrocarbon groupings (nonpolar hydrocarbon chains and rings compatible with nonpolar oil and fat). They are structurally similar to fats, whereas water-soluble vitamins have polar hydroxyl (-OH) and carboxyl (-COOH) groups that are attracted to and form hydrogen bonds with water. One of the most important differences between vitamins is the result of their solubility: Fat-soluble vitamins are stored in the body tissues and organs for relatively long periods, while water-soluble vitamins are eliminated from the body in a relatively fast manner, sometimes in a matter of hours.

Vitamin A (retinol) maintains the health of the eyes, skin, and mucous membranes and is particularly important for good vision in dim light. There are various physiological equivalents to vitamin A, that is, compounds with closely related structures that can be used as the vitamin itself. Beta carotene is a provitamin (a substance that can be easily converted to a vitamin) of vitamin A found in carrots. The vitamin can also be found in liver and liver oils. Lack of vitamin A can cause night or total blindness.

The B vitamins are often considered a B complex group because they work together as coenzymes in biochemical reactions leading to growth and energy production. They are water-soluble and quickly eliminated from food in the cooking process. Members of this group include pyridoxine (B6), involved in at least sixty enzyme reactions (mainly in the metabolism and synthesis of proteins); thiamine (B1), a coenzyme in carbohydrate metabolism and involved in energy production, digestion, and nerve activity; riboflavin (B2), used in obtaining energy from foods; pantothenic acid (B3), needed for proper growth; niacin (B4), needed for the production of healthy tissues; cobalamin (B_{12}), involved in the production and growth of red blood cells; and folic acid (B_9), also involved in the production of red blood cells and metabolism. They are present in various foods, especially meat and dairy products. Deficiency symptoms include anemia, skin disorders, and nervous system disorders.

Vitamin C, or ascorbic acid, is involved in the destruction of invading bacteria, in the synthesis and activity of interferon (which prevents entry of viruses into cells), in decreasing the effect of toxic substances (such as drugs and pollutants), and in the formation of connective tissue. Humans are one of the few animals for which ascorbic acid is a vitamin since other species produce it in their metabolic processes. Deficiency symptoms include the degeneration of tissue and scurvy. Vitamin C is found mostly in citrus fruits.

Vitamin D (calciferol) promotes the absorption of calcium and phosphorus through the intestinal wall and into the bloodstream. Its deficiency induces the disease rickets and, in adults, the malformation of bones. Unlike other vitamins, it forms in the body through the action of the sun's ultraviolet light. As with the vitamin B complex, vitamin D has a set of closely related molecular structures, called "D_1, D_2, D_3," etc. All these structures have the same physiological function. Because of limited sun exposure, copious clothing, and indoor living and working conditions, humans need to add vitamin D to their diet, as in fortified milk, cod liver oil, or vitamin supplements.

Vitamin E (alpha-tocopherol) is an antioxidant of polyunsaturated fatty acids (fatty acids with numerous double bonds). These fatty acids readily form peroxides, which are particularly damaging because they can lead to runaway oxidation in cells. Vitamin E protects the integrity of cell membranes, which contain considerable amounts of fat. It also helps maintain the integrity of the circulatory and central nervous systems; is involved in the functioning of the kidneys, lungs, liver, and genitalia; and detoxifies poisonous materials absorbed by the body. Since aging, in some theories, is considered the cumulative effect of free radicals (reactive atoms) running wild in the body, the antioxidant properties of vitamin E may make it a good candidate for inhibiting aging or preventing premature aging. Its deficiency symptoms in humans are unknown. Vitamin E is present in various foods, especially in grain oils.

Biotin (also called "vitamin H") participates in metabolism by acting as a carboxyl carrier for several enzymes. Its sources are liver, cereals, and egg yolks. Symptoms of deficiency include alopecia (the loss or absence of hair) and skin rashes.

Vitamin K completes the list of vitamins. It participates in the clotting of blood, and its deficiency can cause bleeding and liver damage. This vitamin is commonly found in plants and vegetables.

The term "minerals," when used in a nutritional context, includes all the nutritional, chemical elements of foods obtained from macronutrients, except for carbon, hydrogen, nitrogen, oxygen, and sulfur. This term also refers to metal elements combined with others in compounds such as soluble inorganic salts. It is in this combined form that they serve indispensable functions in the body.

Minerals pass slowly through the body and are excreted in the feces, urine, and sweat. Therefore, they must be replaced and an appropriate balance continuously maintained. Because living beings cannot generate minerals in their bodies, they must obtain them from foods or food supplements. Plants pick up minerals directly from the soil, and animals get them from the plants that they ingest. Unlike vitamins, which plants synthesize, foods do not contain minerals if they were not grown in the soil. Among their many functions, minerals are components of enzymes, are structural components of body parts such as bones, are involved in maintaining the electrolyte balance in body fluids, and transport materials, as hemoglobin does in blood.

There are seventeen known minerals, although many others may exist. Since most of them are present in the body in relatively small amounts, their functions have been determined through the symptoms of various dietary deficiencies. Minerals can be grouped into two classes: the major elements—calcium, phosphorus and magnesium—are required in amounts of 1 gram or more per day. The trace elements include chlorine, chromium, cobalt, copper, fluorine, iodine, iron, manganese, molybdenum, nickel, selenium, sulfur, vanadium, etc. zinc, are needed in milligram or microgram quantities each day.

Calcium, probably the best-known mineral, is present in the body in a greater amount than any other mineral: up to 1.5 or 2.0 percent of total body weight, with 99 percent of it in bones and teeth. In the nervous system, calcium is used to slow down the heartbeat. It is metabolized in the body by a hormone synthesized from calciferol (vitamin D). Excess calcium can give rise to kidney stones. Its deficiency is common in postmenopausal women, who produce less estrogen. This decrease encourages bone dissolution, and when bones are dissolved, calcium is lost. Calcium is found in milk and dairy products, fish, and green vegetables. Phosphorus, the second most common mineral, is a structural component of bones and soft tissue. It is found in nearly all foods.

Sodium and potassium cations (positively charged atoms) are components of many minerals. They work in the conservation of electrolytic balance in cell fluids. Potassium governs the activity of many cellular enzymes. At the same time, sodium keeps the water content of cellular fluids in a healthy balance. For the body to work correctly, it needs the appropriate ratio of sodium to potassium. Potassium ions concentrate inside the cell, while sodium ions concentrate outside the cell. Natural, unprocessed foods have high sodium-to-potassium ratios. However, because sodium and potassium compounds are very soluble in water, they dissolve during processing and cooking and are lost. Sodium is replenished by adding salt to food, but this is not the case with potassium, which is not added to food. Care must be taken in this matter, either by eating more fresh foods or using a specialized table salt containing a mixture of sodium chloride and potassium chloride. Sodium retention leads to water retention and edema (swollen legs and ankles), and high blood pressure in some individuals. Sodium is mainly found in table salt, and potassium is found in meat, dairy products, and fruit.

Magnesium and chloride ions are the most common minerals in cell fluids, as they regulate fluid balances and electrical charges. Magnesium controls the formation of proteins inside the cell and the transmission of electrical signals from cell to cell. Chloride is present in the stomach as hydrochloric acid or stomach acid. Magnesium is found in whole-grain cereals, dried fruits, and leafy green vegetables, and chloride is found in table salt.

Trace elements work in various ways, with most of them incorporated into the structure of enzymes, hormones, and related molecules or acting in conjunction with vitamins. Among the trace elements, one of the more important ones is iron, which is a critical part of the hemoglobin molecule of red blood cells and is involved in oxygen transport. Fluoride, another trace element, helps harden the enamel of teeth to make them resistant to decay; zinc plays a vital role in growth, the healing of wounds, and the development of male sex glands, and manganese is needed for healthy bones and a well-functioning nervous system. Iodine is involved in the proper operation of the thyroid gland, chromium is important in the metabolism of glucose, and cobalt aids in cell function. Copper and selenium are other trace elements needed by the body. Most trace elements are found in fish, meat, fruits, and vegetables.

RELATED DISEASES

Vitamin deficiencies are not common in the United States and other Western countries. A well-balanced diet provides ample vitamins of all kinds. Megadoses of vitamins can create harmful effects; however, a toxic dose exists for many vitamins. For example, vitamin A, when taken in excess, can cause headache, nausea, vomiting, fatigue, swelling, bleeding, pain in the arms and legs, and congenital disabilities. An acute deficiency of the vitamin, however, can impair vision and eventually cause blindness. Consequently, there must be a balance in vitamin intake. This balance can be achieved by following the recommended daily (or dietary) allowances (RDAs).

In the United States, the Food and Nutrition Board of the National Academy of Sciences and the

National Research Council determined the daily needs for some vitamins and minerals. The Food and Drug Administration (FDA) made these findings the basis for its list of RDAs. These allowances are presented in units of grams or milligrams. These amounts are determined using international units of biological activity. (Some vitamins come in several forms, all of which are physiologically equivalent.) RDAs do not cover every vitamin and mineral needed for good health, nor do they cover the more extreme nutritional requirements resulting from illness or unusual genetic makeup. They just serve as general guidelines for healthy individuals. For some substances lacking specific RDAs, such as chromium and a handful of other elements, the FDA lists the daily ranges of these micronutrients that it considers safe and effective. RDAs depend on gender, age, weight, and other conditions. They are usually presented on food labels as percentages of the daily dietary requirement.

In 2005, the US Department of Agriculture (USDA) updated the Food Guide Pyramid. The new pyramid emphasizes the need for physical activity (thirty minutes of moderate or vigorous exercise per day) and the importance of variety in the diet. Consumers can get personalized recommendations based on their age and gender at MyPyramid.gov. Unlike the older food pyramids, the 2005 version suggests food quantities in cups and ounces rather than servings, which was ambiguous and confusing to consumers. For example, for a 2,000 calorie per day diet, the recommendations are six ounces of whole grains, two and a half cups of vegetables, two cups of fruit, three cups of dairy, and five and a half ounces from the meat and bean subgroup.

The main criticisms of the 2005 food recommendations are that they do not mention any specific foods from which to abstain, that people who do not have Web access cannot obtain personalized recommendations, and that the beef and dairy industry lobbies play a role in the USDA's decisions about these matters. Consumers should keep in mind that the primary role of the USDA is to promote agriculture in the United States. Politics are embedded in diet decisions, and recommending that people eat less is not good for business. Some nutritionists and scientists believe that diet matters should be under the auspices of a more neutral party, such as the National Institutes of Health (NIH).

The activity of a vitamin or mineral depends only on its molecular structure, not on its source. Therefore, the synthetic vitamins found in food supplements provide the same nutrients as naturally occurring ones. However, it is crucial to remember that other substances or nutrients are present in the food that is being consumed to obtain the necessary vitamin and mineral requirements. Authentic food often contains additional substances that enhance the absorption and utilization of its nutrients. For example, the calcium naturally present in food is more likely to carry with it any vitamin D or phosphorus that the body might need for its optimum use than is the calcium found in an antacid tablet or a food supplement. A balanced diet provides a diversity of nutrients that no pills can match.

The primary medical use of the vitamins is curing the deficiency diseases—those caused by their absence from the diet. An FDA panel has judged nine vitamins to be safe and effective as over-the-counter drugs. Supplementation is commonly thought of as a means of maintaining nutritional equilibrium in the body.

Many different analytical methods—such as ultraviolet-visible and infrared spectroscopy, paper, thin-layer, and gas-liquid chromatography; and mass spectroscopy—as well as biological assays, have been used for the detection and identification of vitamins. They have greatly helped to explain the complex structures of these compounds. These methods are also used to determine the vitamin content of a particular food item, providing the consumer with valuable nutritional information.

PERSPECTIVE AND PROSPECTS

Vitamin deficiency diseases such as scurvy, beriberi, and pellagra have plagued the world since the existence of written records. The concept of a vitamin or "accessory growth factor" was developed in the early part of the twentieth century. In 1912, Casimir Funk, a Polish biochemist, isolated a dietary growth factor from the outer covering of rice grains and found that it cured the disease when added to the food of those who had beriberi. The factor was an organic compound called an "amine" (that is, a compound containing nitrogen combined with carbon and hydrogen). Funk coined the term "vitamine" (meaning "life-giving amine") for the compound, which is now called "thiamin" or "vitamin B_1." In the next five decades, there was an exciting era of the isolation, identification, and synthesis of vitamins. It was soon found that these compounds were not all amines, and the term was changed to "vitamins." As more information on the structure of vitamins was obtained, names changed from general ones (such as vitamin C) to more specific ones (such as ascorbic acid). These discoveries led to the availability of inexpensive synthetic vitamins and a dramatic reduction in overt vitamin deficiency disease.

Small amounts of vitamins are essential for good health. Still, the benefits of taking megadoses of specific vitamins to prevent or cure certain ailments are often debated. There is evidence that the use of high levels of vitamins can prevent or alleviate several diseases. Improvements in the analytical methods used to detect and identify vitamins have led to better and more sensitive detection limits for these compounds. The result has been increased knowledge of vitamins and minerals and their function.

In 2002, the American Medical Association endorsed the notion that adults should take a multivitamin daily. This reversed the organization's long-standing antivitamin stance that vitamins were a waste of time and money for all people except pregnant women and some people with chronic illnesses. The current recommendation, published in the *Journal of the American Medical Association*, acknowledges that vitamins may prevent some kinds of chronic diseases, such as heart disease, cancer, and osteoporosis. Nevertheless, the efficacy and safety of regular intake of multivitamins remain a subject of debate in the medical community. In a 2013 book entitled *Do You Believe in Magic?: The Sense and Nonsense of Alternative Medicine*, Dr. Paul Offit of the University of Pennsylvania cites several studies suggesting that regular intake of vitamin supplements increases disease risk. However, critics like Dr. Dallas Clouatre of the American College of Nutrition argue that many vitamin studies rely on unscientific data—such as self-reporting through questionnaires.

—Maria Pacheco, PhD, Lisa Levin Sobczak, RNC, and LeAnna DeAngelo, PhD

Further Reading

Balch, James F., and Phyllis A. Balch. *Prescription for Nutritional Healing: A Practical A to Z Reference to Drug-Free Remedies Using Vitamins, Minerals, Herbs, and Food Supplements*. 4th rev. ed., Avery, 2008.

Duyff, Roberta Larson. *American Dietetic Association Complete Food and Nutrition Guide*. 3rd ed., John Wiley & Sons, 2007.

Lieberman, Shari, and Nancy Bruning. *Real Vitamin and Mineral Book*. 4th ed., Avery, 2007.

Murray, Michael. *The Pill Book Guide to Natural Medicines: Vitamins, Minerals, Nutritional Supplements, Herbs, and Other Natural Products*. Bantam, 2002.

Offit, Paul A. *Do You Believe in Magic?: The Sense and Nonsense of Alternative Medicine*. Harper, 2012.

Preidt, Robert. "Too Little Vitamin D May Hasten Disability as You Age." *MedlinePlus*, 17 July 2013.

Shelton, C. D. *Vitamins, Minerals & Supplements: Essential or Over-Hyped?* Amazon Digital Services Inc., 2013.

Nitrogen fixation

Category: Biochemistry
Anatomy or system affected: None

Specialties and related fields: Bacteriology, ecology, geochemical cycles, plant biochemistry, plant physiology, soil science

Definition: the process carried out by certain soil bacteria whereby nitrogen gas from the atmosphere is converted by enzyme-mediated reduction to ammonia

KEY TERMS

ammonification: the conversion of atmospheric nitrogen gas to ammonia and ammonium ion by soil bacteria, particularly *Rhizobium* species

bacteroid: a relatively large, cell-like structure constructed of four to eight individual, but enlarged, *Rhizobium* bacteria cells

nitrification: the conversion of ammonium ion to nitrate ion by soil bacteria, particularly *Rhizobium* species

nitrogen fixation: the overall process of converting atmospheric nitrogen gas to nitrate ion and making it available to plants via their roots

Nitrogen fixation is the process carried out by certain soil bacteria whereby nitrogen gas from the atmosphere is converted by enzyme-mediated reduction to ammonia. This strongly alkaline substance easily accepts a hydrogen ion to form an ammonium ion. Subsequent oxidation, called "nitrification," converts the ammonium ions to nitrate ions, a soluble ionic form of nitrogen available to higher plants in water absorbed through their roots.

NITROGEN IN BIOLOGICAL SYSTEMS

Nitrogen, the fourth most abundant element in most organisms, can account for as much as 4 percent of a plant's dry weight. Most of this nitrogen is present as a constituent of protein structures. Still, it is also a component of numerous other biological compounds, such as chlorophylls and nucleic acids. Thus, for normal plant growth and development, nitrogen must be maintained at fairly high levels in the soil.

Earth's atmosphere is about 79 percent nitrogen gas (N_2). Unfortunately, this form of nitrogen is of no direct value to higher plants; they must acquire their nitrogen in the form of either ammonium or nitrate. These two forms of nitrogen can be supplied to the soil as fertilizer by humans or by nature as the product of microbial action.

Three microbial processes render nitrogen into forms usable by higher plants. These are ammonification, nitrification, and nitrogen fixation. Ammonification occurs when various forms of organic nitrogen, present in the proteins of plant and animal residues and animal wastes (manures), are converted to ammonium. Nitrification is the process by which ammonium is converted to nitrate. Populations of free-living soil microorganisms, like Azotobacter, Beijerinckia, and Clostridium, carry out both these enzyme-mediated processes.

In the nitrogen fixation process, atmospheric nitrogen is converted to ammonium. While some nitrogen fixers are free-living microbes, bacteria that live symbiotically within the roots of several plant species are also responsible for much of this conversion in terrestrial ecosystems.

NITROGEN-FIXING BACTERIA

Nitrogen-fixing bacteria have been shown to coexist with various lower plants, including lichens, liverworts, mosses, and ferns. Among the more advanced seed-bearing plants, nitrogen fixers are associated with some tropical grasses and many shrubs and trees, such as the alders. Agriculturally, the legumes are the most important group of plants coexisting with nitrogen-fixing bacteria. Some 1,500 species of legumes, including peas, beans, clover, and alfalfa, have been shown to live symbiotically with nitrogen-fixing bacteria called *Rhizobium*. A different species of *Rhizobium* infects each different species of legume.

The bacteria penetrate the root by entering the filamentous projections of epidermal cells called "root hairs." The epidermal cells respond to the invasion by enclosing the bacterium in a threadlike structure referred to as an infection thread. The infection thread begins to grow and branch. As it does so, the Rhizobia reproduce numerous times inside the thread. After penetrating several layers of cells, the infection thread eventually reaches the root cortex, rupturing and releasing the encased bacteria.

The release of the bacteria induces the secretion of plant hormones that stimulate specialized root cortical cells to divide several times. As these cells divide, the Rhizobia are encapsulated, and a nodule is formed. Within the cytosol of the nodule cells, the bacteria become nonmotile, increase in size, and accumulate in groups of four to eight bacteroids. These bacteroids are responsible for the biochemical conversion of elemental nitrogen to ammonium.

CHEMISTRY OF NITROGEN FIXATION

Chemically, nitrogen fixation requires that six electrons and eight hydrogen ions be transferred to the atmospheric nitrogen molecule. This reaction is an energy-requiring process; therefore, adenosine triphosphate (ATP), the cell's form of stored energy, must be available for the reaction to take place. The electrons, hydrogen ions, and ATP are supplied by the cellular respiration process in the root cells.

The nitrogenase enzyme transfers the electrons and hydrogen ions to atmospheric nitrogen (N_2). This enzyme consists of two subunits. One subunit takes the electrons and hydrogen ions from the respiratory products and transfers them to the other subunit. The ATP binds with part II of the nitrogenase, and hydrolysis releases the energy stored in the molecule.

This provides the energy required for the reaction to proceed and makes passing the electrons and hydrogen ions onto the second subunit. The nitrogenase transfers the electrons and hydrogen ions to the nitrogen atom in the last step. This final transfer results in the production of ammonium. The ammonium moves out of the bacteroids into the cytosol. It is converted to an organic form of nitrogen that can be transported throughout the plant.

RATES OF FIXATION

Not all soil bacteria species fix nitrogen at the same rate. Several factors can account for these differences. Some plants form nodules much more abundantly than others. Because of their more extensive nodule formation, these plants will fix more nitrogen than those that produce fewer nodules.

The nitrogenase of all rhizobial species tends to transfer electrons to hydrogen ions rather than to nitrogen. As a result, hydrogen gas, which escapes into the atmosphere, is produced. This represents a loss of electrons that could have been used to produce ammonium. However, some Rhizobia species have a second enzyme, called "hydrogenase," which uses hydrogen gas to produce water. ATP is produced as a byproduct of this process. Consequently, these rhizobial species are more efficient because less energy is wasted.

In addition, the fixation rate and the amount of nitrogen fixed will vary with age or the stage of plant development. In most cases, fixation rates are highest when the fruits and seeds are produced. Many plants' seeds, especially legumes, are high in protein. Hence, nitrogen fixation and transport out of the nodules must be higher when the seeds are developing. More than 85 percent of the total nitrogen fixation in legumes occurs at such times.

PLANT-BACTERIA MUTUALISM

Nitrogen-fixing bacteria exist in a symbiotic relationship with their plant hosts. The bacteria supply the plant with much-needed nitrogen, while the plant supplies the bacteria with carbohydrates and other nutrients. Some of the energy derived from the plant-supplied carbohydrates is used in nitrogen fixation. Still, there is ample left over to supply the bacteria with all the energy necessary for survival.

PROVIDING NITROGEN

Plant production throughout the world is limited more by the shortage of nitrogen than any other nutrient. The root zone that encompasses the upper 6-inch (15-centimeter) layer of soil contains from 220 to 13,228 pounds (100 to 6,000 kilograms) of total nitrogen per acre (hectare). This includes all forms of nitrogen, many of which are unavailable to plants. The total nitrogen content is determined by several factors, such as the minerals making up the soil, the kinds of vegetation, and the extent to which these factors are affected over time by climate, topography, and the presence of people.

Ammonium and nitrate are the only forms readily available to plants, being readily soluble in water. In addition to organic nitrogen compounds that can easily be converted to the available forms, these two molecules are the only ones of ecological or agricultural importance. Unfortunately, these forms of nitrogen are continually being removed from the soil. Crop removal, leaching (the removal of minerals as water percolates through the soil), denitrification (the process by which anaerobic microbes convert nitrates to gaseous nitrogen-containing compounds that escape the soil), and erosion account for a total loss of approximately 276 pounds per acre (125 kilograms per hectare) annually. While some of the lost nitrogen can be replaced by available forms falling to the ground in the rain, that amount is much too low to be of value in plant growth. Microbial fixation and the application of fertilizers are the only sources that supply sufficient nitrogen for plant growth.

RESEARCH APPLICATIONS

The nonsymbiotic nitrogen fixers are extremely important, especially in nonagricultural soils. Forest, desert, and prairie ecosystems are dependent on nitrogen fixation by free-living species to replace the annual nitrogen loss. Without it, the growth of many plant species would suffer, and food chains in these ecosystems would soon be disrupted. Several studies have investigated the advantages of incorporating free-living nitrogen fixers into nonlegume crop production, but clear benefits are uncertain.

On the other hand, knowledge of symbiotic nitrogen fixation has resulted in definite improvements in the production of legume crops. When *Rhizobium* is included with seeds as they are planted, increased yields have been observed in nearly every case. There is considerable interest in enhancing the efficiency of the nitrogen-fixing process by increasing nodulation in the roots of some species or by incorporating the hydrogenase system into species that do not have it through bioengineering. An increase in biological nitrogen fixation could enhance the soil's nitrogen content and decrease the dependency on commercial nitrogen fertilizers.

The application of nitrogen fertilizers to nonleguminous crops was once one of a farmer's best investments. Commercial fertilizers, however, have become very expensive because of increased energy costs. In addition, many environmental problems have developed from the accumulation of nitrates from fertilizers finding their way into rivers, ponds, and lakes through agricultural runoff. Consequently, there is renewed interest in symbiotic nitrogen fixation. For years, before the extensive use of nitrogen fertilizers, farmers planted legume crops alternately with other crops. The legume crops were plowed under to supply nitrogen to the soil. There has been a resurgence in this technique of using "green manures" to increase soil fertility. There is considerable interest in growing plants with symbiotic nitrogen fixers in the same fields as plants lacking to improve the natural nitrogen balance in certain environments. Several bioengineering studies are directed at incorporating nitrogen fixation into the gene structures of nonlegume crop plants; this research is difficult, however, because the genetics of the process is very complex.

—D. R. Gossett

Further Reading

California Institute of Technology. "Nitrogen Fixation Research Could Shed Light on Biological Mystery: New Process Could Make Fertilizer Production More Sustainable." *ScienceDaily*, 30 May 2017, www.sciencedaily.com/releases/2017/05/170530140710.htm.

de Bruijn, Frans, editor. *Biological Nitrogen Fixation*. Wiley-Blackwell, 2015.

Legocki, Andrezej, Hermann Bothe, and Alfred Pühler, editors. *Biological Fixation of Nitrogen for Ecology and Sustainable Agriculture.* Springer, 1997.

Salisbury, Frank B., and Cleon W. Ross. *Plant Physiology*. 4th ed., Wadsworth, 1992.

Stacey, Gary, Robert H. Burris, and Harold J. Evans, editors. *Biological Nitrogen Fixation.* Chapman and Hall, 1992.

Tortora, Gerard J., et al. *Microbiology: An Introduction.* 13th ed., Pearson, 2018.

Microbial Genetics

Gene expression is a complicated topic, but the first forays into molecular genetics were with bacteria. Consequently, bacterial gene expression is extremely well understood and exploited for industrial and pharmaceutical purposes. In this section, we discuss microbial genetics and its exploitation by genetic engineering techniques and how microbes exploit their genetics to develop antibiotic resistance. Recombinant DNA technology, plasmids, and antibiotic resistance are among the topics discussed.

Operon

Category: Molecular genetics
Anatomy or system affected: None
Specialties and related fields: Bacteriology, biochemistry, genetics, molecular biology
Definition: a genetic regulatory system found in bacteria and their viruses in which genes coding for functionally related proteins are clustered along the deoxyribonucleic acid (DNA), allowing gene expression to be controlled coordinately in response to the needs of the cell

KEY TERMS

operator: a segment of DNA where the repressor binds, preventing the transcription of specific genes
promoter: a DNA sequence needed to turn a gene on or off where the transcription is initiated
RNA polymerase: an enzyme that is responsible for copying a DNA sequence into a ribonucleic acid (RNA) sequence during the process of transcription
translation: the process in which ribosomes in the cytoplasm or endoplasmic reticulum synthesize proteins after the process of transcription of DNA to RNA in the cell's nucleus
transcription: the process of copying a segment of DNA into RNA

INTRODUCTION

An operon is a cluster of genes with a single promoter. A promoter is an area of genes that starts transcription, or genetic copying, within an organism's deoxyribonucleic acid or DNA. The DNA uses information from genes to direct the replication of cells, determining which cells will have which functions. The genes in the operon are all related to instructing the cells for function or task; they, therefore, share a promoter so that they are copied together. Operons are part of the cellular regulatory system in bacteria and viruses. They allow the cell to produce proteins when needed and restrict their production, helping the cell conserve its resources.

BACKGROUND

Every living thing has cells as its fundamental units of structure. Each cell has a specific function. That function is encoded in genes, which provide instructions to the cells for how to develop. This information is passed on in two ways: from one generation of an organism to another during reproduction or replication and within the organism during processes known as transcription and translation.

Scientists knew that cells shared information for many years, but they did not understand how this happened. The early years of the 1960s were an important time for scientists working to understand genetics. Nearly one hundred years after German monk Gregor Mendel first documented how pea plants inherited specific physical traits, scientists were just beginning to understand how cells used DNA to transcribe the information needed to replicate cells and generate the materials needed for life. Two French scientists working with bacteria DNA went public with a critical discovery regarding how information is passed within cells of an organism in the early years of the decade.

The scientists Francois Jacob and Jacques Monod were working with the bacterium *Escherichia coli*, or *E. coli*, when they discovered it had three genes specifically used to form the proteins needed to break down lactose. Lactose is a milk sugar made up of glucose and galactose. The three genes could turn the production of the proteins off and on, and they were all controlled from one central point or promoter. This group of *E. coli* genes controlled by one promoter was the first operon to be identified.

Jacob and Monod published their findings in 1960 and 1961 in the journal *Proceedings of the French Academy of Sciences.* They named their discovery an operon, based on the French verb *operer,* which means "to effect" or "work on." The French scientists had identified a part of the system that regulates gene function for the first time; they earned a share of the 1965 Nobel Prize in Physiology or Medicine.

OVERVIEW

Operons are part of the transcription process that tells cells when to make proteins. DNA contains the blueprint and instructions needed to build proteins. This information is carried in ribonucleic acid, or RNA, to the cells by messenger RNA, or mRNA. The mRNA carries the instructions to the cells, which then create the protein per the instructions.

This transcription process has three stages known as initiation, elongation, and termination. Operons are part of the initiation of the transcription process. The RNA interacts with the promoter to begin the gene's transcription and transfer the instructions it contains to a new cell. In some cases, multiple genes are attached to a promoter. This is because the genes contain instructions that must be used together to form the correct protein, similar to how a recipe has multiple ingredients and steps that must be used together for it to come out properly. Multiple genes attached to one promoter are what Jacob and Monod identified as an operon.

During the transcription process, the RNA opens the instructions contained in the genes by unwinding their DNA. The RNA begins a process that codes the instructions in the genes into mRNA; this is the elongation step of the transcription. Messenger RNA synthesis continues until it reaches a signal in the instructions that indicate their end. The transcription process is terminated, and the DNA rewinds into its typical double helix configuration.

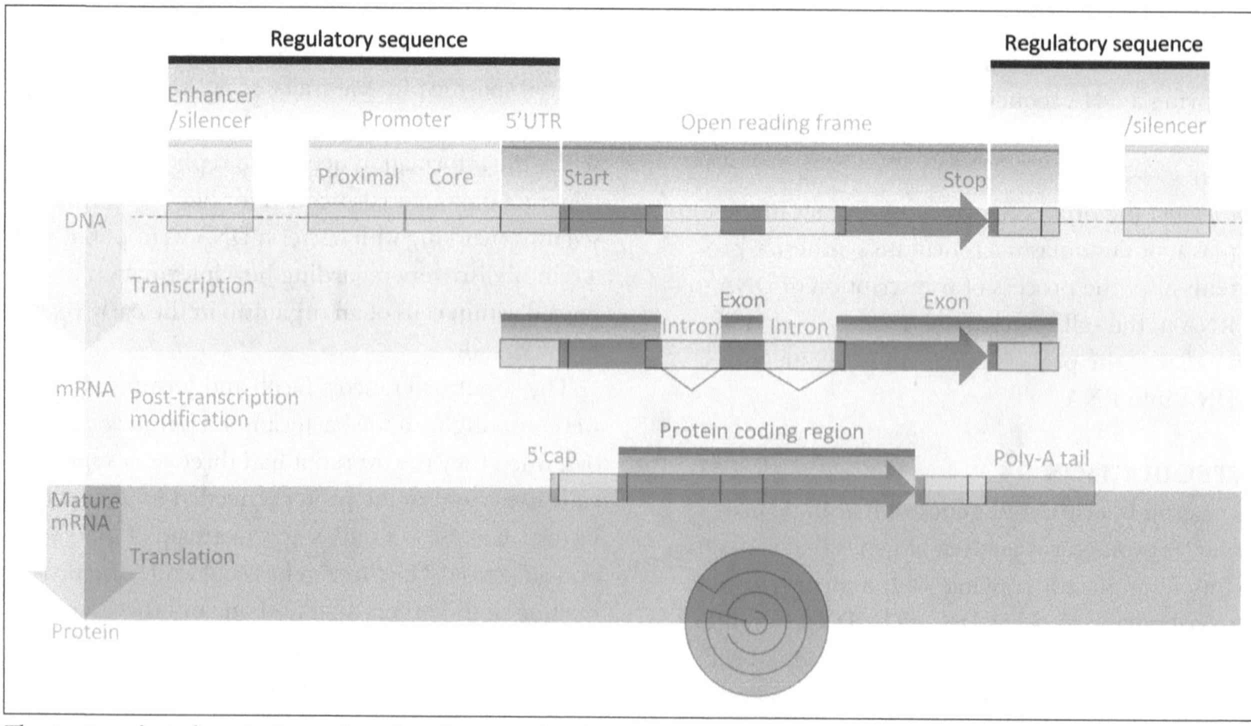

The structure of a prokaryotic operon of protein-coding genes. Regulatory sequences control when expression occurs for the multiple protein-coding regions. Promoter, operator, and enhancer-like regions regulate the transcription of the gene into ana messenger ribonucleic acid (mRNA). The mRNA untranslated regions regulate translation into the final protein products. By Thomas Shafee, WikiJournal of Medicine 4 (1), via Wikimedia Commons.

In the case of operons, the coding provided often triggers the formation of enzymes that initiate the formation of an amino acid. Amino acids are essential to the biosynthesis of many substances used by living organisms, including proteins, hormones, and enzymes. However, the promoter associated with an operon can be regulated by other influences that tell the promoter when to activate the genes associated with it and when to prevent them from activating. For example, in the *E. coli* bacteria initially studied by Jacob and Monod, they observed that the operon that controlled the release of the proteins needed to digest lactose was only triggered when lactose was present. When lactose was absent, these proteins were not needed, and the lactose operon, or lac operon, was not triggered. This means that the operon performs an essential duty in regulating the function of genes.

While operons are important, they have not been found in all living organisms. They are present in bacteria, archaea, viruses, mitochondrial, and chloroplast genomes. Few have been identified in multicell organisms, and none have been found in humans.

Scientists are unsure why operons are not found in more advanced organisms. One theory is that they help simpler organisms eliminate extraneous signals or "noise" from their environment and conserve resources for primary functions. Researchers continue to investigate operons and their functions because they might provide insight into how to provoke cells to perform actions they usually do not, which could help fight diseases.

—*Janine Ungvarsky*

Further Reading

"Genetics of Bacteria and Viruses." *Kean University*, www.kean.edu/~jfasick/docs/Cell%20Biology/chapt18_lecture%20%5BCompatibility%20Mode%5D.pdf. Accessed 24 June 2017.

Holmes, Randall K., and Michael G. Jobling. "Genetics." *Medical Microbiology*, edited by Samuel Baron, University of Texas Medical Branch at Galveston, 1996.

"'I Think I've Just Thought Up Something Important': Francois Jacob (1920-2013)." *National Geographic*, 21 Apr. 2013, phenomena.nationalgeographic.com/2013/04/21/i-think-ive-just-thought-up-something-important-francois-jacob-1920-2013/. Accessed 24 June 2017.

"Mystery of Operon Evolution Probed." *Science Daily*, 30 Aug. 2012, www.sciencedaily.com/releases/2012/08/120830173502.htm. Accessed 24 June 2017.

"The Nobel Prize in Physiology or Medicine 1965: François Jacob, André Lwoff, Jacques Monod: Award Ceremony Speech." *Nobelprize.org*, www.nobelprize.org/nobel_prizes/medicine/laureates/1965/press.html. Accessed 24 June 2017.

Osbourn, Anne E., and Ben Field. "Operons." *Cellular and Molecular Life Sciences*, vol. 66, no. 23, Dec. 2009, pp. 3755-75, www.ncbi.nlm.nih.gov/pmc/articles/PMC2776167/. Accessed 24 June 2017.

Price, Morgan N., et al. "The Life-Cycle of Operons." *PLOS*, 28 June 2006, journals.plos.org/plosgenetics/article?id=10.1371/journal.pgen.0020096. Accessed 24 June 2017.

Semeniuk, Ivan. "Jacques Monod: A Scientist Whose Revolution Is Still Unfolding." *Globe and Mail*, 27 Sept. 2013, www.theglobeandmail.com/technology/science/jacques-monod-a-scientist-whose-revolution-is-still-unfolding/article14572219/. Accessed 24 June 2017.

Lateral Gene Transfer

Category: Molecular genetics
Also known as: Horizontal gene transfer
Anatomy or system affected: N/A
Specialties and related fields: Bacteriology, biochemistry, biotechnology, genetics, microbiology
Definition: the movement of genes between unrelated organisms

KEY TERMS

gene transfer: the movement of fragments of genetic information, whole genes, or groups of genes between organisms

genetically modified organism (GMO): an organism produced by using biotechnology to introduce a new gene or genes, or new regulatory sequences for genes, into it to give the organism a new trait, usually to adapt the organism to a new environment, provide resistance to pest species, or enable the production of new products from the organism

transposons: mobile genetic elements that may be responsible for the movement of genetic material between unrelated organisms

SIGNIFICANCE

Lateral gene transfer is the movement of genes between organisms, and it is also sometimes called "horizontal gene transfer." In contrast, vertical gene transfer is the movement of genes between parents and their offspring. Vertical gene transfer is the basis of the study of transmission genetics. In contrast, lateral gene transfer is important in the study of evolutionary genetics, as well as having important implications in the fields of medicine and agriculture.

GENE TRANSFER IN PROKARYOTES

The fact that genes may move between bacteria has been known since the experiments of Frederick Griffith with pneumonia-causing bacteria in the 1920s. Griffith discovered the process of bacterial transformation. The organism acquires genetic material from its environment and expresses the traits contained in its cells' deoxyribonucleic acid (DNA). Bacteria may also acquire foreign genetic material by the process of transduction. In transduction, a bacteriophage picks up a piece of host DNA from one cell. It delivers it to another cell, where it integrates into the genome. This material may then be expressed in the same manner as any other of the host's genes. A third mechanism, conjugation, allows two bacteria to connect through a cytoplasmic bridge to exchange genetic information.

With the development of molecular biology, evidence has accumulated that supports the lateral movement of genes between prokaryotic species. In the case of *Escherichia coli*, one of the most heavily re-

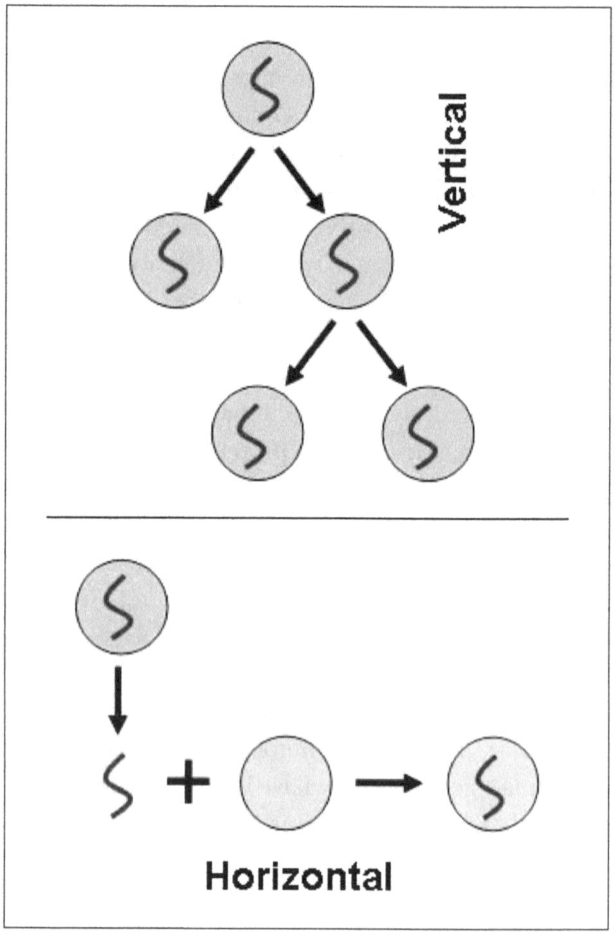

Horizontal and vertical gene transfer. By Gregorius Pilosus (own work), via Wikimedia Commons.

searched bacteria on the planet, there is evidence that as much as 20 percent of the organism's approximately 4,403 genes may have been transferred laterally into the species from other bacteria. This may explain the ability of *E. coli*, and indeed many other prokaryotic species, to adapt to new environments. It may also explain why some members are pathogenic in a given bacterial genus while others are not. Rather than evolving pathogenic traits, bacteria may have acquired genetic sequences from other organisms and then exploited their new abilities.

It is also now possible to screen the genomes of bacteria for similarities in genetic sequences and use this information to reassess previously established phylogenetic relationships. Once again, the majority of this work has been done in prokaryotic organisms, with the primary focus being on the relationship between the domains *Archaea* and *Bacteria*. Several researchers have detected evidence of lateral gene transfer between thermophilic bacteria and *Archaea* prokaryotes. Although the degree of gene transfer between these domains is under contention, there is widespread agreement that the transfer of genes occurred early in their evolutionary history. The fact that there was lateral gene transfer has complicated accurate determinations of divergence time and order.

GENE TRANSFER IN EUKARYOTES

Although not as common as in prokaryotes, there is evidence of gene transfer in eukaryotic organisms. A mechanism by which gene transfer may be possible is the "transposon." Barbara McClintock first proposed the existence of transposons, or mobile genetic elements, in 1948. One of the first examples of a transposon moving laterally between species was discovered in *Drosophila* in the 1950s. A transposon called a "P element" was found to have moved from *D. willistoni* to *D. melanogaster*. Interestingly, these studies are that the movement of the P element was enabled by a parasitic mite common to the two species. This suggests that parasites may play an important role in lateral gene transfer, especially in higher organisms. Furthermore, since the transposon may move parts of the host genome during the transition, it may play a crucial role in gene transfer.

The completion of the Human Genome Project, and the technological advances in genomic processing that it developed, have allowed researchers to compare the human genome with the genomes of other organisms to look for evidence of lateral transfer. It is estimated that between 113 and 223 human genes may not result from vertical gene transfer. Instead, it might have been introduced laterally from bacteria.

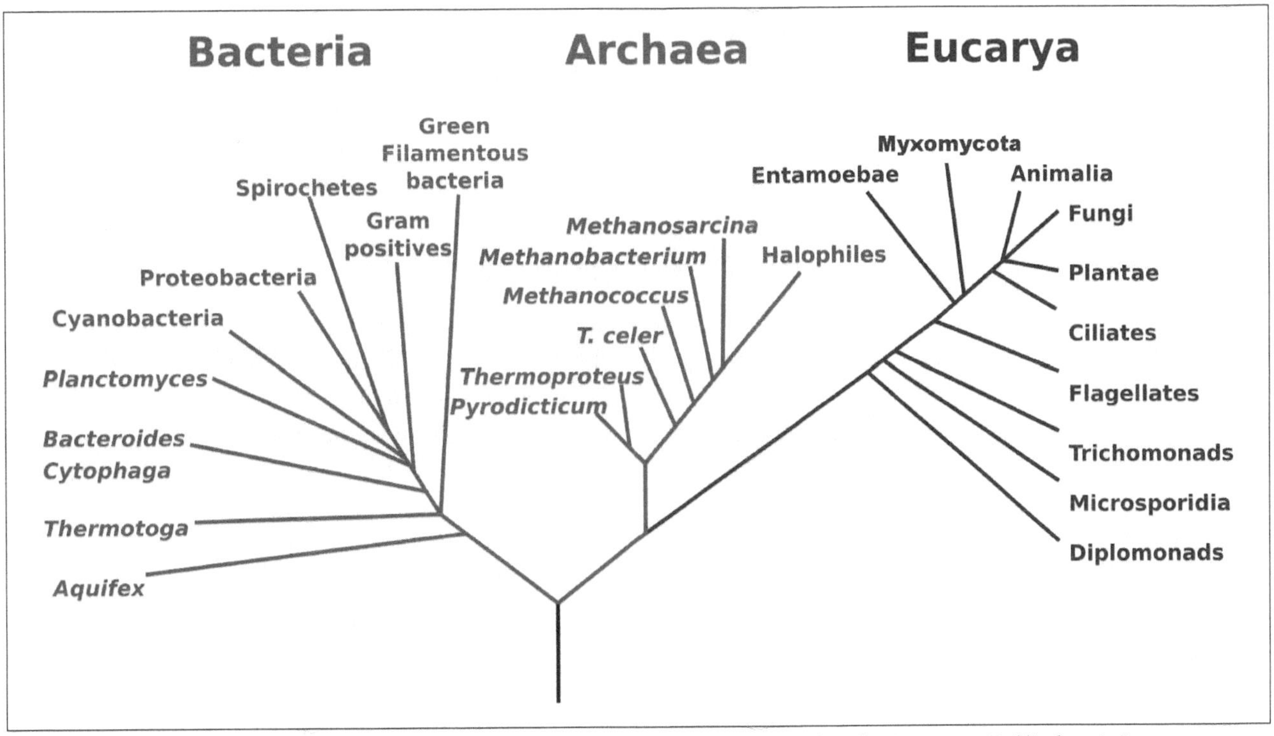

A phylogenetic tree showing the separation of bacteria, archaea, and eukarytes. MPF, via Wikimedia Commons. [Public domain.]

IMPLICATIONS

While the concept of lateral gene transfer may initially seem to be a concern only for evolutionary geneticists in their construction of phylogenetic trees, in reality, the effects of lateral gene transfer pose concerns concerning both medicine and agriculture, specifically in the case of transgenic plants.

Currently, the biggest concern regarding lateral gene transfer is the unintentional movement of genes from genetically modified organisms (GMOs) into other plant species. Such transfer may occur by parasites, as has occurred with *Drosophila* in animals, or by dispersing pollen grains out of the treated field. This second possibility holds particular significance for corn growers, whose crop is wind-pollinated. Genetically modified corn containing the microbial insecticide "Bt," may cross-pollinate with unintentional species, reducing the effectiveness of pest management strategies. In another case, the movement of herbicide-resistant genes from a GMO to a weed species may result in the formation of a superweed.

On the beneficial side, lateral gene transfer may also play a part in medicine as part of gene therapy. Several researchers are examining the possibility of using viruses, transposons, and other systems to move genes, or parts of genes, into target cells in the human body, where they may be therapeutic in treating diseases and disorders.

—Michael Windelspecht, PhD

Further Reading

Bushman, Frederick. *Lateral Gene Transfer: Mechanisms and Consequences*. Cold Spring Harbor Laboratory Press, 2001.

Gogarten, Maria B., Johann Peter Gogarten, and Lorraine C. Olendzenski, editors. *Horizontal Gene Transfer: Genomes in Flux*. Springer, 2009.

Hensel, Michael, and Herbert Schmidt, editors. *Horizontal Gene Transfer in the Evolution of Pathogenesis*. Cambridge UP, 2008.

Rissler, Jane, and Margaret Mellon. *The Ecological Risks of Engineered Crops*. MIT Press, 1996.

Syvanen, Michael, and Clarence Kado. *Horizontal Gene Transfer*. 2nd ed., Academic Press, 2002.

DNA: Recombinant technology

Category: Molecular genetics
Anatomy or system affected: None
Specialties and related fields: Biochemistry, biology, biotechnology, evolutionary biology, genetics
Definition: the production of DNA molecules formed by laboratory methods of genetic recombination that bring together genetic material from multiple sources, creating sequences that would not otherwise be found in the genome

KEY TERMS

in vitro: a term used to indicate reactions and processes that are carried out "in glass," or in the laboratory environment rather than in nature

recombinant DNA: deoxyribonucleic acid (DNA) molecules that have been engineered to include genes from the DNA of different species

restriction enzyme: a protein or protein complex that coordinates to a specific gene or region of the DNA molecule and excises that segment from the molecule

Recombinant deoxyribonucleic acid (DNA) technology uses science's understanding of the molecular structure of DNA, the nucleic acid that encodes genetic information, to alter DNA to manipulate genetic traits. Such technology has immense implications for agriculture, horticulture, and the generation of medicinal compounds from plants.

A MODERN TWIST ON GENETICS

Recombinant DNA technology has been essential for understanding DNA sequences. Because of their large, complex genomes, studying one gene in eukaryotes was difficult. However, recombinant DNA technology has allowed the isolation and amplification of specific DNA fragments facilitating the molecular analysis of genes. In addition, the tools of recombinant DNA technology have been used to create genetically modified plants. Such modifications include introducing resistance to insects, herbicides, viruses, and bacterial and fungal diseases into plants. Plants have also been made to produce antibodies so that plants can serve as edible vaccines.

DNA STRUCTURE

Organisms contain two kinds of nucleic acids: ribonucleic acid, or RNA, and deoxyribonucleic acid, or DNA. DNA is made of a double chain or helix. The structure of one chain, or strand, is a backbone made up of repetitions of the same basic unit. A five-carbon sugar molecule called "2'-deoxyribose" is attached to a phosphate residue. RNA contains ribose sugars instead. Also attached to the sugar part of the backbone are other molecules called "bases." The four bases are adenine (A), guanine (G), cytosine (C), and thymine (T). DNA molecules are double strands held together because each base in one strand is paired to (hydrogen bonds with) a base in the other strand. Adenine always pairs with the base thymine, and guanine always pairs with cytosine. A and T are called "complementary bases." Likewise, G and C are complementary bases. In RNA, the base uracil replaces thymine.

DNA is shaped much like a helical ladder, with the sugar and phosphate backbones being the sides of the ladder and the base pairs that hold the two strands together being the rungs of the ladder. DNA is often

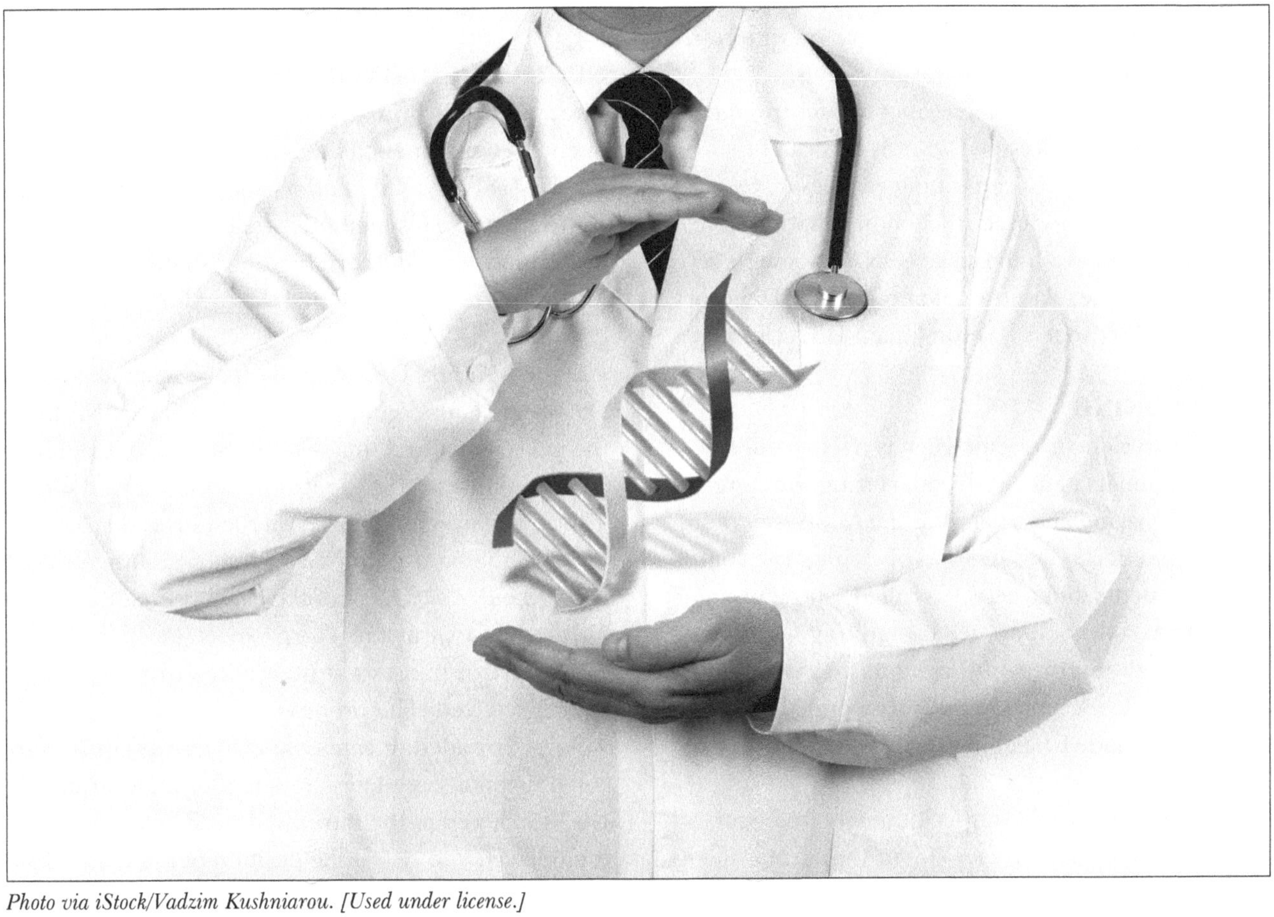

Photo via iStock/Vadzim Kushniarou. [Used under license.]

represented as a string of letters, with each letter representing a base. The order of As, Ts, Gs, and Cs (the ladder's rungs) along a DNA double helix is the sequence of that DNA, which contains the genetic information.

RESTRICTION ENZYMES

In recombinant DNA technology, scientists can use restriction enzymes to make cuts at specific sequences. Some restriction enzymes cut straight across the two strands of DNA in the double helix, creating blunt ends. Other restriction enzymes cut the two strands in a staggered pattern, leaving short, specific single strands at the cut sites. These single-stranded regions, called "sticky ends" or "cohesive ends," can base-pair (hydrogen-bond) with complementary base sequences from other, similarly cut DNAs. These sticky ends allow the joining of DNA from any source cut with restriction enzymes that create the same ends.

Cutting with restriction enzymes creates fragments of DNA with sequence-specific ends that can be spliced into small, self-replicating vehicles or vector molecules and introduced into a host cell where the vector molecules with the added DNA fragments replicate to produce a large amount of specific DNA for analyses. This process is recombinant DNA cloning.

DNA CLONING

One way to clone a specific gene is to clone all the DNA fragments generated from cutting with a restriction enzyme and screen for the clone containing the desired gene. This method of cloning random DNA segments into a vector is called "shotgunning." The entire collection of such cloned fragments, which together represent the organism's entire genome, is called a "gene library." Genomic DNA libraries are made by cloning the total genomic DNA of an organism.

Another way to clone a specific gene is to begin with messenger ribonucleic acid (mRNA) from the organism. (Messenger RNA is a molecule that functions to create complementary copies of DNA strands. At a ribosome, the messenger RNA then determines the order of amino acids joined to make a protein.) Using reverse transcriptase (an enzyme that uses RNA as a template for DNA synthesis), scientists can make a DNA copy of the mRNA. The complementary strand is also synthesized to create a double-stranded DNA called "cDNA" (complementary DNA) complementary to the mRNA. These cDNAs are then cloned to create a complementary DNA library. Individually cloned cDNAs can trap the corresponding mRNA on a nitrocellulose filter. At this point, the mRNA can be used in a cell-free protein synthesis system to allow identification of the protein encoded by that cDNA clone. Alternatively, the cDNA can be used to find sequences complementary to it in a genomic library to obtain a clone of the specific gene.

NUCLEIC ACID HYBRIDIZATION

The ability to hybridize nucleic acids to find sequences complementary to a particular DNA is another essential tool that offers another way to identify cloned genes. This method is called "nucleic acid hybridization" or "Southern blotting" (named after E. M. Southern, who developed the method). In this procedure, DNA is cut with restriction endonucleases, and the resulting DNA fragments are separated by size using agarose gel electrophoresis. The DNA in the gel is denatured (made single-stranded) by high pH and transferred to a nitrocellulose or nylon filter. The DNAs are immobilized on the nitrocellulose or nylon in the same pattern as the gel (a Southern blot). A specific DNA or RNA probe is hybridized to nitrocellulose or nylon. The probe is "labeled" with a radioactive or fluorescent (nonradioactive) tag so it can be detected. The probe is denatured by heat, so it is single-stranded. It can anneal (hybridize) with its complementary sequence among the single-stranded DNAs tethered to the nitrocellulose. The probe is then detected to reveal the position of the DNAs that hybridized with the probe.

POLYMERASE CHAIN REACTIONS

Another molecular biology tool is to use a polymerase chain reaction (PCR) to amplify specific segments of DNA *in vitro* (in the test tube). PCR requires a pair of sequences, called "primers," about twenty base pairs long that are complementary to the ends of the region of DNA to be amplified. High temperature is used to denature the double-stranded DNA. The primers anneal (base-pair) to their complementary sequences at a lower temperature, and a thermal-stable DNA polymerase copies the single-stranded templates. After replicating the segment between the two primers (one cycle), the newly synthesized double-stranded DNA molecules are denatured by high temperature, the temperature is lowered, primers anneal, and the second cycle of replication occurs. The number of DNA molecules produced doubles with each cycle of replication. As a result, a million copies of a single DNA molecule can be produced in only a few hours if the appropriate sequences for the two primers are known. PCR is a very sensitive method: Even a single DNA molecule can be amplified. PCR is much faster than recombinant DNA cloning and can produce a large amount of a specific piece of DNA.

Developing ways to determine DNA sequences (the sequences of adenine, thymine, guanine, and cytosine base pairs on the DNA "rungs") has led to the identification of the complete DNA sequences of the genomes of several organisms, including the model plant *Arabidopsis thaliana* (mouse-ear cress) as well as the much more complex human genome.

—*Susan J. Karcher*

Further Reading

Chaudhuri, Keya. *Recombinant DNA Technology*. The Energy and Resources Institute, 2015.

Hill, Walter E. *Genetic Engineering: A Primer*. Harwood Academic Publishers, 2000.

Kreuzer, Helen, and Adrianne Massey. *Recombinant DNA and Biotechnology: A Guide for Students*. Blackwell Science, 2000.

Old, R. W., and S. B. Primrose. *Principles of Gene Manipulation: An Introduction to Genetic Engineering*. Blackwell Scientific, 1994.

Reece, Jane B., and Neil A. Campbell. *Biology*. 6th ed., Benjamin Cummings, 2002.

Tiwari, Shivangi, and Manisha Sharma. *Recombinant DNA Technology in the Synthesis of Human Insulin*. AP Lambert Academic Publishing, 2018.

Watson, James, Michael Gilman, Jan Witkowski, and Mark Zoller. *Recombinant DNA*. 2nd ed., W. H. Freeman, 1992.

Plasmids

Category: Molecular genetics

Anatomy or system affected: None

Specialties and related fields: Bacteriology, biochemistry, biotechnology, genetics, microbiology, molecular biology

Definition: DNA (deoxyribonucleic acid) molecules that exist separately from the chromosome

KEY TERMS

commensalism: a relationship in which two organisms rely on each other for survival

conjugation: the process by which one bacterium transfers genetic material to another through direct contact

gene: a region of DNA containing instructions for the manufacture of a protein

transposon: a piece of DNA that can copy itself from one location to another

SIGNIFICANCE

Plasmids are deoxyribonucleic acid (DNA) molecules that exist separately from the chromosome. Plasmids exist in a commensal relationship with their host and may provide the host with new abilities. They are used in genetic research as vehicles for carrying genes. In the wild, they promote the exchange of genes and contribute to the problem of antibiotic resistance.

PLASMID STRUCTURE

The structure of plasmids is usually circular, although linear forms do exist. Their size ranges from a few thousand base pairs to hundreds of thousands of base pairs. They are found primarily in bacteria but have also been found in fungi, plants, and even humans.

In its commensal relationship with its host, the plasmid can be thought of as a molecular parasite whose primary function is to maintain itself within its host and to spread itself as widely as possible to other hosts. The majority of genes present on a plasmid will be dedicated to this function. Researchers have discovered that despite the great diversity of plasmids, most of them have similar genes dedicated to this function. This relative simplicity of plasmids makes them ideal models of gene function and useful tools for molecular biology. Genes of interest can be placed on a plasmid, which are easily moved in and out of cells. Using plasmids isolated from the wild, molecular biologists have designed many varieties of artificial plasmids, which have greatly facilitated research in molecular biology.

PLASMID REPLICATION

A plasmid must be able to copy itself or replicate to survive and propagate. The genes that direct this process are known as the replication genes. These genes do not carry out all the replication functions. Instead, they co-opt the host's replication machinery to replicate the plasmid. Replication allows the plasmid to propagate by creating copies of itself that can be passed to each daughter cell when the host divides. In this manner, the plasmid propagates along with the host.

A second function of the replication genes is to control the copy number of the plasmid. The number of plasmid copies that exist inside a host can vary considerably. Plasmids can exist at a very low copy number (one or two copies per cell) or a higher copy number, with dozens of copies per cell. Adjusting the copy number is an important consideration for a plasmid. Plasmid replication is an expensive process that consumes the energy and resources of the host cell. A plasmid with a high copy number can place a significant energy drain on its host cell. In environments where the nutrient supply is low, a plasmid-bearing cell may not be able to compete successfully with other, non-plasmid-containing cells. Wild plasmids often exist at a low copy number or create a high copy number for only a brief time.

PLASMID PARTITIONING

Because the presence of a plasmid is expensive in terms of energy, a cell harboring a plasmid will grow more slowly than a similar cell with no plasmid. This can cause a problem for a plasmid if it fails to partition properly during its host's division. If the plasmid does not partition properly, then one of the host's daughter cells will not contain a plasmid. Since this cell does not have to spend energy replicating a plasmid, it will gain the ability to grow faster, as will all of its offspring. In such a situation, the population of non-plasmid-containing cells could outgrow the population of plasmid-containing cells and use up all the nutrients in the environment. Plasmids have evolved strategies to prevent improper partitioning to avoid this problem. One strategy is for the plasmid to contain partitioning genes. Partitioning genes encode proteins that actively partition plasmids into each daughter cell during the host cell's division. Active partitioning greatly reduces the errors in partitioning that might occur if partitioning were left to chance.

The plasmid addition system is a second strategy that plasmids use to prevent partitioning errors. In this strategy, genes on the plasmid direct the production of both a toxin and an antidote. The antidote protein is very unstable and degrades quickly, but the toxin is quite stable. As long as the plasmid is present, the cell's cytoplasm will be full of toxins and antidotes. Should a daughter cell fail to receive a plasmid during division, the residual antidote and toxin present in the cytoplasm from the mother cell will begin to degrade since there is no longer a plasmid present to direct the synthesis of either toxin or antidote. Since the antidote is very unstable, it will degrade first, leaving only toxin, which will kill the cell.

PLASMID TRANSFER BETWEEN CELLS

Plasmid propagation occurs by spreading plasmids from parent cells to their offspring (referred to as vertical transfer). Still, propagation can also occur between two different cells (horizontal transfer). Many plasmids can transfer themselves from one host to another through conjugation. Conjugal plasmids contain a collection of genes that direct the host cell that contains them to attach to other cells and transfer a copy of the plasmid. In this manner, the plasmid can spread itself to other hosts and is not limited to spreading itself only to the descendants of the original host cell.

One of the first plasmids to be identified was discovered because of its conjugation ability. This plasmid, known as the "F plasmid," or "F factor," is a plasmid found in the bacterium *Escherichia coli*. Cells harboring the F plasmid are designated "F+ cells." They can transfer their plasmid to other *E. coli* cells that do not contain the F plasmid (called "F- cells").

Conjugal plasmids can be specific and transfer only between closely related members of the same species (such as the F plasmid). They can also be very promiscuous and allow transfer between unrelated species. An extreme example of cross-species transfer is the Ti plasmid of the bacterial species *Agrobacterium tumefaciens*. The Ti plasmid can transfer part of itself from *A. tumefaciens* into the cells of dicotyledonous plants. Plant cells that receive parts of the Ti plasmid are induced to grow and form a tumor-like structure, called a "gall," that provides a hospitable environment for *A. tumefaciens*.

HOST BENEFITS FROM PLASMIDS

In most commensal relationships, there is an exchange of benefits between the two partners. The same is true for plasmids and their hosts. In many cases, plasmids provide their host cells with a collection of genes that enhance the ability of the host cell to survive. Enhancements include metabolizing a wider range of materials for food and the ability to survive in hostile environments. One particular hostile environment in which plasmids can provide the ability to survive in the human body. Several pathogenic microorganisms gain their ability to inhabit the human body and thus cause disease from genes contained on plasmids. An example of this is *Bacillus anthracis*, the agent that causes anthrax. Many of the genes that allow this organism to cause disease are contained on one of two plasmids, called "pXO1" and "pXO2." *Yersinia pestis*, the causative agent of bubonic plague, also gains its disease-causing ability from plasmids.

R FACTORS

Another example of plasmids conferring on their hosts the ability to survive in a hostile environment is antibiotic resistance. Plasmids known as R factors contain genes that make their bacterial hosts resistant to antibiotics. These R factors are usually conjugal plasmids to move easily from cell to cell. Because the antibiotic resistance genes they carry are usually parts of transposons, they can readily copy themselves from one piece of DNA to another. Two different R factors that happened to be together in one cell could exchange copies of each other's antibiotic resistance genes. Many R factors exist that contain multiple antibiotic resistance genes. Such plasmids can form "multidrug-resistant" (MDR) strains of pathogenic bacteria, which are difficult to treat. There is much evidence to suggest that the widespread use of antibiotics has contributed to the development of MDR pathogens, which are emerging as an important health concern.

ROLE OF PLASMIDS IN EVOLUTION

Through conjugation, plasmids can transfer genetic information from one species of the bacterial cell to another. During its stay in a particular host, a plasmid may acquire some of the chromosomal genes of the host, which it then carries to a new host by conjugation. These genes can then be transferred from the plasmid to the chromosome of the new host. If the new host and the old host are different species, this gene transfer can introduce new genes and thus new traits into a cell. Bacteria, being asexual, produce daughter cells genetically identical to their parent. The existence of conjugal plasmids, which allow for the transfer of genes between bacterial species, may represent an important mechanism by which bacteria generate diversity and create new species.

GENETIC ENGINEERING OF PLASMIDS

The identification of restriction endonuclease sites within plasmids allowed scientists to manipulate the organization and makeup of these molecules. Researchers were now able to insert genes of interest into these restriction sites within the plasmids and have the recombined plasmid vector taken up by bacteria through transformation. Transcription of the inserted gene by the host bacterium is dependent on an upstream promoter that is active in that bacterial strain. The use of genetic engineering and the creation of artificial plasmid vectors have revolutionized basic research and have led to the creation of modern industrial microbiology. Because bacteria containing the vector can express proteins encoded by the inserted gene(s) at a high level, mass quantities of desired proteins can be produced on an industrial scale. Vectors have been used to express medically important proteins such as insulin, human growth hormone, and human factor IX, a blood-clotting factor. Vectors have also been used to express proteins within eukaryotic cells. Additional DNA elements are necessary for the optimal production of proteins in these cells. Viral or mammalian promoters recognized by host ribonucleic acid (RNA) polymerases and enhancer sequences allow proteins to be expressed from the vector. Poly-adenylation sequences are added downstream of the inserted gene for proper expression of messenger RNA (mRNA).

The ability to express proteins within mammalian cells has enabled the development of DNA vaccines in which antigenic proteins are expressed from plasmids. These plasmids are taken up by cells following introduction into the animal or human subject by injection or alternate means of inoculation. Expression of these proteins in the cell allows for an immunological response by the vaccine. Expression of cytokine or other immunomodulatory molecules from the same plasmid has been explored to enhance the immune response to the coexpressed antigenic protein using DNA vaccines. Plasmid DNA itself can sometimes have immunostimulatory effects due to CpG dinucleotides. Clinical trials using DNA vaccines targeting cancer have demonstrated the immunogenicity of these vaccines in humans. Veterinary applications of DNA vaccines have been approved for use. Plasmids have also been used as delivery vehicles for the expression of double-stranded RNAs (dsRNAs) to suppress specific mRNAs by RNA interference (RNAi). Because of the abbreviated length (about twenty-one nucleotides) and the need for a distinction termination point of these dsRNAs, RNA polymerase III promoters and the accompanying termination signals have often been included on the plasmids to express these transcripts within cells. Delivery of the RNAi expression plasmids to the desired location within the body remains a challenge for developing this technology.

—Douglas H. Brown, PhD and Daniel E. McCallus, PhD

Further Reading

Bower, D. M., and K. L. J. Prather. "Engineering of Bacterial Strains and Vectors for the Production of

Plasmid DNA." *Applied Microbiology and Biotechnology*, vol. 82, no. 5, 2009, pp. 805-13.
Levy, Stuart B. "The Challenge of Antibiotic Resistance." *Scientific American*, vol. 278, 1998, pp. 46-53.
Summers, David K. *The Biology of Plasmids*. Blackwell, 1996.
Thomas, Christopher M. "Paradigms of Plasmid Organization." *Molecular Microbiology*, vol. 37, no. 3, 2000, pp. 485-91.
Van Gaal, E. V. B., W. E. Hennink, D. J. A. Crommelin, and E. Mastrobattista. "Plasmid Engineering for Controlled and Sustained Gene Expression for Nonviral Gene Therapy." *Pharmaceutical Research*, vol. 23, no. 6, 2006, pp. 1053-74.

ANTIBIOTIC RESISTANCE

Category: Treatment
Also known as: Antimicrobial resistance, bacterial resistance, drug resistance
Anatomy or system affected: Cells
Specialties and related fields: Bacteriology, biochemistry, biotechnology, microbiology, pharmacology, preventive medicine, public health
Definition: the ability of bacteria and other microorganisms to resist the effects of an antibiotic they were once sensitive to

KEY TERMS

aminoglycosides: a group of gram-negative antibacterial medications that inhibit protein synthesis and contain as a portion of the molecule an amino-modified glycoside
antibiotics: small molecules that inhibit the growth or directly kill microorganisms
azoles: antifungal agents that are five-membered heterocyclic compounds containing a nitrogen atom and at least one other noncarbon atom as part of the ring
beta-lactam antibiotics: drugs that inhibit bacterial cell wall biosynthesis and contain a beta-lactam ring in their molecular structure; this group of antibiotics includes penicillin derivatives, cephalosporins and cephamycins, monobactams, carbapenems, and carbacephems
beta-lactamase: enzymes produced by bacteria that provide multiresistance to ß-lactam antibiotics such as penicillins, cephalosporins, cephamycins, monobactams, and carbapenems
fluoroquinolones: broad-spectrum antibiotics that share a bicyclic core structure related to the substance 4-quinolone
macrolides: a class of antibiotics that inhibit bacterial protein synthesis that consists of a large macrocyclic lactone ring
porins: protein pore in bacterial membranes that indiscriminately allow the passage of small molecules into bacterial cells

DEVELOPMENT HISTORY

The first antibiotic, the beta-lactam antibiotic penicillin, was discovered in the London laboratory of Alexander Fleming in 1928. It was not used in a human patient until March 14, 1942. The antibiotic sulfonamide was discovered in 1932 by Gerhard Domagk. He made the first sulphonamide—prontosil rubrum in 1935 that was used in human patients that same year. The antibiotic streptomycin was discovered in 1943 by A. I. Schatz, a graduate student in the Rutgers University lab of antibiotic pioneer S. A. Waksman. Streptomycin was isolated from the soil actinobacterium *Streptomyces griseus*. It successfully controlled tuberculosis (*Mycobacterium tuberculosis*) and plague (*Yersinia pestis*).

Soon after the introduction of these antibiotics, pathogenic bacteria adapted quickly to block the drugs' effects. Resistance to beta-lactam antibiotics was noted in 1944 and accounted for more than three-quarters of the hospital (nosocomial) acquired infections in 1950. The long-term risk of resistance was acknowledged as early as 1956.

Excitement about the treatment potential of these early antibiotics, introduced in the 1940s and 1950s,

COMMON BACTERIA RESISTANT TO ANTIBIOTICS, WITH ASSOCIATED INFECTIONS

Organisms	Disease Caused
	Urgent Threats
Carbapenem-resistant *Acinetobacter*	Nosocomial infections
Candida auris (*C. auris*)	Bloodstream infections, wound infections, and ear infections
Clostridioides difficile (*C. difficile*)	Enterocolitis
Carbapenem-resistant Enterobacteriaceae (CRE)	Nosocomial pneumonia, bloodstream infections, urinary tract infections, wound infections, and meningitis
Drug-resistant *Neisseria gonorrhoeae* (*N. gonorrhoeae*)	Sexually transmitted infections
	Serious Threats
Drug-resistant *Campylobacter*	Gastroenteritis
Drug-resistant *Candida*	Vaginal candidiasis, invasive candidiasis, infections of the mouth, throat, and esophagus.
Extended-spectrum beta-lactamase (ESBL)-producing Enterobacteriaceae	Urinary tract infections (UTIs), pneumonia, blood infections, wound infections
Vancomycin-resistant Enterococci (VRE)	Sepsis, endocarditis, and meningitis
Multidrug-resistant *Pseudomonas aeruginosa* (*P. aeruginosa*)	Pneumonia, bloodstream infections, urinary tract infections, and surgical site infections
Drug-resistant nontyphoidal *Salmonella*	Diarrhea (sometimes bloody), fever, and abdominal cramps, sepsis
Drug-resistant *Salmonella* serotype Typhi	Typhoid fever
Drug-resistant *Shigella*	Diarrhea, fever, abdominal pain
Methicillin-resistant *Staphylococcus aureus* (MRSA)	Skin infections, pneumonia, endocarditis, sepsis

contributed to the drugs' rapid, widespread use. Antibiotic use became more promoted and more commonplace in the 1950s and 1960s. It was often prescribed empirically and inappropriately, without regard to long-term resistance effects such as increased virulence and multiple resistance mechanisms.

By the 1960s, methicillin-resistant *Staphylococcus aureus* (MRSA) was identified. MRSA rates continued to increase in the United States into the twenty-first century from just greater than 2 percent in 1975 to nearly 60 percent in 2003. As reported by the Centers for Disease Control and Prevention (CDC) in 2013, however, the MRSA rates in the United States fell between 2005 and 2011 by 31 percent. As MRSA spread through hospital populations and even into animal and community groups, the last resort, the glycopeptide antibiotic vancomycin, was used more frequently. In September 2007, vancomycin-resistant *Staphylococcus aureus* and vancomycin-intermediate *S. aureus* were identified in the United States. Bacteria have developed physical methods and genetic mutations to prevent, reduce, or inactivate antibiotic activity against them.

MECHANISMS OF RESISTANCE

Microbes have four identified mechanisms of antibiotic resistance: (1) antibiotic inactivation or modification, (2) alteration of the antibiotic target, (3) bypassing metabolic pathways, and (4) preventing antibiotic accumulation. More than one mechanism can be used to develop widespread resistance, and different mechanisms are more effective for different antibiotic classes. Once the bacteria develop resistance to an antibiotic, the benefit is passed on to others in the same drug class through genetic mutations in the infectious deoxyribonucleic acid (DNA). Thus, the mutations and resistance spread among people as the bacterial disease is spread. This concept of antibiotic-resistant bacteria in people who have not been directly exposed to the antibiotic supports the urgency of counteracting resistance throughout the human population.

Bacterial resistance develops because of changes to enzymes, target sites, or cell-wall components. Examples of enzyme-mediated resistance are the development of beta-lactamases. These enzymes chemically inactivate beta-lactam antibiotics, as shown below.

There are four different classes of ß-lactamase, class A, B, C, and D. Each distinct class of ß-lactamase has an entirely different three-dimensional structure and amino acid sequence. Class A ß-lactamases include TEM-1 ß-lactamases that confer ampicillin resistance. SHV ß-lactamases are very similar to TEM-1 ß-lactamases, and mutations in TEM-1 or SHV ß-lactamases can produce extended-spectrum ß-lactamases (ESBLs) that can hydrolyze penicillins, cephalosporins, and carbapenems. Therefore, anyone infected with an organism that harbors an ESBL enzyme has limited treatment options. CTX-M ß-lactamases hydrolyze several cephalosporin antibiotics.

To treat infections caused by ß-lactamase-containing microorganisms, drug makers designed ß-lactamase inhibitors. These small molecules bind to ß-lactamases and prevent them from operating. By combining antibiotics with ß-lactamase inhibitors, the spectrum of the antibiotic is significantly expanded. Examples of these antibiotic/ß-lactamase inhibitor combinations include Augmentin (amoxicillin/clavulanic acid), Unasys (ampicillin/sulbactam), Zosyn (piperacillin/tazobactam), and Timentin (ticarcillin/clavulanic acid). Unfortunately, bacteria have countered with inhibitor-resistant ß-lactamases. Examples of such enzymes include Amp-C ß-lactamases (class C) and several class B ß-lactamase (metallo-ß-lactamases) such as IMP-type and VIM ß-lactamases, OXA ß-lactamases (class D), and KPN ß-lactamases (class A).

Chemical modification of aminoglycoside antibiotics confers resistance to these drugs. An enzyme called "aminoglycoside phosphotransferase" adds a phosphate group to aminoglycosides, preventing them from binding to ribosomes and inhibiting protein synthesis.

Reduced bacterial cell-wall permeability, particularly with gram-negative bacteria, is also a common resistance method. Outer membrane-embedded proteins called "porins" allow small molecules to enter bacterial cells. Reduced porin expression or altered porins that fail to allow antibiotics into cells make those bacterial strains resistant to antibiotics. Drug efflux, which occurs when bacterial transport pumps efflux antibiotics from the bacterial cell, is the most common resistance mechanism to tetracycline antibiotics.

Changes to the target site on the bacteria, in which antibiotics cannot recognize the binding site and fail to affect bacteria, are less common with beta-lactams and more common with aminoglycosides, glycopeptides, fluoroquinolones, and macrolides. In some cases, bacteria may otherwise block the target site to prevent antibiotic binding; this occurs against tetracycline antibiotics. Bacteria may increase the number of binding sites on the wall, too, so that antibiotics cannot achieve sufficient intracellular concentrations for activity. This resistance mechanism confers resistance to sulfonamide antibiotics and with glycopeptides such as vancomycin.

Cellular adaptations that help bacteria avoid any interaction with antibiotics and bind antibiotics elsewhere on the bacteria to prevent action on the bacterial cell target also incur drug resistance; the latter method is specific to glycopeptides like vancomycin.

METHODS TO REDUCE RESISTANCE

Early attempts to decrease resistance started in the 1980s when hospitals began instituting guidelines to cycle, or rotate, antibiotic use for certain diseases. Cyclic administration of antibiotics restricts prescribing the most used antibiotics and favors an alternative antibiotic treatment instead.

Research in the late twentieth and early twenty-first centuries has identified no objective evidence of success at minimizing resistance with cycling. Many factors about resistance and efficacy are still unknown. However, restricted antibiotic use in the Netherlands and Scandinavia resulted in decreased hospital occurrences of MRSA, which supports closely monitored antibiotic prescribing to reduce resistance buildup.

A clear correlation between the occurrence of resistance and empiric use of antibiotics, reported in the May 2010 issue of the *British Medical Journal*, has supported the longstanding belief that nonempirical treatment (treatment that is identified based on factual data that shows efficacy, such as a sensitivity analysis) reduces the likelihood of increasing antibiotic resistance because of ineffective antibiotic use. Although occurrence reductions have not been proven, the rate of resistance development is likely to be lower when antibiotics are used correctly.

The prohibition of the use of human antibiotics Antibiotics livestock in animals is debated among health experts. The use of antibiotics in animal husbandry to prevent infections in livestock, such as cattle, pigs, and chickens, can increase the rate of resistant bacteria development by introducing primary antibiotics before they even infect humans. Although animal use of antibiotics began in the 1950s to improve the health and quantity of livestock for food use, the practice is now banned in the European Union and countries worldwide. However, the United States has not banned antibiotic use in animals; the US Department of Agriculture (USDA) has emphasized the importance of reducing antibiotics in meat consumed by humans to reduce drug resistance for the treatment of human infections. However, according to the USDA, over 32 million pounds of antibiotics sold in the United States in 2012 were used for food animals, which was a 16 percent increase since 2009. The Food and Drug Administration (FDA) released two policy documents in 2013 that addressed antibiotics in food animal production. Despite criticisms, many are optimistic that the issue is being addressed.

IMPACT

Drastic changes to bacteria can occur relatively quickly (often within one decade) to reduce antibiotic efficacy, and much needs to be done to identify the means to long-term resistance. To preserve the effectiveness of antibiotics, doctors must prescribe them with more care and attention. Sensitivity analyses, which identify the antibiotics that retain activity against specific microbes in a particular patient, are increasingly used in hospital settings to determine initial antibiotic therapy and monitor continued antibiotic use and infection response.

In 1996, antimicrobial stewardship identified the connection of bacterial resistance with widespread antibiotic use even in persons who had not received treatment with specific antibiotics. Fewer antibiotics are being developed in the twenty-first century because of the high cost of development. These high costs, fewer successful treatment options, and increased resistance mutations (including multidrug or multimechanism patterns) have increased the urgency to improve antibiotic use and find nontraditional methods to suppress bacterial infections.

—Nicole M. Van Hoey, PharmD and Michael A. Buratovich, PhD

Further Reading

Arias, Cesar A., and Barbara E. Murray. "Antibiotic-Resistant Bugs in the Twenty-first Century: A Clinical Super-Challenge." *New England Journal of Medicine*, vol. 360, no. 5, 2009, pp. 439-43.

Brown, Nicholas, M., and Boyan B. Bonev. *Bacterial Resistance to Antibiotics: From Molecules to Man.* Wiley-Blackwell, 2019.

Centers for Disease Control and Prevention. *Antibiotic Resistance Threats in the United States*. US Department of Health and Human Services, CDC, 2019.
"FDA Annual Summary Report on Antimicrobials Sold or Distributed in 2012 for Use in Food-Producing Animals." *FDA*. US Department of Health and Human Services, 2 Oct. 2014. Accessed 30 Nov. 2015.
Forsbeg, Kevin J., et al. "The Shared Antibiotic Resistome of Soil Bacteria and Human Pathogens." *Science*, vol. 337, no. 6098, 2012, pp. 1107-11.
Hauser, Alan R. *Antibiotics for Clinicians*. 3rd ed., Lippincott Williams & Wilkins, 2018.
"New FDA Policies on Antibiotic Use in Food Animal Production." *PEW Trusts*. Pew Charitable Trusts, 10 Mar. 2015. Accessed 30 Nov. 2015.
Polk, Ronald E., and Neil O. Fishman. "Antimicrobial Stewardship." *Mandell, Douglas, and Bennett's Principles and Practice of Infectious Diseases*, edited by Gerald L. Mandell, John F. Bennett, and Raphael Dolin, 7th ed., Churchill Livingstone/Elsevier, 2010.
"Public Gets Early Snapshot of MRSA and *C. difficile* Infections in Individual Hospitals." *CDC*. US Department of Health and Human Services, 12 Dec. 2013. Accessed 30 Nov. 3015.
Schmitz, Franz-Josef, and Ad C. Fluit. "Mechanisms of Antibacterial Resistance." *Cohen and Powderly Infectious Diseases*, edited by Jonathan Cohen, Steven M. Opal, and William G. Powderly, 3rd ed., Mosby/Elsevier, 2010.

MOLECULAR MICROBIOLOGY

Category: Biochemistry
Anatomy or system affected: None
Specialties and related fields: Bacteriology, biochemistry, biotechnology, cell biology, molecular genetics
Definition: the branch of microbiology devoted to the study of the molecular basis of the physiological processes that occur in microorganisms

KEY TERMS

gel electrophoresis: a method for separation and analysis of macromolecules and their fragments, based on their size and charge

next-generation sequencing (NGS): high-throughput approaches deoxyribonucleic acid (DNA) sequencing using the concept of massively parallel processing

polymerase chain reaction (PCR): a method widely used to rapidly make millions to billions of copies of a specific DNA sample, allowing scientists to take a very small sample of DNA and amplify it to a large enough amount to study in detail

pulsed-field gel electrophoresis (PFGE): a laboratory technique used for the separation of large DNA molecules by applying to a gel matrix an electric field that periodically changes direction

INTRODUCTION

Molecular Microbiology is the branch of microbiology devoted to studying the molecular basis of the physiological processes that occur in microorganisms. This field explores the nature of biological phenomena at the molecular level by studying deoxyribonucleic acid (DNA) and ribonucleic acid (RNA), proteins, and other macromolecules involved in genetic information and cell function. Molecular microbiology is characterized by using advanced tools and techniques of separation, manipulation, imaging, and analysis. When applied to the clinical microbiology laboratory, molecular methods such as polymerase chain reaction (PCR) or analysis of proteins and macromolecules can detect and identify microorganisms causing infections.

OVERVIEW

One of the leading areas of development in clinical diagnostics is molecular microbiology. The development of innovative molecular test methods and automated devices for such testing has led to a paradigm shift in clinical microbiology laboratories, decreasing the usage and sole reliance on traditional culture-based methods to detect and characterize microbes. Many new methods are particularly helpful in detecting fastidious, slow-growing, or nonviable isolates that are not identified by conventional culture techniques. The specificity, sensitivity, and turnaround time achieved by molecular techniques have significantly enhanced patients' quality of care. This section will briefly discuss a few of the molecular methods commonly utilized in clinical microbiology laboratories to assist in diagnosing infectious diseases.

EXAMPLES OF FREQUENTLY USED MOLECULAR TECHNIQUES

Polymerase chain reaction. The polymerase chain reaction is one of the most revolutionizing techniques in molecular microbiology. It ranks as one of the most widely used amplification methods in clinical laboratories. The increased sensitivity provided by PCR allows it to be vitally important in the early detection of many infectious agents. Its principal methodology is based on the vital steps: denaturation, annealing, and primer extension. During denaturation, double-stranded DNA is heated to about 95° Celsius, disrupting the hydrogen bonds between double-stranded causing the DNA to be separated into two single-strands. The second phase, annealing, is categorized by binding sequence-specific primers to the DNA template. The reaction temperature during this step depends on the length of the primers, but it usually ranges between 45 to 60° C. The third step, primer extension, is the phase in which the DNA polymerase extends the primer via its DNA polymerizing activity, synthesizing DNA molecules to build a complementary strand. The optimal temperature for this step is about 72° C. While there are many types of polymerases, Taq polymerase—derived from Thermus aquaticus bacteria—is the most widely used due to its thermophilic nature. These three steps are performed under a set number of cycles with different heating and cooling temperatures, which causes an exponential increase in the amplification of the target DNA region.

Gel electrophoresis and pulsed gel electrophoresis. Gel electrophoresis is a standard method used in many clinical microbiology laboratories to separate DNA, ribonucleic acid (RNA), or protein fragments according to their charge and molecular size. In this method, samples are loaded into one end of the gel that contains small pores. An electrical field applied to the opposite pole pushes the sample through it. Based on their size and charge, the molecules will travel through the gel in different directions or at different speeds, allowing them to be separated. For example, the negative charge of DNA molecules is a key factor in the migration of loaded samples toward positive electrodes on the gel in a mass-dependent manner. Since all DNA fragments have the same charge, smaller fragments will migrate through the gel faster than larger fragments within a set time. Standard gel electrophoresis is usually used to identify smaller fragments that range up to 50 kilobases. Larger fragments tend to co-migrate and appear as broad bands at the top of the gel, making the identification of molecules difficult. Clinical laboratories use a method known as pulse-field gel electrophoresis (PFGE) to identify larger molecules.

The principle behind pulse PFGE is similar to standard gel electrophoresis. It requires electrodes that generate an electrical charge through the chamber. However, the electrical field is applied to the gel in "pulses" from multiple directions at different times. By utilizing this method, DNA fragments as large as ten megabases are separated due to their reorientation and movement at different speeds through the gel. As the DNA migrates, it produces a fingerprint analyzed by a reference database to provide a specific identification. PFGE has been vital in rapid genomic analysis and was considered a gold standard due to its reproducibility, cost, availability, and selectivity.

Multiplex immunoassay. A multiplex assay is an immune assay or platform that allows the analysis of multiple analytes of different organisms simultaneously. This methodology uses affinity capture ligands such as antibodies, proteins, or peptides to detect specific analytes within a biological specimen and is incorporated into many new molecular diagnostics tools.

Whole-genome sequencing and next generation sequencing. Whole-genome sequencing (WGS) is a powerful tool used in clinical microbiology laboratories to rapidly evaluate the entire DNA sequence, including the nuclear, mitochondrial, and noncoding sequences found in an organism's genome. WGS has been a vital diagnostic tool for generating accurate reference genomes, microbial identification, epidemiological monitoring of emerging pathogens, and antimicrobial resistance. With time, various sequencing methods have been developed to increase organism identification efficiency, one of which is next-generation sequencing (NGS). NGS can sequence the whole genomes of numerous pathogens simultaneously, significantly revolutionizing molecular microbiology through its broad-range applications by utilizing a multiplex assay.

IMPACT

Molecular microbiology markedly impacts how clinical microbiologists currently detect and identify microorganisms causing infections. Through the development and innovation of many new molecular diagnostic tools, laboratories now can more rapidly obtain vital information on the identity and pathogenicity of microorganisms, provide rapid diagnosis of infectious disease and expand the investigation of pathogenesis and epidemiology of infectious agents. When molecular microbiology techniques are integrated into testing algorithms and conventional microbiology methods, these techniques provide a more rapid and thorough approach to the complementary analysis of the biochemical, phenotypic, and genotypic microbial characteristics and allow a more rapid diagnosis of infectious diseases.

—Thessicar Antoine-Reid, PhD

Further Reading

Babakhani, S., and M. Oloomi. "Transposons: The Agents of Antimicrobial Resistance in Bacteria." *Journal of Basic Microbiology*, vol. 58, 2018, pp. 905-17.

Du, D., X. Wang-Kan, A. Neuberger, et al. "Multidrug Efflux Pumps: Structure, Function, and Regulation." *Nature Reviews Microbiology*, vol. 16, 2018, pp. 523-39.

Fairfax, M. R., M. H. Bluth, and H. Salimnia. "Diagnostic Molecular Microbiology: A 2018 Snapshot." *Clinics in Laboratory Medicine*, vol. 38, 2018, pp. 253-76.

Khoury, G., G. Darcis, M. Y. Lee, et al. "The Molecular Biology of HIV Latency." *Advances in Experimental Medicine and Biology*, vol. 18, no. 1075, 2018, pp. 187-212.

Krump, N. A., and J. You. "Molecular Mechanisms of Viral Oncogenesis in Humans." *Nature Reviews Microbiology*, vol. 16, 2018, pp. 684-98.

Lakhundi, S., and K. Zhang. "Methicillin-Resistant Staphylococcus Aureus: Molecular Characterization, Evolution, and Epidemiology." *Clinical Microbiology Reviews*, vol. 31, 2018.

Ma, Liang, et al. "A Molecular Window into the Biology and Epidemiology of Pneumocystis spp." *Clinical Microbiology Reviews*, vol. 31, no. 3, 2018, p. e00009-18, doi:10.1128/CMR.00009-18.

Otasevic, S., S. Momcilovic, and N. M. Stojanovic. "Non-culture Based Assays for the Detection of Fungal Pathogens." J Mycol Med, vol. 28, 2018, pp. 236-48.

Pierson, T.C., and M. S. Diamond. The Emergence of Zika Virus and Its New Clinical Syndromes." *Nature*, vol. 560, 2018, pp. 573-81.

GEOCHEMICAL CYCLES

In this section, we describe the role microbes play in geochemical cycles. These cycles make specific minerals—such as carbon, nitrogen, phosphorus, and sulfur—available to living organisms as they cycle through different oxidation states.

CARBON CYCLE

Category: Geological processes and formations
Anatomy or system affected: All systems
Specialties and related fields: Bacteriology, biochemistry, biotechnology, chemistry, ecology, microbiology, organic chemistry
Definition: the movement of the element carbon through the Earth's rock and sediment, the aquatic environment, land environments, and the atmosphere; large amounts of organic carbon are found in living organisms and dead organic material

KEY TERMS

biosphere: the regions of the surface, atmosphere, and hydrosphere of the Earth occupied by living organisms
carnivores: meat-eating animals and some plants
decomposers: microorganisms like bacteria and fungi that degrade plant litter and debris and animal carcasses and return complex biological macromolecules to simpler precursors
geosphere: the solid part of the planet, including the crust and mantle of the Earth
herbivores: plant-eating animals
hydrosphere: the total amount of water on a planet that includes water on the surface of the planet, underground, and in the air
pedosphere: the soil mantle of the Earth

BACKGROUND

The surface of the Earth contains an enormous carbon reservoir. Most of this reservoir is in rock and sediment. Since the "turnover" time of such forms of carbon is so long (on the order of thousands of years), the entrance of this material into the carbon cycle is insignificant on the human scale. The carbon cycle represents the movement of this element through the biosphere in a process mediated by photosynthetic plants on land and in the sea. The process involves the fixation of carbon dioxide (CO_2) into organic molecules by photosynthesis. The energy utilized in the process is stored in chemical forms, such as that in carbohydrates (sugars such as glucose). The organic material is eventually oxidized, as occurs when a photosynthetic organism dies; through the process of respiration, the carbon is returned to the atmosphere in the form of carbon dioxide.

PHOTOSYNTHESIS

Organisms that use carbon dioxide as their source of carbon are known as "autotrophs." Many of these organisms also use sunlight as the source of energy to reduce carbon dioxide; hence, they are frequently referred to as "photoautotrophs." This process of carbon dioxide fixation is carried out by phytoplankton in the seas, by land plants, particularly trees, and by many microorganisms. Most of the process is carried out by the land plants. Because photosynthetic plants and microorganisms return carbon from the atmosphere to the biosphere, they are known as "producers."

The following equation can summarize the process of photosynthesis: CO_2 + water + energy $\rightarrow$ carbohydrates + oxygen. The process requires energy from sunlight, which is then stored in the form of the chemical energy in carbohydrates. While most plants produce oxygen in the process—the source of the oxygen in the Earth's atmosphere—some bacteria may produce products other than oxygen. Organisms that carry out carbon dioxide fixation, using photosynthesis to synthesize carbohydrates, are often referred to as producers. Approximately 18 to 27 billion metric tons of carbon are fixed each year by the process—clearly a large amount, but only a tiny proportion of the total carbon found on the Earth. Approximately 410 billion metric tons of carbon are contained within the Earth's forests; some 635 billion metric tons exist in the form of atmospheric carbon dioxide.

Much of the organic carbon on the Earth is found in the form of land plants, including forests and grasslands. When these plants or plant materials die, as when leaves fall to the Earth in autumn, the dead organic material becomes humus. Much of the carbon initially bound during photosynthesis is in the form of humus. Degradation of humus is a slow process that takes decades and mainly occurs in the pedosphere. However, it is the decomposition of humus, mainly through respiration, that returns much of the carbon dioxide to the atmosphere. Thus the carbon cycle represents a dynamic equilibrium between the carbon in the atmosphere and carbon fixed in organic material.

RESPIRATION

Respiration represents the reverse of photosynthesis. All organisms that utilize oxygen, including humans, carry out the process. However, it is primarily humic decomposition by microorganisms that returns most

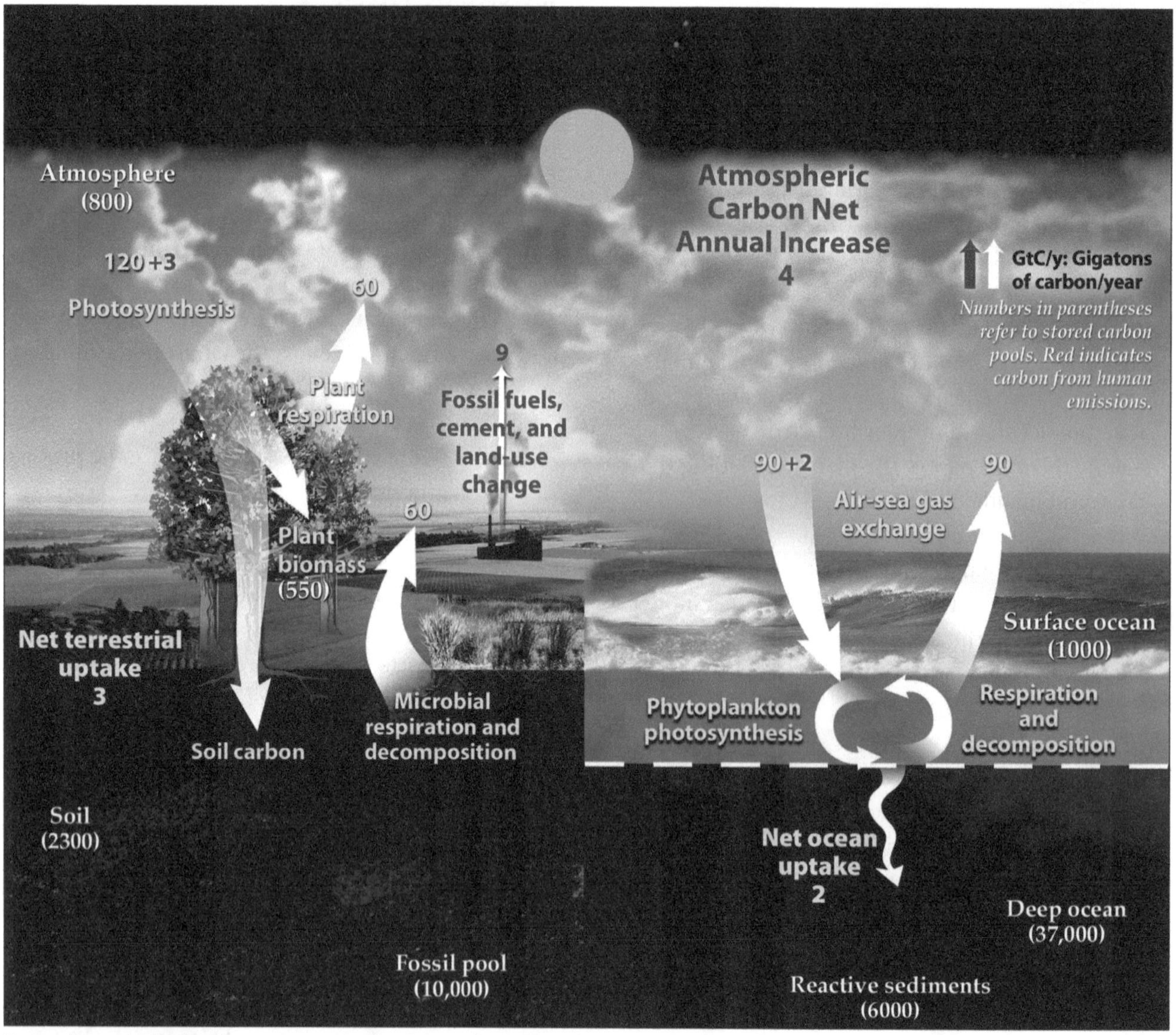

Fast carbon cycle showing the movement of carbon between land, atmosphere, and oceans in billions of tons (gigatons) per year. Image adapted from U.S. Department of Energy, via Wikimedia Commons. [Public domain.]

of the carbon to the atmosphere. Organisms that decompose dead plant and animal material are known as "decomposers." Depending on the particular microorganism, the carbon is either carbon dioxide or methane (CH_4). Respiration is generally represented by the equation carbohydrate + oxygen → carbon dioxide + water + energy. The energy released by the reaction is utilized by the organism (the consumer) to carry out its metabolic processes.

CARBON SEDIMENT

Despite the massive levels of carbon cycling between the atmosphere and living organisms, most carbon is found within carbonate deposits on land and in ocean sediments. Some of this originates in marine ecosystems, where organisms utilize dissolved carbon dioxide to produce carbonate shells (calcium carbonate). As these organisms die, the shells sink and become part of the ocean sediment. Returning carbon to ocean sediment subtracts it from the biosphere and hydrosphere and makes it part of the geosphere. Other organic deposits, such as oil and coal, originate from fossil deposits of dead organic material. The recycling time for such sediments and deposits is generally on the order of thousands of years; hence their contribution to the carbon cycle is negligible on a human timescale. Some sediment is recycled naturally, as when sediment dissolves or acid rain falls on carbonate rock (limestone), releasing carbon dioxide. However, when such deposits are burned as fossil fuels, carbon dioxide levels in the atmosphere may increase at a rapid rate.

ENVIRONMENTAL IMPACT OF HUMAN ACTIVITIES

Carbon dioxide gas is only a small proportion (0.036 percent) of the volume of the atmosphere. However, carbon dioxide acts much like a thermostat because of its ability to trap heat from the Earth. Even small changes in levels of this gas can significantly alter environmental temperatures. Around 1850, humans began burning large quantities of fossil fuels; the use of such fuels accelerated significantly with the invention of the automobile. Between five and six billion metric tons of carbon are released into the atmosphere every year from the burning of fossil carbon. Some of the released carbon probably returns to the Earth through biological carbon fixation, with a possible increase in the land biomass of trees or other plants. (Whether this is so remains a matter of dispute.) Indeed, large-scale deforestation could potentially remove this means by which atmospheric carbon dioxide levels could be controlled naturally.

—Richard Adler

Further Reading

Archer, David. *The Global Carbon Cycle*. Princeton UP, 2010.
Berner, Robert A. *The Phanerozoic Carbon Cycle: CO_2 and O_2*. Oxford UP, 2004.
Field, Christopher B., and Michael R. Raupach, editors. *The Global Carbon Cycle: Integrating Humans, Climate, and the Natural World*. Island Press, 2004.
Hazen, Robert M. *Symphony in C: Carbon and the Evolution of (Almost) Everything*. W. W. Norton & Company, 2019.
Houghton, R. A. "The Contemporary Carbon Cycle." *Biogeochemistry*, edited by W. H. Schlesinger, Elsevier, 2005.
Madigan, Michael, et al. *Brock Biology of Microorganisms*. 15th ed., Pearson/Benjamin Cummings, 2017.
Reichle, David E. *The Global Carbon Cycle and Climate Change: Scaling Ecological Energetics from Organism to the Biosphere*. Elsevier, 2019.
Volk, Tyler. *CO_2 Rising: The World's Greatest Environmental Challenge*. MIT Press, 2008.
Wigley, T. M. L., and D. S. Schimel, editors. *The Carbon Cycle*. Cambridge UP, 2000.

NITROGEN CYCLE

Category: Geochemical processes
Anatomy or system affected: All systems
Specialties and related fields: Bacteriology, biochemistry, geology, microbiology

Definition: the series of processes by which nitrogen and its compounds are interconverted in the environment and living organisms, including nitrogen fixation and decomposition

KEY TERMS

decomposition: the process by which bacteria and fungi break dead organisms into their simple compounds

denitrification: the chemical reduction of soil nitrates or nitrites by denitrifying bacteria leading to gaseous N losses

nitrification: the process that converts ammonia to nitrite and then to nitrate

nitrogen fixation: any natural or industrial process that causes free nitrogen (N_2), which is a relatively inert gas plentiful in air, to combine chemically with other elements to form more-reactive nitrogen compounds such as ammonia, nitrates, or nitrites

INTRODUCTION

Nitrogen (N) is one of the most dynamic elements in the Earth's biosphere; it undergoes transformations that constantly convert it between organic, inorganic, gaseous, and mineral forms. Nitrogen is an essential element in all living things. It is a crucial component of organic molecules such as proteins and nucleic acids.

BACKGROUND

Nitrogen is in high demand in biological systems. However, most nitrogen is not readily available to plants and animals. Although the biosphere contains 300,000 teragrams (billion kilograms) of nitrogen, that amount is far less nitrogen than is in the hydrosphere (23 million teragrams) and much less nitrogen than is in the atmosphere (about 4 billion teragrams). Almost all atmospheric nitrogen is nitrogen gas (N_2), which composes 78 percent of the atmosphere by volume. The most significant reservoir of nitrogen on Earth is the lithosphere (164 billion teragrams). Here the nitrogen is bound up in rocks, minerals, and deep-ocean sediments.

Even though living things exist in a "sea" of nitrogen gas, it does them little good. The bond between the nitrogen atoms is so strong that nitrogen gas is relatively inert. For living things to use nitrogen gas, it must first be converted to an organic or an inorganic form. The nitrogen cycle is the collection of processes driven by microbial activity that converts nitrogen gas into these usable forms and later returns nitrogen gas to the atmosphere. It is considered a cycle because each process can ultimately convert every nitrogen atom. However, that conversion may take a long time. It is estimated, for example, that the average nitrogen molecule spends 625 years in the biosphere before returning to the atmosphere to complete the cycle.

NITROGEN FIXATION

The first step in the nitrogen cycle is nitrogen fixation. Nitrogen fixation is the conversion, by bacteria, of nitrogen gas into ammonium (NH_4^+) and then organic nitrogen (proteins, nucleic acids, and other nitrogen-containing compounds). It is estimated that biological nitrogen fixation adds about 160 billion kilograms of nitrogen to the biosphere each year. This represents about half of the nitrogen taken up by plants and animals. The microorganisms that carry out nitrogen fixation are highly specialized. Each one carries a unique enzyme complex, called "nitrogenase," that allows it to carry out fixation at temperatures and pressures capable of permitting life, something industrial nitrogen fixation does not allow.

Nitrogen-fixing microbes may either be free-living or grow in association with higher organisms such as legumes (in which case the process is called "symbiotic nitrogen fixation"). Symbiotic nitrogen fixation is an important process because legumes are so highly valued as a natural resource. Because they can form these symbiotic associations with nitrogen-fixing bac-

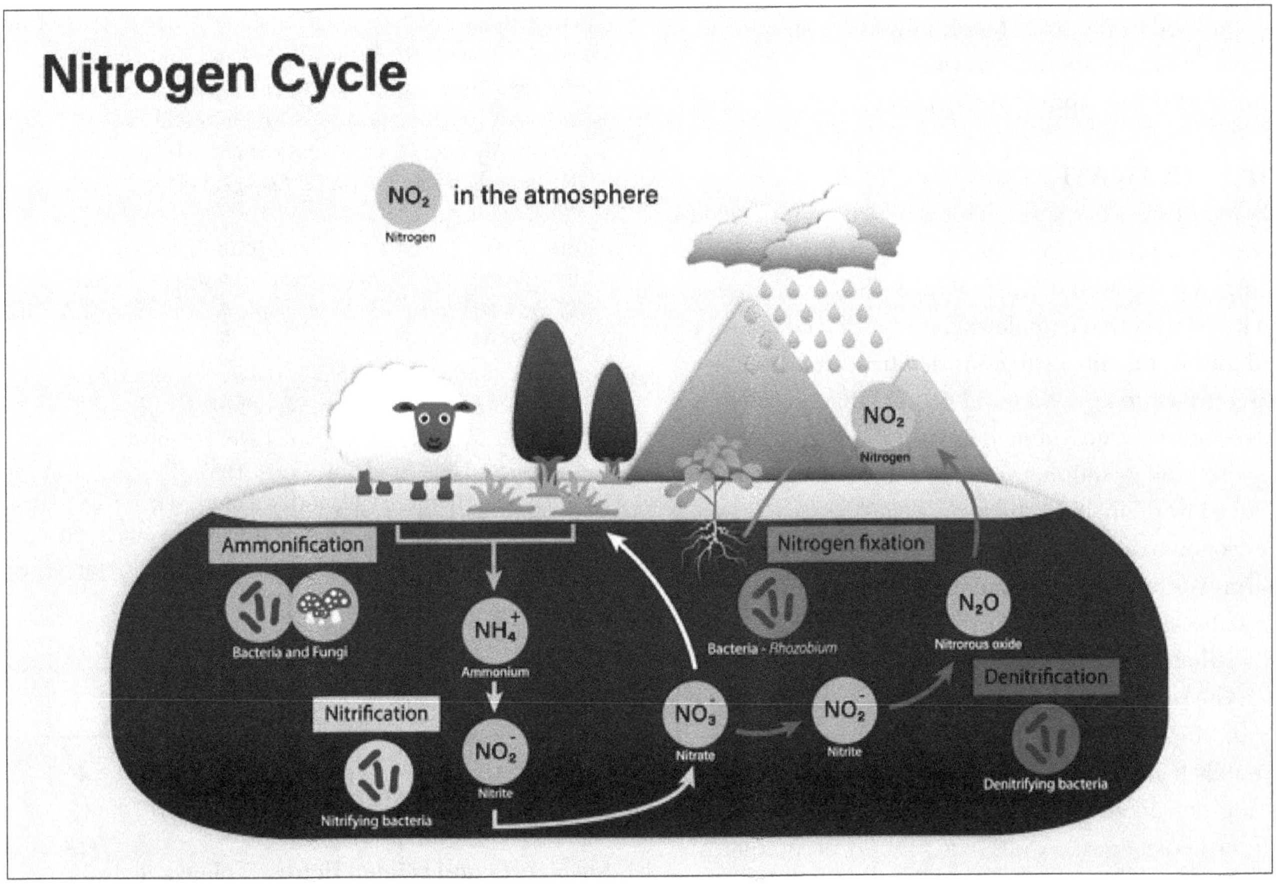

Image via iStock/olando_o. [Used under license.]

teria, legumes can produce seeds and leaves with more nitrogen than other plants. They return much of that nitrogen to the soil when they die, enriching it for future growth.

MINERALIZATION AND NITRIFICATION

When plants and animals die, they undergo a process called "mineralization" (also called "ammonification"). In this stage of the nitrogen cycle, the organic nitrogen in decomposing tissue is converted back into ammonium. Some of the ammonium is taken up by plants as they grow. This process is called "assimilation" or "uptake." Some of the ammonium is taken up by microbes in the soil. In this case, nitrogen is not available for plant growth. If this happens, it is said that the nitrogen is immobilized. Some nitrogen is also incorporated into the clay minerals of the soil. In this case, it is said that the nitrogen is fixed—it is not immediately available for plant and microbial growth. Still, it may become available at a later date.

Ammonium has another potential fate, and this step in the nitrogen cycle is nitrification. In nitrification, the ammonium in the soil is oxidized by bacteria (and some fungi) to nitrate (NO_3-) in a two-step process. First, ammonium is oxidized to nitrite, and next, nitrite is rapidly oxidized to nitrate. Nitrification requires oxygen, so it occurs only in well-aerated environments. Plants and microbes can also take up the nitrate that forms during nitrification. However, unlike ammonium, which is a cation and readily adsorbed by soil, nitrate is an anion and readily leeches or runs off of the soil. Hence, nitrate is a significant water contaminant in areas where excessive fertilization or manure application occurs.

DENITRIFICATION

Some process is responsible for returning nitrogen to the atmosphere; otherwise, organic and inorganic nitrogen forms would accumulate in the environment. The process that completes the nitrogen cycle and replenishes the nitrogen gas is denitrification. Denitrification is a bacterial process that occurs in anaerobic or oxygen-limited environments (e.g., waterlogged soil or sediment). Nitrate and nitrite are reduced by denitrifying bacteria, which can use these nitrogen oxides in place of oxygen for their metabolism. Wetlands are particularly important in this process because at least half of the denitrification in the biosphere occurs in wetlands.

The primary product of denitrification is nitrogen gas, which returns to the atmosphere and approximately balances the amount of nitrogen gas that is biologically fixed each year. In some cases, however, an intermediate gas, nitrous oxide (N_2O), accumulates. Nitrous oxide has serious environmental consequences. Like carbon dioxide, it absorbs infrared radiation, so it contributes to global warming. More important, when nitrous oxide rises to the stratosphere, it contributes to the catalytic destruction of the ozone layer. Besides the potential for fertilizer nitrogen to contribute to nitrate contamination of groundwater, there is the concern that some of it can be denitrified and contribute to ozone destruction.

The nitrogen cycle is a global cycle involving land, sea, and air. It circulates nitrogen through various forms that contribute to life on Earth. When the cycle is disturbed—when an area is deforested, and nitrogen uptake into trees is stopped, or when excessive fertilization is used—nitrogen can become an environmental problem.

—Mark S. Coyne

Further Reading

Chapin, F. Stuart, III, Pamela A. Matson, and Harold A. Mooney. "Internal Cycling of Nitrogen." *Principles of Terrestrial Ecosystem Ecology*. Springer, 2002.

Jacobson, Michael C., et al. *Earth System Science: From Biogeochemical Cycles to Global Change*. 2nd ed., Academic Press, 2000.

Mosier, Arvin, J. Keith Syers, and John R. Freney, editors. *Agriculture and the Nitrogen Cycle: Assessing the Impacts of Fertilizer Use on Food Production and the Environment*. Island Press, 2004.

Nieder, R., and D. K. Benbi. *Carbon and Nitrogen in the Terrestrial Environment*. Springer, 2008.

Schlesinger, William H. *Biogeochemistry: An Analysis of Global Change*. 2nd ed., Academic Press, 1997.

Sigel, Astrid, Helmut Sigel, and Roland K. O. Sigel, editors. *Biogeochemical Cycles of Elements*. Taylor & Francis, 2005.

Sprent, Janet I. *The Ecology of the Nitrogen Cycle*. Cambridge UP, 1987.

PHOSPHORUS CYCLE

Category: Geochemical cycles
Anatomy or system affected: None
Specialties and related fields: Ecology, environmental studies, genetics, geochemistry, plant physiology
Definition: the movement of the element phosphorus as it circulates through the living and nonliving portions of the biosphere

KEY TERMS

eutrophication: a process by which the dissolved oxygen in a body of water is consumed by the decomposition of masses of dead algae, to the point that there is no longer enough dissolved oxygen to support aquatic life

minerals: highly stable materials composed of inorganic materials, generally found as rock but often crystalline in structure

phosphate: chemically, a complex ion consisting of a phosphorus atom to which four oxygen atoms are bonded and carrying two negative charges termed inorganic phosphate, a vital component of certain energy-rich compounds such as adenosine triphosphate (ATP), and a major component of the structures of the nucleic acids

phosphates: collectively refers to all materials in which the phosphate ion is an integral component; usually refers to phosphate-based fertilizers, detergents, and other such environmental pollutants

VITAL ELEMENTS

Many of the chemical elements found on Earth are vital to the processes and systems of living organisms. Unlike oxygen and carbon, phosphorus follows complex pathways. It circulates through Earth's soils, rocks, waters, and atmosphere and through the organisms that inhabit these many ecosystems.

Elements or minerals are stored in discrete parts of Earth's ecosystems called "compartments." Examples of compartments include all the plants in a forest, a certain species of tree, or even the leaves or needles of a tree. Chemical elements reside within the compartments in certain amounts or pools. A basic description of biogeochemical cycles involves following nutrients in minerals or elements from pool to pool through the multitudes of ecosystem compartments.

PHOSPHORUS AND PLANTS

Phosphorus compounds reside primarily in rocks, and phosphorus does not go through an atmospheric phase. Rather, phosphorus-laden rocks release phosphate (PO_4^{3}) into the ecosystem due to weathering and erosion. Phosphorus is a vital nutrient (second only to nitrogen) to plants, which absorb phosphates through their root hairs. Phosphorus then passes on through the food chain when other organisms consume the plants. Phosphorus is an essential component of many biological molecules, including deoxyribonucleic acid (DNA) and ribonucleic acid (RNA). Adenosine triphosphate (ATP), one of the nucleotides consumed to make up DNA and RNA, is also the main energy transfer molecule in the multitude of chemical reactions within organisms.

Because phosphorus is a major plant nutrient, massive amounts of phosphate-based fertilizers are either derived from natural sources (in the form of bat or bird guano or as mined potash) or chemically manufactured for use by agriculture. As late as the early 1970s, phosphates were a major constituent of household detergents until it was discovered that large amounts of phosphates were being released into the environment in the discarded wash water. In aquatic systems such as rivers and lakes, where such runoff eventually appears, an infusion of phosphates can cause algal blooms (rapidly forming, dense populations of algae). When the algae die, they are consumed by bacteria. Decomposition by bacteria requires large amounts of oxygen, which soon depletes the available oxygen in the water. If the process continues unchecked, fish and other organisms die from lack of oxygen. Both phosphates and nitrates contribute to cultural eutrophication.

Phosphates not taken up by plants go into the sedimentary phase, where they are very chemically reactive with other minerals. These reactions produce compounds that effectively remove phosphates from the active nutrient pool. This sedimentary phase is characterized by its long residence time compared to the rapid cycling through the biological phase. Phosphates can remain locked up in rocks for millions of years before being exposed and broken down by weathering, which once again makes them available to plants.

PHOSPHORUS AND THE ENVIRONMENT

Because the phosphorus cycle is so complex, its interactions with other biogeochemical cycles are not completely understood. Studying these interactions is emerging as a vital field among the environmental sciences. Excessive phosphates in a eutrophic lake disrupt the carbon cycle by reacting with bicarbonates, thus increasing the pH. Many freshwater organ-

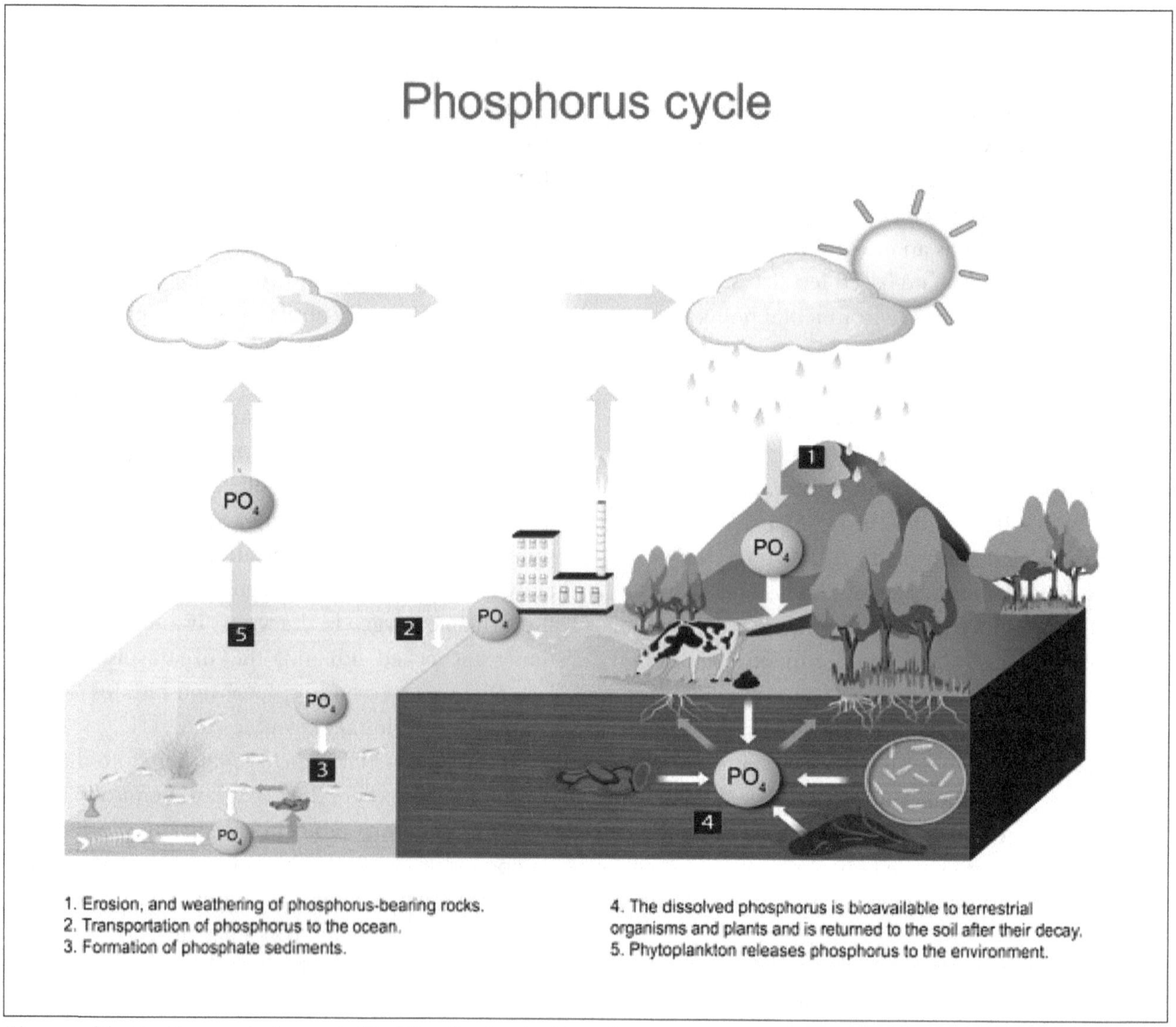
Phosphorus cycle
PO4
PO4
PO4
PO4
PO4
PO4
1
2
3
4
5
1. Erosion, and weathering of phosphorus-bearing rocks.
2. Transportation of phosphorus to the ocean.
3. Formation of phosphate sediments.
4. The dissolved phosphorus is bioavailable to terrestrial organisms and plants and is returned to the soil after their decay.
5. Phytoplankton releases phosphorus to the environment.

Diagram of the phosphorus cycle. Image via iStock/ttsz. [Used under license.]

isms depend on a neutral pH level for their survival. The presence of phosphorus under these oxygen-depleted conditions can also indirectly affect the sulfur cycle, leading to the conversion of sulfate to sulfide. When sulfide combines with hydrogen to form the gas hydrogen sulfide, it takes on the familiar "rotten egg" smell.

One of the keys to preventing environmental degradation by altering global chemical cycles lies in recognizing the effects of such alterations. With the perception of an environmental crisis in the early 1970s, more attention was paid to the role of human activity in these cycles. Test lakes were studied to determine why freshwater fisheries became oxygen-depleted at accelerated rates. Dramatic progress has been made to eliminate algal blooms and oxygen depletion by limiting the phosphorus-laden effluents being discharged into lakes.

—*David M. Schlom*

Further Reading

Frossard, Emmanuel, A. Oberson, and Else K. Bünemann. *Phosphorus in Action: Biological Processes in Soil Phosphorus Cycling*. Springer, 2011.

Krebs, Charles J. *Ecology: The Experimental Analysis of Distribution and Abundance*. 6th ed., Pearson, 2014.

Lasserre, P., and J. M. Martin, editors. *Biogeochemical Processes at the Land-Sea Boundary*. Elsevier, 1986.

Pomeroy, Lawrence, editor. *Cycles of Essential Elements*. Dowden, 1974.

Scholz, Roland W., et al. *Sustainable Phosphorus Management: A Global Transdisciplinary Roadmap*. Springer, 2014.

Tiessen, Holm, editor. *Phosphorus in the Global Environment: Transfers, Cycles, and Management*. Wiley, 1995.

Sulfur cycle

Category: Geological processes and formations
Anatomy or system affected: None
Specialties and related fields: Ecology, environmental chemistry, geochemistry, geology, microbiology
Definition: the transport and fate of sulfur as it moves through sedimentary rocks, the atmosphere, and the oceans

KEY TERMS

oxidation: the loss of electrons by molecules, decreasing its negative charge or increasing its positive charge

reduction: the gain of electrons by molecules, increasing their negative charge or decreasing their positive charge

sulfate: a salt or ester of sulfuric acid, containing the anion SO_4^{2-} or the divalent group $—OSO_2O—$

sulfide: an inorganic anion of sulfur with the chemical formula S^2 or a compound containing one or more S^2 ions

SUMMARY

Sulfur is an important secondary element for proteins and amino acids. In living organisms, sulfur is present in its most reduced state as a hydrosulfide group (SH-). Sulfur can be found in enzymes and large-scale biological structures such as hair and nails.

DEFINITION

The sulfur cycle describes the transport and fate of sulfur as it moves through sedimentary rocks, the atmosphere, and the oceans. The cycle defines the chemical speciation, reactions, and transformation of sulfur compounds introduced by natural and anthropogenic (human-made) sources and their effects on the ecosystem.

OVERVIEW

Most naturally occurring sulfur (also spelled "sulphur" and abbreviated as S) can be found in the inner mantle and core of the Earth. The amount of sulfur on the Earth's surface has greatly increased since the Industrial Revolution. The human-induced sulfur in the atmosphere results from extensive usage of fossil fuels and sulfide ores.

Decomposition of organic sulfur compounds by bacteria produces hydrogen sulfide (H_2S), which has the distinctive smell of rotten eggs. Hydrogen sulfide is either released into the atmosphere as gas or reacts with trace metals in the water or sediments to produce insoluble sulfides. In certain cases, bacteria can also produce elemental sulfur.

Most anthropogenic effects on the sulfur cycle involve increasing the oxidation state of sulfur, resulting in sulfur dioxide (SO_2), sulfur trioxide (SO_3), or sulfates ($SO^2{}_4$-). On the contrary, biological activities tend to reduce the oxidation state of sulfur, producing hydrogen sulfide or dimethylsulfide $(CH_3)_2S$.

Sulfur is discharged from volcanoes and human activities into the atmosphere as sulfur dioxide. In addition, hydrogen sulfide emitted from vegetation and the oceans is oxidized to sulfur dioxide. In the atmosphere, sulfur dioxide is oxidized to sulfur trioxide and sulfates. Subsequently, it is precipitated back to the Earth and oceans. Sulfates are also directly transferred from the oceanic waters to the atmosphere. A large part of sulfur entering the oceans is lost in the deep bottom sediments.

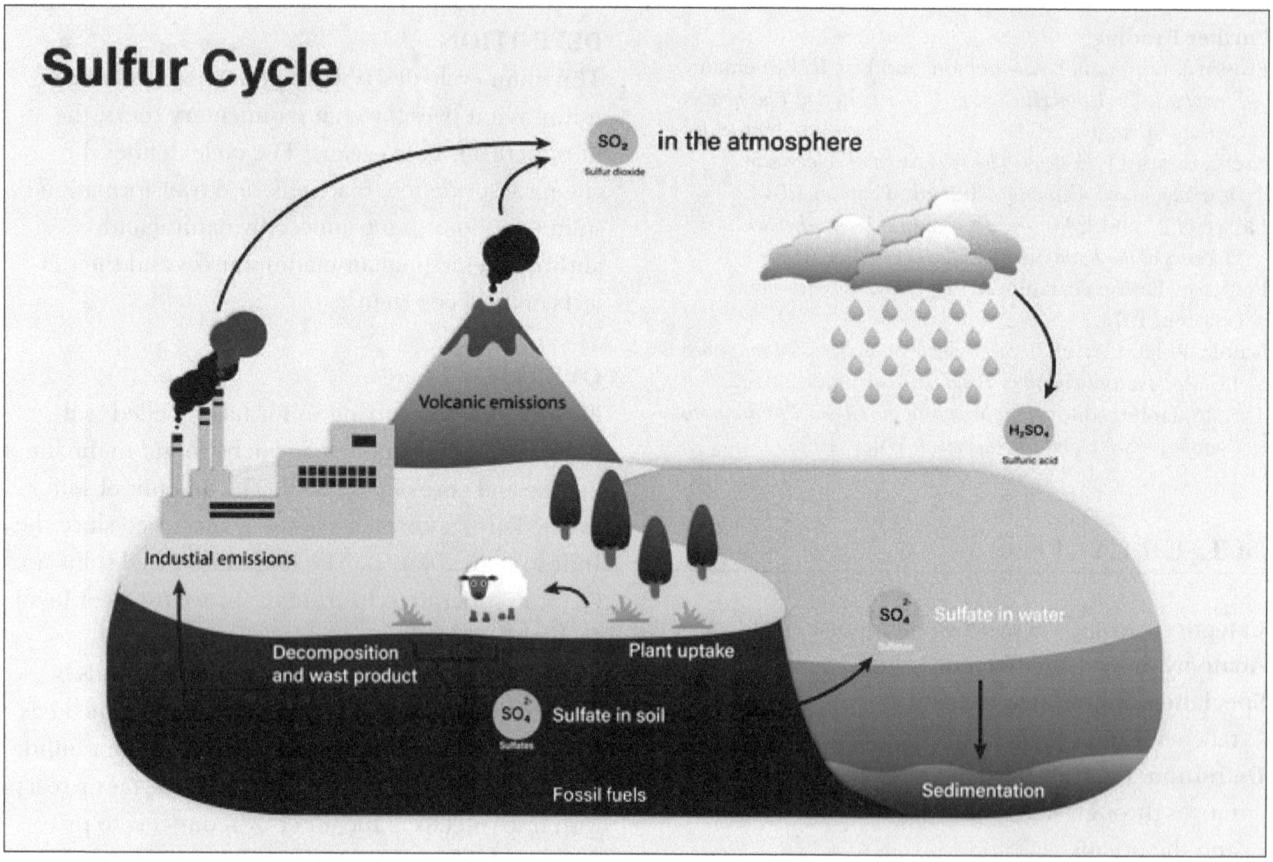

Diagram of the sulphur cycle. Image via iStock/olando_o. [Used under license.]

The most interesting component of the sulfur cycle is atmospheric sulfur dioxide. The amount of sulfur dioxide produced by anthropogenic activities is of the same magnitude as that generated by natural causes. Once sulfur dioxide is in the atmosphere, part of it is oxidized to sulfate. At the same time, the rest is removed by dry deposition. Dry deposition implies the sorption of sulfur compounds by wet solid surfaces and water bodies. Sulfates are primarily removed by wet deposition, a process that involves both the "rained out" sulfates that enter the clouds and the "washed out" sulfates that are intercepted by falling raindrops or snowflakes.

Increased levels of sulfur dioxide and nitrogen oxides in the atmosphere can reduce pH in the rain to less than 5.6, resulting in acid rain conditions. Acid rain can strongly negatively impact flora and fauna; it can also cause deterioration of buildings and other structures. High levels of sulfur dioxide can pose health hazards to humans who suffer from respiratory problems. Remedial actions for reducing anthropogenically induced sulfur compounds involve fuel desulfurization, fuel substitution, and flue-gas desulfurization.

—*Panagiotis D. Scarlatos*

Further Reading

Sagan, Dorion. *The Global Sulfur Cycle*. National Aeronautics and Space Administration, 1985.

Smil, Vaclav. *Carbon-Nitrogen-Sulfur: Human Interference in Grand Biospheric Cycles*. Springer, 2012.

Stevenson, F. J. *Cycles of Soils: Carbon, Nitrogen, Phosphorus, Micronutrients*. 2nd ed., Wiley India Pvt. Ltd., 2015.

Microbial Exploitation

Given the unique metabolic capabilities of microbes, humans have used microbes to make food, drinks, and organic acids; to treat sewage; and to create biosynthetics. Microbes also help degrade large molecules on an industrial level. Entries in this section—which include anaerobic digestion and lactic acid—discuss each of these uses.

ANAEROBIC DIGESTION

Category: Biotechnology, environmentalism
Anatomy or system affected: None
Specialties and related fields: Bacteriology, biochemistry, biotechnology, environmental science, microbiology
Definition: a biological process that converts animal waste or other sewage into biogas and helps manage waste and create an alternative energy source

KEY TERMS

anaerobic: processes that occur in the absence of oxygen
carbon dioxide: a gas composed of molecules with one carbon atom and two oxygen atoms (CO_2) that is the main waste product of biological metabolism
digestion: the degradation of large biological molecules into small precursors
heavy chain: the bigger subunits of an antibody
methane: a chemically simple gas that consists of one carbon atom bound to four hydrogen atoms (CH_4)
organic material: a compilation of molecules made predominantly of carbon but also contains varying quantities of nitrogen, oxygen, and hydrogen

HISTORY

Anaerobic digestion is a process that has been in use for hundreds of years. The first recorded use of biogas took place in Syria during the tenth century BCE. Later, seventeenth-century scientist Jan Baptist van Helmont concluded that decaying organic matter could turn into flammable gases. Count Alessandro Volta expanded on this idea and found a correlation between the amount of matter and gas produced. In 1859, the first anaerobic digestion plant was built in Bombay (now Mumbai), India. Other anaerobic digestion plants were built in the latter half of the nineteenth century in England, where the fuel produced was used to light street lamps. More recently, this technology has been used extensively in India, China, and Nepal. China and India have begun to move away from farm-based facilities, which produce smaller amounts of gas, and toward larger systems that can generate electricity. Biogas production is not, however, limited to these countries. Europe and the United States have also installed hundreds of anaerobic digestion systems since the 1970s. However, it is only since the 1990s that more sophisticated systems have been installed.

HOW IT WORKS

Anaerobic digestion is a chemical process that does not require oxygen. The primary input into the system is usually a type of animal waste; however, any organic material can be processed by anaerobic digestion. Once the process is complete, there are two significant outputs. The first is methane gas, which can be converted for various uses, and the other is digested solids and liquids, which can be converted into fertilizers.

The process is relatively simple. The waste is first collected and put into the digester along with some water. This mixture of waste and water is known as slurry. After the slurry has been mixed, there are three steps to the digestion process. First, microorganisms break down the plant and animal matter into smaller molecules, such as sugar. The resulting mix is then further converted into organic acids by other microorganisms. Finally, this mixture is turned into the outputs of biogas and fertilizer. The entire process needs between fifteen and thirty days to complete.

All three steps utilize microorganisms to break down the slurry into smaller component parts. These microorganisms are easily upset by temperature fluctuations within the biogas digester. Therefore, it is vital to keep the temperature stable for the process to work most effectively. In the United States, this is done by adding extra insulation or heating to the digester. Currently, it is believed that a digester temperature of 72° Fahrenheit nets the maximum production of biogas. Other items that can affect production include the amount of water added, how the manure is mixed, and the range of nitrogen. All of these variables can be easily controlled through frequent monitoring.

The resulting biogas is composed of roughly two-thirds methane and one-third carbon dioxide. Methane is a potent greenhouse gas. It is twenty-one times more effective at trapping heat within the Earth's atmosphere than carbon dioxide, which is the main byproduct of aerobic processes. Methane is also flammable and can be harvested for applications in cooking, heating, or lighting. On a larger scale, methane can generate electricity. A second benefit derived from anaerobic digestion is the leftover slurry. The anaerobic process does not reduce the nutrient content of the slurry, which allows it to remain an excellent fertilizer. The slurry can replace chemical fertilizers on fields or be diverted into a pond to support fish, which would lead to additional income. A final benefit is better waste management. Anaerobic digestion processes the animal waste, reducing odor, the number of fly eggs and weed seeds, and the number of pathogens from land-applied nutrients.

The primary equipment required for anaerobic digestion is the digester. There are four main biogas digester designs. These designs are the flexible vessel, floating drum, fixed dome, and plug-flow digesters. In addition to these four designs, a large number of hybrid models are in use. Each of these designs has strengths and weaknesses. In the United States, more than half of the digesters use a plug-flow design, although each design is in use around the country.

The potential for anaerobic digestion production in the United States is tremendous. Not every farm is appropriate for this technology. Still, it is estimated that 7,000 farms could take advantage of biogas within the United States. To assist those farms, the Department of Agriculture, the Department of Energy, and the Environmental Protection Agency formed AgStar, which helps farmers take advantage of this simple technology.

—Ryan Fogle

Further Reading

Balagurusamy, Nagamani, and Anuj Kumar Chandel. *Biogas Production: From Anaerobic Digestion to a Sustainable Bioenergy Industry*. Springer, 2020.

Balsam, John, and Dave Ryan. "Anaerobic Digestion of Animal Wastes: Factors to Consider." *National Sustainable Agriculture Information Service*, 2006, attra.ncat.org/product/anaerobic-digestion-of-animal-wastes-factors-to-consider/.

Environmental Protection Agency. "How Does Anaerobic Digestion Works?" 22 Jan. 2021, www.epa.gov/agstar/how-does-anaerobic-digestion-work.

Horan, Nigel, Abu Zahrim Yaser, and Newati Wid. *Anaerobic Digestion Processes: Applications and Effluent Treatment*. Springer, 2018.

Pennsylvania State University. "Biogas Production." www.biogas.psu.edu/.

Pullen, Tim. *Anaerobic Digestion—Making Biogas—Making Energy*. Routledge, 2015.

BIOSYNTHETICS

Category: Biotechnology
Anatomy or system affected: Cells
Specialties and related fields: Biochemistry, biotechnology, microbiology
Definition: the process of using small, simple molecules to make larger, more complex molecules, either inside the body or in the laboratory

KEY TERMS

amino acids: the building blocks of proteins
antibody: glycoprotein that binds to and immobilizes a substance that the cell recognizes as foreign
antigen: a substance that triggers an immune response
binding assay: an experimental method for selecting one molecule out of several possibilities by specific binding

DNA (deoxyribonucleic acid): a molecule that contains the genetic code
enzyme: biological catalyst, usually a globular protein
gene: individual unit of inheritance that consists of a sequence of DNA
hormone: a substance produced by endocrine glands and delivered by the bloodstream to target cells, producing the desired effect
hydrophilic: property of tending to dissolve in water
insulin: hormone released from the pancreas
monoclonal antibody: antibody produced from the progeny of a single cell and specific for a single antigen
peptide: molecule formed by linking two to several dozen amino acids
protein: macromolecule formed by polymerization of amino acids

INTRODUCTION

Biosynthesis is the process of using small, simple molecules to make larger, more complex molecules, either inside the body or in the laboratory. Numerous applications for drug development and medicine include synthesizing proteins, hormones, dietary supplements, blood products, and surgical dressings for wounds. Additional techniques to facilitate the diagnosis and treatment of disease include protein biomarkers for immune assays, the development of proteomics to analyze changes in proteins in response to a drug, the development of polyclonal and monoclonal antibodies, immunizations, and various drug delivery systems.

BASIC PRINCIPLES

In general, the term "biosynthetic" refers to any material produced via a biosynthetic process. A biosynthetic process uses enzymes and energetic molecules to transform small molecules into larger molecules within the cells of organisms. The two types of metabolites produced from cellular biosynthetic pathways include the primary metabolites of fatty acids and deoxyribonucleiuc acid (DNA) needed by cells and the secondary metabolites of pheromones, antibiotics, and vitamins that assist the entire organism. Additional small molecules, such as adenosine triphosphate (ATP), provide the energetic driving force for the biosynthetic pathways, and other small molecules, including enzymes, further facilitate the reactions in these pathways. Thus, there have been many possibilities for numerous scientists, including chemists, biochemists, biologists, and geneticists, to create innovations.

The term "biosynthetic" differs from "chemosynthetic" because chemosynthetic indicates the production of materials that cannot occur within a living organism. Scientists generally begin developing a new medical application or dietary supplement by first isolating and characterizing the DNA of the proteins or other small molecules directly involved in the

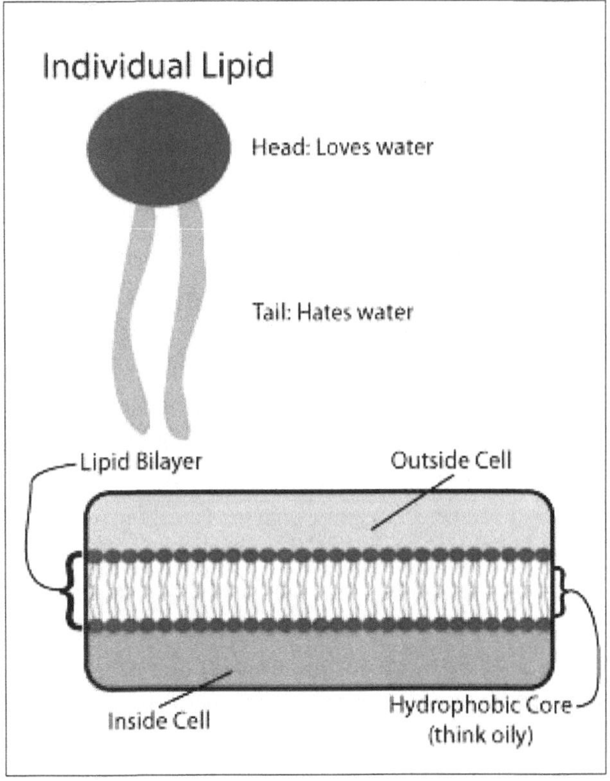

Lipid membrane layer. Image by Bradleyhintze, via Wikimedia Commons.

biological process. They then try to duplicate this naturally occurring biological process to produce massive quantities of the desired material. Ultimately, they combine these naturally occurring processes with chemicals that mimic the process during laboratory manufacturing.

BACKGROUND

The biochemical pharmacologist Hermann Karl Felix "Hugh" Blaschko was a trailblazer whose discoveries in the 1930s initiated the field of biosynthetics. His work elucidated the biosynthetic pathway for adrenaline, which is often called the "fight-and-flight hormone." It encompassed the study of the enzymes necessary for the regulation of this hormone. This work led the way toward the development of syntheses using amino acids for therapeutic applications. Another critical development in biosynthetics was discovering the role of the amino acid L-arginine in synthesizing creatine, a vital biomolecule, by G. L. Foster, Rudolf Schoenheimer, and D. Rittenberg in 1939. Since that time, L-arginine has also been shown to be a precursor to nitrous oxide and nitric oxide, as well as a component of the urea cycle, which is essential for ammonia regulation and thus influences the operation of the kidneys and other organs. Nitric oxide is important in the regulation of blood flow to muscles. These discoveries involving L-arginine have led to dietary supplements useful to bodybuilders who wish to enhance their weight-lifting performance.

Throughout the 1940s, 1950s, and 1960s, scientists made progress toward understanding the genetic composition of organisms, enzymes, and biosynthetic pathways. Researchers made contributions to understanding pyrimidine, galactosidase, *Escherichia coli*, and chlorophyll. Practical biosynthetic applications made possible by these fundamental discoveries began to manifest themselves throughout the 1970s, 1980s, and 1990s, with the development of surgical dressings, therapeutic hormones, and plant supplements for increased nutritional value.

HOW IT WORKS

General process. Often the isolation and characterization of a specific gene responsible for producing an important enzyme or other small molecule is the first step in a lengthy process toward the synthesis of a product that undergoes lengthy clinical trials before the final, approved product is ready for manufacture. Once the gene has been characterized, its DNA is further characterized to facilitate the process of peptide synthesis (the process of producing long peptides is known as protein biosynthesis). The process of peptide synthesis involves the general concepts of antigenicity, hydrophilicity, surface probability, and flexibility indexes. The process involves an analysis of the peptide's characteristics, the use of software and databases to determine hydrophilicity (affinity for water), the study of the antigenicity (capacity to stimulate the production of antibodies) to assist with antibody production, the study of surface probability (which determines the likelihood of inducing the formation of antibodies), the determination of the protein sequence, phosphorylation (a process that activates or deactivates many protein enzymes), and then selection of two to three peptides, followed by a comparison of their homology (similarity of structure).

In a general process called "screening," the efficacy of an antibiotic is first tested using bacterial cultures, followed by injection of the antibiotic into laboratory animals, such as rats, rabbits, or guinea pigs. Clinical trials are conducted according to protocols established by the Food and Drug Administration (FDA). Combinatorial chemistry, a faster screening method, is often used instead. FDA-approved products are then manufactured on a larger scale.

Antibody production. The application of a binding assay is used to isolate the purified protein that is to be the source of an antigen. This antigen is then used as a conjugate to a carrier protein, such as keyhole limpet hemocyanin (KLH), to produce a target peptide with a length of thirteen to twenty amino acids to stimulate the immune system. A carrier protein is a membrane protein that can bind to a substance to facilitate the substance's passive transport into a cell. Injection into a laboratory animal occurs next. The animals undergo a series of four to six immunizations separated by about twenty days. Enzyme-linked immunosorbent assay (ELISA) is used to detect antibodies. ELISA is based on the antibody-antigen binding interaction and often uses color to indicate antibodies' concentration. Purification of antibodies obtained from the antiserum for specific antigen-binding completes the antibody production process.

Antigen preparation. This process is facilitated through bioinformatics analysis to choose the appropriate two to three peptides based on the protein sequence provided by a customer. KLH conjugation is used for immunization, and bovine serum albumin (BSA) conjugation is carried out for screening. A cell can be cryopreserved after immunization protocols, and specific antibodies have been selected during fusion and screening.

Combinatorial chemistry. In combinatorial chemistry synthesis, a high-throughput screening method, the starting small molecule is attached to a type of polymeric resin, followed by different permutations of reagents, to produce large libraries containing hundreds of unique products that can be rapidly screened for enzymatic activity, specific antigen-binding, or protein-protein interactions. Often the process is controlled by a computer and completed through the application of robotics. A customer can specify antigen details. A pharmaceutical company can design a protocol involving the general phases of antigen preparation, immunization, fusion, and screening of assays, and finally selection, purification, and production of antibodies.

EXAMPLES OF APPLICATIONS OF BIOSYNTHETICS

The generation of biosynthetic products has led to successful treatments for many types of cancer, pneumonia, cardiovascular diseases, diabetes, tuberculosis, neurological disorders, strokes, blood disorders, and many other diseases and conditions.

Biosynthetic corneas were used to restore vision in people with keratoconus, a condition that causes corneal scarring. These biosynthetic corneas replaced rejection-prone, scarce cadaver corneas.

The J. Craig Venter Institute synthesized the first self-replicating synthetic bacterial cell in 2010. Synthesis of such cells may aid researchers and help develop new drugs.

Synthetic genomics has made it possible to design and assemble chromosomes and genes and gene pathways, which may create green biofuels, pharmaceuticals, and vaccines.

Scientists at the University of Sheffield are mapping the metabolism of the nitrogen-fixing blue-green bacterium *Nostoc*. This organism not only fixes atmospheric nitrogen to ammonia, but also releases hydrogen gas, which could be used as fuel. Once they understand the metabolic process thoroughly, they hope to genetically engineer an organism that can produce hydrogen more efficiently.

Scientists have identified biosynthetic gene clusters for many aminoglycoside antibiotics, including streptomycin, kanamycin, butirosin, neomycin, and gentamicin. A complete understanding of how these antibiotics work may enable scientists to get around the problem of antibiotic-resistant bacteria.

Mass-produced biosynthetic bovine growth hormone, which when injected into dairy cows raises milk production, has been used in many developing countries. However, its use is controversial as questions have arisen regarding its effects on the health of the cows and the people who drink the milk.

Biosensors. Biosensors are microelectronic devices that use antibodies, enzymes, or other biological molecules to interact with an optical device or electrode to record data electronically. Home health-care providers can operate these devices to transmit data obtained from blood or urine samples, for example, to a clinical laboratory some distance away.

Therapeutic proteins. Plasmids transfer human genes that provide the code for proteins essential for growth hormones, blood clotting, and insulin production to bacterial cells.

Disposable micropumps for drug delivery. Disposable micropumps manufactured by Acuros in Germany can deliver a preset amount of liquid hormones, proteins, antibodies, or other medications. An osmotic microactuator, based on osmotic pressure, regulates the amount of drug delivered, and there are no moving parts or power supply components.

High-throughput screening. High-throughput screening can assay more than twenty thousand potentially useful drugs per week using multiwell plates, standard binding assay methodologies, and robotics.

Protein biomarker assays. NextGen Sciences has developed a mass spectrometry method for protein biomarker assays that do not depend on antibodies but instead use surrogate proteins to develop assays. The mass spectrometer measures the amount of surrogate peptides and applies statistical evaluation to assess each biomarker. This first stage requires that a protein be confirmed; then, only these selected proteins are used for the second stage of validation of these protein biomarkers. The mass spectrometry data are used along with carbon-13 or nitrogen-15 isotopically labeled standards to calculate protein concentrations. Reporting the concentration of protein biomarkers is important to allow batches containing hundreds of samples to be analyzed and validated. This technique uses proteomics (the quantitative analysis of proteins based on a physiological response) to allow for much faster development of assays than immunoassays. A wide range of at least 500 plasma proteins and 3,000 tissue proteins can be analyzed at once.

Gene expression databases. Gene Logic's Bio-Express System is a comprehensive genome-wide gene expression database. The BioExpress System allows patients' cells to be collected and analyzed to develop a useful biomarker profile for comparison with a database sample to indicate a therapeutic target. This process is made possible by using high-throughput gene expression profiling of the mononuclear cell fractions present in a blood sample. The software is capable of mining a database that has access to more than 18,000 samples containing biomarkers for the expression of the gene associated with ovarian cancer. This system is also capable of developing biomarker profiles to help diagnosis autoimmune diseases. Autoimmune diseases include rheumatoid arthritis, Crohn's disease, multiple sclerosis, systemic lupus erythematosus, and psoriasis, which affect about 20 million people in the United States.

Biosynthetic temporary skin substitute. A biosynthetic skin substitute is a good treatment for partial-thickness wounds, including skin tears, burns, and abrasions. After applying a gel to the wound's surface, a semipermeable membrane of biosynthetic skin is used to cover the wound for protection from infection. Before developing biosynthetic skin grafts, a physician had to choose between an allograft, which uses cadaver skin, and a xenograft, which uses tissue from another species. Biosynthetic dressings have also been developed. The dressing called "Hydrofiber" contains ionic silver and has been shown to prevent the spread of bacteria.

Needle-free drug delivery systems. The three types of needle-free drug delivery systems are liquid, powder, and depot injections. Each of these types uses some form of mechanical compression to create enough pressure to force the medication into the skin. Although these needle-free delivery systems cost more initially and require more technical expertise because of their complexity, they also have many advantages. In addition to eliminating pain from needle injections and reducing physician visits, these needle-free delivery systems decrease the frequency of incorrect doses. They are being used to deliver anesthetics, chemotherapy injections, vaccines, and hormones.

Nanoparticles. DNA nanotechnology uses discoveries involving nanoparticles and nanomaterials to manipulate DNA's molecular recognition abilities to build tiny medical robots that mimic bond parts or function within cells.

Messenger ribonucleic acid (mRNA) vaccines consist of synthetic RNAs surrounded by lipid nanoparticles. These effectively deliver genes into living cells for vaccination or, potentially, gene and even anticancer therapy.

Hydrogels. Hydrogels are crosslinked, water-loving polymers that do not dissolve in water but are remarkably absorbent. Despite these characteristics, hydrogels form well-defined structures. These properties make hydrogels ideal for several different applications, particularly in the biomedical area.

Hydrogels have various uses in tissue engineering. To grow organs in the laboratory, tissue engineers make hydrogel scaffolds that assume the form of the organ. After making the hydrogel scaffold, tissue engineers apply progenitor cells that attach, grow, degrade the scaffold, and form the organ. These types of three-dimensional culture systems mimic the environment found in the body. Hydrogels are made synthetically or by microorganisms. Cellulose-based hydrogels made with bacterial-made cellulose are biodegradable and renewable. The quality of the cellulose made by bacteria is better than plant-made cellulose for hydrogel production.

Other uses for hydrogels include coating cell culture wells for tissue culture, environmentally smart gels that sense changes in environmental conditions, drug-release hydrogels that slowly deliver drugs into the bloodstream, and surgical hydrogels that help debride dead cells and dying tissues.

These are but a few of the many potential uses of hydrogels.

FUTURE PROSPECTS

The Human Genome Project has facilitated the mapping of genes, which has been instrumental to the development of vaccines to treat influenza, cervical cancer, and malaria and the creation of new diagnostic tools for analysis. As a result, the pharmaceutical industry in the United States has become a multibillion-dollar industry. The generation of biosynthetic products has enhanced the lives of thousands of people through the development of treatments for many types of cancer, pneumonia, cardiovascular diseases, diabetes, tuberculosis, neurological disorders, strokes, blood disorders, and many other diseases.

Combinatorial chemistry has allowed for rapid screening of potentially successful medications that may enhance and extend many people's lives. Typically, only one out of every 5,000 to 10,000 compounds screened makes it through the multiyear process of clinical trials to become an FDA-approved drug. However, the desire to recoup the money spent during the years of research required to bring a drug to market has caused some pharmaceutical companies to launch a product as early as possible, resulting in serious litigation because some drugs caused harmful side effects. Applying biosynthetic growth hormones for nonmedical applications, such as bodybuilding, has also caused ethical and medical controversy. However, as the global population continues to grow and the percentage of elderly persons increases, the need for the products of biosynthetic research will continue to grow.

—Jeanne L. Kuhler, MS, PhD

Further Reading

Arya, Dev. *Aminoglycoside Antibiotics: From Chemical Biology to Drug Discovery*. Wiley-Interscience, 2007.

Dewick, Paul. *Medicinal Natural Products: A Biosynthetic Approach*. John Wiley & Sons, 2009.

Lazo, John, and Peter Wipf. "Combinatorial Chemistry and Contemporary Pharmacology." *The Journal of*

Pharmacology and Experimental Therapeutics, vol. 293, no. 3, Feb. 2000, pp. 705-9.
Pettit, George. *Biosynthetic Products for Cancer Chemotherapy*. Vol. 5. Elsevier Science, 1985.
Savageau, Michael. *Biochemical Systems Analysis: A Study of Function and Design in Molecular Biology*. CreateSpace, 2010.
Spentzos, Dimitri. "Gene Expression Signature with Independent Prognostic Significance in Epithelial Ovarian Cancer." *Journal of Clinical Oncology*, vol. 22, no. 23, Dec. 2004, pp. 4648-58.
Stanforth, Stephen. *Natural Product Chemistry at a Glance*. Wiley-Blackwell, 2006.
Steinle, Heidrun, et al. "Delivery of Synthetic mRNAs for Tissue Regeneration." *Advanced Drug Delivery Reviews*, vol. 179, no. 114007, 2021, doi:10.1016/j.addr.2021.114007.

SEWAGE TREATMENT AND DISPOSAL

Category: Applied microbiology
Anatomy or system affected: None
Specialties and related fields: Biology, commercial products, ecosystems, ecology, environment, environmentalism, industries, life sciences, technology, and applied science, water resources
Definition: the process that removes most of the contaminants from wastewater or sewage and produces both a liquid effluent suitable for disposal to the natural environment and a sludge

KEY TERMS

activated sludge: aerated sewage containing aerobic microorganisms which help to break it down
biochemical oxygen demand (BOD): the amount of dissolved oxygen that must be present in water for microorganisms to decompose the organic matter in the water, used as a measure of the degree of pollution
sewage: wastewater and excrement conveyed in sewers.
wastewater: water that has been used in the home, in a business, or as part of an industrial process

SUMMARY

The Minoan civilization on the island of Crete near Greece had one of the earliest known sewers in the world (ca. 1600 BCE). A large sewer known as the Cloaca Maxima was built during the sixth century BCE in ancient Rome to drain the Forum. The Romans also reused public bathing water to flush public toilets. London, England, had a drainage system by the thirteenth century, but effluent could not be discharged until 1815. Sewers were constructed in Paris, France, before the sixteenth century. Still, less than five percent of the homes were connected by 1893. The widespread introduction of sewers in densely populated areas did not occur until the mid-nineteenth century.

A typical wastewater disposal system consists of a network of pipes, a treatment plant, and an outfall to the ground or, more commonly, to a stream or the ocean. Older wastewater systems are generally combined; domestic, industrial, and stormwater runoff are conveyed in the same pipes to a treatment plant. Although initially cheaper to build, combined systems are less desirable than separated systems because most effluent must bypass the treatment plant during storms, when street runoff rapidly increases. Newer wastewater systems are designed so that separate pipes handle wastewater and storm runoff.

About 60 to 75 percent of the water supplied to a community will wind up as effluent that must be treated and disposed of. The remaining water is used in industrial processes, lawn sprinkling, and other types of consumptive use. Domestic sewage contains varying proportions of human excrement, paper, soap, dirt, food waste, and other substances. Much of the waste substance is organic and is decayed by bacteria. Accordingly, domestic sewage is biodegradable and capable of producing offensive odors. The composition of industrial waste varies from relatively clean rinse water to effluent that can contain corrosive, toxic, flammable, or even explosive materials.

Therefore, communities usually require the pretreatment of industrial effluent.

Aerobic (oxygen-requiring) bacteria decompose the organic material in sewage. However, the dissolved oxygen (DO) in water can be used up in the process of microbial decomposition. Suppose too much organic waste enters the water body. In that case, the biochemical oxygen demand (BOD) can exhaust the DO in the water, thereby damaging the aquatic ecosystem. Indeed, most fish species die in water in which the DO falls below four milligrams per liter for extended periods.

The function of a wastewater treatment plant is to produce a discharge that is free of odors, suspended solids, and objectionable bacteria. The wastewater treatment processes are categorized as primary, secondary, or tertiary. Primary treatment is mostly mechanical, as it involves the removal of floating and suspended solids through screening and sedimentation in settling basins. This form of treatment can remove 40 to 90 percent of the suspended solids and 25 to 85 percent of the BOD.

Secondary treatment involves biological processing in addition to mechanical treatment. One form of biological processing is a trickling filter, where wastewater is sprayed over the crushed stone and allowed to flow in thin films over biological growths that cover the stone. The organisms in the biologic growths, including bacteria, fungi, and protozoa, decompose the dissolved organic materials in the

Wastewter sludge contains bacteria, fungi, and protozoa, so microbes play an important role in sewage treatment. Photo via iStock/DavidOrr. [Used under license.]

wastewater. These growths eventually slough off and are carried to settling tanks by the wastewater flow. Another type of secondary treatment is the activated sludge process. In this procedure, flocs of bacteria, fungi, and protozoa are stirred in the wastewater with results about the same as those achieved with trickling filters. Depending on the efficiency of the plant and the nature of the incoming wastewater, both types of biological processes can remove 50 to 95 percent of the suspended solids and BOD. The efficiency of the secondary treatment can be seriously lowered if the plant's design capacity is overloaded with excessive effluent coming from stormwater runoff in combined sewers. This is one important reason public health officials favor separate sewers, even though they are more expensive. The biologic processes can also be severely affected by toxic industrial waste, killing the "good" bacteria that are crucial in the treatment process. Accordingly, many communities require the pretreatment of industrial wastes.

Tertiary treatment is the most advanced method and consequently the most expensive. It includes several procedures such as ozone, a strong oxidizing agent, to remove most of the remaining BOD, odor, and taste; adding alum to remove phosphate, and denitrification. The final effluent from any treatment level is usually chlorinated before release.

In areas where population densities are lower than about 1,000 people per square kilometer (2,600 per square mile), the costs of a sewer system and treatment plant are difficult to justify. Accordingly, septic systems are commonly used for wastewater disposal in low-density residential areas. In such a system, household effluent is piped to a buried septic tank, which acts as a small sedimentation basin and anaerobic (without oxygen) sludge-digestion facility. The effluent exits from this tank into a disposal field, where aerobic microorganisms degrade dissolved and solid organic compounds. For a septic system to operate effectively, the soil must be of sufficient depth and permeability so that microbial decomposition can occur before the effluent reaches the water table. The Environmental Protection Agency estimated that about twenty-five percent of the homes in the United States used septic systems as of 2014.

—Robert M. Hordon

Further Reading

Hill, Marquita K. "Water Pollution." *Understanding Environmental Pollution.* 3rd ed., Cambridge UP, 2010.

Lester, J., and D. Edge. "Sewage and Sewage Sludge Treatment." *Pollution: Causes, Effects, and Control,* edited by Roy M. Harrison, 4th ed., Royal Society of Chemistry, 2001.

McGhee, Terrence. *Water Supply and Sewerage.* McGraw-Hill, 1991.

Miller, G. Tyler, Jr., and Scott Spoolman. "Water Pollution." *Living in the Environment: Principles, Connections, and Solutions.* 16th ed., Brooks/Cole, 2009.

Qasim, Syed A. *Wastewater Treatment Plants: Planning, Design, and Operation.* 2nd ed., Technomic, 1999.

Roseland, Mark. "Water and Sewage." *Toward Sustainable Communities: Resources for Citizens and Their Governments.* Rev. ed., New Society, 2005.

Salvato, Joseph A. *Environmental Engineering and Sanitation.* 4th ed., John Wiley & Sons, 1992.

BEER AND WINE MAKING

Category: Applied and industrial microbiology
Also known as: Brewing and vinification
Anatomy or system affected: None
Specialties and related fields: Enology, fermentation science, industrial microbiology, microbiology, viticulture
Definition: a process that begins with selecting the right fruit, fermenting the pulp or juices of the fruit into alcohol with specialized microorganisms, and processing and bottling the finished liquid

KEY TERMS

brewing: beer production by steeping a starch source (usually cereal grains or barley) in water and fermenting it with yeast *ethanol*: also known as ethyl alcohol (C_2H_5OH), is the primary metabolic product of yeast-based fermentation

fermentation: a chemical process carried out by microorganisms in which simple sugars, like glucose, are broken down to alcohols, like ethyl alcohol and other organic acids

finings: chemicals added before alcoholic beverages are bottled that remove undesirable compounds like benzenoid organic compounds or copper ions by precipitating them so that they settle at the bottom of the container

Saccharomyces cerevisiae: also known as "the baker's yeast, the primary yeast species used to ferment the juice of grapes and grains into wine and beer

vinification: the process of winemaking

wort: a solution with extracted sugars from starch sources that are fermented to make beer

HISTORY

Aside from water, beer is thought to be the world's oldest beverage. Some anthropologists believe that beer consumption dates from the Neolithic era of the Stone Age (ca. 9000 BCE) when prehistoric people stumbled upon damp, fermented grains, which they ingested and enjoyed enough to figure out the brewing process.

The earliest farmers' first crops were grains. Fermenting grains requires moisture and yeast, which occur naturally and are likely to be near grain crops. As people began to understand the brewing process better, it became more sophisticated.

By the third millennium BCE, beer had become an essential part of human culture. The *Epic of Gilgamesh*, one of the first known works of literature, describes beer as a symbol of sophistication and intelligence. The master brewers in ancient Mesopotamia were primarily women, reflected in the ancient Sumerians' belief in Ninkasi, the goddess of brewing. Archaeologists have discovered what could be considered the first drinking song, the "Hymn to Ninkasi," which doubled as a brewing recipe.

Early civilizations believed that the intoxicating effect of beer was the work of gods, and drunkenness was thought to be holy. The Babylonians, who succeeded the Sumerians, had as many as twenty different beer recipes. Ancient Egyptians improved on the brewing process, using bread dough and dates to make a better-tasting beer. The preservative power of bottling beer was discovered in the sixteenth century in England. Christopher Columbus introduced beer to America and soon observed the natives making corn-based beer.

The history of wine is a bit foggier. Still, it is believed that the first intentionally made wine dates from around 6000 BCE, coincident with the emergence of pottery. As people figured out how to mold and fire clay into pots and jars, they discovered that the porous surface of the substance was perfect for storing and producing wine.

Wine most likely originated in the Middle East and is an important beverage in the Hebrew Bible. Romans brought wine into Western Europe, and eventually, wine was being produced on all inhabited continents.

Almost all the wine made in the world comes from a single species of grape, *Vitis vinifera*, from which more than 4,000 distinct varieties have been developed. Other species, including *V. labrusca* and *V. rotundifolia*, are occasionally used. However, their sugar content tends to be too low to achieve the necessary alcohol content.

Many factors affect the quality and characteristics of wines, including climate, soil type, and topography. Specific growing techniques used by growers or vineyards can also influence wine quality.

Both beverages have evolved and developed over time. Still, the simple, physical process required for making both beer and wine has remained essentially the same since prehistoric people first observed it.

Other beer-making supplies include:

- Airlock (for the opening on the carboy)
- Stoppers
- Racking tube and siphon hose
- Hydrometer for testing gravity
- Long-handled, nonwooden spoon that will fit in the opening of the carboy and reach the bottom

Winemaking supplies are like those used in brewing beer. An 8-gallon plastic bucket with a lid, marked at the 6-gallon point, should be used for fermentation. Avoid using the same bucket for beer and wine since this affects the taste and fermentation. A 2-gallon carboy is also needed.

Other supplies include a long, narrow tube or jar to hold the wine while its gravity is checked. Hydrometers generally come in tubes, which may be used for this step of the process. A wine thief is a pipette used to remove the developing wine from the carboy. The wine may be poured from the carboy into the other containers or bottles, but using a wine thief will make this process much easier and less messy.

All equipment should be clean, and some items, such as the bucket, should be sterilized at the outset. This helps minimize the risk of other microbes contaminating and ruining the brew.

Additional supplies for winemaking:

- Airlock and bung (rubber stopper with a hole in it) for the carboy
- Hydrometer
- Dairy thermometer
- A long piece (at least 5 feet) of food-grade plastic tubing

Photo via iStock/ArtistGNDphotography. [Used under license.]

- Long-handled spoon
- Bottles and corks

Beer has only four basic ingredients (water, yeast, hops, and grain), which can be altered, substituted, or augmented for different types or flavors of beer. Wine, at its simplest, can be made with nothing more than fruit juice, water, and yeast.

TECHNIQUES

Beer production is called "brewing." Brewing has multiple steps, including malting, milling, mashing, lautering, boiling, fermenting, conditioning, filtering, and packaging.

Malting prepares the grains for brewing by releasing the starches in the grains. The first step steeps the gains in water for about forty hours. Next, the grains are spread on the floor of a germination room for approximately five days. After germination, the grains are gradually heated in a kilning (a process called "kilning") for several hours. Upon completion of kilning, the grains are called "malt."

After malting, the grains are crushed in a mill (milling) to expose the sugar-rich endosperm and cotyledons of the grain. The milled grain is mixed with hot water to form a "mash." This step, mashing, activates naturally occurring enzymes in the grains that degrade the complex carbohydrates, like starch, to simpler sugars (*saccharification*). Mashing produces a sugar-rich fluid called a "wort." The wort is strained to remove larger particles (*lautering*). The mash is mixed with water in a large container called a "mash tun in commercial breweries."

The wort is pumped from the mash tun into a large boiling kettle called a "copper." The wort is boiled with hops and other flavorings in the copper, such as herbs, spices, or sugars. Hops create the bitterness, aroma, and flavor that characterizes beer. Boiling also sterilizes the beers and drives several critical chemical reactions in the wort. After boiling, the wort is called "hopped wort." After boiling, the hopped wort settles in a vessel called a "whirlpool." Here large particulates settle out, and the hopped wort is clarified.

From the whirlpool, the hopped wort passes through a heat exchanger that cools it for fermentation (yeast die at temperatures above 60° Fahrenheit). Once in the fermentation tank, the hopped wort is fermented once yeast are added. Fermentation converts the sugars into ethanol and carbon dioxide. After fermentation, the beer goes to a conditioning tank. The beers ages in the conditioning tank and undesirable flavors diminish, and the overall flavor smoothens. Conditioning usually takes a week to several months. After conditioning, the beer is filtered and, sometimes, carbonated. "Finings" are chemical compounds added to beer and other alcoholic beverages that remove undesirable chemicals, adjust the flavor or aroma, and clarify them. Finings remove these molecules by precipitating them into a sediment at the bottom of the cask or bottle. Materials used for fining include bentonite, gelatin, and other compounds. After fining, the beer is bottled or put into casks.

There are two basic types of beer: lager and ale. Generally, ale is sweet, fruity, and full-bodied, while lager has a crisp, clean, often bitter taste. The difference between the two is determined by the type of yeast used and the amount of time spent fermenting. There are also three types of grain ingredients: grain extract, partial mashes, and all-grain.

Extracts are the easiest type to work with and have the smallest margin of error, so beginners should perfect this method before moving on to the more complicated, labor-intensive techniques. Extracts may be powder or syrup and must be fully dissolved in boiling water before adding the other ingredients.

There are two main types of wine: red and white. The color difference is caused by the material used for fermentation. Red wine is made from the pulp, or must, of red and black grapes. Fermentation during red wine production occurs with the grape skins, which

Photo via iStock/Morsa Images. [Used under license.]

convey its color and flavor to the wine. White wine is made by fermenting the juice of the grapes in the absence of the fruit pulp. White wines made from red grapes use fermented juice extracted from the grapes that have experienced minimal contact with the grape skins. Rosé or, as called in some cases, "blush" wines are usually made from red grapes whose juice stays in contact with the must long enough to acquire the pinkish color. Less commonly, red wines are blended with white wines to achieve a pinkish color.

The primary fermentation can occur naturally with yeast present on the grapes or in the air. Alternatively, specific cultures of specific yeast strains are added to the juice for white wine. The fermentation process takes one to two weeks and converts most of the sugar into ethanol and carbon dioxide. The fermented solution is called the "free-run juice," and wineries pump off this material into tanks. If the skins are pressed to extract any remaining juice, this fermented material is called "press juice." Pressed juice has lower acidity (a higher pH), higher potassium levels, more phenolic compounds, and greater quantities of suspended materials like proteins and natural gum. Phenolic compounds like tannins can give the wine greater body (how heavy or light the wine feels in your mouth) and aroma. However, press juice also contains compounds that make the wine bitter. To achieve the desired flavor and characteristics, winemakers can mix free run juice with press juice at varying ratios.

Red wine production requires an aging process in which bacteria carry out the "malolactic conversion." Lactic acid bacteria use the malolactic enzyme to con-

Photo via iStock/Givaga. [Used under license.]

vert the dicarboxylic acid, malic acid, to the monocarboxylic acid, lactic acid.

Malic acid has a tart taste, whereas lactic acid has a somewhat creamy taste. The malolactic fermentation softens the taste of the wine and reduces its astringency. Red wine sometimes is transferred to oak barrels, which imparts tannins and other aromatic compounds to the wine.

Different grape varieties are used for the different wines, and a wine's name usually comes from the type of grape used. Without several acres of prime cropland in a temperate climate, it is challenging to grow grapes specifically for winemaking. Fortunately, many vineyards sell concentrate from their grapes. Unpasteurized grape juice from the supermarket will also work. Still, it is relatively expensive and will not yield a specific variety of wine. Other kinds of fruits, such as plums and elderberries, can also be used. Several wine-based drinks, such as brandy and sangria, can be made with finished wine.

TRENDS

Since the 1980s and 1990s, there has been a resurgence of microbreweries, particularly in the United States, where microbrews are among the most famous beers in the country. This trend has resulted in more nuance and individuality in commercial beers and grew directly out of the popularity of home brewing during the 1970s.

Today, so-called hybrid or blended wines are becoming more popular among commercial and amateur winemakers. The process involves using more than one type of grape or grape juice to create a unique flavor, not possible with just one kind of grape.

Experimentation with producing other, old-fashioned fermented beverages, such as hard cider (made from the juice of whole apples), perry (from pears), and mead (from honey), has become trendy. Botanical brewing with herbs, mushrooms, or foraged items also took root among homebrewers.

Much of the fun of making both beer and wine is experimentation with ingredients and processes. Making a perfect facsimile of a Napa Valley pinot noir or a Red Hook lager is admirable. Still, the true satisfaction of home brewing comes from concocting something unique or difficult to buy.

In the 2010s, automated countertop brewing machines—akin to fancy coffee makers or bread machines—hit the consumer market. Those largely self-contained devices allow users to select a beer recipe, add ingredients (with varying amounts of customization), and launch the process. At the same time, the machine itself controls the actual brewing and clean-up. Companion mobile applications provide instructions or track progress. Other innovative brewing gadgets include kegerators, cooling equipment, and small conical fermenters.

BEER AND WINE MAKING FOR FUN VS. PROFIT

Making beer or wine can be a rewarding and exciting hobby, either alone or as a social activity with friends or club members. Beer and winemaking are both hobbies meant to be shared; serving up the first batch of homebrew to friends and family is a uniquely gratifying experience.

Unfortunately, since homebrews are not regulated or taxed by the government, selling them is illegal; this makes starting a homebrewing business more complicated than simply printing up a homemade label and building a website. A workaround that can help home brewers decide whether to scale up is to license their recipes to professional craft brewers who can distribute the product legally.

Most microbreweries are started by home brewers looking to expand their hobby. Still, many matters beyond legal concerns need to be considered when starting a brewery or a vineyard. As more and more hobbyists turn their craft into businesses, the market has become highly competitive. The same hurdles involved in starting any business, including market research, real estate, and staffing, are augmented by the specifics of beer or winemaking.

Brewing enthusiasts may also find work in brewing-supply companies or established breweries, or microbreweries. Teaching and writing about homebrewing techniques are other ways to profit from this passion.

—Alex K. Rich and Michael A. Buratovich, PhD

Further Reading

The American Wine Society Presents the Complete Handbook of Winemaking. G.W. Kent, Inc., 1993.

Frederick, Matthew. *Homebrewing for Beginners: A Beginner's Guide to Learning the Supplies, Techniques, and Methods for Brewing Beer at Home*. Independently published, 2019.

Higgins, Patrick, Maura Kate Kilgore, and Paul Hertlein. *The Homebrewer's Recipe Guide*. Simon & Schuster, 1996.

Kania, Leon W. *The Alaskan Bootlegger's Bible: Makin' Beer, Wine, Liqueurs and Moonshine Whiskey: An Old Alaskan Tells How It Is Done*. 2nd ed., Happy Mountain Publications LLC, 2019.

Iverson, Jon. *Home Winemaking Step by Step*. Stonemark Publishing Co., 2002.

Miller, Dave. *Dave Miller's Homebrewing Guide*. Storey Publishing, 1995.

Nachel, Marty. *Homebrewing for Dummies*. Wiley Publishing, Inc., 1997.

Palmer, John J. *How To Brew: Everything You Need to Know to Brew Great Beer Every Time.* 4th ed., Brewers Publications; 2017.

Papazian, Charlie. *Microbrewed Adventures*. HarperCollins Publishers, 2005.

Peragine, John. *The Complete Guide to Making Your Own Wine at Home: Everything You Need to Know Explained Simply*. Atlantic Publishing Group Inc., 2010.

Snyder, Stephen. *The Brewmaster's Bible*. HarperCollins Publishers, 1997.

BREAD

Category: Industrial microbiology

Specialties and related fields: Bacteriology, biochemistry, biotechnology, food science, industrial microbiology,

Definition: food made of flour, water, and yeast or another leavening agent, mixed, and baked

KEY TERMS

dietary fiber: also known as roughage; the portion of plant-derived food that human digestive enzymes cannot completely break down

dough: a thick, malleable mixture of flour and liquid, used for baking into bread or pastry

flour: a powder obtained by grinding grain, typically wheat, and used to make bread, cakes, and pastry

grains: wheat or any other cultivated cereal crop used as food

leaven: a substance, typically yeast, that is used in dough to make it rise

yeast: a microscopic, single-celled fungus consisting of single oval cells that reproduce by budding and convert sugar into alcohols or organic acids and carbon dioxide

Depiction of a bread shop, Tacuinum Sanitatis from Northern Italy, beginning of the 15th century. Image via Wikimedia Commons. [Public domain.]

GRAINS: WHAT WE KNOW

As one of the most ancient, prepared foods consumed by humans, bread has become a primary worldwide food source. There are many methods for preparing bread, but the basic ingredients are flour or meal combined with milk or water to form dough, then baked. Bread is usually leavened, which creates gas bubbles that fluff up bread. The most common microbe used as leavening agents is the baker's yeast Saccharomyces cerevisiae. Different microbes can confer unique tastes and qualities to bread. For example, sourdough bread is made with Lactobacillus cultures as their leavening agent. Lactobacilli are lactic acid bacteria that ferment glucose to lactic acid, which lends the bread its unique sour flavor. Lactic acid bacteria thrive in acidic environments. The carbon dioxide they produce helps the dough rise and give it its structure. Other leavening agents include chemicals such as baking soda or high-pressure aeration, which creates the gas bubbles that fluff up bread. Bread can be made from rye, cornmeal, and other grains but is usually made with wheat flour.

The nutritional value of wheat varies greatly depending on the degree of its refinement. In its unprocessed state (i.e., whole wheat), which includes the bran and germ, wheat contains many valuable nutrients, including vitamins B_1, B_2, B_3, and E, manganese, magnesium, calcium, phosphorus, zinc, copper, iron, and tryptophan. It is also a fantastic source of dietary fiber. However, in the United States, most of the wheat used in bread production has been processed into a 60 percent extraction (i.e., 40 percent of the original wheat, including the bran and germ, has been removed), bleached white flour. This degree of refinement strips the wheat of over half of its nutritional value, which is why, in 1941, the United States decided to "enrich" white flour with vitamins B_1, B_2, B_3, and iron. Even with this refortification, white flour is dramatically inferior to 100 percent whole-wheat in its nutritional contribution. Regular consumption of bread and other foods made from refined grain has been associated with weight gain, increased risk for insulin resistance and diabetes mellitus, type-2 (DM2), and cardiovascular disease. In contrast, the

Various leavened breads. Photo by 3268zauber, via Wikimedia Commons.

consumption of whole-grain bread has proven protective against these conditions. A few of the nutritional benefits of consuming 100 percent whole-wheat bread includes lowering cholesterol and blood pressure, slowing the absorption of glucose and stabilizing blood sugar levels, and supporting bowel regularity.

NUTRIENTS IN BREAD

Whole-wheat bread typically provides 2 grams of dietary fiber/slice. Some benefits of dietary fiber include the following:

- It binds with water and slows the digestive process, thus allowing the body to manage better postprandial (i.e., after eating) glucose and insulin responses;
- It can increase the volume of the intestinal contents, which hinders the absorption of cholesterol. The added bulk also promotes more regular bowel movements, promoting intestinal health.

Wheat contains vitamins B_1, B_2, and B_3, which create energy by aiding the breakdown of carbohydrates, providing cardiovascular protection, maintaining the nervous system, and supporting the production of red blood cells, hormones, and necessary cholesterol.

Whole wheat contains betaine, a metabolite of choline, which reduces inflammation.

Whole wheat has numerous phytonutrients (i.e., beneficial plant-derived chemicals), which serve as antioxidants, have anticancer properties, and reduce inflammation. One important phytonutrient in whole-wheat is the lignan, enterolactone, which has estrogen-like effects. Increasing serum levels of enterolactones may help protect against heart disease and hormone-dependent cancers such as breast and prostate cancers.

Wheat germ is rich in vitamin E, a fat-soluble vitamin that functions primarily as an antioxidant and serves to maintain cell membranes, assist in vitamin K absorption, and contribute to the immune system.

DIETARY INTAKE GUIDELINES

The US Food and Drug Administration (FDA) recommends 25 to 30 gm of dietary fiber intake per day, the amount provided in about 2 cups of 100 percent whole-wheat flour.

RESEARCH FINDINGS

Researchers have found that the dietary fiber in wheat can promote the growth of beneficial bacteria (i.e., flora) in the intestines. This prebiotic action increases the formation of fermentation products, such as the short-chain fatty acids (SCFAs), butyrate, propionate, and acetate, which inhibit the growth and induce death of cancerous cells in the colon. At the same time, these SCFAs serve as an energy source to normal cells, enhancing their survival.

Diets high in simple carbohydrates, such as those made from refined wheat flour, are associated with dyslipidemia (i.e., high levels of cholesterol and triglycerides) and diabetes. Even whole grains have a relatively high glycemic index (i.e., elevated blood sugar after eating), which can cause elevated blood sugar resulting in the increased production of insulin. Researchers report that reformulating refined bread products using composite flours (i.e., flours from other starches such as sweet potato) can reduce the postprandial glycemic response. However, it is still vital to emphasize that diet modification for the prevention of obesity, heart disease, and diabetes should be well-rounded, including unsaturated fats, lean proteins, and fruits and vegetables, as well as whole grains.

SUMMARY

Consumers should become knowledgeable about the physiologic risks and benefits of bread. Whole-wheat bread is a good source of fiber, B vitamins, and phytonutrients. These nutrients promote gastrointestinal health, reduce inflammation, lower blood pressure, and may help prevent cardiovascular disease and type 2 diabetes. Research suggests that diets high in refined wheat flour are associated with a higher risk of high cholesterol and diabetes.

—Cherie Marcel, BS

Further Reading

Bodinham, C. L., et al. "Short-Term Effects of Whole-Grain Wheat on Appetite and Food Intake in Healthy Adults: A Pilot Study." *British Journal of Nutrition*, vol. 106, no. 3, 2011, pp. 327-30, doi:10.1017/S0007114511000225.

Burton, P. M., et al. "Glycemic Impact and Health: New Horizons in White Bread Formulations." *Critical Reviews in Food Science & Nutrition*, vol. 51, no. 10, 2011, pp. 965-82, doi:10.1080/10408398.2010.491584.

German, J. B., C. J. Dillard. "Saturated Fats: What Dietary Intake?" *American Journal of Clinical Nutrition*, vol. 80, no. 3, 2004, pp. 550-59.

Gil, A., et al. "Wholegrain Cereals and Bread: A Duet of the Mediterranean Diet for the Prevention of Chronic Diseases." *Public Health Nutrition*, vol. 14, no. 12, 2011, pp. 2316-22, doi:10.1017/S1368980011002576.

Hamelman, Jeffery. *Bread: A Baker's Book of Techniques and Recipes*. 3rd ed., Wiley, 2021.

Tighe, P., et al. "Effect of Increased Consumption of Whole-Grain Foods on Blood Pressure and Other Cardiovascular Risk Markers in Healthy Middle-Aged Persons: A Randomized Controlled Trial." *American Journal of Clinical Nutrition*, vol. 92, no. 4, 2010, pp. 733-40, doi:10.3945/ ajcn.2010.29417.

Youdim, Adrienne. "Fiber." *Merck Manual Consumer Version*, Jan. 2020, www.merckmanuals.com/home/disorders-of-nutrition/overview-of-nutrition/fiber. Accessed 21 Nov. 2021.

Walton, G. E., et al. "A Randomised, Double-Blind, Placebo Controlled Cross-Over Study to Determine the Gastrointestinal Effects of Consumption of Arabinoxylan-Oligosaccharides Enriched Bread in Healthy Volunteers." *Nutrition Journal*, vol. 11, no. 36, 2012, doi:10.1186/1475-2891-11-36.

INDUSTRIAL FERMENTATION

Category: Biotechnology

Anatomy or system affected: None

Specialties and related fields: Biology, biochemistry, bioprocess engineering, biotechnology, chemical engineering, microbiology, organic chemistry

Definition: the intentional use of fermentation by microorganisms such as bacteria and fungi as well as eukaryotic cells like CHO cells and insect cells, to make products useful to humans

KEY TERMS

antioxidant: a chemical that prevents the oxidation of other chemicals

biomass: a mass of organisms; traditionally, this term refers to the biomass of plants and microorganisms

bioreactor: an apparatus for cell growth with practical purposes under controlled conditions. bioreactors are closed systems and vary in size from the small laboratory scale (5 to 10 milliliters) to the large industrial scale (more than 500,000 liters)

enzymes: biological catalysts made of proteins

fermentation: in biology, the metabolic reactions necessary to generate energy in living (mainly microbial) cells; in industry, any large industrial process based on living things is called "fermentation"

fermenter: a type of traditional bioreactor (stirred or nonstirred tanks) where cell fermentation takes place; fermenters can be operated as continuous or batch-culture systems; in continuous culture, nutrients are continuously fed into the fermentation vessel, allowing the cells to ferment indefinitely

probiotics: microorganisms and substances that promote the development of healthy intestinal microbial communities

substrate: a molecule that is broken down by fermentation

INTRODUCTION

Industrial fermentation is an interdisciplinary science that applies principles associated with biology and engineering. The biological aspect focuses on microbiology and biochemistry, and the engineering aspect applies fluid dynamics and materials engineering. Industrial fermentation is associated primarily with the commercial exploitation of microorganisms on a large scale. The microbes used may be natural species, mutants, or microorganisms that have been genetically engineered. Many products of considerable economic value are derived from industrial fermentation processes. Common products such as antibiotics, cheese, pickles, wine, beer, biofuels, vitamins, amino acids, solvents, and biological insecticides and pesticides are produced via industrial fermentation.

DEFINITION AND BASIC PRINCIPLES

Industrial fermentation uses living organisms (mainly microorganisms), typically on a large scale, to produce commercial products or carry out important chemical transformations. Industrial fermentation aims to improve biochemical or physiological processes that microbes can perform while yielding the highest quality and quantity of a particular product. The development of fermentation processes requires knowledge from microbiology, biochemistry, genetics, chemistry, chemical, bioprocess engineering, mathematics, and computer science. The major microorganisms used in industrial fermentation are fungi (such as yeast) and bacteria. Fermentation is performed in large fermenters or other bioreactors, often of several thousand liters in volume. Industrial fermentation is a part of many industries, including microbiology, food, pharmaceutical, biotechnology, and chemical.

BACKGROUND AND HISTORY

Traditional fermentations such as those for making bread, cheese, yogurt, vinegar, beer, and wine had been used by people for thousands of years before its microbial nature was understood. Brewing beer was one of the first applications of fermentation in ancient Egypt as long as 10,000 years ago. The exact origins of dairy products are unknown—it may have been as early as 8000 BCE. It was probably nomadic Turkish tribes in Central Asia who invented cheese and yogurt making. Traditionally, dairy fermentation was a means of milk preservation. The scientific understanding of fermentation began only in the nineteenth century after French scientist Louis Pasteur published the results of his studies on the microbial nature of winemaking.

The first industrial fermentation bioprocesses based on knowledge of microbes appeared in the early twentieth century. Russian biochemist Chaim Weizmann is considered to be the father of industrial fermentation. Weizmann used the bacterium *Clostridium acetobutylicum* to produce acetone from starch in 1916. Acetone was used to make explosives during World War I.

Significant growth of this field began in the middle of the twentieth century when the fermentation process for the large-scale production of antibiotic penicillin was developed. The goal of industrial-scale production of penicillin during World War II led to the development of fermenters by engineers working together with biologists from the pharmaceutical company Pfizer. The fungus Penicillium grows and produces an antibiotic much more effectively under controlled conditions inside the fermenter. Continuous progress in industrial fermentation technology in the twentieth century has followed the development of genetic engineering. Genetic engineering allows gene transfer between species and creates possibilities to generate new products from genetically modified microorganisms grown in fermenters.

The twenty-first century has been characterized by the introduction of biofuels made by industrial fermentation processes. Once again, past and future developments in fermentation technology require contributions from various disciplines, including microbiology, genetics, biochemistry, chemistry, engineering, mathematics, and computer science.

HOW IT WORKS

Industrial fermentation is based on microbial metabolism. Microbes produce different kinds of substances that they use for the growth and maintenance of their cells. These substances can be helpful to humans. The goal of industrial fermentation technology is to enhance the microbial production of valuable substances.

Process of fermentation. In biology, fermentation is a process of harvesting energy from organic molecules in oxygen-free conditions. Sugars are a prime example of what can be fermented, and however, many other organic molecules can be used. Different fermentations are known and are categorized by the substrate metabolized or the type of the product.

In industry, any extensive microbiological process is called "fermentation." Thus, the term fermentation has a different meaning than in biology, and most industrial fermentations require oxygen.

Industrial fermentation organisms. Different organisms, such as bacteria, fungi, and plant and animal cells, are used in industrial fermentation. An industrial fermentation organism must produce the product of interest in high yield, grow rapidly on inexpensive culture media available in bulk quantities, be open to genetic manipulation, and be nonpathogenic (does not cause any diseases).

Fermentation media. Microorganisms need nutrients (substrates) to make the desired product by fermentation. Nutrients for microbial growth are known

Inside of a mash tun while making whiskey. Photo via iStock/MartinM303. [Used under license.]

as media, and most fermentation requires liquid media or broth. General media components include carbon, nitrogen, oxygen, and hydrogen in organic or inorganic compounds. Other minor or trace elements must also be supplied, for example, iron, phosphorus, or sulfur.

Fermentation systems. Industrial fermentation takes place in fermenters, which are also called "bioreactors." Fermenters are closed vessels (to avoid microbial contamination) that reach vast volumes, as many as several hundred thousand liters. Designed by engineers, the primary purpose of a fermenter is to provide controllable conditions for the growth of microbial cells or other cells. Parameters such as pH, temperature, nutrients, fluid flow, and other variables are controlled. There are two kinds of fermenters, those for anaerobic processes (oxygen-free) and aerobic processes. Aerobic fermentation is the most common in the industry. Anaerobic fermenters can be as simple as stainless-steel tanks or barrels. Aerobic fermenters are more complicated. The most critical part of these systems is aeration. In a large-scale fermenter, the transfer of oxygen is essential. Oxygen transfer and dispersion are provided by stirring with impellers or oxygen (air) sparging.

Fermentation control and monitoring. Industrial fermentation control is crucial to ensure that organisms behave properly. In most cases, computers are used for controlling and monitoring the fermentation process. Computers control temperature, pH, cell density, oxygen concentration, level of nutrients, and product concentration.

APPLICATIONS AND PRODUCTS

There is a wide range of industrial fermentation products and applications.

Food, beverages, food additives, and supplements. Industrial fermentation plays a major role in the production of food. Food products traditionally made by fermentation include dairy products (cheeses, sour cream, yogurt, and kefir); food additives and supplements (flavors, proteins, vitamins, and carotenoids); alcoholic beverages (beer, wines, and distilled spirits); plant products (bread, coffee, soy sauce, tofu, sauerkraut); and fermented meat and fish (pepperoni and salami).

Industrial fermentation. The primary and largest industry revolves around food products. Milk from cows, sheep, goats, and horses has traditionally been used to produce fermented dairy products. These products include cheese, sour cream, kefir, and yogurt. More recently, so-called probiotics appeared and have been marketed as health-food drinks. Dairy products are produced via fermentation using lactic bacteria such as *Lactobacillus acidophilus* and *Bifidobacterium*. Fungi are also involved in making some cheeses. Fermentation produces lactic acid and other flavors and aroma compounds that make dairy products taste good.

Many industrial fermentation products are added into food as flavors, vitamins, colors, preservatives, and antioxidants. These products are more desirable than food additives produced chemically. Many of the vitamins are made by microbial fermentation, including thiamine (vitamin B_1), riboflavin (vitamin B_2), cobalamin (vitamin B_{12}), and vitamin C (ascorbic acid). Vitamin C is a vitamin and is also an important antioxidant that helps prevent heart diseases. Carotenoids are another effective antioxidant, and they are also used as a natural food color for butter and ice cream. Carotenoids are red, orange, and yellow pigments produced by bacteria, algae, and plants.

Food preservatives are yet another product of industrial fermentation. Organic acids, mainly lactic and citric acids, are extensively used as food preservatives. Some of these preservatives (such as citric acid) are used as flavoring agents. A mixture of two bacterial species (*Lactobacillus* and *Streptococcus*) is usually used for the industrial production of lactic acid. The mold *Aspergillus niger* is used for citric acid manufacturing. Another common preservative is the protein nisin. Nisin is produced via fermentation by the bacterium *Lactococcus lactis*. It is employed in the dairy industry, especially for the production of processed cheese.

Antibiotics and other health-care products. Antibiotics are chemicals produced by fungi and bacteria that kill or inhibit the growth of other microbes. They are the second most significant product of industrial fermentation. Most antibiotics are generated by molds or bacteria called "actinomycetes." More than 4,000 antibiotics have been isolated from microorganisms, but only about fifty are produced regularly. Among them, beta-lactams, such as penicillins and tetracyclines, are most common. Penicillin is produced by the mold *Penicillium chrysogenum* via corn fermentation in bioreactors of up to 200,000 liters.

The other primary health-care products produced with the help of industrial fermentation are bacterial vaccines, therapeutic proteins, steroids, and gene therapy vectors. There are two categories of bacterial vaccines: living and inactivated vaccines. Living vaccines consist of weakened, also known as attenuated, bacteria. Examples of live vaccines include diseases such as anthrax, which is caused by *Bacillus anthracis*, and typhoid fever, caused by *Salmonella typhi*. Inactivated vaccines are composed of bacterial cells or their parts that have been inactivated by heat or formaldehyde. Examples of these vaccines are those for meningitis, whooping cough, and cholera. Vaccine production takes place in fermenters no larger than 1,000 liters in volume. It requires highly controlled operations to avoid the release of bacteria into the environment. All exhaust gases pass through sterilization processes.

Therapeutic proteins include growth hormone, insulin, wound-healing factors, and interferon. Previously, such compounds were made from animal tissues and were very expensive to manufacture. Genetic engineering now allows their production by fermentation from bacteria. Human growth hormone is synthesized in the human brain and controls growth. Too little growth hormone can cause some cases of dwarfism. The American company Genentech started producing human growth hormone from genetically modified *Escherichia coli* by fermentation in 1985. Insulin is an animal and human hormone that is involved in the regulation of blood sugar. The body's inability to make sufficient insulin causes diabetes. Insulin extracted from pigs had been used to treat diabetes. However, it has been replaced by insulin produced by industrial fermentation from genetically modified bacteria.

Chemicals. Numerous chemicals, such as amino acids, polymers, organic acids (citric, acetic, and lactic), and bioinsecticides are produced by industrial fermentation. Amino acids are used in foods and animal feed and the pharmaceutical, cosmetic, and chemical industries. Bacteria such as *Kocuria rhizophila* and *Corynebacterium glutamicum* are used for industrial fermentation to produce chemicals. Bacterial toxins are effective against different insects. Since the 1960s, preparations of the bacteria *Bacillus thuringiensis* have been produced by fermentation as a biological insecticide.

Enzymes. Enzymes are used in many industries as catalysts. Microorganisms are the favored source for industrial enzymes. Seventy percent of these enzymes are made from Bacillus bacteria via fermentation. Most commercial microbial enzymes are hydrolases, which break down different organic molecules such as proteins and lipids. The enzyme glucose isomerase is vital in producing fructose syrups from corn and is widely used in the food industry.

Biomass production. During biomass production by fermentation, the cells produced are the products. Biomass is used for four purposes: as a source of protein for human food or animal feed, in the industry as fermentation starter cultures, in agriculture as a pesticide or fertilizer, and as a fuel source.

One major product of this application of industrial fermentation is baker's yeast biomass. Baker's yeast is required for making bread, bakery products, beer, wine, ethanol, microbial media, vitamins, animal feed, and biochemicals for research. Yeast is produced in large aerated fermenters of up to 200,000 liters, and molasses is used as a nutrient source for the cells. Yeast is recovered from fermentation liquid by centrifugation and then dried. It can then be sold as compressed yeast cakes or dry yeast.

Many bacteria have been considered potential sources of protein to fulfill the food needs in some countries of the world. As of 2011, only a few species are cultivated worldwide as a source of food and feed. Among them, cyanobacteria are the most popular. The protein level of the cyanobacterium *Spirulina* can be as high as those found in meat, nuts, and soybeans, from 50 to 70 percent. This cyanobacterium has been used as human food for millennia in Asia, parts of Africa, and Mexico.

Apart from yeast and bacteria, people are also using the biomass of algae. Algae are a source of animal feed, plant fertilizer, chemicals, and biodiesel. Because light is necessary to grow algae, the biomass is produced in open ponds, transparent tubular glass, or plastic bioreactors, called "photobioreactors."

Biofuels. Industrial fermentation is used in the production of biofuels, mainly ethanol and biogas. The action of microorganisms in bioreactors produces these two biofuels. Fermentation can also be used for the generation of biodiesel, butanol, and biohydrogen. Biofuels are considered, by many, as a future substitute for fossil fuels. Pollution from fossil fuels affects public health and causes global climate change due to carbon dioxide (CO_2) release. Using biofuels as an energy source generates fewer pollutants and little or no CO_2.

Production of ethanol is a process based on fungal or bacterial fermentation of a variety of materials. In the United States, most ethanol is produced by yeast (fungal) fermentation of sugar from cornstarch. Sugar is extracted using enzymes, and then yeast cells convert the sugar into ethanol and CO_2. Ethanol is separated from the fermentation broth by distillation. Brazil, the second-largest ethanol producer after the United States, uses sugarcane fermentation to generate ethanol. The Brazilian production of ethanol from sugarcane is more efficient than the American corn-based ethanol.

Biogas is produced during the anaerobic (nonoxygen) fermentation of organic matter by communities of microorganisms (bacteria and Archaea). There are different types of biogas. One type contains a mixture of methane (50 to 75 percent) and CO_2. Another type is composed primarily of nitrogen, hydrogen, and carbon monoxide (CO) with trace amounts of methane. Methane is generated by microorganisms called "Archaea" and is an integral part of their metabolism. Biogas is generated from wastewater, animal waste, and "gas wells" in landfills for practical use.

IMPACT ON INDUSTRY

Industrial fermentation plays a significant role in many multibillion-dollar industries, including food, pharmaceutical, microbiological, biotechnological, chemical, and biofuel. The United States maintains a dominant position in the world in industrial fermentation. The first large-scale industrial fermentation process to produce the antibiotic penicillin was developed in the United States. Many other developed and developing countries use industrial fermentation to produce varieties of products. Some countries have made industrial fermentation a prime national interest. Brazil, for example, is using industrial fermentation for ethanol production. This country has the largest and most successful ethanol for fuel program in the world. As a result of this successful program, Brazil reached complete self-sufficiency in energy supply in 2006.

Government and university research. Governmental agencies such as the National Science Foundation (NSF), the United States Department of Energy (DOE), and the United States Department of Agriculture provide funding for research in the industrial fermentation area. Currently, a vast majority of research is concentrated on biofuel generation by microorganisms and environmental applications.

Industry and business. Scientists in the industry traditionally carry out a significant load of research in the fermentation area. Pfizer and Merck were pioneers in the industrial production of the first antibiotic penicillin produced by fungal fermentation. A significant proportion of industrial fermentation research in industry has been directed to health-care products (such as antibiotics).

Major corporations. Examples of major corporations in the food and beverage industries are Kraft Foods, Dannon, Coors, Guinness, and Anheuser-Busch. In the biofuel industry, major companies are Archer Daniels Midland, Poet Energy, Abengoa Bioenergy, and VeraSun Energy.

CAREERS AND COURSEWORK

There are several career options for people who are interested in being trained in industrial fermentation. Food, biotechnology, microbiology, pharmaceutical, chemical, and biofuel companies are the biggest employers in the area. Students interested in researching industrial fermentation can find jobs in university, government, and industry laboratories.

When choosing a career in industrial fermentation, one should be prepared for interdisciplinary science. Students should obtain skills in microbiology, molecular biology, bioengineering, plant biology, organic chemistry, biochemistry, agriculture, bioprocess engineering, and chemical engineering.

Most professionals in industrial fermentation have a bachelor's degree in biology, microbiology, or biotechnology. Individuals with managerial responsibilities often have a master's or doctorate in biology, microbiology, fermentation, molecular biology, biochemistry, biotechnology, bioprocess or chemical engineering, or genetics. Some universities, including the University of California, Davis, and Oregon State University, offer degrees in fermentation.

A career in industrial fermentation presents various work options such as research, process development, production, technical services, or quality control. Some industrial fermentation specialists may be considered genetic engineers (using deoxyribonucleic acid (DNA) techniques to modify living organisms). In contrast, others are classified as bioprocess or chemical engineers (optimizing bioreactors and biochemical pathways for the desired product).

SOCIAL CONTEXT AND FUTURE PROSPECTS

Industrial fermentation plays a major role in providing food, chemicals, and fuels. End users are consumers, farmers, medical doctors, and industrialists. Industrial fermentation is changing the course of history. People have made food by fermentation for centuries. In the twentieth century, the development of antibiotics and their production by industrial fermentation had the most significant impact on the practice of medicine than any other development. The growth of the industrial fermentation field is continuing rapidly. Since the beginning of the twenty-first century, industrial fermentation has undergone unprecedented growth and expansion due to biofuel introduction. This record growth is evident in the US ethanol industry. In 1980, the US ethanol industry produced 175 million gallons of ethanol by fermentation, and in 2009, 10.6 billion gallons.

The role of industrial fermentation in human society is likely to expand in the future because of increasing requirements for resources.

—Sergei A. Markov, PhD

Further Reading

Bourgaize, David, Thomas R. Jewell, and Rodolfo G. Buiser. *Biotechnology: Demystifying the Concepts.* Benjamin/Cummings, 2000.

Doran, Pauline M. *Bioprocess Engineering Principles.* Academic Press, 1995.

Glazer, Alexander N., and Hiroshi Nikaido. *Microbial Biotechnology: Fundamentals of Applied Microbiology.* 2nd ed., Cambridge UP, 2007.

Lydersen, Bjorn K., Nancy A. D'Elia, and Kim L. Nelson, editors. *Bioprocess Engineering: Systems, Equipment and Facilities.* John Wiley & Sons, 1994.

Stanbury, Peter F., et al. *Principles of Fermentation Technology.* 3rd ed., Butterworth-Heinemann, 2016.

Wright, Richard T., and Dorothy F. Boorse. *Environmental Science: Toward a Sustainable Future.* 11th ed., Benjamin/Cummings, 2011.

LACTIC ACID

Category: Industrial microbiology

Also known as: Lactate

Anatomy or system affected: Blood, cells, circulatory system, liver, skeletal muscles

Specialties and related fields: Bacteriology, biochemistry, biotechnology, hematology, microbiology

Definition: an organic acid ($C_3H_6O_3$) generally found in muscle as a byproduct of anaerobic glycolysis, produced in carbohydrate matter usually by bacterial fermentation, and used especially in food and medicine and in industry

KEY TERMS

constant region: The highly conserved C-terminal portion of the antibody

Fc-fusion proteins: specific proteins of interest fused to the Fc region of the antibody at its C-terminal end

fermentation: a metabolic process that extracts energy from carbohydrates and other organic molecules in the absence of oxygen

lactate dehydrogenase: an enzyme that catalyzes the reduction of pyruvate to lactic acid
lactic acid bacteria: a group of bacteria that ferment glucose to lactic acid
malolactic fermentation: a process in winemaking in which tart-tasting malic acid is converted to the softer tasting lactic acid
pyruvic acid: the product of the metabolic pathway known as glycolysis in which glucose is degraded in a series of enzyme-catalyzed reactions that yields energy, and pyruvic acid

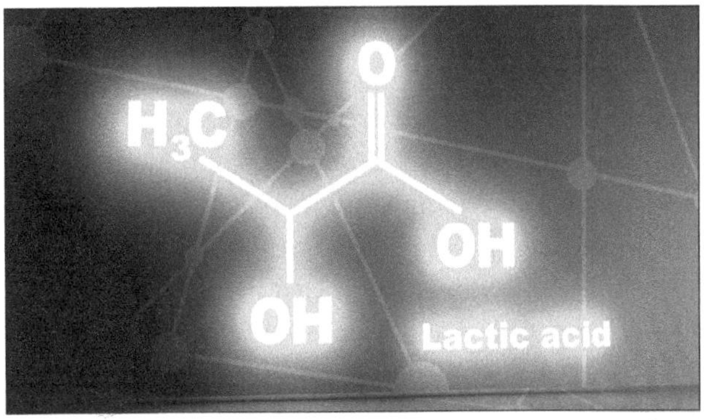

Formula for lactic acid. Image via iStock/Evgeny Gromov. [Used under license.]

INTRODUCTION

Lactic acid is a naturally occurring substance. It is created both as a byproduct of glucose, or sugar, metabolism in animals and by bacteria in dairy products, wine, and other fermented foods. In addition, lactic acid may be reproduced chemically and used as an ingredient in food, cosmetics, and other products.

LACTIC ACID IN METABOLISM

Animals and human beings require both fuel and oxygen to move and function. The process by which fuel and oxygen are used to generate the energy needed to make the body's systems function properly is called "metabolism." As various parts of the body metabolize fuel, a complex series of reactions take place. Some of these processes create lactic acid, a clear or slightly yellowish water-soluble, syrupy liquid.

When the body uses stored glucose as fuel but does not have sufficient oxygen to completely use the glucose—as in anaerobic exercise, which is done in short, intense bursts—glucose converts to lactic acid. Aching muscles following an intense workout can be attributed to lactic acid. The ache eventually diminishes as blood removes excess lactic acid over time.

Some parts of the body, namely the heart, liver, and kidneys, can convert excess lactic acid into carbon dioxide, which the body removes via the respiratory system through exhalation. The kidneys and liver can use lactic acid to produce more glucose through a process known as the Cori cycle. Sometimes, however, the body produces more lactic acid than it can process and remove or neutralize. When this happens, a condition known as lactic acidosis occurs. This condition is rare in otherwise healthy people. Usually, it results from system impairment caused by conditions such as carbon monoxide or cyanide poisoning, diabetes, heart attack, or excessive alcohol consumption.

Blood tests that detect lactic acidosis are often part of the process used to diagnose conditions such as heart failure, diseases of the liver or lungs, and sepsis infections.

LACTIC ACID IN DAIRY, WINE, AND OTHER FOODS

Lactic acid is part of the natural fermentation process that results in yogurt, kefir, cheese, buttermilk, sour cream, some sausages, olives, and fermented vegetable dishes such as sauerkraut and kimchee. It was first identified in 1780 by Swedish chemist Carl Wilhelm Scheele (1742-86). Louis Pasteur (1822-95), the famed French scientist, discovered the first lactobacillus bacteria in 1856. By the late 1800s, researchers were able to produce lactic acid commercially. However, researchers have determined that humans have taken advantage of lactic acid-generating bacteria to culture foods for at least four thousand years and possibly much longer.

Many types of bacteria produce lactic acid as part of their digestive process. Also known as lactobacilli, or LAB, these bacteria are ubiquitous. They are found in many of the mucosal membranes of the human body, such as the mouth and intestines. Lactobacilli are considered probiotics that produce beneficial effects in the digestive system. Lactobacilli are so efficient that people could harness their power to ferment and preserve food long before the bacteria were identified or the process was understood.

One early discovery was the bacteria's ability to convert highly perishable milk to products that could be stored and used longer. In this process, the lactobacilli that occur naturally in the environment around the milk transform the lactose, or milk sugar, into lactic acid. As more and more lactic acid is produced, the original protein in the milk curdles and changes. The product develops a different taste and texture. It is this lactic acid process that gives these dairy products their characteristic tangy taste and thicker texture. The same bacteria serve as probiotics in the human digestive system, enhancing naturally occurring within the body. Probiotics have been proven to help with the treatment of intestinal inflammation and the prevention of urinary tract infections, among other conditions.

Winemakers rely on a process known as malolactic fermentation (MLF) to help wines develop a proper level of acidity. In this process, malic acid—a tart, naturally occurring acid found in grapes—is converted into mellower lactic acid by lactobacilli.

Bacteria also are responsible for generating the lactic acid that plays a key role in fermenting various food products, such as beer, ciders, olives, pickles, sourdough bread, and more. Lactic acid helps prevent the growth of more dangerous organisms that cause food spoilage and lead to food-borne illnesses.

OTHER USES

Lactic acid can be reproduced chemically and is an additive in many products. Creams and lotions containing lactic acid are applied topically to the skin to treat various conditions, such as dry skin, eczema, and keratosis. Lactic acid, along with salicylic acid and urea, are keratolytics, which means they help to break down keratin. Keratin is what holds the top layer of human skin together. This layer is comprised of dead skin cells, which lactic acid and other keratolytics help dissolve and remove, allowing the next layer to absorb water and other emollients that help moisturize the skin.

Other applications include a natural descaling and antisoap scum agent, a detergent, and an antibacterial agent.

—Janine Ungvarsky

Further Reading

Cheryan, Munir. "Lactic Acid." *University of Illinois at Urbana-Champaign*. Munir Cheryan, faculty.fshn.illinois.edu/~mcheryan/lactic.htm. Accessed 22 Jan. 2016.

Ikeda, David M., et al. "Natural Farming: Lactic Acid Bacteria." *Sustainable Agriculture*. College of Tropical Agriculture and Human Resources, University of Hawaii at Manoa, Aug. 2013, www.ctahr.hawaii.edu/oc/freepubs/pdf/sa-8.pdf. Accessed 22 Jan. 2016.

"Is Lactic Acid a Four-Letter Word?" *PBS LearningMedia*. PBS & WGBH Educational Foundation, www.pbslearningmedia.org/resource/tdc02.sci.life.cell.lactic/is-lactic-acid-a-four-letter-word/. Accessed 22 Jan. 2016.

"Lactic Acid." *Calwineries*. Calwineries Inc., www.calwineries.com/learn/wine-chemistry/wine-acids/lactic-acid. Accessed 22 Jan. 2016.

"Lactic Acid Bacteria—Their Uses in Food." *European Food Information Council*. European Food Information Council, www.eufic.org/article/en/artid/lactic-acid-bacteria/. Accessed 22 Jan. 2016.

"Lactic Acid Test." *Medline Plus*. US National Library of Medicine, www.nlm.nih.gov/medlineplus/ency/article/003507.htm. Accessed 22 Jan. 2016.

"Lactic Acid Topical." *WebMD*. WebMD, LLC, www.webmd.com/drugs/2/drug-64136-762/lactic-acid-topical/keratolyticemollients-topical/details. Accessed 22 Jan. 2016.

Zacharof, Myrto P., and Robert W. Lovitt. "*Lactobacilli*: Their Role and Importance in Contemporary Food and Pharmaceutical Industry, Past-Present-Future." *Swansea University*, U.K. PowerPoint. Accessed 22 Jan. 2016.

Microbial Symbioses

Organisms live together in many ways. This section examines some of the more famous examples of symbiosis between microbes and multicellular organisms. Additionally, we discuss the well-known example of lichens in which saprophytic fungi live in a mutualistic relationship with photosynthetic algae. Topics include mycorrhizae, ruminants, and termites.

MYCORRHIZAE

Category: Mycology
Anatomy or system affected: None
Specialties and related fields: Agriculture, evolutionary biology, mycology, silviculture
Definition: a mutualistic, symbiotic relationship between plant roots or other underground organs and fungi; mycorrhizae are among the most abundant symbioses in the world

KEY TERMS

hypha (pl: hyphae): the individual filaments of a fungus
mycelium: the complete structure of the fungal growth
mycobiont: the fungal partner in a plant-fungus symbiosis
symbiont: one of the partnering species in a symbiotic association
symbiosis: a relationship in which two species form a mutually beneficial partnership

FUNGAL ASSOCIATIONS

Mycorrhizal associations have been described in virtually all economically important plant groups. Investigators in Europe detected fungal associations in most European species of flowering plants, all gymnosperms, ferns, and some bryophytes, especially the liverworts. Similar patterns are predicted in other ecosystems. Continuing studies of ecosystems, from boreal forests to temperate grasslands to tropical rain forests and agroecosystems, also suggest that most plant groups are intimately linked to one or more species of fungus.

It is theorized that most of the plants in stable habitats where competition for resources is common probably have some form of mycorrhizal association. Species from all major taxonomic groups of fungi, including the *Ascomycotina*, *Basidiomycotina*, *Deuteromycotina*, and *Zygomycotina*, have been found as partners with plants in mycorrhizae.

Considering the prevalence of mycorrhizae in the world today, botanists theorize that mycorrhizae probably arose early in the development of land plants. Some suggest that mycorrhizae may have been an important factor in the colonization of land.

The fungal partner (or mycobiont) in a mycorrhizal relationship benefits by gaining a source of carbon. Often these mycobionts are poor competitors in the soil environment. Some mycobionts have coevolved to the point that they can no longer live independently of a plant host.

The plant partner in the mycorrhizal relationship benefits from improved nutrient absorption. This may occur in different ways; for example, the mycobiont may directly transfer nutrients to the root. Infected roots experience more branching, thus increasing the volume of soil that the plant can penetrate and exploit. Evidence also suggests that mycorrhizal roots may live longer than roots without these associations. Comparison of the growth of plants without mycorrhizae to those with fungal partners suggests that mycorrhizae enhance overall plant growth.

TYPES OF MYCORRHIZAE

Mycorrhizae may be classified into two broad groups: endomycorrhizae and ectomycorrhizae. Endomycorrhizae enter the cells of the root cortex. Ectomycorrhizae colonize plant roots but do not invade root cortex cells.

The most common form of endomycorrhizae is the vesicular-arbuscular mycorrhizae. The fungi involved are zygomycetes. These mycorrhizae have internal structures called "arbuscules," which are highly branched, thin-walled tubules inside the root cortex cells near the vascular cylinder. It is estimated that 80 percent of all plant species may have vesicular-arbuscular mycorrhizae. This type of mycorrhiza is especially important in tropical trees.

There are several other subtypes of endomycorrhizae. Ericoid mycorrhizae, found in the family Ericaceae and closely related families, supply the host plants with nitrogen. These are usually restricted to nutrient-poor, highly acidic conditions, such as heathlands. Arbutoid mycorrhizae, found in members of the Arbutoideae and related families, share some similarities with ectomycorrhizae. They form more developed structures called the "sheath" and "Hartig net."

Monotropoid mycorrhizae, found in the plant family Monotropaceae, are associated with plants that lack chlorophyll. The host plant is completely dependent on the mycobiont, which also has connections to the roots of a nearby tree. Thus the host, such as *Monotropa*, indirectly parasitizes another plant using the mycobiont as an intermediate. Orchidaceous mycorrhizae are essential for orchid seed germination.

Ectomycorrhizae are common in forest trees and shrubs in the temperate and subarctic zones. Well-developed fungal sheaths characterize these mycorrhizae, along with special structures known as Hartig nets. Basidiomycetes are the usual mycobionts and often form mushrooms or truffles. Ectomycorrhizae help protect the host plant from diseases by forming a physical fungal barrier to infection.

ANATOMY AND DEVELOPMENT

Individual filaments of a fungal body are called "hyphae," and the entire fungal body is called a "mycelium." Root infection may occur from fungal spores that germinate in the soil or from fungal hyphae growing from the body of a nearby mycorrhiza. When infection occurs, hyphae are drawn toward certain chemical secretions from a plant root.

In ectomycorrhizae, root hairs do not develop in roots after infection occurs. Infected roots have a fungal sheath or mantle that ranges from 0.0008 to 0.00016 inches (20 to 40 micrometers) thick. Fungal hyphae penetrate the root by entering between epidermal cells. These hyphae push cells of the outer root cortex apart and continue to grow outside of the cells. This association of hyphal cells and root cortex cells is named a Hartig net. In ectomycorrhizae, the mycobionts never invade plant cells, nor do they penetrate the endodermis or enter the vascular cylinder. Fungi may ensheathe the root tip, but the apical meristem is never invaded. Main roots experience fewer anatomical changes than lateral roots after infection. Lateral roots become thickened, may show the development of characteristic pigments, and grow very slowly. Infected roots also show different branching patterns than those of uninfected roots.

Endomycorrhizae are highly variable in structure. Many endomycorrhizae do not have sheaths or Hartig nets. In all endomycorrhizae, hyphae penetrate root cortex cells, while portions of the mycelium remain in contact with the soil. The hyphae that remain in the soil are important in fungal reproduction and produce large numbers of haploid spores. Fungi do not invade root meristems, vascular cylinders, or chloroplast-containing cells in the plant.

Some of the host cells contain fungal extensions called "vesicles" that are filled with lipids. Vesicles are specialized structures that are often thick-walled and may serve as storage sites or possibly in reproduction. Vesicles are also produced on the hyphae that grow in the soil. Near the vascular cylinder, the hyphae branch dichotomously and form large numbers of thin-walled tubules known as arbuscules that invade host cells. The arbuscules cause the host membranes to fold inward, creating a large surface area of the plant-fungus interface. The arbuscules last for about fourteen days before they break down on their own or are digested by the host cell. Host cells whose fungal arbuscules have broken down may be reinvaded by other hyphae.

—Darrell L. Ray

Further Reading

Deacon, J. W. *Modern Mycology*. 3rd ed., Blackwell Science, 1997.

Harley, J. L., and S. E. Smith. *Mycorrhizal Symbiosis*. Academic Press, 1983.

Raven, Peter H., Ray F. Evert, and Susan E. Eichhorn. *Biology of Plants*. 6th ed., W. H. Freeman/Worth, 1999.

Lichens

Category: Plants and vegetation
Anatomy or system affected: None
Specialties and related fields: Botany, ecology, mycology, phycology
Definition: a complex life form that is a symbiotic partnership of two separate organisms, a fungus, and an alga

KEY TERMS

mutualism: a symbiotic association between organisms of two different species in which each benefit

symbiosis: any close and long-term biological interaction between two different biological organisms, mutualistic, commensalistic, or parasitic.

DEFINITION

Lichens are widespread organisms found in widely diverse habitats, including rock surfaces, trees, and human-made structures. Lichens use structures on which they grow as hosts to support their growth. They are composites composed of two different organisms, an alga, and a fungus, resulting in a symbiotic organism that has a morphology very different from the two original organisms. In lichen symbiosis, the alga provides energy through photosynthesis, and the fungus provides protection and support. Lichens come in three types: crustose, a crusty form that grows tightly on rocks or trees; foliose, which resembles foliage; and fruticose, which has the appearance of "fingers." Lichens provide food for animals, such as reindeer living in arctic regions and habitats for invertebrates. Lichens absorb nutrients from air and rain and are an important part of nutrient recycling. They grow very slowly, with rates less than 5 millimeters per year.

Some species even have growth rates of less than 0.5 millimeters per year. Lichens often live for long periods, with some species in the arctic estimated to be over five thousand years old.

SIGNIFICANCE FOR CLIMATE CHANGE

Because of their sensitivity to environmental factors, including temperature changes, lichens can serve as indicator species (or canaries in the coal mine) to predict environmental changes. Historically, lichens have been important indicators of pollution and climate change. Many long-term climate change studies have used lichen growth to estimate changes in environmental temperature over time. Mapping the distribution of climate-sensitive species indicates current climatic conditions, whereas monitoring over time reveals past climate change effects. Lichen growth studies have been an important part of the debate over greenhouse effects by providing data that support climate change and global warming. Because of global warming, arctic-alpine species have been diminishing, while more tropical species have flourished.

Lichens grow at prolonged rates, increasing in diameter as they grow. The size of individual lichen

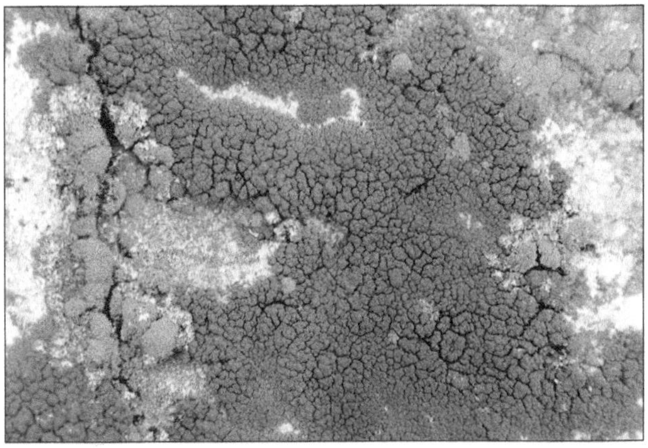

Orange lichens growing on a rock. Photo via iStock/Coica. [Used under license.]

patches on rocks can be used to estimate age. Measuring the diameter of the largest lichen on a rock surface is a method to determine the period during which that rock was exposed. The study of lichen growth to determine the age of or to "date" surfaces is called "lichenometry." The slow growth and longevity of crustose lichens have made them especially useful in lichenometry. Measurement of lichens has been used to document effects of global warming, such as glacial deposits, the former extent of persistent snow cover, and avalanche activity. Most climate change studies using lichens have been conducted using a group of crustose lichens of the genus *Rhizocarpon*, which is abundant in many Arctic environments.

—*C. J. Walsh*

Further Reading

Allen, Jessica L., et al. *Urban Lichens: A Field Guide for Northeastern North America*. Yale UP, 2021.

Brodo, Irwin M., et al. *Keys to Lichens of North America*. Yale UP, 2016.

Nash, Thomas H. *Lichen Biology*. 2nd ed., Cambridge UP, 2008.

Smith, Lorrain Annie. *Lichens*. Library of Alexandria, 2020.

Ruminants

Category: Applied microbiology

Anatomy or system affected: Rumen

Specialties and related fields: Animal science, biochemistry, evolutionary biology, fermentation science, mammalogy, microbial ecology, microbiology

Definition: even-toed ungulate mammals that chew the cud regurgitated from their rumen; the ruminants comprise the cattle, sheep, antelopes, deer, giraffes, and their relatives

KEY TERMS

archaea: a group of single-celled prokaryotic organisms with distinct molecular characteristics separating them from bacteria and eukaryotes

artiodactyl: an herbivore that walks on two toes, which have evolved into hoofs

carnivore: an animal that eats only animal flesh

esophagus: the tube through which food passes from mouth to stomach

gestation: the term of pregnancy

herbivore: an animal that eats only plants

methanogens: methane-producing bacteria, especially archaeans that reduce carbon dioxide to methane

nutrient: a nourishing food ingredient

omnivore: an animal that eats both plants and animals

DEFINING THE RUMINANTS

Ruminants are herbivorous animals that store their food in the first chamber of the stomach, called the "rumen," when it is first swallowed, then after some digestion has taken place, regurgitate it as "cud," which is chewed again and reswallowed into another chamber of the stomach for further digestion. This maximizes the amount of nutrition the animal can derive from hard-to-digest plant food. Wild ruminants tend to eat very quickly, getting as much food mass into their rumens as possible, then retiring to places of safety where they can digest at their leisure. Ruminants include sheep, cows, camels, pronghorns, deer, goats, and antelope. They eat lichens, grass, leaves, and twigs.

The main ruminant suborders are the *Tylopoda* and *Pecora*. Tylopods have three-chambered stomachs. Examples are camels and llamas. Pecorids are sheep, goats, antelope, deer, and cattle. Most pecorids have horns or antlers. These true ruminants have four-chambered stomachs, whose compartments are the rumen, reticulum, psalterium, and abomasum. The abomasum is most similar to the stomachs of nonruminant mammals. At the same time, the other three compartments are developments peculiar to ruminants.

Ruminants chew or grind food between their lower molars and a hard pad in the gums of the upper jaw. The rumen collects partly chewed food when it is first swallowed. The food undergoes digestion in the rumen and passes into the reticulum. There, it is softened by further digestion into cud. Then the reticulum returns the cud to the mouth for rechewing so that it is mixed with more saliva. Swallowing the chewed cud next sends food to the third compartment, the psalterium, for more digestion. The psalterium empties into the abomasum, where food mixes with gastric juices and digestion continues. Finally, the food enters the intestine, which absorbs nutrients carried through the body through blood.

The rumen is a large fermentation chamber filled with a remarkable microbial community that digests the food for ruminants. Ruminants do not produce the enzymes they need to degrade the plant polysaccharides they eat. Instead, the microbes in the rumen do it for them. The rumen microbiome degrades starch and cellulose to simpler sugars (monosaccharides). The rumen microbes ferment the sugars to volatile fatty acids that serve as a nutrient source for ruminants.

Rumen microbes are classified as cellulose-degrading, starch-degrading, and protein-degrading, depending on which macromolecules they degrade. Archaea within the rumen include methane-producing methanogens. These archaea take the hydrogen and carbon dioxide gases from fermenting microbes and convert them to methane. The animal belches out this methane.

The core microbes in the rumen include Prevotella, Butyrivibrio, and Ruminococcus, Lachnospiraceae, Ruminococcaceae, Bacteroidales, and Clostridiales. The main archaea include Methanobrevibacter gottschalkii and Methanobrevibacter ruminantium, Methanosphaera species, and two Methanomassiliicoccaceae-affiliated groups.

DOMESTICATED RUMINANTS

Cattle, sheep, goats, and reindeer are all domesticated ruminants, although there are still wild species. The world population of these ruminants exceeds four billion. Ruminants are useful food sources for humans because they are large mammals, providing a lot of meat and milk, wool, hide, and fuel. Yet, because they eat plants, they are low on the food chain; since 90 percent of the energy from any food source is captured in digestion, domesticated ruminants are relatively efficient transmitters of food energy from plants to humans. However, the large size of these ruminants means that their metabolisms are relatively slow. Thus they can afford the time it takes to digest grasses, leaves, and twigs through rumination. In contrast, smaller herbivores with higher metabolisms, such as rodents, must digest food more quickly and thus eat more nutrient-rich plant food, such as seeds.

SOME WILD RUMINANT SPECIES

Wild ruminants are important to food chains because they eat plants, preventing plant overgrowth. They also are eaten by carnivores and omnivores. Bactrian camels, with three-chambered tylopod stomachs, inhabit the steppes and mountains of the Gobi Desert. These two-humped camels are domesticated as food and draft animals in Afghanistan, Iran, and China. Bactrian camels subsist on a diet of grass, leaves, herbs, twigs, and other plant parts. Their humps contain stored fat. Given the extreme aridity of their native environment, ruminant digestion allows them to derive the maximum nutrition from scarce food supplies.

Chamois goats live in the mountains of Europe and southwestern Asia. Their diet consists of grass and lichens in the summer, while they eat pine needles and bark in winter. Pronghorns live in the open plains and semideserts of the North American West. They are the only living *Antilocapridae*, relatives of antelope. These true ruminants eat herbs, sagebrush, and grasses in the summer and dig under the snow for grass and twigs in the winter. The large reindeer of northern Europe and Asia inhabit forests, grasslands, and mountains. Reindeer eat grass, moss, leaves, twigs, and lichens. Sable antelope live in southeastern

African woodlands and grasslands, where they eat grass, shrub leaves, and twigs.

—*Sanford S. Singer*

Further Reading

Church, D. C., editor. *The Ruminant Animal: Digestive Physiology and Nutrition*. Waveland Press, 1993.

Constantinescu, Gheorghe M., Brian M. Frappier, and Germain Nappert. *Guide to Regional Ruminant Anatomy Based on the Dissection of the Goat*. U of Iowa P, 2001.

Cronje, P. B., E. A. Boomker, and P. H. Henning, editors. *Ruminant Physiology: Digestion, Metabolism, Growth, and Reproduction*. Oxford UP, 2000.

Henderson, G., et al. "Rumen Microbial Community Composition Varies with Diet and Host, but a Core Microbiome Is Found Across a Wide Geographical Range." *Science Reports*, vol. 5, no. 14567, 2015, doi.org/10.1038/srep14567.

Wilson, R. T. *Ecophysiology of the Camelidae and Desert Ruminants*. Springer Verlag, 1990.

TERMITES

Category: Biology

Anatomy or system affected: None

Specialties and related fields: Bacteriology, biochemistry, ecology, entomology, microbiology, protozoology,

Definition: Soil or wood-inhabiting eusocial insects which generally have soft, white bodies and secretive habits

KEY TERMS

alates: recently molted winged adult termites

carton: cardboard-like material composed of wood fragments, saliva, and fecal matter, used for constructing termite nests

cellulose: a fibrous polysaccharide that chiefly constitutes the cell walls of plants

pheromone: a chemical substance produced by an animal that usually elicits certain behavioral responses in other animals of the same species

protozoan: mobile, one-celled animal

reproductives: sexually mature male and female

symbiotic: having a mutually beneficial relationship

CASTE SYSTEM AND NESTS

Termites are notable for their highly organized societies. Because most termites are effectively deaf and blind, they communicate through touch, smell, and taste. Most species are divided into castes of reproductives, workers, and soldiers. Normally, only one reproducing pair is the primary reproductives, or queen and king. Secondary reproductives are also present if the queen or king dies.

The sterile workers and soldiers are of both sexes. Workers care for the eggs and nymphs, provide food

Termites. Photo via Wikimedia Commons. [Public domain.]

for the nymphs, soldiers, and reproductives, and construct, repair, and maintain the nest. Soldiers have evolved modified heads and jaws for defending the nest. The heads are large and hard with powerful, scissor-like mandibles or long, tubular snouts that squirt sticky chemicals. Some soldiers have both formidable jaws and chemical weapons.

Termites are vulnerable to desiccation, changes in temperature, and hungry ants, birds, aardvarks, and other predators. They maintain a moist, temperature-controlled, safe environment by constructing nests. Dry-wood termites never touch the soil but nest in the wood they feed upon, gnawing out tunnels and chambers inside living tree trunks and branches, rotting logs, or furniture and wooden buildings. Subterranean termites must maintain contact with the soil for food sources such as grass and humus or moisture. Many species are master builders, constructing elaborate nests of a carton in trees with covered runways leading to the ground or mounds in the soil complete with ventilation shafts, towering chimneys up to 9 meters (29.5 feet) high, and even fungus gardens to supplement their cellulose diet. Although the architecture varies widely, most nests provide an inner chamber for the egg-laying queen and her king and brood chambers and food storage areas.

Termite Facts

Classification
Kingdom: Animalia
Subkingdom: Bilateria
Phylum: Arthropoda
Subphylum: Hexapoda
Class: Insecta
Subclass: Pterygota
Order: Blattodea
Superfamily: Blattoidea
Families: Hodotermitidae, Kalotermitidae, Rhinotermitidae, Termitidae, Termopsdae

Geographical location
They are located in every continent except Antarctica.

Habitat
The majority of species inhabit tropical rain forests, but many live in temperate and subtropical zones, deserts, or mountains up to an altitude of 2,500 meters (8,200 feet).

Gestational period
The period varies, but averages approximately two weeks.

Life span
The primary queen and king may live to seventy years, although ten or twenty years is more common; workers and soldiers live from two to five years.

Special anatomy
Isoptera, a taxonomic synonym for superfamily Blattoidea, means "equal wings"—adult reproductives have two oval-shaped pairs of overlapping wings of nearly equal length; small to medium-sized, very soft, usually light-colored bodies; thorax fused with the abdomen; workers have specialized mouthparts for chewing wood, while soldiers' mouthparts have been modified for fighting; except for reproductives, termites usually lack eyes.

Termite fossils date from about 130 million years ago. Still, they probably evolved much earlier from a primitive, wood-eating, roachlike ancestor. There are about 1,900 termite species divided among five families. Four families are considered primitive or lower termites because, like their primitive wood-eating roach relatives, they harbor symbiotic protozoa in the hindgut that digest cellulose. Without these protozoans, the termites would starve to death. The higher termite family, Termitidae, is the largest, containing about 75 percent of all termite species. Higher termites may be able to digest cellulose themselves, or bacteria in the gut may secrete enzymes to aid digestion.

LIFE CYCLE

Unlike insects such as butterflies, termites undergo incomplete or gradual metamorphosis. From the time they hatch, immature termites look like pale, wingless, miniature versions of adults. These nymphs molt periodically, shedding the outer skin to allow growth and then eating their outer skin. All nymphs start the same. The correct balance among the castes appears to be maintained by bodily secretions containing hormones that are transferred by licking, but the mechanism is not yet understood.

New colonies are usually formed when alates swarm during certain seasons. Workers prepare by digging tunnels to the surface, with exit holes and sometimes launching platforms. Once the alates leave, soldiers prevent them from returning. They are weak fliers and usually descend within a few hundred meters of the original nest. Wings are shed after landing. Females attract males by raising the abdomen and emitting a pheromone. Before mating, pairs locate a likely site for a nest and seal the entrance with fecal matter. The first batch of eggs is usually small. The king and queen take care of the eggs until enough older nymphs take over. After a few years, the queen's ovaries and abdomen increase in size, and her egg-laying accelerate. She may grow to 11 centimeters (4.3 inches) long and produce up to 36,000 eggs per day.

—*Sue Tarjan*

Further Reading

Behnke, Frances L. *A Natural History of Termites*. Charles Scribner's Sons, 1977.

"Blattoidea." *ITIS*, Integrated Taxonomic Information System, 8 Nov. 2017, www.itis.gov/servlet/SingleRpt/SingleRpt?search_topic=TSN&search_value=666640#null. Accessed 31 Jan. 2018.

Choe, Jae C., and Bernard J. Crespi, editors. *The Evolution of Social Behavior in Insects and Arachnids*. Cambridge UP, 1997.

Harris, W. Victor. *Termites: Their Recognition and Control*. Longmans, 1961.

Krishna, Kumar, and Frances M. Weesner, eds. *Biology of Termites*. 2 vols. Academic Press, 1969-1970.

Telford, Carol, and Rod Theodorou. *Through a Termite City*. Heineman Interactive Library, 1998.

MICROBIOME

Category: Microbial ecology

Anatomy or system affected: Gastrointestinal system, genitourinary system, intestines, mouth, pharynx, reproductive system, skin, stomach, urethra, vagina

Specialties and related fields: Bacteriology, dermatology, gastroenterology, microbiology

Definition: the microorganisms in a particular environment

KEY TERMS

competitive exclusion: the inevitable elimination from a habitat of one of two different species with identical needs for resources

normal flora: a diverse microbial flora is associated with the skin and mucous membranes of every human being from shortly after birth until death

lactic acid bacteria: a group of gram-positive, acid-tolerant, generally nonsporulating, nonrespiring, either rod-shaped or spherical bacteria that commonly ferment sugars to lactic acid

DEFINITION

"Microbiome" generally refers to all the microbes (extremely small organisms that cannot be seen by the naked eye, such as bacteria, fungi, and viruses) that live in a community. Our bodies have several microbiomes; for example, the gut contains a microbiome. The nasal cavity and mouth have a different microbiome, and the skin has an even different microbiome. These communities of microbes help with many bodily functions, including keeping the body healthy and fighting off infectious diseases.

FUNCTIONS

We are born with a certain microbiome that we acquire from our birth environment, both from our mother's womb and from the environment surrounding us at birth. Even the circumstances of our birth individualize our microbiome—babies born vaginally get microbes from the birth canal, while babies born by C-section get more microbes from the skin; babies born at the hospital have a different microbiome profile than those born at home. As we grow and develop, our microbiomes change based on our changing environment, including the foods we eat and the air we breathe. By the time we become an adult, all our microbiomes are very different from those at birth. Even the microbiomes of one individual are very different. For example, the skin microbiome usually contains microbes that do well in dry conditions and can be very different from the mouth or nasal microbiome, which contains microbes that need a moist and damp environment.

However, one common thread through all the different microbiomes is that they contain microbes. Many of these microbes are helpful for us; they help us with many bodily functions, such as digestion, cell division, vitamin production, metabolism, and protection from disease. The human microbiota includes bacteria, archaea, fungi, protists, and viruses.

Sauerkraut, or fermented cabbage, can provide healthy probiotics like lactobacilli *which can increase microbial diversity and promote the growth of gut flora. Photo by Gandydancer via Wikimedia Commons.*

IMPACT

Our different microbiomes help us fight off infectious diseases in many ways. First, our microbiome crowds out pathogens and gets in their way. They outcompete pathogens for resources and prevent them from colonizing our bodies. This is called "competitive exclusion."

Second, helpful microbes can release antimicrobials that keep harmful bacteria, fungi, and viruses away. For example, our skin's microbiome covers us with a protective shield made of helpful microbes. Microbes help keep the skin soft and flexible, creating a strong barrier that microbes cannot penetrate. Harmful microbes must fight through this shield to get into the body to cause disease. Similarly, a healthy colony of microbes in the nasal cavity and mouth can keep harmful bacteria, such as the kind that cause colds or sinus infections, away. Some kinds of bacteria, such as *Streptomyces*, produce natural compounds that have even been used in making medicines. Healthy bacteria can also signal the body to produce other molecules that fight infection.

Another way that our microbiome uses to help fight off infections is by changing the environment in the body. For example, in the vagina, *Lactobacillus* bacteria help to create an extremely acidic environment by releasing lactic acid. Harmful bacteria, yeast, and viruses often cannot survive in an acidic environment. Disrupting the vaginal flora can lead to infections like bacterial vaginosis and candidiasis. The microbiome of our skin also produces byproducts that lower the skin's pH, creating an acidic environment that keeps bacteria from growing and invading our body.

Some helpful bacteria, such as *Bifidobacteria*, keep toxins from passing through the intestinal wall and bloodstream. For example, *E. coli*, a disease-causing bacteria, can release toxins into the bloodstream that damage kidneys and destroy red blood cells, but helpful bacteria from the gut microbiome can prevent these toxins from leaving the gut. Our gut microbiome can even help deactivate toxic molecules that we ingest with our food.

Microbes can also help the body avoid infection by helping to "train" the immune system. Gut and skin microbes stimulate immune tissues and increase antibodies that fight pathogens. The body can learn to recognize and attack harmful pathogens and, at the same time, recognize and encourage healthy microbes.

Our microbiome can also help promote healing when infected with some pathogen. When our gut microbiome is damaged, the bacteria living there release chemicals that signal our cells to begin dividing faster to help us heal.

Sometimes, we need antibiotics to kill infections, and these medications can negatively affect our various microbiomes (and others). Antibiotics are unable to differentiate between helpful and dangerous microbes. As a result, antibiotics can kill some of the helpful bacteria we need to keep our microbiomes healthy. For example, antibiotics can interfere with the vaginal microbiome, so women may be left with an overgrowth of yeast (*Candida*) after taking a course of antibiotics.

To keep our microbiomes healthy, or to restore them after they have been compromised from antibiotics or other medications, diseases, or even from an unhealthy diet, one should eat fiber (whole grains, fruits, and vegetables, especially greens), fermented foods that contain probiotics (sauerkraut, kimchi, tempeh, kefir), and yogurt. If necessary, one can also take prebiotics or probiotics as a supplement. Limiting sugar intake can also help keep the microbiome healthy.

—Marianne M. Madsen, MS

Further Reading

Blaser, Martin J. *Missing Microbes*. Henry Holt and Co., 2014.

Harris, Vanessa C., Bastiaan W. Haak, Michael Boele van Hensbroek, and Willem J. Wiersing. "The Intestinal Microbiome in Infectious Diseases: The Clinical Relevance of a Rapidly Emerging Field." *Open Forum Infectious Diseases*. 8 July 2017, doi:[10.1093/ofid/ofx144].

Honda, Kenya, and Dan R. Littman. "The Microbiome in Infectious Disease and Inflammation." *Annual Review of Immunology*6 Jan. 2021, doi:[10.1146/annurev-immunol-020711-074937].

Ragab, Gaafar, T. Prescott Atkinson, and Matthew L. Stoll, editors. *The Microbiome in Rheumatic Diseases and Infection*. Springer, 2018.

Shreiner, Andrew B., John Y. Kao, and Vincent B. Young. "The Gut Microbiome in Health and in Disease." *Current Opinion in Gastroenterology*, vol. 31, no. 1, 2015, pp. 69-75, doi:[10.1097/MOG.0000000000000139].

Young, Vincent B., Robert A. Britton, and Thomas M. Schmidt. *The Human Microbiome and Infectious Diseases: Beyond Koch. Interdisciplinary Perspectives on Infectious Disease*. Hindawi Publishing Company, 2008, www.hindawi.com/journals/ipid/si/861658/.

The Bacteria

Different bacterial species cause different diseases. These diseases result in significant human suffering and loss. In this section, we explain some of the better-known bacteria and the diseases they cause. Among the topics discussed are food poisoning, tetanus, cholera, typhoid fever, and pneumonia.

DIPHTHERIA

Category: Diseases and conditions
Anatomy or system affected: Throat, tissue, tonsils, upper respiratory tract
Specialties and related fields: Bacteriology, immunology, internal medicine, microbiology, otolaryngology, preventive medicine, public health
Definition: a highly contagious, life-threatening bacterial infection that commonly attacks the mucous membranes associated with the breathing system (the tonsils, throat, and nose) and can also infect the skin

KEY TERMS

beta-bacteriophage: a virus that infects Corynebacterium diphtheriae and transduces the genes to express diphtheria toxin
diphtheria toxin: a powerful inhibitor of protein synthesis in eukaryotic cells produced by some Corynebacterium diphtheria strains
elongation factor 2: also known as EF-2; a critical protein synthesis factor in eukaryotic cells
pseudomembrane: a layer of exudate resembling a membrane that forms on the surface of the skin or mucous membrane.

INTRODUCTION

Diphtheria is a highly contagious and life-threatening infection caused by bacteria. The infection most commonly attacks the mucous membranes associated with the breathing system (the tonsils, throat, and nose) and can also infect the skin. In addition, the infection damages the heart, nerves, kidneys, and brain.

The vaccine for diphtheria is safe and is effective at preventing the disease. A series of shots are given during childhood; booster shots are required every ten years to keep the immunity strong. Before vaccines and medications were available to prevent and treat the disease, nearly one of every ten infected people died. Diphtheria was the leading cause of death among children.

Diphtheria is a medical emergency that requires immediate care from a doctor. Not everyone who gets diphtheria shows signs of illness, though they may be able to infect others. The sooner the infection is treated, the more favorable the outcome.

CAUSES

Diphtheria is caused by the bacterium *Corynebacterium diphtheriae* (*C. diphtheriae*). This bacterial species is an aerobic, club-looking, gram-positive microorganism. There are four subspecies of *C. diphtheriae*, *C. diphtheriae* mitis, *C. diphtheriae* intermedius, *C. diphtheriae* gravis, and *C. diphtheriae* Belfanti. When clustered together, these bacteria seem to resemble Chinese characters.

C. diphtheriae are picky organisms (fastidious) and only grow enriched media with special nutrients. Cystine-tellurite blood agar is the culture media used to grow *C. diphtheriae*. On this medium, the organism forms black colonies.

C. diphtheriae subspecies are either toxigenic or nontoxigenic, depending on if they produce diphtheria toxin. All *C. diphtheriae* strains begin as nontoxigenic until they are infected with a virus called the "beta-bacteriophage." This virus inserts its genome into the bacterial chromosome and confers the ability to produce diphtheria toxin.

Diphtheria toxin has an A and a B subunit. The B larger subunit binds the host cell and brings the toxin into it. Then the A subunit modifies a crucial protein synthesis factor called "elongation factor 2," inactivating it. Shutting down protein synthesis in the host cell kills it.

C. diphtheriae spreads from person to person through contact with droplets of moisture that are coughed or sneezed into the air and breathed. People nearby breathe in these droplets or touch contaminated personal items (fomites) such as tissues or drinking glasses that an infected person used. Also, people can contract diphtheria through infected skin.

RISK FACTORS

Risk factors include having never been immunized against diphtheria, not having had a booster dose in the past ten years, living in crowded or unsanitary conditions, having a compromised immune system, and being undernourished. Visiting resource-poor countries where diphtheria is endemic also increases the risk of diphtheria for unvaccinated individuals.

SYMPTOMS

Signs and symptoms of diphtheria usually begin two to five days after a person is infected. The most telltale sign of diphtheria is a gray covering on the back of the throat called a "pseudomembrane." Pseudomembranes consist of scar tissue, dead bacterial cells, dead white and red blood cells, throat, and epithelial cells. The pseudomembrane can detach and block the airway.

If untreated, diphtheria toxin can spread throughout the body and damage the heart, nerves, and kidneys. Symptoms include sore throat and painful swallowing, fever up to 103° Fahrenheit, swollen glands in the neck, difficulty breathing, difficulty swallowing, weakness, and a gray covering on the back of the throat.

SCREENING AND DIAGNOSIS

A doctor will ask about symptoms and medical history and will perform a physical exam. Diphtheria will be suspected if the throat and tonsils are covered with a gray membrane. Tests to confirm a diagnosis may include taking a sample of the gray membrane that coats the back of the throat and taking a sample of tissue from an infected area of skin.

TREATMENT AND THERAPY

If a doctor suspects diphtheria, the patient's treatment will start immediately, even before the lab results are returned. Treatment options include antitoxin, a substance injected into the body that neutralizes the diphtheria poison that is traveling in the body; antibiotics, a substance injected or given as a pill that kills the diphtheria bacteria in the body and heals the infection (also reduces the length of time a person is contagious); and isolation and bed rest. It takes up to six weeks to recover from diphtheria, especially if the heart is affected. Isolation may be necessary while a person is still contagious.

PREVENTION AND OUTCOMES

There are four types of diphtheria vaccine. DTaP is for children who receive three doses at two, four, and six months of age, their first booster at fifteen to eighteen months, and a second booster at four to six years. Tdap is for preteens who receive this vaccine at age eleven to twelve years. Td or Tdap is for adults who receive a booster every ten years. These vaccines are combinations of pertussis, tetanus, and diphtheria vaccines and are highly effective.

To help reduce the chance of getting diphtheria, persons should get immunized and stay up to date on future immunizations. Suppose a person has been in contact with someone who has diphtheria. In that case, that person should be watched closely for symptoms and should work with a doctor to determine appropriate treatment, if necessary.

—Julie J. Martin and Michael A. Buratovich, PhD

Further Reading

Centers for Disease Control and Prevention. "Diphtheria Vaccination." *Vaccines and Preventable Disease*, 22 Jan. 2020, www.cdc.gov/vaccines/vpd/diphtheria/index.html. Accessed 24 Nov. 2021.

Murray, Patrick R., et al. *Medical Microbiology*. 9th ed., Elsevier, 2020.

Pan American Health Organization. *Control of Diphtheria, Pertussis, Tetanus, "Haemophilus influenzae" Type B, and Hepatitis B Field Guide*. Pan American Health Organization, 2005.

Parker, James N., and Philip M. Parker, editors. *The Official Patient's Sourcebook on Diphtheria*. Icon Health, 2002.

Chlamydia

Category: Diseases and disorders

Anatomy or system affected: Anus, cervix, eyes, lymphatic system, throat, urethra, urogenital system, rectum, reproductive system

Specialties and related fields: Bacteriology, gynecology, immunology, internal medicine, microbiology, neonatology, pathology, pharmacology, preventive medicine, public health, women's health

Definition: a common sexually transmitted disease, caused by bacteria called *Chlamydia trachomatis*, that infects both men and women

KEY TERMS

contact tracing: also known as partner referral; a process that consists of identifying sexual partners of infected patients, informing the partners of their exposure to disease, and offering resources for counseling and treatment

elementary body: a resting form of *Chlamydia trachomitis* that infects host cells

incidence: probability or risk of contracting a disease within a population

prevalence: the proportion of infectious cases in a population at a given time

reticulate body: the metabolically active form of *Chlamydia trachomitis* that the elementary body transforms into after penetrating a host cell that appropriates the host cell resources to divide and form elementary bodies, which culminates in host cell lysis and liberation of elementary bodies for future infective cycles

screening procedures: tests that are carried out in populations, which are usually asymptomatic and at high risk for a disease, to identify those in need of treatment

sexually transmitted infection (STI): an infection caused by organisms transferred through sexual contact (genital-genital, orogenital, or anogenital); transmission of infection occurs through exposure to lesions or secretions which contain the organisms

CAUSES AND SYMPTOMS

Chlamydia infection is caused *by Chlamydia trachomitis* (*C. trachomitis*). This small-gram-negative bacterium is also an obligate intracellular parasite. This organism has two distinct phases in its life cycle: (1) small resting cells called "elementary bodies" that attach to and penetrate cells; (2) within six to eight hours after penetration of host cells, the elementary bodies are transformed into the metabolically active reticulate bodies that use the resources of the host cell to divide, eventually lysing the cell and releasing newly formed elementary bodies. Because of its obligate parasitism, *C. trachomitis* cannot be grown on artificial media but must be grown in tissue culture. Also, because a significant proportion of its life cycle occurs within cells, immunity to this organism is often short-lived, and reinfection or persistent infection is common.

C. trachomitis is the cause of the most commonly reported bacterial infection and the most common sexually-transmitted infection (STI) in the United States.

Information on Chlamydia

Causes: Bacterial infection of mucosa (genital tract, urinary tract, anorectal tract, eyes, throat) through sexual transmission or childbirth

Symptoms: Often none, but may include urinary discomfort, lower abdominal pain, mucopurulent discharge from vagina or urethra, pain upon urination and increased urinary frequency, complications such as pelvic inflammatory disease (PID) and infertility; in infants, eye infection, visual impairment, blindness, respiratory tract infection, proctitis

Duration: Chronic until treated

Treatments: Antibiotics (azithromycin, doxycycline, ofloxacin, levofloxacin)

In 2017, the United States Centers for Disease Control and Prevention (CDC) reported 1,708,569 *C. trachomitis* infections in the United States, an incidence rate of 528.8 cases per 100,000 people. *C. trachomitis* infection in women increased 11.1 percent from 2013 to 2017. The infection rate in women is twice that of men, which may be an artifact of increased screening in women relative to men. Infection prevalence is highest in women fourteen to twenty-four years of age, and prevalence rates are the highest in African American women (14 percent).

Globally, the World Health Organization estimates that in 2008 there were 100 million prevalent cases. An estimated 105.7 million incident cases of *C. trachomitis* infection had occurred worldwide, representing a 4.1 increase in new cases from 2005. However, these increases might result from improved diagnostic and screening methods in resource-limited locations. Recent findings have raised suspicions that the estimated number of *C. trachomitis* cases has been significantly underestimated. For example, in 2006, a new Swedish variant of *C. trachomitis* was identified that was not identified by the available nucleic acid amplification tests (NAATs). Therefore, the number of *C. trachomitis* cases in Sweden was probably grossly underestimated.

C. trachomitis infects mucosal surfaces, such as the genital tract, urinary tract, anorectal tract, eyes, and throat. These infections cause inflammation of the cervix (cervicitis), urethra (urethritis), eyes (conjunctivitis), and rectum (proctitis). In women, symptoms may include urinary discomfort, lower abdominal pain, and abnormal vaginal discharge. Women are commonly asymptomatic in the early stages of the disease.

Since most *C. trachomitis* cases are asymptomatic, most patients are diagnosed by screening procedures, usually NAATs. Nevertheless, asymptomatic patients can infect others and suffer severe consequences of *C. trachomitis* infection. Those at the highest risk for *C. trachomitis* infections include women younger than twenty-five years because of the absence of immunity from previous infections, women who have more than one sex partner in the last three months or report a new sex partner, women who have a history of previous *C. trachomitis* infection, and those whose insistence on condom use is inconsistent. Surprisingly, women who have sex with women show an increased risk of *C. trachomitis* infection. Because *C. trachomitis* often occurs in combination with other STIs, all women with a confirmed STI should also be screened for *C. trachomitis*.

Initial *C. trachomitis* infections usually result in genital infections. The incubation period for cervicitis is usually about one-two weeks. If the woman has cervicitis symptoms, they usually include a puss/mucus-filled discharge and a tender, easily bleeding cervix. In most cases (85 percent), the urethra is also infected and can cause a dysuria-pyuria syndrome in which there is pain upon urination (dysuria), increased frequency of urination, and white blood cells in the urine (pyuria), but no bacterial cells will be present upon culture, however. The ascent of *C. trachomitis* throughout the woman's reproductive organs to involve the uterus, fallopian tubes, and pelvic cavity results in pelvic inflammatory disease (PID). This chronic infection of the upper reproductive tract can lead to scarring of these organs and increases the patient's risk of infertility and ectopic pregnancies. On occasion (5 to 15 percent of cases), PID can extend to the Glisson's capsule surrounding the liver leading to perihepatitis. This condition is called "Fitz-Hugh-Curtis syndrome," and it causes right upper quadrant pain. Rarely, *C. trachomitis* spreads to the regional lymph nodes and causes abscesses; a condition termed lymphogranuloma venereum. This condition is commonly accompanied by systemic symptoms such as fever, chills, and muscle and joint aches.

An infant may contract *C. trachomitis* as it passes through the birth canal of an infected mother. The disease can lead to an eye infection called "trachoma" that results in visual impairment and blindness, and infection of the respiratory tract. Trachoma is the leading infectious cause of blindness worldwide.

TREATMENT AND THERAPY

A patient receives antibiotic therapy if laboratory tests indicate infection with *C. trachomatis*. Antibiotics effectively treat uncomplicated *C. trachomitis* infections. First-line agents for uncomplicated *C. trachomitis* include the tetracycline antibiotic doxycycline (100 mg, twice a day for seven days) and the macrolide antibiotic azithromycin (1 gram, single dose). Because of its long half-life (five to seven days) and excellent intracellular tissue and tissue penetration, azithromycin is the usual drug of choice, although doxycycline costs substantially less. Azithromycin is also safe for pregnant women. Alternative treatments include the quinolone ofloxacin (300 mg twice a day for seven days) and levofloxacin (500 mg once a day for seven days). Still, pregnant or lactating women should not take these drugs, nor should adolescents younger than eighteen years of age because of potentially harmful effects on bone formation.

In cases of PID, drug regimens must contain antibiotics that will treat both *Neisseria gonorrhea* and *C. trachomitis* infections, and drug combinations are preferred. Such drug combinations may consist of intravenous, intramuscular/oral, and primarily oral regimens.

For chlamydial proctitis, a combination of an injection of ceftriaxone (250 mg) plus doxycycline (100 mg twice a day) is recommended. For trachoma (chlamydial conjunctivitis), oral azithromycin (20 mg/kg up to 1 gram maximum) is preferred. However, topical tetracycline (1 percent eye ointment, twice daily for six weeks) is a satisfactory alternative. If available, topical azithromycin (1.5 percent) for three days, applied in the morning and evening, is as effective as oral azithromycin.

Patients with risk factors for STIs or symptoms of the disease may be treated presumptively with antibiotic therapy, even before the results of laboratory tests for chlamydial infection return.

Because *C. trachomitis* infections are associated with other STIs, such as gonorrhea, human immunodeficiency virus (HIV), syphilis, and hepatitis B and C, the patient should be advised to undergo testing for these diseases as well. Approximately 35 to 50 percent of patients with gonorrhea also have *C. trachomitis*. Therefore, an antibiotic against it is given in combination with an antibiotic for gonorrhea unless laboratory tests have declared the patient free of

Drug regimens for PID

IV drug regimens	Intramuscular/oral	Mostly Oral
a) Cefotetan 2 g IV every 12 hours + Doxycycline 100 mg orally or IV every 12 hours b) Cefoxitin 2 g IV every 6 hours + c) Doxycycline 100 mg orally or IV every 12 hours d) Clindamycin 900 mg IV every 8 hours + Gentamicin loading dose IV or IM (2 mg/kg), followed by a maintenance dose (1.5 mg/kg) every 8 hours. Single daily dosing (3-5 mg/kg) can be substituted.	a) Ceftriaxone 250 mg IM in a single dose + Doxycycline 100 mg orally twice a day for 14 days. b) Cefoxitin 2 g IM in a single dose and Probenecid, 1 g orally administered concurrently in a single dose + Doxycycline 100 mg orally twice a day for 14 days. c) Other parenteral third-generation cephalosporins (e.g., ceftizoxime or cefotaxime) + Doxycycline 100 mg orally twice a day for 14 days. Each of these regimens may or may not include Metronidazole 500 mg orally twice a day for 14 days.	a) Azithromycin (500 mg IV daily for 1–2 doses, followed by 250 mg orally daily for 12–14 days) b) Azithromycin as above in combination with metronidazole (500 mg orally twice a day for 14 days) c) Azithromycin (1 g orally once a week for two weeks) + ceftriaxone 250 mg IM single dose

C. trachomitis. A single injection of ceftriaxone (250 mg) cures most uncomplicated gonococcal infections and is safe during pregnancy. Early treatment of *C. trachomitis* and gonorrhea prevents complications such as PID or infertility.

In addition to antibiotics, a key component of treating *C. trachomitis* infections involves counseling to prevent STIs. Patients are encouraged to use barrier methods such as condoms during intercourse and avoid high-risk sexual behaviors to minimize future exposure to C. trachomitis and other STIs.

Another critical component to treating chlamydia and other STIs is contact tracing, which occurs once the infection is confirmed with laboratory testing. With the patient's cooperation, all sexual partners of the patient are notified regarding their exposure to the disease. Partners are encouraged to seek medical attention, even if they have no symptoms themselves, to prevent reinfection of the patient during subsequent sexual encounters or further spread of the disease to other sexual partners.

In the United States, erythromycin eye drops are given prophylactically to all newborns to prevent eye infections with *C. trachomitis* that could lead to visual impairment.

PERSPECTIVE AND PROSPECTS

Diseases caused by *C. trachomatis* have been described as early as ancient Egyptian times. However, it was not until 1907 that the bacterium was identified, and there was some controversy about whether it was an actual bacterium because it requires host cells to live (obligate intracellular parasite). The organism favors cells that are more available on the cervix of young (versus older) women, which is partly why *C. trachomitis rates* are so much higher in the young. In the 1960s, a clinically useful diagnostic test was developed that allowed the screening of many specimens within a few days. The development of a relatively easy test for *C. trachomitis* enabled clinicians to screen a large number of asymptomatic but at-risk patients (such as those under age twenty-five or those with multiple sexual partners).

A promising area of research is the search for a vaccine for *C. trachomatis*. This research focuses on identifying antigens on the bacterium that are important for its function, such as proteins responsible for bacterial attachment to or uptake into cells. The premise is to use these proteins to generate an immune response in patients. When patients are exposed to *C. trachomitis*, their immune systems are prepared to respond against it.

—Clair Kaplan, APRN/MSN;
additional material by Anne Lynn S. Chang, MD;
and updated by Michael A. Buratovich, PhD

Further Reading

Chlamydia-CDC Fact Sheet. Centers for Disease Control and Prevention, 23 Jan. 2014, www.cdc.gov/std/chlamydia/stdfact-chlamydia.htm. Accessed 5 Mar. 2019.

Jameson, J. Larry, et al., editors. *Harrison's Principles of Internal Medicine*. 20th ed., McGraw-Hill, 2018.

Passos, M. R. L., et al., editors. *Atlas of Sexually Transmitted Diseases: Clinical Aspects and Differential Diagnosis*. Springer, 2018.

Ryan, Kenneth J., et al., editors. *Sherris Medical Microbiology: An Introduction to Infectious Diseases*. 7th ed., McGraw-Hill, 2018.

Sexually Transmitted Infections Treatment Guidelines, 2021: Chlamydial Infections. Centers for Disease Control and Prevention, 22 July 2021, www.cdc.gov/std/treatment-guidelines/chlamydia.htm. Accessed 20 Nov. 2021.

TETANUS

Category: Diseases and conditions
Also known as: Lockjaw
Anatomy or system affected: Jaw, mouth, musculoskeletal system, nervous system
Specialties and related fields: Bacteriology, internal medicine, microbiology, neurology, orthopedics,

pharmacology, preventive medicine, public health, toxicology

Definition: a bacterial intoxication that affects the nervous system, resulting in severe muscle spasms

KEY TERMS

motor neurons: a nerve cell forming part of a pathway along which impulses pass from the brain or spinal cord to a muscle or gland.

Renshaw cells: inhibitory interneurons located in the ventral cord and through their localized connections with motor neurons and other interneurons help to ensure a balance between contraction of synergist and antagonist muscles

SNARE proteins: molecular motors that drive the biological fusion of two membranes

tetanospasmin: an extremely potent neurotoxin produced by the vegetative cell of *Clostridium tetani* in anaerobic conditions, causing tetanus

Tetanus is a bacterial intoxication that affects the nervous system. Tetanus bacteria from soil, dust, or manure enter the body through breaks in the skin. The infection may result in severe muscle spasms. Such spasms lead to lockjaw, which prevents the opening or closing of the mouth. Tetanus can be fatal.

CAUSES

Tetanus is caused by a toxin called "tetanospasmin" produced by the bacterium *Clostridium tetani* (*C. tetani*). This microorganism is an obligately anaerobic, gram-positive, endospore-forming bacterium. The endospores look like miniature tennis rackets.

DISEASE PROCESS

C. tetani enter the skin through penetrating skin trauma. However, because these microorganisms are strict anaerobes, they cannot grow unless the wound is devoid of air. Puncture wounds are usually anaerobic and just the right temperature to nurture the growth of *C. tetani*. This organism usually grows

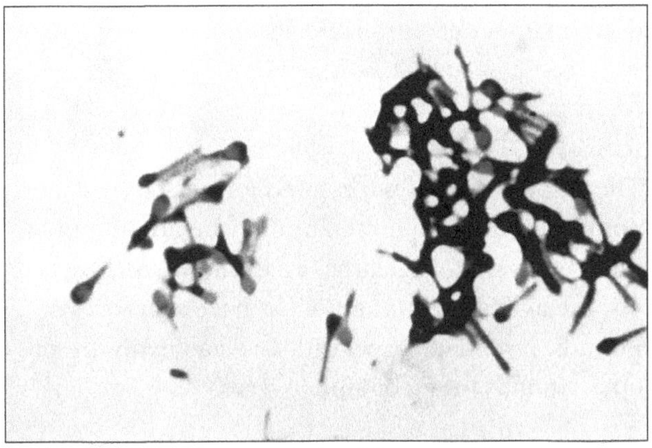

A photomicrograph of a group of Clostridium tetani *bacteria, responsible for causing tetanus in humans. Image by CDC Public Health Image Library. [Public domain.]*

slowly at first, and symptoms do not appear for one to four weeks after the injury.

The bacterium secretes a toxin called "tetanospasmin." This protein enters inhibitory neurons called "Renshaw cells." Renshaw cells are inhibitory interneurons located in the ventral spinal cord. They have connections with motor neurons and other interneurons. They help to balance the contraction of agonist and antagonist muscles. They release the neurotransmitter glycine, which inhibits motor neurons from overstimulating the muscles they innervate.

Neurons synthesize their neurotransmitters and package them into synaptic vesicles. When an action potential reaches the axon terminus, these neurotransmitter-laden vesicles fuse with the terminal axonal membrane. Fusion of the axon terminal membrane with the synaptic vesicles releases the neurotransmitters into the synaptic cleft between connected neurons. Vesicle-membrane fusion is mediated by molecules called "SNARE proteins." Vesicle-specific SNARE proteins engage with membrane-specific SNARE proteins to drive fusion between these two membranes.

Tetanospasmin degrades SNARE proteins within Renshaw cells, preventing them from releasing glycine. Without the fine-tuning of motor neuron activity provided by the Renshaw cells, muscle contraction becomes uncontrollable, leading to painful spasms.

RISK FACTORS

The risk for tetanus is increased for persons who are not immunized against tetanus or who are not updating tetanus shots regularly, are intravenous-drug users, are age fifty years and older, have skin sores or wounds, have had burns, and have had exposure of open wounds to soil or animal feces.

SYMPTOMS

Symptoms of tetanus may include headache; stiff jaw muscles (lockjaw) or neck muscles; drooling or trouble swallowing; muscle spasticity or rigidity; sweating; fever; irritability; pain or tingling at the wound site; high or low blood pressure; seizures; difficult breathing; a heartbeat that is irregular, too fast, or too slow; cardiac arrest; dehydration; and pneumonia (a complication of the infection).

The classical "tetanic" triad includes trismus or lockjaw due to mild to severe spasms of the lower jaw, risus sardonicus, an abnormal grin caused by spasms of facial muscles, and opisthotonos, whole-body muscle spasms that cause arching of the back. Tetanic muscle contractions are so powerful that they can cause muscle tears or bone fractures.

A rare form of tetanus is local tetanus, where spasms are localized around the wound. This condition persists for a few months and then subsides. Local tetanus is lethal in one percent of cases. An even rarer form of tetanus is cephalic tetanus. Only the muscles innervated by the twelve cranial nerves are affected. Skull fractures, eye injuries, middle ear infections, or tooth extractions can lead to cephalic tetanus.

Neonatal tetanus results when the mother is not vaccinated, and the umbilical cord is cut with a nonsterile blade. Immunized mothers transmit passive immunity to their babies, protecting them from neonatal tetanus. Improved hygiene and the use of sterile blades to cut umbilical cords have also reduced the rates of neonatal tetanus.

SCREENING AND DIAGNOSIS

A doctor will ask about symptoms and medical history and perform a physical exam. The diagnosis is based mainly on the patient's medical history. Wound culture rarely yields *C. tetani* because the organism grows very slowly and is sensitive to air.

A positive spatula test is a definitive sign of tetanus. The spatula test involves touching the back wall of the throat with a wooden spatula. This usually makes someone gag, but someone with tetanus will involuntarily bite down hard on the spatula.

TREATMENT AND THERAPY

Treatment may include hospitalization to manage complications of the infection; opening and cleaning of the wound, or sometimes surgical removal of the entire wounded area; antibiotics; a tetanus immunoglobulin (antibodies against tetanus that help neutralize the tetanus toxin); and a tetanus shot, if the patient's tetanus vaccine is not up to date.

Muscle relaxants, pain relievers, and other supportive care measures can help the patient survive while the immune system fights the infection.

Muscle spasms in a person with tetanus. Painting by Sir Charles Bell, 1809, via Wikimedia Commons. [Public domain.]

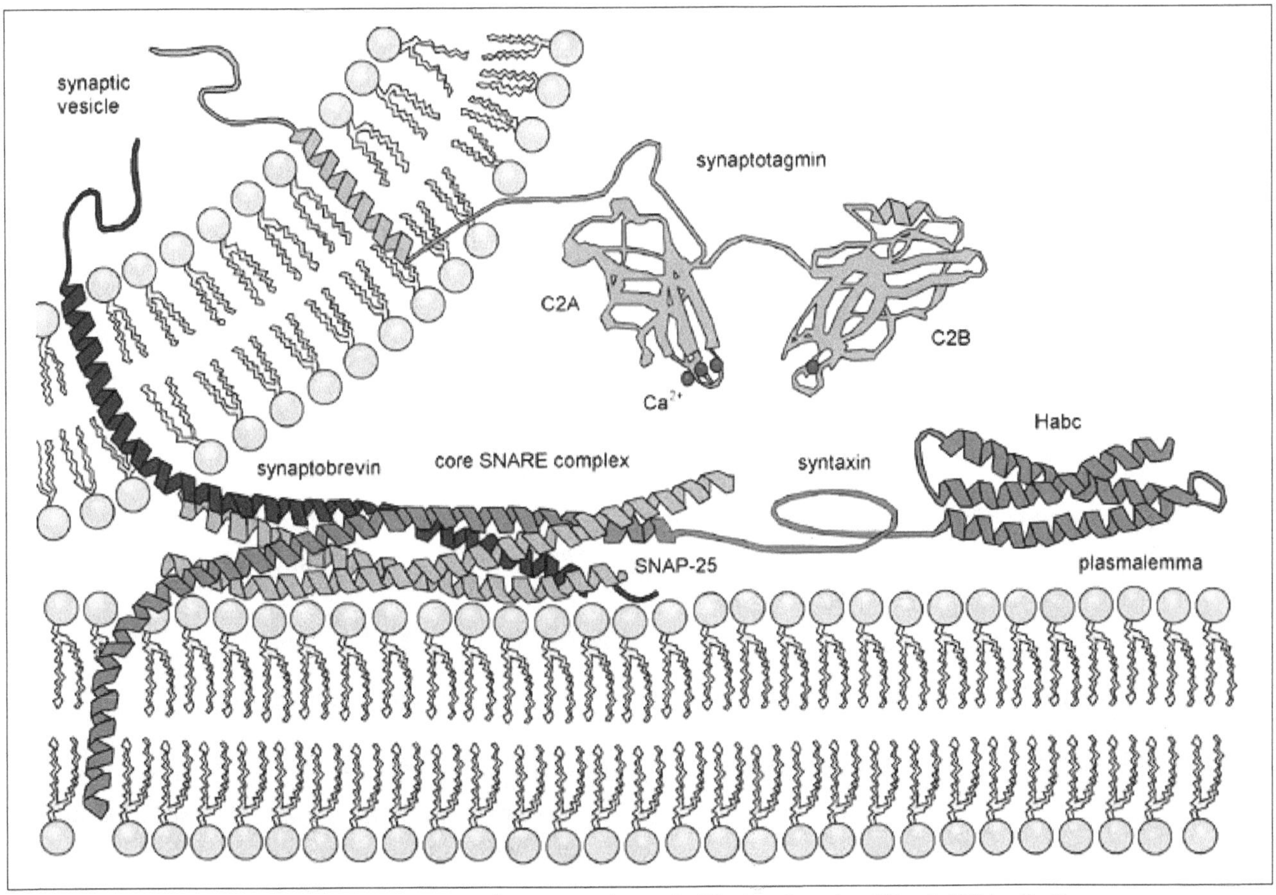

Molecular machinery engaged in synaptic transmitter release. Image by Danko Dimchev Georgiev, MD, via Wikimedia Commons.

For cases in which the patient has trouble breathing or swallowing, a breathing tube may be inserted into the throat to help keep the airway open. A surgical procedure called a "tracheotomy" may be done to provide an open airway in certain situations.

PREVENTION AND OUTCOMES

The best means of prevention is immunization. All children (with a few exceptions) should receive the DTaP vaccine, which protects against diphtheria, tetanus, and pertussis. This vaccination includes a series of five shots and a booster shot. The regular immunization schedule (for children and adults) is as follows: DTaP vaccines at two, four, and six months of age; at fifteen to eighteen months of age; and four to six years of age; and a booster dose of Tdap given at eleven or twelve years of age (for children who have not already had the Td booster).

Children aged thirteen to eighteen years who missed the booster dose or received Td only can receive one dose of Tdap five years after the last dose and can receive a booster of Tdap (a onetime dose for persons age nineteen to sixty-four years) or Td (every ten years) to provide continued protection.

For children, ages four months to six years who have not yet received the vaccination, the Centers for Disease Control and Prevention recommends the following "catch-up" schedule: first and second dose (with a minimum four-week interval between doses); second and third dose (with a minimum four-week interval between doses); third and fourth dose (with a minimum six-month interval between doses); and fourth and fifth dose (with a minimum six-month interval between doses). The fifth dose is not necessary if the fourth dose was administered at age four years or older.

DTaP is not indicated for persons aged seven years or older. Children age seven years and older and adults who have not been vaccinated should also be vaccinated. The choice and timing will vary based on age and prior vaccine exposure.

In addition to the vaccine, one can prevent tetanus by properly caring for wounds, including promptly cleaning all wounds and consulting a doctor for medical care. One should consult a doctor, especially if the patient has not had a tetanus vaccination in the ten years before injury.

—Jenna Hollenstein

Further Reading

Brachman, Philip S., and Elias Abrutyn, editors. *Bacterial Infections of Humans: Epidemiology and Control*. 4th ed., Springer Science, 2009.

Brown, Pamela. *Quick Reference to Wound Care*. Jones and Bartlett, 2009.

"Centers for Disease Control and Prevention Report Finds Tetanus Reaching Younger Adults." *Vaccine Weekly*, 16 July 2003, pp. 21-22.

Centers for Disease Control and Prevention. "Recommended Immunization Schedules for Persons Aged 0-18 Years—United States, 2008." *Morbidity and Mortality Weekly Report*, vol. 57, 2008, pp. Q1-4, www.cdc.gov/mmwr/preview/mmwrhtml/mm5701a8.htm.

———. "Tetanus (Lockjaw) Vaccination." 22 Jan. 2020, www.cdc.gov/vaccines/vpd-vac/tetanus. Accessed 7 Dec. 2021.

Gremillion, Henry A., editor. *Temporomandibular Disorders and Orofacial Pain*. Saunders/Elsevier, 2007.

Pan American Health Organization. World Health Organization. *Control of Diphtheria, Pertussis, Tetanus, "Haemophilus influenzae" Type B, and Hepatitis B Field Guide*. Pan American Health Organization, 2005.

Rodrigo, Chaturaka, et al. "Pharmacological Management of Tetanus: An Evidence-Based Review." *Critical Care*, vol. 18, no. 2, 2014, p. 217, doi:10.1186/cc13797.

CHOLERA

Category: Diseases and conditions

Anatomy or system affected: Gastrointestinal system, intestines

Specialties and related fields: Bacteriology, gastroenterology, immunology, internal medicine, microbiology, public health

Definition: an infection caused by the *Vibrio cholerae* bacterium characterized by acute watery diarrhea with severe dehydration

KEY TERMS

cholera toxin: a six-subunit protein secreted by *V. cholerae* once it attaches to the small intestine wall that drives extensive salt and water excretion by the small intestine mucosae

flagellum: a whiplike tail that extends from the *V. cholerae* and rotates to provide robust motility to the organism

toxin coregulated pilus: a threadlike extension of the *V. cholerae* cell surface that helps the microorganisms attached to the small intestinal wall

INTRODUCTION

Cholera, an infection caused by the *Vibrio cholerae* bacterium, can kill a person within hours if it is left untreated. Contaminated water and food transmit the bacteria. The most dangerous symptom is acute watery diarrhea with severe dehydration.

Vibrio cholerae (*V. cholerae*) is a gram-negative bacterium, which stains pink on a Gram stain. Under the microscope, Gram-stained *V. cholerae* look like little pink commas. This microorganism has a single flagellum that looks like a long tail. The flagellum rapidly rotates to propel *V. cholerae* vigorously through the gastrointestinal tract.

V. cholerae transmission occurs through the fecal-oral route. Consuming untreated sewage water and anything that contacts it, eating undercooked fish, including shellfish, and not washing after a bowel movement effectively transmits cholera. Consequently, it is more common in underdeveloped countries with poor water treatment and sanitation facilities. Africa and South America have high rates of cholera.

When *V. cholerae* enters the stomach, it powers down to survive stomach acid. Once it passes into the small intestine, *V. cholerae* uses its flagella to burrow through the mucous layer of the intestinal wall and attach to the villi in the small intestine. The organism uses a ropelike structure called the "toxin-coregulated pilus" (TCP) to adhere to the small intestinal wall. Once attached, *V. cholerae* multiplies and secretes its potent cholera toxin.

Different *V. cholerae* strains produce different types of toxins, some that cause few symptoms and others that produce severe disease. *V. cholerae* strains that produce cholera enterotoxin, also known as "choleragen," usually cause the most severe disease.

Cholera enterotoxin modifies a highly regulated intracellular protein called the "G protein" so that it is stuck in the activated state. The constitutively activated G protein causes runaway cyclic adenosine monophosphate (cAMP) synthesis. The jump in cAMP levels in the intestinal cells stimulates chloride transport into the small intestine. It also prevents the

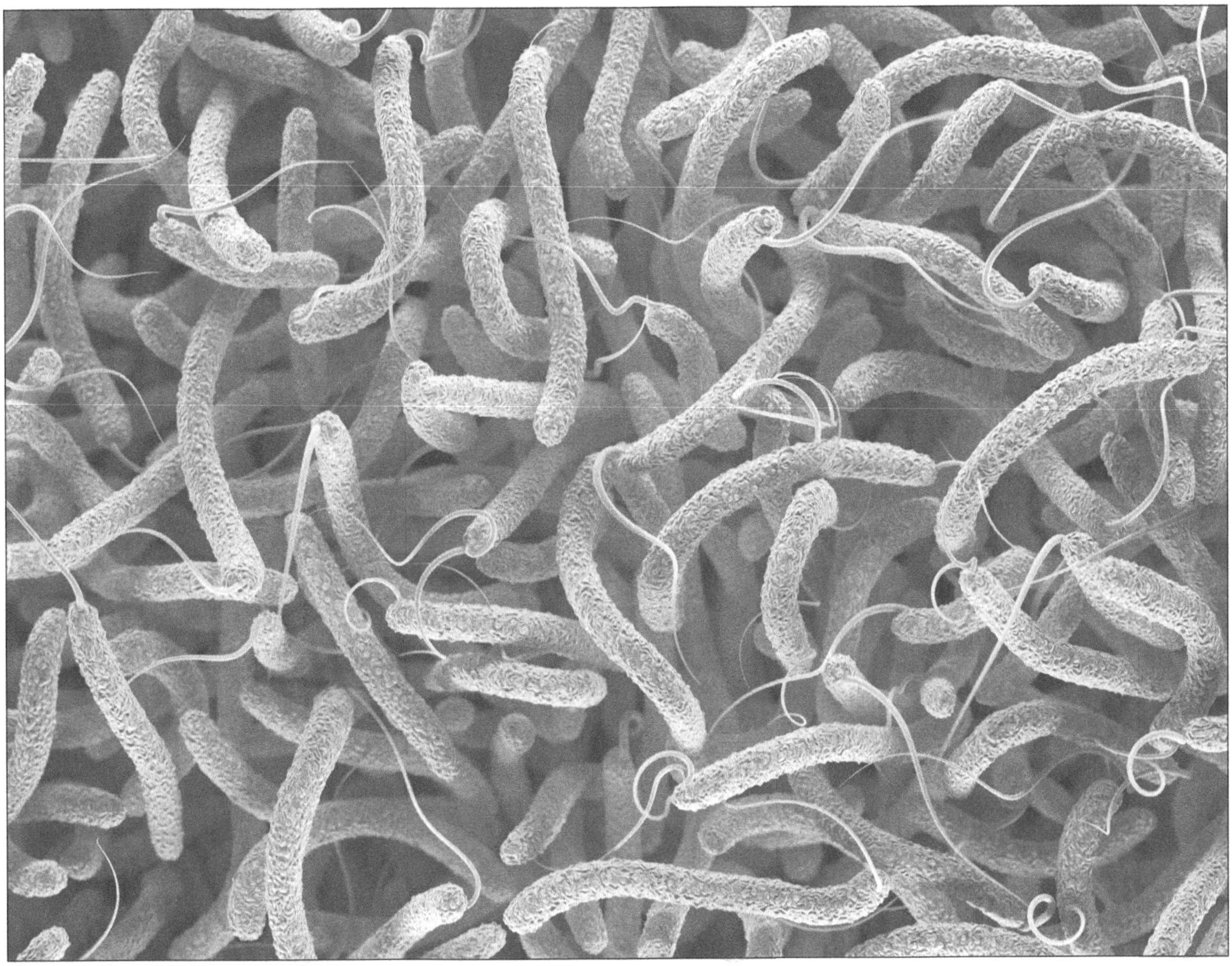

3D rendering of vibrio cholera. *Image via iStock/ktsimage. [Used under license.]*

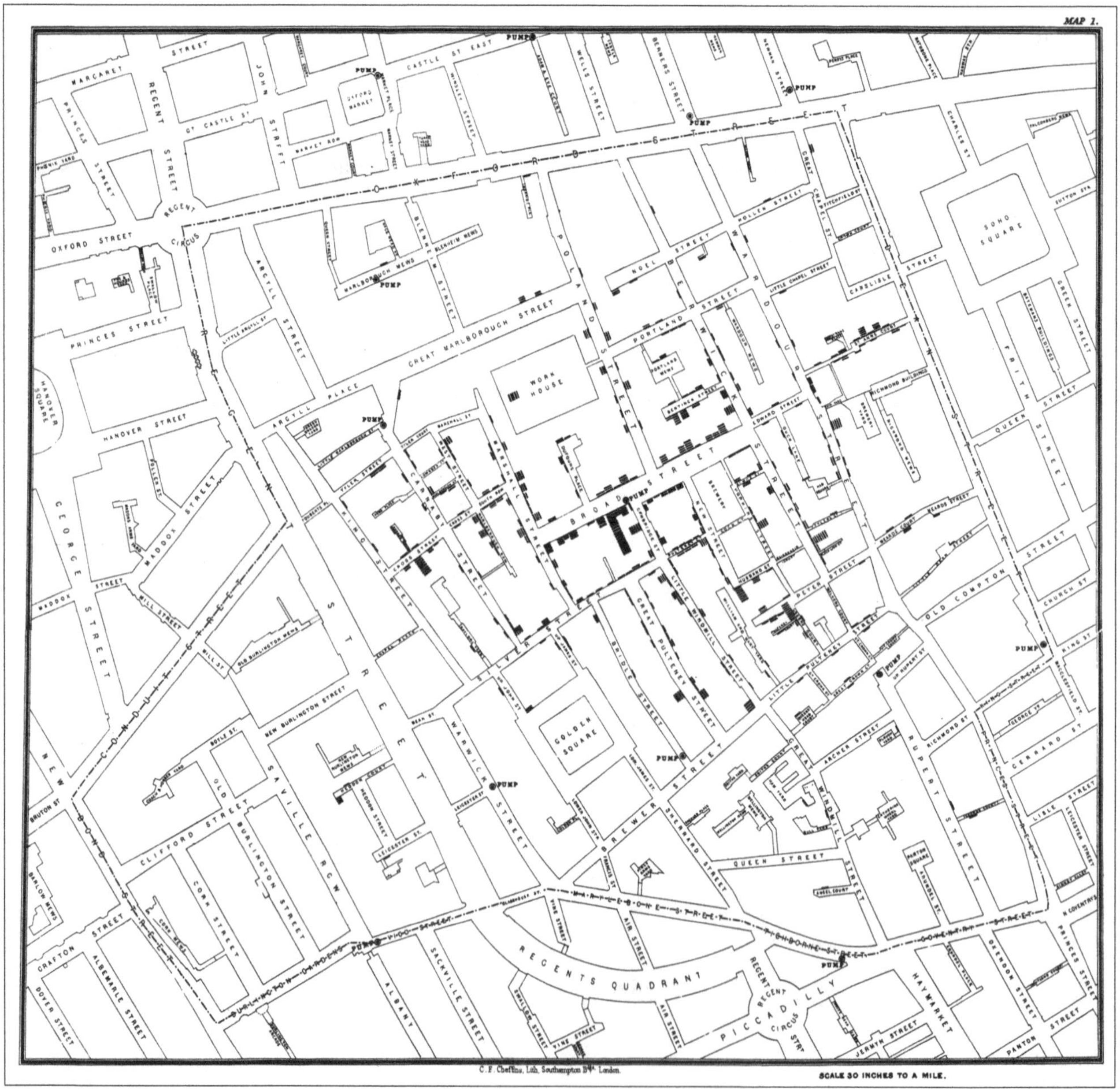

John Snow, an English physician, famously used maps to determine that, in the 1854 Broad Street cholera outbreak, a water pump was the most likely source of bacteria. From John Snow's On the Mode of Communication of Cholera, *2nd Ed, John Churchill, New Burlington Street, London, England, 1855 via Wikimedia Commons.[Public domain.]*

absorption of sodium and chloride. The efflux of chloride also causes water, bicarbonate, and potassium to leave the cells and enter the small intestine. This water-electrolyte movement fills the small intestine with large quantities of water and salts, causing the copious, relentless diarrhea that characterizes this disease. *V. cholerae* bacteria are shed with diarrhea, thus perpetuating the infective cycle. Dehydration and electrolyte imbalances result and, unless corrected, can cause coma and death.

EPIDEMIOLOGY

One of the features of cholera is the ease with which the disease spreads, despite the success of treatment with rehydration and antibiotics. In the twenty-first century, cholera remains a threat in Asia, Africa, and parts of Central America and South America. The incubation period is as little as two to five hours or two to three days. Each year, 3 million to 5 million people become infected, and between 100,000 and 120,000 people die of the disease.

HISTORY

It was challenging to convince people that cholera existed for centuries because acute diarrhea is a component of many illnesses. In 1883, German scientist Robert Koch hypothesized that microbes caused cholera. He obtained fecal specimens during a cholera outbreak in Egypt and isolated the bacterium he named *V. cholerae*.

CHOLERA IN MARINE PLANKTON

Outbreaks of cholera can occur in nonendemic areas when, for example, an infected person travels to another country or when infected water is carried in the ballast of ships to another country. These two processes alone, however, could not explain all of the outbreaks of cholera observed worldwide. In the late 1960s, the bacterium *Vibrio cholerae* was found in the ocean associated with marine plankton. This association, along with climate change, helps to explain the spread of cholera.

Plankton, the tiny organisms suspended in the ocean's upper layers, can be divided into phytoplankton (small plants) and zooplankton (tiny animals). *V. cholerae* is associated with the surface and gut of copepods, which are members of the zooplankton group. These small crustaceans act as a reservoir for the cholera bacteria, allowing them to survive in the ocean for long periods.

A change in weather that causes the ocean temperature to rise could also cause currents that stir up nutrients from the lower layers of the ocean to the upper layers. The numbers of phytoplankton, which live in the upper layers of ocean waters, increase in these periods because of the warmer temperatures and greater availability of nutrients. Zooplankton numbers increase too because of the increase in their primary food source, the phytoplankton. Consequently, the number of cholera bacteria increases to a level that can cause the disease. Thus, climate change can result in an outbreak of cholera in a region where cholera is endemic or if currents move the plankton to other coastal areas in a new, nonendemic region. This scenario is believed to explain the 1991 cholera epidemic in Peru when the oceanic oscillation known as El Niño caused warming of ocean temperature.

Because of the association of *V. cholerae* with plankton, scientists hope to track or identify future epidemics with satellite imagery. Increases in phytoplankton turn the ocean from blue to green. Thus, changes in green areas in the ocean on satellite images show where the phytoplankton and, by association, zooplankton and cholera bacteria are relocating or increasing in number.

The association of cholera with zooplankton also has helped reveal a new way to prevent the disease. People get cholera by ingesting several thousand cholera bacteria at one time. A single copepod can harbor ten thousand bacteria; therefore, the ingestion of one infected copepod can cause disease in a person. Researchers have found a simple and inexpensive way to reduce this risk from copepods dramatically. Filtering water through four layers of fabric used to make saris, which are commonly worn in regions plagued by cholera, removes 99 percent of copepods from water containing high plankton levels.

Now that the entire genetic sequence of *V. cholerae* has been determined, scientists have additional genetic data to elucidate the relationship of the bacterium with copepods, which may help scientists and health experts find more ways of controlling the spread of the disease.

CAUSES

Cholera is caused by drinking water contaminated with feces or by eating food that infected persons have handled. Raw or insufficiently cooked shellfish, such as shrimp, lobster, and crabs, can also transmit *V. cholerae*, which is carried in the chitin of the shell.

RISK FACTORS

The risk factors for cholera include eating contaminated food or fluids, eating raw or undercooked shellfish, living, or traveling in areas where cholera is present, having blood type O (a ninefold increase in risk), having a compromised immune system, and having low levels of stomach acid (e.g., are taking proton pump inhibitors such as omeprazole). Cholera also is most common in areas with poor sanitation and water quality.

People with low immunity, such as malnourished children or persons with human immunodeficiency virus (HIV) infection, are much more likely to die if they are infected. The risk of contracting cholera is greatest in urban slums and camps for refugees and other displaced peoples.

SYMPTOMS

Cholera's most prominent symptom is the acute onset of watery diarrhea, often accompanied by vomiting, dehydration, weakness, and abdominal pain. Of people infected with *V. cholerae* bacteria, about 75 percent never develop symptoms. The bacteria remain in infected patients' feces for seven to fourteen days, are excreted into the environment, and can infect other people.

About 25 percent of persons with the disease will experience cholera symptoms. Of this smaller number, about 80 to 90 percent have mild to moderate symptoms only. The remaining symptomatic patients are at risk of death if they do not receive treatment.

Dehydration and electrolyte imbalances can cause disorientation, swollen tongue, dry mouth, sunken eyes, cold, clammy skin, or shriveled and dry hands and feet. Low bicarbonate levels can cause metabolic acidosis. Metabolic acidosis induces deep, labored breathing (Kussmaul breathing). Low potassium levels can interfere with muscle function. Leg cramps, weakness, and abnormal heart rhythms are symptoms of low blood potassium levels. Low chloride and sodium levels cause headaches, poor balance, disorientation, seizures, and coma. Dehydration leads to hypovolemic shock with low blood pressure (hypotension).

SCREENING AND DIAGNOSIS

You should consult a doctor if you have recently traveled to areas where cholera is common. If cholera is suspected, the doctor will order stool and blood samples for testing. Stool samples are grown on thiosulfate-citrate-bile salts-sucrose (TCBS) agar to isolate *V. cholerae*. In response to a 1991 cholera outbreak in South America, biotechnology companies developed three rapid diagnostic tests. Technicians can perform these tests in the field with little training, and the results are provided within a few minutes. The MARTTM, MedicosTM, and Institut Pasteur cholera dipsticks are very easy to use. Unfortunately, their accuracy is good but not great.

TREATMENT AND THERAPY

The rapid loss of fluids and electrolytes can prove fatal. Hence, cholera treatment focuses on oral rehydration therapy, a method that is successful for about 80 percent of infected persons. This type of supportive therapy is adequate for mild to moderate cholera cases since such infections resolve on their own in about three to seven days. Some persons are so severely dehydrated that they require intravenous fluids. In severe cases, antibiotics may be given to shorten the illness's course and reduce symptoms.

Antibiotic susceptibility varies between different *V. cholerae* strains. Therefore, antibiotic susceptibility tests on the isolated bacteria should direct the choice of antibiotics for treatment. Tetracyclines, ciprofloxacin, ofloxacin, furazolidone, or trimethoprim/sulfamethoxazole are potential treatments.

Four oral cholera vaccines are available: Vaxchora, Dukoral, ShanChol, and Euvichol-Plus/Euvichol. Only Vaxchora is available in the United States. However, as of December 2020, the maker of Vaxchora temporarily ceased making and selling it. Therefore, Vaxchora is currently in limited supply or unavailable. Vaxchora reduces the chance of severe diarrhea in people by 90 percent at ten days after vaccination and by 80 percent at three months after vaccination. It is unknown if Vaxchora is safe for women who are pregnant or breastfeeding women. Nor is it known how long protection lasts beyond three to six months after getting the vaccine.

The other three cholera vaccines are available outside the United States. These vaccines are administered orally and for short-term protection. At the same time, long-term measures must be implemented, such as improvements to local water supplies.

The World Health Organization (WHO) recommends against using antidiarrheal, antiemetic, antispasmodic, cardiotonic, or corticosteroid drugs to treat cholera. WHO also eschews blood transfusions and plasma volume expanders. Infected persons may resume eating a regular diet as soon as vomiting has stopped. Among those who receive quick treatment, the mortality rate is less than 1 percent.

PREVENTION AND OUTCOMES

The most effective prevention against cholera is proper water treatment and sanitation. The fecal wastewater of infected persons and all contaminated materials, such as clothing and bedding, must be sterilized. Hands that have touched infected persons must be disinfected.

Public sewage can be treated with chlorine, ozone, or ultraviolet light and through sterilization or antimicrobial filtration before returning it to water supplies or waterways. The same methods should be applied to all water used for drinking, washing, or cooking.

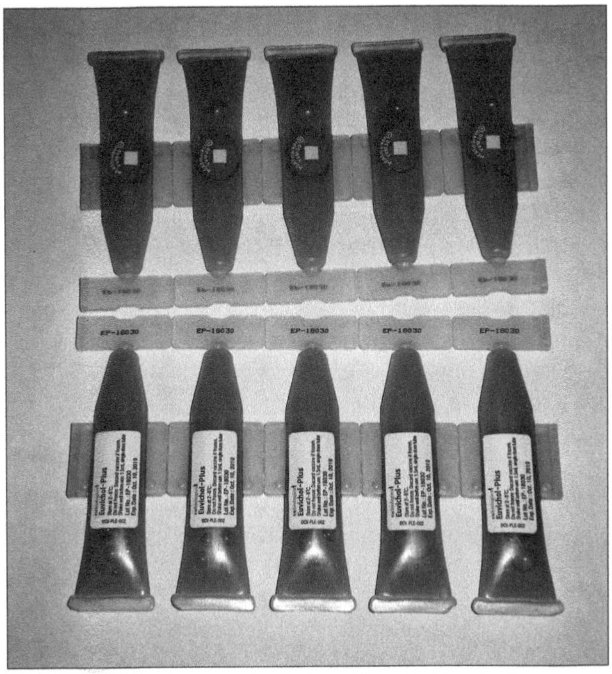

Euvichol-plus oral vaccine for cholera. Photo by Mr. Ibrahem, via Wikimedia Commons.

Food safety measures include avoiding raw foods, except when the peel can be removed; cooking food until it is thoroughly heated; and eating cooked food when it is still hot. One should wash and dry cooking and serving utensils after use and not let cooked food come in contact with uncooked food or unclean utensils. People can also protect themselves with thorough handwashing after defecation, after contacting feces, before preparing food, and before feeding children.

—*Vicki J. Isola, PhD, Merrill Evans, MA and Michael A. Buratovich, PhD*

Further Reading

Fanous, Matthew, and Kevin C. King. "Cholera." *StatPearls*. StatPearls Publishing, 2021.

Feldman, Mark, Lawrence S. Friedman, and Lawrence J. Brandt, editors. *Sleisenger and Fordtran's Gastrointestinal and Liver Disease: Pathophysiology, Diagnosis, Management.* New ed. 2 vols. Saunders/Elsevier, 2010.

Kalluri, P., et al. "Evaluation of Three Rapid Diagnostic Tests for Cholera: Does the Skill Level of the Technician Matter?" *Tropical Medicine in Health*, vol. 11, no. 1, 2006, pp. 49-55.

Ojeda Rodriguez, Jafet A. and Chadi I. Kahwaji. "Vibrio Cholerae." *StatPearls*. StatPearls Publishing, 2021.

Pennisi, Elizabeth. "Infectious Disease: Cholera Strengthened by Trip Through Gut." *Science*, vol. 296, 2002, pp. 1783-84.

Reidl, Joachim, et al. "Vibrio cholerae and Cholera: Out of the Water and into the Host." *FEMS Microbiological Reviews*, vol. 26, 2002, pp. 125-39.

Sampath, Shrikanth et al. "Pandemics Throughout the History." *Cureus*, vol. 13, no. 9, 2021, p. e18136, doi:10.7759/cureus.18136

Sridhar, Saranya. "An Affordable Cholera Vaccine: An Important Step Forward." *The Lancet*, vol. 374, 2009, pp. 1658-60.

FOOD POISONING

Category: Diseases and disorders

Anatomy or system affected: Gastrointestinal system, intestines, stomach

Specialties and related fields: Environmental health, epidemiology, gastroenterology, public health, toxicology

Definition: foodborne illness caused by bacteria, viruses, or parasites consumed in food and resulting in an acute gastrointestinal disturbance that may include diarrhea, nausea, vomiting, and abdominal discomfort

KEY TERMS

contamination: infection of a food item by a pathogen

foodborne infection: a disease caused by eating foods contaminated by infectious microorganisms, with onset occurring within twenty-four hours (e.g., salmonellosis)

foodborne intoxication: a disease caused by eating foods containing microorganisms that produce toxins, with onset occurring within six hours (e.g., botulism)

microorganism: an organism that is too small to be seen with the naked eye

parasite: an organism that lives on another organism (the host) and causes harm to the host while it benefits

pathogen: a disease-causing organism

thermal death point: the lowest temperature that can destroy a foodborne organism

CAUSES AND SYMPTOMS

Often a person feeling the symptoms of nausea, vomiting, diarrhea, and abdominal discomfort assumes that they have contracted influenza. The presence of an actual influenza virus, however, is uncommon. These symptoms are likely caused by eating food that contains undesirable bacteria, viruses, or parasites. This is called "foodborne illness" or "food poisoning." Most foodborne pathogens are colorless, odorless, and tasteless. Fortunately, there are recommendations based on scientific principles to help prevent foodborne illness.

Food poisoning is a global problem. In developing countries, diarrhea is a factor in child malnutrition and is estimated to cause millions of deaths per year. According to the Centers for Disease Control and Prevention (CDC), approximately 49 million Americans are affected by foodborne illnesses every year, resting in 3,000 deaths and 128,000 hospitalizations. Despite advances in modern technology, foodborne illness is a significant problem in developed countries as well.

Certain foods, particularly foods with a high protein and moisture content, provide an ideal environment for the multiplication of pathogens. The foods with high risk in the United States are raw shellfish (especially mollusks), underdone poultry, raw eggs, rare meats, raw milk, and cooked food that another person handled before it was packaged and chilled. In addition to those foods listed, some developing

Image via iStock/Tetiana Lazunova. [Used under license.]

countries could add raw vegetables, raw fruits that cannot be peeled, foods from sidewalk vendors, and tap water or water from unknown and contaminated sources.

Most of the documented foodborne illness cases are caused by only a few bacteria, viruses, and parasites. Bacteria, known as *Salmonella*, are ingested by humans in contaminated foods such as beef, poultry, and eggs; they may also be transmitted by kitchen utensils and the hands of people who have handled infected food or utensils. Once the bacteria are inside the body, the incubation time is from eight to twenty-four hours. Since the bacteria multiply inside the body and attack the gastrointestinal tract, this disease is known as a true food infection. The main symptoms are diarrhea, abdominal cramps, and vomiting. The bacteria are killed by cooking foods to the well-done stage.

The major foodborne intoxication in the United States is caused by eating food contaminated with the toxin of *Staphylococcus* bacteria. Because the toxin or poison has already been produced in the food item that is ingested, the onset of symptoms is usually very rapid (between one-half hour and six hours). Improperly stored or cooked foods (particularly meats, tuna, and potato salad) are the primary carriers of these bacteria. Since this toxin cannot be killed by reheating the food items to a high temperature, foods must be properly stored.

Botulism is a rare food poisoning caused by the toxin of *Clostridium botulinum*. It is anaerobic, meaning that it multiplies in environments without oxygen and is mainly found in improperly home-canned food items. Originally one of the sources of the disease was from eating sausages (the Latin word for which is *botulus*); hence, the term "botulism." A tiny amount of toxin, the size of a grain of salt, could kill hundreds of people within an hour. Danger signs include double vision and difficulty swallowing and breathing.

Though everyone is at risk for foodborne illness, certain groups develop more severe symptoms and are at a greater risk for serious illness and death. Higher-risk groups include pregnant women, very young children, the elderly, and immunocompromised individuals, such as patients with acquired immunodeficiency syndrome (AIDS) and cancer.

Bacteria known as *Listeria* were first documented in 1981 as being transmitted by food. Most people are at low risk of becoming ill after ingesting these bacteria; however, pregnant women are at high risk. *Listeria* infection is rare in the United States, but it does cause serious illness. It is associated with raw (unpasteurized) milk consumption, nonreheated hot dogs, undercooked chicken, and various soft cheeses (Mexican style, feta, Brie, Camembert, and blue-veined cheese), and food purchased from delicatessen counters. *Listeria* causes a short-term illness in pregnant women; however, these bacteria can cause stillbirths and spontaneous abortions. A parasite called *Toxoplasma gondii* is also of particular risk for pregnant women. For this reason, raw or very rare meat should not be eaten. (In addition, since cats may shed these parasites in their feces, it is recommended that pregnant women avoid cleaning cat litter boxes.)

As the protective antibodies from the mother are lost, infants become more susceptible to food poisoning. Botulism generally occurs by ingesting the toxin or poison; however, in infant botulism, it is the spores that germinate and produce the toxin within the infant's intestinal tract. Since honey and corn syrup have been found to contain spores, it is recommended that they not be fed to infants under one year of age, especially those under six months.

Determining whether a foodborne organism causes disease is highly skilled work. The Centers for Disease Control and Prevention (CDC) in Atlanta investigate diseases and their causes. It has been estimated that the true incidence of foodborne illness in the United States is ten to one hundred times greater than that reported to the CDC. The CDC report some of the more interesting cases and outbreaks in narrative form in the *Morbidity and Mortality Weekly Report*.

INFORMATION ON FOOD POISONING

Causes: Bacteria, viruses, or parasites consumed in food

Symptoms: Diarrhea, nausea, vomiting, fever, abdominal discomfort

Duration: One to three days

Treatments: Rest, avoidance of food, extra fluid intake

TREATMENT AND THERAPY

In cases of severe food poisoning marked by vomiting, diarrhea, or collapse—especially in botulism and ingestion of poisonous plant material such as suspicious mushrooms—emergency medical attention should be sought immediately, and, if possible, specimens of the suspected food should be submitted for analysis. Identifying the food source is especially important if that source is a public venue such as a restaurant because stemming a widespread outbreak of food poisoning may be possible. In less severe cases of food poisoning, the victim should rest, eat nothing, but drink fluids that contain some salt and sugar; the person should begin to recover after several hours or one or two days and should see a doctor if not well after two or three days.

Some types of food poisoning can be treated with antibiotics, and there is an antitoxin available for cases of botulism. However, the best "treatment" for food poisoning is prevention. While there is ample information regarding the prevention of food poisoning, many outbreaks still occur due to carelessness in the kitchen. Good food safety is basically good common sense. Yet, it can only make sense when one has learned how foodborne pathogens spread and how to apply food safety steps to prevent foodborne illness. Based on the research literature and the suggestions made by the World Health Organization (WHO) and other groups, the recommendations are to cook foods well, prevent cross-contamination, and keep hot foods hot and cold foods cold.

Cooking foods well means cooking them to a high enough temperature in the slowest-to-heat part and for a long enough time to destroy pathogens that have already gained access to foods. Cooking foods well is only a concern when they have become previously contaminated from other sources or are naturally contaminated. There are several possible sources of contamination of food products.

Coastal water may contaminate seafood. Filter-feeding marine animals (such as clams, scallops, oysters, cockles, and mussels) and some fish (such as anchovies, sardines, and herring) live by pumping in seawater and sifting out organisms that they need for food. Therefore, they can concentrate suspended material by many orders of magnitude. Shellfish grown in contaminated coastal waters are the most frequent carriers of a virus called "hepatitis A."

Contaminated eggs can be another vehicle of foodborne illness. Contamination of eggs can occur from external as well as internal sources. Suppose moist conditions are present and there is a crack in the shell. In that case, the fecal material of hens carrying the microorganism can penetrate the shell and membrane of the egg and can multiply. In the early 1990s, *Salmonella enteritidis* began to appear in the intact egg, particularly in the northeastern part of the United States. It is hypothesized that contamination occurs in the oviduct of the hen before the egg is laid. Food vehicles in which *Salmonella enteritidis* has been reported include sandwiches dipped in eggs and cooked, hollandaise sauce, eggs Benedict, commercial frozen pasta with raw egg and cheese stuffing, Caesar salad dressing, and blended food in which cross-contamination had occurred. Foods such as cookie or cake dough or homemade ice cream made with raw eggs are other possible vehicles of foodborne illness.

The bacterium Bacillus cereus (B. cereus) is an endospore-forming organism that can withstand high temperatures. B. cereus is found in many different foods include bean sprouts, dairy products, rice, and other vegetables. Because of its heat resistance, this organism often survives cooking. Cooked rice is a common source of B. cereus food poisoning.

B. cereus makes two different toxins, a diarrheal toxin and an emetic toxin. The emetic toxin causes symptoms thirty minutes to six hours after ingestion of contaminated food. Symptoms include abdominal cramps, nausea, and vomiting that resolve about six to twenty-four hours later. The diarrheal toxin causes symptoms about eight to sixteen hours after ingestion. Symptoms are abdominal cramping and diarrhea that usually resolve after twenty-four hours. The diarrheal toxin is heat-sensitive, but the emetic toxin is not. Therefore, cooking inactivates diarrheal toxin but not emetic toxin.

Milk, especially raw milk, can be contaminated. Sources of milk contaminants could be an unhealthy cow (such as from mastitis, a significant infection of the mammary gland of the dairy cow) or unclean methods of milking, such as not cleaning the teats well before attaching them to the milker or unclean utensils (milking tanks). If milk is not cooled fast enough, contaminants can multiply.

Modern mechanized milking procedures have reduced but not eliminated foodborne pathogens. Post-pasteurization contamination may occur, especially if bulk tanks or equipment have not been adequately cleaned and sanitized. In 1985, one of the largest salmonellosis outbreaks occurred in Chicago, with the causal food being pasteurized milk. More than sixteen thousand people were infected, and ten died. A small connecting piece in the milk tank that allowed milk and microorganisms to collect was determined to be the source of the contamination. Bulk tanks should be adequately maintained, and piping should be inspected regularly for opportunities for raw milk to contaminate the pasteurized product.

Recommendations for cooking temperatures are based on the temperature required to kill foodborne pathogens and aesthetics and palatability. Generally, a margin of safety is built into the cooking temperature because of the possibility of nonuniform heating. Based on generally accepted temperature requirements, cooking red meat until 71° Celsius (160° Fahrenheit) will reach the thermal death point. Hamburger should be well cooked so that it is medium-brown inside. If pressed, it should feel firm, and the juices that run out should be clear. Cooking poultry to the well-done stage is done for palatability. Tenderness is indicated when there is a flexible hip joint, and juices should run clear and not pink when the meat is pierced with a fork. Fish should be cooked until it loses its translucent appearance and flakes when pierced with a fork. Eggs should be thoroughly cooked until the yolk is thickened and the white is firm, not runny. Cooked or chilled foods that are served hot (that is, leftovers) should be reheated to come to a rolling boil.

Cross-contamination occurs when microorganisms are transmitted from humans, cutting boards, and utensils to food. Contamination between foods, especially from raw meat and poultry to fresh vegetables or other ready-to-eat foods, is significant.

One of the best ways to prevent cross-contamination is simply washing one's hands with soap and water. Twenty seconds is the minimum time that should be spent washing one's hands. The CDC also recommends washing the hands thoroughly and then drying them with a paper towel that is thrown away to prevent the spread of disease. However, thoroughly washed hands can still be a source of bacteria, so one should use tongs and spoons when cooking to prevent contamination.

It is imperative to wash one's hands after certain activities, such as blowing the nose or sneezing, using the lavatory, diapering a baby, smoking, petting animals or pets, and before cooking or handling food.

Other sources of cross-contamination include utensils and cutting surfaces. If people use the same knife and cutting board to cut up raw chicken for a stir-fry and peaches for a fruit salad, they are putting themselves at great risk for foodborne illness. The bacteria on the cutting board and the knife could cross-contaminate the peaches. While the chicken will be cooked until it is well done, the peaches in the salad will not be. In this situation, one could cut the fruit first, then the chicken, and then wash and sanitize the knife and cutting board.

Cleaning and sanitizing is a two-step process. Cleaning involves using soap and water and a scrubber or dishcloth to remove the major debris from the surface. The second step, sanitizing, involves using a diluted chlorine solution to kill bacteria and viruses.

Wooden cutting boards are one of the worst offenders in terms of causing cross-contamination. Since bacteria and viruses are microscopic, they can adhere to and grow in the grooves of a wooden cutting board and spread to other foods when the cutting board is used again. The use of a plastic or acrylic cutting board prevents this problem.

The danger zone in which bacteria can multiply is a range of 4.4° C (40° F) to 60° C (140° F). Room temperature is generally right in the middle of this danger zone. The danger zone is critical because bacteria are increasing in number even though they cannot be seen. They can double and even quadruple in fifteen to thirty minutes. Consequently, perishable foods such as meats, poultry, fish, milk, cooked rice, leftover pizza, hard-cooked eggs, leftover refried beans, and potato salad should not be left in the danger zone for more than two hours. Keeping foods hot means maintaining them at a temperature higher than 60 degrees Celsius. Keeping foods cold means maintaining them at a temperature lower than 4° C (40° F).

Other rules help prevent contamination. When shopping, the grocery store should be the last stop not to store foods in a hot car. When mealtime is over, place leftovers in the refrigerator or freezer as soon as possible. When packing for a picnic, food items should be kept in an ice chest to keep them cold or brought slightly frozen. Many instances of foodborne illnesses and death could be prevented if such food safety rules were followed.

PERSPECTIVE AND PROSPECTS

When the lifestyle of people changed from a hunting-and-gathering society to a more agrarian one, the need to preserve food from spoilage was necessary for survival. As early as 3000 BCE, salt was used as a meat preservative, and the production of cheese had begun in the Near East. The production of wine and the preservation of fish by smoking were also introduced at that time. Even though people had tried many methods throughout history to preserve foods and keep them from spoiling, the relationship between illness and pathogens or toxins in food was not recognized and documented until 1857. It was then that the French chemist Louis Pasteur demonstrated that the microorganisms in raw milk caused spoilage.

Stories from the American Civil War (1860-65) demonstrate the problems of institutional feeding of many people for long periods. Gastrointestinal diseases were rampant during that time. During the first year of the war, of the people who had diarrhea and dysentery, the morbidity rate was 640 per 1,000. It increased to 995 per 1,000 in 1862. More men died of disease and illness than were killed in battle.

Food can be contaminated by disease-causing organisms at any step of the food-handling chain, from the farm to the table. An important role of government and industry is to ensure a safe food supply. In the United States, setting and monitoring of food safety standards are the responsibility of the Food and Drug Administration (FDA) under the auspices of the US Department of Health and Human Services and the Food Safety and Inspection Service (FSIS) under the auspices of the US Department of Agriculture (USDA). The FDA is responsible for the wholesomeness of all food sold in interstate commerce, except meat and poultry. At the same time, the USDA is responsible for inspecting meat and poultry sold in interstate commerce and internationally. Some major food safety laws and policies that have guided the provision of safe food are the Federal Food and Drugs Act in 1906; the Federal Meat Inspection Act in 1906-7; the Food, Drug, and Cosmetic Act in 1938; and the Poultry Products Inspection Act in 1957.

Historically, the diseases of tuberculosis, scarlet fever, strep throat, typhoid fever, and diphtheria have been associated with raw or unpasteurized milk. The reporting of foodborne illness was initiated in the 1920s by the US Public Health Service (USPHS), when annual summaries of outbreaks of milk-borne disease were recorded and reported. Later, reports of waterborne and foodborne diseases were added.

The public's and the FDA's attitudes about what is hazardous in the food supply have often differed. The public generally believes that the safety of additives and chemical contaminants in food is a higher priority than microbiological and nutritional hazards—the exact opposite of the FDA's priorities. (For example, in the mid-1980s, the story about Alar, a chemical used to slow the ripening of apples, represented a very emotional topic. There was particular concern about the risks this chemical might pose to children who ate large amounts of apple products.) As more reliable information is available about both areas of concern, priorities are likely to change.

—Martha M. Henze, MD, RD
and Maria Pacheco, PhD

Further Reading

Carson-DeWitt, Roasalyn. "Food Poisoning." *Health Library*, 22 Mar. 2013.

Cliver, Dean O., and Hans P. Riemann, editors. *Foodborne Diseases*. 2nd ed., Academic Press, 2002.

Gaman, P. M., and K. B. Sherrington. *The Science of Food: An Introduction to Food Science, Nutrition, and Microbiology*. 4th ed., Butterworth-Heinemann/Elsevier, 2008.

Griffith, C. J. "Do Businesses Get the Food Poisoning They Deserve? The Importance of Food Safety Culture." *British Food Journal*, vol. 112, no. 4, 2010, pp. 416-25.

Jay, James M., Martin J. Loessner, and David A. Golden. *Modern Food Microbiology*. 7th ed., Springer, 2005.

Leon, Warren, and Caroline Smith DeWaal. *Is Our Food Safe? A Consumer's Guide to Protecting Your Health and the Environment*. Crown, 2002.

Lew, Kristi. *Food Poisoning: E. Coli and the Food Supply*. Rosen Publishers, 2011.

Longrée, Karla, and Gertrude Armbruster. *Quantity Food Sanitation*. 5th ed., John Wiley & Sons, 1996.

Marriot, Norman G., and Robert B. Gravani. *Principles of Food Sanitation*. 5th ed., Springer, 2006.

Nestle, Marion. *Safe Food: The Politics of Food Safety*. Updated ed., U of California P, 2010.

Ray, Bibek. *Fundamental Food Microbiology*. 4th ed., Taylor & Francis, 2008.

Stenfors Arnesen, Lotte P., et al. "From Soil to Gut: *Bacillus cereus* and Its Food Poisoning Toxins." *FEMS Microbiology Reviews*, vol. 32, no. 4, 2008, pp. 579-606, doi:10.1111/j.1574-6976.2008.00112.x.

Troncoso, Alcides, Cecilia Ramos Clausen, and Jessica Rivas. *Where Can You Catch Botulism Food Poisoning?: Foodborne Botulism*. Lambert Academic Publishing, 2012.

Wilson, Michael, Brian Henderson, and Rod McNab. *Bacterial Disease Mechanisms: An Introduction to Cellular Microbiology*. Cambridge UP, 2002.

BOTULISM

Category: Diseases and conditions

Anatomy or system affected: Nervous system

Specialties and related fields: Bacteriology, biochemistry, biotechnology, immunology, internal medicine, microbiology, neurology, pharmacology, public health

Definition: a potentially deadly illness that is caused by a toxin produced by the bacterium *Clostridium botulinum*

KEY TERMS

botulinum toxin: a potent neurotoxin that prevents nerves from releasing neurotransmitters

clostridia: a group of strictly anaerobic, gram-positive bacteria that cause several diseases and live mainly in soil

endospores: a dormant, tough, and nonreproductive structure produced by some bacteria

neurotransmitter: small, bioactive amines released by a nerve fiber at the arrival of a nerve impulse that diffuses across the synapse or junction and causes the transfer of the impulse to another neuron, muscle, or other structure

isotypes: the different classes of antibodies

paralysis: the loss of the ability to move, or feel anything, in part or most of the body, typically because of illness, poison, or injury

CAUSES

Clostridium botulinum (*C. botulinum*) is a gram-positive, rod-shaped, endospore-forming, obligate anaerobic bacteria. This organism is ubiquitous and occurs on the surfaces of vegetables, fruits, and seafood and exists in soil and marine sediment worldwide and at the bottom of streams and lakes. The intestinal tracts of fish, mammals, crabs, and other shellfish may contain *C. botulinum* and its endospores. *C. botulinum* endospores are heat resistant and can survive 100° Celsius at one atmosphere for five or more hours. Consequently, they can survive in improperly prepared foods. *C. botulinum* endospores are destroyed by heating to 120° C for five minutes. If the proper environmental conditions occur, *C. botulinum* endospores can germinate and grow into toxin-producing vegetative cells. The right conditions for endospore germination include: (a) low oxygen levels (anaerobic or semianaerobic environments; (b) a near-neutral pH (growth is inhibited at pH < 4.6); and (c) temperatures between 25 to 37° C, but some *C. botulinum* strains may grow at even lower temperatures.

Botulinum toxin is perhaps the most potent known poison known to biology. As a comparison, the minimum lethal dose (MLD) in experimental mice of curare and sodium cyanide are 500 and 10,000 mcg/kg, respectively. However, the MLD of botulinum toxin is 0.0003 mcg/kg. One gram of aerosolized botulism toxin could potentially kill at least 1.5 million people. Consequently, contact with even a tiny amount of the botulinum toxin can cause illness. People come in contact with this toxin in one of three ways. First, food can be contaminated with the bacterium and its toxin. The toxin produced by *C. botulinum*—not *C. botulinum* itself—causes botulism in humans. Food contaminated with the toxin include home-canned goods, sausage, other meat products, seafood, canned vegetables, and honey.

Several types of botulism differ in how they are acquired. Infant botulism occurs the infants ingest C. botulinum endospores that colonize their gastrointes-

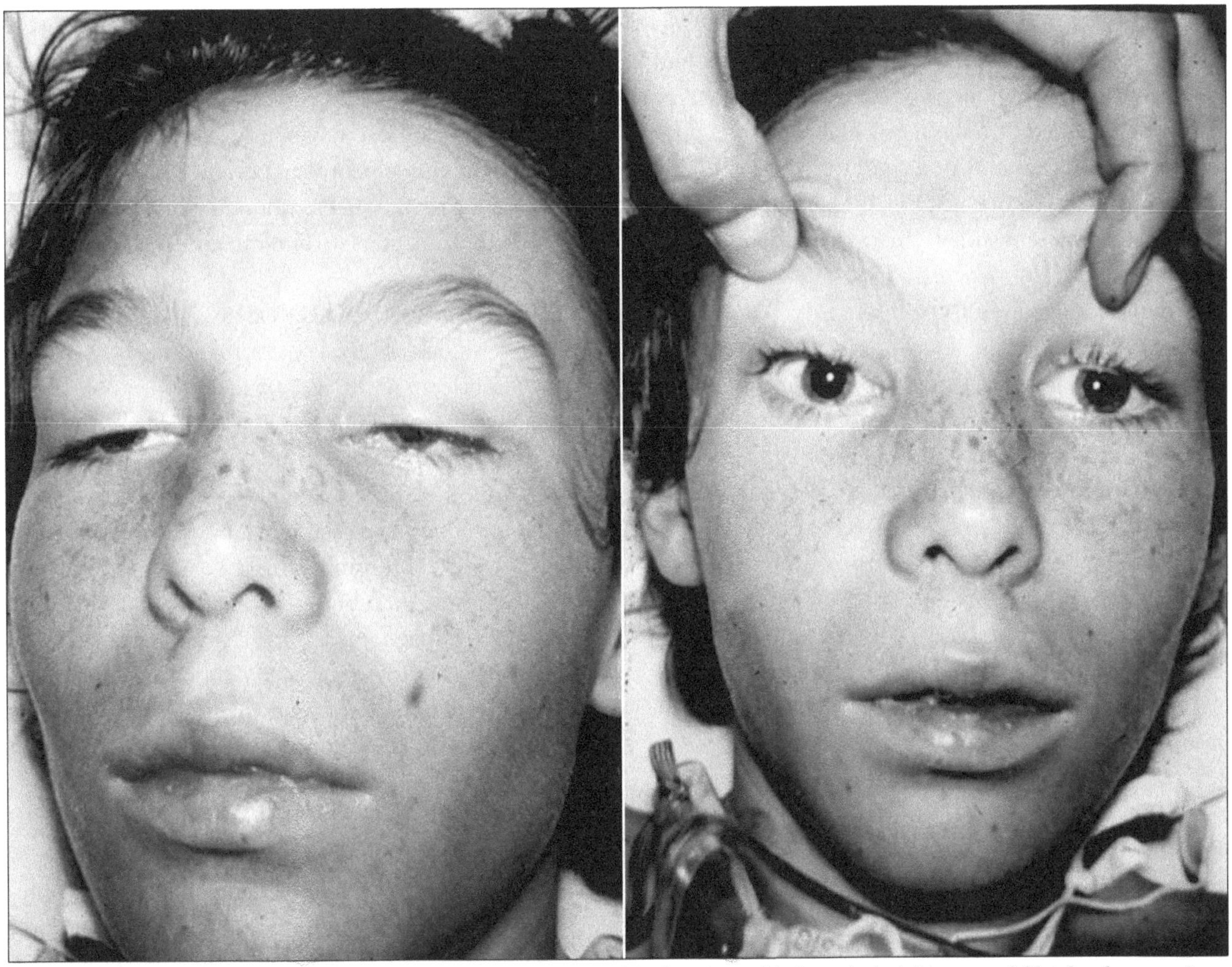

A 14-year-old with botulism, characterised by weakness of the eye muscles and the drooping eyelids shown in the left image, and dilated and non-moving pupils shown in the right image. This youth was fully conscious. Photos by Herbert L. Fred, MD and Hendrik A. van Dijk, via Wikimedia Commons.

tinal tracts. Since infants lack the stable intestinal microflora of adults, the ingested C. botulinum endospores germinate, produce botulinum toxin, and become residents of the infant's bowel. In the United States, most infant botulism cases result from ingestion of environmental dust and soil containing C. botulinum spores. The highest incidence of reported cases occurs in Utah, Pennsylvania, and California, states with the highest soil botulinum spore counts. Ingestion of raw honey also causes infant botulism, but this is a minor cause.

Foodborne botulism is caused by ingesting food contaminated by preformed botulinum toxin. Most foodborne botulism cases involve home-canned fruits, vegetables, and fish. In the United States, the highest rates of foodborne botulism occur among Alaska Natives who eat aged fish, seals, or whale blubber. Moonshine brewed in prisons is also a source of botulinum outbreaks in prisons. In China, home-fermented tofu cause over half the cases of foodborne botulism.

Wound botulism results from wounds contaminated with C. botulinum endospores. Because the endospores require an oxygen-poor environment to germinate, puncture wounds would provide the best type of wound for endospores to germinate and thrive. Wound botulism is associated with injection drug use, particularly with "black tar" heroin. Cocaine users have rarely suffered from wound botulism in their sinuses.

"Iatrogenic" diseases are conditions caused by medical treatments. Iatrogenic botulism occurs in patients who have received botulinum toxin for cosmetic indications.

RISK FACTORS

Risk factors for botulism include eating improperly canned foods and (rarely) using intravenous or intranasal drugs. For infants, consuming honey is a risk factor.

SYMPTOMS

Symptoms begin in the face and eyes and progress down both sides of the body. If the disease is left untreated, muscles in the arms, legs, and torso, and those used in breathing, become paralyzed. Death can occur.

Symptoms in adults can range from mild to severe and include muscle weakness, dizziness, double or blurred vision, droopy eyelids, trouble swallowing, dry mouth, sore throat, slurred speech, difficulty breathing, and constipation. The symptoms include constipation, not eating or sucking, little energy, poor muscle tone, and a feeble cry in babies.

When food is the cause of botulism, symptoms usually start within thirty-six hours of eating the contaminated food. Some people notice symptoms within a few hours, but others may not develop symptoms for several days. Some people experience nausea, vomiting, and diarrhea. When a wound is the cause of botulism, symptoms start within four to fourteen days.

SCREENING AND DIAGNOSIS

A doctor will ask about symptoms and medical history and will perform a physical exam. Blood, stool, and stomach contents will be tested for the toxin. In infants, too, the stool is tested for *C. botulinum*. If available, samples of questionable food may also be tested for the toxin and bacteria. A wound culture will be done if wound botulism is suspected.

Tests to rule out other medical conditions include blood tests, a magnetic resonance imaging (MRI) scan (a scan that uses radio waves and a powerful magnet to produce detailed computer images), spinal fluid analysis, and nerve conduction tests.

TREATMENT AND THERAPY

The most severe complication is respiratory failure. Treatment aims to maintain adequate oxygen supply, which may require a ventilator and close monitoring in an intensive care unit. Feeding through a tube may also be necessary.

If treatment begins early, an antitoxin can stop the paralysis from progressing and may shorten the duration of symptoms. It does not, however, reverse the disease process. Methods to eliminate the toxin include enemas, suctioning of stomach contents, medication to stimulate vomiting, surgery to clean a wound, and antibiotics to treat a wound infection. High temperatures can destroy botulinum toxin.

PREVENTION AND OUTCOMES

Strategies to prevent botulism include the following: Avoid feeding honey to children who are younger than one year of age; refrigerate oils that contain garlic or herbs; bake potatoes without foil (if potatoes are wrapped in foil, keep them hot until served or refrigerate them); avoid tasting foods that appear spoiled; avoid eating food from a can that is bulging; boil home-canned foods for ten to twenty minutes before eating; practice good hygiene when canning; seek medical care for wounds and return to the doctor if the wounds look infected (exhibits redness, warmth, pus, or tenderness); and avoid injecting illicit drugs.

—Debra Wood and Michael A. Buratovich, PhD

Further Reading

Benjamin, Ivor, et al., editors. *Andreoli and Carpenter's Cecil Essentials of Medicine*. 9th ed., Saunders, 2015.

Centers for Disease Control and Prevention. "Botulism." 19 Aug. 2019, www.cdc.gov/botulism/. Accessed 21 Nov. 2021.

Brachman, Philip S., and Elias Abrutyn, editors. *Bacterial Infections of Humans: Epidemiology and Control*. 4th ed., Springer Science, 2009.

Bennett, John E., et al., editors. *Mandell, Douglas, and Bennett's Principles and Practice of Infectious Diseases*. 9th ed., Elsevier, 2019.

Jameson, J. Larry, et al., editors. *Harrison's Principles of Internal Medicine*. 20th ed., McGraw-Hill, 2018.

Kliegman, Robert M., et al., editors. *Nelson Textbook of Pediatrics*. 21st ed., Elsevier; 2019.

Pommerville, Jeffery. *Alcamo's Fundamentals of Microbiology: Body Systems*. 2nd ed., Jones & Bartlett Learning. 2012.

LISTERIOSIS

Category: Diseases and conditions
Anatomy or system affected: Blood, brain, central nervous system, placenta
Specialties and related fields: Bacteriology, food science, gastroenterology, internal medicine, microbiology, neonatology, neurology, obstetrics and gynecology, pediatrics, public health
Definition: a food-borne illness that can lead to death in newborns and persons with compromised immune systems

KEY TERMS

actin: a protein that forms (together with myosin) the contractile filaments of muscle cells, and is also involved in motion in other types of cells

cytoskeleton: a microscopic network of protein filaments and tubules in the cytoplasm of many living cells, giving them shape and coherence

internalin: surface proteins found on *Listeria monocytogenes* that exist in two known forms, InlA and InlB, and are used by the bacteria to invade mammalian cells via cadherins transmembrane proteins and Met receptors, respectively

listeriolysin O: a hemolysin produced by the bacterium *Listeria monocytogenes*

INTRODUCTION

Listeriosis is a food-borne illness that can lead to death in newborns and persons with compromised immune systems. Infants born to women infected with listeriosis may have meningitis (brain infection) or bacteremia (bacterial blood infection). Infected infants who survive may suffer neurological damage and developmental delays. Listeriosis can cause the death of a fetus of an infected pregnant woman.

Up to 65 percent of deaths from food-borne illnesses in the United States are caused by listeriosis. About twenty-five hundred people become ill with listeriosis per year in the United States, and five hundred die. The numbers of the infected may be greater, but such cases have not been identified, likely because symptoms were mild.

CAUSES

Listeriosis is caused by the bacterium *Listeria monocytogenes*, a pathogen that lives in water and soil. It is resistant to refrigeration and is found in ill-prepared or subsequently contaminated meats and vegetables, particularly in luncheon meats, hot dogs, soft cheeses, coleslaw, and unpasteurized milk.

Listeria monocytogenes (*L. monocytogenes*) is a gram-positive rod that is highly motile if grown at 37° Celsius or below. At temperatures above 37° C, it stops expressing the flagellin protein that makes the flagellum and becomes immobile.

In the gastrointestinal tract, *L. monocytogenes* use a cell surface protein called "internalin" to bind host epithelial cells and sneak inside them. Once inside, this bacterium co-opts the cell's actin cytoskeleton to move throughout its cytoplasm and transfer to adjacent cells. *L. monocytogenes* also secretes a toxin called "listeriolysin O" that kills T lymphocytes, numbing the immune system to its presence.

For uncertain reasons, *L. monocytogenes* prefer placental and fetal tissue. *L. monocytogenes* can cross the placenta in infected mothers and infect the fetus.

In people with poorly functioning immune systems, *L. monocytogenes* can cause a liver abscess or move to the brain through the blood-brain barrier and cause meningitis. Meningitis commonly occurs in neonates, the elderly, or adults with diabetes mellitus.

RISK FACTORS

According to the Centers for Disease Control and Prevention, pregnant women have twenty times the risk of developing listeriosis as others, and an estimated one-third of listeriosis cases occur during pregnancy. Other persons at risk are those with compromised immune systems, such as persons with the acquired immune deficiency syndrome (AIDS), who have a three hundred times greater risk for listeriosis than healthy persons. In addition, others at risk include persons with cancer, kidney disease, or diabetes; persons who have had an organ transplant and take immunosuppressant drugs; persons taking corticosteroids; and persons aged sixty years and older.

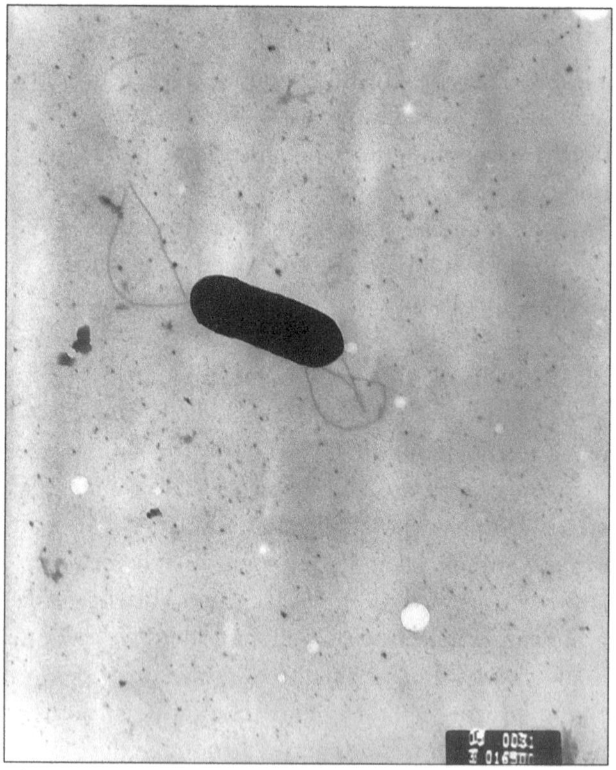

Electron micrograph of a flagellated Listeria monocytogenes *bacterium. Photo via Wikimedia Commons. [Public domain.]*

SYMPTOMS

If listeriosis is limited to the gastrointestinal tract, it causes low-grade fever, diarrhea, and vomiting.

Newborn infants may have jaundice, pneumonia, skin rash, lethargy, and vomiting. Symptoms may occur anytime from two to seventy days from when the individual consumed the contaminated food. Healthy people may have mild symptoms or no symptoms. Pregnant women may have mild symptoms, but her fetus remains at risk for infection.

Recent listeria outbreaks have been linked to cantaloupes and frozen vegetables, processed meats, soft cheeses, and several other foods. Image courtesy of the USDA. Photo by Scott Bauer via Wikimedia Commons. [Public domain.]

SCREENING AND DIAGNOSIS

If listeriosis is suspected, the blood, urine, or feces is screened for *L. monocytogenes*. Also, a spinal fluid test may be used for screening, and the amniotic fluid of a pregnant woman may be tested.

TREATMENT AND THERAPY

If listeriosis remains limited to the gastrointestinal tract, antibiotic treatment is not indicated. Hydration and rest are the best measures in such cases. However, disseminated listeriosis is serious and is treated with intravenous ampicillin plus gentamicin. Pregnant women should not receive gentamicin, and people with a penicillin allergy should not receive ampicillin. In such cases, meropenem is the preferred alternative treatment.

PREVENTION AND OUTCOMES

Active measures can help to avoid infection. One should thoroughly cook all meats, wash all vegetables, and avoid unpasteurized milk products. Everyone should separate uncooked meats from vegetables and other foods during food preparation.

As soon as possible after food preparation, the preparer should wash their hands and any cutting boards and knives used to prepare uncooked foods.

Persons at high risk for listeriosis should avoid soft cheese unless the label on the product indicates the manufacturer made the cheese with pasteurized milk. Pregnant women should avoid deli meat, cold salad, soft cheese, and pâté.

—Christine Adamec, MBA and Michael A. Buratovich, PhD

Further Reading

Centers for Disease Control and Prevention. "Listeria (Listeriosis)." 10 Sept. 2021, www.cdc.gov/listeria/index.html. Accessed 12 Dec. 2021.

Jackson, Kelly A., et al. "Listeriosis Outbreaks Associated with Soft Cheeses, United States, 1998-2014." *Emerging Infectious Diseases,* vol. 24, no. 6, 2018, pp. 1116-18, doi:10.3201/eid2406.171051.

Radoshevich, Lilliana, and Pascale Cossart. "Listeria Monocytogenes: Towards a Complete Picture of its Physiology and Pathogenesis." *Nature Reviews: Microbiology,* vol. 16, no. 1, 2018, pp. 32-46, doi:10.1038/nrmicro.2017.126.

Tack, Danielle M., et al. "Preliminary Incidence and Trends of Infections with Pathogens Transmitted Commonly Through Food—Foodborne Diseases Active Surveillance Network, 10 U.S. Sites, 2016-2019." *Morbidity and Mortality Weekly Report,* vol. 69, no. 17, 2020, pp. 509-14, doi:10.15585/mmwr.mm6917a1.

US National Institutes of Health. "Listeriosis." 30 Nov. 2021, www.nlm.nih.gov/medlineplus/ency/article/001380.htm. Accessed 12 Dec. 2021.

TYPHOID FEVER

Category: Diseases and conditions
Also known as: Enteric fever
Anatomy or system affected: All
Specialties and related fields: Bacteriology, immunology, gastroenterology, internal medicine, hematology, microbiology, preventive medicine, public health
Definition: typhoid fever and paratyphoid fever are serious illnesses caused by *Salmonella* bacteria, either *S. typhi* or *S. paratyphi*

KEY TERMS

capsule: a polysaccharide layer that lies outside the cell envelope
mucosa: the moist, inner lining of some organs and body cavities (such as the nose, mouth, lungs, and stomach)
Peyer's patches: clusters of subepithelial, lymphoid follicles found in the intestine
sepsis: a serious condition resulting from the presence of harmful microorganisms in the blood or other tissues and the body's response to their presence, potentially leading to the malfunctioning of various organs, shock, and death
serotype: a distinct variation within a species of bacteria or virus distinguishable by serological means
submucosa: the layer of areolar connective tissue lying beneath a mucous membrane

Typhoid fever and paratyphoid fever are serious illnesses caused by *Salmonella* bacteria, either *S. typhi* or *S. paratyphi*. The illness occurs most often in developing countries where sanitation is poor. Typhoid fever can be fatal, especially if not treated.

CAUSES

Salmonella is a bacterial genus that belongs to the enteric bacterial group. There are two main species of *Salmonella*: *Salmonella bongori* and *Salmonella enterica*, which has six distinct subspecies. One of these subspecies is enterica, which has over 2,500 serotypes. These serotypes are divided into two main groups based on the clinical symptoms they cause, typhoidal or nontyphoidal *Salmonella*.

The typhoidal group includes an organism designated serotype *Salmonella typhi* (*S. typhi*). This organism specifically infects humans and causes enteric fever, more commonly known as "typhoid fever."

Typhoid fever is caused by eating foods or drinking beverages contaminated with the *Salmonella* bacterium. Contamination can occur from food or drinks handled by someone who is sick or getting sick with typhoid fever, food or drinks handled by someone who has no symptoms but carries the bacteria, water, or food contaminated by sewage, unpasteurized dairy products, and poultry products left unrefrigerated.

DISEASE PROCESS

Generally, *Salmonella* are encapsulated, gram-negative, rod-shaped bacteria. These microorganisms are surrounded by a polysaccharide layer outside their cell wall called a "capsule." They are facultative anaerobes, meaning that they can live and grow in the presence or absence of oxygen. Salmonella also have flagella, which makes them highly motile.

S. typhi is killed by stomach acid. It may require a relatively large inoculum of microorganisms to cause an infection. After ingesting contaminated food or water, *Salmonella* reach the small intestine's distal ileum. They attach to a microfold cell (also known as an "M cell"). This cell engulfs the bacterium and transports it to the underlying Peyer's patches. Peyer's patches are intestinal immune tissue found on the surface of the mucosa but also extends into the submucosa. The infected Peyer's patches respond by releasing small proteins called "cytokines" that summon monocytes and macrophages. These white blood cells damage the small intestine lining, potentially causing ileal perforation and peritonitis.

Macrophages phagocytose *S. typhi* but are unable to kill it. The bacterium uses a small needle-like appendage called a "Type III secretion system" that injects proteins into the macrophage to prevent it from killing the bacteria. The microorganism, therefore, can thrive within the white blood cells that were sent to destroy it.

The infected macrophage carries *S. typhi* to nearby mesenteric enteric lymph nodes and eventually to the bloodstream. This bacterium can infect the liver, spleen, bone marrow, gallbladder, and additional lymph nodes from the bloodstream. The colonization of the blood by these bacteria is called "bacteremia." Bacteremia can progress to sepsis. Sepsis is a life-threatening condition that features systemic vasodilation and significantly reduced perfusion of vital organs (i.e., the organs are not getting enough nutrients and oxygen-rich blood).

Because the spleen plays a significant role in the immune response against encapsulated bacteria, those who have had their spleen removed or have sickle cell disease can develop osteomyelitis, an acute infection of the bones.

Others can become chronically infected with *S. typhi* and become carriers, potentially, for the rest of their lives. The bacterium usually reservoirs in the gallbladder and is shed in feces.

Humans are the only known reservoir for *S. typhi*. Therefore, the transmission route is fecal to oral, even if the contaminated feces come from someone with no symptoms.

RISK FACTORS

Risk factors for typhoid fever include drinking contaminated water, eating raw shellfish, eating fruits and vegetables that are raw or have been washed with contaminated water, and living in, or recent travel, to a country with poor sanitation. This disease is particularly endemic to Asia, Africa, Latin America, and the Caribbean.

SYMPTOMS

The symptoms of typhoid fever typically appear one to two weeks after the initial infection. The disease can take four to six weeks to resolve. Symptoms include a high persistent fever, abdominal pain, constipation followed by diarrhea, and light red-colored spots on the chest and abdomen. Enlargement of the liver (hepatomegaly) and spleen (splenomegaly) occurs in some people as the infection spreads through the bloodstream. Diarrhea can cause dehydration, weakness, headaches, and high fever can cause a confused mental state.

SCREENING AND DIAGNOSIS

A doctor will ask about symptoms and medical history and perform a physical exam. Typhoid fever is usually diagnosed with blood cultures or cultures from intestinal secretions such as vomit or a duodenal aspirate. Stool cultures can identify *S. typhi* during the first week of infection.

Bile cultures work the best for carriers who show no symptoms because stool cultures can give false negatives.

TREATMENT AND THERAPY

First-line treatment involves managing symptoms, including fluid and electrolyte replacement for diarrhea and nonsteroidal anti-inflammatory drugs for pain and fever.

Typhoid fever is highly contagious and is treated with antibiotics, but only if the disease fails to resolve itself. Usually, broad-spectrum antibiotics are used like fluoroquinolones or cephalosporins. Ciprofloxacin and levofloxacin are the preferred fluoroquinolones, or the third-generation cephalosporin ceftriaxone or the macrolide azithromycin are the preferred drugs. However, over 80 percent of S. *typhi* isolates in South Asia are not susceptible to fluoroquinolones. Therefore, fluoroquinolones should be used with caution. Treatment is for five to seven days or fourteen days for those with poorly functioning immune systems.

People may become typhoid carriers even after their illness has subsided in some cases. Chronic carriers can shed contagious *Salmonella* bacteria in their stool or urine. This chronic condition is treated with fluoroquinolone antibiotics and surgical removal of the gallbladder.

PREVENTION AND OUTCOMES

There are two main ways to prevent typhoid fever: vaccination and careful food monitoring. A typhoid vaccine is recommended for persons planning to visit a country where typhoid fever is prevalent. However, the vaccine is not always effective, and careful food monitoring is just as important. When in an area where typhoid fever is prevalent, one should always take the following precautions with food and water: Drink only bottled water or water that has been boiled for a minimum of one minute; eat foods while they are still hot and ensure that they are thoroughly cooked; avoid raw fruits and vegetables that cannot be peeled; avoid raw shellfish, and avoid unpasteurized dairy products.

—Michelle Badash and Michael A. Buratovich, PhD

Further Reading

Bhan, M. K., R. Bahl, and S. Bhatnagar. "Typhoid and Paratyphoid Fever." *The Lancet,* vol. 366 27 Aug.-2 Sept. 2005, pp. 749-62.

Foster, Neil, et al. "Revisiting Persistent *Salmonella* Infection and the Carrier State: What Do We Know?" *Pathogens,* vol. 10, no. 10, 2021, p. 1299, doi:10.3390/pathogens 10101299.

Levine, M. M. "Typhoid Fever." *Bacterial Infections of Humans: Epidemiology and Control,* edited by Philip S. Brachman and Elias Abrutyn, 4th ed., Springer Science, 2009.

Mintz, Eric. "Typhoid and Paratyphoid Fever." *CDC Health Information for International Travel 2010,* wwwnc.cdc.gov/travel/yellowbook/2010/table-of-contents.aspx.

Murray, Patrick R., et al. *Medical Microbiology.* 9th ed., Elsevier, 2020.

Qamar, Farah Naz, et al. "Salmonellosis Including Enteric Fever." *Pediatric Clinics of North America,* vol. 69, no. 1, 2022, pp. 65-77, doi:10.1016/j.pcl.2021.09.007.

Tortora, Gerard J., et al. *Microbiology: An Introduction.* 13th ed., Pearson, 2018.

SHIGELLOSIS

Category: Diseases and conditions

Anatomy or system affected: Bladder, gastrointestinal system, intestines, kidneys, stomach, urinary tract

Specialties and related fields: Bacteriology, critical care, infectious disease, internal medicine, hematology, microbiology, nephrology, public health, urology

Definition: an acute bacterial infection attacks the lining of the intestines, resulting in diarrhea, fever, and stomach cramps

KEY TERMS

endothelial cells: the main type of cell found in the inside lining of blood vessels, lymph vessels, and the heart

gastroenteritis: inflammation of the stomach and intestines, typically resulting from bacterial toxins or viral infection and causing vomiting and diarrhea

ischemia: an inadequate blood supply to an organ or part of the body

mucosae: an epithelial layer rich in mucous glands that lines body cavities and passages (as of the gastrointestinal or respiratory tract) that communicate directly or indirectly with the outside of the body

Shiga toxin: a toxin that inhibits protein synthesis in host cells produced by the bacterium *Shigella* dysenteriae, or a similar one produced by E. coli, that causes dysentery in humans

virulent: The ability of an agent of infection to produce disease

CAUSES

Shigella is a gram-negative, rod-shaped, nonmotile bacterium that is a member of the enteric bacteria. Shigella causes a gastrointestinal disease called "shigellosis." The genus *Shigella* is divided into four species: *Shigella dysenteriae* (*S. dysenteriae*), *S. flexneri*, *S. boydii*, and *S. sonnei*. Each of these species is divided into multiple serotypes. *S. dysenteriae* causes severe epidemic disease in less developed countries, and *S. flexneri* causes disease in developing countries. *S. boydii* is only found on the Indian subcontinent, and *S. sonnei* causes disease in developed and countries between developed and less developed (transitioning) countries.

Shigellosis is spread by fecal-oral contact. Transmission occurs when people ingest food or water contaminated with *Shigella*. Shigellosis can spread when food grows in a field with contaminated water. Flies can spread shigellosis by breeding in infected feces and landing on food or surfaces. Shigellosis can also spread when people drink, swim, or play in contaminated water. Transmission can be spread through sexual contact, particularly through anal and oral sex.

Shigellosis commonly occurs in developing countries with inadequate sanitation. It thrives in areas with overcrowding, poor handwashing techniques, and a lack of safe food and water protocols. The disease also spreads easily in close quarters, such as daycare centers, refugee camps, jails, and prisons.

Shigella can survive stomach acid and needs as few as ten microbes to cause disease. *Shigella*, therefore, is one of the most virulent organisms known to medicine.

RISK FACTORS

About eighteen thousand cases of shigellosis are reported in the United States alone each year. Many milder cases are not diagnosed or reported; public health officials estimate the actual number of cases could be as high as 360,000 annually. The infection is most common in children aged two to four years.

Shigellosis is the most common cause of diarrhea among visitors to developing countries. It infects hundreds of millions of people worldwide each year, resulting in an estimated one million deaths, mostly among children in countries with inadequate medical resources.

DISEASE PROCESS

After entering the gastrointestinal tract, *Shigella* passes through the stomach to the small intestine. After multiplying in the small intestine, it moves to the colon and targets microfold (M) cells. M cells take up the bacteria, but *Shigella* passes through the M cell to the underlying resident macrophage. The macrophage is part of a network of lymphoid tissue called "mucosal-associated lymphoid tissue" or

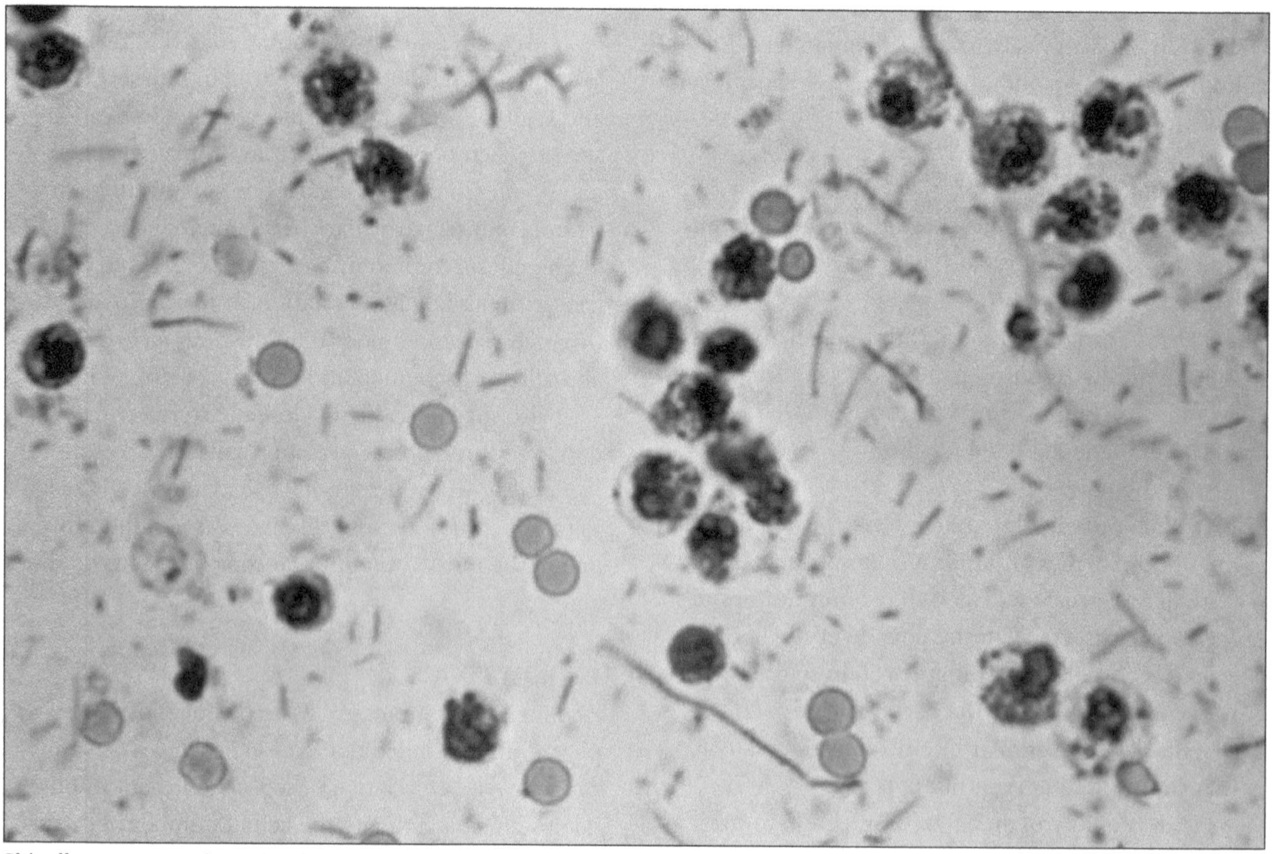
Shigella seen in a stool sample. Photo via CDC/Wikimedia Commons. [Public domain.]

MALT. MALTs extend into the submucosa and are loaded with macrophages and other immune cells.

Macrophages readily phagocytose *Shigella*, but it degrades the phagosome and escapes into the cytoplasm. Once in the cytoplasm, *Shigella* activates the macrophage's programmed cell death pathway, killing it. The dying macrophage releases several cytokines, including interleukin-1ß, that activate a robust inflammatory response. The inflammatory response damages and kills the colon's mucosal cells, resulting in ulcerations and abscesses. The released bacteria enter the mucosal cells from the bottom and spread throughout the mucosal epithelium, damaging it as they go.

S. dysenteriae, serotype 1 produces Shiga toxin, which enters cells and turns off protein synthesis. The extensive damage to the colon wall causes bleeding and allows Shiga toxin to enter the bloodstream. If Shiga toxin travels to the kidney, it enters the epithelial cells of the delicate glomeruli and kills them. Dead glomerular endothelial cells make holes in the kidney's filtration system, and proteins start to leak into the urine (proteinuria). Endothelial cell death also triggers cytokine release, which evokes inflammation and initiates clot formation.

Blood clots within the glomeruli significantly reduce platelet numbers (thrombocytopenia) and form a network of tiny spaces through which red blood cells must pass. These clots act like cheese cutters as red blood cells pass through them, fragmenting them into schistocytes—a process known as microangiopathic hemolysis. Continuous red blood cell destruction causes anemia. The combination of blocked vessels and fewer red blood cells causes kidney ischemia.

Since the ischemic kidney can no longer filter blood, metabolic wastes, like urea accumulate in the blood, causing uremia. The combination of uremia, hemolytic anemia, and thrombocytopenia characterizes hemolytic uremic syndrome (HUS). HUS is a consequence of Shiga toxin entering the bloodstream.

SYMPTOMS

The symptoms of watery diarrhea, fever, vomiting, stomach, and rectal cramps begin from one to three days after a person encounters the bacterium. In some cases, the incubation period is as short as twelve hours. Often, diarrhea contains blood and mucus, and the infected person may develop a fever. Some experience nausea, vomiting, anorexia, or cramping rectal pain. The symptoms usually resolve within five to seven days, even without treatment. However, the infected person may still be contagious for another week or two.

Some people do not experience symptoms at all, but they can pass *Shigella* to others while infected. Sometimes, dehydration occurs, leading to the death of a person who has severe shigellosis.

In developing countries, infected persons may experience prolonged, acute diarrhea, lasting seven to thirteen days, or persistent diarrhea, lasting fourteen days or more; this leads to malnutrition. These long-term conditions have the potential to be life-threatening.

About 3 percent of persons with shigellosis develop Reiter's syndrome, consisting of joint pain (usually the knee), irritation of the eyes, and painful urination. The syndrome can last for months or years, leading to chronic, treatment-resistant arthritis. Another rare complication is hemolytic-uremic syndrome, a form of kidney failure that includes anemia and clotting problems.

Tonic-clonic seizures may occur in severe cases, especially in children. The person loses consciousness, followed by involuntary muscular contractions that cycle from a rigid (tonic) phase to a rhythmic contraction and relaxation (clonic) phase. Neonates and malnourished children have an increased risk of sepsis.

Persons with high fever, confusion, headache with a stiff neck, lethargy, or seizures should seek emergency care. These symptoms are most common in children.

SCREENING AND DIAGNOSIS

A stool culture identifies *Shigella* in the feces of an infected person. Because there are several different strains of *Shigella*, the stool culture will help determine the correct treatment. Fecal samples are usually cultured on MacConkey agar. On this medium, *Shigella* form white, nonlactose fermenting colonies. Staining stools with methylene blue shows polymorphonuclear leukocytes. A blood test also may be done if symptoms are severe or to rule out other causes.

Polymerase chain reaction (PCR) directly identifies *Shigella* deoxyribonucleic (DNA).

TREATMENT AND THERAPY

Persons with mild cases of shigellosis usually recover quickly, without treatment. However, antibiotics can shorten the course of the illness by a few days. Some *Shigella* bacteria have become resistant to antibiotics.

Severe cases require antibiotics, and the standard treatments include trimethoprim/sulfamethoxazole, fluoroquinolones, macrolides, and beta-lactams. However, third-generation cephalosporins like cefixime or ceftriaxone are the preferred treatment for drug-resistant strains.

Infected persons should drink water and electrolyte solutions (such as *Gatorade*) to replace fluids lost by diarrhea. They should follow their normal diet, as much as possible, to ensure nutrition. Doctors recommend that persons with shigellosis avoid foods that are high in fat and sugar, spicy foods, and alcohol and coffee until two days after all symptoms have disappeared.

Because shigellosis is particularly dangerous for children, research continues into *Shigella* vaccines. A person who has recovered from shigellosis is unlikely to be infected with the same strain for several years. Antidiarrheal medicines such as loperamide (Imodium) and atropine (Lomotil) should not be taken by persons with shigellosis, as they can make the illness worse.

PREVENTION AND OUTCOMES

Handwashing with soap can prevent the spread of shigellosis. Because of contagion, infected persons should be separated from those not infected. People with shigellosis should not prepare food or pour water for others. If a child in diapers is infected, all persons who change the diapers should observe strict sanitation and dispose of the diapers properly. Precautions for food and water safety will prevent shigellosis. Travelers should drink only treated or boiled water and eat cooked hot foods and self-peeled fruit.

Government agencies in the United States are working to prevent outbreaks of shigellosis. *Shigella* infections are monitored by the Centers for Disease Control and Prevention and state and local health departments. These agencies investigate shigellosis, track how it is transmitted, and develop methods of controlling the disease. In addition, these agencies research ways to identify and treat *Shigella* infection. The US Food and Drug Administration checks imported foods and advocates for improved food-preparation techniques in restaurants and food-processing plants. The US Environmental Protection Agency checks the safety of drinking water.

—Merrill Evans, MA and Michael A. Buratovich, PhD

Further Reading

Aslam, Aysha, and Chika N. Okafor. "Shigella." *StatPearls*. StatPearls Publishing, 11 Aug. 2021.

Baker, Kate S., et al. "Horizontal Antimicrobial Resistance Transfer Drives Epidemics of Multiple *Shigella* Species." *Nature Communications*, vol. 9, no. 1, 2018, p. 1462, doi:10.1038/s41467-018-03949-8.

Graciaa, Daniel S., et al. "Outbreaks Associated with Untreated Recreational Water—United States, 2000-2014." *Morbidity and Mortality Weekly Report*, vol. 67, no. 25, 2018, pp. 701-6, doi:10.15585/mmwr.mm6725a1.

Kotloff, Karen L., et al. "Shigellosis." *The Lancet*, vol. 391, no. 10122, 2018, pp. 801-12, doi:10.1016/S0140-6736 (17)33296-8.

Muthuramalingam, Meenakumari, et al. "The *Shigella* Type III Secretion System: An Overview from Top to Bottom." *Microorganisms*, vol. 9, no. 2, 2021, p. 451, doi:10.3390/ microorganisms9020451.

Tack, Danielle M., et al. "Preliminary Incidence and Trends of Infections with Pathogens Transmitted Commonly Through Food—Foodborne Diseases Active Surveillance Network, 10 US- Sites, 2016-2019." *Morbidity and Mortality Weekly Report*, vol. 69, no. 17, 2020, pp. 509-14, doi:10.15585/mmwr.mm6917a1.

ESCHERICHIA COLI INFECTION

Category: Diseases and conditions
Also known as: *E. coli* infection, *E. coli* O157: H7
Anatomy or system affected: Gastrointestinal system, intestines, urinary system
Specialties and related fields: Bacteriology, gastroenterology, immunology, internal medicine, microbiology, public health
Definition: a bacterium that normally lives in the intestines of healthy people and animals, but strains of it can cause disease

KEY TERMS

cytoskeleton: a microscopic network of protein filaments and tubules in the cytoplasm of many living cells, giving them shape and coherence

dysentery: infection of the intestines resulting in severe diarrhea with the presence of blood and mucus in the feces

Shiga toxin: a toxin produced by the bacterium *Shigella dysenteriae*, or a similar one produced by *E. coli*, that causes dysentery in humans

thrombocytopenia: deficiency of platelets in the blood. This causes bleeding into the tissues, bruising, and slow blood clotting after injury

uremia: a raised level in the blood of urea and other nitrogenous waste compounds that are normally eliminated by the kidneys

INTRODUCTION

There are six main strains of the *Escherichia coli* bacterium that affect the human gastrointestinal tract, each with distinct qualities. Enterohemorrhagic *E. coli* (EHEC) includes the most lethal strain of *E. coli*, 0157:17, which releases a Shiga toxin. EHEC is extremely virulent and can cause the potentially fatal hemolytic-uremic syndrome in children and postdiarrheal thrombotic purpura in the elderly. Enteropathogenic *E. coli* (EPEC), enterotoxigenic E. coli (ETEC), and enteroaggregative *E. coli* (EAEC) are all causes of infantile diarrhea in developing countries. ETEC is a major cause of travelers' diarrhea. Enteroinvasive *E. coli* (EIEC) causes bacillary dysentery. Uropathogenic *E. coli* (UPEC) is the cause of 90 percent of all urinary tract infections. Gram-negative neonatal meningitis or sepsis caused in neonates.

CAUSES

E. coli is a gram-negative, rod-shaped bacterium that is a member of the enteric group of bacteria. This organism is highly motile, ferments lactose, and grows in the absence or presence of air. This microorganism is a normal inhabitant of almost everyone's gastrointestinal tracts. Normally, this organism does not cause disease but helps with digestion, nutrient absorption, and preventing intestinal pathogens from colonizing the digestive system.

E. coli are surrounded by an external polysaccharide layer called a "capsule." Different *E. coli* strains make distinct capsule polysaccharides, and capsular antigens are called "K" antigens. The outer membrane of *E. coli*, like all gram-negative bacteria, contains a carbohydrate-lipid complex called "lipopolysaccharide" (LPS). Like the capsules, different strains make slightly different LPS molecules, known as "O" antigens. The flagella that help these bacteria move so vigorously are made of flagellin protein. Different *E. coli* strains make varying flagellins called "H" antigens. Finally, *E. coli* have small, hair-like extensions of the cell called "fimbriae" that help them adhere to host cells. Fimbriae vary from strain to strain and are called "F" antigens. *E. coli* strains are named according to the types of H, F, O, and K antigens they possess. For example, *E. coli* O157:H7 is associated with hemorrhagic colitis, hemolytic uremic syndrome, and diarrheal outbreaks. *E. coli* O104:H4 is an enteroaggregative *E. coli* and caused the 2011 *E. coli* O104:H4 outbreak.

In an EHEC infection, *E. coli* attaches to the large intestinal cells and secrete a cytotoxin (or verotoxin) called "Shiga toxin." This toxin is a potent protein synthesis inhibitor that kills intestinal mucosal cells. The sloughed mucosae cause ulcerations, intense inflammation, and bloody diarrhea. Shiga-toxin absorption into the bloodstream causes extensive kidney damage that results in kidney failure, red blood cell damage, anemia, and low platelets. This condition, hemorrhagic uremic syndrome (HUS), causes uremia (high blood urea levels), hemolytic anemia, and thrombocytopenia.

EPEC occurs almost exclusively in children under two years of age. During an EPEC infection, the bacteria invade the intestinal cells and destroy their cytoskeleton. The intestinal mucosae flatten and lose their ability to absorb water, which causes watery diarrhea.

ETEC uses fimbriae to attach to the intestinal mucosae and secrete two types of toxins into the small intestine. One toxin is heat stable, and the other is heat-sensitive, both of which cause watery diarrhea and intestinal inflammation.

EIEC invade intestinal cells and multiply inside them. They evoke extensive inflammation, mucosal destruction, and bloody diarrhea. EAEC is characterized by "stacked brick" aggregated cells that release Shiga toxins and hemolysins that damage intestinal walls. EAEC induces interleukin (IL)-8 release from intestinal cells, causing the mucosal cells to detach from each other.

Most urinary tract infections are caused by the colonization of UPEC from the rectal area in the urethra, leading to an infection of the urethra, kidneys, or prostate. Neonatal meningitis and sepsis are caused by *E. coli* contamination from the maternal genital area during birth or fecal-contaminated persons or equipment, such as respiratory therapy machines in the neonatal nursery.

E. coli is primarily transmitted through the fecal- oral route. It is found in the intestines of cattle and humans. Unsanitary practices at slaughterhouses can cause beef (particularly ground beef) to become infected with fecal matter. Fecal contamination also occurs during milking. *E. coli* can also be transmitted from person to person. *E. coli* can live in water, including water for drinking and bathing and water for swimming.

RISK FACTORS

Children younger than five years of age and the elderly are most at risk of developing serious complications from *E. coli* infections, including hemolytic-uremic syndrome and postdiarrhea thrombotic purpura. Those who suffer from malnutrition are also vulnerable to *E. coli* infections.

Photo via iStock/Manjurul. [Used under license.]

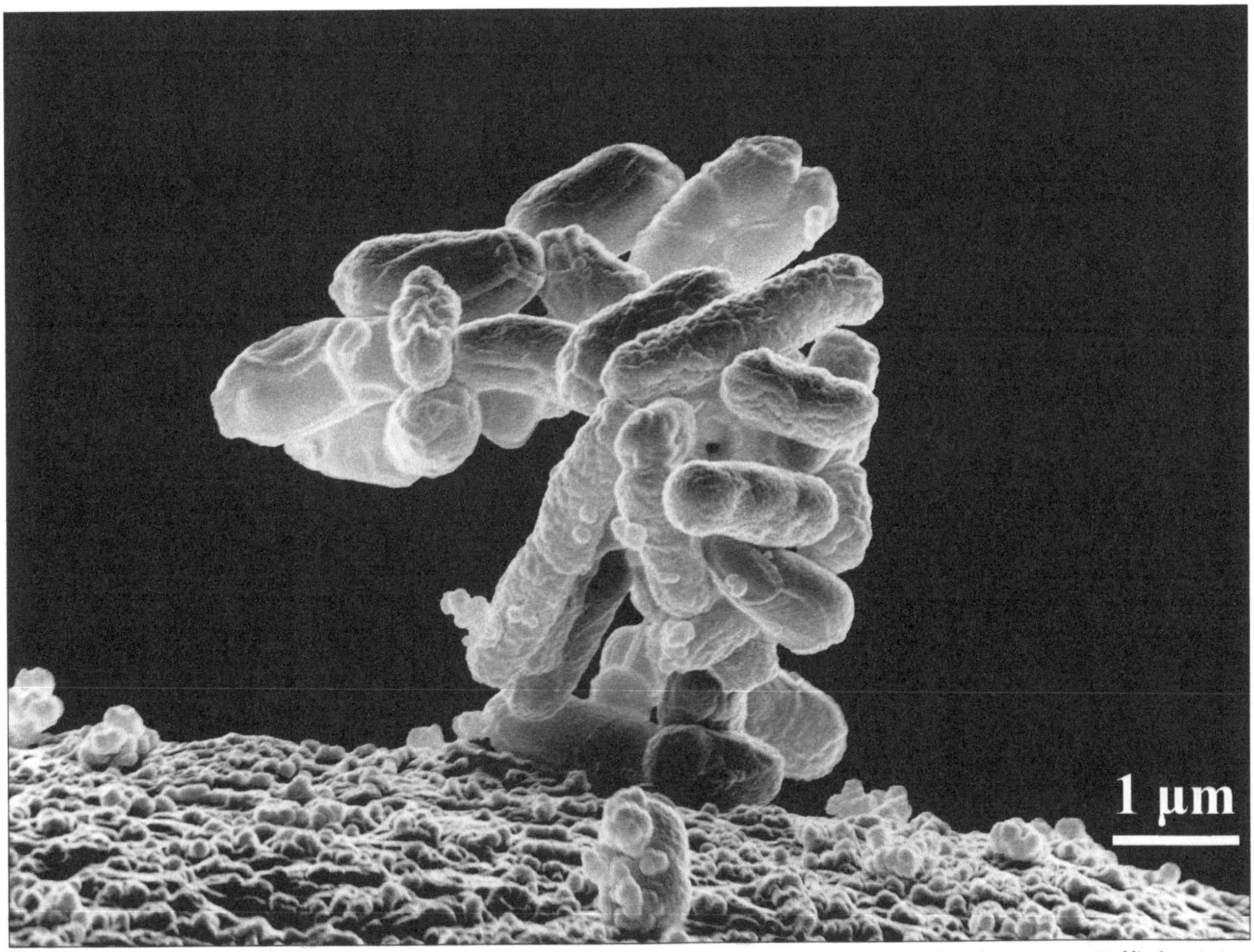

Low-temperature electron micrograph of E. coli. *Photo byfkfkrErbe, digital colorization by Christopher Pooley, via Wikimedia Commons. [Public domain.]*

Women are fourteen times more likely than men to get a UPEC infection. Conditions predisposing women to urinary tract infections include incontinence, postvoid residual urine, sexual intercourse, menopause, diabetes mellitus, catheterization, and pregnancy. Neonatal meningitis or sepsis risk factors include less than thirty-seven weeks gestation, low birth weight, and metabolic abnormalities.

SYMPTOMS

EHEC symptoms are a sudden onset of nonbloody diarrhea that develops into a bloody stool, severe abdominal cramps, and a low-grade fever. EPEC manifests as severe, chronic diarrhea with dehydration. Symptoms of ETEC are watery diarrhea, cramps, and a low fever. EIEC usually manifests with diarrhea with blood or mucus (or both), abdominal cramps, vomiting, chills, high fever, and malaise. Symptoms of EAEC are mucoid and watery diarrhea without fever or vomiting, often lasting more than fourteen days.

A urinary tract infection symptoms include pain with urination, urgency, back pain, cloudy urine, and chills. Neonatal meningitis and sepsis present with fever, grunting respirations, cyanosis, and apnea.

SCREENING AND DIAGNOSIS

The advent of molecular testing of stool samples in diarrhea patients has greatly increased the ease of diagnosis. However, clinicians should interpret the results of such laboratory tests in the context of clinical symptoms.

Stool cultures on MacConkey medium or eosin-methylene blue agar effectively grow *E. coli*. Stool cultures, however, are not specific for strains of *E. coli*, which may be epidemiologically significant. Public health authorities recommend that all stool specimens be cultured for 0157: H7 and reported.

Enzyme-linked immunosorbent assay (ELISA) test kits are commonly used to rapidly screen fecal specimens. Cytotoxic assays are considered the gold standard because of their high specificity and sensitivity; however, they are slow and require special lab facilities. Deoxyribonucleic acid (DNA) based assays are commonly used, such as polymerase chain reaction (PCR).

TREATMENT AND THERAPY

For most healthy persons, *E. coli* infections are self-limiting and do not require treatment other than for dehydration. In cases of EHEC, antibiotics should not be used; the antibiotic kills the bacteria, causing the toxin to be released into the bloodstream and thereby increasing the chance of HUS. Persons with HUS may require dialysis and transfusions. Antimotility agents and opiates should not be used because they prolong the time the toxin remains in the body. EIEC and ETAC are usually treated with azithromycin or fluoroquinolones (e.g., ciprofloxacin or levofloxacin) for nonresistant strains and with rehydration and electrolyte replacement therapy.

Antibiotics do not reduce the symptoms or other complications associated with Shiga toxin-producing *E. coli* (STEC) infections. Consequently, antibiotics are not recommended for STEC infections.

Some urinary tract infections may resolve without antibiotics; however, a three-day course of azithromycin or a fluoroquinolone (three to five days) is standard. Treatment for neonatal meningitis and sepsis is antibiotic therapy with ampicillin and an aminoglycoside for ampicillin-sensitive strains. For ampicillin-resistant strains, the preferred treatment is a combination of an extended-spectrum cephalosporin, like cefotaxime plus an aminoglycoside, usually gentami- cin. Given the effect of aminoglycosides on hearing, gentamicin is discontinued once the cerebrospinal fluid is sterile. The treatment goes for a minimum of twenty-one days.

PREVENTION AND OUTCOMES

General prevention for foodborne transmission of any *E. coli* infections includes cooking meat and poultry until juices run clear (162° Fahrenheit, or 72° Celsius). One should cook hamburger meat until the meat is no longer pink; avoid consuming raw milk or unpasteurized dairy products and juices; wash all fruits and vegetables; wash one's hands after using the toilet, after changing a diaper, and before touching and eating food. Travelers should drink only bottled water; avoid ice, unpeeled fruits, and salad; and eat only foods that are served steaming hot. Taking daily doses of a bismuth subsalicylate preparation can help protect against ETEC. In developing countries, mothers are encouraged to breastfeed their infants to prevent several infantile diarrheas. Cranberry juice is effective in preventing recurring urinary tract infections. Neonates infected with meningitis or sepsis should be isolated to prevent cluster infections.

Rapid diagnosis and reporting of outbreaks enable epidemiological investigation and control measures. Vaccines for individual strains of *E. coli* are being developed for cattle and humans.

—S. M. Willis, MS, MA and
Michael A. Buratovich, PhD

Further Reading

Ameer, Muhammad Atif, et al. "*Escherichia Coli* (*E. Coli* 0157 H7)." *StatPearls*. StatPearls Publishing, 10 July 2021.

Herzig, Carolyn T. A., et al. "Notes from the Field: Enteroinvasive *Escherichia coli* Outbreak Associated with a Potluck Party-North Carolina, June-July 2018." *Morbidity and Mortality Weekly Report,* vol. 68, no. 7, 2019, pp. 183-84, doi:10.15585/mmwr.mm6807a5.

Mueller, Matthew, and Christopher R. Tainter. "*Escherichia Coli.*" *StatPearls*. StatPearls Publishing, 26 July 2021.

Pakbin, Babak, et al. "Virulence Factors of Enteric Pathogenic *Escherichia coli*: A Review." *International Journal of Molecular Sciences,* vol. 22, no. 18, 2021, p. 9922, doi:10.3390/ijms22189922.

Pinaud, Laurie, et al. "Host Cell Targeting by Enteropathogenic Bacteria T3SS Effectors." *Trends in Microbiology,* vol. 26, no. 4, 2018, pp. 266-83, doi:10.1016/j.tim.2018.01.010.

Sora, Valerio M., et al. "Extraintestinal Pathogenic *Escherichia coli*: Virulence Factors and Antibiotic Resistance." *Pathogens,* vol. 10, no. 11, 2021, p. 1355, doi:10.3390/pathogens10111355.

SALMONELLA INFECTION

Category: Diseases and disorders

Anatomy or system affected: Blood, gastrointestinal system

Specialties and related fields: Bacteriology, family medicine, gastroenterology, pediatrics, public health

Definition: a broad spectrum of clinical diseases caused by many types of *Salmonella* bacteria

KEY TERMS

asymptomatic: without symptoms

bacteremia: the presence of bacteria in the blood, which is usually associated with chills and fever

gastroenteritis: infection of the gastrointestinal tract, usually accompanied by nausea, vomiting, diarrhea, and abdominal pains

typhoid fever: a particular disease syndrome most often associated with infection by *Salmonella typhi* but occasionally caused by other types of *Salmonella* bacteria

CAUSES AND SYMPTOMS

Salmonella are a group of bacteria that cause enteric or typhoid fever. All types can cause gastrointestinal, blood, and various local infections. All types of *Salmonella* can be carried in the gastrointestinal tract without symptoms after recovery from infection.

The clinical disease caused by *Salmonella* depends on the type of bacteria, the number of organisms ingested, and the age and immune status of the person infected. Infection with *Salmonella* can occur with the ingestion of one or 100 million organisms. Increasing the dosage of bacteria decreases the incubation period and increases the severity of the resulting disease. After ingestion, the bacteria adhere to and invade the gastrointestinal tract. In the intestinal tract wall, *Salmonella* survive and multiply in immune cells and then enter the bloodstream, where they proceed to any area of the body. Young infants and people with immune deficiencies and hemolytic anemia are at increased risk for severe and complicated infections.

Typhoid fever or enteric fever is very rare in the United States, causing less than five hundred cases per year; it is primarily seen in people coming from developing countries. Classically, this disease is caused by *Salmonella typhi* bacteria, but other types of *Salmonella* can also cause it. During the first week of illness, symptoms include progressively increasing fever with an associated headache, muscle aches, abdominal pains, and lethargy. In the second week, the heart rate decreases, the liver, and spleen enlarge, small red bumps form on the trunk, and the patient enters into a stupor. During the third to fourth week, intestinal bleeding and perforation are common. The fever begins to remit in the fifth to sixth week of illness. Diarrhea usually starts in the first week and resolves within six weeks. Without treatment, death can occur from gastrointestinal bleeding and perforation. Infants tend to have much more severe disease than older children.

Salmonellosis caused by nontyphoid *Salmonella* is more common in the United States, causing about fifty thousand cases per year. The major reservoir of nontyphoid *Salmonella* is the gastrointestinal tract of many animals, including mammals, reptiles, birds, and insects. Farm animals and pet reptiles commonly carry *Salmonella*. Some antibiotic resistance is caused by the use of antibiotics in animal feeds. *Salmonella* can be isolated from 50 percent of chicken, 16 percent of pork, 5 percent of beef, and 40 percent of frozen eggs in retail stores. Contaminated eggs and milk products are common sources of human infection.

Gastroenteritis is the most common disease caused by nontyphoid *Salmonella*. The incubation period for this disease is about one day, ranging from six hours to three days. Symptoms include nausea, vomiting, and abdominal pain. Diarrhea typically contains blood and white cells. Usually, symptoms disappear in less than a week in healthy children. Still, symp-

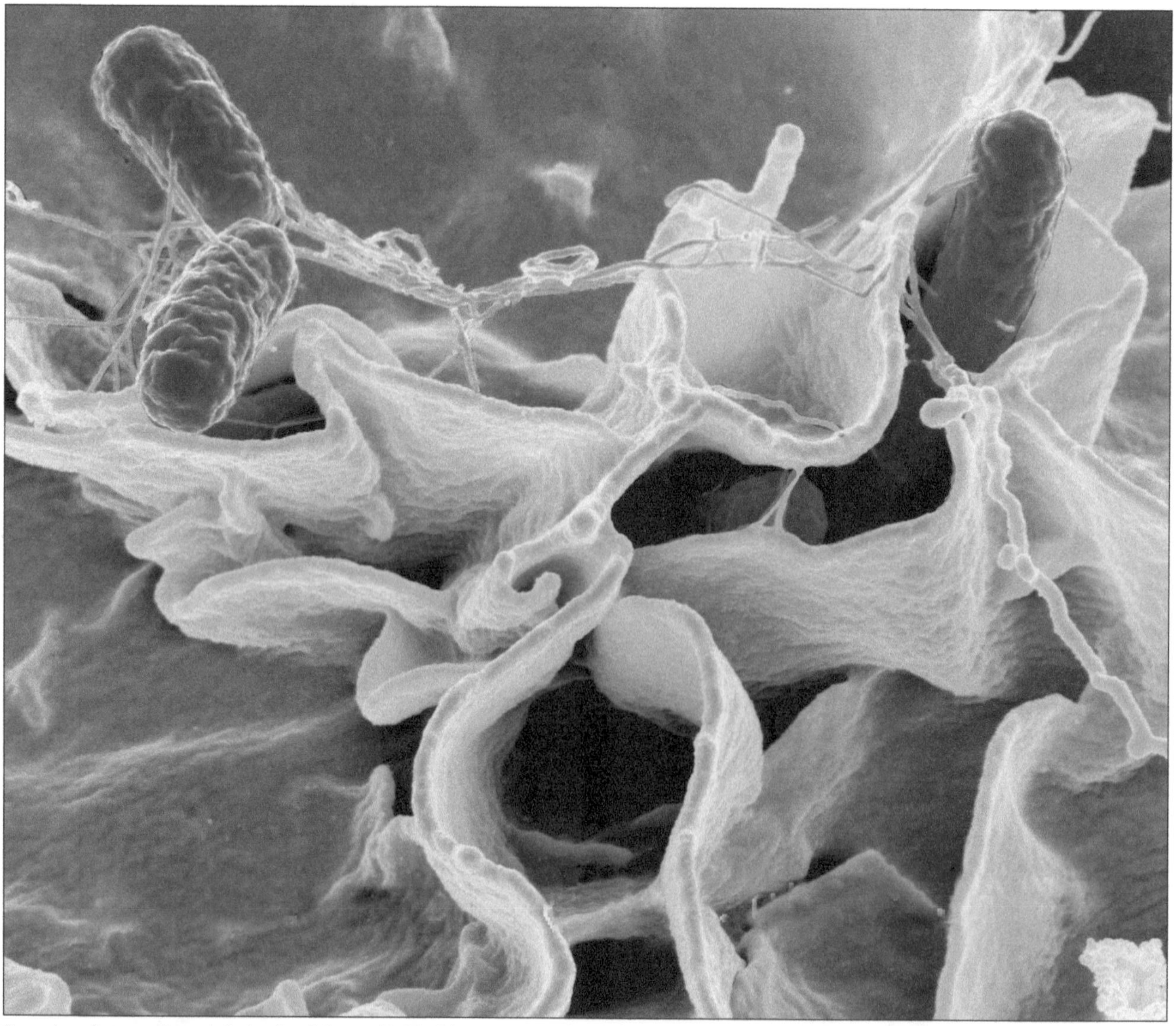

Scanning electron micrograph showing Salmonella Typhimurum *invading cultured human cells. Photo via Wikimedia Commons. [Public domain.]*

toms may persist for several weeks in young infants and children with immune deficiencies.

Bacteremia can occur in 1 to 5 percent of patients with *Salmonella* gastroenteritis. Bacteremia is generally associated with fever, chills, and toxicity in the older child but may be asymptomatic in the infant. Children with an increased risk of bacteremia include those with acquired immunodeficiency syndrome (AIDS) or other immune deficiencies and hemolytic anemias such as sickle cell anemia.

Bacteremia can lead to infection of almost any organ. Children with sickle cell anemia are more prone to bone infections and meningitis. *Salmonella* may localize to areas of the body that have received trauma or contain damaged tissue or a foreign body. Meningitis, inflammation of the covering of the spine and brain, is primarily seen as a complication of bacteremia in infants. Meningitis has a 50 percent death rate, and residual developmental and hearing defects are commonly found in survivors. Patients who have persistent bacteremia should be evaluated for heart infection.

The diagnosis of *Salmonella* infection is best made by culturing stool and blood samples. With enteric fever, it is important to culture multiple sites multiple times. The clinical laboratory must perform antibiotic susceptibility testing routinely to guide therapy. Other bacterial causes of gastroenteritis can be confused with *Salmonella* infection.

TREATMENT AND THERAPY

Treatment for gastroenteritis usually does not require antibiotics. Antibiotics do not speed the resolution of disease but instead lead to prolonged excretion of *Salmonella*. Therapy is primarily focused on correcting fluid and salt abnormalities and on general supportive care. However, if the patient has indications of sepsis, shock, or chills, then antibiotics should be administered. Infants under three months of age and children with immune deficiencies should also be treated with antibiotics. Ampicillin is usually used as the initial treatment in uncomplicated cases, and third-generation cephalosporin antibiotics are used in severe and complicated cases. Over 20 percent of nontyphoid *Salmonella* in the United States is resistant to ampicillin and other antibiotics. Antibiotic treatment should last ten days to two weeks in children with bacteremia and four to six weeks with bone infection or meningitis. Local infections may require surgical drainage.

Typhoid fever is treated for a minimum of two weeks. It is important to perform susceptibility testing for the possibility of resistance so that proper antibiotic therapy can be chosen. Chronic carriers of *Salmonella typhi* should be treated with antibiotics. If eradication is unsuccessful, surgical assessment of the biliary tract should be sought.

Prevention of the spread of *Salmonella* requires several public health procedures. Hand washing is critical to the prevention of transmission. *Salmonella* carriers should be excluded from food preparation and childcare settings. Hospitalized infants and children should be isolated. Proper sewage disposal, water purification, and chlorination are essential public health measures. In developing countries, the promotion of prolonged breastfeeding also reduces the infection rate.

There are two typhoid vaccines commercially available in the United States. The first is an oral, live attenuated vaccine requiring four doses for one week and a booster every five years. The second is a parenteral capsular polysaccharide vaccine given as a single intramuscular injection and a booster every two years. The vaccines are 50 to 80 percent protective. The live attenuated vaccine should not be given to pregnant patients, taking antibiotics, or immunocompromised, such as persons with the human immunodeficiency virus (HIV).

PERSPECTIVE AND PROSPECTS

Salmonella was identified as the cause of typhoid fever in 1880 and was first cultured in 1884. Since 1920, improvements in sanitation, water supplies, and sewage disposal have resulted in a marked decrease in typhoid fever in the United States. In 1920, 36,000 cases were reported; after 1965, the number of cases per year has rarely exceeded 500. Since then, the number of cases has remained fairly constant because of the importation of disease by tourists, immigrants, and migrant laborers. About 62 percent of *Salmonella typhi* infections are acquired through foreign travel. Direct person-to-person transmission is rare except in the homosexual population.

Recent research is focused on public health. Measures to decrease food contamination such as improved cleanliness, decreased use of antibiotics in animal feeds, and food irradiation are being evaluated and used to decrease human transmission. Research into alternate vaccines with fewer side effects and an improved immune response is underway.

Nontyphoid *Salmonella* causes one-half million infections per year in the United States. One-third of these infections are in children less than five years of age, and 40 percent are in adults over thirty years of age.

—Peter D. Reuman, MD, MPH

Further Reading

Bellenir, Karen, and Peter D. Dresser, editors. *Contagious and Noncontagious Infectious Diseases Sourcebook*. Omnigraphics, 1996.

Biddle, Wayne. *A Field Guide to Germs*. 2nd ed., Anchor Books, 2002.

Dodd, Christine, E. R., et al., editors. *Foodborne Diseases*. 3rd ed., Academic Press, 2017.

Jay, James M., Martin J. Loessner, and David A. Golden. *Modern Food Microbiology*. 7th ed., Springer, 2005.

Leon, Warren, and Caroline Smith DeWaal. *Is Our Food Safe? A Consumer's Guide to Protecting Your Health and the Environment*. Crown, 2002.

Parker, James N., and Philip M. Parker, editors. *The Official Patient's Sourcebook on Salmonella Enteritidis Infection*. Icon Health, 2002.

Porter, Robert E., editor. *The Merck Manual of Diagnosis and Therapy*. 20th ed., Merck, 2018.

Klebsiella

Category: Pathogen

Anatomy or system affected: Bladder, blood, brain, circulatory system, heart, immune system, kidneys, lungs, lymphatic system, meninges, nervous system, peritoneum, prostate, urethra, urinary tract

Specialties and related fields: Bacteriology, intensive care, microbiology, neurology, pharmacology, public health, pulmonary medicine, urology

Definition: a bacterial species that is typically acquired in health-care settings such as hospitals, nursing homes, and other long-term care facilities

KEY TERMS

capsule: a gelatinous envelope surrounding a bacterial cell, usually polysaccharide but sometimes polypeptide in nature that lies outside the cell envelope and helps bacteria cause disease

complement: a system of plasma proteins that can be activated directly by pathogens or indirectly by pathogen-bound antibodies, leading to a cascade of reactions that occurs on the surface of pathogens, killing them and generating components with various effector functions

lipopolysaccharide: the major component of the outer membrane of gram-negative bacteria

pili: short, hairlike structures on the cell surface of prokaryotic cells that may assist in movement but are more often involved in adherence to surfaces, which facilitates infection

siderophore: small, high-affinity iron-binding compounds secreted by microorganisms such as bacteria and fungi

urease: an enzyme that catalyzes the hydrolysis of urea, forming ammonia and carbon dioxide

INTRODUCTION

Klebsiella are gram-negative, rod-shaped bacteria that belong to the enteric bacteria group. *Klebsiella* typi-

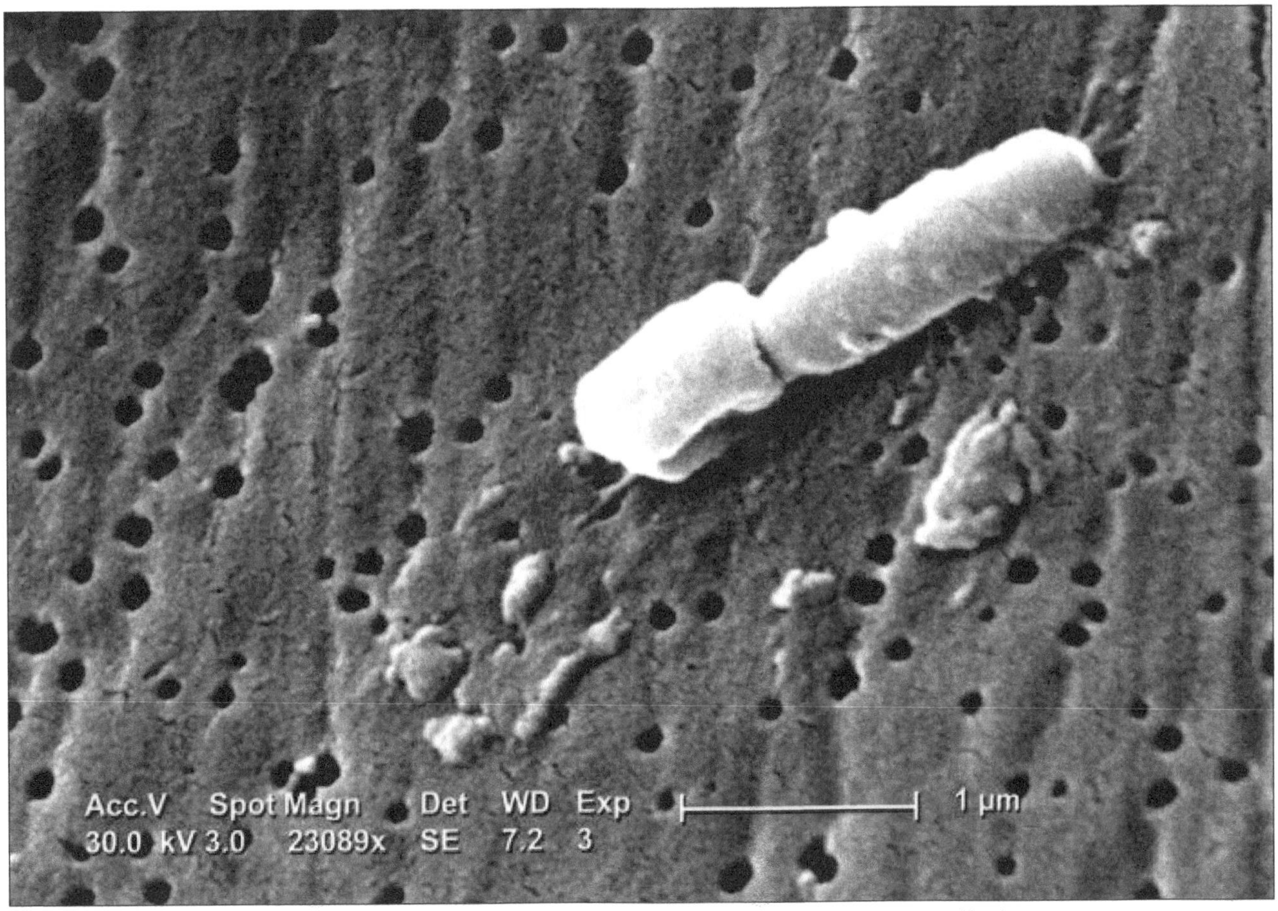

Electron microscopic picture of bacteria Klebsiella pneumoniae. *Photo via CDC/Wikimedia Commons. [Public doman.]*

cally colonize the throat, oral cavity, and gastrointestinal tract. Usually, *Klebsiella* do not cause disease since they are a part of the microbiome of healthy people. However, when people become severely ill, they can acquire pathogenic strains of *Klebsiella* that can cause severe disease. Alternatively, patients might have other bodily sites colonized by their native *Klebsiella* strains.

Klebsiella bacteria are usually acquired in health-care settings such as hospitals, nursing homes, and other long-term care facilities. Species of *Klebsiella* can cause various infections, including pneumonia, urinary tract infections, blood infections, meningitis, and infections of wounds or surgical incisions. They present specific problems for health-care providers because they often attack people who are already weakened by illness or injury and are increasingly resistant to antibiotics.

OVERVIEW

Klebsiella bacteria are named after the German biologist who discovered them in the late nineteenth century, Edwin Klebs (1834-1913). They occur naturally in the human intestines, where they are not dangerous, and in human feces. The bacteria are nonmotile or incapable of moving on their own. They are entirely covered in a capsule made of carbohydrate molecules known as "polysaccharides." This capsule helps to protect the bacterium from attacks by the body's immune system.

This organism has other features that help it cause disease. It makes urease an enzyme that degrades urea, the primary nitrogenous waste in urine, to carbon dioxide and ammonia. Urease helps *Klebsiella* colonize the urinary tract. The polysaccharide capsule helps it escape phagocytosis by white blood cells like macrophages and neutrophils. If these phagocytic cells try to get their pseudopods around *Klebsiella*, its capsule helps it slip from their grasp. Beneath the capsule is the primary molecule of the outer membrane, lipopolysaccharide (LPS). Although LPS is common to the outer membrane of all gram-negative bacteria, *Klebsiella*'s LPS protects it from complement proteins in the blood that seek to destroy it. Iron acquisition is difficult when infecting a human body. *Klebsiella* produces a small molecule that functions as an iron grabber called a "siderophore" to overcome this problem. Siderophores help this organism survive inside you by snatching iron from your cells. Finally, *Klebsiella* have tiny hairlike extensions called "pili" that help them attach to host cells.

Three known species of the genus *Klebsiella* affect humans. These are *Klebsiella pneumoniae, Klebsiella oxytoca,* and *Klebsiella rhinoscleromatis.* They are spread by person-to-person contact. The bacteria can also easily enter the body through medical devices such as catheters or ventilators. While otherwise healthy people can contract a *Klebsiella* infection, they are more common in people whose immune systems are already weakened and those in an inpatient medical setting.

Klebsiella causes a host of hospital-acquired infections in people with underlying illnesses, such as diabetes mellitus, or have indwelling catheters, endotracheal tubes, or urinary catheters. In diabetics and alcoholics, it causes lobar pneumonia. The aspiration of microbes from the gut into the oropharynx to the lungs can cause lung abscesses. In people with urinary catheters, *Klebsiella* causes urinary tract infections that may result in prostatitis (prostate gland infection), pyelonephritis (kidney infection), or cystitis (bladder infection). People with blood vessel catheters may have Klebsiella introduced directly into their blood and suffer from bacteremias that can spread to the heart, cause endocarditis or the brain, and cause meningitis. *Klebsiella* can infect the peritoneal fluid in people with cirrhosis and ascites and cause spontaneous bacterial peritonitis.

The symptoms exhibited by patients depend on what part of the body is affected by the bacteria. For instance, a person who contracts pneumonia from a *Klebsiella* bacteria will have the same symptoms of fever, chills, chest pain, breathing issues, and a productive cough with blood-tinged sputum. Patients with urinary tract infections have pain during urination (dysuria), need to urinate a lot (urinary frequency), and have a strong need to urinate (urinary urgency). Urease production by *Klebsiella* degrades urea to ammonia. Ammonia picks up hydrogen ions from urine, increasing its pH. Alkaline urine promotes phosphate, calcium, and phosphate precipitation. These ions combine with ammonia to form struvite stones that can form large, staghorn kidney stones. Bacteremia patients have a fever, hypotension, and tachycardia, and spontaneous bacterial peritonitis causes fever, chills, and extreme abdominal pain.

Patients who display these symptoms should have their blood, urine, or peritoneum cultured to confirm that *Klebsiella* is the cause of the sickness. Therefore, the medical team needs to order tests to determine the cause and identify the presence of *Klebsiella* as soon as possible.

Klebsiella bacteria are opportunistic organisms, which means they take advantage of various forms of weakness to enter and infect the body. In addition to the risk of being immunocompromised or a hospital or surgical patient, the elderly are also more likely to be affected. In some parts of the world, such as Asia, senior citizens account for most of those who fall victim to a *Klebsiella*-related infection. However, in the United States, about 66 percent of the patients diagnosed with *Klebsiella* infections each year are alcoholics. While a *Klebsiella* infection results in death about 50 percent of the time, that rate soars to nearly 100 percent for patients with alcoholism.

Klebsiella infections are diagnosed by gathering a specimen for testing in the laboratory. They are treated with antibiotics, although the bacteria are resistant to many antibiotics. *Klebsiella* species make enzymes called "beta-lactamases" that make them resistant to ampicillin and amoxicillin. These strains are susceptible to cephalosporins, aminoglycosides, fluoroquinolones, and carbapenems. Other strains produce extended-spectrum beta-lactamases (ESBL) that confer resistance to cephalosporins, aminoglycosides, and fluoroquinolones, and these strains are usually treated with carbapenems. Other strains produce carbapenemases that make the cells resistant to carbapenems. These organisms are treated with colistin, tigecycline, or fosfomycin. A combination product, ceftazidime/avibactam *(Avycaz)*, was approved for use in the United States for the types of infections caused by carbapenemase-producing *Klebsiella* bacteria in 2015.

—Janine Ungvarsky and Michael A. Buratovich, PhD

Further Reading

Bush, Larry M., and Charles E. Schmidt. "Klebsiella, Enterobacter, and Serratia Infections." *Merck Manuals*, Mar. 2021, www.merckmanuals.com/home/infections/bacterial-infections-gram-negative-bacteria/klebsiella-enterobacter-and-serratia-infections. Accessed 1 Dec. 2021.

"Klebsiella Pneumoniae in Health Care Settings." *Centers for Disease Control*, www.cdc.gov/hai/organisms/klebsiella/klebsiella.html. Accessed 1 Dec. 2021.

"Klebsiella Species." *Gov.UK*, 15 July 2008, www.gov.uk/guidance/klebsiella-species. Accessed 1 Dec. 2021.

Murray, Patrick R., et al. *Medical Microbiology*. 9th ed., Elsevier, 2020.

Qureshi, Shahab, et al. "Klebsiella Infections." *Medscape*, 7 Nov. 2016, emedicine.medscape.com/article/219907-overview. Accessed 1 Dec. 2021.

Shaw, Gina. "Breaking News: Deadly Klebsiella Pneumoniae Strain Resistant to Carbapenems." *Emergency Medicine News*, Mar. 2013, journals.lww.com/em-news/Fulltext/2013/03000/Breaking_News__Deadly_Klebsiella_Pneumoniae_Strain.3.aspx. Accessed 1 Dec. 2021.

Srivastava, Roli. "Drug Resistant Bug Klebsiella Causes Worry." *The Hindu*, 23 May 2016, www.thehindu.com/sci-tech/health/medicine-and-research/drugresistant-bug-klebsiella-causes-worry/article7928014.ece. Accessed 1 Dec. 2021.

Pneumonia

Category: Diseases and disorders

Anatomy or system affected: Lungs, respiratory system

Specialties and related fields: Emergency medicine, epidemiology, family medicine, internal medicine, occupational health, public health, pulmonary medicine

Definition: inflammation of one of several possible areas of the respiratory system, mainly in the lungs or bronchial passageways, resulting from bacterial or viral infection

KEY TERMS

biofilm: a microbial colony that secretes an extracellular polysaccharide matrix that protects and adheres them to surfaces

Gram's stain: a laboratory method for tracing the presence of certain bacteria in lung tissue; the procedure involves the observation of different levels of tissue discoloration as specific chemical reactions are induced

methicillin-resistant Staphylococcus aureus: also known as MRSA, a genetically distinct strain of *Staphylococcus aureus* that causes several difficult-to-treat infections in humans

mycoplasma: a group of small typically parasitic bacteria that lack cell walls and sometimes cause diseases

Pneumocystis pneumonia: a form of pneumonia caused by the single-celled parasite *Pneumocystis jirovecii*;

dangerous primarily to persons with impaired immunity mechanisms, particularly victims of acquired immunodeficiency syndrome (AIDS)

pulmonary consolidation: the presence of exudate in the airways and alveoli, usually because of infection

Streptococcus pneumoniae: commonly referred to as pneumococcus; the main bacteria responsible for pneumonia

CAUSES AND SYMPTOMS

Pneumonia is an infection of the lung. When microbes overcome the innate immune systems in the lung, they colonize the bronchioles or alveoli. The infecting pathogens set off an inflammatory response. White blood cells move into the lung to phagocytose the attacking bacteria. The war between the pathogens and the white blood cells produces a fluid made from leaky blood vessels, dead bacteria, dead white blood cells, and proteins. This fluid is called an "exudate," and it fills the alveoli and can be coughed up as sputum.

Just as the causes of pneumonia can vary, the disease itself may take different forms. Pneumonia is divided into "classic" and atypical types of pneumonia. Classic pneumonia causes fever, shortness of breath, and fatigue. Atypical pneumonia or "walking" pneumonia causes a headache, a cough that does not bring up material nonproductive), and, usually, no fever. Other symptoms usually accompany pneumonia are fast heart rate and increased breathing rate. Most classic pneumonia cases are caused by *Streptococcus pneumoniae*, *Haemophilus influenzae*, and *Staphylococcus aureus*. Atypical pneumonia is caused by *Mycoplasma pneumoniae*, *Chlamydia trachomatis*, *Chlamydophila psittaci* and *C. pneumoniae,* and *Legionella pneumophila*. *Legionella* tends to cause more severe pneumonia symptoms that include headaches, mild cough, high fever, confusion, and watery diarrhea. Because these microorganisms are difficult to culture, attempts to isolate the infecting organism will prove futile. Some viruses that cause classic pneumonia are influenza virus, rhinoviruses, parainfluenza viruses, adenoviruses, human metapneumovirus, human bocaviruses, respiratory syncytial virus (RSV), and SARS-CoV-2 (the virus that causes COVID-19).

Fungal pneumonia is relatively uncommon relative to bacterial and viral pneumonia. However, the risk of contracting fungal pneumonia varies between locations. For example, Coccidioides (C. immitis and C. posadasii) cause coccidioidomycosis. These fungi are endemic to desert regions of the Southwestern United States, Mexico, and Central and South America. Coccidioidomycosis, a fungal pneumonia, is asymptomatic in most people but might be severe in others. Histoplasmosis is caused by *Histoplasma capsulatum*, a soil fungus common in the Midwestern and Central United States, along the Ohio and Mississippi River Valleys. This fungus, when inhaled, grows as a yeast in the body and causes bronchopulmonary pneumonia that usually resolves. *Blastomyces* dermatitidis and B. gilchristii cause blastomycosis, an infection that is typically asymptomatic but can produce acute or chronic pneumonia. These fungi occur in the southeastern and south-central states bordering the Mississippi and Ohio River basins, the midwestern states and Canadian provinces bordering the Great Lakes, and along the St. Lawrence and Nelson Rivers. These fungal types of pneumonia are significantly more severe in people with poorly functioning immune systems and can disseminate to other organs.

Upon physical exam, drumming on the chest produces a dull sound since the alveoli are no longer filled with air but pus and fluid, a condition known as "consolidation." Therefore, the chest loses its usual drum-like quality. The chest also shows a phenomenon called "tactile fremitus." Sound and vibrations travel faster through fluid than air. Therefore, when the physician places her hand on the chest or back of the patient, the body wall vibrates vigorously when the patient speaks. When listening to the chest with a stethoscope over the areas of consolidation,

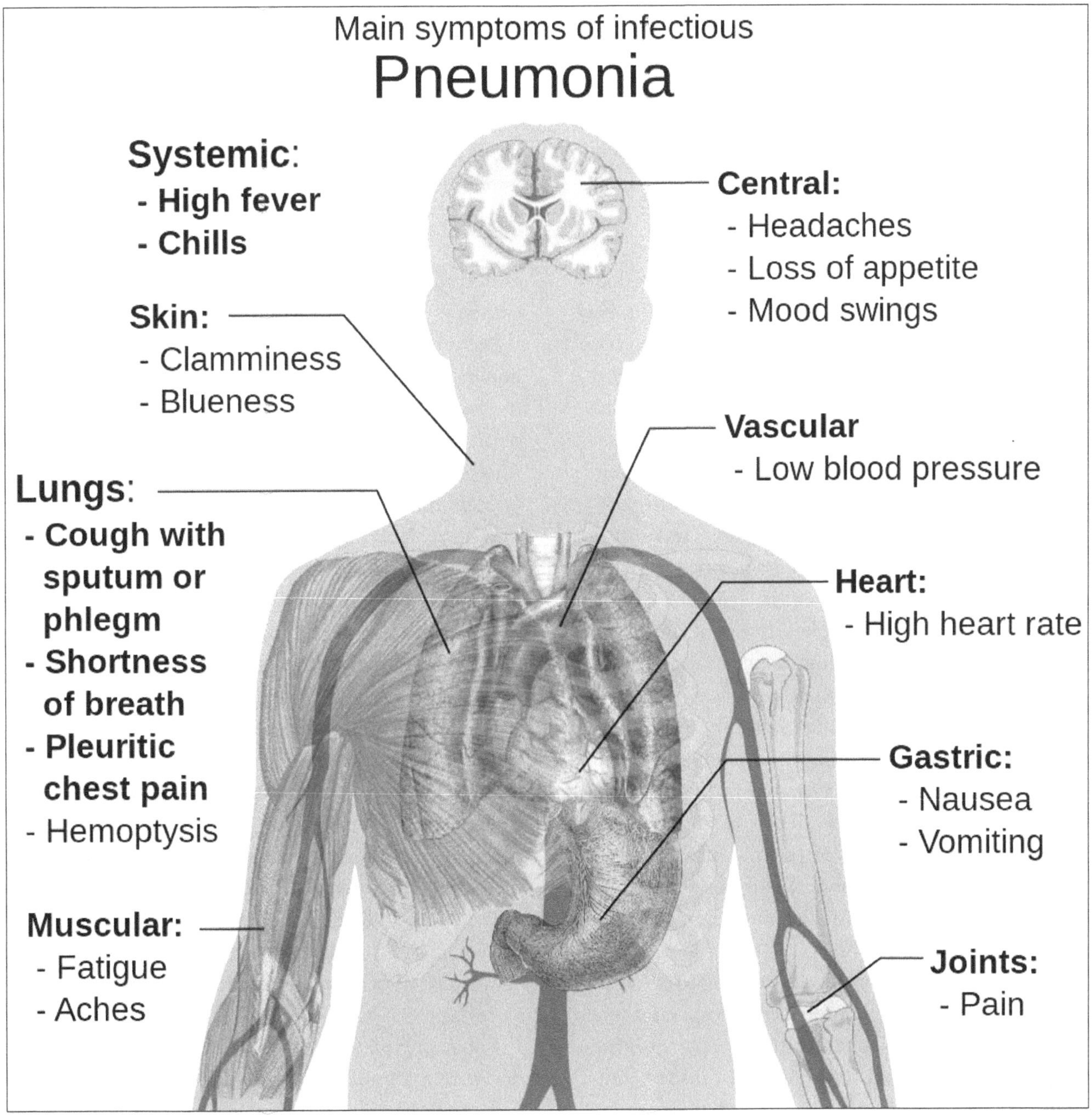

Signs of pneumonia. Image via Wikimedia Commons. [Public domain.]

spoken sounds are louder, a condition called "bronchophony." Also, air moving through fluid generates crackles.

Different age groups are at higher risk for certain pathogens. Pneumonia in neonates or those younger than four weeks is caused by Group B streptococci, *Escherichia coli*, and respiratory syncytial virus. For children four weeks to eighteen years, respiratory syncytial virus, *Chlamydia trachomitis* (more common in those younger than three years), *Chlamydophila pneumoniae*, and *Streptococcus pneumoniae*. In those eighteen to forty years, the common causes are *Mycoplasma pneumoniae*, *Chlamydophila*, *Streptococcus pneumoniae*, and influenza virus. For those over forty, the most common causes are *Streptococcus pneumoniae*, anaerobic bacteria, and viruses, such as SARS-CoV-2.

Pneumonia in intravenous drug users is typically caused by *Staphylococcus aureus* and *Streptococcus pneumoniae*. In people with cystic fibrosis, the primary culprit is *Pseudomonas aeruginosa*. In patients who have poorly functioning immune systems, the main causes of pneumonia are *Staphylococcus aureus*, gram-negative enteric bacteria, and viruses. People with human immunodeficiency virus (HIV) whose T-helper lymphocyte counts drop below 200 cells per microliter can suffer from acquired immunodeficiency syndrome (AIDS)-related opportunistic infections. The most prominent infection is "pneumocystis pneumonia" (PCP), caused by the fungus *Pneumocystis jirovecii*.

Viral pneumonia can weaken someone and damage their lungs, predisposing them to a secondary bacterial infection. *Streptococcus pneumoniae*, *Staphylococcus aureus*, and *Haemophilus influenzae* are the main causes of secondary bacterial pneumonia after primary viral pneumonia. Organ transplantation patients who must take a cocktail of immunosuppressive drugs are at risk for pneumonia with intranuclear and cytoplasmic inclusion bodies. This type of pneumonia in posttransplant patients is caused by cytomegalovirus.

Pneumonia is also classified according to where they are acquired. Community-acquired pneumonia is acquired outside hospitals or health-care settings. *Streptococcus pneumoniae* is the primary cause of community-acquired pneumonia. In contrast, someone has hospital-acquired or nosocomial pneumonia if they contract pneumonia after being in the hospital for at least two days. Methicillin-resistant *Staphylococcus aureus* (MRSA) is asymptomatically carried by hospital staff and is the top cause of hospital-acquired pneumonia. *Legionella pneumophila* is found in hospital water-cooling systems at is another common cause of hospital-acquired hospital. Ventilator-acquired pneumonia is a type of hospital-acquired pneumonia that develops when someone has been intubated for more than forty-eight hours. *Pseudomonas aeruginosa* and *Staphylococcus aureus* form biofilms that stick to the endotracheal tube. Since intubated people cannot cough to clear their airways, any microbes stuck to the endotracheal tubes work their way into the lungs and cause pneumonia.

Aspiration pneumonia results from inhaling food, liquids, or vomited gastric contents. The portion of the lung affected by aspiration pneumonia depends on the patient's position. Anyone lying on their back who breathes in food or drink will develop aspiration pneumonia in the upper lobes of their lungs. Those who do so while standing or sitting up will have pneumonia in their lower lobes. Aspiration pneumonia also occurs in those with defective gag and cough reflexes. Consequently, individuals who have suffered a stroke or brain injury and abuse drugs or alcohol are at higher risk of aspiration pneumonia. One of the main causes of aspiration pneumonia is the enteric bacterium *Klebsiella pneumoniae*.

Another pneumonia classification depends on where within the respiratory system it occurs. Bronchopneumonia is characterized by spotty infection throughout the lungs in the bronchioles and alveoli. Lobar pneumonia causes complete consolidation of the whole lobe of a lung. Interstitial pneumonia is characterized by infection of the tissues between the alveoli.

Lobar pneumonia occurs in stages. The lungs undergo congestion during the first twenty-four to forty-eight hours as the alveoli fill with exudate. The next stage, red hepatization (also known as "consolidation"), occurs over the next forty-eight hours. Red and white blood cells leave the blood vessels and fill the alveoli during this stage, causing them to look like liver. The red blood cells disintegrate for the next three days, and the exudate looks gray and dry. This stage is known as "gray hepatization." Resolution begins approximately eight days after the infection began. Macrophages remove much of the exudate, and the rest is cleared from the lungs to the pharynx, where it is coughed up.

Infection of the pleural membranes that surround the lung may cause pleuritic chest pain, which is a stabbing or burning pain when exhaling or inhaling with a productive cough that brings up yellow sputum.

DIAGNOSIS

One of the most useful tools for diagnosing pneumonia is a chest X-ray. Exudates within the lung appear as dark spots in people with pneumonia. X-rays have the added advantage of locating the site of infection. Bronchopneumonia appears as patchy areas spread throughout the lung. In individuals with lobar pneumonia, the affected area is localized to a single lobe or set of lung lobes. The infiltrate is spread throughout the lungs for those with interstitial pneumonia but is concentrated in the perihilar region and looks weblike.

Sputum cultures can identify the bacteria or fungi causing pneumonia. Gram staining the isolated organism distinguishes between gram-positive and gram-negative bacteria. However, many of the causes of atypical pneumonia are not easily cultured. *Mycoplasma pneumoniae*, for example, requires a special medium for growth in the laboratory called "Eaton agar," which is enriched with cholesterol. *Mycoplasma pneumoniae* infections cause cold autoimmune hemolytic anemia. Antibodies against these bacteria bind to red blood cells in the cold. A serological test called the "Coombs test" detects these autoantibodies and effectively diagnoses *M. pneumoniae* infection.

Legionella pneumophila also does not grow on standard laboratory media. It requires a buffered charcoal yeast extract medium supplemented with the amino acid cysteine and iron to grow. This organism also states poorly with the Gram stain. Large quantities of neutrophils in the sputum with no detectable bacteria on a Gram stain might be a clue that Legionella causes pneumonia.

Chlamydia and Chlamydophila are obligate intracellular parasites and will not grow on standard bacteriological media. Polymerase chain reaction tests effectively identify them. In Giemsa stains of sputum, these microorganisms cause cytoplasmic inclusions inside cells.

Sputum cultures for fungi are grown on Sabouraud dextrose agar. *Coccidioides* fungi grow as round spherules in tissue that are readily identifiable with a microscopic examination of sputum. Treating the sample with lactophenol cotton blue effectively stains fungi for microscopic examination. *Coccidioides* species revert to hyphal growth upon laboratory culture. *Histoplasma capsulatum* grows as a yeast in the lung but as a hyphal organism in culture. *Blastomyces* species are large yeast in lung tissue but revert to hyphal growth in culture. These pathogenic fungi display a phenomenon known as thermal dimorphism. They grow as hyphae below 37° Celsius but as spherules or yeast at 37° C.

Viral infections are identified through serological or polymerase chain reaction tests.

TREATMENT AND THERAPY

Bacterial pneumonia is commonly treated with antibiotics. The antibiotic regimen depends on which organisms are causing the infection. For healthy people younger than sixty-five years old, community-acquired pneumonia is treated with high-dose amoxicillin plus a macrolide (azithromycin or clarithromycin) or doxycycline. If the patient has preexisting health challenges, such as heart, lung, kidney, or liver disease, alcoholism, diabetes mellitus, or is a smoker, pneumonia should be treated with amoxicillin/clavulanate plus a macrolide or doxycycline. Suppose someone has a penicillin allergy and can tolerate cephalosporin antibiotics.

Their pneumonia should be treated with a third-generation cephalosporin, such as cefpodoxime and cefditoren plus a macrolide or doxycycline. An alternative is to treat the patient with the new antibiotic lefamulin (Xenleta). Individuals who cannot tolerate any beta-lactams should be treated with a fluoroquinolone (levofloxacin, moxifloxacin, or gemifloxacin) or lefamulin.

Treatment regimens change for those with hospital-acquired pneumonia. Suppose there is no suspicion that MRSA or *Pseudomonas* causes the infection. In that case, the treatment regimen is the same as those above. If *Pseudomonas* is the cause of pneumonia, then a combination treatment with an antipseudomonal beta-lactam (piperacillin/tazobactam, cefepime, ceftazidime, meropenem, or imipenem) plus an antipseudomonal fluoroquinolone (such as ciprofloxacin or levofloxacin) is recommended. For MRSA pneumonia, the preferred treatment of choice is intravenous vancomycin or linezolid. Ceftaroline is a viable alternative.

For pneumocystis pneumonia, trimethoprim/sulfamethoxazole (TMP-SMZ) is the preferred regimen. Alternative treatments include trimethoprim/dapsone, clindamycin/primaquine, or atovaquone for mild to moderate disease. For severe diseases, intravenous clindamycin/primaquine or pentamidine are viable options. Nevertheless, pentamidine is poorly tolerated and should be reserved for dire cases.

Fungal pneumonia is treated with azole drugs, including oral fluconazole or itraconazole. Severe cases may require an intravenous drug called "amphotericin B."

Drugs for viral infections are sparse, but recent advances have increased the available antiviral drugs. The drugs for influenza infections include oral oseltamivir (Tamiflu), intravenous peramivir, or inhaled zanamivir. All three drugs must be started within forty-eight hours of symptom onset. A newer drug, baloxavir, is given as a one-dose formulation.

Medications for cytomegalovirus infections include intravenous ganciclovir or oral valganciclovir for nonsevere cases. A newer anticytomegalovirus drug is oral letermovir (Prevymis), which is better tolerated than other alternatives.

PERSPECTIVE AND PROSPECTS

Although modern medicine has not been able to reduce substantially or eliminate the number of cases of pneumonia, much has been learned about the disease and its causes. Scientific advances in the campaign to combat the effects of pneumonia in all world areas began with the first isolation of *Streptococcus pneumoniae* in France and the United States in 1880. The French discovery of pneumococci is associated with the laboratory of Louis Pasteur. Simultaneously, George Sternberg was completing work in the medical department of the US Army. In the first decade after the isolation of pneumococci, many researchers contributed to laboratory findings that linked these bacteria to inflammatory infections in the lungs of animals. They extended their research to include the effects on humans.

One of the most important early breakthroughs came in 1884 when the Danish researcher Hans Christian Joachim Gram developed a laboratory method for identifying specific bacteria in tissue specimens. This technique, called "Gram's stain," revealed that different chemical reactions occur when lung tissue samples and secretions from individuals ill with pneumonia and healthy persons are tested. The tissues stain very differently. The next step would lead to research into the phenomenon of phagocytosis, a process within pulmonary tissue that combats inapparent pneumococcus infection in healthy people. This specific discovery became linked with efforts to develop an immunization technology against pneumonia.

Until the 1980s, medical researchers used their knowledge of pneumonia mainly to develop methods of immunization against the disease. They also tried to diversify the drugs used in treating pneumonia. Efforts to produce a vaccine against pneumonia began with experiments by the German researchers George and Felix Klemperer, who tested antiserum in animals in 1891. The Klemperers were able to show that the offspring of adult rabbits that had been immunized were resistant to pneumococcal invasion and infection. Soon after that, they carried out the first injections of immune serum into human patients. This research ultimately led to the finding that there was no actual antitoxin or antibacterial property in the serum. Instead, it promoted phagocytosis, a process of encapsulation around pneumococci that aids in the immunological response of white blood cells in the body. The vaccine stimulates the body to create its defenses.

In 1911 in South Africa, an experimental pneumonia vaccine program was undertaken. Although the specific program was unsuccessful, the British physician and scientist Frederick Lister extended its theory. Unequivocal success with a pneumonia vaccine did not come until the last year of World War II. In 1945, C. M. MacLeod and several colleagues published research findings proving that pneumococcal infection in humans was preventable with vaccines containing as many as fourteen specific antigens. These were termed capsular polysaccharides. The breakthrough that made those findings possible had been pioneered in 1930 when these antigens were injected into human beings for the first time. Previously, they had been used only in experiments with mice.

Pneumonia vaccines are critically important components of programs to prevent disease among older population members. In March 2013, the *American Journal of Medicine* published a study of 1,400 pneumonia patients over fifty. Researchers found a correlation between pneumonia patients requiring hospitalization and an increased risk of decline in their mental abilities. Experts recommend that the elderly receive a pneumonia vaccine each year. The death rate from pneumonia continues to rise, but not as quickly as the percentage of the population that is elderly. Pneumonia is one of the ten leading causes of death in the United States. In 1900, it was the second or third most common killer. Without vaccines, it might easily still be the second or third leading cause of death.

Children under the age of two are also at high risk of catching pneumonia. The World Health Organization reported in November 2021 that pneumonia caused fourteen percent of all death of children under five years old. It killed an estimated 740,180 children in 2019.

—Byron D. Cannon, PhD and
L. Fleming Fallon Jr., MD, PhD, MPH;
updated by Michael A. Buratovich, PhD

Further Reading

Centers for Disease Control and Prevention. "Pneumonia," 9 Mar. 2020, www.cdc.gov/pneumonia/index.html. Accessed 10 Dec. 2021.

Kumar, Vinay, et al., editors. *Robbins & Cotran Pathologic Basis of Disease*. 10th ed., Elsevier, 2020.

Murray, Patrick R., et al. *Medical Microbiology*. 9th ed., Elsevier, 2020.

Niederman, Michael S., George A. Sarosi, and Jeffrey Glassroth. *Respiratory Infections*. 2nd ed., Lippincott Williams & Wilkins, 2001.

Papadakis, Maxine A., et al., editors. *Current Medical Diagnosis and Treatment 2022*. 61st ed., McGraw-Hill Education/Medical, 2021.

Parker, James N., and Philip M. Parker, editors. *The Official Patient's Sourcebook on Streptococcus Pneumoniae Infections*. Icon Health, 2002.

Singh, Sunit K., editor. *Human Respiratory Viral Infections*. CRC Press, 2020.

West, John B., and Andrew M. Luks. *Pulmonary Pathophysiology: The Essentials*. 10th ed., Lippincott Williams & Wilkins, 2021.

World Health Organization. "Pneumonia." 11 Nov. 2021, www.who.int/news-room/fact-sheets/detail/pneumonia. Accessed 10 Dec. 2021.

CAMPYLOBACTER

Category: Pathogen

Anatomy or system affected: Blood, cells, gastrointestinal tract, immune system, joints, lymphatic system

Specialties and related fields: Bacteriology, gastroenterology, immunology, internal medicine, microbiology, public health, rheumatology

Definition: small, curved, highly motile bacteria associated with animals that cause gastroenteritis in humans

KEY TERMS

adhesins: proteins present on the surface of bacteria that help in attaching to surfaces

cytolethal distending toxin: a complex protein toxin made by Campylobacter cells that enters intestinal cells, degrades their nuclear deoxyribonucleic acid (DNA), killing them.

fecal-oral transmission: a mode of disease transmission in which contaminated feces from an infected person or animal are somehow ingested by someone who, subsequently, becomes infected

fimbriae: small, stiff extensions of bacterial cells that help them adhere to surfaces

flagella: a hairlike extension of bacteria that rotates in a circle to propel the organisms through their medium

INTRODUCTION

Campylobacter is a slender, curved rod, highly motile, gram-negative bacterium. The genus *Campylobacter* was first proposed in 1963. At that time, it included only *C. fetus* and *C. bululus* (later renamed *C. sputorum*). Presently, *Campylobacter* is grouped with specialized gram-negative bacteria in "superfamily VI" containing Arcobacter and Helicobacter. Arcobacters are closely related to campylobacters, and some cause intestinal infection in humans. Helicobacter pylori are well known as a cause of stomach diseases in humans. Other Helicobacter species also infect humans.

Campylobacter, the leading cause of bacterial gastroenteritis worldwide, has a corkscrew appearance. The pathogen propels itself with one or two flagella, depending on the subspecies. It thrives best in a nonacidic environment with 3 to 5 percent oxygen and 2 to 10 percent carbon dioxide. It is sometimes found in nonchlorinated bodies of water, such as ponds and streams.

In humans, the primary source of *Campylobacter* infection, or campylobacteriosis, is *C. jejuni*, which accounts for about 90 percent of all *Campylobacter* infections worldwide and up to 99 percent of infections in the United States. Another Campylobacter that causes gastroenteritis in humans is C. coli.

In 2019, Campylobacter infection incidence in the United States was 19.5 infections per 100,000 persons. Therefore, *Campylobacter* has the highest incidence of all pathogens, and *Campylobacter* incidence increased 13 percent from 2016 to 2018. Between states, there is considerable variation in the incidence of Campylobacter infections. Such infections are more common in western states, with California showing an infection rate triple that in Tennessee.

C. jejuni usually resides in the gastrointestinal tracts of birds, but cows are also common carriers. Therefore, the most common mode of infection is the consumption of raw or undercooked poultry or unpasteurized milk. *Campylobacter* is routinely shed by animals and is found in water sources, fresh or saline. In water, *Campylobacter* can survive for many weeks at low temperatures (< 15°C). Water is a direct source of human infection, but eating contaminated food is a more significant cause of *Campylobacter* gastroenteritis.

It is estimated that about two million people experience symptomatic *Campylobacter* infections each year in the United States. The incidence of such infections is as much as six times greater in rural areas. This higher incidence may occur because people in rural locations are more likely to drink unpasteurized (raw) milk than are persons in urban settings.

NATURAL HABITAT AND FEATURES

Campylobacter colonizes the intestinal tract, the urogenital tract, or the oral cavity of healthy and sick animals, particularly chickens. It is also found in the intestinal tract of humans. *C. jejuni* is found in human and bovine (cow) feces, while *C. coli* is commonly found in the feces of pigs, humans, and chickens and contaminated water. *C. helveticus* is found in the feces of cats and dogs. C. upsaliensis is the most important Campylobacter species after C. jejuni and C. coli. C. upsaliensis resides in younger dogs and cats, especially animals fed human food. In Europe, C. upsaliensis comprises about 1 to 2 percent of clinical Campylobacter isolates.

The acidity of the human stomach kills most ingested *Campylobacter*. Nevertheless, some bacteria survive and attach themselves to the intestinal epithelial cells or the mucus on these cells. The motility of *Campylobacter* helps them skirt move through the gastrointestinal tract. The organism is attracted to the mucous that coats the inner layer of the gastrointestinal tract. It also prefers the lower small intestine (ileum) and the large intestine (colon and rectum). Several adhesins (proteins called "FlpA" and "CadF") on the surface of *Campylobacter* help them attach to the intestinal mucosae. These bacteria's high motility and spiral shape help them drill into the mucosal wall,

Campylobacter *infection is a type of stomach flu (gastroenteritis), sometimes called food poisoning. The most common symptoms are diarrhea, vomiting, stomach cramping and fever. Photo via iStock/Moyo Studio. [Used under license.]*

where they secrete a protein toxin called "cytolethal distending toxin" (CDT). CDT has three protein subunits (CdtA, CdtB, & CdtC). Two subunits transport CdtB into intestinal cells, where it enters the nucleus and degrades the cell's chromosomes, killing it.

Usually, *Campylobacter* remains in the intestine of humans. Rarely, however, it migrates to the bloodstream or the lymphatic system. Such a migration is unusual in persons with normally-functioning immune systems.

PATHOGENICITY AND CLINICAL SIGNIFICANCE

Campylobacter infection has an incubation period of one to seven days and lasts up to ten days. Fewer than five hundred organisms are required to cause an infection in the host, approximately equivalent to about one drop of juice from an infected chicken.

People with *Campylobacter* infections typically have a fever, muscle pain, malaise, and headache, followed by crampy abdominal pain and diarrhea that is watery and foul-smelling at first but may turn bloody. Blood diarrhea appears more commonly in children.

Extensive damage to the intestine and its inflammation may cause the colon to dilate, a condition called "toxic megacolon." The symptoms of toxic megacolon include bloating, tachycardia (fast heart rate), and loss of bowel sounds on physical examination.

The immune response to *Campylobacteria* may elicit the production of antiganglioside antibodies that attack the peripheral nerves and cause Guillain-Barré syndrome (GBS). GBS is characterized by ascending paralysis that starts at the feet and extends to the upper body. *Campylobacter* is one of the most common bacterial causes of GBS. An estimated 1 in 1,000 persons infected with *Campylobacter* develop GBS. GBS is a leading cause of acute paralysis in the United States. Most infected persons recover in six to twelve months, but some never recover. According to the Centers for Disease Control and Prevention (CDC), up to 40 percent of all cases of GBS in the United States may be caused by infection with *Campylobacter*. GBS develops within two to four weeks after infection.

The immune response against *Campylobacter* may trigger an autoimmune response in which immune cells attack the joints; a condition called "reactive arthritis." The main symptom of reactive arthritis is pain in the large joints.

Persons with acquired immunodeficiency syndrome (AIDS) have a *Campylobacter* incidence about forty times greater than those without AIDS. Some persons without AIDS have an immune deficiency in immunoglobulin A (IgA), thus increasing their risk for infection with *Campylobacter*. Breast-fed babies have a reduced risk for infection with *Campylobacter*, probably because of the lactating woman's transfer of maternal substances, particularly secretory IgA.

DRUG SUSCEPTIBILITY

In most cases, Campylobacter infections are mild and self-limited. In the United States, the estimated mortality rate from symptomatic Campylobacter infections is 2.4 per 1000 confirmed cases. Maintaining proper hydration and electrolytes are of cardinal importance, and antibiotics are unnecessary for most cases of C. jejuni gastroenteritis. Avoid drugs that decrease gastric motility, such as loperamide, since they can prolong the infection and increase its severity.

In cases of severe disease, 500 mg of oral azithromycin for three to five days is the preferred treatment (10 mg/kg body weight in children). Alternative treatments include levofloxacin (500 mg once a day for three to five days) or ciprofloxacin (500 mg twice a day for three to five days). People with compromised immune systems should undergo treatment for seven to fourteen days.

Increasing worldwide resistance of the *Campylobacter* pathogen to fluoroquinolone drugs has been noted since the late 1990s. Primarily responsible for this resistance is the treatment of animals with fluoroquinolones to promote their growth. There is some resistance to macrolides, but it is much lower than fluoroquinolones such as ciprofloxacin.

Some studies have shown that *Campylobacter* infections acquired during travel are more resistant to antibiotics than those acquired at home. For example, in one study in the Netherlands, resistance to fluoroquinolone antibiotics was 54 percent in travel-related infections. In contrast, the resistance rate was a significantly lower 33 percent in infections in the study subject's native area.

—Christine Adamec, MBA;
Updated by Michael A. Buratovich, PhD

Further Reading

Campylobacter (Campylobacterosis). Centers for Disease Control and Prevention, 14 Apr. 2021, www.cdc.gov/campylobacter/index.html. Accessed 20 Nov. 2021.

Foodborne Diseases Active Surveillance Network. Centers for Disease Control and Prevention, wwwn.cdc.gov/foodnetfast/. Accessed on 20 Nov. 2021.

Geissler, Aimee L., et al. "Increasing Campylobacter Infections, Outbreaks, and Antimicrobial Resistance in the United States, 2004-2012." *Clinical Infectious Diseases,* vol. 65, no. 10, 2017, pp. 1624-31, doi:10.1093/cid/cix624.

Janssen, Riny, et al. "Host-Pathogen Interactions in *Campylobacter* Infections: The Host Perspective." *Clinical Microbiology Reviews,* vol. 21, no. 3, 2008, pp. 505-18.

Konkel Michael E., et al. "Taking Control: Campylobacter jejuni Binding to Fibronectin Sets the Stage for Cellular Adherence and Invasion." *Frontiers in Microbiology,* vol. 11, 2020, p. 564, DOI=10.3389/fmicb.2020.00564.

Nachamkin, Irving, Christine M. Szymanski, and Martin J. Blaser, editors. *Campylobacter.* 3rd ed., ASM, 2008.

Helicobacter pylori infection

Category: Diseases and conditions

Anatomy or system affected: Gastrointestinal system, intestines, stomach

Specialties and related fields: Bacteriology, biochemistry, biotechnology, immunology, gastroenterology, internal medicine, microbiology, oncology, pathology, pharmacology, preventive medicine, public health

Definition: a gram-negative bacillus that causes an infection of the inner mucus lining of the stomach and is the primary cause of gastric ulcer

KEY TERMS

adhesin: cell surface molecules that help bacteria attach to surfaces

CagA: cytotoxin-associated gene A; a protein toxin made by Helicobacter pylori that disrupts cell adhesion junctions between stomach epithelial cells

flagella: a whiplike extension of bacterial cells that rotates to propel them through the medium

gastritis: antibodies derived from a single parent clonal cell

type-IV secretion system: a syringe-like structure that pathogenic bacteria use to inject poisons into cells

peptic ulcer: an erosion of the stomach lining that may bleed or perforate

urease: an enzyme that degrades urea (CH_4N_2O) to ammonium (NH_{4+}), bicarbonate (HCO_3-), and hydroxyl ions (OH-) and helps *Helicobacter pylori* (*H. pylori*) evade the acidity of the stomach

CAUSES

H. pylori infections result from the ingestion of contaminated food or liquids. This microorganism is spread through the fecal-oral route and perhaps by other routes as well. Contact with fecal matter, saliva, or someone else's vomit may contribute to *H. pylori* transmission. *H. pylori* are found in the stomachs of over half the world's population.

The bacterium *H. pylori* is a short, spiral-shaped microorganism with multiple flagella that help it move rapidly. A cadre of adhesins on the cell surface helps *H. pylori* readily attach to the stomach mucosa. The organism survives the stomach's acidity by using

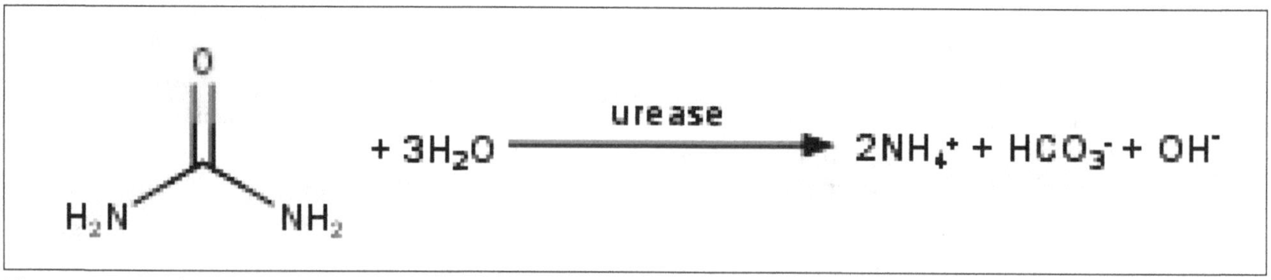

The enzyme urease degrades urea to ammonia and bicarbonate. Illustration by Michael A. Buratovich.

the enzyme urease to degrade urea to ammonia and bicarbonate.

Ammonia and bicarbonate neutralize the strong gastric acidity. The stomach has four regions: the cardia, fundus, body, and pylorus. The pylorus is divided into two main parts: the antrum and the pyloric canal that connects to the first part of the small intestine, the duodenum.

While in the stomach, *H. pylori* use its flagella to move to the stomach lining and a less acidic stomach region, like the antrum with fewer acid-secreting parietal cells. *H. pylori* use a host of cell surface proteins called "adhesins" to attach to foveolar cells (also known as "surface mucus cells") that secrete mucus. Once attached, *H. pylori* secrete virulence factors that damage stomach mucosa and help the bacterial cells flourish.

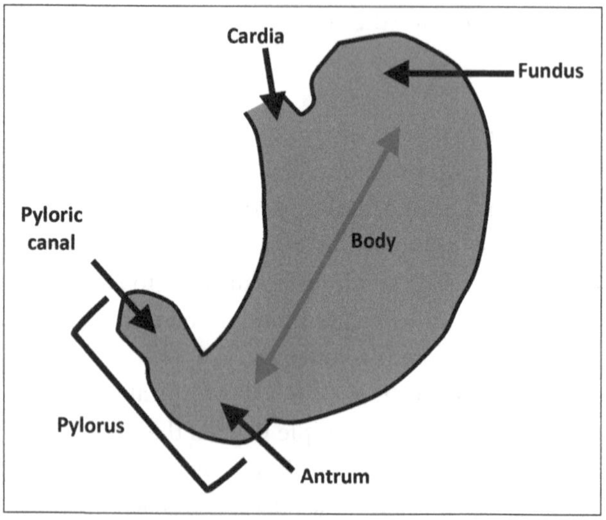

The regions of the stomach. Illustration by Michael A. Buratovich.

Some *H. pylori* make type IV-secretory systems that are syringe-like molecules that inject toxins into host cells. One molecule injected into host cells is a protein called "CagA" (cytotoxin-associated gene A). CagA unzips the attachments between stomach epithelial cells. Acid squeezes between these cells, damaging the underlying tissue and evoking and respectable immune response. Inflammation of the stomach lining is known as gastritis. Another injected toxin is called "vacuolating cytotoxin A" (VacA), which causes epithelial cell death, exposing the underlying mucosal and further exposing it to acid. The bacterial secretions stimulate the formation of inflammatory cytokines, leading to chronic gastritis. The mucus layer is damaged and thinned by *H. pylori* secretions of cytotoxins and enzymes such as proteases and phospholipases. With the loss of the protective mucus layer, the strong acids of the stomach attack and damage the stomach lining, resulting in peptic ulcers. Most peptic ulcer cases in the United States are associated with *H. pylori* infections.

Enzymes secreted by H. pylori damage the mucus layer. The loss of the protective mucus layer causes the strong stomach acid to damage the stomach lining, resulting in peptic ulcers. Most peptic ulcer cases in the United States are associated with *H. pylori* infections.

RISK FACTORS

There is a much greater risk of contracting *H. pylori* infection in developing countries because of unsanitary conditions. Contaminated food and water are primary sources, but other sources include contact with the stool, vomit, or saliva of an infected person.

SYMPTOMS

Most *H. pylori* infections do not cause symptoms. However, *H. pylori* strains that produce CagA and VacA cause stomach inflammation. Gastric inflammation stimulates cells in the stomach lining called "G cells" to increase their secretion of the peptide hormone gastrin. Increased gastrin secretion stimulates stomach parietal cells to produce even more hydrochloric acid. This cycle of inflammation/increased stomach acid accelerates the destruction of the stomach epithelium and the formation of gastric and duodenal ulcers. Antral gastritis leads to duodenal ulcers, but gastritis of the body of the stomach leads to gastritis. Deep ulcers can erode blood vessels or may erode through the stomach lining and cause a perforation. Rarely, chronic duodenal ulcers cause excessive swelling and scarring of the pyloric canal, narrowing it and obstructing food passage.

If someone has symptoms, they will usually experience heartburn, appetite loss, shortness of breath, and abdominal pain just above the stomach that worsens a few hours after eating.

If someone suffers from a peptic ulcer that causes gastric bleeding, they will have blood in their vomit or feces. Erosion of nearby arteries can cause rapid blood loss and shock. Stomach perforation causes air to collect under the diaphragm, which irritates the phrenic nerve, sending referred pain to the shoulder.

Because *H. pylori* constantly sequester iron from their hosts, a chronic infection causes iron deficiency anemia.

Chronic *H. pylori* infections with CagA-positive strains are tightly associated with developing two different gastric cancers; mucosa-associated lymphoid tissue or MALT lymphomas involve B lymphocytes and adenocarcinomas of the gastric glands.

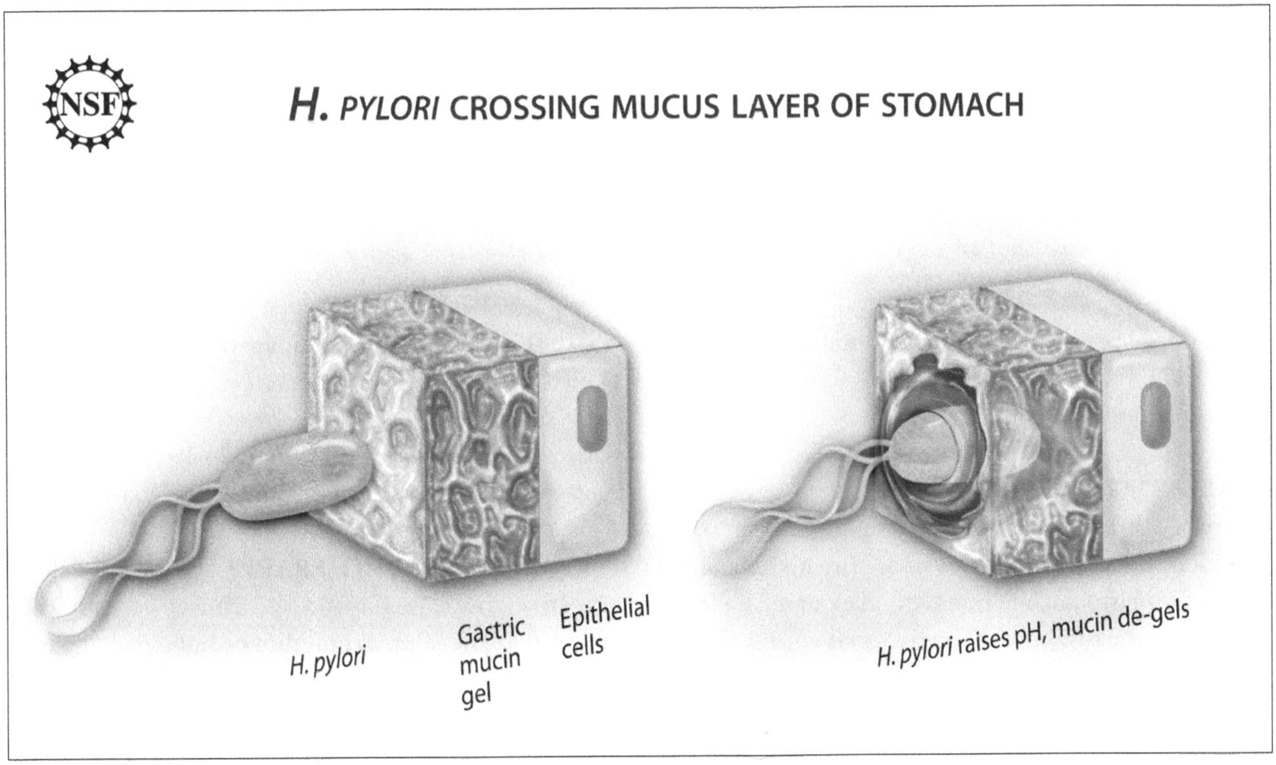

Diagram showing how H. pylori *reaches the epithelium of the stomach. Image via Wikimedia Commons. [Public domain.]*

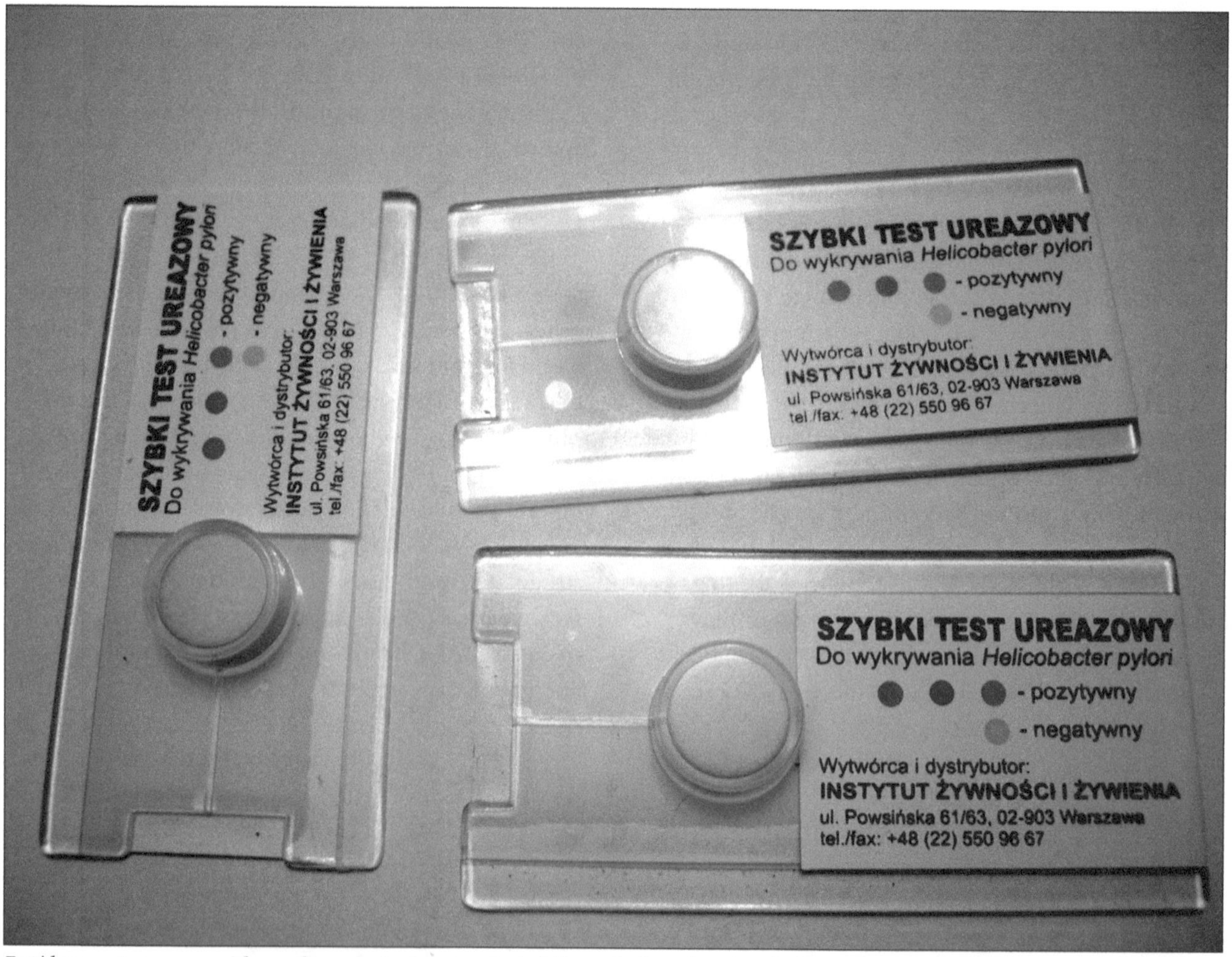

Rapid urease tests are a quicker and more cost-efficient method of diagnosis. Samples taken during gastroscopy are placed into a medium that changes urea to ammonia, thus producing a redder coloration. Photo by Louve.pl via Wikimedia Commons.

Anyone experiencing severe abdominal pain, difficulty swallowing, bloody stools, or vomit should seek immediate medical help.

SCREENING AND DIAGNOSIS

There are three primary ways to diagnosis *H. pylori* infection. In endoscopy, a physician threads a flexible tube into the stomach to remove and examine a tissue sample for the presence of the bacterium. Culturing *H. pylori* from gastric biopsies can help diagnose it. A breath test involves a patient ingesting a test meal containing radioactively labeled urea. *H. pylori* break down the urea, forming radioactive carbon dioxide, which is easily detected. Blood tests detect the presence of antibodies against *H. pylori*, which could indicate a current or prior infection. Finally, a fecal antigen test can detect *H. pylori* antigens in stools.

TREATMENT AND THERAPY

Treatment usually consists of administering multiple drugs that kill *H. pylori* and diminish the damage to the stomach lining. There are several effective treatment regimens for *H. pylori* infections.

The first treatment regimen is designed around the macrolide antibiotic clarithromycin. Clarithromycin plus amoxicillin and a proton pump inhibitor, twice daily for fourteen days is a standard "triple treatment." Alternatives include substituting metronidazole for amoxicillin in people who are allergic to penicillin antibiotics. The cure rate for this treatment regimen is <80 percent. Concomitant therapy uses clarithromycin, amoxicillin, metronidazole, and a proton pump inhibitor for ten to fourteen days. This regimen has a 90 percent cure rate in some studies. Other regimens vary the timing of the antibiotics, and others substitute clarithromycin with levofloxacin. Proton pump inhibitors like omeprazole, esomeprazole, pantoprazole, rabepra- zole, or lansoprazole suppress gastric acid production, promote healing, and improve the effectiveness of the antibiotics.

PREVENTION AND OUTCOMES

Although the mode of transmission of *H. pylori* is not fully understood, what is known is that improved sanitation is an essential preventive measure. A vaccine against this organism is under development.

—David A. Olle, MS and Michael A. Buratovich, PhD

Further Reading

Chey, William D., et al. "ACG Clinical Guideline: Treatment of *Helicobacter pylori* Infection." *The American Journal of Gastroenterology,* vol. 112, no. 2, 2017, pp. 212-39, doi:10.1038/ajg.2016.563.

Crowe, Sheila E. "*Helicobacter pylori* Infection." *The New England Journal of Medicine,* vol. 380, no. 12, 2019, pp. 1158-65, doi:10.1056/NEJMcp1710945.

"Helicobacter Pylori Infections." *MedlinePlus,* 5 May 2021, medlineplus.gov/helicobacterpyloriinfections.html. Accessed 25 Nov. 2021.

Maixner, Frank, et al. "The 5300-year-old Helicobacter pylori Genome of the Iceman." *Science,* 8 Jan. 2016, pp. 162-65.

Öztekin, Merve, et al. "Overview of *Helicobacter pylori* Infection: Clinical Features, Treatment, and Nutritional Aspects." *Diseases (Basel, Switzerland),* vol. 9, no. 4, 2021, p. 66, doi:10.3390/diseases9040066.

Sultan, Mutaz I., et al. "*Helicobacter pylori* Infection." 16 Nov. 2018, http://emedicine.medscape.com/article/929452-overview. Accessed 25 Nov. 2021.

LEGIONNAIRES' DISEASE

Category: Diseases and conditions

Also known as: Legionnaires' pneumonia, legionella disease, legionellosis

Anatomy or system affected: Lungs, respiratory system

Specialties and related fields: Bacteriology, internal medicine, microbiology, public health, pulmonology

Definition: a lung infection and form of pneumonia named for a pneumonia outbreak at the American Legionnaires Convention in Philadelphia in 1976

KEY TERMS

alveolar macrophages: a type of macrophage, a professional phagocyte, found in the airways and at the level of the alveoli in the lungs but separated from their walls

alveoli: tiny air sacs at the end of the bronchioles where the lungs and the blood exchange oxygen and carbon dioxide during the process of breathing in and breathing out

phagocytosis: a cellular process for ingesting and eliminating particles larger than 0.5 μm in diameter, including microorganisms, foreign substances, and apoptotic cells

porins: pore proteins in the outer membrane of gram-negative bacteria that mediate the diffusion of small hydrophilic molecules into cells

type IV secretion system: a secretion protein complex found in gram negative bacteria, gram positive bacteria, and archaea that can transport proteins and deoxyribonucleic acid (DNA) across the cell membrane

CAUSES

Legionnaires' disease is caused by the gram-negative bacterium *Legionella pneumophilia* (*L. pneumophila*). This microorganism is rod-shaped and has a cell wall with a thin layer of peptidoglycan underlying an outer membrane. However, *L. pneumophila* stains poorly with a Gram stain but is best visualized with a silver stain. *L. pneumophila* can be found in sources of standing water, such as cooling towers, hot water tanks, air conditioners, heating, ventilating, air conditioning (HVAC) systems, and soil. *L. pneumophila* and some related bacterial species live as intracellular parasites of free-living amoeba and ciliates.

Legionnaires' disease is contracted by breathing water vapor from a standing water source or soil that contains *Legionella* bacteria into the lungs. The infection is not transmitted from one person to another.

Phagocytosis is a cellular process whereby specific white blood cells engulf something bound to their surfaces and draw it inside. When *L. pneumophila* reaches the lungs, it is phagocytosed by the lung's resident white blood cell, the alveolar macrophage. Alveolar macrophages remove debris and other microbial invaders from the alveoli by phagocytosing them. *L. pneumophila* has several surface structures that bind tightly to alveolar macrophages, such as flagella, pili, and outer membrane proteins called "porins." Phagocytosis is mediated by a surface protein called "Mip" (macrophage infectivity potentiator) expressed on L. pneumophila's cell surface.

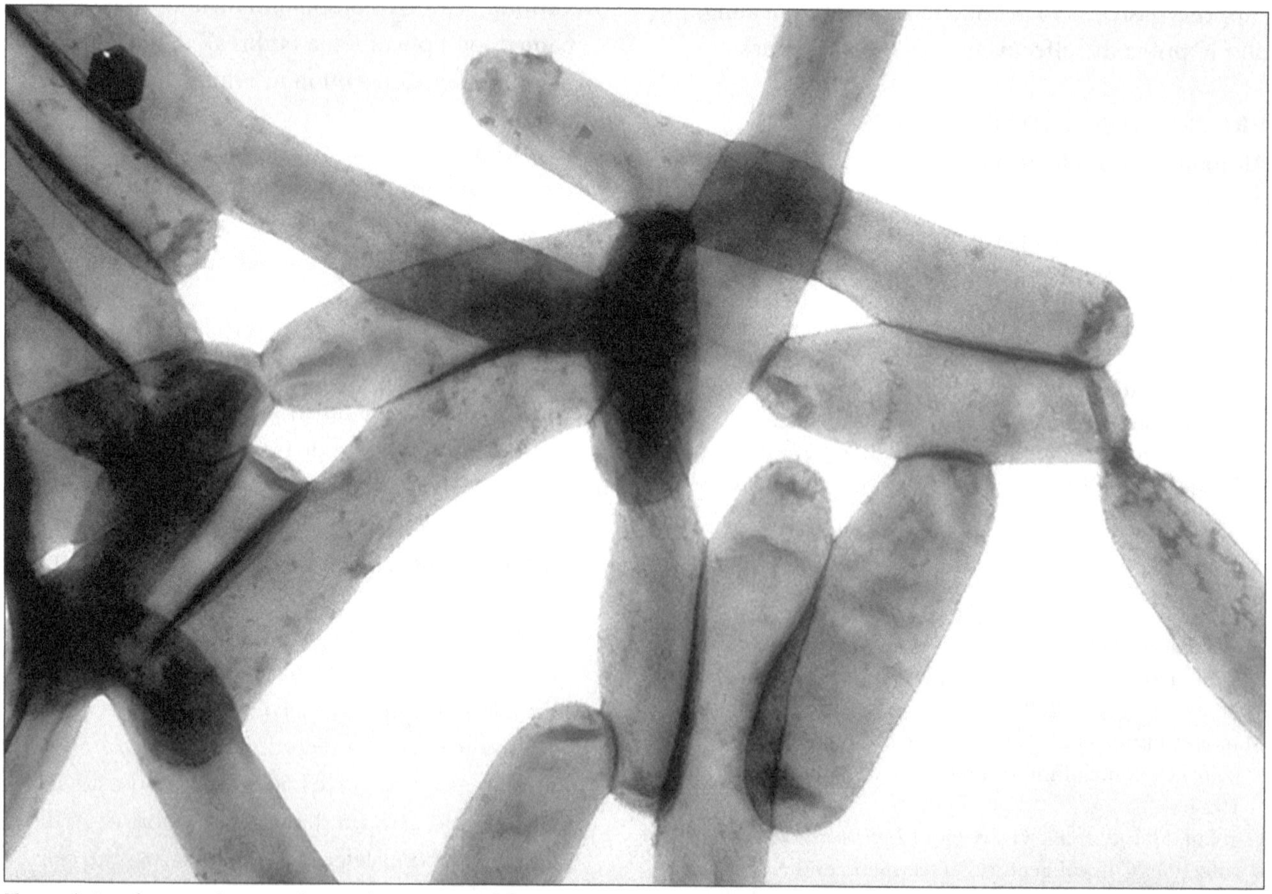

Transmission electron microscopy image of L. pneumophila, *responsible for over 90% of Legionnaires' disease cases. By CDC Public Health Image Library, via Wikimedia Commons. [Public domain.]*

After macrophages phagocytose something, it finds itself in a vesicle called a "phagosome." The macrophage fuses the phagosome with another intracellular vesicle filled with degradative enzymes called a "lysosome" to form a phagolysosome. These lysosomal enzymes drop the pH of the phagolysosome and systematically pick the engulfed microbe apart piece by piece. *L. pneumophila*, however, uses Dot/Icm type IV secretion system to pump a host of proteins into the phagosome. These *L. pneumophila* proteins harden the phagosome and prevent it from fusing with the lysosome. They also transform the phagosome into a nursery for *L. pneumophila* growth and division. The growing bacteria fill the macrophage, and when there are too many for the macrophage to hold, it bursts. *L. pneumophila* goes on to infect other cells.

Macrophage death and bacterial growth cause the release of chemotactic factors. These summon white blood cells called "polymorphonuclear" (PMNs) cells and other macrophages from the blood to the infection site. These white blood cells try to kill the bacteria, filling the alveoli with fibrin, protein, fluid, and dead cell debris. The infected macrophages also can transport the bacteria to sites outside the lung, including the gastrointestinal (GI) tract, central nervous system, kidneys, and heart.

RISK FACTORS

Factors that increase the chance for Legionnaires' disease include advanced age, smoking, excessive alcohol intake, chronic lung disease, weakened immune system (as with acquired immunodeficiency syndrome), kidney failure, diabetes, taking cortisone or other immunosuppressive drugs, organ transplant, and working with soil, especially newly tilled soil, or potting soil. Also, men are at higher risk for the disease.

SYMPTOMS

L. pneumophila causes two types of disease, legionellosis, and Pontiac fever.

Legionellosis causes fatigue, high fevers (104° Fahrenheit), chills and muscle aches, dry cough, chest pain with coughing or breathing, loss of appetite, and headache. Symptoms that develop if the infection becomes serious include shortness of breath; abdominal pain; nausea, vomiting, diarrhea; mental problems; confusion, and memory loss.

Pontiac fever is a much milder disease that causes flu-like symptoms.

SCREENING AND DIAGNOSIS

A doctor will ask about symptoms and medical history and perform a physical exam. Tests may include blood tests to look for high or rising antibodies to *Legionella* bacteria, sputum cultures (which examines mucus from deep inside the lungs to identify the cause of the infection), kidney function tests (poor kidney function is often seen with *Legionella* infection), urine antigen tests to check for *Legionella* lipopolysaccharide antigen in urine, and a chest X-ray to diagnose pneumonia or lung infection. Polymerase chain reaction tests detect *Legionella* deoxyribonucleic acid (DNA).

Sputum cultures are problematic since *L. pneumophila* does not grow on traditional microbiological media like blood or brain-heart infusion agar. Instead, this microorganism requires special nutrients, like cysteine and iron, to grow. Sputum cultures from someone suspected of having legionellosis should be grown on a special medium called "buffered charcoal yeast extract" (BCYE), supplemented with cysteine and iron.

Chest X-rays usually show a patchy infiltrate with consolidation of one lobe. Blood tests might reveal low blood sodium levels (hyponatremia), high white blood cell counts (leukocytosis), and low platelet counts (thrombocytopenia).

TREATMENT AND THERAPY

The preferred treatment for *Legionella* infections is levofloxacin and azithromycin. These drugs kill bacteria, effectively penetrate lung tissue, and are active against all *Legionella* species. Alternative treatments include moxifloxacin or ciprofloxacin, clarithromycin, or doxycycline.

PREVENTION AND OUTCOMES

Proper design, maintenance, and cleaning of high-risk areas can reduce the risk of spreading the disease. This includes any area with standing water. One can reduce the risk of getting Legionnaires' disease by not smoking, by limiting the amount of alcohol intake, by wearing gloves and a mask if working with freshly tilled soil or potting soil, by not inhaling dust from the soil, and by moistening the soil to lower the amount of dust.

—Rick Alan and Michael A. Buratovich, PhD

Further Reading

Brady, Mark F., and Vidya Sundareshan. "Legionnaires' Disease." *StatPearls*. StatPearls Publishing, 2021.

Broaddus, V. Courtney, et al., editors. *Murray & Nadel's Textbook of Respiratory Medicine*. 7th ed., Elsevier, 2021.

Centers for Disease Control and Prevention. "*Legionella* (Legionnaires' Disease and Pontiac Fever." 25 Mar. 2021, www.cdc.gov/legionella/about/index.html. Accessed 11 Dec. 2021.

Levitzky, Michael G. *Pulmonary Physiology*. 9th ed., McGraw-Hill Education/Medical, 2018.

Mason, Robert J., et al., editors. *Murray and Nadel's Textbook of Respiratory Medicine*. 5th ed., Saunders/Elsevier, 2010.

Miyashita, Naoyuki. "Atypical Pneumonia: Pathophysiology, Diagnosis, and Treatment." *Respiratory Investigation*, S2212-5345(21)00177-5, 2021, doi:10.1016/j.resinv.2021.09.009.

Popper, Helmut. *Pathology of Lung Disease: Morphology—Pathogenesis*. 2nd ed., Springer, 2021.

Ryan, Kenneth J., et al., editors. *Sherris Medical Microbiology: An Introduction to Infectious Diseases*. 7th ed., McGraw-Hill, 2018.

BRUCELLOSIS

Category: Diseases and disorders

Also known as: Malta fever, Mediterranean fever, undulant fever

Anatomy or system affected: Brain, heart, joints, liver, musculoskeletal system, nervous system, reproductive system, spine, spleen

Specialties and related fields: Bacteriology, cardiology, epidemiology, microbiology, neurology, occupational health, preventive medicine, public health, rheumatology

Definition: a bacterial infection transmitted to humans from infected livestock and wildlife usually acquired from unpasteurized dairy products

KEY TERMS

antibiotics: drugs that inhibit or kill bacteria and are used to treat bacterial infections

chronic: a disease of long duration, typically several weeks, months, or years

pasteurization: a method first described by the French scientist Louis Pasteur for killing bacteria in food and beverages

relapse: the recurrence of signs and symptoms after they have subsided or ceased

vaccine: a substance, usually injected, that stimulates the immune system to protect the body against a specific infectious disease

zoonotic: an infection of animals that can be transmitted to humans

CAUSES AND SYMPTOMS

Brucellosis is a zoonotic disease, an infection that humans acquire from infected animals. Species of the bacterial genus *Brucella* cause brucellosis. They are transmitted to humans from cattle, sheep, goats, elk, bison, caribou (reindeer), camels, wild pigs, and other animals. The bacteria have been found in porpoises, harbor seals, some types of whales, and dogs. *Brucella* are gram-negative, nonmotile, coccobacilli (short rods) that require oxygen for growth (obligate aerobes).

Most human brucellosis cases are acquired by consuming unpasteurized milk, cheese, other dairy products, or undercooked meat. *Brucella* bacteria can survive up to two days in milk at 46° Fahrenheit (8° Celsius), up to three weeks in frozen meat, and up to three months in goat cheese. People working in microbiology labs, slaughterhouses, and meat-packing plants get brucellosis by inhaling aerosolized particles contaminated with the bacteria. *Brucella* bacteria also enter the body through contact with wounds and mucous membranes (eyes, nose, and mouth). Veterinarians and hunters are exposed to brucellosis in this way. Consequently, brucellosis is an occupational disease in shepherds, slaughterhouse workers, diary-industry professionals, and veterinarians or laboratory personnel handling *Brucella* cultures or preparing *Brucella* vaccines. Person-to-person transmission is rare, as is transmission by breast milk, transplant, transfusion (blood donation), congenital transmission, sexual transmission, and hospital-based transmission, which have all been documented but remain pretty rare.

After ingestion, inhalation, or contact, the bacteria move into the blood. *Brucella* do not remain in the blood for long; they invade various human cells, which protects the bacteria from the immune system. Once in the blood, the bacteria enter white blood cells called "neutrophils." The neutrophils usually seek out, ingest, and destroy bacteria. However, *Brucella* enter neutrophils and remain alive because the bacteria produce substances that prevent neutrophils from killing them. The infected neutrophils enter the lymphatic system, a network of vessels that filters tissue fluids (called "lymph" when inside the vessels) and returns it to the circulatory system. *Brucella* use the lymphatic system to spread throughout the body and invade cells of nearly every organ system. The bacteria infect the liver, spleen, kidneys, joints, breasts, and central nervous system. Reproduction of the bacteria in these organs causes tissue damage and triggers the characteristic signs and symptoms of brucellosis.

Symptoms arise two to four weeks following exposure to *Brucella*. The first symptoms resemble a severe flu-like illness and include fever, headache, fatigue, malaise (a feeling of ill health), anorexia (loss of appetite), and pain in muscles, back, and joints. These symptoms are caused by the immune system's response to the infection. However, the immune response to these bacteria is ineffective because most bacteria reside inside the body's cells. Fever and symptoms often decline during infection. However, bacterial reproduction kills cells, which release the bacteria, inducing another wave of fever and symptoms. Brucellosis has been called "undulant fever" for these fevers that rise and fall like a wave. If untreated, the disease progresses and becomes chronic. A chronic infection lasts many weeks or months, perhaps years.

Recurrent fevers and relapses characterize chronic infection. A relapse is the return of signs and symptoms. Chronic brucellosis is associated with arthritis, inflammation of joints, and spondylitis, inflammation of the joints between bones of the spinal column. Arthritis and spondylitis are painful and disabling. Endocarditis, inflammation of the heart's inner lining, is a dangerous complication of chronic brucellosis and is associated with fever and heart murmurs. In addition, brucellosis affects the nervous system, causing encephalitis, inflammation of the brain, and meningitis, inflammation of membranes surrounding the brain and spinal cord. Encephalitis and meningitis are severe complications of brucellosis. Fever and a sore neck usually accompany meningitis. The bacteria cause swelling in other organs, including the liver, spleen, testes, epididymis (tubes that carry sperm from the testes), and scrotum. Some patients experience chronic fatigue and depression. The severity of the disease depends on the infecting *Brucella* species. *B. canis* causes mild disease, but *B. melitensis* causes severe disease. *B. abortus* main affects the liver and spleen, causing sarcoidosis-like lesions (granulomas). *B. suis* tends to cause liver abscesses. Brucellosis is rarely fatal, resulting in death in probably less than two percent of cases in the United States. Most fatalities are related to endocarditis or severe neurologic disease (meningitis or encephalitis).

Infected animals exhibit no external signs of disease. Infected livestock such as cattle, goats, sheep, and domestic bison may abort spontaneously or produce weak offspring. Reduced milk production is associated with infection but can be caused by other diseases and disorders. Fortunately, brucellosis has been nearly eradicated from domestic livestock in the United States.

TREATMENT AND THERAPY

Most cases of brucellosis turn out well. However, early diagnosis and treatment are essential for a full recovery and preventing serious, disabling complications associated with chronic, relapsing disease. Diagnosis is based on the symptoms, laboratory tests, and a history of recent activities, such as travel abroad that may have resulted in exposure to brucellosis. Brucellosis is well-controlled in US livestock, and outbreaks of brucellosis among humans are unexpected. Most brucellosis cases in the United States are acquired abroad by consuming unpasteurized dairy products, particularly cheese from goats and sheep, in countries where brucellosis is not well controlled. An outbreak in the United States is so unlikely that any brucellosis outbreak is investigated as a possible bioterrorist event. Thus, given a history of exposure and symptoms consistent with brucellosis, laboratory tests confirm the presence of an ongoing infection.

Brucella are very slow-growing bacteria, and colonies can take six to eight weeks of incubation to appear on solid media. *Brucella* isolation from the blood or bone marrow definitively confirms a brucellosis diagnosis. Unfortunately, the slow growth of *Brucella* makes this difficult. However, automated blood culture systems, such as the *Bactec* systems, are much more effective and can isolate *Brucella* in only one week. When cultures fail to provide results in a timely fashion, serological tests like the serum agglutination test, enzyme-linked immunosorbent assay, or complement fixation tests can detect rising levels of antibodies against *Brucella* antigens in the blood. Antibodies are substances produced by B lymphocytes in response to infectious agents such as viruses and bacteria. The body produces antibodies that are specific for each type of infectious agent. Thus, the presence of antibodies for *Brucella* is a reliable sign of brucellosis.

Because *Brucella* travel widely through the body, other body fluids and bone marrow can also be cultured. Cerebrospinal fluid is sampled and studied in the laboratory to determine if meningitis or encephalitis is related to brucellosis. Cerebrospinal fluid surrounds and protects the brain and spinal cord. The fluid is obtained by lumbar puncture, a procedure in which a needle is inserted between the lower back vertebrae to draw fluid into a syringe. Healthy cerebrospinal fluid contains no bacteria. An echocardiogram, an image of the heart's internal structure, detects signs of endocarditis.

Antibiotics are drugs that kill bacteria and provide an effective treatment for brucellosis. Oral doxycycline (100 mg, twice daily) and rifampin (600 to 900 mg/day) are standard treatments for simple brucellosis. Six weeks to several months of antibiotic therapy are required, depending on the severity of the infection. Children younger than eight years who suffer from brucellosis should receive rifampin and trimethoprim/sulfamethoxazole for six weeks. People with chronic brucellosis should receive combinations of three to four different antibiotics. Arthritis and spondylitis are treated with analgesics, medications that reduce pain. Meningitis and encephalitis are treated with drugs that suppress inflammation.

Brucellosis can be prevented by reducing the risk of exposure. Unpasteurized dairy products, including milk and cheese, should not be consumed, especially when traveling abroad. So-called village cheeses, usually made from goat and sheep milk, carry brucellosis. Meat should always be fully cooked before serving. Workers in meat-packing and slaughterhouse facilities should follow their employer's and OSHA (Occupational Safety Health Administration) standards regarding safe handling of meat. These procedures may include using personal protective gear such as gloves and eye protection (to prevent splashes with contaminated fluids). Microbiology laboratory workers who study material contaminated with brucellosis bacteria should also follow similar precautions established by the laboratory employer and OSHA. Hunters should wear gloves to protect their hands when dressing game, especially elk, moose, and wild pigs. Game meat should be stored appropriately and cooked thoroughly before serving.

PERSPECTIVE AND PROSPECTS

A Scottish physician, Sir David Bruce, first described the *Brucella* bacteria in 1887. Bruce made important discoveries about the disease among goats on the Mediterranean island of Malta, which is the source of this disease's former name, Malta fever.

Brucellosis is the most common zoonosis worldwide, and it is a prominent health-care concern in developing countries. Places with endemic brucellosis include countries of the Mediterranean basin, Middle East, Central Asia, China, the Indian subcontinent, sub-Saharan Africa, and parts of Mexico and Central and South America. Globally, approximately 500,000 brucellosis cases are reported each year, and an estimated 2.4 billion people are at risk. International tourism and migration have contributed to recent increases in brucellosis prevalence. In the United States, 100 to 200 brucellosis cases are reported annually. Important sources of infection are imported, unpasteurized dairy products from neighboring countries (mainly Mexico) and domestic sources.

Brucellosis once caused enormous losses in the US livestock industry. In 1952, reduced milk production and livestock losses cost the industry $400 million. Infected cattle and other livestock drove a relatively high rate of human brucellosis in the United States. As of 2013, brucellosis is uncommon in the United States, affecting about 100 people each year. The infection has become uncommon because of the Cooperative State-Federal Brucellosis Eradication Program, which began in 1934. This program has nearly eradicated brucellosis among domestic livestock. In 1934 11.5 percent of cattle herds in the United States tested positive for brucellosis. As of 2000, only six herds tested positive, and they had a low infection rate of less than 0.25 percent.

Continued prevention of human infections depends on effective control of animal brucellosis. The disease is incurable in animals. Thus, control has relied on vaccination and herd screening, followed by removing and killing infected animals. The vaccine is effective at preventing infection in cattle. However, it is less effective at preventing disease in domestic bison. Scientists are seeking an improved vaccine because bison can transmit the infection to cattle. Wildlife such as elk and bison in the Yellowstone area carry and can transmit brucellosis to domestic livestock. Control of brucellosis in wild elk and bison has proven to be difficult because these animals migrate and gather in large numbers near food during winter. Congregating elk and bison are more likely to spread the infection within a herd. More important, wild elk and bison can enter ranch land and expose domestic cattle and bison to infection. To prevent the congregation of wild herds of elk and bison, the United States Animal Health Association recommends that wildlife managers stop the practice of feeding wild herds during the winter. However, most human cases of brucellosis in the United States do not occur in Yellowstone (Montana, Wyoming, and Idaho). Over half of human infections in the United States occur in California, Texas, Arizona, and Florida.

Worldwide, brucellosis is not well controlled in domestic and wild animals, which results in enormous economic losses. Brucellosis is most common in countries with underdeveloped public health or domestic animal health programs and where pasteurization and animal vaccination are not systematically practiced. Considerable effort has been focused on developing more effective vaccines to control the infection in domestic livestock.

Vaccines have been effective in the United States, even though several species of *Brucella* are known. Only a few species cause disease in humans. These include *Brucella melitensis, B. suis, B. abortus,* and *B. canis.* Of these, *B. melitensis* causes the most severe disease in humans. It is transmitted by sheep and goats and has been eradicated in the United States. However, it continues to cause disease worldwide. *B. suis* infects wild pigs and rodents; it occasionally infects hunters. *B. abortus* is transmitted by cattle, but it has been nearly eradicated in the United States thanks to an aggressive animal vaccination program. However, it persists in wild elk and wild bison. *B. canis* infects dogs and is an uncommon cause of infection in veterinarians and dog breeders. *B. ceti* infects porpoises and whales, and *B. pinnipediae* infects harbor seals. These rarely cause disease in humans. Other *Brucella* species include *B. microti,* isolated from wildlife animals, *B. papionis* (isolated from baboons), *B. inopinata,* isolated from a human breast implant wound, and *B. vulpis,* which was isolated from red foxes. Two other *Brucella* species, *B. neotomae* (isolated from desert woodrats) and *B. ovis* (isolated from sheep), do not cause disease in humans.

—Mark Zelman, PhD and Michael A. Buratovich, PhD

Further Reading

Biddle, Wayne. *Field Guide to Germs.* Random House, 2002.

Committee on Revisiting Brucellosis in the Greater Yellowstone Area, et al. *Revisiting Brucellosis in the Greater Yellowstone Area.* National Academies Press, 2021.

Constable, Peter D., et al. *Veterinary Medicine: A Textbook of the Diseases of Cattle, Sheep, Pigs, Goats and Horses.* 11th ed., Saunders, 2017.

El-Sayed, A., and W. Awad. "Brucellosis: Evolution and Expected Comeback." *International Journal of Veterinary Science and Medicine,* vol. 6 (suppl.), 2018, pp. S31-35.

Franco, Maria Pia, Maximilian Mulder, Robert H. Gilman, and Henk L. Smits. "Human Brucellosis." *Lancet Infectious Diseases,* vol. 7, 2007, pp. 775-86.

Jameson, J. Larry, et al., editors. *Harrison's Principles of Internal Medicine.* 20th ed., McGraw-Hill, 2018.

Mackay, Katurah. "Bison Plan Called 'Absurd.'" *National Parks,* vol. 72, nos. 7-8, 1998, pp. 14-15.

Professional Guide to Diseases. 10th ed., Lippincott Williams & Wilkins, 2013.

PSEUDOMONAS INFECTIONS

Category: Diseases and conditions
Also known as: Pseudomonal bacteremia
Anatomy or system affected: All
Specialties and related fields: Bacteriology, dermatology, internal medicine, microbiology, ophthalmology, pathology, pharmacology, pulmonology, urology
Definition: a gram-negative nonfermenting bacillus that is a feared pathogen

KEY TERMS

cystic fibrosis: a hereditary disorder affecting the exocrine glands. It causes the production of abnormally thick mucus, leading to the blockage of the pancreatic ducts, intestines, and bronchi and often resulting in respiratory infection

nosocomial infections: infections acquired during the process of receiving health care that was not present during the time of admission

pseudomonads: a genus of gram-negative, gammaproteobacteria that contain 191 validly described species, have tremendous metabolic diversity, and can colonize a wide range of niches *pyoverdine*: a green pigment made by Pseudomonas aeruginosa that also acts and an iron binder and iron transport molecule

INTRODUCTION

Infections of the skin, blood, bones, eyes, ears, the central nervous system, the heart, the lungs, the gastrointestinal system, wounds, and the urinary tract may all be traced to an infection with the bacterium *Pseudomonas*. Pseudomonas infections range from mild to life-threatening but rarely affect healthy persons and often cause hospital-acquired or nosocomial infections. All infections are potentially curable, but infection with *Pseudomonas* is one of the most difficult types to treat. *Pseudomonas* is present in soil and water and can be found on plants, animals, and healthy persons.

CAUSES

The gram-negative *Pseudomonas* bacterium causes pseudomonas infections. The most prevalent is *P. aeruginosa*. Pseudomonads can infect any body organ or part.

P. aeruginosa makes many different virulence factors that help cause disease. First, its outer membrane is filled with lipopolysaccharide (LPS) made of lipids and polysaccharides. LPS is also called the "bacterial endotoxin" because it is recognized by the immune system and generates an immune response against it that causes extensive tissue damage.

Second, some P. aeruginosa strains synthesize slimy exopolysaccharides that ensconce the bacteria in a sticky, gummy biofilm. Biofilms adhere the bacteria to surfaces and protect them from the immune system and antibiotics.

Third, *P. aeruginosa* produces several toxins and uses a type III secretion system that acts like a tiny, molecular syringe to inject those toxins directly into host epithelial cells.

Fourth, this microorganism releases other toxins straight into the surrounding environment. Phospholipase C, for example, breaks down host cell membranes, destroying them. A second powerful toxin secreted by *P. aeruginosa* is exotoxin A. Even though exotoxin A is secreted outside cells, one of its subunits binds to receptors, and shepherd's the protein into host cells. Inside host cells, exotoxin A is a potent protein synthesis inhibitor, causing cell death.

Fifth, another toxin produced by *P. aeruginosa* is pyocyanin. This blue pigment has a sweet, grape-like smell. However, it generated reactive oxygen species inside cells, killing any competing bacteria and host cells.

Sixth, the organism releases a green pigment called "pyoverdine." This small molecule is a siderophore or iron-binding molecule that steals iron from host cells and transports it into P aeruginosa.

RISK FACTORS

Considered an opportunistic bacterium, *Pseudomonas* attacks debilitated persons, often those who are hospitalized or who have a disorder that weakens the immune system. Any break in the skin or any medical device, such as a urine catheter, may provide an opportunity for the bacterium to enter the body and cause infection.

Persons with human immunodeficiency virus (HIV) infection or cancer, and transplant recipients, are at increased risk because of their weakened immune systems, usually caused by the drugs they take to treat their diseases.

Persons with diabetes or cystic fibrosis, type 2 diabetes mellitus, chronic granulomatous disease, deep wounds, severe burns, or abuse intravenous drugs are at greater risk.

SYMPTOMS

P. aeruginosa may infect a variety of sites in the body, and symptoms depend on the site involved. External otitis, or swimmer's ear, causes pain and a discharge from the ear canal. In contrast, malignant external otitis seen in persons with diabetes has symptoms of fever, loss of hearing, and severe pain. Drainage is often seen in eye infections with *Pseudomonas*. Skin infections cause lesions or develop in open sores and may have green-blue drainage with a fruity odor. The greenish pigment is pyoverdine, an iron-chelator, that binds to iron and transports it into *P. aeruginosa*. Infection of the heart, or endocarditis, comes with a fever, a heart murmur, lesions, and an enlarged spleen.

Diarrhea and dehydration are the most common symptoms of gastrointestinal infections. Pneumonia with fever, difficulty breathing with a rattling sound, and lack of oxygen are symptoms of respiratory system infections. Fever, headache, and confusion are seen in persons with meningitis caused by *Pseudomonas* infection. Bacterial blood infection, or bacteremia, comes with jaundice, fever, rapid breathing, and a rapid heart rate. Lesions may occur with any *Pseudomonas* infection.

SCREENING AND DIAGNOSIS

Blood, wound drainage, body fluids, and tissue are sent to a laboratory for culture to determine the presence of *Pseudomonas* bacteria. The clinical site of the infection is cultured to determine the causative organism for the person's symptoms. Radiology or X-ray studies may show lesions within the body. Still, cultures are the only way to determine the actual organism causing problems.

TREATMENT AND THERAPY

Combinations of antipseudomonal antibiotics are used to treat infections. Antipseudomonal antibiotics include:

- Piperacillin-tazobactam is a beta-lactam/beta-lactamase inhibitor.
- Cephalosporins include ceftazidime, cefepime, cefoperazone, and the newer cephalosporin, cefiderocol.
- Fluoroquinolones are one antibiotic group that is effective against *P. aeruginosa*, except for moxifloxacin.
- Carbapenems like imipenem, doripenem, and meropenem.
- The monobactam aztreonam is active against *P. aeruginosa*.
- Aminoglycosides (gentamicin, tobramycin, amikacin, plazomicin).
- Advanced beta-lactamase inhibitor combinations: ceftazidime-avibactam, ceftolozane-tazobactam, imipenem-cilastatin-relebactam.
- Polymyxins: colistin and polymyxin B.

Supportive therapy, depending on the clinical condition of the infected person, is used. Hospitalization may be needed if symptoms are severe. Respiratory support, including the use of mechanical ventilation, may be indicated. Finally, the revision of wounds and the surgical removal of abscesses are options.

PREVENTION AND OUTCOMES

Good hygiene is the best prevention against *Pseudomonas* infection. Washing food carefully, drinking safe water, not tracking dirt from shoes into living spaces, and handwashing are helpful. In hospital settings, one should avoid the use of catheters, should change bandages often, and should clean equipment (such as ventilators, restrooms, and mops, and other cleaning supplies) where moist conditions are commonly found.

—Patricia Stanfill Edens, RN, PhD, FACHE and Michael A. Buratovich, PhD

Further Reading

Bennett, John E., et al., editors. *Mandell, Douglas, and Bennett's Principles and Practice of Infectious Diseases*. 9th ed., Elsevier, 2019.

de Sousa, Telma, et al. "Genomic and Metabolic Characteristics of the Pathogenicity in Pseudomonas aeruginosa." *International Journal of Molecular Sciences*, vol. 22, no. 23, 2021, p. 12892, doi:10.3390/ijms222312892.

Engleberg, N. Cary, et al. *Schaechter's Mechanisms of Microbial Disease*. 6th ed., Lippincott Williams & Wilkins, 2021.

St. Georgiev, Vassil. *Opportunistic Infections: Treatment and Prophylaxis*. Humana Press, 2003.

Wilson, Benda A., et al. *Bacterial Pathogenesis: A Molecular Approach*. 4th ed., ASM Press, 2019.

Streptococcal infections

Category: Diseases and conditions
Also known as: Strep infections, strep throat
Anatomy or system affected: Arms, back, bladder, brain, chest, circulatory system, ears, feet, genitals, hands, heart, joints, legs, lungs, muscles, neck, skin, throat, vagina
Specialties and related fields: Bacteriology, cardiology, critical care, dermatology, emergency medicine, family medicine, general surgery, microbiology, neurology, obstetrics, pediatrics, pulmonology
Definition: infections caused by bacteria belonging to the genus *Streptococcus*, such as strep throat, scarlet fever, endocarditis, impetigo, cellulitis, pneumonia, rheumatic fever, and necrotizing fasciitis

KEY TERMS

bacteremia: the presence of bacteria in the blood
capsule: a polysaccharide layer that lies outside the bacterial cell envelope
dextran: a complex branched polysaccharide derived from the condensation of glucose
endocarditis: inflammation of the endocarditis
pili: a hairlike appendage found on the surface of many bacteria
protein M: a virulence factor on the surface of certain species of *Streptococcus* pyogenes
virulence factor: cellular structures, molecules, and regulatory systems that enable microbial pathogens to colonize the host, evade its immune system, or obtain nutrition

INTRODUCTION

Streptococcal infections are caused by spherical, gram-positive streptococcal bacteria that reproduce in a twisted, chain-like fashion. These bacteria may be found in humans on the skin and respiratory, genitourinary, and gastrointestinal systems. Under certain conditions, these opportunist bacteria can cause minor to life-threatening infections.

CAUSES

Streptococcus bacteria are classified into Lancefield groups according to their cell wall carbohydrates. The American microbiologist Rebecca Lancefield developed this classification scheme in 1928. Streptococci are grouped into Lancefield groups A, B, C, D, etc. Some *Streptococcus* species do not possess Lancefield carbohydrates, including *Streptococcus pneumoniae* and viridians streptococci.

Several *Streptococcus* species cause human diseases. *Streptococcus pyogenes*, the only Group A *Streptococcus* (GAS), may be part of the normal flora of the throat, skin, and vagina. Scientists have identified more than one hundred different strains of GAS. *S. pyogenes* causes strep throat, tonsillitis, impetigo, cellulitis, and more serious diseases such as scarlet fever, glomerulonephritis, rheumatic heart disease, toxic shock syndrome, and bacteremia.

S. pyogenes have a host of virulence factors to cause disease. First, they are surrounded by a polysaccharide capsule that protects them from white blood cells. Second, on the surface of the capsule are several adhesins or molecules that help them stick to host cell surfaces, like *Streptococcus* fibronectin-binding protein and M protein. Third, they secrete an enzyme called "hyaluronidase" that degrades molecules that sick cells together, increasing their invasiveness. Fourth, they secrete toxins like streptolysin O and S and erythrogenic toxins that damage tissues and release potential food sources. Antibodies against M protein cross-react with heart valve tissue, causing acute rheumatic heart.

Streptococcus agalactiae (*S. agalactiae*) is a Group B *Streptococcus* (GBS). *S. agalactiae* is found in the intestine, vagina, and rectum of up to 35 percent of all healthy women with no symptoms. This organism, however, can be passed to a fetus during delivery and can result in a life-threatening infection in the newborn. *S. agalactiae* are surrounded by a polysaccharide capsule that protects them from white blood cells. These bacteria also have extensions of the cell called "pili" that they use to attach to host cells. They also make a toxin called "beta-hemolysin" that destroys red blood cells and destroys our tissues. *S. agalactiae* causes neonatal pneumonia, sepsis, meningitis, septic arthritis, and chorioamnionitis or cystitis in pregnant females.

Group D *Streptococcus* (GDS), also known as the *Streptococcus bovis* group, contains *S. gallolyticus*. This microorganism is part of the normal flora of the gastrointestinal tract. These bacteria can pass through the intestinal wall during episodes of intestinal bleeding caused by colorectal cancer. Once in the bloodstream, they can cause endocarditis. *S. gallolyticus* endocarditis is a reason to screen for colon cancer.

People who have a history of heart valve disease and have recently had dental work can suffer from viridans streptococcal endocarditis. *Streptococcus sanguinis* is the most common cause of endocarditis in this group. These microorganisms make a dextran capsule that binds tightly to fibrin-platelet aggregates on damaged heart valves. Viridans streptococci do not contain Lancefield carbohydrates, and many of these live in our mouths.

Streptococcus pneumoniae do not have Lancefield carbohydrates. These bacteria colonize the nasal cavities and sinuses of many people. Nevertheless, they can cause disease in people with weakened immune systems. People with no spleen, sickle cell disease, diabetes, cancer, human immunodeficiency virus (HIV) infection, or are very young or very old are at increased risk of *S. pneumoniae* infections. *S. pneumoniae* causes nose, sinus, and middle ear infections, mastoiditis, septic arthritis, bacteremia, endocarditis, and meningitis. *S. pneumoniae* are encapsulated by a polysaccharide layer that protects from white blood cells. They also have pili that they use to attach to host cells. Once attached to a surface, *S. pneumoniae* can secrete exopolysaccharides that cover the cells in the slime layer and form a biofilm. Biofilms hide the cells from the immune system and increase their resistance to antibiotics. These bacteria also secrete an IgA protease that degrades the primary protector of mucosal surfaces. Finally, they secrete an enzyme called "pneumolysin" that activates the complement system in the bloodstream, leading to significant tissue damage.

RISK FACTORS

The risk of streptococcal infections varies based on the strain. Generally, the risk is greater when the host has a weakened immune system or a chronic disease such as diabetes. Certain treatments such as chemotherapy or drugs such as steroids may increase the risk of infection. Teenagers are at risk for strep throat infections, and a higher incidence is seen in the early spring.

Older adults, infants, and persons with a chronic disease are more susceptible to *S. pneumoniae*. A pregnant woman can transmit *S. agalactiae* to her baby during delivery.

SYMPTOMS

Group A strep infections present with varied symptoms, depending on the type of infection. The most common condition is strep throat with symptoms such as an inflamed, red throat with white patches on the tonsils, difficulty swallowing, enlarged lymph nodes with a swollen or tender neck, headache, weakness, a loss of appetite, stomach pain, and a fever. Impetigo, a strep skin rash, develops as weepy red sores. However, serious illnesses, such as scarlet or rheumatic fever, postpartum fever, wound infection, and pneumonia may also result from a strep infection.

A life-threatening response to group A strep that occurs when bacteria enter the blood or lungs is invasive GAS disease. Two uncommon but damaging diseases are toxic shock syndrome and necrotizing fasciitis, which manifests as pain, swelling, fever, and redness at the site of the infection and destroys muscles and skin. Early toxic shock syndrome may occur with dizziness, fever, flu-like symptoms, and confusion, which could progress to shock symptoms, including a drop in blood pressure and a general failure of body organs—the host risks death in about 35 percent of toxic shock syndrome cases.

Endocarditis usually causes a fever and a new heart murmur that results from turbulent blood flow past the damaged heart valves. Vegetations or growths on the heart valves can detach and move through the bloodstream as septic emboli. Septic emboli from the left side of the heart cause strokes, and those from the right side cause pulmonary embolisms. These emboli can block small blood vessels in the fingertips, causing "splinter hemorrhages" under fingernails or flat, red lesions on the palms or soles of the feet called "Janeway lesions." Antibody-antigen complexes also lodge in blood vessels in the fingers and toes and cause painful, raised spots called "Osler's nodules" or in the retina of the eyes and cause Roth spots.

Individuals with meningitis present with the classic triad of fever, neck stiffness, and a headache.

Pneumonia is characterized by difficulty breathing (dyspnea), high heart rate (tachycardia), fatigue, and fever. On physical exam, there are decreased breath sounds and tactile fremitus (the chest tends to rattle when they speak). The chest has a dullness to percussion, suggesting consolidation. When listening to the breath sounds through a stethoscope, the sound of air moving is very clear, with crackling sounds as the air passes through liquid.

SCREENING AND DIAGNOSIS

Screening and diagnosis of streptococcal infections are based on physical examination, the presenting symptoms, and the results of laboratory tests. Strep throat is diagnosed with a rapid strep test or a swab of the infected person's throat mucosa. This simple test takes about twenty minutes to complete and is completed in the doctor's office. If there is a positive result, no further testing is needed. Suppose a negative result occurs in a person who has significant symptoms. In that case, a throat culture is sent to a lab for diagnosis, which occurs within twenty-four to forty-eight hours.

Group B strep is diagnosed by reviewing presenting symptoms and risk factors and a lab examination of blood or spinal fluid, or both. The diagnosis of infection with *S. pneumoniae* is confirmed through a chest X-ray, and a lab culture called an "optochin susceptibility test."

Endocarditis is usually identified with blood cultures. Catching the microorganism with a blood culture may require several attempts. Serological and imaging techniques diagnose pneumonia. Sputum cultures are a definitive way to diagnose pneumonia. Meningitis is diagnosed either from a cerebrospinal fluid culture or polymerase chain reaction tests that identify the organism's genomic deoxyribonucleic acid (DNA).

TREATMENT AND THERAPY

A thorough medical history will allow the health-care provider to assess risk factors when determining effective therapy. Generally, streptococcal infections can be effectively treated with appropriate oral or intravenous antibiotics. The primary classes of drug treatment include penicillin, cephalosporins, and erythromycins. Because of overuse, however, some antibiotics are no longer effective against strep infections. Persons with a sore throat and confirmed strep throat should not return to work or school until a minimum of twenty-four hours have passed after starting antibiotics.

S. pyogenes is still susceptible to penicillin, but oral amoxicillin is the preferred treatment. For those with an allergy to penicillin, cephalosporins like ceftriaxone or macrolide antibiotics like azithromycin are adequate alternatives. Penicillin or ampicillin treats most *S. agalactiae* infections, but penicillin resistance is a concern. In such cases, cefazolin or vancomycin are adequate alternative treatments.

S. pneumoniae has progressively acquired increased resistance to amoxicillin and ampicillin. Preferred treatments for infections with this organism include amoxicillin/clavulanic acid, third-generation cephalosporins like ceftriaxone or cefotaxime, or fluoroquinolones.

For endocarditis, intravenous vancomycin is the treatment of choice.

To avoid spreading the disease, one should wash the dishes and utensils of infected persons in hot, soapy water and keep them separate from other dishes and utensils. After a person has been treated with antibiotics, the patient should replace their toothbrush to prevent contamination. Also, one should encourage infected persons to cover their mouths while coughing or sneezing to minimize the spread of the disease by droplets.

PREVENTION AND OUTCOMES

Prevention of streptococcal infections includes frequent handwashing with soap and warm water or applying antiseptic hand sanitizer. One should minimize skin-to-skin contact with any infected person; wash hands thoroughly before preparing or eating foods and after shaking hands; keep hands away from face and mouth; and avoid sharing personal belongings such as toiletries, towels, or combs and brushes.

Adults age sixty-five years and older and those with a serious chronic disease should be encouraged to take the pneumococcal vaccine. About 30 percent of persons who contract pneumonia may get secondary bacteremia, resulting in death. A pregnant woman who is a carrier of GBS should be tested in the ninth month of her pregnancy to ensure the fetus receives appropriate antibiotics during labor.

To avoid skin infections, one should clean and apply a protective bandage with antibiotic ointment to all cuts and scrapes. Good nutrition, adequate sleep, and regular exercise can help build a resistant immune system.

—Marylane Wade Koch, MSN, RN and Michael A. Buratovich, PhD

Further Reading

Centers for Disease Control and Prevention. "Group A Streptococcal (GAS) Disease." www.cdc.gov/ncidod/dbmd/diseaseinfo/groupastreptococcal.

Newberger, Ryan, and Vikas Gupta. "Streptococcus Group A." *StatPearls*. StatPearls Publishing, 11 Aug. 2021.

Parker, James N., and Philip M. Parker, editors. *The Official Patient's Sourcebook on "Streptococcus pneumoniae" Infections*. Icon Health, 2002.

Pechère, Jean Claude, and Edward L. Kaplan, editors. *Streptococcal Pharyngitis: Optimal Management*. S. Karger, 2004.

Shapira, Raz, et al. "Streptococcus Gallolyticus Endocarditis on a Prosthetic Tricuspid Valve: A Case Report and Review of the Literature." *Journal of Medical Case Reports*, vol. 15, no. 1, 2021, p. 528, doi:10.1186/s13256-021-03125-5.

Vincent, Miriam T. "Sore Throat-Strep Throat? When to Worry." *Pediatrics for Parents*, vol. 21, no. 8, 1 Aug. 2004, pp. 11-12.

Zhu, Yao, and Xin-Zhu Lin. "Updates in Prevention Policies of Early-Onset Group B Streptococcal Infection in Newborns." *Pediatrics and Neonatology*, vol. 62, no. 5, 2021, pp. 465-75, doi:10.1016/j.pedneo.2021.05.007.

STAPHYLOCOCCAL INFECTIONS

Category: Diseases and disorders

Also known as: Staph infections

Anatomy or system affected: Blood, circulatory system, gastrointestinal system, musculoskeletal system, nervous system, respiratory system, urinary system

Specialties and related fields: Bacteriology, cardiology, dermatology, epidemiology, general surgery, internal medicine, microbiology, orthopedics, urology

Definition: infections caused by bacteria from the genus *Staphylococcus*

KEY TERMS

antistaphylococcal penicillins: a group of penicillin antibiotics that kill methicillin-sensitive *Staphylococcus aureus* strains and other staphylococci

beta-lactam antibiotics: antibacterial drugs that inhibit cell wall synthesis in bacteria and contain a beta-lactam ring in their structure

methicillin-resistant Staphylococcus aureus: also known as MRSA

Panton-Valentine leucocidin: a toxin produced by some strains of *Staphylococcus aureus* that destroys white blood cells and damaged tissues

superantigens: a class of antigens that result in excessive activation of the immune system

toxic shock syndrome toxin: a superantigen produced by some *Staphylococcus aureus* and *Streptococcus pyogenes* isolates that causes toxic shock syndrome

CAUSES AND SYMPTOMS

Under the microscope, staphylococci bacteria grow in irregular, grapelike clusters from which they derive their name. Individually, the organisms are spherical and appear purple (positive) when stained with the Gram technique. Staphylococci are found predominantly living on the skin and mucous membranes of mammals and birds. Usually, they exist in a benignly symbiotic relationship with their hosts. When the barrier imposed by either the skin or mucous membranes is breached, however, staphylococci may then cause disease, assisted by various enzymes and toxins that they can manufacture. Many strains of staphylococci, such as *Staphylococcus aureus*, produce coagulase, which catalyzes the formation of a clot from fibrinogen proteins in the blood. Hyaluronidases, lipases, and other proteolytic enzymes carve out a cavity within the clot, covered by a coagulase-generated fibrin coat, and an abscess is formed. Abscess formation is one of the hallmarks of staphylococcal infection and can be found in virtually any organ in the body because of local invasion or spread to the site via the bloodstream. Staphylococcal strains can produce a variety of proteins called "exotoxins" that have significant roles in determining the type of illness that results. Superantigens, such as toxic shock syndrome toxin-1 (TSST-1), target the circulatory system and can markedly lower blood pressure. TSST-1 producing strains have caused disease by growing in vaginal tampons and simultaneously causing postsurgical wound infections. Food poisoning results after ingesting food that is contaminated with staphylococcal strains that produce enterotoxins. The toxins are produced in the food after contamination from the colonized noses or infected skin of food handlers through sneezing or direct contact. The enterotoxins are relatively heat-stable and do not result in any unusual taste, odor, or appearance of the food. Abdominal cramps, nausea, vomiting, and diarrhea occur one to six hours after ingesting the contaminated food containing the toxin. Toxic shock syndrome is not an actual infection, as it is the preformed toxin that produces the illness. Exfoliative toxin is produced by the strains causing scalded skin syndrome. The illness is manifested by fever and reddened skin that subsequently peels off.

Staphylococci can form a variety of cytotoxins that damage the membranes of bodily tissues. These toxins can destroy red and white blood cells as well as organ cells. White blood cell-killing toxins from staphylococci were first reported in 1932, and the Panton-Valentine leukocidin (PVL) was named in honor of these scientists. PVL is produced by strains of *S. aureus* that cause skin and soft tissue infection and pneumonia in the community (outside the hospital). *S. aureus* is the preeminent pathogen causing infection in hospitalized patients. It is the most common cause of surgical wound infections. It is joined by another species, *S. epidermidis*, which can produce slime, enabling the bacteria to adhere to the surfaces of medical devices such as vascular catheters, central nervous system shunts, artificial heart valves, and prosthetic joints. Together, these two species account for many hospital-acquired infections. Infection of skeletal muscle and adjacent tissues is called "pyomyositis" or "necrotizing fasciitis." This type of infection has been common in developing countries with tropical climates. However, since the 1970s, it has been seen more frequently in modern countries with temperate climates. The pathogenesis of pyomyositis is not entirely understood. Some cases follow trauma, and there are several risk factors, such as intravenous drug abuse, human immunodeficiency virus (HIV) infection, and skin diseases. Other cases occur in healthy individuals without any apparent risk factor. Recently, virulent strains of community-acquired methicillin-resistant *S. aureus* (MRSA) have caused pyomyositis in the United States. *S. saprophyticus* is an important cause of urinary tract infections in sexually active young women. Bladder infections caused by this organism usually produce painful urination (dysuria), urinary urgency, pain in the lower abdomen below the navel, and, occasionally, blood in the urine.

TREATMENT AND THERAPY

Specific treatment of a staphylococcal infection hinges on administering an effective antibiotic. Staphylococci are resistant to a host of regularly prescribed antibiotics. Therefore, drug developers discovered a group of antistaphylococcal penicillins to treat staphylococcal infections specifically. These drugs include methicillin (which was removed from human use because it causes kidney damage), nafcillin, oxacillin, and dicloxacillin. Unfortunately, some *Staphylococcus aureus* strains acquired a gene called *mecA* from nonpathogenic staphylococci. The *mecA* gene encodes an abnormally low-affinity cell wall-synthesizing enzyme, PBP2a, that helps the organism grow and divide in the presence of antistaphylococcal penicillins and other antibiotics like them. *Staphylococcus aureus* strains with the *mecA* gene are called "methicillin-resistant *Staphylococcus aureus*" or "MRSA."

MRSA is resistant to beta-lactam antibiotics, including penicillin, methicillin, and other related commonly used antibiotics. Vancomycin, an antibiotic developed in the 1950s, has remained effective for nearly all strains. Still, it must be given intravenously and has some serious potential side effects. Some newer antibiotics, daptomycin, dalfopristin/quinupristin, ceftaroline, linezolid, and tedizolid, are effective against MRSA and used successfully to treat infections.

Antibiotics are not the only measure necessary to cure these infections. Surgical or percutaneous catheter drainage of abscesses, surgical debridement of dead tissue, and the removal of medical devices or prostheses are often necessary. Other medical supportive modalities, such as fluid replacement, vasopressors, or mechanical ventilation, may be required.

PERSPECTIVE AND PROSPECTS

The Centers for Disease Control and Prevention (CDC) collaborates with other medical organizations to develop and promote strategies to reduce the transmission of staphylococci, primarily MRSA, in both health care and community settings. The CDC has also launched a campaign to prevent antimicrobial resistance by educating both the public and health-care providers about unnecessary and inappropriate antibiotic usage. Legislative efforts have been directed at eliminating the use of antibiotics in animal feed to prevent resistance.

Newer antimicrobial agents continue to be developed, but the stream has slowed. Molecular-based treatments directed toward toxins and other virulence factors are being actively pursued.

—H. Bradford Hawley, MD and Michael A. Buratovich, PhD

Futher Reading

Crossley, Kent B., and Gordon L. Archer, editors. *The Staphylococci in Human Disease*. Churchill Livingston, 1997.

Kasper, Dennis L., et al., editors. *Harrison's Principles of Internal Medicine*. 16th ed., McGraw-Hill, 2005.

Koneman, Elmer W. *The Other End of the Microscope: The Bacteria Tell Their Own Story*. ASM Press, 2002.

McCarthy, Matt. *Superbugs: The Race to Stop an Epidemic*. Avery, 2019.

METHICILLIN-RESISTANT STAPH INFECTION

Category: Diseases and conditions

Also known as: CA-MRSA, HA-MRSA, health-care-associated MRSA, methicillin-resistant *Staphylococcus aureus* community-acquired MRSA, methicillin-resistant *Staphylococcus aureus* infection, methicillin-resistant *Staphylococcus aureus* nosocomial MRSA

Anatomy or system affected: Blood, bones, circulatory system, heart, lungs, respiratory system, skin

Specialties and related fields: Bacteriology, cardiology, dermatology, immunology, infectious disease, internal medicine, microbiology, orthopedics, pathology, pharmacology, public health, pulmonology

Definition: a *Staphylococcus aureus* strain that is genetically distinct from other *S. aureus* strains that are responsible for several difficult-to-treat infections in humans

KEY TERMS

beta-lactam antibiotics: antibiotics that contain a beta-lactam ring in their molecular structure and include penicillin derivatives, cephalosporins, and cephamycins, monobactams, carbapenems, and carbacephems

beta-lactamases: enzymes produced by bacteria that provide multi-resistance to ß-lactam antibiotics such as penicillins, cephalosporins, cephamycins, monobactams, and carbapenems

penicillin-binding proteins: enzymes involved in peptidoglycan biosynthesis and contribute essential roles in bacterial cell wall biosynthesis

peptidoglycan: a polymer consisting of sugars and amino acids that forms a mesh-like peptidoglycan layer outside the plasma membrane of most bacteria, forming the cell wall

INTRODUCTION

Staphylococcus aureus is a gram-positive, spherical bacteria that grow in clusters. Gram-positive bacteria have a thick cell wall composed of peptidoglycan, and *Staphylococcus aureus* stains purple in a Gram stain.

Methicillin-resistant *Staphylococcus aureus* (MRSA) is a genetically distinct strain of the *Staphylococcus aureus* (*S. aureus*). The bacterium can affect the skin, blood, bones, heart, or lungs. Some people are colonized with MRSA but show no symptoms. People infected with MRSA show symptoms.

There are two types of MRSA infection: community-acquired (CA) and hospital-acquired (HA). People who have community-acquired MRSA infection were infected outside a hospital setting (such as a dormitory). Nosocomial MRSA infections, usually caused by HA-MRSA, occur in hospital settings.

MRSA has acquired the *mec* gene, which encodes a cell wall synthesizing enzyme called "penicillin-binding protein 2a" (PBP2a) that is not inhibited by most beta-lactam antibiotics. Penicillin-binding proteins are enzymes embedded in the bacterial membrane that catalyze peptidoglycan synthesis, the main component of bacteria cell walls. MRSA also expresses several beta-lactamase enzymes that degrade many beta-lactam antibiotics.

MRSA makes several virulence factors that help them colonize human tissue. First, they make various cytotoxins. Alpha-toxin and Panton-Valentine leucocidin kill neutrophils, which are white blood

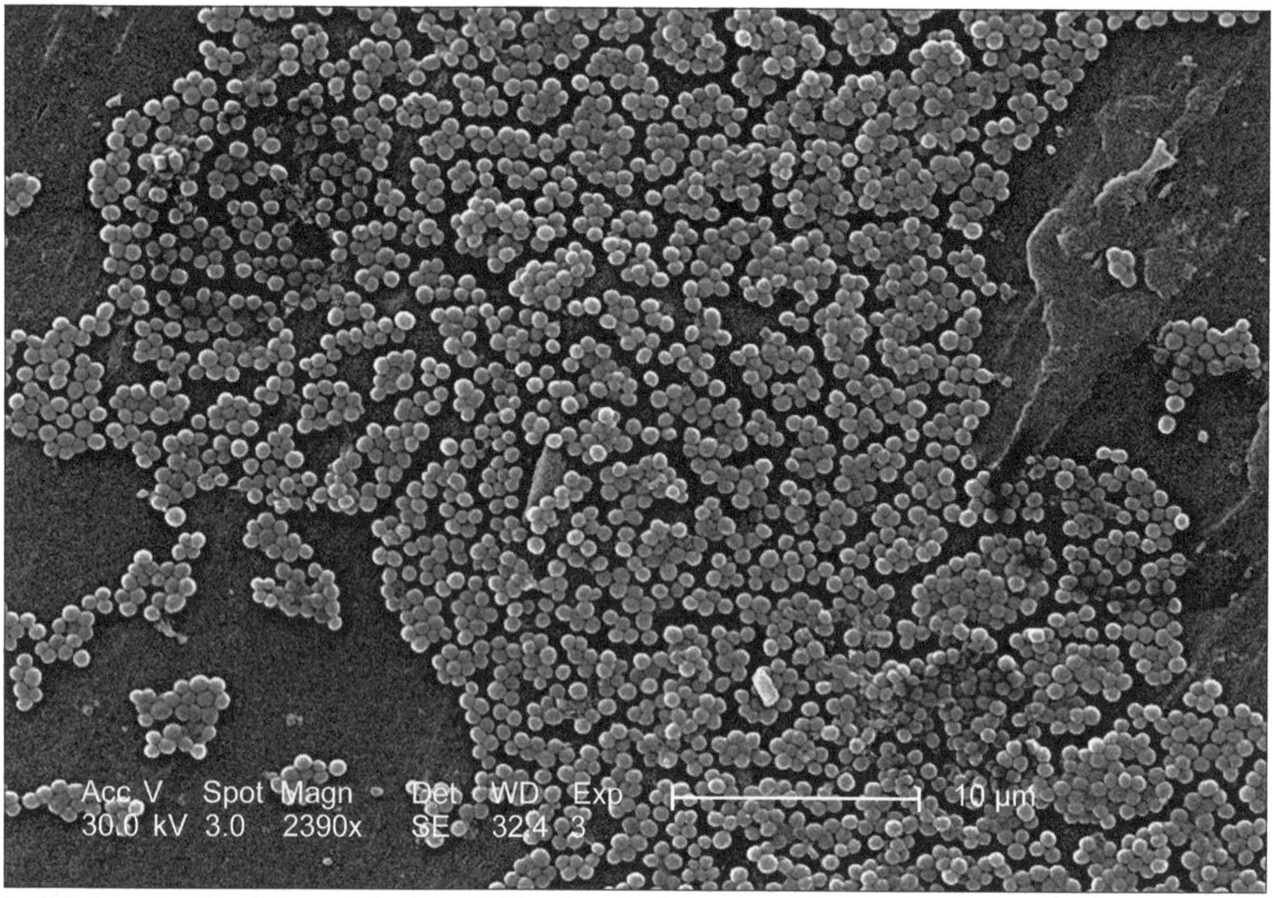

A colorized scanning electron micrograph showing methicillin-resistant Staphylococcus aureus *bacteria at 2390x magnification. Photo by Jeff Hageman, M.H.S./Janice Haney Carr, CDC, via Wikimedia Commons.*

cells that travel to the site of infection to destroy the bacteria. Second, MRSA makes hemolysins that poke holes in red blood cells, macrophages, and lymphocytes. Third, they secrete superantigens like toxic shock syndrome toxin-1 (TSST-1) that stimulate T lymphocytes to release excessive cytokines, causing uncontrolled inflammation. Fourth, some strains produce epidermolytic toxins A and B that degrade the attachments between the epidermis and the dermis. These toxins cause extensive skin blistering. Fifth, they make coagulase, which activates thrombin and activates coagulation. MRSA surrounds itself with fibrin and hides from the immune system. Sixth, MRSA makes a surface protein called "protein A." Protein A binds antibodies and inactivates them, preventing white blood cells from getting their pseudopods around them. Seventh, MRSA can secrete a film of extracellular polysaccharides surrounding the cells with a slime coat called a "biofilm." Biofilms adhere MRSA to surfaces like catheters, implants, and other surfaces, hide them from the immune system and diminish their antibiotic susceptibility.

Some MRSA strains express enterotoxins that cause food poisoning.

CAUSES

An MRSA infection can spread through several mechanisms, including from contaminated surfaces, from person to person, and from one area of the body to another. Skin trauma introduces MRSA into the lower layers of the skin. MRSA can colonize medical implants, such as intravenous lines, PICC (peripherally inserted central catheter) lines, central lines, urinary catheters, and endotracheal tubes and use them to colonize tissues. Open sores are also entry points for MRSA. Intravenous drug use also increases the risk of MRSA endocarditis. Hospital patients on a ventilator for longer than 48 hours are at increased risk of developing MRSA pneumonia.

RISK FACTORS

The following factors increase the chance of community-acquired infection: impaired immunity, sharing crowded spaces (such as dormitories and locker rooms), using intravenous drugs, serious illness, exposure to animals (as pet owners, veterinarians, and pig farmers, for example), using antibiotics, having a chronic skin disorder, and past MRSA infection. Also at higher risk are young children, athletes, prisoners, and military personnel.

For nosocomial infection, the risk factors are impaired immunity, exposure to a hospital or clinical setting, advanced age, chronic illness, using antibiotics, having a wound, living in a long-term care center, and having an indwelling medical device (such as a feeding tube, endotracheal tube, or intravenous catheter). Also, men are at higher risk.

SYMPTOMS

The symptoms of MRSA skin infections usually include erythema (superficial reddening of the skin) and localized swelling and warmth. Hair follicle infections cause folliculitis. If the infection migrates further into the skin, it forms a boil or furuncle. Furuncles may drain pus, blood, or an amber-colored liquid. Furuncles that cluster or fuse are called "carbuncles." Scalded staphylococcal skin syndrome (SSSS) is a skin infection characterized by a fever, rash, and fluid-filled blisters. It results from epidermolytic toxins that detach the epidermis from the dermis. This disease typically occurs in infants and usually resolves within a few weeks.

There are three types of impetigo: (1) nonbullous, (2) bullous, and (3) ecthyma. Nonbullous impetigo is the most common and appears as red bumps that blister—these blisters rupture, ooze, and crust over with a yellow scab. Bullous impetigo forms on the trunk and limbs and are usually not painful. They form large blisters that rupture and scab over with a thin, brown crust. Ecthyma impetigo is a painful, deeper nonbullous impetigo that evolves into yellow scabs. The blisters are often itchy and painful. A fever is rare.

Toxic shock syndrome is caused by the overactivation of T cells by superantigens. The hyperstimulated T cells overexpress cytokines, creating a cytokine storm. The cytokine storm results in fever, a sunburn-like rash, low blood pressure, and diminished blood flow through the organs.

Cellulitis is a skin infection characterized by a swollen, warm, red area with poorly defined borders that spreads quickly. Fever and chills with enlarged lymph nodes also accompany this condition. Abscesses are a collection of subcutaneous pus that is walled off.

MRSA endocarditis causes fever, chills, and fatigue. It also causes heart murmurs, splinter hemorrhages under the fingernails, and Janeway lesions on the palms or soles of the feet.

Staphylococcal food poisoning has a rapid onset and resolution. It causes vomiting, diarrhea vomiting, and abdominal pain.

Staphylococcal pneumonia causes shortness of breath, fast heart rate, and high breath rate.

SCREENING AND DIAGNOSIS

A doctor will ask about symptoms and medical history and perform a physical exam. Tests may include cultures, blood tests, urine tests, and a skin biopsy (removal of a skin sample to test for infection).

Several rapid detection tests are commercially available for MRSA. Chromogenic agar and bacteriophage techniques are two ways to identify MRSA from cultured material rapidly. Chromogenic agars incorporate antibiotics into the agar to identify MRSA from primary isolation plates within twenty-four and forty-eight hours. The KeyPath MRSA/MSSA Blood Culture Test uses MRSA-specific bacteriophages (viruses that attack bacteria) to detect and distinguish between MRSA and methicillin-susceptible S. aureus (MSSA) in blood cultures within five hours.

TREATMENT AND THERAPY

Treatment options include antibiotics prescribed to kill the bacteria and incision and drainage of an abscess. The doctor (but not the patient) opens the abscess and allows the fluid to drain. Another treatment is cleansing the skin. To treat the infection and keep it from spreading, one should wash skin with an antibacterial cleanser, apply an antibiotic, and cover skin with a sterile dressing.

The preferred antibiotic treatment for MRSA skin infections is vancomycin or daptomycin. Both drugs are given intravenously. Oral drugs for MRSA include oxazolidinones (linezolid and tedizolid), the fluoroquinolone delafloxacin, and the tetracycline omadacycline. Intravenous formulations of these antibiotics are also available in hospital settings.

Other intravenous agents that work well if MRSA strains are resistant to vancomycin are the fifth-generation cephalosporins ceftaroline and ceftobiprole. These are short-acting medications. Longer-acting glycopeptide medicines include dalbavancin, oritavancin, and telavancin.

PREVENTION AND OUTCOMES

To help reduce the chance of getting an MRSA infection, one should thoroughly wash hands with soap and water, keep cuts and wounds clean and covered until healed, and avoid contact with other people's wounds and materials contaminated by wounds. Hospitalized persons' visitors and health-care workers should wear special clothing and gloves to prevent spreading the infection to others.

—Krisha McCoy and Michael A. Buratovich, PhD

Further Reading

Ahmad-Mansour, Nour, et al. "*Staphylococcus aureus* Toxins: An Update on Their Pathogenic Properties and Potential Treatments." *Toxins*, vol. 13, no. 10, 2021, p. 677, doi:10.3390/toxins13100677.

Benjamin, Ivor, et al., editors. *Andreoli and Carpenter's Cecil Essentials of Medicine*. 9th ed., Saunders, 2015.

Centers for Disease Control and Prevention. "Methicillin-resistant *Staphylococcus aureus* (MRSA) Infections." 5 Feb. 2019, www.cdc.gov/mrsa/index.html. Accessed 8 Dec. 2021.

Crossley, Kent B., Kimberly K. Jefferson, and Gordon L. Archer, editors. *Staphylococci in Human Disease*. Wiley, 2009.

Singer, Adam J., and David A Talan. "Management of Skin Abscesses in the Era of Methicillin-Resistant *Staphylococcus aureus*." *The New England Journal of Medicine*, vol. 370, no. 11, 2014, pp. 1039-47, doi:10.1056/NEJMra1212788.

Siddiqui, Abdul H., and Janak Koirala. "Methicillin-Resistant *Staphylococcus aureus*." *StatPearls*. StatPearls Publishing, 19 July 2021.

MYCOBACTERIAL INFECTIONS

Category: Diseases and conditions
Also known as: Tuberculosis
Anatomy or system affected: All
Specialties and related fields: Bacteriology, immunology, internal medicine, microbiology, orthopedics, pathology, pharmacology, public health, pulmonary medicine, radiology
Definition: chronic or acute systemic infections that are spread by a common type of bacteria in the environment

KEY TERMS

Addison disease: a condition characterized by adrenal insufficiency *cord factor*: trehalose dimycolate is the primary glycolipid found on the exterior of *Mycobacterium tuberculosis* cells similar species

cytokines: small proteins, such as interferon, interleukin, and growth factors, that are secreted by certain cells of the immune system and influence other cells

Ghon complex: a nonpathognomonic radiographic finding on a chest X-ray that is significant for pulmonary infection of tuberculosis

granuloma: an aggregation of macrophages that forms in response to chronic inflammation

Pott disease: tuberculosis spondylitis; a rare infectious disease of the spine caused by an extraspinal infection that involves multiple vertebrae, causing osteomyelitis and arthritis

Ranke complex: the combination of late fibrocalcific lesions of the lung and lymph node, which evolved from the Ghon complex

DEFINITION

Mycobacterial infections are chronic or acute systemic infections spread by a common type of bacteria in the environment, especially aquatic environments. Mycobacterial infections include tuberculosis; atypical mycobacterial infections include skin, bone, soft tissue, lymph nodes, and gastrointestinal tract; they also include lung disease and septic arthritis.

CAUSES

Mycobacterial infections are caused by one species within the gram-positive, aerobic bacteria family called "Mycobacteriaceae," which belongs to the Actinomycetales order. Specifically, *Mycobacterium tuberculosis* causes tuberculosis, *M. kansasii* causes lung disease, and *M. ulcerans* and *M. marinum* cause skin infections. *M. avium* subspecies *intracellulare* causes lung disease but primarily affects the lungs of those with acquired immunodeficiency syndrome; *M. avium* subspecies *intracellulare* also causes ulcers, diarrhea, fever, pustules, nodules, lesions, and swollen lymph nodes.

Mycobacterium tuberculosis (*M. tuberculosis*) causes tuberculosis, which initially infects the lungs but can disseminate to any organ. The usual means of transmission of *M. tuberculosis* is aerosol. However, the consumption of raw milk was a significant source of infection for the advent of pasteurization.

Mycobacteria are rod-shaped bacteria with thick, exceedingly complex cell walls. Gram-positive bacteria have thick cell walls made of peptidoglycan. Mycobacteria, however, have peptidoglycan with an array of sugar polymers (arabinogalactans), sugars linked to lipids (glycolipids), waxes, fatty acids called "mycolic acids," and lipids. The mycobacterial cell wall protects the organism and influences its interaction with the immune system.

When inhaled, *M. tuberculosis* adheres to the airway epithelium or sojourns to the alveoli of the lung. If the organism adheres, the airway cilia move it to the pharynx, where it is swallowed. *M. tuberculosis* dies in the acid of the stomach. However, if it makes its way to the alveoli, it encounters an alveolar macrophage, which binds it and phagocytoses it.

Inside the alveolar macrophage, *M. tuberculosis* reprograms the white cell so that it cannot kill it. The microorganism lives and grows within the alveolar macrophage. These mycobacteria survive and multiply within the macrophage. About three weeks after the initial infection, a glycolipid on the surface of *M. tuberculosis* called "cord factor" elicits the release of several cytokines that attract macrophages and helper T cells. The macrophages surround the infected macrophage and cordon it off by forming a granuloma. As macrophages present T-helper cells with mycobacterial antigens, they release cytokines that summon even more macrophages. Dead cells, bacteria, and macrophages are located at the center of the granuloma, and surrounding this core are fused macrophages (multinucleated giant or Langerhans cells) and helper T cells. Granulomas in tuberculosis have dead tissue at their cores that looks cheese-like. Therefore, they are called "caseating." Areas in the lung with caseation are called "Ghon focus." If tuberculosis also involves the local lymph nodes and the lung, the affected areas are collectively known as "Ghon complexes." Ghon complexes usually occur in the middle and lower lobes of the lung and between the pleural membranes surrounding the lungs and the body wall (subpleural). Long-lasting Ghon complexes can calcify and scar to form "Ranke complexes." If the granulomas cannot contain the bacteria, the infected macrophages move throughout the lung, further damaging it.

The immune system destroys the mycobacteria, and the lung clears in some cases. However, the mycobacteria are surrounded in other cases, but they remain alive. Should the immune system become weakened, the Ghon complexes can reactivate, burst open, and the cells can disseminate to any organ in the body. Mycobacteria from compromised Ghon complexes usually spread to the upper lobes of the lung since there is so much oxygen there. However, the immune system releases more cytokines to recruit more immune cells to fight the infection, which causes more lung damage. Severe tuberculosis even causes cavity formation within the lung. The infection within the lung erodes blood vessels, causing the infected person to cough up blood (hemoptysis).

Blood vessel erosion allows the mycobacteria to disseminate to any organ in the body. If the mycobacteria go to the kidney, they cause pus in the urine (pyuria). If they spread to the adrenal glands, they cause Addison disease. If they disseminate to the vertebral column, they cause Pott disease. Hematogenous spread to the liver causes hepatitis,

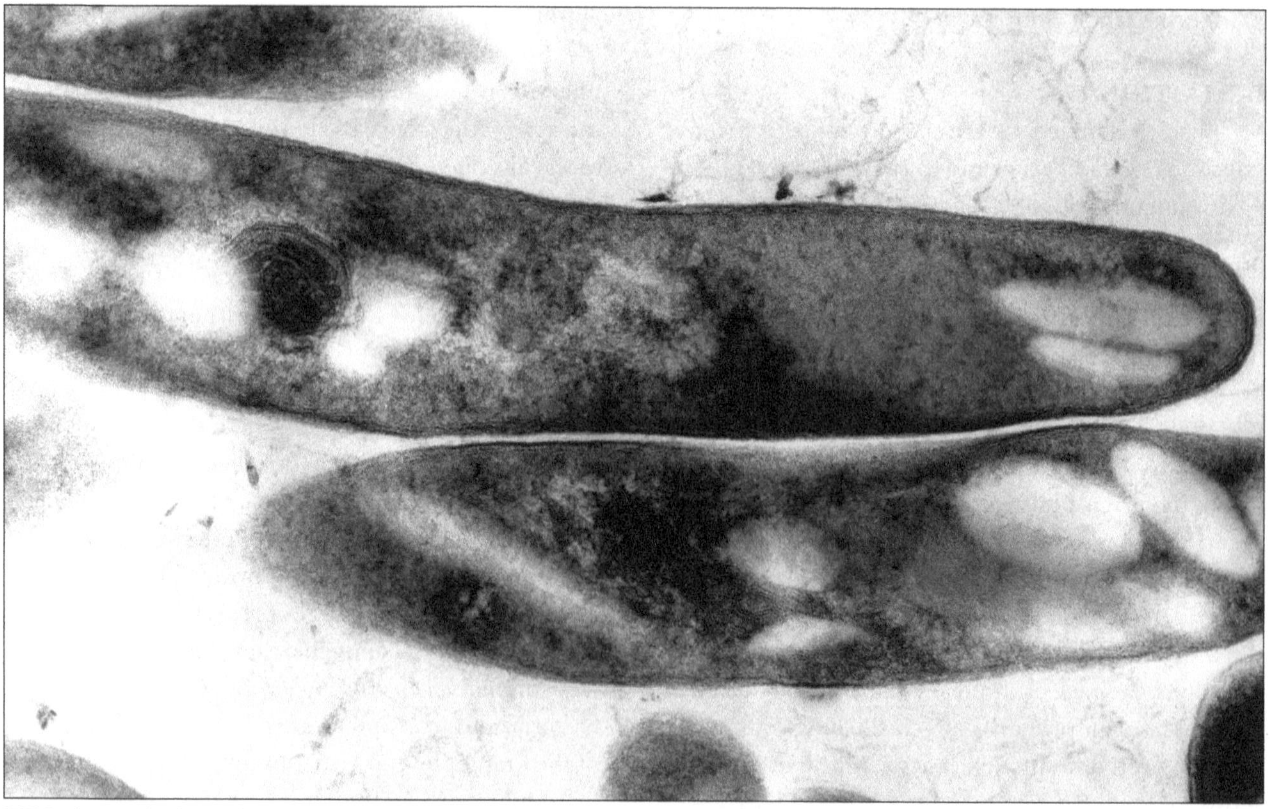

TEM micrograph of M. tuberculosis. *Photo via CDC/Wikimedia Commons. [Public domain.]*

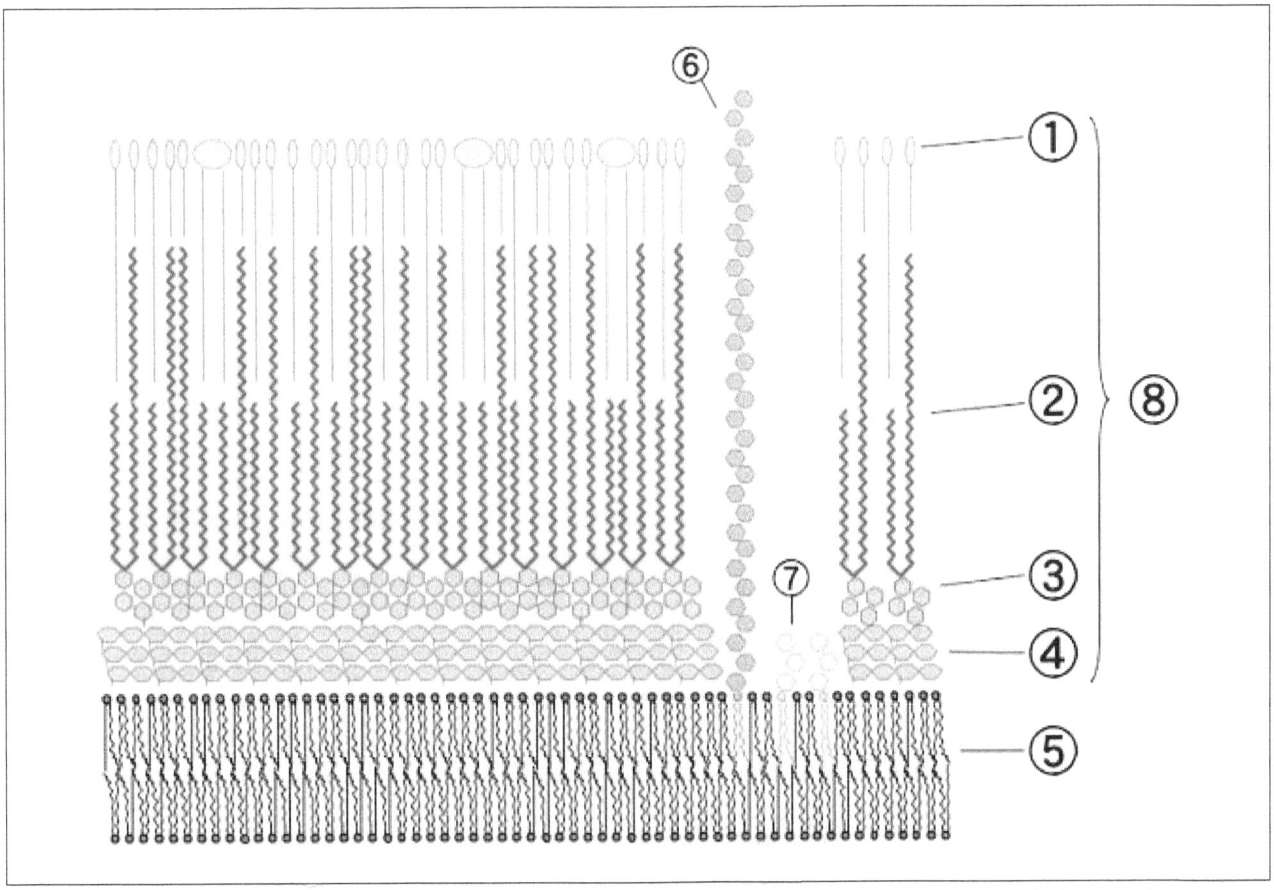

Schematic diagram of Mycobacterial cell wall. 1. outer lipids; 2. mycolic acid; 3. polysaccharides (arabinogalactan); 4. peptidoglycan; 5. plasma membrane; 6. lipoarabinomannan (LAM); 7. phosphatidylinositol mannoside; 8. cell wall skeleton. By Y tambe, own work, via Wikimedia Commons.

and if *M. tuberculosis* spreads to the cervical lymph nodes, it causes lymphadenitis. Mycobacteria can also spread to the brain and cause meningitis, the joints and cause mycobacterial arthritis, and the long bones, causing osteomyelitis.

RISK FACTORS

Exposure to contaminated water sources is a major risk factor for mycobacterial infection. Other risk factors are having a preexisting lung disease, having an impaired immune system, undergoing surgery, and having an organ transplant. Also at higher risk are persons with human immunodeficiency virus (HIV) infection and persons living in unsanitary conditions.

SYMPTOMS

Persons with HIV who have a mycobacterial infection often show a cough, weight loss, chest pain, breathlessness, hemoptysis, night sweats, chills, and fever. Persons with a mycobacterial skin infection often have reddish raised nodules on the elbows, feet, knees, and hands. Pain in the joints, tendons, and bones can be signs of tenosynovitis and infections that could lead to arthritis and osteomyelitis. Enlarged lymph nodes are often a symptom of persons with mycobacterial infection of the lymph nodes. Signs of tuberculosis include fever and chills, rapid breathing, night sweats, pale skin, prolonged coughing that produces bloody sputum, weight loss, loss of appetite, and pleurisy.

SCREENING AND DIAGNOSIS

Screening methods include blood, bone marrow, lymph node, sputum, and stool cultures. Traditional methods of bacteria analysis, including growth rate and pigmentation studies and acid-fast staining, confirm the identity of the bacteria. A bacterial-species-specific polymerase chain reaction analysis for screening assays has been developed. Deoxyribonucleic acid (DNA) fingerprinting and DNA sequencing techniques are often used for bacteria identification. A tissue biopsy is useful for diagnosis, and X-rays or computed tomography scans may be used to detect internal infection sites.

TREATMENT AND THERAPY

Tuberculosis and other mycobacterial diseases are always treated with combination treatments. There are two phases for *M. tuberculosis* treatment: an intensive phase (four drugs are administered for two months), followed by a continuation phase, during which two or three drugs are administered for two to seven months.

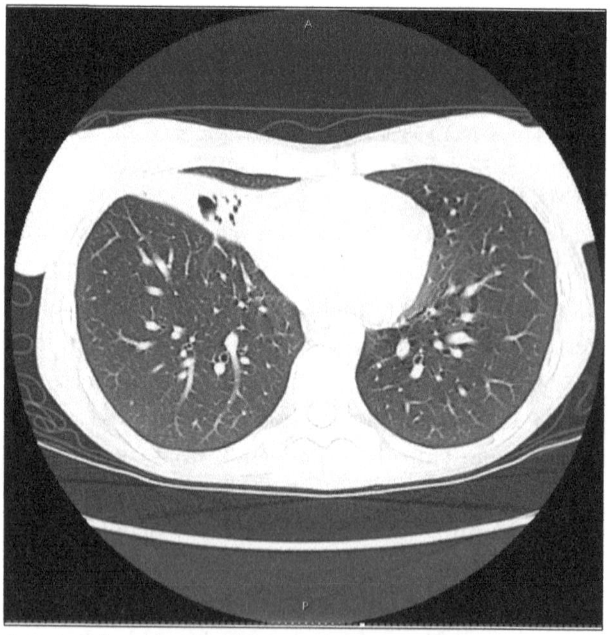

A computed tomography (CT) scan of a patient with Lady Windemere syndrome. Damage can be seen in the right middle lobe atelectasis. Photo by Samir via Wikimedia Commons.

There are two main treatment regimens. The traditional regimen goes for six months. The traditional regimen has an intensive two-month phase and a four-month continuation phase. During the first two months, the patient takes isoniazid, rifampin, pyrazinamide, and ethambutol (RIPE). The continuation phase consists of two drugs, isoniazid, and rifampin, taken for at least four additional months, for a total of at least six months.

The shortened four-month regimen has an intensive eight-week phase and a continuation phase of nine weeks. Clinical trials have established that this shortened regimen is as effective as the six-month regimen. The shortened regimen begins with eight weeks of rifapentine, isoniazid, pyrazinamide, and moxifloxacin once a day. The nine-week continuation phase consists of rifapentine, isoniazid, and moxifloxacin administered once daily.

Multidrug-resistant (MDR) and extremely drug-resistant (XDR) strains require specialized regimens designed by infectious disease specialists.

PREVENTION AND OUTCOMES

To decrease the chance of getting a mycobacterial infection, one should avoid stagnant aquatic environments and should avoid contact with fish and cattle. Chlorination of swimming pools is also an effective prevention method because chlorine kills the bacteria that can cause these infections.

—Jeanne L. Kuhler, PhD and Michael A. Buratovich, PhD

Further Reading

Krishnan, Vidya. *Phantom Plague: How Tuberculosis Shaped History*. PublicAffairs, 2022.

Madigan, Michael T., et al. *Brock Biology of Microorganisms*. 15th ed., Pearson/Prentice Hall, 2017.

Ryan, Frank. *Tuberculosis: The Greatest Story Never Told*. Swift Publishers, 2019.

Schlossberg, David, editor. *Tuberculosis and Nontuberculous Mycobacterial Infections*. 7th ed., ASM Press, 2017.

Shah, Maunank, and Susan E Dorman. "Latent Tuberculosis Infection." *The New England Journal of Medicine*, vol. 385, no 24, 2021, pp. 2271-80, doi:10.1056/NEJMcp2108501.

MYCOPLASMA

Category: Pathogens

Anatomy or system affected: Blood, blood vessels, brain, bronchial tubes, endocardium, heart, lungs, nose, pharynx, respiratory system, sinuses, urogenital system

Specialties and related fields: Bacteriology, cardiology, family practice, internal medicine, microbiology, neurology, pediatrics, respiratory medicine, urology

Definition: *Mycoplasma* is a bacterial genus belonging to the class Mollicutes; several *Mycoplasma* species have been established as human pathogens, including *pneumoniae*, *hominis*, and *genitalium*

KEY TERMS

beta-lactam: antibiotics that contain a beta-lactam ring in their molecular structure, including penicillin derivatives, cephalosporins, and cephamycins, monobactams, carbapenems, and carbacephems

cell wall: a peptidoglycan layer that surrounds the cell, composed of disaccharides and amino acids that gives bacteria structural support

cervicitis: inflammation of the cervix

walking pneumonia: a nonmedical term for a mild case of pneumonia that is also called "atypical pneumonia" and is caused by bacteria or viruses; often by *Mycoplasma pneumonia*

NATURAL HABITAT AND FEATURES

Mycoplasma has been isolated from humans and animals, including cows, dogs, cats, pigs, horses, poultry, sheep, goats, and small rodents. *Mycoplasma* is the smallest bacteria that can live independently. It has a small genome size in the lower complexity limit necessary for self-replicating organisms. *Mycoplasma* can survive in the presence or absence of oxygen. These microbes are transmitted through contaminated blood or inhalation.

Mycoplasma lacks a cell wall, so it does not react in a Gram stain and is not susceptible to antibiotics that target cell walls. It has a specialized organelle, or tip, providing motility and mediating bacterial interactions with its host cells. Adherence proteins allow *Mycoplasma* attachment to cells lining the respiratory and genitourinary tracts, acting as a parasite on the surface of its host cells and using their precursors to produce its genetic material. Some species (*pneumoniae*, *genitalium*, *fermentans*, *penetrans*, and *gallisepticum*, a poultry pathogen) can invade host cells and live intracellularly.

Mycoplasma produces hydrogen peroxide and superoxide, substances that injure the mucosal surface. Mucosal damage activates inflammatory mediators, which is associated with the infectious process. *Mycoplasma* is challenging to grow in culture. Consequently, bacterial identification mainly depends on molecular-biochemical techniques.

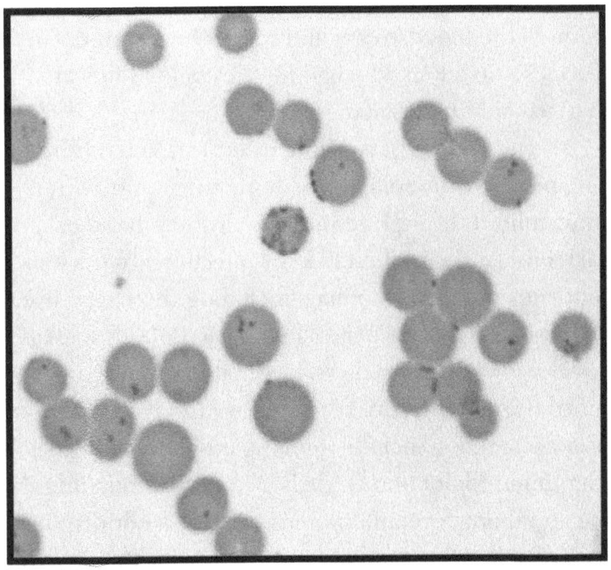

Mycoplasma haemofelis. *Photo by Nr387241, via Wikimedia Commons.*

PATHOGENICITY AND CLINICAL SIGNIFICANCE

Although seven *Mycoplasma* species are detected in the human genitourinary tract, only three species (*genitalium*, *hominis*, and *Ureaplasma* species) are associated with urogenital disease. Nonchlamydial nongonococcal urethritis in men may result from *M. genitalium* and *Ureaplasma* species. *M. genitalium* has also been isolated from the urogenital tract of women with cervicitis and pelvic inflammatory disease. *M. genitalium* and *Ureaplasma* species have also been implicated in extragenital infections.

Hominis and *Ureaplasma* species have been implicated in chorioamnionitis, endometritis, pyelonephritis, postpartum or postabortum fevers, neonatal meningitis, pneumonia, bacteremia, and arthritis (specifically, *hominis* in postpartum women and *Ureaplasma* species in sexually acquired reactive arthritis).

M. hominis has been related to extragenital infections, including sepsis, hematoma infection, vascular and catheter-related infections, sternal wound infections following thoracic surgery, prosthetic valve endocarditis, brain abscesses, and pneumonia. These infections occurred mainly through the spread of bacteria in the bloodstream and mostly in immunocompromised persons who had injuries of anatomical barriers and had polytrauma.

M. *pneumoniae* causes lung infections, often called "atypical pneumonia" or "walking pneumonia." It is transmitted through respiratory droplets between persons. At the highest risk for infection are persons in close contact with others, including those who live, work, or perform activities in crowded places such as schools, homeless shelters, hospitals, prisons, and dormitories. Other risk factors for *Mycoplasma* respiratory infection include smoking and lower preexisting immunoglobulin G levels. *Mycoplasma* pneumonia has pulmonary manifestations (such as a nonproductive cough) and extrapulmonary manifestations (such as cardiologic, neurologic, and dermatologic symptoms). There is no age or gender preference for the disease. Although people of all ages are at risk, infection rarely occurs in children younger than five years.

DRUG SUSCEPTIBILITY

Mycoplasma lack cell walls and are resistant to all antibiotics that target cell wall synthesis, including beta-lactams, cephalosporins, carbapenems, monobactams, and glycopeptides. Sulfa drugs that target folic acid synthesis are also ineffective against *Mycoplasma*.

M. hominis is treated with doxycycline, the drug of choice, usually for ten to fourteen days. However, the duration of treatment is based on observations of symptom resolution and clinical judgment. Resistant strains have been reported, and alternate choices of antibiotics include azithromycin.

Ureaplasma infections are treated with doxycycline or azithromycin. A seven-day course of doxycycline can be used to treat urethritis caused by *Ureaplasma* species. Alternative antimicrobials for *Ureaplasma* include fluoroquinolones (such as levofloxacin and ofloxacin). Clinical observations are important in considering treatment duration.

Neonatal meningitis caused by *M. hominis* and *Ureaplasma* species is often treated with macrolide antibiotics. Lower respiratory infections in newborns are treated with azithromycin or erythromycin. The suggested duration of treatment for *Mycoplasma* infections in newborns is ten to fourteen days.

M. *genitalium* and M. *pneumonia* are treated with macrolides (such as azithromycin, clarithromycin, and erythromycin), fluoroquinolones (such as levofloxacin, and moxifloxacin), and tetracyclines (such as doxycycline). The duration of treatment ranges from five to fourteen days, depending on which antibiotic is used.

—Miriam E. Schwartz, MD, PhD and Shawkat Dhanani, MD, MPH

Further Reading

Blanchard, Alain, and Cecile M. Bebear. "Mycoplasmas of Humans." *Molecular Biology and Pathogenicity of Mycoplasmas*, edited by Shmuel Razin and Richard Herrmann, Kluwer Academic, 2002.

Johannson, Karl-Erik, and Bertil Petterrson. "Taxonomy of Mollicutes." *Molecular Biology and Pathogenicity of Mycoplasmas*, edited by Shmuel Razin and Richard Herrmann, Kluwer Academic, 2002.

Lanao, Andrea E., et al. "Mycoplasma Infections." *StatPearls*. StatPearls Publishing, 10 Aug. 2021.

Mandell, Lionel A., et al. "Infectious Diseases Society of America/American Thoracic Society Consensus Guidelines on the Management of Community-Acquired Pneumonia in Adults." *Clinical Infectious Diseases*, vol. 44, 2007, pp. S27-72.

Miyashita, Naoyuki. "Atypical Pneumonia: Pathophysiology, Diagnosis, and Treatment." *Respiratory Investigation*, S2212-5345(21)00177-5, 5 Nov. 2021, doi:10.1016/j.resinv.2021.09.009.

Ryan, Kenneth J., et al., editors. *Sherris Medical Microbiology: An Introduction to Infectious Diseases*. 7th ed., McGraw-Hill, 2018.

Rickettsia

Category: Pathogens

Anatomy or system affected: Blood vessels, brain, cells, circulatory system, heart, immune system, lungs, lymphatic system, skin

Specialties and related fields: Bacteriology, cardiology, dermatology, entomology, immunology, infectious disease. internal medicine, microbiology, neurology, pathology, public health, pulmonology

Definition: obligate, intracellular, parasitic, gram-negative coccobacilli; humans are usually accidental hosts, while other mammals and arthropods serve as reservoirs; their adenosine triphosphate (ATP) transport system allows them to be energy parasites; and rickettsial-type organisms also have been linked to plant diseases

KEY TERMS

edema: a condition characterized by an excess of watery fluid collecting in the cavities or tissues of the body

endocytosis: a cellular process in which substances are brought into the cell; the material to be internalized is surrounded by an area of the cell membrane, which then buds off inside the cell to form a vesicle containing the ingested material

endosome: a collection of intracellular sorting vesicles in eukaryotic cells

endothelial cells: the main type of cell found in the inside lining of blood vessels, lymph vessels, and the heart

hypovolemia: a decreased volume of circulating blood in the body

lymphohistiocytic vasculitis: an inflammation of blood vessels causes by lymphocytes and macrophages

spotted fever group rickettsioses: a group of diseases caused by closely related bacteria spread to people through the bites of infected ticks and mites

typhus: an infectious disease caused by rickettsiae, characterized by a purple rash, headaches, fever, and usually delirium, and historically a cause of high mortality during wars and famines

NATURAL HABITAT AND FEATURES

Since *Rickettsia* are small, obligate, intracellular parasites, they were originally thought to be viruses. Further studies have shown them to be true bacteria. All *Rickettsia* have a gram-negative-type cell wall, but Gram stains do not efficiently stain them. Giemsa staining effectively dyes rickettsiae for visualization. Giemsa staining turns rickettsiae bluish-purple. Other methods for visualizing *Rickettsia* include the Gimenez and Machiavello stains, which turn these organisms red.

Rickettsia genomes are small (2,100-1,200 kilobases) and incomplete since they lack genes for anaerobic metabolism enzymes and produce most amino acids and nucleotides. They possess the enzymes for aerobic metabolism but normally use a unique adenosine triphosphate (ATP) transport system to steal ATP from their hosts instead of making it themselves, allowing them to be energy parasites. The genome of one of these bacteria, *R. prowazekii*, is the most closely related bacterial genome to the genome of mitochondria.

No *Rickettsia* species can grow on artificial media; instead, they must be cultured in living tissue, usually a chick embryo or cultured cells. In infected humans, *Rickettsia* species usually self-induces phagocytosis by the endothelial cells lining blood vessels. Inside the cells, they escape from the phagosome into the cytoplasm, where they replicate. Many species escape the cell by causing lysis, which destroys the host cell, but other species exit by extrusion through filopodia: finger-like projections on the cell surface.

Serology and deoxyribonucleic acid (DNA) studies have separated these bacteria into two main groups: the typhus group (*R. prowazekii* and *R. typhus*) and the spotted fever group (SFG) (including members like *R. rickettsii* and *R. parkeri*). Another group, formerly called the "*Rickettsia* scrub typhus group," has been separated into the related genus *Orientia*. *Ehrlichia*, *Anaplasma*, and *Coxiella* are similar but only distantly related small intracellular parasites.

The most common reservoirs for *Rickettsia* are ticks, fleas, and mites. Rodents and other mammals also serve as reservoirs. *R. prowazekii*, the causative agent of epidemic typhus, has a human reservoir. It is transmitted from human to human through body lice feces. *R. typhi* is transmitted by rat fleas (*Xenopsylla cheopis*) and *R. rickettsii* by dog tick (*Dermacentor variabilis* and *D. andersoni*).

PATHOGENICITY AND CLINICAL SIGNIFICANCE

Transmission and the disease course are slightly different between the typhus and SFGs. In the typhus group, *R. prowazekii*, the causative agent of epidemic typhus, are deposited on the host's skin in the feces of human body lice. Irritation caused by the louse's saliva causes humans to scratch and sweep louse feces and bacteria to enter through the scratch-abraded skin.

Once inside the body, rickettsiae infect the endothelial cells that line blood vessels. These organisms have lipopolysaccharides, rickettsial outer membrane proteins (rOmps), and surface-exposed proteins (SEPS) on their surfaces. These molecules paste rickettsiae to endothelial cell surfaces. The bound rickettsiae activate the endothelial cells' endocytosis machinery to engulf the rickettsiae. Inside the host cell, rickettsiae secrete enzymes (phospholipase D and hemolysin C) that degrade the endosome, releasing the organisms into the cytoplasm where they replicate.

Rickettsiae damage endothelial cells, inducing blood vessel inflammation and endothelial cell death; a process is called "lymphohistiocytic vasculitis." Damaged blood vessels leak fluid into tissues causing swelling (edema), fluid loss (hypovolemia), low blood pressure (hypotension), and low blood protein levels (hypoalbuminemia). Fluid build-up in the brain causes encephalitis; in the heart, it causes myocarditis, and in the lungs, it causes pulmonary edema. Damaged blood vessels also cause clots to form inside small blood vessels, occluding them.

Symptoms appear suddenly after about eight days of incubation and include fever, chills, headache, tiredness, and muscle and joint pain. One week later, a rash appears in some infected persons. This rash starts on the trunk and spreads toward the extremities but spares the hands and feet. Stupor and delirium may follow. Mortality can be up to 70 percent of those infected, and full recovery can take several months. *R. prowazekii* can remain dormant in people for decades and cause a relapse infection called "Brill-Zinsser disease."

Humans are the main reservoir of the disease; however, other mammals can serve as reservoirs. In the Eastern United States, flying squirrels are important reservoirs. The lice themselves are not reservoirs because they die soon after becoming infected. Crowded conditions are needed for epidemic spread.

R. typhi, the causative agent of endemic typhus, is deposited on humans in the feces of rat or cat fleas. These feces are inhaled or rubbed into the skin. The course of the disease is much like epidemic typhus, but the disease is much milder. Initial symptoms are very flu-like and might be mistaken for a viral illness. The symptoms are nonspecific: myalgias, headache, chills, myalgias, nausea, vomiting, abdominal pain, fever, and diarrhea. After several days a rash appears on the trunk that spreads to the extremities but spares the palms of the hands and the soles of the feet. Humans recover in less than three weeks, even when not treated.

R. rickettsia causes Rocky Mountain spotted fever (RMSF), the most common rickettsial disease in the United States. In the SFG, the bacteria are released into the arthropod's saliva and enter the mammalian host. The arthropods may hatch already infected because there is transovarial transfer of bacteria from the female to her eggs. Uninfected arthropods also may become infected when they take a blood meal from an infected mammal. Three different tick species transmit *R. rickettsii*: (1) *Dermacentor variabilis*, the American dog tick, is found in the Eastern and South-central United States; (2) *Dermacentor andersoni*, the mountain wood tick, lives west of the Mississippi River; (3) *Rhipicephalus sanguineus*, or the common dog tick, resides in the Southwestern United States.

R. rickettsii is not the only *Rickettsia* species that causes RMSF. Instead, closely related organisms form the SFG. All members of the SFG are transmitted by arthropods and cause similar diseases. *R. parkeri* is also transmitted by several tick species, including the dog tick. Ticks must remain attached for some time for disease transmission because the bacteria are in a dormant state and must become active before they can enter the saliva and then the mammal, which may take up to forty-eight hours. The ticks themselves are

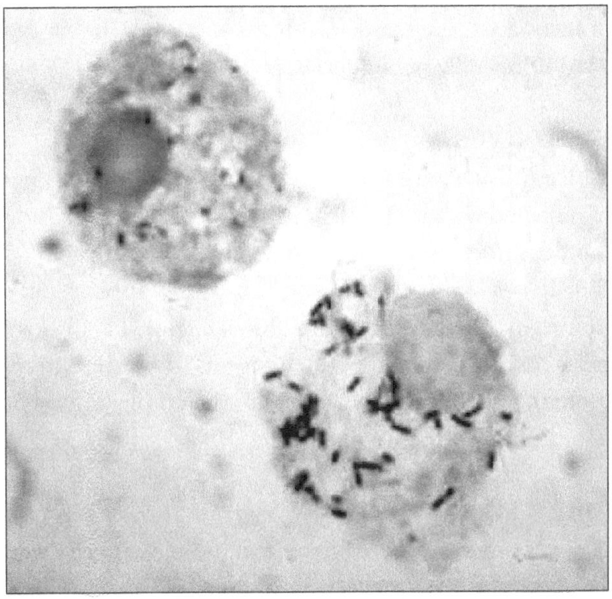

Red-stained Rickettsia rickettsii *visible in cells of an Ixodid vector tick. Photo via CDC/Wikimedia Commons. [Public domain.]*

the main reservoir, while wild rodents serve as secondary reservoirs. Other causes of SFG rickettsioses are Pacific Coast tick fever, caused by R. philipii, and rickettsialpox, caused by *R. akari*. Rickettsialpox differs from other SFG rickettsioses since it is transmitted by the bite of infected mouse mites (Liponyssoides sanguineus). This disease occurs throughout the United States, but most cases appear in the northeastern United States, particularly New York City. Pacific Coast tick fever is transmitted by the bite of infected Pacific Coast ticks, Dermacentor occidentalis, which live along the western coastline in California, Oregon, and Washington.

The onset of symptoms is sudden, two to twelve days after the tick bite, and includes fever, chills, headache, and muscle pain. A rash appears in almost all infected persons two or three days later. This rash begins on the hands and feet, often includes the palms and soles, and spreads toward the trunk. Complications include respiratory and renal failure, seizures, and coma. Mortality is about 20 percent in untreated persons. Other spotted fevers show similar infection patterns and symptoms, although the symptoms may be milder.

DIAGNOSIS

Since *Rickettsia* are not easily cultured, diagnosis is confirmed with serologic testing. For *R. rickettsii*, antibodies appear seven to ten days after illness onset. For R. *typhi* and *R. prowazekii,* antibodies appear ten to twenty-one days after the illness begins. Laboratory tests usually reveal low platelet levels (thrombocytopenia), low sodium (hyponatremia), and prolonged clotting times.

DRUG SUSCEPTIBILITY

Doxycycline, a tetracycline-type antibiotic, is the drug of choice for treating rickettsial diseases.

Doxycycline is taken orally, minimally, for five to seven days and at least three days after the fever subsides. Chloramphenicol is an alternative. The antibiotics can be administered intravenously in severe cases.

—Richard W. Cheney Jr., PhD and Michael A. Buratovich, PhD

Further Reading

Badash, Michelle. "Rocky Mountain Spotted Fever (RMSF)." Reviewed by David L. Horn. *Health Library*, 2016, healthlibrary.epnet.com/GetContent.aspx?token=da29d243-e573-4601-8b42-77cd0ccb14b2&chunkiid=11588. Accessed 15 Nov. 2016.

Paddock, C. D. et al. "Rickettsia parkeri: A Newly Recognized Cause of Spotted Fever Rickettsiosis in the United States." *Clinical Infectious Diseases*, vol. 38, 2004, pp. 805-11.

Didier, Raoult, and Phillipe Parola, editors. *Rickettsial Diseases*. Informa Health Care, 2007.

Hechemy, Karim E., et al., editors. *Rickettsiology and Rickettsial Diseases*. Wiley-Blackwell, 2009.

Madigan, Michael T., and John M. Martinko. *Brock Biology of Microorganisms*. 12th ed., Pearson/Prentice Hall, 2010.

McQuiston, Jennifer. "Rickettsial (Spotted & Typhus Fevers) & Related Infections (Anaplasmosis & Ehrlichiosis)." *CDC Health Information for International Travel*. Oxford UP, 2016. Centers for Disease Control and Prevention, 10 July 2015, wwwnc.cdc.gov/travel/yellowbook/2016/infectious-diseases-related-to-travel/rickettsial-spotted-typhus-fevers-related-infections-anaplasmosis-ehrlichiosis. Accessed 15 Nov. 2016.

"Typhus." *MedlinePlus*, 7 Dec. 2014, medlineplus.gov/ency/article/001363.htm. Accessed 15 Nov. 2016.

Bordetella

Category: Pathogen

Anatomy or system affected: Larynx, lungs, nose, pharynx, respiratory system, throat

Specialties and related fields: Bacteriology, immunology, internal medicine, microbiology, public health, pulmonary medicine

Definition: most *Bordetella* species are obligate respiratory pathogens of animals and humans; *B. pertussis* causes a severe and potentially life-threatening disease (whooping cough, or pertussis) of infants and young children, characterized by repeated and violent coughing spells and the characteristic whooping sound that comes from breathing difficulties

KEY TERMS

adhesin: molecules synthesized by microorganisms that allow them to stick to specific surfaces

agglutinogen: an adhesin used by *Bordetella pertussis* to adhere to the respiratory epithelium

cilia: hairlike extensions of the apical cell surface of epithelial cells that line the proximal parts of the respiratory tract; cilia beat in a concerted, coordinated fashion to clear the airways of foreign debris

filamentous hemagglutinin: a toxin made by *Bordetella pertussis* that aids in its colonizing the upper respiratory tract

lymphocytosis: an increase in the number or proportion of lymphocytes in the blood.

pertactin: an outer membrane protein of *Bordetella pertussis* that promotes adhesion to respiratory epithelial cells

NATURAL HABITAT AND FEATURES

Members of the bacterial genus *Bordetella* are gram-negative coccobacilli (short rods) that stain pink in a Gram stain. There are nine species of *Bordetella*: *B. pertussis B. parapertussis*, *B. bronchiseptica*, *B. avium*, *B. hinzii*, *B. holmesii*, *B. trematum*, *B. ansorpii*, and *B. petrii*. Of these species, *B. pertussis*, *B. parapertussis*, and *B. bronchiseptica* are the most closely related. *B. pertussis* causes whooping cough, and *B. parapertussis* causes a disease like whooping cough in humans. *B. bronchiseptica* infects a range of mammals, including humans, and causes a spectrum of respiratory disorders. All *Bordetella* species (except for *B. petrii*) are obligate respiratory pathogens of animals and humans. *B. pertussis* and *B. parapertussis* cause disease only in humans. The rest, except *B. petrii*, are found naturally in diseased animals, including birds. All *Bordetella* species can, in rare cases, cause disease in immunocompromised humans.

Bordetella species do not ferment their food sources and have a strict requirement for oxygen (are strict aerobes). *B. pertussis* requires enriched media containing charcoal or blood (or both) to grow in the laboratory because of their sensitivity to unsaturated fatty acids and sulfur compounds in regular agar media. When grown on Bordet-Gengou agar, *B. pertussis* forms small (less than 1 millimeter) smooth, transparent, shiny colonies with a circular edge in about five to seven days of incubation at 98.6° Fahrenheit (37° Celsius). *B. parapertussis* forms similar but larger, duller brownish colonies after two days, and *B. bronchiseptica* forms larger, rougher, and pitted colonies in one to two days on this medium. Other species can be grown successfully on less stringent media.

Bordetella species can be differentiated by growth, biochemical, and antigenic characteristics. Molecular methods, including fluorescent antibody and other immunological (serological), polymerase chain reaction (PCR), and 16S rRNA (ribosomal ribonucleic acid) gene sequencing, have been employed to identify and study properties of various species.

PATHOGENICITY AND CLINICAL SIGNIFICANCE

B. pertussis and *B. parapertussis* cause whooping cough. Whooping cough is most severe in infants less than one year, with significant morbidity and mortality rates. Roughly 85 to 90 percent of those exposed get the disease, with the majority being hospitalized. Patients with *B. parapertussis* usually have less severe disease, indistinguishable from a mild upper-respiratory-tract infection.

Bordetella are transmitted from one person to another through sneezes or coughs that release thousands of bacteria-filled droplets that spray out about six feet. Droplets can land in the mouths or noses or get directly inhaled into the lungs of nearby people. Since *Bordetella* can survive several days on dry surfaces, they can infect others when they touch contaminated surfaces, like a contaminated doorknob, and then touch their eyes, nose, or mouth.

Older children, adolescents, and adults can also contract whooping cough. Cases usually are milder because of increased immunity; however, immunocompromised persons can experience severe disease. Research suggests that adolescents and adults can infect susceptible infants and vice versa. Therefore, health authorities recommend giving adolescent siblings, parents, and health-care workers an additional pertussis booster immunization.

B. pertussis has been studied most extensively, so its pathogenesis and the disease-causing roles of its many virulence factors are well understood. The incubation period lasts five to twenty-one days after exposure. The organism employs adhesins and toxins during this time, including filamentous hemagglutinin, pertussis toxin, pertactin, and fimbriae proteins.

Recognizable symptoms occur during the catarrhal stage when the pathogen multiplies rapidly. Because these symptoms resemble a common cold, the organism can be transmitted before patients realize they have severe disease. This stage typically lasts one to two weeks, with persons exhibiting rhinorrhea, mild fever, coryza, and mild cough (although infants can exhibit apnea and respiratory distress).

The paroxysmal stage occurs when numerous toxins, including pertussis toxin, adenylate cyclase, dermonecrotic toxin, and tracheal cytotoxin, cause biochemical abnormalities and tissue destruction that advance the disease process and battle the host's immune defenses.

Pertussis toxin causes an increase in the levels of circulating lymphocytes, specifically dramatic increases in the numbers of T lymphocytes. Pertussis toxin induces T-lymphocyte proliferation and concomitantly causes them to leave the spleen and thymus and enter the bloodstream. However, the toxin also inhibits blood-based lymphocytes from exiting the bloodstream and migrating into tissues. Additionally, Pertussis toxin affects blood vessels by increasing their sensitivity to histamine. Increased histamine sensitivity causes fluid to leak from blood vessels and into airway tissues. These pathophysiological effects cause the airways to swell up, making breathing difficult and causing the classic "whooping" sound during a coughing fit.

The paroxysmal stage usually lasts two to six weeks with characteristic multiple spasms of dry cough, often with projectile vomiting and exhaustion. In infants, the characteristic whoop occurs when they struggle to breathe. These paroxysmal, uninterrupted fits of coughing are followed by inspiratory whooping noises that result from air sliding past a glottis that is partially closed and swollen. The violent force of these coughing fits can cause vomiting, a collapsed lung, broken ribs, and tiny petechiae (pinpoint red spots caused by bleeding) in the face as capillaries burst from the pressure.

Young infants often gasp and cannot breathe (apnea), have a bluish tinge to their skin (cyanosis), and may suffer an "apparent life-threatening event" (ALTE) during this phase. Decreased blood oxygen levels can cause seizures, encephalopathy, and even death.

During the convalescent stage (which lasts two to four weeks), patients have decreasing bouts of coughing and vomiting; however, secondary complications can occur, generally by other pathogens that colonize the host because of the damage that occurred during the infection. These complications include pneumonia, which can lead to seizures, encephalopathy, and death. During this stage, recovery occurs when the host's defenses revive and tissues, especially the ciliated epithelium, regenerate.

Most *Bordetella* species can infect animals (including birds) and immunocompromised humans. *B. bronchiseptica* can establish asymptomatic or severe respiratory infections in various mammals: kennel cough in dogs, atrophic rhinitis in pigs, snuffles in rabbits, and guinea pig bronchopneumonia. *B. avium* causes a potentially fatal respiratory disease of birds, including chickens and turkeys, resulting in significant economic loss. *B. hinzii* is found naturally as a commensal organism in the respiratory tracts of poultry. The least understood species, *B. trematum*, has been found associated with wounds and ear infections. *B. avium*, *B. hinzii*, and *B. petrii* have all been found in the lungs of persons with cystic fibrosis. Any disease-causing role is unclear.

BORDETELLA DIAGNOSIS

The best time to diagnose pertussis is during the catarrhal phase since the antibiotics can kill the bacteria and reduce damage. The best way to identify *Bordetella* is to swab the nasopharynx and grow the bacteria in culture or identify *Bordetella* deoxyribonucleic acid (DNA) by polymerase chain reaction. Another option is to employ direct fluorescent antibodies to detect *Bordetella* antigens. Finally, serological tests can detect antibodies against *Bordetella* antigens in the patient's blood. A vital blood test for infants is a complete blood count (CBC) since the degree of lymphocytosis in infants is an important predictor of the severity of illness.

DRUG SUSCEPTIBILITY

Treatment for whooping cough is primarily supportive; however, early antibiotic therapy can diminish the severity and duration of the disease. It is critical to interfere with transmission to susceptible persons.

Traditionally, erythromycin has been used, but some infants experienced infantile hypertrophic pyloric stenosis. Another macrolide antibiotic, clarithromycin, is not safe for infants. Azithromycin is effective and is preferred for infants younger than one month of age.

Azithromycin or clarithromycin are preferred for people older than one month because they cause fewer side effects and less gastrointestinal upset. Persons older than two months of age who cannot tolerate a macrolide antibiotic can take trimethoprim-sulfamethoxazole (Bactrim) for fourteen days. For persons exposed to clinically diagnosed pertussis cases, prophylaxis for five days with azithromycin or clarithromycin is recommended.

Although the efficacy of the pertussis portion of the diphtheria-tetanus acellular pertussis (DTaP) vaccine is not 100 percent, immunization of infants, children, adolescents, and adults is the most effective way to combat the spread of pertussis.

—Steven A. Kuhl, PhD and Michael A. Buratovich, PhD

Further Reading

Broaddus, V. Courtney, et al., editors. *Murray and Nadel's Textbook of Respiratory Medicine*. 7th ed., Elsevier, 2021.

Lauria, Ashley M., and Christopher P. Zabbo. "Pertussis." *StatPearls*. StatPearls Publishing, 26 June 2021.

Levitzky, Michael G. *Pulmonary Physiology*. 9th ed., McGraw-Hill Medical, 2018.

Kliegman, Robert M., et al. editors. *Nelson Textbook of Pediatrics*. 21st ed., Elsevier, 2019.

Mattoo, Seema, and James D. Cherry. "Molecular Pathogenesis, Epidemiology, and Clinical Manifestations of Respiratory Infections Due to *Bordetella pertussis* and Other *Bordetella* Subspecies." *Clinical Microbiology Reviews*, vol. 18, 2005, pp. 326-82.

Sandora, Thomas J., Courtney A. Gidengil, and Grace M. Lee. "Pertussis Vaccination for Health Care Workers." *Clinical Microbiology Reviews*, vol. 21, 2008, pp.426-34.

Weiss, Alison. "The Genus *Bordetella*." *The Prokaryotes: A Handbook on the Biology of Bacteria*, edited by Martin Dworkin, et al., Vol. 5, Springer, 2006, pp. 602-47.

HAEMOPHILUS

Category: Pathogen

Anatomy or system affected: Blood, brain, bronchi, ear, epiglottis, joints, lungs, nasopharynx, sinuses, skin

Specialties and related fields: Bacteriology, family practice, immunology, internal medicine, microbiology, neurology, orthopedics, otolaryngology, pediatrics, public health, respiratory medicine

Definition: gram-negative, nonmotile, non-spore-forming, pleiomorphic coccobacilli that cause several human infectious diseases

KEY TERMS

biofilms: communities of microorganisms encased in polysaccharide-containing materials secreted by those microorganisms

capsule: a polysaccharide layer that lies outside the cell envelope

lipooligosaccharides: glycolipids found in the outer membrane of some types of gram-negative bacteria, such as Neisseria and Haemophilus species

IgA protease: an enzyme secreted by Haemophilus species that degrades immunoglobulin A, the primary antibody that protects mucosal surfaces

serotypes: a serologically distinguishable strain of a microorganism

untypable: uncapsulated strains of Haemophilus influenzae

DEFINITION

Haemophilus is a genus of gram-negative, nonmotile, non-spore-forming, pleiomorphic coccobacilli. Its name is derived from Greek and means "blood lover." Most strains of *Haemophilus* require hemin (factor X) and NAD (factor V), both of which are naturally found in blood. Strains may be aerobes or facultative anaerobes. Many *Haemophilus* species are normal flora in the upper respiratory and urogenital tracts of humans and other animals.

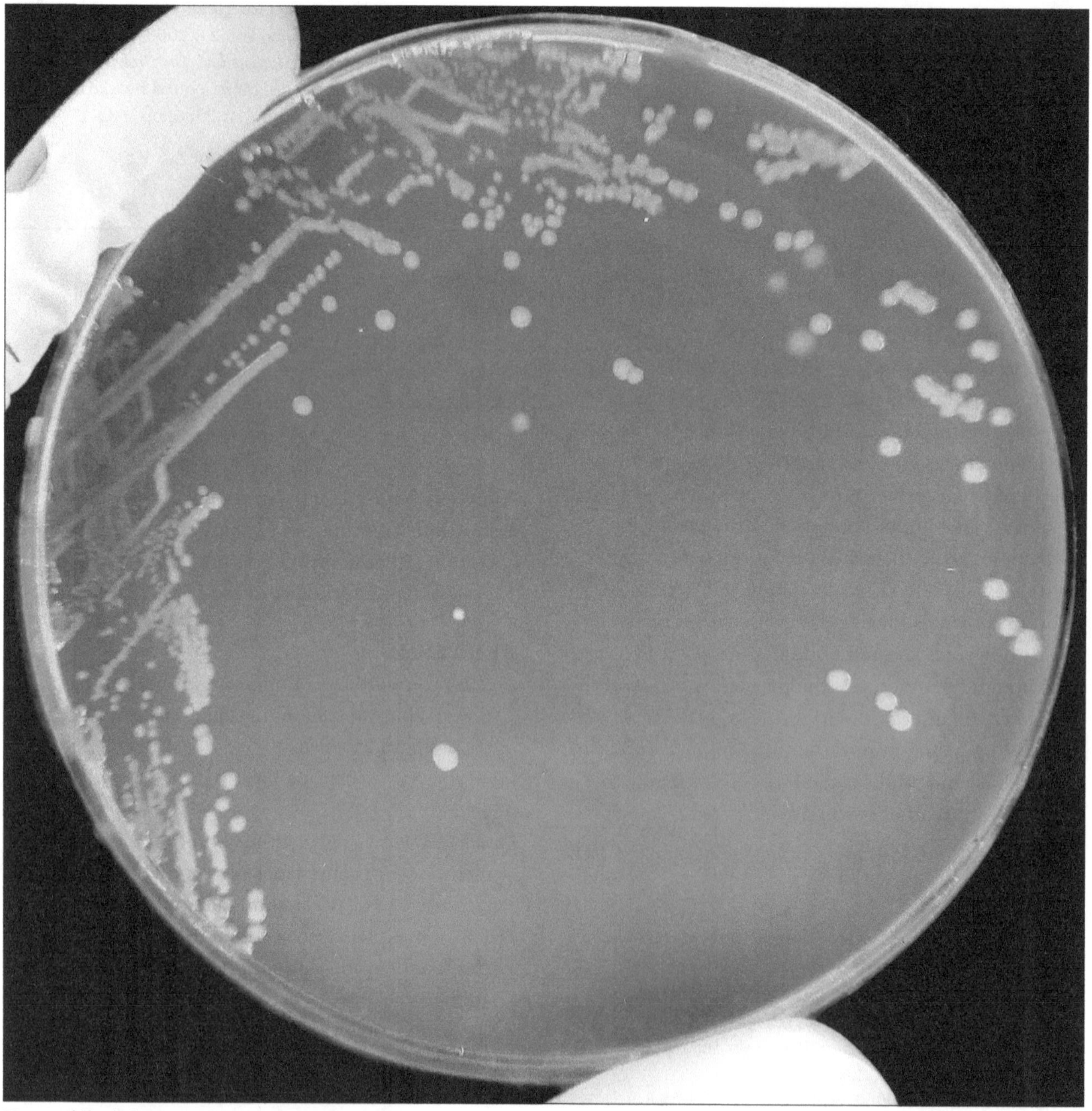

Haemophilus influenzae *colonies growing on the chocolate agar. Colonies shown with reflected light. Photo by Stefan Walkowski, via Wikimedia Commons.*

NATURAL HABITAT AND FEATURES

Although *Haemophilus* species are usually classified as coccobacilli, they are quite pleiomorphic (variable) and can take on various shapes in culture. Most are cultured on chocolate agar, a nutrient-dense agar with added denatured hemoglobin. Additional NAD is usually added to this agar when culturing *Haemophilus* spp. Incubation is best at 98.6° Fahrenheit (37° Celsius), and growth is enhanced in an incubator enriched with carbon dioxide. Most species are commensal in the upper respiratory tract and are opportunistic pathogens with relatively limited host ranges. The major pathogenic species in humans are *H. influenzae* and *H. ducreyi*.

PATHOGENICITY AND CLINICAL SIGNIFICANCE

H. influenzae was named when it was isolated in the 1890s from persons suffering from influenza. It was later shown to be a secondary bacterial infection and not the causative agent of that disease. It is similar to many other members of this genus. *H. aegyptius*, which causes conjunctivitis and Brazilian purpuric fever, has been reclassified as a subtype of *H. influenzae* rather than a separate species. Natural infections occur only in humans, although infection can be artificially induced in a few animal species.

Both encapsulated and nonencapsulated strains of *H. influenzae* exist. Both strains have an outer membrane that contains molecules called "lipooligosaccharides" (LOS). LOS from *H. influenzae* inhibit the cilia that coat the upper respiratory tract. This prevents them from clearing bacteria and other debris from the upper respiratory tract, which helps *Haemophilus influenzae* colonize the respiratory tracts. Both capsulated and unencapsulated strains of *Haemophilus influenzae* make an enzyme called "IgA protease." IgA protease degrades secretory antibodies that protect epithelial surfaces in the respiratory tract; this helps *H. influenzae* evade being killed by white blood cells.

Encapsulated strains are more pathogenic because the capsule offers some protection against the host's immune system. There are six different serotypes of encapsulated H. influenzae strains A-F. *H. influenzae* serotype B (Hib) is the most pathogenic of this group. Before introducing the conjugated Hib vaccine in 1987, approximately 95 percent of all invasive *Haemophilus* infections in children, including 75 percent of meningitis cases, and 50 percent of *Haemophilus* infections in adults, were caused by Hib. In developed countries, Hib infections in children markedly decreased since the early 1990s, when the Hib vaccine became widely used. In the United States, childhood Hib infections decreased 99 percent between 1990 and 2000.

Hib can spread from the upper respiratory tract to the epiglottis and cause epiglottitis. Alternatively, Hib can spread to the skin of the face and cause cellulitis. It can breach the blood capillaries and cause bacteremias. If it causes bacteremia, Hib can spread through the blood to the brain, where it can cause meningitis, the joints, where it can cause septic arthritis, or the bones, causing osteomyelitis.

Nonencapsulated *H. influenzae* strains, also known as untypable strains, cause less invasive infections. They can cause inner ear (otitis media) or sinus infections (sinusitis). Untypable strains can spread to the lower respiratory tract and cause bronchitis or pneumonia. Interestingly, untypable *H. influenzae* strains colonize the nasopharynx of 40 to 80 percent of children and adults, and encapsulated strains colonize the nasopharynx of 3 to 5 percent of all children. In 2006, nonencapsulated strains accounted for almost two-thirds of all *H. influenzae* infections in the United States.

H. influenzae is transmitted, mainly through respiratory droplets and secretions. Smoking, viral infections, chronic lung disease, and immunodeficiency can make untypable infections much more likely. Among vaccinated children, Hib tends to mainly infect children and those who have undergone splenectomy or those with sickle cell disease. Likewise, those with a congenital deficiency of complement components or malignancies are also at higher risk of contracting Hib. Children with an acute viral infection, such as influenza, can suffer from a fatal *Haemophilus influenzae* superinfection, even if their spleen is normal.

Nonencapsulated *H.* strains tend to infect children or adults with poorly functioning immune systems, such as people with diabetes, malignancies, human immunodeficiency virus (HIV) infection, or viral infections (postviral pneumonia). People with respiratory conditions, such as chronic obstructive pulmonary disease or cystic fibrosis, are at risk of *Haemophilus influenzae* infections.

Diagnosis of H. influenzae usually requires isolating the organism from blood, cerebrospinal fluid, synovial fluid from the joints, pleural fluid surrounding the lungs, or sinuses. Alternatively, serological tests, such as enzyme-linked immunosorbent assay, latex agglutination, or coagglutination tests, detect antibodies against *H. influenzae*.

Haemophilus influenzae infections are treated with ceftriaxone. The nontypeable strains are treated with amoxicillin, with or without clavulanate. Second and third-generation cephalosporins, macrolides, and fluoroquinolones are used as alternatives.

H. ducreyi was first isolated in 1899. It is most commonly isolated from the urogenital mucosa of humans, the bacterium's only natural host. Like most members of its genus, *H. ducreyi* is a fastidious bacterium that requires enriched chocolate agar for growth. Genetic testing of *H. ducreyi* has shown it to be genetically related (albeit distantly) to other *Haemophilus* spp. Some bacteriologists have suggested that *H. ducreyi* be placed in its own genus in a family all its own.

H. ducreyi makes a host of molecules to attach to genitourinary epithelia, attracting white blood cells' attention. Phagocytic cells engulf this organism, but *H. ducreyi* makes an enzyme called "copper-zinc superoxide dismutase" that protects it from reactive oxygen species made by phagocytes. Additionally, *H. ducreyi* synthesizes two toxins. The first is HdCTD (*H. ducreyi* cytolethal distending toxin), and the second is cytotoxic hemolysin. These two toxins collaborate to destroy white blood and epithelial cells. The cell destruction leads to erosions that look like raised, red elevations of skin called "papules." The papules fill with pus and become pustules that burst and turn into painful ulcers called "soft chancres."

H. ducreyi infection leads to chancroid (soft chancre), a common sexually transmitted disease in less developed countries in tropical and subtropical regions. The disease causes ulceration of the genitalia and is endemic to sub-Saharan Africa, especially among men who have sex with sex workers, who often are reservoirs for *H. ducreyi*. *H. ducreyi* infection increases the likelihood of HIV transmission ten to one hundred times. Chancroid is uncommon in the United States, with the last major outbreak in the 1980s. Single-dose azithromycin or ceftriaxone are the preferred treatments for *H. ducreyi* infections. Multiple-dose erythromycin or ciprofloxacin are used for severe cases.

Other *Haemophilus* spp. that are commensal in humans only rarely cause opportunistic infections. *H. haemolyticus*, *H. parahaemolyticus*, and *H. parainfluenzae* are commonly found in the nasopharynx and oral cavities. Still, they are seen associated only with pharyngitis and other conditions in debilitated persons. It has been suggested that *H. avium* and *H. agni* be placed within other genera in the Pasteurellaceae family because they are genetically distant from all other *Haemophilus* spp. Other species, such as *H. paracuniculus* and *H. parasuis*, are somewhat genetically closer to the *Haemophilus* spp. that affect humans, but their taxonomy is under scientific review.

—Richard W. Cheney Jr., PhD and Michael A. Buratovich, PhD

Further Reading

Brothwell, Julie A., et al. "Interactions of the Skin Pathogen *Haemophilus ducreyi* with the Human Host." *Frontiers in Immunology*, vol. 11, no. 615402, 2021, doi:10.3389/fimmu.2020.615402.

Garrity, George M., editor. *The Proteobacteria*. Vol. 2 in *Bergey's Manual of Systematic Bacteriology*. 2nd ed., Springer, 2005.

Khattak, Zoia E., and Fatima Anjum. "Haemophilus Influenzae." *StatPearls*, StatPearls Publishing, 14 Sept. 2021.

Madigan, Michael T., et al. *Brock Biology of Microorganisms*. 16th ed., Pearson, 2020.

Spinola, Stanley M., Margaret E. Bauer, and Robert S. Munson, Jr. "Immunopathenogenesis of *Haemophilus ducreyi* Infection (Chancroid)." *Infection and Immunity*, vol. 70, 2002, pp. 1667-76.

Sinusitis

Category: Diseases and disorders

Anatomy or system affected: Nose, respiratory system

Specialties and related fields: Family medicine, internal medicine, otorhinolaryngology

Definition: inflammation, irritation and swelling of the sinuses

KEY TERMS

deviated septum: a condition that causes a shift of the bones and cartilage from the middle of the nose to either side, making one side of the nasal passages much smaller than the other

nasal polyps: noncancerous growths inside the nose; usually associated with allergies or asthma, which can block the sinus drainage tract

orbit: the bones and other tissues that surround the eye, commonly known as the eye socket

CAUSES AND SYMPTOMS

The sinuses are airspaces in the skull's forehead just above the eyes, on either side of the nose below the eyes, and in the area just above the nose and in between the eyes. Mucus and tiny hairs, called "cilia," line the sinuses. Cilia trap inhaled particles and bacteria and move them back out through the nose, eliminating these potential irritants inhaled during normal breathing. The tracts through which the sinuses drain are relatively small and easily blocked by swelling of the area. Sinus blockage impairs drainage and cause the buildup of normal sinus secretions.

The term "sinusitis" refers to irritation or swelling of the sinuses and their membranes. Typical symptoms may include a feeling of congestion or pressure in the nose or face and a runny nose with secretions that may vary in color from clear to yellowish-green to bloody. The facial pressure is often worse when bending forward.

Most often, sinusitis is precipitated by the common cold. Another frequent cause is allergies, with typical symptoms of sneezing, runny nose, and itchy, watery eyes. An allergic patient sensitive to a particular airborne substance (pollen, ragweed, dust, animal dander) has a particularly vigorous response when these particles land in the nose and enter the sinuses. Increased production of mucus and the body's natural immune defenses combine to produce thick and copious nasal secretions that can fill the sinuses in an attempt to eliminate the offending agent.

Another factor predisposing a patient to sinusitis is environmental exposure to smoke or air pollution that irritates the sinuses. Problems that cause a blockage of the sinus drainage system, including nasal polyps, a deviated septum, or pregnancy (which leads to swelling of the nasal membranes due to hormonal changes), can interfere with mucus drainage from the sinuses. Finally, other genetic diseases such as cystic fibrosis or immune system disorders can predispose patients to sinusitis.

Although viruses or allergies cause most cases of sinusitis, these can often lead to infection by bacteria if they do not resolve promptly. Bacterial sinusitis requires treatment with antibiotics to avoid the rare but

Influenza viruses and rhinoviruses are the most common causes of sinusitis. Photo via iStock/Cecilie_Arcurs. [Used under license.]

severe complications of infection of the orbit or the brain and its surrounding tissues.

The distinction between bacterial and other causes of sinusitis depends on the patient's symptoms and a physical examination. A patient is more likely to have bacterial sinusitis if two or three of the following symptoms are present for at least seven days: facial pressure, nasal congestion, discolored nasal mucus, decreased sense of smell, productive or "wet" cough, fever, tooth pain on the upper jaw, or bad breath.

Sinus X-rays, done frequently in the past, are not considered a reliable diagnostic test for sinusitis. Though sinus computed tomography (CT) scans allow detailed visualization of sinus anatomy, they do not reliably distinguish bacterial sinusitis from other forms. They are helpful only in cases of long-standing, refractory symptoms for which sinus surgery is considered.

TREATMENT AND THERAPY

The initial sinusitis treatment involves extra fluids, anti-inflammatory drugs such as ibuprofen, antihistamines, short-term use of nasal decongestant sprays (no longer than three days), and oral decongestants such as pseudoephedrine. Humidified air (e.g., steam from a hot shower) and nasal irrigation with water or saline can offer short-term symptom relief.

If allergies are the cause of sinusitis, then oral or nasal allergy medications are appropriate. Examples are nonprescription antihistamines such as chlorpheniramine, brompheniramine, or diphenhydramine; they can cause drowsiness in some patients. Loratadine, fexofenadine and cetirizine, and other related, newer generation antihistamines are also available over the counter. They offer once-daily dosing and are significantly less sedating. Other nasal sprays such as topical steroids are available by prescription and provide significant relief.

Most sinusitis cases result from viral infections that do not require antibiotics. Symptomatic treatments include saline nasal washes, intranasal corticosteroids, and over-the-counter analgesics (e.g., acetaminophen, ibuprofen, aspirin, etc.). Nasal decongestants may help viral sinusitis cases, but there is no credible evidence that decongestants help with bacterial sinusitis infections.

If symptoms persist longer than ten days, then antibiotic therapy may be necessary, and a health-care provider evaluation is warranted. Other indications that antibiotics are warranted include the presence of pus-filled nasal discharge, facial pain for three to four days, or if symptoms improve after an upper respiratory viral infection and then become worse five to six days later. Many different types of antibiotics are effective for sinusitis, and prescription practices vary. Acute sinusitis treatment also differs from chronic sinusitis treatments. For mild to moderate acute sinusitis, the standard treatment is Augmentin twice a day for fourteen days. The alternative treatment is doxycycline, twice a day, for seven days. If these initial treatments fail, then the patient is given one of three options: (1) high-dose Augmentin (2 grams) twice a day for fourteen days; (2) levofloxacin daily for seven days, or (3) moxifloxacin once daily for seven days. These treatment regimens are for outpatient cases. If the patient suffers from severe sinusitis that requires hospitalization, they are administered intravenous antibiotics.

PERSPECTIVE AND PROSPECTS

Before the antibiotic era, sinusitis treatment involved drainage of the sinuses by extracting a tooth, puncturing the roof of the mouth, or entering the nose and creating a drainage tract to drain the sinuses and subsequently irrigate them with fluid for cleansing. Given the invasiveness of these procedures, they have become uncommon with the development of effective antibiotic therapy.

The development of tiny, high-resolution cameras known as "endoscopes" in the 1950s created a revolution in the understanding of sinus disease. Direct visualization of the nasal passages and sinus drainage tracts allowed a better understanding of the sinus anatomy. It thus led to the use of this equipment to facilitate surgical treatment.

IN THE NEWS: BALLOON SINUPLASTY

In late 2005, the Acclarent Company received permission from the Food and Drug Administration (FDA) to market a device to clear blocked sinuses similar to that used to clear blocked arteries in the heart. A flexible catheter tube inserted into the nostril guides a balloon into the targeted sinus. The balloon is inflated, spreading the bones of the passageway sufficiently to permit accumulated mucus or pus to drain. A minimally invasive outpatient procedure, balloon sinuplasty, is performed under local anesthesia and takes one to two hours. Patients report little or no pain and can often return to regular activity within twenty-four hours. The sinuplasty devices cost from $1,200 to $1,500 and are not reusable. Total costs for the procedure run from $4,000 to $6,800; some private insurance plans cover the procedure.

If the patient has nasal polyps, the surgeon must remove them before they have balloon sinuplasty. Nevertheless, it provides a possible alternative for sinusitis sufferers who do not need or prefer not to undergo surgical procedures involving cutting away bone or other tissue to open the blocked sinus. Surgeons using the devices praised them. The American Rhinologic Society was more cautious in its October 2006 position statement, asserting that the technology had limited indication at the time. In November 2006, the California Blue Cross labeled the procedure investigational and not medically necessary. Since the FDA's clearance was based on the devices' comparability to already approved methods, it did not require safety or effectiveness data submission. The most extensive 2006 clinical trial, which claimed that the procedure was safe and effective, followed 109 patients over twenty-four weeks. Longer-term efficacy remains unknown.

—*Milton Berman, PhD*

Occasionally, patients with recurrent symptoms require surgical removal of infected sinus tissue and enlargement of the natural drainage tracts to minimize sinus obstruction. A specialist in otorhinolaryngology can perform such surgery using an endoscope without the need for general anesthesia. Patients do not typically require hospitalization, and complications are rare.

—*Gregory B. Seymann, MD and Michael A. Buratovich, PhD*

Further Reading

"Acute Sinusitis." *Mayo Clinic*, 27 Aug. 2021, www.mayoclinic.org/diseases-conditions/acute-sinusitis/symptoms-causes/syc-20351671. Accessed 27 Aug. 2021.

Beers, Mark H., et al., editors. *The Merck Manual of Diagnosis and Therapy.* 18th ed., Merck Research Laboratories, 2006.

Brook, Itzhak, editor. *Sinusitis: From Microbiology to Management.* Taylor & Francis, 2006.

Kennedy, David W., and Marilyn Olsen. *Living with Chronic Sinusitis: A Patient's Guide to Sinusitis, Nasal Allergies, Polyps, and Their Treatment Options.* Hatherleigh Press, 2007.

McCaffrey, Thomas. "Functional Endoscopic Sinus Surgery: An Overview." *Mayo Clinic Proceedings*, vol. 68, 1993, pp. 571-77.

Mickelson, Samuel, and Michael Benninger. "The Nose and Paranasal Sinuses." *Textbook of Primary Care Medicine*, edited by John Noble, 3rd ed., Mosby, 2001.

"Sinus Infection (Sinusitis)." *Cleveland Clinic*, 6 Apr. 2020, my.clevelandclinic.org/health/diseases/17701-sinusitis. Accessed 28 Aug. 2021.

Younis, Ramzi T., editor. *Pediatric Sinusitis and Sinus Surgery.* Taylor & Francis, 2006.

PHARYNGITIS

Category: Diseases and disorders

Also known as: Sore throat

Anatomy or system affected: Throat, tonsils

Specialties and related fields: Otorhinolaryngology, pediatrics

Definition: inflammation of the mucous membranes of the pharynx or throat, often caused by a viral infection or bacteria

KEY TERMS

adenovirus: a group of deoxyribonucleic (DNA) viruses first discovered in adenoid tissue that mostly cause respiratory diseases

cytokines: small proteins, such as interferon, interleukin, and growth factors, that are secreted by certain cells of the immune system and influence other cells

group A streptococci: bacteria commonly found in the throat and on the skin that cause strep throat and impetigo

inflammation: a localized physical condition in which part of the body becomes reddened, swollen, hot, and often painful, especially as a reaction to injury or infection

pharynx: the hollow tube inside the neck that starts behind the nose and ends at the top of the trachea (windpipe) and esophagus (the tube that goes to the stomach)

CAUSES AND SYMPTOMS

Sore throat is the chief complaint of pharyngitis. The throat (pharynx) extends from the nasal passages above and behind the mouth to the esophagus in the neck. Viruses or bacteria infect the pharynx and cause it to swell. Goblet cells and submucosal glands produce excessive mucus. White blood cells throng to the site of infection to kill the invading microorganisms and release cytokines. If enough cytokines are released, a fever ensures. The throat often appears red, swollen, or puffy and may have white spots of pus. Fever and cough are also common, and examination may reveal swollen tonsils. Throat scratchiness, pain when swallowing, enlarged lymph nodes in the neck, cough, and irritation are also common in pharyngitis.

Bacteria and viruses that cause pharyngitis enter the body through the nose or mouth. Sneezing or coughing releases thousands of droplets that contain hosts of infective bacteria and viruses. Droplets can spread up to six and a half feet away. These viruses and bacteria can survive on surfaces for hours. People can contract them by toughing contaminated doorknobs, tables, or other objects and subsequently touching the eyes, nose, or mouth.

Viruses that cause the common cold (coronavirus and rhinovirus) or other respiratory diseases may also produce symptoms of pharyngitis. Epstein-Barr virus bypasses the nose and directly attacks the pharynx.

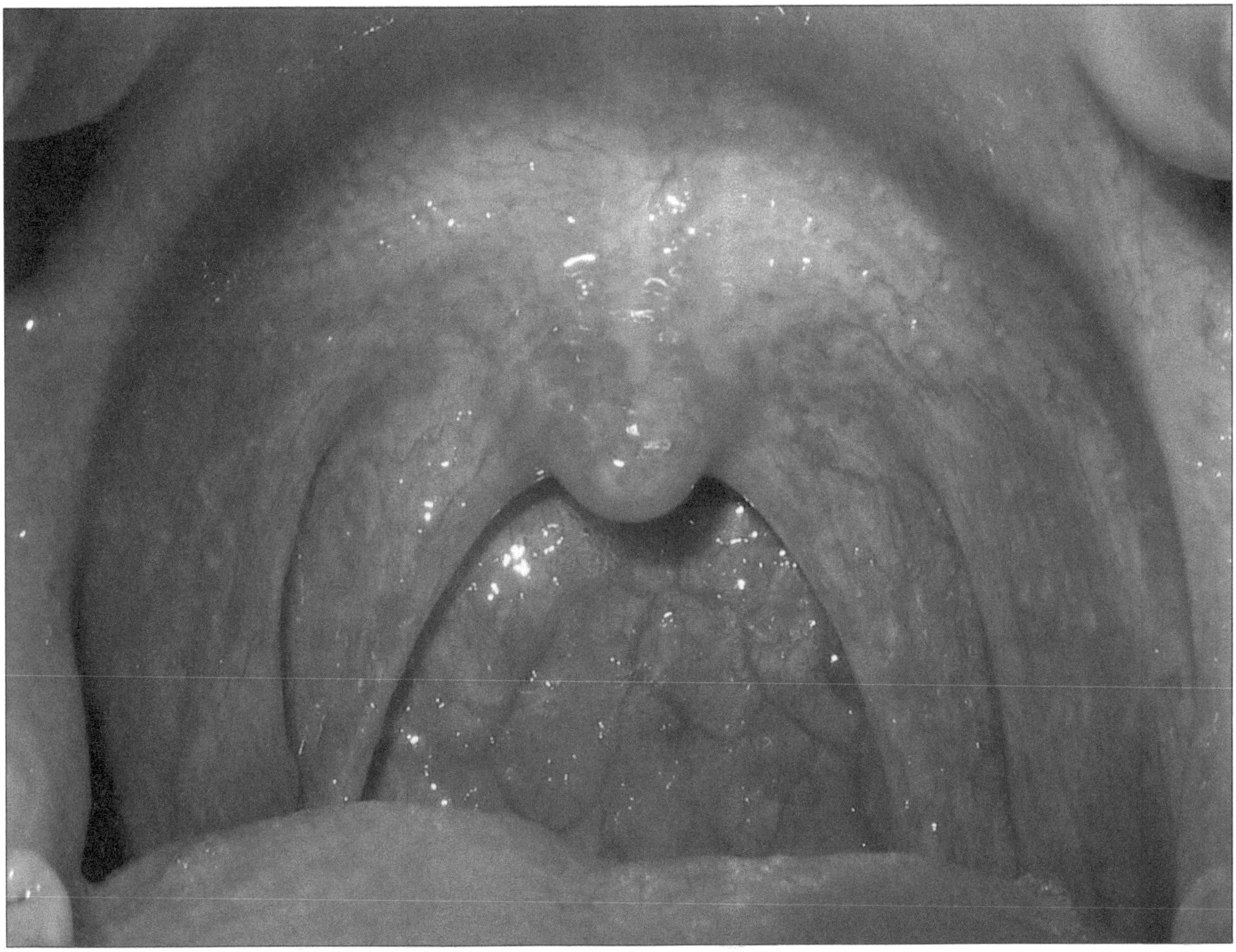

Viral pharyngitis resulting in visible redness. Photo by Dake, via Wikimedia Commons.

Adenovirus is the second most common cause of viral pharyngitis in children in the United States. Herpes Simplex virus-1 (HSV-1), and sometimes HSV-2, can cause painful blisters in the throat and mouth. Additionally, sinusitis and postnasal drip may irritate the pharynx.

Group A *Streptococcus*, which includes *Streptococcus pyogenes* (*S. pyogenes*), causes "strep throat." If the infection spreads to the tonsils, it is called "tonsillitis". Pharyngitis associated with fever and the appearance of pus on the tonsils may indicate streptococcal pharyngitis, which can be diagnosed by a "quick antigen" test and confirmed by a throat culture.

Persistent pharyngitis accompanied by malaise unresponsive to antibiotics may indicate mononucleosis or other nonbacterial causes such as seasonal allergies, inhaling pollutants such as household cleaners or automobile exhaust, and smoking or exposure to second-hand smoke.

TREATMENT AND THERAPY

Viral pharyngitis is treated with acetaminophen or over-the-counter pain remedies and warm salt water gargles. HSV pharyngitis is treated with oral acyclovir or valacyclovir for seven to ten days. Bacterial pharyngitis is treated with a course of antibiotics either orally or by injection. For the most part, *S. pyogenes* strains have remained susceptible to penicillin, and strep throat is treated with oral amoxicillin. Treating strep throat quickly (within ten days) is important to ensure that complications like acute rheumatic heart do not occur.

Proper nutrition is important. Zinc boosts the immune system, has some antiviral activity, and relieves soreness. Vitamin C maintains the immune system and the mucous membranes, and beta-carotene restores the integrity of mucous membranes and supports immune function. If the tonsils are chronically infected, they may require surgical removal (tonsillectomy).

PERSPECTIVE AND PROSPECTS

Viruses cause approximately 40 to 60 percent of cases of pharyngitis, and about 15 percent are associated with streptococcal bacteria. In the United States, children average five sore throats per year and strep infection every four years. The incidence of pharyngitis and strep is highest in children between five and eighteen. Pharyngitis is rare in children below three years of age. Adults experience two sore throats per year and strep infection approximately every eight years. Worldwide, the incidence is higher.

—Marcia J. Weiss, MA, JD

Further Reading

Ferrari, Mario. *PDxMD Ear, Nose, and Throat Disorders*. PDxMD, 2003.

Goldstein, Mark N. "Office Evaluation and Management of the Sore Throat." *Otolaryngologic Clinics of North America*, vol. 25, Aug. 1992, pp. 837-42.

Litin, Scott C., editor. *Mayo Clinic Family Health Book*. 4th ed., HarperResource, 2009.

Morrison, Roger. *Desktop Guide to Keynotes and Confirmatory Symptoms*. Hahnemann Clinic, 1993.

Parker, Philip M., and James N. Parker. *Pharyngitis: A Medical Dictionary, Bibliography, and Annotated Research Guide to Internet References*. ICON Health Publications, 2004.

Pechère, Jean Claude, and Edward L. Kaplan, editors. *Streptococcal Pharyngitis: Optimal Management*. S. Karger, 2004.

Vorvick, Linda J. "Pharyngitis." *MedlinePlus*, 8 Jan. 2012.

NEISSERIAL INFECTIONS

Category: Diseases and conditions

Also known as: Bacterial meningitis, clap, gonococcus, gonorrhea, meningococcus, spinal meningitis

Anatomy or system affected: Anus, adrenal glands, blood, brain, central nervous system, cervix, eyes, fallopian tubes, genitourinary tract, joints, ovaries, peritoneum, skin, spinal cord, throat, urethra

Specialties and related fields: Internal medicine, neurology, orthopedics, otolaryngology, pediatrics, proctology, public health, urology, women's health

Definition: *Neisseria* are gram-negative, bean-shaped cocci that grow in pairs or diplococci; the bacterium infects the genitourinary tract, rectum, throat, conjunctiva, and the tissue covering the brain and spinal cord

KEY TERMS

capsule: a polysaccharide layer that lies outside the cell envelope

gonococci: bacteria that cause gonorrhea

lipooligosaccharide: glycolipids found in the outer membrane of some types of gram-negative bacteria, such as *Neisseria* and *Haemophilus* species

lipopolysaccharide: large molecules consisting of a lipid and a polysaccharide composed of O-antigen, outer core, and inner core joined by a covalent bond; they are found in the outer membrane of gram-negative bacteria

meninges: the three membranes (the dura mater, arachnoid, and pia mater) that line the skull and vertebral canal and enclose the brain and spinal cord.

meningococci: bacteria that cause some forms of meningitis and cerebrospinal infection

neutrophils: the most abundant type of granulocytes that make up 40 to 70 percent of all white blood cells and form an essential part of the innate immune system

phagocytosis: the ingestion of bacteria or other material by phagocytes

virulence factors: bacteria-associated molecules required for them to cause disease

CAUSES

Neisseria are cocci, which means their cells are spherically or nearly spherically shaped. When stained with a Gram stain, gram-negative bacteria stain pink because they fail to retain the crystal violet. Gram-negative bacteria, like *Neisseria*, have a thick peptidoglycan layer outside their cell membrane. Outside the peptidoglycan layer is an outer membrane with unusual lipids in its outer leaf called "lipopolysaccharides" (LPS).

Two *Neisseria* species, *Neisseria meningitidis* (*N. meningitidis*) and *Neisseria gonorrhoeae* (*N. gonorrhoeae*), cause disease in humans. *N. meningitidis* causes meningitis, an infection of the meningeal membranes that surround the brain and spinal cord. *N gonorrhoeae* causes a sexually transmitted infection popularly known as gonorrhea. *N. meningitidis* is typically called "meningococcus," and *N. gonorrhoeae* is usually called "gonococcus."

In the laboratory, *Neisseria* require very rich culture media to grow. They are typically grown on chocolate agar or another rich medium called "Thayer-Martin agar." *N. meningitidis* ferments the disaccharide maltose, but *N. gonorrhoeae* do not. These two organisms grow best in a carbon-dioxide enriched atmosphere.

Meningococci have a thick polysaccharide layer outside their cell walls called a "capsule." Capsules surround these bacteria and protect them from white blood cells and other potentially damaging molecules. Meningococci are grouped based on the composition of their polysaccharide capsule. Meningococci in groups A, B, C, W, X, and Y cause meningitis, and strain B, C, and Y cause most illnesses observed in the United States. However, *N. meningitidis* inhabits the throat of about ten percent of adults without causing disease. Gonococci are not encapsulated.

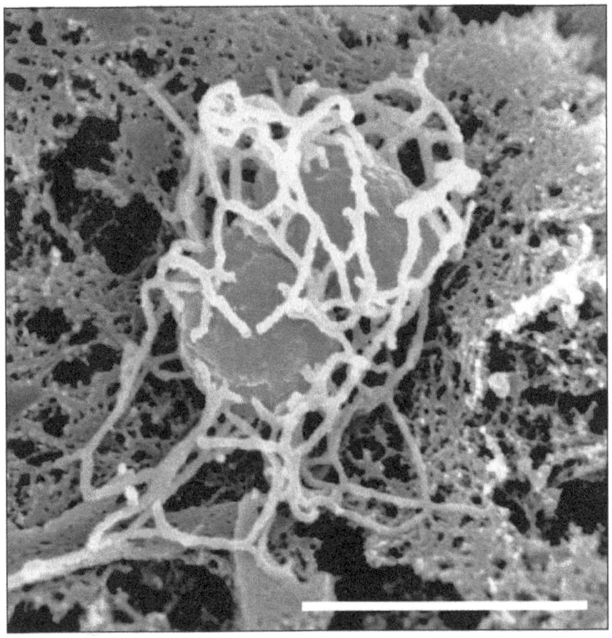

Neisseria meningitidis. *Image by Arthur Charles-Orszag, own work, via Wikimedia Commons.*

MENINGOCOCCAL DISEASE

Meningococci have an armory of virulence factors. The polysaccharide capsule makes them slippery, preventing white blood cells from getting a good grip on them. Second, *N. meningitidis* have small, hairlike extensions called "pili" that help them attach to cells. Third, beneath the capsule, the outer membrane has two proteins called "opacity proteins" (Opa). Opa proteins work with pili to bind meningococci to host cells. Fourth, *N. meningitidis* secretes an enzyme (IgA protease) that degrades secretory antibodies. Immunoglobulin A (IgA) is secreted across mucosal surfaces (respiratory tract, pharynx, nasal cavity, and other places) to neutralize invading organisms. Meningococci take away one of the body's main defenses.

Should meningococci get phagocytosed by defending white blood cells called "neutrophils," they survive inside the neutrophil. Neutrophils use hydrogen peroxide (H_2O_2) to kill cells they have eaten. However, meningococci release the enzyme catalase, breaking down hydrogen peroxide into water and oxygen. The neutrophil takes these bacteria for a ride around the body while they steal its resources to divide inside it. Eventually, the bacteria overwhelm and kill the neutrophil. The dying neutrophil releases meningococci into the bloodstream when it bursts, known as meningococcemia.

In the bloodstream, meningococci have an array of protective mechanisms and secrete toxins that cause blood vessels to leak. The outer membrane of meningococci has a variant on LPS called "lipooligosaccharide" (LOS). This complex sugar-lipid molecule triggers an extensive immune response that dilates blood vessels and activates clotting pathways throughout the body. This phenomenon, disseminated intravascular coagulation (DIC), causes blood pressure to drop precipitously, preventing blood from flowing to vital organs; this is called "septic shock."

Because DIC uses up all the platelets, blood cannot clot normally, and severe bleeding occurs. An organ where copious bleeding occurs is the adrenals glands above the kidneys. Blood pooling in the adrenal glands increases the pressure inside them, pinching their blood vessels shut. This causes the death of adrenal tissue, a condition known as Waterhouse-Friderichsen syndrome. The absence of adrenal hormones like aldosterone and cortisol aggravates the shock.

Meningococcemia can lead to meningitis because the organism uses toxins and other molecules to break through the blood-brain barrier and infect the meningeal membranes surrounding the brain and spinal cord.

People with poorly functioning immune systems (individuals who abuse alcohol, have human immunodeficiency virus (HIV), malignancies, or diabetes) have a higher risk of complications and severe disease. Since meningococci are encapsulated bacteria, and the spleen plays such a seminal role in destroying such microorganisms, people without a functioning spleen (e.g., those whose spleens were surgically removed or have sickle-cell disease) are at high risk of meningococcal infections.

GONOCOCCAL DISEASE

Like their bacterial cousin, gonococci have a respectable armamentarium of virulence factors. Pili help them attach to mucosal surfaces and help gonococci exchange genes. If that isn't bad enough, *N. gonorrhoeae* have genetic mechanisms by which they can vary the content of their pili genes. If one cell comes upon some useful genes, they are passed to the entire gonococcal population. This event, known as phase variation, prevents the immune system from developing antibodies that definitively recognize infecting gonococci. Phase variation also explains where there is no vaccine for gonorrhea.

Like meningococci, gonococci make an IgA protease that protects at mucosal surfaces. Gonococci also make catalase that protects them inside neutrophils. Neutrophils loaded with gonococci transport them to the bloodstream. When the infected neutrophils explode, they colonize the bloodstream, causing gonococcemia. LOS in the gonococcal outer membrane causes sepsis, much like its meningococcal cousin. The most insidious ability of *N. gonorrhoeae* is their tendency to attach the sugar sialic acid to their exteriors (sialylation). Human cells have sialic acid on their exteriors. Thus, gonococci effectively camouflage themselves against the immune system.

N. gonorrhoeae usually enter the body through sexual contact. In males, they cause inflammation of the urethra (urethritis), prostate (prostatitis), or epididymis (epididymitis). In women, it causes inflammation of the urethra, vagina (vaginitis, most common), and cervix (cervicitis). From the cervix, gonococci can spread to the uterus, fallopian tubes, and, on occasion, the ovaries. Inflammation of these reproductive organs is called "pelvic inflammatory disease" (PID). Suppose PID spreads to the peritoneum, the membranes that line the abdominal wall and its organs, and the Glisson capsule that lines the liver. In that case, it causes a complication called "Fitz-High-Curtis syndrome."

During gonococcemia, the bacteria can spread to other organs like the joints or the heart. If gonococci spread to the joints, they can cause septic arthritis. Septic arthritis is more common in sexually active adolescents. If they infect the heart valves, they can cause endocarditis.

Infected pregnant mothers can pass gonococci to their babies during delivery. The bacteria infect the baby's eyes, causing neonatal conjunctivitis, usually appearing two to five days after delivery.

RISK FACTORS

Meningococci are the only organism that causes epidemic meningitis, usually among people living in tight quarters. *N. meningitidis* is transmitted in droplets caused by coughing. Hence, persons in close contact, such as dormitory residents or personnel in military barracks, are at risk for transmission. Tobacco smokers and those exposed to second-hand smoke are more susceptible.

N. gonorrhoeae is transmitted through sexual contact; failing to use condoms and having multiple partners produces the greatest risk.

SYMPTOMS

Meningococcal meningitis is characterized by a rapidly rising fever, headache, and a stiff neck, followed by coma. Common symptoms are a stiff neck disallowing the infected person from touching chin to chest and a spotty rash that does not "bleach" when pressed with a clear glass. Meningococcemia causes low blood pressure and a racing heart rate.

Men with gonorrhea have pain with urination and have a pus-filled discharge. These symptoms usually develop less than one week after sexual contact with an infected partner. If untreated, the gonococci can infect the prostate gland. Sterility results if the sperm ducts are blocked with scar tissue.

The symptoms of gonorrhea are less pronounced in women. Gonococci enter the vagina and then move into the cervix, uterus, and Fallopian tubes. The only symptom of infection is a pus-filled discharge. Sterility, a long-term consequence of gonorrhea in women, occurs when scar tissue is deposited and blocks the Fallopian tubes. Pelvic inflammatory disease may cause lower abdominal pain and lead to loss of fertility.

Anal sex with infected partners can cause proctitis. The symptoms of gonococcal proctitis are rectal bleeding, diarrhea, pain during bowel movements, pain on the left side of the abdomen, rectal pain, and passing mucus during bowel movements.

Those who engage in oral sex with infected partners can suffer from gonococcal pharyngitis (sore throat). The symptoms are pain when swallowing and burning in the throat.

People with gonococcal arthritic have inflammation in multiple joints. There is usually painful swelling in the elbows, wrists, and ankles. Those with gonococcal endocarditis usually show fever, chills, and tiredness.

Infants with neonatal conjunctivitis have swollen eyelids and pussy discharge from the eyes.

SCREENING AND DIAGNOSIS

Definitive diagnosis of meningococcal meningitis is performed by identifying the bacterium in the cerebrospinal fluid, retrieved by a spinal tap. In the laboratory, clinical specimens are applied to a glass slide and stained using the Gram-staining procedure.

Gonococci are identified by culturing vaginal or urethral swabs. The cultures should grow paired, Gram-negative cocci that look like coffee beans.

Gram-negative, bean-shaped diplococci that are visible under the microscope indicate infection. Especially in females, the Gram stain can yield a false-negative result. Polymerase chain reaction analysis to detect bacterial deoxyribonucleic acid (DNA) is performed in many labs for meningococci and gonococci.

TREATMENT AND THERAPY

Rifampin is used to prevent the development of meningitis in asymptomatic persons exposed to infected persons. The drug of choice to treat *N. meningitidis* meningitis is ceftriaxone. Suppose the isolate is susceptible to penicillin G. In that case, the patient is switched to penicillin G. Gonococcal infections are treated with ceftriaxone and doxycycline.

PREVENTION AND OUTCOMES

Meningitis caused by four groups is preventable with a tetravalent glycoconjugate vaccine. *Menactra, Menveo*, and *MenQuadfi* are four available meningococcal vaccines that protect against groups A, C, W, and Y strains. *Bexsero* and *Trumenba* are vaccines against group B meningococci. The Centers for Disease Control and Prevention recommends that all children eleven to twelve years old get the tetravalent vaccine and a booster at sixteen years. Teenagers and young adults should get the vaccine against the B strain if they are at high risk for meningococcal disease.

Suspicion of meningococcal meningitis will cause public health officials to recommend immediate antibiotic treatment for all close contacts of the infected person.

Gonorrhea is preventable with condom use or through sexual abstinence.

—*Kimberly A. Napoli, MS and Michael A. Buratovich, PhD*

Further Reading

Dombrowski, Julia C. "Chlamydia and Gonorrhea." *Annals of Internal Medicine*, vol. 174, no. 10, 2021, p. ITC145-ITC160, doi:10.7326/AITC202110190.

Handsfield, H. H., et al. "*Neisseria gonorrhoeae*." *Mandell, Douglas, and Bennett's Principles and Practice of Infectious Diseases*, edited by Gerald L. Mandell, John F. Bennett, and Raphael Dolin, 7th ed., Churchill Livingstone/Elsevier, 2010.

Nguyen, Nixon, and Derrick Ashong. "Neisseria Meningitidis." *StatPearls*. StatPearls Publishing, 12 Oct. 2021.

Schrier, Robert W., editor. *Diseases of the Kidney and Urinary Tract*. 8th ed., Wolters Kluwer Health/Lippincott Williams & Wilkins, 2007.

Shmaefsky, Brian. *Meningitis*. Rev. ed., Chelsea House, 2010.

GONORRHEA

Category: Diseases and disorders

Also known as: The clap

Anatomy or system affected: Anus, blood, cervix, circulatory system, epididymis, eyes, genitourinary system, fallopian tubes, immune system, joints, lymphatic system, penis, prostate, reproductive system, throat, uterus,

Specialties and related fields: Bacteriology, gynecology, immunology, internal medicine, microbiology, ophthalmology, orthopedics, otolaryngology, pathology, pharmacology, preventive medicine, proctology, public health

Definition: a sexually transmitted infection caused by a gram-negative diplococcus that evades the immune system and can establish long-term infections with significant consequences for reproductive health

KEY TERMS

contact tracing: also known as partner referral; a process that consists of identifying the sexual partners of infected patients, informing these partners of their exposure to disease, and offering resources for counseling and treatment

phase variation: how bacteria deal with rapidly varying environments without random mutation by varying gene expression, often in an on-off fashion, within different parts of a bacterial population.

pili: short, filamentous projections on a bacterial cell, used for motility or adhering to other bacterial cells

screening procedures: tests carried out in populations that are usually asymptomatic and at high risk for a disease to identify those in need of treatment

sexually transmitted disease: an infection caused by organisms transferred through sexual contact (genital-genital, oral-genital, oral-anal, or anal-genital); the transmission of infection occurs through exposure to lesions or secretions that contain the organisms

virulence factors: bacteria-associated molecules required them to cause disease

CAUSES AND SYMPTOMS

Gonorrhea is the second most common bacterial sexually transmitted infection (STI) in the United States, the most common being chlamydia. In the United States, the incidence of gonorrhea has fallen. In 1995, the incidence was about 150 cases out of every 100,000 persons, down from the mid-1970s of more than 400 cases per 100,000 persons. After 1997, the percentage of infections increased slightly, but by 2009, the rate per 100,000 reached an all-time low of 98.1. The following two years saw a slight increase in the rate of infection, reaching 104.2 in 2011. However, overall, from 2007 to 2011, the rate of infection decreased by nearly 12 percent. The highest incidence of gonorrhea is in sexually active men and women under twenty-five years of age; since 2002, the infection rate for women of any age group has been higher than for men (108.9 cases per 100,000 compared with 98.7 for men). In 2018, there were 583,405 reported cases of gonorrhea, but the actual number of cases is likely much higher due to underreporting and asymptomatic infections.

Gonorrhea is caused by the bacterium *Neisseria gonorrhea* (also known as gonococcus), a gram-negative diplococcus. The bacterium infects the mucous membranes with which it comes in contact, most commonly the urethra and the cervix and the throat, rectum, and eyes. Some men will be asymptomatic, but most will experience urinary discomfort and a purulent urethral discharge. Long-term complications of this infection in men include epididymitis, prostatitis, and urethral strictures (scarring). In women, the disease is more likely to be asymptomatic.

N. gonorrhoeae has a cadre of virulence factors that help them cause disease. First, gonococci have tiny, thread-like extensions of the cell surface called "pili." Pili help *N. gonorrhoeae* attach to mucosal surfaces and physically connect to each other. Attaching to epithelial surfaces helps gonococci colonize the reproductive tract, and by attaching to each other, *N. gonorrhoeae* exchange genes.

Another mediator of gonococcal binding to leukocytes, epithelial, and endothelial cells is Opa proteins. Opa proteins mediate gonococcal binding with leukocytes, epithelial, and endothelial cells. Pili and Opa vary with every infection, a phenomenon known as "phase variation." *N. gonorrhoeae* change the genes that encode pili and Opa proteins, which keeps the immune system guessing. Upon infection, the immune system mounts an immune response against the pili and Opa proteins on the surfaces of gonococci and forms memory of it. A second infection by the same organism causes the immune system to respond quickly and specifically against the organism. However, since gonococci change the antigens on their pili and Opa proteins each time it infects a host, the immune system cannot produce a quick, specific immune response. Phase variation explains why there's no effective vaccine against *N. gonorrhoeae*.

Another important virulence factor is an IgA protease. This enzyme degrades the primary protective antibody found in mucosal secretions, like those in the vagina or cervix. Destroying IgA neutralizes the first line of defense at mucosal surfaces.

If gonococci are engulfed by white blood cells called "neutrophils," the bacteria release the enzyme catalase. Catalase breaks down hydrogen peroxide released by neutrophils to kill bacteria. *N. gonorrhoeae* can live inside neutrophils and hide from the immune system.

Finally, in an act of molecular camouflage, *N gonorrhoeae* can attach sialic acid molecules to their surfaces. By wrapping itself in sialic acid, a molecule found in the host's cells, *N. gonorrhoeae* makes itself anonymous to the host's immune system.

Women with symptoms may have purulent vaginal discharge, urinary discomfort, urethral discharge, lower abdominal discomfort, or pain with intercourse. Pelvic inflammatory disease (PID) and its consequences may occur if gonorrheal infection ascends past the cervix into the upper genital tract (uterus, Fallopian tubes, ovaries, and pelvic cavity) in women. Complications of PID include infertility and an increased risk of ectopic pregnancy.

In rare cases, gonorrhea can enter the bloodstream and disseminate throughout the body, causing fever, joint pain, and skin lesions. Gonorrhea can infect the heart valves, pericardium, and meninges as well.

Information on Gonorrhea

Causes: Bacterial infection through intercourse

Symptoms: In men, sometimes urinary discomfort, and discharge, with long-term complications of epididymitis, prostatitis, and urethral scarring; in women, sometimes vaginal discharge, urinary discomfort, urethral discharge, lower abdominal discomfort, and pain with intercourse, with possible pelvic inflammatory disease, infertility, and increased risk of ectopic pregnancy

Duration: Acute

Treatments: Antibiotics, counseling regarding safe sex

When it infects the joints, a condition known as septic arthritis occurs, characterized by pain and swelling and potential destruction of the joints.

An infected mother can transmit gonorrhea to infants through the birth canal, leading to an eye infection that can damage the eye and impair vision. Fortunately, erythromycin eye drops are routinely given to newborns to prevent eye infections. These eye drops are effective against *Neisseria gonorrhea* as well as *Chlamydia trachomatis*.

TREATMENT AND THERAPY

Treatment for gonorrhea consists of the use of antibiotics. With the development of penicillin-resistant strains of gonorrhea, effective therapy relies on antibiotics, such as ceftriaxone, to which gonorrhea remains susceptible. In uncomplicated cases of gonorrheal infection, such as cervicitis or urethritis, a single dose is given.

A patient who has risk factors for STIs (primarily contact with a suspected infected partner) or a clinical picture suggestive of gonorrhea, or both, may receive treatment presumptively before confirmatory laboratory test results for gonorrhea are available. Because many patients with gonorrhea also have chlamydia, patients are treated concomitantly with an antibiotic directed against chlamydia, such as doxycycline. Azithromycin is an alternative treatment and is preferred if the patient is pregnant. Once laboratory test results confirm the diagnosis of gonorrhea, patients should be advised that the Centers for Disease Control and Prevention recommends that they be tested for other STIs, such as human immunodeficiency virus (HIV), hepatitis B and C, and syphilis.

As with all STIs, a key component of therapy includes counseling regarding safer sex. Preventative measures include using barrier contraceptives, such as condoms, and the avoidance of high-risk sexual behaviors. Contact tracing is another crucial element to STI treatment. It notifies the patient's sexual partners of their exposure to gonorrhea or other STIs. Contact tracing also involves offering resources to these partners for medical attention. Contact tracing can prevent the reinfection of the patient through subsequent sexual encounters and the spread of STIs from the patient's partner to their subsequent sexual partners.

PERSPECTIVE AND PROSPECTS

The symptoms of gonorrhea have been described in numerous cultures in the past, including those dating back to the ancient Chinese, Egyptians, and Romans. Albert Neisser first identified the actual gonorrhea bacterium in the 1870s, and it was one of the first bacteria ever discovered. *Neisseria gonorrhea* has continued to be well-studied on both the molecular and epidemiological levels.

Antibiotic therapy, in the form of sulfanilamide, was first used to combat *N. gonorrhea* in the 1930s. By the 1940s, however, gonococcal strains resistant to this antibiotic appeared, and the therapy of choice became penicillin. Over the next several decades, *N. gonorrhea* evolved the ability to resist penicillin, forcing clinicians to use other drugs to combat the bacterium, such as ceftriaxone and ciprofloxacin. In the 1980s, the Centers for Disease Control and Prevention (CDC) instituted surveillance programs to monitor antibiotic resistance patterns in different US cities. Continued success in combating *N. gonorrhea* will depend on the ability to minimize the development of antibiotic resistance.

Finally, since many patients with gonorrhea infection have no symptoms, screening programs of asymptomatic patients in high-risk groups (those younger than twenty-five and/or with multiple sexual partners) play a vital role in decreasing the incidence of *N. gonorrhea* infections.

—Anne Lynn S. Chang, MD and Michael A. Buratovich, PhD

Further Reading

Armed Forces Health Surveillance Center. "Predictive Value of Reportable Medical Events for *Neisseria gonorrhoeae* and *Chlamydia trachomatis*." *MSMR*, vol. 20, no. 2, 2013, pp. 11-14.

Beharry, M. S., T. Shafii, and G. R. Burstein. "Agnosis and Treatment of Chlamydia, Gonorrhea, and Trichomonas in Adolescents." *Pediatric Annals* vol. 42, no. 2, 2013, pp. 26-33.

Centers for Disease Control and Prevention. "Sexually Transmitted Diseases." 17 June 2021, www.cdc.gov/std/default.htm. Accessed 22 Nov. 2021.

Cristaudo, Antonio, and Massimo Giulani, editors. *Sexually Transmitted Infections: Advances in Understanding and Management*. Springer, 2020.

Jameson, J. Larry, et al., editors. *Harrison's Principles of Internal Medicine*. 20th ed., McGraw-Hill, 2018.

Ryan, Kenneth J., et al., editors. *Sherris Medical Microbiology: An Introduction to Infectious Diseases*. 7th ed., McGraw-Hill, 2018.

Sutton, Amy L., editor. *Sexually Transmitted Diseases Sourcebook*. 5th ed., Omnigraphics, 2013.

SYPHILIS

Category: Diseases and disorders

Also known as: "Bad blood," bejel (endemic syphilis)

Anatomy or system affected: Anus, bones, brain, eyes, genitals, heart, joints, kidneys, nervous system, reproductive system

Specialties and related fields: Bacteriology, embryology, epidemiology, gynecology, internal medicine, microbiology, neonatology, neurology, pediatrics, public health, rheumatology

Definition: a sexually transmitted disease caused by the spirochete bacterium *Treponema pallidum* that can progress from a genital lesion to a systemic disorder involving multiple organs

KEY TERMS

chancre: a painless ulcer, particularly one developing on the genitals because of venereal disease *endarteritis*: inflammation of the inner lining of an artery

endoflagella: whip/propeller-like structures found beneath the outer membrane of spirochetes

gumma: a small soft swelling that is characteristic of the late stages of syphilis and occurs in the connective tissue of the liver, brain, testes, and heart

spirochetes: a flexible spirally twisted bacterium, especially one that causes syphilis

CAUSES AND SYMPTOMS

Syphilis is a sexually transmitted disease (STD) resulting from infection by *Treponema pallidum*. *Treponema pallidum* is a spirochete. Spirochetes are gram-negative bacteria that are long, corkscrew-shaped cells. These bacteria are distinguished from other bacteria by endoflagella (also known as axial filaments). Endoflagella are embedded in the cell membrane and project into the periplasmic space between the cell membrane and the cell wall. In other bacteria, flagella extend through the cell wall to the exterior. As endoflagella rotate, the spirochete twists in a corkscrew fashion that helps it burrow through its medium.

The history of syphilis is unclear. Evidence exists that its origin may have been linked with a disease, yaws, found in the Western Hemisphere at the time of explorer Christopher Columbus (1451-1506). Yaws is a relatively mild disease transmitted through contaminated objects or open skin lesions, but not generally through sexual transmission; it results from infection by a subspecies of *Treponema* called *T. pallidum ssp. pertenue*. The theory suggests that this may have been the form of the disease brought back to Europe on one of Columbus's ships. Mutation and sexual transmission in the population of Europe may have produced the more serious form of the disease.

Syphilis is a disease characterized by several distinct stages. Initial exposure to the organism during sexual intercourse results in the formation of a painless skin lesion called a "chancre" at the site of infection (primary syphilis), developing anywhere from a week to months after infection. Spirochete bacteria may be isolated from the lesion and be found inside white blood cells (macrophages and neutrophils) that infiltrate the area. The white cells may be a mechanism for the systemic spread of the organism. The lesion generally heals spontaneously, leaving the impression that the disease has been eliminated.

During the weeks after the formation of the chancre, the spirochetes multiply to large numbers and become disseminated throughout the body. A second stage (secondary syphilis) often appears within two months following regression of the chancre. Symptoms are often flu-like, with malaise, headache, fever, and joint aches. A skin rash often appears, covering most of the body. Sores may develop in the mouth and throat and on many of the mucous membranes in the body. The organism is highly transmissible during this period. The rash and other symptoms generally fade over weeks.

Approximately 10 percent of untreated cases develop a third, or tertiary, stage of syphilis. The organism can infiltrate any organ or system in the body, resulting in soft tumors (gummas) in the eyes, lungs, bone, brain, or other organs. Symptoms are characteristic of the organ infected. For example, infection of the brain or other areas of the central nervous system is described as neurosyphilis or syphilitic dementia, characterized by memory loss, personality changes, and neurodegeneration. Even if tertiary syphilis is treated, the prognosis for the patient at this stage is often poor.

Cardiovascular syphilis occurs when *Treponema pallidum* affects the heart or large vessels that service it. Larger vessels have tiny arterioles that supply blood to their vessel walls called "vasa vasorum." *T. pallidum* inflames the vasa vasorum, a condition called "endarteritis," which deeply damages the aorta. Profound damage to the aorta wall weakens it and increases the risk of aortic aneurysms.

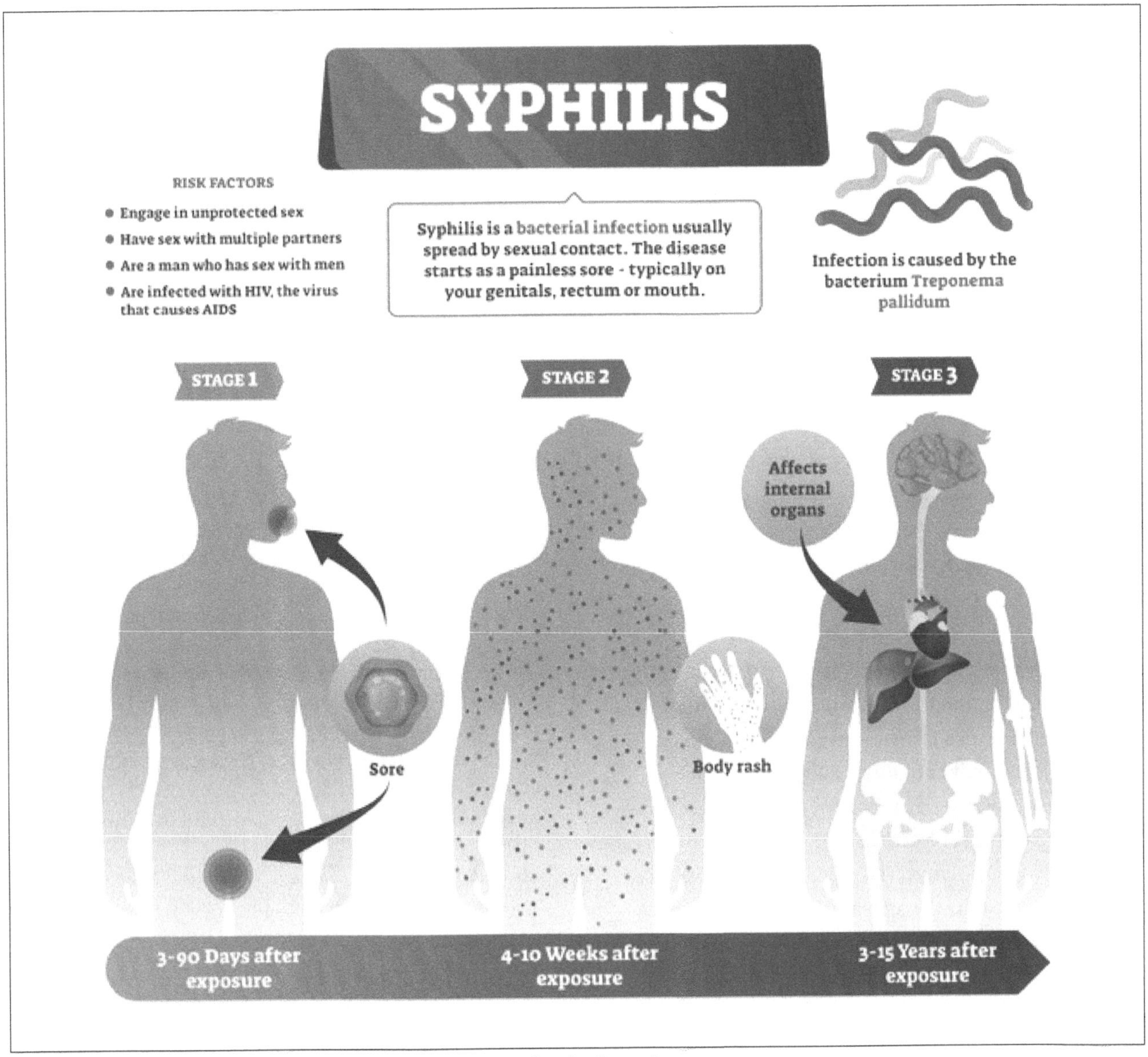

Signs and dangers of syphilis. Image via iStock/VectorMine. [Used under license.]

Treponema can cross the placenta, and the infection of a pregnant woman may result in congenital syphilis or infection of her unborn child. Infection may kill the fetus or cause it to be born with obvious deformities such as blindness or physical abnormalities. The infant may also be asymptomatic. An undiagnosed infection will likely progress, with symptoms appearing within weeks after birth. It is common for a rash to appear, with evidence of tertiary stage neurosyphilis or cardiovascular syphilis.

A diagnosis of syphilis can be made through microscopic examination of lesion exudates, noting the presence of spirochetes. However, *Treponema* is notoriously unstable, and the test must be made shortly after obtaining the specimen. Diagnosis is commonly based on serological testing for serum antibodies against the organism or tissue lipids released from infected or damaged cells.

TREATMENT AND THERAPY

Penicillin is the preferred method of treatment for both primary and secondary syphilis. Suppose the disease has progressed to the tertiary stage. In that case, antibiotic treatment will still eliminate the organism. However, it will not reverse organ damage that may have occurred. Treatment for related organ involvement is symptomatic.

Alternative antibiotics include erythromycin, tetracyclines, and chloramphenicol, if necessary. However, only penicillin is effective during the tertiary stage or for use in pregnant women.

PERSPECTIVE AND PROSPECTS

Despite the long-time existence of effective therapy, penicillin or alternative antibiotics, and the absence of any reservoir for *T. palladium* other than humans, syphilis remains the third most common sexually transmitted bacterial disease in the West. Only gonorrhea and chlamydia are more common.

As a result of effective therapy and the generally obvious symptoms of the disease, tertiary syphilis has largely disappeared. However, sexual practices continue to sustain the spread of the disease, with approximately fifty thousand cases reported each year in the United States. Three factors are primary contributors to the resurgence of the disease: prostitution, the increase in riskier sexual practices among homosexual men, and general apathy toward a relatively easy disease to treat in its early stages. An increase in congenital syphilis also reflects the presence of the disease in women of childbearing years. In the absence of condom use, both unwanted pregnancy and the spread of STDs such as syphilis may result.

No vaccine currently exists for syphilis. The inability to culture the organism in the laboratory has made research related to *Treponema* difficult. The organism does not infect animals other than humans to act as a vaccine production method. However, genetic engineering has resulted in the cloning of several bacterial gene products related to surface proteins and virulence factors, allowing the possibility for a vaccine in the future. For now, the best means of controlling syphilis remains to prevent its spread through education and safer-sex practices and early treatment of those infected.

—Richard Adler, PhD

Further Reading

Centers for Disease Control and Prevention. *The National Plan to Eliminate Syphilis from the United States*. US Department of Health and Human Services, 2006.

Mandell, Gerald L., John E. Bennett, and Raphael Dolin, editors. *Mandell, Douglas, and Bennett's Principles and Practice of Infectious Diseases*. 7th ed., Churchill Livingstone/Elsevier, 2010.

Murray, Patrick R., Ken S. Rosenthal, and Michael A. Pfaller. *Medical Microbiology*. 6th ed., Mosby/Elsevier, 2009.

Parker, James N., and Philip M. Parker, editors. *The Official Patient's Sourcebook on Syphilis*. Icon Health, 2002.

Quetel, Claude. *The History of Syphilis*. Johns Hopkins UP, 1990.

Sutton, Amy L., editor. *Sexually Transmitted Diseases Sourcebook*. 3rd ed., Omnigraphics, 2006.

Tularemia

Category: Diseases and conditions

Also known as: Francis disease, deer-fly fever, rabbit fever, market men disease, water-rat trappers' disease, wild hare disease (yato-byo), and Ohara disease

Anatomy or system affected: All

Specialties and related fields: Bacteriology, cardiology, infectious disease, internal medicine, hematology, microbiology, neurology, ophthalmology, otolaryngology, public health

Definition: a severe infectious bacterial disease of animals transmissible to humans, characterized by ulcers at the site of infection, fever, and loss of weight

KEY TERMS

acid phosphatase: an enzyme that acts to liberate phosphate under acidic conditions

capsule: a polysaccharide layer outside the cell envelope

febrile: having or showing the symptoms of a fever

type IV pili: filaments on the surfaces of many gram-negative bacteria that mediate an extraordinary array of functions, including adhesion, motility, microcolony formation and secretion of proteases and colonization factors

suppuration: the formation or discharge of pus

virulence factor: cellular structures, molecules, and regulatory systems that enable microbial pathogens, be they bacteria, viruses, fungi, or protozoans, to colonize specific niches in the host's body, evade the immune system, and obtain nutrition from the host

INTRODUCTION

Tularemia is a rare bacterial infection that can be deadly. Governments have studied its use as a biological weapon that releases bacteria into the air. The disease occurs naturally through exposure to infected animals or insects or contaminated water or food.

There are different types of the disease, depending on where the exposure and symptoms occur. These types are ulceroglandular (skin), glandular (lymph nodes), oculoglandular (eye), oropharyngeal (mouth and throat), intestinal (bowels), pneumonic (lung), and typhoidal (systemwide disease).

CAUSES

The bacterium *Francisella tularensis* causes tularemia. This organism is a gram-negative coccobacillus (a cross between a spherical coccus and a rod-shaped bacillus). There are four main subspecies of *Francisella tularensis*: F. tularensis subspecies tularensis (also called F. tularensis type A or Francisella neoarctica), F. tularensis subspecies holarctica (also called F. tularensis type B), F. tularensis subspecies novicida, and F. tularensis subspecies mediasiatica. Of these, F. tularensis subspecies tularensis is the most virulent, and subspecies tularensis and holarctica cause the most human and animal diseases. Subspecies holarctica has the widest geographic distribution.

F. tularensis is usually found in small animals, such as mice and rabbits. The germs can survive for weeks in a cool, moist environment. A person can catch the disease if bitten by an infected animal, tick, or deer fly. Infection can also occur through contact with an infected animal's tissues or contaminated water, food, or soil. The bacteria also can enter a person's body through the lungs, eyes, mucous membranes, or skin. This infection is not transmitted from person to person.

F. tularemia has several virulence factors that help it evade the immune system and cause disease. This microorganism is covered by a polysaccharide capsule that protects it from being phagocytosed by white blood cells. Beneath the capsule is the outer membrane that contains a modified lipid-sugar complex called "lipopolysaccharide" (LPS). White blood cells, like macrophages and neutrophils, use a surface receptor called "Toll-like receptor 4" (TLR4) to bind LPS in the outer membrane of gram-negative bacteria and phagocytose them. Bacterial binding stimulated white blood cells to release small proteins called "cytokines" that evoke inflammation. Activated white blood cells also make nitric oxide to damage and destroy the invading bacteria.

Unfortunately, the *F. tularemia*'s LPS is unique and does not bind TLR4. Therefore, this bacterium interacts with white blood cells without activating them. This microbe uses type IV pili to attach to macrophages. Macrophages phagocytose them, but *F. tularemia* produces an enzyme called "acid phosphatase" that prevents phagosome-lysosome fusion. Consequently, *F. tularemia* can live and divide within macrophages. Additionally, these bacteria synthesize an iron-stealing molecule called a "siderophore." These molecules bind to iron so tightly that they can snatch it from host cells.

The infected macrophages can ferry *F. tularensis* to the bloodstream and, from there, to any organ in the body. *F. tularensis* cause bloodstream infections, known as bacteremias or sepsis. It can also infect lymph nodes and fill them with pus (suppuration), infect the liver (hepatitis), kidneys, and cause renal failure, or the lungs and cause pneumonia.

RISK FACTORS

The main risk factor for tularemia is exposure to *F. tularensis*. Exposure can occur through hunting, trapping, or butchering infected animals; working with infected animals or their tissue; working in a laboratory with the bacteria; biological terrorism; eating meat from an infected animal, and bites by an infected mosquito or tick.

SYMPTOMS

Tularemia comes in six different forms: ulceroglandular, glandular, oculoglandular, oropharyngeal, pneumonic, and typhoidal. *F. tularemia* transmitted by ticks and flies causes the ulceroglandular and glandular forms. These forms of tularemia cause a skin and lymph node infection in the ulceroglandular form and just the lymph nodes in the glandular form. Splashing infected material into the eyes or rubbing the eyes with contaminated fingers causes the oculoglandular form. Ingesting contaminated food or water causes the oropharyngeal form, which affects the mouth and throat. Tularemic pneumonia is either primary when someone inhales contaminated droplets or secondary when *F. tularemia* spreads to the lungs through the blood after an initial infection at a different site. The typhoidal form results from entry at any site and causes a systemic febrile illness.

Symptoms usually occur three to five days after exposure, but they can begin earlier or later. Symptoms vary depending on where the bacteria enter the body. Other factors include the number of bacteria, their strength, and the ability of the infected person's immune system to fight the germs.

Pneumonic symptoms include fever, chills, fatigue, headache, body aches, sore throat, cough, and a burning sensation or pain in the chest. Ulceroglandular symptoms include a raised, red bump that continues to swell. The raised area opens, drains pus, and forms an ulcer, and it may form a dark scab. Other symptoms are swollen and tender lymph nodes, a fever, and chills. Glandular symptoms include swollen, tender lymph nodes that are not sore. Oculoglandular symptoms include sensitivity to light; tearing; a puffy eyelid; swelling, redness, and sores in the eye; and swollen lymph nodes. Oropharyngeal symptoms include irritated membranes in the mouth, sore throat, ulcers in the throat or tonsils, and swollen lymph nodes. Intestinal symptoms include fever, abdominal pain, diarrhea, and vomiting.

Typhoidal symptoms include fever, chills, headache, muscle aches, poor appetite, nausea, vomiting, diarrhea, abdominal pain, and cough. Symptoms of progression from other types include swollen lymph nodes, difficulty breathing, bleeding, confusion, coma, organ failure, shock, and death.

SCREENING AND DIAGNOSIS

A doctor will ask about symptoms, medical history, and possible sources of exposure and will perform a physical exam. Tests may include a chest X-ray, examination of body fluids using special techniques and precautions, a skin test to assess immune response, a culture of body fluids to check for bacteria, and a blood test to detect antibodies to the bacteria. Other cases in the infected person's environment would alert health-care workers of the possibility of a bioterrorism attack.

F. tularensis is a very picky (fastidious) bacterium. It does not grow on most microbiological media like blood agar but requires an enriched culture medium. Suppose the attending physician has sufficient reason to suspect that *F. tularensis* is the suspected infecting agent. In that case, cultures should be grown on cysteine enriched chocolate agars, such as buffered charcoal yeast extract (BCYE) and cysteine heart agar with blood (CHAB). Chocolate contains lysed red blood cells, which gives it a deep brown chocolate color. BCYE is a buffered charcoal yeast extract medium enriched with L-cysteine and lysed blood. CHAB is a glucose cysteine agar base enriched with thiamine and lysed blood. *F. tularensis* is a slow-growing microorganism that makes colonies in about 47 to 72 hours. On BCYE, *F. tularensis* forms round, gray-white colonies. On CHAB, it forms greenish-white, round, smooth, mucoid colonies.

TREATMENT AND THERAPY

Antibiotics typically produce a quick response to lung disease. The drugs are injected into a muscle or given through a vein. Later in treatment, some drugs can be given by mouth. Treatment lasts ten to fourteen days. The antibiotics that might be prescribed include streptomycin, with or without chloramphenicol; gentamicin; doxycycline; and quinolone antibiotics, such as ciprofloxacin. In addition to drug therapy, the lymph nodes may require draining. All cases of tularemia are reported to public health officials.

PREVENTION AND OUTCOMES

Antibiotics that effectively treat tularemia include the aminoglycosides streptomycin and gentamicin, tetracyclines, fluoroquinolones, and chloramphenicol. In most cases, the preferred treatment is intravenous gentamicin.

Public health officials may order antibiotics in the event of terrorism exposure. An affected population may be placed on a "fever watch." People will receive drugs after developing a fever or flu-like symptoms. A vaccine exists to help prevent tularemia, but the vaccine is only partially effective. The vaccine is not available in the United States and is not recommended for the general population. However, it is recommended for laboratory workers in regular, close contact with large quantities of the organism.

Measures to prevent the disease from natural causes include not handling sick or dead animals; wearing gloves, a mask, and goggles if skinning or butchering animals; completely cooking game meats; wearing protective clothing if in areas where ticks or deer flies live; using tick repellant; checking skin often for ticks; not touching a tick with one's hands, and following precautions when working in a laboratory.

—David L. Horn, MD, FACP and
Michael A. Buratovich, PhD

Further Reading

American Society for Microbiology (ASM). "Sentinel Level Clinical Laboratory Guidelines for Suspected Agents of Bioterrorism and Emerging Infectious Diseases: *Francisella tularensis*." Revised Oct. 2020, asm.org/ASM/media/Policy-and-Advocacy/Biosafety_Sentinel_Guideline_October_2018_FINAL.pdf. Accessed on 11 Dec. 2021.

Bennett, John E., et al., editors. *Mandell, Douglas, and Bennett's Principles and Practice of Infectious Diseases.* 9th ed., Elsevier, 2019.

Centers for Disease Control and Prevention. "Emergency Preparedness and Response." *Tularemia*, 4 Apr. 2018, emergency.cdc.gov/agent/tularemia/. Accessed 11 Dec. 2021.

Hennebique, Aurélie, et al. "Tularemia as a Waterborne Disease: A Review." *Emerging Microbes & Infections,* vol. 8, no. 1, 2019, pp. 1027-42, doi:10.1080/22221751.2019.1638734.

Maurin, Max. "*Francisella tularensis*, Tularemia and Serological Diagnosis." *Frontiers in Cellular and Infection Microbiology,* vol. 10, no. 512090, 2020, doi:10.3389/fcimb.2020.512090.

Rathjen, Nicholas A., and S. David Shahbodaghi. "Bioterrorism." *American Family Physician*, vol. 104, no. 4, 2021, pp. 376-85.

Snowden, Jessica, and Kari A. Simonsen. "Tularemia." *StatPearls*. StatPearls Publishing, 21 July 2021.

FUNGI

This section discusses those fungi that cause human disease as well as the unicellular fungi, yeast, that help make our beer and bread. Many fungal diseases cause disease in healthy people but are particularly dangerous to those with compromised immune systems. Topics include coccidioidomycosis, mycotoxins, ringworm, and yeasts.

Histoplasmosis

Category: Diseases and conditions

Anatomy or system affected: Adrenal glands, blood, brain, eyes, heart, immune system, intestines, liver, lungs, lymphatics, respiratory system, spleen

Specialties and related fields: Immunology, internal medicine, microbiology, mycology, pathology, public health, pulmonary medicine

Definition: a fungal infection that often causes a respiratory illness

KEY TERMS

Histoplasma capsulatum: a pathogenic fungus that initially infects the lungs but can disseminate to other organs

hyphae: the tubular filaments characteristic of molds

mediastinum: the area between the lungs

microconidia: an asexual resting spore made by the mycelium form of Histoplasma capsulatum

mycelium: a collection of hyphae

mycoses: a fungal infection

thermal dimorphism: the ability of an organism to live and grow in two distinct forms at two different temperature ranges

CAUSES

Histoplasmosis is a pulmonary infection caused by the fungus *Histoplasma capsulatum (H. capsulatum)*. This fungus initially infects the lungs but may spread elsewhere in the body. Humans become infected by exposure to soil or dust contaminated with bird or bat droppings.

Histoplasmosis is globally distributed, and it occurs in parts of Central and South America, Africa, Asia, and Australia. In the United States, endemic areas include the Ohio-Mississippi River valleys and parts of southern Pennsylvania, northern Maryland, central New York, and Texas. Histoplasmosis outbreaks are often associated with bat caves or tree or building removal that mobilizes airborne microconidia at construction sites where birds or bats live.

H. capsulatum shows thermal dimorphism. At temperatures below 35° Celsius, it grows as a mold and produces asexual spores called "microconidia" (2 to 5 μm in diameter). The microconidia are the infectious particles people inhale. Upon inhalation, the fungus transforms from a mold to budding yeast (1 to 5 μm in diameter) at 37° C.

RISK FACTORS

Risk factors for histoplasmosis include:

- outdoor occupations that feature contact with bird or bat droppings, such aviaries, construction, or excavation;
- outdoor activities that put one in contact with bird or bat droppings, such as camping or cave exploration;
- keeping birds as pets;
- living in areas where the fungus is endemic;
- travel to locations where histoplasmosis is endemic; and
- having a medical condition, such as human immunodeficiency virus (HIV) infection, or advanced age that weakens the immune system.

Histoplasma may be able to live in the body for decades, according to a case study published in the journal *BMJ Case Reports* in 2017. The case report described a seventy-year-old man diagnosed with histoplasmosis despite having rarely left his home state of Arizona, a region in which *Histoplasma is* not typically found. The man is thought to have inhaled the spores during a brief visit to North Carolina thirty years before he began displaying symptoms of histoplasmosis.

SYMPTOMS

Fewer than five percent of all individuals exposed to *H. capsulatum* show symptoms. The most common pulmonary form is symptomatic pulmonary histoplasmosis. If patients have symptoms, the most common are fever, chills, headache, sore muscles, appetite loss, cough, and chest pain that develop two to four weeks after exposure but may arise earlier persons inhaled large quantities of microconidia.

Acute diffuse pulmonary histoplasmosis occurs if individuals inhale large amounts of microconidia. In these cases, people experience the same symptoms as those with symptomatic pulmonary histoplasmosis, but the disease is much more severe. This condition can progress to respiratory failure or dissemination to other organs. Patients can recover without treatment, but antifungal agents should be given to all patients. During recovery for these patients, shortness of breath and fatigue can last for months.

Chronic pulmonary histoplasmosis tends to occur in people with preexisting lung diseases. Affected individuals develop a productive cough, shortness of breath, chest pain, fatigue, fevers, and sweats and have scarring in the lungs with cavity formation on chest X-rays.

Broncholithiasis occurs when chronically infected lymph nodes around the lungs calcify or erode into adjacent airways. This causes chronic cough, wheezing, coughing up blood (hemoptysis), fever, chills, and pus-filled sputum production. Patients may report spitting up small stones or gritty material.

The spread of histoplasmosis from the lungs to other organisms results in disseminated histoplasmosis. Several potential clinical syndromes may result from disseminated histoplasmosis, including:

- Adrenal perivasculitis (common)
- Endocarditis
- Mediastinal disease (mediastinal granuloma or mediastinitis)
- Meningitis
- Ocular histoplasmosis
- Intestinal ulcers

SCREENING AND DIAGNOSIS

A doctor will ask about symptoms and medical history and will perform a physical exam. Tests may include blood tests, a blood culture, a sputum culture, a pulmonary function test, skin testing, urine antigen testing, X-rays of chest or abdomen (or both), and bone marrow tests.

Serological tests include detecting antibodies against the fungus or fungal antigens in urine, sputum, or other body fluids.

TREATMENT AND THERAPY

Treatment includes the use of antifungal medications, which may include amphotericin B or itraconazole. Persons with acquired immunodeficiency syndrome may require treatment with an antifungal medication for the rest of their lives to prevent further attacks of histoplasmosis.

Lipid preparations of amphotericin B (AmBisome) are generally preferred because they are less toxic to kidneys. *H. capsulatum* is very susceptible to amphotericin B. However, because of its toxicity, it is reserved for patients with moderate or severe disease. Other antifungal drugs active against *H. capsulatum* include posaconazole, isavuconazole, and voriconazole. Fluconazole and echinocandins are not effective against *H. capsulatum* and should not be used.

PREVENTION AND OUTCOMES

Persons who anticipate being exposed to bird or bat droppings should wear face masks, and persons with weakened immune systems should altogether avoid bird and bat droppings.

—Rosalyn Carson-DeWitt, MD and Michael A. Buratovich, PhD

Further Reading

Azar, Marwan M., et al. "Current Concepts in the Epidemiology, Diagnosis, and Management of Histoplasmosis Syndromes." *Seminars in Respiratory and Critical Care Medicine,* vol. 41, no. 1, 2020, pp. 13-30, doi:10.1055/s-0039-1698429.

Centers for Disease Control and Prevention. "Histoplasmosis." *Fungal Disease*, 29 Dec. 2020, www.cdc.gov/fungal/diseases/histoplasmosis/index.html. Accessed 24 Nov. 2021.
Conover, Michael R., and Rosanna M. Vail. *Human Diseases from Wildlife*. CRC, 2015.
Des Jardins, Terry R., et al. *Clinical Manifestations and Assessment of Respiratory Disease*. 7th ed., Mosby Elsevier, 2016.
Hospenthal, Duane R., and Michael G. Rinaldi. *Diagnosis and Treatment of Fungal Infections*. 2nd ed., Springer, 2015.
Kaufman, C. A. "Histoplasmosis." *Clinics in Chest Medicine*, vol. 30, 2009, p. 217.
Longo, Dan L., et al. "Histoplasmosis." *Harrison's Principles of Internal Medicine*. 19th ed., McGraw-Hill, 2015.
Mason, Robert J., et al., editors. *Murray and Nadel's Textbook of Respiratory Medicine*. 5th ed., Saunders/Elsevier, 2010.
Papadakis, Maxine, et al., editors. *Current Medical Diagnosis and Treatment 2015*. 54th ed., McGraw-Hill, 2015.

Coccidioidomycosis

Category: Diseases and disorders
Also known as: San Joaquin Valley fever, valley fever
Anatomy or system affected: All
Specialties and related fields: Dermatology, family medicine, general surgery, internal medicine, microbiology, mycology, pulmonary medicine
Definition: a fungal infection acquired by inhaling the spores of particular soil-based fungi; it initially attacks the lungs and often resolves without causing symptoms, but it can cause pneumonia and disseminate throughout the body

KEY TERMS

arthroconidia: asexual fungal spores that are made by the segmentation of preexisting fungal hyphae
endospores: tiny, round cells produced by spherules as they divide into smaller and smaller cells that are released upon rupture
erythema nodosum: inflammation of the fatty layer of the skin that results in red, painful bumps, usually located on the front of the legs
hyphae: the long, filamentous, often branching cells of many fungi
mycelium: a body of the fungal organism that consists of a collection of hyphae
mycetoma: a progressive and chronic fungal or bacterial infection that causes overgrowth of the infected tissue and the formation of sinuses filled with the infecting organism
pulmonary nodules: small, round growths on the lung that contain either trapped microorganisms or cancer cells
spherules: a thick-walled, spherical structure that is the tissue-specific form of *Coccidioides* species

CAUSES AND SYMPTOMS

Coccidioidomycosis is an infection caused by the soil-based fungi *Coccidioides immitis* (*C. immitis*) and *Coccidioides posadasii* (*C. posadasii*). These fungi are found only in the Western hemisphere, and they prefer dry, alkaline soils. *C. immitis* and *C. posadasii* are endemic to the southwestern United States (south-central California, Nevada, Arizona, New Mexico, and western Texas), those regions of Mexico that border the western United States, parts of Central America (Guatemala, Honduras, and Nicaragua), and the desert regions of South America (Argentina, Paraguay, and Venezuela).

While in the soil, *Coccidioides*, like most fungi, grows as thin, branching filaments called "hyphae." A collection of hyphae is called a "mycelium." When it rains, the mycelium grows quite rapidly, but it forms resting cells called "arthrospores" once the soil dries out. If disturbed by wind, earthquakes, or soil excavation, these arthrospores become airborne and, if inhaled, can cause coccidioidomycosis.

Once inhaled, the arthrospore transforms into a thick-walled, spherical structure called a "spherule" that divides itself into hundreds of small endospores. When the spherule ruptures, it releases the endospores, which grow into spherules that form more endospores.

About 60 percent of patients show no symptoms, and the disease resolves spontaneously. Patients who show symptoms suffer from fever, sore throat, headache, cough, fatigue, painful bumps on the skin (erythema nodosum), and chest pain approximately one to three weeks after inhaling arthrospores. About 95 percent of symptomatic patients recover without further problems after several weeks. If symptoms persist beyond three months, however, then the patient has chronic progressive coccidioidal pneumonia. Between 5 and 7 percent of patients with coccidioidal pneumonia form pulmonary nodules, which are areas of the lung where the immune system has walled off the organism from the rest of the lung. On an X-ray, these nodules can look exactly like cancerous masses in the lung. A biopsy is often necessary to distinguish between lung cancer and coccidioidal pulmonary nodules. In 5 percent of patients with coccidioidal pulmonary nodules, the nodules enlarge to form pulmonary cavities that can become infected, rupture, and bleed, causing the release of pus between the lungs and the ribs (empyema). Small cavities (less than 2.5 centimeters) can heal after one to two years. However, larger cavities can persist and cause the patient to spit up blood (hemoptysis) and grow fungi throughout the cavity (mycetoma).

A minority of patients develop disseminated coccidioidomycosis, in which the organism penetrates blood vessels, invades the bloodstream, and infects any organ in the body. Disseminated coccidioido-

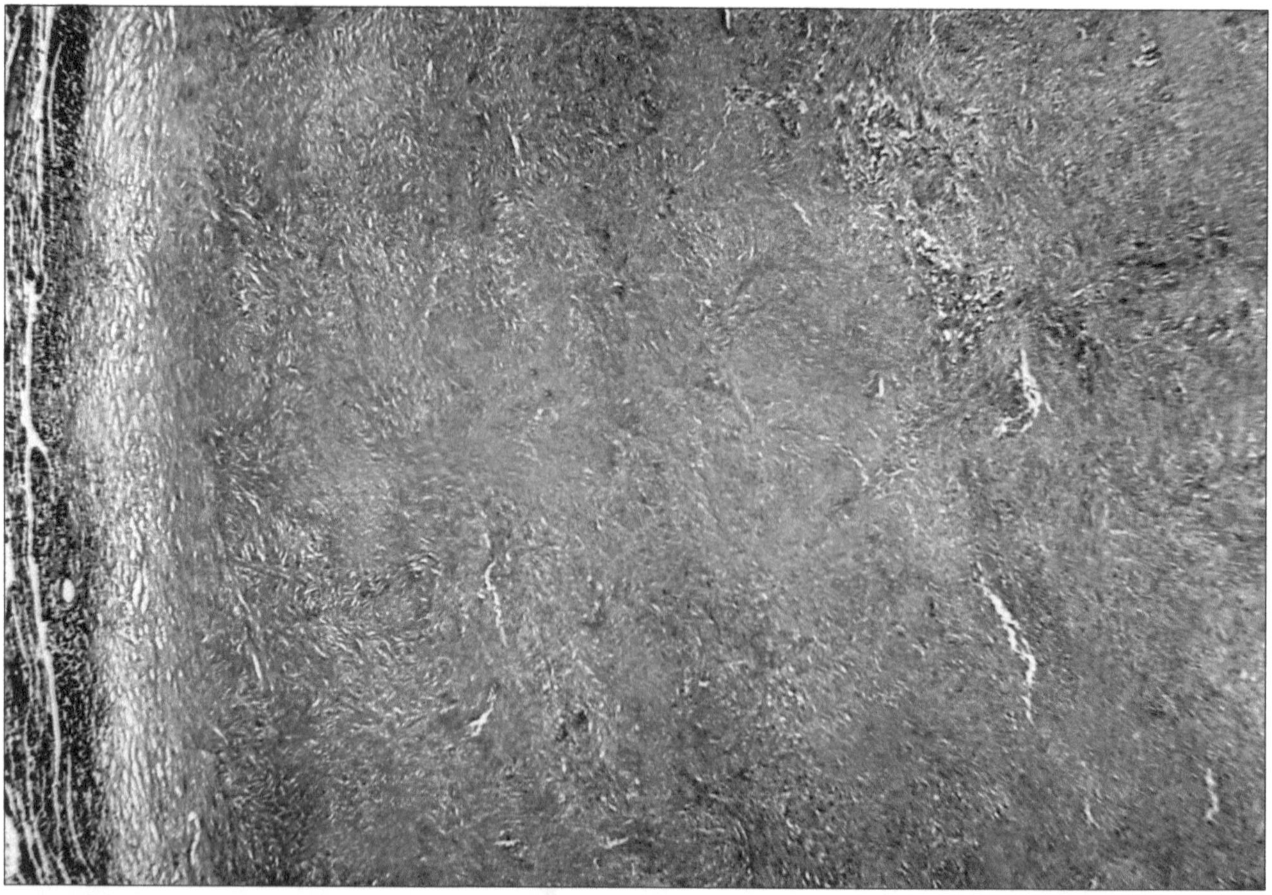

Histo-pathological changes in a case of coccidioidomycosis of the lung showing a large fibrocaseous nodule. Photo by Dr. Martin D. Hicklin/CDC, via Wikimedia Commons. [Public domain.]

mycosis occurs weeks or months after primary pneumonia. It can even develop in cases where there is no previous evidence of respiratory disease. Specific ethnic groups such as Filipinos and African Americans show increased risk of developing disseminated disease, as do pregnant women in the third trimester of their pregnancy, infants younger than one-year-old, diabetics, patients with acquired immunodeficiency syndrome (AIDS), COVID-19, those taking illicit drugs, or suffering from diseases that suppress the immune system.

DIAGNOSIS, TREATMENT, AND THERAPY

Most patients suspected of having coccidioidomycosis are evaluated with serologic testing. Enzyme-linked immunoassays (EIA) and immunodiffusion tests are the serologic tests most used to make a diagnosis. Antibodies, however, take seven-twenty-one days to develop. Therefore, culturing the fungus might be the earliest and, in some instances, the only means of diagnosis. Cultures are usually obtained from hospitalized patients. The most common sources for culture are sputum, cerebrospinal fluid, bone, and soft tissue. *C. immitis* grows on almost any laboratory microbiological media. Growth may appear within three days and definitely within the first week. Commercially available genetic probing kits (*Genprobe*, San Diego) can detect *C. immitis* DNA (deoxyribonucleic acid) and confirm its presence.

Asymptomatic or symptomatic infections are usually self-limited and require little more than supportive care. Patients with coccidioidal pneumonia require fluconazole or itraconazole treatment for at least twelve months and intravenous amphotericin B for stubborn cases. Pulmonary nodules are typically not treated, but they may require surgery. Pulmonary cavities are only treated with antifungal drugs if the patient shows symptoms. Surgical removal might also be warranted if the infection resists treatment. Disseminated coccidioidomycosis requires higher doses of fluconazole. Very sick patients may require amphotericin B or a combination of fluconazole and amphotericin B. Amphotericin B is preferred for pregnant women since other drugs harm the developing fetus.

PERSPECTIVE AND PROSPECTS

Coccidioidomycosis was first described in 1892 by Roberto Johann Wernicke and Alejandro Posadas in South America. The first case in the United States was reported in California in 1894. Two years later, Emmet Rixford and Thomas Caspar Gilchrist reported several clinical infections that were caused by an organism that, they thought, resembled the protozoan *Coccidia*. Therefore they named it *Coccidioides*, which means "*Coccidia*-like." In 1905, William Ophüls described the fungal life cycle and pathology of *C. immitis*. Charles E. Smith studied the epidemiology of coccidioidomycosis in the San Joaquin Valley of California and developed the coccidioidin skin test and serological testing for the disease.

According to the Centers for Disease Control and Prevention (CDC), in 2019, there were 18,407 reported coccidioidomycosis cases. Most were in people who live in Arizona or California. The rates of Valley fever are highest among people aged sixty and older. However, the number of reported Valley fever cases likely underestimates the actual number since coccidioidomycosis testing rates remain low in endemic areas. A 2013 CDC study reported increased coccidioidomycosis cases in the southwestern United States between 1998 and 2011. Cases in Arizona, California, Nevada, New Mexico, and Utah increased from 2,265 reported in 1998 to 22,000 reported in 2011.

New treatments under investigation for coccidioidomycosis include posaconazole, voriconazole, caspofungin, and a new lipid-dispersal formulation of amphotericin B that reduces its kidney toxicity. Nikkomycin Z is another experimental agent that is very active against *Coccidioides* in culture and infected animals.

—Michael A. Buratovich, PhD

Further Reading

Akram, Sami M., and Janak Koirala. "Coccidioidomycosis." *StatPearls*. StatPearls Publishing, 20 Sept. 2021.

Centers for Disease Control and Prevention. "Valley Fever (Coccidioidomycosis) Statistics." www.cdc.gov/fungal/diseases/coccidioidomycosis/statistics.html. Accessed on 21 Nov. 2021.

Frías-De-León, María Guadalupe, et al. "Epidemiology of Systemic Mycoses in the COVID-19 Pandemic." *Journal of Fungi*, vol. 7, no. 7, 2021, p. 556, doi:10.3390/jof7070556.

Galgiani, John N. "Changing Perceptions and Creating Opportunities for Its Control." *Annals of the New York Academy of Sciences*, vol. 1111, 2007, pp. 1-18.

Kibbler, Christopher C., et al., editors. *Oxford Textbook of Medical Mycology*. Oxford UP, 2018.

Parish, James, M., and James E. Blair. "Coccidioidomycosis." *Mayo Clinic Proceedings*, vol. 83, no. 3, 2008, pp. 343-48.

Tsang, Clarisse A., et al. "Increase in Reported Coccidioidomycosis—United States, 1998-2011." *Morbidity and Mortality Report*, vol. 62, no. 12, 2013, pp. 217-21.

BLASTOMYCOSIS

Also known as: Gilchrist's disease
Category: Diseases and conditions
Anatomy or system affected: Bones, central nervous system, genitourinary tract, lungs, respiratory system, skin
Specialties and related fields: immunology, internal medicine, microbiology, mycology, pathology, pharmacology, pulmonology
Definition: an infection caused by *Blastomyces dermatitidis* that primarily affects the lungs but may spread to other parts of the body

KEY TERMS

dimorphic: organisms that exist in two distinct forms under different conditions or during different parts of their life cycle
mold: a fungus that grows in the form of multicellular filaments called "hyphae"
mycelium: a network of tubular, branching hyphae
spores: resting cells made by molds
yeast: a single cell, microscopic fungus consisting of single oval cells that reproduce by budding

CAUSES

Blastomycosis is an infection caused by *Blastomyces dermatitidis*, a fungus typically found in soil. It is endemic to the central and southeastern United States, Canada, and parts of Africa. This type of infection primarily affects the lungs but may spread to other parts of the body. A second *Blastomyces* species, *B. gilchristii*, usually causes chronic pneumonia.

B. dermatitidis is a dimorphic fungus that exists as either a mold or yeast, depending on the environment where it is found. The fungus exists in mold form in moist soil enriched with decaying plant material, wood, or animal waste, often near rivers. Inhalation of fungal spores into the lungs causes a respiratory infection known as pulmonary blastomycosis. At temperatures below 37°Celsius (98.6°Fahrenheit), *Blastomyces* grows as a mold that produces spores. At 37°C (98.6°F), the organism grows as a multinucleate yeast with a thick cell wall. In the lungs, inhaled *Blastomyces* spores transform into yeast that are resistant to phagocytosis by resident white blood cells called "alveolar macrophages." A surface glycoprotein called "BAD-1" lets Blastomyces yeast cells adhere to tissues and immune cells in the lung.

Pulmonary blastomycosis can result in chronic pneumonia that causes few or no symptoms or a fulminant disease that causes lobar pneumonia. Dispersal throughout from the lungs to other organs results in extrapulmonary disseminated blastomycosis. The most common extrapulmonary site of infection is the skin, followed by the bones, the prostate and other genitourinary organs, and the brain.

RISK FACTORS

The disease may affect all ages; however, most reported cases involve healthy males with an outdoor occupation (e.g., farming) or hobby. Recreational exposure to soil also increases the risk of contraction blastomycosis. There is a relatively high prevalence of blastomycosis in North America, particularly in the Ohio and Mississippi river valleys. Persons with diabetes mellitus or weakened immune systems, including organ transplant recipients and those on immunosuppressants, are more likely to have a severe form of the disease. Atypical for fungal infections, blastomycosis is not more likely to appear in persons with acquired immunodeficiency syndrome (AIDS).

SYMPTOMS

Some persons with pulmonary blastomycosis are asymptomatic. Others may have flu-like symptoms, including fever, chills, myalgia, headache, cough, hemoptysis (coughing up blood), chest pain, weight loss, and fatigue. Extrapulmonary blastomycosis of the skin is indicated by verrucous (warty) or ulcerated lesions on the face, neck, and extremities and is a significant indication of the disease. As the disease progresses, pain and lesions may occur on the bones, genitalia, parts of the central nervous system, and organs. Persons with severe disease may show symptoms simulating bacterial pneumonia, tuberculosis, lung cancer, or adult respiratory distress syndrome.

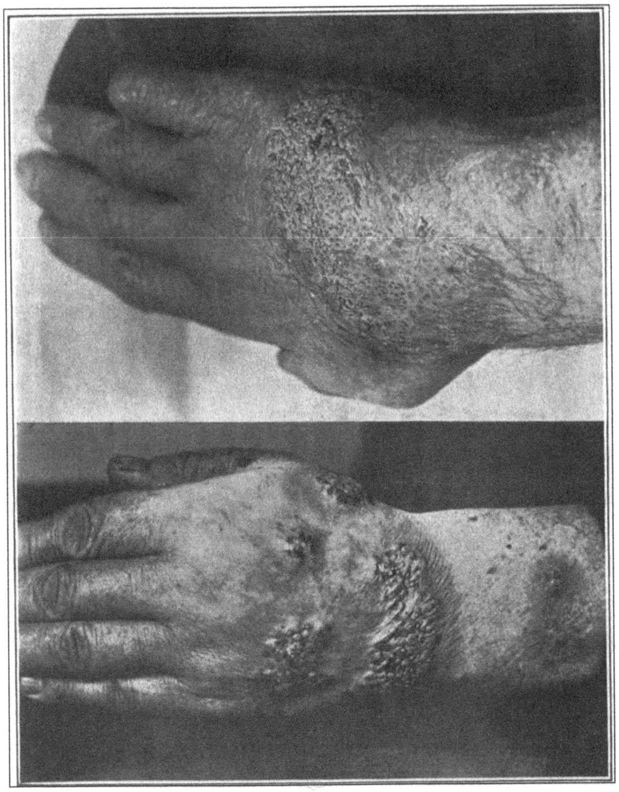

Skin lesions of blastomycosis. Photo via Wikimedia Commons. [Public domain.]

SCREENING AND DIAGNOSIS

Blastomycosis is a rare systemic infection. Primary care physicians often consult with an infectious disease specialist for diagnosis and treatment. Diagnostic tests include blood and urine antigen analyses, tissue biopsy, sputum culture, chest X-ray, and bronchoscopy. Polymerase chain reaction tests effectively identify *Blastomyces* deoxyribonucleic acid (DNA) in sputum, tissues, or blood. Definitive diagnosis of blastomycosis requires culture on Sabouraud dextrose agar and analysis of infected tissue under a microscope. *Blastomyces* yeast are usually 8 to 15 μm in diameter and reproduce by budding. Periodic acid-Schiff stains effectively visualize yeast in biopsied tissue or other fluids.

TREATMENT AND THERAPY

Persons with blastomycosis should be treated based on the extent and severity of the disease. Amphotericin B and itraconazole are the drugs of choice. Oral itraconazole is recommended for persons with pulmonary blastomycosis. Intravenous amphotericin B is recommended for persons with severe disease. A blastomycosis infection has the potential to be fatal if untreated.

PREVENTION AND OUTCOMES

B. dermatitidis is a microscopic airborne fungus. The best form of prevention is to avoid endemic areas where the fungus is prevalent.

—Rose Ciulla Bohling, PhD

Further Reading

Chander, Jagdish. *Textbook of Medical Mycology*. 4th ed., Jaypee Brothers Medical Pub., 2017.

Jameson, J. Larry, et al., editors. *Harrison's Principles of Internal Medicine*. 20th ed., McGraw-Hill, 2018.

Kibbler, Christopher C., et al., editors. *Oxford Textbook of Medical Mycology*. Oxford UP, 2018.

Levitzky, Michael G. *Pulmonary Physiology*. 9th ed., McGraw-Hill Education/Medical, 2018.

Steele, Russell W., and Avinash Shetty. "Blastomycosis." emedicine.medscape.com/article/961731-overview.

Webster, John, and Roland Weber. *Introduction to Fungi*. Cambridge UP, 2007.

SPOROTRICHOSIS

Category: Diseases and conditions

Anatomy or system affected: Brain, cerebrospinal fluid, joints, lungs, lymphatic system, skin, spinal cord

Specialties and related fields: Dermatology, internal medicine, microbiology, mycology, neurology, veterinary medicine

Definition: an infectious disease caused by the soil fungus *Sporothrix schenckii* and *S. brasiliensis* that usually affects the skin but can affect other organs

KEY TERMS

conidia: a type of asexual reproductive spore of fungi usually produced at the tip or side of hyphae

hyphae: a long, branching filamentous structure of a fungus

mycosis: a disease caused by any fungus that invades the tissues, causing superficial, subcutaneous, or systemic disease

thermal dimorphism: a unique group of fungi that respond to shifts in temperature by converting between hyphae (22 to 25° Celsius) and yeast (37° C). This morphologic switch, known as the phase transition, defines the biology and lifestyle of these fungi.

yeast: eukaryotic, single-celled microorganisms classified as members of the fungi

DEFINITION

Sporotrichosis is an infectious disease caused by the soil fungus *Sporothrix schenckii* that usually affects the skin. Sporotrichosis is commonly acquired through cutaneous inoculation. In rare cases, people can inhale it. It is not spread from person to person. Still, zoonotic transmission from infected animals (such as cats and horses) is well documented.

Sporothrix shows a property called "thermal dimorphism." It grows as a filamentous hyphal form at temperatures lower than 37° C. The fungus forms asexual resting structures called "conidia" that can aerosolize and be inhaled in the hyphal form. When introduced into the human body, it transforms into a budding yeast.

Sporotrichosis is the most frequently documented implantation mycosis in Latin America. A distinct *Sporothrix* species called *Sporothrix brasiliensis* (*S. brasiliensis*) is causing an epidemic in cats. The epicenter of this outbreak is in Brazil. Transmission of *S. brasiliensis* from cats to humans occurs routinely. However, the fungus is passed in the yeast phase in this case. Cats are heavily infected, and humans can become infected through cat scratches and bites, or through cat's cough or sneezing, and direct contact between skin and animal secretions.

CAUSES

S. schenckii is widely distributed in the natural environment and can be found on rose thorns and twigs and in sphagnum moss, hay, and soil. The fungus enters the skin through small cuts or punctures, spreading from the initial lesion along lymphatic channels. Hematogenous dissemination is rare in healthy individuals. However, this fungus can spread to the brain, spinal cord, and other organs in those with compromised immune systems.

RISK FACTORS

Persons handling thorny plants, sphagnum moss, or baled hay (such as farmers, nursery workers, landscapers, and gardeners) are at higher risk for the disease. Sporotrichosis resulting from inhalation has been documented in persons with severe chronic obstructive pulmonary disease. Immunosuppressive states and alcoholism predispose to disseminated disease. The disease is most common among adults and slightly more prevalent in males. Sporotrichosis in children may be more common in tropical regions.

In Southern America, cat owners infected with HIV, who take immunosuppressive drugs or cancer treatments are at high risk of contracting *S. brasiliensis*. Likewise, those who work with stray cats, such as animal control officers or veterinarians, are at risk of getting scratched or contracting the fungus from inhalation or animal secretions.

SYMPTOMS

The first symptom is a firm, pink to purple, usually painless skin nodule that resembles an insect bite. It may appear from one to twelve weeks after exposure to the fungus. Over time, the nodule may ulcerate and become chronic. The characteristic infection progresses proximally along lymphatic channels. In most cases, the disease is limited to the skin. The disease can rarely infect the bones, joints, lungs, and brain. Widespread cutaneous lesions and involvement of multiple visceral organs (including eye, prostate, oral

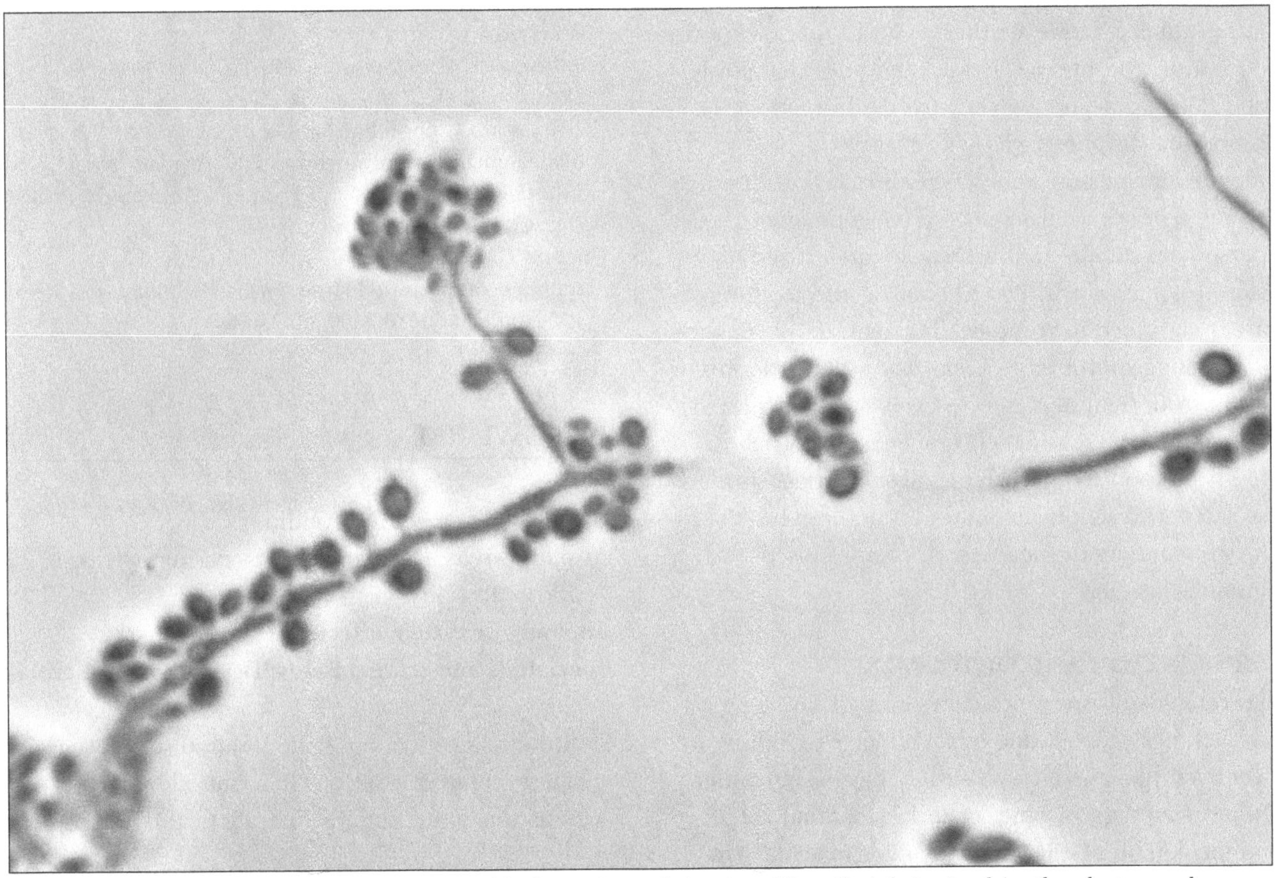

A photomicrograph rof the conidiophores and conidia of the fungus Sporothrix schenckii, *a skin infection involving the subcutaneous layer. CDC via Wikimedia Commons. [Public domain.]*

mucosa, paranasal sinuses, and larynx) predominantly occur in persons with compromised immune systems, such as those with human immunodeficiency virus infection, diabetes, alcoholism, or other disorders of the immune system.

SCREENING AND DIAGNOSIS

Diagnosis is based on a nodule biopsy and laboratory identification of the mold.

TREATMENT AND THERAPY

Oral itraconazole (200 milligrams once or twice daily) is the drug of choice for cutaneous and lymphocutaneous forms of the disease. Itraconazole may also use it to treat bone and joint infections. Posaconazole shows excellent activity against this fungus in culture. However, there is little clinical experience with it beyond a few case studies. Ketoconazole and fluconazole are poorly active against *Sporothrix*. Liposomal amphotericin B is generally recommended for persons with severe disease or with pulmonary, brain, or disseminated infection; once the person has stabilized, itraconazole can be used for step-down therapy. Infected bone or infected lung areas may need to be surgically removed.

Treating sporotrichosis may take several months or years. With treatment, full recovery can be expected. Spontaneous resolution of cutaneous sporotrichosis has been reported. Disseminated sporotrichosis is associated with significant morbidity and can be life-threatening for people with compromised immune systems.

PREVENTION AND OUTCOMES

Preventive measures include wearing gloves, long sleeves, heavy boots, and other protective clothing when handling wires, rose bushes, hay bales, conifer (pine) seedlings, or other materials that may cause minor skin breaks. It is also advisable to avoid skin contact with sphagnum moss, which has been implicated as a source of the fungus in several outbreaks.

Cats in Latin American countries should be regarded as heavily infected with *S. brasiliensis*. People should wear N-95 masks around cats, and anyone should handle stray animals with gloves and a mask. When handling animals with skin lesions, gloves minimize the risk of zoonotic transmission.

—Katia Marazova, MD, PhD and Michael A. Buratovich, PhD

Further Reading

Etchecopaz, Alejandro, et al. "Sporothrix brasiliensis: A Review of an Emerging South American Fungal Pathogen, Its Related Disease, Presentation and Spread in Argentina." *Journal of Fungi*, vol. 7, no. 3, 2021, p. 170, doi:10.3390/jof7030170.

Greenfield, Ronald A. "Sporotrichosis." 2 Mar. 2021, emedicine.medscape.com/article/228723-overview. Accessed 5 Dec. 2021.

Kauffman, Carol A. "Sporotrichosis." *Clinical Infectious Diseases*, vol. 29, 1999, pp. 231-36.

———, et al. "Clinical Practice Guidelines for the Management of Sporotrichosis: 2007 Update by the Infectious Diseases Society of America." *Clinical Infectious Diseases*, vol. 45, 2007, pp. 1255-65.

Queiroz-Telles, Flavio, et al. "Sporotrichosis in Immunocompromised Hosts." *Journal of Fungi*, vol. 5, no. 1, 2019, p. 8, doi:10.3390/jof5010008.

RINGWORM

Category: Diseases and conditions

Also known as: Dermatomycosis, dermatophytosis, tinea infection

Anatomy or system affected: Nails, scalp, skin

Specialties and related fields: Dermatology, microbiology

Definition: a contagious itching skin disease occurring in small circular patches, caused by some fungi that affect chiefly the scalp or the feet

KEY TERMS

dermatophyte: a pathogenic fungus that grows on skin, mucous membranes, hair, nails, feathers, and other body surfaces, causing ringworm and related diseases

epidermis: the surface epithelium of the skin that overlies the dermis

keratin: a fibrous protein forming the main structural constituent of hair, feathers, hoofs, claws, horns

stratum corneum: the horny outer layer of the skin

tinea: a type of ringworm infection

INTRODUCTION

Ringworm is a fungal infection of the skin, including the nails, hands, feet, and scalp. Despite its name, ringworm has nothing to do with worms. Both adults and children can be affected, but ringworm occurs most commonly in children. A fungal infection of the feet is sometimes called "athlete's foot."

CAUSES

Ringworm infections only infect the uppermost layer of the skin, the stratum corneum. The skin consists of an upper epidermis and an underlying dermis. Below the dermis is the subcutaneous fascia, also known as the hypodermis.

The epidermis has four cell types: keratinocytes, melanocytes, Langerhans' cells, and Merkel's cell. Keratinocytes are the most prevalent cell type of the epidermis. They are loaded with the protein keratin and lipids. When they absorb ultraviolet light, cholesterol derivatives are converted into vitamin D precursors. Melanocytes produce the pigment melanin that determines the hue of the skin. Ultraviolet light exposure stimulates melanin production in melanocytes. Melanocytes transfer their melanin granules to keratinocytes (a process called "pigment donation") to color them. Langerhans' cells are dendritic cells that present antigen to immune cells. These cells are in the lower parts of the epidermis. Merkel cells are ovoid sensory neurons found in the lowest parts of the epidermis that sense light touch.

The lowest layer of the epidermis has a stem cell population that constantly rejuvenates it. This stem cell layer, the stratum Basale, is a bed of box-shaped, actively dividing cells that also contain melanocytes. Above the stratum Basale is the stratum spinosum that contains polygonal cells with spiny extensions. Langerhans' cells are in the stratum spinosum. Above the stratum spinosum is a layer called the "stratum granulosum," which is three to five cells thick. The cells in this layer have lots of keratohyalin and lipid granules. The penultimate layer of the epidermis is the stratum lucidum, which is two to three layers thick. This skin layer consists of clear cells filled with eleidin, a compound made from keratohyalin. The top layer of the skin is the stratum corneum. This layer is twenty to thirty cell layers thick. These cells are bags of keratin that have lost their nuclei and all other organelles. This layer thickens in calluses.

Dermatophytes infect the stratum corneum of the skin. Sometimes these infections can dip beneath the stratum corneum and cause inflammation A person can get ringworm from direct skin-to-skin contact with infected people or pets. Ringworm is also transmitted by sharing hats and personal hair-grooming items (such as brushes and combs) and through contact with an infected person's locker room floors, shower stalls, seats, or clothing.

The fungi that cause dermatophyte infections are divided into three groups. The anthropophilic dermatophytes only infect humans and cause mild chronic inflammation. These include *Trichophyton rubrum* (*T. rubrum*), *T. mentagrophytes*, *Epidermophyton floccosum*, and *T. tonsurans* (very common in the USA), *T. interdigitale, Microsporon audouinii (M. audouinii), T. violaceum, M. ferrugineum, T. schoenleinii, T. megninii, T. soudanense, and T. yaoundei*. Zoophilic organisms cause marked inflammation in humans and occur in animals. Zoophilic dermatophytes include *Microsporum canis* (*M. canis*,

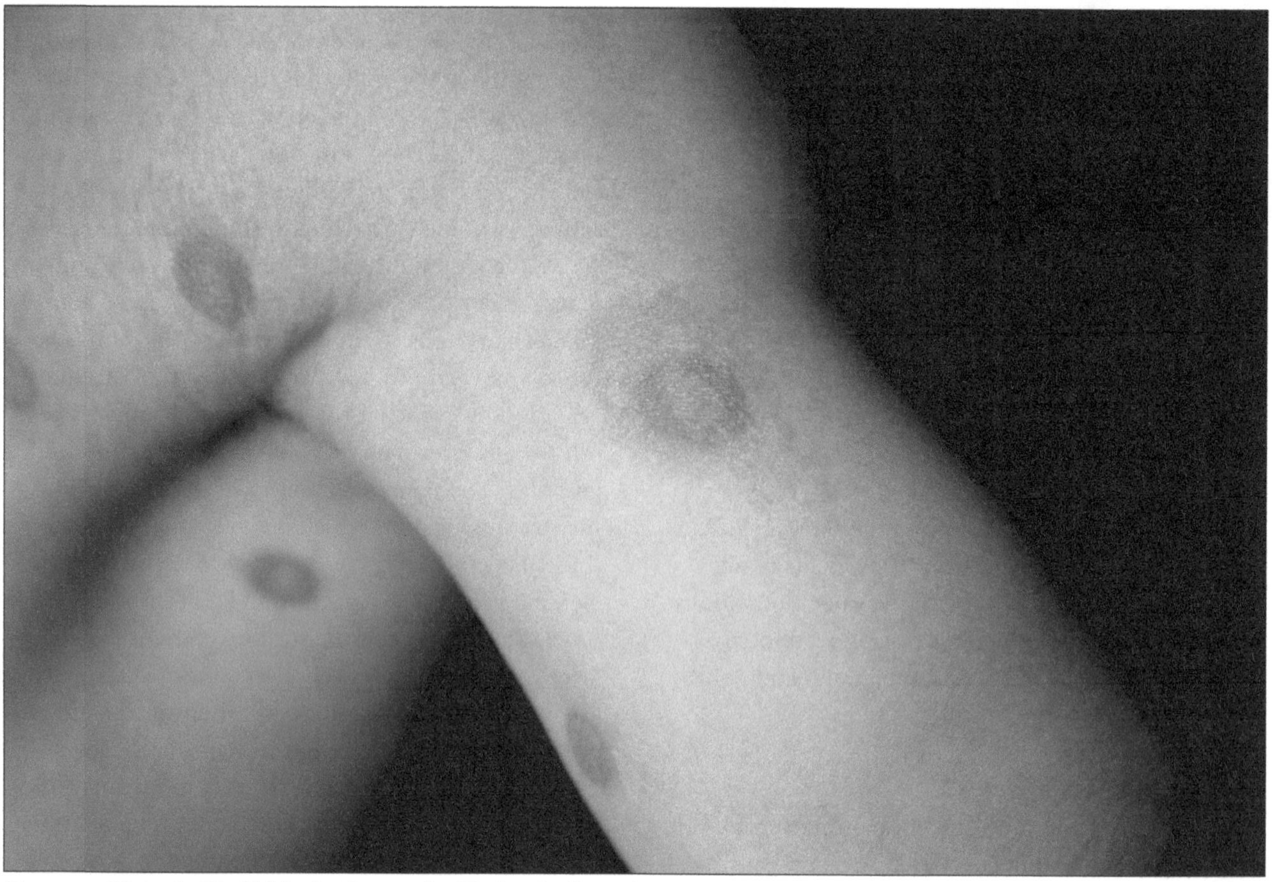

Ringworm. Photo via iStock/alejandrophotography. [Used under license.]

from cats and dogs), *T. equinum* (from horses), *T. erinacei* (from hedgehogs and other animals), *T. verrucosum* (from cattle), *M. nanum* (from pigs), *M. distortum* (a variant of *M. canis*). The third group, the geophilic dermatophytes, cause significant inflammation that can lead to scarring. These fungi can occur in humans and animals but are found in the soil. *M. gypseum* (also called *Nannizzia gypsea*) and *M. fulvum* are two geophilic dermatophytes.

RISK FACTORS

Risk factors for developing ringworm include contact with surfaces (such as seat backs and shower stalls), clothing or personal grooming items used by an infected person; skin-to-skin contact with an infected person or pet; and spending time in nurseries, schools, day-care centers, or locker rooms. At higher risk are children age twelve years or younger. Ringworm of the scalp rarely occurs in children after puberty or in adults.

SYMPTOMS

Tinea infections are one of the most common causes of superficial fungal infections. They are distinguished by the area of the body affected. When ringworm appears on the skin, it makes circular, reddish patches with raised borders. Eventually, the patches grow larger, and the centers of the patches turn clear, giving a ringlike appearance. Symptoms of ringworm vary, depending on the part of the body affected. On the scalp (tinea capitis), the infection begins with small bumps on the head that grow larger and form a circular pattern. Hair may become brittle and break, forming scaly, hairless patches. On the hands (tinea manus), the infection affects the palms and spaces between the fingers. On the feet (tinea pedis, or athlete's foot), the infection may cause scaling between the toes or thickening and scaling on the heels or soles. Infection of the nails (tinea unguium) causes fingernails and toenails to become yellow, thick, and crumbly. Infection of the groin area (tinea cruris, or jock itch) causes a chafed, reddish, itchy, sometimes painful rash. Infection of the skin around the entire body (tinea corporis) produces flat, scaly, round spots. Infection on the face (tinea faciei) produces red, scaly patches. Tinea barbae affects the facial hair follicles of bearded individuals.

Ringworm symptoms on the body usually appear four to ten days after exposure. Scalp symptoms will appear in ten to fourteen days.

SCREENING AND DIAGNOSIS

A doctor will ask about symptoms and medical history and examine the patient's skin. Ringworm is often easily diagnosed by appearance. However, symptoms may be like other conditions. Scraped skin is cultured on sabouraud dextrose medium. The type of spores and how they are presented in the cultured fungus determines its classification. Scraped skin also has fungal hyphae growing throughout it. Wood's lamp examinations or potassium hydroxide preparations make these hyphae easily visible under the microscope.

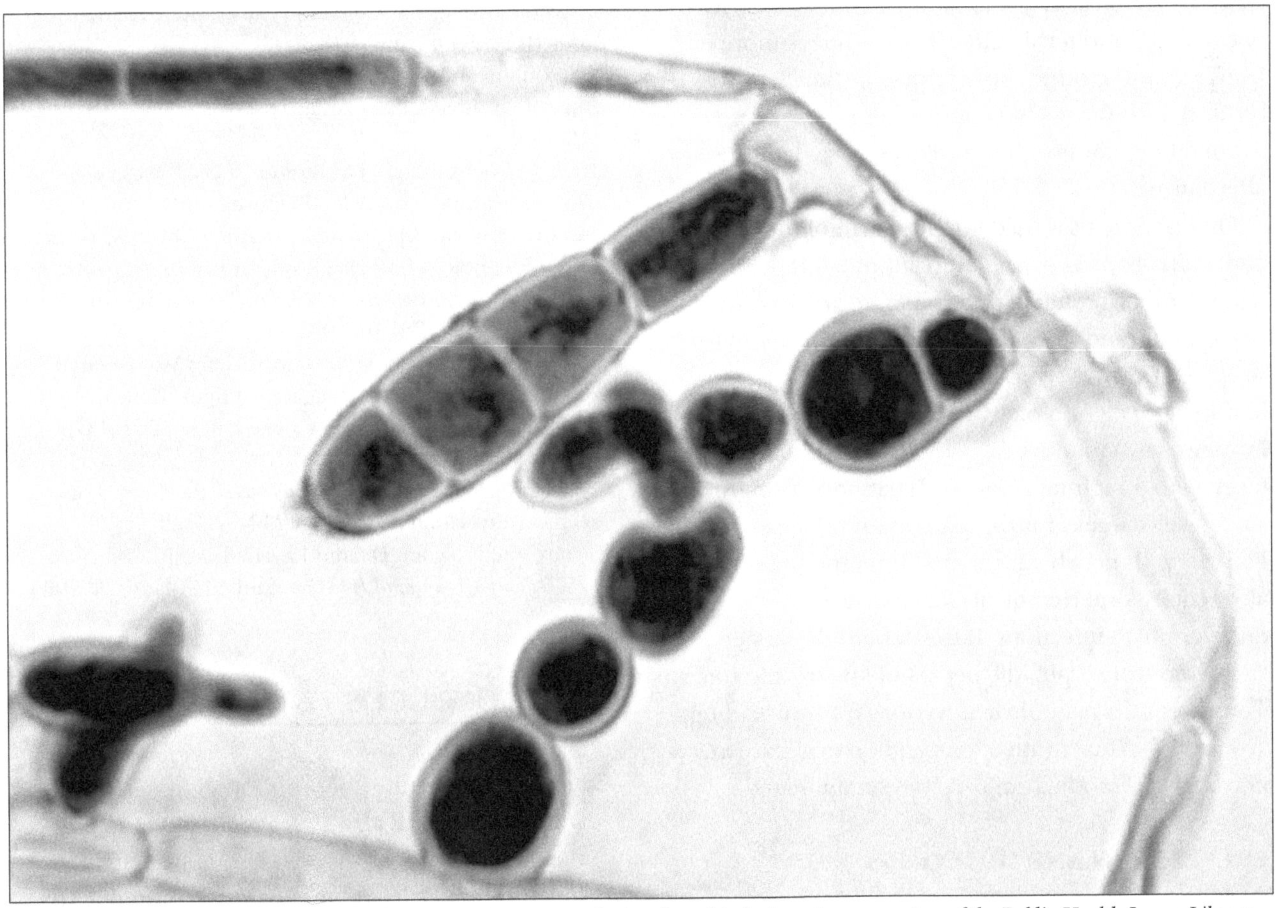

Epidermophyton floccosum, *a fungus that causes ringworm and other fungal infections. Image courtesy of the Public Health Image Library, CDC. [Public domain.]*

Lactophenol cotton blue stains also help with visualization. Skin biopsies can help in uncertain cases.

TREATMENT AND THERAPY

In healthy people, most ringworm infections are not serious. Treatment for ringworm may include a topical treatment. This type of treatment is used for ringworm of the skin or body and includes antifungal creams and powders. It usually takes at least two weeks for the ringworm to clear. After ringworm clears, treatment is usually continued for at least two more weeks.

For ringworm involving the body, hands, or feet, non-prescription treatment is highly effective. The following are some available treatments that can cure ringworm: clotrimazole, tolnaftate, undecylenic acid, miconazole, and terbinafine. Terbinafine is more effective than the other medications. It usually needs to be used for only one week instead of four weeks. Terbinafine, however, is more expensive than the alternatives.

Oral treatment is used for ringworm of the nails and scalp. Topical treatments cannot penetrate the scalp. Oral griseofulvin or terbinafine are used for tinea capitis and tinea unguium. Early treatment for scalp ringworm is critical in preventing permanent hair loss. One gram of griseofulvin is given per day for one to two months. Griseofulvin should always be taken with a high-fat meal. Terbinafine is given for six to twelve weeks for tinea unguium (250 mg/day). Patients with poorly functioning immune systems may also require oral treatments for non-severe dermatophyte infections. If the patient developed ringworm from a pet, the pet should be treated too. To prevent dermatophyte infections from spreading, avoid close contact with others and do not share any personal objects that can become contaminated.

PREVENTION AND OUTCOMES

To help prevent ringworm, one should avoid contact with an infected person, animal, surface, or object; avoid sharing personal hair-grooming items or clothing or shoes; wear sandals in locker room areas; avoid scratching during infection, to prevent ringworm from spreading to other areas; wear clothing that minimizes sweating and moisture buildup; wear breathable shoes or sandals, and keep moisture-prone areas of the body clean and dry.

—Michelle Badash and Michael A. Buratovich, PhD

Further Reading

Al-Janabi, A. A., and F. H. Al-Khikani. "Dermatophytoses: A Short Definition, Pathogenesis, and Treatment." *International Journal of Health & Allied Sciences*, vol. 9, no. 3, 2020, pp. 210-14, doi:10.4103/ijhas.IJHAS_123_19.

Coulibaly, O., C. L'Ollivier, R. Piarroux, and S. Ranque. "Epidemiology of Human Dermatophytoses in Africa." *Medical Mycology*, vol. 56, no. 2, 2017, pp. 145-61, doi:10.1093/mmy/myx048.

"The Cutaneous Mycoses." *Mycology Online*, mycology.adelaide.edu.au/mycoses/cutaneous. Accessed 23 Nov. 2020.

Griffiths, Christopher, et al., editors. *Rook's Textbook of Dermatology*. 9th ed., Wiley-Blackwell, 2016.

Hayette, M.-P., and R. Sacheli. "Dermatophytosis, Trends in Epidemiology and Diagnostic Approach." *Current Fungal Infection Reports*, vol. 9, no. 3, 2015, pp. 164-79, doi:10.1007/s12281-015-0231-4.

Kalsi, A. S. "Tinea unguium. T. violaceum Was Isolated." *ResearchGate*, www.researchgate.net/figure/Tinea-unguium-T-violaceum-was-isolated_fig6_328964379. Accessed 23 Nov. 2020.

Oakley, A. "Tinea capitis." *DermNet NZ*, dermnetnz.org/topics/tinea-capitis/. Accessed 23 Nov. 2020.

Richardson, Malcolm D., and Elizabeth M. Johnson. *The Pocket Guide to Fungal Infection*. 2nd ed., Blackwell, 2006.

CRYPTOCOCCUS

Category: Pathogen

Anatomy or system affected: Blood, brain, circulatory system, cribriform plate, immune system, lungs, lymphatic system, meninges. nasopharynx, nervous system, respiratory system

Specialties and related fields: Immunology, internal medicine, microbiology, mycology, neurology, pathology, pharmacology, public health

Definition: a fungus found worldwide in soil and areas on and around trees that causes cryptococcosis, an invasive meningeal mycosis in humans

KEY TERMS

basidiomycete: a group of higher fungi that have septate hyphae and spores borne on a club-like structure called a "basidium"

capsule: a polysaccharide layer that lies outside the cell wall

meninges: the three membranes (the dura mater, arachnoid, and pia mater) that line the skull and vertebral canal and enclose the brain and spinal cord

phenol oxidase: An enzyme in Cryptococcus species that catalyzes the conversion of phenolic compounds, including catecholamines, such as dopamine and epinephrine, to melanin.

serotypes: a distinct variation within a species of bacteria or virus that is serologically distinguishable

NATURAL HABITAT AND FEATURES

Cryptococcus infections or cryptococcosis is caused by the fungi *Cryptococcus neoformans* (*C. neoformans*) and *Cryptococcus gattii* (*C. gattii*). *C. neoformans* is the most common species found in the United States. C. neoformans is found in soil samples from around the world in areas frequented by birds, especially pigeons and chickens. This fungus is also prevalent in roosting sites of pigeons and is associated with rotting vegetation. *C. gattii*, typically found in tropical and subtropical climates, has also been identified in Canada and the United States. *C. gattii* is not associated with bird feces; instead, it is associated with the bark, leaves, and plant debris of eucalyptus trees and gum trees.

Cryptococci are basidiomycetous, encapsulated yeasts. The basidiomycetes include rusts, smuts, mushrooms, and puffballs. The C. neoformans life cycle involves asexual and sexual stages. In the asexual stage, the organism grows as yeast that reproduce by budding. These haploid, unicellular yeasts are the only forms of C. neoformans that cause human infections. The sexual or perfect stage C. neoformans has been observed in the laboratory. The yeast exists in one of two mating types designated "a" and alpha. Mating between yeast produces the sexual state consisting of dikaryotic hyphae that contain clamp connections characteristic of the basidiomycetes. The sexual state forms specialized structures called "basidia." Meiosis occurs within the terminal portion of the basidia, producing haploid basidiospores. Once forcibly ejected from the basidium, the spores bud as yeast, the life cycle is completed. Initially, the spores have no capsules but quickly develop capsules before budding begins.

Cryptococcus yeast cells are round or oval-shaped, surrounded by a complex cell wall consisting of a filagree of glucans, proteins, melanin, glycoproteins, chitin, and chitosans. The presence of melanin in the cell wall of *C. neoformans* and *C. gattii* is unusual for this genus. Melanin is made by an enzyme called "phenol oxidase," which converts phenolic compounds to melanin. The brain contains a host of neurotransmitters like dopamine and epinephrine, potential substrates for the cryptococcal phenol oxidase. The utilization of neurotransmitters might explain these organisms' preference for the brain. The phenol oxidase enzyme may be an important virulence factor since melanin is an important antioxidant that protects the yeast from white blood cells. Mutant *Cryptococcus* strains that lack phenolic oxidase do not cause disease in laboratory animals and are more susceptible to white blood cell-mediated killing.

Outside the cell wall is a polysaccharide capsule made of mannose, xylose, and glucuronic acid. The capsule prevents white blood cells from phagocytosing it, making it an essential virulence factor for *Cryptococcus*. Mutant cryptococci that make

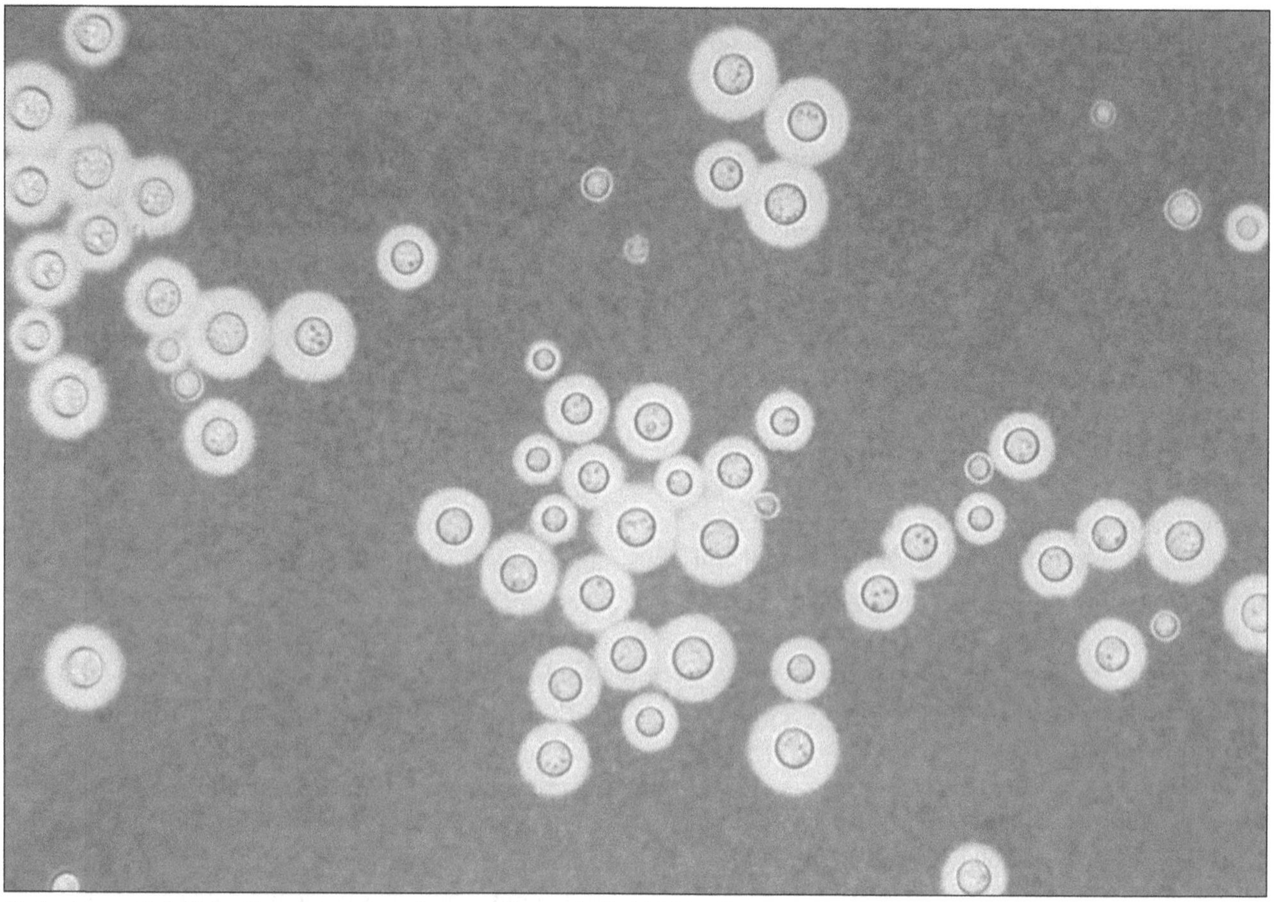

Photomicrograph depicting Cryptococcus neoformans. *Image via CDC/Wikimedia Commons. [Public domain.]*

weak or no capsules are less virulent in animal models and are more easily killed by white blood cells.

C. neoformans and *gattii* are distinct cryptococcal species. Each species has five serotypes based on the antigenic specificity of the capsular polysaccharide. *C. neoformans* includes serotypes A, D, and a hybrid AD. *C. gattii* includes serotypes B and C. Serotype A causes most cryptococcal infections in immunocompromised persons.

C. neoformans and *C gattii* grow as encapsulated yeasts at 98° Fahrenheit (37° Celsius). Their identification is based on its microscopic appearance: smooth, convex, and yellow or tan colonies on solid media at 68° to 98° F (20° to 37° C).

PATHOGENICITY AND CLINICAL SIGNIFICANCE

Cryptococcal infections response mainly depends on the infected person's immune status before infection and the involved sites. Responses include harmless colonization of the airway and asymptomatic infections to meningitis and disseminated disease. This organism's primary infective site is the respiratory system, and human-to-human transmission does not occur. *C. neoformans* can also infect nonhuman animals.

Upon inhalation, the yeast arrive in the lung alveoli. If they survive the neutral to alkaline pH shift and physiologic carbon dioxide concentrations, alveolar macrophages may phagocytose them. *C. neoformans* and *C. gattii* pulmonary infections may cause pneumonia-like symptoms, including shortness of breath, cough, chest pain, and fever. A chest X-ray may reveal focal or diffuse infiltrates and a nodule or mass.

Although it initially infects the lungs, *Cryptococcus* mainly affects the central nervous system. Cryptococcal meningitis and meningoencephalitis are the most common and severe forms of cryptococcal disease affecting the central nervous system. These forms can be fatal if not treated appropriately; death can occur from two weeks to years following the onset of symptoms. The most common symptoms are headaches, altered mental status, confusion, lethargy, obtundation (decreased alertness), seizures, and coma. Other infection-related organ involvement sites are the skin, prostate, bones, eyes, heart (as myocarditis), liver (as hepatitis), and adrenals.

C. neoformans typically infects immunocompromised persons but can also infect persons who are not immunocompromised. Persons with human immunodeficiency virus infection and other immunocompromised persons, including those undergoing organ

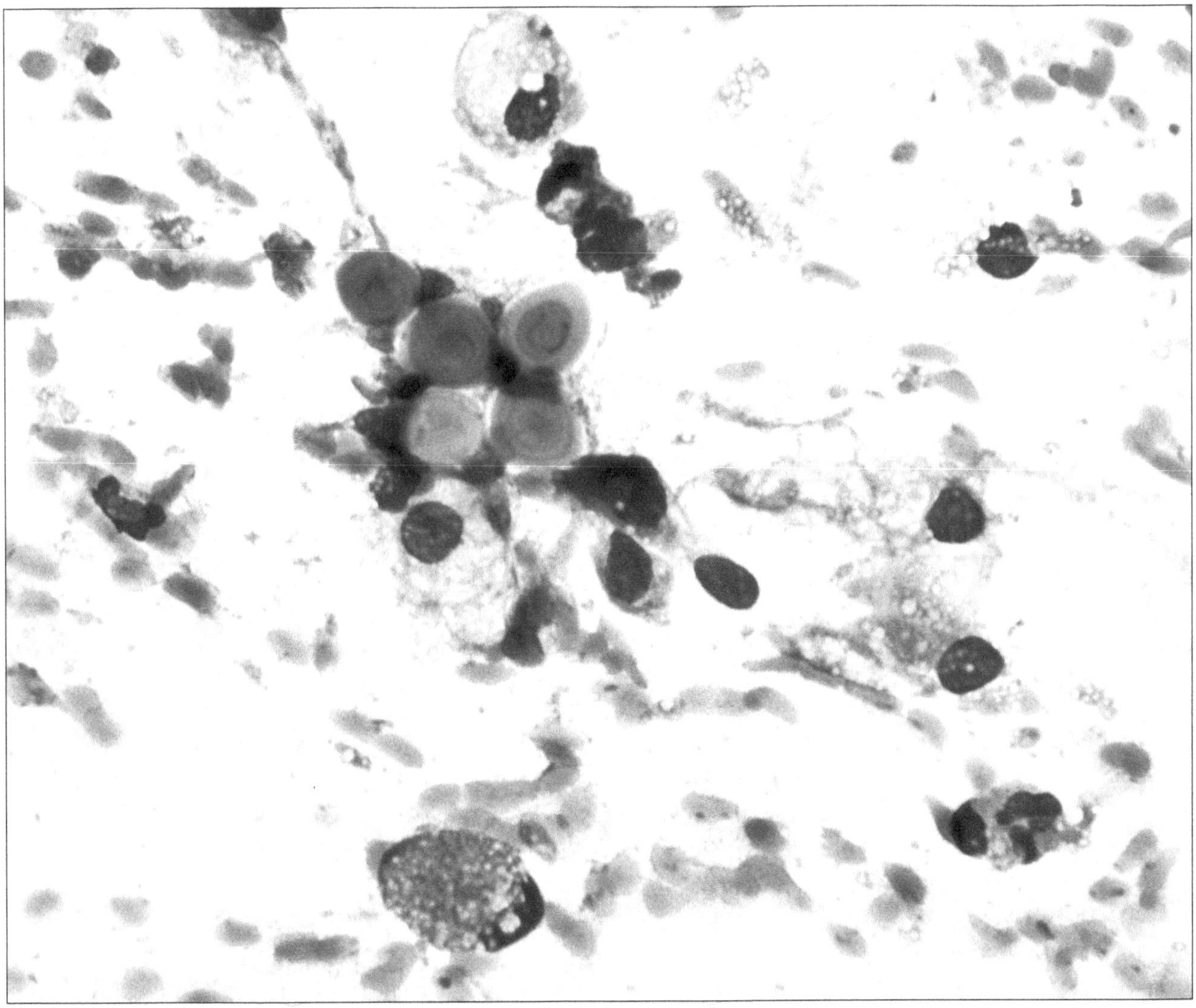

Field stain showing Cryptococcus *species in lung tissue. Photo by Nephron, via Wikimedia Commons.*

transplantation and receiving corticosteroid treatment, are at high risk of developing cryptococcosis. The incubation period for *C. neoformans* is unknown.

C. gattii rarely infects immunocompromised persons. Instead, this organism usually infects persons with healthy immune systems. Persons infected with *C. gattii* may exhibit symptoms two to fourteen months following exposure. *C. gattii* infections respond much slower to treatment than C. *neoformans* infections, which increases the risk of developing significant central nervous system sequela.

Cryptococcal-encapsulated yeast cells can be visualized using an India ink preparation on cerebral spinal fluid. Mucicarmine stains highlight both the yeast form and the capsule and specifically stain Cryptococcus. In addition, blood, urine, tissue, and sputum can also be examined microscopically for the presence of *Cryptococcus*. A rapid cryptococcal antigen test also can use blood or cerebral spinal fluid. The organism must be cultured to determine the infection type definitively, which requires special testing at state health department laboratories or the Centers for Disease Control and Prevention (CDC). Computed tomography scans and magnetic resonance imaging studies may help in distinguishing cryptococcal infections from other symptomology.

DRUG SUSCEPTIBILITY

Meningeal and other serious cryptococcal infections can have a rapid onset of symptoms. Hence, the administration of the appropriate antifungal "cocktail" is critical. Lipid formulations of amphotericin B, combined with oral flucytosine or fluconazole, are first-line drug therapy for cryptococcosis. This combination penetrates the blood-brain barrier more effectively. Amphotericin B has a rapid onset of action, leading to faster clinical improvement. Liposomal amphotericin B is preferred since this preparation of amphotericin spares renal function. Ketoconazole or itraconazole should not be used in the initial treatment of cryptococcosis because they do not adequately penetrate the blood-brain barrier.

Cryptococcosis includes varying degrees of treatment regimens following initial drug therapy. These regimens depend on immune system involvement. An immunocompromised person with a human immunodeficiency virus (HIV) infection will begin initial aggressive treatment with the therapeutic goal of controlling the acute cryptococcal infection, followed by lifelong suppression therapy. The therapeutic goal for persons with cryptococcosis but not HIV-positive is a permanent cure, with no chronic suppressive therapy.

When systemic therapy becomes refractory, intrathecal or intraventricular amphotericin B may be required. Successful therapy is considered after negative cerebral spinal fluid cultures, and the infected person has had significant clinical improvement.

—Stephanie McCallum Blake, MSN and Michael A. Buratovich, PhD

Further Reading

"*Cryptococcus neoformans* Infections." *Red Book: 2009 Report of the Committee on Infectious Diseases*, edited by Larry K. Pickering, et al., 28th ed., American Academy of Pediatrics, 2009.

Kibbler, Christopher C., et al., editors. *Oxford Textbook of Medical Mycology. Oxford UP, 2018.*

King, John W., and Meredith L. DeWitt. "Cryptococcosis." emedicine.medscape.com/article/215354-overview.

Thomas, Nancy J., D. Bruce Hunter, and Carter T. Atkinson, editors. *Infectious Diseases of Wild Birds.* Blackwell, 2007.

Webster, John, and Roland Weber. *Introduction to Fungi.* Cambridge UP, 2007.

Mycotoxins

Category: Toxicology

Anatomy or system affected: All systems

Specialties and related fields: Biochemistry, biotechnology, microbiology, mycology,

Definition: natural secondary byproducts of fungi that can produce toxic effects when ingested or inhaled

KEY TERMS

aflatoxin: a family of toxins produced by certain fungi that are found on crops such as maize (corn), peanuts, cottonseed, and tree nuts; the primary fungi that produce aflatoxins are *Aspergillus flavus* and *Aspergillus parasiticus*, which are abundant in warm and humid regions of the world

trichothecene: any of several mycotoxins that are produced by various fungi (such as genera *Fusarium* and *Trichothecium*) and that include some contaminants of livestock feed, and some held to be found in yellow rain

SIGNIFICANCE

Forensic scientists are sometimes called upon to identify mycotoxins and toxigenic molds, which can cause severe health problems with long-term exposure. Mycotoxins are also a concern to law-enforcement agencies because these agents can be readily isolated and thus have the potential for use as biological weapons.

Mycotoxin production occurs when favorable environmental conditions allow fungi to grow on plants or plant-based materials. Mold and mycotoxin contamination can increase extreme environmental conditions, such as drought, excessive precipitation, floods, sudden frost, and constant high humidity. Most of the world's croplands, forests, and population centers are in temperate zones, prime breeding grounds for toxigenic fungi. Estimates suggest that approximately 25 percent of the world's grain supply is contaminated with mycotoxins at any given time. It has also been estimated that at least 10 percent of all buildings in North America are contaminated with toxigenic molds at levels that pose a health risk.

The severity of mycotoxicosis, or mycotoxin poisoning, depends on the toxicity of the mycotoxin, the extent of exposure, and the age and health of the victim. The human health impacts of mycotoxicosis are multiple, including allergies, chronic bronchitis, skin necrosis, respiratory failure, loss of bone marrow, liver and kidney failure, skin irritation, anorexia, tremors, vasoconstriction, headache, chronic fatigue, cancers, vomiting, gastric and intestinal irritation, hemorrhaging, tachycardia, severe immunodeficiency, neurocognitive dysfunction, anxiety, tremors, fibromyalgia, lupus, ataxia, and reproductive problems. Because of these toxic effects, mycotoxins are part of an ongoing controversy over their use as biological weapons to produce neurological impairment.

Toxigenic fungi associated with animal and human food chains are in three main genera: *Aspergillus*, *Fusarium*, and *Penicillium*. Although *Aspergillus* and *Penicillium* species are important mycotoxin producers—*Penicillium* alone can produce twenty-seven different mycotoxins—*Fusarium* is most significant in their effects on crops, poultry, livestock, and farm workers. *Claviceps* fungi are responsible for historical epidemics of ergotism, and *Stachybotrys chartarum*, considered one of the most poisonous molds on Earth, is commonly found in human dwellings.

Stachybotrys chartarum and *Chaetomium* are fungi often identified within domestic housing and considered the source of "sick house syndrome." The fungi are found in dark, moist indoor environments such as wall cavities, attics, basements, and ventilation systems. The fungi produce black spores resembling soot that grow aggressively on moist drywall and are carried by circulating air. Several high-profile sick house toxic tort cases have been litigated in the United States.

Trichothecene mycotoxin (T-2 mycotoxin), a derivative of *Aspergillus*, *Stachybotrys*, and *Fusarium*, is considered among the most potent naturally occurring toxins. T-2 mycotoxin is the only biologically active toxin effective through inhalation, ingestion, and dermal exposure. Declassified US government documents suggest that T-2 mycotoxins have been identified as biological warfare agents since the mid-1970s in Laos, Afghanistan, Kampuchea, and the Arabian Peninsula. It has been asserted that T-2 mycotoxin exposure is a causal agent of the illness known as Gulf War syndrome.

—*Randall L. Milstein*

Further Reading

DeVries, Jonathan W., Mary W. Trucksess, and Lauren S. Jackson, editors. *Mycotoxins and Food Safety*. Kluwer Academic, 2002.

Diaz, D. E., editor. *The Mycotoxin Blue Book*. Nottingham UP, 2005.

Matossian, Mary Kilbourne. *Poisons of the Past: Molds, Epidemics, and History*. Yale UP, 1989.

Money, Nicholas P. *Carpet Monsters and Killer Spores: A Natural History of Toxic Mold*. Oxford UP, 2004.

Rea, William J., et al. "Effects of Toxic Exposure to Molds and Mycotoxins in Building-Related Illnesses." *Archives of Environmental Health*, vol. 58, no. 7, 2003, pp. 399-405.

Yeasts

Category: Mycology

Anatomy or system affected: Gastrointestinal tract, skin

Specialties and related fields: Economic botany, food, fungi, microorganisms, mycology, plant types, taxonomic groups

Definition: eukaryotic, single-celled microorganisms classified as members of the fungi

Key terms

budding: a form of asexual reproduction in which a new individual develops from some generative anatomical point of the parent organism

pseudomycelium: a cellular association occurring in various higher bacteria and yeasts in which cells cling together in chains resembling small true mycelia

teleomorph: the sexual reproductive stage (morph), typically a fruiting body

INTRODUCTION

Among mycologists, there is some disagreement over what should be called "yeast." Many mycologists use the term to describe any fungus with a unicellular budding form at any time in its life. They often use the term "monomorphic" to describe always unicellular and the term "dimorphic" to describe those that can have both unicellular and filamentous growth. Others, however, reserve the name yeast for those permanently unicellular species and use the term "yeastlike" to describe those fungi that can alternate between mycelial and unicellular forms. Because some species traditionally been called yeasts have later been shown to have a mycelial form, the former broader definition will be used here.

TAXONOMY

Yeasts are found in all three major fungal phyla, *Zygomycota*, *Ascomycota*, and *Basidiomycota*, but the vast majority are ascomycetes. As in many fungi, the placement of some species in the proper phylum is made difficult by the lack of data on sexual reproduction. In others, the sexually reproducing, or teleomorph, form and the asexually reproducing, or anamorph, form have been assigned different names. In addition, dimorphic fungi were often assigned different names for their yeast and mycelial phases. *Mucor indicus* (synonymous with *Mucor rouxii*) is a zygomycete yeast. Basidiomycete genera include *Filobasidiella* (anamorph: *Cryptococcus*), *Rhodospiridium*, and *Ustilago*. Some ascomycete genera are *Saccharomyces*, *Candida*, *Blastomyces*, and *Ajellomyces* (anamorph: *Histoplasma*).

REPRODUCTION

The most common reproductive mechanism seen in yeasts is "budding." During this asexual process, the nucleus divides. A small section of the original cell containing one of the new nuclei begins to bulge from the original cell. The cell and the bud begin to separate by forming a new cell wall called, at this stage, the "cell plate." The bud grows and, usually, separates from the original cell. In some species, buds do not separate and, after they have grown, may produce buds of their own. This pattern leads to a connected group of cells produced by sequential budding called a "pseudomycelium." Yeasts may bud new cells from any part of the original cell (called "multilateral budding") or from just the tips of the cell (called "polar budding"). The release of a bud often leaves a bud scar, and the scarred area can usually not produce another bud.

Other reproduction methods include fission, in which the original cell divides equally, and the production of various kinds of spores occurs both asexually and sexually.

CELLS

Like all fungi, yeasts are eukaryotic organisms that exist in haploid, diploid, and dikaryotic (two haploid nuclei per cell) states. Unlike filamentous fungi, in which the zygote is the only diploid cell, ascomycete yeasts such as *Saccharomyces* can have a prolonged diploid state after the haploid nuclei of the dikaryote fuse. Cell components of yeasts are quite similar to those of filamentous fungi. One exception is that monomorphic yeasts have much lower chitin levels in their cell walls, and the small amount present is found mainly in the bud scars.

USES

Both brewing and bread making, which use various *Saccharomyces* species, have existed for millennia. Four-thousand-year-old tomb paintings in Egypt depict both, but only since the mid-1800s has the involvement of yeast in these processes been studied. Complex carbohydrates are converted to glucose in both, and the yeast ferments glucose, producing ethyl alcohol and carbon dioxide. *Saccharomyces cerevisiae* is the most common baker's yeast, although *Candida milleri* is important in producing sourdough bread. The bottom-fermenting *Saccharomyces carlsbergensis*, which tolerates cold (10° Celsius), is the most used yeast in beer brewing. Wines, a few beers, and most ales use *Saccharomyces cerevisiae*, a top fermenter that requires higher temperatures (20 to 30° C).

Yeasts are also important in developing flavor and texture in certain cheeses. Limburger, Camembert, Brie, and Swiss cheeses all rely on yeast fermentation. Because of their high nutrient content, yeasts themselves are important foods and food additives. Yeasts have also been used to produce various industrial chemicals and biochemicals, including glycerol, ethanol, B vitamins, and polysaccharides. With the advent of modern genetic techniques, yeasts are being engineered to produce many other useful products.

PATHOGENS

Many yeasts can cause disease in plants or animals. *Histoplasma capsulatum*, *Blastomyces dermatiditis*, and *Cryptococcus neoformans* are all dimorphic fungi that can cause systemic infection in humans when the fungi are in the yeast form. *Candida albicans*, also dimorphic, is an opportunistic human pathogen pathogenic in its filamentous form. Most *Candida* infections, such as vaginal yeast infections and thrush, are superficial, but systemic infections can occur in immunocompromised individuals. *Pneumocystis jirovecii*, which causes respiratory infections in patients with acquired immunodeficiency syndrome, was originally classified as a protist but is now thought to be a dimorphic fungus. The dimorphic genera *Ustilago* and *Taphrina* both contain many plant pathogens.

—*Richard W. Cheney Jr.*

Further Reading

Ingold, C. T., and H. J. Hudson. *The Biology of Fungi*. 6th ed., Chapman & Hall, 1993.

Money, Nicholas P. *The Rise of Yeast: How the Sugar Fungus Shaped Civilization*. Oxford UP, 2017.

Phaff, Herman I., et al. *The Life of Yeasts*. 2nd ed., Harvard UP, 1978.

Schaechter, Moselio. *Mechanisms of Microbial Disease*. Lippincott Williams & Wilkins, 1998.

Protozoans

Here, we describe those pathogenic protozoans that cause human and animal diseases. These organisms have remarkably complex life cycles and cause tremendous disease burdens in the Third World. This section includes entries on amoebic dysentery, malaria, trypanosomiasis, toxoplasmosis, and others.

PROTOZOAN DISEASES

Category: Diseases and conditions
Anatomy or system affected: All
Specialties and related fields: Cardiology, gastroenterology, hematology, immunology, internal medicine, microbiology, neurology, travel medicine, parasitology, pathology, protozoology public health, veterinary medicine
Definition: diseases caused by eukaryotic protozoans

KEY TERMS

arthropod vector: mosquitoes, fleas, sand flies, lice, fleas, ticks, and mites that transmit parasites either by injection into the bloodstream directly via their salivary glands, or forcing parasites into a pool of blood that develops when chewing the skin
cyst: resting, dormant protozoan stages with a protective membrane or thickened wall
sporozoite: a minute, elongated cell that arise from the repeated division of the oocyst during sporogony in certain protozoans
trophozoite: the activated, feeding stage in the life cycle of certain protozoa

INTRODUCTION

The protozoa are a large and diverse group of often-pathogenic organisms that can cause many diseases in humans. Traditionally, these organisms have been described as single-celled eukaryotic microorganisms, but newer ultrastructural information challenges this uniform classification. The protozoan *Giardia lamblia*, for example, has been found to lack mitochondria and may be a transitional organism somewhere between the prokaryotic bacteria and eukaryotic protozoa.

Common protozoan diseases include travelers' diarrhea, malaria, trypanosomiasis (African sleeping sickness), and vaginitis. Their transmission mode classifies these diseases, and the most common are enteric, sexual, and arthropodal.

CAUSES

Enteric transmission is generally associated with intestinal illness in humans. Common protozoa that cause intestinal illness include the flagellate *G. lamblia*, the ameba *Entamoeba histolytica*, spore-forming organisms, and ciliates. Diseases caused by these groups begin with the ingestion of contaminated water or by fecal-oral transmission. *Cryptosporidium* also causes gastroenteritis and is transmitted via the fecal-oral route. This organism causes self-limited diarrhea in healthy people but potentially life-threatening infections in those with subfunctional immune systems. *Toxoplasma* shares this group's route of oral-fecal transmission but is not associated with gastroenteritis. Many of these organisms make a thick-walled, resistant structure called a "cyst" that protects them from harsh environments. Cysts are passed into feces where they contaminate water. In tissue, the active form of the protozoan is called the "trophozoite."

Sexually transmitted *Trichomonas vaginalis* infection is the most common type of pathogenic protozoan disease. *T. vaginalis* causes vaginitis in sexually active women who have multiple partners. Infection in men is usually asymptomatic. The organism survives in moist environments. Trichomoniasis may frequently coexist with other sexually transmitted diseases and may increase the risk of human immunodeficiency virus (HIV) transmission.

Arthropod-borne protozoa include the parasitic flagellate *Trypanosoma*, which is transmitted by the tsetse fly and causes trypanosomiasis. *Trypanosoma cruzi* causes Chagas disease, and this protozoan is transmitted by the bite of triatomine bugs ("kissing bugs"). The most common species that transmit Chagas disease are *Triatoma dimidiata*, *Triatoma infestans*, and *Rhodnius prolixus*. *T. cruzi* is also transmitted by contaminated food and water since the organism penetrates mucous membranes. Blood transfusions from infected people can also infect the recipients. Infected mothers can pass Chagas disease to their babies through the placenta.

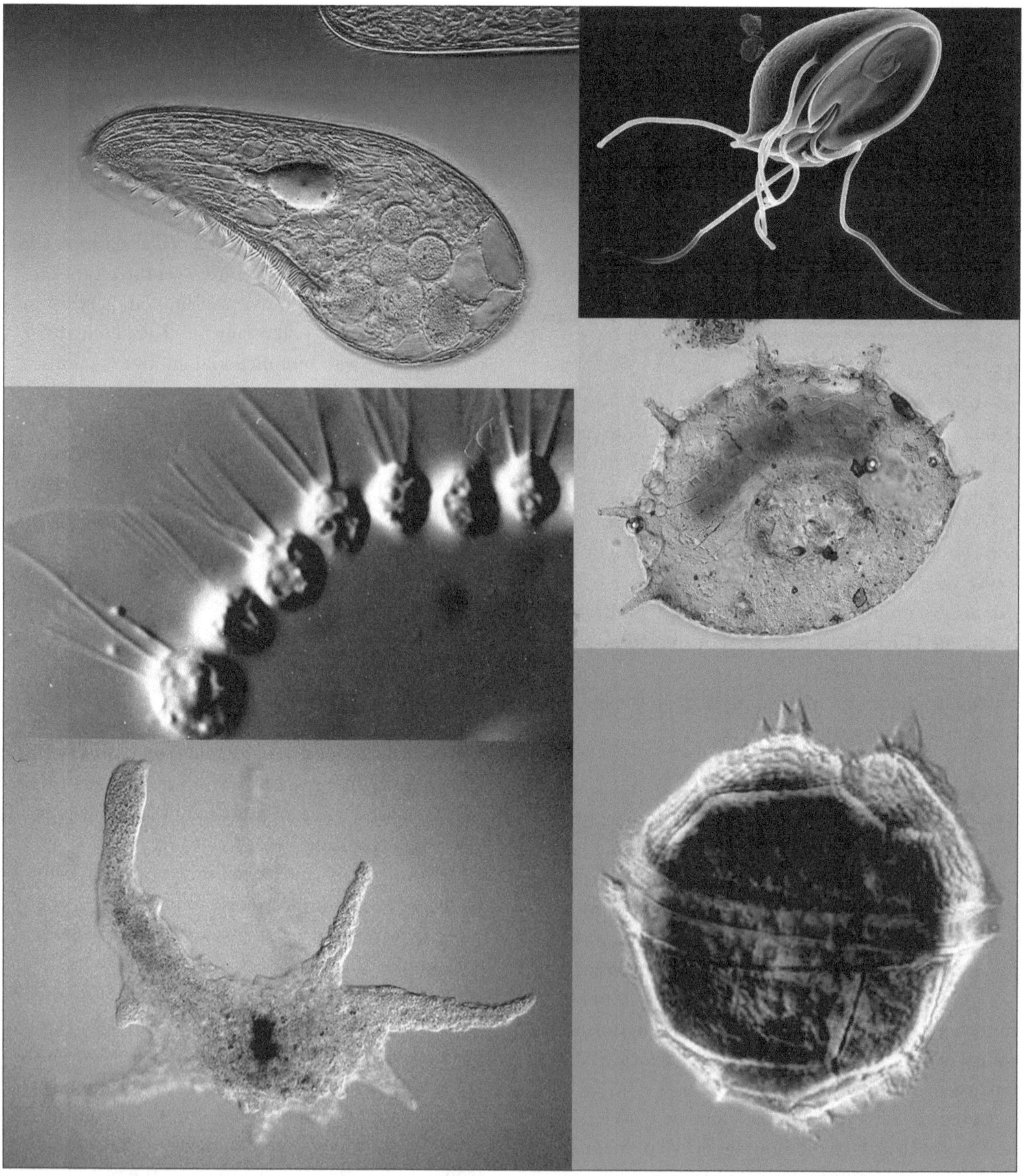

Various protozoa. From top left: Blepharisma japonica, Giardia muris, Centropyxis aculeata, Perdidinium willei, Chaos carolinense, *and* Desmarella moniliformis. *Images courtesy of Frank Fox, Sergey Karpov, CDC/Dr. Stan Erlandsen, Picturepest; Thierry Arnet; Dr. Tsukii Yuuji via Wikimedia Commons.*

Malaria, the leading cause of death in tropical countries, is caused by five *Plasmodium* species transmitted by *Anopheles* mosquitoes. Plasmodia infect the insect's salivary glands, and when they take a blood meal, the salivary glands inject hordes of sporozoites into the blood. Babesiosis is a tick-borne illness caused by *Babesia microti*. Symptoms of babesiosis are similar to malaria. Infection with the protozoan *Leishmania* is caused by the bite of an infected sandfly (*Lutzomyia* or *Phlebotomus*).

The freshwater amoebae *Naegleria fowleri* causes primary amebic meningoencephalitis (PAM). This organism inhabits warm, stagnant water. It infects swimmers when they dive into the water, and the water rushes up their noses. The organism moves through the cribriform plate into the brain, where it feeds on brain tissue.

Another amoeba, *Acanthamoeba*, lives in soil, fresh water, sewage, seawater, swimming pools, bottled water, tap water, hospital air-conditioning units, and contact lens cases. This ubiquitous organism enters the body through breaks in the skin or inhalation. It causes granulomatous amoebic encephalitis, mainly in those with poorly functioning immune systems.

RISK FACTORS

Risk factors for enteric transmission of protozoa are poor sanitary conditions and living or traveling to parts of the world where these conditions are endemic. Elderly persons and children may be at increased risk for these diseases.

Giardiasis is more common in children than in adults and may be concentrated in child day-care centers. Because beavers are an animal reservoir for *Giardia*, campers to drink untreated river water downstream of beaver colonies are at risk for giardiasis.

Pregnant women exposed to cat feces, undercooked meat, or unpasteurized milk are at increased risk for fetal transmission of toxoplasmosis. People with weakened immune systems are at higher risk from all the spore-forming protozoa. These risk factors include acquired immunodeficiency syndrome (AIDS), renal transplantation, cancer, and IgA deficiency.

The risk factor for tick-borne babesiosis is living in areas where ticks are common. In the United States, this includes New England and New Jersey's coastal areas and the Upper Midwest. The risk of being infected with *Plasmodium*, *Leishmania*, or *Trypanosoma* is directly related to living or traveling in tropical or subtropical parts of the world where these organisms are endemic. *T. cruzi* is endemic in the southern United States and Latin America.

SYMPTOMS

Symptoms of protozoan diseases vary greatly and can range from mild to severe. Scientists group them roughly by mode of transmission. Protozoan diseases transmitted by contaminated water or oral-fecal transmission and cause intestinal illness commonly lead to nausea, bloating, anorexia, weight loss, abdominal pain, diarrhea, colitis, and dysentery. Toxoplasmosis infection may cause symptoms of fever, body aches, headache, fatigue, and adenopathy. Infants infected with toxoplasmosis may be born with symptoms including seizures, jaundice, hepato- or splenomegaly, and eye infection.

Common symptoms of arthropod-borne protozoan diseases include fever, chills, sweats, headache, myalgia, fatigue, anorexia, and weight loss. The visceral form of leishmaniasis causes symptoms similar to other arthropod-borne diseases. The cutaneous form of leishmaniasis causes symptoms that include skin bumps or nodules that may ulcerate and scab.

Trichomoniasis causes symptoms of copious, watery vaginal discharge. Vulvovaginal irritation may be accompanied by dysuria, dyspareunia, and abdominal pain. Infection in men is usually asymptomatic but may also result in dysuria.

Chagas disease has an acute and chronic phase. The acute phase is mostly symptomatic but might show tiredness (malaise), fever, appetite loss, headache, a nodule at the infection site (chagoma), enlarged lymph nodes, liver, and spleen. The chronic phase may have cardiac (fatigue, chest pain, swelling, heart murmur, shortness of breath) or gastrointestinal (constipation, abdominal pain, bloating, trouble swallowing, and vomiting) manifestations.

SCREENING AND DIAGNOSIS

The infected person's medical history and a physical examination are important for diagnosis. Still, symptoms of protozoan diseases may mimic many other diseases. Laboratory studies, then, are the most important screening and diagnostic tools. For all protozoa-related intestinal diseases, identifying the organism in a stool sample is the definitive method of confirming the diagnosis.

Diagnosis of arthropod-borne diseases frequently relies on serology, detecting underlying anemia, and identifying protozoa in blood or a blood smear. Other laboratory studies that may aid diagnosis include enzyme-linked immunosorbent assay antibody detection, electron microscopy, and polymerase chain reaction. Histology, imaging studies, and endoscopy may be used in selected cases.

Trichomoniasis is frequently diagnosed by doing a wet mount of vaginal secretions. Immunofluorescence antibody staining and culture are more sensitive, but they could delay diagnosis. Skin scrapings from cutaneous leishmaniasis may be examined microscopically or may be cultured for diagnosis.

TREATMENT AND THERAPY

Treatment for most protozoan diseases requires specific antiprotozoal medication. Intravenous rehydration therapy is an important aspect of treating intestinal protozoan infections that cause dehydration. Nutritional status must also be addressed, especially in newborns and infants. Surgery may play a role in cases of necrotizing colitis or amebic liver abscess.

Some medications used to treat intestinal protozoan infections include iodoquinol, paromomycin, metronidazole, tinidazole, quinacrine, furazolidone, tetracycline, nitazoxanide, and trimethoprim/sulfamethoxazole. Toxoplasmosis may be treated with the antimalarial medication pyrimethamine and the antibiotic sulfadiazine. Trichomoniasis responds to metronidazole and tinidazole. Babesiosis responds to quinine sulfate and clindamycin.

Trypanosome infections require antitrypanosomal medicines. The Chagas disease, benznidazole, and nifurtimox are two agents with proven efficacy. Neither drug is safe for pregnant women. Treating *T. bucei* infections depends on the stage of the disease. For the hemolymphic stage, pentamidine or suramin are the preferred treatments, and eflornithine, with

Rhodnius prolixus *nymphs and adult, via Wikimedia Commons.*

or without nifurtimox, or melarsoprol are the main treatments.

Naegleria fowleri infections are treated with amphotericin B with or without fluconazole or miltefosine. *Acanthamoeba* infections are treated with miltefosine, fluconazole, pentamidine isethionate without or without sulfamethoxazole/trimethoprim, metronidazole, or macrolide antibiotics.

Malaria treatment depends on the Plasmodium species and where malaria was acquired. Chloroquine is the first-line prophylactic treatment for malaria unless you are traveling to an area where chloroquine-resistant malaria is endemic, including Southeast Asia, Ethiopia, and Madagascar (*P. vivax*) or Southeast Asia, Oceania, and South America (*P. falciparum*). Artemisinin-based combination therapies are among the most highly recommended where chloroquine resistance occurs. These combinations include artemether/lumefantrine, artesunate-amodiaquine, artesunate-mefloquine, and artesunate-pyronaridine. Artemisinin-based antimalarials have few side effects, are potent against all blood stages (asexual forms) of malaria, and are cleared from the body the fastest relative to other antimalarial drugs. Primaquine is the best drug for the liver stages of malaria.

PREVENTION AND OUTCOMES

It is not possible to completely prevent the wide spectrum of protozoan diseases. Amebiasis is estimated to infect 10 percent of the world's population. *G. lamblia* is the most commonly isolated parasite globally, infecting up to 40 percent of children in developing countries. The best hope for prevention is through education and public health efforts to provide safe water supplies. Arthropod-borne protozoal diseases may be prevented by avoiding endemic areas, wearing protective clothing, using insecticides, sleeping in screened areas, and taking medications to prevent infection in the case of malaria. Insecticide-laced mosquito nets effectively prevent mosquito bites during the night.

—Christopher Iliades, MD and Michael A. Buratovich, PhD

Further Reading

Bogitsh, Burton J., et al. *Human Parasitology*. 5th ed., Academic Press, 2018.

Chacon-Cruz, Enrique. "Intestinal Protozoal Diseases." 26 Apr. 2017, emedicine.medscape.com/article/999282-overview. Accessed 8 Dec. 2021.

Madigan, Michael T., et al. *Brock Biology of Microorganisms*. 15th ed., Pearson, 2017.

Mullen, Gary T., and Lance A. Durden, editors. *Medical and Veterinary Entomology*. 3rd ed., Academic Press, 2018.

Papadakis, Maxine A., et al., editors. *Current Medical Diagnosis and Treatment 2022*. 61st ed., McGraw-Hill Education/Medical, 2021

LEISHMANIASIS

Category: Diseases and conditions

Also known as: Black fever, espundia, kala-azar, post kala-azar dermal leishmaniasis

Anatomy or system affected: All

Specialties and related fields: Bacteriology, biochemistry, biotechnology, immunology, internal medicine, microbiology, oncology, pathology, pharmacology, preventive medicine, public health

Definition: a heterogeneous group of protozoan parasitic diseases designated as cutaneous, mucocutaneous, or visceral

KEY TERMS

kala-azar: a form of the disease leishmaniasis marked by emaciation, anemia, fever, and enlargement of the liver and spleen

sandflies: a small hairy biting fly of tropical and subtropical regions that transmits several diseases, including leishmaniasis.

trypanosomes: a single-celled parasitic protozoan with a trailing flagellum, infesting the blood

CAUSES

As part of the life cycle of the protozoan genus *Leishmania*, the organisms are injected into the bloodstream of human hosts through the bite of vector sandflies. They proliferate within phagocytes (immune system cells) to continue their existence as obligate parasites. More than twenty disease-causing species of *Leishmania*, borne by some thirty species of blood-feeding sandflies (Phlebotomus), account for the striking epidemiologic and clinical diversity of leishmaniasis.

RISK FACTORS

Perhaps the most significant risk factor for leishmaniasis is extreme poverty, malnutrition, lessened resistance to infection, and poor housing. Cattle and other livestock may increase sandfly density, and sanitation is generally inadequate in endemic areas. Whether economic or war-driven, large-scale migrations expose vulnerable populations to new *Leishmania* strains. On a global scale, urbanization, deforestation, and climate change introduce human migrants to new routes of infection.

EPIDEMIOLOGY

The definite number of leishmaniasis cases remains uncertain. There are an estimated 700,000 to 1.2 million new cases of cutaneous leishmaniasis each year. There were probably fewer than 100,000 new cases of visceral leishmaniasis last year, but this number is equivocal.

Leishmaniasis is found in Asia, the Middle East, North Africa, and southern Europe, but not Australia or the Pacific Islands. In the Western Hemisphere, leishmaniasis occurs in Mexico, Central America, but not Chile or Uruguay. Globally, some ninety countries have leishmaniasis. Most cases result from travel

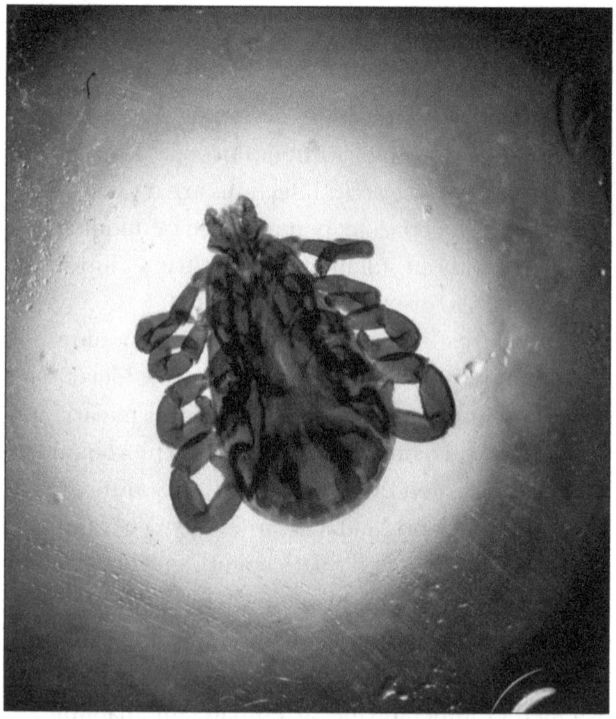

Phlebotomus flym *responsible for the transmission of leishmaniasis. Photo via iStock/piola666. [Used under license.]*

to endemic areas or military personnel who contracted it while on tour in the United States.

SYMPTOMS

The form of infection depends on the locale and the *Leishmania* species encountered. The clinical spectrum varies from self-limiting skin lesions (cutaneous leishmaniasis) to lethal systemic infection (visceral leishmaniasis).

The skin surface nodules or ulcers of cutaneous leishmaniasis often heal without treatment. Once healed, however, the infection can invade facial mucous membranes up to years later, causing the mucocutaneous form of the disease. Mucosal structures of the face and throat may be destroyed, mutilating facial features.

Visceral leishmaniasis, or kala-azar, is the most severe form of the disease. Progressive fever, body wasting, anemia, and enlarged liver and spleen are characteristic. Untreated, the case-fatality rate is more than 90 percent; the rate is about 10 percent with treatment.

Visceral leishmaniasis can also reappear after recovery as post-kala-azar dermal leishmaniasis (PKDL). Between 5 and 60 percent of cured persons develop chronic, unsightly PKDL. Large numbers of parasites in exposed skin of those with PKDL offer ready access to sandflies, contributing to the transmission of visceral leishmaniasis between epidemics.

SCREENING AND DIAGNOSIS

The World Health Organization, among other agencies, is active in leishmaniasis screening and identification. Often an entire village or district in an endemic area is surveyed, ideally with house-to-house screening.

The diagnostic gold standard is the detection of the DNA (deoxyribonucleic acid) of the responsible *Leishmania* species. The molecular techniques are limited to well-equipped laboratories rarely available in endemic countries.

If feasible, the diagnosis of cutaneous leishmaniasis is confirmed by microscopic examination of skin scrapings. Samples that can be analyzed later are collected directly from skin lesions. The characteristic appearance of mucocutaneous leishmaniasis often suffices for diagnosis.

Symptoms of visceral leishmaniasis resemble those of malaria, and coinfection with acquired immunodeficiency syndrome also complicates diagnosis. Accurate and early diagnosis of visceral leishmaniasis has been advanced by developing serologic tests that can be used in real-world field settings. In general, these tests detect antibodies in the blood of those thought to be infected. Another immunologic test can detect antigens in the urine of infected persons.

An important development in serologic testing is using a recombinant antigen that corresponds to the partial sequence of a *Leishmania* protein. A finger-stick blood sample means the test result can be determined in twenty minutes.

TREATMENT AND THERAPY

An effective, inexpensive, and widely available therapy does not exist. Parenteral administration of antimony compounds has been successfully used for decades. It is still the most common treatment for cutaneous and mucocutaneous leishmaniasis. It is, however, toxic and expensive. Oral antifungal agents have met with some recent success. Among the available treatments for cutaneous leishmaniasis that do not require drugs are cauterization, cryotherapy, and topical creams.

Drug options for visceral leishmaniasis are generally toxic, expensive, and difficult to administer. Lengthy treatment and the need for frequent laboratory monitoring add limitations. Spreading parasitic resistance to antimonial drugs is a major problem that antimonial treatment may fail in some locales. Where they are still useful, antimony compounds require thirty days of painful intramuscular injections.

Amphotericin B is also effective, but it is administered intravenously and requires a month's hospitalization. A liposomal formulation is a major advance in that it is well-tolerated. It shortens treatment, but the drug is beyond the financial reach of most endemic countries. However, miltefosine, an anticancer drug, is the first oral drug for visceral leishmaniasis; it cannot be given to pregnant women. Two treatment approaches for visceral leishmaniasis are advocated to counter drug resistance: Combination therapy with available drugs would reduce each dose and shorten the treatment course. Immunotherapy (developing ways to strengthen the host immune response) is being pursued as a research strategy.

PREVENTION AND OUTCOMES

Prevention must consider variables that include the *Leishmania* species, regional geography, and vector biology. A combination of approaches is required:

sandfly control, spraying dwellings and animal shelters, treating sleeping nets with an insecticide and applying insect repellents to skin and fabrics. Active, early case detection and rapid treatment can inhibit infection spread.

Leishmaniasis is one of the few parasitic diseases that vaccination could theoretically control. However, protection achieved with animal models has not carried over into successful field studies. The development of vaccines using recombinant methods continues.

—*Judith Weinblatt, MS, MA*

Further Reading

Clark, David P. *Germs, Genes, and Civilization: How Epidemics Shaped Who We Are Today*. FT Press, 2010.

Cliff, Andrew, Peter Haggett, and Matthew Smallman-Raynor. *World Atlas of Epidemic Diseases*. Oxford UP, 2004.

Marquardt, William, editor. *Biology of Disease Vectors*. 2nd ed., Academic Press/Elsevier, 2005.

Maxfield, Luke, and Jonathan S. Crane. "Leishmaniasis." *StatPearls*. StatPearls Publishing, 18 July 2021.

Parker, James M., and Philip M. Parker, editors. *The Official Patient's Sourcebook on Leishmaniasis: A Revised and Updated Directory for the Internet Age*. Icon Health, 2002.

Pradhan, S., et al. "Treatment Options for Leishmaniasis." *Clinical and Experimental Dermatology*, 4 Sept. 2021, doi:10.1111/ced.14919.

Trypanosomiasis

Category: Diseases and conditions

Also known as: African lethargy, African sleeping sickness, Gambian sleeping sickness, trypanosomosis

Anatomy or system affected: Blood, lymphatic system, nervous system

Specialties and related fields: Hematology, neurology, protozoology, tropical medicine

Definition: a tropical parasitic disease transmitted by the African tsetse fly, which infects the blood, lymphatic system, and nervous system in humans and animals

KEY TERMS

asymptomatic: infected but with no discernable symptoms of a disease

carrier: a person infected by an organism who can transmit that organism to other people but who is asymptomatic

lymph node: immune system structure that filters substances that travel through the lymphatic fluid; contains lymphocytes that help fight infection and disease

tsetse fly: an African blood-sucking fly that bites humans and other mammals; the tsetse fly transmits sleeping sickness and nagana

variant surface glycoprotein: a ~60kDa protein that densely packs the cell surface of protozoan parasites belonging to the genus *Trypanosoma*

CAUSES

A bite from the tsetse fly (Glossina) causes trypanosomiasis. Glossina is only found in sub-Saharan Africa. This insect transmits Trypanosoma parasites. Trypanosomiasis also can be transferred through the placenta during pregnancy. Additionally, laboratory workers have become infected with trypanosomiasis by accidentally pricking their skin with needles infected with Trypanosoma.

Human African trypanosomiasis (HAT) is also known as sleeping sickness. There are two forms of this disease. An acute form of trypanosomiasis is caused by Trypanosoma brucei rhodesiense (rhodesiense HAT). It occurs mainly in East and Southern Africa. Second, a more chronic form of trypanosomiasis occurs mainly in West and Central Africa and is caused by Trypanosoma brucei gambiense (gambiense HAT). These two T. brucei subspecies have identical microscopic morphology, and Glossina transmits both diseases. However, these two forms of HAT differ in epidemiology, clinical presentation, and management.

During T. brucei's lifecycle, it modulates its exposed surface antigens. T. brucei forms metacyclic trypanosomes in the tsetse fly and a distinct bloodstream-form in mammalian and human hosts. These distinct forms have proteins on their surface known as variant surface glycoprotein (VSG). The immune system responds vigorously to VSGs. Trypanosomes are covered with approximately 10 million copies of a single type of VSG. However, T. brucei uses a genetic system to vary the amino acid sequence of its VSG, thus confounding the immune system during the breadth of the infection.

RISK FACTORS

Because the vector of trypanosomiasis, the tsetse fly, is found only in rural areas of sub-Saharan Africa, those who frequent those areas, and villagers, hunters, and fishermen, are at greatest risk. Persons inhabiting rural African woodland and savannah regions, especially near bodies of water and dense vegetation, are most susceptible.

SYMPTOMS

The symptoms of trypanosomiasis include fever, headaches, swollen lymph nodes, extreme fatigue, skin rash, itching, joint and muscle pain, weight loss, confusion, sleepiness, slurred speech, impaired coordination, and altered personality.

SCREENING AND DIAGNOSIS

After conducting a physical examination and questioning the patient about symptoms and medical history, a physician will draw blood and spinal fluid samples. Electron microscopy of spinal fluid is necessary to confirm a trypanosomiasis diagnosis because trypanosomiasis is frequently asymptomatic or manifests mild symptoms in its initial stage and is sometimes difficult to discern in blood.

TREATMENT AND THERAPY

Five drugs are used to treat trypanosomiasis, depending upon the stage of the illness. Suramin and pentamidine are used to treat trypanosomiasis in its initial stage when it is confined to the blood and lymphatic systems. If trypanosomiasis is advanced and has infected the nervous system, then melarsoprol, or eflornithine, sometimes combined with nifurtimox, is administered in a hospital setting. For two years after treatment, the patient's spinal fluid is drawn and tested for trypanosomiasis at six-month intervals. The patient may relapse or become reinfected, requiring further treatment.

PREVENTION AND OUTCOMES

The World Health Organization has greatly reduced trypanosomiasis by treating male tsetse flies with radiation—rendering them sterile—then releasing them back into the environment, thereby lowering the number of tsetse flies. Although no vaccine or drug prevents trypanosomiasis, several steps can reduce the likelihood of infection. One should use netting or screens around tents or other living areas to barricade against insects in endemic areas. All skin, wherever possible, should be covered by medium-weight clothing to protect against insect bites. People in endemic areas should avoid wearing bright and dark colors because tsetse flies are attracted to those colors; one should instead wear light colors.

Tsetse flies bite during the daytime, but they repose in bushes during the hottest part of the day, so people should avoid bushes and dense vegetation. Because tsetse flies are attracted to swirling dust created by moving vehicles on the African savannah, vehicles should be examined carefully for tsetse flies before being entered.

—Mary E. Markland, MA

Further Reading

Büscher, Philippe, et al. "Human African Trypanosomiasis." *The Lancet,* vol. 390, no. 10110, 2017, pp. 2397-2409, doi:10.1016/S0140-6736(17)31510-6.

Dunn, Noel, et al. "African Trypanosomiasis." *StatPearls*. StatPearls Publishing, 11 Aug., 2021.

Fernandes, Vitória de Souza, et al. "Antiprotozoal Agents: How Have They Changed over a Decade?" *Archiv der Pharmazie*, e2100338, 2021, doi:10.1002/ardp.202100338.

Maxfield, Luke, and Rene Bermudez. "Trypanosomiasis." *StatPearls*. StatPearls Publishing, 11 Aug. 2021.

World Health Organization. "Trypanosomiasis, Human African (Sleeping Sickness)" 18 May 2021, www.who.int/news-room/fact-sheets/detail/trypanosomiasis-human-african-(sleeping-sickness). Accessed 13 Dec. 2021.

AMEBIC DYSENTERY

Category: Diseases and conditions

Also known as: Amebiasis

Anatomy or system affected: Gastrointestinal system, intestines, liver, stomach

Specialties and related fields: Bacteriology, gastroenterology, internal medicine, microbiology, parasitology, pathology, pharmacology, preventive medicine, public health

Definition: a treatable, though potentially severe intestinal parasite infection caused by *Entamoeba histolytica*, initially associated with stomach pain, bloody stools, and fever

KEY TERMS

colitis: Inflammation of the large intestine (colon), which usually is associated with bloody diarrhea and fever

diarrhea: Loose or watery stools, usually a decrease in consistency or increase in frequency from an individual baseline

ELISA: enzyme-linked immunosorbent assay, an assay that uses an immobilized enzyme linked to an antibody that binds to the ligand of interest to quantitatively measure the quantity of that ligand in a liquid sample

intestines: The tube connecting the stomach and anus in which nutrients are absorbed from food; divided into the small intestine and the colon, or large intestine

luminal agents: drugs that kill the amoebae but are not absorbed by the gastrointestinal tract and, therefore, kill all unattached parasites

mucosa: The semipermeable layers of cells lining the gut, through which fluid and nutrients are absorbed

peristalsis: The wavelike muscular contractions that move food and waste products through the intestines; problems with peristalsis are called "motility disorders"

stool: The waste products expelled from the body through the anus during defecation; feces

trophozoites: the active, feeding form of the parasite

CAUSES

Amebic dysentery is caused by the parasitic protozoan *Entamoeba histolytica*. A person contracts amebic dysentery by placing something in their mouth that has touched the stool of a person infected with *E. histolytica*, by swallowing water or food that has been contaminated with *E. histolytica*, and by touching cysts (eggs) from *E. histolytica*-contaminated surfaces and bringing those cysts to the mouth. A classic fecal-oral mechanism transmits this parasite.

RISK FACTORS

The risk factors that increase the chance of developing amebic dysentery include living in or traveling to developing countries, places with poor sanitary conditions, tropical or subtropical areas, living in institutions, and having anal intercourse. Travel to endemic areas that increase the risk of *E. histolytica* exposure includes Africa, Southern Asia, and Central America. Malnutrition, poor hygiene, and immunodeficiency, as in the case of acquired immunodeficiency syndrome (AIDS) and cancer patients and others undergoing immunosuppressive therapies, increase of risk of severe disease.

SYMPTOMS

The symptoms of amebic dysentery include loose stools, nausea, weight loss, stomach pain, stomach cramping, bloody diarrhea and mucus in stools, fever, weight loss, and dehydration. If there is rare pulmonary involvement, then a cough is present. If the infection spreads to the liver, which rarely occurs, the patient presents with jaundice and right upper quadrant pain.

SCREENING AND DIAGNOSIS

Routine microscopic examination of stool samples can detect *E. histolytica* cysts and trophozoites. Unfortunately, single examinations can often miss the organism. Examining stool samples multiple times over ten days increases the effectiveness of microscopic examination.

Enzyme-linked immunosorbent assay (ELISA) tests effectively detect *E. histolytica* antigens in fecal samples, and ELISA kits are commercially available. Polymerase chain reaction (PCR) tests effectively detect *E. histolytica* deoxyribonucleic acid (DNA) from patient stool samples. However, few clinics have the laboratory equipment to perform such tests.

Serological tests can detect blood-based antibodies against *E. histolytica* antigens in 70 to 90 percent of cases. ELISA-based antibody tests are extremely helpful in patients with severe disease.

Imaging the lower bowel with colonoscopy or retrosigmoidoscopy not only provides vital material for PCR assays and can differentiate between amebic dysentery and inflammatory bowel diseases. Bowel imaging usually reveals small mucosal ulcers covered with yellowish exudates, and biopsies usually produce trophozoites. Bowel imaging is usually reserved for patients when other tests are negative for *E. histolytica*, but symptoms persist.

In cases of severe disease, chest X-rays, ultrasounds, computer-aided tomography (CAT), or magnetic resonance imaging (MRI) scans can detect liver abscesses. Ultrasound is usually preferred because of its low cost and noninvasive nature. Ultrasound or CAT-guided liver needle aspiration can isolate trophozoites from a liver abscess.

TREATMENT AND THERAPY

Several antibiotics are available to treat amebic dysentery. In asymptomatic or non-invasive symptomatic disease cases, luminal agents that are not well absorbed by the gastrointestinal tract provide the best therapeutic options. Luminal agents include paromomycin, iodoquinol, and diloxanide furoate. Treatment prevents the disease from becoming severe. The treatment also diminishes the shedding of cysts and trophozoites that can infect others from a public health perspective.

For severe disease, metronidazole is the mainstay treatment. Tinidazole is a viable alternative treatment and may even provide better therapeutic effects with few side effects. Liver abscesses less than 10 centimeters are effectively treated with metronidazole, but larger abscesses require surgery. Metronidazole treatment does not kill luminal trophozoites. Therefore, treatment with a luminal agent should always follow metronidazole or tinidazole treatment.

Bacterial coinfection is common in severe disease, and cotreatment with broad-spectrum antibiotics commonly occurs. Other interventions include rehydration to address patient dehydration.

PREVENTION AND OUTCOMES

To help reduce the chance of getting amebic dysentery, one should take the following steps when traveling to a country that has poor sanitation: Drink only bottled water or water that has been boiled for a minimum of one minute; avoid eating fresh fruit or vegetables that another person has peeled; avoid eating or drinking unpasteurized milk, cheese, or dairy products; and avoid eating or drinking items sold by street vendors.

Soap or low concentrations of iodine or chloride do not kill E. histolytic cysts. Fruits and vegetables should be washed with detergent soap and soaked in vinegar for 15 minutes for consumption.

—*Krisha McCoy and Michael A. Buratovich, PhD*

Further Reading

Centers for Disease Control and Prevention. "Amebiasis." 20 July 2015, www.cdc.gov/parasites/amebiasis/general-info.html. Accessed 6 Nov. 2021.

Gonzales, Maria Liza M., et al. "Antiamoebic Drugs for Treating Amoebic Colitis." *The Cochrane Database of Systematic Reviews,* vol. 1, 1, CD006085, 2019, doi:10.1002/14651858.CD006085.pub3.

Jameson, J. Larry, et al., editors. *Harrison's Principles of Internal Medicine*. 20th ed., McGraw-Hill Education, 2018.

Johnson, Leonard R., editor. *Gastrointestinal Physiology*. 7th ed., Mosby/Elsevier, 2007.

Pritt, B. S., and C. Graham Clark. "Amebiasis." Mayo Clinic Proceedings, vol. 83, no. 10, 2008, pp. 1154-60, doi.org/10.4065/83.10.1154.

Shirley, Debbie-Ann T., et al. "Significance of Amebiasis: 10 Reasons Why Neglecting Amebiasis Might Come Back to Bite Us in the Gut." *PLoS Neglected Tropical Diseases*, vol. 13, no. 11, 2019, e0007744, doi:10.1371/journal.pntd.0007744.

TRICHOMONIASIS

Category: Diseases and disorders

Also known as: Trich

Anatomy or system affected: Genitals, reproductive system, urinary system

Specialties and related fields: Epidemiology, gynecology, neonatology, obstetrics, perinatology, public health, urology

Definition: a sexually transmitted disease in which motile *Trichomonas vaginalis* protozoans become established in the genitourinary tract of men and women

KEY TERMS

asymptomatic: infected but with no discernable symptoms of a disease

carrier: a person infected by an organism who can transmit that organism to other people but who is asymptomatic

Centers for Disease Control and Prevention (CDC): a government facility located in Atlanta that coordinates investigations of disease occurrence in the United States

protozoan: a unicellular organism with an organized nucleus

sexually transmitted disease (STD): a disease that is usually transmitted from person to person through contact between the vaginal or urethral discharges from an infected person and the genital mucous membranes of a person susceptible to infection

urethritis: inflammation and infection of the urinary tract

vaginitis: inflammation and infection of the vagina

CAUSES AND SYMPTOMS

Flagellated motile protozoans known as *Trichomonas vaginalis* cause trichomoniasis, one of the most widespread and common sexually transmitted diseases (STDs). The disease is common among people with multiple sex partners, those who engage in unprotected sex, and those who seek services at STD clinics. Trichomoniasis in pregnant women is a leading cause of premature birth in the United States.

Some estimates suggest that 180 million people a year are infected with trichomoniasis worldwide. The most common population found to be infected is females sixteen to thirty-five years old, which is prime childbearing age. This is an important epidemiological group, as trichomoniasis infections are a leading cause of premature rupture of the placenta, premature birth, and low birth weight.

After infection, there is an incubation period of about seven days, ranging from about four to twenty days. Although up to 70 percent of infected women may remain asymptomatic, *T. vaginalis* infections may sometimes produce a frothy yellow or green vaginal discharge. Women's symptoms may also include urethritis, vaginitis, and itching of the vulva. Sometimes, vaginal inspection shows a distinctive "strawberry cervix" (red patches on the cervix) and red spots on the vaginal walls. Men's symptoms sometimes include urethritis, dysuria, a frothy or purulent urethral discharge, and, in rare cases, scrotal pain as the tube connecting the testicle with the vas deferens becomes inflamed.

The symptoms of *T. vaginalis* are of questionable value in diagnosing the infection. Most existing tests, such as microscopic viewing of wet mounts, Pap tests, and polymerase chain reaction (PCR), often fail to show the infectious agent in people with symptoms. In addition, many infected people remain asymptomatic for many years. The culture of vaginal and urethral smears is considered the most effective way of detecting *T. vaginalis* infection. These factors add to the difficulty in reducing infection rates.

TREATMENT AND THERAPY

The CDC's *Sexually Transmitted Diseases Treatment Guidelines 2006*, which includes trichomoniasis, focuses on microbiological cure, alleviation of signs and symptoms, prevention of sequelae, and prevention of transmission.

The infection is treated with a single oral dose of either metronidazole or tinidazole. Any sex partner should be simultaneously treated even if they are asymptomatic. Treatment is successful in 90 to 100 percent of cases. Treatment during pregnancy is controversial, but no case of fetal malformation has been attributed to metronidazole. Studies have shown that trichomoniasis is associated with low infant birth weight, premature rupture of the membranes, and preterm births. However, studies of pregnant women with trichomoniasis who are treated failed to improve preterm deliveries and even trended toward more preterm deliveries; therefore, treatment remains controversial.

PERSPECTIVE AND PROSPECTS

Many men and women infected by the organism remain asymptomatic for years, spreading the disease to other people through sex. Safer sex practices help prevent transmission. People with multiple sex partners should use latex or polyurethane condoms to help curtail the spread of this disease. Sex education programs must emphasize that people with unusual genital symptoms, including urethritis and vaginal discharge, seek medical treatment.

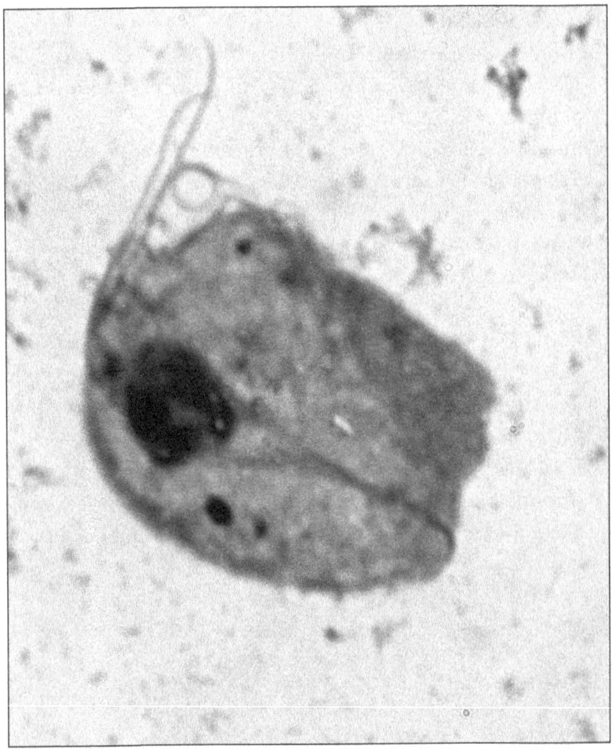

Trichomonas vaginalis *May-Grünwald staining. Photo by Dr Graham Beards, via Wikimedia Commons.*

People infected by *T. vaginalis* may also be infected by other STD organisms, especially the bacterium that causes gonorrhea. Medical professionals believe that infection by the *Trichomonas* protozoan predisposes a person to infection by the human immunodeficiency virus (HIV) upon exposure through unprotected sex with infected partners.

Trichomoniasis in young children may indicate sexual abuse, and health professionals may be obligated to report such infections if local regulations require it.

—*Anita Baker-Blocker, MPH, PhD*

Further Reading

Boston Women's Health Collective. *Our Bodies, Ourselves: A New Edition for a New Era*. 35th-anniversary ed., Simon & Schuster, 2005.

Centers for Disease Control and Prevention. *Sexually Transmitted Diseases Treatment Guidelines*, 22 July 2021, www.cdc.gov/std/treatment. Accessed 5 Dec. 2021.

Heymann, David L., editor. *Control of Communicable Diseases Manual*. 19th ed., American Public Health Association, 2008.

Scharbo-DeHaan, Marianne, and Donna G. Anderson. "The CDC 2002 Guidelines for the Treatment of Sexually Transmitted Diseases: Implications for Women's Health Care." *Journal of Midwifery and Women's Health*, vol. 48, Feb. 2003, pp. 96-104.

Sommers, Michael. *Yeast Infections, Trichomoniasis, and Toxic Shock Syndrome (Girls' Health)*. Rosen Publishing Group, 2007.

Sutton, Amy L., editor. *Sexually Transmitted Diseases Sourcebook*. 5th ed., Omnigraphics, 2013.

TOXOPLASMOSIS

Category: Diseases and conditions
Anatomy or system affected: All
Specialties and related fields: Immunology, internal medicine, microbiology, neonatology, neurology, obstetrics and gynecology, ophthalmology, public health
Definition: a disease caused by toxoplasmas, transmitted chiefly through undercooked meat or in soil or cat feces; symptoms generally pass unremarked in adults, but infection can be dangerous to unborn children

KEY TERMS

bradyzoite: also called "cystozoites," are the life stage of Toxoplasma gondii found in the tissue cyst and replicate slowly

chorioretinitis: inflammation of the choroid, which is a lining of the retina deep in the eye

oocyst: a cyst containing a zygote formed by a parasitic protozoan

sporozoite: a motile spore-like stage in the life cycle of some parasitic sporozoans, that is typically the infective agent introduced into a host

tachyzoite: relatively faster-growing, actively multiplying, and invasive cell type in the life cycle of Toxoplasma gondii

tissue cysts: a closed sac, having a distinct envelope surrounding it within infected tissues

CAUSES

Toxoplasmosis is a parasite infection caused by the protozoan *Toxoplasma gondii* (*T. gondii*). This disease has worldwide distribution. In people with normally functioning immune systems, toxoplasmosis causes no symptoms (asymptomatic). However, in individuals with dysfunctional immune systems, *T. gondii* infections can cause an acute systemic infection or eye disease.

Toxoplasmosis is passed from animals to humans. People can contract it by touching contaminated cat feces or something that has had contact with cat feces, such as soil or insects; by eating undercooked, infected meat; and touching one's mouth after touching contaminated meat. In rare cases, receiving a blood transfusion or an organ transplant can lead to the infection.

Cats acquire *T. gondii* from eating infected rodents. Thirty to fifty percent of the global domestic cat population is chronically infected with *T. gondii*. The protozoan survives stomach acid and infects the intestinal epithelial cells. The parasites undergo sexual development and reproduction in the cat intestine. The infected intestinal epithelial rupture, and the cat sheds oocysts in their feces. These oocysts are resilient and remain infectious for months in cold, dry climates.

When humans ingest oocysts, their stomach acid erodes the oocyst wall, freeing the sporozoites. The sporozoites invade intestinal epithelial cells and develop into highly motile, fast-dividing tachyzoites.

Tachyzoites reproduce inside host cells. Rupture of infected cells releases tachyzoites which can enter the bloodstream and spread to any organ.

The immune response against the tachyzoites drives them to develop into bradyzoites, which are slow-dividing, semidormant cells. Inside infected cells, bradyzoites can form tissue cysts. Tissue cysts can form in any organ but tend to occur in striated muscle (including heart muscle), the brain, or the eyes.

A pregnant woman who gets toxoplasmosis for the first time has a 15 to 60 percent chance of passing it to her fetus. Active infection usually occurs only once in a person's life, although the protozoan remains inactive in the body. Suppose a woman has become immune to the infection before getting pregnant. In that case, she will not pass the condition to her fetus.

RISK FACTORS

People at risk for having symptoms from toxoplasmosis are infants born to women who are first exposed to toxoplasmosis just before becoming pregnant or during pregnancy; people with weakened immune systems from conditions such as human immunodeficiency virus infection, acquired immunodeficiency syndrome, and cancer; and those who have had an organ transplant.

People who consume raw or undercooked meat have a higher risk of contracting toxoplasmosis. Likewise, those who consume water or vegetables contaminated by soil can contract *T. gondii*.

SYMPTOMS

Most people do not have symptoms, but those who do may experience swollen lymph nodes, a fever, fatigue, a sore throat, muscle aches and pains, and a rash. People with weakened immune systems may develop toxoplasmosis infections in multiple organs. Infection is most common in the brain (encephalitis), eye (chorioretinitis), and lung (pneumonitis). Symptoms may include fever, seizures; a headache; visual defects; problems with speech, movement, or thinking; mental illness; and shortness of breath.

In infants, the severity of symptoms depends on when the mother became infected during pregnancy. Suppose infection occurs during the first three months of pregnancy. In that case, the fetus is less likely to become infected. Still, if the fetus does become infected, symptoms will be much more severe. The fetus is more likely to become infected during the last six months of pregnancy, but symptoms will be less serious. Toxoplasmosis can also cause miscarriage or stillbirth.

About one in ten infants born with toxoplasmosis has severe symptoms, including visual defects because of eye infections (chorioretinitis), enlarged liver and spleen, jaundice (yellow skin and eyes), pneumonitis, myocarditis (inflammation of the heart), brain malformations, mental retardation, cerebral palsy, and seizures. Many infants infected with toxoplasmosis who seem healthy at birth may develop problems months or years later. These include visual defects, hearing loss, learning disabilities, and seizures.

SCREENING AND DIAGNOSIS

A doctor will ask about symptoms and medical history and perform a physical exam. Blood tests are done to look for antibodies produced by the body to fight toxoplasmosis. Other lab tests are done to look for the protozoan itself. The Sabin-Feldman dye test is a sensitive and specific diagnostic test for toxoplasmosis. Unfortunately, few laboratories have the reagents available for this test. Pregnant women who are infected will undergo prenatal tests, including ultrasound and amniocentesis, to determine if the fetus is infected.

Polymerase chain reaction (PCR) tests effectively detect *T. gondii* deoxyribonucleic acid (DNA). However, inactive cysts may evade detection with PCR.

TREATMENT AND THERAPY

People who are healthy and not pregnant do not need treatment. Symptoms usually disappear within a few weeks or months. Sulfa drugs, which consist of trimethoprim and sulfamethoxazole, are excellent chemoprophylactic (or preventative) agents.

For people with acute disease, the preferred treatments are pyrimethamine plus sulfadiazine and leucovorin or pyrimethamine plus clindamycin and leucovorin for four to six weeks. Because some localities may not have access to pyrimethamine, an alternative is trimethoprim and sulfamethoxazole. If the patient has a sulfa drug allergy, a viable alternative is atovaquone.

Traditional antitoxoplasmosis medications do not effectively penetrate tissue cysts if the patient has a latent infection. For latent infections, atovaquone with or without clindamycin kills tissue cysts.

People with a weakened immune system are treated with antitoxoplasmosis medicines for several months. If a pregnant woman is diagnosed before eighteen weeks gestation, she is usually given the antibiotic spiramycin. This medicine can decrease the chance of the fetus becoming infected by about 60 percent. If she is diagnosed after eighteen weeks gestation, she is treated with the drug combination pyrimethamine, sulfadiazine, and leucovorin. These drugs can reduce the severity of, but not eliminate, a newborn's symptoms. Once born, the infant will be given pyrimethamine, a sulfonamide, and leucovorin for twelve months.

Cases of toxoplasmic chorioretinitis are sight-threatening infections. Patients are usually treated with pyrimethamine, sulfadiazine, and folinic acid.

PREVENTION AND OUTCOMES

Women who are pregnant or considering becoming pregnant should consult a physician about having a blood test to determine if they are immune to toxoplasmosis (which would indicate a previous exposure). If not immune, women should take the following steps to avoid sources of toxoplasmosis: Avoid eating raw or undercooked meat (if one touches raw meat, avoid touching one's eyes, mouth, or nose); wash one's hands and cutting boards, knives, and sink with soap and warm water; and wash all raw vegetables and fruits.

One should also avoid emptying a cat's litter box; avoid children's sandboxes because cats often use them as a litter box; avoid feeding a cat raw or undercooked meat; and keep one's cat indoors to prevent it from hunting rodents or birds that could be infected. Also, when gardening, one should wear gloves; keep one's hands away from one's eyes, mouth, and nose; and wash one's hands when finished. These steps also apply to persons with weakened immune systems.

—*Laurie Rosenblum and Michael A. Buratovich, PhD*

Further Reading

Ambroise-Thomas, Pierre, and Eskild Petersen, editors. *Congenital Toxoplasmosis: Scientific Background, Clinical Management, and Control*. Springer, 2000.

American Congress of Obstetricians and Gynecologists. "Perinatal Viral and Parasitic Infections." *ACOG Practice Bulletin*, no. 20, 2000, www.acog.org.

Despommier, Dickson D., et al. *Parasitic Diseases*. 5th ed., Apple Tree, 2006.

Joynson, David H. M., and Tim G. Wreghitt, editors. *Toxoplasmosis: A Comprehensive Clinical Guide*. Rev. ed., Cambridge UP, 2005.

Madireddy, Sowmya, et al. "Toxoplasmosis." *StatPearls*. StatPearls Publishing, 28 Sept. 2021.

Martin, Richard J., Avroy A. Fanaroff, and Michele C. Walsh, editors. *Fanaroff and Martin's Neonatal-Perinatal Medicine: Diseases of the Fetus and Infant*. 2 vols. 8th ed., Mosby/Elsevier, 2006.

Parker, James N., and Philip M. Parker, editors. *The Official Patient's Sourcebook on Toxoplasmosis*. Icon Health, 2002.

MALARIA

Category: Diseases and conditions
Anatomy or system affected: All

Specialties and related fields: Infectious disease, internal medicine, microbiology, pharmacology, protozoology, public health, tropical health

Definition: a parasite infection passed to humans through the bite of an infected mosquito

KEY TERMS

gametogony: a stage in the sexual cycle of sporozoans in which gametes are formed, often by schizogony

merozoites: a small amoeboid sporozoan trophozoite

schizogony: asexual reproduction by multiple fission, found in some protozoa, especially parasitic sporozoans

sporozoites: a motile spore-like stage in the life cycle of some parasitic sporozoans (e.g., the malaria organism) that is typically the infective agent introduced into a host

trophozoites: a growing stage in the life cycle of some sporozoan parasites, when they are absorbing nutrients from the host

CAUSES

Malaria is caused by one of the following five parasites: *Plasmodium falciparum*, *P. vivax*, *P. ovale*, *P. malariae*, and *P. knowlesi*. *P. ovale* is found in tropical Western Africa, New Guinea, the eastern parts of Indonesia, and the Philippines. This *Plasmodium* species causes less than one percent of all global malaria cases. *P. malariae* occurs in sub-Saharan Africa, much of southeast Asia and Indonesia, and many Pacific islands. This *Plasmodium* species is also found in parts of the Amazon Basin of South America. Outside Africa, *P. vivax* is the dominant malaria-causing species. It is highly prevalent in the South-East Asian and Western Pacific regions. *P. knowlesi* causes simian malaria in macaques. Human malaria cases caused by *P. knowlesi* occur throughout Southeast Asia, particularly in Malaysia. *P. falciparum* is the most prevalent species of malaria. It occurs throughout Africa, South-East Asia, the Eastern Mediterranean, and the Western Pacific.

An *Anopheles* mosquito becomes infected when it bites someone with malaria. The mosquito passes malaria to a new person through a new bite. Malaria can also be passed from a pregnant girl or woman to her fetus. A blood transfusion from an infected donor can also transmit malaria to the recipient.

P. falciparum is by far the most dangerous of the forms of malaria. In most areas, it is also the most common form.

LIFE CYCLE

The bite of *Anopheles* mosquitoes transmits malaria. *Plasmodium* parasites live in the insect's salivary glands as wormlike sporozoites. When the mosquito takes a blood meal, it inserts its proboscis into the skin and injects salivary gland secretions. Sporozoites spill from the mosquito salivary glands into the bloodstream. Within minutes, the parasites reach the liver and infect hepatocytes (liver cells). In the liver, plasmodia destroy liver cells and divide through schizogony.

The time the parasites spend in the liver differs between *Plasmodium* species. *P. falciparum*, *P. malariae*, and *P. knowlesi* spend one to two weeks in the liver. During their time in the liver, these parasites mature into merozoites, move into the blood, and infect red blood cells. *P. ovale* and *P. vivax*, however, after a few

Anopheles albimanus *mosquito, the sole vector of malaria. Image by James Gathany, CDC Public Health Image Library, via Wikimedia Commons. [Public domain.]*

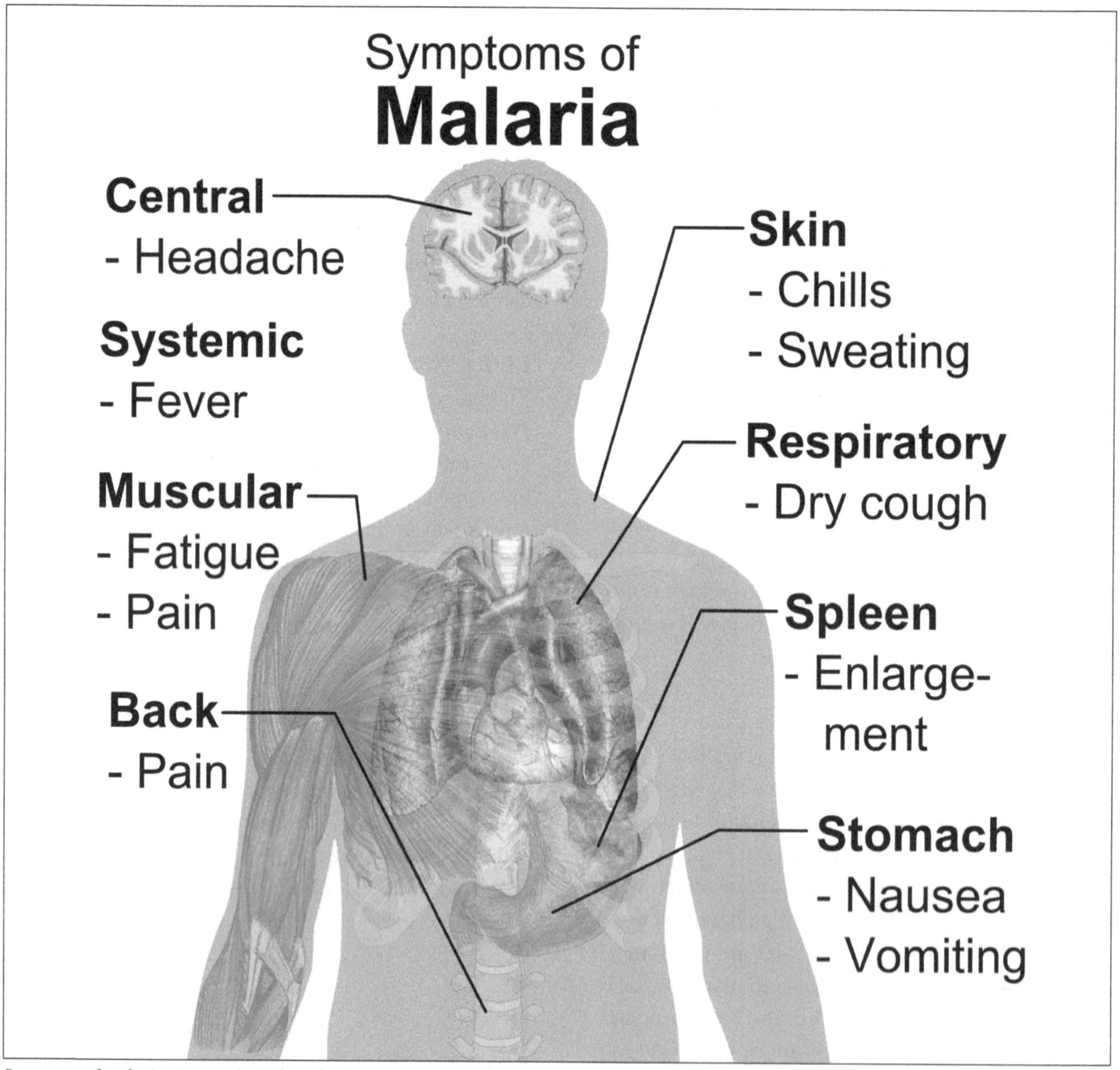

Symptoms of malaria. Image via Wikimedia Commons. [Public domain.]

months or years, enter a dormant phase. The dormant parasites are called "hypnozoites" that do not divide. The liver phase of malaria is called the "exoerythrocytic phase" since it occurs outside red blood cells (erythrocytes) and is usually asymptomatic.

After returning to the bloodstream as merozoites, the parasites infect red blood cells. This begins the erythrocytic stage. *P. ovale* and *P. falciparum* infect red blood cells of any age. *P. vivax* merozoites infect reticulocytes or very young, immature red blood cells. *P. malariae* and *P. knowlesi* prefer older red blood cells. The merozoites divide in the red blood cells and form a tiny ring inside them. The parasites transform from merozoites into trophozoites and digest the red blood cell's hemoglobin. The byproduct of hemoglobin degradation, hemozoin, looks like brown smudges across the cell. The trophozoites divide by mitosis and fill the red blood cell with tiny bodies called "schizonts." The red blood cells rupture and release merozoites that infect other red blood cells.

Red blood cell rupture causes bursts of fever that come in waves every time the red blood cells rupture—the timing when the red blood cells rupture differs between *Plasmodium* species. *P. malariae* causes fevers every seventy-two hours. *P. vivax* and *P. ovale* induce fevers every forty-eight hours. With *P. knowlesi*, the fever comes every twenty-four hours. Finally, *P falciparum* causes fevers that vary from seventy-two hours to twenty-four hours, called "malignant tertian fever."

A selection of trophozoites undergo gametogony and give rise to sausage-shaped gametes. Should the mosquito take another blood meal from the infected person, it will suck up these gametes.

The maturation and proliferation of these parasites in the mosquito is called the "sporogenic cycle." In the mosquito's midgut, the gametes fuse. The zygotes, known as ookinetes, penetrate the midgut wall.

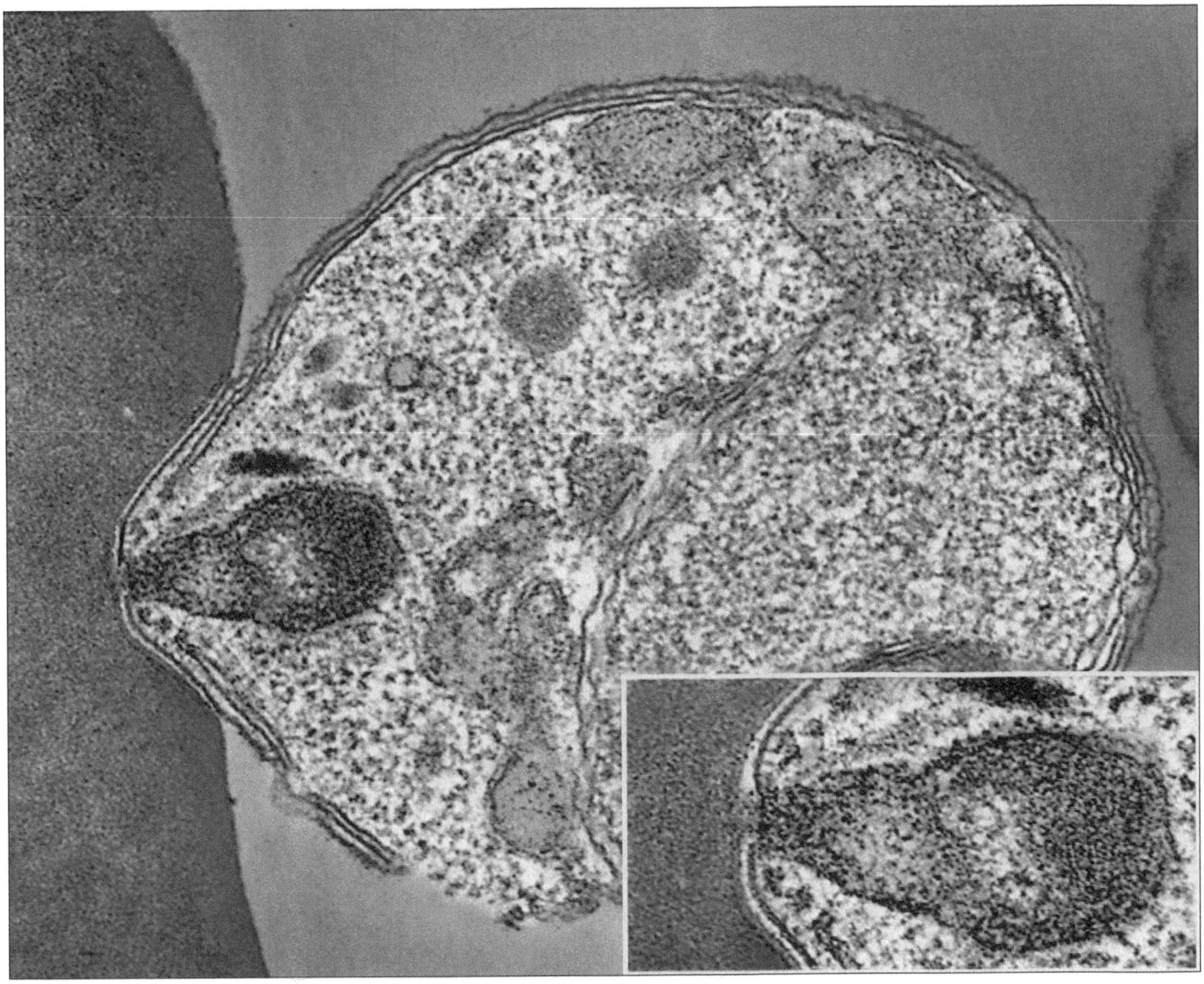

Malaria parasite connecting to a red blood cell. Photo by NIAID, via Wikimedia Commons.

The ookinetes develop into oocysts that grow, rupture, and release sporozoites that invade the mosquito's salivary glands.

RISK FACTORS

Risk factors that increase the chance of getting malaria include living in or traveling to hot, humid climates where *Anopheles* mosquitoes are prevalent; failing to use insect repellants containing N, N-diethyl-meta-toluamide (DEET) when outdoors; failing to use sleeping nets (especially nets treated with permethrin); failing to use medications to prevent malaria infection; and visiting or living in Africa, Asia, or Latin America. Malaria occurs regularly in tourists who fail to follow recommended precautions. The majority of fatal malaria cases seem to be acquired by tourists visiting game parks and other rural areas in East Africa.

SYMPTOMS

Once inside the bloodstream, parasites travel to the liver and multiply there (the hepatic phase). During this phase, the infected person has no symptoms. After several days, the parasites' offspring are released into the bloodstream, where they infect red blood cells. The infected red blood cells burst within forty-eight hours, and the parasites infect more red blood cells. This process leads to recurrent fevers (as high as 106° Fahrenheit), chills, diffuse muscles aches, headaches, nausea and vomiting (or both), di-

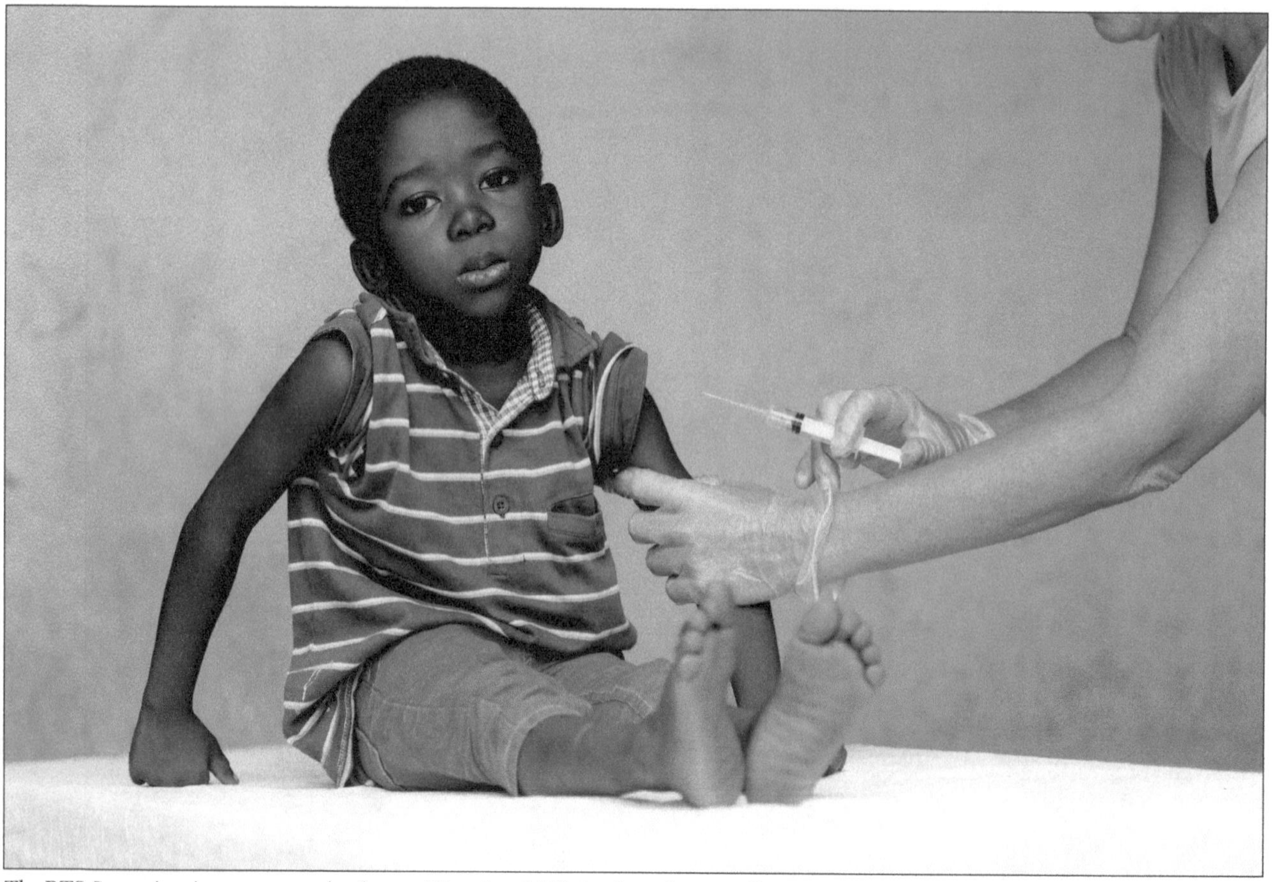

The RTS,S, vaccine, in use across sub-saharan Africa since 2009, has been the most effective so far in battling malaria. Photo via iStock/Riccardo Lennart Niels Mayer. [Used under license.]

arrhea, anemia, and jaundice (yellow coloring of the skin or eyes).

Without treatment, the cycle of red blood cell destruction and fever will continue. This can lead to death. Symptoms usually begin within ten days to four weeks of being bitten by an infected mosquito. *P. malariae* may not produce symptoms for a year or more. *P. falciparum* infections, which cause more severe symptoms, are associated with higher death rates.

Complicated malaria typically occurs with untreated *P. falciparum* malaria. The spleen filters out most *Plasmodium*-infected red blood cells. The exception is *P. falciparum*, which coats the surfaces of its infected red blood cells with a sticky protein that looks like small knobs. These knobs cause cytoadherence, in which the infected red blood cells clump and clog small blood vessels. Blood vessel obstruction prevents adequate blood flow to organs leading to organ damage. The patient suffers from cerebral malaria if the brain is affected, leading to altered mental status, seizures, and coma. Bilious malaria results when the liver is affected, resulting in jaundice, diarrhea, vomiting, and liver failure. Other commonly affected organs include the kidneys, spleen, and lungs, which produce a sepsis-like condition that may lead to death.

SCREENING AND DIAGNOSIS

A doctor will ask about symptoms, medical history, and travel history and perform a physical exam. Blood smears can detect merozoites in red blood cells. Commercially available Rapid Diagnostic kits detect malarial antigens in the blood. These tests are known as "immunochromatographic" tests and use a dipstick or cassette format and provide results in two to fifteen minutes. Polymerase chain reaction tests can determine which *Plasmodium* species is causing the disease.

TREATMENT AND THERAPY

Malaria treatment depends on the stage of the infection. Chemoprophylaxis kills sporozoites before they infect the liver, preventing infection. Travelers to countries where malaria is endemic take these agents. Chemoprophylactic agents include the combination medication atovaquone/proguanil (Malarone), chloroquine, doxycycline, mefloquine, primaquine, and tafenoquine. Chloroquine and doxycycline are started one to two days before leaving and four weeks after returning from malarious areas. Travelers should begin taking Malarone and primaquine one to two days before traveling to a malarious area and continue taking them one week after returning. Mefloquine is started two weeks before leaving for a trip and continued for four weeks after returning. Travelers begin taking tafenoquine three days before leaving and one week after leaving a malarious zone.

In people with malaria, the treatment of choice depends on the type of malaria and where it was acquired. In areas where *Plasmodium* species are resistant to chloroquine, the four treatment options include artemether/lumefantrine (Coartem), Malarone, and quinine combined with doxycycline, tetracycline, or clindamycin.

P. falciparum infections acquired from Central America west of the Panama Canal, Haiti, and the Dominican Republic are treatable with oral chloroquine or hydroxychloroquine.

P. vivax or *P. ovale* infections produce dormant hypnozoites that resist treatment. In such cases, primaquine or tafenoquine effectively kill hypnozoites.

Severe malaria is treated with intravenous artesunate.

PREVENTION AND OUTCOMES

To reduce the chance of getting malaria in an area where malaria is prevalent, one should take antimalarial medication before, during, and after travel. One should use DEET insect repellent (a minimum of 30 to 35 percent DEET) when outside and use proper mosquito netting (sleeping nets) at night. Electronic mosquito repellents, which are supposed to repel mosquitoes by emitting a sound, do not prevent mosquito bites. One should use flying-insect spray in non-air-conditioned rooms while sleeping, wear clothing that covers as much skin as possible, and avoid being outdoors from dusk to dawn when mosquitoes are most prevalent.

—Michelle Badash and Michael A. Buratovich, PhD

Further Reading

Crompton, Peter D., Susan K. Pierce, and Louis H. Miller. "Advances and Challenges in Malaria Vaccine Development." *Journal of Clinical Investigation*, vol. 120, 2010, pp. 4168-78.

Enayati, A., J. Hemingway, and P. Garner. "Electronic Mosquito Repellents for Preventing Mosquito Bites and Malaria Infection." *Cochrane Database of Systematic Reviews* (2009): CD005434, www.ebscohost.com/dynamed.

Jong, Elaine C., and Russell McMullen, editors. *Travel and Tropical Medicine Manual*. 4th ed., Saunders/Elsevier, 2008.

Mandell, Gerald L., John E. Bennett, and Raphael Dolin, editors. *Mandell, Douglas, and Bennett's Principles and Practice of Infectious Diseases*. 7th ed., Churchill Livingstone/Elsevier, 2010.

O'Hanlon, Leslie Harris. "Tinkering with Genes to Fight Insect-Borne Disease: Researchers Create Genetically Modified Bugs to Fight Malaria, Chagas', and Other Diseases." *The Lancet*, vol. 363, 17 Apr. 2004, p. 1288.

Savelkoel, Jelmer, et al. "Abbreviated Atovaquone-Proguanil Prophylaxis Regimens in Travellers After Leaving Malaria-Endemic Areas: A Systematic Review." *Travel Medicine and Infectious Disease*, vol. 21, 2018, pp. 3-20, doi:10.1016/j.tmaid.2017.12.005.

Viruses

This section introduces viruses and examines the primary viral diseases that infect human populations. Some are childhood rites of passage, but others are potentially debilitating or even deadly diseases. This section helps the reader to differentiate between them. Virus types and the structure and life cycle of viruses are discussed, as well as viral genetics, hepatitis B and hepatitis C, retroviruses, polio, influenza, measles, rubella, and rabies.

VIRUSES: STRUCTURE AND LIFE CYCLE

Category: Pathogens
Anatomy or system affected: None
Specialties and related fields: Biochemistry, biotechnology, structural biology, virology
Definition: all viruses contain nucleic acid, either deoxyribonucleic acid (DNA) or ribonucleic acid (RNA) (but not both), and a protein coat that encases the nucleic acid; some viruses are also enclosed by an envelope of lipids and protein molecules

KEY TERMS

capsid: the protein shell of a virus particle surrounding its nucleic acid
minus-sense RNA: a noncoding RNA strand that an RNA-dependent RNA polymerase must copy to produce a translatable mRNA
plus-sense RNA: a single-stranded RNA virus, a plus-strand is one having the same polarity as viral mRNA and containing codon sequences that can be translated into viral protein
virion: the complete, infective form of a virus outside a host cell, with a core of RNA or DNA and a capsid

INTRODUCTION

A virus is a parasitic pathogenic microorganism consisting of a protein coat called a "capsid" that surrounds genetic material, either deoxyribonucleic acid (DNA) or ribonucleic acid (RNA). Each virus type carries out a life cycle tailored to that particular organism. Still, in general, the process begins with the entrance of the virus's genetic material into the host cell, the replication of the viral genome, and the viral genome's packaging within newly produced capsid proteins.

STRUCTURAL CHARACTERISTICS

The individual virus particle, known as a "virion," consists of genetic material, either DNA or RNA, surrounded by a capsid protein coat. Some viruses also include an external lipid envelope generally obtained by budding through a cell membrane during the assembly process. Both the morphology (physical shape) or appearance of the virus and the presence of a viral envelope are genetically determined. The morphology of virus particles encompasses many sizes, ranging from 20 nanometers (nm) in diameter for the smallest viruses, such as those that are the etiological agents for the common cold (rhinoviruses) or certain forms of hepatitis (hepatitis A virus), to the largest and most complex viruses in the 500 nm range (smallpox virus). The poxviruses are large enough to be observed with conventional light microscopes. By comparison, the average-size bacterium is approximately 1 to 2 micrometers (m) in diameter, or roughly two to four times larger than the largest viruses. A blood cell is approximately 20 m in diameter.

Viruses are limited in the quantity of genetic material they carry. Consequently, the most efficient means to encode the proteins used for the capsid is to utilize repeating protein units known as "protomers," which can self-assemble into the subunits, or capsomeres, of the capsid. The result of utilizing repeating units is that the morphological symmetry will be one of two forms: icosahedral (cuboidal) or helical. The only exception is found among the large poxviruses that exhibit a more complex symmetry, reflecting their ability to encode more than two hundred proteins.

Helical capsids resemble long hollow tubes in which the genome is in the center, and capsomeres are arranged in a helical fashion around the core. All known helical viruses contain RNA as the genetic material. Examples of helical viruses include some bacterial viruses, the tobacco mosaic virus (TMV), influenza virus, and measles virus. The helical capsid for some viruses, including measles and influenza, is enclosed within a viral envelope.

Icosahedral-shaped viruses have twenty faces, an equilateral triangle, and twelve corners or vertices. These viruses exhibit what is known as 5:3:2 symmetry, representing the symmetry exhibited by respective axes of the virus. Capsomeres on the faces consist of six protomers (hexons). In contrast, capsomeres that make up the vertices consist of five protomers (pentons). The precise numbers of protomers and the diameter of the virus particle are functions of the size of the genome. Icosahedral viruses include the papilloma (wart) viruses, poliovirus, rhinoviruses (common cold), and herpesviruses. The herpesviruses also contain an external envelope. The largest viruses, including the poxviruses, exhibit a more complex structure that is neither helical nor icosahedral. The poxviruses also contain complex internal structures.

The viral capsid in many viruses is enclosed within a lipid membrane called the "envelope." Except for the poxviruses, the envelope is derived entirely from host cell membranes. Viruses such as the herpesviruses, which replicate in the cell's nucleus, acquire an envelope by budding through the inner nu-

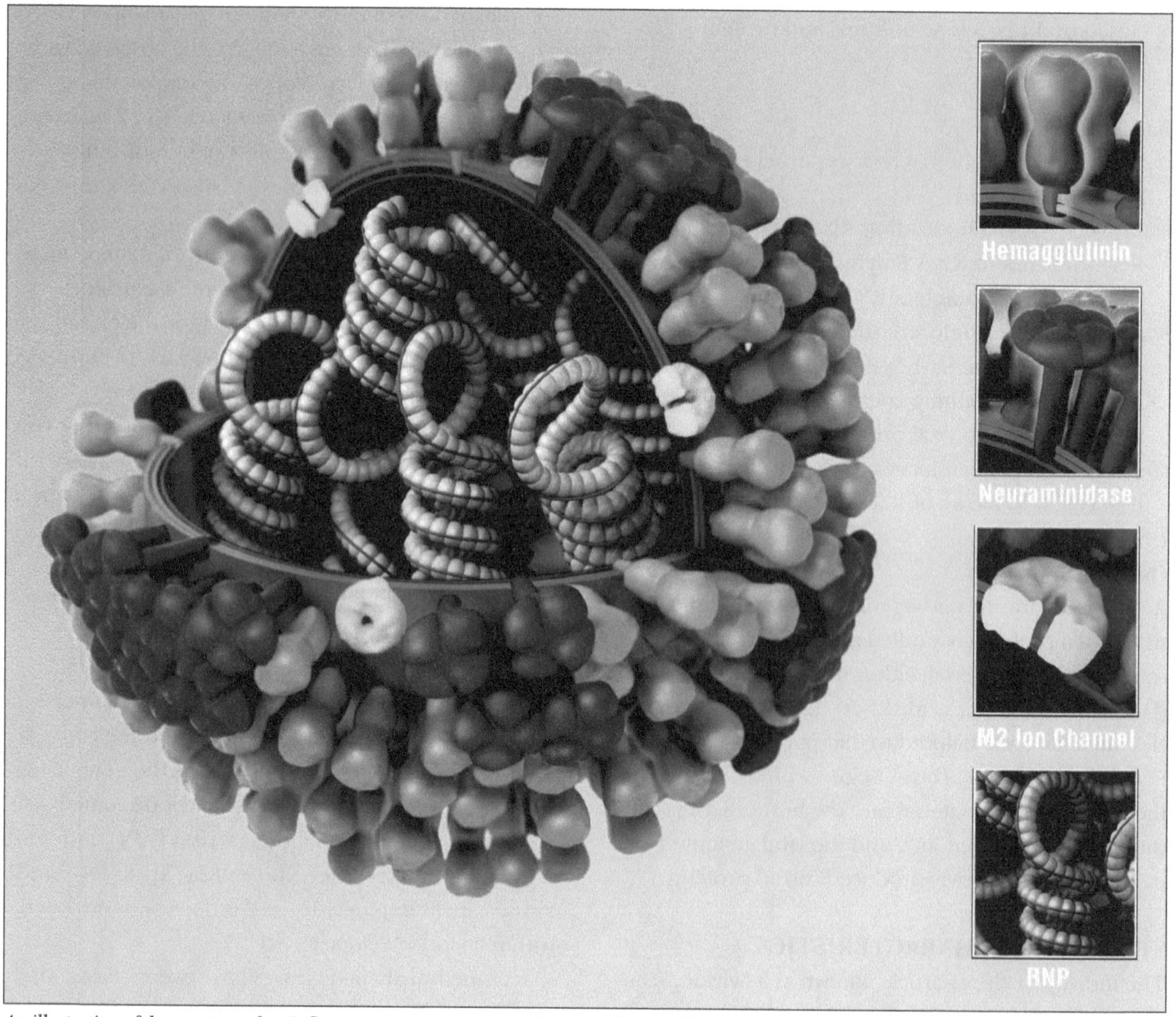

An illustration of the structure of an influenza virion, or virus particle. Image by Dan Higgins, CDC Public Health Image Library.[Public domain.]

clear membrane. Viruses such as influenza and measles obtain their envelope by budding through cytoplasmic membranes. Viral envelopes usually have protein projections, viral encoded spikes or peplomers, on their surface that determine the host range for the virus. For example, influenza viruses have two sets of spikes embedded within their envelopes: the hemagglutinin (H) antigen protein, which attaches the virus to the target cell, and a neuraminidase (N) protein, which is used for release from the cell.

VIRAL GENOMES

The structures of viral genetic molecules encompass numerous categories. Viral genomes may be either single-stranded DNA (parvoviruses, which are associated with fifth disease in humans), double-stranded DNA (adenoviruses and herpesviruses), single-stranded RNA (poliovirus, influenza, measles, rabies, and human immunodeficiency virus), or double-stranded RNA (rotaviruses). The genome in some RNA viruses may consist of a single segment (poliovirus, measles, and rabies) or may consist of multiple individual segments (influenza and rotaviruses). The human immunodeficiency virus (HIV) genome is diploid, consisting of two identical copies of the RNA. The Baltimore classification scheme used to categorize or classify viruses is based upon the type of genome and its replication strategy.

VIRAL INFECTION

Viral infection begins with the adsorption of the particle to the host cell, with a variety of targets referred to as the host range. Attachment depends on the interaction between viral surface molecules and specific receptors on the target cell. That virus cannot infect cells that lack such receptors. For example, HIV infects a class of lymphocytes called "T cells" that express the CD4 receptor protein. The rhinoviruses attach to a molecule on the surface of respiratory mucosal tissues called the "intercellular adhesion molecule-1" (ICAM-1). Influenza H antigen attaches to a class of carbohydrates on the surface of respiratory cells. Most viruses are species-specific, infecting members only within the same species. Influenza virus is an exception. Because many organisms, including humans and birds, express the same carbohydrates on respiratory tissues, influenza has a wide host range that crosses species lines.

The ability of a virus to infect specific tissues is dependent on the expression of receptors by the host cell, leading to the question of why evolution does not select for cells that no longer express those receptor molecules. The answer lies in why cells express such receptors in the first place. The molecule is required for normal functions of the cell, particularly in its interactions with other cells. For example, the CD4 HIV receptor on T lymphocytes is critical for lymphocyte interactions with other classes of white blood cells. ICAM-1 molecules facilitate cell-cell interactions and serve as a signal mediator in immune functions.

Attachment is followed by penetration of the viral capsid into the cell. If the virus has an envelope, the fusion of the viral and cell membranes, analogous to two oil droplets fusing, allows the capsid to enter the cell cytoplasm. Alternatively, suppose the virus lacks an envelope (and sometimes with viruses containing an envelope). In that case, the particle enters through a process called "endocytosis." The cell membrane flows around or "envelops" the attached viral particle. Once inside the cell, the capsid is disassembled, releasing the genome. DNA viruses such as the herpesviruses generally travel to the nucleus for replication. In contrast, RNA viruses such as poliovirus and rhinoviruses replicate in the cytoplasm.

VIRAL MULTIPLICATION

Once the viral capsid has been disassembled, the expression and replication of the viral genome begin. The process used by DNA viruses differs from that in the replication of RNA viruses; cells already contain the basic machinery for the replication and expression of DNA, while no cellular enzymes are present for the replication of RNA.

The proteins necessary for duplication (DNA polymerases, DNA ligases, and other auxiliary molecules) and transcription (RNA polymerase) of viral DNA are already present in the cell's nucleus. The viral DNA replication and expression processes differ little from those that normally take place in the cell. For smaller viruses, cellular proteins are sufficient. Some larger viruses, such as the herpesviruses and poxviruses, have the genetic capacity to encode some of their replication enzymes. For example, both herpesviruses and poxviruses synthesize their own specific DNA polymerases and are not dependent on the cellular enzymes.

Transcription and translation of genetic material immediately following infection results in the production of proteins utilized to duplicate viral DNA. These are referred to as early genes, reflecting the timing of their expression. Genes expressed after DNA duplication are late genes and primarily encode structural or capsid proteins.

Because the cell lacks any machinery for duplication of RNA, RNA viruses must encode their enzymes to duplicate their genetic material. The Baltimore classification scheme for RNA viruses roughly classifies these viruses on the nature or polarity of the RNA genome. Messenger RNA (mRNA), the RNA that is directly translated by cell ribosomes into protein, is defined as having a positive (+) polarity; RNA that is complementary to mRNA is defined as a minus (-) polarity. The + and - symbols here refer only to the molecule's orientation and do not reflect a positive or negative charge. Positive-stranded viruses have a genome identical to mRNA. In contrast, minus- or negative-stranded viruses possess a genome complementary to their mRNA.

Positive-stranded RNA viruses include the rhinoviruses, poliovirus, hepatitis A virus, and mosquito-borne encephalitis viruses. Following the entry into the cell and removal of the capsid, the RNA immediately attaches cell ribosomes and begins the translation of viral proteins. A viral-specific RNA transcriptase is used to replicate the viral genome among the enzymes being synthesized.

The category of negative-stranded RNA viruses includes influenza viruses, measles, mumps, rubella viruses, and rabies virus. Because the RNA is complementary to mRNA, it cannot be directly translated following cell penetration. Negative-stranded viruses incorporate the viral transcriptase directly into the progeny capsids during assembly. Following infection, the viral mRNA is transcribed by the RNA transcriptase, which the virus carries. The transcriptase also copies the positive mRNA into the progeny (or negative) strands for the next generation of viral particles.

HIV, the etiological agent of acquired immunodeficiency syndrome (AIDS), is in an unusual class of viruses called the "retroviruses." Other viruses in this class include the RNA tumor viruses, the agents associated with tumors, and leukemia, primarily in nonhuman animals. The genome in the retrovirus is a diploid plus-stranded RNA. However, the genome does not function directly as the mRNA following infection. The particle carries within its capsid an enzyme referred to as a reverse transcriptase, the function of which is to copy the RNA genome into a double-stranded DNA. The viral DNA integrates within the host cell chromosome. The viral mRNA is transcribed, producing capsid proteins and copies of the reverse transcriptase enzyme for progeny virions.

ASSEMBLY AND RELEASE

Viral assembly is largely nonenzymatic and results from charge interactions among the structural proteins that make up the capsomeres. Animal viruses begin the assembly process in the cell region in which replication of the genome has taken place: RNA viruses in the cytoplasm and DNA viruses in the nucleus of the cell. Expression of the genes for structural proteins occurs primarily after the genome has been replicated. Capsid assembly begins as individual capsomeres commence forming scaffolds around the progeny genomes. The complexity of the process depends upon the coding capacity of the virus; larger and more complex viruses utilize a greater variety of assembly proteins, while smaller viruses may utilize only two or three protein molecules. The assembly of protein capsids is generally associated with a final cleavage step. Capsid precursor proteins are cut to produce the final protein products.

Assembly and release of enveloped particles require an additional step: budding through a cell membrane. Some viruses, such as measles and influenza, encode a matrix or membrane protein (M) and those proteins that make up the spikes. The matrix and spikes are then inserted into the cell membrane. Assembly of the capsid is completed in association with the M protein, followed by reverse endocytosis (exocytosis) as the virus buds through the membrane and acquires the envelope. Budding and releasing influenza virus requires the viral neuraminidase (N protein) activity that separates the envelope proteins on progeny viruses that would otherwise remain attached to carbohydrate residues on the cell surface. Some anti-influenza drugs act by inhibiting this enzyme activity. Maturation of the HIV capsid also requires a final proteolytic step using a viral encoded protease. Certain anti-HIV drugs target this reaction.

The effect of viral infection on the cell itself depends upon two factors: the extent of damage to the cell and whether cell processes are shut down. Productive infection by DNA viruses usually results in cell death. Viral products inhibit both transcription and translation of cell proteins, and the release of progeny virions coincides with cell lysis. Alternatively, herpes virus infections may result in a latent infection in which the virus does not carry out a complete cycle and is retained in a nonreplicative form within the cell. The cell remains functional while the human host carries the virus throughout their life. Enveloped viruses are released from the cell by budding through the cell membrane, which may damage or kill the host cell. The complete replication cycle for RNA viruses generally occurs in twelve to twenty-four hours. In contrast, DNA viruses require a slightly longer time, ranging from twenty-four to forty-eight hours.

IMPACT

Viruses are a class of strict intracellular parasites. Unlike bacteria, viruses are largely devoid of metabolic and enzymatic reactions. Therefore, they are dependent on enzymes and other molecules provided by the host cell. For decades following the discovery of viruses, experts believed the inert nature and dependence of viruses on host functions precluded the development of antibiotics specifically targeting these organisms. However, in 1967, Joseph Kates and Brian McAuslan reported the presence of a viral polymerase in the capsid of poxviruses. In subsequent years, numerous viral encoded enzymes were discovered in infected cells. One of them is reverse transcriptase, an enzyme that copies viral genomic RNA into complementary DNA in cells infected with HIV and RNA tumor viruses. The presence of enzymes and other molecules that are unique to viruses and required for their replication meant that antiviral compounds targeting viruses could be developed.

Because many viruses utilize their encoded polymerases to replicate their genomes, the first generation of antiviral antibiotics targeted these molecules. DNA analogs such as acyclovir and ganciclovir, molecules resembling normal nucleotides but that block genome replication, proved effective in treating herpes virus infections. Amantadine was shown to block influenza virus infection and has proven effective in treating that illness. Two of the drugs targeting influenza, zanamivir and Tamiflu, act at the level of virus release, inhibiting the cleavage reaction involving the viral neuraminidase.

Several generations of drugs have been effective in controlling HIV replication. These include DNA analogs such as zidovudine (or azidothymidine, AZT) and protease inhibitors that block the assembly of the virus. Although viruses do utilize host macromolecules for replication, the production of molecules unique to the virus has provided an opportunity to apply antiviral drugs.

—Richard Adler, PhD

Further Reading

Flint, Jane, et al. *Principles of Virology*. 5th ed., ASM Press, 2020.

Hewlett, Martinez J., et al. *Basic Virology*. 4th ed., Wiley-Blackwell, 2021.

Norkin, Leonard. *Virology: Molecular Biology and Pathogenesis*. ASM Press, 2010.

Strauss, James, and Ellen Strauss. *Viruses and Human Disease*. 2nd ed., Academic Press/Elsevier, 2008.

Tidona, Christian, and Gholamreza Darai, editors. *The Springer Index of Viruses*. Springer, 2002.

Willey, Joanne, et al. *Prescott's Microbiology*. 11th ed., McGraw-Hill Education, 2019.

VIRUS TYPES

Category: Pathogens
Anatomy or system affected: None
Specialties and related fields: Biochemistry, microbiology, virology
Definition: naming viruses and placing them into a taxonomic system

KEY TERMS

deoxyribonucleic acid (DNA): the molecule inside cells that contains the genetic information responsible for the development and function of an organism

envelope: lipoprotein outer layer of some viruses derived from the plasma membrane of the host cell

icosahedron: a polyhedron with twenty faces

ribonucleic acid (RNA): a polymeric molecule essential in various biological roles in coding, decoding, regulation, and expression of genes

virus: a submicroscopic infectious agent that replicates only inside the living cells of an organism

INTRODUCTION

Viruses are intracellular, parasitic, pathogenic organisms consisting of either deoxyribonucleic acid (DNA) or ribonucleic acid (RNA), either single-stranded or double-stranded depending on the virus type. A protein coat surrounds both types called the "capsid"; the combination of genome and capsid is called the "nucleocapsid." Some viruses also have an outer lipid envelope with embedded spikes acquired by budding through a cellular membrane.

CLASSIFICATION

The virus particle's morphology (physical shape) represents one broad category of viral classification. Most viruses are limited in the quantity of genetic material they encode. Consequently, the most efficient means to encode the proteins used for the capsid is to utilize repeating protein units known as protomers, which can self-assemble into the subunits, or capsomeres, of the capsid. The result of utilizing repeating units is that the morphological symmetry will be one of two forms: icosahedral (cuboidal) and helical. The only exception is found among the large poxviruses that can encode more than two hundred proteins, allowing for significantly greater complexity of structure.

Helical capsids resemble long hollow tubes in which the genome is in the center. Capsomeres are arranged in a helical fashion around the core. All known helical viruses contain RNA as the genetic material. The helical nucleocapsid for some viruses, such as rabies, measles, and influenza, is enclosed within a viral envelope. Icosahedral viruses have twenty faces, an equilateral triangle, and twelve corners or vertices. Icosahedral viruses include the papilloma (wart) viruses, poliovirus, rhinovirus (which causes the common cold), and herpesvirus. The herpesviruses also contain an external envelope.

The host range of the virus, or the species the virus can infect, is determined by proteins on the virus's surface (the capsid on nonenveloped viruses or the spikes on enveloped particles) and viral receptors on the surface of the host cell. As a rule, viruses are species-specific: Human rhinoviruses, measles, and herpesviruses infect humans only, for example. In some cases, different species share common receptors, which explains why certain viruses, such as influenza and rabies, may cross species lines. Likewise, viruses may exhibit specific cell tropisms within the species, infecting only those tissues that express certain receptors. Influenza is a respiratory virus and does not infect other tissues; the popular term "stomach flu" is a misnomer because it involves neither the influenza virus nor the stomach.

VIRAL GENOMES

The disadvantage of classifying viruses based on morphology is the failure of this method to consider evolutionary relationships. Viruses that are closely related genetically may produce radically different pathologies; respiratory viruses like the rhinoviruses are genetically related to poliovirus and hepatitis A virus, even though the transmission methods and the sites of infection differ.

In the early 1970s, David Baltimore, then a virologist at the Massachusetts Institute of Technology, proposed a method of classification in which virus families were grouped according to the structure and replication strategy of the genome. Viruses within the same class generally shared a genetic relationship, even though their pathologies differed. Baltimore proposed six classes: double-stranded or single-stranded DNA, classes I and II respectively; double-stranded RNA (class III); single-stranded RNA of positive (+) polarity (class IV); single-stranded RNA of negative (-) polarity (class V); and RNA viruses, which replicate through a DNA intermediate (class VI). Class VII was later added, representing double-stranded DNA viruses that use an RNA intermediate. The positive and negative polarities refer not to any charge but the orientation of the genome concerning messenger RNA (mRNA). mRNA is defined as being a positive strand. Genomes that are positive-stranded are identical to the mRNA, while negative-stranded genomes are complementary to mRNA.

Viruses are grouped in subcategories of families in which the suffix *viridae* denotes a family within each class. For example, all herpesviruses are within the family Herpesviridae. The suffix virus denotes *the viral genus*, as in "rhinovirus" or "herpesvirus." The Baltimore classes are as follows:

Class I: Double-stranded DNA viruses. These include both nonenveloped viruses (Polyomaviridae, Papillomaviridae, and Adenoviridae) and two enveloped families (Herpesviridae and Poxviridae). The polyomaviruses that make up the family are highly species-specific, and most are not associated with human infections. The ability of several family members to cause neoplastic changes, or cancers, in cultured cells has led to extensive research into mechanisms of cell regulation. However, there is no evidence that these viruses pose a threat to humans. In particular, one member of the family, simian virus 40 (SV40), generated concern because it was a contaminant in early poliovirus vaccines grown in rhesus monkey cells; no evidence has been found to suggest the virus poses a threat to humans. However, two variants of the SV40 virus, JC virus, and BK virus, are associated with rare neurological diseases in immunocompromised persons.

The Papillomaviridae are well known as the etiological agents of human warts. More than one hundred serotypes of the human papillomavirus (HPV) are known, the vast majority causing only benign growths called "condylomas" (warts). Because genital HPV is so common, genital warts represent one of the most common types of sexually transmitted diseases. However, about one dozen HPV serotypes are capable of malignant transformation of cervical cells, resulting in cervical cancer, one of the leading causes of cancer deaths in women. Gardasil, the cervical cancer vaccine, is a quadrivalent vaccine directed against the four most common HPV serotypes associated with cervical cancer.

The Adenoviridae, or adenoviruses, are associated with respiratory infections in humans. More than one hundred serotypes are known, about one-half of which are associated with human infections. In most cases, infections appear in children as either a mild infection or an illness associated with sore throats or fever, or both.

The Herpesviridae family includes eight known types of human herpesviruses (HHV). While the illnesses associated with these viruses vary, all exhibit latency. Following recovery, the infected person harbors the virus in a nonreplicative state for the remainder of their life. The virus may periodically become reactivated in some people, resulting in illnesses that may be mild to severe. The best known of these viruses include HHV-1,2, often called "herpes simplex types 1 and 2," associated with cold sores. HHV-3, or varicella-zoster virus, is the agent of chickenpox. Reactivation of the virus from the latent state results in the localized rash known as shingles. HHV-4, the Epstein-Barr virus (EBV), is associated with infectious mononucleosis. EBV has generated significant research, as it is also a potential cancer virus, the etiological agent of Burkitt's lymphoma and nasopharyngeal carcinoma, and possibly the etiological agent of Hodgkin's disease.

The largest group morphologically of double-stranded DNA viruses are the poxviruses. It is estimated that the variola (smallpox) virus killed an estimated 400,000 persons annually in Europe before developing an effective vaccine by English physician Edward Jenner. Smallpox is the only viral disease eradicated from human civilization due to an effective vaccination campaign in the 1970s.

Class II: Single-stranded DNA viruses. The Parvoviridae (*parvo* means "small") contain linear single-stranded DNA genomes. The only parvovirus associated with human disease is B19, the etiological agent for erythema infectiosum, or fifth disease, a common rash in children.

Class III: Double-stranded RNA viruses. The only family of double-stranded RNA animal viruses is the Reoviridae ("reo" stands for "respiratory enteric orphan"). The name originally reflected its isolation from the gastrointestinal and respiratory tracts and the mistaken belief that it was of no clinical significance. The evolution of the original classification of reovirus as an enteric virus to its own family reflected the increasing role of molecular biology in the study of viruses during the 1960s. The virus was later discovered to contain a distinctive genome, both double-stranded and existing in the form of ten to twelve segments. Reoviruses are nonenveloped viruses with a double-layered icosahedral capsid.

Most members of this virus family cause no significant clinical disease. The most important pathogen is rotavirus, arguably one of the most important causes of gastrointestinal disease in young children. Estimates hold that nearly 100 percent of children worldwide are infected early in childhood, with some one-half million deaths caused primarily by the combination of severe diarrhea and poor health care throughout much of the world.

Class IV: Single-stranded, positive-stranded RNA viruses. Four major families are placed in this class, two icosahedral nonenveloped families (Picornaviridae, or picornaviruses, and Caliciviridae) and two on enveloped icosahedral viruses (Togaviridae and Coronaviridae).

The term "picornavirus" refers to a "small" (*pico*) RNA virus. These viruses include the first animal virus discovered (foot-and-mouth-disease virus) and polioviruses, hepatitis A virus, and rhinoviruses, associated with the common cold. The development of the first Salk vaccine and, subsequently, of the Sabin oral vaccine to prevent poliomyelitis has ranked among the most important developments in controlling infectious diseases. From its peak annual incidence of greater than fifty thousand polio cases in the United States in the early 1950s, the disease was largely eradicated worldwide by the twenty-first century, with only a few pockets of infection remaining in developing countries. The rhinoviruses, which include more than 120 serotypes, are the etiological agents for most colds. Many serotypes are the primary reason people average two to three colds each year until well into their adult years.

The term "hepatitis" refers to a clinical condition associated with infection by several different and unrelated types of viruses. Hepatitis A virus, like a poliovirus and several other types of picornaviruses, is transmitted through a fecal-oral route and begins as a gastrointestinal infection.

The caliciviruses (calyx- or cup-shaped structures on the viral surface) include the norovirus (Norwalk virus), among the most common causes of gastroenteritis in adults. The Norwalk virus is frequently the cause of intestinal illnesses on cruise ships, schools, and nursing homes.

The togaviruses, named for the toga or coat appearance of the envelope, include primarily arthropod-borne viruses associated with viral encephalitis, yellow fever, West Nile virus, and hepatitis C virus. Yellow fever was the first viral disease demonstrated to be transmitted by mosquitoes. The building of the Panama Canal during the first decade of the twentieth century was made possible in large part by the Walter Reed Commission's program for control of mosquitoes in Cuba and Panama. Not all togaviruses are arthropod-borne, however. Rubella virus, associated with German measles, is a virus transmitted by respiratory droplets, classified within the togaviruses because of its molecular similarity.

Coronaviruses contain a single RNA genome, the largest known among the RNA viruses. Human infections are relatively common, probably second only to those caused by rhinoviruses. They often result in symptoms resembling those of the common cold. The SARS (severe acute respiratory syndrome) epidemic that appeared in early 2003 represented an unusually virulent virus strain.

Class V: Single-stranded, negative-stranded RNA viruses. The negative-stranded RNA viruses include four major enveloped families: Orthomyxoviridae (influenza viruses), Paramyxoviridae (measles, mumps), Filoviridae (Ebola virus), and Rhabdoviridae (rabies). Only the myxoviruses have segmented genomes.

The myxoviruses (myxa or mucus) include all the influenza viruses. The type of influenza (A, B, and C) refers to the nucleocapsid proteins: Type A is the most common cause of epidemics. The viral envelope includes two types of spikes: the hemagglutinin (H) protein, which is used to attach to the target cell, and the neuraminidase (N), which is used for release from the cell. The particular strain of the virus is indicated by one of the sixteen types of H protein and nine types of N protein. For example, the 2009 swine influenza pandemic was the H1N1 type. Because the genome of influenza viruses is segmented, coinfection of cells by different strains of the virus may result in reassortment of segments, creating an entirely new strain, as happened with the swine influenza virus.

Measles virus is similar to those viruses that cause illnesses in animals: distemper and rinderpest viruses. Genetic analysis has suggested that all three viruses originated from a common ancestor. Humans became infected from cross-species infection and adaptation as animals were domesticated.

The filoviruses include the Marburg virus, discovered in Marburg, Germany, in 1967 when workers were infected from handling monkey tissue. The Ebola virus was discovered following an outbreak near the Ebola River in northern Congo. Both viruses cause life-threatening hemorrhagic fevers.

Class VI: RNA viruses with DNA intermediate. The Retroviridae or retroviruses contain a positive-stranded RNA but replicate through a DNA intermediate. Following infection, the RNA is copied by a viral reverse transcriptase into a double-stranded DNA, then integrated into the host genome. Expression of the viral genes and progeny virus production utilizes only the integrated "provirus."

Three subclasses of retroviruses are known: RNA tumor viruses, originally discovered by Peyton Rous early in the twentieth century; lentiviruses (slow viruses, reflecting the slow progression of the disease) such as human immunodeficiency virus (HIV); and "foamy" viruses, which are not associated with any known human disease.

The RNA tumor viruses were critical in discovering the role of oncogenes in creating cancer cells. Still, with few exceptions, they are not associated with human cancers. Those viruses, such as human T-cell lymphotropic viruses (HTLV-1,2), do not kill the cell but disrupt regulation. HIV, the agent of acquired immunodeficiency syndrome, ultimately kills the infected cell. Because the target cell, the T lymphocyte, is critical to the regulation of the immune response, the result is a complete breakdown of the immune system. HIV likely originated from similar viruses in chimpanzees that jumped species and adapted to humans.

Class VII: DNA retroviruses. The newest members of the Baltimore classification system, the Hepadnaviridae (hepatitis DNA viruses), contain a double-stranded DNA genome that replicates using an RNA intermediate. The RNA is generated using a cellular RNA polymerase that copies the viral genome. In turn, the RNA is copied by a viral reverse transcriptase into progeny DNA.

The most important group member is hepatitis B virus, the primary cause of severe viral liver disease, hepatocellular carcinoma, or liver cancer. An estimated 500 million persons worldwide are believed to carry the virus, which causes nearly two million deaths annually.

IMPACT

The ability to sequence the genomes of an increasing number of viruses has led to an understanding of the phylogenetic relationships among these organisms, despite the seemingly unrelated array of diseases with which they are associated. Viruses within the same family have been shown to share a common ancestry. Among the questions that can be addressed is the origin of human viruses, many of which began as zoonotic diseases in other animals. As human civilization began to creep into new animal habitats and domesticated animals such as dogs and ruminants (cattle and sheep), viruses adapted to new hosts. The process continues, as virus infections associated with newly discovered agents such as the Ebola virus, hantavirus, and even HIV have moved from nonhuman hosts, such as rodents and nonhuman primates, into the human population.

—Richard Adler, PhD

Further Reading

Bishop, Roxanne H. *Influenza and RNA Viruses: Emergence, Classification and Management*. Nova Biomedical, 2014.

Flint, Jane, et al. Principles of Virology. 5th ed., ASM Press, 2020.

Hewlett, Martinez J., et al. *Basic Virology*. 4th ed., Wiley-Blackwell, 2021.

King, Andrew M. Q. *Virus Taxonomy: Classification and Nomenclature of Viruses: Ninth Report of the International Committee on Taxonomy of Viruses*. Academic, 2012.

Norkin, Leonard. *Virology: Molecular Biology and Pathogenesis*. ASM Press, 2010.

Strauss, James, and Ellen Strauss. *Viruses and Human Disease*. 2nd ed., Academic Press/Elsevier, 2008.

Willey, Joanne, et al. *Prescott's Microbiology*. 11th ed., McGraw-Hill Education, 2019.

Zimmer, Carl. *A Planet of Viruses*. 2nd ed., U of Chicago P, 2015.

Viroids and virusoids

Category: Viral genetics
Anatomy or system affected: None
Specialties and related fields: Biochemistry, biotechnology, botany, plant pathology, virology

Definition: viroids are naked strands of ribonucleic acid (RNA), 270 to 380 nucleotides long, that are circular and do not code for any proteins; however, some viroids are catalytic RNAs (ribozymes), cleaving and ligating themselves; despite their simplicity, they can cause disease in susceptible plants, many of them economically important; and virusoids or satellite RNAs, are like viroids, except that they require a helper virus to infect a plant and reproduce

KEY TERMS

RNA polymerase: an enzyme that catalyzes the joining of ribonucleotides to make RNA using deoxyribonucleic acid (DNA) or another RNA strand as a template

RNase: an enzyme that catalyzes the cutting of an RNA molecule

GENERAL CHARACTERISTICS OF VIROIDS AND VIRUSOIDS

Viroids, and some virusoids, are circular, single-stranded ribonucleic acid (RNA) molecules, which normally appear as rods but, when denatured by heating, appear as closed circles. The rod-shaped structure is formed by extensive base pairing within the RNA molecule. The secondary structure is divided into five structural domains. One domain is called the "pathogenicity" (P) domain because differences among variant strains of the same viroid species seem to correlate with differences in pathogenicity. Virusoids may also comprise linear RNA or, rarely, double-stranded circular RNA. The difference between viroids and virusoids is in their mode of transmission. Viroids have no protective covering of any kind. They are no more than the RNA that makes up their genetic material. They depend on breaks in a plant's epidermis or travel with pollen or ovules to gain entry. Virusoids, also known as satellite RNAs, are packaged in the protein coat of other plant viruses, referred to as helpers, and are therefore dependent on the other virus.

Viroids are typically divided into two groups based on the nature of their RNA molecule. Group A is the smallest group, and their RNA can self-cleave. These include the avocado sunblotch and peach latent mosaic viroids. Group B contains all the other viroids, and their RNA cannot self-cleavage. Species in group B include the potato spindle tuber, coconut cadang, tomato plant macho, and citrus bent leaf viroids.

Virusoids are less well studied than viroids and, although more diverse, are most similar to group B viroids in that they cannot self-cleave. Examples include the tomato black ring virus viroid, the peanut stunt virus viroid, and the tobacco ringspot virus viroid. Because so little is known about virusoids, the remainder of this article will focus on viroids.

VIROID PATHOGENESIS

Suppose infected leaves are homogenized in a blender and passed through an "ultrafilter" fine enough to exclude bacteria. In that case, the infection is easily transmitted to another plant by painting some of the filtrates on a leaf. Even billionfold dilutions of the filtrate retain the ability to cause infection, suggesting that it is being replicated. RNase destroys infectivity, suggesting that the genetic material (RNA) is exposed to the medium, unlike viruses, which have a protective protein coat. When isolated from other cell components, an absorbance spectrum shows that viroids are pure nucleic acid, lacking a protein coat.

Although viroids are structurally simple and do not code for any proteins, they still cause disease. Although the molecular mechanisms of viroid pathogenesis are unknown, it is clear that the pathogenesis domain (P domain) is primarily responsible.

Changes in the sequence of nucleotides in the P domain have been correlated with pathogenicity. Some research suggests that the pathogenicity of a viroid strain is related to the resistance of the P domain to heat denaturation, with the stability of this region being inversely related to severity. However, some evidence suggests that this may not be entirely true. In a series of nucleotide substitutions introduced by researchers into the P region of an intermediate strain (that is, intermediate in pathogenicity) of potato spindle tuber viroid (PSTVd), four showed viroid infectivity and pathogenicity that were the same as those of a previously reported severe strain of PSTVd. Altogether, eight different mutant strains were analyzed, and resistance to denaturation and PSTVd pathogenicity were not correlated in all cases.

Research is underway to understand how viroids move from cell to cell and traverse the cytoplasm to the nucleus, where many viroids replicate. There is evidence that a possible interaction might involve viroid RNA activating an RNA-activated protein kinase in response to a nucleotide sequence similar to that of the normal RNA activator. Protein kinases are integral to intracellular signaling pathways that control many aspects of cell metabolism. Once researchers understand the signals that viroids use to get around, they may devise treatments against them. A better understanding of the process may also shed light on normal biochemical communication pathways in plant cells.

Plant pathologist Theodor O. Diener took the scientific world by surprise in 1971 when he discovered the viroid, a plant pathogen 80 times smaller than a virus. By Barry Fitzgerald, via Wikimedia Commons. [Public domain.]

VIROID REPLICATION

Viroids replicate by a rolling circle mechanism, a method also used by some viruses. The original strand is referred to as the "(+) strand," and complementary copies of it are called "(-) strands." Type A and B viroids replicate slightly differently. In type A viroids, the circular (+) strand is replicated by RNA-dependent RNA polymerase to form several linear copies of the RNA (-) strand connected end to end. Site-specific self-cleavage produces individual (-) strands later circularized by a host RNA ligase. Each (-) strand is finally copied by the RNA polymerase to make several linear copies of (+) strand RNA. Cleavage of this last strand makes individual RNA (+) strands, which are then circularized. Self-cleavage in viroids represents one of the cases in which RNA acts as an enzyme. The RNA forms a "hammerhead" structure that enzymatically cleaves the longer RNAs at just the right sites.

Replication of type B viroids is mediated by normal host DNA-dependent RNA polymerase, which mistakes the viroid RNA for DNA. The overall process is similar to what happens with type A viroids, except that the (-) strand is not cleaved but instead copied directly, yielding a (+) strand cleaved by host RNase to form individual copies that are ligated to become circular.

ECONOMIC IMPACT OF VIROIDS

Genetically engineered plants in the future might make proteins that would essentially confer immunity by preventing viroids from entering the nucleus. With no access to the nucleus, a viroid would be incapable of replicating, effectively preventing the damage normally associated with viroid infection. Currently, no transgenic plants exist, and viroids can reduce agricultural productivity if outbreaks are not quickly checked. The typical treatment is to destroy the affected plants, as there is no cure.

Although predominantly negative, viroids may have some potentially positive benefits. They have already been used in unique ways to study plant genetics. They may provide insights into how plant proteins and nucleic acids move in and out of cell nuclei. It may also be possible to harness the benefits of viroid infection for certain agricultural applications, such as dwarfing citrus trees. Considerably more will need to be learned about viroids before they can be adequately controlled or used for human benefit.

—Bryan Ness, PhD

Further Reading

Dalakouras, Athanasios, Elena Dadami, and Michael Wassenegger. "Viroid-Induced DNA Methylation in Plants." *Biomolecular Concepts,* vol. 4, no. 6, 2013, pp. 557-65.

Diener, T. O., R. A. Owens, and R. W. Hammond. "Viroids, the Smallest and Simplest Agents of Infectious Disease: How Do They Make Plants Sick?" *Intervirology,* vol. 35, nos. 1-4, 1993, pp. 186-95.

Duran-Vila, Núria, et al. "Structure and Evolution of Viroids." *Origin and Evolution of Viruses, edited by* Esteban Domingo, Colin R. Parrish, and John J. Holland, 2nd ed., Elsevier/Academic, 2008.

Gómez, Gustavo, and Vicente Pallás. "Viroids: A Light in the Darkness of the lnc RNA-Directed Regulatory Networks in Plants." *New Phytologist,* vol. 198, no. 1, 2013, pp. 10-15.

Hadidi, Ahmed, et al., editors. *Viroids.* Science, 2003.

Hammond, R. W. "Analysis of the Virulence Modulating Region of Potato Spindle Tuber Viroid (PSTVd) by Site-Directed Mutagenesis." *Virology,* vol. 187, no. 2, 1992, pp. 654-62.

Owens, R. A., W. Chen, Y. Hu, and Y-H. Hsu. "Suppression of Potato Spindle Tuber Viroid Replication and Symptom Expression by Mutations, Which Stabilize the Pathogenicity Domain." *Virology,* vol. 208, no. 2, 1995, pp. 554-64.

Shors, Teri. "What About Prions and Viroids?" and "Plant Viruses." *Understanding Viruses.* 2nd ed., Jones, 2013.

Wassenegger, M., et al. "RNA-Directed De Novo Methylation of Genomic Sequences in Plants." *Cell,* vol. 76, no. 3, 1994, pp. 567-76.

Virus-related cancers

Category: Diseases, Symptoms, and Conditions
Anatomy or system affected: All systems
Specialties and related fields: Biochemistry, biotechnology, immunology, microbiology, oncology, virology
Related conditions: Cervical cancer, hepatocarcinoma, Kaposi sarcoma, T-cell lymphomas, B-cell lymphomas
Definition: cancers caused by viruses

KEY TERMS

Kaposi sarcoma: a type of cancer in which lesions (abnormal areas) grow in the skin, lymph nodes, lining of the mouth, nose, and throat, and other tissues of the body

lymphoma: cancer that begins in infection-fighting cells of the immune system, called "lymphocytes"

reverse transcriptase: an enzyme encoded from the genetic material of retroviruses that catalyzes the transcription of retrovirus RNA (ribonucleic acid) into DNA (deoxyribonucleic acid)

tumor: an abnormal mass of tissue that forms when cells grow and divide more than they should or do not die when they should

A PRIMER ON VIRUSES

A virus consists of nucleic acid, either ribonucleic acid (RNA) or deoxyribonucleic acid (DNA), inside a protein coat or capsid, which is enclosed in an outer membrane. On its own, a virus cannot replicate itself.

A virus must first find a way into the individual to infect someone. Usually, this is not through the skin but rather the mouth, lungs, penis, vagina, gastrointestinal tract, or breaks in the skin (wounds, sores). The virus attaches to its host cell and injects genetic material. The cell makes viral DNA or RNA and produces viral proteins instead of its usual products. The host cell creates new viruses, which are released, destroying the host cell. The new viruses find other cells and begin the process anew.

Like the human immunodeficiency virus (HIV), retroviruses have RNA in a protein capsid with a lipid envelope that contains receptor-binding proteins used to attach to the host cell. The retrovirus injects RNA and reverse transcriptase (RTase) into the cell. This allows the cell to make viral DNA, then a complementary strand of DNA. The double-strand copies of DNA become part of the host cell's chromosome.

DEFINITION

Cancer is a malignant disease characterized by the uncontrolled growth of anaplastic cells that invade or spread into sites beyond their origin. While most human cancers are the product of genetic mutations, some cancers result from infection by viruses.

RISK FACTORS

Viruses associated with cancer human immunodeficiency virus (HIV) and hepatitis B and C viruses (HBV, HCV) are most commonly spread through sexual relations or contamination of intravenous fluids. The pooling of blood fluids for isolation of the clotting factor VIII during the early years of the acquired immunodeficiency syndrome (AIDS) epidemic was the primary factor in the infection of hemophiliacs with HIV. Unprotected sexual relations with infected partners and intravenous drug use, in which infected users shared needles, also transmitted HIV and the hepatitis viruses. The human papillomavirus (HPV), the etiological agent for cervical carcinoma, is commonly transmitted by infected sexual partners.

ETIOLOGY AND THE DISEASE PROCESS

Most human cancers are not infectious. They are not the direct result of microbial infection. However, certain groups of viruses have long been known to be associated with malignancies in animals, and several members of these groups have been shown to cause certain human cancers. Viruses can cause genetic changes in oncogenes genes that directly regulate cell division, such as growth factors and their receptors, signal mechanisms, tumor suppressors, and disrupt the regulation of cell growth. However, people can be infected with these viruses and not develop cancer, so how they become activated is unknown.

More than one hundred types of human papillomaviruses have been identified. About half are capable of causing cervical cancer. Some 75 percent of cervical cancer cases are associated with three serotypes of HPV. Three genes encoded by these viruses appear to be linked to the disease, the most important of which is known as E7. The E7 protein inhibits tumor-suppressor proteins such as the retinoblastoma (Rb) molecule, which regulates the steps necessary for cell replication. Infection of cervical cells by these strains of the virus may eventually result in uncontrolled growth and malignancy.

Hepatocellular carcinoma (liver cancer) is among the most common cancers worldwide. It has been shown to result from HBV infection and, less commonly, the HCV. Both HBV and HCV are associated with potentially severe forms of liver disease, and a person with a chronic hepatitis infection or a long-term carrier may eventually develop liver cancer. Little is understood as to how infection results in cancer, as no viral protein having oncogenic ability has been described. However, one protein encoded by HBV, the X gene product, appears to bind the p53 protein. This tumor-suppressor molecule regulates the activation of genes associated with cell division. The core protein of HCV has also been shown to bind p53, similarly suggesting tumor induction.

Kaposi sarcoma is believed to be caused by human herpesvirus 8 (HHV-8), which produces gene products that induce cell replication. Classic Kaposi sarcoma, a rare form of the disease not associated with HIV infection, appears to involve viral and genetic factors. The population at risk for this form of Kaposi sarcoma appears to be limited to men from the Mediterranean, indicating a possible genetic link, but HHV-8 infection may also play a role.

A variety of B-cell and T-cell lymphomas are also associated with viral infections. The human T-cell lymphotropic virus (HTLV) is the etiological agent for certain forms of T-cell disease. T-cell lymphomas resulting from HTLV infection are most commonly found in southern Japan, Africa, and portions of the Caribbean. About 5 percent of persons infected by HTLV develop cancer. The mechanism by which cancer is induced is unclear but appears to involve a disruption of the cell's signaling mechanism, regulating cell replication.

There are numerous non-Hodgkin lymphomas (NHLs), some of which are associated with viral infections. The Epstein-Barr virus (EBV), a member of the herpesvirus family, is the etiological agent behind Burkitt lymphoma, an illness in which malaria is a cofactor, and nasopharyngeal carcinoma, an illness found among those whose ancestry can be traced to southern China.

Hodgkin disease (HD) is likewise a cancer of the lymphocytes. The etiological agent also appears to be the Epstein-Barr virus. The most common form of HD, referred to as classic HD, accounting for 95 percent of all cases, is characterized by Reed-Sternberg (R-S) cells, an unusual type of lymphocyte.

The precise mechanism by which EBV infection results in cancer is unclear but appears to involve a translocation event with the cell chromosome, activating a cellular oncogene. The age of initial exposure seems to be particularly important. The most common illness associated with EBV is infectious mononucleosis, a benign infection of lymphocytes common among teenagers. Infection earlier or later in life seems to be associated with a greater risk of malignancy, reasons for which are unknown.

INCIDENCE

Incidence rates for virus-related cancers vary widely by virus and affected site. Incidence rates for cancers that have viral and nonviral causes are generally not broken down by cause.

Approximately 500,000 cases of cervical cancer are reported annually worldwide. In the United States, cervical cancer accounts for some 5,000 deaths each year. Incidence rates in the United States have been reduced to approximately 12,000 cases per year due to greater use of the cervical cancer vaccine and earlier observation of abnormal cells using the Pap smear.

Liver cancer is estimated to result in 500,000 deaths worldwide each year. Liver cancers account for one-quarter of all cancers in developing countries. In Africa and Southeast Asia, the incidence rate is 20 per 100,000 people, while in the United States, it is about 5 per 100,000 people. Most of these cases result from infection by HBV. However, HCV is increasingly associated with cases of liver cancer. A precise number is unavailable since the extent of HCV in the population is unknown; however, studies from Japan have found HCV in 75 percent of cases of liver cancer. Most cases of liver cancer in the United States are found in immigrant populations from areas in which the disease is endemic.

An increasing number of cases of Kaposi sarcoma began to appear in the late 1980s as a result of the AIDS epidemic. Incidence rates peaked at 9.5 per 100,000 men in 1989, falling to 6 cases per one million by 2014, resulting in better treatment for HIV-positive individuals.

In the United States, non-Hodgkin lymphomas (all types) are the fifth most common form of cancer, with an incidence rate of about 20 per 100,000 people per year. Approximately 9,000 cases of Hodgkin disease are diagnosed annually in the United States, predominately among young adults. The presence of the R-S cell is the diagnostic characteristic of the disease.

SYMPTOMS

Symptoms of virus-related cancers vary depending on the site affected. For cervical cancer, the early stages are generally asymptomatic. As the disease progresses, women experience unusual vaginal bleeding or pain.

For liver cancer, the symptoms include abdominal pain, unexplained weight loss, and a sudden onset of jaundice. Blood tests may reveal the elevation of certain liver proteins such as alpha-fetoprotein.

The first symptoms of Kaposi sarcoma are generally purplish lesions or nodules on or under the skin or mucous membranes.

Symptoms of lymphomas are general, usually manifesting as swollen lymph nodes, fever, or unexplained weight loss.

SCREENING AND DIAGNOSIS

A standard screening test exists for cervical cancer, but other virus-related cancers do not have routine screening procedures.

Observing abnormal cervical cells in a Pap test allows early cervical cancer diagnosis. Although not all abnormal-appearing cells are cancerous, their presence may indicate a precancerous state. A colposcopy, the visual observation of the cervix, may help the physician decide whether to perform a biopsy. As with most cancers, the extent of the disease is the basis for staging.

If a patient's symptoms suggest liver cancer, the physician will investigate further, using a variety of imaging scans. Ultrasound is generally the first choice because it is non-invasive and easily performed. This may be followed up with a computed tomography (CT) scan or magnetic resonance imaging (MRI). Ultimately, a biopsy is necessary to confirm cancer. The staging of liver cancer is based on the size of the tumor and the extent of spread.

Kaposi sarcoma is diagnosed by taking the patient's history and examining the skin for the lesions typical of the disease. Biopsies are performed to confirm the disease, and imaging tests are used to find lesions in the stomach or lungs. Because Kaposi sarcoma is AIDS-related, its staging is based on the extent of the tumor, the state of the immune system, and the amount of systemic illness.

Non-Hodgkin lymphoma is commonly diagnosed by biopsy of an enlarged lymph node that does not respond to antibiotic treatment. If cancer is found in the sample, imaging tests such as X-ray, CT scans, MRI, and positron emission tomography (PET) scans can help determine how much cancer has spread. Staging for non-Hodgkin lymphoma uses the Ann Arbor system based on the degree of spread. In addition, two prognostic systems, one for slow-growing and one for fast-growing lymphomas, have been developed. These systems attempt to describe a patient's risk of dying and help physicians select appropriate courses of treatment.

TREATMENT AND THERAPY

Treatment for virus-related cancers varies depending on the site but generally involves surgery, radiation therapy, chemotherapy, or a combination of therapies.

Treatment of cervical cancer generally starts with surgery to remove as much cancerous tissue as safely possible. Additional treatments depend on the stage of the disease. Radiation therapy may be external or internal using a radioactive implant. Chemotherapy, usually intravenous, is recommended if metastasis has occurred.

A small liver tumor can be removed surgically. Historically, a liver transplant had been recommended for most forms of liver cancer, though later, it was determined that such a radical approach might not be necessary for the absence of extensive involvement. Extensive liver cancer may require chemotherapy, though other approaches have also been useful. Transarterial chemoembolization (TACE), the embolization of tumor blood vessels using a gel or coil; radiofrequency ablation (RFA), the insertion of an electrical probe directly into the tumor, followed by ablation of the tissue; and proton beam therapy have all been shown to be useful in reducing the size of localized tumors.

Treatment of Kaposi sarcoma depends on the extent of the tumor. It can involve either surgical removal or radiation therapy and chemotherapy.

Non-Hodgkin lymphoma in HIV-infected patients is complicated by the patients' low blood cell counts. However, highly active antiretroviral therapy has made it easier for patients to endure chemotherapy.

Treatment of Hodgkin disease generally involves a combination of radiation and chemotherapy, both low and high-dose. The use of monoclonal antibodies directed against the R-S cells has also proven to be of some benefit. The prognosis for HD patients is favorable, particularly if diagnosed early. Some 95 percent of treated patients survive five years or longer, many likely cancer-free.

PROGNOSIS, PREVENTION, AND OUTCOMES

Prognosis and prevention of virus-related cancers vary depending on specific cancer and virus. Prevention generally involves attempting to avoid the spread of viruses that cause cancers. The screening of donated blood has reduced the risk of viral infection through transfusion. Safe-sex practices (such as condom use) help prevent the spread of sexually transmitted viruses such as HPV, HIV, HBV, and HCV. Efforts have been made to educate intravenous drug users about the dangers of sharing needles and spreading viruses. Vaccines have been developed for HBV and HPV. The HBV vaccine can prevent most infections by this virus. Since 80 percent of cases of liver cancer are associated with this virus, immunization should reduce the incidence rate. In 2006, a vaccine for HPV, Gardasil, was approved by the US Food and Drug Administration. While the long-term results of vaccine use are unknown and questions about its efficacy still exist, it is hoped that early immunization for HPV may protect women against most forms of the virus.

The prognosis for patients with cervical cancer depends on the stage of the disease and its response to therapy. Early diagnosis can result in the elimination of the disease. At the same time, once metastasis has taken place, the outcome becomes increasingly poor.

Liver cancer is curable if caught early and the tumor removed. None of the techniques used to treat advanced cancer is effective long-term, and the prognosis for patients with advanced liver cancer remains poor, with death generally resulting within a year of diagnosis.

The prognosis for HIV-related Kaposi sarcoma is improving because of better treatments for AIDS patients. If detected early, the five-year survival rate can reach 90 percent; however, if the disease has reached the lungs, the survival rate drops to 30 percent.

—Richard Adler, PhD

Further Reading

Crawford, Dorothy, et al. *Cancer Virus: The Discovery of the Epstein-Barr Virus.* Oxford UP, 2014.

Grand, J. A., editor. *Viruses, Cell Transformation, and Cancer.* Elsevier, 2001.

Pelengaris, Stella, and Michael Khan. *The Molecular Biology of Cancer.* Blackwell, 2006.

Skloot, Rebecca. *The Immortal Life of Henrietta Lacks.* Random House, 2011.

Tabor, Edward, editor. *Viruses and Liver Cancer.* Elsevier, 2006.

VIRAL GENETICS

Category: Viral genetics
Specialties and related fields: Biochemistry, biotechnology, genetics, microbiology, virology
Definition: the composition and structure of virus genomes and how they replicate, express their genes, and infect cells

KEY TERMS

capsid: the protective protein coating of a virus particle
ribosome: a cytoplasmic organelle that serves as the site for amino acid incorporation during the synthesis of protein
virions: mature infectious virus particles

SIGNIFICANCE

The composition and structures of virus genomes are more varied than any identified in the entire bacterial, botanical, or animal kingdoms. Unlike the genomes of all other cells, which are composed of deoxyribonucleic acid (DNA), virus genomes may contain their genetic information encoded in either DNA or ribonucleic acid (RNA). Viruses cannot replicate on their own but must instead use the reproductive machinery of host cells to reproduce themselves.

WHAT IS A VIRUS?

Viruses are submicroscopic, obligate intracellular parasites. This definition differentiates viruses from all other groups of living organisms. There exists more biological diversity within viruses than in all other known life-forms combined. This results from viruses successfully parasitizing all known groups of living organisms. Viruses have evolved in parallel with other species by capturing and using genes from infected host cells for functions that they require to produce their progeny, enhance their escape from their host's cells and immune system, and survive the intracellular and extracellular environment. At the molecular level, the composition and structures of virus genomes are more varied than any others identified in the entire bacterial, botanical, or animal kingdoms. Unlike the genomes of all other cells composed of DNA, virus genomes may contain their genetic information encoded in either DNA or RNA. The nucleic acid comprising a virus genome may be single-stranded or double-stranded. It may occur in a linear, circular, or segmented configuration.

THE NEED FOR A HOST

Virus particles themselves do not grow or undergo division. Instead, virus particles are produced by assembling preformed components. In contrast, other agents grow from an increase in the integrated sum of their components and reproduce by division. The reason is that viruses lack the genetic information that encodes the apparatus necessary for generating metabolic energy or for protein synthesis (ribosomes). The most critical interaction between a virus and a host cell is the need for the virus for the host's cellular apparatus for nucleic acid and the synthesis of proteins. No known virus has the biochemical or genetic potential to generate the energy necessary for producing all biological processes. Viruses depend totally on a host cell for this function.

Viruses are not living organisms in the traditional sense, but they nevertheless function as living things; they replicate their genes. Inside a host cell, viruses are "alive." In contrast, they are merely a complex assemblage of metabolically inert chemicals outside the host—basically a protein shell. Therefore, while viruses have no inner metabolism and cannot reproduce independently, they carry the means necessary to get into other cells and then use those cells' reproductive machinery to make copies of themselves. Viruses thrive at the host cells' expense.

REPLICATION

The sole goal of a virus is to replicate its genetic information. The type of host cell infected by a virus directly affects the process of replication. For viruses of prokaryotes (bacteria, primarily), reproduction reflects the physical simplicity of the host cell. Reproduction is more complex for viruses with eukaryotic host cells (plants and animals). The coding capacity of the genome forces the virus to choose a reproductive strategy. The strategy might involve near-total reliance on the host cell, resulting in a compact genome encoded for only a few essential proteins (+), or could involve a large, complex virus genome encoded with nearly all the information necessary for replication, relying on the host cell only for energy and ribosomes. Those viruses with an RNA genome plus messenger RNAs (mRNAs) do not need to enter the nucleus of their host cell. However, during replication, many often do. DNA genome viruses mostly replicate in the host cell's nucleus, where host DNA is replicated, and the biochemical apparatus required for this process is located. Some DNA viruses (poxviruses) have evolved to contain the biochemical capacity to replicate in their host's cytoplasm, with a minimal need for the host cell's other functions.

Virus replication involves several stages carried out by all viruses, including the onset of infection, replication, and release of mature virions from an infected host cell. The stages can be defined in eight basic steps: attachment, penetration, uncoating, replica-

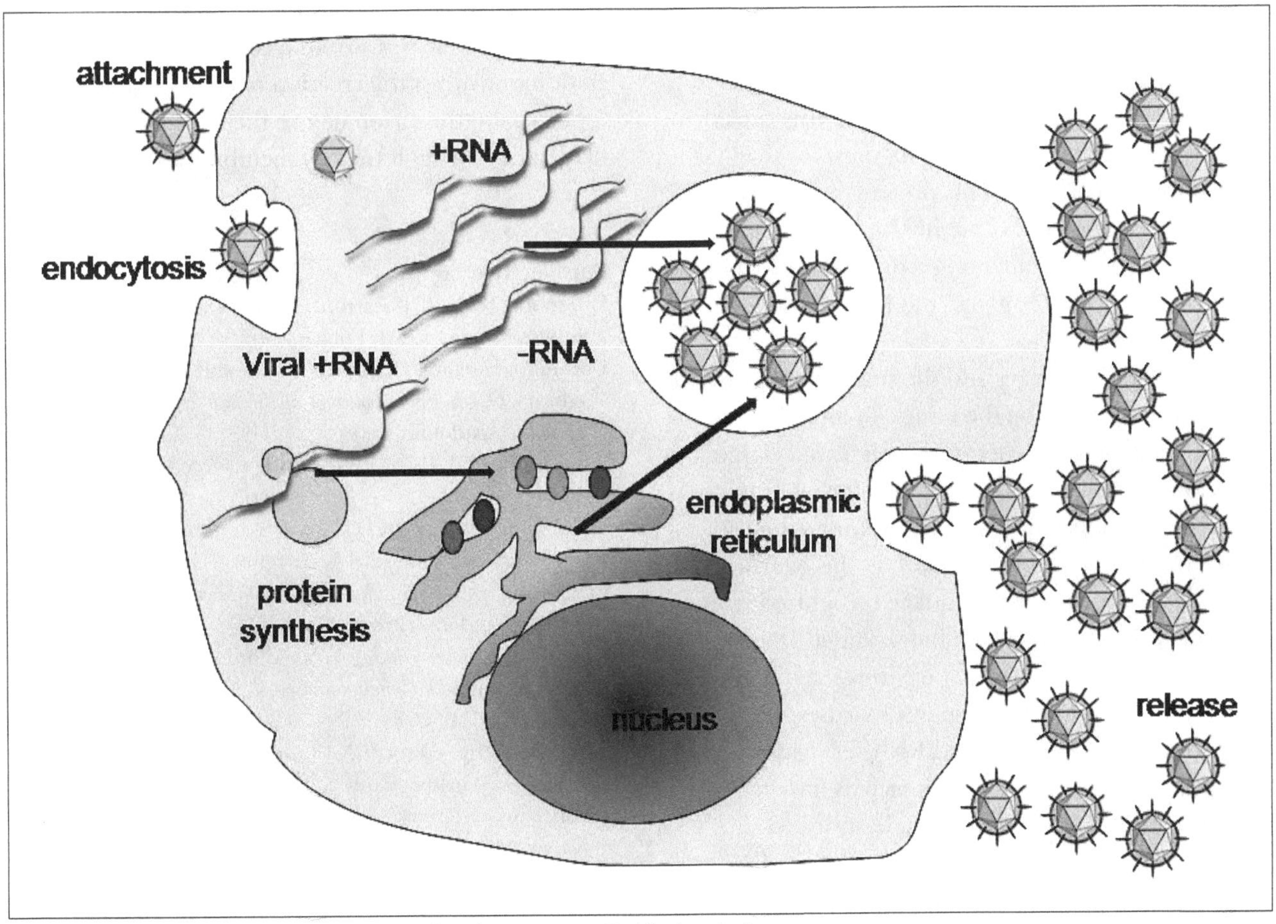

Virus replication cycle. By GrahamColm, via Wikimedia Commons. [Public domain.]

tion, gene expression, assembly, maturation, and release.

The first stage, attachment, occurs when a virus interacts with a host cell and attaches itself—binds with a virus-attachment protein (antireceptor)—to a cellular receptor molecule in the cell membrane. The receptor may be a protein or a carbohydrate residue. Some complex viruses, such as herpes viruses, use more than one receptor and therefore have alternate routes of cellular invasion.

Shortly after attachment, the target cell is penetrated. Cell penetration is usually an energy-dependent process, and the cell must be metabolically active for penetration to occur. The virus bound to the cellular receptor molecule is translocated across the cell membrane by the receptor and is engulfed by the cell's cytoplasm.

Uncoating occurs after penetration and results in the complete or partial removal of the virus capsid and the exposure of the virus genome as a nucleoprotein complex. This protein complex can be a simple RNA genome or highly complex, as in the case of a retrovirus containing a diploid RNA genome responsible for converting a viral RNA genome into a DNA provirus.

How a virus replicates and the resulting expression of its genes depends on the nature of its genetic materials. Control of gene expression is a vital element of virus replication. Viruses use the biochemical apparatus of their infected host cells to express their genetic information as proteins and do this by using the appropriate biochemical language recognized by the host cell. Viruses include double-stranded DNA viruses such as papovaviruses, poxviruses, and herpesviruses; single-stranded sense DNA viruses such as parvoviruses; double-stranded RNA reoviruses; single-stranded sense RNA viruses such as flaviviruses, togaviruses, and caliciviruses; single-stranded antisense RNA such as filoviruses and bunyaviruses; single-stranded sense RNA with DNA intermediate retroviruses; and double-stranded DNA with RNA intermediate-like hepadnaviruses.

During assembly, the basic structure of the virus particle is formed. Virus proteins anchor themselves to the cellular membrane. As virus proteins and genome molecules reach a critical concentration, assembly begins. The result is that a genome is stuffed into a completed protein shell. The maturation process prepares the virus particle for infecting subsequent cells and usually involves the cleavage of proteins to form matured products or conformational structural changes.

For most viruses, the release is simply breaking open the infected cell and exiting. The breakage normally occurs through the physical interaction of proteins against the inner surface of the host cell membrane. A virus may also exit a cell by budding. Budding involves the creation of a lipoprotein envelope around the virion before the virion's being extruded out through the cell membrane.

—Randall L. Milstein, PhD

Further Reading

Dimmock, N. J., A. J. Easton, and K. N. Leppard. *Introduction to Modern Virology*. 6th ed., Blackwell, 2007.

Domingo, Esteban, Colin R. Parrish, and John J. Holland, editors. *Origin and Evolution of Viruses*. 2nd ed., Elsevier/Academic, 2008.

Flint, Jane, et al. *Principles of Virology*. 5th ed., ASM Press, 2020.

Poehlmann, Stefan, and Graham Simmons. *Viral Entry into Host Cells*. Springer, 2013.

Ryu, Wang-Shic. *Molecular Virology of Human Pathogenic Viruses*. Academic Press, 2016.

Shors, Teri. *Understanding Viruses*. 2nd ed., Jones, 2013.

Tibayrenc, Michel. *Genetics and Evolution of Infectious Disease*. Elsevier, 2011.

Yang, Decheng, editor. *RNA Viruses: Host Gene Responses to Infections*. World Scientific, 2009.

Zimmer, Carl. *Parasite Rex: Inside the Bizarre World of Nature's Most Dangerous Creatures*. Free, 2000.

Simian virus 40

Category: Viruses
Also known as: Simian vacuolating virus 40, SV40
Anatomy or system affected: Bone, brain, cells, immune system, lungs, lymphocytes, lymphatic system, nervous system, respiratory system
Specialties and related fields: Microbiology, oncology, pathology, virology
Definition: a polyomavirus of the family Papovaviridae that is found in several species of monkeys

KEY TERMS

lymphomas: cancer of the lymph nodes
mesotheliomas: a cancer of mesothelial tissue in the lungs, associated especially with exposure to asbestos
oncogenes: genes that have the potential to cause cancer
tumor suppressor genes: genes that make proteins that help control cell growth

RELATED CANCERS

Malignant mesothelioma, osteosarcoma, choroid plexus tumors, ependymomas, and non-Hodgkin lymphoma.

EXPOSURE ROUTES

The actual route of exposure of simian virus 40 in humans is under investigation. There is speculation that millions of Americans were exposed to the virus between 1955 and 1963 during the mass immunizations with the original Salk (injectable) and Sabin (oral) polio vaccines. However, some people too young to have received the original polio vaccines have tested positive for exposure to the virus. Therefore, other routes of exposure, such as person to person, may be possible.

WHERE FOUND

As a latent infection in several macaque monkey species, in biomedical research labs where it is used to transform human cells or inoculated into laboratory animals for oncology studies.

AT RISK

People who were vaccinated with the Sabin and Salk polio vaccines between 1955 and 1963; about one hundred army camp men who were inoculated with adenovirus vaccines contaminated with simian virus 40 in the 1950s and 1960s; lab researchers working with the virus.

ETIOLOGY AND SYMPTOMS OF ASSOCIATED CANCERS

Carcinogenesis may be induced by the inactivation of cellular tumor-suppressor proteins (TP53 and RB1). The SV40 genome encodes two oncogenes, large T and small t antigen. These proteins cause cancer in experimental mouse studies.

HISTORY

The virus was discovered in 1960 in the rhesus macaque kidney cells used to amplify the poliovirus for the original Salk and Sabin polio vaccines. In 1961, after learning that inoculated simian virus 40 caused cancer in laboratory animals, the US federal government required that new polio vaccine stocks be free of the virus. Since then, the Salk and Sabin vaccine stocks have been produced using human or African green monkey cell lines extensively screened for viral contaminants.

The National Cancer Institute has reported that forty years of epidemiological studies in the United States and Europe have not shown increased cancer risk in people who may have been exposed to simian virus 40. However, polymerase chain reaction (PCR) testing has revealed traces of simian virus 40 in many malignant mesothelioma tumors and (in one study) 42 percent of non-Hodgkin lymphomas, among others. However, association does not mean causation, and PCR testing techniques for simian virus 40 have not been standardized. Lab contamination could also be a problem. The linkage between simian virus 40 exposure and cancer in humans is still being actively investigated.

—Lisa J. Shientag, VMD

Further Reading

Carbone, Michele, et al. "SV40 and Human Mesothelioma." *Translational Lung Cancer Research,* vol. 9, no. Suppl 1, 2020, pp. S47-59, doi:10.21037/tlcr.2020.02.03.

Colvin, Emily K., et al. "SV40 TAg Mouse Models of Cancer." *Seminars in Cell & Developmental Biology,* vol. 27, 2014, pp. 61-73, doi:10.1016/j.semcdb.2014.02.004.

Sáenz Robles, Maria Teresa, and James M Pipas. "T Antigen Transgenic Mouse Models." *Seminars in Cancer Biology,* vol. 19, no. 4, 2009, pp. 229-35, doi:10.1016/j.semcancer.2009.02.002.

Topalis, D., et al. "The Large Tumor Antigen: A "Swiss Army Knife." Protein Possessing the Functions Required for the Polyomavirus Life Cycle." *Antiviral Research,* vol. 97, no. 2, 2013, pp. 122-36. doi:10.1016/j.antiviral.2012.11.007.

Hepatitis B virus (HBV)

Category: Pathogen

Anatomy or system affected: Blood, blood vessels, bone marrow, kidneys, liver

Specialties and related fields: Family practice, gastroenterology, hematology, hepatology, nephrology, pediatrics, preventative medicine, public health, virology

Definition: the hepatitis B virus causes hepatitis B, a type of liver inflammation mainly spread through contact with blood and blood products and sexual contact with an infected person or carrier

KEY TERMS

aplastic anemia: a condition in which the bone marrow stops producing new blood cells

hepatitis: inflammation of the liver

horizontal transmission: the spread of an infectious agent from one person or group to another, usually through contact with contaminated material, such as sputum or feces

membranous glomerulonephritis (MGN): when the small blood vessels in the kidney (glomeruli) that filter wastes from the blood become damaged and thickened, causing proteins to leak from the damaged blood vessels into the urine (proteinuria)

polyarteritis nodosa: a rare multisystem disorder characterized by widespread inflammation, weakening, and damage to small and medium-sized arteries

vertical transmission: the passage of a disease-causing agent (pathogen) from mother to baby during the period immediately before and after birth

virion: the complete, infective form of a virus outside a host cell, with a core of ribonucleic acid (RNA) or deoxyribonucleic acid (DNA) and a capsid

HEPATITIS B VIRUS

Hepatitis B virus (HBV) is a member of the hepadnavirus group. The surface of this virus has an envelope composed of a phospholipid bilayer that surrounds a protein capsid. Embedded in the envelope are glycoproteins called the "hepatitis B surface antigen."

The protein capsid is made from a protein called the "hepatitis B core antigen." Between the envelope and the capsid is a protein called the "hepatitis B e antigen" (HBeAg). Within the capsid is a partially double-stranded deoxyribonucleic acid (DNA) molecule and a DNA polymerase enzyme. This DNA poly-

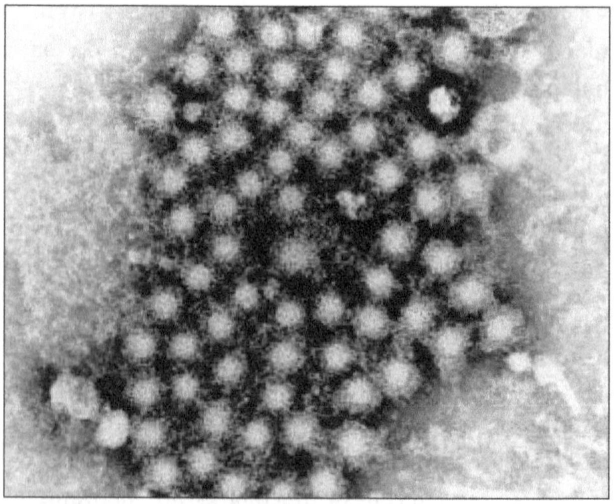

A transmission electron micrograph of hepatitis virions, Centers for Disease Control and Prevention. [Public domain.]

merase is DNA and ribonucleic acid (RNA) dependent, meaning that it can synthesize DNA from a DNA or an RNA template.

INFECTION COURSE

Upon entering host cells, HBV DNA goes to the nucleus. In the nucleus, DNA polymerase converts the partially double-stranded DNA into completely double-stranded circular DNA. Host cell RNA polymerases transcribe the HBV DNA into messenger RNAs sent to the cytoplasm for translation into viral proteins. The viral DNA polymerase reverse transcribes the HBV RNAs into DNA molecules that are packaged with the viral proteins to form new virions. These new virions bud from the host cell and infect nearby cells.

After becoming infected, people experience a long incubation period that lasts 30 to 180 days. A prodromal period follows the incubation period characterized by itchiness, rash, joint pain, fever, tiredness, and swollen lymph nodes.

After the prodromal period, HBV infection proceeds to acute hepatitis with complete resolution or chronic hepatitis that may or may not cause cirrhosis of the liver. People with chronic HBV infection without cirrhosis are carriers of the virus who shed it in their body fluid and can spread the infections to others. Some people may even progress to fulminant hepatitis, characterized by rapid and extensive death of liver cells and loss of liver function. The possibility of HBV infection progressing to chronic hepatitis is less than five percent in adults, twenty to thirty percent in children, and ninety percent in newborns. Consequently, babies born to mothers with active HBV infections should receive antihepatitis B immunoglobulins and an initial dose of the hepatitis B vaccine.

HBV does not destroy the cells it infects. The liver damage results from cytotoxic T cells that destroy infected liver cells that present hepatitis B surface antigen and core antigen.

HBV also causes disease in other body parts besides the liver, specifically the bone marrow, kidneys, and blood vessels. HBV can damage the bone marrow and cause aplastic anemia, a condition in which the bone marrow stops producing new blood cells. Second, HBV sometimes causes membranous glomerulonephritis (MGN). The delicate blood vessels in the kidney, the glomeruli, thicken and stop working correctly. Protein starts to spill into the urine (proteinuria), and the kidneys start to fail. Third, chronic HBV infections may result in inflammation and scarring of the arterial walls, a condition called "polyarteritis nodosa."

EXPOSURE ROUTES

HBV has three main modes of transmission. First, needles or blood are the primary modes of HBV transmission. Intravenous drug users who reuse needles from infected individuals, health-care workers exposed to blood and needlestick accidents, blood transfusion recipients, and dialysis patients all have an increased risk of HBV infection.

Second, bodily fluids like semen, breast milk, tears, sweat, and saliva transmit HBV. Therefore, sexual activity can transmit HBV. Transmission by blood and bodily fluids constitute "horizontal transmission."

Third, mothers can transmit HBV to their babies through the placenta before birth or through blood, breast milk, or other body fluids after birth. HBV transmission from mother to baby is called "vertical transmission."

WHERE FOUND

About a million people in the United States have chronic hepatitis B virus (HBV) infection. Though the incidence of chronic hepatitis B is decreasing in the United States because of a vaccine, HBV infection has increased to one in ten Americans of Asian ancestry.

AT RISK

Patients infected with the hepatitis B virus at greatest risk for liver cancer are those with cirrhosis (scarring of the liver) and a family history of liver cancer. In the

United States, the highest incidence of liver cancer occurs in Asian immigrants; the frequency is lowest in whites, followed by Hispanics and African Americans. Closed environments such as prisons put people at risk. Across the globe, the vast majority of those who develop liver cancer have had the hepatitis B virus for most of their lives. However, in the United States, those with chronic hepatitis B infection mostly have contracted the infection in adulthood in association with other risk factors such as coinfection with the hepatitis C virus (HCV) or chronic alcohol abuse.

ETIOLOGY AND SYMPTOMS OF ASSOCIATED CANCERS

The hepatitis B virus causes hepatitis B, leading to chronic hepatitis B and cirrhosis, followed by liver cancer. The frequency of liver cancer can be correlated with the frequency of hepatitis B infection. It is believed that the inflammatory process involving chronic hepatitis B may be a crucial factor in the development of cancer. Most persons who are hepatitis B carriers and even those with early liver cancer may be asymptomatic. By the time symptoms do appear, liver cancer is usually inoperable. Symptoms may include jaundice, fatigue, abdominal pain and swelling, loss of appetite, nausea, vomiting, and joint pain.

HISTORY

Although hepatitis has been long known, its etiology remained a mystery until a virus found in human blood became suspect; hepatitis B virus was isolated in 1963. The Discovery of a specific antigen linked to the hepatitis B virus led to developing a test to screen blood (1990), significantly reducing the incidence of posttransfusion hepatitis. A vaccine was then developed that protects against the hepatitis B virus and, indirectly, against liver cancer.

PREVENTION AND TREATMENT

The HBV vaccine is a subunit vaccine that contains the hepatitis B surface antigen. Injection of this HBV antigen elicits a robust immune response that protects against HBV infection.

HBV treatment options include interferon or nucleoside/nucleotide analogs. Recombinant interferon-alpha is administered by subcutaneous injection. This medication is for younger patients since it is not well tolerated in older patients. Interferon injections are given weekly for forty-eight weeks and can suppress active hepatitis B infections.

The nucleoside/nucleotide analogs inhibit the HBV DNA polymerase. Available agents include entecavir, tenofovir, lamivudine, adefovir, and telbivudine. Viral suppression with nucleoside/nucleotide analogs takes significantly longer than with interferon.

—Cynthia Racer, MA, MPH and
Michael A. Buratovich, PhD

Further Reading

Lampertico, P., et al. "Review Article: Long-Term Safety of Nucleoside and Nucleotide Analogues in HBV-Monoinfected Patients." *Alimentary Pharmacology & Therapeutics,* vol. 44, no. 1, 2016, pp. 16-34, doi:10.1111/apt.13659.

Lok, Anna S. F., et al. "Antiviral Therapy for Chronic Hepatitis B Viral Infection in Adults: A Systematic Review and Meta-analysis." *Hepatology,* vol. 63, no.1, 2016, pp. 284-306, doi:10.1002/hep.28280.

Mast, Eric E., et al. "A Comprehensive Immunization Strategy to Eliminate Transmission of Hepatitis B Virus Infection in the United States: Recommendations of the Advisory Committee on Immunization Practices (ACIP) Part II: Immunization of Adults." *MMWR. Recommendations and Reports: Morbidity and Mortality Weekly Report,* vol. 55, no. RR-16, 2006, pp. 1-33.

Terrault, Norah A., et al. "Update on Prevention, Diagnosis, and Treatment of Chronic Hepatitis B: AASLD 2018 Hepatitis B Guidance." *Hepatology,* vol. 67, no. 4, 2018, pp. 1560-99, doi:10.1002/hep.29800.

HEPATITIS C VIRUS (HCV)

Category: Pathogen
Anatomy or system affected: Liver

Specialties and related fields: Gastroenterology, hepatology, immunology, internal medicine, microbiology, oncology, pharmacology, virology

Definition: a member of the flavivirus group that infects the liver and causes hepatitis

KEY TERMS

cirrhosis: a chronic disease of the liver marked by degeneration of cells, inflammation, and fibrous thickening of tissue, typically a result of alcoholism or hepatitis

endosome: intracellular sorting organelles in eukaryotic cells

hepatitis: liver inflammation

hepatocellular carcinoma: the most common type of primary liver cancer

hepatocytes: the majority cell type in the liver

proteases: enzymes that catalyze the degradation of proteins into smaller polypeptides

receptor-mediated endocytosis: a process by which cells absorb metabolites, hormones, proteins—and in some cases viruses—by the inward budding of the plasma membrane

INTRODUCTION

The hepatitis C virus (HCV) is a flavivirus that causes hepatitis C that causes liver inflammation. The flaviviruses include viruses that cause yellow fever, Dengue fever, and West Nile virus. HCV is mainly spread via contact with infected blood. Chronic HCV infection also increases the risk of liver cancer. Liver cancer is usually preceded by chronic hepatitis and cirrhosis. HCV is a ribonucleic acid (RNA) virus, and an error-prone RNA replicase replicates its genomic RNA. Therefore, the hepatitis C virus easily mutates, making vaccine development difficult.

HCV is an enveloped virus with a membrane surrounding its genome. HCV viral particles are roughly spherical but vary in size from 40 to 80 nanometers in diameter. There are 7 HCV genotypes and 84 HCV subtypes. The viral envelope is studded with two glycoproteins called "E1" and "E2." These envelop-embedded glycoproteins interact with target proteins on the surfaces of host cells.

Inside the viral envelope is a nucleocapsid that houses the viral single-stranded RNA genome. The HCV genome is a positive-sense RNA, meaning the viral genome is used directly by the host cell ribosomes to synthesize viral proteins.

EXPOSURE ROUTES

HCV is transmitted by blood-to-blood contact with an infected individual or via blood transfusion with infected blood; rarely, maternal transmission may occur. Sexual transmission is also possible in those who have unprotected sex with multiple partners.

HCV is not passed from a mother to her baby during pregnancy. Still, there is a 5 percent chance that the mother will transmit the virus to her baby during delivery.

HCV LIFE CYCLE

After entering the body, HCV circulates through the blood and infects B lymphocytes (antibody-secreting white blood cells) and hepatocytes (liver cells). HCV uses its E1 and E2 surface proteins to bind the low-density lipoprotein (LDL) receptors on the surfaces of liver cells and B lymphocytes. Binding the LDL receptor stimulates the host cell to engulf the HCV particle through "receptor-mediated endocytosis" (RME). RME brings the viral particle into the host cell cytoplasm within a membrane-enclosed vesicle called an "endosome."

Through the activity of E1 and E2, the viral membrane fuses with the endosome membrane, releasing the HCV nucleocapsid into the cytoplasm. The nucleocapsid releases the viral RNA. Host cell ribosomes bind it and translate it into a large viral protein precursor known as a "polyprotein." Enzymes called "proteases" clip the polyprotein into viral structural and nonstructural proteins. Some proteases are from the host cell, but the HCV genome encodes others.

HCV structural proteins are part of mature HCV viral particles. Nonstructured proteins aid the infection process but are not incorporated into mature viral particles.

One of the nonstructural proteins in the NS5B RNA-dependent RNA polymerase makes multiple copies of HCV genomic RNA for incorporation into viral particles. Assembled HCV viral particles (known as virions) bud from the host cell to infect other cells.

DISEASE PROCESS

HCV causes three types of disease. The most common is chronic hepatitis, but others include acute hepatitis and, rarely, fulminant hepatitis that progresses quickly to liver failure.

Chronic hepatitis has no symptoms except, in some people, fatigue. Without treatment, chronic hepatitis progresses to cirrhosis within ten to fifteen years and liver failure after twenty years. Chronic HCV infection also increases the risk of the most common type of liver cancer, hepatocellular carcinoma.

Acute hepatitis usually causes mild symptoms that include nausea, fatigue, diminished appetite, joint and muscle pain. Jaundice, the yellowing of the skin and whites of the eyes, usually does not occur during acute HCV hepatitis.

Over sixty percent of people with HCV develop complications like cryoglobulinemia, in which B lymphocytes make abnormal antibodies that clump together in cold temperatures. These clumped antibodies stick to the wall of small vessels and damage them. Blood vessel damage can result in skin, kidney, thyroid gland, or pancreatic malfunction. These complications arise because HCV infects antibody-secreting lymphocytes, which adversely affects their antibody production.

WHERE FOUND

One hundred fifty million to 200 million people worldwide are infected with HCV. The number of acute cases of hepatitis C reported in the United States increased 20 percent, from 1,778 reported cases in 2012 to 2,138 reported cases in 2013.

WHO IS AT RISK?

People who have a history of injected or inhaled drug use, have been exposed to blood via sexual contact, have received a transfusion of unscreened blood, or have been exposed to contaminated instruments during tattooing, ear and body piercing, and dental procedures are at risk. People who received blood, blood products, or transplanted organs before 1992 (before testing of blood for hepatitis C was begun) are also at risk for HCV infection.

Health-care workers are at risk from needlestick injuries. Coinfection with hepatitis D virus (HBV) or the human immunodeficiency virus (HIV) and alcohol abuse puts people at risk for chronic hepatitis C, cirrhosis (scarring of the liver), and primary liver cancer, as does having a relative with liver cancer. Much less often, HCV transmission occurs among HIV-positive persons. Also at risk are those who were ever incarcerated because of the high prevalence of hepatitis C in the prison population. Patients on long-term hemodialysis are also at risk because of the potential exposure to HCV during hemodialysis.

DIAGNOSIS

Serological tests that detect antibodies against HCV help diagnose this disease. Antibodies against HCV are detectable seven to 31 weeks of infections. A positive antibody test requires confirmation with another test.

Detecting viral antigens in the blood is another possible way to diagnose HCV. Viral antigens are detectable in the blood one to three weeks after exposure and up to four to six months in acute hepatitis cases. Viral antigens are detectable in the blood for longer than ten years in cases of chronic hepatitis.

Polymerase chain reaction (PCR) tests can detect HCV RNA. PCR tests are necessary confirmatory tests for those whose antibody tests are negative.

Liver biopsies can ascertain the level or confirm the presence of liver damage. HCV-infected livers have copious lymphocyte infiltration with large fatty particles in them.

ETIOLOGY AND SYMPTOMS OF ASSOCIATED CANCERS

HCV infection causes an acute illness with a discrete onset of signs or symptoms consistent with acute viral hepatitis. Such signs and symptoms include fever, headache, malaise, anorexia, nausea, vomiting, diarrhea, and abdominal pain, and either (a) jaundice, or (b) elevated serum alanine aminotransferase (ALT) levels >400 IU/L. Most persons infected with HCV are asymptomatic; however, many have chronic liver disease, ranging from mild to severe.

Specific symptoms of liver cancer are usually absent until cirrhosis has occurred. Signs and symptoms of adult primary liver cancer may include a hard lump below the right side of the rib cage; right side upper abdomen discomfort; right shoulder pain, nausea, and unusual fatigue. Signs and symptoms for multiple myeloma may include bone pain and skeletal and spinal fractures; for Burkitt lymphoma, symptoms may include swollen lymph nodes, abdominal pain, tumors, weight loss, and fatigue.

HISTORY

Blood tests were first developed to identify the causative viruses of hepatitis B (1963) and hepatitis A (1973). Still, some posttransfusion blood samples proved negative for both. In the 1980s, scientists identified another virus as the causative agent of "non-A, non-B hepatitis" and called it "hepatitis C virus." In 1990 blood banks began screening donors for the hepatitis C virus, substantially lowering the risk of contracting posttransfusion hepatitis C.

TREATMENT

Traditionally, the primary treatment for HCV infection was recombinant interferon-alpha, pegylated interferon, with or without a drug called "ribavirin." This treatment, unfortunately, suppressed but did not eliminate HCV and had manifold, severe side effects. Today, newer drugs called "direct-acting antivirals" (DAAs) are employed that are much better tolerated and completely rid the body of HCV.

DAAs target HCV proteases and the viral RNA replicase. Since both proteins are unique to HCV, inhibiting them inhibits viral replication without affecting host cell processes. Treatment regimens depend on whether the patient has cirrhosis and which HCV strain has infected the patient.

For example, a recommended treatment regimen includes Mavyret, a combination of the NS3/4A protease glecaprevir (300 milligrams) and the NS5 protein inhibitor pibrentasvir (120 mg) to be taken three times a day with food for eight weeks. An alternative regimen would be Epclusa, a combination of the NS5B RNA-dependent RNA polymerase sofosbuvir (400 mg) and the NS5A protein inhibitor velpatasvir (100 mg) for twelve weeks. Both drug regimens are effective against all genotypes of HCV.

PROGNOSIS, PREVENTION, AND OUTCOMES

Of every 100 people infected with hepatitis C, it is estimated that seventy-five to eighty-five will become chronically infected, liver disease will develop in sixty to seventy, cirrhosis will develop in five to twenty, and one to five will die from complications of liver disease such as cirrhosis or liver cancer. Liver cancer (hepatocellular carcinoma) is associated with cirrhosis due to chronic hepatitis C infection.

Prevention has included programs aimed at avoiding needle sharing among drug addicts. Needle exchange programs and educational interventions have reduced the transmission of hepatitis C infection. However, rates of hepatitis C remain high among addicts (30 percent of younger users). This is primarily because the population of drug addicts is a complex population to reach and intervene.

Among health-care workers, safe needle-usage techniques have been introduced to reduce accidental needlesticks. Also, newer syringes with self-capping needle systems are used to avoid manually replacing a cap after drawing blood, which also reduces the risk of needlesticks. There is no clear way to prevent transmission of hepatitis C from mother to child. Individuals with multiple sexual partners should use barrier precautions such as condoms to limit the risk of hepatitis C and other sexually transmitted diseases. People with hepatitis C infection should avoid sharing razors or toothbrushes with others because of the possibility that these items may be contaminated with blood. Body piercing(s) or tattoo(s) should be obtained at licensed piercing and tattoo shops (facilities) while assuring the body piercing or tattoo shop uses infection-control practices. Screening tests for blood products have almost eliminated the risk of transmission of hepatitis C infection through transfusion.

—Cynthia Racer, MA, MPH; updated by Jeffrey P. Larson, PT, ATC and Michael A. Buratovich, PhD

Further Reading

Celsa, Ciro, et al. "Direct-acting Antiviral Agents and Risk of Hepatocellular Carcinoma: Critical Appraisal of the Evidence." *Annals of Hepatology*, vol. 100568, 2021, doi:10.1016/j.aohep.2021.100568.

Centers for Disease Control and Prevention. "Guidelines for Viral Hepatitis Surveillance and Case Management." *Viral Hepatitis*, 31 May 2015, www.cdc.gov/hepatitis/statistics/surveillanceguidelines.htm. Accessed 28 Nov. 2021.

———. "Recommendations for Prevention and Control of Hepatitis C Virus (HCV) Infection and HCV-related Chronic Disease." *MMWR*, vol. 47, no. RR-19, 1998, pp. 1-54.

Flint, Jane, et al. *Principles of Virology*. 5th ed., ASM Press, 2020.

Yadav, Sanu R., et al. "Hepatitis C: Current State of Treatment in Children." *Pediatric Clinics of North America*, vol. 68, no.6, 2021, pp. 1321-31, doi:10.1016/j.pcl.2021.07.008.

EPSTEIN-BARR VIRUS

Category: Virology
Also known as: EBV, human herpesvirus 4 (HHV-4)
Anatomy or system affected: Blood, cells, circulatory system, immune system, lymphatic system
Specialties and related fields: Family practice, immunology, microbiology, oncology, pediatrics
Definition: a type of human herpesvirus that causes mononucleosis and is associated with a variety of human cancers

KEY TERMS

B lymphocytes: antibody-secreting white blood cells
Burkitt's lymphoma: an aggressive (fast-growing) type of B-cell non-Hodgkin lymphoma that occurs most often in children and young adult
heterophile antibodies: unusual antibodies made by B lymphocytes infected with Epstein-Barr virus that bind to antigens on the surfaces of animal red blood cells
lytic phase: an infective phase of Epstein-Barr virus, usually in oropharyngeal epithelial cells, in which the virus reproduces in the cells and destroys them by lysing them
latent phase: an infective phase of Epstein-Barr virus, usually occurring in B lymphocytes, in which the virus infects the cells but does not destroy them
mononucleosis: an infectious viral disease characterized by swelling of the lymph glands and prolonged fatigue
Monospot test: antibodies derived from a single parent clonal cell
tonsillitis: inflammation of the tonsils

DEFINITION

Epstein-Barr virus is one of the eight known types of human herpes viruses and is a member of the gamma subtype in this group. Like many herpes virus species, the Epstein-Barr virus is a double-stranded DNA vi-

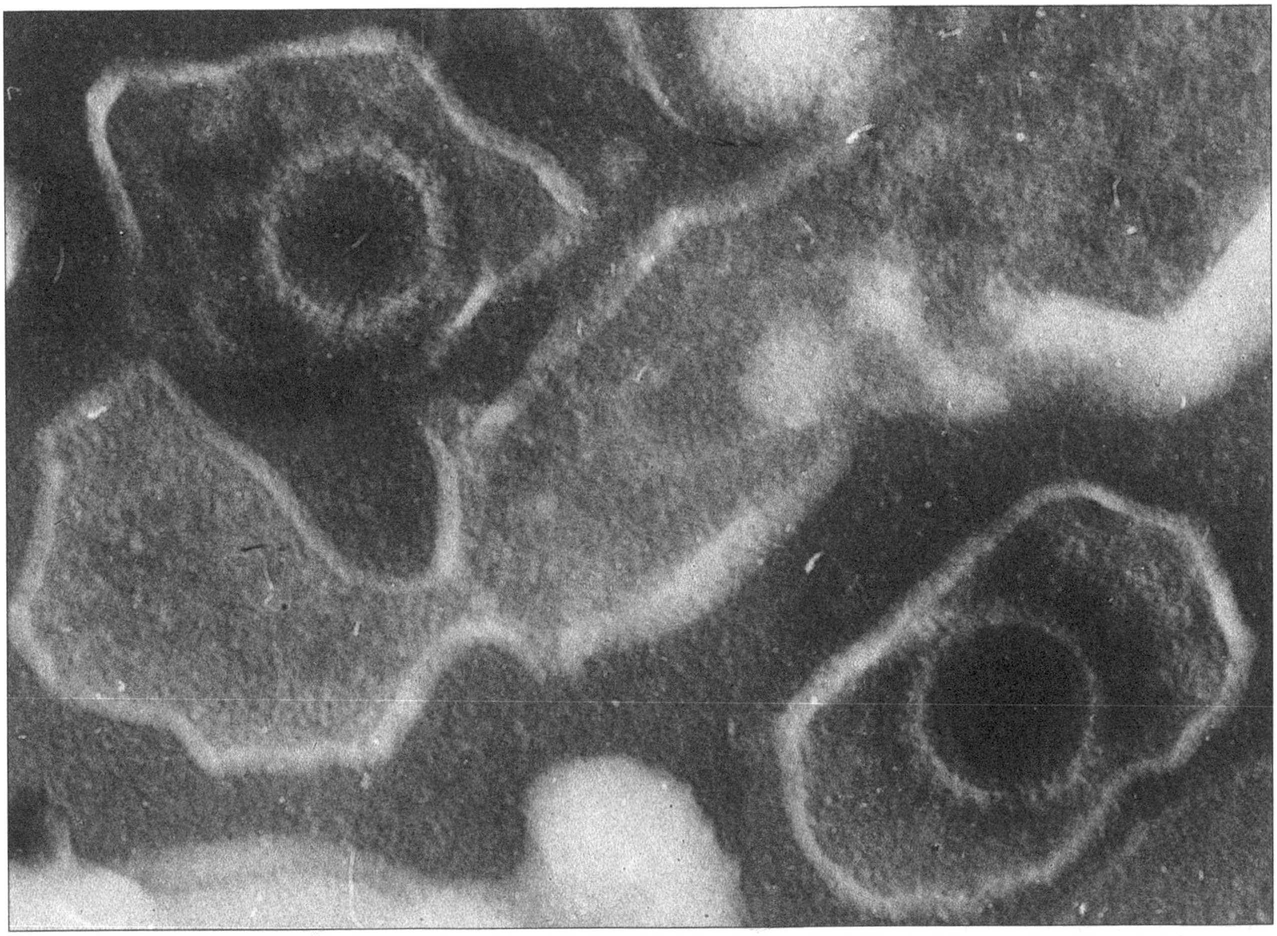

Electron microscopic image of two Epstein Barr Virus virions (viral particles) showing round capsids—protein-encased genetic material—surrounded by the membrane envelope. By Liza Gross, PLoS Biol 3(12) (2005, via Wikimedia Commons.

rus that establishes a lifelong presence in the human body, remaining quiescent for long periods and then inexplicably becoming active. Causally related to mononucleosis, it is also associated with various human cancers, such as Burkitt lymphoma and nasopharyngeal carcinoma, and is therefore considered a carcinogenic virus.

EXPOSURE ROUTES

Humans are the only known reservoir of Epstein-Barr virus, which is present in oropharyngeal secretions and is usually transmitted through saliva; transmission of the virus requires intimate contact with the saliva of an infected person including contact with objects such as shared toothbrushes. Transmission through the air or blood does not usually occur. Still, transmission through blood, semen, and organ transplants has been reported. Most cases of Epstein-Barr virus transfer occur by sharing food or drinks or kissing.

DISEASES

When the Epstein-Barr virus reaches someone's mouth, it infects two types of cells in the oropharynx. First, it infects epithelial cells that line the services of the oropharynx. Second, it infects B lymphocytes, which are the lymphoid cells that create antibodies to fight infections. In epithelial cells, the Epstein-Barr virus undergoes the lytic cycle. Viral deoxyribonucleic acid (DNA) is transcribed during the lytic cycle, and the cellular machinery translates messenger ribonucleic acids (mRNAs). These processes form viral proteins that are packaged into new viruses that leave the host cell after its destruction. These new viruses subsequently infect neighboring epithelial cells.

When the Epstein-Barr virus reaches the lymphoid tissue in the oropharynx, the tonsils, it infects B lymphocytes. The CD21 receptor on the surface of the B lymphocyte is a molecule used by the Epstein-Barr virus to attach and enter the cell. In the infected B cell, the virus enters the latent phase. While in the latent phase, the virus is present in the B lymphocytes but does not kill them. Infected B cells, however, spread the infection to other lymphoid tissues of the body, including the liver, spleen, and other lymph nodes. Typically, the immune response against infection creates antibodies and cytotoxic T lymphocytes that neutralize the virus and kill the infected B lymphocytes, respectively. Therefore, the immune response against the Epstein-Barr virus limits its spread throughout the lymphoid tissues, stopping its spread. In some people, the immune response against the Epstein-Barr virus prevents them from having any symptoms (asymptomatic infection).

Destruction of the oropharyngeal epithelium causes infectious mononucleosis. The most common symptoms of this disease are fever, pharyngitis (inflammation of the throat), and swollen lymph nodes (lymphadenopathy). Throat inflammation results from the destruction of the resident epithelial cells. Swollen lymph nodes result from infected B cells that spread throughout the lymph tissue, causing it to swell. Another common symptom of infectious mononucleosis is fatigue, which can be quite severe and may last for several months.

Other possible symptoms include tonsillitis (inflamed tonsils), palatal petechiae (red spots on the palate), enlargement of the liver (hepatomegaly), and enlargement of the spleen (splenomegaly). Palatal petechiae result from damage to the epithelial cells on the palate by the virus. Hepatomegaly and splenomegaly result from the flood of infected B lymphocytes and T-cytotoxic cells to these organs, causing them to swell.

Splenomegaly, though rare, is of particular concern because it is susceptible to rupture. Splenic rupture can result in excessive bleeding and even death.

Another infrequent symptom of infectious mononucleosis is a rash that consists of pink macules or patches that do not itch and appear on the trunk and arms. The presence of a rash causes infectious mononucleosis to be frequently misdiagnosed as a group A streptococcus infection (strep throat). For some unknown reason, patients with infectious mononucleosis who are prescribed antibiotics like amoxicillin can develop an itchy maculopapular rash. This is not an allergic reaction since people who recover from infectious mononucleosis can take these antibiotics without incident.

Symptoms usually resolve on their own within a few weeks. Treatment is supportive, including rest, nonsteroidal anti-inflammatory drugs, or acetaminophen to reduce fever and alleviate throat pain, avoiding contact with others, and avoiding contact sports release three to four weeks to prevent splenic rupture. Patients with infectious mononucleosis may be contagious for several weeks.

DIAGNOSIS

Infectious mononucleosis is typically suspected based on clinical signs and symptoms such as fever, sore throat, swollen cervical lymph nodes, and fatigue. The posterior cervical lymph nodes in the back of the neck are commonly swollen during Epstein-Barr virus infections because those lymph nodes drain the tonsils where the B lymphocytes are initially infected.

Peripheral blood smears reveal the presence of atypical lymphocytes, most of which are enlarged cytotoxic T lymphocytes. Infected B lymphocytes make unusual antibodies known as "heterophile antibodies." These heterophile or other loving antibodies bind antigens from other animal species. They bind antigens on sheep and horse red blood cells, causing those red blood cells to clump or agglutinate. A so-called Monospot test detects the presence of heterophile antibodies. A positive Monospot test is definitive for Epstein-Barr virus infection. Unfortunately, false negatives of the Monospot test may occur. Early in the infection, heterophile antibodies may not yet have been produced in high enough quantities to be detected. Likewise, in children under the age of four, their B cells do not produce heterophile antibodies.

The presence of Epstein-Barr virus-specific antibodies can confirm an infectious mononucleosis diagnosis. Antibodies against Epstein-Barr virus code protein appear early enough in the infection for detection.

COMPLICATIONS

Latently infected B lymphocytes are usually killed by T-cytotoxic cells. Sometimes, however, cytotoxic T cells are unable to kill all the infected B cells. These leftover, infected B lymphocytes experience an extended latency period. They express viral genes that cause increased B-lymphocyte proliferation. Increased latency in B lymphocytes increases the risk of developing B-cell cancers or B-cell lymphomas. The leading B-cell cancers linked to Epstein-Barr virus are Hodgkin lymphoma and non-Hodgkin lymphoma, especially Burkitt lymphoma and primary central nervous system lymphoma.

Epstein-Barr virus-infected epithelial cells can sometimes enter the latent phase and express viral genes that increase their proliferation. In such cases, a cancer of the upper part of the throat called "nasopharyngeal carcinoma" results.

In individuals with poorly functioning immune systems, such as those with human immunodeficiency virus (HIV), cytotoxic T cells cannot destroy all the infected B cells. Consequently, B-cell cancer development in such individuals is more likely. Individuals with HIV exposed to Epstein-Barr virus may develop a white plaque on the lateral side of the tongue that resists things scraped off. This condition is called "oral hairy leukoplakia."

WHERE FOUND

The Centers for Disease Control and Prevention (CDC) estimates that the Epstein-Barr virus has infected 95 percent of adult Americans between the ages of thirty-five and forty. Still, it is less prevalent in children and teenagers, a pattern observed in the developed world but not in developing regions such as Africa and Asia. In Africa, for example, most children have been infected by the virus by the age of three. Epstein-Barr virus has also been associated with nasopharyngeal cancers in Asia (especially China) and Burkitt lymphoma in equatorial Africa and Papua New Guinea. In tropical regions, Burkitt lymphoma has been shown to coexist with malaria. In the United States, the Epstein-Barr virus has also been associated with nasopharyngeal cancers in immigrants from Asia. The incidence of Burkitt lymphoma has been increasing. Both Hodgkin and non-Hodgkin lymphomas are found in people whose immune systems have been compromised by drug therapy and disease. Epstein-Barr virus has also been associated with approximately 10 percent of gastric carcinomas.

Some, though certainly not all, cases of chronic fatigue syndrome show a clear association with chronic Epstein-Barr virus infection. Chronic fatigue syndrome is a multifaceted illness that has many symptoms and a host of varying clinical presentations. Recently, chronic fatigue syndrome was merged with myalgic encephalomyelitis.

AT RISK

Epstein-Barr virus has been shown to take advantage of those with weakened immune systems; Burkitt lymphoma, a non-Hodgkin lymphoma (NHL), is found in organ transplant patients undergoing immunosuppression therapy, as well as those living with human immunodeficiency virus (HIV) and acquired immunodeficiency syndrome (AIDS) who are immunocompromised by their disease. Because Burkitt lymphoma typically occurs in tropical climates where malaria is endemic, it is believed that the immune systems of those with malaria are altered, resulting in tumor production. The Epstein-Barr virus is also associated with nasopharyngeal carcinoma, which is prevalent in those of Chinese and Southeast Asian ancestry. Environmental/occupational exposure to pesticides and organic solvents shows no significant positive association with Epstein-Barr virus and related cancers.

ETIOLOGY AND SYMPTOMS OF ASSOCIATED CANCERS

Epstein-Barr virus is among the most ubiquitous of viruses. It occurs in nearly all regions of the world, and most people become infected with Epstein-Barr virus sometime during their lifetimes. In the United States, as soon as maternal antibodies dissipate, infants become vulnerable to Epstein-Barr virus infection. In adolescents, infection with the Epstein-Barr virus results in infectious mononucleosis in 35 to 50 percent of cases. Although the symptoms of infectious mononucleosis usually dissipate within weeks to several months, the Epstein-Barr virus lies dormant in a few cells in the throat and blood for the remainder of the person's life. The virus may reactivate from time to time and is often present in the saliva, suppressing the immune system by causing repeated mutations in B cells, which may then proliferate unabated, resulting in tumors.

Epstein-Barr virus establishes a lifelong latent infection in the body's immune system that may later result in the emergence of lymphoma or carcinoma. The initial symptoms of Burkitt lymphoma may include a swollen lymph node in the upper body or abdomen. If the tumor is found in the chest, breathing difficulties may ensue. In other patients, itching, weight loss, fever, and fatigue may be present. Burkitt lymphoma commonly results in the formation of a large tumor mass in the jawbone. Adults with AIDS often develop tumors in various parts of the body. Symptoms of nasopharyngeal cancer may include a lump in the neck or nose, numbness on the side of the face, headaches, ear pain, and difficulty speaking or breathing.

HISTORY

In the latter part of the nineteenth century and the early part of the twentieth century, the medical community in the United States and Europe began to report on a novel syndrome consisting of fever, sore throat, and swollen glands that were later termed mononucleosis and were later found to be causally related to Epstein-Barr virus. Epstein-Barr virus was discovered in the 1960s from a biopsy of a tumor associated with Burkitt lymphoma and was the first virus to be directly linked to human cancer; in 1964, Michael Epstein and Yvonne Barr isolated virus particles from cell lines derived from Burkitt lymphoma, hence the name, Epstein-Barr virus. Subsequently, the Epstein-Barr virus was found to be the leading viral cause of cancer in humans, having an etiological role in Burkitt lymphoma and other B-cell lymphomas as well as nasopharyngeal carcinoma.

—*Cynthia Racer, MA, MPH, Michelle Herdman, and Michael A. Buratovich, PhD*

Further Reading

Centers for Disease Control and Prevention. "Epstein Barr Virus and Infectious Mononucleosis," 28 Sept. 2020, www.cdc.gov/epstein-barr/index.html. Accessed 23 Nov. 2021.

Cui, Xinle, and Clifford M. Snapper. "Epstein Barr Virus: Development of Vaccines and Immune Cell Therapy for EBV-Associated Diseases." *Frontiers in Immunology*, vol. 12, no. 734471, 2021, doi:10.3389/fimmu.2021.734471.

Münz, Christian, editor. *Epstein Barr Virus: One Herpes Virus: Many Diseases* (Current Topics in Microbiology and Immunology, vol. 390). Springer, 2015.

Noor, Nazir, et al. "A Comprehensive Update of the Current Understanding of Chronic Fatigue Syndrome." *Anesthesiology and Pain Medicine*, vol. 11, no. 3, 2021, p. e113629, doi:10.5812/aapm.113629.

Robertson, E. S. *Epstein-Barr virus*. Caister Academic Press, 2005.

Tselis, A., and H. B. Jenson. *Epstein-Barr Virus*. Taylor & Francis, 2006.

Tsao, S., C. M. Tsang, K. To, and K. Lo. "The Role of Epstein-Barr Virus in Epithelial Malignancies." *Journal of Pathology*, vol. 235, 2015, pp. 323-33.

US Department of Health and Human Services, Public Health Service, National Toxicology Program. *Eleventh Report on Carcinogens*. US Department of Health and Human Services, 2005.

Young, L. S., and A. B. Rickinson. "Epstein-Barr Virus: Forty Years On." *Nature Reviews: Cancer*, vol. 4, no. 10, pp. 757-68.

HERPES SIMPLEX VIRUS

Category: Pathogen

Also known as: HSV, HSV1, HSV2

Anatomy or system affected: Lips, pharynx, reproductive system, blood, cells, circulatory system, immune system, lymphatic system

Specialties and related fields: Dermatology, family practice, immunology, internal medicine, microbiology, neurology, obstetrics and gynecology, oncology, ophthalmology, pediatrics, virology

Definition: herpes simplex virus is an enveloped, double-stranded, deoxyribonucleic acid (DNA) virus that causes skin ulcers in infected persons

KEY TERMS

dorsal root ganglia: a collection of cell bodies of the afferent sensory fibers that lie between adjacent vertebrae

genital herpes: a disease characterized by blisters in the genital area, caused by a variety of the herpes simplex virus

herpetic whitlow: herpes simplex infection of the fingertips or nail beds

keratoconjunctivitis: inflammation of the cornea and conjunctiva of the eye

latent cycle: a type of herpes simplex infection in neurons that does not include destruction of the cell but dormancy

lytic cycle: the infective cycle of herpes simplex viruses in epithelial cells that includes the destruction of the host cells

trigeminal ganglion: the large flattened sensory root ganglion of the trigeminal nerve that lies within the skull and behind the orbit

RISK FACTORS

Herpes Simplex Virus-2 (HSV2) is sexually transmitted, making sexually active adolescents and adults at risk for infection. People with weakened immune systems, such as cancer patients, are at an increased risk of recurring HSV infection and disease.

ETIOLOGY AND THE DISEASE PROCESS

In general, Herpes Simplex Virus-1 (HSV1) tends to cause infections above the waist and HSV2 below the waist. However, both viruses can cause both types of infections. HSV1 & 2 infections are most contagious from virus-filled lesions. However, these viruses are also spread by shedding in saliva or genital secretions by people with no symptoms. Therefore, an infected person does not need apparent sores to transmit HSV1 & 2.

When herpes simplex viruses alight on the skin, they enter through cracks in the skin or mucosa and infect epithelial cells. Inside epithelial cells, herpes simplex viruses have their deoxyribonucleic acid (DNA) transcribed into messenger ribonucleic acids (mRNAs) that the cell's ribosomes translate into viral proteins. The viral proteins assemble with viral DNA molecules to form mature virions. This infective cycle culminates in the destruction of epithelial cells and the release of viruses that infect neighboring cells. This is called the "viral lytic cycle."

Eventually, HSV1 & 2 infect nearby sensory neurons and journey up the axon to the neuronal cell body. Sensory neurons of the genitalia have their cell bodies in the dorsal root ganglia of the sacrum. The trigeminal ganglia house the cells bodies of facial sensory neurons. In the neuronal cell body, HSV1 & 2 initiate the latent cycle. Instead of destroying these neurons, they become permanent homes for herpes simplex viruses. Over time, the neurons make a few copies of HSV1 & 2 that travel down the axons and infect overlying epithelial cells, producing localized blisters on the face, lips, or genitals.

Typically, blisters form only on one side of the face or body. A tingling or burning sensation may precede blister appearance. Blisters arise throughout a person's life, and stress, skin damage, viral infections, or other events may trigger them. Oral herpes usually affects children and causes blisters on the gums, lip, face, palate, fever, and enlarged lymph nodes. Blisters usually resolve after a few weeks. Older children and adults usually experience pharyngitis. The common manifestation of herpes is blisters on the lip border on one side of the face. These blisters heal within a week.

Genital herpes, at first, causes ulcers and pustules on the external genitalia of men and women. Reactivation usually causes a few blisters that resolve within a week.

Scratching herpes sores infects the fingertips of nail beds, causing herpetic whitlow. Infected fingertips can infect other regions of the body; a process

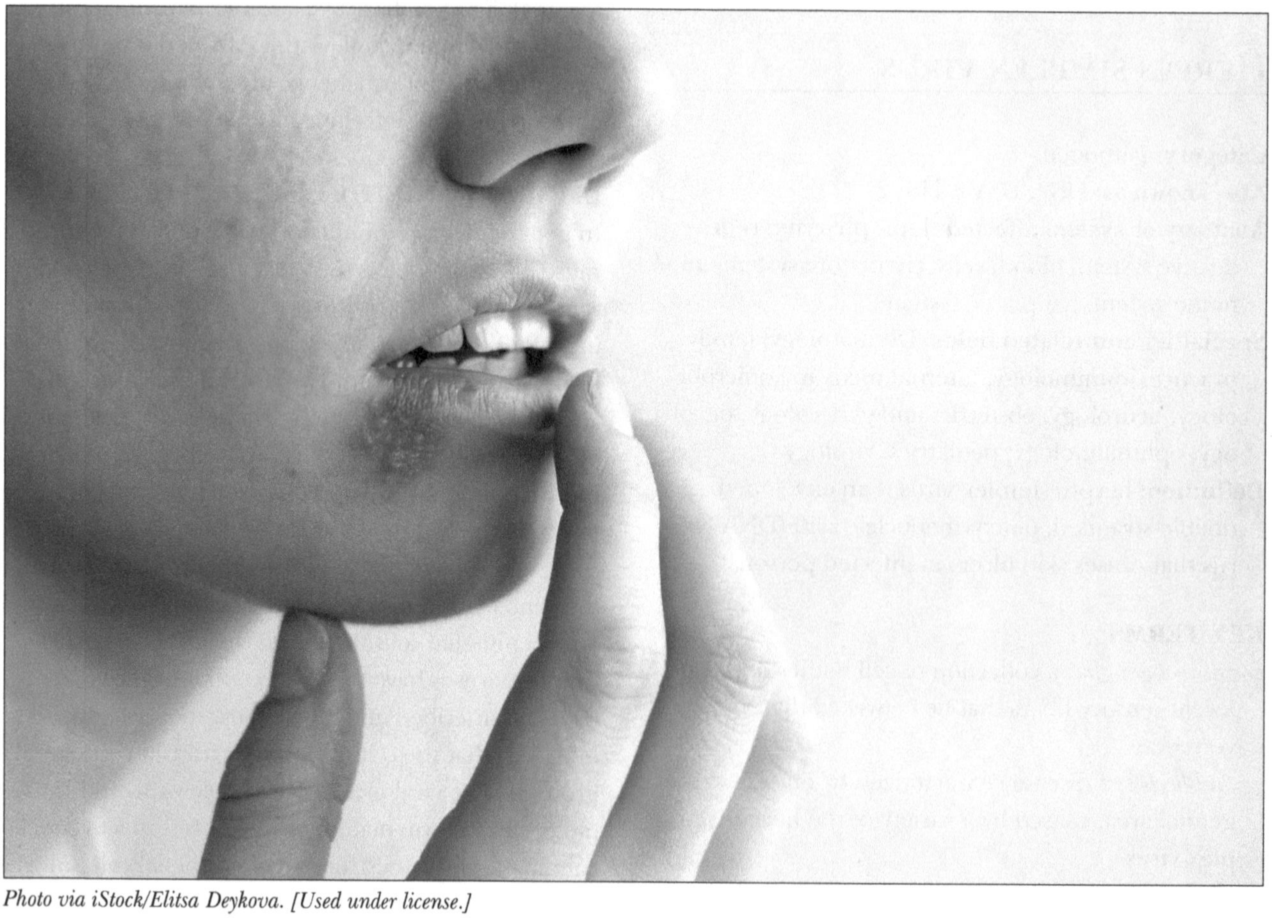

Photo via iStock/Elitsa Deykova. [Used under license.]

called "autoinfection." HSV1 & 2 can infect the trunk, head, or arms, and legs through autoinfection. Wrestlers can infect each other with herpes simplex due to their excessive skin-to-skin contact, a condition called "herpes gladiatorum." Eczema herpeticum occurs when individuals with burn injuries or atopic dermatitis suffer serious herpes infections.

Herpes simplex eye infections cause keratoconjunctivitis or inflammation of the cornea and conjunctiva (whites of the eye). The eyes suffer from pain, redness, sensitivity to light, blurry vision, and tearing. The cornea also has a branching dendritic lesion that is a hallmark of ocular herpes infections.

Rarely, HSV1 & 2 can spread to the central nervous system and cause meningitis or encephalitis. Usually, central nervous system infections result from the entry of the virus into the bloodstream.

Mothers can pass HSV1 & 2 to their babies but usually do so during delivery. Neonatal herpes simplex causes one of three conditions. The first is skin, eye, and mucous membrane infections that cause lesions to form wherever the skin gets damaged. Second is central nervous system involvement that causes irritability, seizures, and lethargy. If not treated, these infections can transition into herpes virus sepsis and organ failure.

Cancer, human immunodeficiency virus (HIV), acquired immunodeficiency syndrome (AIDS), and the use of medications (corticosteroids) that weaken the immune system may also trigger the reappearance of symptoms.

Some studies have suggested that women infected with both the herpes simplex virus and a high-risk type of human papillomavirus (HPV) have a greater likelihood of developing cervical cancer than women with HPV infection alone. However, HSV infection need not be present for cervical cancer to develop.

INCIDENCE

Infections with herpes simplex virus are ubiquitous and are transmitted from person to person whether they have symptoms. Most children will acquire an HSV1 infection during their first few years, usually through contact with infected saliva. In the United States, up to 90 percent of adults have antibodies to HSV1, and up to 30 percent have antibodies to HSV2.

SYMPTOMS

HSV infections in children beyond the neonatal (newborn) period, adolescents, and adults usually have no symptoms. HSV1 may cause fever (especially during the first episode), mouth sores (fever blisters), and enlarged lymph nodes in the neck or groin. HSV2 may cause genital lesions with a burning and tingling sensation, muscle pain, vaginal discharge, and difficulty urinating.

SCREENING AND DIAGNOSIS

HSV infections can be diagnosed by: (1) examination of the physical appearance of the skin lesions; (2) herpes culture; (3) using herpes specific antibodies (direct fluorescent antibody [DFA] test) to detect the virus; (4) screening the patient for HSV antibodies; (5) the detection of herpes DNA by the polymerase chain reaction (PCR); and (6) the Tzanck test (Tzanck smear) where the skin is scraped, fluid collected, the sample stained and microscopically examined for cells characteristic of herpes infection.

TREATMENT AND THERAPY

Mild cases of the disease may not require treatment. For more severe cases, analgesics such as ibuprofen and paracetamol (acetaminophen) can be used to reduce pain and fever. Topical anesthetics, including lidocaine, tetracaine, benzocaine, and prilocaine, can also be used to alleviate pain.

Antiherpes medications consist of guanosine analogs, including acyclovir, and drugs like acyclovir, valacyclovir, penciclovir, famciclovir, ganciclovir, and valganciclovir. After being taken up by cells, these drugs are phosphorylated by enzymes encoded by herpes viruses to monophosphate forms. The monophosphate forms are converted by the cell into triphosphate forms that are incorporated into replication viral DNA replication, terminating it.

Acyclovir is active against HSV 1 & 2 and varicella-zoster virus, which causes chickenpox. It comes in an oral, topical, and intravenous form. Acyclovir is used for mild mucocutaneous and genital lesions and prevents herpes simplex outbreaks in people with compromised immune systems. Intravenous acyclovir treats severe herpes infections. Other drugs, such as famciclovir (Famvir), penciclovir (Denavir, Vectavir, and Fenivir), and valacyclovir (Valtrex), can also be used to treat herpes infections.

PROGNOSIS, PREVENTION, AND OUTCOMES

Herpes has no cure. Recurrences, however, may be milder over time. HSV skin lesions usually heal on their own in seven to ten days, but they may take longer to heal in people with weakened immune systems. Genital HSV infection may be prevented by using condoms and by reducing the number of sexual partners. Condoms, however, do not always cover the whole infected area, and infection may still occur.

There is no licensed vaccine against herpes simplex virus, but several candidate vaccines have been studied.

—Diego Pineda, MS;
updated by Charles L. Vigue, PhD
and Michael A. Buratovich, PhD

Further Reading

Acton, Ashton, editor. *Herpes Simplex Virus: New Insights for the Healthcare Professional*. Scholarly Editions, 2013.

Committee on Infectious Diseases, American Academy of Pediatrics. "Herpes Simplex." *Red Book: 2006 Report of the Committee on Infectious Diseases*, edited by L. K. Pickering, C. J. Baker, S. S. Long, and J. A. McMillan, 27th ed., American Academy of Pediatrics, 2006.

Diefenbach, Russell J., and Cornel Fraefel, editors. *Herpes Simplex Virus: Methods and Protocols* (Methods in Molecular Biology.) Humana Press, 2014.

Flint, Jane, et al. *Principles of Virology*. 5th ed., ASM Press, 2020.

Howley, Peter, M., et al., editors. *Fields Virology: DNA Viruses*. 7th ed., Lippincott Williams & Wilkins, 2021.

Stanberry, Lawrence. *Understanding Herpes*. 2nd ed., UP of Mississippi, 2006.

Retroviruses

Category: Diseases and disorders
Anatomy or system affected: All
Specialties and related fields: Biochemistry, genetics, oncology, pathology, public health, virology
Definition: ribonucleic acid (RNA) viruses that replicate by synthesizing a double-stranded deoxyribonucleic acid (DNA) molecule that integrates into the host genome; they infect virtually all animals and sometimes cause serious diseases, including cancer

KEY TERMS

capsid: a virally encoded protein that surrounds and protects the viral RNA genome

envelope: a lipid bilayer membrane that surrounds the retrovirus particle

glycoprotein: a protein to which is attached one or more sugar molecules

integrase: a virally encoded enzyme that catalyzes the integration of viral double-stranded DNA into the host genome

matrix: the layer of virally encoded protein that surrounds the viral capsid

nucleoprotein: a virally encoded protein that is directly associated with the viral nucleic acid

oncogene: a gene or DNA segment that can cause cancer

oncogenic: having the potential to cause cancer, as in oncogenic retroviruses

positive-sense RNA (+RNA): an RNA molecule that can serve as a template for protein synthesis *reverse transcriptase*: an enzyme that synthesizes double-stranded DNA from single-stranded RNA

src: a gene found in Rous Sarcoma Virus that confers on the virus the ability to transform normal cells into cancer cells

tRNA: an RNA molecule that attaches to an amino acid and interacts with the ribosome during protein synthesis; in retroviruses, it serves as a primer for reverse transcriptase

BIOLOGY OF RETROVIRUSES

Retroviruses are members of the viral family Retroviridae. They are enveloped, positive sense (+) ribonucleic acid (RNA) viruses about 100 nanometers in diameter that replicate within the host's cytoplasm through a double-stranded deoxyribonucleic acid (DNA) intermediate that is integrated into the host genome. In addition to the +RNA, a cellular tRNA hydrogen-bonded to the +RNA serves as a primer for reverse transcriptase.

The viral RNA genome, associated nucleoprotein, reverse transcriptase, and integrase surround a protein capsid. Immediately external to the capsid is the matrix protein. The outer layer of the retrovirus is a lipid bilayer envelope derived from the host's plasma membrane that is acquired as the virus emerges from the host cell. Within the envelope are two glycoproteins encoded by the virus genome and serve as plasma membrane attachment sites during entry into the cell.

The retrovirus genome consists of two, seven kilobase to 11 kilobase +RNA molecules that code for only a few proteins, including *gag*, which codes for the matrix, capsid, and nucleoprotein; *pol*, which codes for reverse transcriptase, RNAse, integrase, and a protease; and *env*, which codes for the envelope glycoproteins.

The retrovirus binds to plasma membrane receptors via the viral envelope glycoproteins. When the retrovirus enters the cell, the viral RNA is released along with its reverse transcriptase. A double-stranded DNA is synthesized from the +RNA using viral reverse transcriptase. Integrase catalyzes the incorporation of the double-stranded DNA molecule into the host genome. When integrated, the viral DNA is referred to as a provirus and replicates with the host genome. Host RNA polymerase transcribes the viral genes, making copies of the viral genome and mRNA molecules translated into viral proteins. Viral RNA and proteins are assembled into new viral particles that emerge from the plasma membrane by budding.

Some retroviruses such as Rous sarcoma virus (RSV), feline leukemia virus (FLV), and mouse mammary tumor virus (MMTV) can induce tumors in their host species. More than twenty-five cancer-causing (oncogenic) retroviruses have been isolated. The retrovirus gains oncogenic potential when it inadvertently acquires a eukaryotic gene during infection. Although the eukaryotic gene may not be oncogenic when first acquired, after several generations, it may mutate or otherwise become altered, transforming it into one that is oncogenic. Retroviruses such as FLV that do not carry an oncogene can still transform by disrupting the function of a normal gene by integrating within it or by integrating next to it so that the neighboring gene can use the viral promoter, resulting in gene overexpression and cellular proliferation.

PERSPECTIVE AND PROSPECTS

The study of retroviruses dates to 1910 with the work of Peyton Rous, who discovered that certain sarcomas in chickens are caused by an agent later identified as a virus. The virus was later named Rous sarcoma virus. In 1970, the laboratories of Howard Temin and David Baltimore independently discovered that certain RNA viruses have an enzyme, now known as reverse transcriptase, that permits the viruses to reverse transcribe their RNA genomes into double-stranded DNA. In the early 1970s, the laboratory of J. Michael Bishop and Harold Varmus demonstrated that the Rous sarcoma virus has a gene, now known as *src*, re-

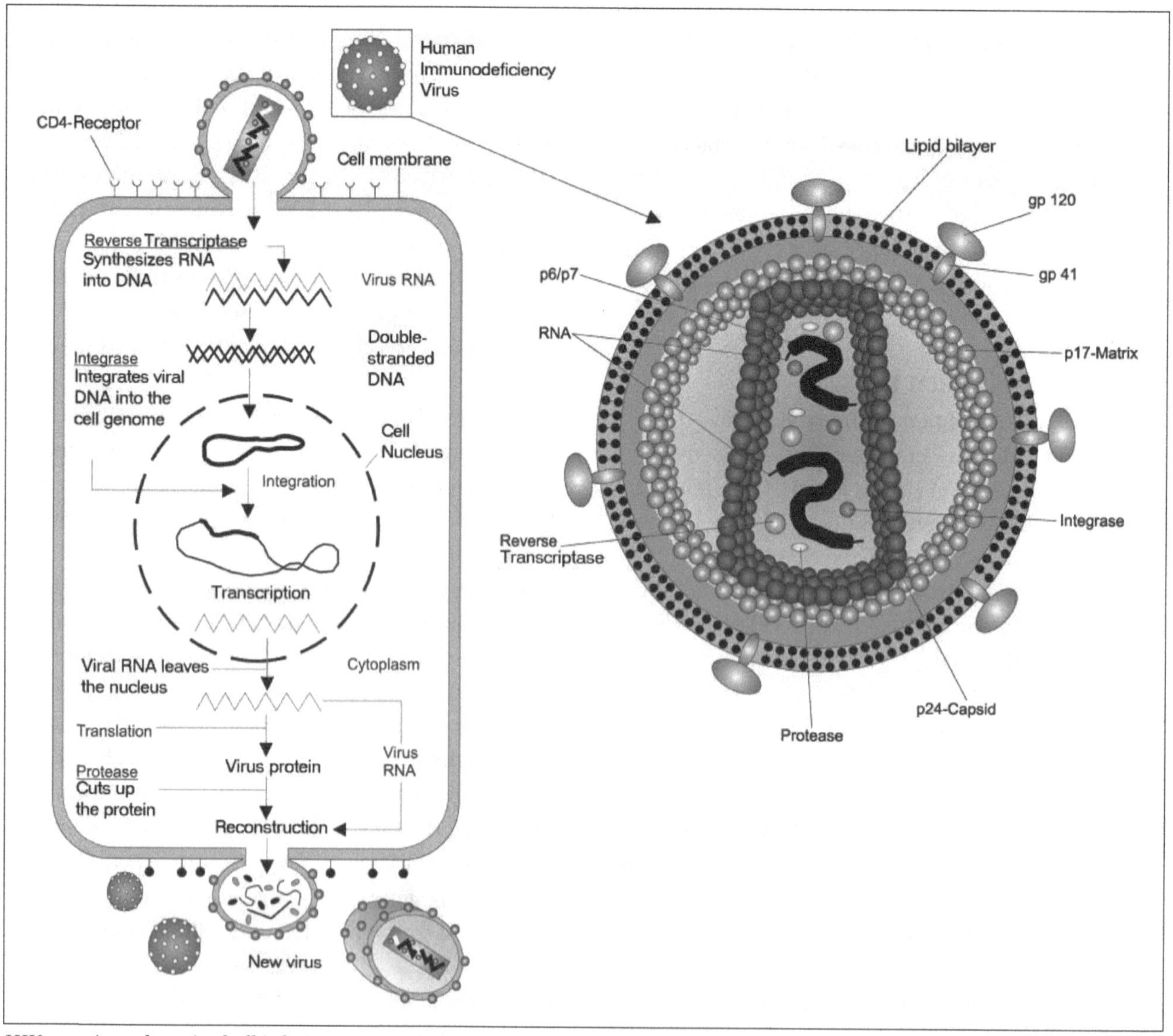

HIV retrovirus schematic of cell infection, virus production and virus structure. Image by Raul654, via Wikimedia Commons.

sponsible for transforming normal cells into tumor cells. Uninfected cells, including human cells, have a normal *src* gene related to the viral *src* gene. In the past, an RSV infected a chicken and incorporated the host *src* gene into its genome. The *src* gene acquired by the virus became altered over time so that it now causes cancer when an RSV infects a chicken cell.

There are many examples of retroviruses, including the human T-cell leukemia virus (HTLV), the first pathogenic human retrovirus discovered in 1980 by Bernard J. Poiesz, Robert Gallo, and their colleagues at the National Institutes of Health and by Mitsuaki Yoshida in Japan. Human immunodeficiency virus (HIV), which causes acquired immunodeficiency syndrome (AIDS), is also a retrovirus; it was discovered in 1983 by Luc Montagnier, Françoise Barré-Sinoussi, and their colleagues at the Pasteur Institute in France.

Since reverse transcriptase does not have the proofreading activities associated with DNA polymerase, retroviruses mutate and evolve more rapidly than DNA viruses, making the development of drugs and vaccines difficult.

Recombinant retroviruses are often used as vectors for genetic engineering. Retroviruses that are modified by removing the genes that make them harmful and replacing them with normal eukaryotic genes can deliver a normal copy of a gene to a defective cell. The DNA copy of the recombinant retrovirus can integrate into the host genome and genetically modify the cell.

—Charles L. Vigue, PhD

Further Reading

Cullen, Bryan R. *Human Retroviruses.* Oxford UP, 1993.

Dudley, Jaquelin. *Retroviruses and Insights into Cancer.* Springer, 2011.

Gallo, Robert. *Virus Hunting: AIDS, Cancer, and the Human Retrovirus-A Story of Scientific Discovery.* Basic Books, 1991

Gallo, Robert C., Dominique Stehelin, and Oliviero E. Varnier. *Retroviruses and Human Pathology.* Humana Press, 1986.

Holmes, Edward C. *The Evolution and Emergence of RNA Viruses.* Oxford UP, 2009.

Kurth, Reinhard, and Norbert Bannert, editors. *Retroviruses: Molecular Biology, Genomics, and Pathogenesis.* Caister Academic Press, 2010.

Singh, Sunit K., and Daniel Ruzek, editors. *Neuroviral Infections: RNA Viruses and Retroviruses.* CRC Press, 2013.

Polio

Category: Pathogen

Also known as: Poliomyelitis

Anatomy or system affected: Central nervous system, gastrointestinal tract, peripheral nerves, spinal cord

Specialties and related fields: Family practice, gastroenterology, immunology, microbiology, neurology, occupational therapy, orthopedics, pediatrics, physical therapy, preventive medicine, public health, virology

Definition: a viral disease that enters the body through the intestines and spreads and causes nerve injury in the spinal cord

KEY TERMS

atrophy: a decrease in the size or functionality of a body part, cell, organ, or other tissue

motor neurons: nerve cells forming part of a pathway along which impulses pass from the brain or spinal cord to a muscle or gland

paralysis: the loss of the ability to move (and sometimes to feel anything) in part or most of the body, typically because of illness, poison, or injury

peripheral nerves: one of two components that make up the nervous system; one being the central nervous system and the other, the peripheral nervous system consisting of the nerves and ganglia outside the brain and spinal cord

postpolio syndrome: a disorder of the nerves and muscles that occurs in some people many years after they have had polio

protease: an enzyme that breaks down proteins and peptides

spinal cord: a column of nerve tissue that runs from the base of the skull down the center of the back

trophic factors: helper molecules that allow a neuron to develop and maintain connections with its neighbors

INTRODUCTION

Polio or poliomyelitis is a viral disease that produced hysteria and fear among North Americans, dreading its transmission during the 1950s.

Before the 1950s, hundreds of thousands of North Americans were paralyzed or killed by polio, whose onset appeared to be an ordinary summer flu. The occurrence of the disease terrified millions of people and kept social activities that could lead to exposure at a minimum. During the decade, the development of the polio vaccines nearly eradicated the disease and increased public confidence in biomedical science.

Poliovirus is a small ribonucleic acid (RNA) virus that is part of the Picornavirus family. A subgroup of the Picornaviruses is the enteroviruses. Enteroviruses infect the gastrointestinal tract of humans and other animals and are divided into four groups based on RNA sequence homology: A, B, C, and D. Poliovirus is included in the human enterovirus C group. There are three serotypes of poliovirus, PV1, PV2, and PV3. PV1 causes most paralytic diseases. Other enteroviruses can also cause paralytic disease.

Poliovirus is a single-stranded, plus-sense RNA virus whose RNA genome is encased in a protein coat called a "capsid." The poliovirus genome is a plus-sense RNA because once the virus enters human cells, its RNA immediately serves as a messenger RNA that is translated into viral proteins. This virus usually infects children under five and is spread by fecal-oral transmission. Poliovirus enters the body through contaminated food and water. Sneezing or coughing can spread thousands of virus-laden droplets that enter the mouth.

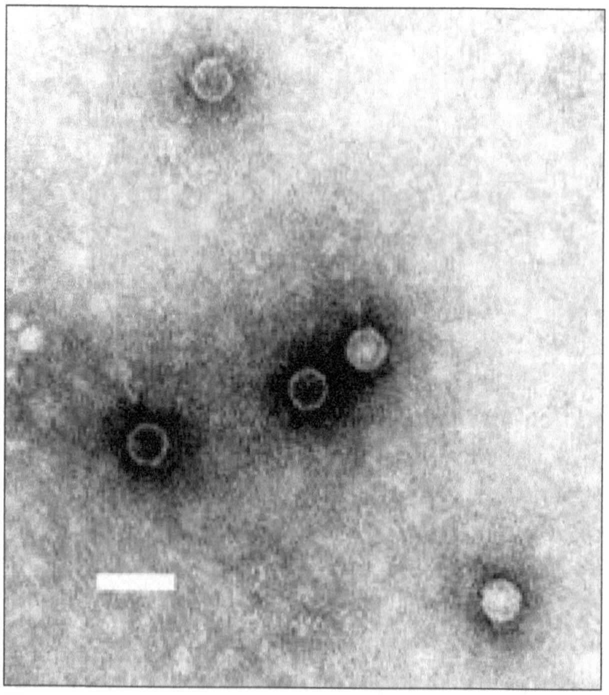

Electron microscopic image of poliovirus, via Wikimedia Commons.

Upon entering the body, poliovirus infects the oropharynx and small intestine mucosal cells. Host cell ribosomes translate the RNA to make a large "polyprotein." This long polypeptide contains all the viral structural and nonstructural proteins fused into one large protein. Internal proteases (enzymes that degrade proteins) clip the individual proteins from the polyprotein. These proteins include viral capsid proteins (VP1, VP2, VP3, and VP4), the RNA-dependent RNA polymerase complex, and proteases. These viral proteins hijack the host cell's machinery and force it to make viral proteins and RNAs. These elements self-assemble into poliovirus virions that fill the host cell, eventually causing it to burst. The newly formed viruses infect nearby lymph nodes and the bloodstream.

Blood-borne polioviruses leave blood vessels for the interstitial tissue of skeletal muscles, where they preferentially infect motor nerves. Initially, the virus infects the axon of the motor neurons innervating the muscle. It travels retrograde (backward) up the axon to the neuronal cell body. Motor neurons are found in the anterior horn of the spinal cord. The infected motor neurons attract white blood cells like neutrophils and macrophages, causing inflammation and damage to the spinal cord. Infected motor neurons ultimately die, and the muscles they innervated fail to receive contraction signals and trophic factors that keep the muscles healthy. The muscles weaken and atrophy.

Rarely, poliovirus infects the brainstem (bulbar polio) and paralyzes the motor nerves that direct swallowing and speaking. Brain stem infections can also destroy the nerves that control the diaphragm and impair breathing.

After the initial infection, some people develop postpolio syndrome (PPS) decades afterward. When poliovirus destroys some motor neurons, unaffected, neighboring neurons form collateral branches to innervate any muscles that lost their innervating motor neurons. Since motor neurons die as the person gets older, larger muscle groups lose their innervation and become paralyzed.

Approximately 90 to 95 percent of poliovirus infections cause symptoms. Less than ten percent of infections cause symptoms, including headache, sore throat, fever, nausea, vomiting, malaise, and fatigue. Poliovirus only causes paralytic disease about one percent of the time. Paralytic disease causes high fever, intense muscle spasms that cause pain and weakness, loss of muscle reflexes that culminate in paralysis. Paralysis develops over a few days, but it usually affects one side more than the other. Paralysis also tends to affect larger muscles in the lower part of the body, like the legs and the thighs. Infants with poliomyelitis develop flaccid paralysis that causes them to go limp, a condition called "floppy baby syndrome." Destruction of the phrenic nerves that innervate the diaphragm can cause death by suffocation.

Initially, physicians thought that polio only affected children and always caused paralysis. However, subsequent investigations demonstrated that people of any age could catch poliovirus and that its victims were not always paralyzed. Historians suggest polio has afflicted humans since ancient times because of biblical depictions of people with withered limbs (e.g., Mephibosheth, 2 Samuel 9; lame man in the temple healed by Peter, Acts 3). In modern times, polio was first described in 1840 by the German physician Jacob von Heine. In 1902, Karl Landsteiner discovered three poliovirus strains: Brunhilde, Lansing, and Leon.

VACCINES

At the beginning of the 1950s, the continuation of worldwide polio epidemics that began during the 1940s became a medical crisis in the United States. In 1950, for example, 3,400 severe cases occurred. In 1951, even more cases developed, and by 1952, there were 58,000 cases. That year, polio epidemics occurred in Europe, China, India, Japan, Korea, and the Philippines.

Man with atrophy and paralysis of the right leg and foot due to polio. Photo via Wikimedia Commons. [Public domain.]

For Americans, the aftermath of polio was painfully clear in photos of President Franklin D. Roosevelt. He contracted the disease in 1921 and had to wear heavy steel braces on his legs. Walking was so difficult for him that he spent most of his time in a wheelchair. During the late 1940s and early 1950s, as the polio epidemic was at its worst, posters advertising the March of Dimes contained heartbreaking pictures of children on crutches and in iron lungs—the cumbersome mechanical aids that enabled polio victims with paralyzed rib cages to breathe. Before a polio

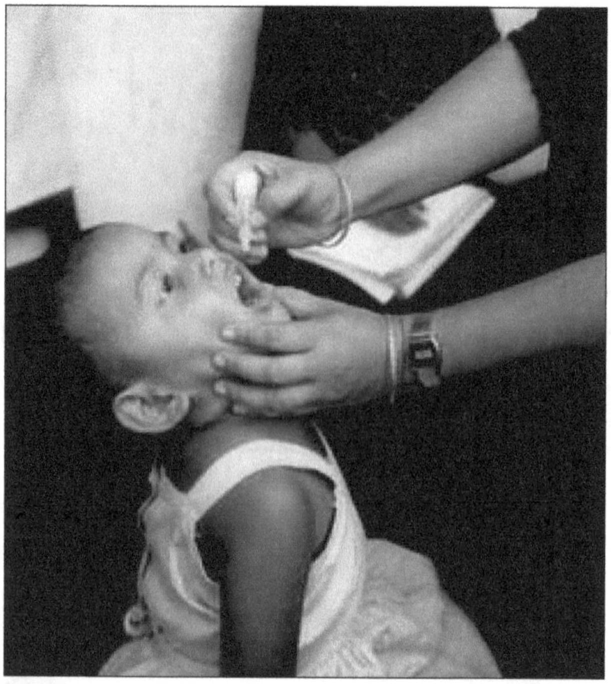

A child receiving an oral polio vaccine. Photo via USAID/Wikimedia Commons. [Public domain.]

vaccine became available, many parents throughout the United States and Canada panicked, keeping their children away from schools and other public facilities, such as movie theaters and swimming pools.

To counter this epidemiological crisis, the National Foundation for Infantile Paralysis funded research that led to the development of vaccines by Jonas E. Salk and Albert B. Sabin during the decade. A major problem that had slowed vaccine development—the inability of polioviruses to grow outside live cells—was obviated by 1949, when Harvard researchers John Enders, Frederick C. Robbins, and Thomas H. Weller developed means to grow the poliovirus outside the body. Using their technique, Salk, a University of Pittsburgh physician, developed an injected vaccine made from killed polioviruses of all three strains.

After 1954, field trials on approximately two million American children—a scope unprecedented in US medical history—began. The results were immediate, and the vaccine was pronounced safe and effective for widespread use. In 1955, mass inoculation began around the United States and the world. At the end of the 1950s, a live and oral form of the vaccine was developed by Sabin to counter problems with the Salk vaccine. The Salk and Sabin vaccines (the latter was not used until the 1960s) protected against all poliovirus types. The number of polio cases declined dramatically after vaccines began being used. In 2021, the World Health Organization (WHO) reported just over 500 cases of polio globally.

IMPACT

The 1950s saw near eradication of this catastrophic disease that had killed or disabled millions of people worldwide. Many historians deem Salk's killed vaccine and Sabin's live vaccine among the most important medical discoveries of the twentieth century. Polio still occurred by the end of the 1950s, although rarely. However, people no longer feared the summer months, when polio infections had been most numerous. Also, the development of the tissue culture methods of Enders, Robbins, and Weller opened the door for later use of mammalian tissue cultures for advances in molecular biology that identified viral causes of cancer and acquired immunodeficiency syndrome (AIDS).

SUBSEQUENT EVENTS

Sabin's oral vaccine, licensed in 1961, quickly replaced Salk's injected vaccine as the standard US polio immunizing agent because of easy administration and the belief that its live viruses would work longer than Salk's killed viruses. Despite the vaccine's effectiveness, data reported in the Morbidity and Mortality Weekly Report showed polio vulnerability of unimmunized people to be quite high throughout the world after the 1950s. In 1979, this was made painfully obvious by epidemics among unvaccinated Amish Americans. Retrospective studies by the WHO, published in 1994, showed that once the vaccines were widely used, the Western Hemisphere quickly became free of naturally occurring polio.

Continued and minor paralytic polio outbreaks after the late 1950s occurred in situations where virus attenuation was incomplete; these outbreaks caused a rethinking of the vaccine use. Reliance on the Sabin vaccine decreased in stages as time went by. In 1997, to reduce vaccine-related polio, the US Centers for Disease Control and Prevention (CDC) recommended that all children get two doses of injected (killed) vaccine, followed by two oral doses of attenuated vaccine. In 2000, to eliminate risks of vaccine-related polio, the CDC revised its recommendation to four injections of the killed virus by age four.

—Sanford S. Singer and Michael A. Buratovich, PhD

Further Reading

Daniel, Thomas M., and Frederick C. Robbins, editors. *Polio*. U of Rochester P, 1997.

Gould, Tony. *A Summer Plague: Polio and Its Survivors*. Yale UP, 1995.

Mbaeyi, Chukwuma, et al. "Progress Toward Poliomyelitis Eradication—Pakistan, Jan. 2020-July 2021." *Morbidity and Mortality Weekly Report*, vol. 70, no. 39, 2021, pp. 1359-64, doi:10.15585/mmwr.mm7039a1.

Sherrow, Victoria. *Jonas Salk*. Facts On File, 1993.

World Health Organization GPEI. "Wild Poliovirus Weekly Update," 30 Nov. 2021, www.polioeradication.org/Dataandmonitoring/Poliothisweek.aspx. Accessed on 7 Dec. 2021.

INFLUENZA

Category: Diseases and conditions

Also known as: The flu, grip, grippe, seasonal flu

Anatomy or system affected: Lungs, muscles, nose, respiratory system, throat

Specialties and related fields: Immunology, microbiology, pediatrics, public health, pulmonary medicine, virology

Definition: a viral disease that affects the respiratory system

KEY TERMS

antigenic drift: the gradual accumulation of point mutations during the circulation of influenza virus. It results from the high error rates associated with ribonucleic acid (RNA)-dependent RNA polymerase during virus replication

antigenic shift: a process by which two or more different strains of a virus combine to form a new strain that has a mixture of the surface antigens of the two or more original strains

hemagglutinin: a surface glycoprotein of the influenza virus that causes red blood cells to clump together.

neuraminidase: an enzyme on the surface of influenza viruses that catalyzes the breakdown of complex sugars that contain neuraminic acid.

orthomyxovirus: a family of negative-sense RNA viruses that infect animal species

Reye's syndrome: a rare but serious condition characterized by the swelling in the liver and brain, caused when children or adolescents take aspirin or other nonsteroidal anti-inflammatory drugs while infected by viruses

INTRODUCTION

Influenza (commonly known as the flu) is a disease that affects the respiratory system. It is caused by a variety of viruses in the Orthomyxovirus family. Influenza infections are not unique to people; they also occur in other animals, most notably birds and pigs. Infection with an influenza virus leads to illness that can be mild or life-threatening, depending on the person's age, general health, and immunity to the particular infecting virus. Every year, the influenza viruses that infect people can differ from those that infected people the previous year.

CAUSES

There are two significant types of influenza viruses: A and B (influenza virus type C causes minor infections). Each influenza A or B virus carries on its outer surface two types of protein: hemagglutinin (H) and neuraminidase (N). Influenza A viruses are classified into subtypes based on the type of HA and NA proteins they carry. There are sixteen types of HA and nine types of NA. When scientists talk about H1N1 influenza, they mean an influenza type A virus that carries HA type 1 and NA type 1 on its surface.

Influenza B viruses, and influenza A subtypes, are further classified into strains. There are hundreds of influenza virus strains, but not all can infect people.

The genes that code for the H and N proteins tend to mutate (change) somewhat each year. The accumulation of these mutations as influenza circulates causes "antigenic drift." Antigenic drift changes the virus enough so that it reduces a person's natural immunity to it. Antigenic drift explains why new flu vaccines must be made each year.

Every few decades or so, an influenza A virus will undergo an "antigenic shift." This significant change in the virus leads to the appearance of a completely new flu virus, against which people have no immunity. The emergence of H1N1 influenza in 2009 probably resulted from such a shift. Viruses that appear because of antigenic shifts may cause pandemics (worldwide epidemics), as did the 2009 H1N1 influenza virus. (The word "pandemic" does not mean "severe illness." It means the infecting microbe can easily cause illness that spreads across the globe.)

Viruses are generally specific to a species. Thus, the bird flu (avian influenza) virus usually cannot cause infection in a human. There have been several cases, however, in which bird flu viruses have infected humans. The best-known avian influenza virus is H5N1, which has caused more than five hundred confirmed cases in humans. Of these cases, 297 were fatal, making H5N1 the deadliest bird flu virus in humans.

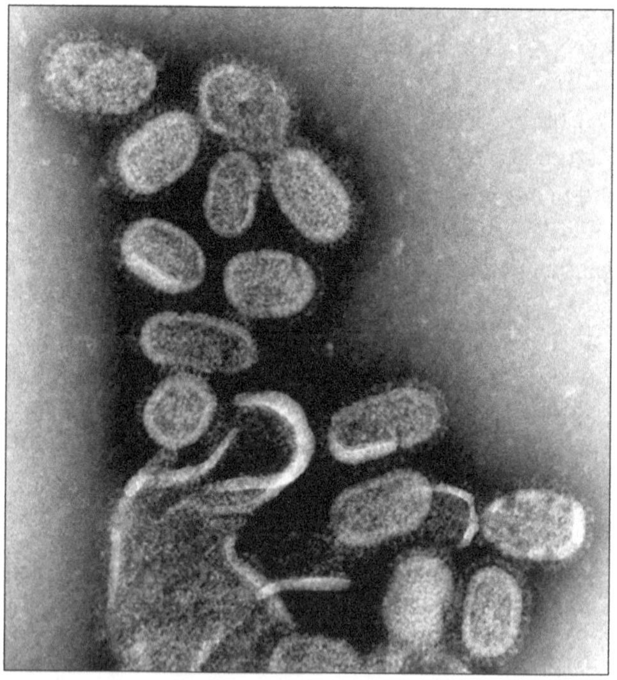

Influenza virus, magnified approximately 100,000 times. Photo via Wikimedia Commons [Public domain.]

The virus is transmitted to humans only by handling sick or uncooked dead birds. Health authorities worldwide remain concerned that if the virus develops the ability to jump among people (instead of, only, from birds to people), it will cause a major pandemic with many deaths.

RISK FACTORS

For the seasonal flu, people younger than age five or older than sixty-five years are most at risk of contracting the flu, as are health-care workers. Crowding increases the risk of virus transmission between people.

In addition, several groups of people are at high risk for complications from the flu. According to the Centers for Disease Control and Prevention (CDC), high-risk groups include pregnant women, people with certain chronic medical conditions (e.g., heart disease or diabetes), people whose immune system is weakened or suppressed, young children, and people older than age fifty years.

H5N1 avian influenza remains a problem in certain parts of the world. People living or traveling in areas where the virus is active are at risk if they handle sick birds or eat uncooked birds infected.

SYMPTOMS

It can take up to four days (in adults) from the infection until symptoms appear. The classic symptoms of the flu are fever and chills, sore throat, cough, runny nose, muscle aches, and headache. The headache can be severe enough to cause sensitivity to light. Muscle aches are most common in the legs, though they can appear anywhere in the body. Extreme fatigue is another common symptom.

Nausea, vomiting, and diarrhea can occur in people with the flu and are especially common in children and people infected with the 2009 H1N1 flu strain. Most flu symptoms disappear in five to six

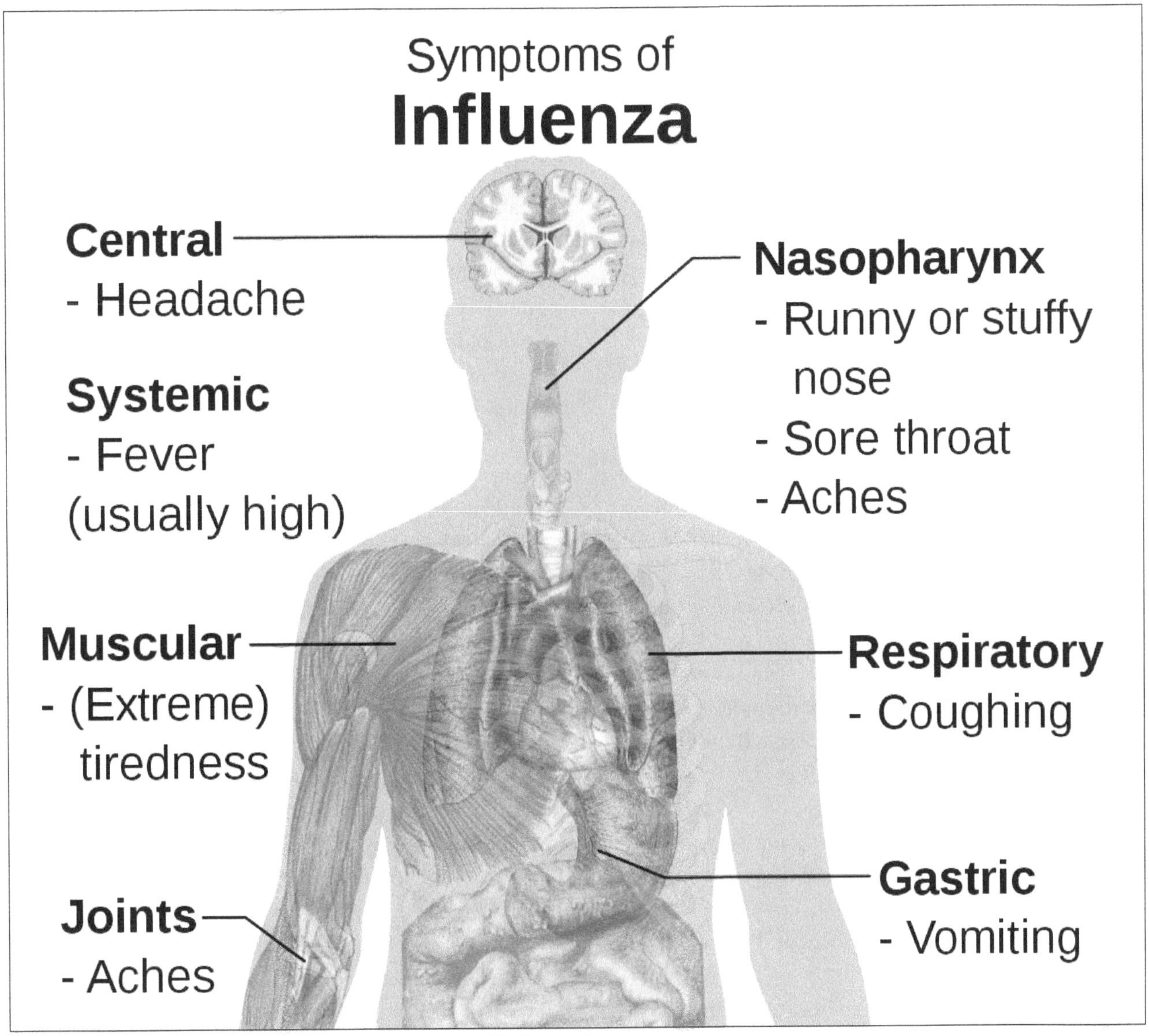

Symptoms of influenza. Image via Wikimedia Commons. [Public domain.]

days, though full recovery takes longer; the fatigue may last several weeks.

Pneumonia is a common complication of influenza. It can be primary (caused by the flu virus) or secondary (caused by another virus or bacteria). Because influenza weakens the body and its immune system, infections by other microbes can colonize a person who is fighting the flu. Symptoms of pneumonia include a cough that gets worse instead of better, difficulty breathing, and, sometimes, bloody phlegm. A person who is recovering from the flu and redevelops fever and cough most likely has bacterial pneumonia.

People with chronic medical conditions should watch for signs that their condition is worsening because of the flu. Severe disease, unfortunately, is not uncommon, especially in people with heart disease or respiratory conditions such as asthma or emphysema.

SCREENING AND DIAGNOSIS

Most of the time, the flu is inferred from the symptoms, and no special testing is required. There are some situations in which knowing the exact subtype of flu virus can influence treatment decisions. Sometimes doctors need to determine if an outbreak of respiratory illness in the population was due to influenza. For that purpose, rapid testing is available.

There are eleven approved rapid tests in the United States. These tests give results in fifteen minutes, but their sensitivity and accuracy vary. Rapid testing is usually done using a swab from nose or throat secretions. (The location of the swab may also affect the test's accuracy in some tests.) Rapid testing can be done only within the first four days of symptom appearance.

The most accurate way of testing for the specific type of flu virus is through reverse transcription-polymerase chain reaction (RT-PCR). Testing with RT-PCR can take up to four hours and is not always available for diagnostic tests.

Medical technologists can culture the influenza virus from swabs taken from affected persons. In a viral culture, the virus obtained from the persons is allowed to multiply in the laboratory, where large quantities allow for typing. Viral cultures are not used to determine treatment because they take three to ten days to grow and provide results. However, viral cultures can determine what type of flu is circulating in a given population.

A test for the presence of the H5 flu virus is available to state and public health authorities. The test, known as influenza A/H5 (Asian lineage) virus real-time RT-PCR primer and probe set, is available when suspected human cases of avian influenza appear in the United States. It takes four hours to get the results. If the H5 virus is detected, further testing is needed to determine if the virus is indeed the H5N1 avian flu virus.

TREATMENT AND THERAPY

For most people who are otherwise healthy, the treatment of influenza consists of treating the symptoms. Treatment includes pain relievers for body aches and headaches and medicine to reduce fever. Many over-the-counter (OTC) multisymptom flu treatments are available. They treat the worst cold symptoms and can bring relief, though they will not cure the flu. OTC products contain a mixture of medications. To avoid overdosing, one should know what medicines the OTCs contain. For example, many OTC products contain acetaminophen, the active ingredient in Tylenol. People who take acetaminophen in addition to multisymptom OTC treatments risk building up a dangerous level of acetaminophen in their bodies.

Children younger than eighteen years of age who might have influenza should not take aspirin. Aspirin in children can cause Reye's syndrome, a potentially fatal disorder that often follows a viral infection. Medications against the flu virus are called "antiviral medications." Two classes of antivirals are available against the flu virus: Neuraminidase inhibitors are effective against influenza A and B. They interfere with the release of the virus from infected cells. Three drugs are available in this class: oseltamivir (Tamiflu), zanamivir (Relenza), and Peramivir (Rapivab). Two drugs are available in this class: amantadine (Symmetrel) and rimantadine (Flumadine). Amantadines are effective against (some) influenza A viruses only, and viral resistance to this class of antivirals is high. Finally, baloxavir (Xofluza) is the first polymerase acidic endonuclease inhibitor class of anti-influenza drugs. This drug inhibits influenza ribonucleic acid (RNA) replication.

Taking these medications within the first forty-eight hours after symptoms appear will reduce the length and severity of the symptoms. Treatment with antiviral drugs is essential in people at high risk for complications. This type of treatment reduces or prevents such complications. Antiviral drugs can also prevent the flu if a person has been exposed to it. However, these medications are not substitutes for influenza vaccines.

Of the neuraminidase inhibitors, zanamivir is given through an inhaler. Because inhaling the medicine can cause intense airway spasms, zanamivir is not recommended for people with certain airway diseases, such as asthma. Use of the inhaler can be problematic for older adults or people with certain physical or mental limitations.

Oseltamivir is Food and Drug Administration (FDA)-approved for treating uncomplicated acute influenza in anyone older than two years and for influenza prophylaxis in patients at least one-year-old. It is the preferred treatment of influenza in pregnant women, hospitalized patients, and outpatients with a severe, complicated, or progressive illness. To mitigate the stomach upset that oseltamivir causes, patients should take this drug with food. Finally, this medication has a bitter taste that puts children off. Mixing the contents of oseltamivir capsules with a thick, sweetened liquid makes it more attractive to children.

Zanamivir is FDA-approved for treating uncomplicated acute influenza in patients at least seven years old and influenza prophylaxis in patients at least five years old. Zanamivir is contraindicated in patients with milk protein allergies. Nor is it recommended for use in patients with severe influenza or underlying airway diseases.

Peramivir is given intravenously and is FDA-approved for treating uncomplicated acute influenza in otherwise healthy patients two years old or older. It is not approved for influenza prevention. Baloxavir is FDA-approved for the treatment of uncomplicated acute influenza in patients at least twelve years old. It also protects those who were recently exposed (within forty-eight hours) to influenza. Taking baloxavir with dairy products, calcium-fortified beverages, or laxatives prevents drug absorption.

Amantadine and rimantadine are approved for the prevention of flu in people one year of age and older. Amantadine is also approved for flu treatment in persons one year and older. Rimantadine is approved for treating persons aged thirteen years and older. Because of the widespread resistance of extant influenza virus strains to these drugs, they are rarely prescribed.

While drug resistance to amantadines has been a growing problem, resistance to oseltamivir is a newer phenomenon. Because oseltamivir is the most used antiviral flu treatment, resistance is a worrisome development. It is, therefore, more important than ever to limit the use of antiviral flu drugs to high-risk groups.

People often ask about taking elderberry for prevention and treatment of influenza. Drug stores and chain supermarkets sell elderberry-containing products that are promoted for relief of cold and flu symptoms and as an immune system booster. However, there is no acceptable evidence that elderberry is effective for prevention or treatment of influenza. Furthermore, its safety is unclear. The bark, stems, leaves, and root of the elderberry plant contain a compound called "sambunigrin," which can release cyanide.

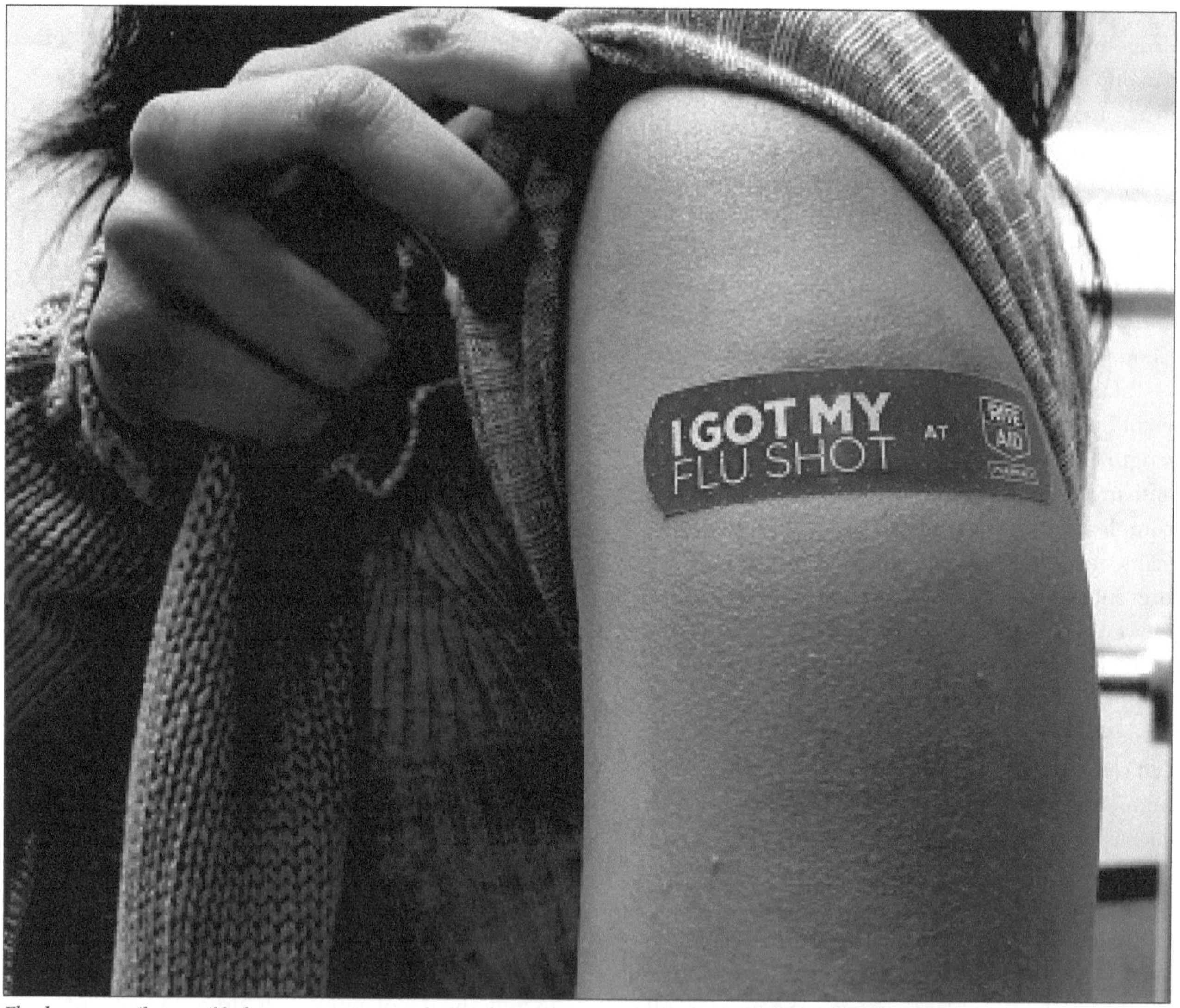

Flu shots are easily accessible for most Americans as they can be administered in pharmacies/drugstores. It is important to immunize annually, as different flu viruses may emerge or be used over time. Photo by Whoisjohngalt, via Wikimedia Commons.

PREVENTION AND OUTCOMES

Vaccination is the best protection against the flu. In early 2010, the CDC's Advisory Committee on Immunization Practices recommended a universal influenza vaccine every year for everyone age six months and older. (The previous recommendation called for yearly vaccinations for children six months to eighteen years of age and certain high-risk groups.)

Because the flu viruses that circulate in the population change every year, getting the flu vaccine each year is crucial. The vaccines change each year according to early testing results that show what virus subtypes are starting to appear. Vaccination is vital in people who are at high risk of severe complications from influenza. It is also important that people who care for or live with a person in any risk group be vaccinated to prevent giving the disease to the high-risk person. Health-care workers are also strongly encouraged to receive the vaccine every year to protect themselves and their patients.

There are four types of influenza vaccines:

- a killed virus vaccine grown in eggs and given by injection
- a live, weakened virus given as a nasal spray
- a killed virus vaccine grown in cultured cells
- and a recombinant influenza vaccine containing three times the amount of antigen used in older people.

The live virus vaccine is given to healthy (nonpregnant) persons between two and forty-nine years. The vaccine is marketed as FluMist or LAIV (live attenuated influenza vaccine). Side effects from the injected vaccine are usually mild and include redness and soreness in the injection area. Allergic reactions to the vaccine may also occur, though they are uncommon. Formerly, health-care workers advised against people allergic to eggs from receiving the injected flu vaccine. However, research has established people with egg allergies can safely receive egg-grown inactivated influenza vaccines. On rare occasions, some people who received the injected flu vaccine developed a paralysis disorder known as "Guillain-Barré syndrome."

Regardless of the type of vaccine, people have no protection against the flu until approximately two weeks after vaccination. People at high risk for flu complications (who receive the injected, killed vaccine) may be given antiviral drugs during the two weeks. The live vaccine can cause mild flu-like symptoms for several days.

Good hygiene is an integral part of protection against the flu. Washing hands frequently or using alcohol-based hand sanitizers will reduce the risk of getting the flu. It is imperative to wash hands before eating and touching areas on the face, especially the nose and mouth. People should be sure to wash their hands after blowing their nose or coughing into their hands. Covering the nose and mouth while coughing or sneezing reduces the risk of spreading influenza virus particles through the air.

—Adi R. Ferrara, BS, ELS and Michael A. Buratovich, PhD

Further Reading

"Antiviral Drugs for Influenza for 2020-2021." *Medical Letter on Drugs and Therapeutics*, vol. 62, no. 1610, 2020, pp. 169-73.

Barry, John M. *The Great Influenza: The Story of the Deadliest Pandemic in History*. Viking Penguin, 2005.

Beigel, John, and Mike Bray. "Current and Future Antiviral Therapy of Severe Seasonal and Avian Influenza." *Antiviral Research*, vol. 78, 2008, pp. 91-102.

"Elderberry for Influenza." *The Medical Letter on Drugs and Therapeutics*, vol. 61, no. 1566, 2019, p. 32.

Flint, Jane, et al. *Principles of Virology, Volume 1: Molecular Biology*. 5th ed., ASM Press, 2020.

"Influenza." *The Merck Manual Home Health Handbook*, edited by Robert S. Porter et al., 3rd ed., Merck Research Laboratories, 2009.

"Influenza Vaccines for 2020-2021." *Medical Letter on Drugs and Therapeutics*, vol. 62, no. 1607, 2020, pp. 145-50.

Strauss, James, and Ellen Strauss. *Viruses and Human Disease*. 2nd ed., Academic Press/Elsevier, 2008.

MEASLES

Also known as: Rubeola

Category: Diseases and conditions

Anatomy or system affected: All

Specialties and related fields: Dermatology, epidemiology, hematology, immunology, internal medicine, microbiology, neurology, ophthalmology, otorhinolaryngology, pulmonary medicine, virology

Definition: a highly contagious viral infection that causes fever, cough, and a rash

KEY TERMS

coryza: runny nose caused by hay fever or infections

exanthem: a widespread rash on the skin that usually occurs in children

Koplik spots: small, white spots on the inside of the cheeks early during measles

macular rash: a rash consisting of flat lesions up to five millimeters in diameter

maculopapular rash: a rash with a mixture of macules and papules

papular rash: a rash consisting of raises bumps up to one millimeter in diameter

rash: an area of irritated or swollen skin characterized by changes in the color, feeling, or texture of skin that might be itchy, red, painful, and irritated

subacute sclerosing panencephalitis (SSPE): a very rare but fatal central nervous system disease caused by measles virus infection

CAUSES

Measles is caused by the measles virus, a ribonucleic acid (RNA) virus that is a member of the morbilliviruses. Morbilliviruses are a part of a larger group of animal viruses called the "paramyxoviruses." Paramyxoviruses are negative-strand RNA viruses. When negative-strand RNA viruses infect cells, the viral RNA is replicated before any viral proteins are synthesized.

The measles virus is highly contagious and spreads by inhaling virus-laden aerosols generated by sneezing, coughing, or touching contaminated surfaces. The measles virus can survive for up to two hours on surfaces or in airborne aerosols. Inhaling aerosolized measles virus or touching contaminated surfaces followed by touching your nose, eyes, or mouth causes infection. Measles is so contagious that ninety percent of unvaccinated people nearby an infected person will also become infected.

INFECTION

When the measles virus enters the respiratory tract of a susceptible person, it infects the epithelial cells in the trachea and bronchi. Measles virus is an enveloped virus with two main surface glycoproteins that help it infect human cells. The H (hemagglutinin) protein binds to the CD46 protein on the surfaces of all nucleated cells. Once bound, the F (fusion) protein drives fusion of the measles virus envelop with the host cell membrane, bringing the virus into the cell. Inside cells, the measles virus uncoats, and its RNA is replicated to make messenger RNAs. The host cell's machinery translates the measles virus mRNAs to form viral proteins. The viral proteins assemble with viral RNAs to form mature viral particles that bud from the host cells to infect other cells.

Days after infection, the measles virus spreads through local tissues to dendritic cells and alveolar macrophages. These two antigen-presenting cells take the virus to local lymph nodes, where the lymphocytes form an immune response to it. The virus continues to spread from respiratory tissues and enters the blood, lung tissue, intestines, and brain.

Ten to fourteen days after the initial infection, the first symptoms appear, and this period is the incubation period.

Symptoms begin with the three Cs: cough, conjunctivitis (redness of the whites of the eyes), and coryza (stuffy nose). One to two days later, rashes appear on

FACTS: MEASLES

Measles is one of the leading causes of death among young children, even though a safe and cost-effective vaccine is available.

In 2008, there were 164,000 measles deaths globally: nearly 450 deaths per day or eighteen deaths per hour.

More than 95 percent of measles deaths occur in developing countries with inadequate health-care infrastructures.

Measles vaccination led to a 78 percent drop in measles deaths between 2000 and 2008 worldwide.

In 2008, about 83 percent of the world's children received one dose of measles vaccine by their first birthday through routine health services, up from 72 percent in 2000.

Source: World Health Organization

the mucous membranes that look like salt granules on a wet background. These small white spots appear most commonly on the inside of the cheeks opposite the molars and are called "Koplik spots." These symptoms are part of the prodromal period.

The exanthem phase follows the prodromal period. A red, blotchy, maculopapular rash spreads from the head to the extremities (cephalocaudal progression), sparing the palms and soles of the feet.

The rash fades after about four days, and the convalescent period begins. Recovery takes about ten to fourteen days, and persistent cough is the last symptom to fade. People are most infectious on the last day of the incubation period and through the

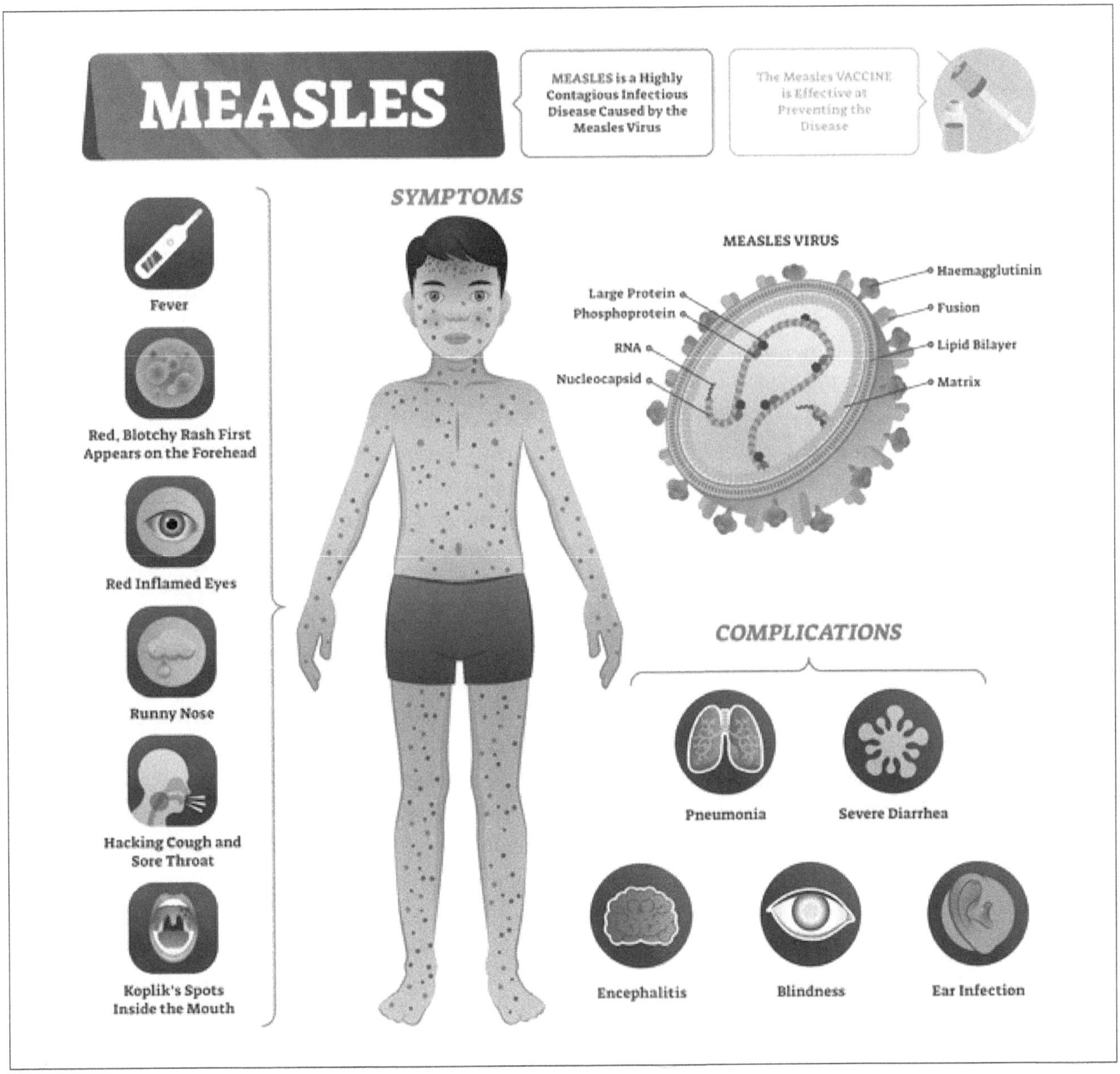

Measles symptoms. Image via iStock/VectorMine. [Used under license.]

prodromal and exanthem phases, which constitutes about eight days.

RISK FACTORS

The factors that increase the chance of developing measles include being unvaccinated or inadequately vaccinated, living in crowded or unsanitary conditions, and traveling to developing countries where measles is common. Also, measles is most common in winter and spring.

Other risk factors include compromised immunity (e.g., untreated human immunodeficiency virus infection [HIV]), even if vaccinated; being born after 1956 and having received no diagnosis of measles; and receiving a vaccine before 1968, without additional vaccination.

SYMPTOMS

Symptoms usually occur eight to twelve days following exposure, including fever (often high), runny nose, red eyes, hacking cough, sore throat, exhaustion, and small spots inside the mouth (two to four days after initial symptoms). Three to five days after initial symptoms appear, a raised, itchy rash will start around the ears, face, and side of the neck and then generally spread to the arms, trunk, and legs over the next two days (and then last about four to six days). Full recovery, without scarring, generally takes seven to ten days from the onset of the rash.

SCREENING AND DIAGNOSIS

Diagnosis is made from the symptoms and the appearance of the rash. Laboratory tests are usually not needed to diagnose measles. Because patients with a subfunctional immune system may display measles infections without classic symptoms, serological tests that detect antibodies against the measles virus effectively diagnose measles.

COMPLICATIONS

Because the measles virus infects a host of organs like the lungs, intestine, and brain, it can cause complications such as pneumonia, diarrhea, and, rarely, encephalitis. These complications, unfortunately, are often severe and can lead to death.

Because it infects immune cells, measles also suppresses the immune response for up to six weeks after the rash fades. Measles increases the risk of middle ear infections (otitis media) and bacterial pneumonia. These complications are worse infants, who have the highest rates of mortality from measles infections.

A severe and fatal complication in children under two years old is subacute sclerosing panencephalitis (SSPE). SSPE occurs seven to ten years after the resolution of measles and results from chronic brain infection by latent measles infections. SSPE symptoms are initially mood changes but become severe later, including seizures, coma, and, if left untreated, death.

TREATMENT AND THERAPY

A virus causes measles, so antibiotics do not treat it. The focus of treatment is on relieving symptoms. Gargling with warm salt water will often relieve the sore throat, and using a humidifier can provide some relief.

Nonaspirin medication, which includes acetaminophen, can treat a high fever. Aspirin is not recommended for children or teens with a current or recent viral infection because of the risk of Reye's syndrome. One should consult the doctor about medicines that are safe for children.

Other treatment includes getting extra rest, drinking increased liquids, and eating a soft, bland diet. Cold sponge baths may also help with symptoms.

In most cases, complications are rare, but persons with severe cases may need to be hospitalized. Complications may include encephalitis (inflammation of the brain) and bacterial pneumonia (a lung infection).

PREVENTION AND OUTCOMES

Getting vaccinated is the best way to prevent measles since the vaccine contains live viruses that can no longer cause disease. The vaccine is usually combined with vaccines against measles, mumps, and rubella (MMR). The MMR vaccine is given twice: at age twelve to fifteen months and age four to six years (or at age eleven to twelve years).

The US Centers for Disease Control and Prevention (CDC) recommends that all children get two doses of MMR vaccine. Children should receive the first dose at twelve to fifteen months of age and the second at four to six years of age. The MMR vaccine protects against measles, mumps, rubella, and varicella (chickenpox). MMR vaccine is licensed for children twelve months to twelve years old.

In some cases, the vaccine is given within three days after exposure to prevent or reduce symptoms. Immunoglobulin is given to certain unvaccinated people within six days of exposure, usually for infants and pregnant women. Treatment is supportive with proper hydration and pain relief to limit the chance of secondary bacterial infections.

In general, one should avoid the vaccine if they have had severe allergic reactions to vaccines or vaccine components, is pregnant (a woman should avoid pregnancy for one to three months after receiving the vaccine), has a weakened immune system, or has a high fever or severe upper respiratory tract infection. If not vaccinated, one should avoid contact with anyone who has measles.

In some cases, providing malnourished children with vitamin A supplements boosts their immune responses against the measles virus and diminishes the risk of complications.

—Rick Alan and Michael A. Buratovich, PhD

Further Reading

Bernstein, David, and Gilbert Schiff. "Viral Exanthems and Localized Skin Infections." *Infectious Diseases*, edited by Sherwood L. Gorbach, John G. Bartlett, and Neil R. Blacklow, Saunders, 2004.

Centers for Disease Control and Prevention. "Vaccine Safety: Measles, Mumps, and Rubella (MMR) Vaccine." www.cdc.gov/vaccinesafety.

Flint, Jane, et al. *Principles of Virology, Volume 1: Molecular Biology*. 5th ed., ASM Press, 2020.

"Measles." *Epidemiology and Prevention of Vaccine-Preventable Diseases*, edited by W. Atkinson et al., 11th ed., Public Health Foundation, 2009.

Peter, G., and P. Gardner. "Standards for Immunization Practice for Vaccines in Children and Adults." *Infectious Disease Clinics of North America*, vol. 15, 2001, pp. 9-19.

Pickering, Larry K., et al., editors. *Red Book: 2009 Report of the Committee on Infectious Diseases*. 28th ed., American Academy of Pediatrics, 2009.

Weedon, David. *Skin Pathology*. 3rd ed., Churchill Livingstone/Elsevier, 2010.

Mumps

Category: Diseases and conditions

Anatomy or system affected: All

Specialties and related fields: Family practice, immunology, internal medicine, microbiology, pediatrics

Definition: a contagious and infectious viral disease that causes swelling of the parotid salivary glands in the face and risk of sterility in adult males

KEY TERMS

glomerulonephritis: acute inflammation of the kidney

glycoprotein: proteins that have carbohydrate groups attached to the polypeptide chain

myocarditis: inflammation of the heart muscle

parotitis: inflammation of a parotid gland, especially

plus-sense RNA: viral genome that is a messenger ribonucleic acid (mRNA) and can be directly translated into viral proteins by the host cell's ribosomes

DEFINITION

Mumps is an acute, systemic, contagious viral infection caused by a single-stranded paramyxovirus whose virion consists of ribonucleic acid (RNA) and seven proteins. The RNA of the virus is surrounded by two surface glycoproteins, the hemagglutinin-neuraminidase, and a hemolysis cell fusion antigen. Mumps is a benign and self-limited disease. Up to one-third of persons contracting the disease have a subclinical infection. As with many viral infections, mumps is commonly more severe in people past puberty than in younger children.

CAUSES

The mumps virus is a member of the paramyxovirus group that includes the measles and parainfluenza viruses. It has an RNA genome packed with a viral polymerase, surrounded by a phospholipid bilayer called the "envelope." The mumps virus envelope is studded with two glycoproteins: the HN (hemagglutinin-neuraminidase) and F (fusion) proteins. The HN protein adheres the mumps virus to cells, and the F protein fuses the viral envelope with the cell membrane. The fusion of these two membranes imports the mumps viral genome into the host cell's cytoplasm.

The mumps viral RNA genome is a plus-sense RNA that host cell ribosomes can immediately translate into viral proteins. The infected cells put lots of F protein on their surfaces which causes cells to fuse, resulting in a multinucleated giant cell (also known as a "syncytium").

Mumps usually causes an acute generalized infection that mostly occurs in school-age children and adolescents. It is transmitted by droplets of saliva and inanimate objects (fomites) that can transfer the pathogen to a host when contaminated with the virus. The virus multiplies in the epithelium of the upper respiratory tract, after which the viral particles enter the bloodstream. This is followed by the infection of one or both parotid glands, the largest salivary glands. This infection is known as "parotitis."

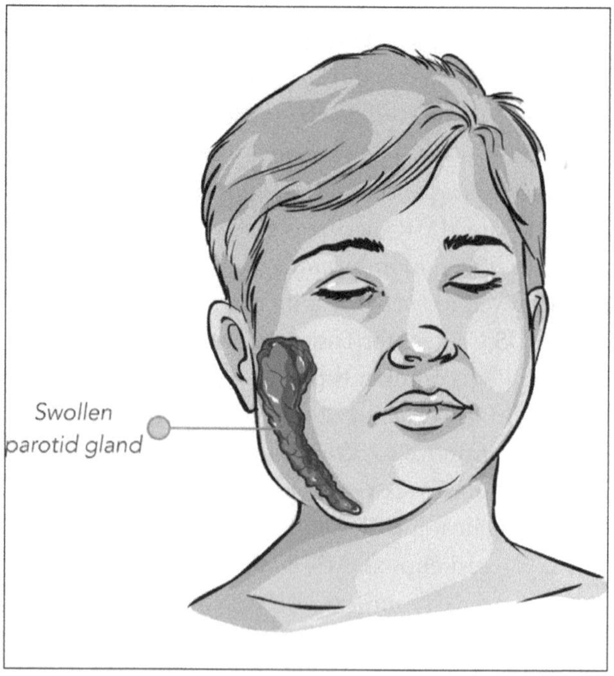

Painful swelling of the parotid glands is the most common symptom of mumps infection. Image via iStock/corbac40. [Used under license.]

Infection of other salivary glands and the meninges, pancreas, and gonads is often seen, but it is not as common. Orchitis, an infection of the testis, is a common (about one in four cases) complication associated with mumps. In rare but severe cases, orchitis may result in sterility. Affected glands show edema and lymphocyte infiltration.

Infrequently, mumps infects the kidney and causes glomerulonephritis, which causes hematuria (blood in the urine) and proteinuria (protein in the urine). Mumps also sometimes affects the joints and causes arthritis in large joints like the shoulder, ankles, hips, and knees. Another organ affected in fifteen percent of mumps cases is the heart (myocarditis). Also, pancreatitis or inflammation of the pancreas is another potential complication of mumps. Unlike rubella, mumps is not passed by pregnant women to their babies.

Long-term immunity is produced with vaccination, and one attack of mumps usually produces lifelong immunity.

RISK FACTORS

Lack of immunization, international travel, and immune deficiencies can make a child more prone to infection by a paramyxovirus. Because the virus is present throughout the world, the risk of exposure to mumps outside the United States may be high, as mumps vaccine is used in only 57 percent of countries that are members of the World Health Organization.

The primary risk factor for contracting mumps is failure to immunize young children. Following the introduction of the mumps vaccine in 1967, the incidence of mumps declined significantly in the United States. At that time, the Advisory Committee on Immunization Practices of the Centers for Disease Control and Prevention recommended that children approaching puberty, and adolescents and adults, be vaccinated. The use of the mumps vaccine in young children was expedited by introducing and extensive use of the measles, mumps, and rubella (MMR) vaccine in 1977.

SYMPTOMS

Parotitis is the classic syndrome of mumps and is evidenced by swelling and inflammation of one or both of the salivary (parotid) glands. Symptoms include low-grade fever, headache, malaise, and anorexia. The incubation period for the disease is fourteen to twenty-one days, and it is communicable from six days before to nine days after facial swelling becomes apparent. However, in 30 percent of infections, no symptoms are observed.

Within twenty-four hours, infected persons experience ear pain near the ear lobe; this pain is made more severe by a chewing movement of the jaw. Acidic foods may exacerbate pain in the parotid gland. After the onset of the disease, one or both parotid glands begin to enlarge; in 70 to 80 percent of cases, the enlargement is bilateral. Pain with pressure is present over the parotid gland. Ordinarily, the parotid gland is not discernible to the touch. Still, in persons with mumps, it quickly swells for several days. Fever diminishes within one week and disappears be-

In the News: Mumps Outbreak in the United States

In December 2005, several students at an unnamed college in eastern Iowa displayed symptoms of illness that included glandular swelling in the salivary region. Antibody testing indicated that the students had active cases of mumps. Several weeks later, an additional case was diagnosed. In the following months, additional cases were reported in the surrounding states of Illinois, Kansas, Minnesota, and Nebraska. Serotyping of isolated viruses indicated that all cases originated from a similar or identical strain. Because not all the cases were directly linked —that is, not all involved known contact—the suspicion among health workers was that portions of the outbreak were maintained through unnoticed infections.

The source of the illness remains unclear, but the initial case may have been contracted in Great Britain. In 2005, some 56,000 cases were diagnosed there, and the strain that first appeared in Iowa appears to be identical. It is likely that a student had either traveled to Great Britain during the outbreak or had contact with someone who had.

By the time the infection had run its course in the summer of 2006, more than 4,700 persons had been diagnosed with mumps, with cases reported in California. Approximately 25 percent of the cases involved college students. Mumps is generally a benign infection. While there were no fatalities, pregnant women and persons with compromised immune systems, such as those who are HIV-positive, may be at risk for severe illness.

An unusual feature of the outbreak was that more than two-thirds of infected persons had already received the recommended two doses of the MMR (measles, mumps, and rubella) vaccine, questioning the long-term effectiveness of current immunization practices. In the light of the outbreak, health authorities recommended that all students be sure of prior immunization against mumps or that they receive an additional two doses of the vaccine.

—Richard Adler, PhD

fore swelling of the parotid gland ceases, which may take up to ten days.

Orchitis, or inflammation of the testis, is the second most common manifestation of mumps. It develops in 20 to 30 percent of postpubertal males with mumps. It is bilateral in one to six of those with testicular involvement. It is uncommon in boys younger than ten years of age. Onset is abrupt, and symptoms include a fever from 102° to 105° Fahrenheit (39° to 41° Celsius), chills, headache, vomiting, and testicular pain. Fever and gonadal swelling usually resolve in one week, but tenderness may persist. The anxiety caused by mumps orchitis is difficult to ease. Still, the psychological fears of sexual impotence and sterility far outweigh the potential debility from testicular atrophy. Sterility is rare, even with bilateral involvement.

SCREENING AND DIAGNOSIS

In most cases, the diagnosis of mumps is made utilizing a history of exposure and evidence of swelling and tenderness of the parotid glands and other classic symptoms of the disease. Although the definitive diagnosis of mumps is dependent on serologic studies or viral isolation, laboratory confirmation of typical mumps is unnecessary.

TREATMENT AND THERAPY

A person with mumps should drink plenty of fluids to promote adequate hydration. Foods and liquids containing acids, such as tomatoes or orange juice, may cause difficulty swallowing. Analgesics, such as ibuprofen, aspirin, or acetaminophen, can relieve a headache or the discomfort of parotitis and can reduce fever. In orchitis, stronger analgesics may be needed. Topical application of warm or cold packs to the parotid glands may relieve discomfort.

Administration of an antiviral agent is not indicated for mumps, as the disease is self-limited. Bed rest is recommended to promote more rapid recovery.

PREVENTION AND OUTCOMES

The most effective way to prevent mumps is to vaccinate susceptible children, adolescents, and adults. This is best achieved in children with the administration of the MMR vaccine. For children, the typical recommended two-dose schedule is administered at age twelve to fifteen months for the first dose and age four to six years for the second dose.

Among the recommendations for managing mumps once the disease has been contracted is to isolate the infected person until the parotid swelling has disappeared. After swelling of the parotid gland is detected, children should be kept out of school or day- care centers for nine days. If an outbreak in these settings should occur, all children involved should be vaccinated. Isolation, however, may be of little value, especially in closed environments such as schools or day-care centers. The virus is present in saliva for several days before parotitis develops. Children with asymptomatic infection can still shed the virus.

—Gerald W. Keister, MA and
Michael A. Baranovichi, PhD

Further Reading

Arumugam, V., et al. "Mumps." *Ferri's Clinical Advisor 2011: Instant Diagnosis and Treatment*, edited by Fred F. Ferry, Mosby/Elsevier, 2011.

Gershon, Anne. "Mumps." *Harrison's Principles of Internal Medicine*, edited by Joan Betterton, 17th ed., McGraw-Hill, 2008.

Gutierrez, K. M. "Mumps Virus." *Principles and Practice of Pediatric Infectious Diseases*, edited by Sarah S. Long, Larry K. Pickering, and Charles G. Prober, 3rd ed., Churchill Livingstone/Elsevier, 2008.

Hviid, A., S. Rubin, and K. Mühlemann. "Mumps." *The Lancet*, vol. 371, Mar. 2008, pp. 932-44.

Litman, Nathan, and Stephen G. Baum. "Mumps Virus." *Mandell, Douglas, and Bennett's Principles and Practice of Infectious Diseases*, edited by Gerald L. Mandell, John F. Bennett, and Raphael Dolin, 7th ed., Churchill Livingstone/Elsevier, 2010.

Peltola, H., et al. "Mumps Outbreaks in Canada and the United States: Time for New Thinking on Mumps Vaccines." *Clinical Infectious Diseases*, vol. 45, Aug. 2007, pp. 459-66.

Rubella

Category: Diseases and conditions
Also known as: German measles, three-day measles
Anatomy or system affected: All
Specialties and related fields: Family practice, immunology, internal medicine, microbiology, obstetrics and gynecology, pathology, preventive medicine, public health, virology
Definition: a contagious but usually mild childhood disease caused by the rubella virus that can lead to congenital disease of newborns if a pregnant woman is exposed to the virus during her first trimester

KEY TERMS

hepatosplenomegaly: enlargement of the liver and spleen
lymph nodes: small bean-shaped structures that are part of the immune system, filter substances that travel through the lymphatic fluid, and they contain lymphocytes (white blood cells) that help the body fight infection and disease
patent ductus arteriosus: an opening between two blood vessels leading from the heart
polymerase chain reaction: a widely used method to rapidly make millions to billions of copies of a specific deoxyribonucleic acid (DNA) sample, allowing scientists to take a very small sample of DNA and amplify it to a large enough amount to study in detail

CAUSES

The rubella virus is a member of the *Rubivirus* genus of the Togaviridae family. Togaviruses are single-stranded ribonucleic acid (RNA) viruses surrounded by a protein capsid within an outer lipid membrane (envelope). These viruses are positive-sense RNA viruses. Their genome serves as a messenger RNA that can be used right away to make protein.

The transmission route is through the respiratory system by direct contact with respiratory droplets discharged from the nose or mouth of an infected person, as might occur during a cough or sneeze. After infecting the pharynx, the rubella virus spreads to the lymph nodes. After infecting the lymph nodes, the virus reenters the bloodstream. From the bloodstream, it enters sundry bodily fluids like urine, synovial fluid in the joints, and cerebrospinal fluid. Rubella virus destroys cells (cytopathic effect) because infected cells commit suicide to prevent viral spread.

Rubella virus also infects unborn children through their mother's placenta. Fetal rubella infection causes congenital rubella syndrome. It seems to damage blood vessels and reduce blood flow to the organs. Reduced blood flow seriously decreases fetal growth and development. When the mother becomes infected during her pregnancy determines the risk to her baby. Congenital disabilities are much more likely if the maternal infection occurs between four to twenty weeks after conception. Infection twenty weeks after conception typically does not cause fetal defects. However, it might lead to intrauterine growth restriction, causing the baby to be smaller than expected.

Rubella infections are usually asymptomatic. But if symptoms do occur, they occur, on average, fourteen days after infection. The contagious period for rubella begins three days after exposure to the virus and lasts for three to four weeks. Humans are the only hosts for this virus.

RISK FACTORS

The main risk factor for rubella is being unvaccinated.

With early immunizations, 99 percent of children never contract rubella. Exposure of nonimmunized children in tight spaces such as schools could pose a risk, but this is unlikely. The main risk is to pregnant women or women of childbearing age who have low rubella titers. The success of the immunization program in the United States has significantly decreased the risk of rubella for pregnant women.

SYMPTOMS

Rubella is usually a mild illness, and most infections cause no symptoms. The affected person may feel tired for a few days. Typical symptoms include malaise with painful enlargement of the lymph nodes behind the ear (postauricular) and neck (suboccipital and cervical lymph nodes). Usually, red macular spots appear on the face; the rash spreads to the trunk and the arms and legs. Some people with rubella do not have a rash; the pattern may vary even in those who develop a rash.

Other symptoms of rubella include a low-grade fever of 101° Fahrenheit or lower. About 10 to 15 percent of older youth and young adults experience joint pain or arthralgia. Cold-like symptoms with congestion and cough may be present. Complications are rare, but extreme cases may result in rubella encephalopathy with headaches and seizures; neuritis, irritated nerves; and orchitis, or inflamed testes.

Rubella contracted during pregnancy increases the risk of miscarriage, fetal death, and stillbirth. Babies born with congenital rubella syndrome commonly have a triad of symptoms, including deafness (sensorineural hearing loss), eye abnormalities (cataracts and retinopathy), and congenital heart disease (such as patent ductus arteriosus). These babies also form blood cells in their skin rather than their bone marrow. This condition, dermal extramedullary hematopoiesis, makes red to purple skin spots called "blueberry muffin" rash.

Congenital rubella syndrome also increases the risk of intellectual disability, developmental delays, behavioral disorders, skin lesions, diabetes mellitus, and enlargement of the liver and spleen (hepatosplenomegaly).

SCREENING AND DIAGNOSIS

The health-care provider will take a medical history and perform a physical examination. Diagnosis will be made by assessing physical symptoms such as enlarged and painful cervical and postauricular lymph nodes, coupled with a low-grade temperature and a macular rash on the face, trunk, and limbs. Most cases are mild and may go undiagnosed. The rubella diagnosis is confirmed by measuring the presence or increased antibody titer of IgM (rubella-specific immunoglobulin M) through blood or culture testing. In the United States, rubella immunizations are mandated for all children before they start attending school, so this disease is rare. Polymerase chain reac-

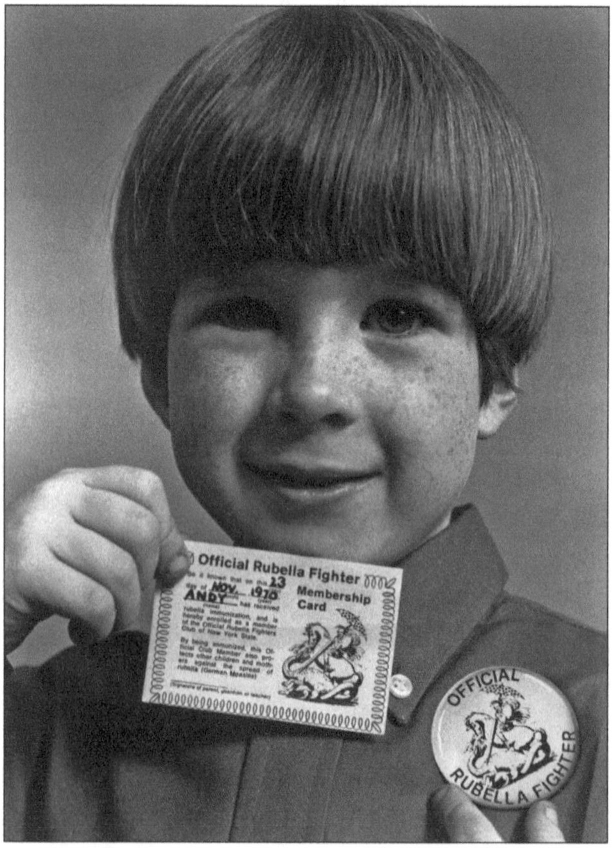

Rubella immunization programs in the United States in the 1960s and 1970s encouraged children's participation in the campaign with membership cards and buttons. Image courtesy of the CDC via the Public Health Image Library.

tion (PCR) tests effectively detect viral RNA. PCR tests of amniotic fluid diagnoses in utero rubella infections.

TREATMENT AND THERAPY

Symptoms are usually mild and need minimal supportive treatment. Acetaminophen or ibuprofen can relieve pain, fever, and joint aches. Maintaining adequate fluid intake is recommended, but no isolation of the infected person is necessary.

PREVENTION AND OUTCOMES

After a rubella outbreak in the mid-1960s, the prolific vaccine researcher Maurice Hilleman developed vaccines against rubella. Getting vaccinated is the best way to prevent rubella. Most states require rubella immunizations in the United States. A live virus is usually given in a combination vaccine for measles, mumps, and rubella (MMR) to infants at twelve to fifteen months of age and through a booster shot at age four to six years (or at age eleven to twelve years). These two immunizations usually provide lifetime immunity to rubella. The vaccine can be given to women of childbearing age.

In general, one should avoid the vaccine if they have had severe allergic reactions to vaccines or vaccine components, is pregnant (a woman should avoid pregnancy for one to three months after receiving the vaccine), has a weakened immune system, or has a high fever or severe upper respiratory tract infection.

Women who are not sure if they have been vaccinated should be tested. This is especially important if they are in occupations, such as health care, teaching, and child care, with a high risk of exposure to rubella. The vaccine offers most people lifelong protection against rubella infection.

Immunizations remain the primary method of prevention of rubella. Risks related to taking the vaccine are minor for most people but can be severe for those who have an allergic reaction. Some parents are concerned that immunization with rubella vaccine may be related to autism. However, extensive epidemiological studies do not confirm this belief.

—Marylane Wade Koch, MSN, RN and Michael A. Buratovich, PhD

Further Reading

Behrman, Richard E., Robert M. Kliegman, and Hal B. Jenson, editors. *Nelson Textbook of Pediatrics*. 18th ed., Saunders/Elsevier, 2007.

Centers for Disease Control and Prevention. "MMR Vaccines: What You Need to Know." www.cdc.gov/vaccines/pubs/vis/downloads/vis-mmr.pdf.

"Congenital Rubella." *The New York Times Health Guide*, health.nytimes.com/health/guides/disease/congenital-rubella/overview.html.

DeStafano, Frank. "Vaccines and Autism: Evidence Does Not Support a Causal Association." *Clinical Pharmacology and Therapeutics*, vol. 82, no. 6, Dec. 2007, pp. 756-59.

Hawkins, Trisha. *The Need to Know Library: Everything You Need to Know About Measles and Rubella*. Rosen, 2001.

Peter, G., and P. Gardner. "Standards for Immunization Practice for Vaccines in Children and Adults." *Infectious Disease Clinics of North America*, vol. 15, 2001, pp. 9-19.

Rabies

Category: Diseases and disorders

Anatomy or system affected: Brain, muscles, musculoskeletal system, nervous system, psychic-emotional system

Specialties and related fields: Epidemiology, neurology, public health, virology

Definition: a virus that attacks the nerve cells and is most often transmitted by the bite of a rabid animal; control of the disease is accomplished through vaccination of pets and immediate immunization of humans if exposed to the disease; once symptoms occur in humans, the disease is nearly always fatal

KEY TERMS

anticoagulant: a chemical that blocks the clotting of blood; some anticoagulants stimulate internal bleeding when ingested by vampire bats and can be used as a method of extermination

epidemiology: the study of the maintenance and spread of disease in a population

fixed virus: a virus that has been repeatedly cultured in the laboratory so that it has lost its natural variation and is more predictable in experiments

passage: one of the culture steps in the production of a fixed virus

replication: the reproduction of a virus; many copies are made within a host cell, then released to infect other host cells

reservoir: the host species in which a parasite is maintained in each area and from which it may infect other species, initiating an epidemic

street virus: a virus derived directly from a natural source; a fixed virus is produced from a street virus by several passages through an artificial culture system

sylvatic rabies: rabies in wild animal populations (as opposed to rabies in domestic animals and pets)

CAUSES AND SYMPTOMS

Rabies is caused by a bullet-shaped virus that attacks warm-blooded animals, especially mammals. The virus can enter many types of mammal cells and cause them to produce and bud off new viruses. However, it is particularly adept at attacking nerve cells and glandular cells. This combination enhances the virus's chance of being transmitted to another host.

The following events occur in an untreated human being after being bitten by a rabid animal. The bite introduces large amounts of saliva, which contains abundant rabies virus because of the virus's efficient growth in salivary glands. The virus enters muscle cells in the vicinity of the bite and replicates there. The new viruses then enter the nerve cells that carry signals from the brain and spinal cord to the muscle cells. They move along these nerve cells to the spinal cord, eventually making their way to the brain. The viruses replicate at certain sites as they ascend the nerve and spinal cord. In the brain, they replicate especially well in the centers that control emotions. Once established in the brain and spinal cord, the virus moves out of these organs along the nerves to most body organs. The salivary glands are favored targets in this migration.

As a result of its extensive migrations, the virus is present in many body tissues. Yet, the critical ones for the pathology and transmission of the disease are the brain and salivary glands, where the virus reproduces especially well. To understand this relationship between transmission and pathology, consider the dog or other animal that bit the human being described above. A sequence of events similar to that described for the human being has occurred in the animal. The viruses attacking the emotional centers of the animal's brain initiated the characteristic aggressive state in which it wandered, attacking anything it encountered. Viruses attacking the brain also stimulated the production and release of copious amounts of saliva, giving rise to another familiar symptom of rabies: frothing at the mouth. The viruses reproducing in the salivary glands and the excessive salivation assured that an abundance of the virus would be chewed into the wound.

A human victim's time sequence and pathology include an incubation period that may range from ten days to a year; in most victims, however, it is between two and eight weeks. During this time, the virus is replicating in cells at the site of the bite and moving to the central nervous system. As the nervous system begins to be involved, generalized symptoms begin. These include fever, headaches, and nausea and last about a week. Neurologic symptoms then develop, including hyperactivity, seizures, and hallucinations. The throat sometimes becomes so sore and prone to spasms that the patient has trouble swallowing and fears choking while drinking. Another common name

Dog bites are a potential source of rabies. Photo via iStock/YuriyGreen. [Used under license.]

for rabies, hydrophobia (literally, fear of water), is based on this aspect of the disease. Paralysis and coma occur about a week after the neurologic symptoms, and death follows a few days later. Once the symptoms begin in a human being, the disease is nearly always fatal.

A similar sequence of events occurs in dogs, though the timing is somewhat different, and dogs occasionally recover. The aggressive stage described above is called the "furious stage" in animals. The gradual development of paralysis is called the "paralytic or dumb stage." Both phases may occur in dogs, or the furious stage may be bypassed. Still, death commonly occurs shortly after symptoms begin.

Rabies in wildlife is called "sylvatic rabies," and the species of wildlife involved differ according to geographic area. Some species act as the virus's reservoir and as the source of rabies epidemics in humans or their pets. In the arctic regions of North America and Eurasia, arctic foxes and wolves are the most important hosts of sylvatic rabies. Red and gray foxes and skunks play important roles in spreading the disease in various parts of eastern Canada and the United States, as do raccoons. Some investigators believe that weasels and their relatives are important carriers in maintaining the virus in nature in many areas. However, they are not particularly important in the direct transfer of the virus to humans. Bats are common sources of rabies throughout the United States, but especially in the southern part of the country. They play a major role in rabies epidemiology in Mexico and Central and South America, where the disease often occurs in vampire bats.

In 2010, there were 6,153 cases in animals, and only 2 cases of rabies in humans reported in the United States and Puerto Rico. Raccoons (36.5 percent), bats (23.2 percent), skunks (23.5 percent), and foxes (7 percent) were the most commonly infected animals. In 2010. 8 percent of rabies reports were for domestic species. Surprisingly, there were more cats than dogs reported with rabies, undoubtedly due to rabies vaccination programs for dogs.

Other mammals, both wild and domestic, are attacked by the virus. Still, they are not important in transmitting rabies between species or acting as reser-

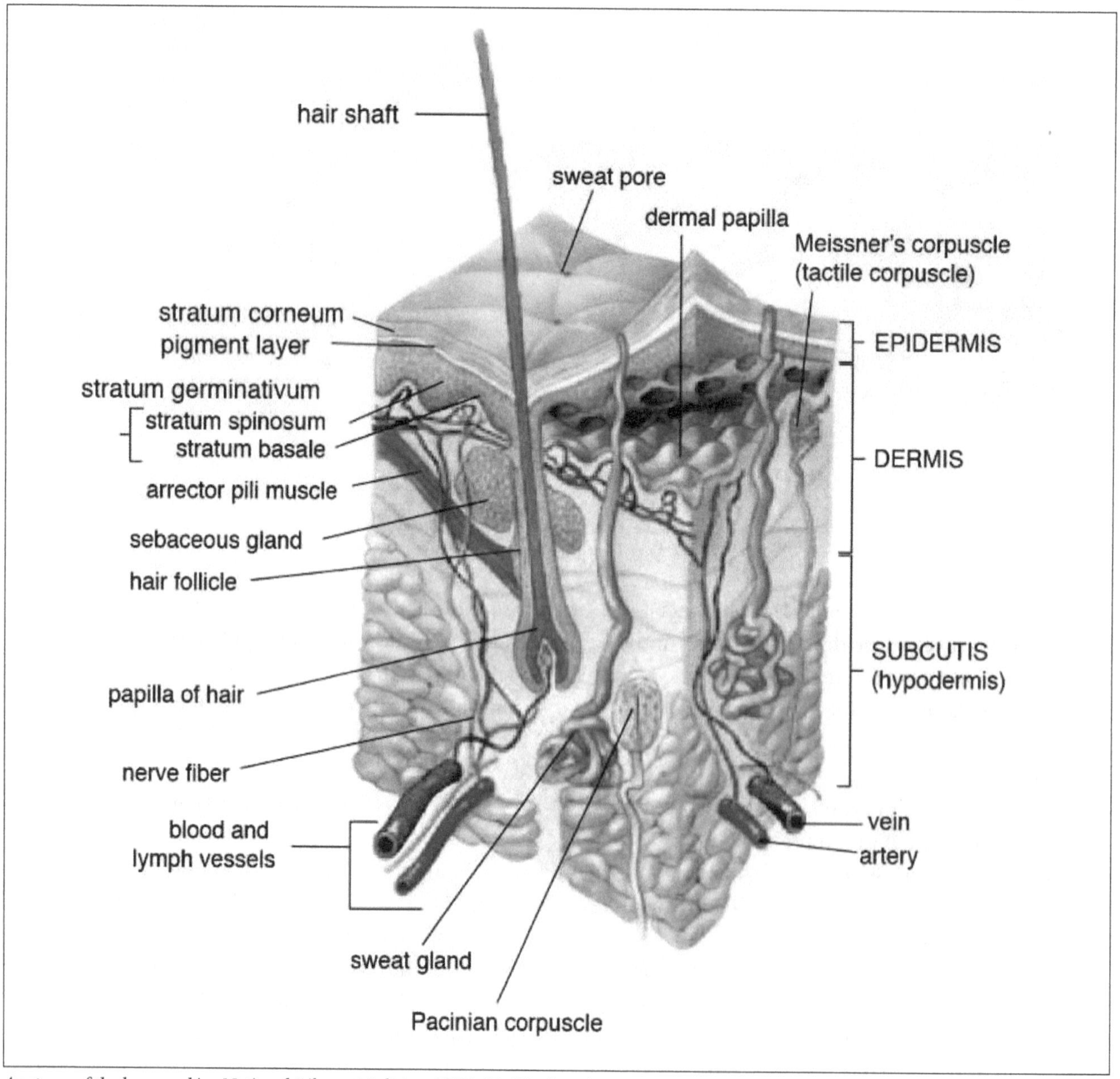

Anatomy of the human skin. National Library Medicine, NIH. [Public domain.]

voirs. Examples of these species are grazing and browsing animals such as cattle and deer, which seldom bite other animals and are not likely to pass the virus to other creatures. However, many of these animals die from rabies, and the economic and ecological impact of these deaths may be great.

The virus attacks these animals in the same manner as humans and dogs, showing similar symptoms. Rabid wild animals do not always suffer a furious stage. Raccoons, for example, often skip the furious stage and go directly into the dumb stage. Early in this stage, they may lose their fear of humans and appear to be friendly. If left alone, they seldom attack, but humans who approach these "friendly" raccoons may be bitten and exposed to rabies. Many bat species also do not go through the furious stage; a bat suffering from dumb rabies is easily caught and may bite if handled, exposing the handler to rabies. Unlike humans, bats, skunks, raccoons, and other wildlife often survive the symptoms of rabies.

In addition to the disease it causes in humans and their pets, the rabies virus has negatively impacted human society. Rabies transmitted to cattle by vampire bats has had a devastating economic effect on the cattle industry in all of Latin America. Rabies in red foxes has occasionally negatively affected Canada's fur industry. In many parts of the world, wildlife populations may be periodically decimated by rabies epidemics, which sometimes reduces the population of a species of recreational importance to humans or one critical to the ecological stability of a region.

IN THE NEWS: RACCOON RABIES

The first death in the United States from the strain of rabies identified with raccoons occurred in northern Virginia in 2003; it was discussed in the *Washington Post* of May 30, 2005, and in more detail in the *Morbidity and Mortality Weekly Report* of November 14, 2003. The victim was a twenty-five-year-old man. Rabies was not diagnosed until after his death, and how he contracted the virus is unknown.

Efforts to fight raccoon rabies rely primarily on a vaccine packaged in a fish meal or some other raccoon-enticing bait. The bait packets are distributed by airdrop from helicopters or light planes in remote areas and hikers in more accessible areas. Raccoons are attracted to the bait, eat the bait and vaccine, and are immunized against rabies; they will neither be sickened by the virus nor pass it on. If a high proportion of the raccoons in each area is immune, then the rabies virus dies out in that area. The bait distribution locations chosen are based on predictions of the future directions of virus spread. Computer models of disease epidemiology often determine these predictions. The specific goals vary according to the region of application. Examples include reducing the prevalence of local infection, restricting the virus to the eastern United States, and retarding its spread to Canada.

These efforts have successfully arrested the spread of rabies in raccoons and other wildlife and in decreasing its prevalence in some hot spots. Still, they probably cannot be used to eliminate the disease, at least in part because no effective method for immunizing bats has been developed. Interaction between bats and wild carnivores (the principal reservoirs of the disease) will probably keep rabies cycling in nature until bats can be included in the vaccination program.

—Carl W. Hoagstrom, PhD

TREATMENT AND THERAPY

Active immunization by vaccination can be used after exposure to rabies because of the relatively long latency period of the virus. If a rabid animal has bitten a person, symptoms usually do not appear for two or more weeks. Prompt vaccination after the bite induces the production of antibodies that attack the virus and neutralize it before it reaches the central nervous system. Two other precautions are often taken. The wound is cleaned and treated with antiviral agents. Passive immunity is often produced by injecting antirabies antiserum into the victim.

The rabies vaccinations of early immunization series were numerous, extremely painful, and not always successful. Since the early twentieth century, it has been possible to determine whether the attacking animal was rabid and thus whether this painful treatment was necessary. The animal was sacrificed, and its brain was sectioned and stained. When treated this way, a rabid animal's brain cells often display Negri bodies, named for the scientist who first described them. They are the sites of production of new virus in the brain cell, and their presence indicates the need for vaccination of the victim. Even though the immunization sequence that was first developed was painful, the certain death that followed the onset of symptoms made immunization imperative if there was any possibility of rabies exposure.

Improved immunization sequences and rabies tests have been developed. These immunization sequences are refined and nearly infallible and require only three inoculations, and are no more painful than most shots. Tests using antibodies are more rapid and reliable at detecting the presence of the rabies virus than the test for Negri bodies.

In contrast to the postexposure immunization described above, preexposure immunization is used to protect persons who might be exposed to rabies in their normal activities and protect pets from contracting rabies. Since the overwhelming majority of human rabies cases come from dog bites, pet vaccination is the most important part of the successful rabies control programs of developed countries. Laws requiring the immunization of pets against rabies and leash laws (which require that pets be controlled and not allowed to wander freely) have effectively reduced human rabies in these countries.

Because wildlife may harbor rabies, attempts have been made to control or eliminate the disease by killing (culling) or immunizing wildlife. Neither approach has been particularly successful, and troublesome side effects accompany the first. The purpose of culling members of host species is to reduce the host population below the point that will sustain the rabies virus's population. This method is based on the idea that each infected host must, on average, infect at least one other susceptible host before it dies for the parasite to persist in the population. The lower the host population, the lower the chance of one host meeting another and thus the lower the probability of an infected host infecting other population members. Yet many of the methods of culling (trapping and poisoning, for example) are not species-specific, and members of other species are killed, sometimes in large numbers. Culling has also been ineffective in many cases. It is most successful in small, isolated areas with a low probability of reinvasion.

Instead of reducing the population size of the host, the goal of immunization is to reduce the number of members that are susceptible to rabies by increasing the number that are immune. Oral immunization by scattering bait containing rabies vaccine has shown promising results in reducing rabies in foxes, coyotes, and raccoons.

An argument against immunization of wildlife to control human rabies is based on the success of the control programs and can be stated as follows. Immunization and regulation of pets, preexposure immunization for humans regularly exposed to rabies, and effective postexposure treatment have already minimized the incidence of human rabies in developed nations. Therefore, wildlife immunization is not necessary for the control of human rabies. In developing nations, where human rabies is still a serious disease, all the potential solutions strain the available resources. However, the most cost-effective solution would be in use in developed countries. There is general agreement, however, that wildlife immunization might be an effective way to increase the population size of a wildlife species that is normally susceptible to rabies, if desirable.

In Latin America, vampire bats carry rabies. These bats deleteriously affect the cattle industry. Culling the bats has been attempted repeatedly, often unsuccessfully, and with serious side effects. For example, other bat species, some of which are important to insect control and the pollination of fruit trees, have been regular victims of indiscriminate attempts at vampire bat control by culling.

A more effective and less ecologically disruptive method for controlling vampire bats employs anticoagulants. These chemicals stimulate bleeding in the digestive tract of vampire bats that swallow them, resulting in death. The anticoagulant can be applied directly to the backs of the vampire bats. It will spread through the bat population when the bats groom one another at the roosting colony. Alternatively, an anticoagulant can be injected into a cow's rumen, the enlarged first chamber of its four-part stomach. The anticoagulant is then absorbed into the animal's blood and spread to vampire bats when they feed on the cattle.

Each method has reduced the number of vampire bat bites in cattle by 90 percent in test areas, but each has drawbacks. Direct application to these bats requires extensive netting or trapping, special equipment, and workers skilled in vampire bat identification. The rumen injection technique requires expensive equipment and workers experienced in handling large numbers of cattle. In developing countries, either combination can be difficult to finance.

Education is another important aspect of rabies control. While dogs are the most common source of human rabies, humans occasionally contract the disease after being bitten by a wild animal. In addition, there are potential avenues of transfer other than bites. These include skinning rabid animals, being licked by a rabid animal on broken skin, and breathing air infested with the rabies virus. All these alternative transmission mechanisms are exceptionally infrequent, but there are documented cases of aerial transmission to humans. For example, two men who were not bitten died of rabies contracted while exploring a bat cave in Texas. The rabies virus must be highly concentrated in the air to be transmitted in this way. These concentrations probably occur only in caves occupied by many bats, and many of them must be carrying the rabies virus.

While educating spelunkers and hunters about these dangers would have a minimal effect on the incidence of rabies, as these transmission mechanisms are so infrequent, such educating could be of the utmost importance to an individual who is spared a rabies infection by the knowledge. Educating people, especially children, to leave animals acting in an unnatural fashion alone would significantly affect rabies incidence. Most wild animals that can be caught or approached closely are sick and may be suffering from dumb or paralytic rabies. They should be avoided and reported to the appropriate authorities, as should any dog or cat that behaves unnaturally. Educating the public about the importance of pet vaccination and pet control is the most important role of education in regulating rabies.

PERSPECTIVE AND PROSPECTS

Rabies in humans and their association with mad dog attacks have been known for more than two thousand years. Even though rabies has never caused epidemics accompanied by mass mortality like smallpox and bubonic plague, its frightful symptoms and ability to turn a loving family pet into a vicious animal have given the disease a terrifying and mysterious aura. As a result, cures and preventions have been sought throughout history.

In the late nineteenth century, Louis Pasteur and his associates performed many experiments. They isolated the rabies virus from a dog and injected it into rabbit brains. Pasteur called this virus a "street" virus because it was isolated directly from dogs in the street. The virus replicated in the rabbit's brain and could be transferred into another rabbit's brain, where it again replicated. The growth of the virus in one of the rabbit brains was called a "passage." A sequence of such passages resulted in a virus that had a more predictable and shorter incubation period. This virus was called a "fixed" virus because of its fixed incubation period. After a hundred such passages, the virus had lost much of its ability to infect dogs.

Pasteur then developed an immunization sequence that protected dogs from the street virus. He air-dried rabbit spinal cord tissue infected with the fixed

Rabies vaccines are often put into bait and distributed in the wild, sometimes via airdrop. Matchbox for scale. Photo by Izvora via Wikimedia Commons.

virus for varying times. Pasteur developed a series of virus solutions, ranging from those that could not infect rabbits through those that could occasionally establish weak infections to those that were maximally infective. He then injected dogs daily for ten days, beginning with the noninfective preparation the first day and increasing the infectivity with each day's injection until, on the tenth day, he was injecting a highly infective virus. The dogs he treated were resistant to experimentally injected street viruses.

Pasteur was still refining his immunization system when a boy who had been attacked by a rabid dog was brought to him. Knowing that the latency period of the virus might allow time for the development of immunity before the symptoms appeared and aware of the almost certain fatal result if nothing was done, Pasteur treated the boy with the sequence that he had used on the dogs. The boy lived with no apparent side effects, and Pasteur's treatment became the standard for rabies. The modern treatment sequence is a refinement of Pasteur's. While the immunization sequence for rabies was not the first to be used successfully—smallpox immunization nearly a century earlier, Pasteur's work established the distinction—the possibility of immunization after exposure to diseases with long incubation periods.

Considerable work has been done on the epidemiology of rabies. Mathematical and computer models have been developed that attempt to predict the characteristics of the disease spread under different conditions and thus suggest means of controlling and preventing rabies epidemics. Arguments over the effectiveness of wildlife vaccination are partially based on such models. The usefulness of these models is not restricted to rabies epidemiology. Instead, it contributes to an understanding of epidemiology in general. Thus, research on rabies continues to enhance the control and prevention of that terrifying disease and add to the general knowledge base of medicine.

—*Carl W. Hoagstrom, PhD*

Further Reading

Badash, Michelle. "Rabies." *Health Library*, 30 Dec. 2011.

Biddle, Wayne. *A Field Guide to Germs.* 2nd ed., Anchor Books, 2002.

Blanton, Jesse D., et al. "Rabies Surveillance in the United States during 2007." *Journal of the American Veterinary Medical Association*, vol. 233, 2008, pp. 884-97.

Centers for Disease Control and Prevention. "Rabies." 23 Sept. 2021, www.cdc.gov/rabies/index.html. Accessed 3 Dec. 2021.

Jackson, Alan C., and William H. Wunner, editors. *Rabies.* Academic Press, 2002.

Kaplan, Colin, G. S. Turner, and D. A. Warrell. *Rabies: The Facts.* 2nd ed., Oxford UP, 1986.

Pace, Brian, and Richard M. Glass. "Rabies." *Journal of the American Medical Association*, vol. 284, no. 8, 30 Aug. 2000, p. 1052.

Parker, James N., and Philip M. Parker, editors. *The Official Patient's Sourcebook on Rabies.* Icon Health, 2002.

"Rabies." *Mayo Clinic*, 28 Jan. 2011.

Rotavirus infection

Category: Diseases and conditions

Also known as: Stomach flu, stomach virus

Anatomy or system affected: Abdomen, gastrointestinal system, intestines, stomach

Specialties and related fields: Family practice, gastroenterology, microbiology, neonatology, pediatrics, preventive medicine, public health, virology

Definition: an intestinal inflammation transmitted by a ribonucleic acid (RNA) virus that results in extreme diarrhea, especially in young children

KEY TERMS

gastroenteritis: inflammation of the stomach and intestines, typically resulting from bacterial toxins or viral infection and causing vomiting and diarrhea

polymerase chain reaction: a method widely used to rapidly make millions to billions of copies of a specific deoxyribonucleic acid (DNA) or RNA sample, allowing scientists to take a very small sample of DNA or RNA and amplify it to a large enough amount to study in detail

rapid antigen tests: point-of-care testing that directly detects the presence or absence of an antigen

reovirus: a group of double-stranded RNA viruses associated with respiratory and enteric infection

serology: the scientific study or diagnostic examination of blood serum, especially regarding the response of the immune system to pathogens or introduced substances

CAUSES

Rotavirus infection is a viral gastroenteritis caused by rotaviruses. Rotaviruses are reoviruses. The prefix "reo" is an abbreviation for respiratory, enteric, and orphan. Reoviruses infect the respiratory system and gastrointestinal tract, but orphan reoviruses do not cause any identified disease. Almost every child has either been vaccinated against rotavirus or infected by rotavirus before the age of five.

Rotaviruses, like other reoviruses, have a double-stranded RNA genome. A spherical protein capsid surrounds the RNA genome of rotaviruses. This virus is "naked," meaning that a lipid envelope does not surround it. The rotavirus genome is divided into ten to twelve segments. There are eight strains of rotavirus, A to H, but most human infections result from infection with group A strains.

Rotavirus is transmitted by the four Fs: fluids, fields, flies, and fingers. The primary mode of rotavirus transmission is ingesting food or water contaminated by fecal material. Depositing infected fecal material on agricultural fields or into the water supply contaminates food or drinking water. Flies can land on infected fecal matter or vomit and transfer the virus, as can touching contaminated surfaces. Rotavirus is very contagious, and people need to contact small amounts of virus to become infected.

RISK FACTORS

Young children are at the greatest risk of contracting rotavirus infection, particularly those between six to twenty-four months, since they have no immunity. Childcare staff and children who attend day-care centers are even more prone to rotavirus infection because they are regularly exposed to children in proximity. The elderly, especially those in nursing homes, are also susceptible to rotavirus infection and those with weakened immune systems. Frequent travelers to developing countries also risk greater exposure because of contaminated food and water sources.

Infection rates increase in the winter months, but infection rates remain the same year-round in tropical regions.

SYMPTOMS

When rotavirus comes to the small intestine, it infects the intestinal enterocytes. It destroys these cells, and the intestine replaces them with immature ones that are less efficient at absorbing nutrients. Thus, rotavirus infections significantly diminish nutrient absorption. Rotavirus also activates the enteric nervous system, which regulates gastrointestinal activity. Hyperactivation of the enteric nervous system causes salt and salt secretion into the intestine, generating the characteristic diarrhea.

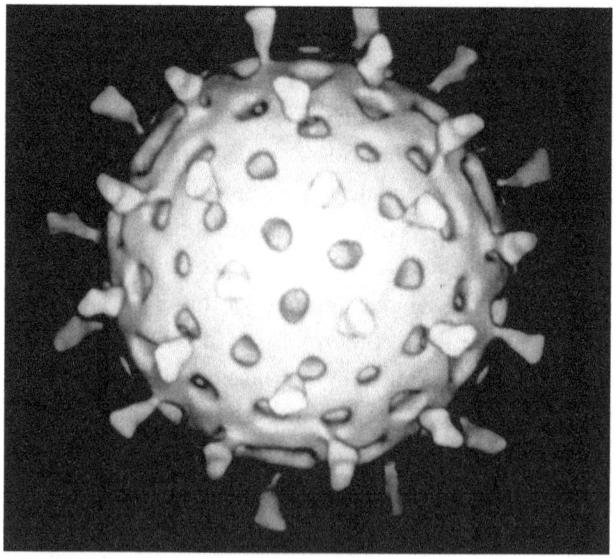

Computer assisted reconstruction of a rotavirus fumaza. Image by Graham Beards, via Wikimedia Commons.

Rotavirus has a short incubation period of twenty-four to forty-eight hours. The symptoms of rotavirus infection include diarrhea, nausea, abdominal pain, vomiting, dehydration, fever, chills, and loss of appetite. Symptoms usually last about eight days.

SCREENING AND DIAGNOSIS

After conducting a physical exam, a physician will question the affected person about their symptoms. Stool tests include immune-based tests, like enzyme-linked immunosorbent assay and latex agglutination test that detect viral antigens. Several commercially available rapid antigen tests can generate results in minutes. Serological tests detect antibodies against rotavirus. Polymerase chain reactions detect rotavirus RNA in the stool.

TREATMENT AND THERAPY

No cure exists for rotavirus infection, but the most serious symptom, dehydration, is treatable by drinking large amounts of liquids, especially liquids that contain electrolytes (such as Gatorade). Severely dehydrated persons will require the intravenous administration of liquids in a hospital setting. Stomach cramps and diarrhea may be slightly mitigated by eating bland food, like soda crackers. Damp cloths or ice packs to the forehead can reduce the fever.

PREVENTION AND OUTCOMES

Completely preventing rotavirus infection is impossible, but one can take steps to reduce the likelihood of infection greatly. First and foremost is vaccination. In 2006, two vaccines, RotaTeq and Rotarix, became available for infants. They are extremely effective in preventing rotavirus infection or lessening the severity of infection if it occurs. Additionally, because tests have shown that the rotavirus survives for several hours on hands, vigilant handwashing is highly effective in reducing transmission. Children should consistently wash their hands, especially after using the toilet and before eating. Childcare workers and all persons associated with children also should practice rigorous handwashing. Dirty diapers should be disposed of immediately after changing, and diaper changing areas should be regularly disinfected. Because rotavirus survives for days on hard surfaces, all toilets, counters, and children's toys should also be cleaned regularly with disinfectant. Persons traveling in developing countries should boil all drinking water before ingesting it.

—Mary E. Markland, MA and Michael A. Buratovich, PhD

Further Reading

Centers for Disease Control and Prevention. "Rotavirus." *CDC*, 26 Mar. 2021, www.cdc.gov/rotavirus/index.html. Accessed 4 Dec. 2021.

Chadwick, Derek, and Jamie A. Goode, editors. *Gastroenteritis Viruses*. John Wiley & Sons, 2001.

Gray, James, and Ulrich Desselberger, editors. *Rotaviruses: Methods and Protocols*. Humana Press, 2000.

Kirschner, Barbara S., and Dennis D. Black. "The Gastrointestinal Tract." *Nelson Essentials of Pediatrics*, edited by Karen J. Marcdante, et al., 6th ed., Saunders/Elsevier, 2011.

Matson, David O. "Rotaviruses." *Principles and Practice of Pediatric Infectious Diseases*, edited by Sarah S. Long, Larry K. Pickering, and Charles G. Prober, 3rd ed., Churchill Livingstone/Elsevier, 2008.

IMMUNOLOGY

The section introduces the reader to the complexities of the human immune system. Our immune system consists of two branches: innate and acquired. The innate system acts quickly, nonspecifically, and has no memory. The acquired immune system acts more slowly, is highly specific, and remembers previous infections. This section discussed the components of each branch and how they interact with microbes. Topics include antibodies, autoimmune disorders, infection control, immunization, sepsis, and steroids.

Antibodies

Category: Biology
Also known as: Gamma globulin proteins, immunoglobulin (Ig)
Anatomy or system affected: Blood, cells, circulatory system, immune system, lymphatic system
Specialties and related fields: Bacteriology, biochemistry, biotechnology, immunology, internal medicine, microbiology, oncology, pathology, pharmacology, preventive medicine, public health
Definition: an antibody is a glycoprotein produced by the body to recognize and eliminate specific foreign antigens, such as bacteria, viruses, or other microorganisms; it is produced by white blood cells called "B lymphocytes"

KEY TERMS

constant region: The highly conserved C-terminal portion of the antibody
Fc-fusion proteins: specific proteins of interest fused to the Fc region of the antibody at its C-terminal end
heavy chain: the bigger subunits of an antibody
isotypes: the different classes of antibodies
light chain: the smaller subunits of an antibody
monoclonal antibodies: antibodies derived from a single parent clonal cell
variable region: the N-terminal portion of the antibody, which possesses antigen-binding sites and has variable amino acid sequences

STRUCTURE AND FUNCTIONS

An antibody is a Y-shaped protein unit, or tetramer, made of four individual subunits. Antibodies consist of two identical smaller subunits called "light chains" and two larger ones called "heavy chains." There are two families of light chains, called "kappa" and "lambda." Each immunoglobulin (Ig) class consists of either one of these two families of light chains and does not change through its life. There are approximately ten different types of heavy chains, and they contribute to the other classes of antibodies or immunoglobulins. The five major types of heavy chains are a, d, e, g, and m types. Each possesses a specific number of amino acids. Depending on the heavy chain present, the antibodies are classified into five major classes or isotypes: IgA, IgD, IgE, IgG, and IgM. Each isotype has a specific function, and more than one Ig isotype is required to eliminate a particular antigen. The presence of specialized proteins called "cytokines" can trigger switching one type of isotype to another.

Each light or heavy chain possesses a constant region conserved in its amino acid composition across distinct Ig types and a variable region that differs between B cells. The constant region is present in the C-terminal portion of the protein, and the variable region is present at the N-terminus. The variable region contains the antigen-binding sites and is thus responsible for binding to the specific antigen. The antibody displays maximum versatility in this region in possessing hundreds of binding sites and, as a result, can recognize and bind hundreds of different antigens. The tips of the variable region are called "fragment antigen-binding" (Fab) regions. The constant region remains identical in all heavy or light chains of the antibodies. It is responsible for activating the complement system and can also initiate phagocytosis. The constant region is also identical between different isotypes. The base of this region consists of the fragment crystallizable (Fc) region. Once the antigen is bound to the antibody surface, the constant region of the antibody elicits an appropriate immune response.

When exposed to the antigen, the B cells transform into plasma cells and begin producing antibodies, utilizing the aid of helper T cells of the immune system. Antibodies, through their antigen-binding sites, bind with the antigen and form antigen-antibody complexes. Each antigen presents a surface that can be recognized and bound by the specific antibody. The antigen-antibody complex is a signal for destruction.

The complex is engulfed by other specialized cells called "macrophages" or "lysed" by triggering a proteolytic pathway called a "complement pathway." This whole response is called the "humoral response" of the immune system.

There are five different antibody isotypes or classes: IgD, IgM, IgG, IgA, and IgE. IgD is an antigen receptor on the surfaces of B lymphocytes that have not been exposed to antigens. This antibody can activate specific white blood cells like mast cells and basophils to produce antimicrobial factors. IgM has a monomer form expressed on B lymphocyte surfaces and a secreted pentameric form that is the first antibody secreted by activated B lymphocytes. IgG comes in four distinct subtypes (IgG1, IgG2, IgG3, and IgG4) that, collectively, are the majority antibody secreted into bodily fluids. IgG can cross the placenta to provide passive immunity to the fetus. IgA is found in mucosal areas such as the gut, respiratory tract, and urogenital tract. It prevents mucosal surface colonization by pathogenic microorganisms. IgA is also found in saliva, tears, and breast milk. IgE binds to allergens, triggers histamine release from basophils and mast cells, and is involved in allergies and protection against parasitic worms.

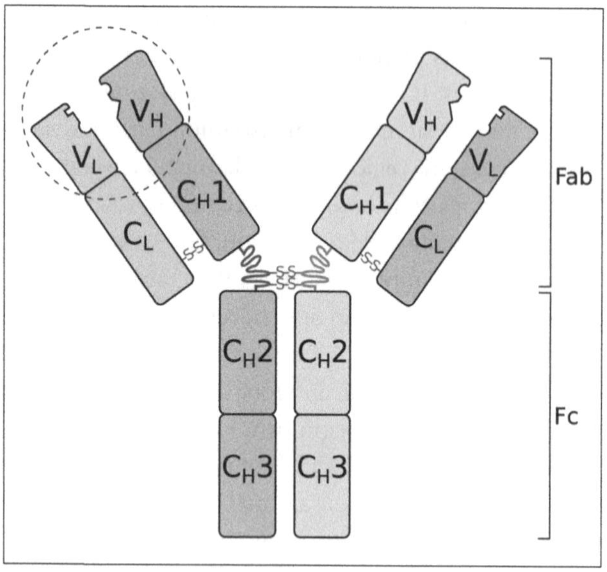

Schematic structure of an antibody: two heavy chains and the two light chains. The antigen binding site is circled. Image by Tokenzero, via Wikimedia Commons.

DISORDERS AND DISEASES

An essential function of an antibody is to distinguish between "self" and "nonself" proteins. An antibody should bind only to nonself proteins. When this recognition capability is compromised, the result is the development of autoimmune diseases or disorders. An autoimmune disease can occur in specific organs, such as the thyroid gland, or the entire body, as with lupus or rheumatoid arthritis. Scleroderma, multiple sclerosis, vitiligo, psoriasis, and alopecia are other autoimmune diseases that can develop because of antibodies attacking the body's proteins.

PERSPECTIVE AND PROSPECTS

The study of antibodies began in the late nineteenth century when scientists discovered a "factor" or antitoxin that could react with antigens or toxins. Emil von Behring and Shibasaburo Kitasato introduced the term "antibody" in 1890. In the early twentieth century, the specificity of antibodies was demonstrated, and the methodology for studying antigen-antibody complexes was deduced. In 1959, Gerald Edelman and Rodney Porter independently published the three-dimensional structure of antibodies. Edelman and Porter received the Nobel Prize for this achievement in 1972. Since then, the field of antibody research has grown, and important discoveries have been made about the structure, functions, and utility of antibodies. Different types of antibodies have been discovered, and their genetic basis inferred.

In addition to affording immunity against pathogens, antibodies are integral agents in clinical diagnostics and treating various human diseases, including cancer and many infectious diseases. Antibody engineering involves the development of monoclonal antibodies directed against specific antigens involved in diseases. Some of these antibodies can be modified to carry drugs or radioactive substances to be delivered specifically to tumor cells in cancer. Several monoclonal antibodies such as Herceptin (trastuzumab), Rituxan (rituximab), and Remicade (infliximab) have been approved by the Food and Drug Administration (FDA) to treat various diseases. Fc-fusion proteins such as Enbrel (etanercept) have also been developed in the last decade to treat multiple diseases. Many monoclonal antibodies and Fc-fusion proteins are being tested in clinical trials. In addition, diagnostic tests may involve the detection of antibodies. Polyclonal antibodies were initially used to prevent viral illnesses and later in the treatment of antibody deficiencies.

Another primary application of antibodies is in scientific research. Some instrumental scientific techniques involve using antibodies, such as enzyme-linked immunosorbent assay (ELISA), Western blotting, flow cytometry, immunohistochemistry, immunohistochemistry, and immunocytochemistry. These techniques are regularly employed in scientific experiments to understand the structure and function of proteins involved in various diseases.

—Geetha Yadav, PhD and Samar Aslam, MD

Further Reading

An, Zhiqiang, editor. *Therapeutic Monoclonal Antibodies: From Bench to Clinic*. John Wiley & Sons, 2009.

Jefferis, Roy, Koicho Kato, and William R. Strohl, editors. *Structure and Function of Antibodies*. Mdpi AG, 2021.

Klein, Christian. *Monoclonal Antibodies*. Mdpi AG, 2018.

Lodish, Harvey, et al. *Molecular Cell Biology*. 9th ed., W. H. Freeman & Co., 2021.

McCullough, Kenneth, and Raymond Spier. *Monoclonal Antibodies in Biotechnology: Theoretical and Practical Aspects*. Cambridge UP, 1990.

ANTIBODIES AND GENETICS

Category: Immunogenetics

Anatomy or system affected: Blood, immune system, lymphatic system

Specialties and related fields: Hematology, immunology

Definition: antibodies provide the mainline defense (immunity) in all vertebrates against infections caused by bacteria, fungi, viruses, or other foreign agents; they are used as therapeutic agents to prevent specific diseases and identify antigens in a wide range of diagnostic procedures; large quantities of antibodies have also been produced in plants for use in human and plant immunotherapy; because of their importance to human and animal health, antibodies are widely studied by geneticists seeking improved antibody production methods

KEY TERMS

B cells: a class of white blood cells (lymphocytes) derived from bone marrow responsible for antibody-directed immunity

B-memory cells: descendants of activated B cells that are long-lived and that synthesize large amounts of antibodies in response to subsequent exposure to the antigen, thus playing an important role in secondary immunity

helper T cells: a class of white blood cells (lymphocytes) derived from bone marrow that prompts the production of antibodies by B cells in the presence of an antigen

lymphocytes: types of white blood cells (including B cells and T cells) that provide immunity

plasma cells: descendants of activated B cells that synthesize and secrete a single antibody type in large quantities and play an essential role in primary immunity

ANTIBODY STRUCTURE

Antibodies, also known as immunoglobulins (Igs), are produced by plasma cells responding to a specific foreign molecule known as an antigen. Most antigens are proteins or proteins conjugated to sugars. Antibodies recognize, bind to, and inactivate antigens that

Classes, Locations, and Functions of Antibodies

Class	Location	Functions
IgG	Blood plasma, tissue Fluid, fetuses	Produces primary and secondary immune responses; protects against bacteria, viruses, and toxins; passes through the placenta and enters fetal bloodstream, thus protecting fetuses.
IgM	Blood plasma	Acts as a B-cell surface receptor for antigens; fights bacteria in primary immune response; powerful agglutinating agent.
IgD	Surface of B cells	Prompts B cells to make antibodies (especially in infants).
IgA	Saliva, milk, urine, tears, respiratory and digestive systems	Protects surface linings of epithelial cells, digestive, respiratory, and urinary systems.
IgE	In secretion with IgA, skin, tonsils, respiratory and digestive systems	Acts as a receptor for antigens causing mast cells (often found in connective tissues surrounding blood vessels) to secrete allergy mediators; excessive production causes allergic reactions (including hay fever and asthma).

have been introduced into an organism by various pathogens such as bacteria, fungi, and viruses.

The simplest form of the antibody molecule is a Y-shaped structure with two identical "heavy chains" and two identical "light chains." These chains are held together by chemical bonds. The lower portion of each chain has a constant region of similar amino acids in all antibody molecules, even among different species. The remaining upper part of each chain, known as the "variable region," differs in its amino acid sequence from other antibodies. The three-dimensional shape of the tips of the variable region (antigen-binding site) allows for the recognition and binding of target molecules (antigens). The high-affinity binding between antibody and antigen results from a combination of hydrophobic, ionic, and van der Waals forces. Antigen-binding sites have specific attachment points on the antigen called "epitopes" or "antigenic determinants."

ANTIBODY DIVERSITY

There are five classes of antibodies (IgG, IgM, IgD, IgA, and IgE), each distinct structure, size, and function. IgG is the principal immunoglobulin and constitutes about 80 percent of all antibodies in the serum. The table below lists the functional distinctions of each antibody subtype.

ANTIBODY GENE REARRANGEMENT AND B-CELL DEVELOPMENT

The variable region of the heavy chain is encoded by a variable or V region, a diversity region or D region, and a J or joining region. The light-chain variable region consists of a V region and a J region. The genes that encode the heavy chain are on chromosome 14, and the light chain has two different gene clusters, kappa, and lambda. The kappa gene cluster is on chromosome two, and the lambda gene cluster is on chromosome 22. The heavy-chain gene cluster consists of 51 different *V* segments, six different *J* segments, 27 distinct *D* segments, and eight constant regions. The kappa light-chain gene cluster consists of 40 different V gene segments, five different J chain clusters, and a single constant region. The lambda gene cluster contains 30 V gene segments and four J gene segments, with four constant regions.

During B-lymphocyte development or lymphopoiesis, the antibody gene clusters are rearranged to create proteins that bind an enormous variety of distinct antigens. The earliest stage of B-lymphocyte development is the early pro-B cell, which forms from the daughter cell after the hematopoietic stem cell divides. The early pro-B cell expresses proteins that rearrange the antibody gene cluster, *RAG-1* & *-2*. The *RAG* proteins join a single D gene segment from one of the heavy-chain gene cluster copies with a single J gene segment. Upon completion of the DJ joining, the early pro-B cell becomes a late pro-B cell. The *RAG-1* & *-2* recombinase joins a single V segment with the DJ segment and the VDJ combination to the μ constant region gene. Successful cutting and pasting of the VDJ-μ gene segments mark the transition of the late pro-B cell to a large pre-B cell. If the cell fails to rearrange its heavy-chain gene segments, the cell dies.

The large pre-B cell expresses the rearranged heavy chain to test it out. The cell traffics the heavy chain to the cell surface. Antibody heavy-chain proteins couple with light-chain proteins, but the large pre-B cell has yet to rearrange the antibody light gene clusters. Therefore, the large pre-B cell couples the heavy chain a "surrogate light chain" to ensure that it properly traffics to the cell surface. The surrogate light chain is made of two proteins, VpreB and lambda 5. The surrogate light chain is a practice light chain for the heavy chain to ensure that it functions properly. Think of the surrogate light chain as a pair of training wheels for the antibody bicycle.

Successful expression of the antibody heavy chain induces the cell to divide and become a small pre-B cell. Next, the small pre-B cell rearranges the light-chain genes, beginning with the kappa genes first. If the cell fails to rearrange its kappa genes, then it moves to the lambda genes. VJ joining of the light-chain gene segments make a functional light chain, and expression of the light chain with the heavy chain places an antibody on the cell's surface, designating it as an immature B cell. If the antibody strongly binds to any self-antigens in the bone marrow, it dies.

The variability in antibody formation is remarkable. The heavy chain alone can encode at least 8262 ($51 \times 6 \times 27$) different heavy-chain variable regions. Suppose a heavy chain is coupled with a kappa light chain. In that case, there are 200 (40 x 5) different light chains and 1.6 x 10^6 (8262 x 200) different possible antibodies that such rearrangements can form.

PRODUCTION OF ANTIBODIES: IMMUNE RESPONSE

Immunity is a state of bodily resistance brought about by the production of antibodies against an invasion by an antigen.

The immune response is mediated by specific white blood cells known as lymphocytes that are made in the bone marrow. There are two types of lymphocytes: T cells, which are formed when lymphocytes migrate to the thymus gland, circulate in the blood, and become associated with lymph nodes and the spleen; and B cells, which are formed in the bone marrow and move directly to the circulatory and the lymph systems. B cells are genetically programmed to produce antibodies. Each B cell synthesizes and secretes only one type of antibody, which can recognize a discrete region (epitope or antigenic determinant) of an antigen with high affinity. Generally, an antigen has several different epitopes. Each B cell produces different antibodies that bind to one of the many epitopes on the same antigen. All of the antibodies in this set, referred to as "polyclonal" antibodies, react with the same antigen.

The adaptive immune response controls infections more effectively than the nonspecific innate response (bodily defenses against infection—such as skin, fever, inflammation, phagocytes, natural killer cells, and some other antimicrobial substances—that are not part of the immune system proper). The immune system has three characteristic responses to antigens: diverse, which effectively neutralizes or destroys various foreign invaders, whether they are microbes, chemicals, dust, or pollen; specific, which effectively differentiates between harmful and harmless antigens; and anamnestic, which has a memory component that remembers and responds faster to a subsequent encounter with an antigen. The primary immune response involves the first combat with antigens. In contrast, the secondary immune response includes the memory component of a first assault. As a result, humans typically get some diseases (such as chickenpox) only once; other infections (such as cold and influenza) often recur because the causative viruses mutate, thus presenting a different antigenic face to the immune system each season.

An antibody-mediated immune response involves several stages: detection of antigens, activation of helper T cells, and antibody production by B cells. White blood cells, known as macrophages, continuously wander through the circulatory system and the interstitial spaces between cells, searching for antigen molecules. Once an antigen is encountered, the invading molecule is engulfed and ingested by a macrophage. Helper T cells become activated by coming in contact with the antigen on the macrophage. In turn, an activated helper T cell identifies and activates a B cell. The activated T cells release cytokines (a class of biochemical signal molecules) that prompt the activated B cell to divide. Immediately, the activated B cell generates two types of daughter cells: antibody-producing cells (each of which synthesizes and releases millions of antibody molecules into the bloodstream in a single day) and B-memory cells (which have a life span of a few months to a year, depending on the immunoglobulin cell from which they derive). The B memory cells are the component of the immune memory system that, in response to a second exposure to the same type of antigen, produces antibodies in larger quantities and at faster rates over a longer time frame than the primary immune response. A similar cascade of events occurs when a macrophage presents an antigen directly to a B cell.

POLYCLONAL AND MONOCLONAL ANTIBODIES

Plasma cells originate from different B cells and manufacture distinct antibody molecules since each B cell was presented with a specific portion of the same antigen by a helper T cell or macrophage. Thus a set of polyclonal antibodies is released in response to an invasion by a foreign agent. Each group of polyclonal antibodies will launch the assault against the foreign agent by recognizing different epitopes of the same antigen. The polyclonal nature of antibodies has been well recognized in the medical field.

In the case of multiple myeloma (a type of cancer), one B cell out of billions in the body proliferates uncontrollably. Eventually, this event compromises the total population of B cells of the body. The immune system will produce vast amounts of IgG originating from the same B cell, which recognizes only one specific epitope of an antigen; therefore, this person's immune system produces a set of antibodies referred to as "monoclonal" antibodies. Monoclonal antibodies form a population of identical antibodies that all specifically recognize one epitope. Thus, someone with this condition may suffer frequent bacterial infections because of a lack of antibody diversity. Indeed, a bacterium whose antigens do not match the antibodies manufactured by the overabundant monoclonal B cells has a selective advantage.

The high-affinity binding capacity of antibodies with antigens has been employed in both therapeutic and diagnostic procedures. However, a manufacturing challenge remains: the effectiveness of commercial preparations of polyclonal antibodies can vary widely from batch to batch. In some instances of immunization, specific epitopes of a particular antigen are potent stimulators of antibody-producing cells. At other times, the immune system responds more vigorously to different epitopes of the same antigen. Thus, one batch of polyclonal antibodies may have a low level of antibody molecules directed against a major epitope and not be as effective as the previous batch. Researchers such as S. K. Rasmussen and others have developed multiple stable cell lines that produce desired monoclonal antibodies to address such inconsistency between batches. Afterward, these multiple batches are combined in a single-batch preparation.

Instead, it may be desirable to produce a cell line that produces monoclonal antibodies with a high affinity for a specific epitope on the antigen for commercial use. Such a cell line would provide a consistent and continual supply of identical (monoclonal) antibodies. Monoclonal antibodies can be produced by hybridoma cells, which are generated by the fusion of cancerous B cells and normal spleen cells obtained from mice immunized with a specific antigen. After the initial selection of hybridomas, monoclonal antibody production is maintained in culture. In addition, the hybridoma cells can be injected into mice to induce tumors that, in turn, will release significant quantities of fluid containing the antibody. This fluid containing monoclonal antibodies can be collected periodically and used immediately or stored for future use. Various systems used to produce monoclonal antibodies include cultured lymphoid cell lines, yeast cells, *Trichoderma reesei* (ascomycetes), insect cells, *Escherichia coli*, *Escherichia coli* monoclonal antibodies, and monkey and Chinese hamster ovary cells. Transgenic organisms and plant cell cultures are potential systems for antibody expression.

IMPACT AND APPLICATIONS

The high-affinity binding capacity of antibodies can potentially inactivate antigens in vivo (within a living organism). Likewise, antibodies may also be employed in many therapeutic and diagnostic applications. In addition, it is a very effective tool in both immunological isolation and detection methods.

Monoclonal antibodies may outnumber all other products being explored by various biotechnology-oriented companies to treat and prevent disease. For example, many strategies for treating cancerous tumors and inhibiting human immunodeficiency virus (HIV) replication employ monoclonal antibodies. HIV is a retrovirus whose genetic material is ribonucleic acid, or RNA, that makes a deoxyribonucleic acid (DNA) copy from an RNA template. HIV causes acquired immunodeficiency syndrome (AIDS). Advances in plant biotechnology have made it possible to use transgenic plants to produce monoclonal antibodies on a large scale for therapeutic or diagnostic use. Indeed, one of the most promising applications of plant-produced antibodies in immunotherapy is passive immunization (e.g., against *Streptococcus mutans*, the most common cause of tooth decay). Large doses of the antibody are required in multiple applications for passive immunotherapy to be effective. Transgenic antibody-producing plants may be one source that can supply vast quantities of antibodies safely and cost-effectively. Hybrid IgA-IgG molecules produced by transgenic plants prevent the colonization of *S. mutans* in culture.

Antibodies expressed in soybeans at a level of 1 percent of total protein may cost approximately one hundred dollars per kilogram of antibody. $100/kg of antibody is relatively inexpensive in comparison with the cost of traditional antibiotics. Transgenic plants can act as bioreactors for the large-scale production of antibodies with no extensive purification schemes. Antibodies have been expressed in transgenic tobacco roots and then accumulated in tobacco seeds. Suppose this technology could obtain a stable accumulation of antibodies in more edible plant organs such as potato tubers. In that case, it could potentially allow for long-term storage and safe and easy delivery of specific antibodies for immunotherapeutic applications. Plant-produced antibodies ("plantibodies") may be more desirable for human use than microbial-produced antibodies; plant-produced antibodies undergo eukaryotic rather than prokaryotic (bacterial) posttranslational modifications. Human glycosylation (a biochemical process whereby sugars are attached to the protein) is more closely related to plants than bacteria.

The potential use of antibody expression in plants for altering existing biochemical pathways has also been demonstrated. For example, germination mediated by a phytochrome (a biochemical produced by plants) has been changed by utilizing plant-produced antibodies. In addition, antibodies expressed in plants have been successfully used to immunize host plants against pathogenic infection; for example, tobacco plants have already been immunized with antibodies against viral attacks. This approach has great potential to replace the traditional methods (use of chemicals) in controlling pathogens.

—Sibdas Ghosh, PhD and Tom E. Scola

Further Reading

"Antibodies." *Genetics & Inherited Conditions.* Salem Press, 2010.

Coico, Richard, and Geoffrey Sunshine. "Antibody Structure and Function." *Immunology: A Short Course.* 6th ed., Wiley-Blackwell, 2009, pp. 41-60.

Diamos Andrew G., et al. "High Level Production of Monoclonal Antibodies Using an Optimized Plant Expression System." *Frontiers in Bioengineering and Biotechnology*, vol. 7, 2020, p. 472, www.frontiersin.org/article/10.3389/fbioe.2019.00472.

Dübel, Stefan, and Janice M. Reichert, editors. *Handbook of Therapeutic Antibodies.* 3 vols. Wiley-VCH, 2014.

Glick, Bernard R., and Jack J. Pasternak, editors. *Molecular Biotechnology: Principles and Applications of Recombinant DNA.* 4th ed., ASM, 2010.

Harlow, Ed, and David Lane, editors. *Using Antibodies: A Laboratory Manual.* Rev. ed., Cold Spring Harbor Laboratory Press, 1999.

Kontermann, Roland, and Stefan Dübel, editors. *Antibody Engineering.* 2nd ed., Springer, 2010.

Mayforth, Ruth D. *Designing Antibodies.* Academic, 1993.

Rasmussen, S. K., et al. "Recombinant Antibody Mixtures: Optimization of Cell Line Generation and Single-Batch Manufacturing Processes." *BMC Proceedings,* vol. 5 (suppl. 8), 2011, p. 2.

Smith, Mathew D. "Antibody Production in Plants." *Biotechnology Advances,* vol. 14, no. 3, 1996, pp. 267-81.

Story, Lachel. "Body Defenses." *Pathophysiology: A Practical Approach.* 2nd ed., Jones, 2015, pp. 31-50.

AUTOIMMUNE DISORDERS

Category: Diseases and disorders
Anatomy or system affected: All tissues
Specialties and related fields: Cardiology, critical care, dermatology, emergency medicine, endocrinology, family medicine, gastroenterology, genetics, hematology, immunology, internal medicine, nephrology, neurology, ophthalmology, orthopedics, pediatrics, pharmacology, psychiatry, pulmonary medicine, rheumatology, vascular medicine
Definition: damage to the tissues or organs of the body caused by failure of the immune system to distinguish between "self" and "nonself," producing autoantibodies called "B cells" and "autoreactive cells" at the request of the T lymphocytes (cells)

KEY TERMS

alleles: one or more variations of a gene that reflects the genes passed from your parents; usually referred to as dominant and recessive genes, while other genes have more than two more alleles

antibody: a molecule of the immune system, produced by B cells and targeted toward eliminating a specific antigen

antigen: a protein or related molecule that is seen as foreign and therefore induces antibody formation in an individual

autoantibody: an antibody that binds to a protein that is a normal part of the human body from which it originates, as opposed to part of a bacteria, virus, or another human being

B cells: also known as B lymphocytes; the antibody-producing cells of the immune system

haplotype: a group of genes within an organism inherited together from a single parent

human leukocyte antigen (HLA): highly polymorphic cell surface proteins that antigen-presenting cells use to present antigen to lymphocytes (class II

HLAs) or are found on the surfaces of every nucleated cell in the body (class I HLAs) and act as identification tags by which the immune system distinguishes between self and nonself; also known as major histocompatibility complex (MHC) proteins

multigenic: referring to a trait or characteristic that requires the product of more than one gene to be expressed

polymorphic: genes that exist in multiple forms or alleles within a population; genes that show extensive variability within a population

selection: the process by which developing immune system cells are either allowed to continue to maturation or destroyed before they can enter the circulation

T cells: also known as T lymphocytes; the immune system cells involved in cellular immunity and regulation of the immune response

tolerance: the ability of the immune system to remain unresponsive to self-antigens

CAUSES OF AUTOIMMUNE DISORDERS

Autoimmunity refers to a group of widely varying diseases or disorders that include familiar examples (type 1 diabetes mellitus, myasthenia gravis, multiple sclerosis, rheumatoid arthritis) and many not as familiar (idiopathic thrombocytopenic purpura, Graves' disease, Felty syndrome, Hashimoto's thyroiditis). The list is long and growing as researchers continue to ferret out the root causes of many disorders that have been known for one hundred years or more. In many cases, environmental triggers or environmentally controlled flare-ups are common. All autoimmune disorders have one thing in common: the failure of the human immune system to distinguish between self (own) and nonself antigens, thus leading the body to attack itself and damage or destroy tissues or organs. Autoimmunity is not a rare event; it occurs in all people and does not necessarily give rise to disease. For instance, aged or damaged cells of the body are usually destroyed by autoantibodies (antibodies directed against the self). However, other autoantibodies, which arise by chance combinations of genes, usually are suppressed during development in the thymus gland. This phenomenon is referred to as selection. Suppose the thymus fails to do its job. In that case, these autoantibodies may be released into the lymph nodes and the bloodstream, seeking out tissue or antigens to attack and destroy.

INFORMATION ON AUTOIMMUNE DISORDERS

Causes: Unknown, possibly hereditary, environmental, or viral

Symptoms: Varies widely; may include pain, fatigue, joint and muscle inflammation, muscle weakness, sleep disturbances, headaches, numbness and tingling, central nervous system disturbances

Duration: Often chronic

Treatments: Alleviation of symptoms, strengthening the immune system

Autoimmune disorders can be classified in several ways. Some diseases affect only one organ system, and some affect multiple systems. Examples of organ-specific disorders are Addison's disease and Graves' disease; non-organ-specific disorders include systemic lupus erythematosus and scleroderma. Alternatively, one can classify autoimmune disorders by the type of immune system cells involved in their onset. Some diseases are caused by the antibody-secreting B cells; they include myasthenia gravis, multiple sclerosis, rheumatic fever, systemic lupus erythematosus, and Graves' disease. Other disorders result from the action of the systemic T cells, including Addison's disease and Hashimoto's thyroiditis.

The onset of an autoimmune disorder hinges on many factors, some of which are still being identified. It is well established, however, that autoimmunity is multifactorial and multigenic. In other words, many environmental factors and many genes are involved in determining susceptibility to autoimmune disorders. In addition, many environmental factors are thought to be involved in controlling remission and flare-ups of autoimmune disorders. Most autoimmune disorders are probably the result of the release of T or B cells that do not correctly distinguish between self and nonself and should have been eliminated or suppressed by the body's immune system.

Autoimmune conditions are caused by hereditary, the environment, bacteria or viruses, fungi, drugs, or some combination of these factors. Most autoimmune diseases have an unknown cause. Some rheumatologists postulate that some gene or group of genes must be activated for autoimmune diseases to develop.

SYMPTOMS

The symptoms of these diseases vary widely and depend on the specific autoimmune disease. Typical symptoms may include pain, fatigue, joint and muscle inflammation, muscle weakness, sleep disturbances, headaches, numbness and tingling, central nervous system disturbances, kidney damage, and liver damage. They can also affect the skin, red blood cells, connective tissue, blood vessels, nerves, and the endocrine glands. Most of them are chronic conditions.

ETIOLOGY OF VARIOUS AUTOIMMUNE DISEASES

Several studies have established genetic links in autoimmune diseases. Still, because most are multigenic, no simple Mendelian inheritance pattern is seen. Nonetheless, there seems to be a clear correlation between specific human leukocyte antigen (HLA) alleles and certain autoimmune disorders. The genes of the major histocompatibility complex (MHC) encode the HLAs. The MHC is a large cluster of tightly linked genes on chromosome six in human beings, but such genes are common to all vertebrates. MHC genes encode cell surface proteins essential for the adaptive immune response. They fall into two main classes, class I and class II MHC protein. Class I MHC proteins are found on the surfaces of all nucleated cells. Viral-infected cells use class I MHC proteins to present viral proteins on their surfaces. The immune system also uses class I MHC proteins to promote "tolerance" or an inability to attack its tissues and cells. Antigen-presenting cells use class II MHCs to present antigen to lymphocytes and other immune cells. Without class II MHC proteins, the immune system cannot recognize foreign substances and mount immune responses to them.

The MHC complex was named from its importance to transplanted tissue compatibility. MHC proteins mediate the interactions of antigen-presenting cells with lymphocytes and other white blood cells. MHC genes are highly polymorphic, meaning that they vary extensively within human populations. The array of alleles within someone's MHC genes determines their donor compatibility for organ transplants and susceptibility to specific autoimmune diseases. Therefore, the correlation between the susceptibility to specific autoimmune disorders and HLA alleles is unsurprising since the class II HLA genes encode proteins that present antigens to lymphocytes. Specific alleles of class II HLA genes encode proteins that have a greater tendency to present self-peptides. Some examples should be illustrative.

For instance, those with HLA allele B27 have a ninetyfold greater risk than the normal population of developing ankylosing spondylitis. Specific HLA-DRB1 alleles, especially those encoding amino acid sequence changes at positions 11 and 13, significantly increase the susceptibility to rheumatoid arthritis. Those with the HLA DR3 haplotype have a twelvefold greater risk of celiac disease and a tenfold greater risk of Sjögren's syndrome. In the case of insulin-dependent diabetes, the relative risk factor is fivefold. If someone has the DR3 and the DR4 haplotype, then the risk increases twentyfold. If the DR3 haplotype and DQw8 subtype of the DR4 haplotype are present, the risk factor is one-hundred-

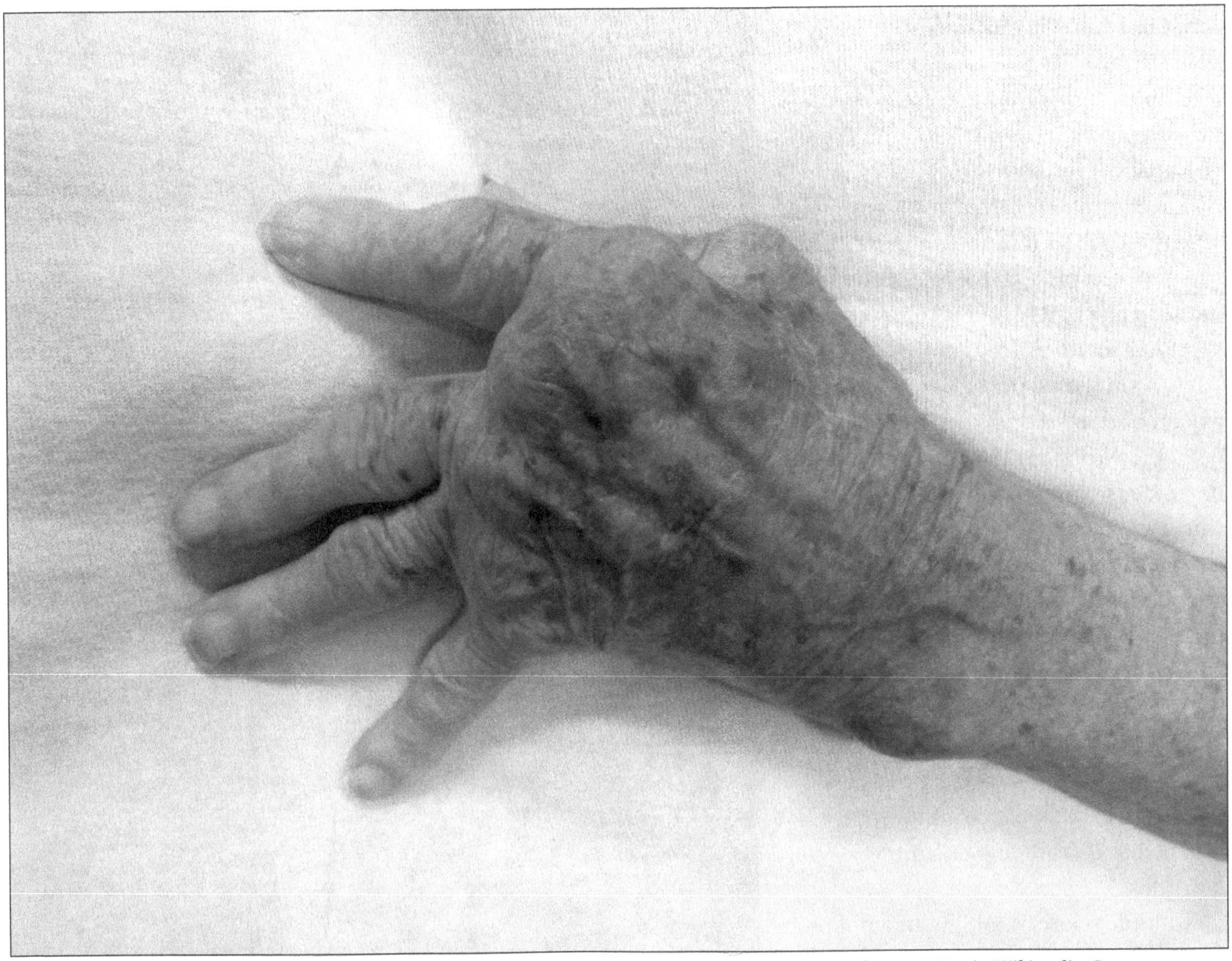

A hand severely affected by rheumatoid arthritis, a common autoimmune disorder. Photo by James Heilman, MD, via Wikimedia Commons.

fold. Yet, these alleles themselves do not automatically cause autoimmune disease, as evidenced by several studies on identical twins in which the rate of disease in the twin of an affected person ranged from 25 to 50 percent. Environmental factors can change susceptibility to autoimmune disease into actual manifestation.

Autoimmune disorders are much more common in women than in men. In addition, they are usually more severe in women. This is likely because estrogen has a role in enhancing the expression of HLA genes and activating macrophages, thus leading to higher tissue destruction. Some autoimmune conditions are noted to flare and subside throughout the menstrual cycle, in conjunction with the rise and fall of estrogen levels. Stress is likely a contributing factor that can cause an autoimmune disorder to flare. This response is likely mediated through the hypothalamus and pituitary glands, which release hormones that directly stimulate the immune system.

The expression of some autoimmune diseases is commonly preceded by infection by a virulent organism. Infections may contribute to autoimmunity in several ways. Some microbes produce antigens that are very close in structure to human antigens. When antibodies are produced against the invading organism, the antibodies also attack self-antigens because of the chemical similarity. Examples of this response are poststreptococcal glomerulonephritis and rheumatic fever. Other invaders may damage human cells and release proteins generally not seen by the immune system (sequestered proteins). These proteins are seen as foreign, and an immune response is set up against these self-antigens. A similar response may be seen when normally sequestered proteins are released through trauma or injury. An example is sympathetic ophthalmia, in which eye lens proteins that are typically not seen in the circulation are released, triggering antibodies that may then attack the opposite (uninjured) eye as well. The symptoms of autoimmune disorders are as varied as the disorders themselves. No one set of symptoms fits all disorders. Symptoms may be systemic or localized, progressive, or stable. Symptoms may also be life-threatening or simply annoying.

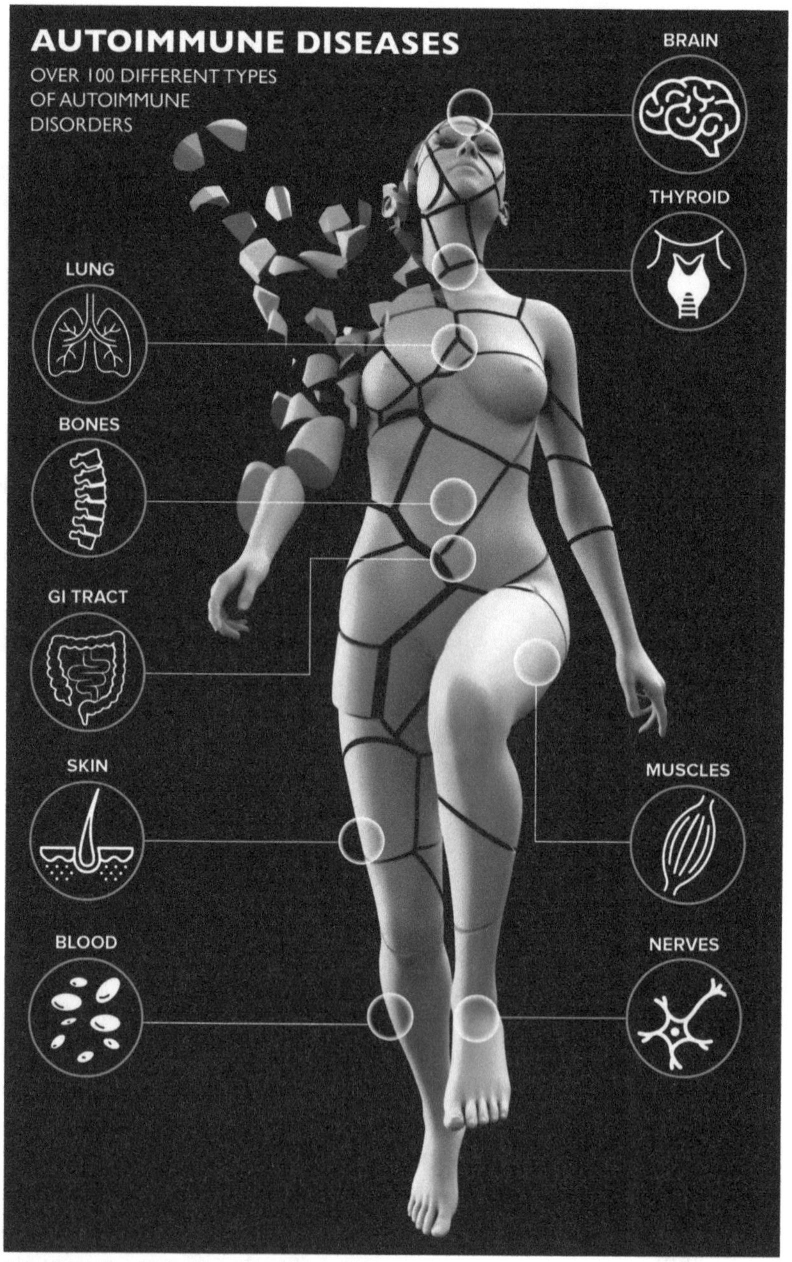

Image via iStock/Naeblys. [Used under license.]

Multiple sclerosis (MS) is an autoimmune disorder involving the central nervous system. Nerve axons of the white matter of the brain are usually surrounded by myelin protein sheaths that protect the nerves and speed the process of transmission. In individuals with MS, these myelin sheath proteins are gradually attacked and destroyed, slowing nerve impulses so that patients develop a loss of control of motor function and vision. This disease often progresses irregularly and unpredictably and is irreversible. It appears to be the result of both B cells producing antibodies against oligodendroglia, the cells that make myelin protein, and T cells, acting against a peptide product from the myelin protein. Although what triggers the initial response is unclear, it has been suggested that onset may follow infection with either Epstein-Barr or hepatitis B viruses. More than 2.1 million people worldwide are affected by MS. It is primarily women who are diagnosed between age twenty and fifty.

Systemic lupus erythematosus (SLE), known simply as lupus, is a generalized disorder that occurs predominantly in women. It is linked to B cells and the production of antibodies against parts of the DNA molecule. These deoxyribonucleic acid (DNA) antibodies bind to free DNA, and a subset of these DNA autoantibodies penetrate cells, translocate to nuclei, inhibit DNA repair, and directly damage DNA. DNA autoantibodies progressively injure cells and tissues and accelerate the accumulated damage associated with the normal aging process. This may form immune complexes deposited in the kidneys and arterioles, leading to tissue destruction and fibrosis, and in the joints, leading to arthritis. Autoantibodies against red blood cells or platelets may also occur in SLE. Antibodies against muscles may be present and contribute to muscle inflammation. In contrast, the presence of antibodies against heart muscle may lead to myocarditis and endocarditis. Antibodies against skin components can lead to a characteristic "butterfly rash" on the bridge of the nose and the area around the eyes seen in many patients with SLE; this rash worsens in the presence of sunlight.

Rheumatoid arthritis is a common, crippling disease. It is controlled by B cells in the joints that are activated to produce several antibodies, including the rheumatoid factor. The result is the formation and deposition of immune complexes in the joint cartilage. Antibodies directed against cartilage may also be seen. The resulting destruction activates chemicals that stimulate T cells to come to the area. In turn, they release destructive enzymes, just as they would if bacteria were invading the joints. All these responses lead to joint damage, inflammation, and pain. The synovia swells and extends into the joints as the disease progresses, causing further pain, discomfort, and joint disfigurement. The cause of antibody activation is unknown and may be quite variable.

Both Hashimoto's thyroiditis and Graves' disease are forms of autoimmune thyroiditis. In Hashimoto's disease, antibodies are formed against a protein within the thyroid cells, leading to immune attack of the thyroid cells and destruction of thyroid tissue. In Graves' disease, the immune system produces antibodies that bind to the receptors for thyroid-stimulating hormone (TSH). This pituitary hormone stimulates the thyroid gland to produce thyroid hormone, and the receptors, in turn, are stimulated. Thus, the thyroid gland is hyperstimulated, and excess thyroid hormone is turned out, a condition known as thyrotoxicosis.

An individual with myasthenia gravis experiences muscle fatigue and extreme weakness with only mild exercise, such as walking short distances. It is caused by autoantibodies that are directed against the acetylcholine (ACh) receptor molecule. In normal neural cells that control large muscles, ACh is stimulated to be released from the neuron and bind to receptors on the muscle fiber endplate. Suppose that receptor is blocked or destroyed by an antibody. In that case, the ACh cannot bind, and therefore the muscle is not stimulated to respond (contract). If a few receptors are blocked, then the muscle may still respond weakly. However, suppose enough antibody is present to block many receptors. In that case, the threshold limit for muscle response will not be achieved. The muscle will not respond even in the presence of repeated stimulation from the neuron. Scleroderma, also known as "progressive systemic sclerosis," predominantly affects middle-aged women and is caused by collagen deposition in various body tissues. Scleroderma may cause the formation of antibodies against the centromere portion of the DNA. Symptoms include calcium deposition in the skin, sensitivity to cold, and decreased esophageal motility. The lungs often experience fibrosis, as do the kidneys.

There is no cure for most autoimmune disorders; they develop into chronic conditions that require a lifetime of care and monitoring. Treatment is quite varied and depends on the underlying cause and etiology of the disease. Overall, the goals of treatment are to reduce the symptoms and to control the disease or disorder while at the same time allowing the immune system to continue fighting the viruses and bacteria affecting the body daily. The drugs used to treat autoimmune disorders are listed below.

Corticosteroids. Prednisone and methylprednisolone are glucocorticoid analogs that suppress inflammatory mediator production and immune effector cells and promote T-lymphocyte apoptosis (death). They are used during acute periods of the disease. Complications arise with high doses and prolonged therapy. Severe adverse effects are bone marrow suppression, gastrointestinal complications, cataracts, and glaucoma.

Disease-modifying antirheumatic drugs (DMARDs). Azathioprine (AZA) is a purine analog (6-mercaptopurine) that inhibits the synthesis of deoxyribonucleic acid (DNA), ribonucleic acid (RNA), and proteins and interferes with purine metabolism and mitosis, suppressing delayed hypersensitivity responses and cell-mediated cytotoxicity. Common adverse effects include leukopenia, pancreatitis, hepatitis, bone marrow suppression, potential malignancies, and pulmonary disease.

Cyclophosphamide. Cyclophosphamide is a nitrogen-derived alkylating agent/cytotoxic immunosuppressant that cross-links DNA and RNA strands, inhibiting cell functions and protein synthesis. It has a dose-dependent effect on the immune system. At high doses, it can induce an aberrant anti-inflammatory, immune effect on lymphocyte activity, affect regulatory T cells, and cause a state of severe immunosuppression that includes significant bone-marrow suppression, leukopenia, anemia, and thrombocytopenia. There can also be adverse effects on the gastrointestinal or renal-genitourinary tracts and the cardiovascular system. There can be an increased risk of malignancy and pulmonary toxicity.

Methotrexate (MTX). MTX is a dihydrofolate reductase inhibitor and antimetabolite approved for Crohn's disease, rheumatoid arthritis, and psoriasis. Despite its efficacy, MTX has severe toxic effects with prolonged use, including liver damage, cytopenias, and several pulmonary diseases; hypersensitivity pneumonitis is most frequently reported.

Hydroxychloroquine. Hydroxychloroquine is an antimalarial agent with immunosuppressant properties and is used to treat systemic lupus erythematosus. The drug has cardiovascular effects that can cause toxic myopathy, cardiomyopathy, and peripheral neuropathy but is usually well tolerated at prescribed doses.

Biological DMARDs. Adalimumab is a monoclonal antibody and tumor necrosis factor (TNF) inhibitor that binds TNF-alpha and blocks its interaction with cell surface receptors. Adverse events include renal-genitourinary effects and dyslipidemia. Other anti-TNF-alpha monoclonal antibodies include golimumab (Simponi) and certolizumab pegol (Cimzia).

Infliximab is a chimeric (part human/part synthetic) monoclonal antibody and anti-TNF agent that binds TNF-alpha and blocks its interaction with cell surface receptors. A twofold risk of infection is the most common adverse event. The risk of developing tuberculosis may be greater than with other anti-TNF agents. Skin and subcutaneous tissue infections and apoptosis-inducing activity (cell death) can occur.

Etanercept is a TNF receptor antagonist that inhibits the binding of TNF-alpha and TNF-beta to cell surface receptors, preventing its interaction with TNF receptors and rendering it biologically inactive. Its use can cause infections of the respiratory tract, skin, or subcutaneous tissue.

Three biological DMARDs (infliximab, adalimumab, and golimumab) are Food and Drug Administration (FDA)-approved for severe ulcerative colitis that does not respond to other medications. Infliximab, adalimumab, and certolizumab pegol are FDA-approved for treating moderate to severe Crohn's disease. Anti-TNF-alpha monoclonal antibodies also treat rheumatoid arthritis, psoriasis, psoriatic psoriasis, and ankylosing spondylitis.

Hormones, proteins, or other substances customarily produced or secreted by the cells or organs damaged in autoimmune disease (such as thyroid hormone or insulin) can usually be supplemented to the point that they are within the proper physiologic range. Sometimes this works well. For instance, to effectively control Graves' disease, a surgeon removes the overactive thyroid gland, and the patient begins taking oral thyroid hormone supplements. Treating type 1 diabetes in children with insulin supplementation, however, is somewhat trickier. Insulin shots cannot duplicate the finely controlled release of insulin from the beta cells of the pancreas. However, with the advent of constant glucose monitors (CGMs) and insulin pumps, diabetic children can experience tight blood glucose control and normal growth and development. Cell phone-based glucose readouts, such as the Dexcom System, further simplify blood glucose monitoring and control.

Many investigators have worked on vaccinations for autoimmune disorders, using animal models of varying types. Some results have been promising, even if the mechanism of action is still mostly unexplained. Vaccinations against autoimmune thyroiditis, encephalitis, and arthritis have been successful in some animal models. Another experimental approach is oral tolerance therapy. Large quantities of the offending autoantigen are given to the patient to induce tolerance to the protein. This approach is like desensitization used with allergy sufferers. Oral doses of myelin, for instance, have shown some success as a treatment for patients with multiple sclerosis.

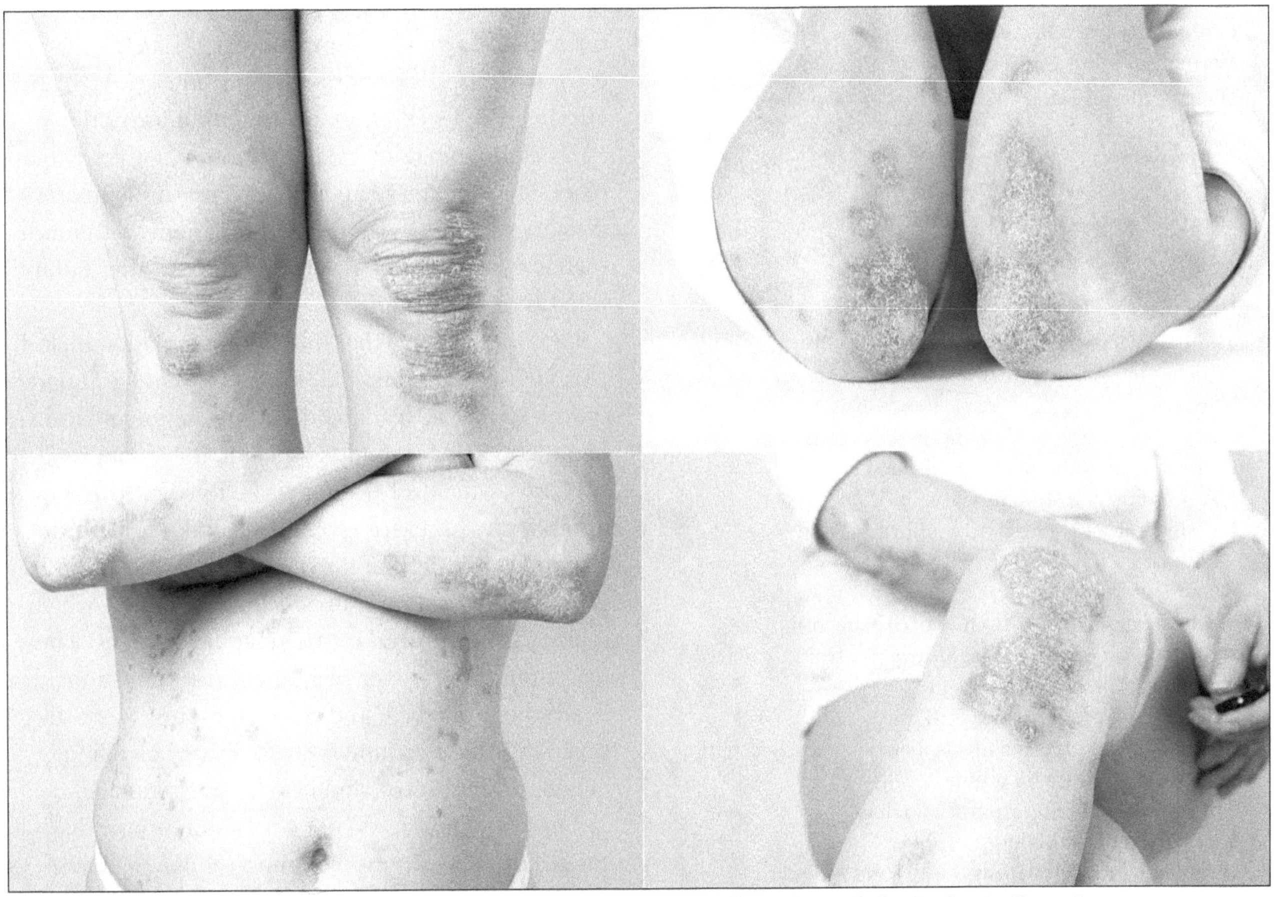

Examples of psoriasis, a frequently occuring autoimmune disorder. Photos via iStock/Natalia Serdyuk. [Used under license.]

Miscellaneous medications. Other medications include rituximab, tocilizumab, and belimumab. Rituximab selectively deletes specific B cells that have a particular marker called "CD20." It treats rheumatoid arthritis, two types of vasculitis, MS, lupus, and other autoimmune diseases. Some patients have hypersensitivity reactions to rituximab. Tocilizumab is a monoclonal antibody that inhibits a cytokine from T cells called "interleukin-6" (IL-6). Tocilizumab treats rheumatoid arthritis, giant cell arteritis, and severe COVID-19. Sarilumab is a monoclonal antibody with similar activity to tocilizumab approved for rheumatoid arthritis. These drugs can cause infusion reactions, high blood pressure, low white blood cell counts, signs of liver damage, and elevated blood lipids.

Belimumab is a monoclonal antibody that inhibits B-cell activating factor and treats lupus and Sjorgen's syndrome. Abatacept (Orencia) is a genetically engineered fusion protein that interferes with T-cell activation. This drug is used with other medicines to treat rheumatoid arthritis. Anakinra (Kineret) is a genetically engineered IL-1 receptor antagonist. It is the least commonly used biologic DMARD for rheumatoid arthritis because it is expensive since daily injections are required. A drug with a similar mechanism of action is a monoclonal antibody called "canakinumab" (Ilaris). This drug is FDA-approved for systemic juvenile idiopathic arthritis and cryopyrin-associated periodic syndromes (CAPS) and effectively relieves gout pain and inflammation.

Ustekinumab (Stelara) is an anti-p40 antibody that blocks interleukin-12 and -23, two pro-inflammatory cytokines. It is FDA-approved for patients with moderately to severely active Crohn's disease after failure with other medications.

Oral medications that inhibit inflammation include JAK kinase inhibitors. JAK kinase is an integral part of lymphocyte activation, and its inhibition quells inflammation. JAK inhibitors include tofacitinib (Xeljanz), which is FDA-approved for the treatment of moderate to severe ulcerative colitis, baricitinib (Olumiant), beneficial for rheumatoid arthritis and severe COVID-19, and upadacitinib (Rinvoq), which is also FDA-approved for rheumatoid arthritis. These drugs can cause upper respiratory tract infections, headaches, diarrhea, and cold symptoms. More severe side effects include gastrointestinal bleeding, fatigue, rashes, sores, and urinary tract damage.

These mediations do not cure autoimmune diseases, but they can put them into remission. These medications are very potent and have significant side

Partial List of Autoimmune Diseases

Disease: Organ/Target
Addison's disease: Adrenal glands
Ankylosing spondylitis: Spine
Antiphospholipid antibody syndrome: Blood clotting
Celiac disease: Small intestine
Crohn's disease: Gastrointestinal tract
Dermatomyositis: Skeletal muscles and skin
Diabetes mellitus type 1: (some forms) Pancreas islet cell
Felty's disease: Joints and spleen
Goodpasture's syndrome: Kidneys and lungs
Graves' disease: Thyroid gland
Guillain-Barré syndrome: Peripheral nervous system
Hashimoto's thyroiditis: Thyroid gland
Hemolytic anemia: Red blood cells, platelets
Multiple sclerosis: Brain and spinal cord
Myasthenia gravis: Junctions between nerves and muscles
Pemphigus vulgaris: Skin and mucus membranes
Pernicious anemia: Stomach parietal cells
Polymyositis: Skeletal muscle
Poststreptococcal glomerulonephritis: Kidneys
Psoriasis: Skin
Rheumatic fever: Heart
Rheumatoid arthritis: Connective tissue and joints
Scleroderma: Heart, lungs, kidney, gastrointestinal tract, and skin
Sjögren's syndrome: saliva and tear glands
Systemic lupus erythematosus: deoxyribonucleic acid (DNA), platelets, all organs
Thrombocytopenic purpura: Platelets
Ulcerative colitis: Colon
Wegener's granulomatosis: Lungs and kidneys

effects. However, they can permit a patient to live a relatively everyday life.

PERSPECTIVE AND PROSPECTS

The history of human understanding of autoimmune disorders is relatively short. Early in the twentieth century, Paul Ehrlich described a condition of "horror autotoxicus," the attack of the human immune system against the body's tissues. His studies set the stage for rapid advancement in the understanding of the human immune system. However, understanding the genetic and molecular basis for autoimmunity, along with the realization that autoimmunity is a normal part of immune system development, began in the 1980s. Even where the cause was well established (such as with insulin-dependent diabetes, established in the 1920s), no significant changes in treatment were made until new genetic tools became available. Indeed, the entire field of immunology, which until the 1970s was in its infancy as a medical field, has grown exponentially as new molecular tools have enabled researchers to elucidate the pathways by which autoimmunity exacts its toll.

There is still plenty to do, both in determining pathways and in developing new therapies aimed at specific targeting of these pathways. With the completion of the Human Genome Project, an incredible amount of new knowledge is available to help researchers produce treatments that are much more targeted and specific than those used in the past. As more is learned about the immune system and inflammation pathways, new therapies may be designed that may prevent the development of most autoimmune diseases. Stem cell treatments with mesenchymal stem cells and mesenchymal stem cell-derived products, while still experimental, may provide relief in several cases for patients who suffer from intractable autoimmune disorders.

—Kerry L. Cheesman, PhD, Christine M. Carroll, RN, BSN, MBA, and Michael A. Buratovich, PhD

Further Reading

Abbas, Abul K., and Andrew H. Lichtman. *Cellular and Molecular Immunology*. 10th ed., Saunders/Elsevier, 2021.

"Autoimmune Conditions." *Medline Plus*, 31 Mar. 2021, https://medlineplus.gov/autoimmunediseases.html. Accessed 20 Aug. 2021.

Chatenoud, Lucienne. "Emerging Biological and Molecular Therapies in Autoimmune Disease." *The Autoimmune Diseases*, edited by Noel R. Rose and Ian R. Mackay, 6th ed., Academic Press, 2020, pp. 1437-57.

Dettmer, Philipp. *Immune: A Journey into the Mysterious System That Keeps You Alive*. Random House, 2021.

"Drugs for Psoriatic Arthritis." *Medical Letter on Drugs and Therapy*, vol. 61, no. 1588, 30 Dec. 2019, pp. 203-10.

Janeway, Charles A., Jr., et al. *Immunobiology: The Immune System in Health and Disease*. 6th ed., Garland Science, 2005.

Kahmini, Fatemeh Rezaei, and Shahab Shahgaldi. "Therapeutic Potential of Mesenchymal Stem Cell-Derived Extracellular Vesicles as Novel Cell-Free Therapy for Treatment of Autoimmune Disorders." *Experimental and Molecular Pathology*, vol. 118, 2011, p. 104566, doi:10.1016/j.yexmp.2020.104566.

Rose, Noel R., and Ian Mackay, editors. *The Autoimmune Diseases*. 6th ed., Academic Press, 2019.

Tsai, Sue, and Pere Santamaria. "MHC Class II Polymorphisms, Autoreactive T-Cells, and Autoimmunity." *Frontiers in Immunology*, vol. 4, no. 321, 2013, pp. 1-7.

Watson, Stephanie. "Autoimmune Diseases: Types, Symptoms, Causes, and More." *Healthline*, 26 Mar. 2019, www.healthline.com/health/autoimmune-disorders?print=true. Accessed 20 Aug. 2021.

Zack, Eric. "Emerging Therapies for Autoimmune Disorders." *Journal of Infusion Nursing*, vol. 37, no. 2, 2012, pp. 109-19.

IMMUNITY AND INFECTIOUS DISEASE

Category: Immune response

Also known as: Acquired (adaptive) immunity, cellular immunity, immune response, inflammatory response, innate immunity

Anatomy or system affected: Blood, cells, circulatory system, immune system, lymphatic system

Specialties and related fields: Bacteriology, immunology, microbiology

Definition: the state, quality, or condition of resistance to pathogens, parasites, and nonliving harmful substances

KEY TERMS

antibodies: secreted glycoproteins produced in response to antigens that bind to and neutralize bacteria, viruses, and foreign substances in the blood

B lymphocytes: white blood cells that secrete antibodies

macrophages: large phagocytic cells found in stationary form in the tissues or as mobile white blood cells, especially at sites of infection

neutrophils: the most abundant type of granulocytes that make up 40 to 70 percent of all white blood cells and form an essential part of the innate immune system

phagocytosis: the ingestion of bacteria or other material by specialized white blood cells known as "phagocytes"

T lymphocytes: white blood cells that are borne in the bone marrow but mature in the thymus and protect the body from infection and cancer

vaccination: treatment with a vaccine to produce immunity against a disease

INNATE AND ACQUIRED IMMUNITY

Ubiquitous pathogens are found on surfaces, on food, and in the air. Innate and acquired immunity confers lifelong protective immunity to the body against foreign substances, including harmful toxins, viruses, and bacteria. Three basic components work closely to protect the body: physical barriers such as the various epithelial surfaces, innate immunity, and acquired immune responses. Inherited genes, environment, lifestyle, and acquired characteristics can influence the state of immunity.

Innate, or natural, immunity is the ability inherent from birth to fight infection without adapting to a specific pathogen. Innate immunity is characterized by physical barriers that defend against harmful agents and by more sophisticated defense mechanisms. Sometimes, physical defenses can be triggered through innate immune responses, such as ciliary action or sneezing from histamine production.

Other defenses include bactericidal enzyme action in secreted bodily fluids and more complex complement proteins. The innate immune system is nonspecific, focusing on conserved pathogen-associated molecular patterns so that many organisms are attacked similarly. Although the quality and efficacy of the initial innate response do not improve after subsequent exposures to the same pathogen, innate immunity includes several other defense mechanisms. Epithelial surfaces, including the genitourinary tract, respiratory tract, skin, and gastrointestinal tract, produce antimicrobial peptides such as defending and cathelicidins that inhibit bacterial and fungal growth. Two nonspecific methods to eliminate microorganisms are phagocytosis and opsonization. In phagocytosis, specialized cells such as neutrophils, monocytes, and macrophages ingest and destroy ingested pathogen particles. In opsonization, phagocytic cells recognize a plasma protein (opsonin) binding to the pathogen's surface, leading to enhanced phagocytosis.

The hallmark of innate immunity is an inflammatory response (inflammation or edema)—proinflammatory mediators such as cytokines, chemokines, and lipid mediators clear the infection. Inflammation, however, is damaging and painful to tissues, and some chronic diseases possess an inflammatory pathology component. Clinically, drugs can be used to control inflammation. The innate immune response is an early defense mechanism against infection. Still, it is also essential in boosting subsequent adaptive immune responses.

During adaptive (acquired) immunity, the immune system develops a defense specific to a particular antigen. It does so with immunological specificity and long-lasting memory beyond the acute infection. An agent evoking a specific immune response is called an "immunogen." Immunogens reacting with antibodies are antigens. Virtually any substance of a specific size, including cell proteins, viral nucleic acids, chemicals, or foreign particles (such as a splinter), can become an antigen. The goal of an acquired immune response is to recognize and destroy substances containing antigens.

Acquired immunity utilizes two sophisticated and flexible mechanisms: cell-mediated immunity and humoral immunity. The cell-mediated response relies on B and T lymphocytes (white blood cells). Following antigen exposure, antigens are taken up and presented to B and T lymphocytes by antigen-presenting cells such as macrophages from the innate system or by dendritic cells from the acquired system. After recognizing their specific matching antigen, B cells differentiate into plasma cells, producing and secreting large amounts of antibodies against the specific antigen.

Likewise, T cells differentiate after antigen recognition into helper T cells (Th) or cytotoxic (killer) T cells (Tc); the T cells release lymphotoxins, causing cell lysis. Th's secrete lymphokines, which further stimulate Tc and B cells to proliferate and divide, attracting neutrophils and improving phagocytes' ability to engulf and kill pathogens. Although innate immunity is available instantly upon infection, acquired immunity takes approximately seven to ten days to mount an initial response. Parts of the innate system, such as complement or phagocytosis,

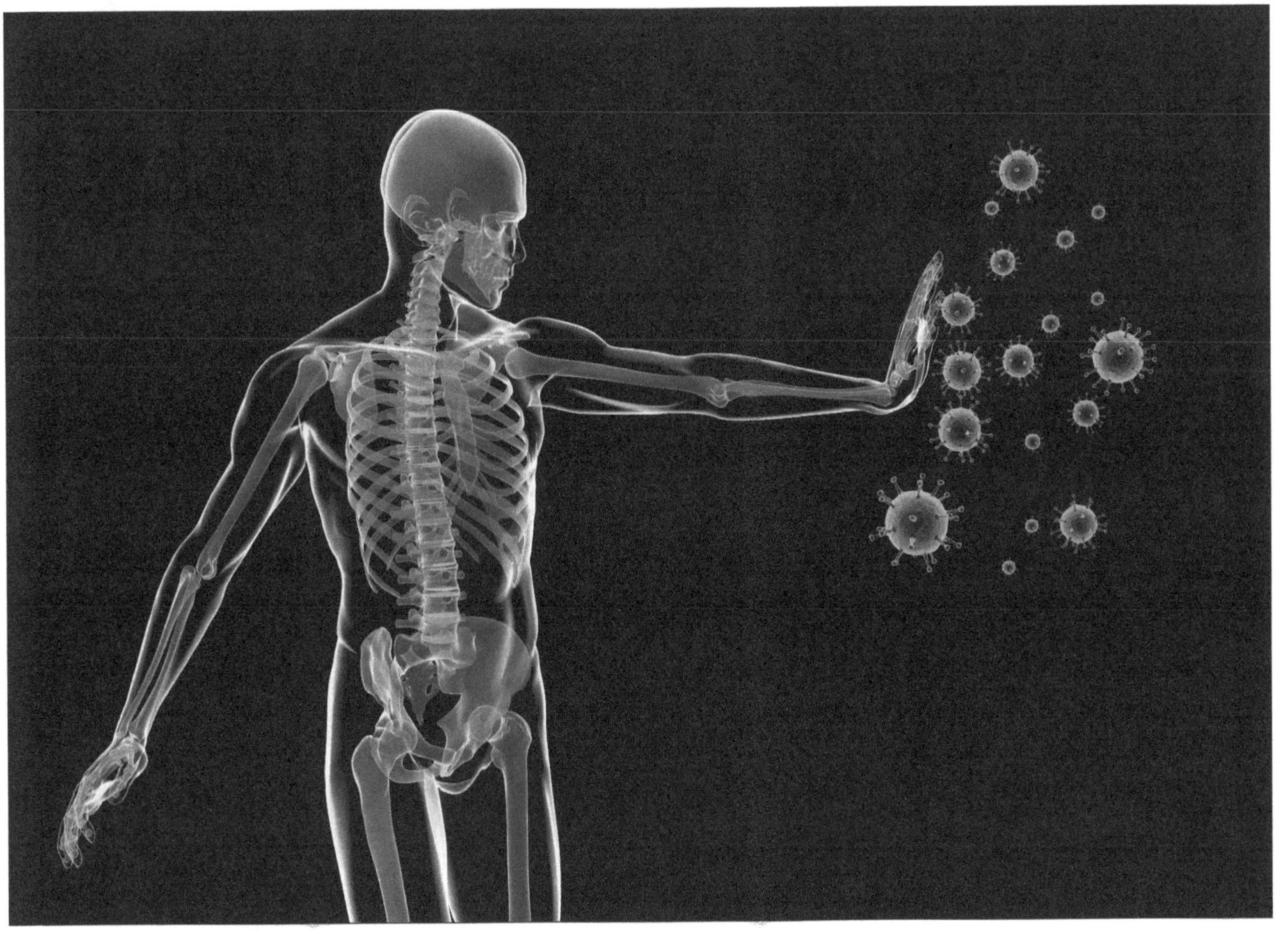

Image via iStock/SciePro. [Used under license.]

can also be activated by the acquired system through antibody mediation.

Immunoglobulin (Ig) is another term for antibody; it binds specifically to antigenic determinants or epitopes. Immunoglobulins inactivate antigens by complement fixation, neutralization, agglutination, and precipitation. Immunoglobulins are made of two identical heavy chains and identical light chains. They are classified based on their heavy chain as IgM, IgG, IgA, IgE, and IgD.

PASSIVE AND ACTIVE IMMUNITY

Antibody-mediated immunity includes passive and active immunity. Exposure to the pathogen/antigen results in active immunity. For example, the antigen is presented (by injection of a weakened, killed, or recombinant pathogenic antigen), resulting in the vaccinated person's body generating a specific immune response against that antigen.

In passive immunity, "natural" or "artificial," the body does not manufacture its antibodies; instead, it gets antibodies from another person. For example, infants undergo natural passive immunity during the transfer of antibodies through the maternal placenta or milk. These infant antibodies disappear between six and twelve months of age with the replacement of breast milk. Passive immunity is short-lived because these antibodies are degraded in the body over time and because no immunological memory exists to produce more antibodies.

Artificial passive immunity involves the transfusion of an antiserum or the injection of antibodies produced by another person or animal. Immediate protection against an antigen is achieved through these antibodies, although it is a short-lived immunity. Examples of passive immunization include tetanus antitoxin and purified human gamma-globulin.

IMMUNITY DISORDERS AND COMPLICATIONS

Sometimes single components of the immune system are inefficient, absent, or excessive. In these cases, the state of protection is not reached adequately. The impaired immune system is considered immunocompromised, and it could leave the host body vulnerable to various opportunistic infections. For example, acquired immunodeficiency syndrome (AIDS) is a result of the depletion of helper T cells after a viral infection. The failure of host defense mechanisms can lead to conditions such as autoimmune diseases, immunodeficiencies, allergies, delayed hypersensitivity states, and transplant rejections. Immune responses in the absence of infection include allergy or hypersensitivity reactions, autoimmunity, and graft rejections. An allergic reaction occurs against innocuous substances, and responses to self-antigens are visible in autoimmune diseases. Immunodeficiencies can be inherited (primary) or acquired (secondary).

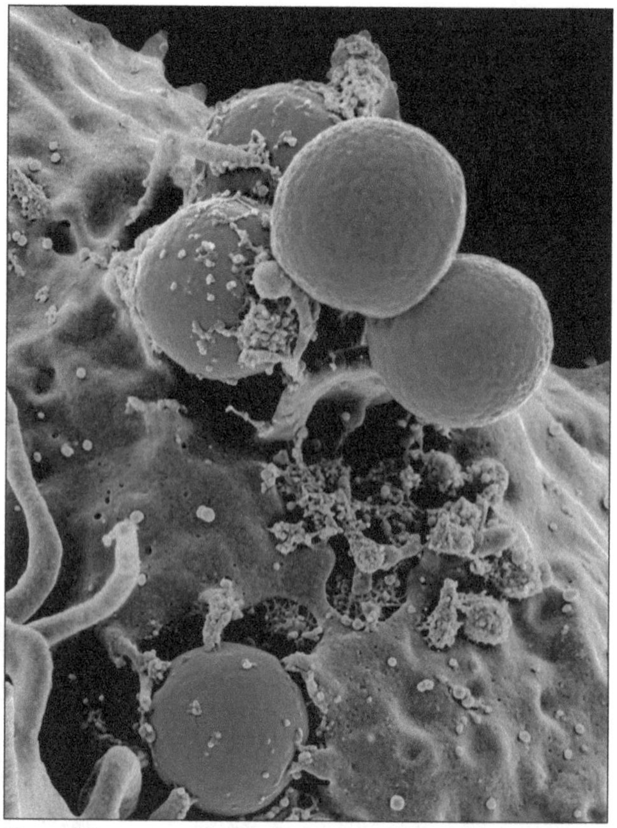

Neutrophils can ingest harmful bacteria like MRSA bacteria. By National Institutes of Health (NIH), via Wikimedia Commons. [Public domain.]

IMPACT

Vaccination is a preventive measure against morbidity and mortality resulting from infectious diseases such as polio, measles, diphtheria, pertussis, rubella, mumps, tetanus, and *Haemophilus influenzae* type B. It is an artificial method of building immunity by deliberately infecting a person so that the body learns self-protection from a pathogen. Controlling and even eradicating infectious diseases reduces frequent doctor's visits, hospitalizations, and deaths, leading to improved public health, reduced disease burdens, and reduced health-care costs—for example, the World Health Organization's immunization campaign from 1967 to 1977 eradicated smallpox.

Another immunization strategy is herd immunity. Immunizations of a high percentage (a herd) of a population protect unvaccinated persons. This type of community immunity tries to break the chain of infection by having large sections of a population immune. It slows infectious disease transmission and can even stop outbreaks.

—Ana Maria Rodriguez-Rojas, MS

Further Reading

Baxter, David. "Active and Passive Immunity, Vaccine Types, Excipients, and Licensing." *Occupational Medicine,* vol. 57, no. 8, 2007, pp. 552-56.

Bonds, M. H., and P. Rohani. "Herd Immunity Acquired Indirectly from Interactions Between the Ecology of Infectious Diseases, Demography, and Economics." *Journal of the Royal Society Interface,* vol. 7, 2010, pp. 541-47.

DeFranco, Anthony, Richard Locksley, and Miranda Robertson. *Immunity: The Immune Response in Infectious and Inflammatory Disease*. Sinauer, 2007.

Keller, M. A., and E. R. Stiehm. "Passive Immunity in Prevention and Treatment of Infectious Diseases." *Clinical Microbiology Reviews,* vol. 13, no. 4, 2000, pp. 602-14.

Murphy, Kenneth, Paul Travers, and Mark Walport. *Janeway's Immunobiology*. 7th ed., Garland Science, 2008.

National Library of Medicine. "Immune System and Disorders." www.nlm.nih.gov/medlineplus/immune systemanddisorders.html.

Strugnell, R. A., and O. L. Wijburg. "The Role of Secretory Antibodies in Infection Immunity." *Nature Reviews Microbiology,* vol. 8, no. 9, 2010, pp. 656-67.

Immune response

Category: Biology

Anatomy or system affected: Blood, bone marrow, cells, circulatory system, glands, liver, lymphatic system, spleen

Specialties and related fields: Biochemistry, cytology, hematology, immunology, microbiology, preventive medicine, serology

Definition: a system that includes the spleen, thymus, lymphatic system, and specialized cells and protects the body from foreign substances

KEY TERMS

antibody: any of the proteins produced in the body during an immune response; recognizes and attacks foreign antigen substances

antigen: a substance within the human body recognized as foreign either by antibodies or by special immune cells; the cause behind the stimulation of the immune response

autoimmunity: an abnormal immune reaction against antigens

immunosuppression: a decrease in the effectiveness of the immune system

pathogen: any disease-causing organism, including a virus, bacterium, protozoan, mold or yeast, or other parasites

STRUCTURE AND FUNCTIONS

The immune system is capable of recognizing and identifying many different substances foreign to the human body. For the immune system to function correctly, it must receive, interpret, and transmit large amounts of information about invaders outside or within the body. These constant and ever-changing

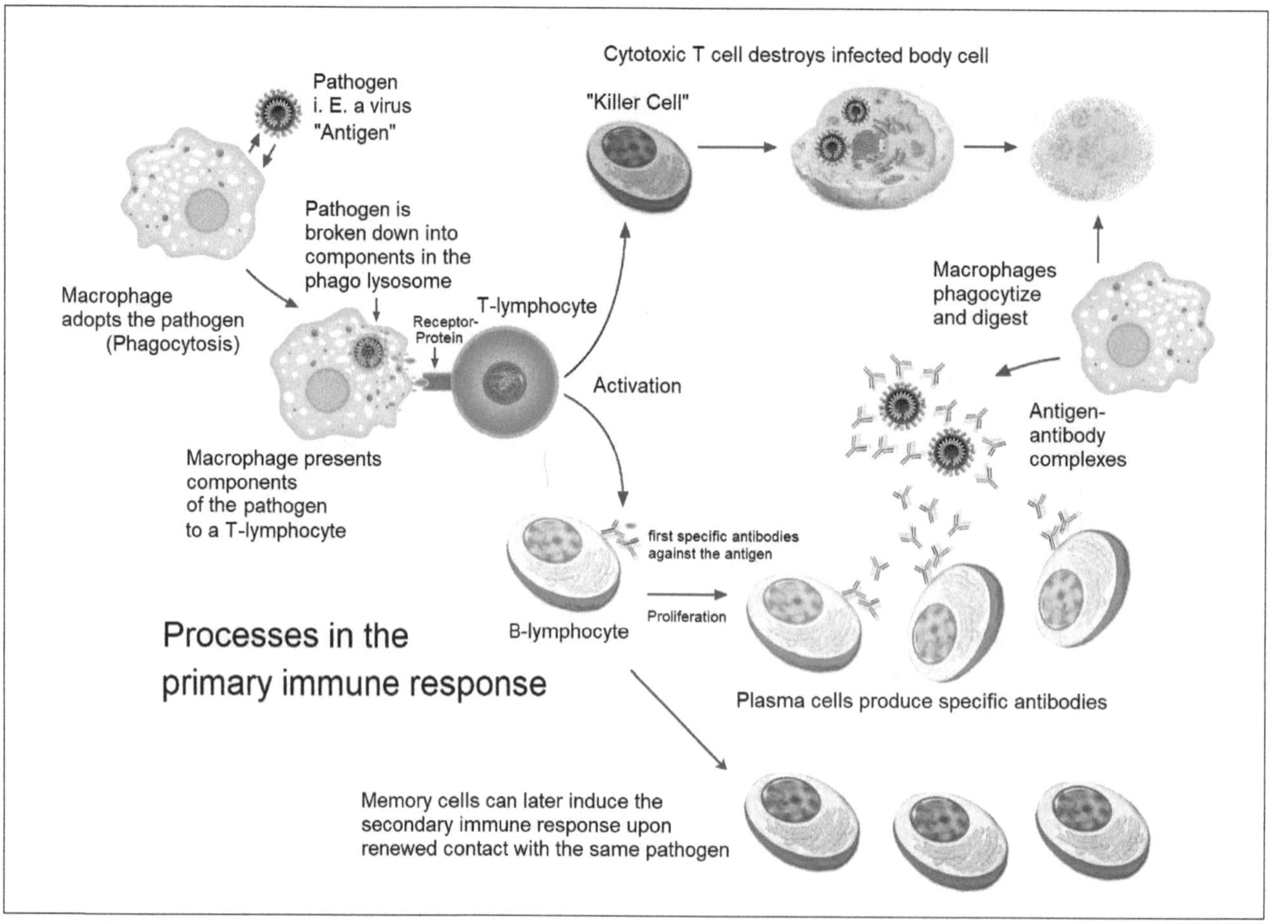

Overview of the processes involved in the primary immune response. Image by Sciencia58/Domdomeg/Petr94/Manu5, via Wikimedia Commons.

threats to the body must be met and destroyed by one complex system—namely, the human immune system. Many organs and parts of the body play a significant role in maintaining resistance; some have more important roles than others, but all parts must work in unison. The circulatory and lymphatic systems, along with specific organs, are of primary importance in the overall workings of the immune system.

Blood. Besides the outer protective layer of the skin and mucous membranes, the first line of defense in the immune system includes the blood in the circulatory system. About 50 percent of human blood is plasma. Plasma contains mainly water, but also proteins, carbohydrates, vitamins, hormones, and cellular waste. The other half of blood is composed of white cells, red cells, and platelets. The red blood cells, called "erythrocytes," are responsible for moving oxygen from the lungs to the other parts of the body. Special cells called "platelets" or "thrombocytes" enable the blood to form clots, thus preventing severe bleeding. An unborn child produces red and white blood cells in the spleen and liver, while a newborn makes blood cells in the center of bones, called the "marrow." After maturity, the bone marrow produces all red and most white blood cells. Although the red cells and platelets are vital, the white cells play a significant role in the immune system.

In a broad sense, white blood cells surround and engulf foreign matter and adjacent dying cells in a process called "phagocytosis." The function is possible since the white blood cells can move, unlike red corpuscles, by pushing their bodies out and pulling forward. Red corpuscles flow with the blood within the circulatory system. White blood cells flow within the bloodstream but move in the lymph vessels, where they work to defend the body against diseases. White blood cells can destroy some of the bacteria and foreign matter they engulf. However, sometimes the corpuscle dies from the toxins produced by the bacteria. The resulting formation of pus is an accumulation of dead white blood cells.

Three major types of white blood cells, known collectively as leukocytes, are involved in immune responses. All three types of leukocytes, granulocytes, monocytes, and lymphocytes, arise from areas in either bone marrow, the spleen, or the liver.

The granulocytes are about twice the size of a red blood cell, originate from red bone marrow, and live only about twelve hours. Under the classification of granulocytes, distinct cells have different structures, sizes, and shapes. These specialized granulocytes include neutrophils, eosinophils, and basophils. None of these cells has a specific memory for future immune responses. Neutrophils eat and digest small foreign matter with the help of special enzymes. Between 40 and 75 percent of the white blood cells in the human body are neutrophils. When these highly mobile neutrophil cells arrive at an injury site, they release their enzymes and degrade the surrounding tissues. Eosinophils are similar to neutrophils but seem specialized in fighting infection caused by parasites because of the array of toxic proteins they secrete. They are also effective against fungal, bacterial, viral, or protozoan infections. Basophils are smaller cells that account for less than 1 percent of the white blood cells found in the blood. After being born in the bone marrow, they travel throughout the body to fight fungal and worm infections and allergies. They also regulate immune responses and prevent coagulation by releasing heparin. Basophils cannot engulf and destroy foreign matter.

The second group of leukocytes includes the monocytes, the largest cells found in the blood. Monocytes are two to three times as large as red cells, yet they are not very numerous, making up 3 to 9 percent of all the leukocytes in the blood. After only a few days in the blood, they move to areas between tissues. Over months or years, the monocytes enlarge ten times in size to specialize in phagocytosis. After this growth, they are called "macrophages." They are also called "terminal cells" since they cannot divide and thus do not reproduce.

The third type of leukocyte, and the most complex white blood cells, are called "lymphocytes" because they come from the lymph system and bone marrow. The T lymphocytes can differentiate into helper, killer, and suppressor cells. Besides recognizing foreign matter precisely, they can live freely in the blood, grow larger and divide, and then change back to their original form after working against the invader. Lymphocytes circulate throughout the body, moving from the bloodstream through the lymph fluid and into the blood. The two major types of lymphocytes are T lymphocytes (also called "T cells") and B lymphocytes (also called "B cells"). Both T and B cells can recognize foreign matter and hook onto it. Some of these special "memory" cells remain in the body for life, preventing a specific invader from causing illness when reencountered in the future. These specialized cells must have a way to travel through the body; one of these transport systems is the lymphatic system.

The lymphatic system. This system is a closed network of vessels that help circulate fluids from the body and return them to the bloodstream. The lymphatic system also defends against disease-causing foreign materials, known as "antigens." The smallest components of the lymph system are the lymphatic capillaries that run parallel to the blood capillaries. The fluid, or lymph, within lymphatic capillaries comes from the liquid that moves across the cell membranes from tissues throughout the body. These capillaries merge into larger lymphatic vessels that fuse into a "lymph node." The lymph fluid is drained into trunks that join one of two collecting ducts. The larger left thoracic duct collects lymph from the lower part of the abdomen, the legs, and the left side of the upper body before emptying into a vein near the neck and shoulder. The right lymphatic duct does the same for the right side of the upper body. After leaving the collecting ducts, the lymph fluid becomes part of the blood plasma in the veins and returns to the heart's right atrium. Lymph does not flow like blood in veins and arteries; instead, it is controlled by muscular activity.

The spleen. This lymphatic organ is in the upper left abdominal cavity, behind the stomach, and under the diaphragm. The hollow spaces within the spleen are filled with blood, making it soft and elastic. The white blood cells in the lining of these open cavities engulf and destroy foreign materials, as well as damaged red blood cells that pass through the spleen.

The thymus. This gland lies between the lungs and above the heart, just behind the upper part of the breast bone. It contains many white cells; some are inactive, but others develop and leave the thymus to become functional in the immune system.

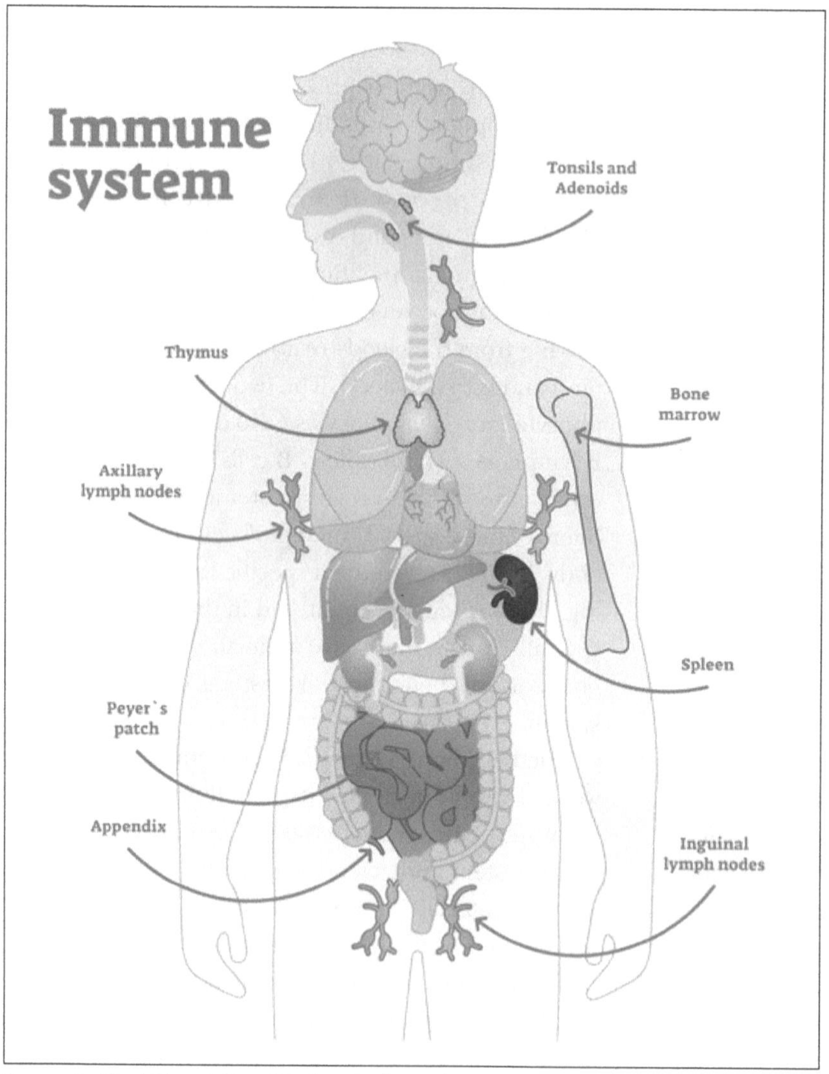

Image via iStock/VectorMine. [Used under license.]

The liver. Well protected by the ribs, the liver sits in the upper right of the abdominal cavity below the diaphragm. Since it is the largest gland in the body, it plays a significant role in metabolism while also aiding the body's ability to clot blood. In addition, various liver cells called "macrophages" help destroy damaged red blood cells. The liver's connection to the immune system is its ability to destroy foreign substances through phagocytosis.

Bone marrow. Marrow is in the center of bones. There are two types of marrow, red or yellow marrow. Red marrow that aids in the formation of white and red blood cells, and the yellow marrow stores fat and does not produce blood cells. Most white blood cells originate from bone marrow hematopoietic stem cells. After their birth in the bone marrow, a subset of lymphocytes travel through the bloodstream to the thymus gland. They undergo special processing that changes them into mature T lymphocytes (the letter *T* indicates that they came from the thymus gland). The other lymphocytes that do not reach the thymus after leaving the bone marrow are named B lymphocytes (*B* because they came from bone marrow). These B lymphocytes are abundant in lymph nodes, the spleen, bone marrow, secretory glands, intestinal lining, and reticuloendothelial tissue.

THE RESPONSES OF THE IMMUNE SYSTEM

Failures of the immune system can lead to devastating diseases, either because the immune system attacks itself or because it fails to defend against outside foreign antigen matter. An antigen can be any substance that stimulates the body to fight, ranging from a bacterial infection to the virus that causes acquired immunodeficiency syndrome (AIDS).

When the body fights against an antigen, the immune system can produce two types of response, either a cellular immune response or a humoral immune response. The cellular response involves specific cells that recognize, attack, and destroy the invading pathogen or antigen. It is the primary response against most viruses, fungi, parasitic organisms, and bacteria (e.g., mycobacteria) and transplanted tissues. The humoral immune response, which consists of complement and antibodies, is the body's primary defense against most other bacteria. The two systems work together, however, communicating by complex chemical mediators.

Another way of looking at how the body fights to keep itself healthy is to separate the immune responses into either primary or secondary responses. The second time a given antigen enters the body, the immune system attacks it with the secondary immune response stored in special immune memory cells. The secondary immune response is faster and more extensive than the primary response. This immune memory is specifically designed for each antigen. It provides immunity to the wide variety of diseases and conditions to which one is exposed daily.

The body begins to build this memory before birth by inventorying all the molecules within the body. Foreign substances not in this memory are considered to be antigens, which will activate an immune response. When the immune system first encounters an antigen, it mounts a primary response that produces lymphocytes sensitized to the invader. Many B lymphocytes can respond to create the appropriate antibody molecules, which are then released into the lymph and transported to the blood. This process may last several weeks. During this primary immune response, the B cells and T cells serve as memory cells. Because immunological memory for the antigen is stored, if the immune system encounters this antigen in the future, the memory cells can react more quickly and effectively. In this secondary immune response, the antibodies are ready to respond by attaching to the surfaces of the antigens. There must be a specific type of antibody produced for every type of antigen. These new antibodies may survive only a few months, but the memory cells live much longer.

There are four main ways that antibodies help neutralize antigens. The antibody can pull together clusters of invading organisms to prevent them from spreading (agglutination). Second, when some antibodies bind to antigens, they activate blood proteins called the "complement system" to punch holes in the invader and destroy it. The antibody can also combine with the antigen, which makes it easier for leukocytes to destroy it. By covering the outside of viruses or toxins, antibodies can neutralize their harmful activity. Given the multifaceted utility of antibodies, it is essential to establish antibody memory. It is this particular memory that leads to future immunity.

These memory cells are responsible for the four different types of immunity. Two are acquired actively, and two are acquired passively. The first type is naturally acquired active immunity, which results after exposure to a live pathogen that causes a disease. The second type, artificially acquired active immunity, is gained after a vaccination. The immune response is triggered after an injection of weakened or dead pathogens is received. However, the body does not suffer the severe symptoms of the disease. An example would be a smallpox vaccination. The third type of immunity is artificially acquired passive immunity, gained through an injection of prepared antibodies. This method is passive since another person made the antibodies (gamma-globulin). This type of immunity usually does not last more than a few weeks, and the person will be susceptible to that pathogen in the future. Naturally acquired passive immunity occurs when the antibodies pass to the fetus from the mother. This type of immunity only includes those antibodies available in the blood of the mother. This process gives an infant certain short-term immunities for the first year of life.

These types of immunity are usually desirable, but there are occasions when an immune response is unwanted, such as after an organ transplant. Tissue or organ transplantation from one person to another may cause the body to reject the foreign tissue, triggering an immune response and possibly destroying the new organ. Consequently, matching the tissue between the recipient and the donor may delay or even prevent the immunological rejection of the transplant. The transplant recipient also receives immuno- suppressive drugs, including steroids, calcineurin inhibitors, and antiproliferative agents that interfere with their ability to form antibodies. Other medications, including daclizumab, basiliximab, rituximab, and antithymocyte globulin, can destroy the lymphocytes that help produce antibodies. Unfortunately, the recipient is often left unprotected against infections since the immune system is not functioning normally.

PERSPECTIVE AND PROSPECTS

In the same way that the discovery of penicillin shocked the world, immunology has created endless possibilities in medicine. When surgeons found that they could transplant an organ from one person to another, the interest in immunology exploded.

This field of medicine has discovered that several factors can diminish the power of the immune system. Improper diet, stress, disease, and excessive physical activity levels can depress the immune system. Other factors that can modify immunity include age, genetics, and metabolic and environmental factors. The susceptibility of the young and the very old to infections illustrate these anatomical, physiological, and microbial factors. For the young, the system is immature, while the aged have suffered a lifetime of assaults from pathogens. The impact of psychological stress is challenging to measure, yet it can negatively affect the immune system.

Unanswered questions remain about how the immune system relates to other body systems. The relationships among the brain and nervous system, hormones, and the respiratory system leave many areas ripe for further study.

Recent research has identified the significant importance of Class I major histocompatibility proteins (IMHCPs) in the cellular immune system. IMHCPs are expressed on the surfaces of all nucleated cells. These proteins are made in the cells in the endoplasmic reticulum, where they bind viral proteins transported into the endoplasmic reticulum by TAP (transporter associated with antigen processing) proteins. Once they bind viral proteins in the endoplasmic reticulum, IMHCPs are trafficked to the cell membrane, where they present the viral antigens to cytotoxic T lymphocytes. Activated T-cytotoxic cells secrete toxins that destroy viral-infected cells. Destruction of viral-infected cells inhibits productive viral infection. Occasionally, IMHCPs can present self-proteins to the immune system, resulting in autoimmune diseases. Genetic factors influence the tendency of the immune system to recognize self-antigens. Conversely, the immune system's efficiency in responding to specific diseases is also subject to genetic variation.

Class II MHCPs (II-MHCPs) are only expressed in antigen-presenting cells, including B lymphocytes, dendritic cells, and macrophages. Like IMHCPs, IIMHCPs are initially made in the endoplasmic reticulum but are later trafficked to endosomes. Endosomes are intracellular vesicles that result from materials brought into the cell by endocytosis. IIMHCPs bind to materials in the endosome, after which they are trafficked to the cell membrane. IIMHCP-antigen complexes are presented to T lymphocytes that recognize the antigen and become activated at the cell surface. Activated T-lymphocytes activate B lymphocytes that recognize the antigen, driving the B cell to proliferate and transform into an antibody-secreting plasma cell. The critical link between the cellular immune system and IMHCPs is illustrated by diseases like AIDS, killing T-helper cells and hamstrings antibody production.

Understanding the genes used in the production of the I-MHCPs and the IIMHCPs has led to methods to control their production, possibilities for the eventual cure of AIDS, emerging cancer treatments, and a better understanding of antibody production.

Additional information is needed on defects in the system, as are explanations for its dysfunctions. With a more profound knowledge of immunology, it may be possible to conquer AIDS, allergies, and asthma and develop birth control methods based on the immune response. Doctors may cure cancer, diabetes, herpes, infertility, multiple sclerosis, and rheumatoid arthritis. The possibilities are endless and could also include perfecting transplants of organs and skin grafts and preventing congenital disabilities and even obesity. Those at risk for genetic disorders could be diagnosed through molecular diagnostic tests. Human gene therapy might even cure those with existing genetic conditions. Genetically engineered drugs and gene replacement therapy could relieve the stress on the human immune system. Until these methods become feasible, however, individuals must protect the natural immunity supplied by their bodies.

—*Maxine M. Urton, PhD*

Further Reading

Adelman, Daniel C., et al., editors. *Manual of Allergy and Immunology*. 5th ed., Lippincott Williams & Wilkins, 2012.

Goering, Richard, et al. *Mims' Medical Microbiology and Immunology*. 6th ed., Elsevier, 2018.

Immune Web, www.*immuneweb*.org.

Janeway, Charles A., Jr., et al. *Immunobiology: The Immune System in Health and Disease*. 7th ed., Garland Science, 2007.

Male, David K., et al., editors. *Immunology*. 8th ed., Elsevier/Saunders, 2018.

Marieb, Elaine N., and Suzanne Keller. *Essentials of Human Anatomy and Physiology*. 12th ed., Pearson/Benjamin Cummings, 2017.

Martin, Seamus J., et al. *Roitt's Essential Immunology*. 13th ed., Blackwell, 2017.

Punt, Jenni, et al., editors. *Kuby Immunology*. 8th ed., W. H. Freeman, 2018.

Tortora, Gerard J., and Bryan Derrickson. *Principles of Anatomy and Physiology*. 16th ed., John Wiley & Sons, 2020.

Infection control

Category: Procedure
Also known as: Infection prevention and control
Anatomy or system affected: All
Specialties and related fields: All
Definition: the practice of reducing infections and other illnesses (primarily in hospitals or other health-care facilities) by employing proper hygiene practices, wearing protective gear, and taking proper precautions during medical procedure

KEY TERMS

contagion: the passage of disease from one person to another by direct or indirect contact

epidemiology: the study and analysis of the causes, patterns, and effects of health and disease in specific populations

immunizations: introducing into a person a weakened form or a small piece of a disease-causing organism to activate their immune system against it and make them immune to that disease

nosocomial infection: an infection acquired at a hospital or that originated at a hospital

INDICATIONS AND PROCEDURES

Hospitals and other health-care facilities are places where people go to find relief from illnesses. Ironically, many kinds of infections may spread in these facilities, seriously endangering the health of patients and health-care personnel. Many people die each year of sicknesses spread in hospitals. Medical staff and patients must be aware of the dangers of *contagions*, or contagious diseases, in the health-care environment and take preventative measures to keep illnesses from occurring and spreading.

Infection control is sometimes overlooked or considered a practical guideline rather than hard science. However, it is an essential part of the health-care system worldwide, necessary for patient health and well-being. It also requires scientific knowledge to be most effectively applied. Practitioners of infection control have to understand the basics of "epidemiology," the study of the spread and control of diseases. People must understand what causes diseases to determine best how to prevent them.

Many kinds of infections are commonly associated with health-care environments and procedures. "Surgical site infections" occur on parts of the body that have undergone surgery. Bloodstream infections can take place if germs are transmitted into the blood through needles used for injections or other proce-

Social distancing is one relatively recent technique for infection control. Photo via iStock/alvarez. [Used under license.]

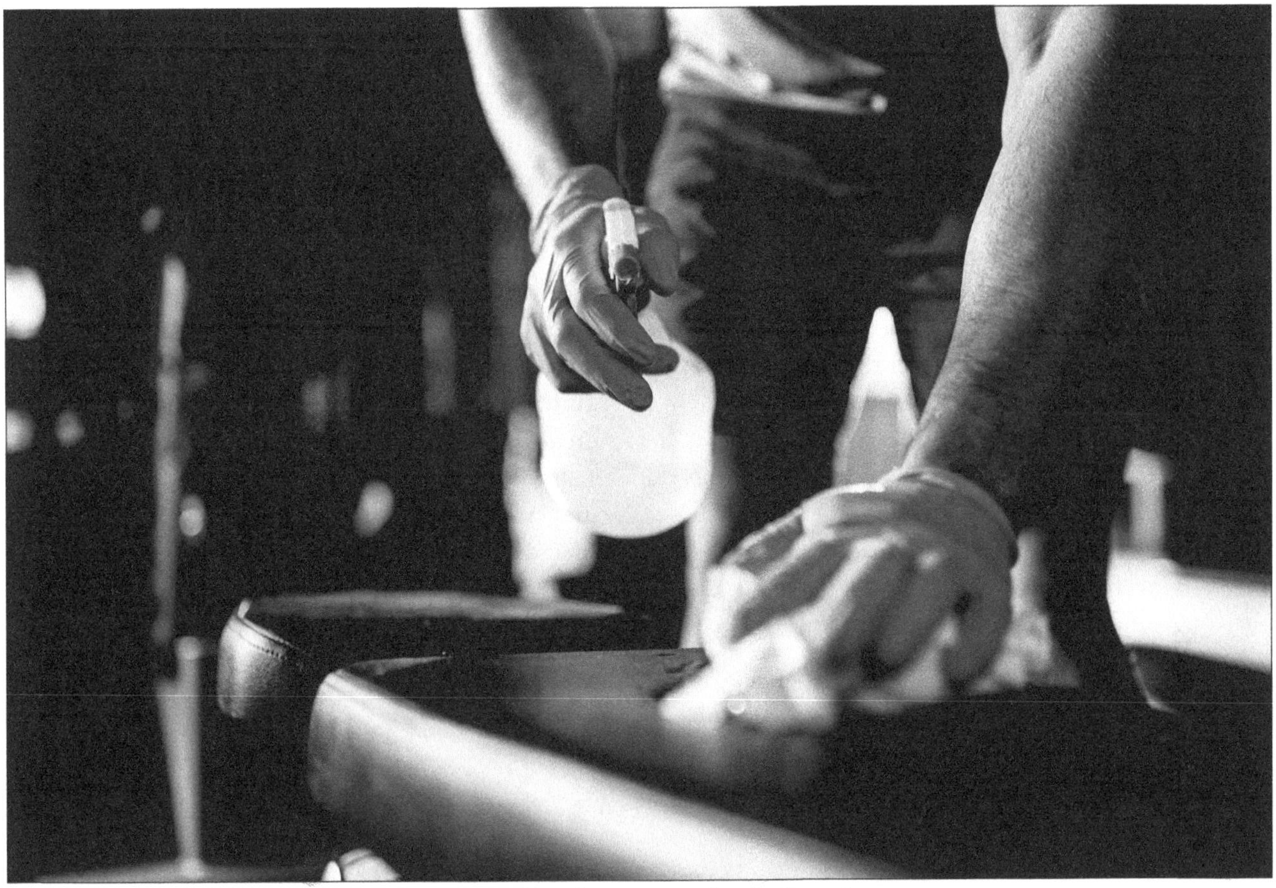

Cleaning common surfaces and equipment. Photo via iStock/PeopleImages. [Used under license.]

dures. Infections of the urinary tract, bladder, or kidneys may result from improper use of "catheters"; thin tubes used to keep passages open and transmit fluids. Pneumonia, a lung infection, is also frequently associated with healthcare-based illness.

Infection in hospitals and other medical facilities may be passed among patients, staff members, and visitors. The problem is widespread and puts all people in the environment at risk. All health-care personnel, patients, and visitors should practice safe behaviors that promote infection control.

Medical personnel must be vigilant in their attention to the infection control protocol. Fortunately, most guidelines for reducing the spread of sickness are relatively simple and relate to proper hygiene. Many are common sense behaviors already practiced by most people in and out of health-care facilities. These actions include hand washing, cough-and-sneeze etiquette, "immunizations" (injections to prevent illnesses), protective gear, safe procedure methods, and management of outbreaks.

Likely the single most effective way to avoid the spread of infection is through proper handwashing. This applies to all people in hospitals as well as the public. People should routinely wash their hands in warm soapy water, eliminating most bacteria and other harmful materials on the skin. Antibacterial lotions and other products may also be helpful if used correctly. Experts believe hand washing can vastly reduce foodborne illnesses and ailments such as the flu and common cold.

Proper use of gloves can also quell the spread of disease. In September 2017, Deborah Patterson Burdsall and her colleagues reported in a study published in the *American Journal of Infection Control* that certified nurse assistants who routinely changed their gloves prevented the spread of disease. Thus, proper glove etiquette seems to be as crucial as proper handwashing hygiene.

Another essential aspect of hygiene is cough-and-sneeze etiquette. It is vital to cover the mouth when coughing or sneezing to prevent the spread of airborne contagion to others. Experts recommend coughing or sneezing into the elbow rather than the hand since coughing or sneezing into the hand transfers harmful materials, which are readily spread through touch.

Medical personnel should keep immunizations and other personal health routines up to date to prevent becoming ill and potentially spreading illnesses to others. Immunization protects people from preventable diseases such as tuberculosis and hepatitis B. The flu and many other sicknesses can also be dodged with such precautions.

Protective gear can shield people from becoming exposed to harmful infectious material. For medical personnel, disposable gloves and smocks are essential protective tools that reduce the chance of contagion spreading by physical contact. Masks reduce the risk of airborne contagions spreading in a health-care facility. According to hospital rules, any object contaminated with blood or other potentially dangerous materials should be washed or disposed of properly.

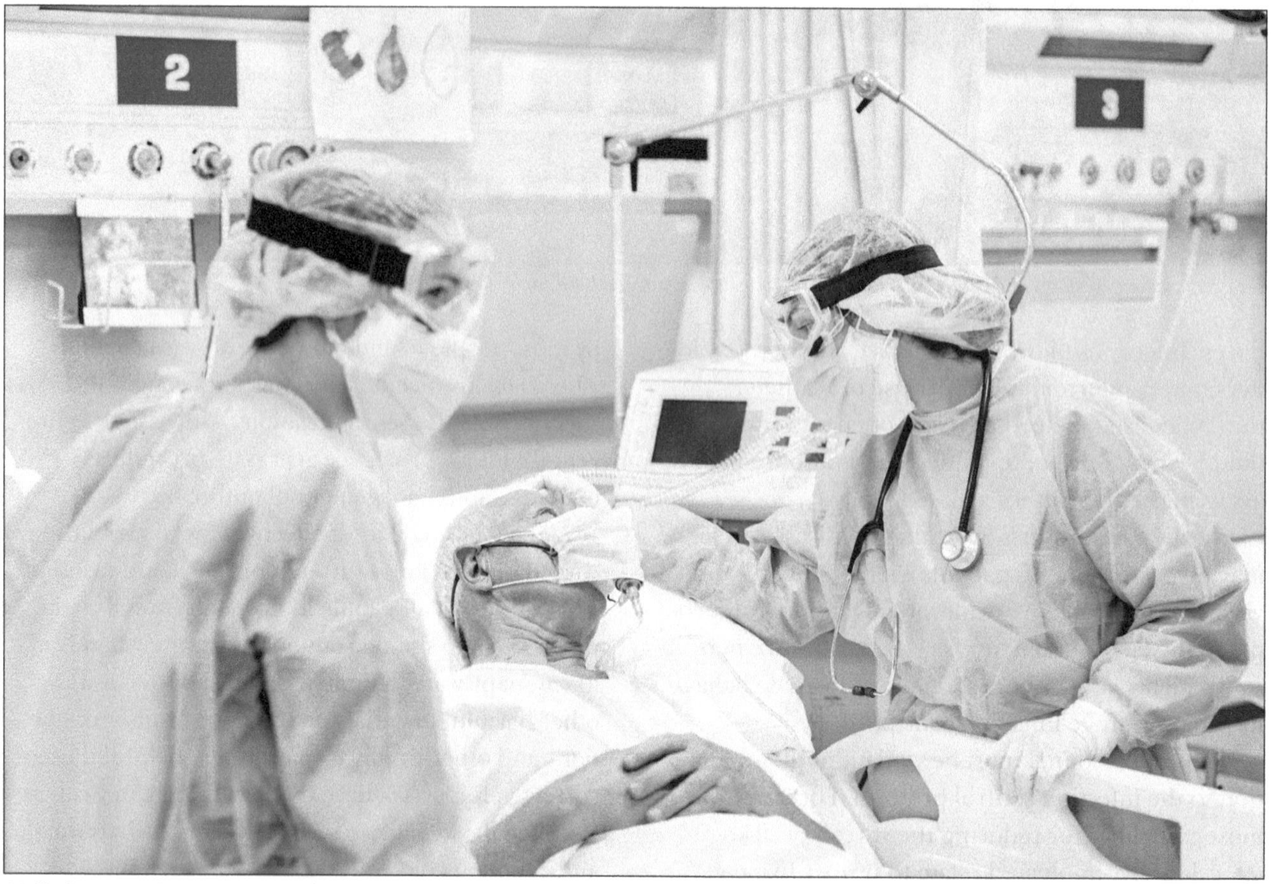

Medical personnel wearing personal protective equipment (PPE). Photo via iStock/Mora Images. [Used under license.]

USES AND COMPLICATIONS

Staff in a hospital or other medical facility should be well trained and prepared to apply specific infection control precautions for different medical procedures. Injections and catheterizations are procedures that involve the most significant risk of causing infection. Medical staff should remember important safety rules such as only using a needle once for an injection. Reusing needles puts patients at extreme risk of infection. Likewise, reusing single-use vials can spread disease to patients. In 2012, outbreaks of invasive *Staphylococcus aureus* infections observed in Delaware and Arizona were definitively linked to injections of pain medications from reused single-use and single-dose vials.

Finally, medical personnel should monitor patients for any signs of sudden health problems. Suppose any indication of an infection outbreak exists among patients. In that case, personnel should act quickly to manage the problem before it spreads. This might involve quarantining infected patients and observing those who came into contact with them. Patients known to be infected or colonized with pathogenic organisms or who might carry pathogens should be placed on contact precaution.

Medical personnel are not the only ones in health care with responsibility for maintaining infection control. It is also vitally important for patients to be active partners in reducing the spread of illness. Patients' families, friends, and other visitors to health-care facilities are also responsible for practicing safe behaviors.

Many patient safeguards are like the commonsense guidelines for medical staff, such as washing hands carefully and practicing cough-and-sneeze etiquette. Patients should also stay informed and observant about their health. They can learn more about safe practices by asking doctors and other medical staff for information. They should carefully monitor themselves for signs of potential trouble, such as loose or dirty bandages or injuries that are not healing properly.

Patients also can help themselves to avoid infections through research into their particular conditions. For example, patients with a high risk of diabetes or those who smoke or are overweight should know these conditions increase the risk of infection. These patients may need to take special cautionary measures to reduce their risk of infection.

The pitfalls of infection control are the time and expense it takes to teach health-care practitioners to master infection control protocols and procedures. Secondly, the equipment and supplies necessary to properly adhere to infection procedure protocols may be expensive and laborious to make available. There is also the tediousness of adhering to these strict protocols. Finally, washing the hands repeatedly each day can dry out the skin on the hands and cause it to chaff and crack. However, these are small prices to pay to ensure that patients receive high-quality care without contracting deadly nosocomial infections while being cared for.

—Mark Dziak and Michael A. Buratovich, PhD

Further Reading

"Infection Control." *MedlinePlus*. US National Library of Medicine, www.nlm.nih.gov/medlineplus/infection control.html. Accessed 11 Feb. 2015.

"Infection Prevention and Control." *Minnesota Department of Health*. Minnesota Department of Health, www.health.state.mn.us/divs/idepc/dtopics/infection control/. Accessed 11 Feb. 2015.

"Infection Prevention and You." *Association for Professionals in Infection Control and Epidemiology*. Association for Professionals in Infection Control and Epidemiology, Inc., www.apic.org/Resource_/TinyMceFileManager/IP_and_You/IPandYou_InfographicPoster_2013.pdf. Accessed 11 Feb. 2015.

"Invasive Staphylococcus aureus Infections Associated with Pain Injections and Reuse of Single-Dose Vials—Arizona and Delaware, 2012." *Monthly Morbidity and Mortality Report*, vol. 61, no. 27, 2012, pp. 501-4, www.cdc.gov/mmwr/preview/mmwrhtml/mm6127a1.htm?s_cid=mm6127a1_w.

LaPorte, Meg. "Glove Use Tied to Better Infection Control." *McKnight's Long-Term Care News*, 6 Oct. 2017,

www.mcknights.com/news/glove-use-tied-to-better infection-control/article/698220/.
"Preventing Infections in the Hospital." *National Patient Safety Foundation*. National Patient Safety Foundation, www.npsf.org/?page=preventinginfections. Accessed 11 Feb. 2015.

B LYMPHOCYTES

Category: Immunology
Also known as: B cells
Anatomy or system affected: Blood, bone marrow, immune system, lymphatic system
Specialties and related fields: Hematology, immunology, microbiology, oncology
Definition: a type of white blood cell that synthesizes and secretes antibodies

KEY TERMS

antibodies: secreted glycoproteins that bind to foreign substances in our bodies and inactivate them, clump them, mark them for destruction, and facilitate their disposal; also known as immunoglobulins

antigens: foreign substances in the body that elicit an immune response

heavy chain: the larger polypeptide chain of an antibody

isotype switching: when activated B cells, by their interaction with T-helper cells change the type of antibody they are secreting to some other antibody subtype

light chain: the smaller polypeptide chain of an antibody

lymphopoiesis: the process of B-lymphocyte development

major histocompatibility complex (MHC) proteins: cell surface proteins that come in two types, class I and II that help antigen-presenting cells present antigen to lymphocytes (class II) and mark every nucleated cell in the body (class I); also called "human leukocyte antigens" (HLAs)

recombinase: the *RAG-1* and *RAG-2* protein complex that cuts and paste gene segments from the antibody gene cluster together to form unique antibodies that bind a wide range of antigens

INTRODUCTION

B lymphocytes are white blood cells that play integral roles in the adaptive immune system. B lymphocytes differentiate into plasma cells and memory B lymphocytes. Plasma cells secrete antibodies to fight infection and destroy abnormal cells such as cancer cells. Memory B lymphocytes are activated by contracting a disease or by vaccination. They confer long-lasting immunity to a specific disease. One subset of B lymphocytes can also infiltrate cancer tumors.

FUNCTION

The immune system is a complex of tissues, cells, and the chemicals these cells produce spread throughout the body. Its function is to rid the body of foreign microbes and damaged or abnormal body cells. The major organs of the immune system are the bone marrow, lymph nodes, thymus, and spleen. Immune system cells and a clear fluid called "lymph" move throughout the body in a series of channels called "lymphatic vessels" that are separate from the circulatory system.

The immune system has two divisions: the innate immune system and the adaptive immune system. The innate system consists of white blood cells (leukocytes) that respond rapidly to a wide range of microbes but cannot confer immunity against disease. The adaptive system has cells that respond more slowly to disease but can create long-lasting immunity. While cells in the innate system can attack many different microbes, each cell of the adaptive system recognizes and attacks only 1 type of microbe or damaged cell. B lymphocytes and T lymphocytes are the main cells of the adaptive immune system.

All cells have specific marker molecules on their surfaces that identify them as "self-cells" or "nonself-cells." Each B lymphocyte and T lymphocyte interacts with only one particular type of surface marker molecule that is displayed on the surfaces of microbes or abnormal cells and identifies them as nonself-cells. These marker molecules (often proteins) that elicit a reaction from B lymphocytes are called "antigens."

DEVELOPMENT

B lymphocytes develop in the bone marrow from hematopoietic stem cells. Hematopoietic stem cells can divide and differentiate into many different types of blood cells. As B lymphocytes mature, each cell develops a unique receptor on its surface that will bind only to a single, specific antigen on a foreign microbe or an abnormal body cell such as a cancer cell. Scientists believe the body can make over one billion unique B lymphocytes. Any B lymphocytes that develop receptors that will bind with marker molecules on healthy self-cells are eliminated while still in the bone marrow so as not to harm the body.

The B-cell receptor is a membrane-bound antibody. Antibodies consist of four polypeptides, two heavy chains, and two light chains. All antibodies have at their front end a variable region that differs from one antibody to another. Behind the variable region is the constant region that is the same for each specific antibody subtype.

The variable region of the heavy chain is encoded by a variable or V region, a diversity region or D region, and a J or joining region. The light-chain variable region consists of a V region and a J region. The genes that encode the heavy chain are on chromosome 14, and the light chain has two different gene clusters, kappa, and lambda. The kappa gene cluster is on chromosome two, and the lambda gene cluster is on chromosome 22. The heavy-chain gene cluster consists of fifty-one different V segments, six different J segments, twenty-seven distinct D segments, and eight different constant regions. The kappa light-chain gene cluster consists of forty different V

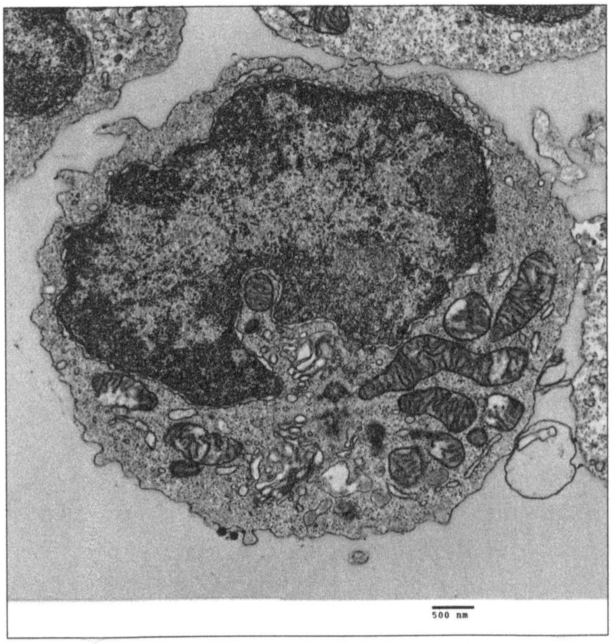

Transmission electron micrograph of a B cell from a human donor. Image by the National Institute of Allergies and Infectious Diseases (NIAID), via Wikimedia Commons.

gene segments, five different J chain clusters, and a single constant region. The lambda gene cluster contains thirty V gene segments and four J gene segments, with four different constant regions.

During B-lymphocyte development or lymphopoiesis, the antibody gene clusters are rearranged to create proteins that bind an enormous variety of distinct antigens. The earliest stage of B-lymphocyte development is the early pro-B cell, which forms from the daughter cell after the hematopoietic stem cell divides. The early pro-B cell expresses proteins that rearrange the antibody gene cluster, *RAG-1* & *-2*. The *RAG* proteins join a single D gene segment from one of the copies of the heavy-chain gene cluster with a single J gene segment. Upon completion of the DJ joining, the early pro-B cell becomes a late pro-B cell. The *RAG-1* & *-2* recombinase further joins a single V segment with the DJ segment and joins the VDJ combination to the μ constant region gene. Successful cutting and pasting of the VDJ-μ gene segments mark the transition of the late pro-B cell to a large pre-B cell. If the cell fails to rearrange its heavy-chain gene segments, the cell dies.

The large pre-B cell expresses the rearranged heavy chain and traffics it to the cell surface. Successful expression of the antibody heavy chain induces the cell to divide and become a small pre-B cell. Next, the small pre-B cell rearranges the light-chain genes, beginning with the kappa genes first. If the cell fails to rearrange its kappa genes, then it moves to the lambda genes. VJ joining of the light-chain gene segments makes a functional light chain, and expression of the light chain with the heavy-chain places an antibody on the surface of the cell and designates the cell as an immature B cell. If the antibody strongly binds to any self-antigens in the bone marrow, it dies.

The variability in antibody formation is remarkable. The heavy chain alone can encode at least 8262 ($51 \times 6 \times 27$) different heavy-chain variable regions. If a heavy chain is coupled with a kappa light chain, then there are 200 (40×5) different light chains and 1.6×10^6 (8262×200) different possible antibodies that such rearrangements can form.

B-CELL ACTIVATION

Immature B lymphocytes can make either immunoglobulin D (IgD) or immunoglobulin M (IgM). The expression of IgD licenses the cell to leave the bone marrow and migrate to lymph nodes. The B cells that have IgM and IgD on their surfaces and have left the bone marrow are called "naïve B cells." Lymph nodes are enlargements of the lymphatic vessels, found in the neck, under the arms, and in the groin, where foreign and abnormal cells are filtered from lymph. When the naïve B lymphocyte enters the lymph node from the blood, they first enter the paracortical region. B cells migrate to the periphery of the lymph node, also known as the cortical region, where they form the primary lymphoid follicles. If a naïve B cell encounters an antigen that binds to its IgM receptor, the lymphocyte is activated and clones itself, producing massive numbers of identical cells. In the lymph node, the proliferating B cell forms a "germinal center," and a follicle with a germinal center is called a "secondary lymphoid follicle."

Most clone cells become plasma cells that secrete IgM antibodies (IgD antibodies are never secreted). Antibodies are large glycoproteins ten times smaller than a virus particle. They bind with any antigen identical to the one that activated the B lymphocyte. An activated B lymphocyte can produce 2000 antibodies per second for up to five days. These antibodies enter the lymphatic and circulatory systems where they attach to the surface of the abnormal cell or invading microbe. This either inactivates the target, causes it to clump with similar cells and be destroyed, or flags the target for destruction by another type of white blood cell. Producing enough antibodies to combat disease takes ten to seventeen days to reach maximal effectiveness, and during that time the individual will show symptoms of the illness.

Activated B lymphocytes also function as antigen-presenting cells that use major histocompatibility complex (MHC) class II proteins to present antigen to T-helper cells. The B lymphocyte presentation of antigen to T-helper cells activates the T cell if it receives two distinct signals. First, the antigen must bind to the T-cell receptor on the surface of the T-helper cells. Second, the T-helper cell has a cell surface protein called "CD40." CD40 must bind to CD40L on the surface of the B lymphocyte. This dual signal system guarantees that only the right cell presenting the right antigen in the proper context activates the T cell. The activated T cell then secretes cytokines that drive the B cell to switch the type of antibody it secretes. This phenomenon, called "isotype switching" requires a second rearrangement of the heavy-chain antibody genes. An enzyme called "activation-induced deaminase," or AID, breaks the intervening deoxyribonucleic acid (DNA) removing all the constant regions between the VDJ and the decided upon antibody. The antibody the B cell eventually secretes depends upon the instructions it receives from the T-helper cell. For example, if the T-helper cell secretes IL-4 and IL-5, then the B cells will become an IgE secreting plasma cell. But if the T-helper cells secrete interferon-gamma, then the B lymphocyte will become an IgG secreting plasma cell.

Not all B lymphocytes become plasma cells. Most B lymphocytes are born and never experience antigen stimulation. A few B lymphocytes, under stimulation by cytokines secreted by helper T lymphocytes, become memory B lymphocytes, or memory B cells. These cells remain in the body for a long time (years to decades). If the same microbe is encountered again, memory B lymphocytes can ramp up antibody production in two to five days, destroying the invading microbe before it can make a person sick. This is the basis for long-term immunity to disease. Vaccines work by introducing a weakened or partial form of a microbe that does not make the individual sick but stimulates the production of plasma cells and memory B lymphocytes.

Another subset of B lymphocytes called "tumor-infiltrating B lymphocytes" (TIL-Bs) appears to migrate to solid cancer tumors and secrete tumor-specific antibodies. Tumor-infiltrating T lymphocytes (TIL-Ts) are also found in solid tumors, and more is known about how they work. TILs have been associated with substantial remission in metastatic melanoma. Experimentally, the presence of TILs also correlates with improved outcomes in ovarian, breast, colorectal, and non-small cell lung cancers.

Why, if the immune system is so effective in preventing illness, does it not wipe out cancer cells and keep the individual cancer-free? One reason is that cancer cells start as normal body cells. At some point, they are transformed because certain genes are inappropriately turned off or on, or are mutated. When this happens, the transformed cell still has many of the surface marker molecules that originally identified it as a self-cell. B lymphocytes that have receptors that would interact with healthy self-cell antigens are killed before they leave the bone marrow to prevent autoimmune diseases. In many cases, transformed cancer cells still look enough like self-cells that B lymphocytes do not recognize them as abnormal and do not mark them for destruction. If B lymphocytes respond to a transformed cell, the response may not be strong enough to be effective since the growth of cancer cells is not limited the way it is in healthy cells. In addition, some cancer cells make chemicals that inactivate or disrupt the functioning of immune system cells.

TREATMENT

The properties of the adaptive immune system can be manipulated in ways to prevent and treat cancer by using vaccines and monoclonal antibodies. Some cancers are primarily caused by viruses. By making a vaccine against the causative virus, it will be rapidly destroyed and cancer can be prevented. As of 2021, two cancer prevention vaccines have been approved.

Many strains of human papillomavirus (HPV) are known to cause cervical, vulvar, anal, penile, and mouth, and colon cancers. The virus is transmitted through sexual activity. Vaccination before puberty against the most common strains of HPV stimulates the adaptive immune system and creates memory B lymphocytes that provide long-term immunity. As of 2021, three HPV vaccines have been approved by the US Food and Drug Administration (FDA). The FDA-approved HPV vaccines include a nonavalent HPV vaccine (Gardasil® 9, 9vHPV), a quadrivalent HPV vaccine (Gardasil®, 4vHPV), and a bivalent HPV vaccine (Cervarix®, 2vHPV). All three HPV vaccines protect against HPV types 16 and 18 that cause most HPV cancers. The distribution of these vaccines has substantially reduced the rate of HPV-caused cancers.

Hepatitis B is a viral infection that can cause liver cancer. Current recommendations in the United States call for children to receive a hepatitis B vaccine at birth. Research is underway to develop vaccines against other microbe-triggered cancers, including various lymphomas (Epstein Barr virus) and stomach cancer (*Helicobacter pylori* bacterium).

Research is also underway to design treatment vaccines to use in individuals with active cancers. As of 2015, only one treatment vaccine has been approved by the FDA. This vaccine, sipuleucel-T, can extend the life of men with metastatic prostate cancer but does not cure the disease. Other current cancer-treating vaccines include live Bacillus of Calmette et Guillen (BCG), which can treat early-stage bladder cancer, and talimogene laherparepvec, which can treat melanoma.

Another B-lymphocyte-related treatment is the use of targeted monoclonal antibodies (mAbs). Scientists first isolate B lymphocytes from laboratory mice that were vaccinated with a specific antigen and fuse them to cancer cells (myelomas). The resultant hybrid cells or "hybridomas" can successfully grow in culture, but also produce a specific type of antibody. After the hybridoma lines are screened to identify which cells produce the antibody of interest, which is a very labor-intensive process, those lines that make the desired antibody are grown in culture. Then the genes that encode the desired antibodies are transferred to a cell line that grows well in culture (e.g., Chinese hamster ovary [CHO] cells) to make small antibody-producing factories. These antibodies are then purified and injected into the body where they attack or mark cancer cells. The effect is short-lived since no memory B cells are created.

The FDA has approved more than a dozen monoclonal antibody biologics for use with other cancer treatments. The mAb drugs can be recognized because their generic name ends in "mab" (e.g., alemtuzumab, trastuzumab). Some mAbs, called "conjugated mAbs," have a chemotherapy drug or radioactive isotope attached to the antibody to increase its effectiveness (e.g., brentuximab vedotin, ibritumomab tiuxetan). Although mAbs have side effects, these are less harsh than those of chemotherapy drugs.

—Tish Davidson, AM and Michael A. Buratovich, PhD

Further Reading

Finn, O. J. "Immuno-Oncology: Understanding the Function and Dysfunction of the Immune System in Cancer." *Annals of Oncology*, vol. 23(Suppl. 8), 2012, pp. iiv6-9, annonc.oxfordjournals.org/content/23/suppl_8/viii6.full.

Melero, I., G. Gaudemack, W. Gerritsen, et al. "Therapeutic Vaccines for Cancer: An Overview of Clinical Trials." *Nature Reviews Clinical Oncology*, vol. 11, 2014, pp. 509-24.

Scott, Andrew M., Jedd D. Wolchok, and Lloyd J. Old. "Antibody Therapy of Cancer." *Nature Reviews Cancer*, vol. 12, 2012, pp. 278-87, www.nature.com/nrc/journal/v12/n4/full/nrc3236.html.

LYMPHOCYTE

Category: Immune system

Anatomy or system affected: Blood, bone marrow, cells, circulatory system, immune system, lymphatic system

Specialties and related fields: Bacteriology, biochemistry, biotechnology, cell biology, immunology, microbiology

Definition: a white blood cell with a single round nucleus, occurring especially in the lymphatic system

KEY TERMS

acquired immunity: a subsystem of the immune system that is composed of specialized, systemic cells and processes that eliminate pathogens or prevent their growth

antigen: a toxin or other foreign substance which induces an immune response in the body, especially the production of antibodies

B lymphocyte: a type of white blood cell that makes antibodies

lymphocytosis: an increase in the concentration of lymphocytes in the blood

T lymphocyte: a type of lymphocyte that, though born in the bone marrow, matures in the thymus

INTRODUCTION

A lymphocyte is a white blood cell, an important part of the immune system that helps the body defend against invading organisms. Lymphocytes originate in the soft inside of the bones and migrate through the body. They become specialized into one of two types. B lymphocytes, or B cells, produce an antibody protein that directs other cells to attack foreign invaders, such as bacteria or viruses. T lymphocytes, or T cells, attack the body's cells if they have been infected by an invader or have become cancerous. Some lymphocytes can "remember" the target of an attack and produce antibodies that can make the body immune to future illnesses from those same invaders.

BACKGROUND

Blood is a fluid that courses through the body via a network of blood vessels called "arteries" and "veins." It has several vital functions, including carrying oxygen to the body's cells and defending against infection. Blood is made up of four components. One of those, plasma, is mostly water-based and acts as a medium to transport blood cells throughout the body. Red blood cells are flat disk-shaped cells that contain a protein called "hemoglobin." Hemoglobin absorbs oxygen from the lungs and carries it to the other cells in the body. It also removes carbon dioxide waste and transports it back to the lungs, where it can be exhaled. Platelets are cell-like structures that clump together to form tiny clots in the blood to stop bleeding.

White blood cells are the body's natural infection fighters. These cells, also called "leukocytes," defend the body against foreign invaders such as bacteria, viruses, fungi, or parasites. White blood cells make up about 1 percent of the body's blood, and they are divided into five types. Monocytes are the largest white blood cells and act as cell-devouring scavengers. They engulf and digest foreign organisms and even other dead white blood cells. Eosinophils release toxins that attack and kill bacteria, viruses, or parasites, such as intestinal worms. High eosinophil counts are linked to allergic and asthmatic reactions. Basophils are the least common type of white blood cell and act as a warning system to alert the body to invading organisms. They do this by releasing an organic compound called "histamine", which widens the blood vessels and allows other white blood cells to reach the site of the infection. Like eosinophils, they are also associated with allergies. Neutrophils are the most common type of infection fighter, comprising about 60 percent of the body's white blood cells. They function as natural assassins, killing their targets through various methods.

OVERVIEW

Lymphocytes, which make up about 20 to 40 percent of the body's white blood cells, are the fifth type. They originate in the soft, springy inside of the bones called the "marrow." From there, they migrate out through the body through the lymphatic system, a network of tissues and organs that includes the lymph nodes, spleen, and thymus. The lymphatic system helps filter out toxins, bacteria, and other foreign substances from the body. Depending on their destination, the lymphocytes will become specialized into one of two types—B lymphocytes or T lymphocytes—both of which have the same goal of defending the body against foreign invaders. When the body senses an invading organism, it recognizes it by its antigen, the shape of the molecules on its surface. The immune system then produces specific lymphocytes with specialized receptors on surfaces designed to battle a specific antigen.

About three-quarters of the lymphocytes travel to the thymus to become mature T cells. The T cells get their name from the thymus, a gland located in the center of the chest between the lungs. The T cells themselves are also specialized into two types. Helper T cells recognize an invading antigen and send out chemical signals that trigger B cells and killer T cells into action. When a killer T cell encounters the specific antigen it recognizes, the helper T cell instructs the killer cell to make copies of itself. The killer T cells use the receptors on their surface to identify body cells that have been infected by an invader or have become cancerous. If an antigen is detected, the killer T cells destroy the cell. Some T cells remain in the body long after the immune system has finished responding to the infection if needed to deal with a recurrence.

B cells make up about a quarter of the body's lymphocytes. They do not attack invading organisms. Instead, they attach themselves to an invader when a matching antigen is detected, awaiting instructions from helper T cells. When they receive the signal, the B cells begin dividing, creating plasma cells and B memory cells. Plasma cells produce special proteins called "antibodies." These antibodies are tailor-made to target the same antigen as their parent cell. They coat an infected cell or invading organism and mark it for destruction, letting killer T cells or infection-devouring cells called "macrophages" know which cells to target. Antibodies can also attach themselves to several antigens at once, clumping together groups of invaders and making them easier to devour.

B memory cells have a long life span and can retain a memory of the receptor pattern of the antigen they were programmed to fight. If the same foreign organism enters the body, the B memory cells remain activated to fight the invader. This function of lymphocytes is responsible for developing immunity against certain illnesses.

An abnormally high number of lymphocytes in the blood can lead to a condition known as lymphocytosis. Because lymphocytes are part of the body's natural defense system, it is not unusual for their count to be elevated while fighting an infection. However, if the count remains high, it may signal the presence of a more serious illness, such as leukemia, human immunodeficiency virus/acquired immunodeficiency syndrome (HIV/AIDS), cancer of the lymphatic system, or a severe bacterial infection.

—*Richard Sheposh*

Further Reading

"The Immune System: Information about Lymphocytes, Dendritic Cells, Macrophages, and White Blood Cells." *Chemocare*, www.real-world-physics-problems.com/gyroscope-physics.html. Accessed 27 Mar. 2017.

"The Immune System—In More Detail." *Nobelprize.org*, 2014, www.nobelprize.org/educational/medicine/immunity/immune-detail.html. Accessed 27 Mar. 2017.

"Learn about the Five Common White Blood Cells." *University of Wisconsin Oshkosh*, www.uwosh.edu/med_tech/what-is-elementary-hematology/white-blood-cells. Accessed 28 Mar. 2017.

"Lymphocytosis (High Lymphocyte Count)." *Mayo Clinic*, www.mayoclinic.org/symptoms/lymphocytosis/basics/definition/sym-20050660. Accessed 28 Mar. 2017.

Morris, Susan York. "Everything You Should Know about Lymphocytes." *Healthline*, 30 Jan. 2017, www.healthline.com/health/lymphocytes#overview1. Accessed 27 Mar. 2017.

Murphy, Kenneth, and Casey Weaver. *Janeway's Immunobiology*. 9th ed., Garland Science, 2016.

Parham, Peter. *The Immune System*. 4th ed., Garland Science, 2015.

Parry, Dr. Nicola. "Functional Difference between T Cells & B Cells." *Seattle Post-Intelligencer*, education.seattlepi.com/functional-difference-between-t-cells-b-cells-4573.html. Accessed 27 Mar. 2017.

HAND HYGIENE COMPLIANCE

Category: Procedure

Anatomy or system affected: Arms, elbows, fingers, hands, skin, wrists

Specialties and related fields: All fields

Definition: a practice of cleaning and cleansing the hands with soap, water, or any other method to prevent the unintentional passage of pathogenic microorganisms while caring for patients

KEY TERMS

disinfection: cleaning surfaces or objects with chemicals to kill microorganisms and significantly reduce their numbers.

sterilization: subjecting objects or surfaces to chemicals or processes that rid them of any living microorganisms

INDICATIONS AND PROCEDURES

Hospitals and health-care services must define the organization's hand hygiene practices to ensure patient well-being. These practices create a sterile, clean environment for surgeries and daily examinations—if they are correctly observed. Three of the most common methods of measuring hand hygiene compliance are direct observation, self-reporting of health-care workers, and indirect measurements of hygiene product (e.g., liquid soap, hand sanitizer, and paper towels) usage from refilling frequency and amount. Observing employee behavior, self-reporting, and peer monitoring allows administrators and managers to determine the most relevant approach for employees based on the gathered baseline data.

Assessing hand hygiene techniques is essential in generating methods to increase compliance from health-care workers and visitors. However, it is essential to measure the frequency and conditions with which employees practice proper hand hygiene. Health-care administrators and managers must observe and question employees to assess if sanitary guidelines are being followed to determine whether employees are using the correct quantity of liquid soap or alcohol-based hand rub per wash, spending a satisfactory amount of time using hand hygiene products, using paper towels when turning off faucets to minimize recontamination, and wearing and removing gloves correctly to avoid contamination. Likewise, further action may be required per wash due to personal articles and features such as long fingernails, artificial nails, and jewelry that can trap additional contaminants.

Hand hygiene methods include washing with soap and water or alcohol-based sanitizers. Alcohol-based hand sanitizers are preferred in most situations since they are the best germ killers. During routine patient care, health-care workers should use an alcohol-based sanitizer: (1) immediately before touching any patient or handling any invasive medical device, (2) if moving from a soiled site of the body to a clean site, (3) after touching a patient or anything in their immediate environment, (4) after touching any bodily fluids or contaminated surfaces, or (5) right after glove removal. Soap and water are best reserved for cases when the hands are visibly dirty, after caring for someone with diarrhea, or after exposure to endospore-producing microorganisms.

To wash your hands with an alcohol-based sanitizer put the sanitizer on your hands and make sure that all surfaces are covered for about 20 seconds until the skin is dry. To wash your hands with soap and water, first wet the hands with water and apply the soap. Vigorously rub the hands together for approximately 15 seconds, ensuring that you cover all areas of the hands. Then rinse the hands with cold or tepid water (hot water dries out the hands) and dry them with a clean paper towel. Use the towel to turn off the faucet.

Health-care administrators and managers often have difficulty monitoring hand hygiene compliance because patient contact between the individual and the environment occurs at varying locations across the health-care setting. Furthermore, hand hygiene opportunities for both clinical and nonclinical staff occur around the clock. Clinicians require different frequencies and intensities of hand hygiene depending on the type of care they have provided and the patient factors. For example, clinicians preparing for surgery have a much more rigorous hand-cleaning protocol to follow than a nurse preparing to insert a nasogastric tube. Otherwise, the same hand hygiene procedures are followed each time consistently. Monitoring hand hygiene compliance is time-consuming; managers may not always be able to divert healthcare specialists from their other duties to monitor hand hygiene methods, especially infection prevention specialists, quality improvement staff, nurses, and respi-

Hand washing is an effective and simple way to prevent infections. Photo via iStock/Maridav. [Used under license.]

ratory therapists. Moreover, close monitoring of hygiene practices may stimulate the Hawthorne effect, where subjects alter behavior and actions when observed in close proximities.

Hand hygiene compliance is a global interest as it can affect any person in any hospital. In 2002, the US Centers for Disease Control and Prevention (CDC) provided restructured and updated hand hygiene guidelines for the United States. In 2004, the World Health Organization (WHO) launched the World Alliance for Patient Safety, a program to emphasize and endorse hand hygiene compliance in health care as a means to eliminate nosocomial infections. The CDC and WHO advocate constant monitoring of all health-care settings to maintain and improve hand hygiene compliance. The WHO has recommended other instances where hand hygiene compliance is necessary: before contact with patients (before approaching the patient), before all aseptic (sterile environment) tasks, after exposure to bodily fluids, after glove removal, after contact with the patient, and after contact with the patient's environment. The nonprofit organizations National Quality Forum and the Joint Commission (formerly the Joint Commission on Accreditation of Healthcare Organizations and Joint Commission on Accreditation of Hospitals) both endorse the CDC and WHO guidelines and have suggested other means of administrator monitoring, such as during patient observations and the use of hand hygiene compliance systems that electronically monitor product volume or count the number of times hand hygiene apparatus is used. It was suggested that implementing single-level and multilevel hand hygiene interventions—adding more dispensers, education, and involvement of patients and staff, feedback initiatives, social marketing, and cultural and organizational change—can alter compliance rates based on social and cultural aspects.

Leadership support can elevate hand hygiene compliance in hospitals and health-care settings. The Association for Professionals in Infection Control and Epidemiology (APIC) has suggested the following eight guidelines for promoting hand hygiene among health-care employees: (1) The health-care organization's leadership must fully support hand hygiene compliance initiatives. (2) Continuous monitoring and response must be available for tracking current and developing endemic infections and drug-resistant pathogens. (3) Senior administrators must create a multidisciplinary design and response team dedicated to hand hygiene compliance. (4) Health-care facilities must provide education and training for staff, patients, families, caregivers, and visitors. (5) Hand hygiene resources, such as hand sanitizer stations and disinfectant wipes, must be readily available in all hallways, entrances, and exit pathways. (6) Administrators can emphasize accountability for good hygiene by recognizing and rewarding hand hygiene compliance among employees. (7) Visual resources with relevant facts and statistics, such as instructive reports, posters, pocket cards, and brochures, can serve as constant reminders. Health-care organizations should direct observers, coworkers, patients, and visitors to give health-care workers feedback in real time. (8) Monitoring and feedback programs can inform employees of the hospital's monthly compliance data and encourage employees to discuss these findings with colleagues.

Successful hand hygiene compliance can be accomplished from a continuous emphasis on hand hygiene messages, administrator acknowledgment of clinical staff's perception of hand hygiene, supervision of hand hygiene exercises, educational tools, and having senior staff role models. In addition, patient engagement was reported to increase hand hygiene compliance. Patients were encouraged to ask clinical staff if they had washed their hands. Successful hand hygiene compliance requires multimodal participation and health-care community involvement to minimize the spread of preventable nosocomial infections in and around health-care settings.

—Pamela Rose V. Samonte, Rhea U. Vallente, PhD, Mindy Rice, MSN, RN, PhD, and Michael A. Buratovich, PhD

Further Reading

APIC. *APIC Implementation Guide: Guide to Hand Hygiene Programs for Infection Prevention.* APIC, June 2015.

Bromwich, Jonah Engel. "You've Been Washing Your Hands Wrong." *New York Times*, 20 Apr. 2016. Accessed 19 Aug. 2016.

Centers for Disease Control and Prevention. "Hand Hygiene in Healthcare Settings." 29 Apr. 2019, www.cdc.gov/handhygiene/index.html. Accessed 25 Nov. 2021.

Centers for Disease Control and Prevention. "Guideline for Hand Hygiene in Health-Care Settings: Recommendations of the Healthcare Infection Control Practices Advisory Committee and the HICPAC/SHEA/APIC/IDSA Hand Hygiene Task Force." *Morbidity and Mortality Weekly Report*, 51.RR-16, 2002.

Pfoh, Elizabeth, Sydney Dy, and Cyrus Engineer. "Chapter 8: Interventions to Improve Hand Hygiene Compliance: Brief Update Review." *Making Health Care Safer II: An Updated Critical Analysis of the Evidence for Patient Safety Practices.* Agency for Healthcare Research and Quality, Mar. 2013.

World Health Organization. "Clean Hands Protect Against Infection." *WHO*, 2016. Accessed 9 Aug. 2016.

World Health Organization. "Patient Safety: World Alliance for Patient Safety." *WHO*, 27 Oct. 2004. Accessed 9 Aug. 2016.

World Health Organization. "The Evidence for Clean Hands." *WHO*, n.d. Accessed 19 Aug. 2016.

HERD IMMUNITY

Category: Public health
Also known as: Community immunity
Specialties and related fields: Epidemiology, immunology, infectious diseases
Definition: a form of protection against the spread of infectious diseases that requires many individuals within a community to be immune to the disease, usually through vaccination

KEY TERMS

immunity: the capability of organisms to resist harmful infections by pathogens
pathogen: a disease-causing agent such as a virus, bacterium, fungus, or parasite

HOW HERD IMMUNITY WORKS

Infectious diseases spread when an infected person directly or indirectly is exposed to an uninfected person. Diseases can spread through person-to-person contact, airborne, waterborne, or foodborne transfer of pathogens, or physical contact with an infected surface. Contracting a contagious disease can be avoided by vaccinations, past exposure to similar pathogens, and, occasionally, by inherited resistance to the disease. The most common of these methods is vaccination.

Photo via iStock/VV Shots. [Used under license.]

Herd immunity occurs when the immunity of a large segment of the population prevents contagious diseases from affecting those who are not immune. The percentage of individuals who must be immune for this to happen varies with the disease's infectiousness. Measles, for example, is so contagious that if one person with measles is in a room for one hour with 100 nonimmune individuals, 90 of them will contract measles. For herd immunity against measles to be effective, about 93 percent of individuals in the community must be immune to the disease. When a large percentage of a community is immune to a disease, the percentage of individuals who can bring that disease into the community is reduced. Suppose the percentage of immune individuals is high enough. In that case, infected individuals may never interact with other vulnerable individuals, thus stopping the disease from spreading through the community or herd.

THREATS TO HERD IMMUNITY

Potentially deadly diseases such as measles, mumps, and polio have been mostly eliminated through mass vaccination campaigns in developed countries. Because a large percentage of the population continues to be vaccinated, some antivaccinationists will refuse vaccination based on their belief that herd immunity will protect them. As of 2017, about 2 percent of parents rejected all vaccines for their children, and about 30 percent rejected at least one vaccine or delayed its administration. Vaccine refusal or delay reduces the percentage of immune individuals in the population, making the entire population more vulnerable to disease and decreasing the effectiveness of herd immunity. Herd immunity is essential to individuals who cannot be vaccinated for medical reasons. Many immunocompromised people (e.g., human immunodeficiency virus (HIV) positive, receiving chemotherapy or immunosuppressant drugs) cannot safely receive certain vaccines. They must depend on herd immunity which is often weak.

Antivaccination movements threaten the effectiveness of herd immunity because when fewer individuals in the community are vaccinated, diseases such as measles or pertussis that have essentially been eliminated can make a resurgence. Often these diseases are brought into a country by travelers from an area where the disease is common. These travelers can be infectious but not yet show symptoms of the disease. In 2015, the United States experienced a measles outbreak when an infected traveler visited a California amusement park from outside the country. One hundred and eighty-nine people in twenty-four states contracted measles, which had been eliminated in the United States in 2000. Most of the cases in the 2015 outbreak were among people who were not vaccinated. In 2016, 6,353 cases of mumps, another disease that has effectively been eliminated, were reported in the United States. Again, most cases were in unvaccinated individuals who failed to be protected by herd immunity.

—Tyler Biscontini and Tish Davidson, AM

Further Reading

College of Physicians of Philadelphia. "Herd Immunity." *History of Vaccines,* www.historyofvaccines.org/content/herd-immunity-0. Accessed 31 July 2017.

Davidson, Tish. *Vaccines: History, Issues, and Science.* Greenwood, 2017.

Kim, Tae Hyong, et al. "Vaccine Herd Effect." *Scandinavian Journal of Infectious Diseases,* vol. 43, no. 9, Sept. 2011, pp. 683-89, www.ncbi.nlm.nih.gov/pmc/articles/PMC3171704. Accessed 31 July 2017.

McGinty, Jo Craven. "How Anti-Vaccination Trends Vex Herd Immunity." *Wall Street Journal.* Dow Jones, Feb. 2015, www.wsj.com/articles/how-anti-vaccination-trends-vex-herdimmunity-1423241871. Accessed 31 July 2017.

HYPERSENSITIVITY REACTION

Category: Allergies
Anatomy or system affected: All systems
Specialties and related fields: Allergists, family medicine, gastroenterology, immunology, internal medicine, microbiology, oncology, ophthalmology, otolaryngology, pathology, pediatrics, pharmacology, pulmonary medicine, rheumatology
Definition: extreme physical sensitivity to particular substances or conditions

KEY TERMS

atopy: the genetic tendency to develop allergic diseases such as allergic rhinitis, asthma, and atopic dermatitis

IgE: antibodies that mediate the release of chemicals that cause an allergic reaction.

inflammation: a protective reaction to injury, disease, or irritation of tissues, characterized by pain, swelling, heat, and pain

mast cells: A type of white blood cell found in connective tissues all through the body, especially under the skin, near blood vessels and lymph vessels, in nerves, and the lungs and intestines

T lymphocytes: a type of white blood cell that is borne in the bone marrow and matures in the thymus

INTRODUCTION

A hypersensitivity reaction is a detrimental chemical reaction inside the body caused by an overactive or, less commonly, underactive immune system. When the body enters into a hypersensitive state due to an overactive immune system, the body is overcompensating in its efforts to fight off foreign substances that invade the body. Although these responses are designed to protect the body from harm, an overreaction of an immune system response can cause the body harm. When the immune system fails to respond adequately, or at all, immune responses do not protect the body as they should. This leaves the body vulnerable to harm and can cause a hypersensitive reaction. Hypersensitivity reactions are divided into two categories: immediate and delayed. Immediate hypersensitivity reactions occur when the body releases antibodies or proteins that resist disease to combat foreign substances. Delayed hypersensitive reactions occur when the body releases a type of white blood cell called a "T lymphocyte" in response to the presence of foreign substances.

BACKGROUND

The body's immune system is responsible for recognizing toxins and other foreign substances that are not normal parts of the human body. Once these substances are recognized, the immune system initiates various processes that fight off the invaders, so the body does not come to any harm. Foreign substances can enter the body in many ways, such as through the mouth, nose, ears, and skin. These substances can also enter the body through cuts and other injuries and can be prevented by keeping wounds clean and covered.

When substances such as bacteria or viruses enter the bloodstream, they can cause disease. Although the immune system cannot always prevent disease, it can fight disease-causing substances until they are no longer inside the body. For example, when a person contracts the cold or flu virus, they often experience fatigue, sore throat, runny nose, and coughing in response to the virus. The immune system releases proteins and white blood cells that specifically fight this virus and eliminate it from the body, leading to symptom relief and recovery.

The immune system plays a crucial role in fighting disease. Still, sometimes the immune system's normal functions react too much, too little, or inappropriately. When the immune system experiences these reactions to a foreign substance, a hypersensitivity reaction can develop. These reactions can cause injury and sometimes death. When antibodies are involved in hypersensitivity reactions, it is called an "immediate hypersensitive reaction." When white blood cells called "T lymphocytes" are involved in a hypersensitivity reaction, it is called a "delayed hypersensitive reaction."

OVERVIEW

Four main types of hypersensitivity reactions exist. Immediate hypersensitivity reactions account for Types I, II & III, while delayed hypersensitivity reactions account for Type IV. Some professionals feel it is necessary to distinguish certain Type II disorders that bind to cell surface receptors instead of the cell surface itself and therefore include a Type V category, but this type of designation is uncommon.

Immediate hypersensitive reactions are initiated when the immune system releases antibodies to combat a foreign substance or antigen. The reaction depends upon what type of foreign substance enters the body, how pervasive the substance is, and what type of antibody reacts with the substance. When the body is first exposed to an antigen, it is called the "sensitiz-

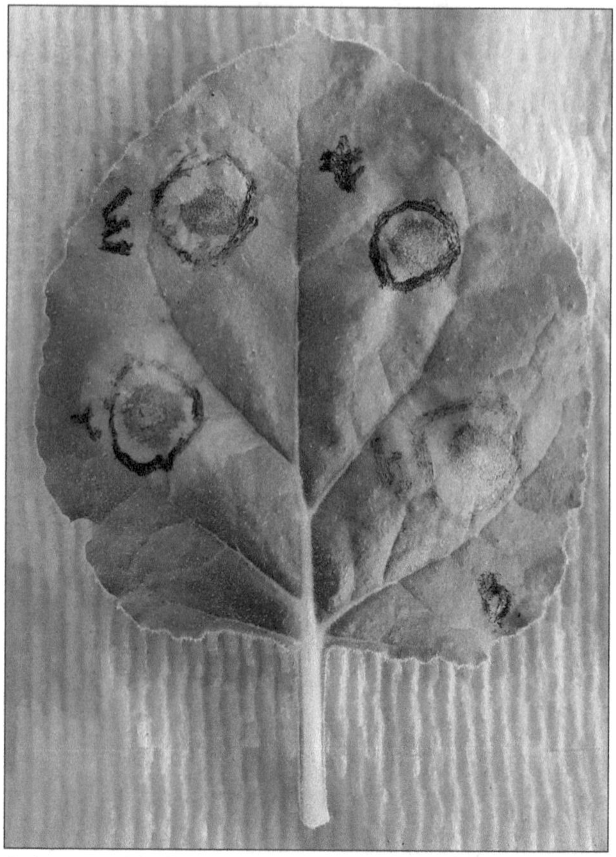

Lesions caused by the plant hypersensitive response. Photo by DanieliusKa, via Wikimedia Commons.

ing dose." The antigen then enters a latent period in which it does not react inside the body. When the same antigen enters the body at a later time, it is called a "shocking dose." The shocking dose causes a hypersensitivity reaction and can lead to several issues.

Type I reactions involve allergic reactions mediated by the antibody immunoglobulin E (IgE). Immediate hypersensitive reactions to allergens occur within minutes. Examples of disorders caused by Type I reactions include anaphylaxis, asthma, atopy, and Churg-Strauss syndrome.

Type II reactions involve antibody-dependent cell-mediated cytotoxicity (ADCC). Cytotoxicity refers to immune system responses that destroy cells. Immunoglobulin M (IgM) and immunoglobulin G (IgG) mediate cytotoxicity reactions by binding to cell surface antigens. In these instances, antigens bind to normal cells that, consequently, are targeted for destruction by the immune system. This can lead to some harmful consequences, such as the development of autoimmune diseases. Examples of disorders related to Type II reactions include autoimmune hemolytic anemia, rheumatic heart disease, Graves' disease, and Goodpasture syndrome.

Although white blood cells play a role in Type III reactions, IgG is the primary mediator of this type of immediate hypersensitivity reaction. Type III is also referred to as immune-complex reactions because they involve circulating antigen-antibody complexes. An antigen-antibody complex is a molecule formed when an antibody binds to an antigen and deposits into blood vessel walls of specific areas of the body. This initiates an inflammatory reaction that can lead to issues such as joint pain and kidney disease. Disorders that commonly occur because of a Type III reaction include serum sickness, rheumatoid arthritis, reactive arthritis, hypersensitivity pneumonitis, and systemic lupus erythematosus.

Delayed hypersensitivity reactions (also called "Type IV reactions") take a day or more to develop. In delayed hypersensitivity reactions, white blood cells known as T lymphocytes, or T cells, mediate the immune response to antigens. Typically, T cells maintain cellular immunity by releasing substances called "lymphokines," which trigger the release of macrophages or cells that protect against infection. Macrophages ingest harmful foreign substances and remove them from the body. A delayed hypersensitivity reaction occurs due to an overreaction of this process that causes the macrophages to act on healthy cells, initiating an inflammatory response that can lead to tissue damage. Examples of disorders caused by Type IV reactions include contact dermatitis, multiple sclerosis, and celiac disease.

Some experts believe this categorization of hypersensitivity reactions is too generalized and does not account for the complex connection between immune system responses and hypersensitivity reactions. An updated classification system was proposed in the late 1990s that separates hypersensitivity reactions into seven categories as follows:

- inactivation/activation antibody reactions
- cytotoxic antibody reactions
- immune-complex reactions
- allergic reactions
- T-cell cytotoxic reactions
- delayed hypersensitivity reactions
- granulomatous reactions (inflammation that produces a mass of connective tissue)

—*Cait Caffrey*

Further Reading

Abramson, Stuart L. "Delayed Hypersensitivity Reactions." *Medscape*, 7 May 2018, emedicine.medscape.com/article/136118-overview. Accessed 5 Nov. 2018.

Buelow, Becky. "Immediate Hypersensitivity Reactions." *Medscape*, 9 Feb. 2015, emedicine.medscape.com/article/136217-overview. Accessed 5 Nov. 2018.

"Hypersensitivity Reactions." *Centers for Disease Control and Prevention*, www.cdc.gov/travel-training/local/HistoryEpidemiologyandVaccination/page27396.html. Accessed 5 Nov. 2018.

"Hypersensitivity Reactions." *Centers for Disease Control and Prevention*, www.cdc.gov/travel-training/local/History EpidemiologyandVaccination/page36674.html. Accessed 5 Nov. 2018.

"Hypersensitivity Reactions." *CliffsNotes*, www.cliffsnotes.com/study-guides/biology/microbiology/disorders-of-the-immune-system/hypersensitivity-reactions. Accessed 5 Nov. 2018.

"Hypersensitivity Reactions (Types I, II, III, IV)?" *Rutgers, The State University of New Jersey*, 9 Apr. 2009, njms.rutgers.edu/gsbs/olc/mci/prot/2009/Hypersensitivities09.pdf. Accessed 27 Nov. 2021.

Janeway, Charles A., et al., editors. "Hypersensitivity Disease." *Immunobiology: The Immune System in Health and Disease*. Garland Science, 2001.

Klatt, Edward C., and Vinay Kumar. *Robbins and Cotran Review of Pathology*. 5th ed., Elsevier, 2021.

IMMUNIZATION AND VACCINATION

Category: Procedure

Anatomy or system affected: Blood, cells, immune system

Specialties and related fields: Immunology, microbiology, preventive medicine, public health

Definition: immunization is the process by which exposure to an infectious agent or chemical confers an organism with resistance to that agent; vaccination involves the injection of a killed or attenuated microorganism to induce immunity

KEY TERMS

active immunity: immunity resulting from antibody production following exposure to an antigen

antibody: a protein secreted by lymphocytes in response to antigen stimuli, such as bacteria or viruses; also referred to as immunoglobulin

antigen: any chemical substance that stimulates the production of antibodies

attenuation: the weakening or elimination of the pathogenic properties of a microorganism; ideally, the organism is rendered harmless

passive immunity: immunity resulting from the introduction of preformed antibodies

serotype: a subgroup member within a larger species that is similar but not identical to other members of the species

toxoid: a toxin that has been chemically treated to eliminate its toxic properties but that retains the same antigens as the original

vaccinia: a virus that causes a pox-like illness in cattle (cowpox) and serves as a smallpox vaccine in humans because of its similarity to the smallpox virus

THE FUNDAMENTALS OF IMMUNIZATION

The major day-to-day function of the immune response is to protect the body from infection. Exposure to foreign antigens such as infectious agents results in the stimulation of either of two components of the immune system: the humoral (or antibody) immune response or the cellular immune response. Although no clear division exists between these two facets of the immune system, the antibody response deals primarily with bacteria that live outside the cell. The cellular response deals primarily with microbes that live within a cell, such as intracellular bacteria or viruses.

A specialized class of white cells called "B lymphocytes" carries out the production of antibodies. Stimulation of these cells results from a complicated interaction between various cells, including antigen-presenting cells (macrophage and dendritic cells) and both T and B lymphocytes. The response is specific in that each type of T or B cell can interact with only a single antigen. The B cell that produces antibodies against a particular characteristic or shape on the surface of a bacterium reacts only with that particular determinant. In turn, the antibodies secreted by that B cell can interact only with specific determinants.

Antibodies secreted by B cells are themselves inert proteins. A variety of effects can result, however, when an antibody binds to an antigen. The specific results depend on the nature of the antigen. For example, the binding of an antibody to a toxin results in the neutralization of that substance. Suppose the antigen is on the surface of a bacterial cell. In that case, the antibody can act as a flag that attracts other chemicals circulating in the blood. The technical term for an antibody bound to a bacterium is *opsonin*. The antibody-bacterium complex becomes much more likely to be ingested and destroyed by a specialized cell called a "phagocyte" than if the antibody were not present. Likewise, suppose the antigen is an extracellular virus particle. In that case, binding of the antibody may inhibit the ability of the virus to infect a cell.

The cellular immune system also reacts in a specific manner. A subclass of T lymphocytes called "cytotoxic T cells" reacts with specific antigenic determinants on the surface of infected cells. When the T cells bind to a target, they locally release toxic chemicals that ultimately kill the target.

The development of vaccines against specific infectious microbial agents resulted in the control or elimination of many diseases caused by these agents. The first formal vaccine developed for the prevention of disease was that used by Edward Jenner against smallpox during the 1790s. Another century passed before the molecular basis for vaccine function began to be understood.

Immunity to an antigen or disease may be induced using either of two methods. If preformed antibodies produced in another human or animal are inoculated into an individual, the result is passive immunity. Passive immunity can be advantageous in that the recipi-

Photo via iStock/filadendron. [Used under license.]

ent achieves immunity in a short period. For example, suppose a person has been exposed to a toxin or has contacted an infectious agent. In that case, passive immunity can provide rapid, short-term protection. However, no long-range protection is achieved because the individual does not generate the capacity to produce that antibody, and the preformed antibodies are gradually removed from the body.

Using antigens introduced through vaccination to activate antibody production results in active immunity. The development of effective active immunity requires a period of several days to several weeks. The immunity is long-term, however, often lasting for the life span of the individual. Furthermore, each additional exposure to that same antigen, either through a vaccine booster or natural exposure, results in a more rapid, greater response than previously achieved. This increased rate of reaction is referred to as an anamnestic response.

The actual material used in a vaccine is variable, depending on the form of antigen. The earliest vaccine used by Jenner against smallpox consisted of a virus that caused disease in cattle called "cowpox." The word "vaccination" is derived from this use; *vacca* is the Latin word for cow. While cowpox is distinct from smallpox, the viruses that cause the two diseases contain similar antigenic determinants. Jenner made this observation and exploited that exposure to the cowpox virus results in active immunization against smallpox.

The use of attenuated strains of bacteria or viruses applies the same principle of cross-reaction. Attenuated organisms are mutants that have lost the ability to cause disease but retain the virulent strain's antigenic character. The most notable application of attenuation is the Sabin oral poliovirus vaccine (OPV). By testing hundreds of virus isolates for the ability to cause polio in monkeys, Albert Sabin was able to isolate certain strains that did not cause disease. These strains formed the basis for his vaccine. Similar testing resulted in the development of attenuated virus vaccines against a wide variety of agents, including measles, mumps, and rubella. Likewise, the Bacillus Calmette-Guérin (BCG) strain of *Mycobacterium tuberculosis* serves as a vaccine against the agent that causes tuberculosis. Unfortunately, this vaccine does not always result in immunity for the recipient. It generally is not used in the United States but is widely used in countries where tuberculosis is common.

In some cases, the isolation of attenuated strains of microorganisms has proved difficult. For this reason, inactivated or killed microorganisms often serve as the basis for vaccine production. The Salk inactivated poliovirus vaccine represents the best-known example. By treating poliovirus with a solution of the chemical formalin, Jonas Salk could inactivate the organism. The virus retained its antigenic potential and served as an effective vaccine. A similar process has resulted in vaccines protecting against other bacterial diseases, such as bubonic plague, cholera, pertussis (whooping cough), and viral influenza.

In some cases, the vaccine is directed not against the etiological agent itself but toxic materials produced by the agent. This is the case with diphtheria and tetanus. The vaccines are produced by treating diphtheria and tetanus toxins secreted by these bacteria with formalin. The toxoids that result are antigenically similar to the actual toxins and so can induce immunity. They are incapable, however, of causing the harmful effects of the respective diseases.

Only those determinants of a virus or bacterium that stimulate neutralizing antibodies are necessary for most vaccines. For this reason, the use of genetically engineered vaccines was begun in the 1980s. The first example was the production of a vaccine against the hepatitis B virus (HBV). The gene that encodes the surface antigen of HBV was isolated and inserted into a piece of genetic material within the yeast *Saccharomyces*. The HBV antigen produced by the yeast was purified and subsequently found to be as effective in a vaccine as the whole virus. Since no live virus is involved, there is no danger of an attenuated strain reverting to its virulent parent. Recently, similar technology has been applied to produce vaccines that protect against chickenpox and hepatitis A virus.

HISTORY AND MAJOR SUCCESSES

Since the first use of vaccination by Jenner in the 1790s for the prevention of smallpox, immunization techniques have been developed for protection against most major infectious illnesses. The term *vaccination* was initially applied to immunization against smallpox. However, its definition has long been expanded to include most immunization techniques. The terms "vaccination" and "immunization" are used interchangeably, although there are technical differences in their definitions.

The nineteenth-century improvements in public health measures, combined with the passage of laws for compulsory vaccination, resulted in a steady decrease in the number of smallpox cases in the United States and most countries of Europe. Even as late as 1930, however, approximately 49,000 cases were reported in the United States. In the 1950s, large numbers of cases were still being reported in areas of Africa and Asia. At that time, the World Health Organization (WHO) of the United Nations decided on a plan to eliminate smallpox based on the fact that humans served as the sole reservoir for the smallpox virus; animals are not naturally infected with smallpox. Through mass immunization techniques, the plan was to isolate areas of infection into smaller and smaller pockets.

WHO's plan to eliminate smallpox proved entirely successful. There are two different forms of smallpox. The last known natural case of variola major was reported in Bangladesh in 1974. The last known case of variola minor was reported in Somalia in 1976. Although an outbreak of smallpox resulting from a laboratory accident was reported in Great Britain, there were no additional naturally caused cases of smallpox. In 1978, WHO declared the world to be free of smallpox.

Ironically, the origins of the vaccine in use during this successful campaign are unknown. The original strain of cowpox used by Jenner was lost sometime during the nineteenth century. The strain used in the vaccine during the twentieth century, called "vaccinia virus," may have originated from an isolate obtained during the Franco-Prussian War in the 1870s.

Although vaccination was effective in immunizing most persons against smallpox, the use of the vaccine itself had some risks. Serious complications were rare but did occasionally occur. With the disappearance of the disease, the need for routine immunization lessened, and, in 1971, compulsory vaccination of children in the United States was discontinued. In 1976, the routine vaccination of hospital employees was discontinued as well. By the 1990s, the only four freezers in the world had known, declared smallpox virus stocks in their freezers. The September 11, 2001, terrorist attacks at the World Trade Center in New York City and the Pentagon introduced a new era in which the fear of terrorists' use of biological agents has prompted renewed discussions about the reintroduction of the smallpox vaccine.

The use of vaccines for the elimination of poliomyelitis represents another success story. Although sporadic outbreaks of polio occurred in earlier centuries and probably as long ago as ancient Egypt, the first epidemics appeared in the late nineteenth century. Ironically, this increase in the incidence of polio was caused by improvements in public health. Poliovirus is easily transmitted through a fecal-oral route. Still, the majority of cases, particularly in young children, are without symptoms or asymptomatic. With improvements in sanitation, the first exposure to polio was often delayed until later childhood or, as in the case of President Franklin D. Roosevelt, in the adult years. Under these circumstances, the disease is often more severe.

In 1955, the inactivated poliovirus vaccine developed by Salk was introduced for general use; in 1961, Sabin's oral poliovirus vaccine was licensed for use. By

Recommended Child and Adolescent Immunization Schedule for ages 18 years or younger

UNITED STATES 2021

Vaccines in the Child and Adolescent Immunization Schedule*

Vaccines	Abbreviations	Trade names
Diphtheria, tetanus, and acellular pertussis vaccine	DTaP	Daptacel® Infanrix®
Diphtheria, tetanus vaccine	DT	No trade name
Haemophilus influenzae type b vaccine	Hib (PRP-T)	ActHIB® Hiberix®
	Hib (PRP-OMP)	PedvaxHIB®
Hepatitis A vaccine	HepA	Havrix® Vaqta®
Hepatitis B vaccine	HepB	Engerix-B® Recombivax HB®
Human papillomavirus vaccine	HPV	Gardasil 9®
Influenza vaccine (inactivated)	IIV	Multiple
Influenza vaccine (live, attenuated)	LAIV4	FluMist® Quadrivalent
Measles, mumps, and rubella vaccine	MMR	M-M-R II®
Meningococcal serogroups A, C, W, Y vaccine	MenACWY-D	Menactra®
	MenACWY-CRM	Menveo®
	MenACWY-TT	MenQuadfi®
Meningococcal serogroup B vaccine	MenB-4C	Bexsero®
	MenB-FHbp	Trumenba®
Pneumococcal 13-valent conjugate vaccine	PCV13	Prevnar 13®
Pneumococcal 23-valent polysaccharide vaccine	PPSV23	Pneumovax 23®
Poliovirus vaccine (inactivated)	IPV	IPOL®
Rotavirus vaccine	RV1 RV5	Rotarix® RotaTeq®
Tetanus, diphtheria, and acellular pertussis vaccine	Tdap	Adacel® Boostrix®
Tetanus and diphtheria vaccine	Td	Tenivac® Tdvax™
Varicella vaccine	VAR	Varivax®
Combination vaccines *(use combination vaccines instead of separate injections when appropriate)*		
DTaP, hepatitis B, and inactivated poliovirus vaccine	DTaP-HepB-IPV	Pediarix®
DTaP, inactivated poliovirus, and *Haemophilus influenzae* type b vaccine	DTaP-IPV/Hib	Pentacel®
DTaP and inactivated poliovirus vaccine	DTaP-IPV	Kinrix® Quadracel®
DTaP, inactivated poliovirus, *Haemophilus influenzae* type b, and hepatitis B vaccine	DTaP-IPV-Hib-HepB	Vaxelis®
Measles, mumps, rubella, and varicella vaccine	MMRV	ProQuad®

*Administer recommended vaccines if immunization history is incomplete or unknown. Do not restart or add doses to vaccine series for extended intervals between doses. When a vaccine is not administered at the recommended age, administer at a subsequent visit. The use of trade names is for identification purposes only and does not imply endorsement by the ACIP or CDC.

How to use the child/adolescent immunization schedule

1 Determine recommended vaccine by age **(Table 1)**

2 Determine recommended interval for catch-up vaccination **(Table 2)**

3 Assess need for additional recommended vaccines by medical condition and other indications **(Table 3)**

4 Review vaccine types, frequencies, intervals, and considerations for special situations **(Notes)**

Recommended by the Advisory Committee on Immunization Practices (www.cdc.gov/vaccines/acip) and approved by the Centers for Disease Control and Prevention (www.cdc.gov), American Academy of Pediatrics (www.aap.org), American Academy of Family Physicians (www.aafp.org), American College of Obstetricians and Gynecologists (www.acog.org), American College of Nurse-Midwives (www.midwife.org), American Academy of Physician Assistants (www.aapa.org), and National Association of Pediatric Nurse Practitioners (www.napnap.org).

Report

- Suspected cases of reportable vaccine-preventable diseases or outbreaks to your state or local health department
- Clinically significant adverse events to the Vaccine Adverse Event Reporting System (VAERS) at www.vaers.hhs.gov or 800-822-7967

Download the CDC Vaccine Schedules App for providers at www.cdc.gov/vaccines/schedules/hcp/schedule-app.html.

Helpful information

- Complete ACIP recommendations: www.cdc.gov/vaccines/hcp/acip-recs/index.html
- *General Best Practice Guidelines for Immunization*: www.cdc.gov/vaccines/hcp/acip-recs/general-recs/index.html
- Outbreak information (including case identification and outbreak response), see Manual for the Surveillance of Vaccine-Preventable Diseases: www.cdc.gov/vaccines/pubs/surv-manual
- ACIP Shared Clinical Decision-Making Recommendations www.cdc.gov/vaccines/acip/acip-scdm-faqs.html

CDC

U.S. Department of Health and Human Services
Centers for Disease Control and Prevention

the 1990s, polio had been eliminated from the Western Hemisphere and developed countries elsewhere. In 2003, WHO developed the Global Polio Eradication Initiative, the purpose of which is to monitor outbreaks of polio and address the means of preventing its transmission. Despite setbacks in portions of Asia and Africa, the number of reported polio cases worldwide was reduced to 223 by 2012, according to WHO data. Since polio eradication is not complete, the American Academy of Pediatrics (AAP) recommends that children receive immunizations at intervals during their first two years of life and again before starting school. Adults who plan to travel to areas of the world in which polio is found should also be immunized.

The most significant advancement in twentieth-century health care in the United States has been eliminating most major childhood diseases. In addition to poliovirus immunization, children routinely receive a variety of early immunizations. Measles, mumps, and rubella (MMR) vaccines are first administered in a single preparation at twelve to fifteen months of age. All three contain live attenuated viruses. The measles vaccine was first introduced in 1966. It resulted in a decline in reported measles cases of nearly 99 percent by the 1980s. Beginning about 1986, however, increasing numbers of measles cases were reported among young adults who had been previously immunized. For this reason, the AAP recommends that children receive a second dose of MMR vaccine before entering school, at approximately four to six years of age. A series of the diphtheria, pertussis, and tetanus (DTaP) vaccine is administered at two, four, six, and fifteen to eighteen months, with tetanus and diphtheria boosters recommended at ten-year intervals throughout the remainder of a person's life.

By eliminating most other major childhood illnesses, *Hemophilus influenzae* type B infections became among the most significant causes of illness and death among young children. In 1985, a vaccine developed from the outer coat of the bacterium was licensed for use. The vaccine worked poorly in children under the age of two, the major population at risk. Consequently, an improved vaccine was developed and licensed in 1987. The second vaccine consisted of a portion of the bacterial coat of *H. influenzae* joined to diphtheria toxoid. Immunization with the vaccine is recommended at two, four, six, and twelve to fifteen months.

Understandings of immunity and immunization grew, and vaccine technology continued to evolve. For example, in 1986, a genetically engineered hepatitis B virus vaccine was developed and licensed. The gene that encodes the virus surface antigen was placed in a small piece of deoxyribonucleic acid (DNA), a plasmid, and inserted into the common baker's yeast *Saccharomyces cerevisiae*. The antigen produced is used in a three-dose series to immunize individuals at risk for the disease: health-care workers, institutional staff, and anyone else who is likely to contact the virus. It is also recommended for all children to eradicate the disease through universal immunization.

Other vaccines developed during the late twentieth century include the pneumococcal vaccine, meningococcal vaccine, hepatitis A vaccine, rotavirus vaccine, and varicella (chickenpox) vaccine. All of these have become part of routine immunization in children and adolescents.

Additionally, preparations of some long-standing vaccines have been improved through new technology. For example, the live poliovirus vaccine (Sabin) proved more effective than the original injectable vaccine (Salk). However, even the attenuated virus rarely causes disease in some vaccine recipients. Successful efforts were undertaken to improve the effectiveness of the vaccine made from killed (inactivated) virus; this enhanced inactivated preparation is now used in the United States. Likewise, the pertussis vaccine, which was poorly tolerated in a significant number of children and adults, has been improved, and the new acellular pertussis vaccine preparation is now recommended for both children and adults.

Although routine vaccination of children has virtually eliminated the most health-threatening infectious disease from that population, immunization of adults against preventable diseases has not been as successful. According to estimates, in the late twentieth century, between fifty thousand and seventy thousand adults died yearly from preventable diseases through immunization, such as pneumococcal pneumonia, influenza, and hepatitis B.

Historically, pneumococcal pneumonia, caused by the bacterium *Streptococcus pneumonia*, has been a killer of adults. In the 1990s, an estimated forty thousand persons, primarily the elderly, died from this disease. The available vaccine is a polysaccharide vaccine, representing serotypes for twenty-three of the major strains of the bacterium. When administered by the age of fifty, the vaccine provides a significant degree of protection against the organism.

Between 1957 and 2005, seasonal influenza outbreaks averaged between ten thousand to twenty-five thousand deaths per year in the United States alone. Two of the epidemics each resulted in more than forty thousand deaths. Most of these deaths were in elderly adults. Although the usefulness of vaccination among the elderly is limited, immunization against influenza will often lessen the severity of the disease, even if it fails to prevent infection. Furthermore, vaccination of large segments of the population, including health-care workers, limits the spread of disease within the population and reduces the exposure to susceptible elderly persons and others at risk for complications from influenza.

Shingles is a blistering skin condition caused by a reactivation of the chickenpox virus (varicella-zoster, or herpes zoster) that occurs primarily among the elderly. Although it rarely causes death, it can be extremely painful, and the pain may persist long after the lesions heal. When the condition involves the area around the eye, blindness may result. A vaccine against this disease has been developed.

Some specialized vaccines are recommended only for international travelers. Both killed injectable and live attenuated oral vaccines against typhoid are licensed. The ease of administration of an oral vaccine has made this form the preferred choice. In addition, vaccines against cholera and yellow fever may be used in appropriate circumstances.

Although active immunization in most circumstances remains the preferred method of protection through vaccination, there are situations in which passive immunization may provide temporary protection. Individuals may be immunosuppressed or lack a functional immune system. This condition may result from infection (acquired immunodeficiency syndrome, or AIDS), medical intervention (chemotherapy), or congenital reasons (severe combined immunodeficiency disease, or SCID). Whatever the cause, active immunization does not develop. In addition, there are circumstances in which the necessary time for the development of immunity through active immunization is not available, such as with exposure to tetanus, hepatitis A or B, or rabies. For passive immunization or replacement therapy in immunodeficiency disorders, immunoglobulin is usually prepared from pools of plasma obtained from large numbers of blood donors. Specific immunoglobulins, directed against specific targets such as rabies or tetanus, are prepared from plasma containing high concentrations of these antibodies.

PERSPECTIVE AND PROSPECTS

The elimination of smallpox represents the classic example in which the efficacy of a vaccine resulted in the eradication of the disease. Smallpox was an ancient disease with origins as early as the twelfth century BCE. It appeared in the Middle East in the sixth century CE, with subsequent dissemination into northern Africa and southern Europe due to the Arab invasions from the sixth to the eighth centuries. The disease spread throughout Europe during the Crusades of the eleventh and twelfth centuries. It reached the Americas early in the colonial period. It has been estimated that smallpox killed 400,000 persons each year and caused more than one-third of all cases of blindness at its peak during the eighteenth century. It has also been estimated that smallpox or other diseases that traveled to the Americas with settlers killed approximately 85 percent of the American Indians who died during colonial periods.

The principle of immunization in prevention did not originate with Jenner, the English physician credited with developing the smallpox vaccine in the 1790s. Variolation was well known in China and parts of the Middle East for centuries prior to Jenner. Variolation consisted of the inhalation of dried crust prepared from the pocks obtained from individuals suffering from mild cases of smallpox. A variation involved removing small amounts of fluid from an active smallpox pustule and scratching the liquid into children's skin. Lady Mary Wortley Montagu, the wife of the British ambassador to the Ottoman Empire, introduced the practice of variolation into Great Britain during the early eighteenth century. The use of variolation was empirical; the practice was often successful. The possibility remained, however, that immunization might introduce the disease.

Born in 1749, Jenner first became aware of the protective effects of cowpox from the story of a local dairymaid who had been exposed to the disease. After years of study and observation, he became convinced of the story's validity. In 1796, he immunized an eight-year-old boy with material from a cowpox lesion, and no ill effects were seen. Further immunizations supported the theory that cowpox protected against smallpox. Jenner called this material "variolae vaccinae." Richard Dunning, a Plymouth physician, in an 1800 analysis of the procedure, was the first to use the term "vaccination."

Wider application of the principle of vaccination followed Louis Pasteur's studies during the 1870s and 1880s. With his attenuation of the bacterium that caused chicken cholera, Pasteur demonstrated that one could manipulate the virulence of a microorganism. This practice soon led to his development of vaccines against both anthrax and rabies.

In the News: Combination Vaccine Pediarix for Infants

In December 2002, the US Food and Drug Administration (FDA) approved a multicomponent vaccine containing diphtheria and tetanus toxoids, acellular pertussis adsorbed (DTaP), hepatitis B (HepB; recombinant), and inactivated poliovirus (IPV) components. This vaccine, manufactured by GlaxoSmithKline Biologicals in Belgium, is called "DTaP-HepB-IPV combined vaccine," or "Pediarix." It is intended to be given to infants in three injections, one each at two, four, and six months of age. Pediarix protects against diphtheria, tetanus, pertussis (whooping cough), hepatitis B, and poliomyelitis in a series of only three injections instead of the nine injections previously given to infants. Thus, it provides benefits for patients and practitioners alike, such as saved office visits and decreased costs. Most important, the shorter injection series makes it much easier for parents, increasing the likelihood that parents will have their children immunized against these serious diseases.

The Pediarix vaccine has been subjected to numerous worldwide clinical trials to evaluate its safety and immunogenicity or impact on the immune system. Data indicate that the vaccine is both safe and effective, with no significant adverse effects being reported. In these trials, the immunogenicity of each of the components in the combined vaccine was not shown to differ significantly from the immunogenicity of the components administered separately. These studies, involving more than seven thousand infants, reported minor adverse effects, including injection-site soreness (pain, redness, or swelling), fever, and irritability. The Pediarix vaccine has been associated with a greater frequency of fever in comparison to the separately administered vaccine components. Infants who are hypersensitive to any vaccine component, including yeast, neomycin, and polymyxin B, should not be given the vaccine. The Advisory Committee on Immunization Practices (ACIP), the Committee on Infectious Diseases of the American Academy of Pediatrics, and the American Academy of Family Physicians have recommended this vaccine for routine use.

—Steven A. Kuhl, PhD

The twentieth century saw the development of effective vaccines against most major childhood diseases. The use of the DPT toxoid became routine in the United States about 1945. The development of the oral Sabin vaccine and inactivated Salk vaccines during the 1950s resulted in the complete elimination of poliomyelitis from the Western Hemisphere by the 1990s. Genetic engineering, in which only the genes necessary to synthesize specific antigens are used, was first applied to the hepatitis B vaccine. It has also been applied successfully to create vaccines against chickenpox and the hepatitis A virus. A vaccine against the hepatitis C virus is under development. This technology provides the potential for manufacturing vaccine "cocktails," or combinations of such genes from various infectious agents in a single vaccine.

Some vaccines have a more wide-ranging impact than infection prevention, as some long-standing viral infections have been shown to cause cancer. This is true of both hepatitis B and hepatitis C, which are implicated in developing primary hepatocellular carcinoma, the most common form of cancer worldwide.

Another such vaccine is directed against infection by certain strains of the human papillomavirus (HPV). HPV is most commonly associated with genital warts and is usually transmitted sexually. However, it may be spread by hand contact as well. Infection by some strains may lead to the development of cervical cancer. According to the WHO, more than 500,000 new cases of cervical cancer are diagnosed annually, and more than 250,000 women die of the disease each year. In 2006, a vaccine called "Gardasil," directed against the four most common types of HPV, was licensed. A three-dose regimen is recommended for girls and women between nine and twenty-six years old. Some experts recommend that the vaccine be extended to boys and young men to interrupt the most common chain of infection more effectively.

Several infections that cause a considerable burden of illness and death worldwide have so far eluded efforts to create a successful vaccine. Malaria is one example. Attempts to develop a vaccine against AIDS also have proven largely unsuccessful. The mechanism that has been exploited by other vaccines, namely the stimulation of T cells to produce antibodies that protect a recipient, is nonfunctional in AIDS. The successful development of a vaccine against AIDS will require scientific ingenuity.

New infections also have attracted the attention of vaccine researchers. The appearance of a new outbreak of avian influenza in 2003, usually referred to simply as the bird flu, raised new concerns about particularly virulent strains of influenza. The emergence in 2009 of a new strain of influenza, the novel H1N1 strain, raised fears of a pandemic (a worldwide epidemic) and challenged the capacity of the pharmaceutical industry to develop an effective vaccine and to produce enough of it in a short time to immunize the world's population against an imminent threat. The emergence of new infections, the threat of bioterrorism, and the unique difficulties posed by some microorganisms remain areas of active investigation in the field of vaccine research and development.

—Richard Adler, PhD,
L. Fleming Fallon Jr., MD, PhD, MPH,
and Margaret Trexler Hessen, MD

Further Reading

Behbehani, Abbas. *The Smallpox Story in Words and Pictures.* University of Kansas Medical Center, 1988.

Brock, Thomas D., editor. *Microorganisms: From Smallpox to Lyme Disease.* W. H. Freeman, 1990.

Delves, Peter J., et al. *Roitt's Essential Immunology.* 13th ed., Blackwell, 2017.

Flajnik, Martin, et al. *Paul's Fundamental Immunology.* 8th ed., Lippincott Williams & Wilkins, 2022.

Grandi, Guido, editor. *Genomics, Proteomics, and Vaccines.* John Wiley & Sons, 2004.

National Institute of Allergy and Infectious Diseases. "How Vaccines Work." *National Institutes of Health,* 19 Apr. 2011.

Playfair, J. H. L., and B. M. Chain. *Immunology at a Glance.* 10th ed., Wiley-Blackwell, 2012.

Plotkin, Stanley A., et al., editors. *Vaccines.* 7th ed., Elsevier, 2017.
Rosario, Diane. *Immunization Resource Guide: Where to Find Answers to All Your Questions About Childhood Vaccinations.* Patter, 2001.

Immunodeficiency disorders

Category: Diseases and disorders
Anatomy or system affected: Immune system
Specialties and related fields: Genetics, immunology
Definition: genetic or acquired disorders that result from disturbances in the normal functioning of the immune system

KEY TERMS

antibody: protein immunoglobulin secreted by B lymphocytes; the production of antibodies is induced by specific foreign invaders, and they combine with and destroy only those invaders

B lymphocytes: also referred to as B cells; white cells of the immune system that produce antibodies; produced within the bone marrow

phagocytes: white cells of the immune system that destroy invading foreign bodies by engulfing and digesting them in a nonspecific immune response; include macrophages and neutrophils

stem cells: multipotential precursor cells within the bone marrow that develop into white cell populations, including lymphocytes and phagocytic cells

T lymphocyte: a type of immune cell that kills host cells infected by bacteria or viruses and secretes chemicals (interleukins) that regulate the immune response

CAUSES AND SYMPTOMS

The defense of the body against foreign invaders is provided by the immune system. In nonspecific immunity, phagocytic cells engulf and destroy invading particles. Specific immunity consists of very specialized cell types that are synthesized in response to a particular type of foreign invader. Self-replicating stem cells within the bone marrow give rise to lymphocytes, which mediate specific immunity. Lymphocytes establish self-replacing colonies within the thymus, spleen, and lymph nodes. The various categories of T lymphocytes are derived from the thymus colonies, while B lymphocytes develop and mature within the bone marrow. B lymphocytes secrete highly specific antibodies that attack bacteria and some viruses.

T lymphocytes do not secrete antibodies. Cytotoxic T cells directly attack body cells that have been infected with a bacterium or virus, while helper T cells regulate the immune response, either by directly interacting with other lymphocytes or by secreting chemicals, called "interleukins," that regulate those cells. In immunodeficiency disorders, some or all of these defenses are compromised, which can have life-threatening consequences. Immunodeficiency diseases are generally the result of genetic abnormalities and are present from birth; others may be acquired through infection or exposure to damaging drug or radiation treatments. Depending upon the specific defect, the result may range from limited defects involving a class of cells to an entire shutdown of the immune system; prognosis depends on the severity of the defect. Since these defects generally involve recessive traits, expression of immune deficiencies usually results from mutations on the X chromosome; sex-linked traits are generally observed only in males, since males carry a single X chromosome.

The most severe immunodeficiency disorder is attributable to the absence of stem cells, which results in a total lack of both B and T lymphocytes. This rare genetic condition is referred to as severe combined immunodeficiency syndrome (SCID). Affected infants show a failure to thrive from birth and can easily die from common bacterial or viral infections. The term SCID encompasses a variety of genetic deficiencies. Certain forms are sex-linked, while other types may be autosomal (non-sex-linked). The most common autosomal form is a deficiency in the enzyme adenosine deaminase (ADA), resulting in disruption of deoxyribonucleic acid (DNA) synthesis in the stem cells.

Major syndromes that involve defects specific to the T-lymphocyte population are characterized by recurrent viral and fungal infections. DiGeorge syndrome results from improper development of the thymus, which in turn results in insufficient production of T lymphocytes, often accompanied by other structural abnormalities in the infant. In severe cases, death results in early childhood from overwhelming viral infections.

The most common disorders affecting B lymphocytes are forms of hypogammaglobulinemia. This condition is characterized by insufficient levels of antibody. The cause is generally associated with increased rates of antibody breakdown or loss in the urine secondary to kidney malfunction. Bruton's agammaglobulinemia is a rare, sex-linked form of the condition, in which B cells fail to mature properly. Severe bacterial infections are the most common symptom. When the disorder is left untreated, infants generally die of severe pneumonia prior to six months of age.

Several immunodeficiency disorders may be the result of partial defects in the production and/or function of B and T lymphocytes. Wiskott-Aldrich syndrome is a genetically inherited disease manifested by recurrent infections and an itchy, scaly inflammation of the skin. Certain classes of antibodies are absent or scarce. Chronic mucocutaneous candidiasis is characterized by chronic fungal infection of the skin and mucous membranes; reduced levels of T cells are responsible for this disfiguring disorder.

Immunodeficiency disorders may also be the result of defects in phagocytic cells; the underlying cause of most of these disorders is ill-defined but often involves deficiencies in hydrolytic enzymes. In chronic granulomatosis, an inherited enzyme deficiency prevents the immune system from destroying bacteria that have been phagocytized. Infants affected by this disorder develop severe infections and chronic inflammations of internal organs and bones. The bacteria responsible for these infections are generally common flora that are not considered pathogens in healthy individuals.

Immune disorders may involve defects in antibody production. However, certain forms of inherited disorders involve another group of proteins called "complement." Complement actually represents a group of serum proteins that interact with each other in a series. The pathway may be activated either specifically from antibody-target interactions or nonspecifically by surface components of certain bacteria. Intermediates in the complement pathway attract or stimulate phagocytosis (opsonins), induce inflammation, and play roles in cell destruction. Some of the intermediates are enzymes that regulate the activation of complement components. The most important intermediate in the pathway is the C3 component. C3 is a protein that acts as an opsonin and at the same time is involved in activating later steps in the sequence. Defects in C3 result in increased susceptibility to infection. Similar immune problems may result from defects in other complement intermediates.

Most of the disorders that affect the immune system are not inherited but develop sometime during the person's life. They are either the result of an infection or a consequence of another disease or its treatment. The use of corticosteroids to treat inflammations, or the illicit use of them in muscle-building, can interfere with the proper production and function of T lymphocytes. Other immunosuppressive drugs used to diminish the possibilities of graft or transplant rejection, or in the treatment of autoimmune diseases, can severely depress antibody production. Chemotherapeutic agents used in the treatment of cancer can affect DNA replication and severely compromise the entire immune system. Whole-body radiation can damage or destroy bone marrow stem cells.

Acquired immunodeficiency syndrome (AIDS) is caused by the human immunodeficiency virus (HIV). HIV is transmitted primarily through unprotected sexual contact, sharing of needles for intravenous drug use, transfusion with contaminated blood products, or contact with contaminated body fluids. HIV specifically infects one type of regulatory T lymphocyte, the helper T cell, resulting in severe immune depression. The virus may be harbored in an inapparent form for years. Initial symptoms may be quite mild, but they generally progress so that the affected individual becomes susceptible to a host of opportunistic bacterial and fungal infections. A rare form of cancer called "Kaposi's sarcoma" is associated with infection by a particular human herpesvirus, HHV8, in persons with AIDS. AIDS also produces neurological damage in about one-third of infected individuals.

TREATMENT AND THERAPY

Most treatment of immunodeficient individuals is palliative. Infections are treated with antibiotics whenever possible. Individuals are counseled to avoid situations in which they may be exposed to contagious agents.

Treatment of immunodeficiency disorders may also target the source of the deficiency. For example, in DiGeorge syndrome, characterized by the congenital absence of the thymus, fetal thymus transplants may correct the problem, with improvement in lymphocyte levels seen within hours after the transplants. The use of thymus extracts has also been beneficial. Syndromes such as hypogammaglobulinemia can be managed by injection with mixtures of antibodies. Drug therapy to substitute for some immune components absent in Wiskott-Aldrich syndrome has been shown to have variable effects. The most effective treatment for chronic mucocutaneous candidiasis is aggressive antifungal medication to eradicate the causative organism; treatment must continue for several months because fungal infections are slow to respond to therapy and frequently recur. Chronic granulomatosis is notoriously difficult to treat, and the most effective therapy has been antibiotic and antifungal agents used aggressively during an overt infection.

Because of the magnitude of the defects, many inherited immunodeficiency disorders are difficult to treat successfully and are commonly fatal early in life. Chronic granulomatosis is usually fatal within the first few years of life, and only about 20 percent of patients reach the age of twenty. SCID is a serious disorder in which affected infants can die before a proper diagnosis is made. For individuals with these and other serious immunodeficiency disorders, maintenance in an environment free of bacteria, viruses, and fungi, such as a sterile "bubble," has been the best means to prevent life-threatening infections. Such an approach, however, precludes the possibility of a normal life. The most effective treatment for individuals with severely compromised immune systems is bone marrow transplantation. In this procedure, bone marrow from a compatible individual is introduced into the bone marrow of the patient. If the procedure works—and the success rate is high—in approxi-

Information on Immunodeficiency Disorders

Causes: Genetic disorders, infections, damage from drug or radiation treatments, environmental factors

Symptoms: Vary; can include recurrent infections, scaly inflammation of skin, chronic inflammations of internal organs and bones, fever, fatigue

Duration: Often chronic

Treatments: Typically targeted at source of deficiency; can include antibody injections, antifungal medications, bone marrow transplantation, drug cocktails, alternative medicine (acupuncture, herbal medicine, meditation, homeopathy)

mately one to six months the transplant recipient's immune system will be reconstituted and functional; full recovery make take up to one year. Bone marrow transplantation is a permanent cure for these disorders, since the transplanted marrow will contain stem cells that produce all the cell types of the immune system. The difficulties in transplantation include finding a compatible donor and preventing infections during the period after the transplant.

Drug therapy for AIDS utilizes treatments that interfere with replication of the virus. The first drug to be approved for use was zidovudine (formerly azidothymidine [AZT]), a DNA analogue, but its success was somewhat limited, as it was associated with severe side effects and the creation of resistant virus. More recent treatments utilize drug "cocktails," combinations of drugs that act at different stages of viral replication. Vaccines and antibiotic therapy are used to prevent or treat the opportunistic illnesses that accompany AIDS. Various drugs may also help to ease symptoms of AIDS such as appetite disturbances, nausea, pain, insomnia, anxiety, depression, fever, and diarrhea. A combination of therapies has been shown to increase life expectancy in AIDS patients. Many patients choose to participate in clinical trials of experimental drugs not approved for general use in the hope that the new drug will be more effective at alleviating the disease. Others seek out alternative or nontraditional medical treatments that have a long history of use in Western cultures. These treatments include acupuncture, herbology, meditation, and homeopathy. An important aspect of therapy for AIDS patients is maintaining mental health through support groups and supportive caregivers.

Illicit use of corticosteroids can seriously compromise the immune system and may lead to permanent damage. The best therapy for this type of acquired immunodeficiency is prevention—that is, to not misuse the drugs. In their supervised use to control inflammation or other disease symptoms, normal immune function will return after treatment has been completed. A huge risk to cancer patients who are being treated with chemotherapy and/or radiation therapy is the depression of the immune system, which can lead to a host of infections being contracted and not easily fought off by the body's compromised immune system. These individuals should avoid exposure to infectious agents when possible and be attentive to lifestyle modifications that can strengthen the immune system and encourage its speedy recovery, including a nutritious diet, plenty of rest, and avoidance of stress. Close monitoring for any signs of infection facilitates rapid antibiotic therapy, which can prevent serious complications.

PERSPECTIVE AND PROSPECTS

Prior to the gains in scientific knowledge about the mechanics of the immune system, individuals with genetic immunodeficiency disorders would die of serious infections during their first few years of life. Even when it was finally realized that these individuals suffered from defects of the immune system, little could be done for most of the disorders, except to treat infections as they developed and to avoid contact with potential disease-causing organisms—a near impossibility if one is to lead a normal life. Housing persons with SCID in sterile bubbles was uncommon because of the expense and impracticality. During the 1970s, bone marrow transplants were first developed; by the 1990s they had progressed to a greater than 80 percent success rate. As a result of improved transplant-rejection drugs, transplants from donors with less-than-perfect tissue matches are now possible. Bone marrow transplantation has been a source of cure for many individuals with immune disorders.

Bone marrow transplantation is not suitable or possible in every case of immunodeficiency disorder, and scientists have long sought a means to cure the genetic defects themselves. In 1992, French Anderson of the National Institutes of Health (NIH) conducted the first gene therapy trial on a young girl suffering from SCID. Some of the girl's bone marrow cells were removed from her body and exposed to an inactivated virus containing a normal gene for ADA, the defective enzyme. Some of the stem cells in the marrow incorporated the healthy gene, and the engineered cells were returned to her body. The cells lodged in her bone marrow, where they produced healthy immune cells. The procedure was repeated successfully in three other children shortly afterward.

Among the exciting applications of research into the molecular biology of immunodeficiencies has been the identification of specific genetic defects. Bone marrow stem cells can now be isolated and identified. In the future, such cells may be engineered such that the defective gene associated with the deficiency may be replaced by a normal copy. Since the cells are those from the same individual, transplantation problems can be avoided. However, bone marrow or cord blood transplants remain the only cure for SCID.

—*Karen E. Kalumuck, PhD and Richard Adler, PhD*

Further Reading

Abbas, Abul K., and Andrew H. Lichtman. *Basic Immunology: Functions and Disorders of the Immune System.* 6th ed., Elsevier, 2019.

Bartlett, John G., and Ann K. Finkbeiner. *The Guide to Living with HIV Infection.* 6th ed., Johns Hopkins UP, 2007.

Berliner, Nancy, editor. *Immunodeficiency, Infection, and Stem Cell Transplantation.* Saunders, 2011.

Delves, Peter J., et al. *Roitt's Essential Immunology.* 13th ed., Blackwell, 2017.

Fehervari, Zoltan, and Shiman Sakaguchi. "Peacekeepers of the Immune System." *Scientific American,* Oct. 2006, pp. 56-63.

Frank, Steven A. *Immunology and Evolution of Infectious Disease.* Princeton UP, 2002.

Geha Raif S., and Luigi Notarangelo. *Case Studies in Immunology: A Clinical Companion.* 6th ed., Garland Science, 2012.

Immune Web, immuneweb.org.

Parker, James N., and Philip M. Parker, editors. *The Official Parent's Sourcebook on Primary Immunodeficiency.* Icon Health, 2002.

Sticherling, Michael, et al., editors. *Treatment of Autoimmune Disorders.* Springer, 2012.

Stine, Gerald J. *AIDS Update 2014.* McGraw-Hill, 2013.

Vickers, Peter S. *Severe Combined Immune Deficiency: Early Hospitalization and Isolation.* John Wiley & Sons, 2009.

IMMUNOLOGY

Category: Specialty

Anatomy or system affected: Blood, cells, immune system

Specialties and related fields: Cytology, hematology, microbiology, preventive medicine, serology

Definition: the study of the immune system, its protection of the body from foreign agents, and its malfunction in autoimmune diseases, in which the body's defenses react against the body's cells or tissues

KEY TERMS

antibody: a protein produced by lymphocytes in response to an antigen; binds only to a specific antigen

antigen: any substance perceived by immunological defenses to be foreign and against which antibody is produced; generally, a protein

autoantibody: an antibody produced against tissue antigens within a host—that is, self-antigens

complement: a series of about twenty serum proteins that, when sequentially activated by immune complexes, may trigger cell damage

determinant: a region on the surface of an antigen capable of creating an immune response or of combining with an antibody produced by an immune response

Hashimoto's disease: thyroiditis; among the earliest characterized autoimmune diseases

lymphocyte: a small white blood cell constituting about 25 percent of all blood cells; two basic types are B cells (antibody production) and T cells (cellular immunity) *systemic lupus erythematosus (SLE)*: commonly called "lupus"; a chronic inflammatory disease characterized by an arthritic condition and a rash

tolerance: the state in which an organism does not normally react against its tissue

SCIENCE AND PROFESSION

The field of immunology deals with the ability of the immune system to react against an enormous repertoire of stimulation by antigens. In most instances, these antigens are foreign infectious agents such as viruses or bacteria. Inherent in this process is the ability to react against nearly any known determinant, whether natural or artificially produced. The most reactive antigenic determinants are proteins, though, to a lesser degree, other substances such as carbohydrates (sugars), lipids (fats), and nucleic acids may also stimulate a response.

In general, the body exhibits tolerance during constant exposure to its tissue. The precise reasons behind tolerance are vague, but the basis for the lack of response lies in two central mechanisms: the elimination during the development of immunological cells capable of responding to the body's tissue and the active prevention of existing reactive cells from responding to self-antigens. When this regulation fails, autoimmune disease may result.

There are two major types of immunological defense: humoral immunity and cell-mediated immunity. Humoral immunity refers to the soluble substances in blood serum, primarily antibodies, and complement. In contrast, cellular immunity refers to

Photo via iStock/Toshe_O. [Used under license.]

the portion of the immune response directly mediated by cells. Though these processes are sometimes categorized separately, they do interact with and regulate each other.

Antibodies are produced by cells called "B lymphocytes" in response to foreign antigens. These proteins bind to the antigen in a specific manner, resulting in a complex that can be removed readily by phagocytic white blood cells. More important, in the context of autoimmunity, antibody-antigen complexes also activate the complement pathway, a series of some twenty enzymes and serum proteins. The result of activation is the lysis of the antigenic targets. In general, the targets are bacteria; in autoimmune diseases, the target may be any cell in the body.

The cellular response utilizes any of several types of cytotoxic cells. These can include a specialized lymphocyte called the "T cell" (because of its development in the thymus) or another unusual large granular lymphocyte called the "natural killer" (NK) cell. NK and cytotoxic T cells function similarly—by binding to the target and releasing toxic granules in apposition to its cell membrane.

Though autoimmune diseases differ in scope, they do tend to exhibit certain common factors. The pathologies associated with most of these illnesses result in part from producing autoantibodies, antibodies produced against the body's cells or tissues. If the antibody binds to tissue in a particular organ, complement is activated in the tissue, destroying those organs. For example, Goodpasture's syndrome is characterized by the deposition of autoantibodies directed against the glomerulus membrane in the kidneys. Complement activation can result in severe organ pathology and subsequent kidney failure.

Suppose the autoantibody binds to soluble material in blood serum. In that case, the resultant antibody-antigen complexes are carried along in the circulation. There is the possibility that they will lodge in various areas of the body. For example, systemic lupus erythematosus (SLE) produces autoantibodies against soluble nucleoprotein released from cells as they undergo normal death and lysis. The immune complexes frequently lodge in the kidney, where they can cause renal failure.

However, all autoimmune diseases do not result solely from autoantibody production. Though a precise role for either cytotoxic T cells or NK cells in human autoimmune disease has not been fully confirmed, several observations make such an association likely. First, large T cells are found in certain organ-specific diseases, including thyroiditis and pernicious anemia. Second, animal models of similar diseases show a specific role for such cells in the pathology of these diseases. Thus, it is likely that these cells do participate in organ destruction.

Autoimmune disorders are categorized in the form of a disease spectrum. At one end of the spectrum, one can place organ-specific diseases. For example, Hashimoto's disease is an autoimmune thyroid disorder characterized by autoantibodies against thyroid antigens. The extensive infiltration and proliferation of lymphocytes are observed (although, as described above, their roles are unproved), along with the subsequent destruction of follicular tissue.

Likewise, diabetes mellitus, type 1 (formerly called "juvenile-onset diabetes"), is an organ-specific autoimmune disease. In this case, however, autoantibodies are directed against the beta cells of the pancreas, which produce insulin. In pernicious (or megaloblastic) anemia, antibodies are produced against intrinsic factor, a molecule necessary for uptake of vitamin B_{12}. People with pernicious (or megaloblastic) anemia suffer from a lack of absorption of the vitamin. Addison's disease, from which US president John F. Kennedy suffered, is a potentially life-threatening condition resulting from antibody production against the adrenal cortex. Myasthenia gravis (MG) is characterized by severe heart or skeletal muscle weakness caused by antibodies directed against neurotransmitter receptors on the muscle.

Cells from any organ may be potential targets for the production of an autoantibody.

Certain organ-specific autoimmune diseases in the spectrum are characterized not by antibodies directed against any specific organ but by cellular infiltration triggered in some manner by less specific autoantibodies. For example, biliary cirrhosis, an inflammatory condition of the liver, is characterized by the obstruction of bile flow through the liver ductules. Though extensive cellular infiltration is observed, serum antibodies are directed against mitochondrial antigens found within all cells. Certain types of chronic hepatitis also exhibit an analogous situation.

In some cases, antibodies may be directed against circulatory cells. Antibodies directed against red blood cells may cause subsequent lysis of the cells, leading to hemolytic anemia. Antibodies directed against blood platelets can cause a reduction in the number of those cells, resulting in thrombocytopenia purpura. Often, these are temporary conditions that have resulted from the binding of a pharmacologic chemical such as an antibiotic to the cell's surface, which triggers an immune response. A more severe condition is hemolytic disease of the newborn (HDN), one example being hemolytic disease of the newborn (formerly erythroblastosis fetalis) or Rh disease. In this case, a mother lacking the Rh protein on her blood cells may produce an immune response against that protein, which is present in the blood of the fetus she is carrying during pregnancy. Before 1967, when an effective preventive measure became available, HDN was a severe problem for many pregnancies. An analogous situation occurs with other cell types.

At the other end of the autoimmune spectrum are those diseases that are not cell- or organ-specific but result in widespread lesions in various body parts. Lupus received its name from the butterfly rash often seen on patients' faces, which resembles a wolf bite (*lupus* is Latin for "wolf"). However, pathologic changes can be found at various sites in the body, including the kidneys, joints, and blood vessels. Likewise, rheumatoid arthritis is characterized by the production of rheumatoid factor, an antibody molecule directed against other antibodies in blood serum. The resultant immune complexes lodge in joints, causing joint pain and destruction associated with severe arthritis.

In most cases, the specific reason for the production of autoantibodies is unknown. Genetic factors are certainly involved since some autoimmune diseases run in families. Bacterial or viral infections may trigger some. Viral antigens may be expressed on the surfaces of specific cells, or the virus itself may be attached to the cell. Heart muscle appears to express antigenic determinants in common with certain streptococcal bacteria. A mild "strep throat" may be followed several weeks later by severe rheumatic fever.

The binding of drugs to cell surfaces may trigger an immune response. For example, penicillin may bind to the surfaces of red blood cells, triggering hemolytic anemia. Likewise, the hypnotic/sedative apronal (Sedormid) may bind to the membrane of platelets.

Most cases of autoimmune disease, however, are triggered by no apparent cause. They may "simply" involve a breakdown of the normal regulatory mechanisms associated with the immune response.

DIAGNOSTIC AND TREATMENT TECHNIQUES

The regulation of self-reactive lymphocytes is necessary for maintaining tolerance by the immune system. When regulation breaks down or is otherwise defective, either humoral or cellular immunity is generated against the cells or tissues. The resultant pathology may be simply a painful nuisance or may have potentially fatal consequences. The difference relates to the extent of damage to particular organs, in the case of organ-specific autoimmune reactions, or to the level of tissue damage in systemic disease.

Despite differences in pathology, the mechanisms of tissue damage are similar in most autoimmune diseases. Most involve the formation of immune complexes. Either antibodies bind to cell surfaces, or immune complexes form in the circulation. In either case, the result is complement activation. Components of the complement pathway, in turn, can either directly damage cell membranes or trigger the infiltration of a variety of cytotoxic cells.

Because the damage associated with most autoimmune diseases results from parallel processes, treatment methods vary little in theory from one illness to another. Most involve the treatment of resultant symptoms; for example, the use of aspirin to reduce minor inflammation and, when necessary, the use of steroids to reduce the level of the immune response. Recently, the focus has shifted from treating symptoms only to attacking the underlying disease mechanism with disease-modifying drugs. Some of these drugs include methotrexate, azathioprine, cyclosporine, and hydroxychloroquine. Newer immune modulators (such as infliximab and etanercept) and monoclonal antibodies (such as rituximab) are used in some autoimmune conditions that are refractory to other measures.

The treatment of autoimmune diseases does not eliminate the problem. The disease remains, but under ideal conditions, it is held under control. At the same time, there exists the danger of side effects of treatment. For example, most methods that reduce the level of the immune response are nonspecific; reducing the severity of the autoimmune disease may cause the patient to become more susceptible to infections by bacteria or viruses.

Specific approaches have been successful in the palliative treatment of some forms of autoimmune disease. For example, patients with MG exhibit significant muscle weakness. An MG patient may have difficulty breathing and may experience extreme fatigue, in severe cases being unable to open their mouth or eyelids. Associated with the disease are autoantibodies produced against the receptor for the neurotransmitter acetylcholine (ACh), the chemical utilized by nerves in regulating movement by the muscle. By blocking the ACh receptor, these antibodies inhibit the ability of nerves to control muscle movement. In effect, the patient loses control of the muscles.

Patients with MG often exhibit abnormalities of the thymus, the gland associated with T-cell production. In addition, there is evidence that the thymus contains ACh receptors that are particularly antigenic (perhaps exacerbating the illness). Removal of the thymus, even in adults, often aids in reducing the symptoms of the disease. Though not superfluous in adults, the thymus carries out its primary functions during the early years of life through adolescence. Thus, its removal generally has few significant implications.

Often, MG will respond to more conventional forms of treatment. Steroid treatment will often reduce symptoms. Metabolic controls may also aid in reducing symptoms. For example, during normal neural function that involves ACh, the enzyme cholinesterase degrades ACh, thereby regulating muscle movement. Anticholinesterase drugs that prolong the presence of ACh at the site of the receptor on the muscle have also been of benefit to some patients.

SLE is among the most common systemic autoimmune diseases. The disease usually strikes women in the prime of life, between the ages of twenty and forty. It is characterized by a butterfly rash over the facial region and weakness, fatigue, and fever. In many respects, the symptoms are those of severe arthritis. As the disease progresses, tissue or organ degradation may occur in the kidney or heart.

The specific cause of the symptomology is the formation of immune complexes, which consist of antibodies against cell components such as deoxyribonucleic acid (DNA) or nucleoprotein. Complexes in the kidney have been large enough to observe with the electron microscope, mainly when the complexes contain cell nuclei. Regions of the skin characterized by inflammation and a rash have similar complexes. The immune complexes are sometimes ingested (phagocytized) by scavenger neutrophils, which make up the most significant proportion (65 percent) of white blood cells. The presence of these so-called lupus erythematosus (LE) cells, white cells with ingested antibody-bound nuclei, was at one time used for the diagnosis of lupus.

As is true for many autoimmune diseases, the control of lupus involves the use of steroids and other immunosuppressive drugs. These have included drugs such as cyclosporin, which blocks T-cell function, and antimitotic drugs such as azathioprine or methotrexate, which block the proliferation of immune cells, as well as immune modulators such as rituximab. Generalized immunosuppression as a side effect is a concern. Often, using combinations of steroids and immunosuppressives makes it possible to use lower concentrations of each, increasing the drugs' effectiveness and reducing the danger of toxicity.

Other palliative treatments of symptomology can increase patient comfort. For example, aspirin may be used to reduce inflammation or joint pain. Topical steroids can reduce the rash. Since lupus may significantly increase the photosensitivity of the skin, staying out of direct sunlight, or at least covering the surface of the skin, may reduce skin lesions. It should be emphasized again that these treatments deal only with symptoms; none will cure the disease.

Since some systemic diseases result from immune complex disorders, a reduction of such complexes benefits some patients. Treatment involves a process called "plasmapheresis." Plasma, the liquid portion of the blood, is removed from the patient (a small proportion at a time). The immune complexes are separated from the plasma. Though a temporary measure, since other complexes continue to form, the process does prove useful.

Rheumatoid arthritis is another common autoimmune disorder. As is true of most autoimmune diseases, rheumatoid arthritis is primarily a disease of women. Symptomatology results from the lodging of immune complexes in joints, resulting in the inflammation of those joints. Many cases result from the formation of antibodies directed against other antibody molecules—a case of the immune system turning against itself. Pathology results from complement activation and the infiltration of various cells into the joint; the result is damage to both cartilage and bone.

Medical treatment usually begins with aspirin or other nonsteroidal anti-inflammatory agents. Other common treatments increase patient comfort: rest, proper exercise, and weight loss, if necessary. In severe cases, steroids, immune modulators, or monoclonal therapy may be necessary.

In general, autoimmune diseases are characterized by alternating periods of symptomatology and remission. Since the precise origin of most of these disorders is unknown, prevention remains difficult. Treatments are generally similar in reducing inflammation as the first line of intervention, with immunosuppression as a last resort.

PERSPECTIVE AND PROSPECTS

During the 1950s, F. Macfarlane Burnet published his theory of clonal selection. Burnet believed that antibody specificity was predetermined in the B cell as it underwent development and maturation. Selection of the cell by the appropriate antigen resulted in the proliferation of that specific cell, a process of clonal selection.

However, Burnet also had to account for tolerance—the inability of immune cells to respond against their antigens. Burnet theorized that exposure to self-antigens, or determinants, resulted in the ablation of any self-reactive cells during prenatal development. Only those self-reactive immune cells directed against sequestered antigens survived.

Though Burnet's theories have reached the level of dogma in the field of immunology, they fail to account for certain autoimmune disorders. Although not appreciated during Burnet's time, W. W. Gull and others recognized autoimmune disorders as early as 1866. In that year, W. W. Gull demonstrated the link between chilling, and a syndrome called "paroxysmal hemoglobinuria." When external tissue such as skin is exposed to cold, large amounts of hemoglobin are discharged into the urine. In the "correct" circumstances, the body does react against itself. In 1904, Karl Landsteiner established the autoimmune basis for the disease by demonstrating the role of complement in the lysis of red blood cells, causing the release of hemoglobin and the symptomatology of the disorder. Furthermore, he demonstrated that one could cause the lysis of normal cells by mixing them with sera from patients who had hemoglobin in their urine.

Hashimoto's disease was among the first organ-specific autoimmune diseases to be described. The disease was first described in 1912 by Hakaru Hashimoto, a Japanese surgeon. The immune basis for the disease was established independently by Ernest Witebsky and Noel Rose in the United States, and by Deborah Doniach and Ivan Roitt in Great Britain, in 1957.

Since the 1950s, dozens of autoimmune disorders have been described. Treatment of these disorders remains, for the most part, nonspecific. Research in the area has examined the precise triggers for autoimmune diseases and developed ways to suppress specifically those immune reactions responsible for the symptomatology. Successes have been associated with vaccines directed against components involved with the reactions under investigation. For example, since the production of autoantibodies is the basis for some forms of the disease, the generation of additional antibody molecules directed against determinants on the autoantibodies at fault could neutralize those components' effects. This procedure could be likened to a police department that arrests its dishonest officers. There is a precedent for such an operation. Newborn children of mothers suffering from MG synthesize just such antibodies against the inappropriate MG antibodies that have crossed the placenta. Synthesis does seem to alleviate the symptoms of the disease.

There is no question that autoimmune disorders represent an aberrant form of the immune response. Nevertheless, understanding the underlying mechanism will shed light on exactly how the immune system is regulated. For example, it remains unclear how antibody production is controlled following a normal immune response. In the presence of an antigen, antibody levels increase for days to weeks, reach a plateau, and then slowly decrease as additional production comes to a halt. How the shutdown takes place remains nebulous.

Tolerance does not result solely from an absence of T or B cells that respond to antigens; it involves an active suppression of the process. A more detailed understanding of the process will generally lead to a more thorough understanding of the immune system.

—Richard Adler, PhD

Further Reading

Delves, Peter J., et al. *Roitt's Essential Immunology*. 13th ed., Blackwell, 2017.

Flint, Jane, et al. *Principles of Virology, Volume 1: Molecular Biology*. 5th ed. ASM Press, 2020.

Frank, Steven A. *Immunology and Evolution of Infectious Disease*. Princeton UP, 2002.

Murphy, Kenneth M., et al. *Janeway's Immunobiology*. 10th ed., W. W. Norton & Company, 2022.

Parham, Peter. *The Immune System*. 3rd ed., Garland Science, 2009.

Punt, Jenni, et al. *Kuby Immunology*. 8th ed., W. H. Freeman, 2018.

Rose, Noel R., and Ian. R. Mackay, editors. *The Autoimmune Diseases*. 6th ed., Academic Press/Elsevier, 2019.

IMMUNOEDITING

Category: Immunology

Anatomy or system affected: Blood, cells, circulatory system, immune system, lymphatic system

Specialties and related fields: Immunology, oncology

Definition: the process by which the immune system and tumor cells interact to create immune-resistant tumor growths

KEY TERMS

B lymphocytes: any of the lymphocytes with antigen-binding antibody molecules on the surface that comprise the antibody-secreting plasma cells when mature and that in mammals differentiate in the bone marrow

cytokines: substances, such as interferon, interleukin, and growth factors, secreted by specific cells of the immune system that influence other cells

immune surveillance: the monitoring process by which cells of the immune system (such as natural killer cells, cytotoxic T cells, or macrophages) detect and destroy premalignant or malignant cells in the body

T lymphocytes: any of several lymphocytes (such as a helper T cell) that differentiate in the thymus, possess particular cell-surface antigen receptors and include some that control the initiation or suppression of cell-mediated and humoral immunity (as by the regulation of T- and B-cell maturation and proliferation) and others that lyse antigen-bearing cells

tumors: abnormal benign or malignant new growths of tissue that possesses no physiological function and arises from uncontrolled, usually rapid cellular proliferation

INTRODUCTION

Immunoediting is the process by which the immune system and tumor cells interact to create immune-resistant tumor growths. A tumor also referred to as a neoplasm, is an uncontrolled growth of tissue in the body. Tumors can spread throughout the body and cause diseases such as cancer. Tumors can also remain at one site in the body and not cause harm. Immunoediting involves the body's innate, or immediate, immune system responses and adaptive, or specialized, immune system responses. It consists of three phases: elimination, equilibrium, and escape. During these phases, the body's immune system recognizes a tumor cell as a foreign agent and attempts to remove it from the system. Tumor cells that elude elimination are placed in a dormant state. They can change and create new immune system responses. If the immune system cannot keep the tumor cell in its dormant phase, the cell escapes and grows, leading to disease.

BACKGROUND

Researchers have been studying the immune system's effect on tumor growth in organisms for many decades. Scientific arguments centered on the nature of the immune system's effectiveness, specifically whether it affected tumor growth positively, negatively, or neutrally. Immunologists knew that the immune system could detect the invasion of pathogens, or viruses, in the body. Still, knowledge of the immune system's effect on tumors remained untouched until the early twentieth century. In 1909, German scientist Paul Ehrlich proposed that the immune system monitors the host of abnormal tissue growth in the body. The idea was broached again fifty years later by immunologists Frank Macfarlane Burnet and Lewis Thomas. Burnet and Lewis believed the immune system was capable of recognizing cells experiencing new and abnormal changes in tissue growth and could work to eliminate the cells before the tumor growth spread, a concept they termed "immune surveillance" or "immunosurveillance."

However, the concept of immune surveillance was only one facet of the immune system's role in tumor growth. Future studies refined and elaborated on this concept, with many researchers theorizing the immune system was capable of more than just serving as the body's lookout for abnormal cells. Direct evidence of immune surveillance and its implications was not formally demonstrated in experiments until 2001 when immunologist Robert D. Schreiber and his col-

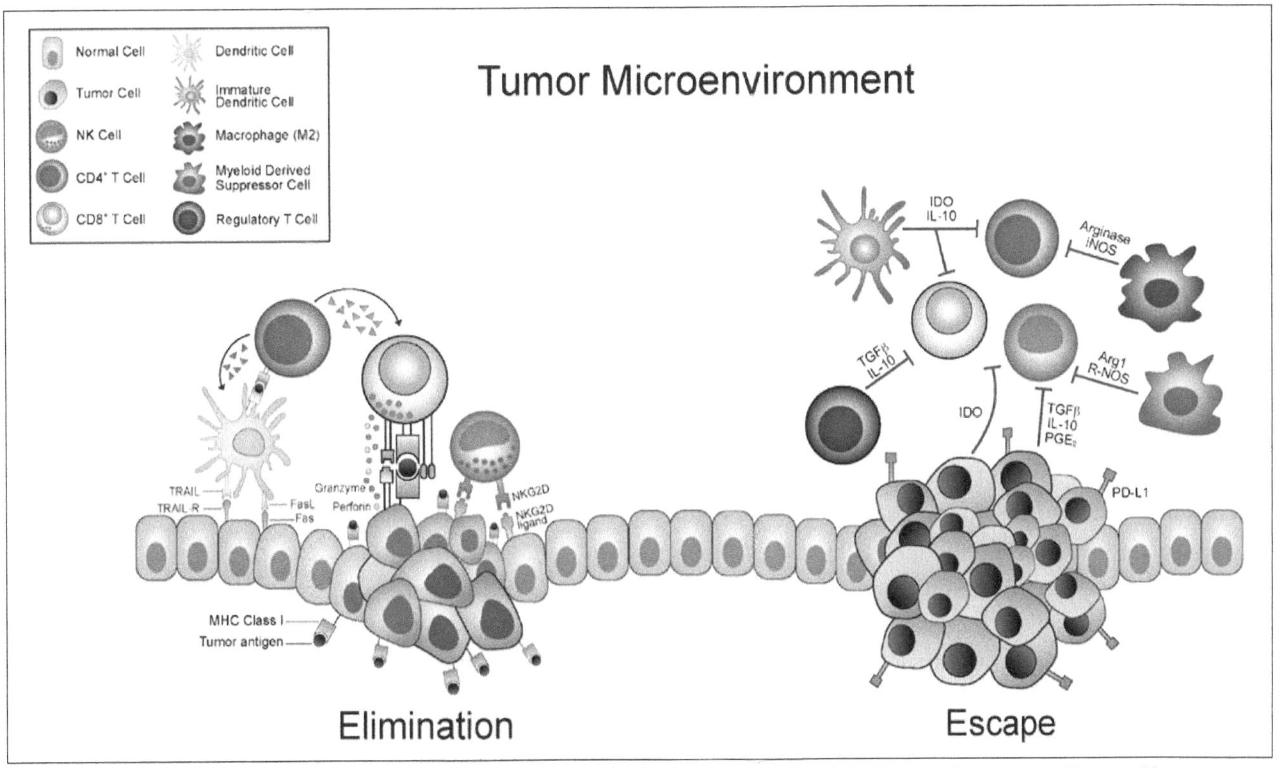

During the elimination phase, immune effector cells such as natural killer cells, with the help of dendritic and CD4+ T-cells, are able to recognize and eliminate tumor cells (left). As a result of heterogeneity, however, tumor cells which are less immunogenic are able to escape immunosurveillance (right). Image by Frontiers in Oncology, via Wikimedia Commons.

leagues compared the incidence of benign and malignant tumors in aging wild-type mice to mice without a recombination-activating gene 2 (*RAG2*). The *RAG2* gene is responsible for variable (V), diversity (D), and joining (J) gene recombination, a process that generates B cells and T cells, which play a significant role in the body's response to harmful foreign materials such as viruses, bacteria, and parasites. B cells and T cells generate antibodies—proteins that fight infections—to fight foreign materials and help the body build an adaptive immune system. The mice without the *RAG2* gene could not generate these processes and therefore had no adaptive immune response.

Schreiber's experiment showed that mice without an adaptive immune response had a higher incidence of cancerous growth and were more susceptible to carcinogens or cancer-causing agents. The wild-type mice, however, had far fewer tumor growths and were less susceptible to carcinogens. This showed that some of the tumor cells present in the *RAG2*-deficient mice were being identified and eliminated in the wild-type mice with an intact immune system. Tumors that did grow in the wild-type mice were less immunogenic or less able to produce an immune response and grew in mice with and without an intact immune system. Schreiber used the term "immunoediting" to describe the process by which the body's immune system interacts with tumors.

OVERVIEW

When engaging in immunoediting, the body is protecting itself from cancerous growths and shaping the makeup of developing tumors. Scientists divided immunoediting into three phases: elimination, equilibrium, and escape to organize this complex process for continued research better. Evidence showing the elimination phase of immunoediting primarily exists within scientific studies on mice. In contrast, the equilibrium and escape phases have been inferred from both mice and human trials.

The elimination phase of immunoediting can be described as immune surveillance. This phase utilizes innate and adaptive immune systems to sense and destroy tumors before they grow into visible masses. With the innate immune system, the presence of tumor cells produces inflammatory signals. This initiates the activation of T cells and natural killer cells, which flood to the site of the tumor cells. These cells invade a tumor cell and produce secreted proteins called "cytokines," which kill the tumor cells through several proteins or oxygen-related tumor cell-killing mechanisms. While tumor cell death is in process, the body begins to produce chemokines, which block new blood vessel formation so the tumor cells cannot travel further. Immune system cells known as "dendritic cells" absorb what remains of the tumor cell following its destruction. The dendritic cells then travel to the lymph nodes for drainage. Tumor-saturated dendritic cells activate the development of T cells, which destroy any remaining immunogenic tumor cells.

The second step in immunoediting is the equilibrium phase. In the equilibrium phase, tumor cells that did not activate an innate immune response and escaped the elimination phase continue to grow until the adaptive immune system responds. The immune system holds the tumor in a dormant yet functioning state, allowing it to change. This is the editing phase of immunoediting, as it is when changes to immunogenicity occur. During this phase, tumor cells can evolve and elude immune system recognition, leading to immunosuppression or suppressing the body's healthy immune response. Cytokines also play a role in the equilibrium phase as the immune system attempts to balance the anti-tumor and tumor-supporting cytokines. Suppose the immune system cannot eliminate the tumor cells because of immunogenic edits. In that case, the equilibrium phase can last for years, and in some instances, for the entirety of the host's life.

Suppose the immune system is no longer able to hold tumor cells in a state of dormancy. In that case, the body enters the escape phase of immunoediting. The immune system no longer controls tumor outgrowth during this phase, and the tumor escapes into the body. Escape is made possible due to several factors, such as the immune system's inability to produce antitumor T cells and natural killer cells. Immunosuppression can also lead to tumor-cell escape. Without immune system regulation, the tumor can grow and spread, leading to diseases such as cancer.

—Cait Caffrey

Further Reading

Corthay, A. "Does the Immune System Naturally Protect Against Cancer?" *Frontiers in Immunology*, vol. 5, no. 197, 2012.

Dunn, Gavin P., et al. "Cancer Immunoediting: From Immunosurveillance to Tumor Escape." *Nature Immunology*, vol. 3, 2002, pp. 991-98.

Dunn, Gavin P., et al. "The Three Es of Cancer Immunoediting." *Annual Review of Immunology*, vol. 23, 2013, pp. 329-60.

Gupta, Ravi G., et al. "Exploiting Tumor Neoantigens to Target Cancer Evolution: Current Challenges and Promising Therapeutic Approaches." *Cancer Discovery*, vol. 11, no. 5, 2021, pp. 1024-39, doi:10.1158/2159-8290.CD-20-1575.

Mittal, Deepak, et al. "New Insights into Cancer Immunoediting and Its Three Component Phases—Elimination, Equilibrium and Escape." *Current Opinion in Immunology*, vol. 27, 2014, pp. 16-25.

Paterson, Yvonne. "Immunoediting." *Encyclopedia of Cancer*, edited by Manfred Schwab, Springer, 2011.

Prendergast, George C., and Elizabeth M. Jaffee. *Cancer Immunotherapy: Immune Suppression and Tumor Growth*. Academic Press, 2013.

Schreiber, Robert D., et al. "Cancer Immunoediting: Integrating Immunity's Roles in Cancer Suppression and Promotion." *Science*, vol. 331, no. 6024, pp. 1565-70.

Teng, Michele W. L., et al. "From Mice to Humans: Developments in Cancer Immunoediting." *Journal of Clinical Investigation*, vol. 25, no. 9, 2015, pp. 3338-46.

IMMUNOGENETICS

Category: Immunology
Anatomy or system affected: All
Specialties and related fields: Allergist, genetics, hematology, immunology, oncology, pathology, pediatrics, rheumatology, serology, transplant surgery
Definition: a study of the major histocompatibility (MHC) genes that identify self-tissues, the genes in B lymphocytes that direct antibody synthesis, and the genes that direct the synthesis of T-lymphocyte receptors

KEY TERMS

apoptosis: cell death that is programmed as a natural consequence of growth and development through normal cellular pathways or signals from neighboring cells
cytokines: soluble intercellular molecules produced by cells such as lymphocytes that can influence the immune response
downstream: describes the left-to-right direction of deoxyribonucleic acid (DNA) whose nucleotides are arranged in sequence with the 5' carbon on the left and the 3' on the right; the direction of ribonucleic acid (RNA) transcription of a genetic message with the beginning of a gene on the left and the end on the right
gene: a unit of heredity transferred from a parent to offspring and that determines some characteristic of the offspring
haplotype: a sequential set of genes on a single chromosome inherited together from one parent; the other parent provides a matching chromosome with a different set of genes
monoclonal antibodies: antibodies with one particular target that have been generated in large quantities from a single hybrid parent cell formed in a laboratory
transposon: a sequence of nucleotides flanked by inverted repeats capable of being removed or inserted within a genome

GENES, B CELLS, AND ANTIBODIES

The fundamental question that led to the development of immunogenetics is how scientists can make the thousands of specific antibodies that protect people from the thousands of organisms they encounter. Frank Macfarlane Burnet proposed the clonal selection theory, which states that an antigen (i.e., anything not-self, such as an invading microorganism) selects, from the thousands of different B cells, the receptor on a particular B cell that fits it like a key fitting a lock. That cell is activated to make a clone of plasma cells, producing millions of soluble antibodies with attachment sites identical to the receptor on that B-cell surface. The problem facing scientists interested in a genetic explanation for this capability was the need for more genes than the number found in the entire human genome.

Susumu Tonegawa first recognized that many antibodies produced in the lifetime of a human did not have to have the equivalent number of physical genes on their chromosomes. From his work, Tonegawa determined that the genes responsible for antibody synthesis are arranged in tandem segments on specific chromosomes relating to specific parts of antibody structure. The amino acids that form the two light polypeptide chains and the two heavy polypeptide chains making up the IgG class of antibodies are programmed by nucleotide sequences of deoxyribonucleic acid (DNA) that exist on three different chromosomes. Light-chain genes are found on chromosomes 2 and 22. The specific nucleotide sequences code for light polypeptide chains. Half the chain has a constant amino acid sequence, and the other half has a variable sequence. The amino acid sequences of the heavy polypeptide chains are constant over three- quarters of their length, with five primary sequences identifying five classes of human immunoglobulins: IgG, IgM, IgD, IgA, and IgE. The other quarter length has a variable sequence that, together with the variable sequence of the light chain, forms the antigen-binding site. The nucleotide sequence coding for the heavy chain is part of chromosome 14.

The actual light-chain locus is organized into sequences of nucleotides designated variable (V), joining (J), and constant (C) segments. The multiple options for the different V and J segments and mixing the different V and J segments cause many different DNA light-chain nucleotide sequences and the synthesis of different antibodies. The same rearrangement occurs between various nucleotide sequences related to the V, diversity (D), and J segments of the heavy-chain locus. The recombination of segments appears to be genetically regulated by recombination signal sequences downstream from the variable segments and recombination activating genes that function during B-cell development. Genetic recombination is complete with the immature B cell committed to producing one kind of antibody. The diversity of antibody molecules is explained by the messenger ribonucleic acid (mRNA) transcript coding for either the light polypeptide chain or the heavy polypeptide chain is formed containing exons transcribed from recombined gene segments during B-cell differentiation. The unique antigen receptor-binding site is formed when the variable regions of one heavy and one light chain come together during the formation of the completed antibody in the endoplasmic reticulum of the mature B cell. The B-cell antigen receptor is an attached surface antibody of the IgM class. The binding of the antigen to the specific B cell activates its cell division and the formation of a clone of plasma cells that produce a unique antibody. If this circulating B cell does not contact its specific antigen within a few weeks, it will die by apoptosis. During plasma cell formation, the class of antibody protein produced switches typically from IgM to IgG by forming an mRNA transcript containing the exon nucleotide sequence made from the IgG heavy-chain C segment rather than the heavy-chain C segment for IgM. The intervening nucleotide sequence of the IgM constant segment is deleted from the chromosome as an excised circle reminiscent of the transposon or plasmid excision process. This switch forms an IgG antibody having the same antigen specificity as the IgM antibody because the variable regions of the light and heavy polypeptide chains remain the same. Although the activation and development of B cells by some antigens may not need T-cell involvement, it is believed that T-cell cytokines influence class switching and most B-cell activity.

MAJOR HISTOCOMPATIBILITY GENES

In humans, the major histocompatibility (MHC) genes encoding "self-antigens" are also called the "HLA complex" and are located on chromosome 6. The nucleotides that compose this DNA complex encode for two sets of cell surface molecules designated MHC Class I and MHC Class II antigens. The Class I region contains loci *A*, *B*, and *C*, which encode for MHC Class I A, B, and C glycoproteins on every nucleated cell in the body. Because the *A*, *B*, and *C* loci comprise highly variable nucleotide sequences, numerous kinds of A, B, and C glycoproteins characterize humans. All people inherit MHC Class I *A*, *B*, and *C* genes as a haplotype from their parents. Children will have tissues with half of their Class I A, B, and C antigens like their mother and half like their father's. Siblings could have tissue antigens that are identical or dissimilar based on their MHC I glycoproteins. Body surveillance by T lymphocytes involves T cells recognizing self-glycoproteins. Cellular invasion by a virus or any other parasite results in processing an antigen and its display in the MHC Class I glycoprotein cleft. T-cytotoxic lymphocytes with T-cell receptors specific to the antigen-MHC I complex will attach to the antigen and activate the clonal selection. Infected host cells are killed when activated cytotoxic T cells bind to the surface and release perforins, causing apoptosis.

MHC Class II genes are designated *HLA-DPA1* and *HLA-DPB1*, *HLA-DQA1* and *HLA-DQB1*, and *HLA-DRA* and *HLA-DRB1*. These genes encode glycoprotein molecules that attach to the cell surface in α and ß pairs. A child will inherit the six genes as a group or haplotype, three α and ß glycoprotein gene pairs from each parent. During glycoprotein synthesis, the child will also have glycoprotein molecules made from the maternal and paternal α and ß pairings.

The Class II MHC molecules are on the membranes of macrophages, B cells, and dendritic cells. These specialized cells capture antigens and attach antigen peptides to the three-dimensional grooves formed by combined α and ß glycoprotein pairs. The antigen attached to the Class II groove is presented to the T-helper cell, with the receptor recognizing the specific antigen while bound to the Class II self-antigen. The specific T-helper cell forms a specific clone of effector cells and memory cells.

GENES, T-HELPER CELLS, AND T-CYTOTOXIC CELLS

The thousands of specific T-cell receptors (TCR) available to any specific antigen one might encounter in a lifetime are formed from progenitor T cells in the human embryonic thymus. The TCR comprises two dissimilar polypeptide chains designated α and ß or γ and δ. They are similar in structure to immunoglobulins and MHC molecules. They have regions of variable amino acid sequences and constant amino acid sequences arranged in loops called "domains." This basic structural configuration places all three molecules in a chemically similar grouping designated the immunoglobulin superfamily. The genes of these molecules are derived from a primordial supergene that encoded the basic domain structure.

The exons encoding the α and γ polypeptides are designated V, J, and C gene segments in sequence and associate with recombination signal sequences similar to the immunoglobulin light-chain gene. The ß and δ polypeptide genes are designated VDJ and C exon segments in sequence associating with recombination signal sequences similar to the immunoglobulin heavy-chain genes. Just as there are multiple forms for each immunoglobulin variable gene segment, there are multiple forms for the variable TCR gene segments. Thymocytes, T-cell precursors in the thymus, undergo chance recombinations of gene segments. These genetic recombinations, as well as the chance combination of a completed α polypeptide with a completed ß polypeptide, provide thousands of completed specific TCRs ready to be chosen by an invading antigen and to form a clone of either T-helper cells or T-cytotoxic cells.

IMMUNOGENETIC DISEASE

The HLA genes of the major histocompatibility complex identify every human being as distinct from all other things, including other human beings, because of the MHC Class I and Class II antigens. Surveillance of self involves B- and T-cell antigen recognition because of MHC self-recognition. How well individual human beings recognize self and their response to antigen in an adaptive immune response are determined by MHC haplotypes and the genes that make immunoglobulins and T-cell receptors. These same genes can explain various disease states, such as autoimmunity, allergy, and immunodeficiency.

Because immunoglobulin structure and T-cell receptor formation are based on a chance mechanism, problems involving self-recognition may occur. Thymocytes with completed T-cell receptors are protected from apoptosis when they demonstrate self-MHC molecule recognition. Alternatively, the thymic epithelial cells present thymocytes with self-antigens processed by specialized macrophages bearing MHC Class I and Class II molecules. Thymocytes reacting with high-affinity receptors to processed self-antigens undergo apoptosis. There is also a negative selection process within the bone marrow that actively eliminates immature B cells with membrane-bound autoantibodies that react with self-antigens. Despite these selective activities, autoreactive T cells and B cells can be part of circulating surveillance, causing autoimmune disease of either single organs or multiple tissues.

Autoimmune diseases occur in families, and there is growing evidence that an individual with a particular HLA haplotype has a greater risk for developing a particular disease. For example, ankylosing spondylitis develops more often in individuals with *HLA-B27* than those with another *HLA-B* allele, and rheumatoid arthritis is associated with several common *HLA-DRB1* alleles. Myasthenia gravis and multiple sclerosis are two neurological diseases caused by autoantibodies. There is evidence that they are related to the restricted expression of T-cell variable genes. Genomic studies provide evidence for the possibility that autoimmune induction occurs because of molecular mimicry between human host proteins and microbial antigens. Among the cross-reacting antigens that have been implicated are papillomavirus E2 and the insulin receptor, and poliovirus VP2 and the acetylcholine receptor.

The genetics of immunity also involves the study of defective genes that cause primary immunodeficiency infectious disease. The deficiency can decrease an adaptive immune response involving B cells, T cells, or both, as is the case with severe combined immunodeficiency disorder (SCID). There is evidence that SCID can demonstrate either autosomal recessive or X-linked inheritance. One such defect is on the short arm of chromosome 11. It involves a mutation of recombination-activating genes that are necessary for the rearrangement of immunoglobulin gene segments and the T-cell receptor gene segments. The inability to recombine the VD and J variable segments prevents the development of active B cells and T cells with various antigen receptors. SCID is essentially incompatible with life and characterized by severe opportunistic infections caused by even usually benign organisms.

Tissue rejection after transplantation remains an integral cause of transplant failure that markedly affects morbidity. Cytokines play a significant role in transplant rejection. Dutch researchers discovered that specific alleles of the genes that encode IL-6 and the IL-6 receptor in the recipient donor but not in the transplant recipient could predict the recipient's susceptibility to transplant rejection.

Allergies have a genetic component, and atopy, an abnormal IgE response, is common to certain families. There is evidence that children have a 30 percent chance of developing an allergic disease if one parent is allergic. In comparison, those children with two allergic parents have a 50 percent chance. The genetic control of IgE production can be related to T_{H2} lymphocyte cytokine stimulation of class switching from the constant segment of IgG to the constant segment of IgE on chromosome 14 in an antigen-selected cell undergoing clonal selection.

IMPACT

Understanding the genetic basis for immune reactions results in novel approaches to protection against disease and improvements in health. Researchers are pursuing therapeutics to control B-cell responses in autoimmune diseases and IgE responses in allergic reactions. Clinical laboratories provide detailed histocompatibility and immunogenetics testing for solid organ and stem cell transplantation and blood and platelet transfusions to reduce graft-versus-host the incidence and severity of graft-versus-host disease.

Immunotherapy capitalizes on a person's immune system to fight cancers or infectious diseases, either by actively stimulating the production of natural antibodies or by passively introducing antibodies engineered in a laboratory. With active stimulation, specific immunity may be induced with vaccines, or nonspecific immunity may be induced with interferons or interleukins. Monoclonal antibodies that target specific cell-surface antigens achieve passive immunity. Several therapeutic monoclonal antibodies have been approved for use in humans by the

US Food and Drug Administration (FDA), particularly in treating colorectal cancer, non-Hodgkin lymphoma, and some types of leukemia. Conversely, other therapeutic monoclonal antibodies have been produced and marketed to suppress immune responses in rheumatoid arthritis and allergic asthma. Researchers who employ related technologies are trying to develop biomarkers that will track the progression of immune disorders and measure their response to various treatment modalities.

Immunogenetics has led to new fields of study in public health, such as medical anthropology, which includes determining how people of certain races or ethnicities are genetically predisposed to certain diseases. Another new field of study is the immunology of aging, which includes attempting to determine the effect of genetic variation on the natural aging process. One crucial issue in this field is how to boost the immune response in the elderly to vaccines, especially those for influenza and pneumonia.

—Patrick J. DeLuca, PhD and Bethany Thivierge, MPH

Further Reading

Abbas, Abul K., Andrew H. Lichtman, and Shiv Pillai. *Basic Immunology: Functions and Disorders of the Immune System.* 6th ed., Elsevier, 2019.

Flajnik, Martin F., Nevil J. Singh, and Steven M. Holland. *Paul's Fundamental Immunology*. Lippincott Williams & Wilkins, 2022.

Genetics Home Reference. "HLA-DRB1." *Genetics Home Reference*. US NLM, 28 July 2014. Accessed. 4 Aug. 2014.

Male, David. *Immunology: An Illustrated Outline*. 6th ed., CRC Press, 2021.

McKusick, Victor A., and Paul J. Converse. "*142857 Major Histocompatibility Complex, Class II, DR Beta-1; HLA-DRB1." *OMIM.org*. Johns Hopkins U, 25 June 2014. Accessed 4 Aug. 2014.

Oksenberg, Jorge R., and David Brassat, editors. *Immunogenetics of Autoimmune Disease*. Springer, 2006.

Owen, Judith A., Janis Kuby, Jenni Punt, and Sharon A. Stranford. *Immunology*. 7th ed., Macmillan, 2013.

Pines, Maya, editor. *Arousing the Fury of the Immune System.* Howard Hughes Medical Institute, 1998.

Poppelaars, Felix, et al. "Donor Genetic Variants in Interleukin-6 and Interleukin-6 Receptor Associate with Biopsy-Proven Rejection Following Kidney Transplantation." Scientific Reports, vol. 11, 2021, p. 16483, https://doi.org/10.1038/s41598-021-95714-z.

INNATE IMMUNITY

Category: Immunity

Also known as: Natural immunity or genetic immunity

Anatomy or system affected: Blood, cells, circulatory system, immune system, lymphatic system

Specialties and related fields: Immunology

Definition: the first line of defense against harmful substances that enter the body

KEY TERMS

lysosome: a membrane-bound cell organelle involved with various cell processes that contains digestive enzymes

lysozyme: an enzyme that catalyzes the destruction of the cell walls of certain bacteria

myeloperoxidase: a peroxidase enzyme that is most abundant in neutrophil granulocytes and produces hypohalous acids to carry out their antimicrobial activity

phagocyte: a type of cell within the body capable of engulfing and absorbing bacteria and other small cells and particles

phagolysosome: or endolysosome, is a cytoplasmic body formed by the fusion of a phagosome with a lysosome in a process that occurs during phagocytosis

phagosome: a vacuole in the cytoplasm of a cell, containing a phagocytosed particle enclosed within a part of the cell membrane

Innate immunity is the first line of defense against harmful substances that enter the body. It is also known as natural immunity or genetic immunity. The innate immune system detects invaders known as pathogens and releases special cells that attack them. Innate immunity is present at birth, does not need to learn, and can be very effective. The innate immune system does not identify what type of attack is present. It reacts in much the same way regardless of the substance present. It is still an essential and valuable part of protecting against disease and illness caused by foreign substances.

OVERVIEW

The immune system is the body's way of protecting against diseases caused by pathogens. Pathogens are organisms such as viruses, bacteria, parasites, and fungi. The immune system also protects against the effects of toxins such as poisons.

Immunity works in two ways: through innate immunity and adaptive immunity. The innate immune system is present at birth and is the first line of defense. The innate immune system activates the adaptive immune system. Also known as acquired immunity, the adaptive immune system develops and remembers a response to specific pathogens. Both innate and adaptive immunity are important to protect a person's overall health.

Innate immunity is coded into a person's genes and provides protection from illness beginning at birth. The innate immune system begins to work almost immediately after exposure to a pathogen or toxin.

The innate immune response can come in physical barriers, chemical barriers, or cellular defenses against the invading pathogen or toxin.

Physical barriers such as skin, eyelashes, and the hairs inside the nose interfere with the ability of a pathogen or toxin to enter the body. The coughing reflex creates a physical barrier by expelling potentially harmful substances.

Chemical barriers include tears and nasal secretions, and they help wash pathogens or toxins from the body. Stomach acid is another chemical barrier; it kills many types of pathogens that enter through the mouth into the digestive system. Eccrine sweat glands release sweat on our skin to cool us down and prevent overheating. As the sweat evaporates, it leaves salt, making the skin salty. High salt concentrations prevent the growth of many types of bacteria. Sebaceous glands also secrete an oily substance called "sebum." Bacteria on the skin, whose growth is promoted by the high salt concentration, degrade the fats in the sebum to glycerol and fatty acids. These free fatty acids lower the pH of the skin, further discouraging the growth of pathogenic bacteria.

Sweat, mucus, tears, saliva, and breastmilk contain proteins, like the enzyme lysozyme and other small molecules that kill some bacteria and discourage or prevent the growth of others. Lysozyme degrades the cell walls of bacteria, killing them. Urine and vaginal secretions are too acidic for many pathogens to tolerate. Semen contains zinc, which inhibits many bacteria and viruses. Antimicrobial proteins called "defensins" are produced by white blood cells and the epithelial linings of the gastrointestinal, genitourinary tracts, and respiratory tracts, and skin. Defensins kill microbes by disrupting their bacterial cell membranes. In the stomach, acid and digestive enzymes kill most of the pathogens that venture into it.

Cellular defenses in the white blood cells generate a variety of other cells that can attack pathogens. They include dendritic cells, mast cells, macrophages, neutrophils, basophils, and eosinophils. These cells, in turn, activate other cells that weaken or kill invading pathogens.

Some white blood cells, in particular macrophages and neutrophils, are "professional phagocytes." Phagocytes gobble up, or phagocytose, bacteria, fungi, and viruses and destroy them. These cells are armed with a host of receptors called "pathogen recognition receptors" or PRRs. They recognize pathogens, bind them, and swallow them into a vesicle called a "phagosome." Inside the phagocyte, vesicles called "lysosomes" fuse with the phagosome to form a phagolysosome. Lysosomes are filled with enzymes that pick microbes apart bit by bit. Phagocytes use something called an "oxidative burst" that fills the phagolysosome with reactive oxygen species (more popularly known as "free radicals") that kill microbes. Neutrophils are especially good at making reactive oxygen species because they have an enzyme called "myeloperoxidase." Neutrophils are usually the first cells to arrive at the injury site, and they clean up invading bacteria and cell debris.

The innate immune system works best in the first seven days after the introduction of a pathogen. It helps neutralize or kill pathogens and helps attract cellular defenses to the part of the body where they are most needed. It also triggers the acquired immune system to develop specialized cells called "antibodies" to attack the pathogens and helps the body eliminate dead cells as the immune system works.

The innate immune system is part of a healthy and efficient immune response to protect the body from illness. However, it can sometimes result in undesired responses. In some cases, the system will react to substances that are generally not harmful. When this unnecessary response is excessive, it results in allergies.

—Janine Ungvarsky and Michael A. Buratovich, PhD

Further Reading

Cain, Phil. "How Does Your Immune System Work?" *World Economic Forum*, 7 Apr. 2020, www.weforum.org/agenda/2020/04/immune-system-fight-off-disease-coronavirus-covid19-pandemic/. Accessed 10 Feb. 2021.

Delves, Peter J. "Innate Immunity." *Merck Manual*, Apr. 2020, www.merckmanuals.com/home/immune-disorders/biology-of-the-immune-system/innate-immunity. Accessed 10 Feb. 2021.

Gleichmann, Nicole. "Innate Vs. Adaptive Immunity." *Immunology and Microbiology*, 20 May 2020, www.technologynetworks.com/immunology/articles/innate-vs-adaptive-immunity-335116. Accessed 10 Feb. 2021.

"Immune Response." *Medline Plus*, 8 Feb. 2021, medlineplus.gov/ency/article/000821.htm. Accessed 10 Feb. 2021.

"Innate and Adaptive Immunity." *American Society for Radiation Oncology*, 2021, www.astro.org/Patient-Care-and-Research/Research/Professional-Development/Research-Primers/Innate-and-Adaptive-Immunity. Accessed 10 Feb. 2021.

"Introduction to Immunology Tutorial." *The Biology Project, University of Arizona*, 24 May 2000, www.biology.arizona.edu/immunology/tutorials/immunology/page3.html. Accessed 10 Feb. 2021.

Shoman, Mary. "How the Immune System Works." *VeryWell Health*, 4 July 2020, www.verywellhealth.com/how-does-the-immune-system-work-3232652. Accessed 10 Feb. 2021.

"What Is Innate Immunity?" *Center for Innate Immunity and Immune Diseases*, ciiid.washington.edu/content/what-innate-immunity. Accessed 10 Feb. 2021.

MONOCLONAL ANTIBODIES

Category: Chemotherapy and other drugs
Anatomy or system affected: All
Specialties and related fields: Bacteriology, biochemistry, biotechnology, immunology, internal medicine, microbiology, oncology, pathology, pharmacology, preventive medicine, public health
Definition: antibodies that recognize only one antigen and are mass-produced in the laboratory from a single clone of a B cell, the type of immune system cell that makes antibodies

KEY TERMS

antibody: a protein made by plasma cells in response to an antigen

antigen: any substance that causes the body to make an immune response against that substance.

B lymphocyte: a type of white blood cell that antibodies

conjugated antibody: an antibody to which a substrate such as an enzyme, toxin, or inorganic compound has been attached

hybridoma: a culture of hybrid cells that results from the fusion of B cells and myeloma cells *infusion*: a method of putting fluids, including drugs, into the bloodstream; also called "intravenous infusion"

DISEASES TREATED

Lymphoma, leukemia, breast cancer, head and neck cancers, colorectal cancer, lung cancer, and autoimmune diseases, including Crohn disease, ulcerative colitis, psoriasis, psoriatic arthritis, rheumatoid arthritis, ankylosing spondylitis, hidradenitis suppurativa, and noninfectious uveitis.

SUBCLASSES OF THIS GROUP

Murine (composed entirely of mouse sequences), chimeric (composed of approximately one-third mouse and two-thirds human sequences), humanized (composed of at least 90 percent human sequences), and human (antibodies that are fully human in composition).

DELIVERY ROUTES

Due to their molecular size and susceptibility to enzymatic digestion in the gut if administered orally, monoclonal antibodies must usually be administered by intravenous (IV) infusion or subcutaneous (SC) injections.

HOW THESE DRUGS WORK

Antibodies are proteins that bind to a specific site, or epitope, on a specific target molecule. In response to infection or immunization with a foreign agent, the immune system generates many different antibodies that bind to the foreign molecules. This pool of polyclonal antibodies contains a mixture of different antibody molecules, each of which binds to a specific epitope. Isolation of a single antibody from a polyclonal antibody pool would yield a highly specific molecular tool with the ability to bind to a single epitope. Georges Köhler, César Milstein, and Niels Kaj Jerne invented the process of producing monoclonal antibodies in 1975 and shared the 1984 Nobel Prize in Physiology or Medicine for their discovery. Since then, monoclonal antibodies have become an important tool in biological research and medicine.

Producing monoclonal antibodies involves fusing an individual B cell, which produces a single antibody with a single specificity but has a finite life span, with a long-lived myeloma tumor cell. The B cell is taken from the spleen or lymph nodes of an animal that has been challenged with the antigen of interest. Combining the B cell and the myeloma cell produces a hybridoma cell, a kind of perpetual antibody-producing factory. The hybridoma cell produces a single specific antibody. It can be grown indefinitely, allowing large amounts of monoclonal antibodies to be produced. Monoclonal antibodies are potentially more effective than conventional drugs in treating cancer since conventional drugs attack cancer cells and normal cells. Monoclonal antibodies attach only to the specific target molecule. Since monoclonal antibodies are specific for a particular antigen, one designed to bind to ovarian cancer cells, for instance, will not bind to colorectal cancer cells.

Georges Köhler and César Milstein made the first monoclonal antibodies from mouse B cells. When administered to humans, mouse antibodies are recognized by the human immune system as foreign (because they are from a different species). They can elicit an immune response against them, causing allergic-type reactions. Researchers have since learned how to replace some portions of the mouse antibody sequences with human antibody sequences. Genetic engineering techniques have allowed the production of chimeric, humanized, and, more recently, fully human monoclonal antibodies.

An antibody molecule consists of two heavy polypeptide chains and two light polypeptide chains. Both heavy and light chains are composed of a region that varies from antibody to antibody, the variable region, and a constant region that is conserved. By combining human sequences for the constant region

Common Monoclonal Antibodies

Drug	Brands	Subclass	Delivery Mode	Cancers Treated
Alemtuzumab	Campath	Humanized	IV	Chronic lymphocytic leukemia
Bevacizumab	Avastin	Humanized	IV	Lung cancer, colorectal cancer
ntblCetuximab	Erbitux	Chimeric	IV	Head and neck cancers, colorectal cancer
Gemtuzumab ozogamicin	Mylotarg	Humanized	IV	Acute myelogenous leukemia
Ibritumomab tiuxetan	Zevalin	Murine	IV	B-cell lymphoma
Lym-1	Oncolym	Murine	IV	Lymphoma
Panitumumab	Vectibix	Human	IV	Colorectal cancer
Rituximab	Rituxan	Chimeric	IV	B-cell lymphoma
Tositumomab	Bexxar	Murine	IV	B-cell lymphoma
Trastuzumab	Herceptin	Humanized	IV	Breast cancer
Bezlotoxumab	Zinplava	Human	IV	Clostridioides difficile infection
Adalimumab	Humira	Human	SC	Autoimmune diseases
Certolizumab	Cimzia	Humanized	SC	Autoimmune diseases
Golimumab	Simponi	Human	SC	Autoimmune disease
Etanercept (fusion protein)	Enbrel	Human	SC	Autoimmune disease
Infliximab	Remicade	Chimeric	SC	Autoimmune diseases
Omalizumab	Xolair	Humanized	SC	Severe, persistent, allergic asthma
Benralizumab	Fasenra	Humanized	SC	Eosinophilic asthma
Mepolizumab	Nucala	Humanized	SC	Eosinophilic asthma
Reslizumab	Cinqair	Humanized	SC	Eosinophilic asthma
Dupilumab	Dupixent	Human	SC	Intractable asthma

with murine sequences for portions of the variable region, the amount of murine sequence can be decreased. Depending on how much murine sequence is left, the result is either a chimeric (with approximately one-third murine and two-thirds human sequence) or a humanized (with at least 90 percent human sequence) monoclonal antibody. Genetically engineered mice strains are now available that contain a large portion of human deoxyribonucleic acid (DNA) that codes for the antibody heavy and light chains, with the mouse's own heavy and light chain genes, inactivated. Using these mice to produce B cells for the construction of hybridomas allows the generation of fully human antibodies, which are likely to be safer and might be more effective than the previous generation of monoclonal antibodies.

One potential treatment for cancer involves using monoclonal antibodies that bind only to a cancer cell-specific component of interest and induce an immunological response against the target cancer cell (referred to as "naked" monoclonal antibodies). Monoclonal antibodies can also be designed for the delivery of another (nonspecific) agent, such as a toxin, radioisotope, or cytokine, to the cancer cell to kill it (referred to as "conjugated" monoclonal antibodies).

Some naked monoclonal antibodies bind to cancer cells and exert their action by marking the cells to help the body's immune system destroy them. Rituxan (rituximab) and Campath (alemtuzumab) are examples of such monoclonal antibodies. Rituximab binds to the CD20 antigen, a protein found on B cells, and is used to treat B-cell non-Hodgkin lymphoma.

Alemtuzumab binds to the CD52 antigen, another protein present on B and T cells. It is used to treat some patients with B-cell chronic lymphocytic leukemia.

Some naked monoclonal antibodies bind to functional parts of cancer cells or other cells that help cancer cells grow and act by interfering with the cancer cell's ability to grow. Herceptin (trastuzumab), Erbitux (cetuximab), and Avastin (bevacizumab) are examples of this type of monoclonal antibody. Trastuzumab binds to the HER2/neu protein, a protein present in large numbers on tumor cells in some cancers that, when activated, helps these cells grow. Trastuzumab acts by inactivating these proteins. It is used to treat some breast cancers. Cetuximab binds to the epidermal growth factor receptor (EGFR) protein, which, when present in high levels on cancer cells, helps them grow. Cetuximab blocks the activation of EGFR and is used to treat some advanced colorectal cancers and some head and neck cancers. Bevacizumab binds to the vascular endothelial growth factor (VEGF), a protein that cancer cells produce to attract the new blood vessels they need for growth. Bevacizumab prevents VEGF from functioning and is used to treat some colorectal, lung, and breast cancers.

These monoclonal antibodies have been used in cancer treatment for many years. At first, oncologists used them mainly after other treatments had failed. However, in response to data from detailed clinical studies, the trend is to use them earlier in the course of cancer treatment.

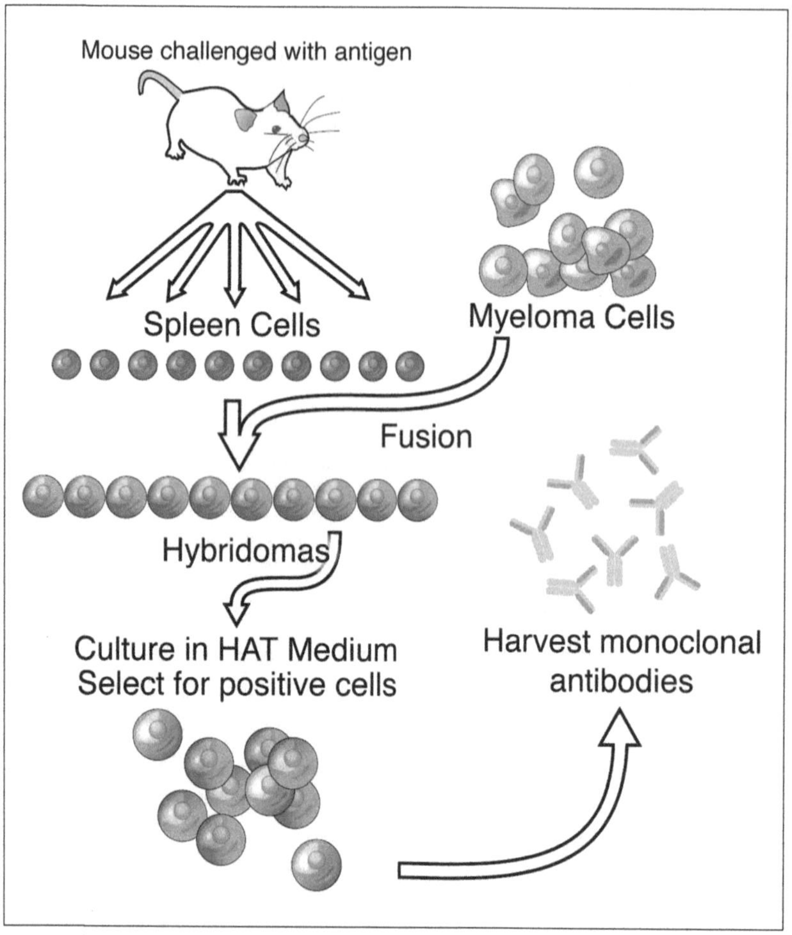

A general representation of the method used to produce monoclonal antibodies. Image by Adenosine, via Wikimedia Commons.

Conjugated monoclonal antibodies (also called "tagged" or "loaded" monoclonal antibodies) are attached to anticancer (chemotherapy) drugs, toxins, or radioactive substances and used as vehicles to deliver these toxic agents directly to cancer cells. Radiolabeled monoclonal antibodies are attached to radioactive substances; treatment with such agents is called "radioimmunotherapy." Chemolabeled monoclonal antibodies are attached to anticancer drugs, and immunotoxins are monoclonal antibodies attached to toxins. Zevalin (ibritumomab tiuxetan) and Bexxar (tositumomab) are examples of radiolabeled monoclonal antibodies. Both bind to an antigen on cancerous B lymphocytes and are used to treat some B cell non-Hodgkin lymphomas. Mylotarg (gentuzumab ozogamicin) is an example of an immunotoxin. It contains the toxin calicheamicin attached to a monoclonal antibody that binds to CD33, a protein antigen on most leukemia cells. Mylotarg treats some acute myelogenous leukemias.

Clinical trials of monoclonal antibody therapy are in progress for patients with almost every type of cancer. As more cancer-associated antigens have been identified and studied, it has been possible for researchers to make monoclonal antibodies against more types of cancer.

SIDE EFFECTS

Antibodies that contain murine sequences can be recognized by the human immune system as foreign, causing systemic inflammatory effects such as fever, chills, weakness, headaches, nausea, vomiting, and diarrhea.

Some monoclonal antibodies also have side effects associated with the target antigen. For example, some monoclonal antibodies can affect the bone marrow's ability to produce blood cells. Diminished bone marrow activity increases the risk of bleeding or infection in some patients. Another example is bevacizumab, which inhibits a growth factor called VEGF. Cancer cells use this protein to build new blood vessels that feed the tumor and facilitate its spread. VEGF inhibition inhibits the maintenance and healing of blood vessels, leading to bleeding.

Therapeutic monoclonal antibodies that treat autoimmune diseases suppress specific arms of the immune response. Therefore, these medications can increase the risk of upper respiratory infections and pneumonia, tuberculosis, and cancer. Consequently, such drugs are usually used when other first-line treatments have failed or become intolerable (e.g., corticosteroids).

—Jill Ferguson, PhD and Michael A. Buratovich, PhD

Further Reading

George, Andrew J. T., and Catherine E. Urch, editors. *Diagnostic and Therapeutic Antibodies*. Humana Press, 2000.

Haur, Harleen, and Dietmar Reusch, editors. *Monoclonal Antibodies: Physicochemical Analysis*. Academic Press, 2021.

Jess, Sowmya. *Understanding Monoclonal Antibodies. Xlibris US, 2020.*

Melero, I., et al. "Immunostimulatory Monoclonal Antibodies for Cancer Therapy." *Nature Reviews Cancer*, vol. 7, 2007, pp. 95-106.

Reichert, J. M., and V. E. Valge-Archer. "Development Trends for Monoclonal Antibody Cancer Therapeutics." *Nature Reviews Drug Discovery*, vol. 6, 2007, pp. 349-56.

Zafir-Lavie, I., Y. Michaeli, and Y. Reiter. "Novel Antibodies as Anticancer Agents." *Oncogene*, vol. 28, 2007, pp. 3714-33.

PHAGOCYTOSIS

Category: Immunology
Anatomy or system affected: Blood, cells, circulatory system, immune system, lymphatic system
Specialties and related fields: Bacteriology, biochemistry, cell biology, immunology, microbiology
Definition: the ingestion of bacteria or other material by phagocytes and amoeboid protozoans

KEY TERMS

macrophage: a type of white blood cell that surrounds and kills microorganisms, removes dead cells and stimulates the action of other immune system cells
monocyte: a large phagocytic white blood cell with a simple oval nucleus and clear, grayish cytoplasm
neutrophil: the most abundant type of granulocytes that make up 40 to 70 percent of all white blood cells in humans and form an essential part of the innate immune system
pseudopod: a temporary arm-like projection of a eukaryotic cell membrane that is developed in the direction of movement

INTRODUCTION

Phagocytosis is the process by which a cell engulfs and consumes other cells or solid particles. In single-celled organisms such as the amoeba, phagocytosis is a feeding method. The process is a vital part of the immune system in higher organisms. It is carried out by specialized cells called "macrophages" and certain white blood cells. These cells, known as "phagocytes," absorb and digest invading organisms such as bacteria, viruses, or other foreign bodies. The term "phagocytosis" comes from the Greek words *phagein*, meaning "to eat," and *cytos*, meaning "cell."

BACKGROUND

Cells are covered by an outer plasma membrane that protects the interior and regulates the substances that enter the cell. Smaller molecules can easily pass through the cell membrane. However, to take in larger molecules, cells use endocytosis. The cellular membrane moves inward during endocytosis, curving around the molecule and surrounding it. Phagocytosis is a type of endocytosis in which a cell ingests solid particles. Pinocytosis, or cellular drinking, occurs when the cell consumes liquids.

By the nineteenth century, scientists had identified white blood cells and observed them in the process of surrounding bacteria. However, the prevailing idea at the time was that these cells carried the foreign invaders and spread them throughout the body. In the 1880s, Russian microbiologist Elie Metchnikoff observed the action of "wandering" cells in transparent starfish larvae and noticed that the cells seemed to clump near the site of an invading organism as if they were attacking it. He realized that the cells were consuming the foreign bodies, not spreading them. He coined the term phagocytosis to describe his discovery, for which he shared the Nobel Prize in 1908.

OVERVIEW

Microscopic organisms that consist of a single cell sustain themselves through the process of phagotrophic nutrition. When an amoeba, for example, detects the presence of a bacterium, it crawls toward its target. The amoeba sends out "arms" called "pseudopodia" formed from its cell membrane when it makes contact. Some single-celled organisms use hair-like structures called" cilia" to attach themselves to their prey. The pseudopodia surround the bacterium and encase it in an internal cavity called a "vesicle." Before the bacterium can reproduce, the amoeba releases other vesicles packed with digestive enzymes. These proteins produce a chemical reaction in organic substances. These vesicles attach themselves to the food, break down the bacterium, and absorb the nutrients.

In the human body, phagocytosis occurs in many cells. Still, the most common are white blood cells found in the immune system. These "professional phagocytes" cells make up about 1 percent of the body's blood and act as natural infection fighters. Monocytes and neutrophils are the most accomplished phagocytes of the five types of white blood cells. Monocytes are the largest white blood cells created in the bone marrow, a soft substance in the center of the bones. After monocytes are released into the bloodstream, they travel to the site of an infection and enter the body's tissue. They then develop into specialized infection-devouring cells known as macrophages. The large size of the cells makes it easier for them to consume foreign invaders.

Since macrophages must encounter an infectious organism before consuming it, the body must first find a way to let the macrophage know which objects to target. One way is for other white blood cells to produce proteins called "antibodies" that attach themselves to certain molecular patterns on the surface of the invading cell. These antibodies act as a homing beacon for the macrophages, marking a particular cell for destruction. Some macrophages have specialized receptors called "Fc receptors" on their surface designed to search for specific antibodies and bind with them. Others have receptors designed to seek out and bind with certain molecules produced by the bacteria called "pattern recognition receptors" (PRRs).

When the receptors on the surface of the macrophage bind with the target, the macrophage begins drawing the cell inward, encircling it, and closing around it. The target becomes enclosed inside an internal bubble called a "phagosome." This structure then fuses with an enzyme-filled bubble inside the cell called a "lysosome." Once formed, the merged structure, known as a phagolysosome, reduces its internal pH level. A pH level measures acidity; the lower the pH level, the more acidic the environment. The low pH in the phagolysosome activates groups of enzymes called "acid hydrolases." Acid hydrolases are active at low pH and inactive at higher pH. These enzymes degrade the material in the phagolysosome piece by piece.

Additionally, the phagolysosomes of neutrophils and macrophages contain enzymes that produce reactive oxygen species (also known as free radicals). Reactive oxygen species include hydrogen peroxide (H_2O_2), superoxide radicals (O_2^-), and other harmful molecules. These enzymes and reactive oxygen species kill or neutralize the target before it has a chance to reproduce, allowing the macrophage's enzymes to break it down for absorption. The resulting waste product is later expelled from the macrophage.

Some viruses can trick a macrophage and hijack it to replicate inside the cell. The influenza virus, for example, uses the drop in pH levels to escape from the phagolysosome and gain access into the macrophage's interior. When a macrophage becomes infected by a virus, the body's other phagocytes will then mark it as a target.

Neutrophils are the most numerous white blood cells, making up about 60 percent of the body's total white blood cells. Neutrophils are the first line of defense to arrive at the site of an infection. They act as natural assassins, using several methods to kill their targets. They are also produced in the bone marrow and live for only a few hours. Their primary method of attack is similar to the macrophage; they engulf an invader and digest it through enzymes released from granules, or grain-like shapes, within their structures. Neutrophils have an enzyme complex in their phagolysosomes called "myeloperoxidase." They can emit a burst of highly reactive oxygen molecules, like hypochlorous acid (HOCl), which damages and destroys invading cells. Neutrophils and other phagocytes can also kill their engulfed prey by releasing antimicrobial proteins specifically designed to destroy bacteria or produce binding proteins that interfere with a bacterium's reproduction ability.

—Richard Sheposh

Further Reading

Dale, David C., Laurence Boxer, and W. Conrad Liles. "The Phagocytes: Neutrophils and Monocytes." *Blood,* 2008, www.bloodjournal.org/content/112/4/935?sso-checked=true. Accessed 16 May 2017.

Gordon, Siamon. "Phagocytosis: An Immunobiologic Process." *Science Direct,* 15 Mar. 2016, www.sciencedirect.com/science/article/pii/S1074761316300656. Accessed 16 May 2017.

Lim, Daniel V. *Microbiology.* 3rd ed., Kendall/Hunt Publishing, 2003.

Mandal, Ananya. "What Is a Macrophage?" *News-Medical.Net,* 14 Jan. 2014, www.news-medical.net/life-sciences/What-is-a-Macrophage.aspx. Accessed 16 May 2017.

Mayadas, Tanya N., Xavier Cullere, and Clifford A. Lowell. "The Multifaceted Functions of Neutrophils." *Annual Review of Pathology,* vol. 9, Jan. 2014, pp. 181-218, www.ncbi.nlm.nih.gov/pmc/articles/PMC4277181/. Accessed 16 May 2017.

"Phagocytosis." *Khan Academy,* www.khanacademy.org/test-prep/mcat/cells/transport-across-a-cell-membrane/a/phagocytosis. Accessed 16 May 2017.

Tan, S. Y., and M. K. Dee. "Elie Metchnikoff (1845-1916): Discoverer of Phagocytosis." *Singapore Medical Journal,* 2009, webext.pasteur.fr/biblio/ressources/histoire/textes_integraux/metchnikoff/smjmetabio2009tan.pdf. Accessed 16 May 2017.

"What Are White Blood Cells?" *University of Rochester Medical Center,* www.urmc.rochester.edu/encyclopedia/content.aspx?ContentTypeID=160&ContentID=35. Accessed 16 May 2017.

SEPSIS

Category: Diseases and conditions
Anatomy or system affected: All
Specialties and related fields: Bacteriology, critical care, emergency medicine, hematology, immunology, internal medicine, microbiology, mycology, nephrology, pharmacology, virology
Definition: a whole-body inflammatory response to an infection

KEY TERMS

apoptosis: the death of cells that occurs as a normal and controlled part of an organism's growth or development
bacteremia: the presence of bacteria in the blood
interleukins: a class of glycoproteins produced by leukocytes for regulating immune responses
inflammation: a localized physical condition in which part of the body becomes reddened, swollen, hot, and often painful, especially as a reaction to injury or infection

INTRODUCTION

Sepsis is a systemic inflammatory response to infection. In the United States, as of 2008, more than 1.1 million persons develop sepsis each year, and between one-quarter and half of those with sepsis die from the infection, according to the US Centers for Disease Control and Prevention (CDC). The number of cases has been rising due to an aging population, higher rates of illness and medical procedures, and improved diagnostics and reporting. In the past, the term "septicemia" (or "blood poisoning") was often used interchangeably with sepsis. Yet, that practice has fallen out of favor because the disease description, "blood poisoning," is considered imprecise.

CAUSES

Sepsis often begins when there is an infection in the body, whether bacterial, viral, fungal, or parasitic. The infection provokes a massive immune response that releases overly large quantities of pro-inflammatory molecules like tumor necrosis factor-alpha, interleukin-1, -6, and -12. In this situation, global blood vessel dilation prevents the body from effectively delivering oxygen to all the organs and cells that need it. The lungs, abdomen, urinary tract, skin, brain, and bone are common starting points for sepsis. Sepsis can also affect the intestine, where bacteria thrive, and already-infected areas after surgery. A foreign object (such as a catheter or drainage tube) inserted into the body can cause sepsis.

RISK FACTORS

Sepsis has become more common, especially among hospitalized persons. People at risk include the elderly, neonatal patients, immunocompromised persons, and persons who use injectable drugs. The widespread use of antibiotics encourages the growth of drug-resistant microorganisms. There is a higher incidence of sepsis when a person is already weakened by a condition such as malnutrition, alcoholism, liver disease, diabetes, a malignant neoplasm (cancer), organ transplantation, bone marrow transplantation, or human immunodeficiency virus (HIV) infection.

Of persons with end-stage renal disease, 75 percent will die of sepsis. Sepsis also causes high mortality rates in persons undergoing dialysis and in renal transplant recipients. Systemic inflammatory response syndrome and acute respiratory distress syndrome are closely related to sepsis.

Men are more susceptible than women to developing sepsis. Minorities appear to be at greater risk of developing sepsis as well. Among persons who already have sepsis, blacks are more likely to die than are whites. Preliminary studies have identified socioeconomic status, educational level, genetics, the number of other chronic diseases a person has, tobacco or alcohol use, nutritional status, and when and where a patient develops sepsis (i.e., before, during, or after hospitalization) as areas for further research into what effect race has on the disease progression and mortality of people with sepsis. Similar factors may affect risk by gender as well.

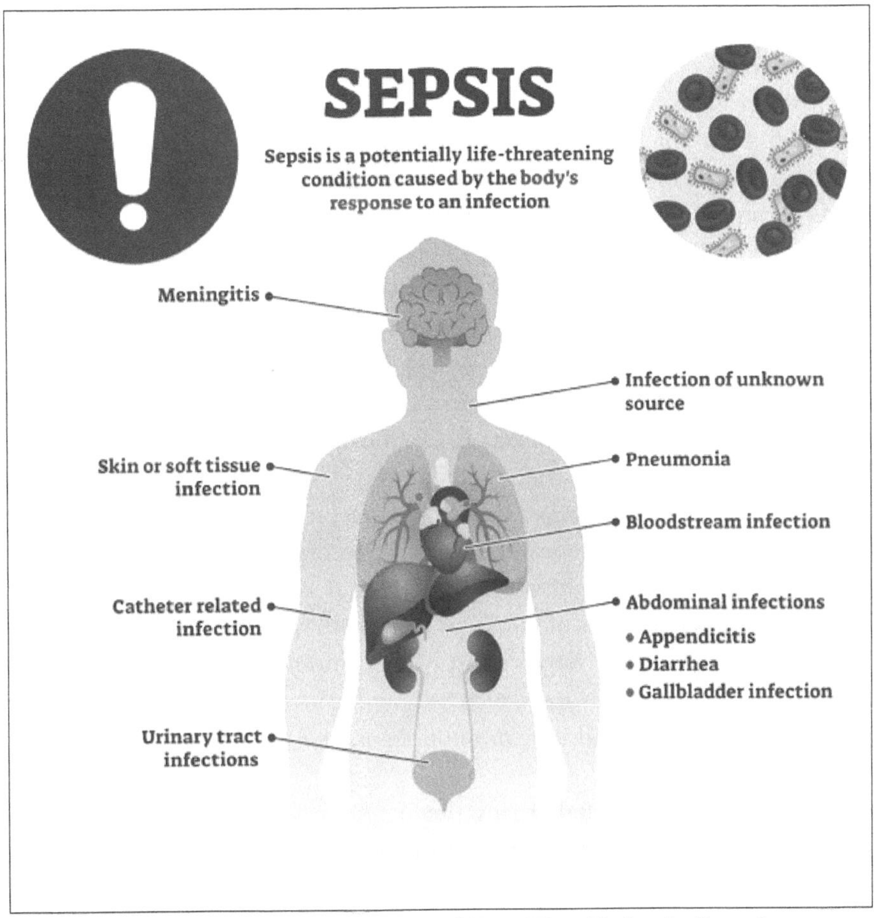

Sepsis dangers and warning signs. Image via iStock/VectorMine. [Used under license.]

SYMPTOMS

Common signs and symptoms include fever, increased heart rate, increased breathing rate, low blood pressure, and confusion. Other symptoms accompanying sepsis include shaking, chills, weakness, decreased urine output, nausea, vomiting, and diarrhea. There are also infection-specific symptoms such as cough with pneumonia or pain during urination with a urinary tract infection. However, very young or older adults or those with poorly functioning immune systems may display no signs of infection or fever.

Sepsis can cause infections to spread uncontrollably and attack crucial body systems, such as the lining of the brain, the sac around the heart, the bones, or the large joints. Sepsis can also bring about impaired intestinal function. The patient may have decreased bowel sounds and ileus.

Sepsis attacks the endothelium, the thin layer of cells within the blood vessels, which affects the circulation, the heart, and, ultimately, the body's organs. Endothelial dysfunction leads to increased vascular permeability, edema (swelling), and clotting abnormalities. Multiple organ failure is a common effect of sepsis. Apoptosis, also known as cell suicide, is closely linked to multiple organ failure and sepsis.

Severe sepsis causes reduced organ blood flow resulting in poor organ function. Poor blood flow is signaled by abnormally low blood pressure, high blood lactate, and low urine output. Septic shock is low blood pressure that fails to correct after giving appreciable intravenous fluids.

The status of a sepsis patient is determined with a qSOFA or quick sequential organ failure assessment score. The qSOFA score is calculated this way:

- A respiratory rate of twenty-two breaths per minute or more = 1
- A decrease in consciousness—Glasgow Coma Scale $< 15 = 1$
- Systolic Blood Pressure ≤ 100 mmHg = 1

Patients with a qSOFA score ≥ 2 have an increased mortality risk.

SCREENING AND DIAGNOSIS

Because sepsis is so lethal, early diagnosis is crucial. Some of the signs are a temperature above 101° or below 96° Fahrenheit, a heart rate above ninety beats per minute, or a breathing rate faster than twenty beats per minute. Additional signs include having a white blood cell count greater than 12,000 cubic millimeters or having pus-forming or other pathogenic organisms. The white cell count will show a "left shift," meaning a predominance of young, immature white cells in circulation. Blood cultures are drawn to determine the source of the infection. Diagnostic tests may also be performed on wound secretions or cerebrospinal fluid. Imaging scans may be done, too.

Several factors can complicate diagnosis. Doctors often do not see persons with sepsis until those persons are in the later stages of illness and tend to have several complex diseases. Sepsis may be one component of a larger disease process, such as systemic inflammatory response syndrome or multiple organ dysfunction syndrome.

If there is damage to vital organs, the diagnosis becomes severe sepsis. The most serious form of sepsis is septic shock, with the complication of low blood pressure (hypotension) that does not respond to standard treatment.

TREATMENT AND THERAPY

Because sepsis spreads so quickly, treatment may start before the results of blood cultures are available. More potent antibiotics are available, covering a broader spectrum, and antifungal agents may be used if the infection is thought to be fungal, rather than bacterial, in origin. Once blood cultures identify the pathogen, narrower-spectrum antimicrobials can be used. Immunosuppressive agents may also be used to decrease inflammation. Other treatments include insulin, painkillers, sedatives, and surgery. One strategy is to attempt invasive treatment of inflammatory, infectious, and neoplastic diseases. A 2015 *Cochrane Review* meta-analysis of clinical trials also shows that low-dose corticosteroids over an extended period appear to reduce mortality and improve the odds of recovery from septic shock.

Respiratory failure is treated with gas exchange and oxygen. Beta-2 receptor stimulators help treat liver failure. For cardiac dysfunction, the patient is treated with volume therapy and vasopressors. Ventilator support is used for neurological problems. Proton pump inhibitors are sometimes administered to prevent stress ulcer complications. Deep vein thrombosis prophylaxis is another necessary treatment to prevent complications from blood clots.

PREVENTION AND OUTCOMES

The best protection against sepsis is frequent handwashing, staying current on immunizations, and seeking prompt care for infections. A doctor should examine skin that has redness, swelling, or pus. In hospitals, the best prevention is identifying sepsis early and treating it with the correct antibiotic. This protocol will help to reduce organ dysfunction. In many cases, however, sepsis strikes persons who are already vulnerable.

Those who survive sepsis or septic shock may experience temporary depression, anxiety, confusion, loss of appetite, aches and pains, fatigue, weight loss, insomnia, or shortness of breath. Most sepsis survivors regain renal function over time; however, those with preexisting renal problems may need ongoing dialysis. More rarely, survivors experience long-term neurocognitive impairments, struggle with insomnia, have ongoing organ dysfunction, or require amputation of a limb.

—Merrill Evans, MA

Further Reading

Angus, Derek C., and Tom van der Poll. "Sepsis and Septic Shock." *New England Journal of Medicine,* vol. 369, 2013, pp. 840-51. Accessed 30 Dec. 2015.

Dellinger, R. Phillip, et al. "Surviving Sepsis Campaign: International Guidelines for Management of Severe Sepsis and Septic Shock: 2008." *Critical Care Medicine,* vol. 36, 2008, pp. 296-327.

Folstad, Steven G. "Soft Tissue Infections." *Emergency Medicine: A Comprehensive Study Guide*, edited by Judith E. Tintinalli, 6th ed., McGraw-Hill, 2004.

Hill, Kathleen "Shock, Sepsis, and Multiple Organ Dysfunction Syndrome." *Introduction to Critical Care Nursing*, edited by Mary Lou Sole, Deborah G. Klein, and Marthe J. Moseley, 6th ed. Saunders/Elsevier, 2013.

Mayr, Florian B., et al. "Infection Rate and Acute Organ Dysfunction Risk as Explanations for Racial Difference in Severe Sepsis." *Journal of the American Medical Association*, vol. 24, 2010, pp. 2495-2503.

National Center for Emerging and Zoonotic Infectious Diseases (NCEZID), Division of Healthcare Quality Promotion (DHQP). "Sepsis Questions and Answers." *CDC*. Centers for Disease Control and Prevention, 5 Oct. 2015. Accessed 30 Dec. 2015.

Valley, Thomas S., and Colin R. Cooke. "The Epidemiology of Sepsis: Questioning Our Understanding of the Role of Race." *Critical Care*, vol. 19, 2015, p. 347. Accessed 30 Dec. 2015.

Zucker-Franklin, D., et al. *Atlas of Blood Cells: Function and Pathology*. 3rd ed., Lea & Febiger, 2003.

STEROIDS

Category: Treatment

Anatomy or system affected: All systems

Specialties and related fields: Biochemistry, biotechnology, endocrinology, immunology, pharmacology substance use, treatment, and therapy

Definition: a class of natural or synthetic organic compounds characterized by a molecular structure of 17 carbon atoms arranged in four rings

KEY TERMS

endocrine glands: glands that release hormones directly into the bloodstream

gonads: a collective term referring to the testes and ovaries

hormone: a class of signaling molecules in multicellular organisms that are transported to distant organs to regulate physiology and behavior

semisynthetic: about compounds that are obtained by altering or augmenting the molecular structure of a compound obtained from a natural source

synthetic: about compounds that are produced entirely by synthesis reactions carried out in the laboratory from simple starting materials

ABSTRACT

Steroids are natural and synthetic hormones that enhance specific body activities and can be used as therapeutic agents to treat many clinical disorders. They are distinguished by a unique molecular arrangement of seventeen carbon atoms conjoined in four adjacent rings called the "steroid nucleus," which is common to all steroid compounds.

STRUCTURE AND FUNCTIONS

Steroids are a group of organic compounds distinguished by a unique molecular arrangement of seventeen carbon atoms conjoined in four adjacent rings. These four rings are referred to as the steroid nucleus and are common to all steroid compounds. Three of these rings are hexagonal six-carbon rings arranged in a bent-line fashion to form a phenanthrene group. The fourth group or ring contains only five carbon atoms. Steroids vary with the nature of the attached groups, the position of a given attached group, or some alteration to the configuration of the steroid nucleus. Small chemical differences in the structure of steroids can reflect very great differences in specific biological effects. Steroids are included in the lipid category of biological molecules because they are nonpolar and insoluble in water.

Any steroid that contains a hydroxyl group (-OH) is called a "sterol." This term comes from a Greek word meaning "solid." Sterols were so named because they were among the earliest compounds found to be solid at room temperature. Once chemical structures were determined, other compounds with similar structures were given the name steroid, which means "sterol-like." The suffix *-oid* comes from Greek and means "similar to."

Chemists have isolated hundreds of steroids from plants and animals; additionally, thousands have been manufactured by chemically modifying natural steroids or synthesizing the entire molecule. The parent compound for steroids is acetic acid. Assisted by various enzymes, acetic acid is altered and transformed into several other compounds before cholesterol is formed. Cholesterol serves as the parent, or precursor compound, for bile acids and the biologically important steroids to the body.

Cholesterol is the most common steroid in the human body, as it is a structural component of cellular membranes. The prefix *chole-* came from the Greek word for liver bile, a digestive fluid manufactured by

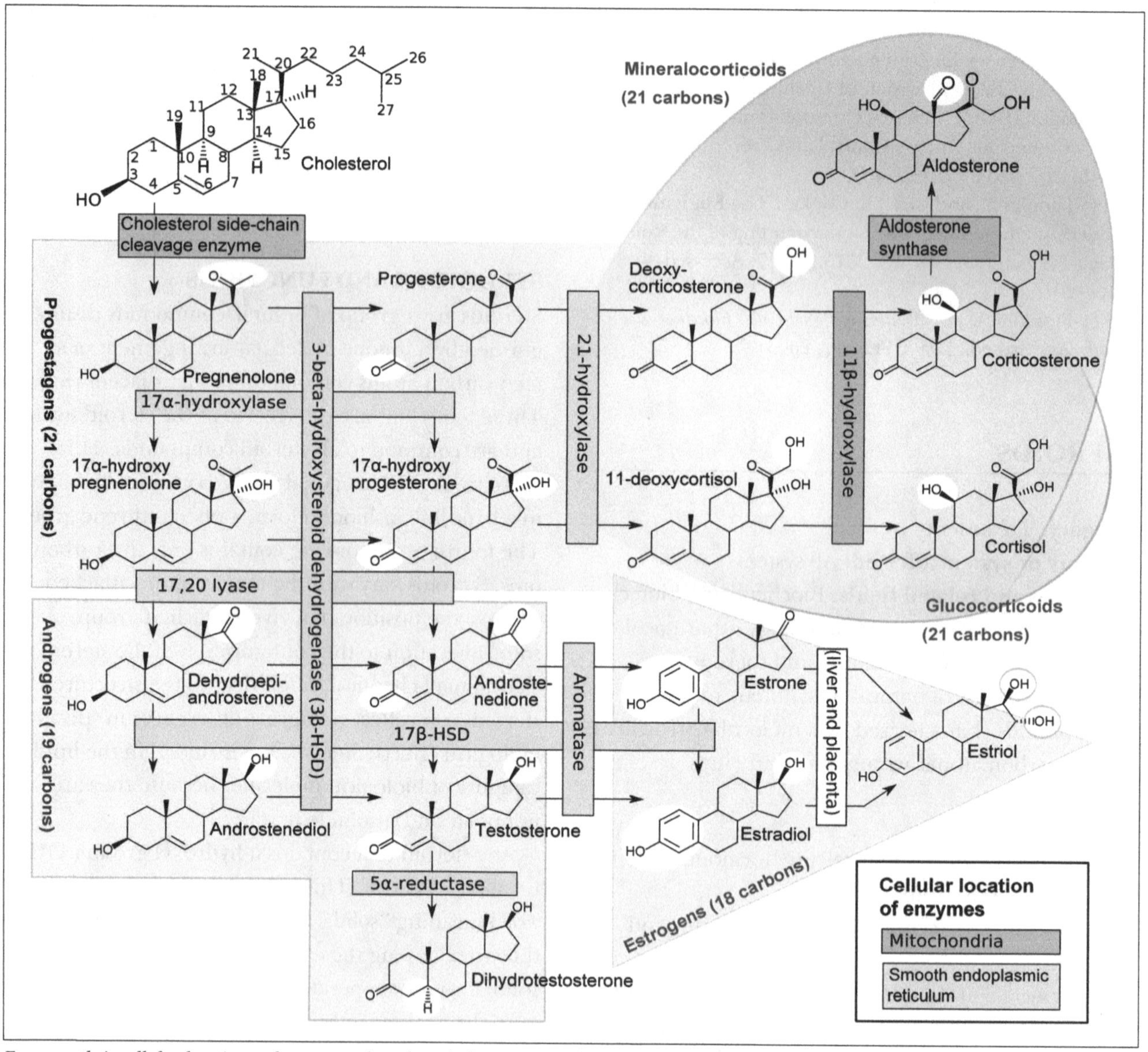

Enzymes, their cellular location, substrates, and products in human steroidogenesis. Image by David Richfield and Mikael Häggström, via Wikimedia Commons.

the liver and secreted into the intestines. The name is appropriate since the bile contains a considerable amount of cholesterol. Bile is stored in the gallbladder and becomes concentrated there. Cholesterol is not very soluble, and if it accumulates in great enough quantities, it will form small crystals in the bile. These crystals may join together to form larger particles that can block the narrow duct that leads from the gallbladder to the intestines. These aggregations of particles are called "gallstones" and are composed of almost pure cholesterol. The blockage can result in a buildup of pressure and cause much pain. Often, surgery is required to remove the obstruction.

Cholesterol is also an important molecule because it is the precursor or parent molecule for the steroid hormones produced by the gonads and the adrenal cortex. The gonads, a collective term referring to the testes and ovaries, secrete the sex steroids. These sex steroids include estradiol and progesterone from the ovaries and testosterone from the testes. The adrenal cortex secretes the corticosteroids, including cortisol and aldosterone, and other steroid compounds. Most steroid hormones are specialized in their function and do not produce general effects on metabolism. Sex hormones influence reproduction by acting on sexual organs to stimulate their development and function, influencing sexual behavior, and stimulating the development of secondary sex characteristics. Some of the steroids secreted by the adrenal cortex have more general effects on the metabolism of carbohydrates and proteins in many tissues.

Steroid hormones combine with specific receptors located in the cytoplasm of the responsive tissues. Since these hormones are lipid-soluble, they pass readily through cell membranes, largely composed of lipids. Inside the cytoplasm, a steroid-specific protein receptor will bind to the hormone. Upon binding, the hormone-receptor complex becomes activated or transformed and translocates to the nucleus. In the nucleus, the activated steroid receptor complex binds to the chromatin, or genetic material, causing an activation of a certain set of genes. Gene activation produces messenger molecules that induce the production of specific proteins that are either used by the cell or secreted elsewhere.

CATEGORIES OF STEROID HORMONES

The steroid hormones of the adrenal cortex fall into three categories, each having separate actions and sites of action. Aldosterone is the principal mineralocorticoid and plays an important role in regulating body levels of sodium and potassium. Cortisol, the major glucocorticoid, regulates carbohydrate metabolism. Adrenal androgens are also produced, but they have only weak activity and play a minor physiological role under most conditions. All adrenal steroids are derived from cholesterol.

Mineralocorticoids. Mineralocorticoids are adrenal steroids that regulate potassium and sodium levels in the body. The most potent mineralocorticoid, aldosterone, is secreted by the adrenal cortex at the rate of about 0.1 milligrams per day. Mineralocorticoids affect the distal tubules of the kidney by stimulating the excretion of potassium and the reabsorption of sodium. The net effect of these actions is to increase the volume of body fluids.

Glucocorticoids. Glucocorticoids are the adrenal steroids that regulate glucose metabolism. In humans, cortisol is responsible for most of the glucocorticoid activity. It is secreted by the adrenal cortex at the rate of about 20 milligrams per day and metabolically affects tissues throughout the body. Cortisol is regulated by the central nervous system and by the body's permissive or stimulatory messenger molecules. Generally, glucocorticoids stimulate glucose production and enhance the use of fat and protein as energy sources.

Androgens. Androgens are steroid hormones secreted primarily by the testes, adrenal glands, and ovaries. Testosterone is the principal androgen secreted by the testes; it regulates the development and function of male sex accessory organs. Increased testosterone secretion during puberty is required for seminal vesicle and prostate growth. Removal of androgens by castration results in these organs undergoing atrophy.

Androgens stimulate the growth of the larynx and cause the lowering of the voice. They increase hemoglobin synthesis, which is higher in males than females, and affect bone growth by causing the conversion of cartilage to bone. Androgens also promote protein synthesis or anabolic activity in skeletal muscle, bone, and kidneys. As a class of compounds, androgens are reasonably safe drugs since they have a limited and relatively predictable set of side effects. In human males, testosterone is synthesized by the testes at the rate of about 8 milligrams per day.

Estrogens and progesterones. Estrogens and progesterones are primarily produced in the ovaries of nonpregnant adult women. In pregnancy, the placenta is the major estrogen and progesterone production site. Smaller estrogen synthesis involves the liver, kidney, skeletal muscle, and testes. Estrogens cause the growth of the female reproductive organs. They are responsible for expressing female secondary sex characteristics, such as breast enlargement, female body contours, skin texture, and distribution of body hair. Estrogens are thought to protect against atherosclerosis and heart attacks since the occurrence of these health problems in mature women is much lower than in males of similar ages.

USES AND COMPLICATIONS

When cortisone was initially discovered, it was labeled a "wonder drug" and thought to possess overall effectiveness in many areas of medicine. Although these expectations have not been realized, various steroids are found to be effective in medical practice and treatment. Steroids are commonly prescribed for those persons whose bodies cannot produce specific steroid hormones in adequate quantities. Steroids are effective as anti-inflammatory agents, reducing inflammatory reactions in various body tissues. They are also prescribed for patients who have undergone organ transplantation or have highly sensitive allergies because they inhibit the immune system's responsiveness.

The primary therapeutic use of androgens is for testicular deficiency. The induction and maintenance of male secondary sex characteristics are desired. In these cases, supplemental doses of androgens are given to stimulate and enhance the development of sexual and accessory sex characteristics. Androgens are also effective in treating some anemias when persons have reduced levels of red blood cells. Androgens are used to treat osteoporosis, which is a decrease in bone or skeletal mass. Androgens are given to women in breast cancer treatment and are effective about 20 percent of the time. They are used to treat the abnormal growth of endometrial tissue in the peritoneal cavity of women, a disease called "endometriosis," and are effective in that role.

Steroids also have anabolic activities that are manifested by stimulating increases in protein production, enhancing the uptake of amino acids into cells, and inhibiting the glucocorticoids from breaking down proteins. They influence embryonic development, especially the differentiation of the central nervous system and the male reproductive tract. The excitatory function of androgens occurs at puberty, during which the reproductive organs are activated to produce sex cells. Androgens also maintain the body's sexual characteristics in the adult. Thus, in cases of androgen deficiency, there is a regression of male sexual behavior, libido, and reproductive function; this regression is reversible with treatment.

It should be noted that anabolic steroids are frequently abused because of these kinds of effects. Athletes and bodybuilders searching for accelerated muscle building or physical definition are two groups in which such abuse has been seen. One complication is that the stimulatory effects of the anabolic steroids make the drug administrations reinforcing and therefore loaded with addiction potential. As such, frequent users of anabolic steroids should be aware of conditions such as dependence, withdrawal, and other problematic side effects of heavy steroid use. Such effects may include increased periods of sleep disturbance, paranoia, anger and agitation, mood swings or instability of mood, violence or other impulsive behavior, and concentration and memory disturbance. Physical problems may include severe acne, jaundice, excess water retention, decreased sperm count, high cholesterol, liver problems, and difficulty with blood sugar control.

The condition known as Addison's disease is caused by a failure of the adrenal gland to secrete adequate amounts of both glucocorticoids and mineralocorticoids. The symptoms of this disease are imbalances in body levels of sodium and potassium, dehydration, reduced blood pressure, rapid weight loss, and generalized weakness. A person with Addison's disease will die if not treated with corticosteroids because of the severe electrolyte imbalance and dehydration.

Another disorder, Cushing's syndrome, results when the adrenal gland secretes corticosteroids in excessive quantities. Symptoms include high blood pressure, protein and carbohydrate metabolism alterations, high blood sugar concentrations, and muscular weakness. A tumor often causes this syndrome in the adrenal gland that promotes the secretion of corticosteroids. Surgical intervention is often used to remove the portion of the gland that is malfunctioning. Symptoms similar to those seen in Cushing's syndrome are found in people with inflammatory diseases who receive lengthy treatments with corticosteroids to reduce the inflammation.

Adrenogenital syndrome results from an excessive level of sex steroids, and adrenal gland hyperactivity is usually the main cause. Androgen is the major sex steroid involved in this clinical condition, which causes premature puberty and enlarged genital sex organs in young children. Other characteristics are increased amounts of body and facial hair and deepening of the voice.

The greatest use of estrogens as therapeutic agents is oral contraception or birth control pills. This method is convenient, reversible, and relatively inexpensive; its use is worldwide and includes approximately 16 percent of American women of childbearing age and 25 percent of all contraceptive users. Most oral contraceptives are active combinations of estrogen and progesterone. Users take a daily pill containing both steroids for twenty or twenty-one days of the menstrual cycle and then a placebo for seven or eight days. Withdrawal bleeding occurs two to three days after discontinuing the pill. The mechanism for the effectiveness of these steroids involves inhibiting the release of hormones that would normally stimulate ovulation. Hormones were also once the primary treatment for menopausal symptoms. However, the Women's Health Initiative Study showed that while hormone therapy had some benefits, it also increased the risk for blood clots, breast cancer, heart attacks, and strokes.

Antiandrogens are substances that prevent or depress the action of androgens, or testosterone, on the body. They are of value in managing patients whose bodies are producing abnormally high levels of androgens, who are undergoing premature puberty, or who are affected with acne, hirsutism (excessive hairiness, especially in women), and certain tumors or neoplasms. Potentially, these drugs can be utilized to cause sterility in males.

Natural body androgens stimulate the growth of the prostate gland in males and enhance the proliferation of many prostate cancers. Treatment of abnormal growths, malignant cancers, or benign tumors in the male prostate gland has frequently used natural or synthetic steroids. Estradiol, a form of the female sex steroid estrogen, is used to control the advancement of prostate carcinoma in some males and can induce remission in 50 to 80 percent of the cases of prostate tumors. Estrogens exert their effect by interfering with androgen production or inhibiting androgen-responsive tissues' function. Thus, in some cases, estrogens inhibit abnormal cellular growth. A manufactured synthetic drug, cyproterone acetate, is also used to treat benign prostatic enlargement in men. Cyproterone acetate is very effective in treating prostate cancers and tumors. It does not have the feminizing side effects of the estrogens; however, it does cause inhibition of sperm production and loss of sexual drive.

PERSPECTIVE AND PROSPECTS

The use of steroids as therapeutic agents began in the early 1930s. At that time, Philip Showalter Hench, who was working in the Mayo Clinic, noticed that the symptoms of arthritic women were alleviated when they became pregnant. He suggested that increased secretions from the adrenal cortex might be the responsible agents. Later, clinical trials were conducted to test the role of corticosteroids in treating acute arthritis. With the use of adequate dosages, the clinical response was impressive. The 1950 Nobel Prize in Physiology or Medicine was awarded to Hench and his coworkers to find that cortisone was effective in treating arthritis.

Pharmaceutical firms have manufactured numerous steroid derivatives, all of which have different effectiveness levels as glucocorticoids, mineralocorticoids, or sex steroids. Organic chemists synthesize adrenal steroid analogs to create compounds that produce heightened biological effects with minimal or no side effects. As a consequence, hundreds of different steroids are available. Most of these are characterized according to their biological effectiveness, such as reducing inflammation or inhibiting the immune system. When determining a course of treatment, a physician chooses a particular steroid that enhances desired effects and has minimal effects in related areas. The skin very poorly absorbs some pharmaceutical derivatives of adrenal steroids. These derivatives are especially useful to apply to the skin when a maximal local effect is desired without a generalized effect on other body regions.

Steroids, however, can also have negative and sometimes dangerous effects on the body, whether they are ingested (cholesterol) or injected (anabolic steroids). Evidence indicates that high blood cholesterol levels are associated with an increased risk of atherosclerosis, a clinical condition in which localized plaques (or atheromas) build up in the walls of arteries, reducing blood flow. Atheromas serve as locations for blood clot formation, further blocking the blood supply to a vital organ such as the heart, brain, or lung. High blood cholesterol may result from a diet rich in cholesterol and saturated fat, or it may result from an inherited condition in which affected individuals have extremely high cholesterol concentrations regardless of their diet. In the latter case, affected persons usually suffer heart attacks during childhood.

Cholesterol is found in foods that are based on animal products. Cholesterol-rich foods include most meats, eggs, and dairy products such as cheese, cream, and butter. Humans readily absorb cholesterol from dietary sources. Most Western diets contain 400 to 600 milligrams of cholesterol per day. About 75 percent is readily absorbed into the bloodstream from the dietary tract. Cholesterol is carried to the arteries by proteins in the blood plasma called "low-density lipoproteins" (LDLs). A given cell may engulf the LDLs and use the cholesterol for different purposes. The LDLs in a given location may stimulate other cells to secrete growth factors that either begin or contribute to atheroma development. Thus, the risk of atherosclerosis is greatly increased. Most people can significantly lower their blood cholesterol levels through controlled exercise and diet. Since saturated fat raises blood cholesterol levels, many health experts recommend limiting their intake of foods such as fatty meat, egg yolk, and liver. Fat should contribute less than 30 percent to the total calories, though others have challenged this assumption.

The use and abuse of anabolic steroids to increase muscle mass and strength is widespread in both amateur and professional sports. Although those promoting steroid use claim increases in muscle mass, strength, and endurance, controlled clinical trials show minimal, if any, enhancement of muscle mass and strength. Testosterone may also enhance training efforts by promoting aggressive behavior. The use of these compounds poses ethical questions. It increases the risk of serious toxicity because of the extremely high doses administered, often as much as one hundred times the usual therapeutic dosages.

—Sharon W. Stark, RN, APRN, DNSc,
Nancy A. Piotrowski, PhD, and
Roman J. Miller, PhD

Further Reading

"Anabolic Steroid (Oral Route, Parenteral Route)." *Mayo Clinic*. Mayo Foundation for Medical Education and Research, 1 Dec. 2015. Accessed 11 May 2016.

"Anabolic Steroids." *MedlinePlus*. National Library of Medicine, National Institutes of Health, 9 Mar. 2016. Accessed 11 May 2016.

Craig, Charles R., and Robert E. Stitzel, editors. *Modern Pharmacology with Clinical Applications*. 6th ed., Lippincott, 2004.

Guyton, Arthur C., and John E. Hall. *Human Physiology and Mechanisms of Disease*. 6th ed., W. B. Saunders, 1997.

Henry, Helen L., and Anthony W. Norman, editors. *Encyclopedia of Hormones*. 3 vols. Academic Press, 2003.

Melmed, Schlomo, et al., editors. *Williams Textbook of Endocrinology*. 14th ed., Elsevier, 2019.

Montgomery, Rex, et al. *Biochemistry: A Case-Oriented Approach*. 6th ed., Mosby, 1996.

"Steroids." *MedlinePlus*. National Library of Medicine, National Institutes of Health, 25 Mar. 2016. Accessed 11 May. 2016.

Tortora, Gerard J., and Bryan Derrickson. *Principles of Anatomy and Physiology*. 16th ed., John Wiley and Sons, 2020.

Zelman, Mark, et al. *Human Diseases: A Systemic Approach*. 7th ed., Pearson, 2010.

Synthetic Antibodies

Category: Immunology
Anatomy or system affected: Blood, cells, circulatory system, immune system, lymphatic system
Specialties and related fields: Biochemistry, biotechnology, immunology, pharmacology
Definition: synthetic antibodies are artificially produced replacements for natural human antibodies; they are used to treat various illnesses and promise to continue to be an important part of medical technology in the future

KEY TERMS

antibody: a protein molecule that binds to a substance to remove, destroy, or deactivate it
antigen: a toxin or other foreign substance which induces an immune response in the body, especially the production of antibodies
B cells: white blood cells that produce antibodies
monoclonal antibodies: identical antibodies produced by identical B cells

THE DEVELOPMENT OF ANTIBODY THERAPY

Natural antibodies are protein molecules produced by white blood cells known as B cells in response to the presence of foreign substances. A specific antibody binds to a specific substance, known as an antigen, in a way that renders it harmless or allows it to be removed from the body or destroyed. A person will produce antibodies naturally upon exposure to harmless versions of an antigen, a process known as active immunization. Active immunization was the first form of antibody therapy to be developed and is used to prevent diseases such as measles and polio.

The oldest method of producing therapeutic antibodies outside the human body is passive immunization. This process involves exposing an animal to an antigen to develop antibodies to it. The antibodies are separated from the animal's blood and administered to a patient. Passive immunization is used to treat diseases such as rabies and diphtheria. A disadvantage of antibodies derived from animal blood is the possibility that the patient may develop an allergic reaction. Because the animal's antibodies are foreign substances, the patient's antibodies may treat them as antigens, leading to fever, rash, itching, joint pain, swollen tissues, and other symptoms. Antibodies derived from human blood are much less likely to cause allergic reactions than antibodies from the blood of other animals. This led researchers to seek a way to develop synthetic human antibodies.

A breakthrough in the search for synthetic antibodies was made in 1975 by Cesar Milstein and Georges

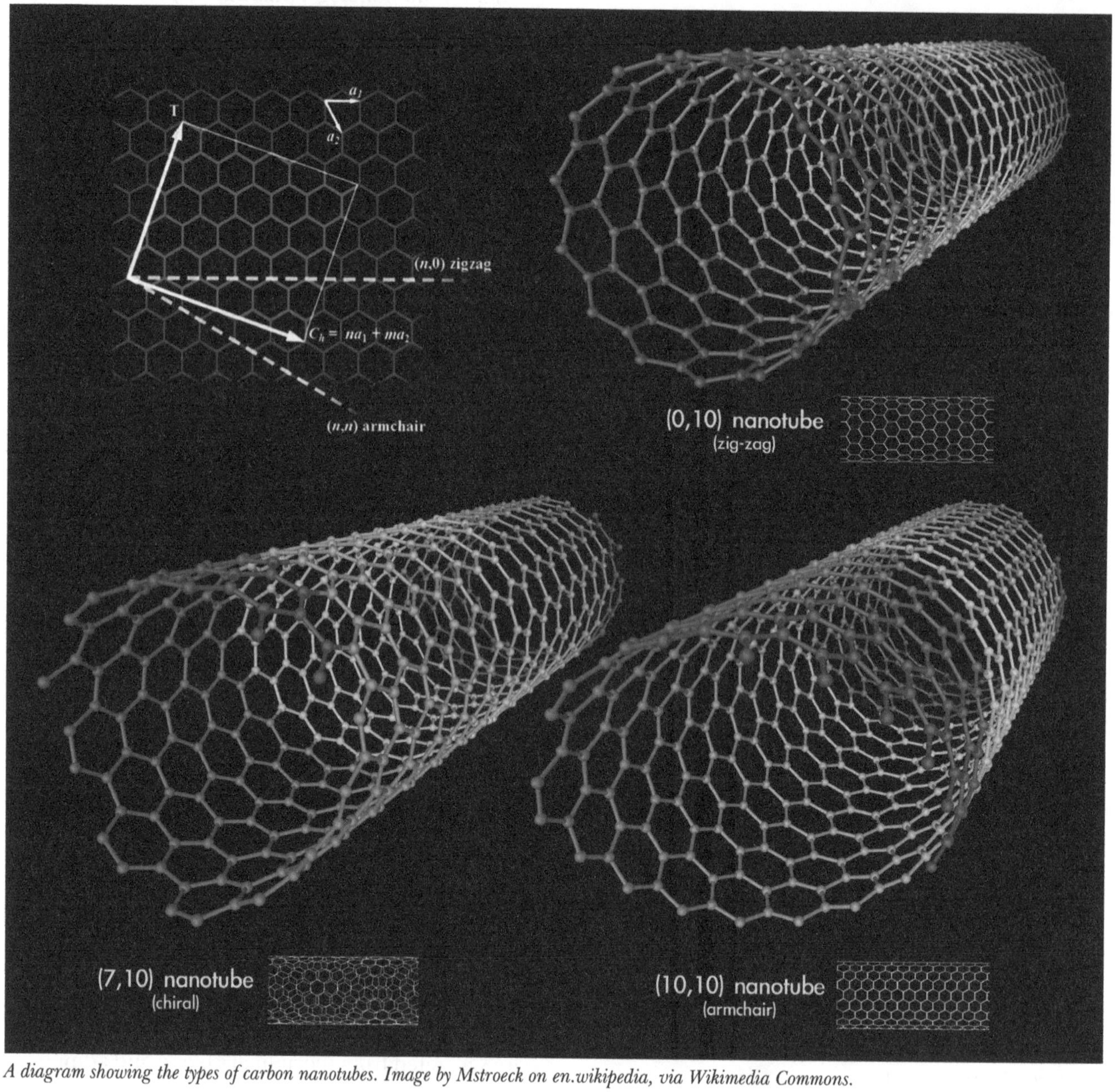

A diagram showing the types of carbon nanotubes. Image by Mstroeck on en.wikipedia, via Wikimedia Commons.

Köhler. They developed a technique that allowed them to produce a specific antibody outside the body of a living animal. This method involved exposing an animal to an antigen, causing it to produce antibodies. Instead of obtaining the antibodies from the animal's blood, they obtained B cells from its spleen. These cells are then combined with abnormal B cells known as myeloma cells. Unlike normal B cells, myeloma cells can reproduce identical copies of themselves unlimited times. The normal B cells and the myeloma cells fuse to form cells known as hybridoma cells. Hybridoma cells can reproduce an unlimited number of times and can produce the same antibodies as the B cells. Those hybridoma cells that produce the desired antibody are separated from the others and allowed to reproduce. The antibodies produced this way are known as monoclonal antibodies.

Because human B cells do not normally form stable hybridoma cells with myeloma cells, B cells from mice have been used. Because mouse antibodies are not identical to human antibodies, they may be treated as antigens by the patient's antibodies, leading to allergic reactions. During the 1980s and 1990s, researchers began to develop methods of producing synthetic antibodies that were similar or identical to human antibodies. An antibody consists of a variable region, which binds to the antigen, and a constant region. The risk of allergic reactions can be reduced by combining variable regions derived from mouse hybridoma cells with constant regions from human cells. The risk can be further reduced by identifying the exact sites on the mouse variable region necessary for binding and integrating these sites into human variable regions. This method produces synthetic antibodies that are very similar to human antibodies.

Other methods exist to produce synthetic antibodies that are identical to human antibodies. A species of virus known as the Epstein-Barr virus can be used to change human B cells so that they will fuse with myeloma cells to form stable hybridoma cells that produce human antibodies. Another method involves genetic engineering to produce mice with B cells that produce human antibodies rather than mouse antibodies. One of the most promising techniques has been the creation of libraries of synthetic human antibodies. This process uses the polymerase chain reaction (PCR) to produce multiple copies of the genetic material within B cells. This genetic material contains the information that results in proteins that come together to form antibodies. Researchers can produce millions of different antibodies by causing these proteins to be produced and allowing them to combine at random. The antibodies are then tested to detect those that bind to selected antigens.

RECOMBINANT ANTIBODIES

Advances in molecular genetic techniques and the characterization of the genes for the variable and constant regions of antibody molecules have made it possible to produce new forms of monoclonal antibodies. The generation of recombinant antibodies does not depend on immunizing animals. Instead, it utilizes combinations of antibody genes to make libraries of recombinant antibodies that are displayed on screening platforms. These screen platforms include bacteriophage, yeast, and in vitro translation systems.

Geneticists discovered that genes inserted into the genes for fibers expressed on the surface of bacterial viruses called "bacteriophages" are expressed and detectable as new protein sequences on the surface of the bacteriophage. Investigators working with antibody genes found that they could produce populations of bacteriophage expressing combinations of antibody-variable genes. Molecular genetic methods have made it possible to generate populations of bacteriophage expressing different combinations of antibody-variable genes with frequencies approaching the number present in an individual mouse or human immune system. The bacteriophage population can be screened for binding to an antigen of interest. The bacteriophage expressing combinations of variable regions binding to the antigen can be multiplied and used to generate recombinant antibody molecules in culture.

Advances in phage display technology generated even more valuable antibodies. However, protein chemists discovered that random mutagenesis of the isolated antibody gene could also be used to derive a panel of mutant binding sites with higher affinity binding than the antibody detected in the initial screening.

Recombinant deoxyribonucleic acid (DNA) technology has also made it possible to modify the procedures for immunization and production of human monoclonal bodies. A process referred to as DNA immunization involves introducing the gene for the target antigen in a form that results in the expression of the protein and an immune response against it. Also, mice with their immunoglobulin genes replaced by the corresponding human genes can be immunized to produce human monoclonal antibodies.

Researchers have also experimented with introducing antibody genes into plants, resulting in plants that produce quantities of the specific antibodies. Hybridomas or bacteriophages expressing specific antibodies of interest may be a potential source of the antibody gene sequences introduced into these plant antibody factories.

MONOCLONAL ANTIBODIES IN PROTEOMICS

Coincident with the development of genomic methods for the determination of gene expression at the ribonucleic acid (RNA) level has been interested in detecting relative protein expression levels. The incorporation of monoclonal antibodies into microarrays that compare the expression of proteins from different cells or tissues has since been developed. As this technology advances, it will likely become an integral part of basic research and clinical assays.

IMPACT AND APPLICATIONS

Some synthetic antibodies are used to help prevent the rejection of transplanted organs. An antibody (Digibind) that binds to the heart drug digoxin can be used to treat overdoses of that drug. Antibodies attached to radioactive isotopes are used in certain diagnostic procedures. Synthetic antibodies have also been used in patients undergoing a heart procedure known as a percutaneous transluminal coronary angioplasty (PTCA). The use of a particular synthetic antibody has been shown to reduce the risk of having one of the blood vessels that supply blood to the heart shut down during or after a PTCA. Researchers also hope to develop synthetic antibodies to treat acquired immunodeficiency syndrome (AIDS) and septic shock, a syndrome caused by toxic substances released by certain bacteria.

The most active area of research involving synthetic antibodies in the 1990s was cancer treatment. On November 26, 1997, the US Food and Drug Administration approved a synthetic antibody for use in non-Hodgkin's lymphoma, a cancer of the white blood cells. It was the first synthetic antibody approved for use in cancer therapy. Today, a host of synthetic antibodies treat many different diseases. As a few examples, alemtuzumab (Campath) treats some patients with chronic lymphocytic leukemia (CLL). Alemtuzumab binds to a cell surface protein called "CD52," found on lymphocytes and leukemia lymphocytes. After it binds to the cancer cells, alemtuzumab attracts immune cells to destroy them. Another synthetic antibody is trastuzumab (Herceptin). This synthetic antibody binds to the HER2 protein, a mutant version of the epidermal growth factor receptor that drives uncontrollable cell growth and proliferation when activated. Breast and stomach cancer cells express large amounts of this protein on their surfaces. When activated, HER2 drives these cells to grow. When trastuzumab binds to HER2, it stops this receptor from becoming active, squelching cell growth. These are but a few of many examples of synthetic antibodies that treat various diseases.

Antibodies are important for immune system protection against pathogens. They are also important tools in biological research, as with antibodies to help determine the structure of cellular proteins. There may be a role for synthetic antibodies in determining the structure of RNA as well. RNA is the genetic material that works with DNA, and DNA is the master genetic code, but the role of RNA is a crucial one. More information on RNA structure may unveil more information on RNA function.

New approaches to therapeutic synthetic antibodies are in development. One approach involves the production of symphobodies. Symphobodies are made up of several different synthetic antibodies attacking a variety of antigens on the same target. These targets can be tumor or cancer cells or whole organisms such as viruses. Multiple synthetic antibody attacks on the same pathogen could result in more effective treatment.

—*Rose Secrest, MD, Richard P. Capriccioso, MD, and Michael A. Buratovich, PhD*

Further Reading

Doerr, Allison. "RNA Antibodies: Upping the Ante." *Nature Methods,* vol. 5, no. 2, 2008, p. 220.

Inbar, Noa Harel, and Itai Benhar. "Selection of Antibodies from Synthetic Antibody Libraries." *Archives of Biochemistry and Biophysics,* vol. 526, no. 2 (special issue), n.d., pp. 87-98. *Biological Abstracts*. Accessed 4 Sept. 2014.

Miersch, S., and S. S. Sidhu. "Synthetic Antibodies: Concepts, Potential and Practical Considerations." *Methods*, vol. 57, no. 4, 2012, pp. 486-98. *Biological Abstracts*. Accessed 4 Sept. 2014.

"The Next Generation of Antibody Therapeutics?" *Pharmaceutical & Diagnostic Innovation,* vol. 4, no. 11, 2006, pp. 13-14.

Sidhu, Sachdev, and Clarence Ronald Geyer. *Phage Display in Biotechnology and Drug Discovery*. 2nd ed., CRC Press, 2015.

Tiller, Tom, editor. *Synthetic Antibodies: Methods and Protocols*. Humana, 2017.

Weber, Marcel, et al. "A Highly Functional Synthetic Phage Display Library Containing over Forty Billion Human Antibody Clones." *Plos ONE,* vol. 9, no. 6, 2014, pp. 1-9. *Academic Search Complete*. Accessed 4 Sept. 2014.

T LYMPHOCYTES

Category: Immune response

Anatomy or system affected: Blood, bone marrow, immune system, lymphatic system, thymus

Specialties and related fields: Bacteriology, hematology, immunology, oncology, virology

Definition: specialized white blood cells that are essential components of the immune system

KEY TERMS

antigen: a toxin or other foreign substance that induces an immune response

antigen presentation: a process by which antigen-presenting cells introduce protein antigens to lymphocytes in the form of short peptide fragments

antigen-presenting cell: a varied collection of immune cells including dendritic cells, macrophages, and B lymphocytes that process and present antigens for recognition by T lymphocytes

histocompatibility leukocyte antigens: cell surface molecules expressed by antigen-presenting cells that present antigenic peptides to T lymphocytes; also known as major histocompatibility complex (MHC) proteins

lymphocyte: a type of white blood cell that has a single round nucleus, and resides, mostly, in the lymphatic system.

T-cell receptor: a T-lymphocyte-specific cell surface protein that recognizes and binds specific antigens

T-cytotoxic cell: a subtype of T lymphocyte that has the CD8 cell surface glycoprotein, and attacks and destroys tumor cells and viral-infected cells

T-helper cell: a subtype of T lymphocyte that secretes signaling molecules called "lymphokines" that induce the maturation and activation of B lymphocytes, macrophages, and T-cytotoxic cells

thymus: a lymphoid organ that lies just over the upper part of the heart and produces mature T cells for the immune system

INTRODUCTION

T lymphocytes are white blood cells of the adaptive immune system that help fight disease. There are several subgroups of T lymphocytes with different immune system functions: cytotoxic T lymphocytes (CTLs), helper T lymphocytes, memory T lymphocytes, regulatory T lymphocytes, and tumor-infiltrating T lymphocytes (TILs). Laboratory manipulation of TILs makes them useful for cancer treatments.

FUNCTION

Some background on the immune system is helpful in understanding how T lymphocytes function and the role they play in fighting infections. The immune system is a complex of tissues, cells, and chemicals produced by these cells that spread throughout the body. Its function is to rid the body of foreign microbes and damaged or abnormal body cells (such as cancer cells). The major organs of the immune system are the bone marrow, lymph nodes, thymus, and spleen. Immune system cells and a clear fluid called "lymph" move throughout the body in a series of channels called "lymphatic vessels" that are separate from the blood-based circulatory system.

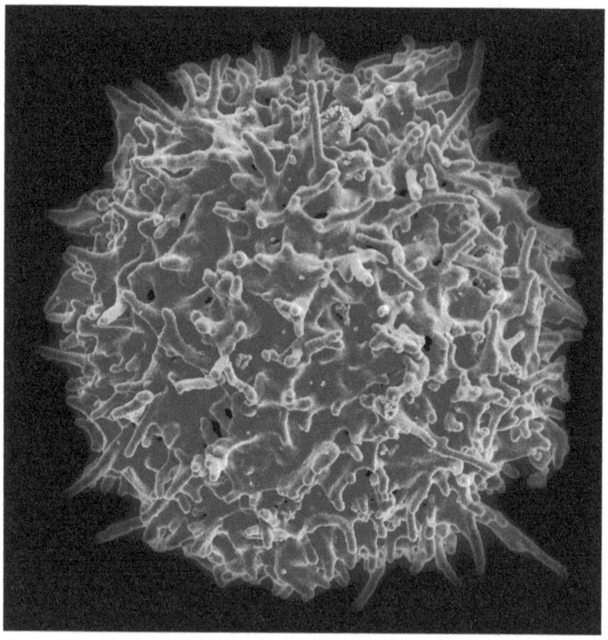

Scanning electron micrograph of a human T lymphocyte (also called a T cell) from the immune system. Photo via NIAID/Wikimedia Commons. [Public domain.]

The immune system has two divisions, the innate immune system, and the adaptive immune system. The innate system consists of white blood cells (leukocytes) that respond rapidly to a wide range of microbes but that cannot confer immunity against disease. While a cell in the innate system can attack many different kinds of microbes, each cell of the adaptive system recognizes and attacks only one type of microbe or damaged cell. T lymphocytes and B lymphocytes are the major cells of the adaptive immune system. Adaptive system cells respond more slowly when first encountering a microbe but can "remember" exposure to that microbe in a way that creates long-lasting immunity to a particular disease.

All cells have marker molecules on their surface that act as a "uniform" that identifies them as "self cells" or "nonself cells." Self markers do not cause an immune response in a healthy person. Nonself marker molecules, also called "antigens," stimulate an immune system reaction. As a distinctive feature of the adaptive immune system, every T lymphocyte or B lymphocyte develops a unique receptor on its surface that will interact only with one highly specific nonself antigen.

DEVELOPMENT

T lymphocytes originate in bone marrow from hematopoietic stem cells, which are cells that have the potential to differentiate into many kinds of blood cells. Immature T lymphocytes leave the bone marrow and travel to the thymus, an organ behind the breastbone. Here they undergo a complicated maturation process that results in 2 major types of T lymphocytes. One type has the CD8 protein on its surface and is called a "$CD8^+$ cell." The other carries the CD4 protein and is called a "$CD4^+$ cell." In addition to these proteins, every one of these cells has a different T-cell-specific antigen receptor on its surface.

While in the thymus, the pre-T lymphocytes rearrange their gene segments for the T-cell receptor. The T-cell receptor consists of two polypeptides, an alpha, and a beta chain. The genes that encode the alpha protein are on chromosome 14 and those that encode the beta chain are on chromosome seven. Early in its sojourn in the thymus, the pre-T lymphocyte rearranges its genes for the beta protein. The beta chain gene cluster consists of approximately 52 variable or V gene segments, two D or diversity gene segments, and about 13 J or joining gene segments. In response to interleukin-2 (IL-2) and interleukin-7 (IL-7), secreted by the thymic epithelium, the pre-T lymphocyte synthesizes two proteins, RAG-1 & -2, that cut and paste a specific D segment with a single J segment. Then the pre-T lymphocyte makes another enzyme called the "V(D)J recombinase" that attaches a single V gene segment to the joined DJ segments. The joining of the VDJ segment to the constant region (μ segment) produces a mature beta-chain gene.

Once a mature beta chain of the T-cell receptor appears on the cell surface, the pre-T lymphocyte expresses the CD4 and CD8 proteins on its surface. It is called a "double positive" or DP cell at this point in its development. DP cells begin to rearrange the genes for the alpha subunit of the T-cell receptor. The alpha gene cluster contains between 70 to 80 copies of the V segment, no D segments, and about 61 copies of the J segment. The joining of a single V segment with a single J segment and a constant region produces a mature alpha chain of the T-cell receptor. The pairing of the mature alpha and beta chains of the T-cell receptor constitutes a functional T-cell receptor. The joining of multiple V with multiple J regions allows for extensive mixing and matching of distinct gene segments. Such wide variability allows T cells to deploy a cadre of receptor proteins that can recognize a kaleidoscope of different antigens.

Any pre-T lymphocytes whose T-cell receptors to molecules on healthy tissue cells die before they leave the thymus so as not to harm the body—a process called "negative selection." Likewise, any pre-T lymphocytes that fail to rearrange their T-cell receptor genes and produce a functional T-cell receptor also undergo programmed cell death. However, all T-cell receptors must interact with human leukocyte antigens (HLAs) on the surfaces of dendritic cells in the thymus. If they fail to bind to HLAs, the pre-T lymphocytes also die, an example of positive selection. The pre-T lymphocyte then stops expressing either CD8 or CD4 and becomes a CD4-positive T-helper cell or a CD8-expression T-cytotoxic cell. Mature T lymphocytes exit the thymus and move to the peripheral circulation or lymphatic system.

ACTIVATION

To activate T lymphocytes, dendritic cells, a type of antigen-presenting cell that constantly circulates through the body, present antigen to the T lymphocyte. The dendritic cell finds a foreign microbe or an abnormal cell, engulfs or "eats" the cell, and takes some of the engulfed cell's proteins, and traffics them to their surfaces, bound to an MHC protein. This process is called "antigen presentation." The first step of activation occurs if the T-cell receptor on the T lymphocyte surface fits the antigen presented on the surface of the antigen-presenting cell. However, T lymphocyte activation requires a second signal. T lymphocytes also have on their surfaces a protein called "CD28." CD28 must bind to the B7 protein complex on the surface of the antigen-presenting cell. This dual signal tells the T lymphocyte that the right cell has presented the antigen and has approved T lymphocyte activation. In the absence of either signal, the T lymphocyte remains inactivated.

The activated T lymphocyte produces IL-2 and a fully functional IL-2 receptor. IL-2 is a potent T-lymphocyte growth factor, and the T lymphocyte divides vigorously, a phenomenon called "clonal expansion." IL-2 expression also activates nearby CD8+ CTLs that recognize the same antigen. Dendritic cells can, likewise, present antigens to CD8+ cells and activate them. The CD8+ cell undergoes clonal expansion, dividing rapidly to produce a large number of identical cells.

After leaving the thymus, CTLs move through the body, looking for microbes or abnormal cells that match their T-cell receptors. When a CTL finds a match, it attaches to the cell and releases chemicals that kill this rogue cell. CTLs play an integral role in fighting viral infections.

A few CD8+ cells become memory T lymphocytes instead of CTLs. Memory cells remain in the body for many years and are the basis for long-term immunity. If the same microbe invades the body again, memory T lymphocytes and memory B lymphocytes respond quickly to control the invader before the individual becomes sick.

When a dendritic cell presents an antigen to a CD4+ T lymphocyte whose receptor matches its foreign surface protein, the CD4+ cell is activated and becomes either a helper T lymphocyte or a regulatory T lymphocyte. There are at least three subgroups of helper T lymphocytes: Th1, Th2, and Th17. Helper cells cannot kill other cells directly. Instead, they secrete chemicals that coordinate the immune system response. One subset of helper cells, Th2, secretes cytokines that stimulate B lymphocytes to produce antibodies against foreign or abnormal cells. Another, Th1, helps CTLs become more effective killers. In addition, some T-helper cells, so-called T follicular helper or Tfh cells, establish memory B cells. Other T-helper cells become memory T lymphocytes.

Some CD4+cells are exposed to a cytokine called "transforming growth factor-beta" (TGF-ß) and become regulatory T lymphocytes. Regulatory T lymphocytes monitor the activity of other T lymphocytes, help prevent immune response against normal body cells, and wind down the immune response once a microbe has been eliminated.

Both CD4+ and CD8+ cells can become tumor-infiltrating T lymphocytes (TILs). Researchers are still learning about TILs and how they can be used to fight cancer.

TREATMENT

The body is exposed to thousands of different microbes each day, and a healthy immune system effectively eliminates most of them before an individual becomes sick. The immune system, however, has challenges in eliminating cancer cells. Cancer cells start as normal body cells with all the surface proteins that identify them as self cells. At some point, these normal cells transform into cancer cells, but they often continue to have enough of the surface proteins that originally marked them as self cells to escape detection by dendritic cells and T lymphocytes. If the immune system does respond to transformed cells, the response may not be effective since the growth of cancer cells is not limited the way it is in healthy cells. In addition, some cancer cells make chemicals that turn off or disrupt the functioning of immune system cells.

As of 2015, the US Food and Drug Administration (FDA) has approved only 1 T-lymphocyte cancer treatment vaccine. Unlike preventative vaccines, patients receive treatment vaccines after they already have cancer. The drug, called "Sipuleucel-T," is used to treat advanced prostate cancer. Scientists take cells from the patient's immune system and expose them in the laboratory to chemicals that turn them into dendritic cells with a special molecule artificially attached to stimulate a strong immune response against prostate cancer cells. The drug does not cure the cancer but can prolong life.

Another approach to treating cancer uses TILs and manipulates them in the laboratory to fight certain cancers. The process is called "adoptive cell therapy" (ACT), and it was first used to successfully treat metastatic melanoma. A surgeon excises a portion of the patient's melanoma tumor that contains TILs. In the laboratory, the tumor is minced and grown in culture with interleukin 2 (IL-2), a cytokine that stimulates

T-cell growth. In a few weeks, the T lymphocytes destroy the tumor leaving a culture of billions of TIL cells—many more than the body could make—that is reintroduced into the patient. Since these cells are clones of self TILs, the immune system does not reject them, and they can infiltrate and successfully attack metastatic melanoma tumors.

Originally, ACT worked only with melanoma tumors, but researchers are working on perfecting ways to use a harmless virus to genetically engineer the patient's own TIL cells to make them more effective against difficult to treat cancers such as acute lymphoblastic leukemia. The cells are taken from the patient, manipulated and cloned in the laboratory, and then re-introduced into the patient. This process is called "chimeric antigen receptor" (CAR) T-cell therapy. The results are promising. The FDA has approved several CAR T-cell treatments, including Abecma (idecabtagene vicleucel), Breyanzi (lisocabtagene maraleucel), Kymriah (tisagenlecleucel), Tecartus (brexucabtagene autoleucel), and Yescarta (axicabtagene ciloleucel). Kymriah is approved for the treatment of acute lymphoblastic leukemia. Yescarta, Kymriah, and Breyanzi are approved for the treatment of B-cell lymphoma. Yescarta is also approved for the treatment of follicular lymphoma. Tecartus is approved for mantle cell lymphoma, and Abecma is approved for the treatment of multiple myeloma. CAR T-cell treatments are not without their caveats. Patients can suffer from high fevers and dangerously low blood pressure, brain swelling, confusion, seizures, and severe headaches. CAR T-cell therapy is being tested in clinical trials against many kinds of cancer. A searchable list of current clinical trials can be found at www.clinicaltrials.gov.

—Tish Davidson, AM and Michael A. Buratovich, PhD

Further Reading

Abbas, Abul K., Andrew H. Lichtman, and Shiv Pillai. *Cellular and Molecular Immunology*. 10th ed., Elsevier, 2021.

Burton, Thomas. "Immunotherapy Treatments for Cancer Gain Momentum." *Wall Street Journal*, 12 Oct. 2017, www.wsj.com/articles/immunotherapy-treatments-for-cancer-gain-momentum-1507825152?mod=searchresults_pos17&page=1.

Kershaw, Michael H., J. A. Westwood, C. Y. Slaney, and P. K. Darcy. "Clinical Application of Genetically Modified T Cells in Cancer Therapy." *Clinical and Translational Immunology*, vol. 3, 2014, p. e16., www.nature.com/cti/journal/v3/n5/full/cti20147a.html.

Man, Yang-gao, et al. "Tumor-Infiltrating Immune Cells Promoting Tumor Invasion and Metastasis: Existing Theories." *Journal of Cancer*, vol. 4, no. 1, 2013, pp. 84-95.

Martin, Seamus, Dennis R. Burton, Ivan M. Roitt, and Peter J. Delves. Roitt's Essential Immunology. 13th ed., Wiley-Blackwell, 2017.

Pardoll, Drew. "T Cells Take Aim at Cancer." *Proceedings of the National Academy of Sciences*, vol. 99, no. 25, 2012, pp. 15840-42, www.pnas.org/content/99/25/15840.full.

VACCINES

The final section of this volume discusses the vaccines that protect us from childhood and adult diseases. Vaccines prevent untold human suffering. This section contains entries that discuss commonly available vaccines as well as a few experimental ones. Among the vaccines covered are those for cancer, chickenpox, COVID-19, hepatitis, typhoid, and rabies.

Adenovirus and Adenovirus-Based Vaccines

Category: Prevention

Anatomy or system affected: Immune system, lymphatic system

Specialties and related fields: Biotechnology, immunology, preventive medicine, public health, virology

Definition: vaccines made from live, attenuated (weakened) adenoviruses that prevent infection with some types of adenoviruses, or genetically engineered vaccines made from recombinant adenoviruses that deliver

KEY TERMS

adenovirus vaccine: a nonpathogenic form of an adenovirus that stimulates the formation of a memory immune response to adenovirus infection

adenoviruses: medium-sized, nonenveloped, icosahedral, double-stranded deoxyribonucleic acid (DNA) viruses originally isolated from human adenoid tissue

antigens: a toxin or other foreign substance that induces an immune response in the body, especially the production of antibodies and T lymphocytes

serotype: also known as a "serovar," a distinct variation within a species of bacteria or virus that elicits a peculiar immune response

viral strain: genetic variant of specific viruses that possess distinct characteristics

INTRODUCTION

Adenoviruses usually cause mild illnesses that include colds, sore throats, acute bronchitis (so-called chest colds), diarrhea, and conjunctivitis ("pink eye"). Infrequently, adenoviruses cause serious diseases, and people with weakened immune systems or existing lung or heart disease have an increased risk of developing severe adenovirus-based illnesses. Severe adenovirus infections include severe pneumonia or infections of the brain or spinal cord and may require hospitalization and even cause death.

Adenoviruses spread easily from one person to another. Close personal contact (e.g., touching or shaking hands), aerosols (coughing and sneezing), fomites (handling contaminated objects), handling infected stools (during diaper changing), or contacting contaminated water (e.g., swimming pools) are all possible means of adenovirus transmission. Certain adenovirus strains, Type 4 and Type 7, have caused severe outbreaks of respiratory illness among military recruits. There are eighty-eight different human adenovirus strains.

Adenovirus vaccine prevents infection with some adenovirus strains. This vaccine consists of live human Adenovirus type 4 and type 7 pelleted into a tablet. The adenoviruses in this vaccine are grown in cultured human cells, isolated, and pelleted into a tablet. The final vaccine comprises two enteric-coated tablets, one tablet of Adenovirus Type 4 and one tablet of Adenovirus Type 7. These two tablets constitute a single vaccine dose. The patient must swallow the tablets whole without chewing them to prevent releasing the virus into the upper respiratory tract. The adenovirus vaccine is designed to pass intact through the stomach and release the live virus in the intestine. Therefore, health-care providers should never give it to patients who cannot swallow.

From 1971 to 1996, the US military routinely administered oral adenovirus vaccines to new recruits. Because military recruits live and work in tight spaces, Adenovirus types 4 and 7 readily spread within this population, causing widespread, and sometimes severe, pneumonia. The adenovirus vaccine induces a strong immune response because it presents B cells with viral antigens in their natural conformation. Infection of epithelial cells in the small intestine allowed viral replication but was asymptomatic. The vaccine was unavailable to civilians because most adenovirus infections are mild and self-limiting.

Adenovirus-based vaccines. Ebola virus is a hemorrhagic fever virus. It is spread between people by contact with bodily fluids such as blood. Initial symptoms include fever, headache, muscle pain, and chills. Later, a person may experience internal bleeding resulting in vomiting or coughing blood. This virus interferes with blood clotting, leading to internal bleeding, inflammation, and tissue damage. It kills up to 90 percent of everyone it infects.

Ad26.ZEBOV/MVA-BN-Filo vaccine is a recombinant adenovirus-based Ebola vaccine. It utilizes a Chimpanzee adenovirus type 3 (ChAd3) vector that encodes the Ebola virus Zaire (EBOV-Z) glycoprotein. This vaccine infects human cells but does not replicate in them. Instead, the infected cells make the Ebola virus surface glycoprotein on their surfaces, presenting it to the immune system. The immune system mounts a response against this Ebola protein, protecting against this dangerous virus. Ad26.ZEBOV/MVA-BN-Filo is given as a two-dose vaccine and was approved in 2020 and later put into use. This vaccine with the vesicular stomatitis virus-based *Ervebo*® vaccine (rVSV-ZEBOV) helped end the Ebola epidemic in June 2020.

In 2019, the COVID-19 pandemic began in Wuhan, China, and soon spread worldwide. Three adenovirus-based vaccines designed to protect against (severe acute respiratory syndrome) SARS-CoV-2, the etiologic agent of COVID-19, were developed in Russia, the United States, and the United Kingdom.

The Johnson and Johnson vaccine uses a replication-incompetent recombinant adenovirus type 26 (Ad26) vector. This vector genetically engineers cells to express the SARS-CoV-2 spike protein to elicit an immune response against it. According to data presented at an October 15, 2021, meeting of the FDA Vaccines and Related Biological Products Advisory Committee, a single dose of the Johnson and Johnson vaccine generated durable protection against COVID-19-induced severe disease and hospitalization for at least six months.

The Oxford/AstraZeneca COVID-19 vaccine is marketed under the trade names Covishield and Vaxzevria. This vaccine was developed by a collaboration between Oxford University and the British-Swedish company AstraZeneca. This vaccine employs a modified chimpanzee adenovirus ChAdOx1 vector to engineer cells to express the SARS-CoV-2 spike protein. According to a phase III clinical trial conducted in the United States by AstraZeneca in 2021, Covishield is 74 percent effective at preventing symptomatic COVID-19 and 100 percent efficacy at preventing severe disease and hospitalization.

The Gam-COVID-Vac vaccine was formerly known as the Sputnik V vaccine. It was developed by the Gamaleya Research Institute of Epidemiology and Microbiology in Russia. This vaccine uses two replication-deficient human adenoviruses, Ad26 and Ad5. To prevent an immune response against the vector from preventing its efficacy, each of the two inoculations is with different recombinant adenovirus strains. A large clinical trial in Buenos Aires, Argentina, examined the efficacy of this vaccine. The trial went from December 29, 2020, to May 15, 2021, and enlisted 663,602 participants aged 60 and older. This trial compared Gam-COVID-Vac against Covishield and the Chinese Sinopharm BIBP vaccines. According to the results of this study, Gam-COVID-Vac had an overall 98 percent efficacy against COVID-19-related deaths.

Several adenovirus-based vaccines are in human clinical trials. Adenovirus-based vaccines against HIV, Ebola, influenza, Mycobacterium tuberculosis and Plasmodium falciparum are currently under human clinical trials. Additionally, vaccines in preclinical trials include those against rabies virus, dengue virus, and Middle East respiratory syndrome coronavirus.

T cell vaccine. Many people carry adenovirus without symptoms. Immunodeficiency, induced so that a person can receive a tissue or organ transplant, allows the virus to spread in the infected person and cause severe and potentially lethal disseminated infection.

An adenovirus T-cell vaccine gives those persons cytotoxic T cells (or killer T cells), which recognize and destroy cells infected by adenovirus, or gives them helper T cells, which help cytotoxic T cells perform those functions.

A vaccine can be prepared by selecting and growing donated helper T cells that naturally recognize antigenic fragments of the adenovirus capsid hexon protein. Or a vaccine can be prepared using donated monocytes to train donated naïve cytotoxic T cells to recognize antigenic fragments of all adenovirus capsid proteins. The monocytes are modified by infection with recombinant adenovirus to produce the antigenic fragments. They present the antigenic fragments to the cytotoxic T cells.

IMPACT

The B-cell vaccine previously used against human adenovirus types 4 and 7 decreased adenovirus respiratory disease in military recruits by 82 to 95 percent. The economic value of that health benefit is estimated to be worth $22 million annually.

Clinical trials suggest that T-cell vaccines may decrease the significant risk of illness or death caused by severe disseminated adenovirus infection. Clinical trials will determine if recombinant vaccines can be suitable for the clinic.

Recombinant adenovirus-based vaccines have saved countless lives in Africa during the Ebola outbreak and globally during the COVID-19 pandemic. Further vaccine development with these highly versatile and adaptable vaccine vectors holds remarkable potential.

—David Caldwell, PhD and Michael A. Buratovich, PhD

Further Reading

Chatziandreou, Ilenia, et al. "Capture and Generation of Adenovirus Specific T Cells for Adoptive Immunotherapy." *British Journal of Haematology*, vol. 132, 2006, pp. 117-26.

Food and Drug Administration. "Vaccines and Related Biological Products Advisory Committee October 14-15, 2021 Meeting Announcement." 14-15 Oct. 2021, www.fda.gov/advisory-committees/advisory-committee-calendar/vaccines-and-related-biological-products-advisory-committee-october-14-15-2021-meeting-announcement#event-information. Accessed 10 Dec. 2021.

García-Montero, Cielo, et al. "An Updated Review of SARS-CoV-2 Vaccines and the Importance of Effective Vaccination Programs in Pandemic Times." *Vaccines*, vol. 9, no. 5, 2021, p. 433, doi:10.3390/vaccines9050433.

Heymann, David L., editor. *Control of Communicable Diseases Manual*. 19th ed., American Public Health Association, 2008.

Howley, Peter M., et al., editors. *Fields' Virology*. 7th ed., Lippincott Williams & Wilkins, 2021.

Ledgerwood, Julie E., et al. "Chimpanzee Adenovirus Vector Ebola Vaccine." *New England Journal of Medicine*, vol. 376, 2017, pp. 928-38, doi:10.1056/NEJMoa1410863.

Lundstrom, Kenneth. "Viral Vectors for COVID-19 Vaccine Development." *Viruses*, vol. 13, no. 2, 2021, p. 317, doi:10.3390/v13020317.

Plotkin, Stanley A., et al., editors. *Vaccines*. 7th ed., Elsevier, 2017.

Russell, Kevin L., et al. "Vaccine-Preventable Adenoviral Respiratory Illness in U.S. Military Recruits, 1999-2004." *Vaccine*, vol. 24, 2006, pp. 2835-42.

ANTHRAX VACCINE

Category: Prevention

Anatomy or system affected: Blood, cells, circulatory system, immune system, lymphatic system

Specialties and related fields: Bacteriology, biochemistry, biotechnology, immunology, preventive medicine, public health

Definition: a vaccine that prevents infection with anthrax, a serious and sometimes fatal disease caused by the bacterium *Bacillus anthracis*; the vaccine protects against cutaneous anthrax (the most common form) and inhalation anthrax

KEY TERMS

anthrax: a zoonotic infection caused by *Bacillus anthracis* that comes in an inhaled, cutaneous, and gastrointestinal form

dose: a quantity of a medicine, drug, or vaccine taken at a particular time.

zoonotic infections: infections by microorganisms that typically colonize animals and are transmitted to humans by human-animal contact

HISTORY AND DEVELOPMENT

The anthrax vaccine was licensed in 1970 in the United States after a successful clinical trial of a precursor formulation in mills that processed imported animal hair. Based on this research, experts estimate that the vaccine is 92.5 percent effective. Only one anthrax vaccine is licensed for use in the United States (Anthrax Vaccine Adsorbed, or BioThrax), but new vaccine formulations have been in development.

RECOMMENDATIONS

The Centers for Disease Control and Prevention (CDC) recommends vaccination against anthrax for persons between the age of eighteen and sixty-five years who could be exposed to large amounts of *B. anthracis* as part of their jobs. At-risk jobs include certain types of laboratory or remediation work, work with animals or animal products, and work in specific US Department of Defense-designated occupations (including certain military and associated personnel).

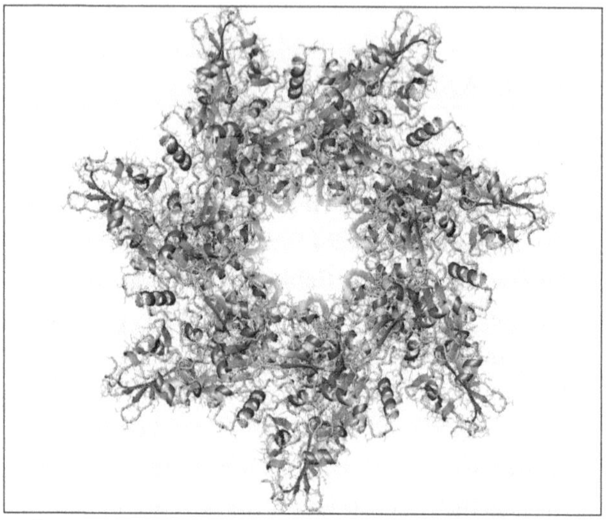

Anthrax toxin protective antigen (fragment) heptamer, Bacillus anthracis. *Image by Astrojan, via Wikimedia Commons.*

Women who are nursing can safely receive the vaccine, and it may be recommended to pregnant women exposed to inhalation anthrax. However, persons with a history of a severe allergic reaction to the vaccine or any other vaccine or vaccine component should not get the vaccine; health-care providers may recommend against vaccination in sick persons and persons with Guillain-Barré syndrome.

ADMINISTRATION

The vaccine is administered in the muscle in a recommended five doses (each 0.5 milliliters [mL]) the first when the risk for exposure is identified. The four follow-up doses are scheduled for week four, and months six, twelve, and eighteen after the first dose. Annual booster vaccinations are recommended to preserve immunity.

The vaccine also can be given to people who have already been exposed to anthrax. In these cases, the vaccine is given under the skin and in a recommended three doses only, each at 0.5 mL (after first exposure, then two and four weeks after the first dose).

SIDE EFFECTS

Minimal risk of serious harm comes with the anthrax vaccine. Severe allergic reactions are infrequent, and when they occur, they appear within the first hour of administration. Signs of a serious reaction that require immediate medical attention include trouble breathing, wheezing, rapid heartbeat, swelling, hives, hoarseness, dizziness, weakness, and paleness. Other potential problems are mild and include tenderness, redness, itching, a lump or bruise at the injection site; muscle aches or limited movability of the arm following injection; headaches; and fatigue.

IMPACT

The use of the anthrax vaccine in persons at risk for anthrax infection is likely to prevent fatalities, which would otherwise be expected in up to 20 percent of cutaneous anthrax cases and the large majority of cases of inhalation anthrax.

—Katherine Hauswirth, MSN, RN

Further Reading

Centers for Disease Control and Prevention. "Anthrax," 20 Nov. 2020, www.cdc.gov/anthrax/index.html.

———. "Anthrax Vaccination: What Everyone Should Know." Nov. 2016, www.cdc.gov/vaccines/vpd/anthrax/public/index.html.

———. "Bioterrorism-Related Anthrax." *Emerging Infectious Diseases*, vol. 8, 2002, pp. 1013-1183.

———. "Use of Anthrax Vaccine in the United States: Recommendations of the Advisory Committee on Immunization Practices." *Morbidity and Mortality Weekly Report*, vol. 49, 2000, pp. 1-20.

Dixon, Terry C., et al. "Anthrax." *New England Journal of Medicine*, vol. 341, 1999, pp. 815-26.

Friedlander, A. M., and S. F. Little. "Advances in the Development of Next-Generation Anthrax Vaccines." *Vaccine*, vol. 27, suppl. 4, 2009, pp. D61-64.

Institute of Medicine. *An Assessment of the CDC Anthrax Vaccine Safety and Efficacy Research Program*. National Academy Press, 2003.

Plotkin, Stanley A., et al., editors. *Vaccines*. 7th ed., Elsevier, 2017.

Antivaccination Movement

Category: Social movements

Anatomy or system affected: Blood, cells, circulatory system, immune system, lymphatic system

Specialties and related fields: Bacteriology, biochemistry, biotechnology, immunology, microbiology, preventive medicine, public health, virology

Definition: a socioreligious movement that opposes vaccination

KEY TERMS

autism spectrum disorder (*ASD*): a neurodevelopmental disorder that starts in early childhood. Symptoms include difficulty interacting with others in normative ways

conspiracy theories: attempts to explain social—often random—events as the secret acts of powerful, evil forces. They tend to include mistrust of government or corporate entities

immunization: a process by which an individual is made resistant to an infectious disease, such as by vaccination. Individuals may also become immune to disease after suffering contagion

opt-outs: express instructions provided to refuse to participate in a service, program, process, or contract

thimerosal: also known as merthiolate; an antiseptic and antifungal agent used in vaccines that was erroneously implicated as a possible cause of ASD in vaccinated children; childhood vaccines have not contained thimerosal since 2001

vaccine: concoctions made from the same organisms that cause a specific disease; they help a person develop immunity to the disease by causing the immune system to respond to a dead or weakened form of a virus without producing full-blown symptoms

INTRODUCTION

The development of vaccines is one of the most significant advances in modern medicine. However, the antivaccine movement has challenged the prevalent assumption that vaccines are a panacea or universal remedy. Concerned antivaccination supporters argue that vaccinations are overused and may be harmful in excess or even in any dosage; many believe conspiracies exist between the pharmaceutical industry, doctors, and governments to hoodwink the public. Landmark court cases from 2008 to 2010 have further nuanced the debate.

OVERVIEW

The inception of modern vaccines dates to the late 1700s, with the development of antismallpox vaccination by Edward Jenner. Since then, there have been those who opposed vaccination, chiefly out of concern that they might cause ills worse than those they purported to cure. Most vaccination opponents in the twenty-first century are parents concerned for their children. They run the gamut from people who want to delay or space out vaccination to those who reject only some types of inoculation to those who refuse any type of vaccination. A rise in the number of people who refuse to vaccinate their children worldwide has caused grave medical and public health concerns. Experts point to the popular spread of antivaccination conspiracy theories as one of the leading causes for declining immunization rates in some societies.

In 1998, *The Lancet,* a respected medical journal, published an article authored by Andrew Wakefield that raised the possibility of a causative link between the Measles, Mumps, and Rubella (MMR) vaccination and autism spectrum disorder (ASD). In a series of investigative reports from 2004-10 in the *Sunday Times,* British investigative journalist Brian Deer established that Wakefield had multiple, undeclared commercial conflicts of interest and had falsified his data. In re-

A protest against COVID-19 vaccination in London, United Kingdom, 2021. Photo via Wikimedia Commons. [Public domain.]

sponse the Deer's reports, the *British Medical Journal* called Wakefield's research article "an elaborate fraud." These revelations led to the longest inquiry ever undertaken by the UK General Medical Council that lasted 217 days. In their verdict, the Council judged Wakefield as "dishonest," "unethical," and "callous." *The Lancet* issued a partial retraction of the Wakefield paper in 2004 and a full retraction in 2010, thus wholly discrediting its research. On May 2, 2010, Andrew Wakefield was "struck off" the UK Medical Registry. Nevertheless, this bogus article was one of the main causes of a decline in vaccinations worldwide. Consequently, according to experts, diseases such as measles, which had been halted in most regions, have become endemic in some areas again. For example, the United Kingdom declared in 2008 that measles had reappeared in its population.

Conspiracy theories—elaborate plots based on unfounded suspicions that nevertheless find common cause with adherents who find them compelling and believable—tend to be based on fear and mistrust of governments, corporations, and other special interests, such as the scientific and medical communities. Antivaccination movement detractors argue that proponents base their concerns on conspiracy theories, namely, that pharmaceutical corporations cover up detrimental information about vaccines or exaggerate their benefits to protect and increase their profits. Proponents can present real historical examples to bolster their position that cover-ups are nothing novel. In the 1940s, the US government experimented with inoculating unsuspecting people in Guatemala with venereal diseases to run a clinical trial. The Tuskegee Syphilis study ran from 1932 to the early 1970s, following the progress of syphilis in hundreds of African Americans, from whom treatment was deliberately withheld even after it became available so that the study might continue. Incontrovertible scientific evidence, however, exists that vaccines are reasonably safe and effective. Vaccines are also necessary for the long-term maintenance of public health.

Since parents face a heavy schedule of routine vaccinations for their young children, they are often more likely to seek information about vaccines on the internet and other media outlets than from medical personnel. Internet sites that disseminate antivaccine theories have been especially influential in dissuading parents from protecting their children from serious childhood diseases with vaccinations.

Pediatrics researchers, on the other hand, urge parents who refuse vaccination for their children to reconsider. In a 2010 pediatrics study showing that unvaccinated children played a role in the spread of a whooping cough epidemic, the researchers argued that vaccination is one of the most outstanding public health achievements in medical history, playing an indispensable role in eradicating smallpox and controlling polio, measles, rubella, and other infectious diseases.

FURTHER INSIGHTS

Vaccines protect inoculated people from various infectious diseases caused by bacteria and viruses. Widespread vaccination led to the eradication of smallpox, one of the deadliest diseases in the world, and has led to significant decreases in rates of many other diseases that were once common. The more vaccinated people in a community, the smaller the risk of infection for the whole community. Community immunity

Image via iStock/kbeis. [Used under license.]

relies on safety in numbers, provided the numbers of vaccinated people are high enough. There are always some unvaccinated individuals in a community for several reasons. For example, vaccination may be potentially dangerous to specific individuals because of personal health issues, and infants may be too young to vaccinate. Suppose the numbers of unvaccinated individuals in a community are low enough. In that case, the unvaccinated may, in a sense, "ride free" on the backs of the vaccinated. Since most people in a community are vaccinated, the spread of disease is halted or controlled. A small number of unvaccinated individuals are unlikely to encounter the disease or spread it to others if they do contract it. When vaccination rates decline from the ideal rate, the safety in numbers effect weakens. In consequence, diseases spread.

Not all communities can reach an ideal immunization rate. Lack of access to immunization is one of the main reasons for mortality in children younger than five in developing countries. Close to 100 percent of these deaths occur in low-income countries. About 70 percent are caused by diseases that are preventable by vaccines widely available in other countries. Moreover, with widespread international travel, new and old diseases can be contracted in one location and spread worldwide by travelers, making immunization and vaccination critical in preventing the reestablishment of formerly common serious illnesses.

Many experts argue that antivaccine movements occur almost exclusively in advanced countries, among parents who probably have never seen a child harmed or killed by preventable diseases such as polio, measles, or meningitis. Many of these diseases have become rare in advanced countries, and the chances of encountering them in a primarily vaccinated country may seem to be very small. A misconception also persists that because many of these diseases were once common, they are relatively harmless.

Until the late twentieth century, measles was a common childhood disease. It can lead to grave complications such as brain swelling, seizures, pneumonia, and hearing loss in close to 30 percent of cases. Measles is highly contagious but was overwhelmingly reduced after developing an effective measles vaccine in the 1960s. The MMR vaccine was developed in the early 1970s, leading to a significant decline in mumps, measles, and rubella worldwide. In fact, according to the US Centers for Disease Control and Prevention (CDC), vaccination of children with MMR vaccine led to a 75 percent decrease in measles around the world at the inception of the twenty-first century. Measles was eliminated in the United States at that time.

Because the MMR vaccine is not widespread in some countries, measles continues to cause close to 150,000 annual deaths. Despite the decline and disappearance of measles in many advanced nations, several of these suffered measles outbreaks in the second decade of the twenty-first century. The United States, for example, suffered a new measles outbreak in seventeen states, with nearly 180 cases. These cases were linked to the decline in vaccination rates. Measles is so contagious that it is necessary to maintain a 95 percent immunization rate to maintain community immunity. Nevertheless, measles immunization rates in the United States are steadily declining, and experts explain that they are no longer high enough to protect vulnerable members of the population in some communities.

Parents who have chosen not to vaccinate their children tend to live in clusters. Mapping of infectious disease outbreaks illustrates this since several measles outbreaks occurred in California, a state with a robust antivaccination community. Some of the schools in California have a vaccination rate of less than 50 percent. In 2010 California had the highest whooping cough epidemic since the 1940s, and scientists found that unvaccinated children were linked to the spread. Outbreaks have also occurred in New York and other states.

In the United States, all states require parents or guardians to comply with vaccine regulations as established by the American Academy of Pediatrics before their children attend public schools. Most states require the same for daycare centers. However, many states provide legal opt-out options, which run the gamut from children who cannot be immunized for health reasons—for example, their immune system is compromised—to waivers for religious or even philosophical reasons.

The tide may be turning for the antivaccination movement. In 2015, California legislators signed a mandate for all children enrolling in daycare or school to be vaccinated, including those who have philosophical or religious objections. Other states, such as Vermont, allowed some religious opt-outs based along strictly established parameters. In contrast, opt-outs for philosophical or personal reasons continued to be allowed in eighteen states.

Parents who refuse to vaccinate their children argue that they protect their children from health risks they believe are posed by vaccination. Supporters of the antivaccination movement are usually well-educated and affluent people who have engaged in research. Some physicians support these parents. For example, many parents and some doctors argue that it might be better to acquire immunization the natural way by experiencing and coming through the disease through contagion.

On the other hand, most doctors and other experts argue that insufficient evidence exists for these claims and that parents are placing their children at risk for diseases spread by their peers and travelers from abroad. They also argue that antivaccination activists are influenced by unsound theories and do not fully comprehend the risks of terrible infectious diseases.

VIEWPOINTS

In 1957, there were close to 60,000 cases of polio in the United States. By 1961, six years after the polio vaccine became widespread, polio cases had fallen 98 percent. There were fewer than 200 cases reported in the United States in the last two decades of the twentieth century and none since 1999. Similarly, in the 1950s, up to 4 million were infected with measles every year, with close to 50,000 requiring hospitalization. By 2012, 55 cases were reported. Nevertheless, outbreaks of nearly-eradicated infectious diseases in the United States have occasionally been reported, such as measles and whooping cough in California, mumps in Ohio, and measles in New York.

Even vaccination advocates acknowledge that some of the concerns of antivaccine activists are legitimate. For example, some vaccines pose risks for some people, and the live polio vaccine has a small risk of inadvertently causing the disease. However, vaccinations against polio and rotavirus were made safer, and other vaccinations have been proven to be reasonably safe. Nevertheless, many parents are more scared of vaccines' potential or perceived risks than the more certain risks posed by the diseases the vaccines are meant to prevent.

Vaccines, throughout their history, have raised fears and suspicions by skeptics who view vaccinations as, at best, overused and, at worst, toxic. Antivaccina- tion adherents claim that vaccines cause many health problems, including ASD, bipolar disorder, and attention deficit disorders. Some doctors support the antivaccine movement and state that vaccination should be elective. Others argue that for a healthy child, childhood diseases that used to be commonplace should not be a severe risk and would naturally immunize the child. The CDC have adamantly refuted the latter position.

Associated assertions commonly made by antivaccination advocates include that pharmaceutical corporations are aware of the severe risks posed by vaccines and are covering them up. Moreover, some argue that the medical field and the government aid and abet the pharmaceutical industry out of greed, expediency, and other ulterior interests.

Scientists argue that the antivaccination movement relies on unfounded arguments. They mention, for example, a prevalent rumor that 97 percent of people who have suffered mumps were vaccinated for it; hence, it was pointless to get vaccinated against the disease. The CDC counters that, while some people will contract mumps even after vaccination, those who received two doses of the MMR vaccine are nine times less likely to get mumps if exposed and that those who do get sick suffer milder symptoms. In response to the assertion that vaccines cause ASD, the symptoms of this condition usually appear when children start their vaccinations, and the apparent correlation is a matter of timing rather than cause and effect.

Interestingly, most antivaccination supporters are better educated and more affluent than the average person. Possibly, people with higher education levels believe that if all variables are controlled, risks can be eliminated if all variables are under control. In other words, rather than believing in the randomness of adversity, they trust nothing terrible will happen to them, and if it does, they will have the resources to handle it. Nevertheless, it is essential to note that antivaccination supporters are a diverse group that spans a broad social, cultural, and ideological spectrum.

The link-to-ASD controversy had widespread effects. In 2001, over 5,000 families filed a claim arguing that MMR vaccines triggered ASD in their children, precisely due to the thimerosal ingredients in some of the vaccines, possibly in addition to other factors. Evidentiary hearings were conducted in 2007. Thousands of pages of evidence were analyzed, and the courts heard the testimony of dozens of experts. In 2009, the courts ruled that the MMR vaccine, whether alone or in combination with other vaccines, was not a cause in the development of ASD or any of the ASD spectrum disorders (ASD). The second set of rulings were issued in 2010, finding that thimerosal in vaccines was not a causal factor in the development of ASD or ASD.

The ruling was also critical of doctors who, in the opinion of the judges, peddled hope to parents rather than sound opinions based on science. Public health officials also expressed hope that the rulings would reassure parents that vaccination was safe and put to rest the rumors linking vaccinations to ASD and related disorders.

Moreover, a scientific study published in the journal *Pediatrics* analyzed 20,000 scientific reports published between the years 2010 to 2013. It determined that vaccinations do not trigger ASD and that they present very low-risk factors. The study found no link between immunization and allergies or leukemia. Nevertheless, an analysis by the RAND Corporation acknowledges that some rare side effects may occur, including fever or seizures. The MMR vaccine is not associated with ASD.

In 2008, however, the US federal government settled a case in federal vaccine court before trial. The government agreed that vaccines aggravated an underlying mitochondrial disorder suffered by the claimant, Hannah Poling. In the statement, the government claims that aggravating her then-undiagnosed condition resulted in ASD even though vaccines did not cause it. It is not known how many children may have similar undiagnosed conditions.

Scientists and policymakers continue to work to eliminate the rare risks related to vaccinations and reassure the public of vaccination's efficiency, need, and safety. They claim some inroads have been made, in particular by being explicit about the specific consequences of contagion with infectious diseases. For example, mumps can cause hearing loss and infertility, measles and meningitis can have serious lifelong consequences, and most of the diseases at issue pose a risk of death.

By 2015, the movement seemed to be slowly losing traction. Parents who refused to vaccinate complained of becoming outcasts in their communities, feared by other parents who saw their unvaccinated children as risks to their own. Doctors and public health officials hope that the availability of up-to-date sound scientific information will slowly turn vaccine opponents around.

—*Trudy Mercadal, PhD*

Further Reading

Barrett, S. "Omnibus Court Rules Against Autism-Vaccine Link." *Autism Watch*, 6 Oct. 2010, www.autism-watch.org/omnibus/overview.shtml. Accessed 15 Nov. 2021.

Blume, S. "Anti-Vaccination Movements and Their Interpretations." *Social Science & Medicine*, vol. 62, no. 3, 2006, pp. 628-42.

Camargo Jr., K., and R. Grant. "Public Health, Science, and Policy Debate: Being Right Is Not Enough." *American Journal of Public Health*, vol. 105, no. 2, 2015, pp. 232-35.

Deer, Brian. *The Doctor Who Fooled the World: Science, Deception, and the War on Vaccines*. Johns Hopkins UP, 2020.

Hellman, M. "Study: Measles, Mumps, and Rubella not Associated with Autism." 1 July 2014, time.com/2943945/study-measles-mumps-and-rubella-vaccines-not-associated-with-autism/. Accessed 15 Nov. 2021.

Jolley, D., and K. M. Douglas. "The Effects of Anti-Vaccine Conspiracy Theories on Vaccination Intentions." *PloS ONE*, vol. 9, no. 2, 2014, pp. 1-9.

Kluger, J. "Who's Afraid of a Little Vaccine?" *Time*, vol. 184, no. 13, 2014, pp. 40-43.

Offit, P. *Deadly Choices: How the Anti-Vaccine Movement Threatens Us All*. Basic Books, 2012.

Poland, G., R. M. Jacobson, and I. G. Ovsyannikova. "Trends Affecting the Future of Vaccine Development and Delivery: The Role of Demographics, Regulatory Science, the Anti-Vaccine Movement, and Vaccinomics." *Vaccine*, vol. 27, nos. 25/26, 2009, pp. 3240-44.

Sears, R. W. *The Vaccine Book: Making the Right Decision for your Child*. Little, Brown, and Company, 2011.

Youngdahl, K., et al. *The History of Vaccines*. College of Physicians of Philadelphia, 2013.

Brucellosis vaccine

Category: Prevention
Also known as: *Brucella abortus* RB51 vaccine
Anatomy or system affected: Immune system
Specialties and related fields: Bacteriology, biotechnology, immunology, microbiology, preventive medicine, public health, veterinary medicine
Definition: an attenuated live bacterial vaccine for cattle (brucellosis vaccine exists for humans)

KEY TERMS

antibodies: secreted glycoproteins produced by B lymphocytes in response to and that counteract specific antigens.

attenuated strains: weakened strains of disease-causing bacteria and viruses that are often used as vaccines because they stimulate a protective immune response while causing no or only mild disease

brucellosis: a bacterial disease caused by members of the genus *Brucella* that typically affect cattle and buffalo and cause undulant fever in humans

vaccine: a substance that stimulates the production of antibodies and provides immunity against one or several diseases

INTRODUCTION

The brucellosis vaccine is an attenuated live bacterial vaccine for cattle licensed conditionally for cattle by the US Food and Drug Administration (FDA) in 1996. RB51 is a strain of live bacterium. RB51 is preferred because it is less likely to cause severe disease in cattle or humans than are other strains of *B. abortus*. *B. abortus* distinguishes serologically vaccinated animals from infected animals. It does not cause false-positive reactions on standard brucellosis serologic tests.

IMMUNIZATION

Cattle immunizations against brucellosis started in 1941. The RB51 immunization denotes a safer immunization both for cattle and for the veterinarians administering it. The vaccine received full approval in 2003. *B. abortus* RB51 vaccine is used in forty-nine states and Puerto Rico, and the US Virgin Islands.

Other *Brucella* vaccines also in use include *B. abortus* S19, SR82, and *B. melitensis* Rev1. *Brucella abortus* strain S19 was initially isolated from milk in 1923 and accidentally attenuated after being kept at room temperature for a year. Cattle were vaccinated with *B. abortus* S19 in 1941, but it is mainly used in calves since adult cows do not tolerate it. *B. abortus* SR82 was developed in the former Soviet Union and was first applied as a vaccine for cattle in 1988. It is as efficient as vaccine S19. Strain *B. melitensis* Rev1 is an attenuated strain that is given to goats and sheep.

Vaccination with live *Brucella* strains has proven to have highly protective against brucellosis in cattle and wildlife animals. These vaccines promote an immune response mediated by T lymphocytes against components of the pathogen.

PATHOLOGY

Brucellosis is a zoonotic infectious disease caused by the bacteria of the genus *Brucella*. Although it is primarily a disease among livestock, it can be transmitted from animals to humans through human ingestion of undercooked meat and unpasteurized dairy products from infected animals and handling infected animal tissue. Four species of *Brucella* cause the most concern: *B. abortus*, principally affecting cattle and bison; *B. suis*, principally affecting swine and reindeer but also cattle and bison; *B. melitensis*, principally affecting goats but not present in the United States; and *B. canis*, principally affecting dogs.

PATHOGENICITY

Brucellae are aerobic gram-negative coccobacilli that produce urease and that reduce nitrate to nitrite. They have a lipopolysaccharide coat that is much less pyrogenic than other gram-negative organisms, which accounts for the rare presence of high fever in brucellosis. Brucellae can enter the human body through breaks in the skin and mucous membranes, conjunctiva, and the respiratory and gastrointestinal tracts. Infection most often occurs by way of contact with or ingesting unpasteurized milk; meat products often have a low bacterial load. Percutaneous needle-stick exposure, conjunctival exposure through eye splash, and inhalation are the most common transmission routes in the United States.

Various *Brucella* species affect sheep, goats, cattle, deer, elk, pigs, dogs, and others. Humans become infected by encountering animals or animal products contaminated with *Brucella*.

IMPACT

The RB51 vaccine was developed as a less pathogenic strain, but it retains pathogenicity for humans; exposure can still pose a human health risk. Identified forms of exposure include needle sticks, eye and wound splashes, and exposure to infected material. In a series of exposures reported to the Centers for Disease Control and Prevention (CDC), persons developed local symptoms of brucellosis infection; those who became ill exhibited some systemic symptoms.

Routine serologic testing for brucellosis is not effective in monitoring for infection. Broader symptoms resulting from exposure to RB51 should be passively monitored for six months from the last exposure.

Acute symptoms of infection include fever, chills, headache, low back pain, joint pain, malaise, and occasional diarrhea. Subacute symptoms include malaise, muscle pain, headache, neck pain, fever, and sweating. Chronic symptoms include anorexia, weight loss, abdominal pain, joint pain, headache, backache, weakness, irritability, insomnia, depression, and constipation. Persons who believe they have been exposed to RB51 and who develop symptoms should consult a doctor or other health-care provider.

—Camillia King, MPH

Further Reading

Ashford, David A., et al. "Adverse Events in Humans Associated with Accidental Exposure to the Livestock

Brucellosis Vaccine RB51." *Vaccine*, vol. 3, no. 22, 2004, pp. 3435-39.
Berkelman, Ruth L. "Human Illness Associated with Use of Veterinary Vaccines." *Clinical Infectious Diseases*, vol. 37, no. 3, 2003, pp. 407-14.
Centers for Disease Control and Prevention. "Brucellosis." Division of Bacterial and Mycotic Diseases, www.cdc.gov/brucellosis/index.html.
Ferrero, Mariana C., et al. "Pathogenesis and Immune Response in Brucella Infection Acquired by the Respiratory Route." *Microbes and Infection*, vol. 22, no. 9, 2020, pp. 407-15, doi:10.1016/j.micinf.2020.06.001.
Goodwin, Zakia I., and David W. Pascual. "Brucellosis Vaccines for Livestock." *Veterinary Immunology and Immunopathology*, vol. 181, 2016, pp. 51-58, doi:10.1016/j.vetimm.2016.03.011.
Hayoun, Michael A., et al. "Brucellosis." *StatPearls*. StatPearls Publishing, 18 Sept. 2021.
López-Santiago, Rubén, et al. "Immune Response to Mucosal *Brucella* Infection." *Frontiers in Immunology*, vol. 10, no. 1759, 2019, doi:10.3389/fimmu.2019.01759.

CANCER VACCINES

Category: Prevention
Anatomy or system affected: Blood, cells, circulatory system, immune system, lymphatic system
Specialties and related fields: Biochemistry, biotechnology, immunology, internal medicine, oncology, pathology, pharmacology, preventive medicine
Definition: vaccines that sensitize the immune system against cancers prevent them from forming or treating already existing cancers

KEY TERMS

cytotoxic T cells: a type of lymphocyte that kills foreign cells, cancer cells, and virus-infected cells
dendritic cells: a special type of immune cell that is found in tissues presents antigens on its surface to other cells of the immune system
immunotherapy: a treatment that uses a person's immune system to fight cancer
metastasis: the spread of cancers from their site of origin to other places in the body
monocytes: a large phagocytic white blood cell with a simple oval nucleus and clear, grayish cytoplasm that, when activated, becomes a macrophage

IMMUNOTHERAPY

Vaccines are commonly known for their benefits in preventing or fighting infectious diseases such as polio, tetanus, or measles. As a form of immunotherapy, vaccines promote immunity, the body's defense against pathogens and injured or abnormal cells, such as cancer cells. The immune system, which can deliver its effector components to different locations in the body, is such a highly specific system that it can isolate one cancer cell from many other healthy cells and destroy that cancer cell.

Utilizing basic principles of infectious disease vaccines, a new type of vaccine is being developed to target one of the most critical public health concerns: cancer. Although some advances have been made, cancer is still the leading cause of death in persons younger than age eighty-five years in the United States.

Cancer is a group of diseases characterized by abnormal and uncontrolled cell growth, invasion, and sometimes metastasis. In a healthy body, cells grow, die, and are replaced in a regulated fashion. Damage or change in the genetic material of cells by internal or environmental factors sometimes results in immortal cells, which continue to multiply until a mass of cancer cells, or a tumor, develops. Most cancer-related deaths are caused by metastasis, in which malignant cells make their way into the bloodstream and establish colonies in other parts of the body. Cancer immunotherapy manipulates the immune system to overcome self-tolerance and to recognize cancer cells.

Like the traditional vaccines that present inactivated, attenuated, or subunit pathogens to the immune system, cancer vaccines present the right cancer antigen in combination with the right adjuvant to

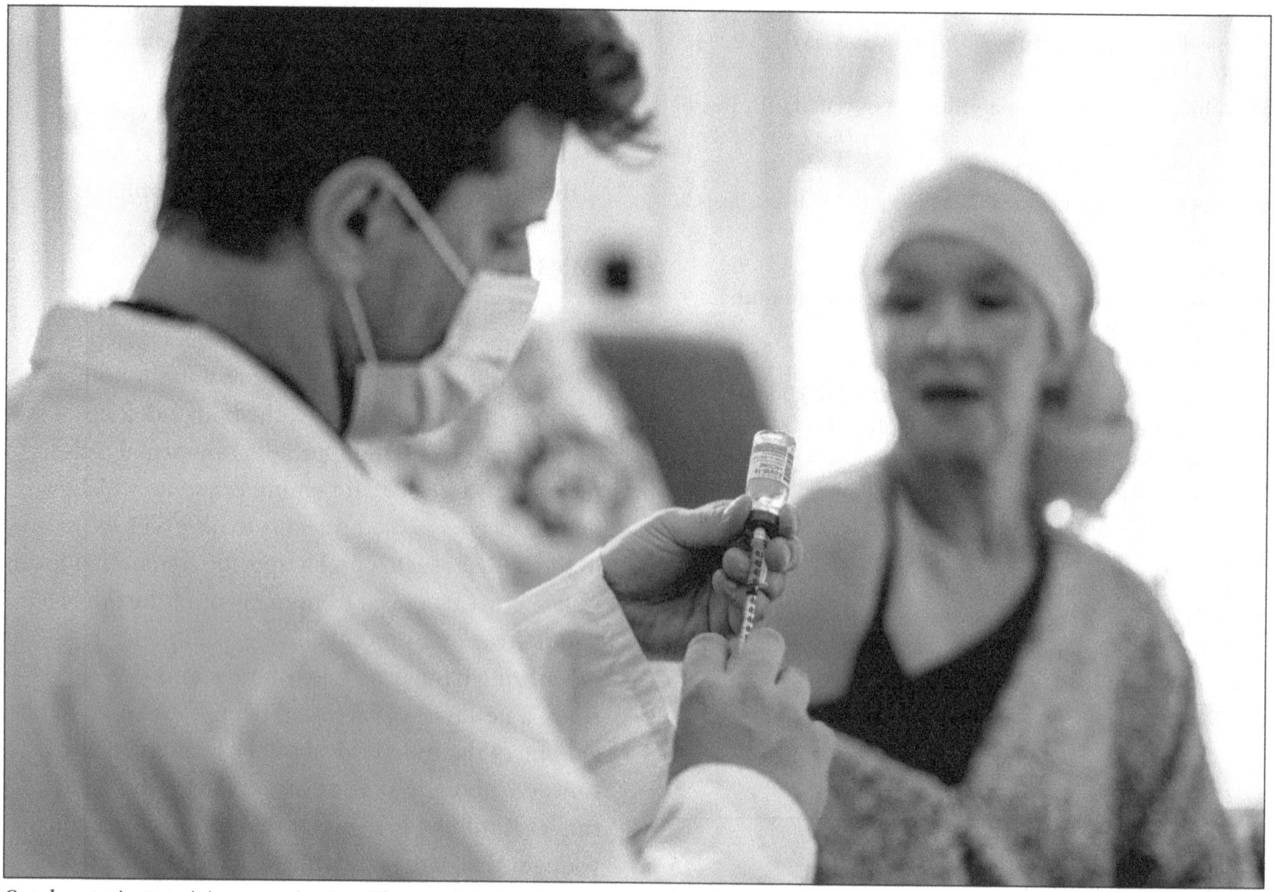

Oncology patient receiving a vaccination. Photo via iStock/FatCamera. [Used under license.]

generate the right type of immune response. This response, whether humoral or cellular, ideally should destroy cancer only and leave healthy cells untouched. Cancer cells are different from normal healthy cells. As such, they are recognized by the immune system as being different. Proteins expressed by cancer cells are different from normal proteins or are absent in normal differentiated cells. These proteins can be immunogenic when presented in the context of a cancer vaccine.

The vaccine is made from cancer-specific proteins or proteins that are found predominantly in cancer cells. Because of the associated immunologic memory, the risk of recurrence is reduced compared with traditional treatments. Rather than compromise the immune system, as many chemotherapy treatments do, cancer vaccines train the immune system to target those specific malignant cells. Consequently, some cancer vaccines are safer and do not have the traditional side effects associated with chemotherapy or radiation therapy. Depending on the specific vaccine, cancer vaccines might be stand-alone therapies or may be used with other conventional cancer therapies.

Every cancer, and its vaccine, is different. Personalized medicine is critical to the development of vaccines that must be tailor-made to each person.

PASSIVE AND ACTIVE IMMUNOTHERAPY

Cancer vaccines are characterized as either active or passive immunotherapies. While the active type aims to elicit the host immune system to fight the disease, the passive type does not depend on the body's defenses to start the attack. Instead, it uses administered medicines (antibodies or T-cell therapy) to destroy the tumor. Passive immunotherapy has no immunologic memory associated with the treatment. Any of these therapies can be targeted to one type of tumor cell or antigen (specific immunotherapies) or can generally stimulate the immune system (nonspecific immunotherapies).

Cancer vaccines are either therapeutic or preventive. Therapeutic vaccines treat persons at the early stages of the disease or with minimal residual disease after removing the primary tumor. In some cases, advanced disease may be treated with a vaccine. Preventive vaccines include the human papillomavirus (HPV) vaccine, preventing cervical, vaginal, and vulvar cancers. The hepatitis B virus (HBV) vaccine lowers the risk of developing liver cancer. The *Helicobacter pylori* vaccine targets the bacterium *H. pylori*, which is associated with stomach cancer. Hence, the HPV, HBV, and *H. pylori* vaccines do not target cancer cells; instead, they are specific to the viruses or bacteria that cause these cancers.

VACCINE STRATEGIES

Cancer vaccines target malignancies such as melanoma, leukemia, non-Hodgkin's lymphoma, and cancers of the lung, breast, kidney, ovary, pancreas, prostate, and colorectal area. The unique complex strategies used in cancer vaccine design depend on various considerations, specific to the specific cancer process, the optimum level of immunity that can potentially be achieved, and a person's health status.

In whole cancer-cell vaccines, cancer cells are irradiated before being returned to the treated person's body through injection. These vaccines contain thousands of potential antigens expressed in the whole tumor. Antigen vaccines, however, use only one antigen (or a few), whereas peptide vaccines present short fragments of the tumor protein.

Preventative Cancer Vaccines

Vaccine	Protective effects
Cervarix	Protects against two strains of HPV, types 16 and 18, that cause most cervical cancers
Gardasil	Protects against infection by HPV types 16, 18, 6, and 11
Gardasil-9	Protects against HPV types 16, 18, 31, 33, 45, 52, and 58
Hepatitis B (HBV) vaccine (HEPLISAV-B®)	Protects against infection by the hepatitis B virus and prevents the development of HBV-related liver cancer

Dendritic cell vaccines use specialized antigen-presenting cells that efficiently present tumor antigens and tumor peptides to the immune system. Dendritic cells break down cancer proteins into small fragments and then present these antigens to T cells, thus improving immunologic antigen recognition and, eventually, cancer destruction. Nucleic acid vaccines use the genetic code that codes for cancer protein antigens. The host cells continuously make the cancer antigen while keeping the immune response stimulated and strong.

Viral, bacterial, and vector-based vaccines can deliver antigens or genes encoding the tumor proteins or peptides to make the host's immune system more apt to respond. Because bacterial and viral components on these vector vaccines represent pathogen danger signals, they may trigger additional immune responses that might benefit the overall response, making it more robust and longer-lasting. Messenger ribonucleic acid (mRNA) vaccines may prove even more robust in their ability to sensitize the immune system to cancer-specific antigens.

Anti-idiotype vaccines can act passively against B-cell lymphomas or actively by mimicking cancer antigens. In the latter case, these vaccines work through antibody cascades. Some of these vaccines contain adjuvants to amplify the humoral or the cell-mediated (or both) immune responses to an antigen and break self-tolerance. Adjuvants have been developed to enhance immunogenicity when mixed with proteins, peptides, or deoxyribonucleic acid (DNA). Tumor peptide-MHC (major histocompatibility complex) complexes are essential for recognizing tumor cells by the immune system because tumor peptides are recognized if joined to the MHC complex. Cytotoxic T cells are the killer cells that recognize the peptide-MHC complexes on the tumor cells and destroy the cancer cells.

Some cancer vaccines protect against viruses that cause infections that can cause cancer. These antiviral vaccines are "protective vaccines" since they protect patients from infections by these cancer-causing viruses.

HPV vaccines help prevent HPV-related anal, cervical, head and neck, penile, vulvar, and vaginal cancers. HPV infection is superficial, and the body does not form a robust immune response against it. HPV vaccines, however, generate excellent, wide-ranging immunity against HPV. This vaccine-generated immune response prevents infection by the most cancer-causing HPV strains and the cancers that they cause.

Therapeutic cancer vaccines include Bacillus Calmette-Guérin (BCG) and Sipuleucel-T (Provenge). BCG stimulates the cellular immune response against early-stage bladder cancers. Sipuleucel-T stimulates antigen-presenting cells in patients with prostate cancer to generate T cells that attack and destroy the tumor. In randomized trials, sipuleucel-T prolongs survival compared with placebo dendritic cells and is safe for patients.

IMPACT

Cancer vaccines have the potential to treat cancers in line with treatments such as surgery or radiation therapy. Cancer vaccines are primarily experimental, although some have already entered the drug market after receiving approval from the US Food and Drug Administration (FDA). Some vaccines have shown promise in clinical trials, while others have advanced through late-stage clinical studies.

After removing the primary tumor by traditional means, using cancer vaccines helps activate the immune system to destroy any remaining cancer cells and target metastasis. Immunotherapy has the potential to strengthen the body's natural defenses, despite cancers that might have already developed. It can prevent new growth of existing cancers, hamper recurrence of treated cancers, and destroy cancer cells not previously eliminated by other treatments.

When cancer is controlled or cured, cachexia usually stops. During cachexia, there is a wasting of adipose and skeletal muscle. Persons with pancreatic and gastric cancer, for example, suffer from acute cachexia. Those with cachexia suffer from poor functional performance, depressed chemotherapy response, and more significant mortality. Therefore, the success of cancer vaccine development may benefit persons with cachexia enormously.

Immunotherapies themselves are costly, but in the long term, they reduce overall medical costs by reducing fees for patient care, management, hospitalization, and death. The pursuit and development of safe and effective cancer vaccines can greatly benefit immunologists, oncologists, molecular biologists, chemists, public health workers, and society. Above all, they help persons with cancer.

—Ana Maria Rodriguez Rojas, MS

Further Reading

Addeo, Alfredo, et al. "A New Generation of Vaccines in the Age of Immunotherapy." *Current Oncology Reports,* vol. 23, no. 12, 2021, p. 137, doi:10.1007/s11912-021-01130-x.

American Cancer Society, www.cancer.org/. Accessed 24 Nov. 2021.

Bidram, Maryam, et al. "mRNA-Based Cancer Vaccines: A Therapeutic Strategy for the Treatment of Melanoma Patients." *Vaccines,* vol. 9, no. 10, 2021, p. 1060, doi:10.3390/vaccines9101060.

Haque, Sakib, et al. "RNA-Based Therapeutics: Current Developments in Targeted Molecular Therapy of

Triple-Negative Breast Cancer." *Pharmaceutics*, vol. 13, no. 10, 2021, p. 1694, doi:10.3390/pharmaceutics 13101694.
Jemal, A., et al. "Cancer Statistics, 2010." *CA: A Cancer Journal for Clinicians*, vol. 60, no. 5, 2010, pp. 277-300.
Murphy, J. F. "Trends in Cancer Immunotherapy." *Clinical Medicine Insights: Oncology*, vol. 4, 2010, pp. 67-80.
Plotkin, Stanley A., et al., editors. *Vaccines*. 7th ed., Elsevier, 2017.
Sonpavde, G., et al. "Emerging Vaccine Therapy Approaches for Prostate Cancer." *Reviews in Urology*, vol. 12, no. 1, 2010, pp. 25-34.

Chickenpox vaccine

Category: Prevention
Also known as: Varicella-zoster vaccine
Anatomy or system affected: Blood, immune system, lymphatic system
Specialties and related fields: Biotechnology, immunology, microbiology, pediatrics, preventive medicine, public health, virology
Definition: a live, attenuated vaccine that produces CD4 and CD8 effector- and antibody-based immunity and immunological memory to the varicella-zoster virus (VZV), the cause of chickenpox

KEY TERMS

animal passage: when human viruses are grown in cultured animal cells, which decreases their capacity to cause disease in humans, making them suitable vaccine candidates
herpes viruses: a large family of double-stranded deoxyribonucleic acid (DNA) viruses that cause infections and specific diseases in animals, including humans
live attenuated vaccines: bacterial and viral strains that have been weakened so that, upon injection into a person with a healthy immune system, these strains elicit robust and protective immune responses against them
varicella: a highly contagious viral infection causing an itchy, blister-like rash on the skin
zoster: also known as shingles; a painful rash that may appear as a stripe of blisters on the trunk of the body

PATHOGENICITY AND CLINICAL SIGNIFICANCE

Varicella is a highly contagious viral illness caused by varicella-zoster virus (VZV), a human herpesvirus of the Alphaherpesvirinae subfamily. Transmission is by respiratory droplets or by direct contact with the virus-containing vesicle fluid. Household transmission rates approach 90 percent.

During the ensuing week, the virus spreads to various body parts, including the skin, liver, central nervous system, lymphatic system, and spleen. Most affected persons have symptoms that include fever, malaise, and inflamed, pruritic vesicles, which resolve in two to three weeks.

Approximately 1 in 50 persons exhibit complications that include encephalitis, pneumonia, and hepatitis. Secondary bacterial skin infections can occur as open skin lesions provide an entry portal. The varicella virus can be transmitted through the placenta to the fetus if the pregnant girl or woman acquires the disease during pregnancy. The fetus may be born with congenital varicella syndrome and demonstrate skin, extremity, ocular, and brain abnormalities.

Herpesvirus remains dormant in the spinal and cranial sensory ganglia. It reactivates typically in later life as the person's antibody level wanes, or the person experiences immune suppression, like that seen in cancer. The reemergence of the herpes virus is called "zoster," or "shingles." It can lead to extremely painful postherpetic neuralgia, which lasts from weeks to years.

DISEASE PREVENTION

The vaccine Varivax was licensed in the United States in 1995. In 1996, the Advisory Committee on Immu-

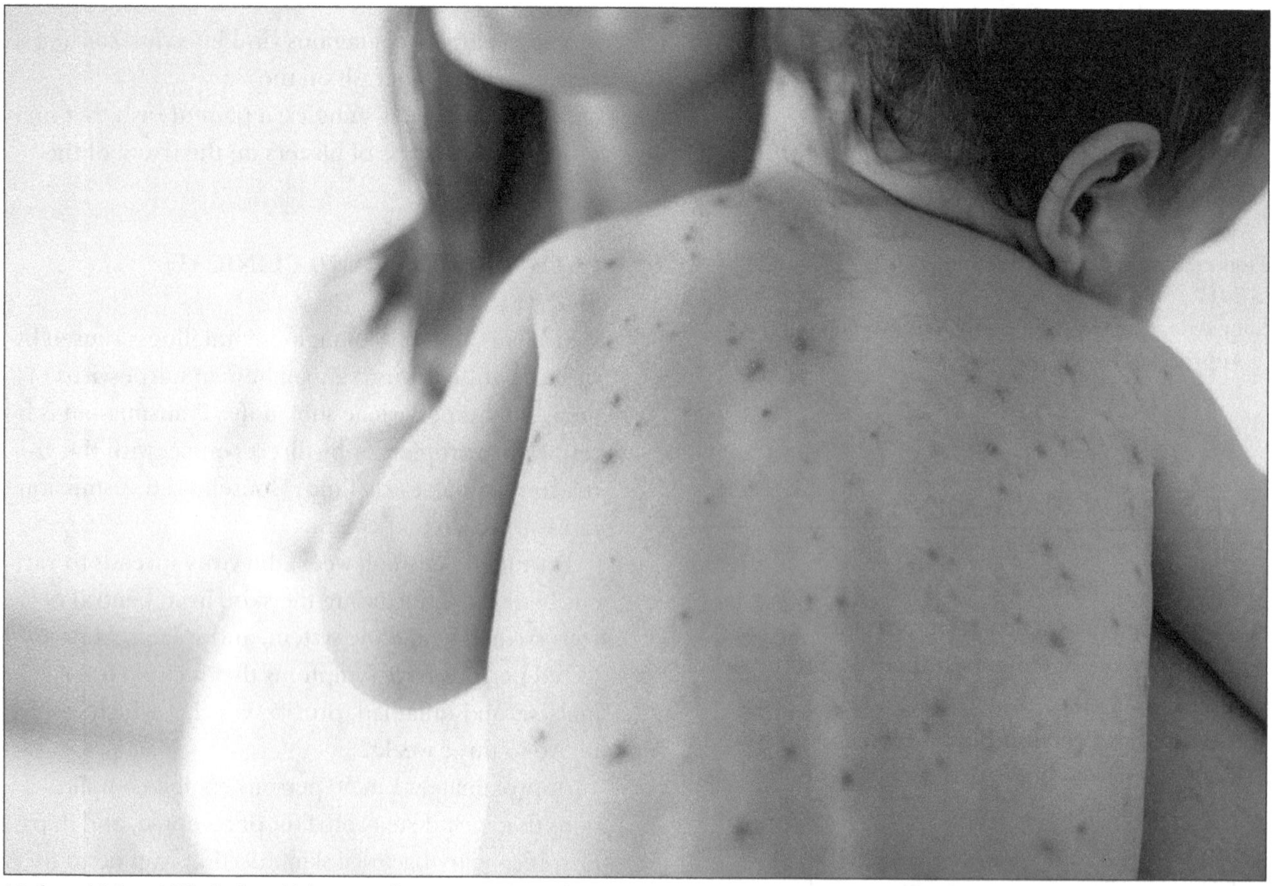

Mother with her child who has chickenpox. Photo via iStock/South_agency. [Used under license.]

nization Practices of the Centers for Disease Control and Prevention (CDC) recommended Varivax as part of routine childhood immunizations. *Varivax* contains a preparation of the Oka/Merck strain of live, attenuated varicella-zoster virus that was initially isolated from a child who had chickenpox. This virus was grown, first, in human embryonic lung cell cultures and later in cultured embryonic guinea pig cells. Animal passage of cultured viruses that cause human disease diminishes their capacity to cause disease in humans. After further propagation in cultured human cells, the virus was freeze-dried in a preparation with sucrose, phosphate, glutamate, processed gelatin, and urea as stabilizers. *Varivax* is administered subcutaneously as a 0.5-mL dose. Children should receive their first dose at twelve through fifteen months.

Though the chickenpox vaccine initially consisted of a single dose, in 2006, the CDC added a second dose. The combination vaccine ProQuad, which contains live, attenuated mumps, measles, rubella, and varicella viruses, was approved in 2005. Children who have had Varivax should have ProQuad at age four through six years. ProQuad is not recommended for children older than 13 years old, who, instead, should have two doses (0.5 ml each) of Varivax four to eight weeks apart.

The Shingrix vaccine is effective in boosting cell-mediated immunity (antibody production) against the varicella-zoster virus. Two doses of Shingrix are more than 90 percent effective at preventing shingles and postherpetic neuralgia. Protection stays above 85 percent for at least the first four years after vaccination.

Healthy adults fifty years and older should get two doses of Shingrix, separated by two to six months. There is no maximum age for Shingrix. The former shingles vaccine, Zostavax, is no longer available in the United States as of November 18, 2020. Zostavax was discontinued because of the demonstrable superiority of the Shingrix vaccine.

The most common side effects of the varicella vaccine include fever, injection-site complaints, and a varicella-like rash. The vaccine is not recommended for persons with hypersensitivity to its ingredients, including gelatin and neomycin; for persons with immunosuppression or active tuberculosis; or pregnant women or girls.

POSTEXPOSURE VACCINE

Postvaricella-exposure vaccination in children has shown some effectiveness in preventing disease if administered within three days of exposure. Protection has not been demonstrated in adolescents and adults.

IMPACT

Before developing a chickenpox vaccine, four million people in the United States acquired varicella annually, leading to ten thousand hospitalizations and one hundred deaths. After the development of a vaccine, these numbers decreased by 85 to 90 percent. The initial vaccine dose reduced varicella infection by 64 percent, and the second dose further reduced infection by 90 percent. Research has shown that the administration of the varicella vaccine in childhood reduces the incidence of herpes zoster in adulthood.

—Wanda Bradshaw, MSN, RN, NNP-BC, PNP, CCRN and Michael A. Buratovich, PhD

Further Reading

Bennet, John E., et al., editors. *Mandell, Douglas, and Bennett's Principles and Practice of Infectious Diseases*. 9th ed., Elsevier; 2019.

Centers for Disease Control and Prevention. "Recommended Immunization Schedules for Persons Aged 0-18 Years—United States, 2008." *Morbidity and Mortality Weekly Report*, vol. 57, 2008, pp. Q1-4, www.cdc.gov/mmwr/preview/mmwrhtml/mm5701a8.htm.

———. "Varicella (Chickenpox) Vaccination." www.cdc.gov/vaccines/vpd-vac/varicella.

Exumé, Myriam. "Product Discontinuation Notice: Zostavax(r) (Zoster Vaccine Live)." www.merckvaccines.com/wp-content/uploads/sites/8/2020/06/US-CIN-00033.pdf. Accessed 20 Nov. 2021.

Macartney, Kristine, and Peter McIntyre. "Vaccines for Post-Exposure Prophylaxis Against Varicella (Chickenpox) in Children and Adults." *The Cochrane Database of Systematic Reviews*, vol. 3, no. CD001833, 2008, doi:10.1002/14651858.CD001833.pub2.

Marin, Mona, et al. "Varicella Prevention in the United States: A Review of Successes and Challenges." *Pediatrics*, vol. 122, no. 3, 2008, pp. e744-51, doi:10.1542/peds.2008-0567.

Roush, Sandra, et al. "Historical Comparisons of Morbidity and Mortality for Vaccine-Preventable Diseases in the United States." *Journal of the American Medical Association*, vol. 298, no. 18, 2007, pp. 2155-63.

Singh, Grisuna, et al. "Recombinant Zoster Vaccine (Shingrix(r)): A New Option for the Prevention of Herpes Zoster and Postherpetic Neuralgia." *The Korean Journal of Pain*, vol. 33, no. 3, 2020, pp. 201-7, doi:10.3344/kjp.2020.33.3.201.

Cholera vaccine

Category: Prevention

Anatomy or system affected: Gastrointestinal tract, immune system, lymphatic system

Specialties and related fields: Bacteriology, immunology, microbiology, preventive medicine, public health

Definition: a vaccine that protects against the potentially lethal gastrointestinal disease cholera

KEY TERMS

cholera: a gastrointestinal disease characterized by severe, watery diarrhea that causes extensive dehydration and, potentially, death *vaccine*: a substance used to stimulate the production of antibodies and provide immunity against one or several diseases

Vibrio cholera: the bacterial species that causes the gastrointestinal disease cholera

INTRODUCTION

Cholera is a gastrointestinal illness caused by infection with the gram-negative bacterium Vibrio cholerae (*V. cholera*) Cholera is characterized by severe watery diarrhea that can rapidly lead to dehydration. If not treated, severe dehydration can lead to death. Cholera epidemics have occurred around the world, although infection among travelers is not common. Cholera is caused by eating contaminated food or water.

Vaxchora is a live attenuated oral vaccine approved by the US Food and Drug Administration (FDA) to prevent cholera caused by *V. cholera* serogroup O1. People two through 64 years of age can receive Vaxchora. It is usually prescribed to people traveling to cholera-affected areas. The Food and Drug Administration (FDA) approved Vaxchora in the United States in June 2016. It is the only cholera vaccine available in the United States to date.

Global cholera cases are almost certainly vastly underreported. Consequently, there are no precise measurements of cholera morbidity and mortality. There are an estimated 3 million cases of diarrheal disease and approximately 100,000 deaths worldwide caused by V. cholerae each year. Cholera is endemic in approximately 50 countries (i.e., those that reported cholera cases in at least three of the five past years), mainly in Africa and Asia. Cholera tends to occur in areas with inadequate access to clean water and sanitation. Epidemics have occurred throughout Africa, Asia, the Middle East, South and Central America, and the Caribbean.

Cholera vaccines have been available for more than twenty years. However, many countries have not used them because of high cost, limited supply, and logistical problems in providing two doses.

VACCINE ADMINISTRATION

Vaxchora is administered orally. The vaccine comes as a powder with 100 ml of sterile buffer. The health-care provider mixes the powder with the buffer, and the patient drinks it. Vaxchora should be taken at least ten days before traveling to a cholera-affected area. Anyone taking Vaxchora should avoid eating and drinking for one hour before and after vaccine administration. No one should take Vaxchora within two weeks of systemic antibiotic treatments or ten days of treatment with the antimalarial medication chloroquine. Women who are pregnant or breastfeeding or anyone who is immunosuppressed should not take this vaccine.

Outside the United States, the are three available cholera vaccines: Durokal, Shanchol, and Euvichol.

Durokal is a killed whole-cell oral vaccine. Durokal is licensed in more than sixty countries. It provides short-term protection against *V. cholerae* for up to 90 percent of all age groups, including infants. Additionally, Durokal provides limited (<50 percent) protection against infection with enterotoxigenic Escherichia coli.

Shanchol is a killed whole-cell oral cholera vaccine. It provides longer-term protection and is pending World Health Organization (WHO) prequalification. Its formula is particularly effective in children younger than five years of age.

Euvichol is a killed whole-cell oral cholera vaccine. It has received World Health Organization prequalification and is being stockpiled with Dukorkal and Shanchol by the WHO for emergency use.

The Durokal, Shanchol, and Euvichol vaccines are administered orally, offering ease of administration and freedom from the risks of needle-borne infection. The vaccines must be administered in two doses, between seven days and six weeks apart. Persons in areas of the world in which cholera is prevalent should receive booster doses every six months. Mucosal vaccines are also under development for the treatment of cholera.

IMPACT

Vaccines are one facet of cholera eradication. They provide temporary protection while safe water and improved sanitation are secured. Community education on safe practices is also necessary. Persons who survive infection by *V. cholerae* develop protective immunity.

—Merrill Evans, MA and Michael A, Buratovich, PhD

Further Reading

Ali, Mohammad, et al. "Updated Global Burden of Cholera in Endemic Countries." *PLoS Neglected Tropical Diseases,* vol. 9, no. 6, 2015, p. e0003832, doi:10.1371/journal.pntd.0003832.

Chatterjee, Patralekha. "High Hopes for Oral Cholera Vaccine." *Bulletin of the World Health Organization,* vol. 88, 2010, pp. 165-66.

"Cholera Outbreak—Haiti, October, 2010." *Morbidity and Mortality Weekly,* Report 59, 2010, p. 1411.

Freedman, David O. "Re-born in the USA: Another Cholera Vaccine for Travellers." *Travel Medicine and Infectious Disease,* vol. 14, no. 4, 2016, pp. 295-96, doi:10.1016/j.tmaid.2016.07.008.

Plotkin, Stanley A., et al., editors. *Vaccines.* 7th ed., Elsevier, 2017.

Wong, Karen K., et al. "Recommendations of the Advisory Committee on Immunization Practices for Use of Cholera Vaccine." *Morbidity and Mortality Weekly Report,* vol. 66, no. 18, 2017, pp. 482-85, doi:10.15585/mmwr.mm6618a6.

World Health Organization. "Cholera." Fact Sheet No. 107, www.who.int/mediacentre/factsheets/fs107.

COVID-19 VACCINE

Category: Prevention

Anatomy or system affected: Blood, cells, circulatory system, immune system, lymphatic system

Specialties and related fields: Biochemistry, biotechnology, immunology, preventive medicine, public health, virology

Definition: a vaccine that decreases the chances of acquiring, transmitting, and suffering from severe COVID-19

KEY TERMS

COVID-19: coronavirus infectious disease discovered in 2019; the infectious disease caused by (severe acute respiratory syndrome) SARS-CoV-2

lipid nanoparticle: a synthetic

light chain: the smaller subunits of an antibody

mRNA: messenger ribonucleic acid; an RNA molecule translated by ribosomes to synthesize a protein

SARS-CoV-2: a coronavirus that infects animals and humans and causes COVID-19

spike protein: the main glycoprotein embedded in the envelope of SARS-CoV-2 that binds to the ACE2 protein on host cells and initiates infection

In response to a worldwide viral pandemic that infected millions in 2020, scientists began working on a vaccine that could protect people against coronavirus disease 2019 (COVID-19), the disease caused by the virus SARS-CoV-2. The virus was discovered in China in late 2019, and by early 2020 had spread rapidly across the globe. While normal vaccine development takes years, scientific teams used relatively new forms of research to produce vaccines in record time. Several vaccines passed through testing and showed significant success in combating COVID-19 in less than a year, with a few approved in certain countries on a limited or emergency basis within just months. More widespread vaccination campaigns were rolled out in 2021.

BACKGROUND

In December 2019, health officials in China began investigating reports of a serious respiratory illness in Wuhan, a city of 11 million people in central China. By January 2020, Chinese scientists had identified the cause of the illness as a new type of coronavirus. Coronaviruses are a family of viruses that can cause illnesses as mild as the common cold or more serious, potentially fatal respiratory diseases, such as severe acute respiratory syndrome (SARS), which resulted in a significant epidemic in 2002 to 2004. These viruses are made up of a strand of ribonucleic acid (RNA) with a covering of protein spikes that resemble a crown. After identifying the new coronavirus, scientists named it severe acute respiratory syndrome coronavirus 2 (SARS-CoV-2). They named the disease it caused coronavirus disease 2019, or COVID-19.

The World Health Organization (WHO) declared a public health emergency in late January 2020. By March, it declared the outbreak a pandemic, with cases of COVID-19 spreading around the world. Despite efforts to control the pandemic through lockdowns, travel bans, and other preventative measures, the disease spread rapidly, with major epicenters emerging in Italy and Iran by the end of February. By late spring, the United States had become the epicenter of SARS-CoV-2 transmission, passing 100,000 deaths in May and 2 million cases in June. Once the fall and winter approached, cases of COVID-19 spiked even higher in the United States and many other countries.

OVERVIEW

Within weeks after discovering SARS-CoV-2, scientists recognized the severity of the new virus and began efforts to combat it. In January 2020, Chinese scientists who had mapped the genetic sequence of SARS-

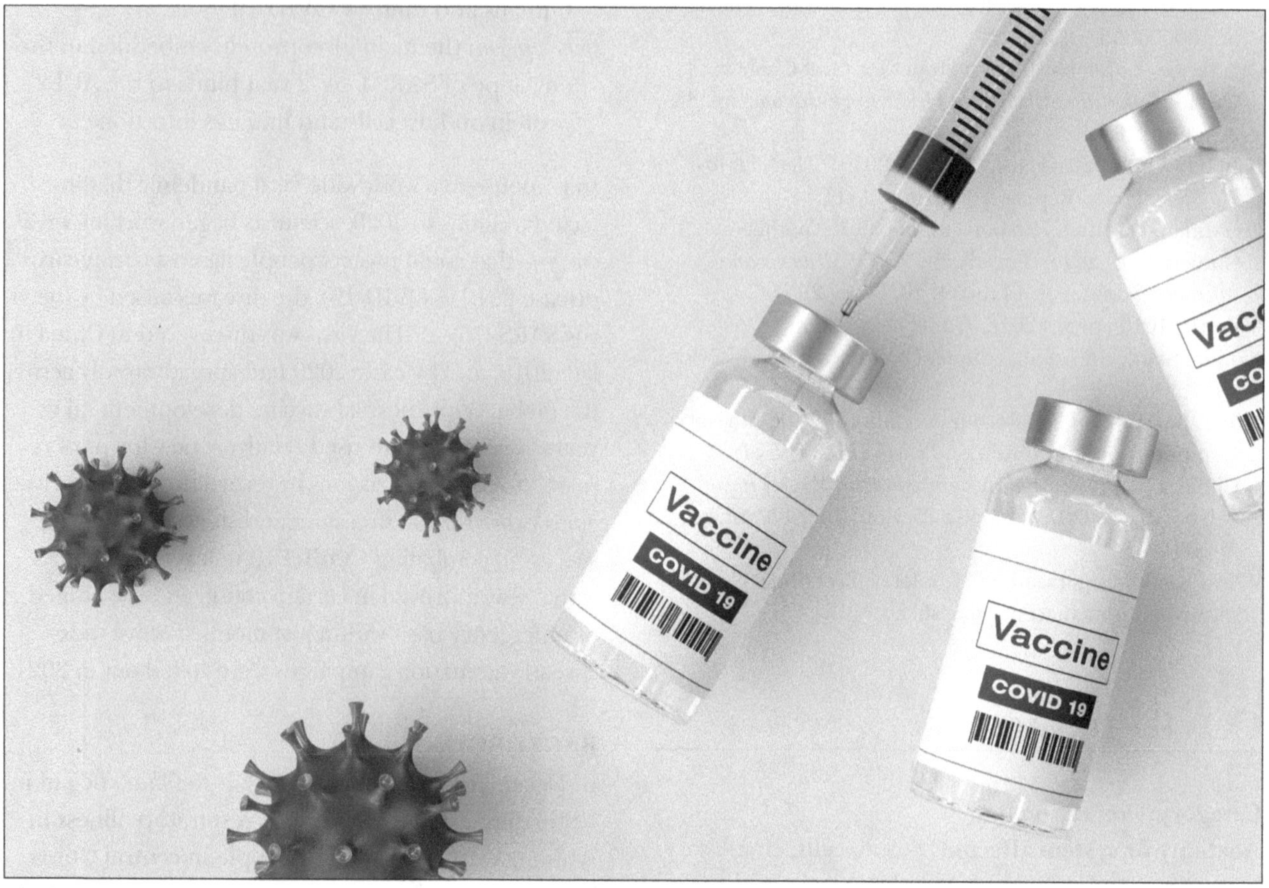

The COVID-19 vaccine was widely distributed throughout the world in 2021. Photo via iStock/K_E_N. [Used under license.]

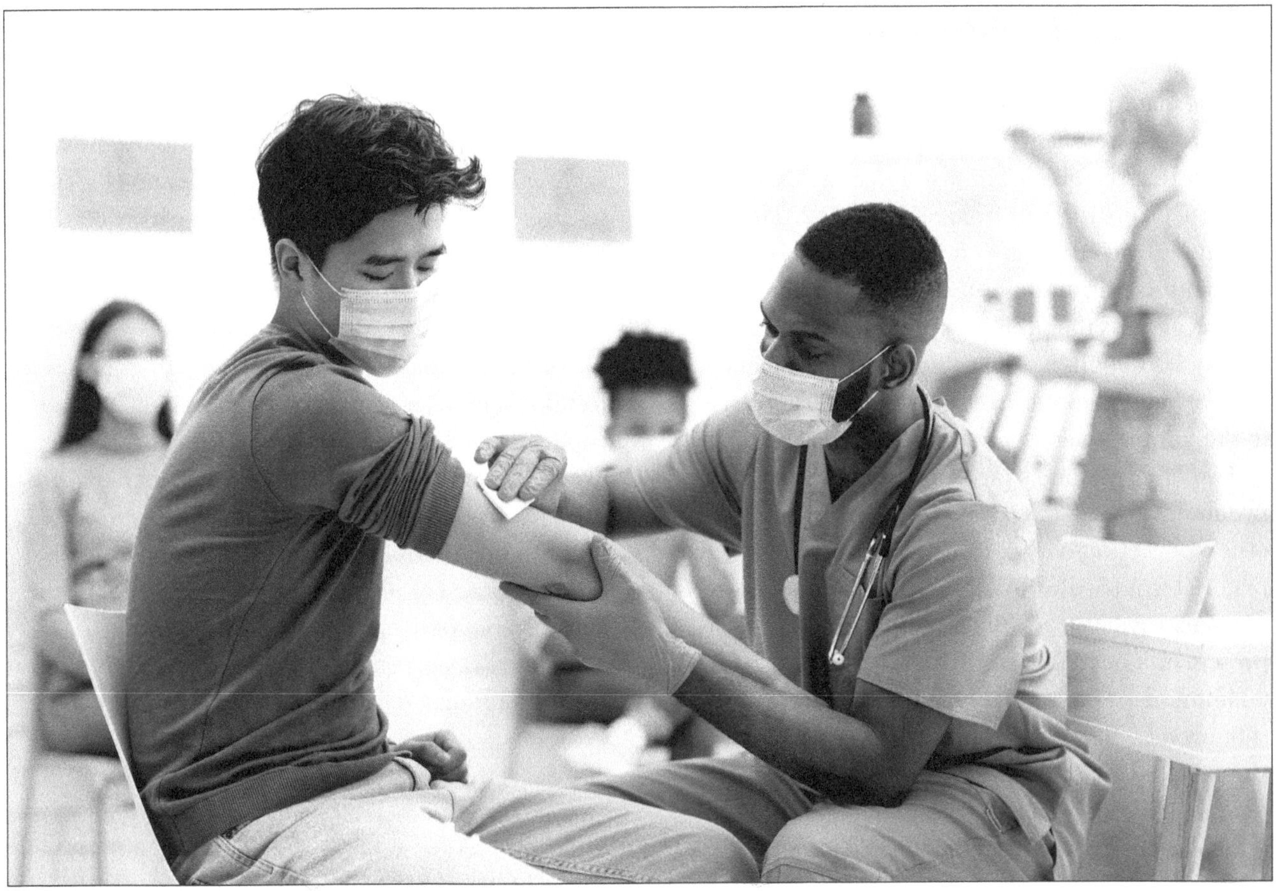

Photo via iStock/Prostock-Studio. [Used under license.]

CoV-2 made that information public. Research teams worldwide began using that data to begin the development of experimental vaccines.

Typically, vaccines take years to develop, test, and gain approval for public use. On average, this process takes about ten to fifteen years, with the fastest approval before 2020 taking four years. However, scientists working on the COVID-19 vaccines had two significant advantages that sped up the process. First, many nations provided significant government funding to develop vaccines and streamline distribution. The second advantage involved new medical research that allowed scientists to work on a new vaccine type by using the virus's genetic code.

In the past, most vaccines were made using a weakened or dead form of a virus or bacteria. These weakened or dead forms were injected into the body, triggering the immune system to respond to the foreign invaders. The immune system uses white blood cells to create proteins called "antibodies" that attack and destroy the invader. These antibodies can recognize and target a specific infection. Other white blood cells remember the invading germ and can respond to produce more antibodies in the future should it return. While this method has proven to be very effective, developing a vaccine in this manner is time-consuming.

To develop a COVID-19 vaccine, several research teams began looking at different ways to prompt the body into creating an immune response. Three approaches emerged as the most successful in the early months of vaccine development. One was mRNA vaccines, which use messenger ribonucleic acid (mRNA) to trigger an immune response. mRNA is a strand of genetic material that carries instructions from a cell's deoxyribonucleic acid (DNA) on creating specific proteins. Scientists can encode a strand of mRNA with the genetic instructions from SARS-CoV-2. The instructions teach the body's immune cells to copy a protein spike from the virus. When the immune system recognizes the protein as an invader, it creates antibodies to fight it. The protein spike is harmless, but the antibodies it triggers remain and can respond in the future should the actual virus enter the body.

Two other vaccine types also showed considerable promise. Vector vaccines use a weakened form of another, nonserious virus as a carrier and have genetic material from SARS-CoV-2 inserted into it. The genetic material instructs the body's cells to create a protein unique to the COVID-19 virus, which triggers an immune response. The third approach, protein subunit vaccines, uses harmless pieces of proteins from SARS-CoV-2 to alert the immune system.

Many pharmaceutical companies, medical researchers, and government health organizations worldwide took up the vaccine creation process. Hundreds of potential vaccines were studied and progressed through the various development and testing stages at different rates. In the summer of 2020, a few countries officially approved certain vaccines on an emergency or limited-use basis after just a few months of development. The first approvals came from China on June 25, 2020, Russia on August 11, and China on August 31. However, some international experts worried that the process may have been too rushed in these cases. Other early vaccines received limited approval in Russia on October 14 and the United Arab Emirates on September 14.

Development and testing of potential vaccines continued throughout 2020. In December, the US Food and Drug Administration (FDA) made its first approvals for emergency use in adults. US pharmaceutical company Pfizer developed the first US-approved vaccine in collaboration with German biotech company BioNTech. The mRNA vaccine, which was administered in two doses, was found to have a success rate of 95 percent. Soon after, another mRNA vaccine from US pharmaceutical company Moderna received FDA approval. According to early data, the Moderna vaccine, also delivered in two doses, was 94.5 percent effective.

In February 2021, the FDA approved a single-dose vector vaccine developed by US company Johnson & Johnson's Janssen Pharmaceutical for individuals eighteen and older. While it proved to be less effective in testing—showing to be 66 percent effective, as reported by the US Centers for Disease Control and Prevention (CDC)—it did have high success against preventing severe illness, hospitalization, and death, and required only one shot. However, on April 13, 2021, the FDA paused the distribution of the Johnson & Johnson/Janssen vaccine after it reported that six women between the ages of eighteen and forty-eight had developed a cerebral venous sinus thrombosis (a rare and severe blood clot) after receiving the vaccine. The pause was soon lifted after the CDC and FDA reviewed the cases and found extremely low risks.

More than two hundred vaccines remained in various stages of development and testing by early 2021, and several other vaccines were approved by countries other than the United States. Notably, British company AstraZeneca developed a vector-based vaccine that used a virus found in chimpanzees as a carrier. In testing, the two-dose vaccine was 62 to 90 percent effective and was shown to reduce the person-to-person spread of the virus. By late February 2021, the AstraZeneca vaccine had been approved for use in several jurisdictions, including the European Union, the United Kingdom, and India. However, concerns over rare occurrences of blot clots after receiving that vaccine led to pauses and restrictions in some countries. Meanwhile, a two-dose protein subunit vaccine developed by US-based Novavax was shown to be 89.3 percent effective.

The first vaccines were distributed in the United States to frontline health-care workers and long-term care residents in December 2020. Further vaccines were distributed by individual states as available in phases, prioritizing those over the age of sixty-five, people with underlying medical conditions, and other essential workers. By April 30, all adults in the United States were declared eligible for a vaccine. On May 10, 2021, the FDA authorized the use of the Pfizer-BioNTech vaccine for twelve- to fifteen-year-olds, followed two days later by a CDC endorsement of the vaccine for children in that age group.

In early May 2021, US president Joe Biden announced that he supported giving other countries access to COVID vaccine patents by waiving intellectual property rights held by the vaccine developers for a limited time. In doing so, Biden reversed previous US policy and acceded to the pleas of Democratic lawmakers in the United States and officials in over one hundred other countries, who had argued that increased international access to the patents would help save lives and reduce the risk that vaccine-resistant COVID strains would emerge. The announcement angered pharmaceutical companies, who argued that the move would make it harder to fund research and development for future vaccines and that poorer countries would not be able to make the vaccines because they lacked the necessary facilities and materials. The Biden administration also announced it would send tens and eventually hundreds of millions of vaccine doses to other countries to help combat the pandemic on the global scale as the situation in the United States improved.

According to the CDC, by June 27, 2021, over 179.26 million people (54 percent of the population) in the United States had received at least one dose of a vaccine, while over 153.02 million people (46.1 percent of the total population) were fully vaccinated. Of the latter, more than 80.6 million people had received the Pfizer vaccine, more than 60.2 million had received the Moderna vaccine, and over 12.1 million people had received the Johnson & Johnson/Janssen vaccine.

While vaccination rates continued to rise in the United States, hesitancy to receive the vaccine persisted among specific populations. Experts warned of the public health challenge posed by the antivaccination movement, and debate ensued regarding the ethics of potentially providing incentives to vaccinated individuals or requiring proof of vaccination, such as "vaccine passports." Meanwhile, company executives, college administrators, and political leaders made difficult decisions regarding requiring vaccination before allowing entry to various establishments or activities.

—*Richard Sheposh*

Further Reading

Brothers, Will. "A Timeline of COVID-19 Vaccine Development." *BioSpace*, 20 Dec. 2020, www.biospace.com/article/a-timeline-of-covid-19-vaccine-development/. Accessed 4 Feb. 2021.

"COVID-19 Dashboard by the Center for Systems Science and Engineering (CSSE) at Johns Hopkins University (JHU)." *Johns Hopkins University & Medicine*, 14 June 2021, coronavirus.jhu.edu/map.html. Accessed 14 June 2021.

"COVID-19 Vaccines: Get the Facts." *Mayo Clinic*, 12 June 2021, www.mayoclinic.org/diseases-conditions/coronavirus/in-depth/coronavirus-vaccine/art-20484859. Accessed 14 June 2021.

"COVID-19 Vaccinations in the United States." *COVID Data Tracker*, Centers for Disease Control and Prevention, 27 June 2021, covid.cdc.gov/covid-data-tracker/#vaccinations. Accessed 28 June 2021.

"Frequently Asked Questions." *Centers for Disease Control and Prevention*, 25 May 2021, www.cdc.gov/coronavirus/2019-ncov/faq.html. Accessed 14 June 2021.

Herper, Matthew. "J&J One-Dose Covid Vaccine Is 66% Effective, a Weapon but Not a Knockout Punch." *Stat*, 29 Jan. 2021, www.statnews.com/2021/01/29/jj-one-dose-covid-vaccine-is-66-effective-a-weapon-but-not-a-knockout-punch/. Accessed 4 Feb. 2021.

Marks, Peter. "Joint CDC and FDA Statement of Johnson & Johnson COVID-19 Vaccine." *US Food and Drug Administration*, 13 Apr. 2021, www.fda.gov/news-events/

press-announcements/joint-cdc-and-fda-statement-johnson-johnson-covid-19-vaccine. Accessed 14 Apr. 2021.

Mullard, Asher. "How COVID Vaccines Are Being Divvied Up Around the World." *Nature*, 30 Nov. 2020, www.nature.com/articles/d41586-020-03370-6. Accessed 4 Feb. 2021.

Santora, Marc, and Rebecca Robbins. "The AstraZeneca Vaccine Is Shown to Cut Transmission of the Virus." *New York Times*, 3 Feb. 2021, www.nytimes.com/2021/02/03/us/astrazeneca-coronavirus-vaccine.html. Accessed 4 Feb. 2021.

Shalal, Andrea, Jeff Mason, and David Lawder. "US Reverses Stance, Backs Giving Poorer Countries Access to COVID vaccine patents." *REUTERS*, 6 May 2021, www.reuters.com/business/healthcare-pharmaceuticals/biden-says-plans-back-wto-waiver-vaccines-2021-05-05/. Accessed 7 May. 2021.

"Understanding How COVID-19 Vaccines Work." *Centers for Disease Control and Prevention*, 9 Mar. 2021, www.cdc.gov/coronavirus/2019-ncov/vaccines/different-vaccines/how-they-work.html. Accessed 19 Apr. 2021.

Zimmer, Carl, Jonathan Corum, and Sui-Lee Wee. "Coronavirus Vaccine Tracker." *The New York Times*, 10 June 2021, www.nytimes.com/interactive/2020/science/coronavirus-vaccine-tracker.html. Accessed 14 June 2021.

DTaP vaccine

Category: Epidemiology

Also known as: Diphtheria, tetanus, and acellular pertussis vaccine

Anatomy or system affected: Blood, cells, circulatory system, immune system, lymphatic system

Specialties and related fields: Bacteriology, biochemistry, biotechnology, immunology, preventive medicine, public health

Definition: a vaccine that protects against three different bacterial illnesses: diphtheria, tetanus, and pertussis

KEY TERMS

antigen: a foreign substance in the body that is recognized and responded to by the immune system

diphtheria: an upper respiratory infection caused by the bacterium *Corynebacterium diphtheriae*

DT: diphtheria and tetanus vaccine that is given to children

DTaP: diphtheria, tetanus, and acellular pertussis vaccine that is given to children

pertussis: whooping cough, a disease caused by the bacterium *Bordetella pertussis*

Td: tetanus and diphtheria vaccine that is given to adolescents and adults

Tdap: combined tetanus, diphtheria, and acellular pertussis vaccine that is given to adolescents and adults

tetanus: an infection caused by the bacterium *Clostridium tetani*

INTRODUCTION

The DTaP vaccine protects against three different bacterial illnesses. The first disease, diphtheria, is caused by the bacterium *Corynebacterium diphtheriae*. Infection with this bacterium causes a severe sore throat and difficulty breathing and swallowing. The second disease, tetanus, is caused by the bacterium *Clostridium tetani* and leads to what is commonly referred to as lockjaw. This disease causes intense muscle contractions and can interfere with breathing. The last disease, pertussis, or whooping cough, is caused by the bacterium *Bordetella pertussis.* This bacterium produces a severe persistent cough with a characteristic whooping sound on inspiration between coughing fits and can lead to respiratory failure.

MECHANISM OF ACTION

The vaccine incorporates the three toxins produced by the bacteria in their inactivated forms (known as toxoids). Administration of these toxoids leads to an immune response without causing the disease that protects the vaccine recipient from future illness.

HISTORY

Individual vaccines against diphtheria, tetanus, and pertussis were first developed in the late nineteenth and early twentieth centuries. The combination vaccine that incorporated all three was first licensed in 1948. The vaccine was further modified in 1991 in response to a high rate of side effects caused by the original whole-cell pertussis component. A new, acellular pertussis element was developed at that time and has resulted in a significant decrease in the side-effect profile of the vaccine. There are seven pediatric DTaP vaccines licensed and currently used in the United States: Daptacel®, Infanrix®, Kinrix®, Pediarix®, Pentacel®, Quadracel®, and Vaxelis™. Pediarix (FDA approved in 2002) is DTaP combined with hepatitis B and inactivated poliovirus, and Pentacel (FDA approved in 2008) combines DTaP with Haemophilus influenzae type B and inactivated poliovirus. Quadracel (FDA approved in 2015) and Kinrix (FDA approved in 2008) combined DTaP with inactivated poliovirus. The FDA approved Daptacel in 2002 and Infanrix in 1997.

ADMINISTRATION

Health experts recommended that children receive the DTaP vaccine at ages two, four, six, and fifteen to eighteen months and again between four and six years. Adolescents and adults should then receive one administration of the Tdap vaccine, which differs from the DTaP. It contains less of the diphtheria and acellular pertussis components. After the Tdap, adults should receive the Td booster immunization against tetanus and diphtheria every ten years.

IMPACT

The impact of the DTaP vaccine on public health has been enormous. Diphtheria has been nearly eradicated in the United States, and the incidence of tetanus and pertussis has been dramatically reduced. However, of the three diseases, pertussis affects many adults and children in the United States, with morbidity and mortality rates rising among infants.

—Jennifer Birkhauser, MD

Further Reading

Advisory Committee on Immunization Practices. "Recommended Adult Immunization Schedule: United States, 2010." *Annals of Internal Medicine*, vol. 152, 2010, pp. 36-39.

Centers for Disease Control and Prevention. "Recommended Immunization Schedules for Persons Aged 0-18 Years—United States, 2008." *Morbidity and Mortality Weekly Report*, vol. 57, 2008, pp. Q1-4, www.cdc.gov/mmwr/preview/mmwrhtml/mm5701a8.htm.

Harvey, Richard A., Pamela C. Champe, and Bruce D. Fisher. *Lippincott's Illustrated Reviews: Microbiology*. 2nd ed., Lippincott Williams and Wilkins, 2006.

Loehr, Jamie. *The Vaccine Answer Book: Two Hundred Essential Answers to Help You Make the Right Decisions for Your Child*. Sourcebooks, 2010.

Miller, Neil Z. *Vaccines Are They Really Safe and Effective?* New Atlantean Press, 2015.

Pan American Health Organization. World Health Organization. *Control of Diphtheria, Pertussis, Tetanus, "Haemophilus influenzae" Type B, and Hepatitis B Field Guide*. Author, 2005.

Playfair, J. H. L., and B. M. Chain. *Immunology at a Glance*. 9th ed., Wiley-Blackwell, 2009.

Plotkin, Stanley A., et al., editors. *Vaccines*. 7th ed., Elsevier, 2017.

Hepatitis vaccines

Category: Prevention

Anatomy or system affected: Blood, cells, circulatory system, immune system, lymphatic system

Specialties and related fields: Immunology, internal medicine, microbiology, oncology, preventive medicine, public health, virology

Definition: vaccinations that prevent infection with hepatitis A and B viruses

KEY TERMS

hepatitis: inflammation of the liver

hepatitis A virus: a ribonucleic acid (RNA) virus that causes inflammation of the liver and is transmitted through ingestion of contaminated food and water or direct contact with an infectious person

hepatitis B virus: a deoxyribonucleic acid (DNA) virus that causes hepatitis (inflammation of the liver) and is carried and passed to others through the blood and other body fluid.

INTRODUCTION

Hepatitis is inflammation of the liver caused by a viral infection. There are five types of hepatitis infection: A, B, C, D, and E. Not all of these types of hepatitis, however, can be prevented by vaccination

PREVENTION

Hepatitis A and B are the types of viral hepatitis that a vaccine can prevent. Havrix and Vaqta are licensed vaccines for Hepatitis A. Engerix-B and Recombivax HB can prevent infections with Hepatitis B virus. Combination vaccines that prevent hepatitis A and hepatitis B virus infection include Comvax, Pediarix, and Twinrix. A vaccine for hepatitis E is being tested but has not been approved by the US Food and Drug Administration.

Hepatitis A vaccine is available for people in high-risk groups, such as day-care and nursing-home staff, laboratory staff, and those traveling to parts of the world where hepatitis is common. Routine childhood immunization against hepatitis A is also recommended.

Hepatitis B vaccine is given to all infants and unvaccinated children. The vaccine is available for adults at high risk, such as health-care professionals, intravenous-drug users, and those who do not practice safer sex.

REQUISITE DOSAGES

Dosages are administered at intervals, and no vaccine series should be restarted. Licensed combination vaccines may be used when any component is indicated and when its other component (or components) is not contraindicated. The use of licensed combination vaccines is preferred over separate injections of their equivalent component vaccines. Engerix-B or Recombivax HB should be used for the hepatitis B vaccine birth dose.

IMPACT

Hepatitis A and B are highly contagious. Hepatitis A is spread readily in locations with poor sanitary conditions, and hepatitis B is spread through contact with infected persons' blood or body fluids. However, hepatitis A and hepatitis E are typically caused by the ingestion of contaminated food or water. Of the many persons at risk of being infected with these diseases, those at higher risk include people who work or travel in areas with high rates of infection and all children older than one year.

—*Margaret Ring Gillock, MS*

Further Reading

Centers for Disease Control and Prevention. "Global Routine Vaccination Coverage, 2009." *MMWR: Morbidity and Mortality Weekly Report,* vol. 59, 2010, pp. 1367-71.

Dienstag, J. L. "Hepatitis B Virus Infection." *New England Journal of Medicine,* vol. 359, 2008, pp. 1486-1500.

Jou, J. H., and A. J. Muir. "In the Clinic: Hepatitis C." *Annals of Internal Medicine,* vol. 148, 2008, ITC6-1-ITC6-16.

Plotkin, Stanley A., et al., editors. *Vaccines*. 7th ed., Elsevier, 2017.

Sjogren, M. H. "Hepatitis A." In *Sleisenger and Fordtran's Gastrointestinal and Liver Disease: Pathophysiology, Diagnosis, Management,* edited by Mark Feldman, Lawrence S. Friedman, and Lawrence J. Brandt, New ed., 2 vols., Saunders/Elsevier, 2010.

HIB VACCINE

Category: Prevention

Also known as: *Haemophilus influenzae* type B vaccine

Anatomy or system affected: Blood, cells, circulatory system, immune system, lymphatic system

Specialties and related fields: Bacteriology, biotechnology, family practice, immunology, microbiology, pediatrics. preventive medicine, public health

Definition: a vaccine protects against disease caused by the bacterium *Haemophilus influenzae* type B

KEY TERMS

capsule: a polysaccharide shell that surrounds some bacterial strains and protects them

H. influenzae: a human pathogen that causes meningitis and other significant infections in children and adults

nontypeable: *H. influenzae* strains that do not make a capsule

INTRODUCTION

The Hib vaccine protects against disease caused by the bacterium *Haemophilus influenzae* type B. This bacterium (also called "Hib") can lead to infection of the coverings of the spinal cord and brain (meningitis) and infections of the epiglottis (epiglottitis) and blood (bacteremia), among other areas of the body. These infections are dangerous and can be fatal, even with adequate treatment.

Other strains of *H. influenzae* exist and are commonly referred to as nontypeable *H. influenzae*. These strains can cause infection, though these diseases are much less virulent than those caused by *H. influenzae*. These infections, common in the ear, sinuses, and lower respiratory tract, rarely spread to the bloodstream and rarely cause meningitis.

MECHANISM

The Hib vaccine is made by taking the capsule (the polysaccharide coating) of the Hib bacterium and linking it to another protein. Injection of this safe combination incites the body to produce an immune response against Hib bacteria coating without causing the disease, thus protecting against future infection.

HISTORY

The first version of the Hib vaccine was released in 1985. It was placed on the recommended pediatric immunization schedule starting in 1989. The vaccine eventually was combined with the DTaP (diphtheria, tetanus, and pertussis) vaccine in 1996 as TriHIBit (diphtheria, tetanus, pertussis, and *Haemophilus influenzae* type B) and later with the DTaP and inactivated poliovirus vaccines as Pentacel.

As of 2021, there are five Hib vaccines licensed for use in the United States. Three of them are monovalent conjugate Hib vaccines. Monovalent means that these vaccines contain only Hib antigens without anything else. These vaccines are "conjugate" because the capsular polysaccharides are attached or "conjugated" to a carrier protein. Conjugation significantly increases the immune response generated against the polysaccharides. These three vaccines are given to infants as young as six weeks of age and include ActHIB, Hiberix, and PedvaxHIB.

The other two Hib vaccines are combination vaccines. Pentacel® contains freeze-dried ActHIB® that is dissolved in a DTaP/Inactivated Polio Virus solution. Children can receive Pentacel for all three Hib doses. Vaxelis(tm) contains hepatitis B antigens, plus DTaP and inactivated poliovirus in addition to PedvaxHIB. Neither Pentacel nor Vaxelis™ should be used for the booster dose.

ADMINISTRATION

Children should receive the Hib vaccine at two, four, six, and twelve to fifteen months of age. The vaccine is commonly administered in combination with DTaP and poliovirus in the combination vaccine Pentacel. The booster shots given between twelve to fifteen months consist only of Hib.

IMPACT

The Hib vaccine is highly effective at preventing the diseases commonly caused by the bacterium *H. influenzae*. Before the development of this vaccine,

Hib was the leading cause of meningitis in children. It is estimated that the mortality rate among infants and children who contracted this illness was 5 percent, with an even greater incidence of permanent brain damage or hearing loss, or both, among survivors. It is important to note that other bacterial causes of meningitis still exist. Still, the incidence of meningitis overall has dramatically declined since the Hib vaccine was added to the immunization schedule.

Epiglottitis, a severe disease commonly caused by Hib, was widespread before Hib vaccination became standard. Epiglottitis has virtually disappeared as a disease, and many pediatricians have learned of this illness only by anecdote.

—Jennifer Birkhauser, MD

Further Reading

Kliegman, Robert M., et al., editors. *Nelson Textbook of Pediatrics*. 21st ed., Elsevier, 2019.

Loehr, Jamie. *The Vaccine Answer Book: Two Hundred Essential Answers to Help You Make the Right Decisions for Your Child*. Sourcebooks, 2010.

Plotkin, Stanley A., et al., editors. *Vaccines*. 7th ed., Elsevier, 2017.

Stearns, Jennifer, et al. *Microbiology for Dummies*. For Dummies, 2019.

Human papillomavirus (HPV) vaccine

Category: Prevention

Anatomy or system affected: Blood, cells, circulatory system, immune system, lymphatic system

Specialties and related fields: Family medicine, immunology, internal medicine, microbiology, oncology, pediatrics, preventive medicine, public health, virology

Definition: a vaccination that activates protective immunity against human papillomavirus strains that cause cervical cancer

KEY TERMS

cervical cancer: a type of cancer that occurs in the cells of the cervix—the lower part of the uterus that connects to the vagina

human papillomavirus: a virus with subtypes that cause diseases in humans ranging from common warts to cervical cancer

papilloma: a small wartlike growth on the skin or a mucous membrane, derived from the epidermis and usually benign

DEFINITION

Three brands of human papillomavirus have been licensed by the US Food and Drug Administration: 9-valent HPV vaccine (Gardasil 9), quadrivalent HPV vaccine (Gardasil), and bivalent HPV vaccine (Cervarix). Both brands can prevent most cases of cervical cancer if the vaccine is given before HPV exposure. Gardasil can also prevent genital warts in both females and males. Since late 2016, only Gardasil-9 is distributed in the United States. Gardasil-9 protects against nine HPV strains (6, 11,

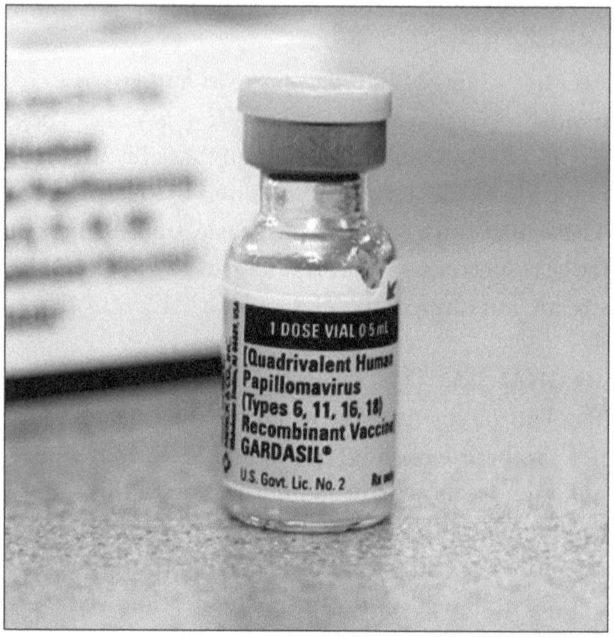

Bottle of the Gardasil vaccine for human pappilomavirus. Photo by Jan Christian, via Wikimedia Commons.

16, 18, 31, 33, 45, 52, and 58) that cause cervical cancer.

More than forty types of HPV can infect the genital areas of both males and females. Most HPV types cause no symptoms and resolve on their own. Some types of HPV, however, cause cervical cancer and other, less common, genital cancers (of the penis, anus, vagina, and vulva). The Centers for Disease Control and Prevention (CDC) estimate that 17,500 women and 9,300 men are affected by cancers caused by HPV each year. Some types of HPV can cause genital warts. Because the HPV vaccine does not prevent all kinds of cervical cancer, females who receive the HPV vaccine still need to have regular Pap tests.

CANDIDATES FOR VACCINATION

The HPV vaccine should be given before beginning sexual activity with another person. The vaccine is most effective in persons who have not been exposed to HPV.

The vaccine is recommended for children aged eleven and twelve. However, the vaccines can be administered to children as young as nine years of age. Also, people through age twenty-six years can receive the vaccine if they did not receive any or all the shots when they were younger.

Vaccination is not recommended for anyone older than twenty-six years old who has not previously received the HPV vaccine. People in this category have already been exposed to HPV and, consequently, HPV vaccination provides less benefit. Approximately 85 percent of people will get an HPV infection in their lifetime. However, anyone between twenty-seven to forty-five years old who has an increased risk of contracting HPV can receive the HPV vaccine upon their doctor's recommendation.

DOSAGE

Children who receive their first HPV vaccination before their fifteenth birthday need only two doses. Anyone who begins their HPV vaccination series after their fifteenth birthday needs three doses. Likewise, anyone with a compromised immune system needs three doses even if they begin HPV vaccination before their fifteenth birthday.

Each dose is 0.5 milliliters, administered intramuscularly, preferably in a deltoid muscle. It is best to use the same vaccine brand for all three doses. The minimum time between doses one and two of the vaccine is four weeks; between doses two and three is twelve weeks. The minimum time between dose one and dose three is twenty-four weeks. Doses received after a shorter-than-recommended time interval should be given again.

RISKS

Generally, the HPV vaccine is very safe, but mild to moderate reactions have been reported. Reactions include pain, redness, itching, bruising, or swelling at the injection site; mild to moderate fever; headache; nausea; vomiting; dizziness; and fainting. Persons who are allergic to the ingredients of the vaccines, including yeast, should not receive the vaccine, nor should pregnant persons.

IMPACT

The HPV vaccine is the first preventive cancer vaccine. Initially, the vaccine was controversial because some parents and religious groups claimed it would make casual sex acceptable, especially among girls. However, studies by Merck (the manufacturer of Gardasil) and independent researchers show no link between receiving the vaccine and increased sexual activity. Lawmakers are debating whether to make this vaccine mandatory; as of December 2015, it was mandatory only in Rhode Island, Virginia, and Washington, DC. Many teenagers were still not receiving it. According to the CDC, in 2014, 60 percent of adolescent girls and 42 percent of boys had received at least one dose of the vaccine, an increase over the previous year but still a lower percentage than those receiving other vaccines recommended for eleven- and twelve-year-olds, such as the Tdap (tetanus, diphtheria, and pertussis) and meningitis vaccines.

—Claudia Daileader Ruland, MA

Further Reading

Centers for Disease Control and Prevention. "FDA Licensure of Bivalent Human Papillomavirus Vaccine (HPV2, Cervarix) for Use in Females: Recommendations of the Advisory Committee on Immunization Practices (ACIP)." *Morbidity and Mortality Weekly Report*, 28 May 2010, pp. 626-29.

———. "FDA Licensure of Quadrivalent Human Papillomavirus Vaccine (HPV4, Gardasil) for Use in Males: Recommendations of the Advisory Committee on Immunization Practices (ACIP)." *Morbidity and Mortality Weekly Report*, 28 May 2010, pp. 630-32.

Centers for Disease Control and Prevention. "HPV Vaccine Information for Young Women." *Centers for Disease Control and Prevention*. Department of Health and Human Services, 26 Mar. 2015. Accessed 30 Dec. 2015.

Daniel, Jennifer. "Good Talks Needed to Combat HPV Vaccine Myth." *New York Times*. New York Times, 9 Nov. 2015. Accessed 30 Dec. 2015.

Dunne, E. F., and L. E. Markowitz. "Genital Human Papillomavirus Infection." *Clinical Infectious Diseases*, vol. 43, 2006, p. 624.

"Human Papillomavirus (HPV) Vaccines." *National Cancer Institute*. National Institutes of Health, 19 Feb. 2015. Accessed 30 Dec. 2015.

Larsen, Laura. *Sexually Transmitted Diseases Sourcebook*. Omnigraphics, 2009.

McCance, Dennis J., editor. *Human Papilloma Viruses*. Elsevier Science, 2002.

Plotkin, Stanley A., et al., editors. *Vaccines*. 7th ed., Elsevier, 2017.

Thompson, Dennis. "CDC Says Too Few US Teens Getting HPV Vaccine." *CBS News*. CBS, 30 July 2015. Accessed 30 Dec. 2015.

INFLUENZA VACCINE

Category: Prevention

Also known as: Flu vaccine

Anatomy or system affected: Blood, cells, circulatory system, immune system, lymphatic system

Specialties and related fields: Biochemistry, biotechnology, family practice, gerontology, immunology, microbiology, pediatrics, preventive medicine, public health, virology

Definition: vaccines that protect against infection by influenza viruses

KEY TERMS

antigenic drift: a kind of genetic variation in viruses, arising from the accumulation of mutations in the virus genes that code for virus-surface proteins that host antibodies recognize

antigenic shift: the process by which two or more different strains of a virus, or strains of two or more different viruses, combine to form a new subtype having a mixture of the surface antigens of the two or more original strain

glycoproteins: a class of proteins that have carbohydrate groups attached to the polypeptide chain

orthomyxoviruses: a family of single-stranded ribonucleic acid (RNA) viruses that have a spherical or filamentous virion with numerous surface projections of glycoprotein and include the causative agents of influenza

INTRODUCTION

The influenza vaccine helps to protect against infection with the influenza virus. Influenza is an acute viral respiratory illness with an abrupt onset. It is spread primarily by respiratory droplets from person to person (mainly through inhalation of virus-containing droplets). Influenza is caused by a group of viruses of the Orthomyxoviridae family, separated into three strain types (A, B, and C) according to their nuclear material.

Vaccination is the most effective protection against influenza. The vaccine may be administered to anyone over six months old wishing to reduce the risk of influenza, and the US Centers for Disease Control (CDC) suggests universal vaccination to provide the best possible protection. It is recommended especially for unhealthy persons and persons likely to transmit influenza to unhealthy persons in a given community. However, some people should not receive the vaccine due to other conditions. Anyone with life-threatening allergies to any vaccine element should avoid vaccination or talk to their doctor about what is suitable. People with egg allergies can usually receive the vaccine. Still, they should consult their doctor, especially if they have had reactions other than hives to eggs. Those with a history of Guillain-Barré Syndrome (GBS) may also be advised not to receive the flu vaccine. Finally, people who are already sick should consider waiting to be vaccinated when they are feeling better.

Influenza vaccines are designed to trigger an immune response to hemagglutinin and neuraminidase, the two proteins found on the surface of the influenza virus. These proteins are constantly changing (mutating), so every year (antigenic drift), seasonal influenza vaccines have to be reformulated with the three strains that are likely to be more effective in fighting new influenza strains.

The World Health Organization's Global Influenza Programme monitors the influenza viruses circulating among humans worldwide and quickly identifies new strains to make new, appropriate vaccines for a particular year.

Predicting which influenza strains will dominate patients depends on which flu strains are circulating during the winter on the other side of the globe. Countries in the northern hemisphere, like Canada and the United States experience winter, and flu season, at the opposite time than countries in the southern hemisphere, like Australia and South America. The flu strains that cause the most cases in countries in the southern hemisphere during their flu season provide the guidelines for those strains that influenza vaccines must include.

INFLUENZA VIRUS STRAINS

Influenza belongs to the orthomyxovirus group, and its ribonucleic acid (RNA) genome is broken into seven different segments. The most common type of influenza virus is type A. Type A influenza viruses are further divided based on the two glycoproteins embedded in their envelopes, the H or hemagglutinin, and the N or neuraminidase glycoproteins. The H and the N glycoproteins vary slightly in their structure. Therefore, different versions of these glycoproteins are designated by a number. For example, a subtype of the type A influenza virus, H3N2, has hemagglutinin number 3 and neuraminidase number 2 on its surface. Type A influenza viruses can infect animals, which allows them to swap RNA segments and different H and N glycoproteins. This phenomenon, antigenic shift, makes influenza viruses with novel combinations of different types of H and N glycoproteins.

To give an influenza strain its full name, you must list the type, the virus' original host, the location where the virus was first identified (usually a city), the strain number, the year it was recognized, and the subtype according to the virus' H and N surface glycoproteins. For example, the type A influenza strain H1N1 arose in ducks from Alberta, Canada, and was the thirty-fifth strain discovered in 1976. Therefore, influenza H1N1 would be A/duck/Alberta/35/76 (H1N1).

Type B influenza only infects humans and does not vary as much as type A. Type B influenza only has a few types of H and N glycoproteins. Type A influenza also infects animals which aid in making new viral strains. Type B flu strains are named with the same conventions as type A, but there is no need to specify its origin since it only infects humans. For example, type B influenza found in Yamagata, Japan was the sixteenth strain identified in 1988 is B/Yamagata/16/88.

Type C influenza is the least common influenza virus strain and the least likely to vary. Influenza type C usually causes mild disease in children. However, unlike type B influenza viruses, they can infect humans and pigs. Type C influenza viruses do not have H and N glycoproteins on their surfaces. Instead, they have a hemagglutinin-esterase-fusion protein to enter and exit cells. Therefore, type influenza viruses are listed without the NH subtype. An example of a type C influenza virus is one found in São Paulo, Brazil, and was the thirty-seventh strain found in 1982. Thus, this virus is called "C/Sao Paulo/37/82."

The influenza vaccines for 2021-2022 contain the four following influenza strains:

- A/Victoria/2570/2019 (H1N1) pdm09-like
- A/Cambodia/e0826360/2020 (H3N2)-like
- B/Washington/02/2019 (Victoria lineage)-like
- B/Phuket/3073/2013 (Yamagata lineage)-like

TYPES OF INFLUENZA VACCINES

The trivalent inactivated influenza vaccine (TIV) has been available since the mid-twentieth century. However, it is not available in the United States for the 2021-22 influenza season. For the 2021-22 flu season, all influenza vaccinations are quadrivalent. TIV and quadrivalent inactivated vaccines (QIV) are administered by intramuscular routes. The four standard-dose, inactivated, quadrivalent influenza vaccines include *Afluria Quadrivalent, Fluarix Quadrivalent, FluLaval Quadrivalent,* and *Fluzone Quadrivalent.*

The influenza vaccine viruses are grown in chicken eggs, and the final product contains trace amounts of egg protein. Regardless, numerous studies have demonstrated that those with egg allergies are not at increased risk for a reaction to any influenza vaccine. Flu vaccines are available in both pediatric- and adult-dose formulations. They can be preservative-free in a single vial or a multidose vial with thimerosal as a preservative.

For those who cannot tolerate egg proteins at any concentration, two completely egg-free vaccines are available. *Flucelvax Quadrivalent* is made from inactivated influenza viruses grown in tissue culture. This vaccine is US Food and Drug Administration (FDA)-approved for children four years and older and completely egg-free. *Flublok Quadrivalent* is a recombinant influenza vaccine for those eighteen years and older, and this vaccine is also egg-free.

The live attenuated influenza vaccine (LAIV) that contains the same four influenza viruses as the QIV is administered by the intranasal route. *FluMist* Quadrivalent is an intranasal live-attenuated influenza vaccine that is FDA-approved for use in healthy nonpregnant persons 2-49 years old. LAIV viruses are also grown in chicken eggs. LAIV is preservative-free and is provided in a single-dose sprayer unit with one-half the dosage sprayed into each nostril. LAIV is not recommended to be used for some flu seasons, depending on the severity of the strain and the effectiveness of the vaccine.

Older adults (65-82 years old) have weaker immune responses to influenza vaccines than younger adults and children. Likewise, the antibody levels elicited by the influenza vaccine diminish more rapidly in older individuals, decreasing vaccine effectiveness. To address this shortcoming of the standard dose, inactivated influenza vaccines, older adults have several options. Flublok Quadrivalent is a recombinant influenza vaccine produced without whole influenza viruses or chicken eggs. These recombinant vaccines contain three times the amount of antigen in the standard-dose, inactivated vaccines. Fluzone High-Dose Quadrivalent is an inactivated vaccine containing four times the amount of antigen in standard-dose inactivated influenza vaccines. This vaccine is FDA-licensed for use in persons ≥ 65 years old. Another option for older adults is an adjuvanted, inactivated influenza vaccine called "Fluad Quadrivalent." This vaccine is FDA-licensed for use in persons ≥ 65 years old. It contains an oily compound called "MF59" (an oil-in-water emulsion of squalene) that increases the immune response to the vaccine. Adjuvants recruit antigen-presenting cells to the injection site and promote influenza antigen uptake by those cells.

Improved technology and innovation have enabled improved methods of administering influenza vaccines, including a reduced-dose injectable made possible by adding adjuvants and using a cell culture vaccine. Scientists are also exploring new routes of administration, such as intradermal (with or without needle) and transcutaneous. A patch delivers the vaccine through microneedles that may barely penetrate the skin before dissolving and releasing the vaccine.

EFFICACY AND FURTHER RESEARCH

Most vaccinated persons develop postvaccination hemagglutination inhibition antibody titers. These antibodies are protective against illness caused by strains similar to those in the vaccine or related variants that may emerge during outbreaks. However, while vaccines are considered the best protection against influenza, they are not 100 percent effective. The constantly mutating nature of the influenza virus means that vaccine makers are continually trying to keep up with the latest strain. The prevalent method of growing the vaccine in eggs means that it may be prone to mutate into a less effective form. Even the most successful vaccines usually only protect about 60 percent of those receiving them. When the vaccine is poorly matched to the prevalent strain, the efficacy can be as low as 10 percent protection. Vaccination is further complicated by individual immune systems' complexity and unique character, which can interact differently with even vaccines made from the correct seasonal strain.

Amid debates over the effectiveness of vaccinations, concern also always existed over the high number of elderly adults who would still succumb to influenza and possibly lose their lives despite being vaccinated because of their naturally weaker immune systems. Therefore, in late 2015, the FDA approved an influenza vaccine booster that had previously been used in several other countries. The vaccine with the adjuvant, which helps stimulate the immune system to make the vaccine more effective, is known as *Fluad* and was first made available in the United States in 2016.

Additionally, scientists have continued to experiment on whether a universal vaccine could be produced to help the immune system fight groups of viruses rather than a specific strain. Research suggests that new methods of vaccine development, such as genetic engineering instead of cultivation in eggs, may help improve vaccine efficacy. However, it has been suggested that many scientists are not inclined to openly discuss issues with existing vaccines due to fears of stoking the arguments of the antivaccination movement, which is disregarded as pseudoscience.

—Oladayo Oyelola, PhD, SC (ASCP)
and Michael A. Buratovich, PhD

Further Reading

Bennett, John E., et al., editors. *Mandell, Douglas, and Bennett's Principles and Practice of Infectious Diseases*. 9th ed., Elsevier, 2019.

Cohen, John. "Why Flu Vaccines So Often Fail." *Science*, 20 Sept. 2017, www.sciencemag.org/news/2017/09/why-flu-vaccines-so-often-fail. Accessed 30 Nov. 2021.

Delves, Peter J., et al. *Roitt's Essential Immunology*. 13th ed., Wiley-Blackwell, 2017.

"Flu (Influenza)." *Vaccines.gov*. US Department of Health and Human Services, 8 Sept. 2021, www.vaccines.gov/diseases/flu/index.html. Accessed 30 Nov. 2021.

Hak, E., et al. "Influence of High-Risk Medical Conditions on the Effectiveness of Influenza Vaccination Among Elderly Members of Three Large Managed-Care Organizations." *Clinical Infectious Diseases*, vol. 35, 2002, pp. 370-77.

"Influenza Vaccine for 2021-2022." *Medical Letter on Drugs and Therapeutics*, vol. 63, no. 1634, 2021, pp. 153-57.

"Key Facts About Seasonal Flu Vaccine." *Centers for Disease Control and Prevention*. US Department of Health and Human Services, 18 Nov. 2021, https://www.cdc.gov/flu/prevent/keyfacts.htm. Accessed 20 Nov. 2021.

Plotkin, Stanley A., et al. *Vaccines*. 7th ed., Elsevier, 2017.

Zhang, Sarah. "Scientists Get One Step Closer to a Universal Flu Vaccine." *Wired*. Condé Nast, 24 Aug. 2015. Accessed 30 Dec. 2015.

MALARIA VACCINE

Category: Prevention

Anatomy or system affected: Blood, cells, circulatory system, immune system, lymphatic system

Specialties and related fields: Biochemistry, biotechnology, immunology, internal medicine, microbiology, preventive medicine, protozoology, public health

Definition: a preparation designed to provide immunity against infection by the parasite *Plasmodium*, which leads to malaria

KEY TERMS

malaria: an infectious disease caused by protozoan parasites from the *Plasmodium* family that can be transmitted by the bite of the Anopheles mosquito or by a contaminated needle or transfusion

Plasmodium: a parasitic protozoan of a genus that includes those causing malaria

BACKGROUND

Four species of *Plasmodium* cause malaria: *P. falciparum*, *P. vivax*, *P. malariae*, and *P. ovale*. These four species cause about 90 percent of malaria cases and are responsible for the most deaths, particularly in Africa. The World Health Organization estimated in 2015 that 3.2 billion people were at risk for malaria exposure. About 214 million people contracted malaria worldwide in 2014, and nearly a half-million people died from the disease.

The *Anopheles* mosquito transmits plasmodia; the incubation period lasts between seven and thirty days, depending on the *Plasmodium* species transmitted. Symptoms of malaria include shivering, fever, headache, vomiting, and sweating. Severe malaria can involve such symptoms as impaired consciousness, seizures, coma, anemia, pulmonary edema, and cardiovascular collapse.

VACCINE STATUS

Preventing malaria infection is a top priority for many health and research organizations worldwide, as they are trying to establish vaccines to protect against the disease. Despite decades of research on the topic, no commercially available vaccine for malaria exists. Many researchers focus on developing vaccines against *P. falciparum*, while a few groups are working on a vaccine for *P. vivax*. The life cycle of *P. falciparum* is quite complex, as it provides several stages on which to focus vaccine development.

The most advanced vaccine is RTS, S/AS01, studied in phase-three trials in several countries in sub-Saharan Africa since 2009; the phase-two trial for this drug showed 30 to 50 percent efficacy in reducing malaria in infants and children. Based on these results, the vaccine will only partially protect those immunized.

Another promising vaccine is FMP2.1/AS02A, which has shown efficacy in children in Mali. Numerous clinical trials have attempted to select safe, effective vaccines. Because of the complexity of the parasite's life cycle, multiple types of vaccines will likely be necessary to interrupt that life cycle.

IMPACT

A viable, disease-preventing malaria vaccine has the potential to save millions of lives by protecting against *Plasmodium* infection.

—*Dawn M. Bielawski, PhD*

Further Reading

Bonam, Srinivasa Reddy, et al. "*Plasmodium falciparum* Malaria Vaccines and Vaccine Adjuvants." *Vaccines,* vol. 9, no. 10, 2021, p. 1072, 24 Sept. 2021, doi:10.3390/vaccines9101072.

Crompton, Peter D., Susan K. Pierce, and Louis H. Miller. "Advances and Challenges in Malaria Vaccine Development." *Journal of Clinical Investigation,* vol. 120, 2010, pp. 4168-78.

Enayati, A., and J. Hemingway. "Malaria Management: Past, Present, and Future." *Annual Review of Entomology*, vol. 55, 2010, pp. 569-91.
Mahamadou, A. Thera, et al. "Safety and Immunogenicity of an AMA1 Malaria Vaccine in Malian Children." *PLoS*, vol. 5, 2010, p. e9041.
Sherman, Irwin W. *The Elusive Malaria Vaccine: Miracle or Mirage?* ASM Press, 2009.
"10 Facts on Malaria." *World Health Organization*, Nov. 2015, www.who.int/features/factfiles/malaria/en. Accessed 16 Nov. 2016.

MMR VACCINE

Category: Prevention
Anatomy or system affected: Blood, cells, circulatory system, immune system, lymphatic system
Specialties and related fields: Immunology, microbiology, preventive medicine, public health, virology
Definition: a vaccine against measles, mumps, and rubella

KEY TERMS

measles: an infectious viral disease that causes fever and a red rash on the skin, typically occurring in childhood
mumps: a contagious and infectious viral disease that causes swelling of the parotid salivary glands in the face and risk of sterility in adult males
rubella: a contagious viral disease with symptoms like mild measles; it can cause fetal malformation if contracted in early pregnancy

INTRODUCTION

The MMR vaccine combines immunizations for three diseases (measles, mumps, and rubella) into a single series of injections. Each of these childhood diseases is caused by a different virus. Measles leads to rash, fever, cough, and irritated eyes. It may lead to pneumonia, seizures, and (in severe cases) brain damage and death—mumps results in characteristic swollen glands in the neck, accompanied by a fever and headaches. Mumps may lead to meningitis, deafness, painful and damaging swelling of the testes, and (in severe cases) death. Rubella, or German measles, causes a rash with a mild fever and arthritis. Rubella infection in pregnant women can cause miscarriage or congenital disabilities and disorders.

According to the Centers for Disease Control and Prevention (CDC), all children should receive two doses of the MMR vaccine. Children should receive their first dose at twelve to fifteen months and their second at four to six years. After receiving the first dose, children can receive their second dose twenty-eight days after their first dose.

An alternative is the MMRV vaccine that protects against measles, mumps, rubella, and chickenpox. The MMRV vaccine is licensed in children who are twelve months through twelve years of age.

BENEFITS OF VACCINATION

The combined MMR vaccine protects children and adults against measles, mumps, and rubella altogether. Before the vaccine was developed, these highly contagious diseases were prevalent, and virtually all children became infected at some point. In the 1960s, vaccines were developed for each disease individually, and in 1971 the separate vaccines were combined into the MMR vaccine. In 1993 doctors began recommending a booster shot to increase children's protection against the diseases. In 2005 a version known as MMRV was made available that combined the standard MMR vaccine with the vaccine for chickenpox, or varicella.

The vaccination program was highly successful at reducing all three diseases in the United States. Outbreaks of measles, mumps, and rubella only typically continue to occur in areas with clusters of nonimmunized children, such as in religious communities that avoid immunization or in families in which a parent or parents fear that the MMR vaccine has harmful side effects and has a link to autism.

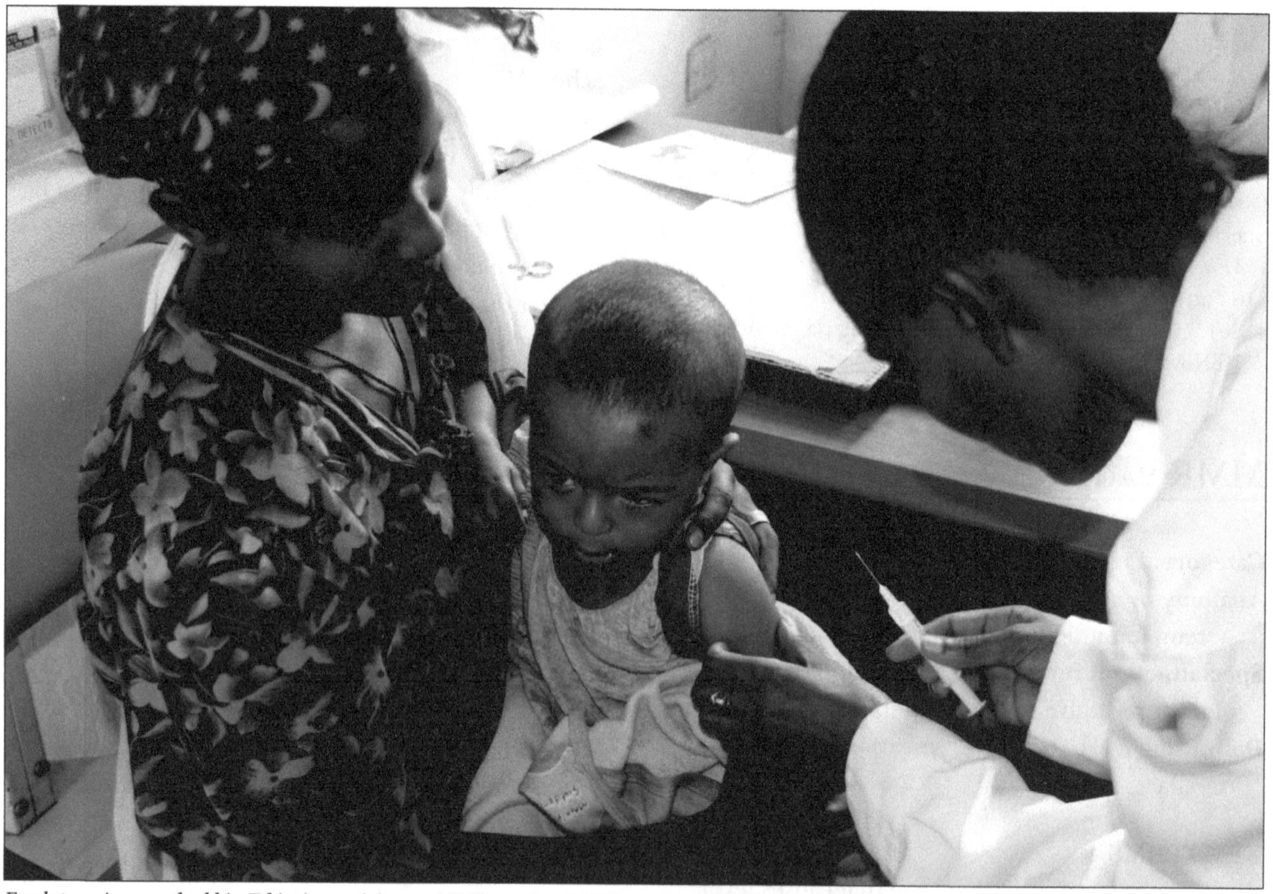

Emabet, a nine-month-old in Ethiopia, receiving an MMR vaccine. MMR vaccines are typically administered between nine and fifteen months. Then a second dose is given at least four weeks later—image courtesy of the DFID UK Department for International Development via Wikimedia Commons.

MMR vaccine and autism. A controversial study published in 1998 by the journal *The Lancet* suggested a link between the MMR vaccine and rising rates of autism. The article soon led to widespread fear among parents of the vaccine's safety, and some parents refused the vaccine for their children. Pockets of nonimmunized children contributed to renewed measles, mumps, and rubella outbreaks in the United States and the United Kingdom, and other European countries. However, the original study was flawed, and *The Lancet* officially retracted the report in February 2010. The article, authored by the discredited British researcher Andrew Wakefield and coauthors, had erroneous conclusions. Additional research attempting to replicate Wakefield's findings did not support his results. Rather, a further study found no evidence of a link between the MMR vaccine and autism, supporting vaccination safety. Still, the negative publicity generated by the report helped sustain a vocal minority of antivaccination advocates despite widespread scientific consensus that failing to vaccinate children hurts public health.

Side effects. The MMR vaccine is associated with mild side effects, including fever, mild rash, and swollen glands. Less common side effects include seizures and temporary joint pain. Rarely, allergic reactions or serious side effects such as deafness, long-term seizures, and brain damage may occur.

IMPACT

According to the Centers for Disease Control of Prevention (CDC), the MMR vaccine has reduced the incidence of measles, mumps, and rubella by more than 99 percent. The vaccine's success in dramatically reducing the spread of these diseases has enabled the US government's Childhood Immunization Initiative to set a goal of eradicating native measles, mumps, and rubella in the United States. This goal acknowledges that the viruses may be brought to the United States by infected people in other countries.

The vaccine leads to lifelong immunity. Children receive the dose between twelve and fifteen months of age and get a booster shot between four and six years of age. After two doses, the vaccine protects 99 percent of the children immunized. In some cases, adults may be recommended to receive the vaccine.

—Cheryl Pokalo Jones

Further Reading

Centers for Disease Control and Prevention. "Vaccine Safety: Measles, Mumps, and Rubella (MMR) Vaccine." www.cdc.gov/vaccinesafety.

Centers for Disease Control and Prevention. "Measles, Mumps, and Rubella (MMR) Vaccination: What Everyone Should Know." *Vaccines and Preventable Diseases*, 26 Jan. 2021, www.cdc.gov/vaccines/vpd/mmr/public/index.html. Accessed 2 Dec. 2021.

Editors of *The Lancet*. "Retraction: Ileal-Lymphoid-Nodular Hyperplasia, Non-specific Colitis, and Pervasive Developmental Disorder in Children." *The Lancet*, vol. 375, 2020, p. 445.

Griffin, Diane E., and Michael B. A. Oldstone, editors. *Measles: History and Basic Biology*. Springer, 2009.

Hawkins, Trisha. *Everything You Need to Know About Measles and Rubella*. Rosen, 2001.

Institute of Medicine. *Immunization Safety Review: Vaccines and Autism*. National Academies Press, 2004.

"Measles: Questions and Answers." *Immunization Action Coalition*. Immunization Action Coalition, n.d. Accessed 23 Dec. 2015.

"MMR Vaccine Does Not Cause Autism." *Immunization Action Coalition*. Immunization Action Coalition, n.d. Accessed 23 Dec. 2015.

mRNA vaccines

Category: Prevention

Anatomy or system affected: Blood, cells, circulatory system, immune system, lymphatic system

Specialties and related fields: Biochemistry, biotechnology, immunology, pharmacology, preventive medicine, public health, virology

Definition: a type of vaccine that uses a copy of a molecule called "messenger ribonucleic acid" (mRNA) to produce an immune response

KEY TERMS

antibodies: a large, Y-shaped protein used by the immune system to identify and neutralize foreign objects such as pathogenic bacteria and viruses

antigen: a molecule or molecular structure that can bind to a specific antibody or T-cell receptor

lymphocytes: a form of small leukocyte (white blood cell) with a single round nucleus, occurring especially in the lymphatic system

nanoparticles: a particle of matter that is between 1 and 100 nanometers in diameter

INTRODUCTION

A messenger ribonucleic acid (mRNA) vaccine uses a piece of genetic material named messenger RNA (mRNA) to instruct the body to trigger an immune response. Previously, many vaccines introduced a weakened or dead piece of a virus or bacterium into the body to jumpstart the immune system. However, mRNA vaccines do not contain any trace of an infectious agent. Instead, they carry genetic instructions that teach the body's immune cells to make a special protein that mimics an infectious agent. The body's immune system begins mounting a defense against the targeted virus or bacterium. Scientists researched mRNA vaccines for decades before achieving breakthroughs in technology in the twenty-first century. That progress proved beneficial in 2020 when several companies used mRNA technology to develop vaccines against the COVID-19 pandemic, which, as of February 2021, had already infected tens of millions and claimed over two million lives.

BACKGROUND

The body's immune system consists of cells, organs, and tissues that work together to fight off infections. One of the chief weapons used by the immune system is white blood cells, which perform various functions in the fight against invading viruses and bacteria. Macrophages are a type of white blood cell specifically designed to seek out and digest foreign invaders. Macrophages leave behind pieces of the invader known as antigens. The immune system sees these antigens as a threat. It begins producing proteins known as antibodies to target those specific antigens.

The body also produces two specialized white blood cells called "lymphocytes" to battle the infection. B-lymphocytes produce antibodies that mark a specific antigen and target it for destruction. T-lymphocytes attack infected cells in the body. One type of T-lymphocyte called a "killer T-cell" targets the invader marked by the B-lymphocyte and destroys it. Some lymphocytes can remember a specific antigen and warn the body if the antigen is detected again. This re-exposure prompts the immune system to produce more antibodies to destroy the invader. In this way, the body can develop immunity to a specific virus or bacteria.

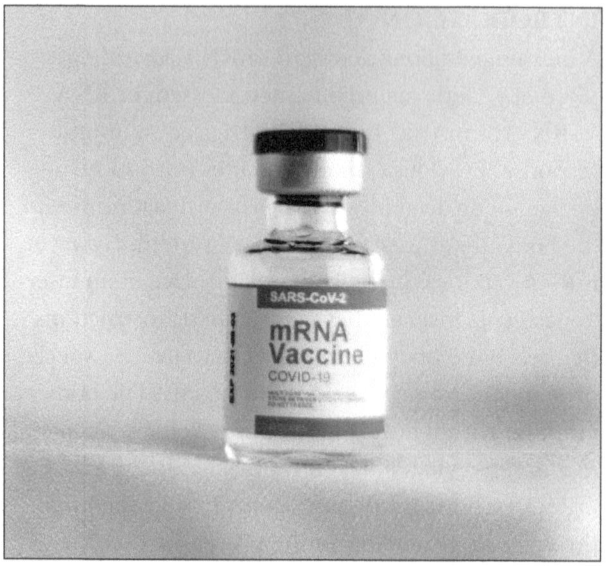

An example of the COVID-19 mRNA vaccines. Photo by Spencerbdavis, via Wikimedia Commons.

OVERVIEW

Vaccines use the immune system's ability to remember antigens to protect the body against disease. Typically, vaccines introduce a weakened or dead form of a virus or bacterium to prompt the immune system to begin making antibodies. The infectious agent's weakened or dead form does not cause serious illness. Still, it carries the specific antigen that can be targeted by the white blood cells. As the immune system reacts to the vaccine, it produces lymphocytes that will remember that antigen and respond to the live virus or bacterium if it enters the body.

While these vaccines have proven highly effective in fighting disease, they do come with some drawbacks. Chief among these is that a piece of a sometimes-serious biological agent is introduced into the body. Even if the virus or bacteria is weakened, it can still cause some symptoms of the illness. Furthermore, determining how weakened an infection can be and still prove effective is often difficult.

In the late twentieth century, scientists began to look at a new way to fight disease using messenger RNA (mRNA). Messenger RNA is a single strand of ribonucleic acid (RNA) that carries instructions from the cell's deoxyribonucleic acid (DNA). The DNA in the cellular nucleus gives an mRNA strand a copy of instructions on making a specific protein. The mRNA travels out into the cell's cytoplasm, transferring that information to complex molecular machines called "ribosomes." These ribosomes read the instructions and follow them to make the protein.

Scientists believed that if they could find a way to introduce specially coded mRNA into the body, it could instruct the cells to make specific proteins. These proteins could be targeted to act as disease-fighting antibodies or enzymes that could reverse illness or repair or grow tissue. Researchers began studying the process in the 1960s, and by 1990 had found a way to introduce mRNA into mice. However, using the method on humans proved extremely difficult. The body's immune system would recognize the laboratory-made mRNA as a foreign invader and destroy it before it could do its job.

In 2005, scientists discovered a way to alter the molecules that made up the mRNA strand to fool the body into accepting it. Soon, they began to view mRNA technology as a potentially revolutionary weapon in the fight against the disease. Researchers at several pharmaceutical companies worldwide began working with mRNA on potential vaccines for diseases such as cancer, rabies, and influenza. They solved the problem of the immune system attacking the mRNA by further altering its genetic makeup or encasing it in a lipid coating to protect it. A lipid is a fatty acid that does not dissolve in water.

For an mRNA vaccine to work, scientists do not need a sample of an infectious agent such as a virus or bacterium. They only need to know the genetic mRNA sequence the virus or bacteria uses to make its protein coat. When the vaccine enters the body, the immune cells consume the mRNA code. Once inside, the mRNA instructs the cell to begin making its unique protein. The body's immune system recognizes the protein antigen and begins producing antibodies to destroy it. After the mRNA has done its job, it is broken down and destroyed by the cell. As a bonus, mRNA vaccines have also been found to trigger the production of killer T-cells, doubling their effectiveness in combatting the disease.

Researchers had made significant progress on mRNA vaccines by January 2020 when the world faced a global health crisis brought on by the rapid spread of a virus. The virus, SARS-CoV-2, caused the disease COVID-19 that infected more than one hundred million people by the end of 2020 and killed more than two million. Scientists at two pharmaceutical companies—Germany's BioNTech and the US-based Moderna—recognized their mRNA research was well-suited to the quick production. Once they received the genetic code of SARS-CoV-2, they were able to create an mRNA strand that mimicked the protein spikes on the virus's surface. The immune system responded to the protein spikes by producing antibodies to fight COVID-19 without patients ever having to have the virus present in their bodies. By the end of 2020, the companies had produced vaccines that passed through a rapid testing process and were distributed in several countries worldwide.

—Richard Sheposh

Further Reading

Fiore, Kristina. "Want to Know More About mRNA Before Your COVID Jab?" *MedPage Today*, 3 Dec. 2020, www.medpagetoday.com/infectiousdisease/covid19/89998. Accessed 3 Feb. 2021.

Garde, Damian, and Jonathan Saltzman. "The Story of mRNA: How a Once-Dismissed Idea Became a Leading Technology in the Covid Vaccine Race." *Stat*, 10 Nov. 2020, www.statnews.com/2020/11/10/the-story-of-mrna-how-a-once-dismissed-idea-became-a-leading-technology-in-the-covid-vaccine-race/. Accessed 3 Feb. 2021.

"History of Vaccine Development." *World Health Organization*, 2021, vaccine-safety-training.org/history-of-vaccine-development.html. Accessed 3 Feb. 2021.

Komaroff, Anthony. "Why Are mRNA Vaccines So Exciting?" *Harvard Health Publishing*, 18 Dec. 2020, www.health.harvard.edu/blog/why-are-mrna-vaccines-so-exciting-2020121021599. Accessed 3 Feb. 2021.

Pardi, Norbert, et al. "mRNA Vaccines—A New Era in Vaccinology." *Nature Reviews Drug Discovery*, vol. 17, 12 Jan. 2018, pp. 261-279, www.nature.com/articles/nrd.2017.243. Accessed 3 Feb. 2021.

Riegelman, Richard. "Population Prevention and COVID-19." *COVID-19 Global Lessons Learned: Interactive Case Studies*. Jones & Bartlett Learning, 2021, pp. 31-42.

"Understanding How COVID-19 Vaccines Work." *Centers for Disease Control and Prevention*, 13 Jan. 2021, www.cdc.gov/coronavirus/2019-ncov/vaccines/different-vaccines/how-they-work.html. Accessed 3 Feb. 2021.

"What Does mRNA Do? mRNA Produces Instructions to Make Proteins That May Treat or Prevent Disease." *Moderna*, 2020, www.modernatx.com/mrna-technology/science-and-fundamentals-mrna-technology. Accessed 3 Feb. 2021.

PNEUMOCOCCAL VACCINE

Category: Prevention

Also known as: Pneumococcal polysaccharide vaccine, pneumococcal conjugate vaccine, *Streptococcus pneumoniae* vaccine, Prevnar 13, Pneumovax 23, PVC13, PPV23

Anatomy or system affected: Blood, cells, circulatory system, immune system, lymphatic system

Specialties and related fields: Bacteriology, biochemistry, biotechnology, family practice, gerontology, immunology, internal medicine, microbiology, pediatrics, preventive medicine, public health

Definition: a vaccine that prevents pneumococcal infections

KEY TERMS

antigen: any substance that causes the body to make an immune response against that substance

capsule: an extracellular polymeric substance that surrounds bacteria and coats them

conjugated vaccines: a type of subunit vaccine which combines a weak antigen with a strong antigen as a carrier so that the immune system has a stronger response to the weak antigen

polysaccharide: a large carbohydrate molecule made of many smaller monosaccharides

DEFINITION

The pneumococcal vaccine prevents disease caused by various types of *Streptococcus pneumoniae* bacteria (also known as pneumococcus), depending on the type of immunization administered. These diseases include pneumonia, middle-ear infection (otitis media), and sinusitis. If untreated, pneumococcal disease can spread quickly to the blood and spinal cord, resulting in bacteremia and meningitis, respectively, which can be devastating. More serious cases can result in death.

VACCINATION TYPES

The two pneumococcal vaccines that are most used to prevent infection by *S. pneumonia* are the pneumococcal polysaccharide vaccine (PPV23) and the pneumococcal conjugate vaccine (PCV13). Pneumococcal polysaccharide vaccines contain purified capsular polysaccharide antigens. Pneumococcal conjugate vaccines contain *S. pneumoniae* serotypes conjugated to CRM197.

PCV13 contains thirteen serotypes of *S. pneumoniae*, including serotype 19A, a leading cause of invasive pneumococci infections in children. PPV23 contains twenty-three *S. pneumoniae* serotypes, including all the serotypes found in PCV13 (except for serotype 6A).

MECHANISM OF ACTION

The vaccine is made by taking the shell, or polysaccharide coating, of the *S. pneumoniae* bacterium and linking it to another protein. Injection of this safe combination incites the body to produce an immune response against this bacterial coating without actually causing the disease, thus protecting against future infection.

Approximately ninety different serotypes of *S. pneumoniae* bacteria exist. The polysaccharide coatings from the thirteen most dangerous types to children are those found in the Prevnar 13 vaccine. The coatings from the twenty-three most commonly encountered types of *S. pneumoniae* are used for the adult version of the vaccine, Pneumovax.

VACCINE HISTORY

The first pneumococcal vaccine was licensed in 1977 and protected against fourteen different types of *S. pneumoniae*. Merck Sharp & Dohme Corp released the

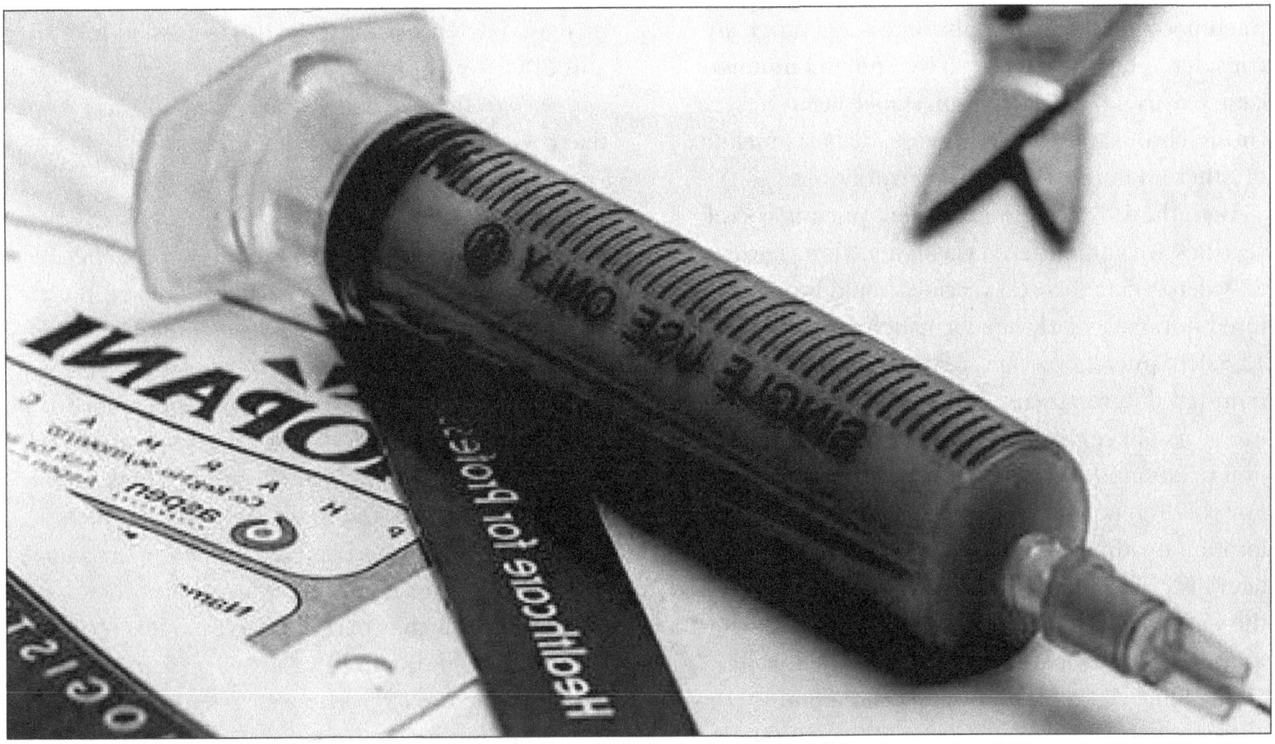

A dose of Prevenar, the most commonly used pneumococcal vaccine in the United States for children between 2 and 14 months of age. Photo by Jusnuel, via Wikimedia Commons.

most recent 23-valent form of the vaccine (PPV23) in 1983 under Pneumovax. As PPV23 is ineffective in children under two years of age, in 2000, the 7-valent pediatric form of the pneumococcal vaccine (PCV7) was licensed under Prevnar, and routine administration to all children was recommended. Researchers further improved the pediatric vaccine to provide broader coverage against pneumococcal disease, especially the serotype 19A isolate, with the 13-valent form of Prevnar (PCV13) released in 2010.

ADMINISTRATION

Medical experts recommend that all children under two years of age and adults sixty-five years or older receive a pneumococcal vaccine. Children should receive initial injections of the 13-valent pneumococcal vaccine at two, four, and six months. An additional booster is recommended between twelve to fifteen months of age. PPV23 is not effective in children under two years of age. Children aged two years and older who are at high risk of developing pneumococcal disease should be given the 23-valent pneumococcal vaccine. Children are considered high-risk if they have conditions that cause weakened immune systems or have heart, lung, or liver disease.

Medical experts recommend that adults sixty-five years and older receive doses of PCV13 and PPV23 as a vaccination series. The Advisory Committee on Immunization Practices (ACIP) recommends administering a dose of PCV13 first, followed by a dose of PPV23 later, usually not exceeding a year from the administration of the initial PCV13 dose. Additional doses of PPV23 can be administered if there is a significant risk of infection. PCV13 and PPSV23 should not be administered simultaneously, as this reduces the efficacy of the 23-valent pneumococcal vaccine. Adults between nineteen and sixty-five should receive a vaccination if they are considered high-risk for pneumococcal disease. Adults in this age range are considered high risk if they have human immunodeficiency virus (HIV), alcoholism, smoke cigarettes, cirrhosis, chronic pulmonary disease, diabetes mellitus, or other immunocompromising conditions.

Both the 13-valent and 23-valent pneumococcal vaccines are administered via an injection. The 23-valent pneumococcal vaccine should be administered intramuscularly or subcutaneously, while the 13-valent pneumococcal vaccine should only be administered intramuscularly. For older children (over two years of age) and adults, the recommended location to administer the injection is the deltoid muscle. For children two years or younger, the recommended location to administer the injection is the vastus lateralis muscle. Both children and adults who receive the vaccination may experience the following symptoms: fever; muscle soreness; and redness, swelling, and itching where the vaccination was administered.

Persons should not be vaccinated with PVC13 if they are allergic to vaccines containing diphtheria toxoid or have had a prior allergic reaction to PCV7 or PVC13. Persons should not get PPV23 if they have had a prior allergic reaction or are allergic to any of the vaccine's components.

IMPACT

Before developing the pneumococcal vaccine, diseases caused by pneumococcus were rapidly becoming resistant to the antibiotics available, rendering them more virulent and difficult to treat. The introduction of the pneumococcal vaccine helped prevent these diseases, making antibiotic resistance less of an issue. However, these bacteria continue to be resistant, making prevention the primary focus of public health efforts.

New pneumococcal vaccines, with increased protection against the different types of *S. pneumoniae*, are under development. In early 2010, a form of the pediatric vaccine with an extended spectrum of thirteen pneumococcus subtypes was released, giving children increased defense against the disease. As both PVC13 and PPV23 only protect against certain serotypes of *S. pneumoniae*, there is a significant need for vaccines that cover a wider range of pneumococcus.

EFFICACY

Pneumococcal vaccines are not 100 percent effective at preventing disease caused by *S. pneumonia*. Both PVC13 and PVP23 protect against a limited number of the approximately ninety serotypes that can cause pneumococcal disease. The 23-valent pneumococcal vaccine effectively protects against invasive pneumococcal disease (IPD). Still, adults are at a high risk of nonbacteremic pneumococcal pneumonia, which PPV23 does not protect against. PVC13 offers some protection against nonbacteremic pneumococcal pneumonia, which is part of the reason it is recommended in addition to PPV23.

One of the ways *S. pneumoniae* is spread is by children transmitting the bacteria from children to adults. Pneumococcus is carried in the nasopharynx of children, and the PVC13 vaccine reduces the number of the thirteen serotypes of pneumococcus contained in the vaccine. Vaccination, therefore, promotes serotype replacement, in which other serotypes of *S. pneumoniae* have room to spread in children. These serotypes are not protected against by either PVC13 or PVP23 and leave adults susceptible to the pneumococcal diseases they cause. Serotype replacement has led medical professionals to question the long-term effects of using PVC13.

—Jennifer Birkhauser, MD

Further Reading

Behrman, Richard E., et al., editors. *Nelson Textbook of Pediatrics*. 2 vols. 20th ed., Elsevier, 2016.

Cafiero-Fonseca, E., et al. "The Full Benefits of Adult Pneumococcal Vaccination: A Systematic Review." *PLOS ONE*, vol. 12, no. 10, 2017, p. e0186903.

Edwards, J., Jennings, M., Apicella, M., and Seib, K. "Is Gonococcal Disease Preventable? The Importance of Understanding Immunity and Pathogenesis in Vaccine

Development." *Critical Reviews in Microbiology*, vol. 42, no. 6, 2016, pp. 928-41.
Fisher, Margaret C. *Immunizations and Infectious Diseases: An Informed Parent's Guide*. American Academy of Pediatrics, 2006.
Harvey, Richard A., et al. *Lippincott's Illustrated Reviews: Microbiology*. 3rd ed., Lippincott Williams and Wilkins, 2013.
Loehr, Jamie. *The Vaccine Answer Book: Two Hundred Essential Answers to Help You Make the Right Decisions for Your Child*. Sourcebooks, 2010.
Plotkin, Stanley A., et al., editors. *Vaccines*. 7th ed., Elsevier, 2017.
"Pneumococcal Polysaccharide Vaccine." *MedlinePlus*. National Library of Medicine, 1 Aug. 2010, medlineplus.gov/druginfo/meds/a607022.html. Accessed 17 Nov. 2016.
Principi, N., and Esposito, S. "Prevention of Community-Acquired Pneumonia with Available Pneumococcal Vaccines." *International Journal of Molecular Sciences*, vol 18, no. 1, 2016, p. 30.
Weinberger, D., Harboe, Z. and Shapiro, E. "Developing Better Pneumococcal Vaccines for Adults." *JAMA Internal Medicine*, vol. 177, no. 3, 2017, p. 303.

Polio vaccine

Category: Prevention
Anatomy or system affected: Blood, cells, circulatory system, immune system, lymphatic system
Specialties and related fields: Family practice, immunology, microbiology, pediatrics, preventive medicine, public health, virology
Definition: a vaccination that prevents people from getting polio

KEY TERMS

poliovirus: the causative agent of polio, is a serotype of the species Enterovirus C in the family Picornaviridae

poliomyelitis: an infectious viral disease that affects the central nervous system and can cause temporary or permanent paralysis

DEFINITION

There are two types of polio vaccine: inactivated and oral, first available in 1955 and 1962, respectively. The vaccines provide immunity to poliomyelitis, or polio, a viral disease that damages nerve cells. The virus enters through the mouth and replicates in the intestines. It then enters the bloodstream and crosses into the central nervous system, attacking the nerve cells.

The first signs of polio are fatigue, headache, nausea, neck stiffness, and fever. The arms and legs are affected first, and in serious cases, the chest muscles are affected, resulting in respiratory failure. Eventually, the nerves no longer send out electrical impulses to move muscles, and the body can become paralyzed; paralysis, however, is uncommon.

TYPES

The inactivated polio vaccine (IPV), developed by Jonas Salk in the early 1950s, was the first polio vaccine available (1955). Salk based his vaccine, which is injected, on a then-new premise: that only the virus's outer shell (capsid) was needed to confer immunity. At the time, all vaccines were manufactured from live but weakened viruses.

In the late 1950s, Albert Sabin produced an oral form of the polio vaccine. Decades earlier, Sabin had shown that polio resides in the intestines rather than the nervous system, laying the theoretical groundwork for an orally administered vaccine. Sabin's oral polio vaccine (OPV), first administered in 1962, used a weakened form of the live poliovirus to stimulate antibody production. Once introduced, OPV quickly became the dominant polio vaccine because it was so easy to administer, and it quickly conferred immunity.

The unique advantage of OPV is the use of live poliovirus. The virus, although weakened, is shed in feces from recently vaccinated persons. An unvaccinated person who encounters the shed virus from a recently vaccinated person, for example, a parent who recently changed a baby's diaper, may contract the weakened poliovirus and thus become passively vaccinated. This ability of OPV to confer immunity to persons not directly vaccinated helped spread immunity and helped eliminate polio outbreaks.

Since 2000, the IPV has been the only polio vaccine given in the United States. This vaccine is given in the leg or arm, depending on the child's age. The Centers for Disease Control and Prevention (CDC) recommends that children receive four doses of the IPV: the first at two months, the second at four months, the third from six to eighteen months, and the final dose at four to six years.

SIDE EFFECTS

Although the live virus in OPV is weakened, it is still a live virus that can cause infection. In rare cases, OPV causes vaccine-associated paralytic poliomyelitis or VAPP. People vaccinated with OPV shed the weakened poliovirus up to six weeks after each dose. Caregivers or others with a weakened immune system, such as those who have had organ transplants or have human immunodeficiency virus (HIV) infection, may develop VAPP if they come in close contact with newly vaccinated children.

The most common adverse event associated with IPV is soreness at the injection site. Allergic reactions, including respiratory difficulties, increased heart rate, hives, dizziness, or throat swelling, are rare.

IMPACT

Polio has no cure and can be prevented only through vaccination. Together, IPV and OPV eradicated polio from most of the world. Polio has become so rare in the United States that the small risk of VAPP associated with OPV is now greater than the benefit of passive immunization. IPV is now the recommended vaccine for all children. Recommendations require three injections for infants at two, four, and six months of age, between six and eighteen months of age, and booster shots between four and six years of age.

—Cheryl Pokalo Jones

Further Reading

Bruno, Richard L. *The Polio Paradox: Understanding and Treating "Post-Polio Syndrome" and Chronic Fatigue*. Warner, 2002.

Hewlett, Martinez J., et al. *Basic Virology*. 4th ed., Wiley-Blackwell, 2021.

Naden, Corinne J., and Rose Blue. *Jonas Salk: Polio Pioneer*. Millbrook, 2001.

Offit, Paul A. *The Cutter Incident: How America's First Polio Vaccine Led to the Growing Vaccine Crisis*. Yale UP, 2005.

Strauss, James, and Ellen Strauss. *Viruses and Human Disease*. 2nd ed., Academic Press/Elsevier, 2008.

RABIES VACCINE

Category: Prevention

Anatomy or system affected: Blood, circulatory system, immune system, lymphatic system

Specialties and related fields: Biochemistry, biotechnology, immunology, microbiology, neurology, preventive medicine, public health, virology

Definition: a vaccine used to prevent rabies

KEY TERMS

immunoglobulin: a class of proteins present in the serum and cells of the immune system, which function as antibodies

rabies: a contagious and fatal viral disease of dogs and other mammals that causes madness and convulsions, transmissible through the saliva to humans

zoonoses: a disease that can be transmitted to humans from animals

INTRODUCTION

Rabies vaccines are made from the killed rabies virus and are administered as a series of shots as soon as possible after potential exposure. Rabies is a serious viral infection of the central nervous system. The virus is transmitted through the saliva of an infected animal through a bite or scratch. Licking alone rarely transmits the disease unless the infected saliva enters an open sore or a mucous membrane. All mammals, such as raccoons, skunks, ferrets, dogs, and cats, are susceptible to infection. However, bats are the most commonly infected animals in the United States.

Although earlier administration of the rabies vaccine is preferred, a physician may give it at any time during the incubation phase. Once symptoms begin, however, it is too late for vaccination.

DISEASE COURSE

Rabies infection begins slowly with an incubation period of one to three months. The virus travels from the bite site through the nerves to the brain, where it replicates. The first symptoms are mild and vague, consisting of headaches, fatigue, and fever. The disease then progresses rapidly. Symptoms of advanced rabies infection include the characteristic hydrophobia, or fear of water, where the presence or even the thought of water causes muscle spasms in the throat. The infected person may become hyperactive and aggressive. The person becomes completely paralyzed and dies as the disease progresses, often from respiratory failure.

TYPES

After exposure to rabies, the infected person will receive two types of vaccine. First, rabies immunoglobulin (RIg) is given at a dose based on the weight of the infected person. Part of the RIg is delivered at the site of the bite, if possible, and the remainder is injected into a muscle. The amount of RIg delivered to the wound depends on the size of the wound. Next, five shots of either human-diploid-cell rabies vaccine (HDCV) or purified chick embryo cell vaccine (PCEC) are administered intramuscularly immediately and then again at three, seven, fourteen, and twenty-eight days. All doses must be administered without interruption.

Preexposure vaccines are given to those at high risk for rabies exposure, such as veterinarians or anyone who frequently contacts wild animals. These consist of either HDCV or PCEC delivered intramuscularly for the initial dose, then again at seven, fourteen, twenty-one, and twenty-eight days. After a rabies exposure, two doses of HDCV or PCEC are still required, but RIg is unnecessary.

The vaccines are also recommended for pregnant women who may have been exposed to the rabies virus. Infants and children receive the vaccines on the same schedule as adults, although the dose of RIg is proportionately smaller.

SIDE EFFECTS

The most common vaccine side effects are swelling, redness, and itching at the vaccine site and headaches, nausea, abdominal pain, dizziness, or muscle aches. In rare cases, the person may develop hives or malaise.

IMPACT

Rabies cannot be treated, but it can be prevented with vaccination. The rabies vaccine is highly effective when administered as soon as possible after a possible exposure to the rabies virus. The disease is always fatal in unvaccinated people. No case of rabies has occurred in any person who has received the vaccine after exposure to animals proven to be rabid.

—Cheryl Pokalo Jones

Further Reading

Atkinson, W., et al., editors. *Epidemiology and Prevention of Vaccine-Preventable Diseases*. 11th ed., Public Health Foundation, 2009.

Hankins, D. G., and J. A. Rosekrans. "Overview, Prevention, and Treatment of Rabies." *Mayo Clinic Proceedings*, 79, no. 5, May 2004, pp. 671-76.

Jackson, Alan C., and William H. Wunner, editors. *Rabies*. 2nd ed., Academic Press, 2007.

Kienzle, Thomas E. *Rabies*. Chelsea House, 2006.

Klosterman, Lorrie. *Rabies*. Marshall Cavendish Benchmark, 2008.

Krauss, Hartmut, et al. *Zoonoses: Infectious Diseases Transmissible from Animals to Humans*. 3rd ed., ASM Press, 2003.

Parker, James N., and Philip M. Parker, editors. *The Official Patient's Sourcebook on Rabies*. Icon Health, 2002.
Playfair, J. H. L., and B. M. Chain. *Immunology at a Glance*. 9th ed., Wiley-Blackwell, 2009.

ROTAVIRUS VACCINE

Category: Prevention
Anatomy or system affected: Blood, circulatory system, gastrointestinal tract, immune system, lymphatic system
Specialties and related fields: Family practice, immunology, internal medicine, microbiology, neonatology, pediatrics, preventive medicine, public health, virology
Definition: a vaccine used to protect against rotavirus infections, which are the leading cause of severe diarrhea among young children

KEY TERMS

rotavirus: a group of ribonucleic acid (RNA) viruses, some of which cause acute enteritis in humans
vaccine: a biological preparation that provides active acquired immunity to a particular infectious disease

DEFINITION

The rotavirus vaccine prevents infection with rotavirus, a pathogen that invades the gastrointestinal system and can cause severe disease accompanied by vomiting, diarrhea, and fever. Many children acquire rotavirus and manifest only mild vomiting and diarrhea. However, often, affected children require hospitalization to manage the resultant dehydration.

MECHANISM OF ACTION

The mechanism of action of the rotavirus vaccine depends upon the brand administered. The RotaTeq vaccine is a combination of a bovine strain of the virus that does not cause disease in humans and a component of the human rotavirus that cannot cause an active infection. These components are then administered together in an oral dose and elicit an immune response without actually causing the disease, therefore protecting from future illness.

The Rotarix brand of the vaccine is derived from a strain of human rotavirus that has been weakened enough not to cause active disease while still eliciting an immune response from the patient.

HISTORY

The vaccine against rotavirus was first licensed in 1998. In 1999, the US Advisory Committee on Immunization Practices withdrew the recommendation that the rotavirus vaccine be administered to all children because of reports of an association with intussusception. This illness causes one segment of the bowel to telescope into another, sometimes requiring surgical repair. In 2006 and 2008, the US Food and Drug Administration licensed new, safer forms of the vaccine under the names RotaTeq and Rotarix, respectively.

ADMINISTRATION

It is recommended that the rotavirus vaccine be administered to all children in two or three doses, depending on which brand of vaccine is given. RotaTeq is the three-dose form of the vaccine and is given at two, four, and six months of age. Rotarix is the two-dose form given at two and four months of age. Both forms of the vaccine are oral and do not require an injection for administration.

IMPACT

Rotavirus is the most common cause of acute gastrointestinal disease worldwide, with increased mortality in developing countries. Since the rotavirus vaccine was developed, concentrated efforts have been made by public health organizations to immunize the children of developing countries. In the United States, uniform administration of the vaccine has led to a greatly decreased incidence of rotavirus disease.

—Jennifer Birkhauser, MD

Further Reading

Behrman, Richard E., Robert M. Kliegman, and Hal B. Jenson, editors. *Nelson Textbook of Pediatrics*. 18th ed., Saunders/Elsevier, 2007.

Centers for Disease Control and Prevention. "Rotavirus Vaccine." 26 Mar. 2021, www.cdc.gov/rotavirus/vaccination.html. Accessed 4 Dec. 2021.

Loehr, Jamie. *The Vaccine Answer Book: Two Hundred Essential Answers to Help You Make the Right Decisions for Your Child*. Sourcebooks, 2010.

Matson, David O. "Rotaviruses." *Principles and Practice of Pediatric Infectious Diseases*, edited by Sarah S. Long, Larry K. Pickering, and Charles G. Prober, 3rd ed., Churchill Livingstone/Elsevier, 2008.

Sears, Robert. *The Vaccine Book: Making the Right Decision for Your Child*. Little, Brown, 2007.

TUBERCULOSIS (TB) VACCINE

Category: Prevention

Also known as: Bacille Calmette-Guérin vaccine, Bacillus Calmette-Guérin vaccine

Anatomy or system affected: Immune system, lymphatic system, skin

Specialties and related fields: Bacteriology, immunology, microbiology, preventive medicine, public health

Definition: a vaccine primarily used against tuberculosis

KEY TERMS

attenuated strains: a strain of a virus whose pathogenicity has been reduced so that it will initiate the immune response without producing the specific disease

subunit vaccine: a vaccine that contains purified parts of the pathogen that are antigenic or necessary to elicit a protective immune response

whole-cell vaccines: vaccines that contain whole cells of the pathogen that have been genetically modified in the laboratory

DEFINITION

The tuberculosis (TB) vaccine is a weakened strain of live bacteria that infect cattle. Albert Calmette and Camille Guérin developed the vaccine to prevent TB, an infectious disease of humans and animals caused by various strains of bacteria of the genus *Mycobacterium*. The weakened strain (*M. bovis*) was obtained by repeatedly growing it in ox bile media until a strain was produced that would not kill experimental animals nor revert to an infectious state. The vaccine, also known as BCG (for its developers), was first used as a vaccine in 1921 after thirteen years of development.

VACCINE ADMINISTRATION

BCG is administered through the skin either by injection or by multiple punctures. Localized skin reactions can occur after vaccination. If drainage occurs, the wound must be covered to prevent transmission of the weakened live bacteria. Serious side effects may include bone infection and disseminated disease, especially in persons who have compromised immune systems.

VACCINE EFFICACY

Studies have shown that BCG protects against tuberculous meningitis and miliary (disseminated) TB in children but provides inadequate prevention against pulmonary TB in adults. Studies also conflict on the duration of protection, ranging from ten to fifteen years to fifty to sixty years. Study designs, geographical location, and statistical factors may have influenced these variable outcomes.

Several factors may influence vaccine efficacy, including the immune status of vaccinated persons. For example, although persons exposed to *Mycobacterium* that is endemic to their environment have some inherent protection against *Mycobacterium* infections, their immune response to BCG is not as pronounced as in persons who have not been exposed to *Mycobacterium*, such as newborns, infants, and those who live in nonendemic areas.

OFFICIAL RECOMMENDATIONS

BCG is not generally recommended in the United States because of the low prevalence of TB and variable vaccine efficacy and interference of BCG with the tuberculin skin test (TST). Selective use of BCG is recommended in some persons, such as children, with negative TST and who are continually exposed to either adults with untreated TB or persons infected by strains resistant to isoniazid and rifampin. Health care workers should be considered for BCG vaccination in specific situations.

BCG vaccination should not be given to immunocompromised persons, such as those with cancer, those with viral infections such as human immunodeficiency virus (HIV), and those taking medications (such as steroids) that cause immune suppression. Pregnant women should not be vaccinated because of live bacteria in the vaccine.

FUTURE VACCINES

Several TB vaccines are under investigation because of drug-resistant strains, the threat of TB in immunocompromised persons, the easy spread of the disease through the air, and the increasing number of infections relative to population growth. These newer vaccines include genetically modified BCG strains, *M. tuberculosis* mutants, *M. tuberculosis* antigens introduced by viruses, and substances included in vaccine modifiers (adjuvants).

Subunit vaccines contain bits and pieces of the pathogen that the immune system recognizes and against which the immune system mounts a response. The most advanced subunit vaccine for TB, M72/AS01E, was developed by GlaxoSmithKline and was taken over by the Bill and Melinda Gates Medical Research Institute. In phase IIb clinical trials, this subunit vaccine protected half the subjects who received it from progressing to active tuberculosis over three years.

Whole-cell vaccines for tuberculosis include MTBVAC and VPM1002. MTBVAC is a genetically modified *Mycobacterium tuberculosis* strain that has a deletion of groups of genes required to produce infection. This vaccine reached phase II clinical trials. VPM1002 is the most advanced live vaccine in clinical trials to date, having successfully completed three phase III trials. This vaccine candidate is safer and more efficacious than BCG.

IMPACT

The BCG vaccine is the most commonly used vaccine globally; more than three billion people have been immunized. Though with variable efficacy, the vaccine confers protection against different manifestations of tuberculosis. A more effective vaccine against TB is needed. It is a contagious disease that infects two billion people, approximately one-third of the world's population.

—Miriam E. Schwartz, MD, PhD,
Shawkat Dhanani, MD, MPH,
and Michael A. Buratovich, PhD

Further Reading

Aronson, N. E., et al. "Long-Term Efficacy of BCG Vaccine in American Indians and Alaska Natives: A Sixty-Year Follow-up Study." *Journal of the American Medical Association*, vol. 291, 2004, pp. 2086-91.

Dockrell, Hazel M., and Ying Zhang. "A Courageous Step Down the Road Toward a New Tuberculosis Vaccine." *American Journal of Respiratory and Critical Care Medicine*, vol. 179, 2009, pp. 628-29.

Hoft, D. F. "Tuberculosis Vaccine Development: Goals, Immunological Design, and Evaluation." *The Lancet*, vol. 372, 2008, pp. 164-75.

Kaufmann, Stefan H. E. "Vaccine Development Against Tuberculosis Over the Last 140 Years: Failure as Part of Success." *Frontiers in Microbiology*, vol. 12, 2021, p. 50124, doi:10.3389/fmicb.2021.750124.

Tait, Dereck R., et al. "Final Analysis of a Trial of M72/AS01E Vaccine to Prevent Tuberculosis." *New England Journal of Medicine*, vol. 381, 2019, pp. 2429-39, doi:10.1056/NEJMoa1909953.
West, John B., and Andrew M. Luks. *Pulmonary Pathophysiology: The Essentials*. 10th ed., Lippincott Williams & Wilkins, 2021.

Typhoid vaccine

Category: Prevention
Anatomy or system affected: Blood, cells, circulatory system, immune system, lymphatic system
Specialties and related fields: Bacteriology, immunology, microbiology, preventive medicine, public health
Definition: vaccines that prevent infection caused by *Salmonella* bacteria found in areas of poor sanitation worldwide

KEY TERMS

Salmonella: a genus of rod-shaped gram-negative bacteria of the family Enterobacteriaceae
typhoid fever: an infection that causes diarrhea and a rash and is most commonly caused by a bacteria called "Salmonella typhi"

PATHOGEN AND DISEASE CHARACTERISTICS

Typhoid fever, an acute illness of fever, rash, and malaise caused by *S. enterica*, serotype typhi (commonly known as *S. typhi*), is distinguished from typhus by its intestinal symptoms. Humans are the only source, and bacteria are spread through fecal contamination of food and water sources. An estimated 22 million cases of typhoid and approximately 200,000 deaths occur each year worldwide. However, only 400 cases occur in the United States (primarily in travelers). Approximately 2 to 4 percent of people with acute fever become chronic carriers.

VACCINE DEVELOPMENT AND DESCRIPTIONS

Early typhoid vaccines had numerous adverse effects, poor efficacy, or low potency. An inactivated injection and an oral attenuated version are now available and are active against strains of *S. typhi*, but even these are not 100 percent effective; approximately 50 to 80 percent of recipients are protected.

Typhim VI is an inactivated cell surface polysaccharide vaccine of *S. typhi*, Ty2 strain for intramuscular administration; it contains 0.25 percent phenol preservative and is safe for ages two years and older and should be administered a minimum of two weeks before possible typhoid exposure. Boosters are recommended every two years if necessary.

Vivotif, an oral vaccine against typhoid, contains live, attenuated cells of the Ty21a strain. The four-capsule regimen should be administered a minimum of one week before possible exposure; one capsule is taken every other day. *Vivotif* is safe for ages six years and older and should be swallowed whole with a cool liquid one hour before a meal. Boosters should be given every five years. Both products require refrigeration at 2° to 8° Celsius (36° to 46° Fahrenheit) before use.

VACCINATION SETTINGS AND RISK GROUPS

Vaccination is recommended for travelers to areas without proper sanitation, for persons who have contact with carriers of *S. typhi*, and laboratory staff who work with pure typhoid culture. (Higher rates of fever are noted in these technicians.)

S. typhi is spread through unclean water sources in Africa, Asia, Central America, and South America. Disease risk is greatest for travelers to South Asia, for longer travel durations, and travelers visiting friends or family.

ADVERSE EFFECTS AND CONTRAINDICATIONS

Typhoid vaccines are not recommended for common use in local populations of risk areas or treatment of chronic carriers.

Vaccine side effects are mild and typically resolve within forty-eight hours. Adverse effects of fever, headache, injection site reaction (with intramuscular administration), and gastrointestinal upset causing reduced capsule absorption (with oral vaccine) occur in less than 10 percent of cases. Immune diseases may increase the risk of any adverse events.

Vaccine contraindications to the vaccine are acute febrile illness and previous allergic reactions (such as hoarseness, wheezing, and anaphylaxis). Live vaccine should not be given to persons with a weakened immune system (such as from human immunodeficiency virus infection), to persons with cancer or who are undergoing cancer treatments, to persons receiving immunomodulating drug treatment or corticosteroid treatment for more than two weeks, and persons receiving certain antibiotics, such as sulfonamides, within one day of a planned vaccination.

IMPACT

Typhoid incidence decreased dramatically with vaccine use and improved sanitation and prevention measures. Because the vaccine is not 100 percent effective, vaccinated travelers should practice preventive measures, such as boiling raw foods, avoiding drinks with ice, and using only boiled or bottled liquids when traveling in countries without adequate sanitation.

—Nicole M. Van Hoey, PharmD

Further Reading

Centers for Disease Control and Prevention. "Typhoid Fever and Paratyphoid Fever." 18 Nov. 2020, www.cdc.gov/typhoid-fever/typhoid-vaccination.html. Accessed 6 Dec. 2020.

"Immunization: Typhoid Vaccine." *Mandell, Douglas, and Bennett's Principles and Practice of Infectious Diseases*, edited by Gerald L. Mandell, John E. Bennett, and Raphael Dolin, 7th ed., Churchill Livingstone/Elsevier, 2010.

Levine, Myron M. "Typhoid Fever." *Bacterial Infections of Humans: Epidemiology and Control*, edited by Philip S. Brachman and Elias Abrutyn, 4th ed., Springer Science, 2009.

———. "Typhoid Fever Vaccines." *Vaccines*, edited by Stanley A. Plotkin, et al., 7th ed., Elsevier, 2017.

Mintz, Eric. "Typhoid and Paratyphoid Fever." *CDC Health Information for International Travel 2010*, wwwnc.cdc.gov/travel/yellowbook/2010/table-of-contents.aspx. (See chapter 2.)

VACCINE SAFETY: OVERVIEW

Category: Prevention

Anatomy or system affected: Blood, cells, circulatory system, immune system, lymphatic system

Specialties and related fields: Bacteriology, biochemistry, biotechnology, immunology, microbiology, preventive medicine, public health

Definition: while no vaccine is 100 percent safe or harmless, the benefits conferred by vaccines far outweigh their potential risks

KEY TERMS

autism spectrum disorder (ASD): a collection of neurologically-based developmental disorders in which individuals have impairments in social interaction and communication skills and a tendency to have repetitive behaviors and/or narrow, often-obsessive interests

clinical trial: a scientific evaluation of a drug or vaccine to determine whether it is effective and safe, with several phases involving greater numbers of test subjects in each phase; early phase trials are typically done with lab animals and later ones with human volunteers

endemic: describes a disease or pathogen always present or usually prevalent in a particular population or geographical area

epidemiology: the branch of medicine that deals with studying the causes, distribution, and control of disease in groups of people *herd immunity*: refers to a significant proportion of a population becoming immune to an infectious disease because of exposure or vaccination, decreasing the likelihood of spread

thimerosal: a mercury-based preservative used in some vaccines beginning in the 1930s

vaccine: a preparation containing an antigen, typically consisting of whole disease-causing organisms (killed or weakened) or parts of such organisms, used to confer immunity against the disease that the antigen causes

vaccine hesitancy: the belief that vaccines may not be effective, have unknown long-term effects and create serious adverse reactions

INTRODUCTION

The mainstream American medical establishment has long contended that the public health benefits of vaccines—to prevent such diseases as diphtheria, tetanus (lockjaw), pertussis (whooping cough), polio, rubella, measles, mumps, hepatitis B, varicella (chickenpox), and influenza—heavily outweigh the relatively small risks associated with such preventive measures. Officials from the Centers for Disease Control and Prevention (CDC) and the Food and Drug Administration (FDA), most major medical associations, and most practicing clinicians have invested significant energy in a campaign to convince people to keep their immunizations up to date. The CDC has also encouraged adults to receive select vaccinations at specific life stages and under certain circumstances, such as before international travel.

However, for a small but persistent number of parents and advocates, the wisdom of routine vaccination against certain common diseases remains suspect. According to these critics, vaccines may cause serious side effects, long-term medical conditions, or even prove fatal. They, therefore, argue that vaccination should be a personal, considered decision and that mandatory vaccination infringes their rights.

In 2020 the coronavirus disease 2019 (COVID-19) pandemic renewed discussions over the need for vaccines to control the spread of disease, how vaccine safety is assessed, and who should be vaccinated.

HISTORY

Controversies over vaccine safety have raged for as long as vaccines have existed. Until the eighteenth century, treatment for smallpox, an often-deadly disease, was limited to an eleventh-century Chinese technique called "variolation," which involved rubbing pus obtained from a smallpox pustule into a scratch on an otherwise healthy individual so that person would develop immunity to the virus. In the early eighteenth century, Lady Mary Wortley Montagu introduced variolation to Western Europe from the Ottoman Empire, where it was common practice. Although Montagu faced bitter opposition from English doctors on religious, medical, economic, and sexist grounds, she eventually succeeded in convincing the royal family of the effectiveness of variolation. In 1796 the English doctor Edward Jenner developed the world's first vaccination, for smallpox, after observing that milkmaids who had contracted cowpox, a milder, related condition, were not susceptible to smallpox.

Deliberately exposing a patient to an infected fluid to confer immunity worked, but the notion of deliberately introducing diseased matter into a healthy body provoked fear, skepticism, and sometimes even violence. Although Jenner was confident enough in the safety of his vaccine to inoculate his infant son, in 1853, more than half a century after his landmark discovery, the English government passed a law making vaccination mandatory for all its citizens. So persistent were the doubts about the safety and necessity of vaccination, however, that in 1898, the government dropped this requirement, although it continued to encourage vaccination.

Between Jenner's discovery of the smallpox vaccine in 1796 and 1980, when the World Health Organization (WHO) officially declared smallpox eradicated, new vaccines emerged for many other infectious diseases that have caused enormous suffering throughout human history. In the early 1950s, for instance, Dr. Jonas Salk (1914-95) of New York famously developed an effective vaccination for another disease of children: polio. That debilitating disease often left patients partially paralyzed (as in the case of President Franklin Delano Roosevelt), unable to walk, and sometimes unable to breathe. Others included vaccines against rabies, plague, cholera, typhoid fever, diphtheria, pertussis, tuberculosis, tetanus, yellow fever, influenza, mumps, measles, rubella, and anthrax.

Ironically, the very success of vaccination efforts in causing once-dreaded diseases to fade into distant memories may have contributed to increasing concerns over vaccine safety. As fewer and fewer Americans witnessed firsthand the ravages of epidemic diseases such as measles (which claimed 120 deaths in 1989-91) and rubella (which, between 1964 and 1965 alone, caused deafness, blindness, heart disease, or cognitive impairment in about 20,000 newborns whose mothers had been infected with the virus during pregnancy), many began focusing more on the risks of vaccines than on the horrors of the diseases they prevented.

In response to growing concerns over vaccine safety, Congress passed in 1986 the National Childhood Vaccine Injury Act, which made mandatory the reporting of adverse health events following specific vaccinations and established a no-fault compensation system for those injured by vaccines. In 1990 the Vaccine Adverse Reporting System (VAERS) was established to monitor the safety of all vaccines approved for use in the United States.

With national vaccine safety protocols in place, public health officials launched an aggressive campaign to promote vaccinations. In 1994 the Vaccines for Children (VFC) program was established to provide free vaccines for eligible children, and in 1995 several major pediatric and medical organizations endorsed a uniform childhood immunization schedule. By 1994 officials were able to certify the elimination of polio in the Americas. By 2003 they declared measles no longer endemic in the Americas, and by 2005 they declared rubella no longer endemic in the United States. Compliance with recommended vaccination improved dramatically with school-entry mandates for immunization.

These triumphs, however, did not end the controversy over vaccine safety. In 1999 the first rotavirus vaccine, which was licensed only a year earlier, was pulled from the market because of its adverse side effects, specifically, intestinal blockage reported through VAERS. Public confidence in vaccine safety controls was further shaken in 2004 when the first Lyme disease vaccine, which the FDA had approved in 1998, was also withdrawn from the market amid a flurry of lawsuits against its manufacturer and increasing concerns of untoward side effects like the actual symptoms of Lyme disease.

However, the greatest controversy in the vaccine safety debate has swirled around belief in a link between the measles, mumps, and rubella (MMR) vaccine and autism spectrum disorder (ASD), an idea first proposed in 1998. The journal retracted the article after the author's conflict of interest was revealed. According to epidemiologic evidence gathered for numerous subsequent studies, the weight of scientific evidence does not support the hypothesis that the MMR vaccine causes ASD. Researchers have suggested that because the vaccine is administered to young children around the same age that ASD symptoms typically emerge, observers have attributed a causal link to what is more likely coincidental timing.

Beginning in 1999, when questions began surfacing about the rotavirus, Lyme disease, and MMR vaccines, consumer advocacy groups mounted a campaign to ban thimerosal, a mercury-based additive, from vaccines. That year, the Public Health Service and the American Academy of Pediatrics recommended the removal of thimerosal from childhood vaccines as a precautionary measure. Because mercury is a neurotoxin harmful to fetuses and infants' developing central nervous system, some groups believed that thimerosal exposure from childhood vaccines might cause attention-deficit/ hyperactivity disorder (ADHD), speech or language delays, and ASD. In 2004 the Institute of Medicine's Immunization Safety Review Committee examined epidemiological data from the United States, Denmark, Sweden, and the United Kingdom before concluding that evidence does not support a causal relationship between thimerosal-containing vaccines and autism.

Three key developments in the early twenty-first century shaped attitudes toward vaccine safety. First, the September 11, 2001 terrorist attacks and the subsequent anthrax scare increased public fears of bioterrorism. Concerns over vaccine safety became overshadowed by worries of inadequate national supplies of anthrax and smallpox vaccine. Second, government officials and scientists issued repeated, dire warnings about the possibility that the virus responsible for creating a fast-spreading epidemic of avian influenza ("bird flu") might cross the species barrier to infect humans. Faced with the prospect of a flu epidemic that could claim millions of human lives, safety concerns took a back seat to the urgent quest to create an effective vaccine to prevent such a global catastrophe. Finally, a 2005 study of more than 30,000 Japanese children definitively debunked, in the eyes of most scientists, the claim that the MMR vaccine could be responsible for rising rates of ASD worldwide. When the number of ASD cases continued to increase after the MMR vaccine was replaced with single vaccines, researchers concluded that the combined vaccine could not be responsible for triggering the disorder on a large scale. A 2014 meta-analysis involving 1.2 million children also found no causation or association between vaccination and ASD. Some skeptics continued to refuse to accept the scientific consensus, however.

The US government also took several actions on vaccine safety. In 2005 Congress saw the reintroduction of a bill that would amend the Federal Food, Drug, and Cosmetic Act to ban mercury in vaccines. The Mercury-Free Vaccines Act stalled in the House Subcommittee on Health; reintroduced again in 2007 and 2009, each bill met the same fate as the earlier version. In 2012 and 2013, the CDC's Advisory Committee on Immunization Practices (ACIP) implemented several revisions, including recommending that the flu vaccine is safe for all persons six months and older. The ACIP also recommended that eleven- and twelve-year-old males receive the human papillomavirus (HPV4) vaccine and that a single dose of the meningococcal vaccination should be administered to military recruits.

Meanwhile, vaccine skepticism surged again in the 2010s, boosted by social media. Conspiracy theories proliferated amid a general atmosphere of increasing public backlash against government, media, and scientific authority. In January 2017, President-elect Donald Trump, who believed in a link between ASD and vaccination, appointed vaccine skeptic Robert F. Kennedy Jr. to head a commission to investigate vaccine safety.

VACCINE SAFETY TODAY

Based on CDC data for 1998 to 2015, vaccination against MMR, diphtheria, tetanus, pertussis, polio, hepatitis B, varicella, and pneumococcal conjugate dipped around 2012-13 before climbing again. In 2017, over 70 percent of children between the ages of a year and a half and three years old had received all their ACIP recommended vaccines, with the lowest rates being found among uninsured, lived-in rural areas, or were black or Indigenous. Declines in MMR vaccination were linked to a measles outbreak in California in 2014-15 and another in New York in 2018-19. It may have contributed to the 150 outbreaks of mumps observed in 2016 and 2017, with more than six thousand cases reported each of those years compared to the few hundred cases per year seen from 2000 to 2005. In the wake of those outbreaks, about a dozen state legislatures attempted to repeal personal, moral, or philosophical exemptions from mandatory vaccination before school entry and were met with protest and backlash. California, New York, and Maine removed all exemptions except for underlying medical conditions.

A novel coronavirus was identified in late 2019 and quickly spread through countries worldwide in early 2020. At the time, there were few medical treatments for it, and more than 2 million people died of COVID-19 within about a year. The COVID-19 pandemic thus brought with it renewed interest in and fears overvaccination. An unprecedented international effort was launched to develop a vaccine in record time. More than 150 potential vaccines for COVID-19 were in some phase of clinical trials in less than a year. Although some of the concerns expressed were old, such as fears about potential side effects and adverse reactions, others were newer. Conspiracy theories and false rumors regarding the origins of the virus and the purpose of vaccination spread rapidly. Some Americans expressed concern about the unusually rapid development and testing process. Public health officials maintained that candidate vaccines were thoroughly and rigorously vetted and credited with improved communication and collaboration among the private and public sectors, unparalleled funding, and ease of volunteer recruitment for trials. Another challenge was the fact that, despite experiencing higher rates of infection and death from COVID-19, black Americans and American Indians expressed significantly higher levels of vaccine hesitancy than members of other races; many observers attributed this reluctance to a long history of medical racism in the United States, including past exploitation for medical research. Other considerations included concerns over cost and health insurance coverage.

While a few vaccines earned emergency approval in certain countries within just a few months, the rigorous testing process in the United States took somewhat longer. In December 2020, drug manufacturers Pfizer and Moderna announced that they had developed vaccines that were more than 90 percent effective at preventing patients from developing COVID-19. Both companies received emergency-use authorizations from the FDA for their vaccines for use in adults who had no history of vaccine allergy and whose immune systems were not compromised by medications or chronic medical conditions. Because of the limited doses available initially, the CDC issued nonbinding recommendations that health-care staff, frontline workers, and people aged seventy-four and older be among the first to receive the vaccine; nonetheless, different states took different approaches and equitable access to the vaccine remained a challenge. As other vaccines also entered wide use, some safety concerns also emerged. A vaccine developed by British company AstraZeneca was put on hold or restricted by some countries in early 2021 due to rare occurrences of blot clots in people receiving the vaccine. Similar concerns developed over a vaccine by Johnson & Johnson (Janssen) approved by the FDA in February 2021, causing the FDA to pause distribution that April briefly. However, after an extensive review, the FDA and the CDC determined that the Johnson & Johnson vaccine was safe and effective, with the known and potential benefits considerably outweighing the known and potential risks.

Despite early challenges with distribution, by April 19, 2020, all adults in the United States were declared eligible for a vaccine. By May 11, the CDC reported that over 46 percent of the total population had received at least one dose of a vaccine, while over 35 percent were fully vaccinated. However, questions remained over whether vaccine hesitancy would prevent the country from reaching herd immunity and fully curbing the pandemic. Notably, research indicated significant political division overvaccination, with Republicans—especially rural males—generally more hesitant to receive a COVID-19 vaccine. Another concern was whether newly emerging virus variants would evade the vaccine and thus how long vaccines would confer immunity.

—Beverly Ballarlo and Nancy Sprague

Further Reading

"Adult Immunization Schedule by Vaccine and Age Group." *Immunization Schedules*. Centers for Disease Control and Prevention, US Department of Health and Human Services, 3 Feb. 2020, www.cdc.gov/vaccines/schedules/hcp/imz/adult.html. Accessed 28 Jan. 2021.

Brophy Marcus, Mary. "States with the Highest Child Vaccine Rates." *CBS News*. CBS Interactive, 25 Apr. 2017, www.cbsnews.com/news/states-child-vaccination-rates-mmr-vaccine-dtap-whooping-cough-chickenpox. Accessed 22 Sept. 2017.

Centers for Disease Control and Prevention. *Health, United States, 2018: With Chartbook on Long-Term Trends in Health*. National Center for Health Statistics. US Department of Health and Human Services, 2019 , *CDC*, www.cdc.gov/nchs/data/hus/hus18.pdf. Accessed 28 Jan. 2021.

"FDA and CDC Lift Recommended Pause on Johnson & Johnson (Janssen) COVID-19 Vaccine Use Following Thorough Safety Review." *FDA*, 23 Apr. 2021, www.fda.gov/news-events/press-announcements/fda-and-cdc-lift-recommended-pause-johnson-johnson-janssen-covid-19-vaccine-use-following-thorough. Accessed 12 May 2021.

Hager, Thomas. "How One Daring Woman Introduced the Idea of Smallpox Inoculation to England." *Time*, 5 Mar. 2019, time.com/5542895/mary-montagu-smallpox/. Accessed 21 Jan. 2021.

Institute of Medicine. *Immunization Safety Review: Vaccines and Autism*. National Academies Press, 2004.

———. *Vaccine Safety Research, Data Access and Public Trust*. National Academies Press, 2005.

Kum, Dezimey. "Fueled by a History of Mistreatment, Black Americans Distrust the New COVID-19 Vaccines." *Time*, 28 Dec. 2020, time.com/5925074/black-americans-covid-19-vaccine-distrust/. Accessed 20 Jan. 2021.

Link, Kurt. *The Vaccine Controversy: The History, Use, and Safety of Vaccinations*. Praeger Publishers, 2005.

Motta, Matt, and Timothy Callaghan. "Why the Next Major Hurdle to Ending the Pandemic Will Be about Persuading People to Get Vaccinated." *The Conversation*, 27 Jan. 2021, theconversation.com/why-the-next-major-hurdle-to-ending-the-pandemic-will-be-about-persuading-people-to-get-vaccinated-153847. Accessed 28 Jan. 2021.

"New Meta-analysis Confirms: No Association between Autism and Vaccines." *Autism Speaks*, 19 May 2014, www.autismspeaks.org/science-news/new-meta-analysis-confirms-no-association-between-autism-and-vaccines. Accessed 28 Jan. 2021.

Rabin, Roni Caryn. "Eager to Limit Exemptions to Vaccination, States Face Staunch Resistance." *The New York Times*, 14 June 2019, www.nytimes.com/2019/06/14/health/vaccine-exemption-health.html. Accessed 28 Jan. 2021.

"Recommended Child and Adolescent Immunization Schedule for Ages 18 years or Younger." *Centers for Disease Control and Prevention*, US Department of Health and Human Services, 29 Jan. 2020, www.cdc.gov/vaccines/schedules/downloads/child/0-18yrs-child-combined-schedule.pdf. Accessed 28 Jan. 2021.

"Republican Men Are Vaccine-Hesitant, But There's Little Focus on Them." *Pew*, 23 Apr. 2021, www.pewtrusts.org/en/research-and-analysis/blogs/stateline/2021/04/23/republican-men-are-vaccine-hesitant-but-theres-little-focus-on-them. Accessed 12 May. 2021.

Vaccine Safety. Centers for Disease Control and Prevention. US Department of Health and Human Services, 12 Jan. 2021, www.cdc.gov/vaccinesafety/index.html. Accessed 28 Jan. 2021.

Yang, Y. Tony, and Vicky Debold. "A Longitudinal Analysis of the Effect of Nonmedical Exemption Law and Vaccine Uptake on Vaccine-Targeted Disease Rates." *American Journal of Public Health*, vol. 104, no. 2, 2014, pp. 371-77, search.ebscohost.com/login.aspx?direct=true&db=a9h&AN=93721908. Accessed 12 Mar. 2014.

Vaccine types

Category: Prevention

Also known as: Immunization

Anatomy or system affected: Blood, immune system, lymphatic system

Specialties and related fields: Allergist, epidemiology, hematology, immunology, microbiology, pharmacology, public health, virology

Definition: the injection, ingestion, or inhalation of suspension of immunogens (molecules that produce an immune response) or a vehicle that causes the synthesis of immunogens that elicit a protective immune response against a specific infection

KEY TERMS

adjuvant: a substance given with a vaccine that enhances the immune response elicited by it

antigen: a foreign substance that elicits an immune response when inoculated into a living organism

attenuated: weakened or partial organisms

conjugate: the attachment of molecules that tend to elicit weak immune responses, such as carbohydrates, to large molecules that enhance the immunogenicity of the attached molecules

excipient: an inactive substance that serves as the vehicle or medium for a drug, vaccine, or other material

immunity: the ability of an immune system to recognize, neutralize, and destroy an infecting organism, protecting the individual from infection

inoculation: the introduction of a substance or group of substances into a living organism

lipid nanoparticles: a shell of cholesterol and other fat-soluble molecules that house an internal messenger ribonucleic acid (mRNA) core and fuse with host cells to deliver the mRNA payload

VACCINES

A vaccine is a suspension of immunogens (molecules that produce an immune response or stimulate the production of antibodies), such as weakened or dead pathogenic (disease-causing) cells or cellular components. The act of administering a vaccine, or immunization, is called "vaccination." Persons who receive a vaccine are considered immunized against a particular pathogen. Vaccines may contain a pathogen, suspending fluid, adjuvants, excipients, and preservatives.

Several types of vaccines are given to humans. These types include live attenuated, inactivated, component or subunit, toxoid, conjugate, deoxyribonucleic acid (DNA), and recombinant vector vaccines. Live attenuated vaccines contain living but altered bacteria or viruses that do not cause disease. Inactivated or killed vaccines contain killed bacteria or inactivated viruses that do not cause disease. Component or subunit vaccines contain parts of the whole bacteria or viruses. Toxoid vaccines contain toxins (or poisons) produced by the pathogen that have been made harmless. Conjugate vaccines allow the immune system to recognize certain bacteria disguised by a polysaccharide outer coating and respond. DNA and recombinant vector vaccines are in the experimental stage, and both use genetic material to stimulate an antibody response.

Some vaccines are combinations of pathogens for different diseases, such as measles, mumps, and rubella (or MMR vaccine). Most vaccines are administered by injection into the muscle (intramuscular); however, some may be given into the skin (subcutaneous), by mouth, or into the nose (intranasal).

Active immunity is classified as natural (after pathogen exposure and infection) or acquired (after vaccination). Passive immunity is also classified as natural (across the placenta during pregnancy) or acquired (injection of antibodies or immunoglobulins pooled from several donors). Immunoglobulins are prepared antibodies given to a person who has already been infected or is at risk of acquiring an infection, thereby providing passive immunization. In this case, the immune system does not need to produce antibodies protecting the body.

Herd immunity occurs when most of, but not all, the people in a given population are immune to a pathogen. If there is an outbreak or exposure to a pathogen, those who are immune will sometimes naturally protect those who are not immune from getting the disease; however, those who are not immune are still more likely to get the disease and spread it to others.

MECHANISMS OF ACTION

A vaccine is given to intentionally expose the immune system to a pathogen in a safe, controlled manner so that the immune system can react and develop antibodies to that pathogen or antigen. Antibodies are large proteins that help fight infection and control disease. Many antibodies disappear after destroying the invading antigens, but the cells involved in antibody production remain and become memory cells. Memory cells "remember" the original antigen and then defend against it if the antigen attempts to reinfect a person. This protection is called "immunity." Therefore, after sufficient antibodies have been developed, reexposing the immune system to that pathogen elicits a fast reaction against it (within minutes to hours). The pathogen is destroyed before a full-fledged infection and organ damage can occur. B cells are a type of lymphocyte (white blood cell) that makes antibodies. B cells use antibodies to identify, inactivate, and help destroy these pathogens.

Vaccines, which protect from the disease without the severe symptoms, have a high effectiveness rate (usually 95 to 99 percent). Vaccine failure, meaning that the vaccine administration did not result in antibody production, is uncommon. Several factors can lead to vaccine failure, including having an already compromised immune system and the inadequate storage or administration of the vaccine. The immune response to a pathogen may decrease over time, so vaccines known as boosters restore antibodies. Protective immunity lasts longer with boosters.

A suspending fluid (such as sterile water or saline) is needed to allow the vaccine to be administered. Preservatives and stabilizers, such as albumin, phenols, and glycine, keep the vaccine from being changed. Adjuvants, or enhancers, help the vaccine work. Adjuvants help promote an earlier, more potent response and a more persistent immune response to the vaccine. Antibiotics prevent the growth of bacteria during the production and storage of the vaccine. Eggs are used to grow the pathogen, and egg protein is found in influenza and yellow fever vaccines. Formaldehyde is used to inactivate bacterial products for toxoid vaccines and kill unwanted viruses and bacteria that might contaminate the vaccine during production. Monosodium glutamate and 2-phenoxy-ethanol are preservatives that help the vaccine remain unchanged during exposure to heat, light, acidity, or humidity. Thimerosal is a mercury-containing preservative that helps prevent contamination and the growth of bacteria.

Most vaccines are given to prevent disease and are effective only if administered to the person before exposure to the pathogen or disease; most vaccines must be given by a certain age to ensure effectiveness. Also, most vaccine-preventable diseases can cause serious or life-threatening infections in infants and young children. For example, exposure and infection with polio can occur at a very young age and cause paralysis, so the vaccine should be given to infants as soon as possible. A pregnant woman can transfer immunity to some pathogens to her fetus. However, this immunity wanes once the newborn is older than six months of age. Breastfeeding can also help extend immunity to some diseases, but even this is limited.

Certain vaccines (such as pneumococcal or hepatitis B vaccines) are given once in a lifetime unless a booster is needed. However, the seasonal influenza vaccine is given annually because hundreds of influenza-like viruses exist; also, the seasonal variations or types of viruses that are prevalent change every year. Vaccination schedules have been developed for children, adolescents, and adults that indicate when these persons should receive doses of required vaccinations or boosters.

TYPES OF VACCINES

The selection of the type of vaccine depends on basic information or factors about the pathogen. These factors include how the pathogen infects cells and how the immune system responds to it. Practical considerations include the regions of the world where the vaccine would be used. Pros and cons are associated with each type of vaccine.

Live attenuated vaccines. Live attenuated vaccines are usually created from the naturally occurring pathogen. The pathogen's ability to cause severe infection is attenuated or weakened by manipulating the virus or bacteria in a laboratory environment. However, these vaccines can still induce antibody production or protective immune responses. The pathogen's attenuation is usually done by "passing" or growing the virus or bacteria from culture to culture before it is formulated into a vaccine. Live attenuated vaccines elicit strong cellular and antibody responses and often confer lifelong immunity with only one or two doses. Not everyone can safely receive live attenuated vaccines, however. People with weakened immune systems cannot be given live vaccines because of the risk of developing disease symptoms.

These types of vaccines usually need to be refrigerated to stay potent. Proper storage then becomes critical in maintaining vaccine efficacy. Examples of live attenuated vaccines include measles, mumps, and rubella (MMR vaccine), oral polio vaccine (OPV), the nasal form of influenza (flu) vaccine, and the varicella vaccine (chickenpox vaccine).

Inactivated vaccines. Inactivated vaccines contain a killed pathogen that cannot cause the disease but can stimulate antibody production. Pathogens are inactivated with chemicals such as formaldehyde. Inactivated vaccines are more stable and safer than live vaccines. These vaccines usually do not require refrigeration and are easily stored and transported in freeze-dried form, making them useful in situations requiring long transportation or with less-developed medical infrastructure. Most inactivated vaccines, however, produce a weaker immune response than

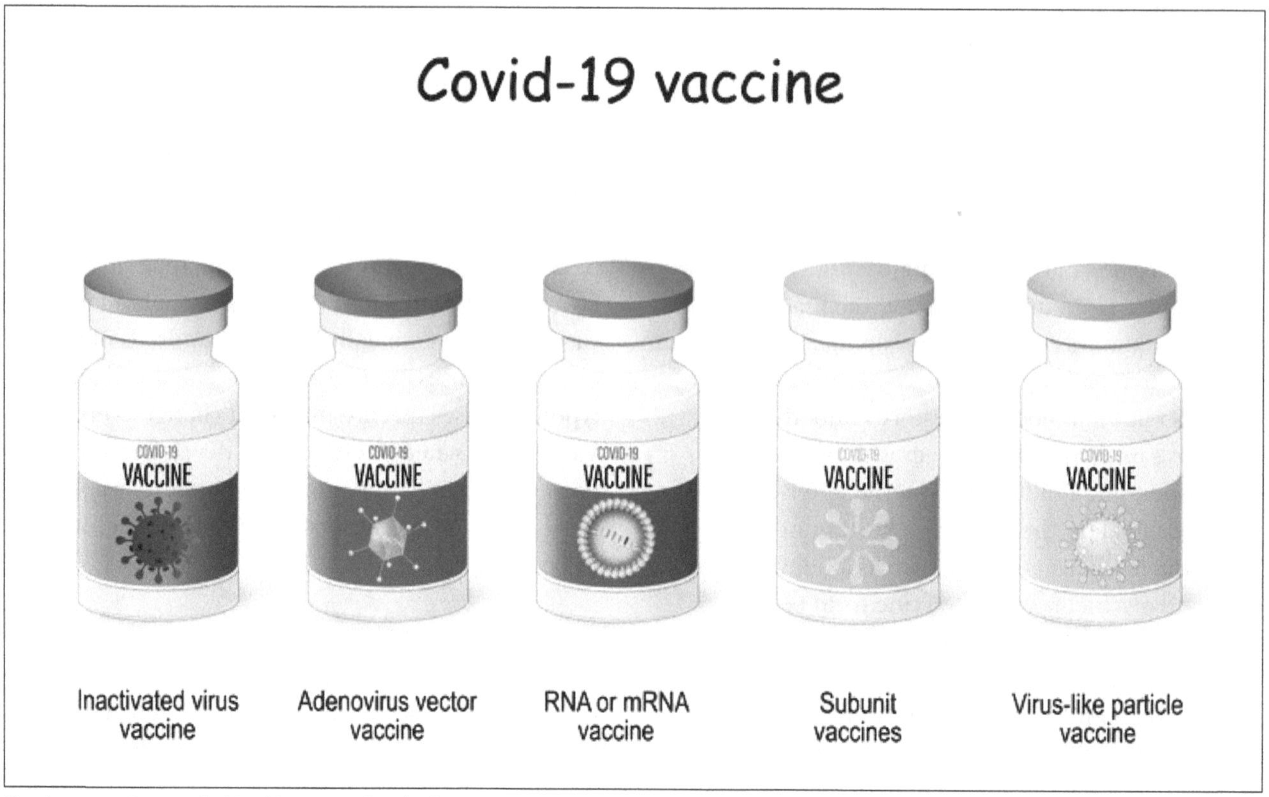

Vaccine types. Image via iStock/ttsz. [Used under license.]

do live vaccines. Several additional doses or booster shots, therefore, are needed to maintain immunity. Examples of inactivated vaccines include inactivated polio vaccine (IPV) and inactivated (injectable form) influenza vaccine.

Component or subunit vaccines. Component or subunit vaccines are made by using only parts of the pathogen. These vaccines cannot cause disease, but they can stimulate the body to produce an immune response against the disease. Component vaccines contain only the essential antigens, but not all the other molecules, of the pathogen, so the chance of an adverse reaction to the vaccine is lessened.

These vaccines can contain anywhere from one to twenty or more antigens. Identifying what antigens best stimulate the immune system can be a tricky, time-consuming process. A recombinant component vaccine has been created for the hepatitis B virus. Hepatitis B genes that code for essential antigens were inserted into common baker's yeast. The yeast then produced the antigens, which were collected and purified for use in the vaccine.

A conjugate vaccine is another type of component vaccine developed for bacteria that possesses an outer coating of sugar molecules called "polysaccharides." The polysaccharide coating disguises the internal antigens of the bacterium so that the immune system does not recognize or respond to it. Vaccines help the immune system link the polysaccharide coating to the bacterium and, therefore, allow antibodies to produce immunity to that pathogen. Examples of component vaccines include *the Haemophilus influenzae* type B (Hib) vaccine, hepatitis B (Hep B) vaccine, hepatitis A (Hep A) vaccine, and pneumococcal conjugate vaccine.

Toxoid vaccines. Toxoid vaccines are made by treating the toxin produced by the pathogen with heat or chemicals, such as formalin (a solution of formaldehyde and sterilized water). A toxoid vaccine may be used for pathogens that secrete toxins or harmful chemicals when the toxoid is the leading cause of illness. Toxins are inactivated and do not produce disease, and detoxified toxins are called "toxoids." After vaccination with a toxoid vaccine, the immune system produces antibodies that block the toxin. Examples of toxoid vaccines include those against diphtheria and tetanus.

DNA vaccines. Experimental DNA vaccines contain the genes that code for antigens. The development of DNA vaccines requires that the genes from the pathogen be isolated and analyzed. DNA vaccines would stimulate an immune response to the free-floating antigen secreted by cells and against the antigens displayed on cell surfaces. DNA vaccines would contain copies of a few of the pathogen's genes, so the vaccine would not cause disease.

DNA vaccines are relatively easy and inexpensive to design and produce. Naked DNA vaccines, which consist of DNA administered directly into the body, could be mixed with molecules that facilitate its uptake by the body's cells. Naked DNA vaccines for influenza and herpes viruses are under investigation.

mRNA vaccines. In 1990, Agnes Jani and Phillip Felgner and their colleagues injected messenger ribonucleic acid (mRNA) into the muscles of laboratory animals and observed protein synthesis. In 1992, Bloom and colleagues from the Scripps Research Institute in La Jolla, California, reversed diabetes insipidus in Brattleboro rats by injecting vasopressin mRNA into their hypothalamus. These two studies demonstrated the ability of injected RNA to induce the expression of specific proteins. However, these studies also showed the caveats of mRNA injection, including its short half-life and the tendency for the immune system to recognize mRNA and destroy it.

Further research revealed that efficient delivery of mRNAs by encasing them in lipid nanoparticles increased mRNA uptake by cells and prevented recognition of the RNA by the immune system. Using modified bases during the synthesis of mRNAs not only increases their half-lives but significantly decreases their immunogenicity. The production of mRNA vaccines is rapid, inexpensive, and highly scalable.

mRNA vaccines made by Pfizer-BioNTech (BNT162b2) and Moderna (mRNA-1273) were granted Emergency Use Authorization (EUA) by the US Food and Drug Administration (FDA) on December 11, 2020 and December 18, 2020, respectively, for the COVID-19 pandemic. Both vaccines encode the SARS-CoV-2 spike protein and induce excellent immune responses against SARS-CoV-2. Other mRNA vaccines for Zika virus, influenza, respiratory syncytial virus, human immunodeficiency virus, and several types of cancer are in development.

Recombinant vector vaccines. Recombinant vector vaccines, also experimental, use an attenuated pathogen to introduce DNA to cells. A vector, in this case, is a harmless virus or bacterium used as a carrier. Certain harmless or attenuated viruses are used to carry portions of the genetic material from other microbes. The carrier viruses then ferry the microbial DNA to cells and display the pathogen's antigens on the cell's surface. The harmless organism mimics a pathogen and provokes an immune response. Recombinant vector vaccines closely mimic a natural infection, effectively stimulating the immune system. Recombinant vector vaccines for human immunodeficiency virus (HIV), rabies, and measles are under investigation.

Adenovirus-based vaccines for COVID-19 include the Oxford-AstraZeneca (ChAdOx1), Johnson & Johnson (JNJ-78436735), and Sputnik V (Gam-COVID-Vac) vaccines. The FDA awarded the Johnson & Johnson vaccine EUA on February 27, 2021. The Oxford-AstraZeneca vaccine has been approved for use in the European Union, Vietnam, Argentina, Bangladesh, Brazil, the Dominican Republic, El Salvador, India, Malaysia, Mexico, Nepal, Pakistan, the Philippines, Sri Lanka, South Korea, and Taiwan. Adenovirus-based vaccines are also available for Ebola.

CONTROVERSY

State laws in the United States mandate that children in daycare and students be immunized against certain diseases. Some exceptions are allowed. Still, many parents refuse to immunize their children for fear of a link between autism, for example, and the use of vaccines containing thimerosal, an ethylmercury-based preservative. Scientific evidence does not support this link. In 2004, the Institute of Medicine conducted a scientific review of thimerosal. It concluded that "the evidence favors rejection of a causal relationship between thimerosal-containing vaccines and autism." Since then, nine additional studies have found no link between thimerosal-containing vaccines and autism spectrum disorder. Nevertheless, thimerosal is no longer used in the production of most single-shot vaccines in the United States.

The link between vaccines and autism comes from a fraudulent paper published in 1997 in the prestigious medical journal *The Lancet* by a British physician named Andrew Wakefield. Not only has no one been able to replicate Wakefield's results, but an intensive investigation by the British journalist Brian Deer established that Wakefield's research project contained serious procedural errors, undisclosed financial conflicts of interest, and ethical violations. The authors retracted their Lancet paper. Wakefield lost his medical license for endangering children and failing to disclose significant financial interests connected to the project. A follow-up study led by Brent Taylor at the Royal Free Hospital Medical School, London, published in 2002, examined 498 autistic children and failed to establish any link between vaccination and either inflammatory bowel disease or autism. Further work has corroborated Taylor's findings. A 2011 Institute of Medicine examined side effect data of eight vaccines given to children and adults and found no link with autism spectrum disorder. A 2013 Centers for Disease Control and Prevention (CDC) study of the antigens in vaccines established that increased exposure to these antigens is not the cause of autism spectrum disorder.

Information sheets alert persons to adverse effects associated with vaccine administration. These sheets also educate parents and others about what to expect after receiving a vaccine. Everyone seeking a vaccine is provided an information sheet before they can be vaccinated.

IMPACT

Disease prevention is the key to public health, and it is always better to prevent a disease than to have to treat it. Vaccination is considered one of the most important medical discoveries in all of human history. Diseases can cause suffering, permanent disability, and death. Vaccines prevent disease in those who get vaccinated and protect those who come into contact with unvaccinated persons. Vaccination has controlled many infectious diseases that were once common, including polio, measles, diphtheria, pertussis (whooping cough), rubella (German measles), mumps, tetanus, and influenza. It even led to the complete eradication of smallpox from the human population.

Not all countries have the same level of vaccination requirements as the United States. Given the current global nature of travel and business, exposure to many diseases is likely. Vaccination minimizes the risk of developing a disease and its associated complications. When persons travel outside the United States, they may require additional vaccinations. One should consult a physician within a minimum of four weeks of traveling to determine what vaccines, if any, are needed.

—Beatriz Manzor Mitrzyk, PharmD and Michael A. Buratovich, PhD

Further Reading

Centers for Disease Control and Prevention. "Autism and Vaccines." www.cdc.gov/vaccinesafety/concerns/autism.html.

Centers for Disease Control and Prevention. "Immunization Schedules: Resources for Parents." www.cdc.gov/vaccines/schedules/parents-adults/resources-parents.html.

Centers for Disease Control and Prevention. "Vaccines and Immunizations." www.cdc.gov/vaccines/.

Deer, Brian. *The Doctor Who Fooled the World*. Johns Hopkins UP, 2020.

DeStefano, Frank, Cristofer S. Price, and Eric S. Weintraub. "Increasing Exposure to Antibody-Stimulating Proteins and Polysaccharides in Vaccines Is Not Associated with Risk of Autism." *Journal of Pediatrics*, vol. 163, no. 2, pp. 561-67.

Merino, Noël. *Vaccines*. Greenhaven, 2015.

Pardi, Norbert, et al. "mRNA Vaccines—A New Era in Vaccinology." *Nature Reviews Drug Discovery*, vol. 17, 2018, pp. 261-79.

Plotkin, Stanley A., et al., editors. *Vaccines*. 7th ed., Elsevier, 2017.

Shoenfeld, Yehuda, and Nancy Agmon-Levin. *Vaccines and Autoimmunity*. Wiley, 2015.

"Vaccines." *National Institute of Allergy and Infectious Disease*. National Institutes of Health, 13 Aug. 2020, www.niaid.nih.gov/research/vaccines.

Yellow fever vaccine

Category: Prevention
Anatomy or system affected: Blood, cells, circulatory system, immune system, lymphatic system
Specialties and related fields: Immunology, internal medicine, microbiology, preventive medicine, public health, virology
Definition: a vaccine that protects individuals from yellow fever

KEY TERMS

flaviviruses: a group of insect-borne, positive-strand ribonucleic acid (RNA) viruses that includes West Nile virus, Zika virus, dengue virus, and yellow fever virus, among others

meningoencephalitis: inflammation of the membranes of the brain and the adjoining cerebral tissue *yellow fever*: a typically short-lived viral infection caused by a flavivirus that can, infrequently, cause liver damage, leading to leading to jaundice

DEFINITION

The yellow fever vaccine was developed to fight yellow fever, an acute infectious disease transmitted by mosquitoes and caused by a flavivirus. Yellow fever remains endemic to parts of South America and in Africa. When in an endemic area, the reported risk of contracting yellow fever is approximately 1 in 267; of those infected, up to 40 percent will die. No antiviral treatment is effective against the yellow fever virus, so a vaccine was developed to prevent people from contracting the disease. The vaccine is prepared from the 17D strain of the disease, a live but attenuated (weaker). More than four hundred million doses of yellow fever vaccine have been administered worldwide.

HISTORY

As was first thought, yellow fever was conclusively identified as a virus rather than as a bacterium in 1928. Max Theiler, a South African-born virologist working at New York's Rockefeller Foundation, developed the yellow fever vaccine in 1937. He initially passed the virus through laboratory mice. He found that the weakened form of the virus provided immunity to Rhesus monkeys. During his work with the virus, Theiler contracted yellow fever but survived and consequently developed immunity. Theiler was awarded the Nobel Prize in Physiology or Medicine in 1951 for developing the yellow fever vaccine.

ADMINISTRATION

Persons traveling to or planning to live in areas where yellow fever is endemic should receive the vaccine. People routinely exposed to yellow fever virus, such as researchers and laboratory staff, are encouraged to receive the vaccine. The vaccine, however, is not recommended for newborns younger than four months of age or women during their first trimester of pregnancy. The yellow fever vaccine is administered in a single injection by a health-care professional. Effective protection from the virus begins after ten days, and the protection lasts a minimum of ten years.

DOCUMENTATION

To legally enter some countries, people must carry internationally recognized proof of receiving the yellow fever vaccine. This proof is established with a stamped document, the International Certificate of Vaccination Against Yellow Fever.

SIDE EFFECTS

The yellow fever vaccine is safe. As with any drug or vaccine, strict regulations are enforced during its development and manufacturing. Common physical reactions to the vaccine include soreness and tenderness or redness at the site of the injection. Also, a slight headache, low-grade fever, or aching muscles can occur five to ten days after receiving the vaccine.

—April Ingram, BS

Further Reading

Bloom, Barry R., and Paul-Henri Lambert, editors. *The Vaccine Book*. Academic Press, 2002.

Centers for Disease Control and Prevention. "Yellow Fever Vaccine." www.cdc.gov/ncidod/dvbid/yellowfever/vaccine.

Frierson, J. Gordon. "The Yellow Fever Vaccine: A History." *Yale Journal of Biology and Medicine,* vol. 83, no. 2, June 2010, pp. 77-85.

Jong, Elaine C., and Russell McMullen, editors. *Travel and Tropical Medicine Manual*. 4th ed., Saunders/Elsevier, 2008.

Norrby, Erling. "Yellow Fever and Max Theiler: The Only Nobel Prize for a Virus Vaccine." *Journal of Experimental Medicine*, vol. 204, no. 12, 26 Nov. 2007, pp. 2779-84.

World Health Organization. "Yellow Fever." www.who.int/mediacentre/factsheets/fs100.

Bibliography

"10 Facts on Malaria." *World Health Organization*, Nov. 2015, www.who.int/features/factfiles/malaria/en. Accessed 16 Nov. 2016.

Abbas, Abul K., and Andrew H. Lichtman, and Shiv Pillai. *Basic Immunology: Functions and Disorders of the Immune System*. 6th ed., Elsevier, 2019.

———. *Cellular and Molecular Immunology*. 10th ed., Saunders/Elsevier, 2021.

"About Porphyria." *American Porphyria Foundation*, 2015, www.porphyriafoundation.com/about-porphyria. Accessed 9 May 2017.

Abramson, Stuart L. "Delayed Hypersensitivity Reactions." *Medscape*, 7 May 2018, emedicine.medscape.com/article/136118-overview. Accessed 5 Nov. 2018.

Acton, Ashton, editor. *Herpes Simplex Virus: New Insights for the Healthcare Professional*. Scholarly Editions, 2013.

"Acute Sinusitis." *Mayo Clinic*, 27 Aug. 2021, www.mayoclinic.org/diseases-conditions/acute-sinusitis/symptoms-causes/syc-20351671. Accessed 27 Aug. 2021.

Addeo, Alfredo, et al. "A New Generation of Vaccines in the Age of Immunotherapy." *Current Oncology Reports*, vol. 23, no. 12, 2021, p. 137, doi:10.1007/s11912-021-01130-x.

Adelman, Daniel C., et al., editors. *Manual of Allergy and Immunology*. 5th ed., Lippincott Williams & Wilkins, 2012.

"Adult Immunization Schedule by Vaccine and Age Group." *Immunization Schedules*. Centers for Disease Control and Prevention, US Department of Health and Human Services, 3 Feb. 2020, www.cdc.gov/vaccines/schedules/hcp/imz/adult.html. Accessed 28 Jan. 2021.

Advisory Committee on Immunization Practices. "Recommended Adult Immunization Schedule: United States, 2010." *Annals of Internal Medicine*, vol. 152, 2010, pp. 36-39.

Ahmad-Mansour, Nour, et al. "*Staphylococcus aureus* Toxins: An Update on Their Pathogenic Properties and Potential Treatments." *Toxins*, vol. 13, no. 10, 2021, p. 677, doi:10.3390/toxins13100677.

Akram, Sami M., and Janak Koirala. "Coccidioidomycosis." *StatPearls*. StatPearls Publishing, 20 Sept. 2021.

Albers, Sonja, et al. "Archaea." *Essentials of Glycobiology*, edited by Ajit Varki et al., 3rd ed., Cold Spring Harbor Laboratory Press, 2017, pp. 283-292, doi:10.1101/glycobiology.3e.022.

Ali, Mohammad, et al. "Updated Global Burden of Cholera in Endemic Countries." *PLoS Neglected Tropical Diseases*, vol. 9, no. 6, 2015, p. e0003832, doi:10.1371/journal.pntd.0003832.

Al-Janabi, A. A., and F. H. Al-Khikani. "Dermatophytoses: A Short Definition, Pathogenesis, and Treatment." *International Journal of Health & Allied Sciences*, vol. 9, no. 3, 2020, pp. 210-14, doi:10.4103/ijhas.IJHAS_123_19.

Allen, Jessica L., et al. *Urban Lichens: A Field Guide for Northeastern North America*. ?Yale UP, 2021.

Al-Nafussi, Awatif. *Tumor Diagnosis: Practical Approach and Pattern Analysis*. Oxford UP, 2005.

Ambroise-Thomas, Pierre, and Eskild Petersen, editors. *Congenital Toxoplasmosis: Scientific Background, Clinical Management, and Control*. Springer, 2000.

Ameer, Muhammad Atif, et al. "*Escherichia Coli* (*E. Coli* 0157 H7)." *StatPearls*. StatPearls Publishing, 10 July 2021.

American Cancer Society, www.cancer.org/. Accessed 24 Nov. 2021.

American Congress of Obstetricians and Gynecologists. "Perinatal Viral and Parasitic Infections." *ACOG Practice Bulletin*, no. 20, 2000, www.acog.org.

American Society for Microbiology (ASM). "Sentinel Level Clinical Laboratory Guidelines for Suspected Agents of Bioterrorism and Emerging Infectious Diseases: *Francisella tularensis*." Revised Oct. 2020, asm.org/ASM/media/Policy-and-Advocacy/Biosafety_Sentinel_Guideline_October_2018_FINAL.pdf. Accessed on 11 Dec. 2021.

The American Wine Society Presents the Complete Handbook of Winemaking. G.W. Kent, Inc., 1993.

Amos, W. Bradshaw, and J. G. Duckett, editors. *Prokaryotic and Eukaryotic Flagella*. Cambridge UP, 1982.

An, Zhiqiang, editor. *Therapeutic Monoclonal Antibodies: From Bench to Clinic*. John Wiley & Sons, 2009.

"Anabolic Steroid (Oral Route, Parenteral Route)." *Mayo Clinic*. Mayo Foundation for Medical Education and Research, 1 Dec. 2015. Accessed 11 May 2016.

"Anabolic Steroids." *MedlinePlus*. National Library of Medicine, National Institutes of Health, 9 Mar. 2016. Accessed 11 May 2016.

Anderson, Denise G., et al. *ISE Nester's Microbiology: A Human Perspective*. 10th ed., McGraw-Hill Education, 2021.

Angus, Derek C., and Tom van der Poll. "Sepsis and Septic Shock." *New England Journal of Medicine*, vol. 369, 2013, pp. 840-51. Accessed 30 Dec. 2015.

"Antibodies." *Genetics & Inherited Conditions*. Salem Press, 2010.

"Antiviral Drugs for Influenza for 2020-2021." *Medical Letter on Drugs and Therapeutics*, vol. 62, no. 1610, 2020, pp. 169-73.

APIC. *APIC Implementation Guide: Guide to Hand Hygiene Programs for Infection Prevention*. APIC, June 2015.

Archer, David. *The Global Carbon Cycle*. Princeton UP, 2010.

Arias, Cesar A., and Barbara E. Murray. "Antibiotic-Resistant Bugs in the Twenty-first Century: A Clinical Super-Challenge." *New England Journal of Medicine*, vol. 360, no. 5, 2009, pp. 439-43.

Armed Forces Health Surveillance Center. "Predictive Value of Reportable Medical Events for *Neisseria gonorrhoeae* and *Chlamydia trachomatis*." *MSMR*, vol. 20, no. 2, 2013, pp. 11-14.

Aronson, N. E., et al. "Long-Term Efficacy of BCG Vaccine in American Indians and Alaska Natives: A Sixty-Year Follow-up Study." *Journal of the American Medical Association*, vol. 291, 2004, pp. 2086-91.

Arumugam, V., et al. "Mumps." *Ferri's Clinical Advisor 2011: Instant Diagnosis and Treatment*, edited by Fred F. Ferry, Mosby/Elsevier, 2011.

Arya, Dev. *Aminoglycoside Antibiotics: From Chemical Biology to Drug Discovery*. Wiley-Interscience, 2007.

Ashford, David A., et al. "Adverse Events in Humans Associated with Accidental Exposure to the Livestock Brucellosis Vaccine RB51." *Vaccine*, vol. 3, no. 22, 2004, pp. 3435-39.

Aslam, Aysha, and Chika N. Okafor. "Shigella." *StatPearls*. StatPearls Publishing, 11 Aug. 2021.

Aspinall, Gerald O. *Polysaccharides*. Pergamon Press, 1970.

Atkins, John F., Raymond F. Gesteland, and Thomas R. Cech, editors. *RNA Worlds: From Life's Origins to Diversity in Gene Regulation*. Cold Spring Harbor Laboratory, 2011.

Atkinson, W., et al., editors. *Epidemiology and Prevention of Vaccine-Preventable Diseases*. 11th ed., Public Health Foundation, 2009.

"Autoimmune Conditions." *Medline Plus*, 31 Mar. 2021, https://medlineplus.gov/autoimmunediseases.html. Accessed 20 Aug. 2021.

Azar, Marwan M., et al. "Current Concepts in the Epidemiology, Diagnosis, and Management of Histoplasmosis Syndromes." *Seminars in Respiratory and Critical Care Medicine*, vol. 41, no. 1, 2020, pp. 13-30, doi:10.1055/s-0039-1698429.

Babakhani, S., and M. Oloomi. "Transposons: The Agents of Antimicrobial Resistance in Bacteria." *Journal of Basic Microbiology*, vol. 58, 2018, pp. 905-17.

Badash, Michelle. "Rabies." *Health Library*, 30 Dec. 2011.

———. "Rocky Mountain Spotted Fever (RMSF)." Reviewed by David L. Horn. *Health Library*, 2016, healthlibrary.epnet.com/GetContent.aspx?token=da29d243-e573-4601-8b42-77cd0ccb14b2&chunkiid=11588. Accessed 15 Nov. 2016.

Baker, Kate S., et al. "Horizontal Antimicrobial Resistance Transfer Drives Epidemics of Multiple *Shigella* Species." *Nature Communications*, vol. 9, no. 1, 2018, p. 1462, doi:10.1038/s41467-018-03949-8.

Balagurusamy, Nagamani, and Anuj Kumar Chandel. *Biogas Production: From Anaerobic Digestion to a Sustainable Bioenergy Industry*. Springer, 2020.

Balch, James F., and Phyllis A. Balch. *Prescription for Nutritional Healing: A Practical A to Z Reference to Drug-Free Remedies Using Vitamins, Minerals, Herbs, and Food Supplements*. 4th rev. ed., Avery, 2008.

Balsam, John, and Dave Ryan. "Anaerobic Digestion of Animal Wastes: Factors to Consider." *National Sustainable Agriculture Information Service*, 2006, attra.ncat.org/product/anaerobic-digestion-of-animal-wastes-factors-to-consider/.

Banschbach, Valerie S., and Robert Letovsky. "The Use of Corn versus Sugarcane to Produce Ethanol Fuel: A Fermentation Experiment for Environmental Studies." *American Biology Teacher*, vol. 72, no. 1, 2010, pp. 31-36.

Barker, David M. *Archaea: Salt-Lovers, Methane-Makers, Thermophiles, and Other Archaeans*. Crabtree Pub. Co., 2010.

Barrett, S. "Omnibus Court Rules Against Autism-Vaccine Link." *Autism Watch*, 6 Oct. 2010, www.autism-watch.org/omnibus/overview.shtml. Accessed 15 Nov. 2021.

Barry, John M. *The Great Influenza: The Story of the Deadliest Pandemic in History*. Viking Penguin, 2005.

Barsanti, Laura, and Paolo Gualtieri. *Algae: Anatomy, Biochemistry, and Biotechnology*. 2nd ed., CRC Press, 2014.

Bartlett, John G., and Ann K. Finkbeiner. *The Guide to Living with HIV Infection*. 6th ed., Johns Hopkins UP, 2007.

Baxter, David. "Active and Passive Immunity, Vaccine Types, Excipients, and Licensing." *Occupational Medicine*, vol. 57, no. 8, 2007, pp. 552-56.

Beers, Mark H., et al., editors. *The Merck Manual of Diagnosis and Therapy*. 18th ed., Merck Research Laboratories, 2006.

Beharry, M. S., T. Shafii, and G. R. Burstein. "Agnosis and Treatment of Chlamydia, Gonorrhea, and Trichomonas in Adolescents." *Pediatric Annals* vol. 42, no. 2, 2013, pp. 26-33.

Behbehani, Abbas. *The Smallpox Story in Words and Pictures.* University of Kansas Medical Center, 1988.

Behme, Stefan. *Manufacturing of Pharmaceutical Proteins: From Technology to Economy*. Wiley, 2009.

Behnke, Frances L. *A Natural History of Termites*. Charles Scribner's Sons, 1977.

Behrman, Richard E., et al., editors. *Nelson Textbook of Pediatrics*. 2 vols. 20th ed., Elsevier, 2016.

Beigel, John, and Mike Bray. "Current and Future Antiviral Therapy of Severe Seasonal and Avian Influenza." *Antiviral Research*, vol. 78, 2008, pp. 91-102.

Bellenir, Karen, and Peter D. Dresser, editors. *Contagious and Noncontagious Infectious Diseases Sourcebook*. Omnigraphics, 1996.

Bellinger, Edward G., and David C. Singee. *Freshwater Algae: Identification and Use as Bioindicators.* Wiley, 2010.

Benjamin, Ivor, et al., editors. *Andreoli and Carpenter's Cecil Essentials of Medicine*. 9th ed., Saunders, 2015.

Bennett, John E., et al., editors. *Mandell, Douglas, and Bennett's Principles and Practice of Infectious Diseases*. 9th ed., Elsevier, 2019.

Berg, Jeremy M., et al. *Biochemistry.* 9th ed., W. H. Freeman, 2019.

Berkelman, Ruth L. "Human Illness Associated with Use of Veterinary Vaccines." *Clinical Infectious Diseases*, vol. 37, no. 3, 2003, pp. 407-14.

Berliner, Nancy, editor. *Immunodeficiency, Infection, and Stem Cell Transplantation.* Saunders, 2011.

Berner, Robert A. *The Phanerozoic Carbon Cycle: CO2and O2*. Oxford UP, 2004.

Bernstein, David, and Gilbert Schiff. "Viral Exanthems and Localized Skin Infections." *Infectious Diseases*, edited by Sherwood L. Gorbach, John G. Bartlett, and Neil R. Blacklow, Saunders, 2004.

Bethaz, Carlo, and Vito Li Puma, editors. *New Research on Protein Synthesis*. Nova, 2014.

Betsy, Tom, and Jim Keogh. *Microbiology DeMYSTiFieD*. 2nd ed., McGraw-Hill Education, 2012.

Bettelheim, Frederick A., et al. *Introduction to General, Organic, and Biochemistry.* 10th ed., Brooks/Cole Cengage Learning, 2013.

Bhan, M. K., R. Bahl, and S. Bhatnagar. "Typhoid and Paratyphoid Fever." *The Lancet,* vol. 366 27 Aug.-2 Sept. 2005, pp. 749-62.

Biddle, Wayne. *A Field Guide to Germs*. 2nd ed., Anchor Books, 2002.

Bidram, Maryam, et al. "mRNA-Based Cancer Vaccines: A Therapeutic Strategy for the Treatment of Melanoma Patients." *Vaccines,* vol. 9, no. 10, 2021, p. 1060, doi:10.3390/vaccines9101060.

"Biofilm Basics." *Montana State University Center for Biofilm Engineering*, www.biofilm.montana.edu/biofilm-basics.html. Accessed 4 Feb. 2016,

Bishop, Roxanne H. *Influenza and RNA Viruses: Emergence, Classification and Management*. Nova Biomedical, 2014.

Bjarnshot, T. "The Role of Bacterial Biofilms in Chronic Infections." *Acta Pathologica, Microbiologica et Immunologica Scandinavica*, May 2013, www.ncbi.nlm.nih.gov/pubmed/23635385. Accessed 4 Feb. 2016.

Bjelakovic, G., L. L. Gluud, D. Nikolova, et al. "Antioxidant Supplements for Liver Diseases." *Cochrane Database of Systematic Reviews*, vol. 3, 2011, CD007749.

Black, R. E., L. H. Allen, Z. A. Bhutta, et al. "Maternal and Child Undernutrition: Global and Regional Exposures and Health Consequences." *The Lancet*, vol. 371, no. 9608, 2008, pp. 243-60.

Blanchard, Alain, and Cecile M. Bebear. "Mycoplasmas of Humans." *Molecular Biology and Pathogenicity of Mycoplasmas*, edited by Shmuel Razin and Richard Herrmann, Kluwer Academic, 2002.

Blankenship, Robert E. *Molecular Mechanisms of Photosynthesis*. 2nd ed., Wiley-Blackwell, 2014.

Blanton, Jesse D., et al. "Rabies Surveillance in the United States during 2007." *Journal of the American Veterinary Medical Association*, vol. 233, 2008, pp. 884-97.

Blaser, Martin J. *Missing Microbes*. Henry Holt and Co., 2014.

"Blattoidea." *ITIS*, Integrated Taxonomic Information System, 8 Nov. 2017, www.itis.gov/servlet/SingleRpt/SingleRpt?search_topic=TSN&search_value=666640#null. Accessed 31 Jan. 2018.

Bloom, Barry R., and Paul-Henri Lambert, editors. *The Vaccine Book.* Academic Press, 2002.

Bloomfield, Molly M., and Lawrence J. Stephens. *Chemistry and the Living Organism.* 6th ed., John Wiley, 1996.

Blume, S. "Anti-Vaccination Movements and Their Interpretations." *Social Science & Medicine*, vol. 62, no. 3, 2006, pp. 628-42.

Bodinham, C. L., et al. "Short-Term Effects of Whole-Grain Wheat on Appetite and Food Intake in Healthy Adults: A Pilot Study." *British Journal of Nutrition*, vol. 106, no. 3, 2011, pp. 327-30, doi:10.1017/S0007114511000225.

Bogitsh, Burton J., et al. *Human Parasitology*. 5th ed., Academic Press, 2018.

Bolton-Smith, C., et al. "A Two-Year Randomized Controlled Trial of Vitamin K1 (Phylloquinone) and Vitamin D3 Plus Calcium on the Bone Health of Older Women." *Journal of Bone and Mineral Research,* vol. 22, no. 4, 2007, pp. 509-19.

Bonam, Srinivasa Reddy, et al. "*Plasmodium falciparum* Malaria Vaccines and Vaccine Adjuvants." *Vaccines,* vol. 9, no. 10, 2021, p. 1072, 24 Sept. 2021, doi:10.3390/vaccines9101072.

Bonds, M. H., and P. Rohani. "Herd Immunity Acquired Indirectly from Interactions Between the Ecology of Infectious Diseases, Demography, and Economics." *Journal of the Royal Society Interface,* vol. 7, 2010, pp. 541-47.

Booth, S. L., et al. "Dietary Vitamin K Intakes Are Associated with Hip Fracture but Not with Bone Mineral Density in Elderly Men and Women." *American Journal of Clinical Nutrition,* vol. 71, 2000, pp. 1201-8.

———. "Effect of Vitamin K Supplementation on Bone Loss in Elderly Men and Women." *Journal of Clinical Endocrinology and Metabolism,* vol. 93, no. 4, 2008, pp. 1217-23.

Boray Tek, F., A. G. Dempster, and I. Kale. "Computer Vision for Microscopy Diagnosis of Malaria." *Malaria Journal,* vol. 8, 2009, p. 153.

Boston Women's Health Collective. *Our Bodies, Ourselves: A New Edition for a New Era.* 35th-anniversary ed., Simon & Schuster, 2005.

Boulton, Christopher, and David Quain. *Brewing Yeast and Fermentation.* Blackwell, 2001.

Bourgaize, David, Thomas R. Jewell, and Rodolfo G. Buiser. *Biotechnology: Demystifying the Concepts.* Benjamin/Cummings, 2000.

Bower, D. M., and K. L. J. Prather. "Engineering of Bacterial Strains and Vectors for the Production of Plasmid DNA." *Applied Microbiology and Biotechnology,* vol. 82, no. 5, 2009, pp. 805-13.

Braam, L. A., et al. "Vitamin K1 Supplementation Retards Bone Loss in Postmenopausal Women Between Fifty and Sixty Years of Age." *Calcified Tissue International,* vol. 73, 2003, pp. 21-26.

Brachman, Philip S., and Elias Abrutyn, editors. *Bacterial Infections of Humans: Epidemiology and Control.* 4th ed., Springer Science, 2009.

Brady, Mark F., and Vidya Sundareshan. "Legionnaires' Disease." *StatPearls.* StatPearls Publishing, 2021.

Broaddus, V. Courtney, et al., editors. *Murray and Nadel's Textbook of Respiratory Medicine.* 7th ed., Elsevier, 2021.

Broadhead, T. W., editor. *Fossil Prokaryotes and Protists: Notes for a Short Course.* U of Tennessee, 1987.

Brock, Thomas D., editor. *Microorganisms: From Smallpox to Lyme Disease.* W. H. Freeman, 1990.

Brodo, Irwin M., et al. *Keys to Lichens of North America.* Yale UP, 2016.

Bromwich, Jonah Engel. "You've Been Washing Your Hands Wrong." *New York Times,* 20 Apr. 2016. Accessed 19 Aug. 2016.

Brook, Itzhak, editor. *Sinusitis: From Microbiology to Management.* Taylor & Francis, 2006.

Brooker, Robert J., et al. *Biology.* 5th ed., McGraw-Hill Higher Education, 2019.

Brophy Marcus, Mary. "States with the Highest Child Vaccine Rates." *CBS News.* CBS Interactive, 25 Apr. 2017, www.cbsnews.com/news/states-child-vaccination-rates-mmr-vaccine-dtap-whooping-cough-chickenpox. Accessed 22 Sept. 2017.

Brothers, Will. "A Timeline of COVID-19 Vaccine Development." *BioSpace,* 20 Dec. 2020, www.biospace.com/article/a-timeline-of-covid-19-vaccine-development/. Accessed 4 Feb. 2021.

Brothwell, Julie A., et al. "Interactions of the Skin Pathogen *Haemophilus ducreyi* with the Human Host." *Frontiers in Immunology,* vol. 11, no. 615402, 2021, doi:10.3389/fimmu.2020.615402.

Brown, Michael S., and Joseph L. Goldstein. "How LDL Receptors Influence Cholesterol and Atherosclerosis." *Scientific American,* vol. 251, Nov. 1984, pp. 58-66.

Brown, Nicholas, M., and Boyan B. Bonev. *Bacterial Resistance to Antibiotics: From Molecules to Man.* Wiley-Blackwell, 2019.

Brown, Pamela. *Quick Reference to Wound Care.* Jones and Bartlett, 2009.

Bruno, Richard L. *The Polio Paradox: Understanding and Treating "Post-Polio Syndrome" and Chronic Fatigue.* Warner, 2002.

Bryer, S. C., and A. H. Goldfarb. "Effect of High Dose Vitamin C Supplementation on Muscle Soreness, Damage, Function, and Oxidative Stress to Eccentric Exercise." *International Journal of Sport Nutrition and Exercise Metabolism,* vol. 16, 2006, pp. 270-80.

Brzostowski, Joseph, and Haewon Sohn, editors. *Confocal Microscopy: Methods and Protocols* (Methods in Molecular Biology, 2304). Humana, 2021.

Buchanan, Bob B., et al. *Biochemistry and Molecular Biology of Plants.* American Society of Plant Physiologists, 2000.

Buelow, Becky. "Immediate Hypersensitivity Reactions." *Medscape,* 9 Feb. 2015, emedicine.medscape.com/article/136217-overview. Accessed 5 Nov. 2018.

Burton, P. M., et al. "Glycemic Impact and Health: New Horizons in White Bread Formulations." *Critical Reviews in Food Science & Nutrition*, vol. 51, no. 10, 2011, pp. 965-82, doi:10.1080/10408398.2010.491584.

Burton, Thomas. "Immunotherapy Treatments for Cancer Gain Momentum." *Wall Street Journal*, 12 Oct. 2017, www.wsj.com/articles/immunotherapy-treatments-for-cancer-gain-momentum-1507825152?mod=searchresults_pos17&page=1.

Büscher, Philippe, et al. "Human African Trypanosomiasis." *The Lancet*, vol. 390, no. 10110, 2017, pp. 2397-2409, doi:10.1016/S0140-6736(17)31510-6.

Bush, Larry M., and Charles E. Schmidt. "Klebsiella, Enterobacter, and Serratia Infections." *Merck Manuals*, Mar. 2021, www.merckmanuals.com/home/infections/bacterial-infections-gram-negative-bacteria/klebsiella-enterobacter-and-serratia-infections. Accessed 1 Dec. 2021.

Bushman, Frederick. *Lateral Gene Transfer: Mechanisms and Consequences*. Cold Spring Harbor Laboratory Press, 2001.

Byrd, Allyson L., and Julia A. Segre. "Infectious Disease: Adapting Koch's Postulates." *Science*, vol. 351, no. 6270, 2016, pp. 224-26, doi:10.1126/science.aad6753.

Cafiero-Fonseca, E., et al. "The Full Benefits of Adult Pneumococcal Vaccination: A Systematic Review." *PLOS ONE*, vol. 12, no. 10, 2017, p. e0186903.

Cain, Phil. "How Does Your Immune System Work?" *World Economic Forum*, 7 Apr. 2020, www.weforum.org/agenda/2020/04/immune-system-fight-off-disease-coronavirus-covid19-pandemic/. Accessed 10 Feb. 2021.

California Institute of Technology. "Nitrogen Fixation Research Could Shed Light on Biological Mystery: New Process Could Make Fertilizer Production More Sustainable." *ScienceDaily*, 30 May 2017, www.sciencedaily.com/releases/2017/05/170530140710.htm.

Camargo Jr., K., and R. Grant. "Public Health, Science, and Policy Debate: Being Right Is Not Enough." *American Journal of Public Health*, vol. 105, no. 2, 2015, pp. 232-35.

Campylobacter (Campylobacterosis). Centers for Disease Control and Prevention, 14 Apr. 2021, www.cdc.gov/campylobacter/index.html. Accessed 20 Nov. 2021.

"Carbohydrates." *Royal Society of Chemistry*, www.rsc.org/Education/Teachers/Resources/cfb/Carbohydrates.htm. Accessed 19 Apr. 2017.

Carbone, Michele, et al. "SV40 and Human Mesothelioma." *Translational Lung Cancer Research*, vol. 9, no. Suppl 1, 2020, pp. S47-59, doi:10.21037/tlcr.2020.02.03.

Carlson, Emily. "The Big, Fat World of Lipids." *NIH National Institute of General Medical Sciences: Inside Life Science*, 9 Aug. 2012.

Carson-DeWitt, Roasalyn. "Food Poisoning." *Health Library*, 22 Mar. 2013.

Cavicchioii, Richard, editor. *Archaea: Molecular and Cellular Biology*. ASM Press, 2007.

Celsa, Ciro, et al. "Direct-acting Antiviral Agents and Risk of Hepatocellular Carcinoma: Critical Appraisal of the Evidence." *Annals of Hepatology*, vol. 100568, 2021, doi:10.1016/j.aohep.2021.100568.

Centers for Disease Control and Prevention. "Amebiasis." 20 July 2015, www.cdc.gov/parasites/amebiasis/general-info.html. Accessed 6 Nov. 2021.

———. "Anthrax." 20 Nov. 2020, www.cdc.gov/anthrax/index.html.

———. "Anthrax Vaccination: What Everyone Should Know." 22 Nov. 2016, www.cdc.gov/vaccines/vpd/anthrax/public/index.html.

———. *Antibiotic Resistance Threats in the United States*. US Department of Health and Human Services, CDC, 2019.

———. "Autism and Vaccines." www.cdc.gov/vaccinesafety/concerns/autism.html.

———. "Bioterrorism-Related Anthrax." *Emerging Infectious Diseases*, vol. 8, 2002, pp. 1013-1183.

———. "Botulism." 19 Aug. 2019, www.cdc.gov/botulism/. Accessed 21 Nov. 2021.

———. "Brucellosis." Division of Bacterial and Mycotic Diseases, www.cdc.gov/brucellosis/index.html.

———. "Diphtheria Vaccination." *Vaccines and Preventable Disease*, 22 Jan. 2020, www.cdc.gov/vaccines/vpd/diphtheria/index.html. Accessed 24 Nov. 2021.

———. "Emergency Preparedness and Response." *Tularemia*, 4 Apr. 2018, emergency.cdc.gov/agent/tularemia/. Accessed 11 Dec. 2021.

———. "Epstein Barr Virus and Infectious Mononucleosis," 28 Sept. 2020, www.cdc.gov/epstein-barr/index.html. Accessed 23 Nov. 2021.

———. "FDA Licensure of Bivalent Human Papillomavirus Vaccine (HPV2, Cervarix) for Use in Females: Recommendations of the Advisory Committee on Immunization Practices (ACIP)." *Morbidity and Mortality Weekly Report*, 28 May 2010, pp. 626-29.

———. "FDA Licensure of Quadrivalent Human Papillomavirus Vaccine (HPV4, Gardasil) for Use in Males: Recommendations of the Advisory Committee on Immunization Practices (ACIP)." *Morbidity and Mortality Weekly Report*, 28 May 2010, pp. 630-32.

———. "Global Routine Vaccination Coverage, 2009." *MMWR: Morbidity and Mortality Weekly Report,* vol. 59, 2010, pp. 1367-71.

———. "Group A Streptococcal (GAS) Disease." www.cdc.gov/ncidod/dbmd/diseaseinfo/groupastreptococcal.

———. "Guideline for Hand Hygiene in Health-Care Settings: Recommendations of the Healthcare Infection Control Practices Advisory Committee and the HICPAC/SHEA/APIC/IDSA Hand Hygiene Task Force." *Morbidity and Mortality Weekly Report,* 51.RR-16, 2002.

———. "Guidelines for Viral Hepatitis Surveillance and Case Management." *Viral Hepatitis,* 31 May 2015, www.cdc.gov/hepatitis/statistics/surveillanceguidelines.htm. Accessed 28 Nov. 2021.

———. "Hand Hygiene in Healthcare Settings." 29 Apr. 2019, www.cdc.gov/handhygiene/index.html. Accessed 25 Nov. 2021.

———. *Health, United States, 2018: With Chartbook on Long-Term Trends in Health.* National Center for Health Statistics. US Department of Health and Human Services, 2019 , *CDC,* www.cdc.gov/nchs/data/hus/hus18.pdf. Accessed 28 Jan. 2021.

———. "Histoplasmosis." *Fungal Disease,* 29 Dec. 2020, www.cdc.gov/fungal/diseases/histoplasmosis/index.html. Accessed 24 Nov. 2021.

———. "HPV Vaccine Information for Young Women." *Centers for Disease Control and Prevention.* Department of Health and Human Services, 26 Mar. 2015. Accessed 30 Dec. 2015.

———. "Immunization Schedules: Resources for Parents." www.cdc.gov/vaccines/schedules/parents-adults/resources-parents.html.

———. "*Legionella* (Legionnaires' Disease and Pontiac Fever." 25 Mar. 2021, www.cdc.gov/legionella/about/index.html. Accessed 11 Dec. 2021.

———. "Listeria (Listeriosis)." 10 Sept. 2021, www.cdc.gov/listeria/index.html. Accessed 12 Dec. 2021.

———. "Measles, Mumps, and Rubella (MMR) Vaccination: What Everyone Should Know." *Vaccines and Preventable Diseases,* 26 Jan. 2021, www.cdc.gov/vaccines/vpd/mmr/public/index.html. Accessed 2 Dec. 2021.

———. "Methicillin-resistant *Staphylococcus aureus* (MRSA) Infections." 5 Feb. 2019, www.cdc.gov/mrsa/index.html. Accessed 8 Dec. 2021.

———. "MMR Vaccines: What You Need to Know." www.cdc.gov/vaccines/pubs/vis/downloads/vis-mmr.pdf.

———. *The National Plan to Eliminate Syphilis from the United States.* US Department of Health and Human Services, 2006.

———. "Pneumonia," 9 Mar. 2020, www.cdc.gov/pneumonia/index.html. Accessed 10 Dec. 2021.

———. "Rabies." 23 Sept. 2021, www.cdc.gov/rabies/index.html. Accessed 3 Dec. 2021.

———. "Recommendations for Prevention and Control of Hepatitis C Virus (HCV) Infection and HCV-related Chronic Disease." *MMWR,* vol. 47, no. RR-19, 1998, pp. 1-54.

———. "Recommended Immunization Schedules for Persons Aged 0-18 Years-United States, 2008." *Morbidity and Mortality Weekly Report,* vol. 57, 2008, pp. Q1-4, www.cdc.gov/mmwr/preview/mmwrhtml/mm5701a8.htm.

———. "Rotavirus." *CDC,* 26 Mar. 2021, www.cdc.gov/rotavirus/index.html. Accessed 4 Dec. 2021.

———. "Sexually Transmitted Diseases." 17 June 2021, www.cdc.gov/std/default.htm. Accessed 22 Nov. 2021.

———. *Sexually Transmitted Diseases Treatment Guidelines,* 22 July 2021www.cdc.gov/std/treatment. Accessed 5 Dec. 2021.

———. "Tetanus (Lockjaw) Vaccination." 22 Jan. 2020, www.cdc.gov/vaccines/vpd-vac/tetanus. Accessed 7 Dec. 2021.

———. "Typhoid Fever and Paratyphoid Fever." 18 Nov. 2020, www.cdc.gov/typhoid-fever/typhoid-vaccination.html. Accessed 6 Dec. 2020.

———. "Use of Anthrax Vaccine in the United States: Recommendations of the Advisory Committee on Immunization Practices." *Morbidity and Mortality Weekly Report,* vol. 49, 2000, pp. 1-20.

———. "Vaccine Safety: Measles, Mumps, and Rubella (MMR) Vaccine." www.cdc.gov/vaccinesafety.

———. "Varicella (Chickenpox) Vaccination." www.cdc.gov/vaccines/vpd-vac/varicella.

———. "Vaccines and Immunizations." www.cdc.gov/vaccines/.

———. "Valley Fever (Coccidioidomycosis) Statistics." www.cdc.gov/fungal/diseases/coccidioidomycosis/statistics.html. Accessed on 21 Nov. 2021.

———. "Yellow Fever Vaccine." www.cdc.gov/ncidod/dvbid/yellowfever/vaccine.

"Centers for Disease Control and Prevention Report Finds Tetanus Reaching Younger Adults." *Vaccine Weekly,* 16 July 2003, pp. 21-22.

Chacon-Cruz, Enrique. "Intestinal Protozoal Diseases." 26 Apr. 2017, emedicine.medscape.com/article/999282-overview. Accessed 8 Dec. 2021.

Chadwick, Derek, and Jamie A. Goode, editors. *Gastroenteritis Viruses*. John Wiley & Sons, 2001.

Chander, Jagdish. *Textbook of Medical Mycology*. 4th ed., Jaypee Brothers Medical Pub., 2017.

Chapin, F. Stuart, III, Pamela A. Matson, and Harold A. Mooney. "Internal Cycling of Nitrogen." *Principles of Terrestrial Ecosystem Ecology*. Springer, 2002.

Chatenoud, Lucienne. "Emerging Biological and Molecular Therapies in Autoimmune Disease." *The Autoimmune Diseases*, edited by Noel R. Rose and Ian R. Mackay, 6th ed., Academic Press, 2020, pp. 1437-57.

Chatterjee, Patralekha. "High Hopes for Oral Cholera Vaccine." *Bulletin of the World Health Organization*, vol. 88, 2010, pp. 165-66.

Chatziandreou, Ilenia, et al. "Capture and Generation of Adenovirus Specific T Cells for Adoptive Immunotherapy." *British Journal of Haematology*, vol. 132, 2006, pp. 117-26.

Chaudhuri, Keya. *Recombinant DNA Technology*. The Energy and Resources Institute, 2015.

"Chemistry I: Atoms and Molecules." *Estrella Mountain*, www2.estrellamountain.edu/faculty/farabee/biobk/BioBookCHEM1.html. Accessed 9 May 2017.

"Chemistry of Porphyrins." *The Museum of Organic Chemistry*, www.org-chem.org/yuuki/porphyrin/porphyrin.html. Accessed 9 May 2017.

Cheryan, Munir. "Lactic Acid." *University of Illinois at Urbana-Champaign*. Munir Cheryan, faculty.fshn.illinois.edu/~mcheryan/lactic.htm. Accessed 22 Jan. 2016.

Chey, William D., et al. "ACG Clinical Guideline: Treatment of *Helicobacter pylori* Infection." *The American Journal of Gastroenterology*, vol. 112, no. 2, 2017, pp. 212-39, doi:10.1038/ajg.2016.563.

Chlamydia-CDC Fact Sheet. Centers for Disease Control and Prevention, 23 Jan. 2014, www.cdc.gov/std/chlamydia/stdfact-chlamydia.htm. Accessed 5 Mar. 2019.

Chlebowski, R. T., et al. "Calcium Plus Vitamin D Supplementation and the Risk of Breast Cancer." *Journal of the National Cancer Institute*, vol. 100, 2008, pp. 1581-91

Choe, Jae C., and Bernard J. Crespi, editors. *The Evolution of Social Behavior in Insects and Arachnids*. Cambridge UP, 1997.

"Cholera Outbreak-Haiti, October, 2010." *Morbidity and Mortality Weekly*, Report 59, 2010, p. 1411.

Christen, W. G., et al. "Vitamin E and Age-Related Cataract in a Randomized Trial of Women." *Ophthalmology*, vol. 115, 2008, pp. 822-29..

Christian, Janet L., and Janet L. Greger. *Nutrition for Living*. 4th ed., Benjamin/Cummings, 1994.

Chuang, C. H., et al. "Adjuvant Effect of Vitamin C on Omeprazole-Amoxicillin-Clarithromycin Triple Therapy for *Helicobacter pylori* Eradication." *Hepatogastroenterology*, vol. 54, 2007, pp. 320-24.

Church, D. C., editor. *The Ruminant Animal: Digestive Physiology and Nutrition*. Waveland Press, 1993.

Clark, David P. *Germs, Genes, and Civilization: How Epidemics Shaped Who We Are Today*. FT Press, 2010.

Cliff, Andrew, Peter Haggett, and Matthew Smallman-Raynor. *World Atlas of Epidemic Diseases*. Oxford UP, 2004.

Cliver, Dean O., and Hans P. Riemann, editors. *Foodborne Diseases*. 2nd ed., Academic Press, 2002.

Cockayne, S., et al. "Vitamin K and the Prevention of Fractures." *Archives of Internal Medicine*, vol. 166, 2006, pp. 1256-61.

Cohen, John. "Why Flu Vaccines So Often Fail." *Science*, 20 Sept. 2017, www.sciencemag.org/news/2017/09/why-flu-vaccines-so-often-fail. Accessed 30 Nov. 2021.

Coico, Richard, and Geoffrey Sunshine. "Antibody Structure and Function." *Immunology: A Short Course*. 6th ed., Wiley-Blackwell, 2009, pp. 41-60.

College of Physicians of Philadelphia. "Herd Immunity." *History of Vaccines*, www.historyofvaccines.org/content/herd-immunity-0. Accessed 31 July 2017.

Colvin, Emily K., et al. "SV40 TAg Mouse Models of Cancer." *Seminars in Cell & Developmental Biology*, vol. 27, 2014, pp. 61-73, doi:10.1016/j.semcdb.2014.02.004.

Committee on Infectious Diseases, American Academy of Pediatrics. "Herpes Simplex." *Red Book: 2006 Report of the Committee on Infectious Diseases*, edited by L. K. Pickering, C. J. Baker, S. S. Long, and J. A. McMillan, 27th ed., American Academy of Pediatrics, 2006.

Committee on Revisiting Brucellosis in the Greater Yellowstone Area, et al. *Revisiting Brucellosis in the Greater Yellowstone Area*. National Academies Press, 2021.

"Congenital Rubella." *The New York Times Health Guide*, health.nytimes.com/health/guides/disease/congenital-rubella/overview.html.

Connolly, D. A., et al. "The Effects of Vitamin C Supplementation on Symptoms of Delayed Onset Muscle Soreness." *Journal of Sports Medicine and Physical Fitness*, vol. 46, 2006, pp. 462-67.

Conover, Michael R., and Rosanna M. Vail. *Human Diseases from Wildlife*. CRC, 2015.

Constable, Peter D., et al. *Veterinary Medicine: A Textbook of the Diseases of Cattle, Sheep, Pigs, Goats and Horses*. 11th ed., Saunders, 2017.

Constantinescu, Gheorghe M., Brian M. Frappier, and Germain Nappert. *Guide to Regional Ruminant Anatomy Based on the Dissection of the Goat*. U of Iowa P, 2001.

Cook, N. R., et al. "A Randomized Factorial Trial of Vitamins C and E and Beta Carotene in the Secondary Prevention of Cardiovascular Events in Women: Results from the Women's Antioxidant Cardiovascular Study." *Archives of Internal Medicine*, vol. 167, 2007, pp. 1610-18.

Cornatzer, W. E. *Role of Nutrition in Health and Disease*. Thomas, 1989.

Corthay, A. "Does the Immune System Naturally Protect Against Cancer?" *Frontiers in Immunology*, vol. 5, no. 197, 2012.

Coulibaly, O., C. L'Ollivier, R. Piarroux, and S. Ranque. "Epidemiology of Human Dermatophytoses in Africa." *Medical Mycology*, vol. 56, no. 2, 2017, pp. 145-61, doi:10.1093/mmy/myx048.

"COVID-19 Dashboard by the Center for Systems Science and Engineering (CSSE) at Johns Hopkins University (JHU)." *Johns Hopkins University & Medicine*, 14 June 2021, coronavirus.jhu.edu/map.html. Accessed 14 June 2021.

"COVID-19 Vaccinations in the United States." *COVID Data Tracker*, Centers for Disease Control and Prevention, 27 June 2021, covid.cdc.gov/covid-data-tracker/#vaccinations. Accessed 28 June 2021.

"COVID-19 Vaccines: Get the Facts." *Mayo Clinic*, 12 June 2021, www.mayoclinic.org/diseases-conditions/coronavirus/in-depth/coronavirus-vaccine/art-20484859. Accessed 14 June 2021.

Craig, Charles R., and Robert E. Stitzel, editors. *Modern Pharmacology with Clinical Applications*. 6th ed., Lippincott, 2004.

Crawford, Dorothy, et al. *Cancer Virus: The Discovery of the Epstein-Barr Virus*. Oxford UP, 2014.

Crick, Francis H. C. "The Genetic Code: III." *Scientific American*, Oct. 1966, pp. 55-62.

Cristaudo, Antonio, and Massimo Giulani, editors. *Sexually Transmitted Infections: Advances in Understanding and Management*. Springer, 2020.

Croft, William J. *Under the Microscope: A Brief History of Microscopy*. World Scientific, 2006.

Crompton, Peter D., Susan K. Pierce, and Louis H. Miller. "Advances and Challenges in Malaria Vaccine Development." *Journal of Clinical Investigation*, vol. 120, 2010, pp. 4168-78.

Cronje, P. B., E. A. Boomker, and P. H. Henning, editors. *Ruminant Physiology: Digestion, Metabolism, Growth, and Reproduction*. Oxford UP, 2000.

Crossley, Kent B., Kimberly K. Jefferson, and Gordon L. Archer, editors. *Staphylococci in Human Disease*. Wiley, 2009.

Crowe, Sheila E. "*Helicobacter pylori* Infection." *The New England Journal of Medicine*, vol. 380, no. 12, 2019, pp. 1158-65, doi:10.1056/NEJMcp1710945.

"*Cryptococcus neoformans* Infections." *Red Book: 2009 Report of the Committee on Infectious Diseases*, edited by Larry K. Pickering, et al., 28th ed., American Academy of Pediatrics, 2009.

Cui, Xinle, and Clifford M. Snapper. "Epstein Barr Virus: Development of Vaccines and Immune Cell Therapy for EBV-Associated Diseases." *Frontiers in Immunology*, vol. 12, no. 734471, 2021, doi:10.3389/fimmu.2021.734471.

Cullen, Bryan R. *Human Retroviruses*. Oxford UP, 1993.

"The Cutaneous Mycoses." *Mycology Online*, mycology.adelaide.edu.au/mycoses/cutaneous. Accessed 23 Nov. 2020.

Da Poian, Andrea T., and Miguel A. R. B. Castanho. *Integrative Human Biochemistry: A Textbook for Medical Biochemistry*. 2nd ed., Springer, 2021.

Dabbs, David J., editor. *Diagnostic Immunohistochemistry: Theranostic and Genomic Applications*. 5th ed., Elsevier, 2018.

Dalakouras, Athanasios, Elena Dadami, and Michael Wassenegger. "Viroid-Induced DNA Methylation in Plants." *Biomolecular Concepts*, vol. 4, no. 6, 2013, pp. 557-65.

Dale, David C., Laurence Boxer, and W. Conrad Liles. "The Phagocytes: Neutrophils and Monocytes." *Blood*, 2008, www.bloodjournal.org/content/112/4/935?sso-checked=true. Accessed 16 May 2017.

Daniel, Jennifer. "Good Talks Needed to Combat HPV Vaccine Myth." *New York Times*. New York Times, 9 Nov. 2015. Accessed 30 Dec. 2015.

Daniel, Thomas M., and Frederick C. Robbins, editors. *Polio*. U of Rochester P, 1997.

Daniel, Wayne W. *Biostatistics: A Foundation for Analysis in the Health Sciences*. 9th ed., John Wiley & Sons, 2009.

Davidson, Tish. *Vaccines: History, Issues, and Science*. Greenwood, 2017.

de Bruijn, Frans, editor. *Biological Nitrogen Fixation*. Wiley-Blackwell, 2015.

de Sousa, Telma, et al. "Genomic and Metabolic Characteristics of the Pathogenicity in Pseudomonas aeruginosa." *International Journal of Molecular Sciences*, vol. 22, no. 23, 2021, p. 12892, doi:10.3390/ijms222312892.

Deacon, J. W. *Modern Mycology*. 3rd ed., Blackwell Science, 1997.

Deer, Brian. *The Doctor Who Fooled the World: Science, Deception, and the War on Vaccines*. Johns Hopkins UP, 2020.

DeFranco, Anthony, Richard Locksley, and Miranda Robertson. *Immunity: The Immune Response in Infectious and Inflammatory Disease*. Sinauer, 2007.

Dellinger, R. Phillip, et al. "Surviving Sepsis Campaign: International Guidelines for Management of Severe Sepsis and Septic Shock: 2008." *Critical Care Medicine*, vol. 36, 2008, pp. 296-327.

Delves, Peter J. "Innate Immunity." *Merck Manual*, Apr. 2020, www.merckmanuals.com/home/immune-disorders/biology-of-the-immune-system/innate-immunity. Accessed 10 Feb. 2021.

Delves, Peter J., et al. *Roitt's Essential Immunology*. 13th ed., Blackwell, 2017.

Denniston, Katherine, et al. *General, Organic, and Biochemistry*. 10th ed., McGraw-Hill Education, 2019.

Des Jardins, Terry R., et al. *Clinical Manifestations and Assessment of Respiratory Disease*. 7th ed., Mosby Elsevier, 2016.

Despommier, Dickson D., et al. *Parasitic Diseases*. 5th ed., Apple Tree, 2006.

DeStafano, Frank. "Vaccines and Autism: Evidence Does Not Support a Causal Association." *Clinical Pharmacology and Therapeutics*, vol. 82, no. 6, Dec. 2007, pp. 756-59.

DeStefano, Frank, Cristofer S. Price, and Eric S. Weintraub. "Increasing Exposure to Antibody-Stimulating Proteins and Polysaccharides in Vaccines Is Not Associated with Risk of Autism." *Journal of Pediatrics*, vol. 163, no. 2, pp. 561-67.

Dettmer, Philipp. *Immune: A Journey into the Mysterious System That Keeps You Alive*. Random House, 2021.

DeVries, Jonathan W., Mary W. Trucksess, and Lauren S. Jackson, editors. *Mycotoxins and Food Safety*. Kluwer Academic, 2002.

Dewick, Paul. *Medicinal Natural Products: A Biosynthetic Approach*. John Wiley & Sons, 2009.

Diamos Andrew G., et al. "High Level Production of Monoclonal Antibodies Using an Optimized Plant Expression System." *Frontiers in Bioengineering and Biotechnology*, vol. 7, 2020, p. 472, www.frontiersin.org/article/10.3389/fbioe.2019.00472.

Diaz, D. E., editor. *The Mycotoxin Blue Book*. Nottingham UP, 2005.

Didier, Raoult, and Phillipe Parola, editors. *Rickettsial Diseases*. Informa Health Care, 2007.

Diefenbach, Russell J., and Cornel Fraefel, editors. *Herpes Simplex Virus: Methods and Protocols* (Methods in Molecular Biology.) Humana Press, 2014.

Diener, T. O., R. A. Owens, and R. W. Hammond. "Viroids, the Smallest and Simplest Agents of Infectious Disease: How Do They Make Plants Sick?" *Intervirology*, vol. 35, nos. 1-4, 1993, pp. 186-95.

Dienstag, J. L. "Hepatitis B Virus Infection." *New England Journal of Medicine*, vol. 359, 2008, pp. 1486-1500.

Dimmock, N. J., A. J. Easton, and K. N. Leppard. *Introduction to Modern Virology*. 6th ed., Blackwell, 2007.

"Discovery of a Glowing Millipede in California and the Gradual Evolution of Bioluminescence in Diplopoda." *Crossmark*, 2015, www.pnas.org/content/112/20/6419.full.pdf. Accessed 9 May 2017.

Dixon, Terry C., et al. "Anthrax." *New England Journal of Medicine*, vol. 341, 1999, pp. 815-26.

Dockrell, Hazel M., and Ying Zhang. "A Courageous Step Down the Road Toward a New Tuberculosis Vaccine." *American Journal of Respiratory and Critical Care Medicine*, vol. 179, 2009, pp. 628-29.

Dodd, Christine, E. R., et al., editors. *Foodborne Diseases*. 3rd ed., Academic Press, 2017.

Doerr, Allison. "RNA Antibodies: Upping the Ante." *Nature Methods*, vol. 5, no. 2, 2008, p. 220.

Dombrowski, Julia C. "Chlamydia and Gonorrhea." *Annals of Internal Medicine*, vol. 174, no. 10, 2021, p. ITC145-ITC160, doi:10.7326/AITC202110190.

Domingo, Esteban, Colin R. Parrish, and John J. Holland, editors. *Origin and Evolution of Viruses*. 2nd ed., Elsevier/Academic, 2008.

Donlan, Rodney M. "Biofilms: Microbial Life on Surface." *Emerging Infectious Diseases*. US Centers for Disease Control, vol. 8, no. 9, Sept. 2002, wwwnc.cdc.gov/eid/article/8/9/02-0063_article. Accessed 4 Feb. 2016.

Doran, Pauline M. *Bioprocess Engineering Principles*. Academic Press, 1995.

Dosedel, Martin et al. "Vitamin C-Sources, Physiological Role, Kinetics, Deficiency, Use, Toxicity, and Determination." *Nutrients*, vol. 13, no. 2, 2021, p. 615, doi:10.3390/nu13020615.

"Drugs for Psoriatic Arthritis." *Medical Letter on Drugs and Therapy*, vol. 61, no. 1588, 30 Dec. 2019, pp. 203-10.

Du, D., X. Wang-Kan, A. Neuberger, et al. "Multidrug Efflux Pumps: Structure, Function, and Regulation." *Nature Reviews Microbiology*, vol. 16, 2018, pp. 523-39.

Dübel, Stefan, and Janice M. Reichert, editors. *Handbook of Therapeutic Antibodies*. 3 vols. Wiley-VCH, 2014.

Dudley, Jaquelin. *Retroviruses and Insights into Cancer*. Springer, 2011.

Dunn, Gavin P., et al. "Cancer Immunoediting: From Immunosurveillance to Tumor Escape." *Nature Immunology*, vol. 3, 2002, pp. 991-98.

———. "The Three Es of Cancer Immunoediting." *Annual Review of Immunology*, vol. 23, 2013, pp. 329-60.

Dunn, Noel, et al. "African Trypanosomiasis." *StatPearls*. StatPearls Publishing, 11 Aug., 2021.

Dunne, E. F., and L. E. Markowitz. "Genital Human Papillomavirus Infection." *Clinical Infectious Diseases*, vol. 43, 2006, p. 624.

Duran-Vila, Núria, et al. "Structure and Evolution of Viroids." *Origin and Evolution of Viruses, edited by* Esteban Domingo, Colin R. Parrish, and John J. Holland, 2nd ed., Elsevier/Academic, 2008.

Duyff, Roberta Larson. *American Dietetic Association Complete Food and Nutrition Guide*. 3rd ed., John Wiley & Sons, 2007.

Eckert, M. J., J. T. Perry, V. Y. Sohn, et al. "Incidence of Low Vitamin A Levels and Ocular Symptoms after Roux-en-Y Gastric Bypass." *Surgery for Obesity and Related Diseases*, vol. 6, no. 6, 2010, pp. 653-57.

Editors of *The Lancet*. "Retraction: Ileal-Lymphoid-Nodular Hyperplasia, Non-specific Colitis, and Pervasive Developmental Disorder in Children." *The Lancet*, vol. 375, 2020, p. 445.

Edwards, J., Jennings, M., Apicella, M., and Seib, K. "Is Gonococcal Disease Preventable? The Importance of Understanding Immunity and Pathogenesis in Vaccine Development." *Critical Reviews in Microbiology*, vol. 42, no. 6, 2016, pp. 928-41.

"Elderberry for Influenza." *The Medical Letter on Drugs and Therapeutics*, vol. 61, no. 1566, 2019, p. 32.

El-Mansi, E. M. T., et al., editors. *Fermentation Microbiology and Biotechnology*. 3rd ed., CRC, 2012.

El-Sayed, A., and W. Awad. "Brucellosis: Evolution and Expected Comeback." *International Journal of Veterinary Science and Medicine*, vol. 6 (suppl.), 2018, pp. S31-35.

Enayati, A., and J. Hemingway. "Malaria Management: Past, Present, and Future." *Annual Review of Entomology*, vol. 55, 2010, pp. 569-91.

Enayati, A., J. Hemingway, and P. Garner. "Electronic Mosquito Repellents for Preventing Mosquito Bites and Malaria Infection." *Cochrane Database of Systematic Reviews* (2009): CD005434, www.ebscohost.com/dynamed.

Engelkirk, Paul G., and Gwendolyn R. W. Burton. *Burton's Microbiology for the Health Sciences, 11th ed.* 8th ed., ? Jones & Bartlett Learning, 2020.

Engleberg, N. Cary, et al. *Schaechter's Mechanisms of Microbial Disease*. 6th ed., Lippincott Williams & Wilkins, 2021.

Environmental Protection Agency. "How Does Anaerobic Digestion Works?" 22 Jan. 2021, www.epa.gov/agstar/how-does-anaerobic-digestion-work.

Etchecopaz, Alejandro, et al. "Sporothrix brasiliensis: A Review of an Emerging South American Fungal Pathogen, Its Related Disease, Presentation and Spread in Argentina." *Journal of Fungi*, vol. 7, no. 3, 2021, p. 170, doi:10.3390/jof7030170.

Exumé, Myriam. "Product Discontinuation Notice: Zostavax(r) (Zoster Vaccine Live)." www.merckvaccines.com/wp-content/uploads/sites/8/2020/06/US-CIN-00033.pdf. Accessed 20 Nov. 2021.

Fairfax, M. R., M. H. Bluth, and H. Salimnia. "Diagnostic Molecular Microbiology: A 2018 Snapshot." *Clinics in Laboratory Medicine*, vol. 38, 2018, pp. 253-76.

Fanous, Matthew, and Kevin C. King. "Cholera." *StatPearls*. StatPearls Publishing, 2021.

"FDA and CDC Lift Recommended Pause on Johnson & Johnson (Janssen) COVID-19 Vaccine Use Following Thorough Safety Review." *FDA*, 23 Apr. 2021, www.fda.gov/news-events/press-announcements/fda-and-cdc-lift-recommended-pause-johnson-johnson-janssen-covid-19-vaccine-use-following-thorough. Accessed 12 May 2021.

"FDA Annual Summary Report on Antimicrobials Sold or Distributed in 2012 for Use in Food-Producing Animals." *FDA*. US Department of Health and Human Services, 2 Oct. 2014. Accessed 30 Nov. 2015.

Fedonkin, Mikhail A., James G. Gehling, Kathleen Grey, et al. *The Rise of Animals: Evolution and Diversification of the Kingdom Animalia*. Johns Hopkins UP, 2007.

Fehervari, Zoltan, and Shiman Sakaguchi. "Peacekeepers of the Immune System." *Scientific American*, Oct. 2006, pp. 56-63.

Feldman, Mark, Lawrence S. Friedman, and Lawrence J. Brandt, editors. *Sleisenger and Fordtran's Gastrointestinal and Liver Disease: Pathophysiology, Diagnosis, Management*. New ed. 2 vols. Saunders/Elsevier, 2010.

Fernandes, Vitória de Souza, et al. "Antiprotozoal Agents: How Have They Changed over a Decade?" *Archiv der Pharmazie*, e2100338, 2021, doi:10.1002/ardp.202100338.

Ferrari, Mario. *PDxMD Ear, Nose, and Throat Disorders*. PDxMD, 2003.

Ferrero, Mariana C., et al. "Pathogenesis and Immune Response in Brucella Infection Acquired by the

Respiratory Route." *Microbes and Infection*, vol. 22, no. 9, 2020, pp. 407-15, doi:10.1016/j.micinf.2020.06.001.

Field, Christopher B., and Michael R. Raupach, editors. *The Global Carbon Cycle: Integrating Humans, Climate, and the Natural World*. Island Press, 2004.

Finn, O. J. "Immuno-Oncology: Understanding the Function and Dysfunction of the Immune System in Cancer." *Annals of Oncology*, vol. 23(Suppl. 8), 2012, pp. iiv6-9, annonc.oxfordjournals.org/content/23/suppl_8/viii6.full.

Fiore, Kristina. "Want to Know More About mRNA Before Your COVID Jab?" *MedPage Today*, 3 Dec. 2020, www.medpagetoday.com/infectiousdisease/covid19/89998. Accessed 3 Feb. 2021.

Fisher, Margaret C. *Immunizations and Infectious Diseases: An Informed Parent's Guide*. American Academy of Pediatrics, 2006.

Flajnik, Martin F., Nevil J. Singh, and Steven M. Holland. 8th ed., *Paul's Fundamental Immunology*. Lippincott Williams & Wilkins, 2022.

Flint, Jane, et al. *Principles of Virology*. 5th ed., ASM Press, 2020.

"Flu (Influenza)." *Vaccines.gov*. US Department of Health and Human Services, 8 Sept. 2021, www.vaccines.gov/diseases/flu/index.html. Accessed 30 Nov. 2021.

Folstad, Steven G. "Soft Tissue Infections." *Emergency Medicine: A Comprehensive Study Guide*, edited by Judith E. Tintinalli, 6th ed., McGraw-Hill, 2004.

Foodborne Diseases Active Surveillance Network. Centers for Disease Control and Prevention, wwwn.cdc.gov/foodnetfast/. Accessed on 20 Nov. 2021.

Forsbeg, Kevin J., et al. "The Shared Antibiotic Resistome of Soil Bacteria and Human Pathogens." *Science*, vol. 337, no. 6098, 2012, pp. 1107-11.

Forterre, Patrick. *Microbes from Hell*. Translated by Teresa Lavender Fagan. U of Chicago P, 2016.

Fosnight, S. M., W. J. Zafirau, and S. E. Hazelett. "Vitamin D Supplementation to Prevent Falls in the Elderly: Evidence and Practical Considerations." *Pharmacotherapy*, vol. 28, 2008, pp. 225-34.

Foster, Neil, et al. "Revisiting Persistent *Salmonella* Infection and the Carrier State: What Do We Know?" *Pathogens*, vol. 10, no. 10, 2021, p. 1299, doi:10.3390/pathogens10101299.

Fox, Stuart Ira, and Krista Rompolski. *Human Physiology*. 16th ed., McGraw-Hill, 2021.

Franco, Maria Pia, Maximilian Mulder, Robert H. Gilman, and Henk L. Smits. "Human Brucellosis." *Lancet Infectious Diseases*, vol. 7, 2007, pp. 775-86.

Frank, Steven A. *Immunology and Evolution of Infectious Disease*. Princeton UP, 2002.

Frederick, Matthew. *Homebrewing for Beginners: A Beginner's Guide to Learning the Supplies, Techniques, and Methods for Brewing Beer at Home*. ?Independently published, 2019.

Freedman, David O. "Re-born in the USA: Another Cholera Vaccine for Travellers." *Travel Medicine and Infectious Disease*, vol. 14, no. 4, 2016, pp. 295-96, doi:10.1016/j.tmaid.2016.07.008.

"Frequently Asked Questions." *Centers for Disease Control and Prevention*, 25 May 2021, www.cdc.gov/coronavirus/2019-ncov/faq.html. Accessed 14 June 2021.

Frías-De-León, María Guadalupe, et al. "Epidemiology of Systemic Mycoses in the COVID-19 Pandemic." *Journal of Fungi*, vol. 7, no. 7, 2021, p. 556, doi:10.3390/jof7070556.

Friedlander, A. M., and S. F. Little. "Advances in the Development of Next-Generation Anthrax Vaccines." *Vaccine*, vol. 27, suppl. 4, 2009, pp. D61-64.

Frierson, J. Gordon. "The Yellow Fever Vaccine: A History." *Yale Journal of Biology and Medicine*, vol. 83, no. 2, June 2010, pp. 77-85.

Frossard, Emmanuel, A. Oberson, and Else K. Bünemann. *Phosphorus in Action: Biological Processes in Soil Phosphorus Cycling*. Springer, 2011.

Galgiani, John N. "Changing Perceptions and Creating Opportunities for Its Control." *Annals of the New York Academy of Sciences*, vol. 1111, 2007, pp. 1-18.

Gallo, Robert. *Virus Hunting: AIDS, Cancer, and the Human Retrovirus-A Story of Scientific Discovery*. Basic Books, 1991

Gallo, Robert C., Dominique Stehelin, and Oliviero E. Varnier. *Retroviruses and Human Pathology*. Humana Press, 1986.

Gaman, P. M., and K. B. Sherrington. *The Science of Food: An Introduction to Food Science, Nutrition, and Microbiology*. 4th ed., Butterworth-Heinemann/Elsevier, 2008.

García-Montero, Cielo, et al. "An Updated Review of SARS-CoV-2 Vaccines and the Importance of Effective Vaccination Programs in Pandemic Times." *Vaccines*, vol. 9, no. 5, 2021, p. 433, doi:10.3390/vaccines9050433.

Garde, Damian, and Jonathan Saltzman. "The Story of mRNA: How a Once-Dismissed Idea Became a Leading Technology in the Covid Vaccine Race." *Stat*, 10 Nov. 2020, www.statnews.com/2020/11/10/the-story-of-mrna-how-a-once-dismissed-idea-became-a-leading-technology-in-the-covid-vaccine-race/. Accessed 3 Feb. 2021.

Garrett, Roger A., and Hans-Peter Klenk, editors. *Archaea: Evolution, Physiology, and Molecular Biology*. Wiley-Blackwell, 2008.

Garrity, George M., editor. *The Proteobacteria*. Vol. 2 in *Bergey's Manual of Systematic Bacteriology*. 2nd ed., Springer, 2005.

Geha Raif S., and Luigi Notarangelo. *Case Studies in Immunology: A Clinical Companion*. 6th ed., Garland Science, 2012.

Geissler, Aimee L., et al. "Increasing Campylobacter Infections, Outbreaks, and Antimicrobial Resistance in the United States, 2004-2012." *Clinical Infectious Diseases,* vol. 65, no. 10, 2017, pp. 1624-31, doi:10.1093/cid/cix624.

Genetics Home Reference. "HLA-DRB1." *Genetics Home Reference*. US NLM, 28 July 2014. Accessed. 4 Aug. 2014.

"Genetics of Bacteria and Viruses." *Kean University*, www.kean.edu/~jfasick/docs/Cell%20Biology/chapt18_lecture%20%5BCompatibility%20Mode%5D.pdf. Accessed 24 June 2017.

George, Andrew J. T., and Catherine E. Urch, editors. *Diagnostic and Therapeutic Antibodies.* Humana Press, 2000.

George, Helga. "What Are Polysaccharides?" *WiseGeek,* 12 Apr. 2017, www.wisegeek.org/what-are-polysaccharides.htm. Accessed 19 Apr. 2017.

Gerhardt, Philipp, et al., editors. *Methods for General Bacteriology*. American Society for Microbiology, 1994.

German, J. B., C. J. Dillard. "Saturated Fats: What Dietary Intake?" *American Journal of Clinical Nutrition*, vol. 80, no. 3, 2004, pp. 550-59.

Gershon, Anne. "Mumps." *Harrison's Principles of Internal Medicine*, edited by Joan Betterton, 17th ed., McGraw-Hill, 2008.

Gil, A., et al. "Wholegrain Cereals and Bread: A Duet of the Mediterranean Diet for the Prevention of Chronic Diseases." *Public Health Nutrition*, vol. 14, no. 12, 2011, pp. 2316-22, doi:10.1017/S1368980011002576.

Gladwin, Mark T., et al. Clinical Microbiology Made Ridiculously Simple. 8th ed., ?MedMaster, 2021.

Glazer, Alexander N., and Hiroshi Nikaido. *Microbial Biotechnology: Fundamentals of Applied Microbiology.* 2nd ed., Cambridge UP, 2007.

Gleichmann, Nicole. "Innate Vs. Adaptive Immunity." *Immunology and Microbiology*, 20 May 2020, www.technologynetworks.com/immunology/articles/innate-vs-adaptive-immunity-335116. Accessed 10 Feb. 2021.

Glick, Bernard R., and Jack J. Pasternak, editors. *Molecular Biotechnology: Principles and Applications of Recombinant DNA.* 4th ed., ASM, 2010.

Gnanam, Chelin Rani. Introduction To Mycology. MJP Publishers, 2013.

Goering, Richard, et al. *Mims' Medical Microbiology and Immunology*. 6th ed., Elsevier, 2018.

Gogarten, Maria B., Johann Peter Gogarten, and Lorraine C. Olendzenski, editors. *Horizontal Gene Transfer: Genomes in Flux*. Springer, 2009.

Goldstein, Mark N. "Office Evaluation and Management of the Sore Throat." *Otolaryngologic Clinics of North America*, vol. 25, Aug. 1992, pp. 837-42.

Gómez, Gustavo, and Vicente Pallás. "Viroids: A Light in the Darkness of the lnc RNA-Directed Regulatory Networks in Plants." *New Phytologist,* vol. 198, no. 1, 2013, pp. 10-15.

Gonzales, Maria Liza M., et al. "Antiamoebic Drugs for Treating Amoebic Colitis." *The Cochrane Database of Systematic Reviews,* vol. 1, 1, CD006085, 2019, doi:10.1002/14651858.CD006085.pub3.

Goodwin, Zakia I., and David W. Pascual. "Brucellosis Vaccines for Livestock." *Veterinary Immunology and Immunopathology,* vol. 181, 2016, pp. 51-58, doi:10.1016/j.vetimm.2016.03.011.

Gordon, Siamon. "Phagocytosis: An Immunobiologic Process." *Science Direct,* 15 Mar. 2016, www.sciencedirect.com/science/article/pii/S1074761316300656. Accessed 16 May 2017.

Gould, Tony. *A Summer Plague: Polio and Its Survivors*. Yale UP, 1995.

Graciaa, Daniel S., et al. "Outbreaks Associated with Untreated Recreational Water-United States, 2000-2014." *Morbidity and Mortality Weekly Report,* vol. 67, no. 25, 2018, pp. 701-6, doi:10.15585/mmwr.mm6725a1.

Graham, Linda E., and Lee W. Wilcox. *Algae.* 2nd ed., Benjamin Cummings, 2009.

Grand, J. A., editor. *Viruses, Cell Transformation, and Cancer.* Elsevier, 2001.

Grandi, Guido, editor. *Genomics, Proteomics, and Vaccines.* John Wiley & Sons, 2004.

Gray, James, and Ulrich Desselberger, editors. *Rotaviruses: Methods and Protocols*. Humana Press, 2000.

Gray, Michael W. "Lynn Margulis and the Endosymbiont Hypothesis: 50 Years Later." *Molecular Biology of the Cell,* vol. 28, no.10, 2017, pp. 1285-87, doi:10.1091/mbc.E16-07-0509.

Greenfield, Ronald A. "Sporotrichosis." 2 Mar. 2021, emedicine.medscape.com/article/228723-overview. Accessed 5 Dec. 2021.

Gremillion, Henry A., editor. *Temporomandibular Disorders and Orofacial Pain.* Saunders/Elsevier, 2007.

Griffin, Diane E., and Michael B. A. Oldstone, editors. *Measles: History and Basic Biology*. Springer, 2009.

Griffith, C. J. "Do Businesses Get the Food Poisoning They Deserve? The Importance of Food Safety Culture." *British Food Journal*, vol. 112, no. 4, 2010, pp. 416-25.

Griffiths, Christopher, et al., editors. *Rook's Textbook of Dermatology*. 9th ed., Wiley-Blackwell, 2016.

Grognot, Marianne, and Katja M. Taute. "More Than Propellers: How Flagella Shape Bacterial Motility Behaviors." *Current Opinion in Microbiology*, vol. 61, 2021, pp. 73-81, doi:10.1016/j.mib.2021.02.005.

Group, Edward. "Understanding Your Nutrition: What Are Polysaccharides?" *Global Healing Center*, 18 Feb. 2016, www.globalhealingcenter.com/natural-health/understanding-nutrition-polysaccharides. Accessed 19 Apr. 2017.

Gudas, L. J., and J. A. Wagner. "Retinoids Regulate Stem Cell Differentiation." *Journal of Cellular Physiology*, vol. 226, no. 2, 2011, pp. 322-30.

Gunde-Cimerman, Nina, Aharon Oren, and Ana Plemenita. *Adaptation to Life at High Salt Concentrations in Archaea, Bacteria, and Eukarya*. Springer, 2011.

Gunners, Kris. "Vitamin D 101-A Detailed Beginner's Guide." *Healthline*, 6 Mar. 2019, www.healthline.com/nutrition/vitamin-d-101.

Gupta, Ravi G., et al. "Exploiting Tumor Neoantigens to Target Cancer Evolution: Current Challenges and Promising Therapeutic Approaches." *Cancer Discovery*, vol. 11, no. 5, 2021, pp. 1024-39, doi:10.1158/2159-8290.CD-20-1575.

Gutierrez, K. M. "Mumps Virus." *Principles and Practice of Pediatric Infectious Diseases*, edited by Sarah S. Long, Larry K. Pickering, and Charles G. Prober, 3rd ed., Churchill Livingstone/Elsevier, 2008.

Guyton, Arthur C., and John E. Hall. *Human Physiology and Mechanisms of Disease*. 6th ed., W. B. Saunders, 1997.

Habu, D., et al. "Role of Vitamin K2 in the Development of Hepatocellular Carcinoma in Women with Viral Cirrhosis of the Liver." *JAMA: The Journal of the American Medical Association*, vol. 292, 2004, pp. 358-61.

Hadidi, Ahmed, et al., editors. *Viroids*. Science, 2003.

Hager, Thomas. "How One Daring Woman Introduced the Idea of Smallpox Inoculation to England." *Time*, 5 Mar. 2019, time.com/5542895/mary-montagu-smallpox/. Accessed 21 Jan. 2021.

Hak, E., et al. "Influence of High-Risk Medical Conditions on the Effectiveness of Influenza Vaccination Among Elderly Members of Three Large Managed-Care Organizations." *Clinical Infectious Diseases*, vol. 35, 2002, pp. 370-77.

Hamelman, Jeffery. *Bread: A Baker's Book of Techniques and Recipes*. 3rd ed., ?Wiley, 2021.

Hammond, R. W. "Analysis of the Virulence Modulating Region of Potato Spindle Tuber Viroid (PSTVd) by Site-Directed Mutagenesis." *Virology*, vol. 187, no. 2, 1992, pp. 654-62.

Handsfield, H. H., et al. "*Neisseria gonorrhoeae*." *Mandell, Douglas, and Bennett's Principles and Practice of Infectious Diseases*, edited by Gerald L. Mandell, John F. Bennett, and Raphael Dolin, 7th ed., Churchill Livingstone/Elsevier, 2010.

Hankins, D. G., and J. A. Rosekrans. "Overview, Prevention, and Treatment of Rabies." *Mayo Clinic Proceedings*, 79, no. 5, May 2004, pp. 671-76.

Haque, Sakib, et al. "RNA-Based Therapeutics: Current Developments in Targeted Molecular Therapy of Triple-Negative Breast Cancer." *Pharmaceutics*, vol. 13, no. 10, 2021, p. 1694, doi:10.3390/pharmaceutics13101694.

Hardy, Simon P. *Human Microbiology*. Taylor and Francis, 2003.

Harley, J. L., and S. E. Smith. *Mycorrhizal Symbiosis*. Academic Press, 1983.

Harlow, Ed, and David Lane, editors. *Using Antibodies: A Laboratory Manual*. Rev. ed., Cold Spring Harbor Laboratory Press, 1999.

Harris, Vanessa C., Bastiaan W. Haak, Michael Boele van Hensbroek, and Willem J. Wiersing. "The Intestinal Microbiome in Infectious Diseases: The Clinical Relevance of a Rapidly Emerging Field." *Open Forum Infectious Diseases*. 8 July 2017, doi:[10.1093/ofid/ofx144].

Harris, W. Victor. *Termites: Their Recognition and Control*. Longmans, 1961.

Harvey, Richard A., et al. *Lippincott's Illustrated Reviews: Microbiology*. 3rd ed., Lippincott Williams and Wilkins, 2013.

Haur, Harleen, and Dietmar Reusch, editors. *Monoclonal Antibodies: Physicochemical Analysis*. Academic Press, 2021.

Hauser, Alan R. *Antibiotics for Clinicians*. 3rd ed., Lippincott Williams & Wilkins, 2018.

Hawkins, Trisha. *The Need to Know Library: Everything You Need to Know About Measles and Rubella*. Rosen, 2001.

Hayette, M. P., and R. Sacheli. "Dermatophytosis, Trends in Epidemiology and Diagnostic Approach." *Current Fungal Infection Reports*, vol. 9, no. 3, 2015, pp. 164-79, doi:10.1007/s12281-015-0231-4.

Hayoun, Michael A., et al. "Brucellosis." *StatPearls*. StatPearls Publishing, 18 Sept. 2021.

Hazen, Robert M. *Symphony in C: Carbon and the Evolution of (Almost) Everything*. W. W. Norton & Company, 2019.

Hechemy, Karim E., et al., editors. *Rickettsiology and Rickettsial Diseases*. Wiley-Blackwell, 2009.

"Helicobacter Pylori Infections." *MedlinePlus*, 5 May 2021, medlineplus.gov/helicobacterpyloriinfections.html. Accessed 25 Nov. 2021.

Hellman, M. "Study: Measles, Mumps, and Rubella not Associated with Autism." 1 July 2014, time.com/2943945/study-measles-mumps-and-rubella-vaccines-not-associated-with-autism/. Accessed 15 Nov. 2021.

"Heme and Bilirubin Metabolism." *The Medical Biochemistry Page*, 2021, themedicalbiochemistrypage.org/heme-and-bilirubin-metabolism/. Accessed 1 Dec 2021.

Hemila, H., E. Chalker, and B. Douglas. "Vitamin C for Preventing and Treating the Common Cold." *Cochrane Database of Systematic Reviews*, 2010, CD000980.

Henderson, G., et al. "Rumen Microbial Community Composition Varies with Diet and Host, but a Core Microbiome Is Found Across a Wide Geographical Range." *Science Reports*, vol. 5, no. 14567, 2015, doi.org/10.1038/srep14567.

Hennebique, Aurélie, et al. "Tularemia as a Waterborne Disease: A Review." *Emerging Microbes & Infections*, vol. 8, no. 1, 2019, pp. 1027-42, doi:10.1080/22221751.2019.1638734.

Henry, Helen L., and Anthony W. Norman, editors. *Encyclopedia of Hormones*. 3 vols. Academic Press, 2003.

Hensel, Michael, and Herbert Schmidt, editors. *Horizontal Gene Transfer in the Evolution of Pathogenesis*. Cambridge UP, 2008.

Herper, Matthew. "J&J One-Dose Covid Vaccine Is 66% Effective, a Weapon but Not a Knockout Punch." *Stat*, 29 Jan. 2021, www.statnews.com/2021/01/29/jj-one-dose-covid-vaccine-is-66-effective-a-weapon-but-not-a-knockout-punch/. Accessed 4 Feb. 2021.

Herzig, Carolyn T. A., et al. "Notes from the Field: Enteroinvasive *Escherichia coli* Outbreak Associated with a Potluck Party-North Carolina, June-July 2018." *Morbidity and Mortality Weekly Report,* vol. 68, no. 7, 2019, pp. 183-84, doi:10.15585/mmwr.mm6807a5.

Hewlett, Martinez J., et al. *Basic Virology*. 4th ed., Wiley-Blackwell, 2021.

Heymann, David L., editor. *Control of Communicable Diseases Manual*. 19th ed., American Public Health Association, 2008.

Hibbet, David S., et al. "A Higher-Level Phylogenetic Classification of the Fungi." *Mycological Research,* vol. 111, 2007, pp. 509-47.

Higgins, Patrick, Maura Kate Kilgore, and Paul Hertlein. *The Homebrewer's Recipe Guide*. Simon & Schuster, 1996.

Hill, Kathleen "Shock, Sepsis, and Multiple Organ Dysfunction Syndrome." *Introduction to Critical Care Nursing*, edited by Mary Lou Sole, Deborah G. Klein, and Marthe J. Moseley, 6th ed. Saunders/Elsevier, 2013.

Hill, Marquita K. "Water Pollution." *Understanding Environmental Pollution*. 3rd ed., Cambridge UP, 2010.

Hill, Walter E. *Genetic Engineering: A Primer*. Harwood Academic Publishers, 2000.

"History of Vaccine Development." *World Health Organization*, 2021, vaccine-safety-training.org/history-of-vaccine-development.html. Accessed 3 Feb. 2021.

Hoenigl, Martin, and Alida Fe Talento, editors. Fungal Infections Complicating COVID-19. ? Mdpi AG, 2021.

Hoft, D. F. "Tuberculosis Vaccine Development: Goals, Immunological Design, and Evaluation." *The Lancet*, vol. 372, 2008, pp. 164-75.

Holmes, Edward C. *The Evolution and Emergence of RNA Viruses*. Oxford UP, 2009.

Holmes, Randall K., and Michael G. Jobling. "Genetics." *Medical Microbiology*, edited by Samuel Baron, University of Texas Medical Branch at Galveston, 1996.

Honda, Kenya, and Dan R. Littman. "The Microbiome in Infectious Disease and Inflammation." *Annual Review of Immunology*6 Jan. 2021, doi: [10.1146/annurev-immunol-020711-074937].

Hopkins, William G., and Normal P. A. Hüner. *Introduction to Plant Physiology*. 4th ed., John Wiley & Sons, 2008.

Horan, Nigel, Abu Zahrim Yaser, and Newati Wid. *Anaerobic Digestion Processes: Applications and Effluent Treatment*. Springer, 2018.

Hospenthal, Duane R., and Michael G. Rinaldi. *Diagnosis and Treatment of Fungal Infections*. 2nd ed., Springer, 2015.

Houghton, R. A. "The Contemporary Carbon Cycle." *Biogeochemistry*, edited by W. H. Schlesinger, Elsevier, 2005.

Howley, Peter M., et al., editors. *Fields' Virology*. 7th ed., Lippincott Williams & Wilkins, 2021.

"The Human Genome." *Nature, vol.* 409, no. 6822, 2001, pp. 813-958.

"Human Papillomavirus (HPV) Vaccines." *National Cancer Institute*. National Institutes of Health, 19 Feb. 2015. Accessed 30 Dec. 2015.

Hvas, A. M., et al. "No Effect of Vitamin B12 Treatment on Cognitive Function and Depression." *Journal of Affective Disorders,* vol. 81, 2004, pp. 269-73.

———. "Vitamin B12 Treatment Has Limited Effect on Health-Related Quality of Life Among Individuals with

Elevated Plasma Methylmalonic Acid." *Journal of Internal Medicine,* vol. 253, 2003, pp. 146-52.

Hviid, A., S. Rubin, and K. Mühlemann. "Mumps." *The Lancet*, vol. 371, Mar. 2008, pp. 932-44.

"Hypersensitivity Reactions (Types I, II, III, IV)?" *Rutgers, The State University of New Jersey*, 9 Apr. 2009, njms.rutgers.edu/gsbs/olc/mci/prot/2009/Hypersensitivities09.pdf. Accessed 27 Nov. 2021.

"Hypersensitivity Reactions." *Centers for Disease Control and Prevention*, www.cdc.gov/travel-training/local/HistoryEpidemiologyandVaccination/page27396.html. Accessed 5 Nov. 2018.

"Hypersensitivity Reactions." *CliffsNotes,* www.cliffsnotes.com/study-guides/biology/microbiology/disorders-of-the-immune-system/hypersensitivity-reactions. Accessed 5 Nov. 2018.

"'I Think I've Just Thought Up Something Important': Francois Jacob (1920-2013)." *National Geographic*, 21 Apr. 2013, phenomena.nationalgeographic.com/2013/04/21/i-think-ive-just-thought-up-something-important-francois-jacob-1920-2013/. Accessed 24 June 2017.

Ikeda, David M., et al. "Natural Farming: Lactic Acid Bacteria." *Sustainable Agriculture.* College of Tropical Agriculture and Human Resources, University of Hawaii at Manoa, Aug. 2013, www.ctahr.hawaii.edu/oc/freepubs/pdf/sa-8.pdf. Accessed 22 Jan. 2016.

"Immune Response." *Medline Plus*, 8 Feb. 2021, medlineplus.gov/ency/article/000821.htm. Accessed 10 Feb. 2021.

"The Immune System: Information about Lymphocytes, Dendritic Cells, Macrophages, and White Blood Cells." *Chemocare*, www.real-world-physics-problems.com/gyroscope-physics.html. Accessed 27 Mar. 2017.

"The Immune System-In More Detail." *Nobelprize.org,* 2014, www.nobelprize.org/educational/medicine/immunity/immune-detail.html. Accessed 27 Mar. 2017.

Immune Web, immuneweb.org.

"Immunization: Typhoid Vaccine." *Mandell, Douglas, and Bennett's Principles and Practice of Infectious Diseases*, edited by Gerald L. Mandell, John E. Bennett, and Raphael Dolin, 7th ed., Churchill Livingstone/Elsevier, 2010.

Inbar, Noa Harel, and Itai Benhar. "Selection of Antibodies from Synthetic Antibody Libraries." *Archives of Biochemistry and Biophysics*, vol. 526, no. 2 (special issue), n.d., pp. 87-98. *Biological Abstracts*. Accessed 4 Sept. 2014.

"Infection Control." *MedlinePlus*. US National Library of Medicine, www.nlm.nih.gov/medlineplus/infectioncontrol.html. Accessed 11 Feb. 2015.

"Infection Prevention and You." *Association for Professionals in Infection Control and Epidemiology*. Association for Professionals in Infection Control and Epidemiology, Inc., www.apic.org/Resource_/TinyMceFileManager/IP_and_You/IPandYou_InfographicPoster_2013.pdf. Accessed 11 Feb. 2015.

"Infection Prevention and Control." *Minnesota Department of Health*. Minnesota Department of Health, www.health.state.mn.us/divs/idepc/dtopics/infectioncontrol/. Accessed 11 Feb. 2015.

"Influenza." *The Merck Manual Home Health Handbook*, edited by Robert S. Porter et al., 3rd ed., Merck Research Laboratories, 2009.

"Influenza Vaccine for 2021-2022." *Medical Letter on Drugs and Therapeutics*, vol. 63, no. 1634, 2021, pp. 153-57.

Ingold, C. T., and H. J. Hudson. *The Biology of Fungi*. 6th ed., Chapman & Hall, 1993.

"Innate and Adaptive Immunity." *American Society for Radiation Oncology*, 2021, www.astro.org/Patient-Care-and-Research/Research/Professional-Development/Research-Primers/Innate-and-Adaptive-Immunity. Accessed 10 Feb. 2021.

Institute of Medicine. *An Assessment of the CDC Anthrax Vaccine Safety and Efficacy Research Program*. National Academy Press, 2003.

———. *Immunization Safety Review: Vaccines and Autism*. National Academies Press, 2004.

———. *Vaccine Safety Research, Data Access and Public Trust*. National Academies Press, 2005.

"Introduction to Immunology Tutorial." *The Biology Project, University of Arizona*, 24 May 2000, www.biology.arizona.edu/immunology/tutorials/immunology/page3.html. Accessed 10 Feb. 2021.

"Invasive Staphylococcus aureus Infections Associated with Pain Injections and Reuse of Single-Dose Vials-Arizona and Delaware, 2012." *Monthly Morbidity and Mortality Report*, vol. 61, no. 27, 2012, pp. 501-4, www.cdc.gov/mmwr/preview/mmwrhtml/mm6127a1.htm?s_cid=mm6127a1_w.

Investigative Ophthalmology & Visual Science, vol. 41, no. 6, pp. 1513-22.

"Is Lactic Acid a Four-Letter Word?" *PBS LearningMedia*. PBS & WGBH Educational Foundation, www.pbslearningmedia.org/resource/tdc02.sci.life.cell.lactic/is-lactic-acid-a-four-letter-word/. Accessed 22 Jan. 2016.

Iverson, Jon. *Home Winemaking Step by Step*. Stonemark Publishing Co., 2002.

Jackson, Alan C., and William H. Wunner, editors. *Rabies*. 2nd ed., Academic Press, 2007.

Jackson, Kelly A., et al. "Listeriosis Outbreaks Associated with Soft Cheeses, United States, 1998-2014." *Emerging Infectious Diseases,* vol. 24, no. 6, 2018, pp. 1116-18, doi:10.3201/eid2406.171051.

Jacobson, Michael C., et al. *Earth System Science: From Biogeochemical Cycles to Global Change*. 2nd ed., Academic Press, 2000.

Jameson, J. Larry, et al., editors. *Harrison's Principles of Internal Medicine*. 20th ed., McGraw-Hill, 2018.

Janeway, Charles A., Jr., et al. *Immunobiology: The Immune System in Health and Disease*. 7th ed., Garland Science, 2007.

Janssen, Riny, et al. "Host-Pathogen Interactions in *Campylobacter* Infections: The Host Perspective." *Clinical Microbiology Reviews*, vol. 21, no. 3, 2008, pp. 505-18.

Jay, James M., Martin J. Loessner, and David A. Golden. *Modern Food Microbiology*. 7th ed., Springer, 2005.

Jefferis, Roy, Koicho Kato, and William R. Strohl, editors. *Structure and Function of Antibodies*. Mdpi AG, 2021.

Jemal, A., et al. "Cancer Statistics, 2010." *CA: A Cancer Journal for Clinicians*, vol. 60, no. 5, 2010, pp. 277-300.

Jess, Sowmya. *Understanding Monoclonal Antibodies*. Xlibris US, 2020.

Johannson, Karl-Erik, and Bertil Petterrson. "Taxonomy of Mollicutes." *Molecular Biology and Pathogenicity of Mycoplasmas*, edited by Shmuel Razin and Richard Herrmann, Kluwer Academic, 2002.

Johnson, Leonard R., editor. *Gastrointestinal Physiology*. 7th ed., Mosby/Elsevier, 2007.

Jolley, D., and K. M. Douglas. "The Effects of Anti-Vaccine Conspiracy Theories on Vaccination Intentions." *PloS ONE*, vol. 9, no. 2, 2014, pp. 1-9.

Jong, Elaine C., and Russell McMullen, editors. *Travel and Tropical Medicine Manual*. 4th ed., Saunders/Elsevier, 2008.

Jou, J. H., and A. J. Muir. "In the Clinic: Hepatitis C." *Annals of Internal Medicine,* vol. 148, 2008, ITC6-1-ITC6-16.

Joynson, David H. M., and Tim G. Wreghitt, editors. *Toxoplasmosis: A Comprehensive Clinical Guide*. Rev. ed., Cambridge UP, 2005.

Kahmini, Fatemeh Rezaei, and Shahab Shahgaldi. "Therapeutic Potential of Mesenchymal Stem Cell-Derived Extracellular Vesicles as Novel Cell-Free Therapy for Treatment of Autoimmune Disorders." *Experimental and Molecular Pathology,* vol. 118, 2011, p. 104566, doi:10.1016/j.yexmp.2020.104566.

Kalluri, P., et al. "Evaluation of Three Rapid Diagnostic Tests for Cholera: Does the Skill Level of the Technician Matter?" *Tropical Medicine in Health*, vol. 11, no. 1, 2006, pp. 49-55.

Kalsi, A. S. "Tinea unguium. T. violaceum Was Isolated." *ResearchGate*, www.researchgate.net/figure/Tinea-unguium-T-violaceum-was-isolated_fig6_328964379. Accessed 23 Nov. 2020.

Kang, J. H., et al. "A Randomized Trial of Vitamin E Supplementation and Cognitive Function in Women." *Archives of Internal Medicine,* vol. 166, 2006, pp. 2462-68.

Kania, Leon W. *The Alaskan Bootlegger's Bible: Makin' Beer, Wine, Liqueurs and Moonshine Whiskey: An Old Alaskan Tells How It Is Done*. 2nd ed., Happy Mountain Publications LLC, 2019.

Kaplan, Colin, G. S. Turner, and D. A. Warrell. *Rabies: The Facts.* 2nd ed., Oxford UP, 1986.

Karlson, E. W., et al. "Vitamin E in the Primary Prevention of Rheumatoid Arthritis." *Arthritis and Rheumatism,* vol. 59, 2008, pp. 1589-95.

Kasper, Dennis L., et al., editors. *Harrison's Principles of Internal Medicine*. 16th ed., McGraw-Hill, 2005.

Kataja-Tuomola, M., et al. "Effect of Alpha-tocopherol and Beta-carotene Supplementation on Macrovascular Complications and Total Mortality from Diabetes." *Annals of Medicine,* vol. 42, 2010, pp. 178-86.

Katz, Sandor Ellix. *The Art of Fermentation: An In-Depth Exploration of Essential Concepts and Processes from around the World*. Chelsea Green, 2012.

Kauffman, Carol A. "Sporotrichosis." *Clinical Infectious Diseases*, vol. 29, 1999, pp. 231-36.

———, et al. "Clinical Practice Guidelines for the Management of Sporotrichosis: 2007 Update by the Infectious Diseases Society of America." *Clinical Infectious Diseases*, vol. 45, 2007, pp. 1255-65.

Kaufman, C. A. "Histoplasmosis." *Clinics in Chest Medicine*, vol. 30, 2009, p. 217.

Kaufmann, Stefan H. E. "Vaccine Development Against Tuberculosis Over the Last 140 Years: Failure as Part of Success." *Frontiers in Microbiology*, vol. 12, 2021, p. 50124, doi:10.3389/fmicb.2021.750124.

Keiler, Kenneth C., editor. *Bacterial Regulatory RNA: Methods and Protocols.* Humana, 2012.

Keith, M. E., et al. "A Controlled Clinical Trial of Vitamin E Supplementation in Patients with Congestive Heart Failure." *American Journal of Clinical Nutrition,* vol. 73. 2001, pp. 219-24.

Keller, M. A., and E. R. Stiehm. "Passive Immunity in Prevention and Treatment of Infectious Diseases."

Clinical Microbiology Reviews, vol. 13, no. 4, 2000, pp. 602-14.

Kennedy, David W., and Marilyn Olsen. *Living with Chronic Sinusitis: A Patient's Guide to Sinusitis, Nasal Allergies, Polyps, and Their Treatment Options*. Hatherleigh Press, 2007.

Kershaw, Michael H., J. A. Westwood, C. Y. Slaney, and P. K. Darcy. "Clinical Application of Genetically Modified T Cells in Cancer Therapy." *Clinical and Translational Immunology*, vol. 3, 2014, p. e16., www.nature.com/cti/journal/v3/n5/full/cti20147a.html.

"Key Facts About Seasonal Flu Vaccine." *Centers for Disease Control and Prevention*. US Department of Health and Human Services, 18 Nov. 2021, www.cdc.gov/flu/prevent/keyfacts.htm. Accessed 20 Nov. 2021.

Khattak, Zoia E., and Fatima Anjum. "Haemophilus Influenzae." *StatPearls*, StatPearls Publishing, 14 Sept. 2021.

Khoury, G., G. Darcis, M. Y. Lee, et al. "The Molecular Biology of HIV Latency." *Advances in Experimental Medicine and Biology*, vol. 18, no. 1075, 2018, pp. 187-212.

Kibbler, Christopher C., et al., editors. *Oxford Textbook of Medical Mycology*. Oxford UP, 2018.

Kienzle, Thomas E. *Rabies*. Chelsea House, 2006.

Kim, Tae Hyong, et al. "Vaccine Herd Effect." *Scandinavian Journal of Infectious Diseases,* vol. 43, no. 9, Sept. 2011, pp. 683-89, www.ncbi.nlm.nih.gov/pmc/articles/PMC3171704. Accessed 31 July 2017.

King, Andrew M. Q. *Virus Taxonomy: Classification and Nomenclature of Viruses: Ninth Report of the International Committee on Taxonomy of Viruses*. Academic, 2012.

King, John W., and Meredith L. DeWitt. "Cryptococcosis." emedicine.medscape.com/article/215354-overview.

Kirschner, Barbara S., and Dennis D. Black. "The Gastrointestinal Tract." *Nelson Essentials of Pediatrics*, edited by Karen J. Marcdante, et al., 6th ed., Saunders/Elsevier, 2011.

Klatt, Edward C., and Vinay Kumar. *Robbins and Cotran Review of Pathology*. 5th ed., ? Elsevier, 2021.

"Klebsiella Pneumoniae in Health Care Settings." *Centers for Disease Control*, www.cdc.gov/hai/organisms/klebsiella/klebsiella.html. Accessed 1 Dec. 2021.

"Klebsiella Species." *Gov.UK*, 15 July 2008, www.gov.uk/guidance/klebsiella-species. Accessed 1 Dec. 2021.

Klein, Christian. *Monoclonal Antibodies*. Mdpi AG, 2018.

Klein, David R. *Organic Chemistry and a Second Language*. 5th ed., Wiley, 2019.

Kliegman, Robert M., et al., editors. *Nelson Textbook of Pediatrics*. 21st ed., Elsevier, 2019.

Klosterman, Lorrie. *Rabies*. Marshall Cavendish Benchmark, 2008.

Kluger, J. "Who's Afraid of a Little Vaccine?" *Time*, vol. 184, no. 13, 2014, pp. 40-43.

Knapen, M. H., L. J. Schurgers, and C. Vermeer. "Vitamin K2 Supplementation Improves Hip Bone Geometry and Bone Strength Indices in Postmenopausal Women." *Osteoporosis International,* vol. 18, no. 7, 2007, pp. 963-72.

Koch, Arthur L. *The Bacteria: Their Origin, Structure, Function, and Antibiosis*. Springer, 2006.

Komaroff, Anthony. "Why Are mRNA Vaccines So Exciting?" *Harvard Health Publishing*, 18 Dec. 2020, www.health.harvard.edu/blog/why-are-mrna-vaccines-so-exciting-2020121021599. Accessed 3 Feb. 2021.

Koneman, Elmer W. *The Other End of the Microscope: The Bacteria Tell Their Own Story*. ASM Press, 2002.

Konkel Michael E., et al. "Taking Control: Campylobacter jejuni Binding to Fibronectin Sets the Stage for Cellular Adherence and Invasion." *Frontiers in Microbiology*, vol. 11, 2020, p. 564, DOI=10.3389/fmicb.2020.00564.

Kontermann, Roland, and Stefan Dübel, editors. *Antibody Engineering*. 2nd ed., Springer, 2010.

Kotloff, Karen L., et al. "Shigellosis." *The Lancet,* vol. 391, no. 10122, 2018, pp. 801-12, doi:10.1016/S0140-6736(17)33296-8.

Krauss, Hartmut, et al. *Zoonoses: Infectious Diseases Transmissible from Animals to Humans*. 3rd ed., ASM Press, 2003.

Krebs, Charles J. *Ecology: The Experimental Analysis of Distribution and Abundance*. 6th ed., Pearson, 2014.

Krebs, Jocelyn E., Elliott S. Goldstein, and Stephen T. Kilpatrick, eds. *Lewin's Genes XI*. 11th ed., Jones, 2014.

Kreuzer, Helen, and Adrianne Massey. *Recombinant DNA and Biotechnology: A Guide for Students*. Blackwell Science, 2000.

Krishna, Kumar, and Frances M. Weesner, eds. *Biology of Termites*. 2 vols. Academic Press, 1969-1970.

Krishnan, Vidya. *Phantom Plague: How Tuberculosis Shaped History*. PublicAffairs, 2022.

Kroll, Jess. "The Differences between Monosaccharides & Polysaccharides." *Sciencing,* sciencing.com/differences-between-monosaccharides-polysaccharides-8319130.html. Accessed 19 Apr. 2017.

Krump, N. A., and J. You. "Molecular Mechanisms of Viral Oncogenesis in Humans." *Nature Reviews Microbiology*, vol. 16, 2018, pp. 684-98.

Kum, Dezimey. "Fueled by a History of Mistreatment, Black Americans Distrust the New COVID-19 Vaccines." *Time*, 28 Dec. 2020, time.com/5925074/black-americans-covid-19-vaccine-distrust/. Accessed 20 Jan. 2021.

Kumar, Vinay, et al., editors. *Robbins & Cotran Pathologic Basis of Disease*. 10th ed., Elsevier, 2020.

Kurth, Reinhard, and Norbert Bannert, editors. *Retroviruses: Molecular Biology, Genomics, and Pathogenesis*. Caister Academic Press, 2010.

"Lactic Acid." *Calwineries*. Calwineries Inc., www.calwineries.com/learn/wine-chemistry/wine-acids/lactic-acid. Accessed 22 Jan. 2016.

"Lactic Acid Bacteria-Their Uses in Food." *European Food Information Council*. European Food Information Council, www.eufic.org/article/en/artid/lactic-acid-bacteria/. Accessed 22 Jan. 2016.

"Lactic Acid Test." *Medline Plus*. US National Library of Medicine, www.nlm.nih.gov/medlineplus/ency/article/003507.htm. Accessed 22 Jan. 2016.

"Lactic Acid Topical." *WebMD*. WebMD, LLC, www.webmd.com/drugs/2/drug-64136-762/lactic-acid-topical/keratolyticemollients-topical/details. Accessed 22 Jan. 2016.

Lafferty, Peter, and Julian Rowe, editors. *The Hutchinson Dictionary of Science*. 2nd ed., Helicon, 1998.

Lake, James A. "The Ribosome." *Scientific American*, Aug. 1981, pp. 84-97.

Lakhundi, S., and K. Zhang. "Methicillin-Resistant Staphylococcus Aureus: Molecular Characterization, Evolution, and Epidemiology." *Clinical Microbiology Reviews*, vol. 31, 2018.

Lanao, Andrea E., et al. "Mycoplasma Infections." *StatPearls*. StatPearls Publishing, 10 Aug. 2021.

LaPorte, Meg. "Glove Use Tied to Better Infection Control." *McKnight's Long-Term Care News*, 6 Oct. 2017, www.mcknights.com/news/glove-use-tied-to-betterinfection-control/article/698220/.

Larsen, Laura. *Sexually Transmitted Diseases Sourcebook*. Omnigraphics, 2009.

Lasserre, P., and J. M. Martin, editors. *Biogeochemical Processes at the Land-Sea Boundary*. Elsevier, 1986.

Lauria, Ashley M., and Christopher P. Zabbo. "Pertussis." *StatPearls*. StatPearls Publishing, 26 June 2021.

Lazo, John, and Peter Wipf. "Combinatorial Chemistry and Contemporary Pharmacology." *The Journal of Pharmacology and Experimental Therapeutics*, vol. 293, no. 3, Feb. 2000, pp. 705-9.

Le Dévédec, S. E., et al. "Systems Microscopy Approaches to Understand Cancer Cell Migration and Metastasis." *Cellular and Molecular Life Sciences*, vol. 67, 2010, pp. 3219-40.

"Learn about the Five Common White Blood Cells." *University of Wisconsin Oshkosh*, www.uwosh.edu/med_tech/what-is-elementary-hematology/white-blood-cells. Accessed 28 Mar. 2017.

Ledgerwood, Julie E., et al. "Chimpanzee Adenovirus Vector Ebola Vaccine." *New England Journal of Medicine*, vol. 376, 2017, pp. 928-38, doi:10.1056/NEJMoa1410863.

Lee, Robert Edward. *Phycology*. 5th ed., Cambridge UP, 2018.

Legocki, Andrezej, Hermann Bothe, and Alfred Pühler, editors. *Biological Fixation of Nitrogen for Ecology and Sustainable Agriculture*. Springer, 1997.

Leon, Warren, and Caroline Smith DeWaal. *Is Our Food Safe? A Consumer's Guide to Protecting Your Health and the Environment*. Crown, 2002.

Leong A. S.-Y., and T. Y.-M. Leong. "Newer Developments in Immunohistology." *Journal of Clinical Pathology*, vol. 59, 2006, pp. 1117-26.

Leong, Anthony S.-Y., Kumarasen Cooper, and F. Joel W.-M. Leong. *Manual of Diagnostic Antibodies for Immunohistology*. 2nd ed., Greenwich Medical Media, 2002.

Lester, J., and D. Edge. "Sewage and Sewage Sludge Treatment." *Pollution: Causes, Effects, and Control*, edited by Roy M. Harrison, 4th ed., Royal Society of Chemistry, 2001.

Levine, Myron M. "Typhoid Fever." *Bacterial Infections of Humans: Epidemiology and Control*, edited by Philip S. Brachman and Elias Abrutyn, 4th ed., Springer Science, 2009.

———. "Typhoid Fever Vaccines." *Vaccines*, edited by Stanley A. Plotkin, Walter A. Orenstein, and Paul A. Offit, 5th ed., Saunders/Elsevier, 2008.

Levitzky, Michael G. *Pulmonary Physiology*. 9th ed., McGraw-Hill Medical, 2018.

Levy, Stuart B. "The Challenge of Antibiotic Resistance." *Scientific American*, vol. 278, 1998, pp. 46-53.

Lew, Kristi. *Food Poisoning: E. Coli and the Food Supply*. Rosen Publishers, 2011.

Lieberman, Shari, and Nancy Bruning. *Real Vitamin and Mineral Book*. 4th ed., Avery, 2007.

Liljas, Anders, and Måns Ehrenberg. *Structural Aspects of Protein Synthesis*. 2nd ed., World Scientific, 2013.

Lim, Daniel V. *Microbiology*. 3rd ed., Kendall/Hunt Publishing, 2003.

Link, Kurt. *The Vaccine Controversy: The History, Use, and Safety of Vaccinations*. Praeger Publishers, 2005.

Litin, Scott C., editor. *Mayo Clinic Family Health Book*. 4th ed., HarperResource, 2009.

Litman, Nathan, and Stephen G. Baum. "Mumps Virus." *Mandell, Douglas, and Bennett's Principles and Practice of Infectious Diseases*, edited by Gerald L. Mandell, John F. Bennett, and Raphael Dolin, 7th ed., Churchill Livingstone/Elsevier, 2010.

Lodish, Harvey, et al. *Molecular Cell Biology*. 9th ed., W. H. Freeman, 2021.

Loehr, Jamie. *The Vaccine Answer Book: Two Hundred Essential Answers to Help You Make the Right Decisions for Your Child.* Sourcebooks, 2010.

Lok, Anna S. F., et al. "Antiviral Therapy for Chronic Hepatitis B Viral Infection in Adults: A Systematic Review and Meta-analysis." *Hepatology*, vol. 63, no.1, 2016, pp. 284-306, doi:10.1002/hep.28280.

Longo, Dan L., et al. "Histoplasmosis." *Harrison's Principles of Internal Medicine*. 19th ed., McGraw-Hill, 2015.

Longrée, Karla, and Gertrude Armbruster. *Quantity Food Sanitation*. 5th ed., John Wiley & Sons, 1996.

López-Santiago, Rubén, et al. "Immune Response to Mucosal *Brucella* Infection." *Frontiers in Immunology*, vol. 10, no. 1759, 2019, doi:10.3389/fimmu.2019.01759.

Lundstrom, Kenneth. "Viral Vectors for COVID-19 Vaccine Development." *Viruses*, vol. 13, no. 2, 2021, p. 317, doi:10.3390/v13020317.

Lydersen, Bjorn K., Nancy A. D'Elia, and Kim L. Nelson, editors. *Bioprocess Engineering: Systems, Equipment and Facilities*. John Wiley & Sons, 1994.

"Lymphocytosis (High Lymphocyte Count)." *Mayo Clinic*, www.mayoclinic.org/symptoms/lymphocytosis/basics/definition/sym-20050660. Accessed 28 Mar. 2017.

Ma, Liang, et al. "A Molecular Window into the Biology and Epidemiology of Pneumocystis spp." *Clinical Microbiology Reviews*, vol. 31, no. 3, 2018, p. e00009-18, doi:10.1128/CMR.00009-18.

Macartney, Kristine, and Peter McIntyre. "Vaccines for Post-Exposure Prophylaxis Against Varicella (Chickenpox) in Children and Adults." *The Cochrane Database of Systematic Reviews*, vol. 3, no. CD001833, 2008, doi:10.1002/14651858.CD001833.pub2.

Mackay, Katurah. "Bison Plan Called 'Absurd.'" *National Parks*, vol. 72, nos. 7-8, 1998, pp. 14-15.

Mactier, H., M. M. Mokaya, L. Farrell, and C. A. Edwards. "Vitamin A Provision for Preterm Infants: Are We Meeting Current Guidelines?" *Archives of Disease in Childhood: Fetal and Neonatal Edition*, vol. 96, no. 4, 2011, pp. F286-89, doi:10.1136/adc.2010.190017.

Madigan, Michael T., et al. *Brock Biology of Microorganisms*. 16th ed., Pearson, 2020.

Madireddy, Sowmya, et al. "Toxoplasmosis." *StatPearls*. StatPearls Publishing, 28 Sept. 2021.

Mahamadou, A. Thera, et al. "Safety and Immunogenicity of an AMA1 Malaria Vaccine in Malian Children." *PLoS*, vol. 5, 2010, p. e9041.

Maixner, Frank, et al. "The 5300-year-old Helicobacter pylori Genome of the Iceman." *Science*, 8 Jan. 2016, pp. 162-65.

Male, David. *Immunology: An Illustrated Outline*. 6th ed., CRC Press, 2021.

Malouf, R., and J. Grimley Evans. "Folic Acid with or Without Vitamin B12 for the Prevention and Treatment of Healthy Elderly and Demented People." *Cochrane Database of Systematic Reviews*, vol. 4, 2008, CD004514.

Man, Yang-gao, et al. "Tumor-Infiltrating Immune Cells Promoting Tumor Invasion and Metastasis: Existing Theories." *Journal of Cancer*, vol. 4, no. 1, 2013, pp. 84-95.

Mandal, Ananya. "What Is a Macrophage?" *News-Medical.Net*, 14 Jan. 2014, www.news-medical.net/life-sciences/What-is-a-Macrophage.aspx. Accessed 16 May 2017.

Mandell, Gerald L., John E. Bennett, and Raphael Dolin, editors. *Mandell, Douglas, and Bennett's Principles and Practice of Infectious Diseases*. 7th ed., Churchill Livingstone/Elsevier, 2010.

Mandell, Lionel A., et al. "Infectious Diseases Society of America/American Thoracic Society Consensus Guidelines on the Management of Community-Acquired Pneumonia in Adults." *Clinical Infectious Diseases*, vol. 44, 2007, pp. S27-72.

Manning, P. J., et al. "Effect of High-Dose Vitamin E on Insulin Resistance and Associated Parameters in Overweight Subjects." *Diabetes Care*, vol. 27, 2004, pp. 2166-71.

Margolis, K. L., et al. "Effect of Calcium and Vitamin D Supplementation on Blood Pressure: The Women's Health Initiative Randomized Trial." *Hypertension*, vol. 52, 2008, pp. 847-55.

Margulis, Lynn, and Michael J. Chapman. *Kingdoms and Domains: An Illustrated Guide to the Phyla of Life on Earth*. 4th ed., Elsevier, 2009.

Marieb, Elaine N., and Suzanne Keller. *Essentials of Human Anatomy and Physiology*. 12th ed., Pearson/Benjamin Cummings, 2017.

Marin, Mona, et al. "Varicella Prevention in the United States: A Review of Successes and Challenges." *Pediatrics*, vol. 122, no. 3, 2008, pp. e744-51, doi:10.1542/peds.2008-0567.

Mark, M., N. B. Ghyselinck, and P. Chambon. "Function of Retinoic Acid Receptors during Embryonic Development." *Nuclear Receptor Signaling Atlas,* vol. 7, 2009, p. e002.

Marks, Peter. "Joint CDC and FDA Statement of Johnson & Johnson COVID-19 Vaccine." *US Food and Drug Administration,* 13 Apr. 2021, www.fda.gov/news-events/press-announcements/joint-cdc-and-fda-statement-johnson-johnson-covid-19-vaccine. Accessed 14 Apr. 2021.

Marquardt, William, editor. *Biology of Disease Vectors.* 2nd ed., Academic Press/Elsevier, 2005.

Marriot, Norman G., and Robert B. Gravani. *Principles of Food Sanitation.* 5th ed., Springer, 2006.

Martin, Richard J., Avroy A. Fanaroff, and Michele C. Walsh, editors. *Fanaroff and Martin's Neonatal-Perinatal Medicine: Diseases of the Fetus and Infant.* 2 vols. 8th ed., Mosby/Elsevier, 2006.

Martin, Seamus, Dennis R. Burton, Ivan M. Roitt, and Peter J. Delves. *Roitt's Essential Immunology.* 13th ed., Wiley-Blackwell, 2017.

Martini, L. A., et al. "Dietary Phylloquinone Depletion and Repletion in Postmenopausal Women: Effects on Bone and Mineral Metabolism." *Osteoporosis International,* vol. 17, no. 6, 2006, pp. 929-35.

Mason, Robert J., et al., editors. *Murray and Nadel's Textbook of Respiratory Medicine.* 5th ed., Saunders/Elsevier, 2010.

Mast, Eric E., et al. "A Comprehensive Immunization Strategy to Eliminate Transmission of Hepatitis B Virus Infection in the United States: Recommendations of the Advisory Committee on Immunization Practices (ACIP) Part II: Immunization of Adults." *MMWR. Recommendations and Reports: Morbidity and Mortality Weekly Report,* vol. 55, no. RR-16, 2006, pp. 1-33.

Matossian, Mary Kilbourne. *Poisons of the Past: Molds, Epidemics, and History.* Yale UP, 1989.

Matson, David O. "Rotaviruses." *Principles and Practice of Pediatric Infectious Diseases,* edited by Sarah S. Long, Larry K. Pickering, and Charles G. Prober, 3rd ed., Churchill Livingstone/Elsevier, 2008.

Matsumoto, Brian, editor. *Cell Biological Applications of Confocal Microscopy.* 2nd ed., Academic Press, 2002.

Mattoo, Seema, and James D. Cherry. "Molecular Pathogenesis, Epidemiology, and Clinical Manifestations of Respiratory Infections Due to *Bordetella pertussis* and Other *Bordetella* Subspecies." *Clinical Microbiology Reviews,* vol. 18, 2005, pp. 326-82.

Maurin, Max. "*Francisella tularensis,* Tularemia and Serological Diagnosis." *Frontiers in Cellular and Infection Microbiology,* vol. 10, no. 512090, 2020, doi:10.3389/fcimb.2020.512090.

Maxfield, Luke, and Rene Bermudez. "Trypanosomiasis." *StatPearls.* StatPearls Publishing, 11 Aug. 2021.

Maxfield, Luke, and Jonathan S. Crane. "Leishmaniasis." *StatPearls.* StatPearls Publishing, 18 July 2021.

Mayadas, Tanya N., Xavier Cullere, and Clifford A. Lowell. "The Multifaceted Functions of Neutrophils." *Annual Review of Pathology,* vol. 9, Jan. 2014, pp. 181-218, www.ncbi.nlm.nih.gov/pmc/articles/PMC4277181/. Accessed 16 May 2017.

Mayforth, Ruth D. *Designing Antibodies.* Academic, 1993.

The Mayo Staff. "Vitamin D." *The Mayo Clinic,* 18 Oct. 2017, www.mayoclinic.org/drugs-supplements-vitamin-d/art-20363792.

Mayr, Florian B., et al. "Infection Rate and Acute Organ Dysfunction Risk as Explanations for Racial Difference in Severe Sepsis." *Journal of the American Medical Association,* vol. 24, 2010, pp. 2495-2503.

Mbaeyi, Chukwuma, et al. "Progress Toward Poliomyelitis Eradication-Pakistan, Jan. 2020-July 2021." *Morbidity and Mortality Weekly Report,* vol. 70, no. 39, 2021, pp. 1359-64, doi:10.15585/mmwr.mm7039a1.

McCaffrey, Thomas. "Functional Endoscopic Sinus Surgery: An Overview." *Mayo Clinic Proceedings,* vol. 68, 1993, pp. 571-77.

McCance, Dennis J., editor. *Human Papilloma Viruses.* Elsevier Science, 2002.

McCarthy, Matt. *Superbugs: The Race to Stop an Epidemic.* Avery, 2019.

McCullough, Kenneth, and Raymond Spier. *Monoclonal Antibodies in Biotechnology: Theoretical and Practical Aspects.* Cambridge UP, 1990.

McGhee, Terrence. *Water Supply and Sewerage.* McGraw-Hill, 1991.

McGinty, Jo Craven. "How Anti-Vaccination Trends Vex Herd Immunity." *Wall Street Journal.* Dow Jones, 6 Feb. 2015, www.wsj.com/articles/how-anti-vaccination-trends-vex-herdimmunity-1423241871. Accessed 31 July 2017.

McKusick, Victor A., and Paul J. Converse. "*142857 Major Histocompatibility Complex, Class II, DR Beta-1; HLA-DRB1." *OMIM.org.* Johns Hopkins U, 25 June 2014. Accessed 4 Aug. 2014.

McMenamin, Mark. *Discovering the First Complex Life: The Garden of the Ediacara.* Columbia UP, 1998.

McQuiston, Jennifer. "Rickettsial (Spotted & Typhus Fevers) & Related Infections (Anaplasmosis & Ehrlichiosis)." *CDC Health Information for International Travel.* Oxford UP, 2016. Centers for Disease Control and Prevention, 10 July 2015, wwwnc.cdc.gov/travel/yellowbook/2016/infectious-diseases-related-to-travel/rickettsial-spotted-typhus-fevers-related-infections-anaplasmosis-ehrlichiosis. Accessed 15 Nov. 2016.

"Measles." *Epidemiology and Prevention of Vaccine-Preventable Diseases*, edited by W. Atkinson et al., 11th ed., Public Health Foundation, 2009.

"Measles: Questions and Answers." *Immunization Action Coalition*. Immunization Action Coalition, n.d. Accessed 23 Dec. 2015.

MedlinePlus. "Dietary Fats." *MedlinePlus*, 28 June 2013.

Melero, I., et al. "Immunostimulatory Monoclonal Antibodies for Cancer Therapy." *Nature Reviews Cancer*, vol. 7, 2007, pp. 95-106.

Melero, I., G. Gaudemack, W. Gerritsen, et al. "Therapeutic Vaccines for Cancer: An Overview of Clinical Trials." *Nature Reviews Clinical Oncology*, vol. 11, 2014, pp. 509-24.

Melmed, Schlomo, et al., editors. *Williams Textbook of Endocrinology*. 14th ed., Elsevier, 2019.

Merino, Noël. *Vaccines*. Greenhaven, 2015.

Meydani, S. N., et al. "Vitamin E and Respiratory Infection in the Elderly." *Annals of the New York Academy of Sciences*, vol. 1031, 2005, pp. 214-22.

Mickelson, Samuel, and Michael Benninger. "The Nose and Paranasal Sinuses." *Textbook of Primary Care Medicine*, edited by John Noble, 3rd ed., Mosby, 2001.

Miersch, S., and S. S. Sidhu. "Synthetic Antibodies: Concepts, Potential and Practical Considerations." *Methods*, vol. 57, no. 4, 2012, pp. 486-98. *Biological Abstracts*. Accessed 4 Sept. 2014.

Miller, Dave. *Dave Miller's Homebrewing Guide*. Storey Publishing, 1995.

Miller, G. Tyler, Jr., and Scott Spoolman. "Water Pollution." *Living in the Environment: Principles, Connections, and Solutions*. 16th ed., Brooks/Cole, 2009.

Miller, Neil Z. *Vaccines Are They Really Safe and Effective?* New Atlantean Press, 2015.

Mintz, Eric. "Typhoid and Paratyphoid Fever." *CDC Health Information for International Travel 2010*, wwwnc.cdc.gov/travel/yellowbook/2010/table-of-contents.aspx.

Mitchell, David A., et al., editors. *Solid-State Fermentation Bioreactors: Fundamentals of Design and Operation*. Springer, 2006.

Mitra, Sandhya. *Genetic Engineering: Principles and Practice*. 2nd ed., McGraw Hill Education, 2015.

Mittal, Deepak, et al. "New Insights into Cancer Immunoediting and Its Three Component Phases-Elimination, Equilibrium and Escape." *Current Opinion in Immunology*, vol. 27, 2014, pp. 16-25.

Miyashita, Naoyuki. "Atypical Pneumonia: Pathophysiology, Diagnosis, and Treatment." *Respiratory Investigation*, S2212-5345(21)00177-5, 2021, doi:10.1016/j.resinv.2021.09.009.

"MMR Vaccine Does Not Cause Autism." *Immunization Action Coalition*. Immunization Action Coalition, n.d. Accessed 23 Dec. 2015.

Money, Nicholas P. *Carpet Monsters and Killer Spores: A Natural History of Toxic Mold*. Oxford UP, 2004.

———. *The Rise of Yeast: How the Sugar Fungus Shaped Civilization*. Oxford UP, 2017.

Monoinfected Patients." *Alimentary Pharmacology & Therapeutics*, vol. 44, no. 1, 2016, pp. 16-34, doi:10.1111/apt.13659.

Montgomery, Rex, et al. *Biochemistry: A Case-Oriented Approach*. 6th ed., Mosby, 1996.

Morris, Susan York. "Everything You Should Know about Lymphocytes." *Healthline*, 30 Jan. 2017, www.healthline.com/health/lymphocytes#overview1. Accessed 27 Mar. 2017.

Morrison, Roger. *Desktop Guide to Keynotes and Confirmatory Symptoms*. Hahnemann Clinic, 1993.

Mosier, Arvin, J. Keith Syers, and John R. Freney, editors. *Agriculture and the Nitrogen Cycle: Assessing the Impacts of Fertilizer Use on Food Production and the Environment*. Island Press, 2004.

Motta, Matt, and Timothy Callaghan. "Why the Next Major Hurdle to Ending the Pandemic Will Be about Persuading People to Get Vaccinated." *The Conversation*, 27 Jan. 2021, theconversation.com/why-the-next-major-hurdle-to-ending-the-pandemic-will-be-about-persuading-people-to-get-vaccinated-153847. Accessed 28 Jan. 2021.

Mueller, Matthew, and Christopher R. Tainter. "*Escherichia Coli*." *StatPearls*. StatPearls Publishing, 26 July 2021.

Mullard, Asher. "How COVID Vaccines Are Being Divvied Up Around the World." *Nature*, 30 Nov. 2020, www.nature.com/articles/d41586-020-03370-6. Accessed 4 Feb. 2021.

Mullen, Gary T., and Lance A. Durden, editors. *Medical and Veterinary Entomology*. 3rd ed., Academic Press, 2018.

Münz, Christian, editor. *Epstein Barr Virus: One Herpes Virus: Many Diseases* (Current Topics in Microbiology and Immunology, vol. 390). Springer, 2015.

Murphy, J. F. "Trends in Cancer Immunotherapy." *Clinical Medicine Insights: Oncology*, vol. 4, 2010, pp. 67-80.

Murphy, Kenneth M., et al. *Janeway's Immunobiology*. 10th ed., W. W. Norton & Company, 2022.

Murray, Michael. *The Pill Book Guide to Natural Medicines: Vitamins, Minerals, Nutritional Supplements, Herbs, and Other Natural Products*. Bantam, 2002.

Murray, Patrick R., et al. *Medical Microbiology*. 9th ed., ? Elsevier, 2020.

Muthuramalingam, Meenakumari, et al. "The *Shigella* Type III Secretion System: An Overview from Top to Bottom." *Microorganisms,* vol. 9, no. 2, 2021, p. 451, doi:10.3390/microorganisms9020451.

"Mystery of Operon Evolution Probed." *Science Daily*, 30 Aug. 2012, www.sciencedaily.com/releases/2012/08/120830173502.htm. Accessed 24 June 2017.

Nachamkin, Irving, Christine M. Szymanski, and Martin J. Blaser, editors. *Campylobacter*. 3rd ed., ASM, 2008.

Nachel, Marty. *Homebrewing for Dummies*. Wiley Publishing, Inc., 1997.

Naden, Corinne J., and Rose Blue. *Jonas Salk: Polio Pioneer*. Millbrook, 2001.

Nall, Rachel. "In Which Foods Are Polysaccharides Found?" *Livestrong,* 22 June 2011, www.livestrong.com/article/477021-polysaccharides-are-found-in-which-foods. Accessed 19 Apr. 2017.

Nash, Thomas H. *Lichen Biology*. 2nd ed., Cambridge UP, 2008.

National Center for Complementary and Integrative Health: Fact Sheet Vitamin D. https://ods.od.nih.gov/search.aspx?zoom_query=Vitamin%20D.

National Center for Emerging and Zoonotic Infectious Diseases (NCEZID), Division of Healthcare Quality Promotion (DHQP). "Sepsis Questions and Answers." *CDC*. Centers for Disease Control and Prevention, 5 Oct. 2015. Accessed 30 Dec. 2015.

———. "Vitamin A." *Medline Plus*, 12 Jan. 2015, www.nlm.nih.gov/medlineplus/ency/article/002400.htm. Accessed 3 Feb. 2015.

Nelson, David L, and Michael M. Cox. *Lehninger Principles of Biochemistry*. 8th ed., W.H. Freeman, 2021.

Nestle, Marion. *Safe Food: The Politics of Food Safety*. Updated ed., U of California P, 2010.

"New FDA Policies on Antibiotic Use in Food Animal Production." *PEW Trusts*. Pew Charitable Trusts, 10 Mar. 2015. Accessed 30 Nov. 2015.

"New Meta-analysis Confirms: No Association between Autism and Vaccines." *Autism Speaks*, 19 May 2014, www.autismspeaks.org/science-news/new-meta-analysis-confirms-no-association-between-autism-and-vaccines. Accessed 28 Jan. 2021.

Newberger, Ryan, and Vikas Gupta. "Streptococcus Group A." *StatPearls*. StatPearls Publishing, 11 Aug. 2021.

"The Next Generation of Antibody Therapeutics?" *Pharmaceutical & Diagnostic Innovation*, vol. 4, no. 11, 2006, pp. 13-14.

Nguyen, Nixon, and Derrick Ashong. "Neisseria Meningitidis." *StatPearls*. StatPearls Publishing, 12 Oct. 2021.

Nieder, R., and D. K. Benbi. *Carbon and Nitrogen in the Terrestrial Environment*. Springer, 2008.

Niederman, Michael S., George A. Sarosi, and Jeffrey Glassroth. *Respiratory Infections*. 2nd ed., Lippincott Williams & Wilkins, 2001.

Nisbet, Evan G. *The Young Earth: An Introduction to Archean Geology*. Unwin Hyman, 1987.

"The Nobel Prize in Physiology or Medicine 1965: François Jacob, André Lwoff, Jacques Monod: Award Ceremony Speech." *Nobelprize.org*, www.nobelprize.org/nobel_prizes/medicine/laureates/1965/press.html. Accessed 24 June 2017.

Noor, Nazir, et al. "A Comprehensive Update of the Current Understanding of Chronic Fatigue Syndrome." *Anesthesiology and Pain Medicine,* vol. 11, no. 3, 2021, p. e113629, doi:10.5812/aapm.113629.

Norkin, Leonard. *Virology: Molecular Biology and Pathogenesis*. ASM Press, 2010.

Norrby, Erling. "Yellow Fever and Max Theiler: The Only Nobel Prize for a Virus Vaccine." *Journal of Experimental Medicine*, vol. 204, no. 12, 26 Nov. 2007, pp. 2779-84.

O'Hanlon, Leslie Harris. "Tinkering with Genes to Fight Insect-Borne Disease: Researchers Create Genetically Modified Bugs to Fight Malaria, Chagas', and Other Diseases." *The Lancet*, vol. 363, 17 Apr. 2004, p. 1288.

Oakley, A. "Tinea capitis." *DermNet NZ*, dermnetnz.org/topics/tinea-capitis/. Accessed 23 Nov. 2020.

"Off Switch for Biofilm Formation Discovered." *Science Daily, 24 Aug. 2015,* www.sciencedaily.com/releases/2015/08/150824163001.htm. Accessed 4 Feb. 2016.

Offit, P. *Deadly Choices: How the Anti-Vaccine Movement Threatens Us All*. Basic Books, 2012.

Offit, Paul A. *The Cutter Incident: How America's First Polio Vaccine Led to the Growing Vaccine Crisis*. Yale UP, 2005.

———. *Do You Believe in Magic?: The Sense and Nonsense of Alternative Medicine*. Harper, 2012.

Ojeda Rodriguez, Jafet A. and Chadi I. Kahwaji. "Vibrio Cholerae." *StatPearls*. StatPearls Publishing, 2021.

Oksenberg, Jorge R., and David Brassat, editors. *Immunogenetics of Autoimmune Disease*. Springer, 2006.

Old, R. W., and S. B. Primrose. *Principles of Gene Manipulation: An Introduction to Genetic Engineering*. Blackwell Scientific, 1994.

Olsen, Gary, and Carl R. Woese. "Archaeal Genomics: An Overview." *Cell*, vol. 89, 1997, pp. 991-94.

Olson, J. A. "Vitamin A, Retinoids, and Carotenoids." *Modern Nutrition in Health and Disease*, edited by M. E. Shils, J. A. Olson, and M. Shike, 8th ed., Lea & Febiger, pp. 287-307.

Oregon State University. "Vitamin A." *Linus Pauling Institute*. Micronutrient Information Center, 2007, lpi.oregonstate.edu/infocenter/vitamins/vitaminA/. Accessed 3 Feb. 2015.

Osbourn, Anne E., and Ben Field. "Operons." *Cellular and Molecular Life Sciences*, vol. 66, no. 23, Dec. 2009, pp. 3755-75, www.ncbi.nlm.nih.gov/pmc/articles/PMC2776167/. Accessed 24 June 2017.

Otasevic, S., S. Momcilovic, and N. M. Stojanovic. "Non-culture Based Assays for the Detection of Fungal Pathogens." J Mycol Med, vol. 28, 2018, pp. 236-48.

Owen, Judith A., Janis Kuby, Jenni Punt, and Sharon A. Stranford. *Immunology*. 7th ed., Macmillan, 2013.

Owens, R. A., W. Chen, Y. Hu, and Y-H. Hsu. "Suppression of Potato Spindle Tuber Viroid Replication and Symptom Expression by Mutations, Which Stabilize the Pathogenicity Domain." *Virology*, vol. 208, no. 2, 1995, pp. 554-64.

"Oxidative Phosphorylation." *Science Direct*. Accessed 18 May 2020.

Öztekin, Merve, et al. "Overview of *Helicobacter pylori* Infection: Clinical Features, Treatment, and Nutritional Aspects." *Diseases (Basel, Switzerland)*, vol. 9, no. 4, 2021, p. 66, doi:10.3390/diseases9040066.

Pace, A., et al. "Vitamin E Neuroprotection for Cisplatin Neuropathy." *Neurology*, vol. 74, 2010, pp. 762-66.

Pace, Brian, and Richard M. Glass. "Rabies." *Journal of the American Medical Association*, vol. 284, no. 8, 30 Aug. 2000, p. 1052.

Paddock, C. D. et al. "Rickettsia parkeri: A Newly Recognized Cause of Spotted Fever Rickettsiosis in the United States." *Clinical Infectious Diseases*, vol. 38, 2004, pp. 805-11.

Paddock, Stephen W., editor. *Confocal Microscopy Methods and Protocols*. Humana Press, 1999.

Pakbin, Babak, et al. "Virulence Factors of Enteric Pathogenic *Escherichia coli*: A Review." *International Journal of Molecular Sciences*, vol. 22, no. 18, 2021, p. 9922, doi:10.3390/ijms22189922.

Palmer, John J. *How To Brew: Everything You Need to Know to Brew Great Beer Every Time*. 4th ed., Brewers Publications; 2017.

Pan American Health Organization. World Health Organization. *Control of Diphtheria, Pertussis, Tetanus, "Haemophilus influenzae" Type B, and Hepatitis B Field Guide*. Author, 2005.

Papadakis, Maxine A., et al., editors. *Current Medical Diagnosis and Treatment 2022*. 61st ed., McGraw-Hill Education/Medical, 2021

Papazian, Charlie. *Microbrewed Adventures*. HarperCollins Publishers, 2005.

Pardi, Norbert, et al. "mRNA Vaccines-A New Era in Vaccinology." *Nature Reviews Drug Discovery*, vol. 17, 12 Jan. 2018, pp. 261-279, www.nature.com/articles/nrd.2017.243. Accessed 3 Feb. 2021.

Pardoll, Drew. "T Cells Take Aim at Cancer." *Proceedings of the National Academy of Sciences*, vol. 99, no. 25, 2012, pp. 15840-42, www.pnas.org/content/99/25/15840.full.

Parham, Peter. *The Immune System*. 4th ed., Garland Science, 2015.

Parish, James, M., and James E. Blair. "Coccidioidomycosis." *Mayo Clinic Proceedings*, vol. 83, no. 3, 2008, pp. 343-48.

Parker, James N., and Philip M. Parker, editors. *The Official Patient's Sourcebook on Leishmaniasis: A Revised and Updated Directory for the Internet Age*. Icon Health, 2002.

———. *The Official Parent's Sourcebook on Primary Immunodeficiency*. Icon Health, 2002.

———. *The Official Patient's Sourcebook on Rabies*. Icon Health, 2002.

———. *The Official Patient's Sourcebook on Salmonella Enteritidis Infection*. Icon Health, 2002.

———. *The Official Patient's Sourcebook on Streptococcus Pneumoniae Infections*. Icon Health, 2002.

———. *The Official Patient's Sourcebook on Syphilis*. Icon Health, 2002.

———. *The Official Patient's Sourcebook on Toxoplasmosis*. Icon Health, 2002.

Parker, Philip M., and James N. Parker. *Pharyngitis: A Medical Dictionary, Bibliography, and Annotated Research Guide to Internet References*. ICON Health Publications, 2004.

Parry, Dr. Nicola. "Functional Difference between T Cells & B Cells." *Seattle Post-Intelligencer*, education.seattlepi.com/functional-difference-between-t-cells-b-cells-4573.html. Accessed 27 Mar. 2017.

Passos, M. R. L., et al., editors. *Atlas of Sexually Transmitted Diseases: Clinical Aspects and Differential Diagnosis*. Springer, 2018.

Paterson, Yvonne. "Immunoediting." *Encyclopedia of Cancer*, edited by Manfred Schwab, Springer, 2011.

Pawley, James B., editor. *Handbook of Biological Confocal Microscopy*. 3rd ed., Springer, 2006.

Pechère, Jean Claude, and Edward L. Kaplan, editors. *Streptococcal Pharyngitis: Optimal Management*. S. Karger, 2004.

Pelczar, Michael J., Jr., E. C. S. Chan, and Noel R. Krieg. *Microbiology: Concepts and Applications*. McGraw, 1993.

Pelengaris, Stella, and Michael Khan. *The Molecular Biology of Cancer*. Blackwell, 2006.

Peltola, H., et al. "Mumps Outbreaks in Canada and the United States: Time for New Thinking on Mumps Vaccines." *Clinical Infectious Diseases*, vol. 45, Aug. 2007, pp. 459-66.

Pennisi, Elizabeth. "Infectious Disease: Cholera Strengthened by Trip Through Gut." *Science*, vol. 296, 2002, pp. 1783-84.

Pennsylvania State University. "Biogas Production." www.biogas.psu.edu/.

Peragine, John. *The Complete Guide to Making Your Own Wine at Home: Everything You Need to Know Explained Simply*. Atlantic Publishing Group Inc., 2010.

Pereira, Leonel, and Joao Magalhaes Neto, editors. *Marine Algae: Biodiversity, Taxonomy, Environmental Assessment, and Biotechnology*. CRC Press, 2014.

Peter, G., and P. Gardner. "Standards for Immunization Practice for Vaccines in Children and Adults." *Infectious Disease Clinics of North America*, vol. 15, 2001, pp. 9-19.

Peters, U., et al. "Vitamin E and Selenium Supplementation and Risk of Prostate Cancer in the Vitamins and Lifestyle (VITAL) Study Cohort." *Cancer Causes and Control*, vol. 19, 2008, pp. 75-87.

Pettit, George. *Biosynthetic Products for Cancer Chemotherapy*. Vol. 5. Elsevier Science, 1985.

Pfoh, Elizabeth, Sydney Dy, and Cyrus Engineer. "Chapter 8: Interventions to Improve Hand Hygiene Compliance: Brief Update Review." *Making Health Care Safer II: An Updated Critical Analysis of the Evidence for Patient Safety Practices*. Agency for Healthcare Research and Quality, Mar. 2013.

Phaff, Herman I., et al. *The Life of Yeasts*. 2nd ed., Harvard UP, 1978.

"Phagocytosis." *Khan Academy*, www.khanacademy.org/test-prep/mcat/cells/transport-across-a-cell-membrane/a/phagocytosis. Accessed 16 May 2017.

Phillips, Julie A. *The Lives of Seaweeds: A Natural History of Our Planet's Seaweeds and Other Algae*. Princeton UP, 2022.

Pickering, Larry K., et al., editors. *Red Book: 2009 Report of the Committee on Infectious Diseases*. 28th ed., American Academy of Pediatrics, 2009.

Pierson, T.C., and M. S. Diamond. The Emergence of Zika Virus and Its New Clinical Syndromes." *Nature*, vol. 560, 2018, pp. 573-81.

Piganeau, Gwenaël. *Genomic Insights into the Biology of Algae*. Academic, 2012.

Pinaud, Laurie, et al. "Host Cell Targeting by Enteropathogenic Bacteria T3SS Effectors." *Trends in Microbiology*, vol. 26, no. 4, 2018, pp. 266-83, doi:10.1016/j.tim.2018.01.010.

Pines, Maya, editor. *Arousing the Fury of the Immune System*. Howard Hughes Medical Institute, 1998.

Playfair, J. H. L., and B. M. Chain. *Immunology at a Glance*. 10th ed., Wiley-Blackwell, 2012.

Plotkin, Stanley A., et al., editors. *Vaccines*. 7th ed., Elsevier, 2017.

"Pneumococcal Polysaccharide Vaccine." *MedlinePlus*, 15 Nov. 2016, medlineplus.gov/druginfo/meds/a607022.html. Accessed 19 Apr. 2017.

Poehlmann, Stefan, and Graham Simmons. *Viral Entry into Host Cells*. Springer, 2013.

Poland, G., R. M. Jacobson, and I. G. Ovsyannikova. "Trends Affecting the Future of Vaccine Development and Delivery: The Role of Demographics, Regulatory Science, the Anti-Vaccine Movement, and Vaccinomics." *Vaccine*, vol. 27, nos. 25/26, 2009, pp. 3240-44.

Polk, Ronald E., and Neil O. Fishman. "Antimicrobial Stewardship." *Mandell, Douglas, and Bennett's Principles and Practice of Infectious Diseases*, edited by Gerald L. Mandell, John F. Bennett, and Raphael Dolin, 7th ed., Churchill Livingstone/Elsevier, 2010.

Pomeroy, Lawrence, editor. *Cycles of Essential Elements*. Dowden, 1974.

Pommerville, Jeffery. *Alcamo's Fundamentals of Microbiology: Body Systems*. 2nd ed., Jones & Bartlett Learning. 2012.

Poppelaars, Felix, et al. "Donor Genetic Variants in Interleukin-6 and Interleukin-6 Receptor Associate with Biopsy-Proven Rejection Following Kidney Transplantation." *Scientific Reports*, vol. 11, 2021, p. 16483, https://doi.org/10.1038/s41598-021-95714-z.

Popper, Helmut. *Pathology of Lung Disease: Morphology-Pathogenesis*. 2nd ed., Springer, 2021.

"Porphyrins." *TCI America*, www.tcichemicals.com/eshop/en/us/category_index/10825/. Accessed 9 May 2017.

"Porphyrins & Porphyria Diagnosis." *American Porphyria Foundation*, 2015, www.porphyriafoundation.com/

testing-and-treatment/testing-for-porphyria/porphyrins-and-porphyria-diagnosis. Accessed 9 May 2017.
"Porphyrins: One Ring in the Colors of Life." *American Scientist*, May/June 2011, www.americanscientist.org/issues/pub/porphyrins-one-ring-in-the-colors-of-life/2. Accessed 9 May 2017.
Porter, Robert E., editor. *The Merck Manual of Diagnosis and Therapy*. 20th ed., Merck, 2018.
Pradhan, S., et al. "Treatment Options for Leishmaniasis." *Clinical and Experimental Dermatology*, 4 Sept. 2021, doi:10.1111/ced.14919.
Preidt, Robert. "Too Little Vitamin D May Hasten Disability as You Age." *MedlinePlus*, 17 July 2013.
Prendergast, George C., and Elizabeth M. Jaffee. *Cancer Immunotherapy: Immune Suppression and Tumor Growth*. Academic Press, 2013.
Prescott, Lansing, John P. Harley, and Donald A. Klein. *Microbiology*. 5th ed., McGraw-Hill, 2002.
"Preventing Infections in the Hospital." *National Patient Safety Foundation*. National Patient Safety Foundation, www.npsf.org/?page=preventinginfections. Accessed 11 Feb. 2015.
Price, Morgan N., et al. "The Life-Cycle of Operons." *PLOS*, 28 June 2006, journals.plos.org/plosgenetics/article?id=10.1371/journal.pgen.0020096. Accessed 24 June 2017.
Priest, Fergus G., and Michael Goodfellow, editors. *Applied Microbial Systematics*. Kluwer Academic, 2000.
Principi, N., and Esposito, S. "Prevention of Community-Acquired Pneumonia with Available Pneumococcal Vaccines." *International Journal of Molecular Sciences*, vol 18, no. 1, 2016, p. 30.
Pritt, B. S., and C. Graham Clark. "Amebiasis." *Mayo Clinic Proceedings*, vol. 83, no. 10, 2008, pp. 1154-60, doi.org/10.4065/83.10.1154.
Professional Guide to Diseases. 10th ed., Lippincott Williams & Wilkins, 2013.
Pruthi, S., et al. "Vitamin E and Evening Primrose Oil for Management of Cyclical Mastalgia." *Alternative Medicine Review*, vol. 15, 2010, pp. 59-67.
"Public Gets Early Snapshot of MRSA and *C. difficile* Infections in Individual Hospitals." *CDC*. US Department of Health and Human Services, 12 Dec. 2013. Accessed 30 Nov. 3015.
Pullen, Tim. *Anaerobic Digestion-Making Biogas-Making Energy*. Routledge, 2015.
Punt, Jenni, et al., editors. *Kuby Immunology*. 8th ed., W. H. Freeman, 2018.
Purwosunu, Y., et al. "Vitamin K Treatment for Postmenopausal Osteoporosis in Indonesia." *Journal of Obstetrics and Gynaecology Research*, vol. 32, 2006, pp. 230-34.
Qamar, Farah Naz, et al. "Salmonellosis Including Enteric Fever." *Pediatric Clinics of North America*, vol. 69, no. 1, 2022, pp. 65-77, doi:10.1016/j.pcl.2021.09.007.
Qasim, Syed A. *Wastewater Treatment Plants: Planning, Design, and Operation*. 2nd ed., Technomic, 1999.
Queiroz-Telles, Flavio, et al. "Sporotrichosis in Immunocompromised Hosts." *Journal of Fungi*, vol. 5, no. 1, 2019, p. 8, doi:10.3390/jof5010008.
Quetel, Claude. *The History of Syphilis*. Johns Hopkins UP, 1990.
Qureshi, Shahab, et al. "Klebsiella Infections." *Medscape*, 7 Nov. 2016, emedicine.medscape.com/article/219907-overview. Accessed 1 Dec. 2021.
"Rabies." *Mayo Clinic*, 28 Jan. 2011.
Rabin, Roni Caryn. "Eager to Limit Exemptions to Vaccination, States Face Staunch Resistance." *The New York Times*, 14 June 2019, www.nytimes.com/2019/06/14/health/vaccine-exemption-health.html. Accessed 28 Jan. 2021.
Rabins, Peter V. *The Why of Things: Causality in Science, Medicine, and Life*. Columbia UP, 2013.
Radoshevich, Lilliana, and Pascale Cossart. "Listeria Monocytogenes: Towards a Complete Picture of its Physiology and Pathogenesis." *Nature Reviews: Microbiology*, vol. 16, no. 1, 2018, pp. 32-46, doi:10.1038/nrmicro.2017.126.
Ragab, Gaafar, T. Prescott Atkinson, and Matthew L. Stoll, editors. *The Microbiome in Rheumatic Diseases and Infection*. Springer, 2018.
Rao, K. K., and D. O. Hall. *Photosynthesis*. 6th ed., Cambridge UP, 1999.
Rasmussen, S. K., et al. "Recombinant Antibody Mixtures: Optimization of Cell Line Generation and Single-Batch Manufacturing Processes." *BMC Proceedings*, vol. 5 (suppl. 8), 2011, p. 2.
Rathjen, Nicholas A., and S. David Shahbodaghi. "Bioterrorism." *American Family Physician*, vol. 104, no. 4, 2021, pp. 376-85.
Raven, Peter H., Ray F. Evert, and Susan E. Eichhorn. *Biology of Plants*. 8th ed., W. H. Freeman, 2013.
Ray, Bibek. *Fundamental Food Microbiology*. 4th ed., Taylor & Francis, 2008.
Rea, William J., et al. "Effects of Toxic Exposure to Molds and Mycotoxins in Building-Related Illnesses." *Archives of Environmental Health*, vol. 58, no. 7, 2003, pp. 399-405.
"Recommended Child and Adolescent Immunization Schedule for Ages 18 years or Younger." *Centers for*

Disease Control and Prevention, US Department of Health and Human Services, 29 Jan. 2020, www.cdc.gov/vaccines/schedules/downloads/child/0-18yrs-child-combined-schedule.pdf. Accessed 28 Jan. 2021.

Reece, Jane B., et al. *Campbell Biology: Concepts & Connections*. 8th ed., Pearson, 2020.

Reichert, J. M., and V. E. Valge-Archer. "Development Trends for Monoclonal Antibody Cancer Therapeutics." *Nature Reviews Drug Discovery*, vol. 6, 2007, pp. 349-56.

Reichle, David E. *The Global Carbon Cycle and Climate Change: Scaling Ecological Energetics from Organism to the Biosphere*. Elsevier, 2019.

Reidl, Joachim, et al. "Vibrio cholerae and Cholera: Out of the Water and into the Host." *FEMS Microbiological Reviews*, vol. 26, 2002, pp. 125-39.

"Republican Men Are Vaccine-Hesitant, But There's Little Focus on Them." *Pew*, 23 Apr. 2021, www.pewtrusts.org/en/research-and-analysis/blogs/stateline/2021/04/23/republican-men-are-vaccine-hesitant-but-theres-little-focus-on-them. Accessed 12 May. 2021.

Reznikoff-Etievant, M. F., et al. "Low Vitamin B12 Level as a Risk Factor for Very Early Recurrent Abortion." *European Journal of Obstetrics, Gynecology, and Reproductive Biology*, vol. 104, 2002, pp. 156-59.

Rich, Alexander, and Sung Hou Kim. "The Three-Dimensional Structure of Transfer RNA." *Scientific American*, Jan. 1978, pp. 52-62.

Richardson, Malcolm D., and Elizabeth M. Johnson. *The Pocket Guide to Fungal Infection*. 2nd ed., Blackwell, 2006.

Riedel, Stefan, et al., Jawetz Melnick & Adelbergs Medical Microbiology, 28th ed. McGraw-Hill Education/Medical, 2019.

Riegelman, Richard. "Population Prevention and COVID-19." *COVID-19 Global Lessons Learned: Interactive Case Studies*. Jones & Bartlett Learning, 2021, pp. 31-42.

Rissler, Jane, and Margaret Mellon. *The Ecological Risks of Engineered Crops*. MIT Press, 1996.

Robertson, E. S. *Epstein-Barr virus*. Caister Academic Press, 2005.

Rock, C. L. "Carotenoid Update." *Journal of the American Dietetic Association*, vol. 103, no. 4, 2003, pp. 423-25.

Rodrigo, Chaturaka, et al. "Pharmacological Management of Tetanus: An Evidence-Based Review." *Critical Care*, vol. 18, no. 2, 2014, p. 217, doi:10.1186/cc13797.

Rombouts, E. K., F. R. Rosendaal, and F. J. van der Meer. "Daily Vitamin K Supplementation Improves Anticoagulant Stability." *Journal of Thrombosis and Haemostasis*, vol. 5, no. 10, 2007, pp. 2043-48.

Rosario, Diane. *Immunization Resource Guide: Where to Find Answers to All Your Questions About Childhood Vaccinations*. Patter, 2001.

Rose, Noel R., and Ian. R. Mackay, editors. *The Autoimmune Diseases*. 6th ed., Academic Press/Elsevier, 2019.

Roseland, Mark. "Water and Sewage." *Toward Sustainable Communities: Resources for Citizens and Their Governments*. Rev. ed., New Society, 2005.

Roush, Sandra, et al. "Historical Comparisons of Morbidity and Mortality for Vaccine-Preventable Diseases in the United States." *Journal of the American Medical Association*, vol. 298, no. 18, 2007, pp. 2155-63.

Ruggiero, Michael A., et al. "A Higher Level Classification of All Living Organisms." *PloS One*, vol. 10, no. 4, 2015, p. e0119248, doi:10.1371/journal.pone.0119248.

Rumbold, A. R., et al. "Vitamins C and E and the Risks of Preeclampsia and Perinatal Complications." *New England Journal of Medicine*, vol. 354, 2006, pp. 1796-1806.

Russell, Kevin L., et al. "Vaccine-Preventable Adenoviral Respiratory Illness in U.S. Military Recruits, 1999-2004." *Vaccine*, vol. 24, 2006, pp. 2835-42.

Ryan, Frank. *Tuberculosis: The Greatest Story Never Told*. Swift Publishers, 2019.

Ryan, Kenneth J., et al., editors. *Sherris Medical Microbiology: An Introduction to Infectious Diseases*. 7th ed., McGraw-Hill, 2018.

Ryu, Wang-Shic. *Molecular Virology of Human Pathogenic Viruses*. Academic Press, 2016.

Sáenz Robles, Maria Teresa, and James M Pipas. "T Antigen Transgenic Mouse Models." *Seminars in Cancer Biology*, vol. 19, no. 4, 2009, pp. 229-35, doi:10.1016/j.semcancer.2009.02.002.

Sagan, Dorion. *The Global Sulfur Cycle*. National Aeronautics and Space Administration, 1985.

Salisbury, Frank B., and Cleon W. Ross. *Plant Physiology*. 4th ed., Wadsworth, 1992.

Salvato, Joseph A. *Environmental Engineering and Sanitation*. 4th ed., John Wiley & Sons, 1992.

Sampath, Shrikanth et al. "Pandemics Throughout the History." *Cureus*, vol. 13, no. 9, 2021, p. e18136, doi:10.7759/cureus.18136

Sanders, Robert. "Discovery Opens Door to Attacking Biofilms that Cause Chronic Infections." *Berkeley News*. University of Berkeley, 12 July 2012, news.berkeley.edu/2012/07/12/discovery-opens-door-to-attacking-biofilms-that-cause-chronic-infections/. Accessed 4 Feb. 2016.

Sandora, Thomas J., Courtney A. Gidengil, and Grace M. Lee. "Pertussis Vaccination for Health Care Workers." *Clinical Microbiology Reviews*, vol. 21, 2008, pp.426-34.

Santora, Marc, and Rebecca Robbins. "The AstraZeneca Vaccine Is Shown to Cut Transmission of the Virus." *New York Times*, 3 Feb. 2021, www.nytimes.com/2021/02/03/us/astrazeneca-coronavirus-vaccine.html. Accessed 4 Feb. 2021.

Sarin, S. K., et al. "High Dose Vitamin K3 Infusion in Advanced Hepatocellular Carcinoma." *Journal of Gastroenterology and Hepatology*, vol. 21, 2006, pp. 1478-82.

Sathe, M. N., and A. S. Patel. "Update in Pediatrics: Focus on Fat-Soluble Vitamins." *Nutrition in Clinical Practice*, vol. 25, no. 4, 2010, pp. 340-46, doi:10.1177/0884533610374198.

Satir, Peter. *Structure and Function in Cilia and Flagella*. Springer, 1965.

Sato, Y., et al. "Amelioration by Mecobalamin of Subclinical Carpal Tunnel Syndrome Involving Unaffected Limbs in Stroke Patients." *Journal of Neurological Sciences*, vol. 231, 2005, pp. 13-18.

Savageau, Michael. *Biochemical Systems Analysis: A Study of Function and Design in Molecular Biology*. CreateSpace, 2010.

Savelkoel, Jelmer, et al. "Abbreviated Atovaquone-Proguanil Prophylaxis Regimens in Travellers After Leaving Malaria-Endemic Areas: A Systematic Review." *Travel Medicine and Infectious Disease*, vol. 21, 2018, pp. 3-20, doi:10.1016/j.tmaid.2017.12.005.

Schaechter, Moselio. *Mechanisms of Microbial Disease*. Lippincott Williams & Wilkins, 1998.

Scharbo-DeHaan, Marianne, and Donna G. Anderson. "The CDC 2002 Guidelines for the Treatment of Sexually Transmitted Diseases: Implications for Women's Health Care." *Journal of Midwifery and Women's Health*, vol. 48, Feb. 2003, pp. 96-104.

Scheckenbach, Frank, et al. "Large-Scale Patterns in Biodiversity of Microbial Eukaryotes from the Abyssal Sea Floor." *Proceedings of the National Academy of Sciences of the United States of America*, vol. 107, 2010, pp. 115-20.

Schlesinger, William H. *Biogeochemistry: An Analysis of Global Change*. 2nd ed., Academic Press, 1997.

Schlossberg, David, editor. *Tuberculosis and Nontuberculous Mycobacterial Infections*. 7th ed., ASM Press, 2017.

Schmitz, Franz-Josef, and Ad C. Fluit. "Mechanisms of Antibacterial Resistance." *Cohen and Powderly Infectious Diseases*, edited by Jonathan Cohen, Steven M. Opal, and William G. Powderly, 3rd ed., Mosby/Elsevier, 2010.

Scholz, Roland W., et al. *Sustainable Phosphorus Management: A Global Transdisciplinary Roadmap*. Springer, 2014.

Schopf, J. William, editor. *Major Events in the History of Life*. Jones and Bartlett, 1992.

Schreiber, Robert D., et al. "Cancer Immunoediting: Integrating Immunity's Roles in Cancer Suppression and Promotion." *Science*, vol. 331, no. 6024, pp. 1565-70.

Schrier, Robert W., editor. *Diseases of the Kidney and Urinary Tract*. 8th ed., Wolters Kluwer Health/Lippincott Williams & Wilkins, 2007.

Scott, Andrew M., Jedd D. Wolchok, and Lloyd J. Old. "Antibody Therapy of Cancer." *Nature Reviews Cancer*, vol. 12, 2012, pp. 278-87, www.nature.com/nrc/journal/v12/n4/full/nrc3236.html.

Sears, R. W. *The Vaccine Book: Making the Right Decision for your Child*. Little, Brown, and Company, 2011.

Semeniuk, Ivan. "Jacques Monod: A Scientist Whose Revolution Is Still Unfolding." *Globe and Mail*, 27 Sept. 2013, www.theglobeandmail.com/technology/science/jacques-monod-a-scientist-whose-revolution-is-still-unfolding/article14572219/. Accessed 24 June 2017.

Sesso, H. D., et al. "Vitamins E and C in the Prevention of Cardiovascular Disease in Men." *Journal of the American Medical Association*, vol. 300, 2008, pp. 2123-33.

Seussen, S. J., et al. "Oral Cyanocobalamin Supplementation in Older People with Vitamin B12 Deficiency." *Archives of Internal Medicine*, vol. 165, 2005, pp. 1167-72.

Sexually Transmitted Infections Treatment Guidelines, 2021: Chlamydial Infections. Centers for Disease Control and Prevention, 22 July 2021, www.cdc.gov/std/treatment-guidelines/chlamydia.htm. Accessed 20 Nov. 2021.

Shah, Maunank, and Susan E Dorman. "Latent Tuberculosis Infection." *The New England Journal of Medicine*, vol. 385, no 24, 2021, pp. 2271-80, doi:10.1056/NEJMcp2108501.

Shahar, E., G. Hassoun, and S. Pollack. "Effect of Vitamin E Supplementation on the Regular Treatment of Seasonal Allergic Rhinitis." *Annals of Allergy, Asthma, and Immunology*, vol. 92, 2004, pp. 654-58.

Shalal, Andrea, Jeff Mason, and David Lawder. "US Reverses Stance, Backs Giving Poorer Countries Access to COVID vaccine patents." *REUTERS*, 6 May 2021, www.reuters.com/business/healthcare-pharmaceuticals/biden-says-plans-back-wto-waiver-vaccines-2021-05-05/. Accessed 7 May. 2021.

Shapira, Raz, et al. "Streptococcus Gallolyticus Endocarditis on a Prosthetic Tricuspid Valve: A Case

Report and Review of the Literature." *Journal of Medical Case Reports*, vol. 15, no. 1, 2021, p. 528, doi:10.1186/s13256-021-03125-5.

Shaw, Gina. "Breaking News: Deadly Klebsiella Pneumoniae Strain Resistant to Carbapenems." *Emergency Medicine News*, Mar. 2013, journals.lww.com/em-news/Fulltext/2013/03000/Breaking_News__Deadly_Klebsiella_Pneumoniae_Strain.3.aspx. Accessed 1 Dec. 2021.

Shelton, C. D. *Vitamins, Minerals & Supplements: Essential or Over-Hyped?* Amazon Digital Services Inc., 2013.

Sheppard, John, editor. *Introduction to Brewing and Fermentation Science: Essential Knowledge for Those Dedicated to Brewing Better Beer*. World Scientific, 2021.

Sherman, Irwin W. *The Elusive Malaria Vaccine: Miracle or Mirage?* ASM Press, 2009.

Sherrow, Victoria. *Jonas Salk*. Facts On File, 1993.

Shirley, Debbie-Ann T., et al. "Significance of Amebiasis: 10 Reasons Why Neglecting Amebiasis Might Come Back to Bite Us in the Gut." *PLoS Neglected Tropical Diseases*, vol. 13, no. 11, 2019, e0007744, doi:10.1371/journal.pntd.0007744.

Shmaefsky, Brian. *Meningitis*. Rev. ed., Chelsea House, 2010.

Shoenfeld, Yehuda, and Nancy Agmon-Levin. *Vaccines and Autoimmunity*. Wiley, 2015.

Shoman, Mary. "How the Immune System Works." *VeryWell Health*, 4 July 2020, www.verywellhealth.com/how-does-the-immune-system-work-3232652. Accessed 10 Feb. 2021.

Shors, Teri. *Understanding Viruses*. 2nd ed., Jones, 2013.

Shreiner, Andrew B., John Y. Kao, and Vincent B. Young. "The Gut Microbiome in Health and in Disease." *Current Opinion in Gastroenterology*, vol. 31, no. 1, 2015, pp. 69-75, doi: [10.1097/MOG.0000000000000139].

Siddiqui, Abdul H., and Janak Koirala. "Methicillin-Resistant *Staphylococcus aureus*." *StatPearls*. StatPearls Publishing, 19 July 2021.

Sidhu, Sachdev, and Clarence Ronald Geyer. *Phage Display in Biotechnology and Drug Discovery*. 2nd ed., CRC Press, 2015.

Sigel, Astrid, Helmut Sigel, and Roland K. O. Sigel, editors. *Biogeochemical Cycles of Elements*. Taylor & Francis, 2005.

Sikorski, Zdzislaw E., and Anna Kolakowska, editors. *Chemical and Functional Properties of Food Lipids*. CRC Press, 2002.

Singer, Adam J., and David A Talan. "Management of Skin Abscesses in the Era of Methicillin-Resistant *Staphylococcus aureus*." *The New England Journal of Medicine*, vol. 370, no. 11, 2014, pp. 1039-47, doi:10.1056/NEJMra1212788.

Singh, Grisuna, et al. "Recombinant Zoster Vaccine (Shingrix(r)): A New Option for the Prevention of Herpes Zoster and Postherpetic Neuralgia." *The Korean Journal of Pain*, vol. 33, no. 3, 2020, pp. 201-7, doi:10.3344/kjp.2020.33.3.201.

Singh, Sunit K., editor. *Human Respiratory Viral Infections*. CRC Press, 2020.

Singh, Sunit K., and Daniel Ruzek, editors. *Neuroviral Infections: RNA Viruses and Retroviruses*. CRC Press, 2013.

"Sinus Infection (Sinusitis)." *Cleveland Clinic*, 6 Apr. 2020, my.clevelandclinic.org/health/diseases/17701-sinusitis. Accessed 28 Aug. 2021.

Sjogren, M. H. "Hepatitis A." In *Sleisenger and Fordtran's Gastrointestinal and Liver Disease: Pathophysiology, Diagnosis, Management*, edited by Mark Feldman, Lawrence S. Friedman, and Lawrence J. Brandt, New ed., 2 vols., Saunders/Elsevier, 2010.

Skloot, Rebecca. *The Immortal Life of Henrietta Lacks*. Random House, 2011.

Sleigh, Michael A., editor. *Cilia and Flagella*. Academic Press, 1974.

Smil, Vaclav. *Carbon-Nitrogen-Sulfur: Human Interference in Grand Biospheric Cycles*. Springer, 2012.

Smith, Lorrain Annie. *Lichens*. ?Library of Alexandria, 2020.

Smith, Mathew D. "Antibody Production in Plants." *Biotechnology Advances*, vol. 14, no. 3, 1996, pp. 267-81.

Snowden, Jessica, and Kari A. Simonsen. "Tularemia." *StatPearls*. StatPearls Publishing, 21 July 2021.

Snyder, Stephen. *The Brewmaster's Bible*. HarperCollins Publishers, 1997.

Sommers, Michael. *Yeast Infections, Trichomoniasis, and Toxic Shock Syndrome (Girls' Health)*. Rosen Publishing Group, 2007.

Sonpavde, G., et al. "Emerging Vaccine Therapy Approaches for Prostate Cancer." *Reviews in Urology*, vol. 12, no. 1, 2010, pp. 25-34.

Sora, Valerio M., et al. "Extraintestinal Pathogenic *Escherichia coli*: Virulence Factors and Antibiotic Resistance." *Pathogens*, vol. 10, no. 11, 2021, p. 1355, doi:10.3390/pathogens10111355.

Spellman, Frank R., and Melissa L. Stoudt. *Environmental Science: Principles and Practices*. Scarecrow, 2013.

Spentzos, Dimitri. "Gene Expression Signature with Independent Prognostic Significance in Epithelial Ovarian Cancer." *Journal of Clinical Oncology*, vol. 22, no. 23, Dec. 2004, pp. 4648-58.

Spinola, Stanley M., Margaret E. Bauer, and Robert S. Munson, Jr. "Immunopathenogenesis of *Haemophilus ducreyi* Infection (Chancroid)." *Infection and Immunity*, vol. 70, 2002, pp. 1667-76.

Sprent, Janet I. *The Ecology of the Nitrogen Cycle*. Cambridge UP, 1987.

Sridhar, Saranya. "An Affordable Cholera Vaccine: An Important Step Forward." *The Lancet*, vol. 374, 2009, pp. 1658-60.

Srivastava, Roli. "Drug Resistant Bug Klebsiella Causes Worry." *The Hindu*, 23 May 2016, www.thehindu.com/sci-tech/health/medicine-and-research/drugresistant-bug-klebsiella-causes-worry/article7928014.ece. Accessed 1 Dec. 2021.

St. Georgiev, Vassil. *Opportunistic Infections: Treatment and Prophylaxis*. Humana Press, 2003.

Stacey, Gary, Robert H. Burris, and Harold J. Evans, editors. *Biological Nitrogen Fixation*. Chapman and Hall, 1992.

Stanberry, Lawrence. *Understanding Herpes*. 2nd ed., UP of Mississippi, 2006.

Stanbury, Peter F., et al. *Principles of Fermentation Technology*. 3rd ed., Butterworth-Heinemann, 2016.

Stanforth, Stephen. *Natural Product Chemistry at a Glance*. Wiley-Blackwell, 2006.

Starr, Cecie, et al. *Biology: The Unity and Diversity of Life*. 15th ed., Cengage Learning, 2018.

Stearns, Jennifer, et al. Microbiology for Dummies. For Dummies, 2019.

Steele, Russell W., and Avinash Shetty. "Blastomycosis." emedicine.medscape.com/article/961731-overview.

Steinle, Heidrun, et al. "Delivery of Synthetic mRNAs for Tissue Regeneration." *Advanced Drug Delivery Reviews*, vol. 179, no. 114007, 2021, doi:10.1016/j.addr.2021.114007.

Stenfors Arnesen, Lotte P., et al. "From Soil to Gut: *Bacillus cereus* and Its Food Poisoning Toxins." *FEMS Microbiology Reviews*, vol. 32, no. 4, 2008, pp. 579-606, doi:10.1111/j.1574-6976.2008.00112.x.

"Steroids." *MedlinePlus*. National Library of Medicine, National Institutes of Health, 25 Mar. 2016. Accessed 11 May. 2016.

Stevenson, F. J. *Cycles of Soils: Carbon, Nitrogen, Phosphorus, Micronutrients*. 2nd ed., Wiley India Pvt. Ltd., 2015.

Sticherling, Michael, et al., editors. *Treatment of Autoimmune Disorders*. Springer, 2012.

Stine, Gerald J. *AIDS Update 2014*. McGraw-Hill, 2013.

Story, Lachel. "Body Defenses." *Pathophysiology: A Practical Approach*. 2nd ed., Jones, 2015, pp. 31-50.

Straus, Eugene, and Alex Straus. *Medical Marvels: The One Hundred Greatest Advances in Medicine*. Prometheus Books, 2006.

Strauss, James, and Ellen Strauss. *Viruses and Human Disease*. 2nd ed., Academic Press/Elsevier, 2008.

Strugnell, R. A., and O. L. Wijburg. "The Role of Secretory Antibodies in Infection Immunity." *Nature Reviews Microbiology*, vol. 8, no. 9, 2010, pp. 656-67.

Stucker, M., et al. "Topical Vitamin B, a New Therapeutic Approach in Atopic Dermatitis: Evaluation of Efficacy and Tolerability in a Randomized Placebo-Controlled Multicentre Clinical Trial." *British Journal of Dermatology*, vol. 150, 2005, pp. 977-83.

Sultan, Mutaz I., et al. "*Helicobacter pylori* Infection." 16 Nov. 2018, http://emedicine.medscape.com/article/929452-overview. Accessed 25 Nov. 2021.

Summers, David K. *The Biology of Plasmids*. Blackwell, 1996.

Sutton, Amy L., editor. *Sexually Transmitted Diseases Sourcebook*. 5th ed., Omnigraphics, 2013.

Syvanen, Michael, and Clarence Kado. *Horizontal Gene Transfer*. 2nd ed., Academic Press, 2002.

Tabor, Edward, editor. *Viruses and Liver Cancer*. Elsevier, 2006.

Tack, Danielle M., et al. "Preliminary Incidence and Trends of Infections with Pathogens Transmitted Commonly Through Food-Foodborne Diseases Active Surveillance Network, 10 U.S. Sites, 2016-2019." *Morbidity and Mortality Weekly Report*, vol. 69, no. 17, 2020, pp. 509-14, doi:10.15585/mmwr.mm6917a1.

Tait, Dereck R., et al. "Final Analysis of a Trial of M72/AS01E Vaccine to Prevent Tuberculosis." *New England Journal of Medicine*, vol. 381, 2019, pp. 2429-39, doi:10.1056/NEJMoa1909953.

Tan, S. Y., and M. K. Dee. "Elie Metchnikoff (1845-1916): Discoverer of Phagocytosis." *Singapore Medical Journal*, 2009, webext.pasteur.fr/biblio/ressources/histoire/textes_integraux/metchnikoff/smjmetabio2009tan.pdf. Accessed 16 May 2017.

Tecklenburg, S. L., et al. "Ascorbic Acid Supplementation Attenuates Exercise-Induced Bronchoconstriction in Patients with Asthma." *Respiratory Medicine*, vol. 101, 2007, pp. 1770-78.

Telford, Carol, and Rod Theodorou. *Through a Termite City*. Heineman Interactive Library, 1998.

Teng, Michele W. L., et al. "From Mice to Humans: Developments in Cancer Immunoediting." *Journal of Clinical Investigation*, vol. 25, no. 9, 2015, pp. 3338-46.

The Teratology Society. "Teratology Society Position Paper: Recommendations for Vitamin A Use during Pregnancy." *Teratology*, vol. 35, no. 2, 1987, pp. 269-75.

Terrault, Norah A., et al. "Update on Prevention, Diagnosis, and Treatment of Chronic Hepatitis B: AASLD 2018 Hepatitis B Guidance." *Hepatology*, vol. 67, no. 4, 2018, pp. 1560-99, doi:10.1002/hep.29800.

Thomas, Christopher M. "Paradigms of Plasmid Organization." *Molecular Microbiology*, vol. 37, no. 3, 2000, pp. 485-91.

Thomas, Nancy J., D. Bruce Hunter, and Carter T. Atkinson, editors. *Infectious Diseases of Wild Birds*. Blackwell, 2007.

Thompson, Dennis. "CDC Says Too Few US Teens Getting HPV Vaccine." *CBS News*. CBS, 30 July 2015. Accessed 30 Dec. 2015.

Tibayrenc, Michel. *Genetics and Evolution of Infectious Disease*. Elsevier, 2011.

Tidona, Christian, and Gholamreza Darai, editors. *The Springer Index of Viruses*. Springer, 2002.

Tiessen, Holm, editor. *Phosphorus in the Global Environment: Transfers, Cycles, and Management*. Wiley, 1995.

Tighe, P., et al. "Effect of Increased Consumption of Whole-Grain Foods on Blood Pressure and Other Cardiovascular Risk Markers in Healthy Middle-Aged Persons: A Randomized Controlled Trial." *American Journal of Clinical Nutrition*, vol. 92, no. 4, 2010, pp. 733-40, doi:10.3945/ ajcn.2010.29417.

Tiller, Tom, editor. *Synthetic Antibodies: Methods and Protocols*. Humana, 2017.

Ting, R. Z., et al. "Risk Factors of Vitamin B12 Deficiency in Patients Receiving Metformin." *Archives of Internal Medicine*, vol. 166, 2006, pp. 1975-99.

Tiwari, Shivangi, and Manisha Sharma. *Recombinant DNA Technology in the Synthesis of Human Insulin*. AP Lambert Academic Publishing, 2018.

Topalis, D., et al. "The Large Tumor Antigen: A "Swiss Army Knife." Protein Possessing the Functions Required for the Polyomavirus Life Cycle." *Antiviral Research*, vol. 97, no. 2, 2013, pp. 122-36. doi:10.1016/j.antiviral.2012.11.007.

Tortora, Gerard J., and Bryan Derrickson. *Principles of Anatomy and Physiology*. 16th ed., John Wiley & Sons, 2020.

Tortora, Gerard J., Berdell R. Funke, and Christine L. Case. *Microbiology: An Introduction*. 10th ed., Benjamin Cummings, 2010.

Tortora, Gerard J., et al. *Microbiology: An Introduction*. 13th ed., Pearson, 2018.

Troncoso, Alcides, Cecilia Ramos Clausen, and Jessica Rivas. *Where Can You Catch Botulism Food Poisoning?: Foodborne Botulism*. Lambert Academic Publishing, 2012.

Tropp, Burton E. *Molecular Biology: Genes to Proteins*. 4th ed., Jones, 2012.

Tsai, Sue, and Pere Santamaria. "MHC Class II Polymorphisms, Autoreactive T-Cells, and Autoimmunity." *Frontiers in Immunology*, vol. 4, no. 321, 2013, pp. 1-7.

Tsang, Clarisse A., et al. "Increase in Reported Coccidioidomycosis-United States, 1998-2011." *Morbidity and Mortality Report*, vol. 62, no. 12, 2013, pp. 217-21.

Tsao, S., C. M. Tsang, K. To, and K. Lo. "The Role of Epstein-Barr Virus in Epithelial Malignancies." *Journal of Pathology*, vol. 235, 2015, pp. 323-33.

Tselis, A., and H. B. Jenson. *Epstein-Barr Virus*. Taylor & Francis, 2006.

"Typhus." *MedlinePlus*, 7 Dec. 2014, medlineplus.gov/ency/article/001363.htm. Accessed 15 Nov. 2016.

"Understanding How COVID-19 Vaccines Work." *Centers for Disease Control and Prevention*, 9 Mar. 2021, www.cdc.gov/coronavirus/2019-ncov/vaccines/different-vaccines/how-they-work.html. Accessed 19 Apr. 2021.

US Department of Health and Human Services, Public Health Service, National Toxicology Program. *Eleventh Report on Carcinogens*. US Department of Health and Human Services, 2005.

US Food and Drug Administration. News & Events. "FDA Warns Consumers to Stop Using Soladek Vitamin Solution." *FDA, News & Events*, 28 Mar. 2011, www.fda.gov/NewsEvents/Newsroom/PressAnnouncements/ucm248588.htm. Accessed 3 Feb. 2015.

———. "Vaccines and Related Biological Products Advisory Committee October 14-15, 2021 Meeting Announcement." 14-15 Oct. 2021, www.fda.gov/advisory-committees/advisory-committee-calendar/vaccines-and-related-biological-products-advisory-committee-october-14-15-2021-meeting-announcement#event-information. Accessed 10 Dec. 2021.

US National Institute of Allergy and Infectious Diseases. "How Vaccines Work." *National Institutes of Health*, 19 Apr. 2011.

US National Institute of General Medical Sciences. "You Are What You Eat." *NIH National Institute of General Medical Sciences: ChemHealthWeb*, 9 Aug. 2012.

US National Institutes of Health. "Dietary Supplement Fact Sheet: Vitamin A." *Office of Dietary Supplements*, 5 June 2013, ods.od.nih.gov/factsheets/VitaminA-HealthProfessional/. Accessed 3 Feb. 2015.

———. "Listeriosis." 30 Nov. 2021, www.nlm.nih.gov/medlineplus/ency/article/001380.htm. Accessed 12 Dec. 2021.

US National Library of Medicine. "Immune System and Disorders." www.nlm.nih.gov/medlineplus/immunesystemanddisorders.html.

USDA National Nutrient Database for Standard Reference, Release 24. "Vitamin A, IU Content of Selected Foods Per Common Measure, Sorted by Nutrient Content." *USDA*, 31 May 2013, www.docstoc.com/docs/30903437/USDA-National-Nutrient-Database-for-Standard-Reference_-Release-22. Accessed 3 Feb. 2015.

Vaccine Safety. Centers for Disease Control and Prevention. US Department of Health and Human Services, 12 Jan. 2021, www.cdc.gov/vaccinesafety/index.html. Accessed 28 Jan. 2021.

"Vaccines." *National Institute of Allergy and Infectious Disease*. National Institutes of Health, 13 Aug. 2020, www.niaid.nih.gov/research/vaccines.

Valley, Thomas S., and Colin R. Cooke. "The Epidemiology of Sepsis: Questioning Our Understanding of the Role of Race." *Critical Care*, vol. 19, 2015, p. 347. Accessed 30 Dec. 2015.

Van Gaal, E. V. B., W. E. Hennink, D. J. A. Crommelin, and E. Mastrobattista. "Plasmid Engineering for Controlled and Sustained Gene Expression for Nonviral Gene Therapy." *Pharmaceutical Research*, vol. 23, no. 6, 2006, pp. 1053-74.

Vance, Dennis E., and Jean E. Vance, editors. *Biochemistry of Lipids, Lipoproteins, and Membranes*. 5th ed., Elsevier, 2008.

Vickers, Peter S. *Severe Combined Immune Deficiency: Early Hospitalization and Isolation*. John Wiley & Sons, 2009.

Vickery, Karen, editor. *Microbial Biofilms in Healthcare: Formation, Prevention and Treatment*. Mdpi AG, 2020.

Vincent, Miriam T. "Sore Throat-Strep Throat? When to Worry." *Pediatrics for Parents*, vol. 21, no. 8, 1 Aug. 2004, pp. 11-12.

"Vitamin D: The Nutrition Source." *Harvard T. H. Chan, School of Public Health*, Mar. 2020, www.hsph.harvard.edu/nutritionsource/vitamin-d.

Volk, Tyler. *CO2 Rising: The World's Greatest Environmental Challenge*. MIT Press, 2008.

Vorvick, Linda J. "Pharyngitis." *MedlinePlus*, 8 Jan. 2012.

Walker, J. G., et al. "Mental Health Literacy, Folic Acid and Vitamin B12, and Physical Activity for the Prevention of Depression in Older Adults." *British Journal of Psychiatry*, vol. 197, no. 1, 2020, pp. 45-54.

Walsh, Thomas J., et al. Larone's Medically Important Fungi: A Guide to Identification. 6th ed., ASM Press, 2018.

Walton, G. E., et al. "A Randomised, Double-Blind, Placebo Controlled Cross-Over Study to Determine the Gastrointestinal Effects of Consumption of Arabinoxylan-Oligosaccharides Enriched Bread in Healthy Volunteers." *Nutrition Journal*, vol. 11, no. 36, 2012, doi:10.1186/1475-2891-11-36.

Wassenegger, M., et al. "RNA-Directed De Novo Methylation of Genomic Sequences in Plants." *Cell*, vol. 76, no. 3, 1994, pp. 567-76.

Watson, James D., et al. *Molecular Biology of the Gene*. 7th ed., Pearson, 2013.

Watson, James, Michael Gilman, Jan Witkowski, and Mark Zoller. *Recombinant DNA*. 2nd ed., W. H. Freeman, 1992.

Watson, Stephanie. "Autoimmune Diseases: Types, Symptoms, Causes, and More." *Healthline*, 26 Mar. 2019, www.healthline.com/health/autoimmune-disorders?print=true. Accessed 20 Aug. 2021.

Weber, Marcel, et al. "A Highly Functional Synthetic Phage Display Library Containing over Forty Billion Human Antibody Clones." *Plos ONE*, vol. 9, no. 6, 2014, pp. 1-9. *Academic Search Complete*. Accessed 4 Sept. 2014.

Webster, John, and Roland Weber. *Introduction to Fungi*. Cambridge UP, 2007.

Weedon, David. *Skin Pathology*. 3rd ed., Churchill Livingstone/Elsevier, 2010.

Weinberger, D., Harboe, Z. and Shapiro, E. "Developing Better Pneumococcal Vaccines for Adults." *JAMA Internal Medicine*, vol. 177, no. 3, 2017, p. 303.

Weiss, Alison. "The Genus *Bordetella*." *The Prokaryotes: A Handbook on the Biology of Bacteria*, edited by Martin Dworkin, et al., Vol. 5, Springer, 2006, pp. 602-47.

Wernegreen, Jennifer J. "In It for the Long Haul: Evolutionary Consequences of Persistent Endosymbiosis." *Current Opinion in Genetics & Development*, vol. 47, 2017, pp. 83-90, doi:10.1016/j.gde.2017.08.006.

West, John B., and Andrew M. Luks. *Pulmonary Pathophysiology: The Essentials*. 10th ed., Lippincott Williams & Wilkins, 2021.

"What Are the Different Polysaccharides?" *InnovateUS*, www.innovateus.net/science/what-are-different-polysaccharides. Accessed 19 Apr. 2017.

"What Are White Blood Cells?" *University of Rochester Medical Center*, www.urmc.rochester.edu/encyclopedia/content.aspx?Co

ntentTypeID=160&ContentID=35. Accessed 16 May 2017.

"What Does mRNA Do? mRNA Produces Instructions to Make Proteins That May Treat or Prevent Disease." *Moderna*, 2020, www.modernatx.com/mrna-technology/science-and-fundamentals-mrna-technology. Accessed 3 Feb. 2021.

"What Is Innate Immunity?" *Center for Innate Immunity and Immune Diseases*, ciiid.washington.edu/content/what-innate-immunity. Accessed 10 Feb. 2021.

Whitford, David. *Proteins: Structure and Function.* Wiley, 2005.

Wigley, T. M. L., and D. S. Schimel, editors. *The Carbon Cycle*. Cambridge UP, 2000.

Willey, Joanne, et al. *Prescott's Microbiology*. 11th ed., McGraw-Hill Education, 2019.

Wilson, Benda A., et al. *Bacterial Pathogenesis: A Molecular Approach*. 4th ed., ASM Press, 2019.

Wilson, Leslie, William Dentler, and Paul T. Matsudaira, editors. *Cilia and Flagella*. Methods in Cell Biology 47. Academic Press, 1995.

Wilson, Michael, Brian Henderson, and Rod McNab. *Bacterial Disease Mechanisms: An Introduction to Cellular Microbiology*. Cambridge UP, 2002.

Wilson, R. T. *Ecophysiology of the Camelidae and Desert Ruminants*. Springer Verlag, 1990.

Winter, Arthur. *Organic Chemistry for Dummies*. 2nd ed., For Dummies, 2016.

Woese, Carl R. "Archaebacteria." *Scientific American,* vol. 244, 1981, pp. 98-122.

Wong, Karen K., et al. "Recommendations of the Advisory Committee on Immunization Practices for Use of Cholera Vaccine." *Morbidity and Mortality Weekly Report,* vol. 66, no. 18, 2017, pp. 482-85, doi:10.15585/mmwr.mm6618a6.

World Health Organization. "Cholera." Fact Sheet No. 107, www.who.int/mediacentre/factsheets/fs107.

———. "Clean Hands Protect Against Infection." *WHO*, 2016. Accessed 9 Aug. 2016.

———. "The Evidence for Clean Hands." *WHO*, n.d. Accessed 19 Aug. 2016.

———. "Global Prevalence of Vitamin A Deficiency in Populations at Risk 1995-2005." *Global Database on Vitamin A Deficiency, WHO*, 2009.

———. "Patient Safety: World Alliance for Patient Safety." *WHO*, 27 Oct. 2004. Accessed 9 Aug. 2016.

———. "Pneumonia." 11 Nov. 2021, www.who.int/news-room/fact-sheets/detail/pneumonia. Accessed 10 Dec. 2021.

———. "Trypanosomiasis, Human African (Sleeping Sickness)" 18 May 2021, https://www.who.int/news-room/fact-sheets/detail/trypanosomiasis-human-african-(sleeping-sickness). Accessed 13 Dec. 2021.

———. GPEI. "Wild Poliovirus Weekly Update," 30 Nov. 2021, www.polioeradication.org/Dataandmonitoring/Poliothisweek.aspx . Accessed on 7 Dec. 2021.

———. "Yellow Fever." www.who.int/mediacentre/factsheets/fs100.

Wright, Richard T., and Dorothy F. Boorse. *Environmental Science: Toward a Sustainable Future*. 11th ed., Benjamin/Cummings, 2011.

Wu, Chaodong, et al. "Regulation of Glycolysis: The Role of Insulin." *Experimental Gerontology*, vol. 40, 2005, pp. 894-99.

Yaacov, Davidov, and Eduard Jurkevitch. "Predation Between Prokaryotes and the Origin of Eukaryotes." *BioEssays*, vol. 31, 2009, pp. 738-57.

Yadav, Sanu R., et al. "Hepatitis C: Current State of Treatment in Children." *Pediatric Clinics of North America,* vol. 68, no.6, 2021, pp. 1321-31, doi:10.1016/j.pcl.2021.07.008.

Yang, Decheng, editor. *RNA Viruses: Host Gene Responses to Infections.* World Scientific, 2009.

Yang, Y. Tony, and Vicky Debold. "A Longitudinal Analysis of the Effect of Nonmedical Exemption Law and Vaccine Uptake on Vaccine-Targeted Disease Rates." *American Journal of Public Health*, vol. 104, no. 2, 2014, pp. 371-77, search.ebscohost.com/login.aspx?direct=true&db=a9h&AN=93721908. Accessed 12 Mar. 2014.

Yong, Ed. I Contain Multitudes: The Microbes Within Us and a Grander View of Life. Ecco, 2016.

Youdim, Adrienne. "Fiber." *Merck Manual Consumer Version*, Jan. 2020, www.merckmanuals.com/home/disorders-of-nutrition/overview-of-nutrition/fiber. Accessed 21 Nov. 2021.

Young, L. S., and A. B. Rickinson. "Epstein-Barr Virus: Forty Years On." *Nature Reviews: Cancer*, vol. 4, no. 10, pp. 757-68.

Young, Vincent B., Robert A. Britton, and Thomas M. Schmidt. *The Human Microbiome and Infectious Diseases: Beyond Koch. Interdisciplinary Perspectives on Infectious Disease*. Hindawi Publishing Company, 2008, www.hindawi.com/journals/ipid/si/861658/.

Youngdahl, K., et al. *The History of Vaccines*. College of Physicians of Philadelphia, 2013.

Younis, Ramzi T., editor. *Pediatric Sinusitis and Sinus Surgery.* Taylor & Francis, 2006.

Zacharof, Myrto P., and Robert W. Lovitt. "*Lactobacilli*: Their Role and Importance in Contemporary Food and Pharmaceutical Industry, Past-Present-Future." *Swansea University*, U.K. PowerPoint. Accessed 22 Jan. 2016.

Zack, Eric. "Emerging Therapies for Autoimmune Disorders." *Journal of Infusion Nursing*, vol. 37, no. 2, 2012, pp. 109-19.

Zafir-Lavie, I., Y. Michaeli, and Y. Reiter. "Novel Antibodies as Anticancer Agents." *Oncogene*, vol. 28, 2007, pp. 3714-33.

Zariwala, Maimoona A., et al. "Primary Ciliary Dyskinesia." *GeneReviews*, edited by Margaret P. Adam, et al., University of Washington, Seattle, 24 Jan. 2007.

Zelman, Mark, et al. *Human Diseases: A Systemic Approach*. 7th ed., Pearson, 2010.

Zhang, Sarah. "Scientists Get One Step Closer to a Universal Flu Vaccine." *Wired*. Condé Nast, 24 Aug. 2015. Accessed 30 Dec. 2015.

Zhu, Yao, and Xin-Zhu Lin. "Updates in Prevention Policies of Early-Onset Group B Streptococcal Infection in Newborns." *Pediatrics and Neonatology*, vol. 62, no. 5, 2021, pp. 465-75, doi:10.1016/j.pedneo.2021.05.007.

Ziaei, S., A. Kazemnejad, and M. Zareai. "The Effect of Vitamin E on Hot Flashes in Menopausal Women." *Gynecologic and Obstetric Investigation*, vol. 64, 2007, pp. 204-7.

Ziaei, S., et al. "A Randomised Controlled Trial of Vitamin E in the Treatment of Primary Dysmenorrhoea." *BJOG: An International Journal of Obstetrics and Gynaecology*, vol. 112, 2005, pp. 466-69.

Zimmer, Carl. *Parasite Rex: Inside the Bizarre World of Nature's Most Dangerous Creatures*. Free, 2000.

———. *A Planet of Viruses*. 2nd ed., U of Chicago P, 2015.

Zimmer, Carl, Jonathan Corum, and Sui-Lee Wee. "Coronavirus Vaccine Tracker." *The New York Times*, 10 June 2021, www.nytimes.com/interactive/2020/science/coronavirus-vaccine-tracker.html. Accessed 14 June 2021.

Zimmerman, Jerry J., et al. "Cellular Respiration." *Science Direct*. Accessed 18 May 2020.

Zollinger, P. E., et al. "Can Vitamin C Prevent Complex Regional Pain Syndrome in Patients with Wrist Fractures?" *Journal of Bone and Joint Surgery: American Volume*, 89, 2007, pp. 1424-31.

Zucker-Franklin, D., et al. *Atlas of Blood Cells: Function and Pathology*. 3rd ed., Lea & Febiger, 2003.

Glossary

11-cis-retinal: a cofactor of opsins in the retina that are light activated

acid phosphatase: an enzyme that acts to liberate phosphate under acidic conditions

acquired immunity: a subsystem of the immune system that is composed of specialized, systemic cells and processes that eliminate pathogens or prevent their growth

actin: a protein that forms (together with myosin) the contractile filaments of muscle cells, and is also involved in motion in other types of cells

activated sludge: aerated sewage containing aerobic microorganisms which help to break it down

active immunity: immunity resulting from antibody production following exposure to an antigen

Addison disease: a condition characterized by adrenal insufficiency

adenosine triphosphate (ATP): an important biological molecule that represents the energy currency of the cell; the energy in a special high-energy bond in ATP is used to drive almost all cellular processes that require energy

adenovirus vaccine: a nonpathogenic form of an adenovirus that stimulates the formation of a memory immune response to adenovirus infection

adenoviruses: medium-sized, nonenveloped, icosahedral, double-stranded deoxyribonucleic acid (DNA) viruses originally isolated from human adenoid tissue

adhesin: molecules synthesized by microorganisms that allow them to stick to specific surfaces

adjuvant: a substance given with a vaccine that enhances the immune response elicited by it

aerobic: occurring in the presence of oxygen

aflatoxin: a family of toxins produced by certain fungi that are found on crops such as maize (corn), peanuts, cottonseed, and tree nuts; the primary fungi that produce aflatoxins are *Aspergillus flavus* and *Aspergillus parasiticus*, which are abundant in warm and humid regions of the world

agglutinogen: an adhesin used by *Bordetella pertussis* to adhere to the respiratory epithelium

alates: recently molted winged adult termites

alcohol: an organic compound containing a hydroxyl (-OH) group attached to a carbon atom

alleles: one or more variations of a gene that reflects the genes passed from your parents; usually referred to as dominant and recessive genes, while other genes have more than two more alleles

alpha-tocopherol: another name for vitamin E

alveolar macrophages: a type of macrophage, a professional phagocyte, found in the airways and at the level of the alveoli in the lungs but separated from their walls

alveoli: tiny air sacs at the end of the bronchioles where the lungs and the blood exchange oxygen and carbon dioxide during the process of breathing in and breathing out

amino acid: the basic subunit of a protein; there are twenty commonly occurring amino acids, any of which are joined by chemical bonds to form a complex protein molecule

amino group: a functional group containing a nitrogen atom bonded to two hydrogen atoms (-NH2)

aminoglycosides: a group of gram-negative antibacterial medications that inhibit protein synthesis and contain as a portion of the molecule an amino-modified glycoside

ammonification: the conversion of atmospheric nitrogen gas to ammonia and ammonium ion by soil bacteria, particularly *Rhizobium* species

anaerobic: processes that occur in the absence of oxygen

animal passage: when human viruses are grown in cultured animal cells, which decreases their capacity to cause disease in humans, making them suitable vaccine candidates

anthrax: a zoonotic infection caused by *Bacillus anthracis* that comes in an inhaled, cutaneous, and gastrointestinal form

antibiotics: drugs that inhibit or kill bacteria and are used to treat bacterial infections

antibodies: secreted glycoproteins that bind to foreign substances in our bodies and inactivate them, clump them, mark them for destruction, and facilitate their disposal; also known as immunoglobulins

anticoagulant: a chemical that blocks the clotting of blood; some anticoagulants stimulate internal bleeding when ingested by vampire bats and can be used as a method of extermination

antigen presentation: a process by which antigen-presenting cells introduce protein antigens to lymphocytes in the form of short peptide fragments

antigen: a toxin or other foreign substance that induces an immune response in the body, especially the production of antibodies and T lymphocytes

antigenic drift: the gradual accumulation of point mutations during the circulation of influenza virus. It results from the high error rates associated with ribonucleic acid (RNA)-dependent RNA polymerase during virus replication

antigenic shift: the process by which two or more different strains of a virus, or strains of two or more different viruses, combine to form a new subtype having a mixture of the surface antigens of the two or more original strain

antigen-presenting cell: a varied collection of immune cells including dendritic cells, macrophages, and B lymphocytes that process and present antigens for recognition by T lymphocytes

antioxidant: a chemical that prevents the oxidation of other chemicals

antistaphylococcal penicillins: a group of penicillin antibiotics that kill methicillin-sensitive *Staphylococcus aureus* strains and other staphylococci

aplastic anemia: a condition in which the bone marrow stops producing new blood cells

apoptosis: cell death that is programmed as a natural consequence of growth and development through normal cellular pathways or signals from neighboring cells

Archaea: a group of single-celled prokaryotic organisms with distinct molecular characteristics separating them from bacteria and eukaryotes

arthroconidia: asexual fungal spores that are made by the segmentation of preexisting fungal hyphae

arthropod vector: mosquitoes, fleas, sand flies, lice, fleas, ticks, and mites that transmit parasites either by injection into the bloodstream directly via their salivary glands, or forcing parasites into a pool of blood that develops when chewing the skin

artiodactyl: an herbivore that walks on two toes, which have evolved into hoofs

ascorbic acid: a vitamin found particularly in citrus fruits and green vegetables; it is essential in maintaining healthy connective tissue and is also thought to act as an antioxidant; and severe deficiency causes scurvy

ascus: a cylindrical sac in which the spores of ascomycete fungi develop

asexual reproduction: organismal production used by many eukaryotes and all prokaryotes that involves a single parent, resulting in genetically identical offspring; includes binary fission, fragmentation, and budding

asymptomatic: infected but with no discernable symptoms of a disease

atopy: the genetic tendency to develop allergic diseases such as allergic rhinitis, asthma, and atopic dermatitis

ATP: Adenosine 5'-triphosphate, or ATP, is the principal molecule for storing and transferring energy in cells

ATP synthase: an enzyme that catalyzes the synthesis of ADP or makes a phosphoanhydride bond to form ATP from ADP and inorganic phosphate.

atrophy: a decrease in the size or functionality of a body part, cell, organ, or other tissue

attenuated strains: weakened strains of disease-causing bacteria and viruses that are often used as vaccines because they stimulate a protective immune response while causing no or only mild disease

attenuation: the weakening or elimination of the pathogenic properties of a microorganism; ideally, the organism is rendered harmless

autism spectrum disorder (ASD): a collection of neurologically-based developmental disorders in which individuals have impairments in social interaction and communication skills and a tendency to have repetitive behaviors and/or narrow, often-obsessive interests

autoantibody: an antibody produced against tissue antigens within a host-that is, self-antigens

autoimmunity: an abnormal immune reaction against antigens

azoles: antifungal agents that are five-membered heterocyclic compounds containing a nitrogen atom and at least one other noncarbon atom as part of the ring

B lymphocytes: any of the lymphocytes with antigen-binding antibody molecules on the surface that comprise the antibody-secreting plasma cells when mature and that in mammals differentiate in the bone marrow

bacteremia: the presence of bacteria in the blood

bacteria: ubiquitous, mostly free-living organisms often consisting of one biological cell that constitute a large domain of prokaryotic microorganisms; typically, a few micrometers in length, bacteria were among the first life-forms to appear on Earth and are present in most of its habitat

bacteroid: a relatively large, cell-like structure constructed of four to eight individual, albeit enlarged, *Rhizobium* bacteria cells

basidiomycete: a group of higher fungi that have septate hyphae and spores borne on a club-like structure called a "basidium"

basidium: a microscopic club-shaped spore-bearing structure produced by certain fungi

beta-bacteriophage: a virus that infects *Corynebacterium diphtheriae* and transduces the genes to express diphtheria toxin

beta-carotene: an organic, intensely colored red-orange pigment abundant in fungi, plants, and fruits

beta-lactam antibiotics: drugs that inhibit bacterial cell wall biosynthesis and contain a beta-lactam ring in their molecular structure; this group of antibiotics includes penicillin derivatives, cephalosporins and cephamycins, monobactams, carbapenems, and carbacephems

beta-lactamases: enzymes produced by bacteria that provide multi-resistance to ß-lactam antibiotics such as penicillins, cephalosporins, cephamycins, monobactams, and carbapenems

binding assay: an experimental method for selecting one molecule out of several possibilities by specific binding

biochemical oxygen demand *(BOD)*: the amount of dissolved oxygen that must be present in water for microorganisms to decompose the organic matter in the water, used as a measure of the degree of pollution

biofilms: communities of microorganisms encased in polysaccharide-containing materials secreted by those microorganisms

biomass: a mass of organisms; traditionally, this term refers to the biomass of plants and microorganisms

biopsy: an examination of tissue removed from a living body to discover the presence, cause, or extent of a disease

bioreactor: an apparatus for cell growth with practical purposes under controlled conditions. bioreactors are closed systems and vary in size from the small laboratory scale (5 to 10 milliliters) to the large industrial scale (more than 500,000 liters)

biosphere: the regions of the surface, atmosphere, and hydrosphere of the Earth occupied by living organisms

blood clotting: the coagulation of the blood to prevent blood loss after damage to a blood vessel

B-memory cells: descendants of activated B cells that are long-lived and that synthesize large amounts of antibodies in response to subsequent exposure to the antigen, thus playing an important role in secondary immunity

botulinum toxin: a potent neurotoxin that prevents nerves from releasing neurotransmitters

bradyzoite: also called "cystozoites," are the life stage of Toxoplasma gondii found in the tissue cyst and replicate slowly

brewing: beer production by steeping a starch source (usually cereal grains or barley) in water and fermenting it with yeast

brucellosis: a bacterial disease caused by members of the genus *Brucella* that typically affect cattle and buffalo and cause undulant fever in humans

budding: a form of asexual reproduction in which a new individual develops from some generative anatomical point of the parent organism

Burkitt's lymphoma: an aggressive (fast-growing) type of B-cell non-Hodgkin lymphoma that occurs most often in children and young adult

CagA: cytotoxin-associated gene A; a protein toxin made by *Helicobacter pylori* that disrupts cell adhesion junctions between stomach epithelial cells

calcium: An essential element for the proper functioning of the body, including bone formation, muscle and heart contraction, and nervous system functioning

capsid: a virally encoded protein that surrounds and protects the viral RNA or DNA genome

capsule: an outer layer common to many bacteria, made of polysaccharides outside the cell envelope. It is regarded as part of the outer envelope of bacterial cells, is a well-organized layer, not easily washed off, and contributes to disease causation in many pathogenic bacteria

carbon dioxide: a gas composed of molecules with one carbon atom and two oxygen atoms (CO2) that is the main waste product of biological metabolism

carboxyl group: a functional group containing a carbon atom double-bonded to an oxygen atom and single bonded to a hydroxyl group (-OH); has the formula CO2H, typically written -COOH

carboxylic acid: an organic compound that contains the carboxyl (-CO2H) group

carnivore: an animal that eats only animal flesh

carrier: a person infected by an organism who can transmit that organism to other people but who is asymptomatic

carton: cardboard-like material composed of wood fragments, saliva, and fecal matter, used for constructing termite nests

catalyst: a chemical species that initiates or speeds up a chemical reaction but is not consumed in the reaction

cell membrane: a fatty layer made mostly of phospholipids that surrounds cells and delimits them

cell wall: a peptidoglycan layer that surrounds the cell, composed of disaccharides and amino acids that gives bacteria structural support; a rigid structure that surrounds cells and gives them their shape

cellular respiration: a complex series of chemical reactions by which chemical energy stored in the bonds of food molecules is released and used to form ATP

cellulose: a fibrous polysaccharide that chiefly constitutes the cell walls of plants

Centers for Disease Control and Prevention (CDC): a government facility located in Atlanta that coordinates investigations of disease occurrence in the United States

cervical cancer: a type of cancer that occurs in the cells of the cervix-the lower part of the uterus that connects to the vagina

cervicitis: inflammation of the cervix

chancre: a painless ulcer, particularly one developing on the genitals because of venereal disease

chemical energy: the energy locked up in the chemical bonds that hold the atoms of a molecule together; food molecules, such as glucose, contain much energy in their bonds

chemoattractants: molecules that attract cells towards them

chemorepellents: molecules that drive cells away from them

chlorophyll: a green pigment, present in all green plants and in cyanobacteria, responsible for absorbing light to provide energy for photosynthesis with a magnesium atom held in a porphyrin ring.

chloroplast: the organelle in eukaryotes in which photosynthesis is performed by algae and green plants

cholecalciferol: another name for the mature form of vitamin D

cholera toxin: a six-subunit protein secreted by *V. cholerae* once it attaches to the small intestine wall that drives extensive salt and water excretion by the small intestine mucosae

cholera: a gastrointestinal disease characterized by severe, watery diarrhea that causes extensive dehydration and, potentially, death

chorioretinitis: inflammation of the choroid, which is a lining of the retina deep in the eye

chronic: a disease of long duration, typically several weeks, months, or years

cilia: short, microscopic hairlike vibrating structures found in large numbers on the surface of specific cells, either causing currents in the surrounding fluid or, in some protozoans and other small organisms, providing propulsion

cirrhosis: a chronic disease of the liver marked by degeneration of cells, inflammation, and fibrous thickening of tissue, typically a result of alcoholism or hepatitis

clade: a group of organisms that probably evolved from a common ancestor

clinical trial: a scientific evaluation of a drug or vaccine to determine whether it is effective and safe, with several phases involving greater numbers of test subjects in each phase; early phase trials are typically done with lab animals and later ones with human volunteers

clostridia: a group of strictly anaerobic, gram-positive bacteria that cause several diseases and live mainly in soil

clotting factors: several plasma components (such as fibrinogen, prothrombin, thromboplastin, and factor VIII) involved in the clotting of blood

cobalamin: a water-soluble vitamin involved in metabolism that acts as a cofactor in deoxyribonucleic acid (DNA) synthesis, in both fatty acid and amino acid metabolism

colitis: Inflammation of the large intestine (colon), which usually is associated with bloody diarrhea and fever

commensalism: a relationship in which two organisms rely on each other for survival

competitive exclusion: the inevitable elimination from a habitat of one of two different species with identical needs for resources

complement: a system of plasma proteins that can be activated directly by pathogens or indirectly by pathogen-bound antibodies, leading to a cascade of reactions that occurs on the surface of pathogens, killing them and generating components with various effector functions

complementary strand: one of the two strands of nucleotides that make up a DNA molecule, with each nucleotide in one strand corresponding to the position of its complementary nucleotide (cytosine for guanine, adenine for thymine, and vice versa) in the other.

compound microscope: a microscope that uses multiple lenses to enlarge the image of a sample

conidia: a type of asexual reproductive spore of fungi usually produced at the tip or side of hyphae

conjugate: the attachment of molecules that tend to elicit weak immune responses, such as carbohydrates, to large molecules that enhance the immunogenicity of the attached molecules

conjugated antibody: an antibody to which a substrate such as an enzyme, toxin, or inorganic compound has been attached

conjugated vaccines: a type of subunit vaccine which combines a weak antigen with a strong antigen as a carrier so that the immune system has a stronger response to the weak antigen

conjugation: the process by which one bacterium transfers genetic material to another through direct contact

conspiracy theories: attempts to explain social-often random-events as the secret acts of powerful, evil forces. They tend to include mistrust of government or corporate entities

constant region: The highly conserved C-terminal portion of the antibody

contact tracing: also known as partner referral; a process that consists of identifying sexual partners of infected patients, informing the partners of their exposure to disease, and offering resources for counseling and treatment

contagion: the passage of disease from one person to another by direct or indirect contact

contamination: infection of a food item by a pathogen

cord factor: trehalose dimycolate is the primary glycolipid found on the exterior of *Mycobacterium tuberculosis* cells similar species

coryza: runny nose caused by hay fever or infections

COVID-19: coronavirus infectious disease discovered in 2019; the infectious disease caused by (severe acute respiratory syndrome) SARS-CoV-2

creatine phosphate: an energy-containing molecule present in significant quantities in muscle tissue; energy is stored in a high-energy bond like that of ATP

cyanide poisoning: poisoning from exposure to any form of cyanide

cyst: resting, dormant protozoan stages with a protective membrane or thickened wall

cystic fibrosis: a hereditary disorder affecting the exocrine glands. It causes the production of abnormally thick mucus, leading to the blockage of the pancreatic ducts, intestines, and bronchi and often resulting in respiratory infection

cytochrome: compounds consisting of heme bonded to a protein that function as electron transfer agents in many metabolic pathways, especially cellular respiration

cytokines: small proteins, such as interferon, interleukin, and growth factors, that are secreted by certain cells of the immune system and influence other cells

cytolethal distending toxin: a complex protein toxin made by Campylobacter cells that enters intestinal cells, degrades their nuclear deoxyribonucleic acid (DNA), killing them.

cytoskeleton: a microscopic network of protein filaments and tubules in the cytoplasm of many living cells, giving them shape and coherence

cytotoxic T cells: a type of lymphocyte that kills foreign cells, cancer cells, and virus-infected cells

decomposers: microorganisms like bacteria and fungi that degrade plant litter and debris and animal carcasses and return complex biological macromolecules to simpler precursors

decomposition: the process by which bacteria and fungi break dead organisms into their simple compounds

dendritic cells: a special type of immune cell that is found in tissues presents antigens on its surface to other cells of the immune system

denitrification: the chemical reduction of soil nitrates or nitrites by denitrifying bacteria leading to gaseous N losses

deoxyribonucleic acid (DNA): a large molecule formed by two complementary strands of nucleotides that encode all living organisms' genetic information

dermatophyte: a pathogenic fungus that grows on skin, mucous membranes, hair, nails, feathers, and other body surfaces, causing ringworm and related diseases

determinant: a region on the surface of an antigen capable of creating an immune response or of combining with an antibody produced by an immune response

deviated septum: a condition that causes a shift of the bones and cartilage from the middle of the nose to either side, making one side of the nasal passages much smaller than the other

dextran: a complex branched polysaccharide derived from the condensation of glucose

diarrhea: Loose or watery stools, usually a decrease in consistency or increase in frequency from an individual baseline

dietary fiber: also known as roughage; the portion of plant-derived food that human digestive enzymes cannot completely break down

digestion: the degradation of large biological molecules into small precursors

dimorphic: organisms that exist in two distinct forms under different conditions or during different parts of their life cycle

diphtheria toxin: a powerful inhibitor of protein synthesis in eukaryotic cells produced by some *Corynebacterium diphtheria* strains

diphtheria: an upper respiratory infection caused by the bacterium *Corynebacterium diphtheriae*

disinfection: cleaning surfaces or objects with chemicals to kill microorganisms and significantly reduce their numbers.

DNA (deoxyribonucleic acid): *see* deoxyribonucleic acid

domain: the highest-level division of life, sometimes called a "superkingdom"

dorsal root ganglia: a collection of cell bodies of the afferent sensory fibers that lie between adjacent vertebrae

dosage: the size or frequency of a dose of a medicine or drug

dose: a quantity of a medicine, drug, or vaccine taken at a particular time.

dough: a thick, malleable mixture of flour and liquid, used for baking into bread or pastry

downstream: describes the left-to-right direction of deoxyribonucleic acid (DNA) whose nucleotides are arranged in sequence with the 5' carbon on the left and the 3' on the right; the direction of ribonucleic acid (RNA) transcription of a genetic message with the beginning of a gene on the left and the end on the right

DT: diphtheria and tetanus vaccine that is given to children

DTaP: diphtheria, tetanus, and acellular pertussis vaccine that is given to children

dynein: a family of cytoskeletal motor proteins that move along microtubules in cells

dysentery: infection of the intestines resulting in severe diarrhea with the presence of blood and mucus in the feces

edema: a condition characterized by an excess of watery fluid collecting in the cavities or tissues of the body

electron transport chain (ETC): a series of protein complexes that transfer electrons from electron donors to electron acceptors via redox reactions and couples this electron transfer with the transfer of protons across a membrane

elementary body: a resting form of *Chlamydia trachomitis* that infects host cells

ELISA: enzyme-linked immunosorbent assay, an assay that uses an immobilized enzyme linked to an antibody that binds to the ligand of interest to quantitatively measure the quantity of that ligand in a liquid sample

elongation factor 2: also known as EF-2; a critical protein synthesis factor in eukaryotic cells

endarteritis: inflammation of the inner lining of an artery

endemic: describes a disease or pathogen always present or usually prevalent in a particular population or geographical area

endocarditis: inflammation of the endocarditis

endocrine glands: glands that release hormones directly into the bloodstream

endocytosis: a cellular process in which substances are brought into the cell; the material to be internalized is surrounded by an area of the cell membrane, which then buds off inside the cell to form a vesicle containing the ingested material

endoflagella: whip/propeller-like structures found beneath the outer membrane of spirochetes

endosome: a collection of intracellular sorting vesicles in eukaryotic cells

endospores: a dormant, tough, and nonreproductive structure produced by some bacteria

endosymbiosis: a type of symbiotic relationship between two organisms in one organism lives inside the cells of the other organism, usually in a mutualistic relationship that benefits both organisms

endosymbiotic theory: the concept that eukaryotes arose from prokaryotes by incorporating free-living microbes into symbiotic relationships

endothelial cells: the main type of cell found in the inside lining of blood vessels, lymph vessels, and the heart

envelope: lipoprotein outer layer of some viruses derived from the plasma *membrane* of the host cell

enzyme: a biological catalyst that speeds up a chemical reaction without itself being used up; enzymes are made of protein, and a single enzyme can usually only catalyze a single chemical reaction

epidemiology: the branch of medicine that deals with studying the causes, distribution, and control of disease in groups of people

epidermis: the surface epithelium of the skin that overlies the dermis

erythema nodosum: inflammation of the fatty layer of the skin that results in red, painful bumps, usually located on the front of the legs

esophagus: the tube through which food passes from mouth to stomach

ester: the relatively non-water-soluble compound formed when an alcohol reacts with a carboxylic acid

ethanol: also known as ethyl alcohol (C2H5OH), is the primary metabolic product of yeast-based fermentation

eukaryote: organisms made of cells with extensive internal compartmentalization

eukaryotic cell: a cell that contains a nucleus and other membrane-bounded organelles

eutrophication: a process by which the dissolved oxygen in a body of water is consumed by the decomposition of masses of dead algae, to the point that there is no longer enough dissolved oxygen to support aquatic life

exanthem: a widespread rash on the skin that usually occurs in children

excipient: an inactive substance that serves as the vehicle or medium for a drug, vaccine, or other material

extracellular polymer: a polymer like a protein, polysaccharide, glycoprotein, lipids, or polynucleotides secreted by microbes that take up large quantities of water and form a slime layer outside the cell

extreme halophiles: microorganisms that require extremely high salt concentrations for optimal growth

fat-soluble vitamins: vitamins that, because of their structure and solubility, migrate to fatty tissues in the body, where they are stored

fatty acid: an organic compound that is composed of a long hydrocarbon chain with a carboxyl group at one end

Fc-fusion proteins: specific proteins of interest fused to the Fc region of the antibody at its C-terminal end

febrile: having or showing the symptoms of a fever

fecal-oral transmission: a mode of disease transmission in which contaminated *feces* from an infected person or animal are somehow ingested by someone who, subsequently, becomes infected

fermentation: in biology, the metabolic reactions necessary to generate energy in living (mainly microbial) cells; in industry, any large industrial process based on living things is called "fermentation;" a metabolic process that extracts energy from carbohydrates and other organic molecules in the absence of oxygen

fermenter: a type of traditional bioreactor (stirred or nonstirred tanks) where cell fermentation takes place; fermenters can be operated as continuous or batch-culture systems; in continuous culture, nutrients are continuously fed into the fermentation vessel, allowing the cells to ferment indefinitely

filamentous hemagglutinin: a toxin made by *Bordetella pertussis* that aids in its colonizing the upper respiratory tract

fimbriae: small, stiff extensions of bacterial cells that help them adhere to surfaces

finings: chemicals added before alcoholic beverages are bottled that remove undesirable compounds like benzenoid organic compounds or copper ions by precipitating them so that they settle at the bottom of the container

fixed virus: a virus that has been repeatedly cultured in the laboratory so that it has lost its natural variation and is more predictable in experiments

flagellin: the structural protein of bacterial flagella

flagellum (pl: *flagella*): relatively long, delicate, whiplike structures on the surface of cells, used to drive motion

flaviviruses: a group of insect-borne, positive-strand ribonucleic acid (RNA) viruses that includes West Nile virus, Zika virus, dengue virus, and yellow fever virus, among others

flour: a powder obtained by grinding grain, typically wheat, and used to make bread, cakes, and pastry

fluorescence: the visible or invisible radiation emitted by certain substances because of incident radiation of a shorter wavelength such as X-rays or ultraviolet light

fluoroquinolones: broad-spectrum antibiotics that share a bicyclic core structure related to the substance 4-quinolone

focus: the adjustment of a lens to produce a clear image

foodborne infection: a disease caused by eating foods contaminated by infectious microorganisms, with onset occurring within twenty-four hours (e.g., salmonellosis)

foodborne intoxication: a disease caused by eating foods containing microorganisms that produce toxins, with onset occurring within six hours (e.g., botulism)

gametogony: a stage in the sexual cycle of sporozoans in which gametes are formed, often by schizogony

gastric bypass surgery: surgical procedures performed on the stomach or intestines to induce weight loss.

gastritis: inflammation of the lining of the stomach

gastroenteritis: inflammation of the stomach and intestines, typically resulting from bacterial toxins or viral infection and causing vomiting and diarrhea

gel electrophoresis: a method for separation and analysis of macromolecules and their fragments, based on their size and charge

gene: a unit of heredity transferred from a parent to offspring and that determines some characteristic of the offspring; a region of DNA containing instructions for the manufacture of a protein

gene expression: the process by which RNA copies genes, which are specific segments of the DNA molecule, and uses the information to synthesize either proteins or other types of RNA

gene transfer: the movement of fragments of genetic information, whole genes, or groups of genes between organisms

genetically modified organism (GMO): an organism produced by using biotechnology to introduce a new gene or genes, or new regulatory sequences for genes, into it to give the organism a new trait, usually to adapt the organism to a new environment, provide resistance to pest species, or enable the production of new products from the organism

genital herpes: a disease characterized by blisters in the genital area, caused by a variety of the herpes simplex virus

geosphere: the solid part of the planet, including the crust and mantle of the Earth

gestation: the term of pregnancy

Ghon complex: a nonpathognomonic radiographic finding on a chest X-ray that is significant for pulmonary infection of tuberculosis

glomerulonephritis: acute inflammation of the kidney

glycerol: three-carbon alcohol that has one hydroxyl compound on each carbon atom

glycocalyx: a collective term for the biofilm in which the microorganisms are embedded

glycogen: a substance deposited in bodily tissues as a store of carbohydrates; a polysaccharide forms glucose on hydrolysis

glycolysis: the breakdown of glucose by enzymes, releasing energy in the form of ATP and pyruvic acid.

glycoproteins: a class of proteins that have carbohydrate groups attached to the polypeptide chain

gonads: a collective term referring to the testes and ovaries

gonococci: bacteria that cause gonorrhea

grains: wheat or any other cultivated cereal crop used as food

Gram's stain: a laboratory method for tracing the presence of certain bacteria in lung tissue; the procedure involves the observation of different levels of tissue discoloration as specific chemical reactions are induced

gram-negative: a group of bacteria that have a thin peptidoglycan cell wall with an outer membrane extender to it, and stain pink with a Gram stain

gram-positive: a group of bacteria that have a thick peptidoglycan cell wall without an outer membrane and stain purple with a Gram stain

granuloma: an aggregation of macrophages that forms in response to chronic inflammation

group A streptococci: *Streptococcus pyogenes*, a bacterial species commonly found in the throat and on the skin that cause strep throat and impetigo

gumma: a small soft swelling that is characteristic of the late stages of syphilis and occurs in the connective tissue of the liver, brain, testes, and heart

H. influenzae: *Haemophilus influenzae*; a human pathogen that causes meningitis and other significant infections in children and adults

haplotype: a sequential set of genes on a single chromosome inherited together from one parent; the other parent provides a matching chromosome with a different set of genes

Hashimoto's disease: thyroiditis; among the earliest characterized autoimmune diseases

heavy chain: the larger polypeptide chain of an antibody

helper T cells: a class of white blood cells (lymphocytes) derived from bone marrow but matures in the thymus that prompts the production of antibodies by B cells in the presence of an antigen

hemagglutinin: a surface glycoprotein of the influenza virus that causes red blood cells to clump together

hepatitis: liver inflammation

hepatitis A virus: a ribonucleic acid (RNA) virus that causes inflammation of the liver and is transmitted through ingestion of contaminated food and water or direct contact with an infectious person

hepatitis B virus: a deoxyribonucleic acid (DNA) virus that causes hepatitis (inflammation of the liver) and is carried and passed to others through the blood and other body fluid.

hepatocellular carcinoma: the most common type of primary liver cancer

hepatocytes: the majority cell type in the liver

hepatosplenomegaly: enlargement of the liver and spleen

herbivore: an animal that eats only plants

herd immunity: refers to a significant proportion of a population becoming immune to an infectious disease because of exposure or vaccination, decreasing the likelihood of spread

herpes viruses: a large family of double-stranded deoxyribonucleic acid (DNA) viruses that cause infections and specific diseases in animals, including humans

herpetic whitlow: herpes simplex infection of the fingertips or nail beds

heterophile antibodies: unusual antibodies made by B lymphocytes infected with Epstein-Barr virus that bind to antigens on the surfaces of animal red blood cells

heterosaccharide: any saccharide composed of more than one simple sugar

heterotrophic: organisms that cannot produce their food but must acquire it from other organic carbon sources, such as bacterial, plant, or animal matter

histocompatibility leukocyte antigens: cell surface molecules expressed by antigen-presenting cells that present antigenic peptides to T lymphocytes; also known as major histocompatibility complex (MHC) proteins

Histoplasma capsulatum: a pathogenic fungus that initially infects the lungs but can disseminate to other organs

homosaccharide: any saccharide composed of one simple sugar

horizontal transmission: the spread of an infectious agent from one person or group to another, usually through contact with contaminated material, such as sputum or feces

hormone: a substance produced by endocrine glands and delivered by the bloodstream to target cells, producing the desired effect

horseradish peroxidase (HRP): an enzyme found in the roots of horseradish used extensively in biochemistry applications

human leukocyte antigen (HLA): highly polymorphic cell surface proteins that antigen-presenting cells use to present antigen to lymphocytes (class II HLAs) or are found on the surfaces of every nucleated cell in the body (class I HLAs) and act as identification tags by which the immune system distinguishes between self and nonself; also known as major histocompatibility complex (MHC) proteins

human papillomavirus: a virus with subtypes that cause diseases in humans ranging from common warts to cervical cancer

hybridoma: a culture of hybrid cells that results from the fusion of B cells and myeloma cells

hydrocarbon: an organic compound composed of only hydrogen and carbon atoms that does not dissolve in water (water-insoluble)

hydrophilic: "water-loving" or "water-attracting"; a term given to molecules or regions of molecules that interact favorably with water

hydrophobic: "water-hating" or "water-repelling"; a term given to molecules or regions of molecules that do not interact favorably with water

hydrosphere: the total amount of water on a planet that includes water on the surface of the planet, underground, and in the air

hypha (pl: hyphae): a long, branching filamentous structure of a fungus

hypothesis: a proposition made as a basis for reasoning, without any assumption of its truth

hypovolemia: a decreased volume of circulating blood in the body

icosahedron: a polyhedron with twenty faces

IgA protease: an enzyme secreted by *Haemophilus* and *Neisseria* species that degrades immunoglobulin A, the primary antibody that protects mucosal surfaces

IgE: antibodies that mediate the release of chemicals that cause an allergic reaction.

immune surveillance: the monitoring process by which cells of the immune system (such as natural killer cells, cytotoxic T cells, or macrophages) detect and destroy premalignant or malignant cells in the body

immunity: the ability of an immune system to recognize, neutralize, and destroy an infecting organism, protecting the individual from infection

immunizations: introducing into a person a weakened form or a small piece of a disease-causing organism to activate their immune system against it and make them immune to that disease

immunoglobulin: a class of proteins present in the serum and cells of the immune system, which function as antibodies

immunosuppression: a decrease in the effectiveness of the immune system

immunotherapy: a treatment that uses a person's immune system to fight cancer

in vitro: a term used to indicate reactions and processes that are carried out "in glass," or in the laboratory environment rather than in nature

incidence: probability or risk of contracting a disease within a population

inflammation: a localized physical condition in which part of the body becomes reddened, swollen, hot, and often painful, especially as a reaction to injury or infection

infusion: a method of putting fluids, including drugs, into the bloodstream; also called "intravenous infusion"

inoculation: the introduction of a substance or group of substances into a living organism

insertion sequence: a small, independently transposable genetic element

insulin: hormone released from the pancreas

integrase: a virally encoded enzyme that catalyzes the integration of viral double-stranded DNA into the host genome

interleukins: a class of glycoproteins produced by leukocytes for regulating immune responses

internalin: surface proteins found on *Listeria monocytogenes* that exist in two known forms, InlA and InlB, and are used by the bacteria to invade mammalian cells via cadherins transmembrane proteins and Met receptors, respectively

intestines: The tube connecting the stomach and anus in which nutrients are absorbed from food; divided into the small intestine and the colon, or large intestine

ischemia: an inadequate blood supply to an organ or part of the body

isotype switching: when activated B cells, by their interaction with T-helper cells change the type of antibody they are secreting to some other antibody subtype

isotypes: the different classes of antibodies

kala-azar: a form of the disease leishmaniasis marked by emaciation, anemia, fever, and enlargement of the liver and spleen

Kaposi sarcoma: a type of cancer in which lesions (abnormal areas) grow in the skin, lymph nodes, lining of the mouth, nose, and throat, and other tissues of the body

keratin: a fibrous protein forming the main structural constituent of hair, feathers, hoofs, claws, horns

keratoconjunctivitis: inflammation of the cornea and conjunctiva of the eye

Koplik spots: small, white spots on the inside of the cheeks early during measles

lactate dehydrogenase: an enzyme that catalyzes the reduction of pyruvate to lactic acid

lactic acid bacteria: a group of gram-positive, acid-tolerant, generally nonsporulating, nonrespiring, either rod-shaped or spherical bacteria that commonly ferment sugars to lactic acid

laser: a device that generates an intense beam of coherent monochromatic light (or other electromagnetic radiation) by stimulated emission of photons from excited atoms or molecules

latent cycle: a type of herpes simplex infection in neurons that does not include destruction of the cell but dormancy

latent phase: an infective phase of Epstein-Barr virus, usually occurring in B lymphocytes, in which the virus infects the cells but does not destroy them

leaven: a substance, typically yeast, that is used in dough to make it rise

light chain: the smaller subunits of an antibody

lipid nanoparticles: a shell of cholesterol and other fat-soluble molecules that house an internal messenger ribonucleic acid (mRNA) core and fuse with host cells to deliver the mRNA payload

lipooligosaccharides: glycolipids found in the outer membrane of some types of gram-negative bacteria, such as *Neisseria* and *Haemophilus* species

lipopolysaccharide: large molecules consisting of a lipid and a polysaccharide composed of O-antigen, outer core, and inner core joined by a covalent bond; they are found in the outer membrane of gram-negative bacteria

listeriolysin O: a hemolysin produced by the bacterium *Listeria monocytogenes*

live attenuated vaccines: bacterial and viral strains that have been weakened so that, upon injection into a person with a healthy immune system, these strains elicit robust and protective immune responses against them

luminal agents: drugs that kill the amoebae but are not absorbed by the gastrointestinal tract and, therefore, kill all unattached parasites

lymph nodes: small bean-shaped structures that are part of the immune system, filter substances that travel through the lymphatic fluid, and they contain lymphocytes (white blood cells) that help the body fight infection and disease

lymphocyte: a small white blood cell constituting about 25 percent of all blood cells; two basic types are B cells (antibody production) and T cells (cellular immunity)

lymphocytosis: an increase in the number or proportion of lymphocytes in the blood.

lymphohistiocytic vasculitis: an inflammation of blood vessels caused by lymphocytes and macrophages

lymphoma: cancer that begins in infection-fighting cells of the immune system, called "lymphocytes"

lymphopoiesis: the process of B-lymphocyte development

lysosome: a membrane-bound cell organelle involved with various cell processes that contains digestive enzymes

lysozyme: an enzyme that catalyzes the destruction of the cell walls of certain bacteria

lytic cycle: the infective cycle of herpes simplex viruses in epithelial cells that includes the destruction of the host cells

lytic phase: an infective phase of Epstein-Barr virus, usually in oropharyngeal epithelial cells, in which the virus reproduces in the cells and destroys them by lysing them

macrolides: a class of antibiotics that inhibit bacterial protein synthesis that consists of a large macrocyclic lactone ring

macronutrients: materials ingested in large amounts to supply the energy and materials for physical bodies

macrophages: large phagocytic cells found in stationary form in the tissues or as mobile white blood cells, especially at sites of infection

macular rash: a rash consisting of flat lesions up to five millimeters in diameter

maculopapular rash: a rash with a mixture of macules and papules

major histocompatibility complex (MHC) proteins: cell surface proteins that come in two types, class I and II that help antigen-presenting cells present antigen to lymphocytes (class II) and mark every nucleated cell in the body (class I); also called "human leukocyte antigens" (HLAs)

malaria: an infectious disease caused by protozoan parasites from the *Plasmodium* family that can be transmitted by the bite of the Anopheles mosquito or by a contaminated needle or transfusion

malolactic fermentation: a process in winemaking in which tart-tasting malic acid is converted to the softer tasting lactic acid

mast cells: A type of white blood cell found in connective tissues all through the body, especially under the skin, near blood vessels and lymph vessels, in nerves, and the lungs and intestines

matrix: the layer of virally encoded protein that surrounds the viral capsid

measles: an infectious viral disease that causes fever and a red rash on the skin, typically occurring in childhood

mediastinum: the area between the lungs

megadose: ten or more times the recommended daily allowance of a nutrient

megaloblastic anemia: a condition in which the bone marrow produces unusually large, structurally abnormal, immature red blood cells called "megaloblasts"

meiosis: a type of cell division that results in four daughter cells, each with half the number of chromosomes of the parent cell, as in the production of gametes and plant spores

membranous glomerulonephritis (MGN): when the small blood vessels in the kidney (glomeruli) that filter wastes from the blood become damaged and thickened, causing proteins to leak from the damaged blood vessels into the urine (proteinuria)

meninges: the three membranes (the dura mater, arachnoid, and pia mater) that line the skull and vertebral canal and enclose the brain and spinal cord

meningococci: bacteria that cause some forms of meningitis and cerebrospinal infection

meningoencephalitis: inflammation of the membranes of the brain and the adjoining cerebral tissue

merozoites: a small amoeboid sporozoan trophozoite

mesotheliomas: a cancer of mesothelial tissue in the lungs, associated especially with exposure to asbestos

metastasis: the spread of cancers from their site of origin to other places in the body

methane: a chemically simple gas that consists of one carbon atom bound to four hydrogen atoms (CH4)

methanogens: methane-producing bacteria, especially archaeans that reduce carbon dioxide to methane

methicillin-resistant Staphylococcus aureus: also known as MRSA, a genetically distinct strain of *Staphylococcus aureus* that causes several difficult-to-treat infections in humans

methylation: the attachment of methyl (-CH3) groups to molecules, usually proteins.

microconidia: an asexual resting spore made by the mycelium form of Histoplasma capsulatum

micronutrients: substances of which only milligrams are needed in the daily diet, such as vitamins and minerals

microorganism: an organism that is too small to be seen with the naked eye

microtubules: a microscopic tubular structure present in numbers in the cytoplasm of cells, sometimes aggregating to form more complex structures

minerals: highly stable materials composed of inorganic materials, generally found as rock but often crystalline in structure

minus-sense RNA: a noncoding RNA strand that an *RNA*-dependent *RNA* polymerase must copy to produce a translatable *mRNA*

mitochondria: an organelle found in large numbers in most cells, in which the biochemical processes of respiration and energy production occur. It has a double membrane; the inner layer being folded inward to form layers (cristae)

mold: a fungus that grows in the form of multicellular filaments called "hyphae"

Monera: a biological kingdom that is made up of prokaryotes

monoclonal antibodies: antibodies with one particular target that have been generated in large quantities from a single hybrid parent cell formed in a laboratory

monocytes: a large phagocytic white blood cell with a simple oval nucleus and clear, grayish cytoplasm that, when activated, becomes a macrophage

mononucleosis: an infectious viral disease characterized by swelling of the lymph glands and prolonged fatigue

monosaccharide: any class of sugars (e.g., glucose) that cannot be hydrolyzed to give a simpler sugar

Monospot test: a rapid test for infectious mononucleosis due to Epstein-Barr virus that detects heterophilic antibodies

motor neurons: a nerve cell forming part of a pathway along which impulses pass from the brain or spinal cord to a muscle or gland.

mRNA: messenger ribonucleic acid; an RNA molecule translated by ribosomes to synthesize a protein

mucosa: the moist, inner lining of some organs and body cavities (such as the nose, mouth, lungs, and stomach)

multigenic: referring to a trait or characteristic that requires the product of more than one gene to be expressed

mumps: a contagious and infectious viral disease that causes swelling of the parotid salivary glands in the face and risk of sterility in adult males

mutualism: a symbiotic association between organisms of two different species in which each benefit

mycelium: a body of the fungal organism that consists of a collection of hyphae

mycetoma: a progressive and chronic fungal or bacterial infection that causes overgrowth of the infected tissue and the formation of sinuses filled with the infecting organism

mycobiont: the fungal partner in a plant-fungus symbiosis

mycoplasma: a group of small typically parasitic bacteria that lack cell walls and sometimes cause diseases

mycosis: a disease caused by any fungus that invades the tissues, causing superficial, subcutaneous, or systemic disease

myeloperoxidase: a peroxidase enzyme that is most abundant in neutrophil granulocytes and produces hypohalous acids to carry out their antimicrobial activity

myocarditis: inflammation of the heart muscle

nanoparticles: a particle of matter that is between 1 and 100 nanometers in diameter

nasal polyps: noncancerous growths inside the nose; usually associated with allergies or asthma, which can block the sinus drainage tract

neuraminidase: an enzyme on the surface of influenza viruses that catalyzes the breakdown of complex sugars that contain neuraminic acid.

neurotransmitter: small, bioactive amines released by a nerve fiber at the arrival of a nerve impulse that diffuses across the synapse or junction and causes the transfer of the impulse to another neuron, muscle, or other structure

neutrophil: the most abundant type of granulocytes that make up 40 to 70 percent of all white blood cells in humans and form an essential part of the innate immune system

next-generation sequencing (NGS): high-throughput approaches deoxyribonucleic acid (DNA) sequencing using the concept of massively parallel processing

nicotinamide adenine dinucleotide (NAD): a molecule used to hold pairs of electrons when they have been removed from a molecule by some biological process; the empty molecule is denoted by NAD+, while it is denoted as NADH when it is carrying electrons

nitrification: the process that converts ammonia to nitrite and then to nitrate

nitrogen fixation: any natural or industrial process that causes free nitrogen (N2), which is a relatively inert gas plentiful in air, to combine chemically with other elements to form more-reactive nitrogen compounds such as ammonia, nitrates, or nitrites

nontypeable: *H. influenzae* strains that do not make a capsule

normal flora: a diverse microbial flora is associated with the skin and mucous membranes of every human being from shortly after birth until death

nosocomial infections: infections acquired during the process of receiving health care that was not present during the time of admission

nosocomial infection: an infection acquired at a hospital or that originated at a hospital

nucleoprotein: a virally encoded protein that is directly associated with the viral nucleic acid

nucleotide: the fundamental structural component of DNA and RNA, consisting of a ribose (in RNA) or deoxyribose (in DNA) sugar molecule bonded to a phosphate group and one of five nucleobases: cytosine, adenine, guanine, thymine (DNA only), or uracil (RNA only)

nutrient: a nourishing food ingredient

objective lens: the lens on the microscope closest to the sample

omnivore: an animal that eats both plants and animals

oncogenes: genes that have the potential to cause cancer if overexpressed or mutated

oncogenic: having the potential to cause cancer, as in oncogenic retroviruses

oocyst: a cyst containing a zygote formed by a parasitic protozoan

operator: a segment of DNA where the repressor binds, preventing the transcription of specific genes

opt-outs: express instructions provided to refuse to participate in a service, program, process, or contract

orbit: the bones and other tissues that surround the eye, commonly known as the eye socket

organelles: subcellular membrane-bounded compartments that perform specific functions within the eukaryotic cell

organic material: a compilation of molecules made predominantly of carbon but also contains varying quantities of nitrogen, oxygen, and hydrogen

orthomyxoviruses: a family of single-stranded ribonucleic acid (RNA) viruses that have a spherical or filamentous virion with numerous surface projections of glycoprotein and include the causative agents of influenza

osteoporosis: a medical condition in which the bones become brittle and fragile from loss of tissue, typically because of hormonal changes, or deficiency of calcium or vitamin D

oxidation: the loss of electrons by molecules, decreasing its negative charge or increasing its positive charge

Panton-Valentine leucocidin: a toxin produced by some strains of *Staphylococcus aureus* that destroys white blood cells and damaged tissues

papilloma: a small wartlike growth on the skin or a mucous membrane, derived from the epidermis and usually benign

papular rash: a rash consisting of raises bumps up to one millimeter in diameter

paralysis: the loss of the ability to move (and sometimes to feel anything) in part or most of the body, typically because of illness, poison, or injury

parasite: an organism that lives on another organism (the host) and causes harm to the host while it benefits

parasitism: the relationship between two species of plants or animals in which one benefits at the expense of the other

parathyroid gland: a gland next to the thyroid which secretes a hormone (parathyroid hormone) that regulates calcium levels in a person's body

parotitis: inflammation of a parotid gland, especially

passage: one of the culture steps in the production of a fixed virus

passive immunity: immunity resulting from the introduction of preformed antibodies

pasteurization: a method first described by the French scientist Louis Pasteur for killing bacteria in food and beverages

patent ductus arteriosus: an opening between two blood vessels leading from the heart

pathogen: any disease-causing organism, including a virus, bacterium, protozoan, mold or yeast, or other parasites

pathogenic: able to cause disease in living organisms

pedosphere: the soil mantle of the Earth

penicillin-binding proteins: enzymes involved in peptidoglycan biosynthesis and contribute essential roles in bacterial cell wall biosynthesis

peptic ulcer: an erosion of the stomach lining that may bleed or perforate

peptide: molecule formed by linking two to several dozen amino acids

peptide bond: a covalent bond that links the carboxyl group of one amino acid to the amine group of another, enabling the formation of proteins and other polypeptides

peptidoglycan: a polymer consisting of sugars and amino acids that forms a mesh-like peptidoglycan layer outside the plasma membrane of most bacteria that forms the cell wall

peripheral nerves: one of two components that make up the nervous system; one being the central nervous system and the other, the peripheral nervous system consisting of the nerves and ganglia outside the brain and spinal cord

peristalsis: The wavelike muscular contractions that move food and waste products through the intestines; problems with peristalsis are called "motility disorders"

pertactin: an outer membrane protein of *Bordetella pertussis* that promotes adhesion to respiratory epithelial cells

pertussis: whooping cough, a disease caused by the bacterium *Bordetella pertussis*
Peyer's patches: clusters of subepithelial, lymphoid follicles found in the intestine
phagocytes: white cells of the immune system that destroy invading foreign bodies by engulfing and digesting them in a nonspecific immune response; include macrophages and neutrophils

phagocytosis: a cellular process for ingesting and eliminating particles larger than 0.5 μm in diameter, including microorganisms, foreign substances, and apoptotic cells

phagolysosome: or endolysosome, is a cytoplasmic body formed by the fusion of a phagosome with a lysosome in a process that occurs during phagocytosis

phagosome: a vacuole in the cytoplasm of a cell, containing a phagocytosed particle enclosed within a part of the cell membrane

pharynx: the hollow tube inside the neck that starts behind the nose and ends at the top of the trachea (windpipe) and esophagus (the tube that goes to the stomach)

phase variation: how bacteria deal with rapidly varying environments without random mutation by varying gene expression, often in an on-off fashion, within different parts of a bacterial population.

phenol oxidase: An enzyme in *Cryptococcus* species that catalyzes the conversion of phenolic compounds, including catecholamines, such as dopamine and epinephrine, to melanin.

pheromone: a chemical substance produced by an animal that usually elicits certain behavioral responses in other animals of the same species

phosphate: chemically, a complex ion consisting of a phosphorus atom to which four oxygen atoms are bonded and carrying two negative charges termed inorganic phosphate, a vital component of certain energy-rich compounds such as adenosine triphosphate (ATP), and a major component of the structures of the nucleic acids

phospholipids: lipid molecules with phosphate in their structure that are the main component of cell membranes

photoautotrophy: the ability of algae, plants, and other microorganisms to derive energy directly from sunlight through the process of photosynthesis and assimilate carbon from atmospheric carbon dioxide

photolysis: the splitting of molecules into component atoms and parts of molecules through the action of light energy

pili: short, hairlike structures on the cell surface of prokaryotic cells that may assist in movement but are more often involved in adherence to surfaces, which facilitates infection

plasma cells: descendants of activated B cells that synthesize and secrete a single antibody type in large quantities and play an essential role in primary immunity

Plasmodium: a parasitic protozoan of a genus that includes those causing malaria

plus-sense RNA: a single-stranded RNA virus, a plus-strand is one having the same polarity as viral mRNA and containing codon sequences that can be translated into viral protein

Pneumocystis pneumonia: a form of pneumonia caused by the single-celled parasite *Pneumocystis jirovecii*; dangerous primarily to persons with impaired immunity mechanisms, particularly victims of acquired immunodeficiency syndrome (AIDS)

poliomyelitis: an infectious viral disease that affects the central nervous system and can cause temporary or permanent paralysis

poliovirus: the causative agent of polio, is a serotype of the species Enterovirus C in the family Picornaviridae

polyarteritis nodosa: a rare multisystem disorder characterized by widespread inflammation, weakening, and damage to small and medium-sized arteries

polymerase chain reaction: a widely used method to rapidly make millions to billions of copies of a specific deoxyribonucleic acid (DNA) sample, allowing scientists to take a very small sample of DNA and amplify it to a large enough amount to study in detail

polymorphic: genes that exist in multiple forms or alleles within a population; genes that show extensive variability within a population

polypeptide: a linear molecule composed of amino acids joined together by peptide bonds; all proteins are functional polypeptides

polysaccharide: a large carbohydrate molecule made of many smaller monosaccharides

porins: pore proteins in the outer membrane of gram-negative bacteria that mediate the diffusion of small hydrophilic molecules into cells

positive-sense RNA (+RNA): an RNA molecule that can serve as a template for protein synthesis

postpolio syndrome: a disorder of the nerves and muscles that occurs in some people many years after they have had polio

postulate: a thing suggested or assumed as true as the basis for reasoning, discussion, or belief

Pott disease: tuberculosis spondylitis; a rare infectious disease of the spine caused by an extraspinal infection that involves multiple vertebrae, causing osteomyelitis and arthritis

Precambrian: the interval of geologic time from the formation of the earth (4.6 billion years ago) to the beginning of the Cambrian period (544 million years ago)

prevalence: the proportion of infectious cases in a population at a given time

primary antibody: an antibody that binds directly to the target; the variable region of the primary antibody recognizes an epitope on the target; it is produced by a host organism that is of a different species than the specimen

probiotics: microorganisms and substances that promote the development of healthy intestinal microbial communities

prokaryotes: microorganisms that lack internal compartmentalization (i.e., their cells lack an organized nucleus and other internal organelles)

promoter: a DNA sequence needed to turn a gene on or off where the transcription is initiated

protease: an enzyme that breaks down proteins and peptides

protein M: a virulence factor on the surface of certain species of *Streptococcus* pyogenes

protein: a biological polymer consisting of one or more long chains of amino acids linked by peptide bonds in a sequence specified by an organism's deoxyribonucleic acid (DNA)

protozoan: single-celled eukaryotes, either free-living or parasitic, that feed on organic matter such as other microorganisms or organic tissues and debris

provitamin: a substance that may be converted within the body to a vitamin

pseudomembrane: a layer of exudate resembling a membrane that forms on the surface of the skin or mucous membrane.

pseudomonads: a genus of gram-negative, gamma-proteobacteria that contain 191 validly described species, have tremendous metabolic diversity, and can colonize a wide range of niches

pseudomycelium: a cellular association occurring in various higher bacteria and yeasts in which cells cling together in chains resembling small true mycelia

pseudopod: a temporary arm-like projection of a eukaryotic cell membrane that is developed in the direction of movement

pulmonary consolidation: the presence of exudate in the airways and alveoli, usually because of infection

pulmonary nodules: small, round growths on the lung that contain either trapped microorganisms or cancer cells

pulsed-field gel electrophoresis (PFGE): a laboratory technique used for the separation of large DNA molecules by applying to a gel matrix an electric field that periodically changes direction

pyoverdine: a green pigment made by *Pseudomonas aeruginosa* that also acts and an iron binder and iron transport molecule

pyruvic acid: the product of the metabolic pathway known as glycolysis in which glucose is degraded in a series of enzyme-catalyzed reactions that yields energy, and pyruvic acid

quorum-sensing: the regulation of gene expression in response to fluctuations in cell-population density

rabies: a contagious and fatal viral disease of dogs and other mammals that causes madness and convulsions, transmissible through the saliva to humans

Ranke complex: the combination of late fibrocalcific lesions of the lung and lymph node, which evolved from the Ghon complex

rapid antigen tests: point-of-care testing that directly detects the presence or absence of an antigen

rash: an area of irritated or swollen *skin* characterized by changes in the color, feeling, or texture of *skin* that might be itchy, red, painful, and irritated

receptor-mediated endocytosis: a process by which cells absorb metabolites, hormones, proteins-and in some cases viruses-by the inward budding of the plasma membrane

recombinant DNA: deoxyribonucleic acid (DNA) molecules that have been engineered to include genes from the DNA of different species

recombinase: the *RAG-1* and *RAG-2* protein complex that cuts and paste gene segments from the antibody gene cluster together to form unique antibodies that bind a wide range of antigens

recommended daily (or dietary) allowance (RDA): the in-take levels of the essential nutrients that are considered adequate to meet the known nutritional needs of most healthy persons

reduction: the gain of electrons by molecules, increasing their negative charge or decreasing their positive charge

relapse: the recurrence of signs and symptoms after they have subsided or ceased

Renshaw cells: inhibitory interneurons located in the ventral cord and through their localized connections with motor neurons and other interneurons help to ensure a balance between contraction of synergist and antagonist muscles

reovirus: a group of double-stranded RNA viruses associated with respiratory and enteric infection

replication: the reproduction of a virus; many copies are made within a host cell, then released to infect other host cells

reproductives: sexually mature male and female

reservoir: the host species in which a parasite is maintained in each area and from which it may infect other species, initiating an epidemic

resolution: the minimum distance at which two distinct points of a specimen can still be seen as separate entities; the degree of detail visible in a microscopic image

restriction enzyme: a protein or protein complex that coordinates to a specific gene or region of the DNA molecule and excises that segment from the molecule

reticulate body: the metabolically active form of *Chlamydia trachomitis* that the elementary body transforms into after penetrating a host cell that appropriates the host cell resources to divide and form elementary bodies, which culminates in host cell lysis and liberation of elementary bodies for future infective cycles

reverse transcriptase: an enzyme encoded from the genetic material of retroviruses that catalyzes the transcription of retrovirus RNA (ribonucleic acid) into DNA (deoxyribonucleic acid)

Reye's syndrome: a rare but serious condition characterized by the swelling in the liver and brain, caused when children or adolescents take aspirin or other nonsteroidal anti-inflammatory drugs while infected by viruses

ribonucleic acid (RNA): a nucleic acid (chain of nucleotides) that serves various functions concerning deoxyribonucleic acid (DNA), including protein synthesis and gene expression, and regulation

ribosome: a cytoplasmic organelle that serves as the site for amino acid incorporation during the synthesis of protein

RNA polymerase: an enzyme that catalyzes the joining of ribonucleotides to make RNA using deoxyribonucleic acid (DNA) or another RNA strand as a template

RNase: an enzyme that catalyzes the cutting of an RNA molecule

rotavirus: a group of ribonucleic acid (RNA) viruses, some of which cause acute enteritis in humans

rubella: a contagious viral disease with symptoms like mild measles; it can cause fetal malformation if contracted in early pregnancy

rubisco: a short-form name for the enzyme ribulose bisphosphate carboxylase/oxygenase, reportedly the most abundant protein on earth

Saccharomyces cerevisiae: also known as "the baker's yeast, the primary yeast species used to ferment the juice of grapes and grains into wine and beer

S-adenosylmethionine: an essential cofactor in the body made from the amino acid methionine and adenosine that participates in single carbon transfer reactions

Salmonella: a genus of rod-shaped gram-negative bacteria of the family Enterobacteriaceae

sandflies: a small hairy biting fly of tropical and subtropical regions that transmits several diseases, including leishmaniasis.

saponification: a reaction in which a strong basic solution splits a molecule into a carboxylic acid unit and an alcohol unit

saprophytes: an organism that feeds on nonliving organic matter known as detritus at a microscopic level

SARS-CoV-2: a coronavirus that infects animals and humans and causes COVID-19

schizogony: asexual reproduction by multiple fission, found in some protozoa, especially parasitic sporozoans

screening procedures: tests carried out in populations that are usually asymptomatic and at high risk for a disease to identify those in need of treatment

secondary antibody: an antibody that binds to the primary antibody-target complex to capture the complex and to deliver a means of detecting the complex

selection: the process by which developing immune system cells are either allowed to continue to maturation or destroyed before they can enter the circulation

semisynthetic: about compounds that are obtained by altering or augmenting the molecular structure of a compound obtained from a natural source

sepsis: a serious condition resulting from the presence of harmful microorganisms in the blood or other tissues and the body's response to their presence, potentially leading to the malfunctioning of various organs, shock, and death

serotypes: a distinct variation within a species of bacteria or virus that is serologically distinguishable

sewage: wastewater and excrement conveyed in sewers.

sexually transmitted disease: an infection caused by organisms transferred through sexual contact (genital-genital, oral-genital, oral-anal, or anal-genital); the transmission of infection occurs through exposure to lesions or secretions that contain the organisms

Shiga toxin: a toxin that inhibits protein synthesis in host cells produced by the bacterium *Shigella dysenteriae*, or a similar one produced by *E. coli*, that causes dysentery in humans

siderophore: small, high-affinity iron-binding compounds secreted by microorganisms such as bacteria and fungi

simple microscope: essentially a magnifying glass made of a single convex lens with a short focal length that magnifies the object through angular magnification

small subunit ribosomal RNA (ssu rRNA): the ribonucleic acid (RNA) molecule found in the small subunit of the ribosome; also called 16S rRNA (in prokaryotes) or 18S rRNA (in eukaryotes)

SNARE proteins: molecular motors that drive the biological fusion of two membranes

spherules: a thick-walled, spherical structure that is the tissue-specific form of *Coccidioides* species

spike protein: the main glycoprotein embedded in the envelope of SARS-CoV-2 that binds to the ACE2 protein on host cells and initiates infection

spinal cord: a column of nerve tissue that runs from the base of the skull down the center of the back

spirochetes: a flexible spirally twisted bacterium, especially one that causes syphilis

spores: a reproductive cell capable of developing into a new individual without fusion with another reproductive cell; resting cells made by molds

sporozoites: a motile spore-like stage in the life cycle of some parasitic sporozoans (e.g., the malaria organism) that is typically the infective agent introduced into a host

spotted fever group rickettsioses: a group of diseases caused by closely related bacteria spread to people through the bites of infected ticks and mites

src: a gene found in Rous Sarcoma Virus that confers on the virus the ability to transform normal cells into cancer cells

starch: an odorless, tasteless white substance that occurs widely in plant tissue and is obtained chiefly from cereals and potatoes; it is a polysaccharide that functions as a carbohydrate store and is an important constituent of the human diet

stem cells: multipotential precursor cells within the bone marrow that develop into white cell populations, including lymphocytes and phagocytic cells

sterilization: subjecting objects or surfaces to chemicals or processes that rid them of any living microorganisms

stool: The waste products expelled from the body through the anus during defecation; feces

stratum corneum: the horny outer layer of the skin

street virus: a virus derived directly from a natural source; a fixed virus is produced from a street virus by several passages through an artificial culture system

Streptococcus pneumoniae: commonly referred to as pneumococcus; the main bacteria responsible for pneumonia

subacute sclerosing panencephalitis (SSPE): a very rare but fatal central nervous system disease caused by measles virus infection

submucosa: the layer of areolar connective tissue lying beneath a mucous membrane

substrate: a molecule that is broken down by fermentation

subunit vaccine: a vaccine that contains purified parts of the pathogen that are antigenic or necessary to elicit a protective immune response

sulfate: a salt or ester of sulfuric acid, containing the anion SO42- or the divalent group -OSO2O-

sulfide: an inorganic anion of sulfur with the chemical formula $S^{2?}$ or a compound containing one or more $S^{2?}$ ions

superantigens: a class of antigens that result in excessive activation of the immune system

suppuration: the formation or discharge of pus

sylvatic rabies: rabies in wild animal populations (as opposed to rabies in domestic animals and pets)

symbiont: one of the partnering species in a symbiotic association

symbiosis: any of several living arrangements between members of two different species, including mutualism, commensalism, and parasitism

synthetic: about compounds that are produced entirely by synthesis reactions carried out in the laboratory from simple starting materials

systemic lupus erythematosus (SLE): commonly called "lupus"; a chronic inflammatory disease characterized by an arthritic condition and a rash

T cells: *see* T lymphocytes

T lymphocytes: any of several lymphocytes (such as a helper T cell) that differentiate in the thymus, possess particular cell-surface antigen receptors and include some that control the initiation or suppression of cell-mediated and humoral immunity (as by the regulation of T- and B-cell maturation and proliferation) and others that lyse antigen-bearing cells

tachyzoite: relatively faster-growing, actively multiplying, and invasive cell type in the life cycle of Toxoplasma gondii

T-cell receptor: a T-lymphocyte-specific cell surface protein that recognizes and binds specific antigens

T-cytotoxic cell: a subtype of T lymphocyte that has the CD8 cell surface glycoprotein, and attacks and destroys tumor cells and viral-infected cells

Td: tetanus and diphtheria vaccine that is given to adolescents and adults

Tdap: combined tetanus, diphtheria, and acellular pertussis vaccine that is given to adolescents and adults

teleomorph: the sexual reproductive stage (morph), typically a fruiting body

tetanospasmin: an extremely potent neurotoxin produced by the vegetative cell of *Clostridium tetani* in anaerobic conditions, causing tetanus

tetanus: an infection caused by the bacterium *Clostridium tetani*

T-helper cell: a subtype of T lymphocyte that secretes signaling molecules called "lymphokines" that induce the maturation and activation of B lymphocytes, macrophages, and T-cytotoxic cells

thermal death point: the lowest temperature that can destroy a foodborne organism

thermal dimorphism: a unique group of fungi that respond to shifts in temperature by converting between hyphae (22 to 25° Celsius) and yeast (37° C). This morphologic switch, known as the phase transition, defines the biology and lifestyle of these fungi.

thimerosal: a mercury-based preservative used in some vaccines beginning in the 1930s

thrombocytopenia: deficiency of platelets in the blood. This causes bleeding into the tissues, bruising, and slow blood clotting after injury

thymus: a lymphoid organ that lies just over the upper part of the heart and produces mature T cells for the immune system

tinea: a type of ringworm infection

tissue cysts: a closed sac, having a distinct envelope surrounding it within infected tissues

tissue sectioning: cleanly and consistently cutting paraffin-embedded or frozen *tissue* into thin slices

tolerance: the state in which an organism does not normally react against its tissue

tonsillitis: inflammation of the tonsils

toxic shock syndrome toxin: a superantigen produced by some *Staphylococcus aureus* and *Streptococcus pyogenes* isolates that causes toxic shock syndrome

toxin coregulated pilus: a threadlike extension of the *V. cholerae* cell surface that helps the microorganisms attached to the small intestinal wall

toxoid: a toxin that has been chemically treated to eliminate its toxic properties but that retains the same antigens as the original

trace elements: elements needed in the diet at levels of less than 100 milligrams per day

transcription: the process of copying a segment of DNA into RNA

translation: the process of forming proteins according to instructions contained in an RNA molecule

transposons: mobile genetic elements that may be responsible for the movement of genetic material between unrelated organisms

trichothecene: any of several mycotoxins that are produced by various fungi (such as genera *Fusarium* and *Trichothecium*) and that include some contaminants of livestock feed, and some held to be found in yellow rain

trigeminal ganglion: the large flattened sensory root ganglion of the trigeminal nerve that lies within the skull and behind the orbit

tRNA: an RNA molecule that attaches to an amino acid and interacts with the ribosome during protein synthesis; in retroviruses, it serves as a primer for reverse transcriptase

trophic factors: helper molecules that allow a neuron to develop and maintain connections with its neighbors

trophozoite: the activated, feeding stage in the life cycle of certain protozoa

trypanosomes: a single-celled parasitic protozoan with a trailing flagellum, infesting the blood

tsetse fly: an African blood-sucking fly that bites humans and other mammals; the tsetse fly transmits sleeping sickness and nagana

tumor suppressor genes: genes that make proteins that help control cell growth

tumors: abnormal benign or malignant new growths of tissue that possesses no physiological function and arises from uncontrolled, usually rapid cellular proliferation

type IV pili: filaments on the surfaces of many gram-negative bacteria that mediate an extraordinary array of functions, including adhesion, motility, microcolony formation and secretion of proteases and colonization factors

type IV secretion system: a secretion protein complex found in gram negative bacteria, gram positive bacteria, and archaea that can transport proteins and deoxyribonucleic acid (DNA) across the cell membrane

typhoid fever: an infection that causes diarrhea and a rash and is most commonly caused by a bacteria called "Salmonella typhi"

typhus: an infectious disease caused by rickettsiae, characterized by a purple rash, headaches, fever, and usually delirium, and historically a cause of high mortality during wars and famines

untypable: uncapsulated strains *of Haemophilus influenzae*

urease: an enzyme that catalyzes the hydrolysis of urea, forming ammonia and carbon dioxide

uremia: a raised level in the blood of urea and other nitrogenous waste compounds that are normally eliminated by the kidneys

urethritis: inflammation and infection of the urinary tract

vaccination: treatment with a vaccine to produce immunity against a disease

vaccine hesitancy: the belief that vaccines may not be effective, have unknown long-term effects and create serious adverse reactions

vaccine: a preparation containing an antigen, typically consisting of whole disease-causing organisms (killed or weakened) or parts of such organisms, used to confer immunity against the disease that the antigen causes

vaccinia: a virus that causes a pox-like illness in cattle (cowpox) and serves as a smallpox vaccine in humans because of its similarity to the smallpox virus

vaginitis: inflammation and infection of the vagina

variable region: the N-terminal portion of the antibody, which possesses antigen-binding sites and has variable amino acid sequences

variant surface glycoprotein: a ~60kDa protein that densely packs the cell surface of protozoan parasites belonging to the genus *Trypanosoma*

varicella: a highly contagious viral infection causing an itchy, blister-like rash on the skin

vertical transmission: the passage of a disease-causing agent (pathogen) from mother to baby during the period immediately before and after birth

Vibrio cholera: the bacterial species that causes the gastrointestinal disease cholera

vinification: the process of winemaking

viral strain: genetic variant of specific viruses that possess distinct characteristics

virion: the complete, infective form of a virus outside a host cell, with a core of RNA or DNA and a capsid

virulence factor: cellular structures, molecules, and regulatory systems that enable microbial pathogens to colonize the host, evade its immune system, or obtain nutrition

virulent: The ability of an agent of infection to produce disease

virus: a submicroscopic infectious agent that replicates only inside the living cells of an organism

vitamin: any of a group of organic compounds which are essential for normal growth and nutrition and are required in small quantities in the diet because the body cannot synthesize them

walking pneumonia: a nonmedical term for a mild case of pneumonia that is also called "atypical pneumonia" and is caused by bacteria or viruses; often by *Mycoplasma pneumonia*

warfarin: a drug used to treat blood clots (such as in deep vein thrombosis-DVT or pulmonary embolus-PE) and prevent new clots from forming

wastewater: water that has been used in the home, in a business, or as part of an industrial process

water-soluble vitamins: vitamins that, because of their structure, show strong solubility in water; they usually pass through the body in a relatively short time

whole-cell vaccines: vaccines that contain whole cells of the pathogen that have been genetically modified in the laboratory

wort: a solution with extracted sugars from starch sources that are fermented to make beer

yeast: a microscopic, single-celled fungus consisting of single oval cells that reproduce by budding and convert sugar into alcohols or organic acids and carbon dioxide

yellow fever: a typically short-lived viral infection caused by a flavivirus that can, infrequently, cause liver damage, leading to leading to jaundice

zoonotic infections: infections by microorganisms that typically colonize animals and are transmitted to humans by human-animal contact

zoster: also known as shingles; a painful rash that may appear as a stripe of blisters on the trunk of the body

zygospores: the thick-walled resting cell of certain fungi and algae, arising from the fusion of two similar gametes

Organizations

American Phytopathological Society (APS)
3352 Sherman Ct.
Ste. 202
St. Paul, MN 55121
651-454-7250
www.apsnet.org

American Society for Microbiology (ASM)
1752 N St. NW
Washington, DC 20036
202-737-3600
asm.org

American Society for Virology (ASV)
University of Michigan Medical School
Department of Microbiology and Immunology
1150 West Medical Center Drive
5635 Med Sci II
Ann Arbor, MI 48109-5620
734-764-9686
asv.org

American Society of Tropical Medicine and Hygiene
241 18th St. South
Ste. 501
Arlington, VA 22202
571-351-5409
info@astmh.org

American Type Culture Collection (ATCC)
www.atcc.org

National Institute of Allergy and Infectious Diseases (NIAID)
National Institutes of Health (NIH)
5601 Fishers Ln
Rockville, MD 20852
301-496-5717
niaid.nih.gov

The Big Picture Book of Viruses
www.virology.net/big_virology

Biotechnology Innovation Organization
1201 Maryland Ave. SW
Ste. 900
Washington, DC 20024
202-962-9200
www.bio.org

British Mycological Society
Charles Darwin House
12 Roger St.
London WC1N2JU
44(0)330-1330002
www.britmycolsoc.org.uk

Center for Biologics Evaluation and Research (CBER)
US Food and Drug Administration (FDA)
10903 New Hampshire Ave.
Silver Spring, MD 20993
www.fda.gov/about-fda/center-biologics-evaluation-and-research-cber/center-biologics-evaluation-and-research

Centers for Disease Control and Prevention (CDC)
Division of Foodborne, Bacterial, and Mycotic Diseases
1600 Clifton Rd.
Atlanta, GA 3032800-232-4636
www.cdc.gov

Center for Food Safety and Applied Nutrition
US Food and Drug Administration
5001 Campus Dr.
College Park, MD 20740
www.fda.gov/about-fda/fda-organization/center-food-safety-and-applied-nutrition-cfsan

Disease Outbreak News (DONs)
World Health Organization (WHO)
www.who.int/emergencies/disease-outbreak-news

Division of Microbiology and Infectious Diseases (DMID)
NIH/NIAID
5601 Fishers Ln
Rockville, MD 20852
301-496-5717
www.niaid.nih.gov/about/dmid

Human Biome Project
National Institutes of Health
9000 Rockville Pike
Bethesda, MD 20892
www.niaid.nih.gov/about/vrc
commonfund.nih.gov/hmp

Infectious Diseases Society of America (IDSA)
4040 Wilson Blvd.
Ste. 300
Arlington, VA 22203
703-299-0200
www.idsociety.org

Microbiology Society
14-16 Meredith St.
London EC1R 0AB
UK
microbiologysociety.org

Microscopy Society of America
11130 Sunrise Valley Dr.
Ste. 350
Reston, VA 20191
703-234-4115
www.microscopy.org

Mycological Society of America (MSA)
2424 American Ln
Madison, WI 53704
608-441-1060
msafungi.org

National Center for Biotechnology
US National Library of Medicine
8600 Rockville Pike
Bethesda, MD 20894
www.ncbi.nlm.nih.gov/

National Center for Emerging and Zoonotic Infectious Diseases
Centers for Disease Control
1600 Clifton Rd.
Atlanta GA 30329-5027
800-232-4636
www.cdc.gov/ncezid

National Oceanic and Atmospheric Administration
Climate Program Office
1401 Constitution Ave. NW
Room 5128
Washington, DC 20230
301-713-1208
www.noaa.gov/

National Science Foundation (NSF)
2415 Eisenhower Ave.
Alexandria, VA 22314
703-292-5111
nsf.gov

Partnership for Food Safety Education
2345 Crystal Dr.
Ste. 800
Arlington, VA 22202
www.fightbac.org/

Pathology Microbiology & Immunology
University of South Carolina
School of Medicine
Columbia, SC 29208
803-216-3400
pmi@uscmed.sc.edu
pathmicro.med.sc.edu/book/immunol-sta.htm

Public Health Agency of Canada
130 Colonnade Rd S
Nepean, ON K1a 0K9
Canada
844-280-5020
www.phac-aspc.gc.ca

Society for Applied Microbiology
Salisbury House
Station Road
Cambridge CB1 2LA
UK
sfam.org.uk

Society for Industrial Microbiology (SIMB)
3929 Old Lee Highway
Ste. 92A
Fairfax, VA 22030
703-691-3357
www.simbhq.org

Todar's Online Textbook of Bacteriology
www.textbookofbacteriology.net/

US Department of Agriculture
Biotechnology #5071
1400 Independence Ave. SW
Washington, CD 20250
202-720-2791
asusda@usda.gov
www.usda.gov/topics/biotechnology

US Geological Survey
US Department of the Interior
USGS Carbon Cycle Research
12201 Sunrise Valley Dr.
Reston, VA
888-275-8747
geochange.er.usgs.gov/carbon

Vaccine Research Center
National Institute of Allergy and Infectious Diseases
National Institutes of Health
9000 Rockville Pike
Bethesda, MD 20892
www.niaid.nih.gov/about/vrc

Subject Index